MECHATRONICS

An Integrated Approach

Clarence W. de Silva

CRC PRESS

Boca Raton London New York Washington, D.C.

Library of Congress Cataloging-in-Publication Data

De Silva, Clarence W.
 Mechatronics : an integrated approach/Clarence de Silva.
    p. cm.
 Includes bibliographical references and index.
 ISBN 0-8493-1274-4
 1. Mechatronics. I. Title.

TJ163.12.D45 2004
621.3—dc22   2004050339

2005 CRC Press LLC

No claim to original U.S. Government works
International Standard Book Number 0-8493-1274-4
Library of Congress Card Number 2004050339
Printed in the United States of America   3 4 5 6 7 8 9 0
Printed on acid-free paper

Dedication

To my parents and teachers who brought me to my destination, to my family who accompanied me supporting and without complaining, to my friends and colleagues who provided the motivation for the journey, and to my students who provided innovative goals and reasons for the trip.

"Nothing tends so much to the advancement of knowledge as the application of a new instrument. The native intellectual powers of men in different times are not so much the causes of the different success of their labours, as the peculiar nature of the means and artificial resources in their possession."

Sir Humphrey Davy (1778–1829)

"The difficulty lies, not in the new ideas, but in escaping the old ones, which ramify, for those brought up as most of us have been, into every corner of our minds."

John Maynard Keynes (1883–1946)

Preface

This is an introductory book on the subject of Mechatronics. It will serve as both a textbook and a reference book for engineering students and practicing professionals. Mechatronics concerns synergistic and concurrent use of mechanics, electronics, computer engineering, and intelligent control systems in modeling, analyzing, designing, developing, and implementing smart electromechanical products. As the modern machinery and electromechanical devices are typically being controlled using analog and digital electronics and computers, the technologies of mechanical engineering in such a system can no longer be isolated from those of electronic and computer engineering. For example, in a robot system or a micromachine, mechanical components are integrated with analog and digital electronic components to provide single functional units or products. Similarly, devices with embedded and integrated sensing, actuation, signal processing, and control have many practical advantages. In the framework of Mechatronics, a unified approach is taken to integrate different types of components and functions, both mechanical and electrical, in modeling, analysis, design, and implementation, with the objective of harmonious operation that meets a desired set of performance specifications.

In the mechatronic approach, a mixed system consisting of subsystems that have primarily mechanical (including fluid and thermal) or primarily electrical character, is treated using integrated engineering concepts. In particular, electromechanical analogies, consistent energy transfer (e.g., kinetic, potential, thermal, fluid, electrostatic, and electromagnetic energies) through energy ports, and integrated design methodologies may be incorporated using innovative concepts such as Mechatronic Design Quotient (MDQ), resulting in benefits with regard to performance, efficiency, reliability, and cost.

Mechatronics has emerged as a bona fide field of practice, research, and development, and simultaneously as an academic discipline in engineering. Historically, the approach taken in learning a new field of engineering has been to first concentrate on a single branch of engineering such as electrical, mechanical, civil, chemical, or aerospace engineering in an undergraduate program and then learn the new concepts and tools during practice or research. Since the discipline of Mechatronics involves electronic and electrical engineering, mechanical and materials engineering, and control and computer engineering, a more appropriate approach would be to acquire a strong foundation in the necessary fundamentals from these various branches of engineering in an integrated manner in a single and unified undergraduate curriculum. In fact many universities in the United States, Canada, Europe, Asia, and Australia have established both undergraduate and graduate programs in Mechatronics. This book is geared toward this focus on integrated education and practice as related to electromechanical systems. The book will be useful as both a textbook at undergraduate and introductory graduate levels and a reference book for engineers, researchers, project managers, and other practicing professionals.

Scope of the Book

Mechatronics is a multidisciplinary field that concerns the integrated modeling, analysis, design, manufacture, and control of smart electromechanical products and systems. The study of Mechatronics requires a good foundation of such core subjects as electrical components and analysis, mechanical components and analysis, sensors and instrumentation, drives and actuators, control including intelligent control, digital processing and hardware, communication and interfacing, software engineering, modeling, and design. A conventional undergraduate curriculum in engineering does not provide such a broad and multidisciplinary foundation. Furthermore, since Mechatronics involves a synergistic combination of these core areas, a unified approach is needed for learning the subject, particularly in relation to integrated modeling, analysis, design, and prototyping of mechatronic systems. The book represents an effort towards this goal.

The book consists of 13 chapters and 2 appendices. The chapters are devoted to presenting the fundamentals in electrical and electronic engineering, mechanical engineering, control engineering, and computer engineering that are necessary for forming the foundation of Mechatronics. In particular, they cover mechanical components, modeling, analysis, instrumentation, sensors, transducers, signal processing, actuators, control, and system design and integration. The book uniformly incorporates the underlying fundamentals into analytical methods, modeling approaches, and design techniques in a systematic manner throughout the main chapters. The practical application of the concepts, approaches, and tools presented in the introductory chapters is demonstrated through a wide range of practical examples and a comprehensive set of case studies. Useful information on software tools and transforms which are not directly useful in the presentation of the fundamentals of Mechatronics are given in a concise manner in the appendices.

The book is an outgrowth of the author's experience in integrating key components of Mechatronics into senior-level courses for engineering students, and in teaching graduate and professional courses in Mechatronics and related topics. Consequently, the main emphasis of the book is for use as an engineering textbook. But, in view of the practical considerations, design issues, computer tools, and industrial techniques that are presented throughout the book, and in view of the simplified and snap-shot style presentation of more advanced theory and concepts, the book will also serve as a useful reference tool for engineers, technicians, project managers, and other practicing professionals in industry and in research laboratories.

To maintain clarity and the focus and to maximize the usefulness of the book, the material is presented in a manner that will be convenient and useful to anyone with a basic engineering background, be it electrical, mechanical, aerospace, control, or computer engineering. Case studies, detail worked examples, and exercises are provided throughout the book. Complete solutions to the end-of-chapter problems are presented in a solutions manual, which will be available to instructors who adopt the book.

Main Features of the Book

- The material is presented in a progressive manner, first giving introductory material and then systematically leading to more advanced concepts and applications in each chapter.

- The material is presented in an integrated and unified manner so that users with a variety of engineering backgrounds (mechanical, electrical, computer, control, aerospace, and material) will be able to follow and equally benefit from it.

- Practical applications and tools are introduced in the very beginning and then uniformly integrated throughout the book.

- Key issues presented in the book are summarized in point form at various places in each chapter for easy reference, recollection, and presentation as viewgraphs.

- Many worked examples and case studies are included throughout the book.

- Numerous problems and exercises, most of which are based on practical situations and applications, are given at the end of each chapter.

- Commercial software tools for analysis, design, and implementation of mechatronic systems are described and illustrated using suitable examples. Only the industry-standard and state-of-the-art software tools are presented.
- References and reading suggestions are given for further information and study.
- Useful material that cannot be conveniently integrated into the chapters is presented in a concise form as separate appendices at the end of the book.

A solutions manual has been developed for the convenience of instructors.

A Note to Instructors

A curriculum for a four-year Bachelor's degree in Mechatronics is given in the solutions manual available from the publisher. MECHATRONICS—An Integrated Approach, is suitable as the text book for several courses in such a curriculum. Several appropriate courses are listed below.

Mechatronics
Mechanical Components
Actuators and Drive Systems
Automatic Control
Electro-mechanical Systems
Mechatronic Product Design
Sensors and Transducers
System Modeling and Simulation
Computer Control Systems

The book is also suitable for introductory graduate-level courses on such subjects as

Control Sensors and Actuators
Instrumentation and Design of Control Systems
Mechatronics
Control Engineering

Clarence W. de Silva
Vancouver, Canada

The Author

Clarence W. de Silva, P.Eng., Fellow ASME and Fellow IEEE, is Professor of Mechanical Engineering at the University of British Columbia, Vancouver, Canada, and has occupied the NSERC Research Chair in Industrial Automation since 1988. He has earned Ph.D. degrees from Massachusetts Institute of Technology (1978) and the University of Cambridge, England (1998). De Silva has also occupied the Mobil Endowed Chair Professorship in the Department of Electrical and Computer Engineering at the National University of Singapore. He has served as a consultant to several companies including IBM and Westinghouse in the U.S., and has led the development of six industrial machines.

He is recipient of the Killam Research Prize, Outstanding Engineering Educator Award of IEEE Canada, Education Award of the Dynamic Systems and Control Division of the American Society of Mechanical Engineers (ASME), Lifetime Achievement Award of the World Automation Congress, IEEE Third Millennium Medal, Meritorious Achievement Award of the Association of Professional Engineers of BC, and the Outstanding Contribution Award of the Systems, Man, and Cybernetics Society of the Institute of Electrical and Electronics Engineers (IEEE).

He has authored or co-authored 16 technical books, 12 edited volumes, about 175 journal papers, and about 200 conference papers and book chapters. He has served on the editorial boards of 12 international journals, in particular as the Editor-in-Chief of the *International Journal of Control and Intelligent Systems*, Editor-in-Chief of the *International Journal of Knowledge-Based Intelligent Engineering Systems*, Senior Technical Editor of Measurements and Control, and Regional Editor, North America, of *Engineering Applications of Artificial Intelligence—the International Journal of Intelligent Real-Time Automation*. He is a Lilly Fellow, Senior Fulbright Fellow to Cambridge University, Fellow of the Advanced Systems Institute of British Columbia, and a Killam Fellow.

Acknowledgments

Many individuals have assisted in the preparation of this book, but it is not practical to acknowledge all such assistance here. First, I wish to recognize the contributions, both direct and indirect, of my graduate students, research associates, and technical staff. Particular mention should be made of Jian Zhang, my research engineer; Poi Loon Tang, my laboratory manager; and Yan Cao, Rick McCourt, and Ken Wong, my graduate research assistants.

I am particularly grateful to Cindy Renee Carelli, Acquisitions Editor–Engineering, CRC Press, for her interest, enthusiasm, support, advice and patience, as usual, throughout the project. Other staff of CRC Press and its affiliates, particularly Jessica Vakili and Priyanka Negi, deserve special mention here. Stephen McLane, Marketing Administration Manager of Aerotech, was very helpful in providing motor data. Finally, I wish to acknowledge the advice and support of various authorities in the field—particularly, Prof. Devendra Garg of Duke University, Prof. Mo Jamshidi of the University of New Mexico, Prof. Tong-Heng Lee of the National University of Singapore, Prof. Arthur Murphy (DuPont Fellow Emeritus), Prof. Grantham Pang of the University of Hong Kong, Prof. Jim A.N. Poo of the National University of Singapore, Dr. Daniel Repperger of U.S. Air Force Research Laboratory, Prof. P.D. Sarath Chandra of the Open University of Sri Lanka, and Prof. David N. Wormley of the Pennsylvania State University.

Source Credits

Figure 1.1 A servomotor is a mechatronic device. (Danaher Motion. With permission.)

Figure 1.2(a) A humanoid robot is a complex and "intelligent" mechatronic system. (American Honda Motor Co. With permission.)

Figure 1.2(b) Components of a humanoid robot. (American Honda Motor Co. With permission.)

Figure 3.2(b) A commercial ball screw unit (Deutsche Star GmbH. With permission.)

Figure 6.8(a) LVDT: A commercial unit (Scheavitz Sensors, Measurement Specialties, Inc. With permission.)

Figure 6.80 A commercial RTD unit (RdF Corp. With permission.)

Figure 7.1(b) Components of a commercial incremental encoder (BEI Electronics, Inc. With permission.)

Figure 8.1 A commercial two-stack stepper motor (Danaher Motion. With permission.)

Figure 8.46 Stepper motor performance curves (Aerotech, Inc. With permission.)

Figure 9.33(b) Speed–torque characteristics of a commercial brushless DC servomotor with a matching amplifier (Aerotech, Inc. With permission.)

Table 8.2 Stepper motor data (Aerotech, Inc. With permission.)

Windows and Word are software products of Microsoft Corporation.

MATLAB and SIMULINK are registered trademarks and products of The MathWorks, Inc.

LabVIEW is a product of National Instruments, Inc. The associated figures are reproduced with permission.

These software tools have been used by the author in teaching and in the development of the present book.

Table of Contents

1

Mechatronic Engineering

The field of Mechatronics concerns the synergistic application of mechanics, electronics, controls, and computer engineering in the development of electromechanical products and systems, through an integrated design approach. A mechatronic system requires a multidisciplinary approach for its design, development, and implementation. In the traditional development of an electromechanical system, the mechanical components and electrical components are designed or selected separately and then integrated, possibly with other components, hardware, and software. In contrast, in the mechatronic approach, the entire electromechanical system is treated concurrently in an integrated manner by a multidisciplinary team of engineers and other professionals. Naturally, a system formed by interconnecting a set of independently designed and manufactured components will not provide the same level of performance as a mechatronic system, that employs an integrated approach for design, development, and implementation. The main reason is straightforward. The best match and compatibility between component functions can be achieved through an integrated and unified approach to design and development, and best operation is possible through an integrated implementation. Generally, a mechatronic product will be more efficient and cost effective, precise and accurate, reliable, flexible and functional, and mechanically less complex, compared to a nonmechatronic product needing a similar level of effort in its development. Performance of a nonmechatronic system can be improved through sophisticated control, but this is achieved at an additional cost of sensors, instrumentation, and control hardware and software, and with added complexity. Mechatronic products and systems include modern automobiles and aircraft, smart household appliances, medical robots, space vehicles, and office automation devices.

1.1 Mechatronic Systems

A typical mechatronic system consists of a mechanical skeleton, actuators, sensors, controllers, signal conditioning/modification devices, computer/digital hardware and software, interface devices, and power sources. Different types of sensing, information acquisition and transfer are involved among all these various types of components. For example, a servomotor (see Figure 1.1), which is a motor with the capability of sensory feedback for accurate generation of complex motions, consists of mechanical, electrical, and electronic components. The main mechanical components are the rotor and the stator. The electrical components include the circuitry for the field windings and rotor windings (if present), and circuitry for power transmission and commutation (if needed). Electronic components include those needed for sensing (e.g., optical encoder for displacement and speed sensing and tachometer for speed sensing). The overall design of a servomotor can be improved by taking a mechatronic approach. The humanoid robot

FIGURE 1.1
A servomotor is a mechatronic device. (Danaher Motion, Rockford, IL. With permission).

shown in Figure 1.2(a) is a more complex and "intelligent" mechatronic system. It may involve many servomotors and a variety of mechatronic components, as is clear from the sketch in Figure 1.2(b). A mechatronic approach can greatly benefit the design and development of a complex electromechanical system of this nature.

In a true mechatronic sense, the design of a mixed multi-component system will require simultaneous consideration and integration and design of all its components, as indicated in Figure 1.3. Such an integrated and "concurrent" design will call for a fresh look at the design process itself, and also a formal consideration of information and energy transfer between components within the system. It is expected that the mechatronic approach will result in higher quality of products and services, improved performance, and increased reliability, approaching some form of optimality. This will enable the development and production of electromechanical systems efficiently, rapidly, and economically. A study of mechatronic engineering should consider all stages of design, development, integration, instrumentation, control, testing, operation, and maintenance of a mechatronic system.

When performing an integrated design of a mechatronic system, the concepts of energy/power present a unifying thread. The reasons are clear. First, in an electromechanical system, ports of power/energy exist that link electrical dynamics and mechanical dynamics. Hence, modeling, analysis, and optimization of a mechatronic system can be carried out using a hybrid-system (or, mixed-system) formulation (a model) that integrates mechanical aspects and electrical aspects of the system. Second, an optimal design will aim for minimal energy dissipation and maximum energy efficiency. There are related implications, for example, greater dissipation of energy will mean reduced

FIGURE 1.2(a)
A humanoid robot is a complex and "intelligent" mechatronic system. (American Honda Motor Co., Torrance, CA, With permission).

overall efficiency and increased thermal problems, noise, vibration, malfunctions, wear and tear. Again, a hybrid model that presents an accurate picture of energy/power flow within the system will present an appropriate framework for the mechatronic design. (Note: Refer to bond graph models and linear graph models in particular, as discussed in Chapter 2).

By definition, a mechatronic design should result in an optimal final product. In particular, a mechatronic design in view of its unified and synergistic treatment of components and functionalities, with respect to a suitable performance index (single or multiple-objective), should be "better" than a traditional design where the electrical design and the mechanical design are carried out separately and sequentially. The mechatronic approach should certainly be better than a simple interconnection of components that can do the intended task.

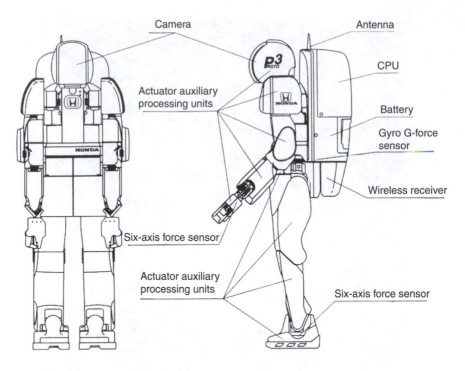

FIGURE 1.2(b)
Components of a humanoid robot. (American Honda Motor Co. With permission).

1.2   Modeling and Design

A design may use excessive safety factors and worst-case specifications (e.g., for mechanical loads and electrical loads). This will not provide an optimal design or may not lead to the most efficient performance. Design for optimal performance may not necessarily lead to the most economical (least costly) design, however. When arriving at a truly optimal design, an objective function that takes into account all important factors (performance, quality, cost, speed, ease of operation, safety, environmental impact, etc.) has to be optimized. A complete design process should incorporate the necessary details of a system for its construction or assembly. Of course, in the beginning of the design process, the desired system does not exist. In this context, a model of the anticipated system can be very useful. In view of the complexity of a design process, particularly when striving for an optimal design, it is useful to incorporate system modeling as a tool for design iteration.

Modeling and design can go hand in hand, in an iterative manner. In the beginning, by knowing some information about the system (e.g., intended functions, performance specifications, past experience and knowledge of related systems) and using the design objectives, it will be possible to develop a model of sufficient (low to moderate) detail and complexity. By analyzing and carrying out computer simulations of the model it will be possible to generate useful information that will guide the design process (e.g., generation of a preliminary design). In this manner design decisions can be made, and the model can be refined using the available (improved) design. This iterative link between modeling and design is schematically shown in Figure 1.4.

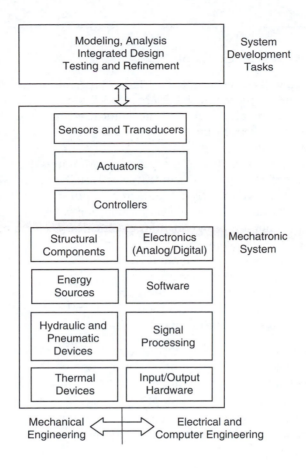

FIGURE 1.3
Concepts of a mechatronic system.

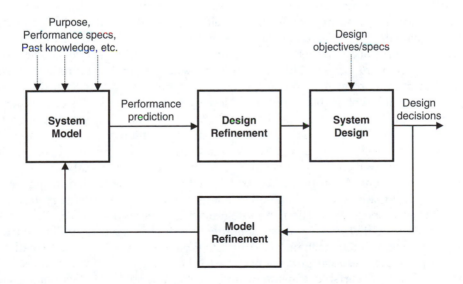

FIGURE 1.4
Link between modeling and design.

1.3   Mechatronic Design Concept

A mechatronic system will consist of many different types of interconnected components and elements. As a result there will be energy conversion from one form to another, particularly between electrical energy and mechanical energy. This enables one to use energy as the unifying concept in the analysis and design of a mechatronic system. Let us explore this idea further.

In an electromechanical system an interaction (or, coupling) exists between electrical dynamics and mechanical dynamics. Specifically, electrical dynamics affect the mechanical dynamics and vice versa. Traditionally, a "sequential" approach has been adopted for the design of mixed systems such as electromechanical systems. For example, the mechanical and structural components are designed first, electrical and electronic components are selected or developed and interconnected next, and a computer is selected and interfaced with the system next, and so on. The dynamic coupling between various components of a system dictates, however, that an accurate design of the system should consider the entire system as a whole rather than designing the electrical/electronic aspects and the mechanical aspects separately and sequentially. When independently designed components are interconnected, several problems can arise:

1. When two independently designed components are interconnected, the original characteristics and operating conditions of the two will change due to loading or dynamic interactions (see Chapter 4).

2. Perfect matching of two independently designed and developed components will be practically impossible. As a result a component can be considerably underutilized or overloaded, in the interconnected system, both conditions being inefficient and undesirable.

3. Some of the external variables in the components will become internal and "hidden" due to interconnection, which can result in potential problems that cannot be explicitly monitored through sensing and cannot be directly controlled.

The need for an integrated and concurrent design for electromechanical systems can be identified as a primary motivation for the development of the field of Mechatronics.

Design objectives for a system are expressed in terms of the desired performance specifications. By definition, a "better" design is where the design objectives (design specifications) are met more closely. The "principle of synergy" in Mechatronics means, an integrated and concurrent design should result in a better product than one obtained through an uncoupled or sequential design. Note that an uncoupled design is where each subsystem is designed separately (and sequentially), while keeping the interactions with the other subsystems constant (i.e., ignoring the dynamic interactions).

The concept of mechatronic design can be illustrated using an example of an electromechanical system, which can be treated as a coupling of an electrical subsystem and a mechanical subsystem. An appropriate model for the system is shown in Figure 1.5(a). Note that the two subsystems are coupled using a loss-free (pure) energy transformer while the losses (energy dissipation) are integral with the subsystems (see Chapter 2). In this system, assume that under normal operating conditions the energy flow is from the electrical subsystem to the mechanical subsystem (i.e., it behaves like a motor rather than a generator). At the electrical port connecting to the energy transformer, there exists a current i (a "through" variable) flowing in, and a voltage v (an "across" variable) with the shown polarity (The concepts of through and across variables and the related terminology are explained in Chapter 2). The product vi is the electrical power, which is positive out

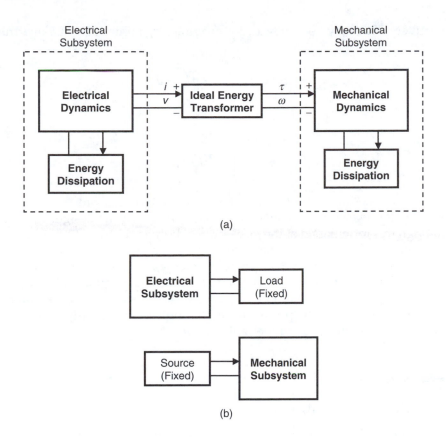

FIGURE 1.5
(a) An electromechanical system; (b) Conventional design.

of the electrical subsystem and into the transformer. Similarly, at the mechanical port coming out of the energy transformer, there exists a torque τ (a through variable) and an angular speed ω (an across variable) with the sign convention given in Figure 1.5(a). Accordingly, a positive mechanical power $\omega\tau$ flows out of the transformer and into the mechanical subsystem. The ideal transformer implies:

$$vi = \omega\tau \qquad (1.1)$$

In a conventional uncoupled design of the system, the electrical subsystem is designed by treating the effects of the mechanical subsystem as a fixed load, and the mechanical subsystem is designed by treating the electrical subsystem as a fixed energy source, as indicated in Figure 1.5(b). Suppose that, in this manner the electrical subsystem achieves a "design index" of I_{ue} and the mechanical subsystem achieves a design index of I_{um}. Note here that the design index is a measure of the degree to which the particular design satisfies the design specifications (design objectives).

When the two uncoupled designs (subsystems) are interconnected, there will be dynamic interactions. As a result, neither the electrical design objectives nor the mechanical design objectives will be satisfied at the levels dictated by I_{ue} and I_{um}, respectively. Instead, they will be satisfied at the lower levels given by the design indices I_e, and I_m. A truly mechatronic design will attempt to bring I_e, and I_m as close as possible to I_{ue} and

I_{um}, respectively. This may be achieved, for example, by minimizing the quadratic cost function

$$J = \alpha_e (I_{ue} - I_e)^2 + \alpha_m (I_{um} - I_m)^2 \tag{1.2}$$

subject to

$$\begin{bmatrix} I_e \\ I_m \end{bmatrix} = D(p) \tag{1.3}$$

where D denotes the transformation that represents the design process, and p denotes information including system parameters that is available for the design.

Even though this formulation of the mechatronic design problem appears rather simple and straightforward, the reality is otherwise. In particular, the design process, as denoted by the transformation D, can be quite complex and typically nonanalytic. Furthermore, minimization of the cost function J or the "mechatronic design quotient" *MDQ* is by and large an iterative practical scheme, and undoubtedly a knowledge-based and nonanalytic procedure. This complicates the process of mechatronic design. In any event, the design process will need the information represented by p.

1.4   Evolution of Mechatronics

Mechanical engineering products and systems that employ some form of electrical engineering principles and devices have been developed and used since the early part of the 20th century. These systems included the automobile, electric typewriter, aircraft, and elevator. Some of the power sources used in these systems were not necessarily electrical, but there were batteries and/or the conversion of thermal power into electricity through generators. These "electromechanical" systems were not "mechatronic" systems, because they did not use the integrated approach characterizing Mechatronics for their analysis, design, development, and implementation.

Rapid advances in electromechanical devices and systems were possible particularly due to developments in control engineering, which began for the most part in the early 1950s, and still more rapid advances in digital computer and communication as a result of integrated circuit (IC) and microprocessor technologies, starting from the late 1960s. With these advances, engineers and scientists felt the need for a multidisciplinary approach to design and hence a "mechatronic" approach. Yasakawa Electric in Japan was the first to coin the term "Mechatronics," for which the company obtained a trademark in 1972. Subsequently, in 1982, the company has released the trademark rights. Even though a need for Mechatronics was felt even in those early times, no formal discipline and educational programs existed for the engineers to be educated and trained in this area. The research and development activities mainly in automated transit systems and robotics, in the 1970s and 1980s undoubtedly paved the way for the evolution of the field of Mechatronics. With today's sophisticated technologies of mechanics and materials, analog and digital electronics, sensors, actuators, controllers, electromechanical design, and microelectromechanical systems (MEMS) with embedded sensors, actuators, and microcontrollers, the field of Mechatronics has attained a good degree of maturity. Now many universities around the world offer undergraduate and graduate programs in mechatronic engineering, which have become highly effective and popular among students, instructors, employees, and employers alike.

1.5   Application Areas

Application areas of Mechatronics are numerous, and involve those that concern mixed systems and particularly electromechanical systems. These applications may involve:

1. Modifications and improvements to conventional designs, by using a mechatronic approach.
2. Development and implementation of original and innovative mechatronic systems.

In either category, the applications will employ sensing, actuation, control, signal conditioning, component interconnection and interfacing, and communication, generally using tools of mechanical, electrical and electronic, computer, and control engineering. Some important areas of application are indicated below.

Transportation is a broad area in which mechatronic engineering has numerous applications. In ground transportation in particular, automobiles, trains, and automated transit systems use mechatronic devices. They include airbag deployment systems, antilock braking systems (ABS), cruise control systems, active suspension systems, and various devices for monitoring, toll collection, navigation, warning, and control in intelligent vehicular highway systems (IVHS). In air transportation, modern aircraft designs with advanced materials, structures, electronics, and control benefit from the concurrent and integrated approach of Mechatronics to develop improved designs of flight simulators, flight control systems, navigation systems, landing gear mechanisms, traveler comfort aids, and the like.

Manufacturing and production engineering is another broad field that uses mechatronic technologies and systems. Factory robots (for welding, spray painting, assembly, inspection, etc.), automated guided vehicles (AGVs), modern computer-numerical control (CNC) machine tools, machining centers, rapid (and virtual) prototyping systems, and micromachining systems are examples of mechatronic applications.

In medical and healthcare applications, robotic technologies for examination, surgery, rehabilitation, drug dispensing, and general patient care are being developed and used. Mechatronic technologies are being applied for patient transit devices, various diagnostic probes and scanners, beds, and exercise machines.

In a modern office environment, automated filing systems, multifunctional copying machines (copying, scanning, printing, FAX, etc.), food dispensers, multimedia presentation and meeting rooms, and climate control systems incorporate mechatronic technologies.

In household applications, home security systems and robots, vacuum cleaners and robots, washers, dryers, dishwashers, garage door openers, and entertainment centers use mechatronic devices and technologies.

In the computer industry, hard disk drives (HDD), disk retrieval, access, and ejection devices, and other electromechanical components can considerably benefit from Mechatronics. The impact goes further because digital computers are integrated into a vast variety of other devices and applications.

In civil engineering applications, cranes, excavators, and other machinery for building, earth removal, mixing and so on, will improve their performance by adopting a mechatronic design approach.

In space applications, mobile robots such as NASA's Mars exploration Rover, space-station robots, and space vehicles are fundamentally mechatronic systems.

It is noted that there is no end to the type of devices and applications that can incorporate Mechatronics. In view of this, the traditional boundaries between engineering disciplines will become increasingly fuzzy, and the field of Mechatronics will grow and evolve further through such merging of disciplines.

1.6 Study of Mechatronics

Due to the interdisciplinary nature of the field of Mechatronics, one should not use a "compartmentalized: approach in studying this discipline. Specifically, rather than using a conventional approach to learning such standard subjects as mechanics, electronics, modeling, control, computer engineering, and signal processing, separately in a disjointed manner, the components need to be integrated into a common "mechatronics" framework, along with other specialized subjects such as sensors, actuators, intelligent control, interface hardware, testing, performance evaluation, and cost-benefit analysis. This integration should be achieved through the common thread of concurrent and mixed-system design. Curricula in Mechatronics have been developed based on this understanding. In any event, in a single and cohesive program of study it may not be feasible to cover all the fundamentals of science and engineering that are needed for mechatronic engineering. A more realistic approach would be to follow a traditional engineering curriculum in the first 2 years of a 4-year undergraduate program, and then get into an integrated mechatronic curriculum in the next 2 years. This assumption has been made in developing the present book.

1.7 Organization of the Book

Mechatronics is a multidisciplinary field, which concerns the integrated modeling, analysis, design, manufacture, control, testing, and operation of smart electromechanical products and systems. The study of Mechatronics requires a good foundation of such core subjects as mechanics, electronics, modeling, control, signal processing and conditioning, communication and computer engineering, and specialized subjects like electrical components, mechanical components, sensors and transducers, instrumentation, drives and actuators, intelligent control, and interfacing hardware and software. In Mechatronics, all these subjects are unified through an integrated approach of modeling, analysis, design, and implementation for mixed systems. A traditional undergraduate curriculum in engineering does not provide such a broad and multidisciplinary foundation. Furthermore, since Mechatronics involves a synergistic combination of many subjects, a unified approach is needed for learning as well, particularly in integrated modeling, analysis, design, and prototyping. It is not feasible, however, to cover all the needed subjects in a single degree program of Mechatronics. In fact, a great deal of the foundation material is covered in the first two years of a standard four-year curriculum in engineering. What is presented in this book is the necessary material in Mechatronics that is not traditionally covered in the first 2 years of an undergraduate engineering program.

The book consists of 13 chapters and 2 appendices. The chapters are devoted to presenting the fundamentals in electrical and electronic engineering, mechanical engineering, control engineering, and computer engineering, which are necessary for forming the core of Mechatronics. In particular, they cover modeling, analysis, mechanics, electronics, instrumentation, sensors, transducers, signal processing, actuators, drive systems, computer engineering, control, and system design and integration. The book uniformly incorporates the underlying fundamentals into analytical methods, modeling approaches, and design techniques in a systematic manner throughout the main chapters. The practical application of the concepts, approaches, and tools presented in the introductory chapters are demonstrated through numerous illustrative examples and a comprehensive set of case studies. The background theory and techniques that are not directly useful to present the fundamentals of Mechatronics are given in a concise manner in the appendices.

This chapter introduces the field of Mechatronics. The evolution of the field is discussed, and the underlying design philosophy of Mechatronics is described. This introductory chapter sets the tone for the study, which spans the remaining chapters. Relevant publications in the field are listed.

Chapter 2 deals with modeling and analysis of dynamic systems. Mechanical, electrical, fluid, and thermal systems, and mixed systems such as electromechanical systems are studied. Several techniques of modeling are presented, while emphasizing those methods that are particularly appropriate for mechatronic systems. Analysis in both time domain and frequency domain is introduced, particularly discussing response analysis and computer simulation.

Chapter 3 concerns mechanical components, which are important constituents of a mechatronic system. Robotic devices, motion transmission devices, object handling devices, underlying phenomena, and analytical methods are presented.

Chapter 4 discusses component interconnection and signal conditioning, which is in fact a significant unifying subject within Mechatronics. Impedance considerations of component interconnection and matching are studied. Amplification, filtering, analog-to-digital conversion, digital-to-analog conversion, bridge circuits, and other signal conversion and conditioning techniques and devices are discussed.

Chapter 5 covers performance analysis of a mechatronic device or component. Methods of performance specification are addressed, both in time domain and frequency domain. Common instrument ratings that are used in industry and generally in the engineering practice are discussed. Related analytical methods are given. Instrument bandwidth considerations are highlighted, and a design approach based on component bandwidth is presented. Errors in digital devices, particularly resulting from signal sampling, are discussed from analytical and practical points of view.

Chapter 6 presents important types, characteristics, and operating principles of analog sensors. Particular attention is given to sensors that are commonly used in mechatronic systems. Motion sensors, force, torque and tactile sensors, optical sensors, ultrasonic sensors, temperature sensors, pressure sensors, and flow sensors are discussed. Analytical basis, selection criteria, and application areas are indicated.

Chapter 7 discusses common types of digital transducers. Unlike analog sensors, digital transducers generate pulses or digital outputs. These devices have clear advantages, particularly when used in computer-based, digital systems. They do possess quantization errors, which are unavoidable in a digital representation of an analog quantity. Related issues of accuracy and resolution are addressed.

Chapter 8 studies stepper motors, which are an important class of actuators. These actuators produce incremental motions. Under satisfactory operating conditions, they have the advantage of being able to generate a specified motion profile in an open-loop manner without requiring motion sensing and feedback control. But, under some conditions of loading and motion, motion steps will be missed. Consequently, it is appropriate to use sensing and feedback control when complex motion trajectories need to be followed under nonuniform and extreme loading conditions.

Chapter 9 outlines continuous-drive actuators such as dc motors, ac motors, hydraulic actuators, and pneumatic actuators. Common varieties of actuators under each category are discussed. Operating principles, analytical methods, design considerations, selection methods, drive systems, and control techniques are described. Advantages and drawbacks of various types of actuators on the basis of the nature and the needs of an application are discussed. Practical examples are given.

Chapter 10 covers the subject of digital logic and hardware, which falls into the area of electronic and computer engineering. Logic devices and integrated circuits are widely used in mechatronic systems, for such purposes as sensing, signal conditioning, and control. Basic principles of digital components and circuits are presented in the chapter. Types and applications of logic devices are discussed. The technology of integrated circuits is introduced.

Chapter 11 addresses another important topic in computer engineering and control. Specifically microprocessors, digital computers, and programmable logic controllers (PLCs) are studied in this chapter. The microprocessor has become a standard component in a large variety of mechatronic devices. A microprocessor, together with memory and software and interface hardware, provides an effective and economical miniature digital computer in mechatronic applications. Smart sensors, actuators, controllers, and other essential components of a mechatronic system can immensely benefit from the programmability, flexibility and the processing power of a microcontroller. PLCs are discrete control devices, which are particularly applicable in a coordinated operation of several mechatronic devices, to achieve a common goal. Considerations of networking and communication, and the compatibility of interconnected (or, networked) components become paramount here. These issues are discussed in the chapter.

Chapter 12 deals with conventional control of mechatronic systems. Both time-domain techniques and frequency-domain techniques of control are covered. In particular, conventional digital control is presented. Underlying analytical methods are described. Tuning and design methods of controllers and compensators in mechatronic applications are treated.

Chapter 13 concludes the main body of the book by presenting the design approach of Mechatronics and by giving extensive case studies of practical mechatronic systems. The techniques covered in the previous chapters come together and are consolidated in these case studies. Several practical projects are given, which may be attempted as exercises in Mechatronics.

Appendix A gives useful techniques of Laplace transform and Fourier transform. Appendix B presents several useful software tools. In particular, SIMULINK©, and MATLAB© toolbox of control systems are outlined. The LabVIEW® program development environment, which is an efficient tool for laboratory experimentation (particularly, data acquisition and control), is also described.

1.8   Problem

You are a mechatronic engineer who has been assigned the task of designing and instrumenting a mechatronic system. In the final project report you will have to describe the steps of establishing the design/performance specifications for the system, selecting and sizing sensors, transducers, actuators, drive systems, controllers, signal conditioning and interface hardware, and software for the instrumentation and component integration of this system. Keeping this in mind, write a project proposal giving the following information:

1. Select a process (plant) as the system to be developed. Describe the plant indicating the purpose of the plant, how the plant operates, what is the system boundary (physical or imaginary), what are important inputs (e.g., voltages, torques, heat transfer rates, flow rates), response variables (e.g., displacements, velocities, temperatures, pressures, currents, voltages), and what are important plant parameters (e.g., mass, stiffness, resistance, inductance, conductivity, fluid capacity). You may use sketches.

2. Indicate the performance requirements (or, operating specifications) for the plant (i.e., how the plant should behave under control). You may use any available information on such requirements as accuracy, resolution, speed, linearity, stability, and operating bandwidth.

3. Give any constraints related to cost, size, weight, environment (e.g., operating temperature, humidity, dust-free or clean room conditions, lighting, wash-down needs), etc.

4. Indicate the type and the nature of the sensors and transducers present in the plant and what additional sensors and transducers might be needed for properly operating and controlling the system.

5. Indicate the type and the nature of the actuators and drive systems present in the plant and which of these actuators have to be controlled. If you need to add new actuators (including control actuators) and drive systems, indicate such requirements in sufficient detail.

6. Mention what types of signal modification and interfacing hardware would be needed (i.e., filters, amplifiers, modulators, demodulators, ADC, DAC, and other data acquisition and control needs). Describe the purpose of these devices. Indicate any software (e.g., driver software) that may be needed along with this hardware.

7. Indicate the nature and operation of the controllers in the system. State whether these controllers are adequate for your system. If you intend to add new controllers briefly give their nature, characteristics, objectives, etc. (e.g., analog, digital, linear, nonlinear, hardware, software, control bandwidth).

8. Describe how the users and/or operators interact with the system, and the nature of the user interface requirements (e.g., graphic user interface or GUI).

The following plants/systems may be considered:

1. A hybrid electric vehicle
2. A household robot
3. A smart camera
4. A smart airbag system for an automobile
5. Rover mobile robot for Mars exploration, developed by NASA
6. An automated guided vehicle (AGV) for a manufacturing plant
7. A flight simulator
8. A hard disk drive for a personal computer
9. A packaging and labeling system for a grocery item
10. A vibration testing system (electrodynamic or hydraulic)
11. An active orthotic device to be worn by a person to assist a disabled or weak hand (which has some sensation, but not fully functional).

2

Dynamic Models and Analogies

Design, development, modification, and control of a mechatronic system require an understanding and a suitable "representation" of the system; specifically, a "model" of the system is required. Any model is an idealization of the actual system. Properties established and results derived are associated with the model rather than the actual system, whereas the excitations are applied to and the output responses are *measured* from the actual system. This distinction is very important particularly in the context of the present chapter. A mechatronic system may consist of several different types of components, and it is termed a *mixed system*. It is useful then to use analogous procedures for modeling such components. In this manner the component models can be conveniently integrated to obtain the overall model. In particular, analytical models may be developed for mechanical, electrical, fluid, and thermal systems in a rather analogous manner, because some clear analogies are present among these four types of systems. In view of the *analogy*, then, a unified approach may be adopted in the analysis, design, and control of mechatronic systems.

2.1 Terminology

Each interacted component or element of a mechatronic system will possess an *input-output* (or *cause-effect*, or *causal*) relationship. A *dynamic system* is one whose response variables are functions of time, with non-negligible "rates" of changes. Also, its present output depends not only on the present input, but also on some historical information (e.g., previous input or output). A more formal mathematical definition can be given, but it is adequate to state here that a typical mechatronic system, which needs to be controlled, is a dynamic system. A model is some form of representation of a practical system. An analytical model (or mathematical model) comprises equations (e.g., differential equations) or an equivalent set of information, which represents the system to some degree of accuracy. Sometimes, a set of curves, digital data (table) stored in a computer, and other numerical data—rather than a set of equations—might be termed an analytical model if such data can represent the system of interest.

2.1.1 Model Types

One way to analyze a system is to impose disturbances (inputs) on the system and analyze the reaction (outputs) of the system. This is known as "experimental modeling" or *model identification*. A model that is developed by exciting the actual system and measuring its response, is called an experimental model. Another way is to analyze the system using an analytical model of the system. In effect, we represent the system with a model, such as a *state space model*, a *linear graph*, a *bond graph*, a *transfer function model* or a *frequency-domain model*. Since disturbing a physical system often is less economical or practical than analyzing its analytical model, analytical models are commonly used in practical applications. Systems for experimental modeling (exciters, measuring devices and analyzers) are commercially available, and experimental modeling is done, if less often than analytical modeling.

In general, models may be grouped into the following categories:

1. Physical models (prototypes)
2. Analytical models
3. Computer (numerical) models
4. Experimental models (using input/output experimental data).

Normally, mathematical definitions for a dynamic system are given with reference to an analytical model of the system, for example, a state-space model. In that context the system and its analytical model are synonymous. In reality, however, an analytical model, or any model for that manner, is an idealization of the actual system. Analytical properties that are established and results that are derived, would be associated with the model rather than the actual system, whereas the excitations are applied to and the output responses are measured from the actual system. This distinction should be clearly recognized.

Analytical models are very useful in predicting the dynamic behavior (response) of a system when it is subjected to a certain excitation (input). For example, vibration is a dynamic phenomenon and its analysis, practical utilization, and effective control require a good understanding of the vibrating system. A recommended way to control a dynamic system is through the use of a suitable model of the system. A model may be employed for designing a mechatronic system for proper performance. In the context of *product testing*, for example, analytical models are commonly used to develop test specifications and the input signal applied to the exciter, and to study dynamic effects and interactions in the test object, the excitation system, and their interfaces. In product qualification by analysis, a suitable analytical model of the product replaces the test specimen. In process control, a dynamic model of the actual process may be employed to develop the necessary control schemes. This is known as *model-based control*.

2.1.1.1 System Response

The response of an analytical model to an imposed disturbance can be expressed in either the *time-domain* (response value versus time) or in the *frequency domain* (amplitude and phase versus frequency). The time-domain response generally involves the solution of a set of differential equations. The frequency domain analysis is done with a set of transfer functions, that is the ratio output/input in the *Laplace transform* ("s") form. We shall see that *mobility, admittance, impedance,* and *transmissibility* are convenient transfer-function representations. For example, transmissibility is important in vibration isolation, and mechanical impedance is useful in tasks such as cutting, joining, and assembly that employ robots.

Experimental determination of transfer function (i.e., frequency-domain experimental modeling) is often used in *modal testing*—i.e., testing for natural "modes" of response—of a mechanical system. This requires imposing forces on the system and measuring its response (motion). The system must be designed to limit both the forces transmitted from the system to the foundation, and the motions transmitted from the support structure to the main system. In these cases, the vibration isolation characteristics of the system can be expressed as transfer functions for *force transmissibility* and *motion transmissibility*. We will see that these two transmissibility functions are identical for a given mechanical system and suspension.

In this chapter we will analyze the following modeling techniques for response analysis and design of a mechatronic system:

1. *State models*, using state variables representing the state of the system in terms of system variables, such as position and velocity of lumped masses, force and displacement in springs, current through an inductor, and voltage across a capacitor. These are time-domain models, with the independent variable t (time).

2. *Linear graphs*—a model using a graphic representation. This is particularly useful as a tool in developing a state model. The linear graph uses *through variables* (e.g., forces or currents) and *across variables* (e.g., velocities or voltages) for each branch (path of energy flow) in the model.

3. *Bond graphs*—another graphical model (like the linear graph), but using branches called bonds to represent power flow. The bond graph uses *flow variables* (e.g., velocities or currents) and *effort variables* (e.g., forces or voltages). A state model can be developed from the bond-graph representation as well.

4. *Transfer-function models*—a very common model type. Uses output/input ratio in the Laplace transform form (i.e., in the "s-domain"). Here the Laplace variable s is the independent variable.

5. *Frequency-domain models*—a special case of (4) above. Here we use the Fourier transform instead of the Laplace transform. Simply stated, $s = j\omega$ in the frequency domain. Here, frequency ω is the independent variable.

2.1.2 Model Development

Development of a suitable analytical model for a large and complex system requires a systematic approach. Tools are available to aid this process. The process of modeling can be made simple by following a systematic sequence of steps. The main steps are summarized below:

1. Identify the system of interest by defining its *purpose* and the system *boundary*.

2. Identify or specify the *variables* of interest. These include inputs (forcing functions or excitations) and outputs (response).

3. Approximate (or model) various segments (components or processes or phenomena) in the system by *ideal elements*, which are suitably interconnected.

4. Draw a *free-body diagram* for the system with isolated/separated elements, as appropriate.

5. Write *constitutive equations* (physical laws) for the elements.

6. Write *continuity* (or conservation) equations for through variables (equilibrium of forces at joints; current balance at nodes, fluid flow balance, etc.)

7. Write *compatibility* equations for across (potential or path) variables. These are loop equations for velocities (geometric connectivity), voltage (potential balance), pressure drop, etc.

8. Eliminate *auxiliary* variables that are redundant and not needed to define the model.

9. Express system *boundary conditions* and response *initial conditions* using system variables.

These steps should be self-explanatory, and should be integral with the particular modeling technique that is used.

2.1.2.1 *Lumped Model of a Distributed System*

There are two broad categories of models for dynamic systems: lumped-parameter models and continuous-parameter models. In a lumped-parameter model, various characteristics of the system are lumped into representative elements located at a discrete set of points in a geometric space. The corresponding analytical models are ordinary differential equations. Most physical systems have distributed-parameter (or continuous) components, which need spatial coordinates (e.g., x, y, z) for their representation. These dynamic systems have time (t) and space coordinates as the independent variables. The corresponding analytical models are partial differential equations. For analytical convenience, we may attempt to approximate such distributed-parameter models into lumped-parameter ones. Lumped-parameter models are more commonly employed than continuous-parameter models, but continuous-parameter elements sometimes are included in otherwise lumped-parameter models in order to improve the model accuracy. Let us address some pertinent issues by considering the case of a heavy spring.

A coil spring has a mass, an elastic (spring) effect, and an energy-dissipation characteristic, each of which is distributed over the entire coil. In an analytical model, however, these individual distributed characteristics can be approximated by a separate mass element, a spring element, and a damper element, which are interconnected in some parallel-series configuration, thereby producing a lumped-parameter model. Since a heavy spring has its mass continuously distributed throughout its body, it has an infinite number of degrees of freedom. A single coordinate cannot represent its motion. But, for many practical purposes, a lumped-parameter approximation with just one lumped mass to represent the inertial characteristics of the spring, would be sufficient. Such an approximation may be obtained by using one of several approaches. One is the energy approach. Another approach is equivalence of natural frequency. Let us consider the energy approach first. Here we represent the spring by a lumped-parameter "model" such that the original spring and the model have the same net kinetic energy and same potential energy. This energy equivalence is used in deriving a lumped mass parameter for the model. Even though damping (energy dissipation) is neglected in the present analysis, it is not difficult to incorporate that as well in the model.

2.1.2.2 *Kinetic Energy Equivalence*

Consider the uniform, heavy spring shown in Figure 2.1, with one end fixed and the other end moving at velocity v. Note that:

m_s = mass of spring

k = stiffness of spring

l = length of spring

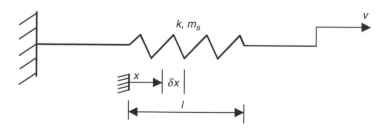

FIGURE 2.1
A uniform heavy spring.

Local speed of element δx of the spring is given by $\frac{x}{l} v$. Element mass $= \frac{m_s}{l} \delta x$. Hence, element kinetic energy KE $= \frac{1}{2} \frac{m_s}{l} \delta x (\frac{x}{l} v)^2$. In the limit, we have $\delta x \to dx$. Then,

$$\text{Total KE} = \int_0^l \frac{1}{2} \frac{m_s}{l} dx \left(\frac{x}{l} v\right)^2 = \frac{1}{2} \frac{m_s v^2}{l^3} \int_0^l x^2 \, dx = \frac{1}{2} \frac{m_s v^2}{3}$$

Hence

Equivalent lumped mass concentrated at the free end $= \frac{1}{3} \times$ spring mass.   (2.1)

NOTE   This derivation assumes that one end of the spring is fixed and, furthermore, the conditions are uniform along the spring.

An example of utilizing this result is shown in Figure 2.2. Here a system with a heavy spring and a lumped mass, is approximated by a light spring (having the same stiffness) and a lumped mass.

2.1.2.3   *Natural Frequency Equivalence*

Now consider the approach of natural frequency equivalence. Here we derive an equivalent lumped-parameter model by equating the fundamental (lowest) natural frequency of the distributed-parameter system to the natural frequency of the lumped-parameter model (in the one-degree-of-freedom case). The method can be easily extended to multidegree-of-freedom lumped parameter models as well. We will illustrate our approach by using an example.

A heavy spring of mass m_s and stiffness k_s with one end fixed and the other end attached to a sliding mass m, is shown in Figure 2.3(a). If the mass m is sufficiently larger than m_s, then at relatively high frequencies the mass will virtually stand still. Under these conditions we have the configuration shown in Figure 2.3(b), where the two ends of the spring are fixed. Also, approximate the distributed mass by an equivalent mass m_e at the mid

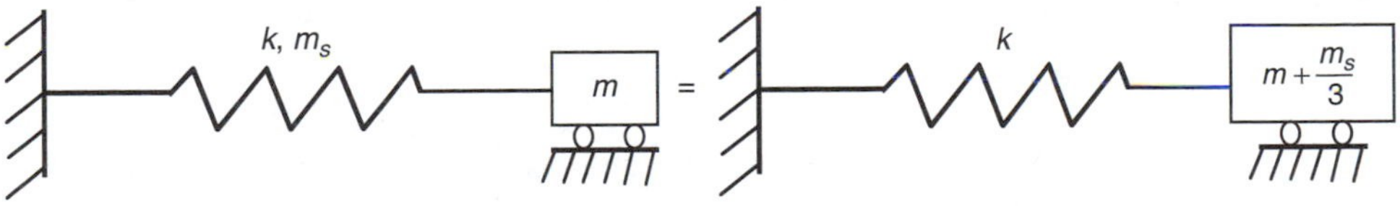

FIGURE 2.2
Lumped-parameter approximation for an oscillator with heavy spring.

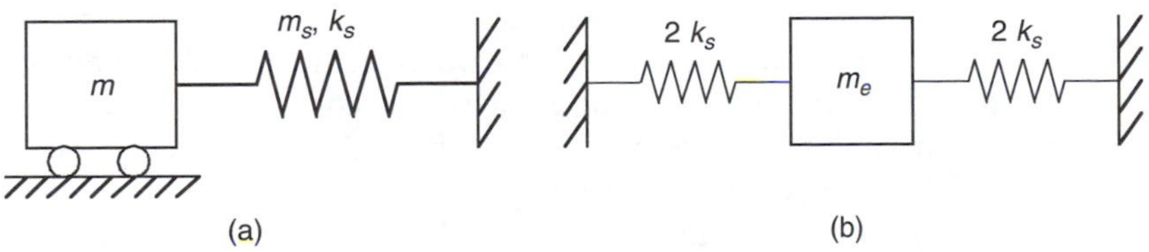

FIGURE 2.3
(a) A lumped mass with a distributed-parameter system, (b) A lumped-parameter model of the system.

point of the spring: each spring segment has double the stiffness of the original spring. Hence the overall stiffness is $4k_s$. The natural frequency of the lumped-model is

$$\omega_e = \sqrt{\frac{4k_s}{m_e}} \tag{2.2}$$

It is known from a complete analysis of a heavy spring that the natural frequency for the fixed-fixed configuration is

$$\omega_s = \pi n \sqrt{\frac{k_s}{m_s}} \tag{2.3}$$

where n is the mode number. Then, for the fundamental (first) mode (i.e., $n = 1$), the natural frequency equivalence gives

$$\sqrt{\frac{4k_s}{m_e}} = \pi \sqrt{\frac{k_s}{m_s}}$$

or,

$$m_e = \frac{4}{\pi^2} m_s \approx 0.4 m_s \tag{2.4}$$

Note that since the effect of inertia decreases with frequency, it is not necessary to consider the case of low frequencies.

The natural frequency equivalence may be generalized as an *eigenvalue* equivalence (*pole equivalence*) for any dynamic system. In this case, the eigenvalues of the lumped parameter model are equated to the corresponding eignevalues of the distributed-parameter system, and the model parameters are determined accordingly.

2.2   Analogies

A system may possess various physical characteristics incorporating, for example, mechanical, electrical, thermal, and fluid components. The procedure of model development will be facilitated if we understand the similarities of the characteristics of different types of components. This issue is addressed in the present section.

Analogies exist among mechanical, electrical, hydraulic, and thermal systems. The basic system elements can be divided into two groups: energy-storage elements and energy-dissipation elements. Table 2.1 gives the linear relationships, which describe the behavior of translatory-mechanical, electrical, thermal, and fluid elements. These relationships are known as constitutive relations. In particular, Newton's second law is considered the constitutive relation for a mass element. The analogy used in Table 2.1 between mechanical and electrical elements is known as the force-current analogy. This analogy appears more logical than a force-voltage analogy, as is clear from Table 2.2. This follows from the fact that both force and current are *through variables*, which are analogous to fluid flow through a pipe, and furthermore, both velocity and voltage are *across variables*, which vary across the flow direction, as in the case of fluid pressure along a pipe. The correspondence between the

TABLE 2.1

Some Linear Constitutive Relations

	Constitutive Relation for		
	Energy Storage Elements		**Energy Dissipating Elements**
System Type	**A-type (Across) Element**	**T-type (Through) Element**	**D-type (Dissipative) Element**
Translatory-Mechanical	*Mass*	*Spring*	*Viscous Damper*
v = velocity f = force	$m\dfrac{dv}{dt} = f$ (Newton's second law) m = mass	$\dfrac{df}{dt} = kv$ (Hooke's law) k = stiffness	$f = bv$ b = damping constant
Electrical	*Capacitor*	*Inductor*	*Resistor*
v = voltage i = current	$C\dfrac{dv}{dt} = i$ C = capacitance	$L\dfrac{di}{dt} = v$ L = inductance	$Ri = v$ R = resistance
Thermal	*Thermal Capacitor*	*None*	*Thermal Resistor*
T = temperature difference Q = heat transfer rate	$C_t\dfrac{dT}{dt} = Q$ C_t = thermal capacitance		$R_t Q = T$ R_t = thermal resistance
Fluid	*Fluid Capacitor*	*Fluid Inertor*	*Fluid Resistor*
P = pressure difference Q = volume flow rate	$C_f\dfrac{dP}{dt} = Q$ C_f = fluid capacitance	$I_f\dfrac{dQ}{dt} = P$ I_f = inertance	$R_f Q = P$ R_f = fluid resistance

TABLE 2.2

Force-Current Analogy

System type	Mechanical	Electrical
System-response variables:		
Through-variables	Force f	Current i
Across-variables	Velocity v	Voltage v
System parameters	m	C
	k	$1/L$
	b	$1/R$

parameter pairs given in Table 2.2 follows from the relations in Table 2.1. A rotational (rotatory) mechanical element possesses constitutive relations between torque and angular velocity, which can be treated as a generalized force and a generalized velocity, respectively (compare this with a rectilinear or translatory mechanical element as listed in Table 2.1). In fluid systems as well, basic elements corresponding to capacitance (capacity), inductance (fluid inertia), and resistance (fluid friction) exist. Constitutive relations between pressure difference and mass flow rate can be written for these elements. In thermal systems, generally only two elements—capacitance and resistance—can be identified. In this case constitutive relations exist between temperature difference and heat transfer rate.

Proper selection of system variables is crucial in developing an analytical model for a dynamic system. A general approach that may be adopted is to use across variables of the *A*-type (or, across-type) energy storage elements and the through variables of the *T*-type (or, through-type) energy storage element as system variables (state variables). Note that if any two elements are not independent (e.g., if two spring elements are directly connected in series or parallel) then only a single state variable should be used to represent both elements. Independent variables are not needed for *D*-type (dissipative) elements because their response can be represented in terms of the state variables of the energy storage elements (*A*-type and *T*-type). State-space models and associated variables will be discussed in more detail in a later section. Now we will discuss various types of physical elements and their analogies.

2.2.1   Mechanical Elements

For mechanical elements we use the velocity (across variable) of each independent mass (*A*-type element) and the force (through variable) of each independent spring (*T*-type element) as the system variables (*state variables*). The corresponding constitutive equations form the "shell" for an analytical model. These equations will directly lead to a *state-space model* of the system, as we will illustrate in subsequent sections.

2.2.1.1   Mass (Inertia) Element

The constitutive equation (Newton's second law) is

$$m \frac{dv}{dt} = f \tag{2.5}$$

Since power $= fv =$ rate of change of energy, by substituting Equation 2.5, the energy of the element may be expressed as

$$E = \int fv \, dt = \int m \frac{dv}{dt} v \, dt = \int mv \, dv$$

or,

$$\text{Energy } E = \frac{1}{2} mv^2 \tag{2.6}$$

This is the well-known *kinetic energy.* Now by integrating Equation 2.5, we have

$$v(t) = v(0^-) + \frac{1}{m} \int_{0^-}^{t} f \, dt \tag{2.7}$$

By setting $t = 0^+$ in Equation 2.7, we see that

$$v(0^+) = v(0^-) \tag{2.8}$$

unless an infinite force is applied to the mass element. Note that 0^- denotes the instant just before $t = 0$ and 0^+ denotes the instant just after $t = 0$. In view of these observations, we may state the following:

1. Velocity can represent the state of an inertia element. This is justified first because, from Equation 2.7, the velocity at any time t can be completely determined with the knowledge of the initial velocity and the applied force, and because, from Equation 2.6, the energy of an inertia element can be represented in terms of v alone.

2. Velocity across an inertia element cannot change instantaneously unless an infinite force/torque is applied to it.

3. A finite force cannot cause an infinite acceleration in an inertia element. A finite instantaneous change (step) in velocity will need an infinite force. Hence, v is a natural output (or state) variable and f is a natural input variable for an inertia element.

2.2.1.2 Spring (Stiffness) Element

The constitutive equation (Hooke's law) is

$$\frac{df}{dt} = kv \tag{2.9}$$

Note that we have differentiated the familiar force-deflection Hooke's law, in order to be consistent with the response/state variable (velocity) that is used for the inertia element.

Now following the same steps as for the inertia element, the energy of a spring element may be expressed as

$$E = \int fv \, dt = \int f \frac{1}{k} \frac{df}{dt} dt = \int \frac{1}{k} f \, df$$

or,

$$\text{Energy} \quad E = \frac{1}{2} \frac{f^2}{k} \tag{2.10}$$

This is the well-known (elastic) *potential energy.*

Also,

$$f(t) = f(0^-) + k \int_{0^-}^{t} v \, dt \tag{2.11}$$

and

$$f(0^+) = f(0^-) \tag{2.12}$$

unless an infinite velocity is applied to the spring element. In summary, we have

1. Force can represent the state of a stiffness (spring) element. This is justified because the force of a spring at any general time t may be completely determined with the knowledge of the initial force and the applied velocity, and also because the energy of a spring element can be represented in terms of f alone.

2. Force through a stiffness element cannot change instantaneously unless an infinite velocity is applied to it.

3. Force f is a natural output (state) variable and v is a natural input variable for a stiffness element.

2.2.2 Electrical Elements

Here we use the voltage (across variable) of each independent capacitor (*A*-type element) and the current (through variable) of each independent inductor (*T*-type element) as system (state) variables.

2.2.2.1 *Capacitor Element*

The constitutive equation is:

$$C\frac{dv}{dt} = i \tag{2.13}$$

Since power is given by iv, by substituting Equation 2.13, the energy in a capacitor may be expressed as

$$E = \int iv \, dt = \int C\frac{dv}{dt} v \, dt = \int Cv \, dv$$

or,

$$\text{Energy} \quad E = \frac{1}{2}C \, v^2 \tag{2.14}$$

This is the *electrostatic energy* of a capacitor.
  Also,

$$v(t) = v(0^-) + \frac{1}{C}\int_{0^-}^{t} i \, dt \tag{2.15}$$

Hence, for a capacitor,

$$v(0^+) = v(0^-) \tag{2.16}$$

unless an infinite current is applied to a capacitor. We summarize:

1. Voltage is an appropriate response variable (or state variable) for a capacitor element.

2. Voltage across a capacitor cannot change instantaneously unless an infinite current is applied.

3. Voltage is a natural output variable and current is a natural input variable for a capacitor.

2.2.2.2   *Inductor Element*

The constitutive equation is

$$L\frac{di}{dt} = v \qquad (2.17)$$

Energy

$$E = \frac{1}{2}Li^2 \qquad (2.18)$$

This is the *electromagnetic energy* of an inductor.

Also,

$$i(t) = i(0^-) + \frac{1}{L}\int_{0^-}^{t} v\,dt \qquad (2.19)$$

Hence, for an inductor,

$$i(0^+) = i(0^-) \qquad (2.20)$$

unless an infinite voltage is applied. We summarize:

1. Current is an appropriate response variable (or state variable) for an inductor.
2. Current through an inductor cannot change instantaneously unless an infinite voltage is applied.
3. Current is a natural output variable and voltage is a natural input variable for an inductor.

2.2.3   Thermal Elements

Here the across variable is temperature (T) and the through variable is the heat transfer rate (Q). The thermal capacitor is an *A*-type element. There is no *T*-type element in a thermal system. The reason is clear. There is only one type of energy (thermal energy) in a thermal system, whereas there are two types of energy in mechanical and electrical systems.

2.2.3.1   *Thermal Capacitor*

Consider a thermal volume v of fluid with, density ρ, and specific heat c. Then, for a net heat transfer rate Q into the control volume we have

$$Q = \rho v c \frac{dT}{dt} \qquad (2.21)$$

or,

$$C_t \frac{dT}{dt} = Q \qquad (2.22)$$

where, $C_t = \rho v c$ is the thermal capacitance of the control volume.

2.2.3.2 Thermal Resistance

There are three basic processes of heat transfer:

1. Conduction
2. Convection
3. Radiation

There is a thermal resistance associated with each process, given by its constitutive relation, as indicated below.

Conduction: $$Q = \frac{kA}{\Delta x} T \qquad (2.23)$$

where
 k = conductivity
 A = area of cross section of the heat conduction element
 Δx = length of heat conduction that has a temperature drop of T

The conductive resistance

$$R_k = \frac{\Delta x}{kA} \qquad (2.24)$$

Convection: $$Q = h_c A T \qquad (2.25)$$

where
 h_c = convection heat transfer coefficient
 A = area of heat convection surface with a temperature drop of T

The conductive resistance

$$R_c = \frac{1}{h_c A} \qquad (2.26)$$

Radiation: $$Q = \sigma F_E F_A \left(T_1^4 - T_2^4 \right) \qquad (2.27)$$

where
 σ = Stefan-Boltzman constant
 F_E = effective emmisivity of the radiation source (of temperature T_1)
 F_A = shape factor of the radiation receiver (of temperature T_2)
 A = effective surface area of the receiver

This corresponds to a nonlinear thermal resistor.

2.2.4 Fluid Elements

Here we use pressure (across variable) of each independent fluid capacitor (*A*-type element) and volume flow rate (through variable) of each independent fluid inertor (*T*-type element) as system (state) variables.

2.2.4.1   Fluid Capacitor

We have,

$$C_f \frac{dP}{dt} = Q \tag{2.28}$$

Note that a fluid capacitor stores potential energy (a "fluid spring") unlike the mechanical *A*-type element (inertia), which stores kinetic energy.

For a liquid control volume *V* of bulk modulus β we have the fluid capacitance

$$C_{bulk} = \frac{V}{\beta} \tag{2.29}$$

For an isothermal (constant temperature, slow-process) gas of volume *V* and pressure *P* we have the fluid capacitance

$$C_{comp} = \frac{V}{P} \tag{2.30}$$

For an adiabatic (zero heat transfer, fast-process) gas we have the capacitance

$$C_{comp} = \frac{V}{kP} \tag{2.31}$$

where

$$k = \frac{c_p}{c_v} \tag{2.32}$$

which is the ratio of specific heats at constant pressure and constant volume.

For an incompressible fluid contained in a flexible vessel of area *A* and stiffness *k*, we have the capacitance

$$C_{elastic} = \frac{A^2}{k} \tag{2.33}$$

NOTE   For a fluid with bulk modulus, the equivalent capacitance would be

$$C_{bulk} + C_{elastic}$$

For an incompressible fluid column of an area of cross-section *A* and density ρ, we have the capacitance

$$C_{grav} = \frac{A}{\rho g} \tag{2.34}$$

2.2.4.2 Fluid Inertor

We have

$$I_f \frac{dQ}{dt} = P \tag{2.35}$$

This is a T-type element. But, it stores kinetic energy, unlike the mechanical T-type element (spring), which stores potential energy. For a flow with uniform velocity distribution across an area A and over a length segment Δx we have the fluid inertance

$$I_f = \rho \frac{\Delta x}{A} \tag{2.36}$$

For a nonuniform velocity distribution, we have

$$I_f = \alpha \rho \frac{\Delta x}{A} \tag{2.37}$$

where a correction factor α has been introduced. For a flow of circular cross-section with a parabolic velocity distribution, we use $\alpha = 2.0$.

2.2.4.3 Fluid Resistance

In the approximate, linear case we have

$$P = R_f Q \tag{2.38}$$

The more general, nonlinear case is given by

$$P = K_R Q^n \tag{2.39}$$

where K_R and n are parameters of nonlinearity. For viscous flow through a uniform pipe we have, for a circular cross-section of diameter d:

$$R_f = 128 \, \mu \frac{\Delta x}{\pi d^4} \tag{2.40}$$

and for a rectangular cross-section of height b which is much smaller than its width w:

$$R_f = 12\mu \frac{\Delta x}{wb^3} \tag{2.41}$$

Also, μ is the absolute viscosity (or, dynamic viscosity) of the fluid, and is related to the kinematic viscosity v through

$$\mu = v\rho \tag{2.42}$$

2.2.5 Natural Oscillations

Mechanical systems can produce natural (free) oscillatory responses (or, free vibrations) because they can possess two types of energy (kinetic and potential). When one type of stored energy is converted into the other type, repeatedly back and forth, the resulting response is oscillatory. Of course, some of the energy will dissipate (through the dissipative mechanism of a damper) and the free natural oscillations will decay as a result. Similarly, electrical circuits and fluid systems can exhibit free, natural oscillatory responses due to the presence of two types of energy storage mechanism, where energy can "flow" back and forth repeatedly between the two types of elements. But, thermal systems have only one type of energy storage element (*A*-type) with only one type of energy (thermal energy). Hence, purely thermal systems cannot naturally produce oscillatory responses unless forced by external means, or integrated with other types of systems (e.g., fluid systems).

2.3 State-Space Representation

More than one variable might be needed to represent the response of a dynamic system. There also could be more than one input variable in a system. Then we have a *multi-variable* system. A time-domain analytical model is a set of differential equations relating the response variables to the input variables. This set of system equations generally is coupled, so that more than one response variable appears in each differential equation. A particularly useful time-domain representation for a dynamic system is a state-space model. State equations define the dynamic state of a system. In the state–space representation, an *n*th-order system is represented by *n* first-order differential equations, which generally are coupled. An entire set of state equations is reduced to one vector-matrix State Equation.

2.3.1 State Space

The word "state" refers to the dynamic status or condition of a system. A complete description of the state will require all the variables that are associated with the time-evolution of the system response (both "magnitude" and "direction" of the response trajectory with respect to time). The state is a *vector*, which traces out a trajectory in the *state space*. The analytical development requires a definition of the "state space." A second-order system requires a two-dimensional or plane space, a third-order system requires a three-dimensional space, and so on.

2.3.1.1 State Equations

A common form of state equations for an *n*th-order linear, unforced (free, no input) system is

$$\dot{x}_1 = a_{11}x_1 + a_{12}x_2 + a_{13}x_3 + \cdots + a_{1n}x_n$$

$$\dot{x}_2 = a_{21}x_1 + a_{22}x_2 + a_{23}x_3 + \cdots + a_{2n}x_n$$

$$\vdots$$
$$\vdots$$

$$\dot{x}_n = a_{n1}x_1 + a_{n2}x_2 + a_{n3}x_3 + \cdots + a_{nn}x_n$$

(2.43)

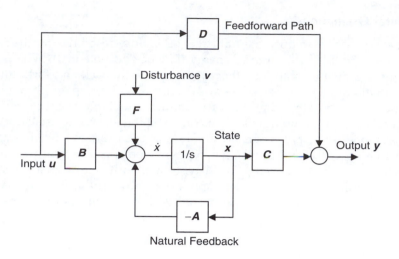

FIGURE 2.4
A multivariable linear control system.

where $\dot{x} = dx/dt$; and x_1, x_2, etc. are the *state variables*. Equation 2.43 simply says that a change in any of the n variables of the system may affect the rate of change of any given variable. This set of equations can be rewritten in the vector-matrix form as

$$\dot{x} = Ax \tag{2.44a}$$

The bold-type upper-case letter indicates that the variable is a *matrix*; a bold-type lower-case letter indicates a *vector*, typically a column vector. Specifically,

$$x = \begin{bmatrix} x_1 \\ x_2 \\ \vdots \\ x_n \end{bmatrix}, \quad \dot{x} = \begin{bmatrix} \dot{x}_1 \\ \dot{x}_2 \\ \vdots \\ \dot{x}_n \end{bmatrix}, \quad A = \begin{bmatrix} a_{11} & a_{12} & \cdots & a_{1n} \\ a_{21} & a_{22} & \cdots & a_{2n} \\ \vdots & \vdots & & \vdots \\ a_{n1} & a_{n2} & \cdots & a_{nn} \end{bmatrix}$$

A generalized linear control system with inputs (i.e., under forced conditions), is shown by the block diagram in Figure 2.4. The state space representation of this system is

$$\dot{x} = Ax + Bu + Fv \tag{2.44b}$$

$$y = Cx + Du \tag{2.45}$$

where
$x = [x_1 \quad x_2 \quad \ldots \quad x_n]^T$ = state vector (nth order)
$u = [u_1 \quad u_2 \quad \ldots \quad u_r]^T$ = input vector (rth order)
$y = [y_1 \quad y_2 \quad \ldots \quad y_m]^T$ = output vector (mth order)
v = disturbance input vector
A = system matrix ($n \times n$)
B = input distribution matrix ($n \times r$)
C = output (or measurement) gain matrix ($m \times n$)
D = feedforward gain matrix ($m \times r$)
F = disturbance input distribution matrix

Note that [ ]^T denotes the transpose of a matrix or vector. You may verify that the block diagram of this system is shown in Figure 2.4.

In Equation 2.44a the disturbance term Fv may be dropped since it can be absorbed into the regular input term Bu. Then we have

$$\dot{x} = Ax + Bu \tag{2.46}$$

The system matrix A tells us how the system responds naturally without any external input, and B tells us how the input u is amplified when reaching the system.

2.3.2 State Models

A state vector x is a column vector, which contains a minimum set of state variables (x_1, $x_2, \ldots, x_n$) which completely determine the state of the dynamic system. The number of states variables (n), is the *order* of the system.

Property 1
The state vector $x(t_0)$ at time t_0 and the input (forcing excitation) $u[t_0, t_1]$ over the time interval $[t_0, t_1]$, will uniquely determine the state vector $x(t_1)$ any future time t_1. In other words, a transformation g can be defined such that

$$x(t_1) = g(t_0, t_1, x(t_0), u[t_0, t_1]) \tag{2.47}$$

Note that by the *causality* property of a dynamic system, future states can be determined if all inputs up to that future time are known. This means that the transformation g is *nonanticipative* (i.e., inputs beyond t_1 are not needed to determine $x(t_1)$. Each forcing function $u[t_0, t_1]$ defines a state trajectory. As mentioned before, the n-dimensional vector space formed by all possible state trajectories is the *state space*.

Property 2
The state $x(t_1)$ and the input $u(t_1)$ at any time t_1 will uniquely determine the system output or response vector $y(t_1)$ at that time. This can be expressed as

$$y(t_1) = h(t_1, x(t_1), u(t_1)) \tag{2.48}$$

This says that the system response (output) at time t_1 depends on the time, the input, and the state vector.

The transformation h has no *memory*—the response at a previous time cannot be determined through the knowledge of the present state and input. Note also that, in general, system outputs (y) are not identical to the states (x) even though the former can be uniquely determined by the latter.

A state model consists of a set of n first-order ordinary differential equations (time-domain) that are coupled (inter-related). In vector form, this is expressed as

$$\dot{x} = f(x, u, t) \tag{2.49}$$

$$y = h(x, u, t) \tag{2.50}$$

Equation 2.49 represents the *n state* (differential) *equations* and Equation 2.50 represents the *algebraic output equations*. If f is a nonlinear vector function, then the state model is nonlinear.

2.3.3   Input-Output Models

Suppose that the Equations 2.50 are substituted into Equation 2.49 to eliminate x and $\dot{x}$, and a set of differential equations for y are obtained (with u and its derivatives present). Then we have an input-output model. If these input-output differential equations are nonlinear, then the system (or strictly, the input-output model) is nonlinear.

Example 2.1

The concepts of state, output, and order of a system, and the importance of the system's initial state, can be shown using a simple example. Consider the rectilinear motion of a particle of mass m subject to an input force $u(t)$. By Newton's second law, its position x can be expressed as the second-order differential equation:

$$m\frac{d^2x}{dt^2} = u(t) \quad \text{or} \quad m\ddot{x} - u = 0 \tag{i}$$

If the output is the position, then Equation i is indeed the input–output equation. If the output is velocity, we can define a state:

$$x_1 = \frac{dx}{dt}$$

and write the state model as

$$\dot{x}_1 = \frac{1}{m}u(t) \tag{ii}$$

with the algebraic output equation

$$y = x_1 \tag{iii}$$

The model in Equations ii through iii represents a "first-order" system with velocity as the output.

If the output is position, we will need two state variables, for example:

$$x_1 = x, \quad x_2 = \frac{dx}{dt}$$

The corresponding state model is now the two equations:

$$\dot{x}_1 = x_2$$

$$\dot{x}_2 = \frac{1}{m}u(t) \tag{iv}$$

with the algebraic output equation (now giving position):

$$y = x_1 \tag{v}$$

which represents the response of a "second-order" system.

If we consider both position and velocity as outputs, an appropriate state model would be Equation iv together with the algebraic output equations:

$$\begin{bmatrix} y_1 \\ y_2 \end{bmatrix} = \begin{bmatrix} 1 & 0 \\ 0 & 1 \end{bmatrix} \begin{bmatrix} x_1 \\ x_2 \end{bmatrix} \tag{v*}$$

It should be noted that in this example, the three variables x, $\dot{x}$, and $\ddot{x}$ do not form the system state because this is not a minimal set. In particular, $\ddot{x}$ is redundant as it is completely known from u.

Another important aspect can be observed when deriving the system response by directly integrating the system equation. When the output is velocity, just one initial condition $\dot{x}(0)$ would be adequate, whereas if the output is position, two initial conditions $x(0)$ and $\dot{x}(0)$ would be needed. In the latter case, just one initial state does not uniquely generate a state trajectory corresponding to a given forcing input. This intuitively clear fact, nevertheless, constitutes an important property of the state of a system y—the number of initial conditions needed = order of the system.

Finally, it is also important to understand the nonuniqueness of the choice of state variables. For instance, an alternative state model for the case where the output is position would be:

$$\dot{x}_1 = -2x_2$$
$$\dot{x}_2 = \frac{3}{m} u(t) \tag{vi}$$

with the output equation:

$$y = -\frac{1}{6} x_1 \tag{vii}$$

Summarizing, state vector of a dynamic system is a least (minimal) set of variables that is required to completely determine the dynamic state of the system at all instants of time. They may or may not have a physical interpretation. State vector is not unique; many choices are possible for a given system. Output (response) variables of a system can be completely determined from any such choice of state variables. Since state vector is a least set, a given state variable cannot be expressed as a linear combination of the remaining state variables in that state vector.

2.3.2.1 Time-Invariant Systems

If in Equation 2.49 and Equation 2.50, there is no explicit dependence on time in the functions f and h, the dynamic system is said to be *time-invariant, or stationary,* or *autonomous.* In this case, the system behavior is not a function of the time origin for a given initial state and input function. In particular, a linear system is time-invariant if the matrices A, B, C, and D (in Equation 2.45 and Equation 2.46) are constant.

2.3.2.2   *Principle of Superposition*

A system is linear if and only if the principle of superposition is satisfied. This principle states that, if y_1 is the output when the input is u_1, and y_2 is the output when the input is u_2, then $\alpha_1 y_1 + \alpha_2 y_2$ is the output for the input $\alpha_1 u_1 + \alpha_2 u_2$, where α_1 and α_2 are any real constants.

Example 2.2

A torsional dynamic model of a pipeline segment is shown in Figure 2.5(a). The free-body diagram in Figure 2.5(b) shows the internal torques acting at sectioned inertia junctions, for free motion. A state model is obtained using the generalized velocities (angular velocities Ω_i) of the independent inertial elements and the generalized forces (torques T_i) of the independent elastic (torsional spring) elements as state variables. A minimum set of states, which is required for a complete representation determines the system order.

There are two inertia elements and three spring elements—a total of five energy-storage elements. The three springs are not independent, however. The motion of any two springs completely determines the motion of the third. This indicates that the system is a fourth-order system. We obtain the state-space model as follows:

Newton's second law gives

$$I_1 \dot{\Omega}_1 = -T_1 + T_2$$
$$I_2 \dot{\Omega}_2 = -T_2 - T_3 \tag{i}$$

Hooke's law gives

$$\dot{T}_1 = k_1 \Omega_1$$
$$\dot{T}_2 = k_2 (\Omega_2 - \Omega_1) \tag{ii}$$

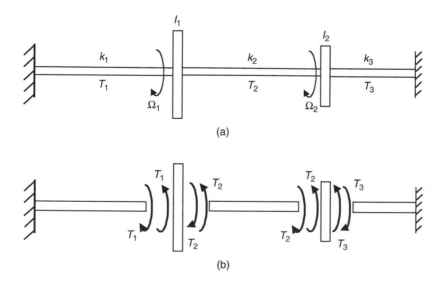

(a)

(b)

FIGURE 2.5
(a) Dynamic model of a pipeline segment, (b) Free body diagram.

Torque T_3 in Equation i is substituted in terms of T_1 and T_2, using the displacement relation (compatibility) for the inertial I_2:

$$\frac{T_1}{k_1} + \frac{T_2}{k_2} = \frac{T_3}{k_3} \tag{iii}$$

The state vector is chosen as

$$x = [\Omega_1 \quad \Omega_2 \quad T_1 \quad T_2]^T$$

The corresponding system matrix is

$$A = \begin{bmatrix} 0 & 0 & -\dfrac{1}{I_1} & \dfrac{1}{I_1} \\[2ex] 0 & 0 & -\dfrac{1}{I_2}\left(\dfrac{k_3}{k_1}\right) & -\dfrac{1}{I_2}\left(1+\dfrac{k_3}{k_2}\right) \\[2ex] k_1 & 0 & 0 & 0 \\[2ex] -k_2 & k_2 & 0 & 0 \end{bmatrix} \tag{iv}$$

The output (displacement) vector is

$$y = \left[\frac{T_1}{k_1}, \quad \frac{T_1}{k_1} + \frac{T_2}{k_2}\right]^T \tag{v}$$

which corresponds to the following output-gain matrix:

$$C = \begin{bmatrix} 0 & 0 & \dfrac{1}{k_1} & 0 \\[2ex] 0 & 0 & \dfrac{1}{k_1} & \dfrac{1}{k_2} \end{bmatrix} \tag{vi}$$

Example 2.3

The rigid output shaft of a diesel engine prime mover is running at known angular velocity $\Omega(t)$. It is connected through a friction clutch to a flexible shaft, which in turn drives a hydraulic pump (see Figure 2.6(a)).

A linear model for this system is shown schematically in Figure 2.6(b). The clutch is represented by a viscous rotatory damper of damping constant B_1 (units: torque/angular velocity). The stiffness of the flexible shaft is K (units: torque/rotation). The pump is represented by a wheel of moment of inertia J (units: torque/angular acceleration) and viscous damping constant B_2.

a. Write down the two state equations relating the state variables T and ω to the input Ω, where T is the torque in flexible shaft and ω is the pump speed.

   HINTS

   1. Free body diagram for the shaft is given in Figure 2.6(c), where ω_1 is the angular speed at the left end of the shaft.
   2. Write down the "torque balance" and "constitutive" relations for the shaft, and eliminate ω_1.

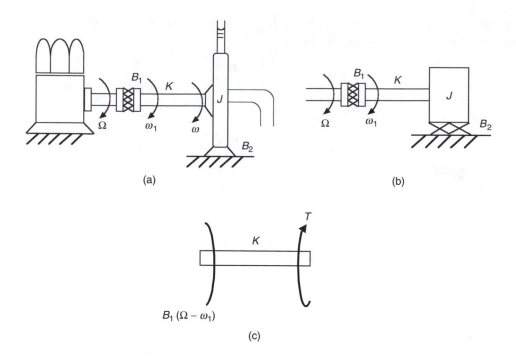

FIGURE 2.6
(a) Diesel engine; (b) Linear model; (c) Free body diagram of the shaft.

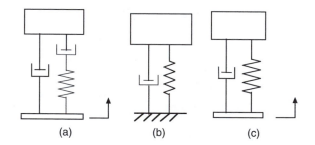

FIGURE 2.7
Three translatory mechanical systems.

      3. Draw the free body diagram for the wheel *J* and use D'Alembert's principle.

  b. Express the state equations in the vector-matrix form.

  c. Which one of the translatory systems given in Figure 2.7 is the system in Figure 2.6(b) analogous to?

SOLUTION
  a. For *K*:

      Constitutive relation

$$\frac{dT}{dt} = K(\omega_1 - \omega) \tag{i}$$

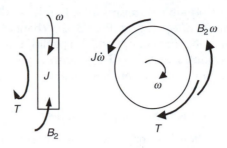

FIGURE 2.8
Free body diagram for the wheel.

Torque balance

$$B_1(\Omega - \omega_1) - T = 0 \qquad \text{(ii)}$$

Substitute Equation ii into Equation i:

$$\frac{dT}{dt} = -\frac{K}{B_1}T - K\omega + K\Omega \qquad \text{(iii)}$$

This is one state equation.
For J (see Figure 2.8):
D'Alembert's principle gives

$$J\dot{\omega} + B_2\omega - T = 0$$

that is,
$$\frac{d\omega}{dt} = -\frac{B_2}{J}\omega + \frac{1}{J}T \qquad \text{(iv)}$$

This is the second state equation.

b. Vector-matrix form of the state-space model:

$$\begin{bmatrix} \dfrac{dT}{dt} \\ \dfrac{d\omega}{dt} \end{bmatrix} = \begin{bmatrix} \dfrac{-K}{B_1} & -K \\ \dfrac{1}{J} & -\dfrac{B_2}{J} \end{bmatrix}\begin{bmatrix} T \\ \omega \end{bmatrix} + \begin{bmatrix} K \\ 0 \end{bmatrix}\Omega$$

c. The translatory system in Figure 2.7(a) is analogous to the given rotatory system.

2.4 Model Linearization

Real systems are nonlinear and they are represented by nonlinear analytical models consisting of nonlinear differential equations (see Equation 2.49 and Equation 2.50). Linear systems (models) are in fact idealized representations, and are represented by linear differential equations (see Equation 2.44 and Equation 2.45). Clearly, it is far more convenient to analyze linear systems. For this reason, nonlinear systems are often approximated by linear models.

It is not possible to represent a highly nonlinear system by a single linear model in its entire range of operation. For small "changes" in the system response, however, a linear model may be developed, which is valid in the neighborhood of an operating point of the system about which small response changes take place. In this section we will study linearization of nonlinear models about an operating point.

2.4.1 Nonlinear State-Space Models

Consider a general nonlinear, time-variant, nth-order system represented by n first-order differential equations, which generally are coupled, as given by

$$\frac{dq_1}{dt} = f_1(q_1, q_2, \ldots, q_n, r_1, r_2, \ldots, r_m, t)$$

$$\frac{dq_2}{dt} = f_2(q_1, q_2, \ldots, q_n, r_1, r_2, \ldots, r_m, t)$$

$$\vdots \tag{2.51}$$

$$\frac{dq_n}{dt} = f_n(q_1, q_2, \ldots, q_n, r_1, r_2, \ldots, r_m, t)$$

The state vector is

$$q = [q_1, q_2, \ldots, q_n]^T \tag{2.52}$$

and the input vector is

$$r = [r_1, r_2, \ldots, r_m]^T \tag{2.53}$$

Equation 2.51 may be written in the vector notation,

$$\dot{q} = f(q, r, t) \tag{2.54a}$$

2.4.2 Linearization

Equilibrium states of the dynamic system given by Equation 2.51, correspond to

$$\dot{q} = 0 \tag{2.55}$$

This is true because in equilibrium (i.e., at an operating point) the system response remains steady and hence its rate of change is zero. Consequently, the equilibrium states $\bar{q}$ are obtained by solving the set of n algebraic equations

$$f(q, r, t) = 0 \tag{2.56}$$

for a particular steady input $\bar{r}$. Usually a system operates in the neighborhood of one of its equilibrium states. This state is known as its *operating point*. The steady state of a dynamic system is also an *equilibrium state*.

Suppose that a slight excitation is given to a dynamic system that is operating at an equilibrium state. If the system response builds up and deviates further from the equilibrium state, the equilibrium state is said to be *unstable*. If the system returns to the original operating point, the equilibrium state is stable. If it remains at the new state without either returning to the equilibrium state or building up the response, the equilibrium state is said to be *neutral*.

To study the stability of various equilibrium states of a nonlinear dynamic system, it is first necessary to linearize the system model about these equilibrium states. Linear models are also useful in analyzing nonlinear systems when it is known that the variations of the system response about the system operating point are small in comparison to the maximum allowable variation (dynamic range). Equation 2.56 can be linearized for small variations δq and δr about an equilibrium point $(\bar{q}, \bar{r})$ by employing up to only the first derivative term in the Taylor series expansion of the nonlinear function f. The higher-order terms are negligible for small δq and δr. This method yields

$$\delta \dot{q} = \frac{\partial f}{\partial q}(\bar{q}, \bar{r}, t)\delta q + \frac{\partial f}{\partial r}(\bar{q}, \bar{r}, t)\delta r \tag{2.57}$$

The state vector and the input vector for the linearized system are denoted by

$$\delta q = x = [x_1, x_2, \ldots, x_n]^T \tag{2.58}$$

$$\delta r = u = [u_1, u_2, \ldots, u_m]^T \tag{2.59}$$

The linear system matrix $A(t)$ and the input gain matrix $B(t)$ are given by

$$A(t) = \frac{\partial f}{\partial q}(\bar{q}, \bar{r}, t) \tag{2.60}$$

$$B(t) = \frac{\partial f}{\partial r}(\bar{q}, \bar{r}, t) \tag{2.61}$$

This gives the linear model 2.44. If the dynamic system is a constant-parameter system, or if it can be assumed as such for the time period of interest, then A and B become constant matrices.

2.4.3 Illustrative Examples

Now we will illustrate model linearization and operating point analysis using several examples.

Example 2.4

The robotic spray-painting system of an automobile assembly plant employs an induction motor and pump combination to supply paint at an overall peak rate of 15 gal/min to a cluster of spray-paint heads in several painting booths. The painting booths are an integral part of the production line in the plant. The pumping and filtering stations are in the ground level of the building and the painting booths are in an upper level. Not all booths or painting heads operate at a given time. The pressure in the paint supply line is maintained at a desired

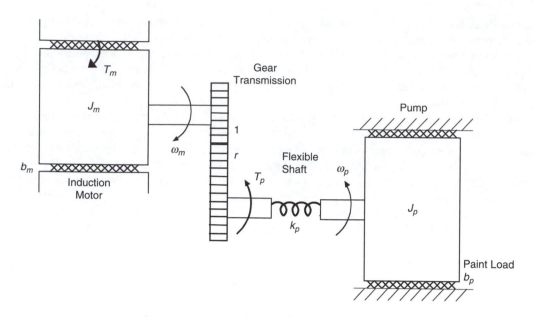

FIGURE 2.9
A model for a paint pumping system in an automobile assembly plant.

level (approximately 275 psi) by controlling the speed of the pump, which is achieved through a combination of voltage control and frequency control of the induction motor. An approximate model for the paint pumping system is shown in Figure 2.9.

The induction motor is linked to the pump through a gear transmission of efficiency η and speed ratio $1{:}r$ and a flexible shaft of torsional stiffness k_p. The moments of inertia of the motor rotor and the pump impeller are denoted by J_m and J_p, respectively. The gear inertia is neglected (or lumped with J_m). The mechanical dissipation in the motor and its bearings is modeled as linear viscous damping of damping constant b_m. The load on the pump (the paint load plus any mechanical dissipation) is also modeled as viscous damping, and the equivalent damping constant is b_p. The magnetic torque T_m generated by the induction motor is given by

$$T_m = \frac{T_0 q \omega_0 (\omega_0 - \omega_m)}{\left(q \omega_0^2 - \omega_m^2\right)} \qquad (2.62)$$

in which ω_m is the motor speed. The parameter T_0 depends directly (quadratically) on the phase voltage supplied to the motor. The second parameter ω_0 is directly proportional to the line frequency of the ac supply. The third parameter q is positive and greater than unity, and this parameter is assumed constant in the control system.

a. Comment about the accuracy of the model shown in Figure 2.9.
b. Taking the motor speed ω_m, the pump-shaft torque T_p, and the pump speed ω_p as the state variables, systematically derive the three state equations for this (nonlinear) model. Clearly explain all steps involved in the derivation. What are the inputs to the system?
c. What do the motor parameters ω_0 and T_0 represent, with regard to motor behavior? Obtain the partial derivatives $\frac{\partial T_m}{\partial \omega_m}$, $\frac{\partial T_m}{\partial T_0}$, and $\frac{\partial T_m}{\partial \omega_0}$ and verify that the first of these three expressions is negative and the other two are positive. Note that under normal operating conditions $0 < \omega_m < \omega_0$.

d. Consider the steady-state operating point where motor speed is steady at $\bar{\omega}_m$. Obtain expressions for the ω_p, T_p, and T_0 at this operating point, in terms of $\bar{\omega}_m$ and $\bar{\omega}_0$.

e. Suppose that $\frac{\partial T_m}{\partial \omega_m} = -b$, $\frac{\partial T_m}{\partial T_0} = \beta_1$, and $\frac{\partial T_m}{\partial \omega_0} = \beta_2$ at the operating point given in Part (d). Note that voltage control is achieved by varying T_0 and frequency control by varying ω_0. Linearize the state model obtained in Part (b) about the operating point and express it in terms of the incremental variables $\hat{\omega}_m$, $\hat{T}_p$, $\hat{\omega}_p$, $\hat{T}_0$, and $\hat{\omega}_0$. Suppose that the (incremental) output variables are the incremental pump speed $\hat{\omega}_p$ and the incremental angle of twist of the pump shaft. Express the state space model in the usual form Equation 2.45 and Equation 2.46, and obtain the matrices A, B, C, and D.

f. For the case of frequency control only (i.e., $\hat{T}_0 = 0$) obtain the input-output differential equation relating $\hat{\omega}_p$ and $\hat{\omega}_0$. Using this equation show that if $\hat{\omega}_0$ is suddenly changed by a step of $\Delta\hat{\omega}_0$ then $\frac{d^3\hat{\omega}_p}{dt^3}$ will simultaneously change by a step of $\frac{\beta_2 k_p}{r J_m J_p} \Delta\hat{\omega}_0$, but the lower derivatives of $\hat{\omega}_p$ will not change instantaneously.

SOLUTION

a. • Backlash and inertia of the gear transmission have been neglected in the model shown. This is not accurate in general. Also, the gear efficiency η, which is assumed constant here, usually varies with the gear speed.

• Usually there is some flexibility in the shaft (coupling), which connects the gear to the drive motor.

• Energy dissipation (in the pump load and in various bearings) has been lumped into a single linear viscous-damping element. In practice, this energy dissipation is nonlinear and distributed.

b. Motor speed
$$\omega_m = \frac{d\theta_m}{dt}$$

Load (pump) speed
$$\omega_p = \frac{d\theta_p}{dt}$$

where
θ_m = motor rotation
θ_p = pump rotation.
Let T_g = torque transmitted by the motor to the gear. By definition, gear efficiency is given by

$$\eta = \frac{T_p r \omega_m}{T_g \omega_m} = \frac{\text{output power}}{\text{input power}}$$

Note that since r is the gear ratio, $r\omega_m$ is the output speed of the gear. Also power = torque × speed. We have

$$T_g = \frac{r}{\eta} T_p \qquad\qquad (i)$$

Newton's second law (Torque = inertia × angular acceleration) for the motor:

$$T_m - T_g - b_m \omega_m = J_m \dot{\omega}_m \tag{ii}$$

Newton's second law for the pump:

$$T_p - b_p \omega_p = J_p \dot{\omega}_p \tag{iii}$$

Hooke's law (Torque = torsional stiffness × angle of twist) for the flexible shaft:

$$T_p = k_p \left(\frac{\theta_m}{r} - \theta_p \right) \tag{iv}$$

Equations ii through iii, and the derivative of Equation iv are the three state equations. Specifically, substitute Equation i into Equation ii:

$$J_m \dot{\omega}_m = T_m - b_m \omega_m - \frac{r}{\eta} T_p \tag{v}$$

Differentiate Equation iv:

$$\dot{T}_p = k_p \left(\frac{\omega_m}{r} - \omega_p \right) \tag{vi}$$

Equation iii:

$$J_p \dot{\omega}_p = T_p - b_p \omega_p \tag{vii}$$

Equations v through vii are the three state equations. Note that this is a nonlinear model with the state vector $[\omega_m \quad T_p \quad \omega_p]^T$. The input is T_m. Strictly,

$$T_m = \frac{T_0 q \omega_0 (\omega_0 - \omega_m)}{\left(q \omega_0^2 - \omega_m^2 \right)} \tag{viii}$$

There are two inputs: ω_0 (the speed of the rotating magnetic field, which is proportional to the line frequency) and T_0 which depends quadratically on the phase voltage.

c. When $\omega_m = 0$ we note from Equation viii that $T_m = T_0$. Hence $T_0 =$ starting torque of the motor.

Also, from Equation viii we see that when $T_m = 0$, we have $\omega_m = \omega_0$. Hence, $\omega_0 =$ no-load speed.

This is the synchronous speed—Under no-load conditions, there is no slip in the induction motor (i.e., actual speed of the motor is equal to the speed ω_0 of the rotating magnetic field).

Differentiate Equation viii with respect to T_0, ω_0, and ω_m. We have

$$\frac{\partial T_m}{\partial T_0} = \frac{q \omega_0 (\omega_0 - \omega_m)}{\left(q \omega_0^2 - \omega_m^2 \right)} = \beta_1 \text{ (say)} \tag{ix}$$

Note that β_1 is positive.

$$\frac{\partial T_m}{\partial \omega_0} = \frac{T_0 q\left[\left(q\omega_0^2 - \omega_m^2\right)(2\omega_0 - \omega_m) - \omega_0(\omega_0 - \omega_m)2q\omega_0\right]}{\left(q\omega_0^2 - \omega_m^2\right)^2}$$

$$= \frac{T_0 q\omega_m\left[(\omega_0 - \omega_m)^2 + (q-1)\omega_0^2\right]}{\left(q\omega_0^2 - \omega_m^2\right)^2} = \beta_2 \text{ (say)}$$

(x)

Note that β_2 is positive.

$$\frac{\partial T_m}{\partial \omega_m} = \frac{T_0 q\omega_0\left[\left(q\omega_0^2 - \omega_m^2\right)(-1) - (\omega_0 - \omega_m)(-2\omega_m)\right]}{\left(q\omega_0^2 - \omega_m^2\right)^2}$$

$$= -\frac{T_0 q\omega_0\left[(q-1)\omega_0^2 + (\omega_0 - \omega_m)^2\right]}{\left(q\omega_0^2 - \omega_m^2\right)^2}$$

(xi)

$$= -b \text{ (say)}$$

Note that b is positive.

d. For a steady-state operating point, the rates of changes of the state variables will be zero. Hence set $\dot{\omega}_m = 0 = \dot{T}_p = \dot{\omega}_p$ in Equations v through vii. We get

$$0 = \overline{T}_m - b_m\overline{\omega}_m - \frac{r}{\eta}\overline{T}_p$$

$$0 = k_p\left(\frac{\overline{\omega}_m}{r} - \overline{\omega}_p\right)$$

$$0 = \overline{T}_p - b_p\overline{\omega}_p$$

Hence,

$$\overline{\omega}_p = \overline{\omega}_m/r$$

(xii)

$$\overline{T}_p = b_p\overline{\omega}_m/r$$

(xiii)

$$\overline{T}_m = b_m\overline{\omega}_m + b_p\overline{\omega}_m/\eta$$

$$= \frac{T_0 q\overline{\omega}_0(\overline{\omega}_0 - \overline{\omega}_m)}{\left(q\overline{\omega}_0^2 - \overline{\omega}_m^2\right)} \text{ (from (2.58))}$$

or,

$$\overline{T}_0 = \frac{\overline{\omega}_m(b_m + b_p/\eta)\left(q\overline{\omega}_0^2 - \overline{\omega}_m^2\right)}{q\overline{\omega}_0(\overline{\omega}_0 - \overline{\omega}_m)}$$

(xiv)

e. Take the increments of the state Equation v, Equation vi, and Equation vii. We get

$$J_m \dot{\hat{\omega}}_m = \hat{T}_m - b_m \hat{\omega}_m - \frac{r}{\eta} \hat{T}_p - b \hat{\omega}_m \qquad \text{(xv)}$$

$$\dot{\hat{T}}_p = k_p \left(\frac{\hat{\omega}_m}{r} - \hat{\omega}_p \right) \qquad \text{(xvi)}$$

$$J_p \hat{\omega}_p = \hat{T}_p - b_p \hat{\omega}_p \qquad \text{(xvii)}$$

where

$$\hat{T}_m = \left[\frac{\partial T_m}{\partial T_0} \right] \hat{T}_0 + \left[\frac{\partial T_m}{\partial \omega_0} \right] \hat{\omega}_0$$

$$= \beta_1 \hat{T}_0 + \beta_2 \hat{\omega}_0 \qquad \text{(xviii)}$$

Equations xv through xvii subject to Equation xiii are the three linearized state equations. Then, defining the linear:

State vector

$$x = [\hat{\omega}_m \quad \hat{T}_p \quad \hat{\omega}_p]^T$$

Input vector

$$u = [\hat{T}_0 \quad \hat{\omega}_0]^T$$

Output vector

$$y = \left[\hat{\omega}_p \quad \hat{T}_p / k_p \right]^T$$

we have

$$A = \begin{bmatrix} -(b + b_m)/J_m & -r/(\eta J_m) & 0 \\ k_p/r & 0 & -k_p \\ 0 & 1/J_p & -b_p/J_p \end{bmatrix}, \quad B = \begin{bmatrix} \beta_1/J_m & \beta_2/J_m \\ 0 & 0 \\ 0 & 0 \end{bmatrix}$$

$$C = \begin{bmatrix} 0 & 0 & 1 \\ 0 & 1/k_p & 0 \end{bmatrix}, \quad D = 0$$

f. For frequency control, $\hat{T}_0 = 0$.

Substitute Equation xvi into Equation xv in order to eliminate $\hat{\omega}_m$. Then substitute Equation xvii into the result in order to eliminate $\hat{T}_p$. On simplification we get the input-output equation

$$rJ_m J_p \frac{d^3 \hat{\omega}_p}{dt^3} + r[J_m b_p + J_p (b_m + b)] \frac{d^2 \hat{\omega}_p}{dt^2} + r \left[k_p \left(J_m + \frac{J_p}{\eta} \right) + b_p (b_m + b) \right] \frac{d \hat{\omega}_p}{dt}$$

$$+ r k_p \left(\frac{b_p}{\eta} + b_m + b \right) \hat{\omega}_p = \beta_2 k_p \hat{\omega}_0 \qquad \text{(xix)}$$

This is a third-order differential equation, as expected, since the system is third order. Also, as we have seen, the state-space model is also third order.

We can observe the following from Equation xix: Suppose that $\hat{\omega}_0$ is changed by a "finite" step of $\Delta\hat{\omega}_0$. Then the RHS of Equation xix will be finite. If, as a result, $\frac{d^2\hat{\omega}_p}{dt^2}$ or the lower derivatives also change by a finite step, then $\frac{d^3\hat{\omega}_m}{dt^3}$ should change by an infinite value (because, the derivative of a step is an impulse which is infinite at the instant of change). But the LHS of Equation xix cannot become infinite because the RHS is finite. Hence, $\frac{d^2\hat{\omega}_p}{dt^2}$, $\frac{d\hat{\omega}_p}{dt}$, and $\hat{\omega}_p$ will not change instantaneously. Only $\frac{d^3\hat{\omega}_p}{dt^3}$ will change instantaneously by a finite value due to the finite step change of $\hat{\omega}_0$. From Equation xix, the resulting change of $\frac{d^3\hat{\omega}_p}{dt^3}$ is $\frac{\beta_2 k_p}{r J_m J_p}\Delta\hat{\omega}_0$.

Example 2.5

An automated wood cutting system contains a cutting unit, which consists of a dc motor and a cutting blade, linked by a flexible shaft and a coupling. The purpose of the flexible shaft is to locate the blade unit at any desirable configuration, away from the motor itself. The coupling unit helps with the shaft alignment. A simplified, lumped-parameter, dynamic model of the cutting device is shown in Figure 2.10.

The following parameters and variables are shown in the figure:

J_m = axial moment of inertia of the motor rotor

b_m = equivalent viscous damping constant of the motor bearings

k = torsional stiffness of the flexible shaft

J_c = axial moment of inertia of the cutter blade

b_c = equivalent viscous damping constant of the cutter bearings

T_m = magnetic torque of the motor

ω_m = motor speed

T_k = torque transmitted through the flexible shaft.

ω_c = cutter speed

T_L = load torque on the cutter from the workpiece (wood)

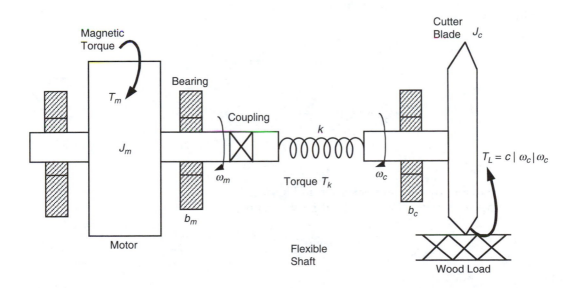

FIGURE 2.10
A wood cutting machine.

In comparison with the flexible shaft, the coupling unit is assumed rigid, and is also assumed light. The cutting load is given by

$$T_L = c |\omega_c| \omega_c \tag{2.63}$$

The parameter c, which depends on factors such as the depth of cut and the material properties of the workpiece, is assumed to be constant in the present analysis.

a. Using T_m as the input, T_L as the output, and $[\omega_m \quad T_k \quad \omega_c]^T$ as the state vector, develop a complete (nonlinear) state model for the system shown in Figure 2.10. What is the order of the system?

b. Using the state model derived in Part (a), obtain a single input-output differential equation for the system, with T_m as the input and ω_c as the output.

c. Consider the steady operating conditions, where $T_m = \overline{T}_m$, $\omega_m = \overline{\omega}_m$, $T_k = \overline{T}_k$, $\omega_c = \overline{\omega}_c$, $T_L = \overline{T}_L$ are all constants. Express the operating point values $\overline{\omega}_m$, $\overline{T}_k$, $\overline{\omega}_c$, and $\overline{T}_L$ in terms of $\overline{T}_m$ and model parameters only. You must consider both cases, $\overline{T}_m > 0$ and $\overline{T}_m < 0$.

d. Now consider an incremental change $\hat{T}_m$ in the motor torque and the corresponding changes $\hat{\omega}_m$, $\hat{T}_k$, $\hat{\omega}_c$, and $\hat{T}_L$ in the system variables. Determine a linear state model (A, B, C, D) for the incremental dynamics of the system in this case, using $x = [\hat{\omega}_m \quad \hat{T}_k \quad \hat{\omega}_c]^T$ as the state vector, $u = [\hat{T}_m]$ as the input and $y = [\hat{T}_L]$ as the output.

e. In the incremental model (see Part(a)), if the twist angle of the flexible shaft (i.e., $\theta_m - \theta_c$) is used as the output what would be a suitable state model? What is the system order then?

f. In the incremental model, if the angular position θ_c of the cutter blade is used as the output variable, explain how the state model obtained in Part (a) should be modified. What is the system order in this case?

Hint for Part (b):

$$\frac{d}{dt}(|\omega_c| \omega_c) = 2 |\omega_c| \dot{\omega}_c \tag{2.64}$$

$$\frac{d^2}{dt^2}(|\omega_c| \omega_c) = 2 |\omega_c| \ddot{\omega}_c + 2\omega_c^2 \, \text{sgn}(\omega_c) \tag{2.65}$$

SOLUTION

a. The free-body diagram is shown in Figure 2.11.
   Constitutive equations for the three elements give:

$$J_m \dot{\omega}_m = T_m - b_m \omega_m - T_k \tag{i}$$

$$\dot{T}_k = k(\omega_m - \omega_c) \tag{ii}$$

$$J_c \dot{\omega}_c = T_k - b_c \omega_c - c |\omega_c| \omega_c \tag{iii}$$

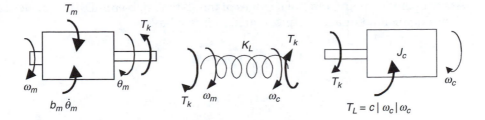

FIGURE 2.11
Free body diagram of the system.

These are the state equations, with

$$\text{State vector} = [\omega_m \quad T_k \quad \omega_c]^T$$

$$\text{Input vector} = [T_m]$$

$$\text{Output vector} = [T_L] = c|\omega_c|\omega_c$$

This is a third-order system (three state equations).

b. Substitute Equation ii in Equation i, to eliminate ω_m:

$$J_m[\ddot{T}_k/k + \dot{\omega}_c] = T_m - b_m[\dot{T}_k/k + \omega_c] - T_k$$

Now substitute Equation iii in this equation, to eliminate T_k, using the fact that (see Equation 2.64 and Equation 2.65)

$$\dot{T}_k = J_c\ddot{\omega}_c + b_c\dot{\omega}_c + 2c|\omega_c|\dot{\omega}_c$$

and

$$\ddot{T}_k = J_c\dddot{\omega}_c + b_c\ddot{\omega}_c + 2c|\omega_c|\ddot{\omega}_c + 2c\dot{\omega}_c^2\,\text{sgn}(\omega_c)$$

We get

$$\frac{J_m}{k}\left[J_c\dddot{\omega}_c + b_c\ddot{\omega}_c + 2c|\omega_c|\ddot{\omega}_c + 2c\dot{\omega}_c^2\,\text{sgn}(\omega_c)\right] + J_m\dot{\omega}_c$$

$$= T_m - \frac{b_m}{k}[J_c\ddot{\omega}_c + b_c\dot{\omega}_c + 2c|\omega_c|\dot{\omega}_c] - b_m\omega_c - [J_c\dot{\omega}_c + b_c\omega_c + c|\omega_c|\omega_c]$$

which can be expressed as

$$J_mJ_c\frac{d^3\omega_c}{dt^3} + (J_mb_c + J_cb_m)\frac{d^2\omega_c}{dt^2} + 2cJ_m|\omega_c|\frac{d^2\omega_c}{dt^2} + (J_mk + J_ck + b_mb_c)\frac{d\omega_c}{dt}$$

$$+ 2b_mc|\omega_c|\frac{d\omega_c}{dt} + 2cJ_m\,\text{sgn}(\omega_c)\left(\frac{d\omega_c}{dt}\right)^2 + k(b_m + b_c)\omega_c + kc|\omega_c|\omega_c = kT_m$$

c. At the operating point, rates of changes of the state variables will be zero. Hence, from Equation i, Equation ii and Equation iii we have

$$0 = \overline{T}_m - b_m \overline{\omega}_m - \overline{T}_k \tag{iv}$$

$$0 = k(\overline{\omega}_m - \overline{\omega}_c) \tag{v}$$

$$0 = \overline{T}_k - b_c \overline{\omega}_c - c \,|\overline{\omega}_c| \overline{\omega}_c \tag{vi}$$

Case 1: $\overline{T}_m > 0 \Rightarrow \overline{\omega}_c > 0$

Eliminate $\overline{T}_k$ using Equation iv and Equation vi

$$0 = \overline{T}_m - b_m \overline{\omega}_m - b_c \overline{\omega}_c - c \overline{\omega}_c^2$$

But $\overline{\omega}_m = \overline{\omega}_c$.
Hence,

$$c \overline{\omega}_c^2 + (b_m + b_c)\overline{\omega}_c - \overline{T}_m = 0$$

or,

$$\overline{\omega}_c = \frac{-(b_m + b_c) \pm \sqrt{(b_m + b_c)^2 + 4c\overline{T}_m}}{2c}$$

Take the positive root.

$$\overline{\omega}_c = \frac{\sqrt{(b_m + b_c)^2 + 4c\overline{T}_m}}{2c} - \frac{(b_m + b_c)}{2c}$$

$$= \overline{\omega}_m$$

From Equation iv:

$$\overline{T}_k = \overline{T}_m - b_m \left[\frac{\sqrt{(b_m + b_c)^2 + 4c\overline{T}_m}}{2c} - \frac{(b_m + b_c)}{2c} \right]$$

From Equation vi:

$$\overline{T}_L = c\,|\overline{\omega}_c|\overline{\omega}_c = \overline{T}_m - (b_m + b_c) \left[\frac{\sqrt{(b_m + b_c)^2 + 4c\overline{T}_m}}{2c} - \frac{(b_m + b_c)}{2c} \right]$$

Case 2: $\overline{T}_m < 0 \Rightarrow \overline{\omega}_c < 0$

Then

$$0 = \overline{T}_m - b_m\overline{\omega}_c - b_c\overline{\omega}_c + c\overline{\omega}_c^2$$

or,

$$c\overline{\omega}_c^2 - (b_m + b_c)\overline{\omega}_c + \overline{T}_m = 0$$

$$\overline{\omega}_c = \frac{(b_m + b_c) \pm \sqrt{(b_m + b_c)^2 - 4c\overline{T}_m}}{2c}$$

Note that $\overline{T}_m < 0$. Use the negative root

$$\overline{\omega}_c = \frac{(b_m + b_c)}{2c} - \frac{\sqrt{(b_m + b_c)^2 - 4c\overline{T}_m}}{2c}$$

The rest will follow as before.

d. Linearize:

$$J_m\dot{\hat{\omega}}_m = \hat{T}_m - b_m\hat{\omega}_m - \hat{T}_k$$

$$\dot{\hat{T}}_k = k(\hat{\omega}_m - \hat{\omega}_c)$$

$$J_c\dot{\hat{\omega}}_c = \hat{T}_k - b_c\hat{\omega}_c - 2c\,|\overline{\omega}_c|\,\hat{\omega}_c$$

with $x = [\hat{\omega}_m \quad \hat{T}_k \quad \hat{\omega}_c]^T$ and $u = [\hat{T}_m]$

$$A = \begin{bmatrix} -b_m/J_m & -1/J_m & 0 \\ k & 0 & -k \\ 0 & 1/J_c & -(b_c + 2c\,|\overline{\omega}_c|)/J_c \end{bmatrix}, \quad B = \begin{bmatrix} 1/J_m \\ 0 \\ 0 \end{bmatrix}$$

with

$$y = [\hat{T}_L] = [2c\,|\overline{\omega}_c|\,\hat{\omega}_c]$$

$$C = [0 \quad 0 \quad 2c\,|\overline{\omega}_c|] \quad \text{and} \quad D = [0]$$

e. $y = \theta_m - \theta_c = \dfrac{T_k}{k}$

Hence, exactly the same state equations are applicable, along with this new output equation.

System order = 3

f. $y = \theta_c$

Here, θ_c cannot be expressed as an algebraic equation of the three previous state variables. A new state variable θ_c has to be defined, along with the additional state equation

$$\frac{d\theta_c}{dt} = \omega_c$$

Note that the system order becomes 4 in this case.

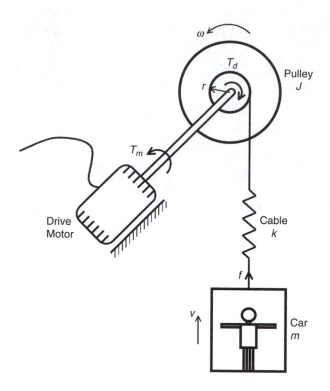

FIGURE 2.12
A simplified model of an elevator.

Example 2.6

A simplified model of an elevator is shown in Figure 2.12.
  Note that
  J = moment of inertia of the cable pulley
  r = radius of the pulley
  k = stiffness of the cable
  m = mass of the car and occupants

  a. Which system parameters are variable? Explain.
  b. Suppose that the damping torque $T_d(\omega)$ at the bearing of the pulley is a nonlinear function of the angular speed ω of the pulley. Taking the state vector x as

$$x = [\omega \quad f \quad v]^T$$

  in which
  f = tension force in the cable
  v = velocity of the car (taken positive upwards),
  the input vector as $u = [T_m]^T$
  in which

$$T_m = \text{torque applied by the motor to the pulley}$$

(positive in the direction indicated in Figure 2.12)

and, the output vector as

$$y = [v]$$

obtain a complete, nonlinear, state-space model for the system.

c. With T_m as the input and v as the output, convert the state-space model into a nonlinear input-output differential equation model. What is the order of the system?

d. Give an equation whose solution provides the steady-state operating speed $\bar{v}$ of the elevator car.

e. Linearize the nonlinear input/output differential-equation model obtained in Part (c), for small changes $\hat{T}_m$ of the input and $\hat{v}$ of the output, about an operating point.

   NOTE $\bar{T}_m$ = steady-state operating-point torque of the motor (assumed to be known).

   HINT Denote $\dfrac{dT_d}{d\omega}$ as $b(\omega)$.

f. Linearize the state-space model obtained in Part (b) and give the model matrices A, B, C, and D in the usual notation. Obtain the linear input/output differential equation from this state-space model and verify that it is identical to what was obtained in Part (e).

SOLUTION

a. The parameter r is a variable due to winding/unwinding of the cable around the pulley. The parameter m is a variable because the car occupancy changes.

b. The state equations are obtained simply by applying Newton's second law to the two inertia elements and Hooke's law to the spring element:

$$J\dot{\omega} = T_m - rf - T_d(\omega) \tag{i}$$

$$\dot{f} = k(r\omega - v) \tag{ii}$$

$$m\dot{v} = f - mg \tag{iii}$$

Output $y = v$

c. Eliminate f by substituting Equation iii into Equation i and Equation ii:

$$J\dot{\omega} = T_m - rm(\dot{v} + g) - T_d(\omega) \tag{iv}$$

$$m\ddot{v} = k(r\omega - v) \tag{v}$$

From Equation v we have

$$\omega = \frac{1}{r}\left(\frac{m}{k}\ddot{v} + v\right)$$

Hence,

$$\dot{\omega} = \frac{1}{r}\left(\frac{m}{k}\dddot{v} + \dot{v}\right)$$

Substitute these into Equation iv, to eliminate ω:

$$\frac{J}{r}\left(\frac{m}{k}\dddot{v} + \dot{v}\right) = T_m - rm(\dot{v} + g) - T_d\left(\frac{1}{r}\left(\frac{m}{k}\ddot{v} + v\right)\right) \qquad \text{(vi)}$$

This is a third-order model (The highest derivative in Equation vi is third order).

d. At steady state $\dot{v} = 0$. Hence $\ddot{v} = 0$ and $\dddot{v} = 0$ as well. Substitute into Equation vi, to get the steady-state equation:

$$\bar{T}_m - rmg - T_d\left(\frac{\bar{v}}{r}\right) = 0$$

where $\bar{T}_m$ = steady-state value of the input T_m.

The solution of this nonlinear equation will give the steady-state operating speed $\bar{v}$ of the elevator.

NOTE   The same result may be obtained from the state Equations i through iii under steady-state conditions:

$$0 = \bar{T}_m - r\bar{f} - T_d(\bar{\omega})$$

$$0 = k(r\bar{\omega} - \bar{v})$$

$$0 = \bar{f} - mg$$

This can be converted into a single equation, by eliminating $\bar{f}$ and $\bar{\omega}$.

e. Linearize Equation vi:

$$\frac{J}{r}\left(\frac{m}{k}\dddot{\hat{v}} + \dot{\hat{v}}\right) = \hat{T}_m - rm\dot{\hat{v}} - b(\bar{\omega})\hat{\omega} \qquad \text{(vii)}$$

where

$$b(\bar{\omega}) = \frac{dT_d(\omega)}{d\omega}\bigg|_{\omega = \bar{\omega}}$$

Now from Equation v:

$$m\ddot{\hat{v}} = k(r\hat{\omega} - \hat{v}) \qquad \text{(viii)}$$

Substitute Equation viii into Equation vii, to eliminate $\hat{\omega}$. We get

$$\frac{J}{r}\left(\frac{m}{k}\dddot{\hat{v}}+\dot{\hat{v}}\right)=\hat{T}_m-r\,m\dot{\hat{v}}-b(\overline{\omega})\frac{1}{r}\left(\frac{m}{k}\ddot{\hat{v}}-\hat{v}\right)$$

or,

$$\frac{J}{r}\frac{m}{k}\dddot{\hat{v}}+\frac{b(\overline{\omega})m}{rk}\ddot{\hat{v}}+\left(\frac{J}{r}+rm\right)\dot{\hat{v}}+\frac{b(\overline{\omega})}{r}\hat{v}=\hat{T}_m$$

(f) Linearize Equation i through iii:

$$J\dot{\hat{\omega}}=\hat{T}_m-r\hat{f}-b(\overline{\omega})\hat{\omega} \tag{ix}$$

$$\dot{\hat{f}}=k\,(r\hat{\omega}-\hat{v}) \tag{x}$$

$$m\dot{\hat{v}}=\hat{f} \tag{xi}$$

Output $y=\hat{v}$, input $u=\hat{T}_m$, state vector $x=[\hat{\omega}\quad\hat{f}\quad\hat{v}]^T$. Hence,

$$A=\begin{bmatrix} -\dfrac{b(\overline{\omega})}{J} & -\dfrac{r}{J} & 0 \\[2mm] rk & 0 & -k \\[2mm] 0 & \dfrac{1}{m} & 0 \end{bmatrix},\quad B=\begin{bmatrix} 1/J \\ 0 \\ 0 \end{bmatrix}$$

$$C=[0\quad 0\quad 1],\qquad D=0$$

Substitute Equation xi into Equation ix and Equation x, to eliminate $\hat{f}$:

$$J\dot{\hat{\omega}}=\hat{T}_m-rm\dot{\hat{v}}-b(\overline{\omega})\hat{\omega}$$

$$m\ddot{\hat{v}}=k\,(r\hat{\omega}-\hat{v})$$

Now eliminate $\hat{\omega}$. We get the same result as before, for the input/output equation.

Example 2.7

A rocket-propelled spacecraft of mass m is fired vertically up (in the Y-direction) from the earth's surface (see Figure 2.13). The vertical distance of the centroid of the spacecraft, measured from the earth's surface, is denoted by y. The upward thrust force of the rocket is $f(t)$. The gravitational pull on the spacecraft is given by $mg[\frac{R}{R+y}]^2$, where g is the acceleration

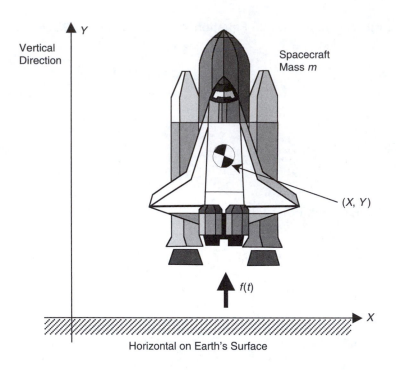

FIGURE 2.13
Coordinate system for the spacecraft problem.

due to gravity at the earth's surface and R is the "average" radius of earth (about 6370 km). The magnitude of the aerodynamic drag force resisting the motion of the spacecraft is approximated by $k\dot{y}^2 e^{-y/r}$ where k and r are positive and constant parameters, and $\dot{y} = \frac{dy}{dt}$. Here, the exponential term represents the loss of air density at higher elevations.

a. Derive the input-output differential equation for the system, treating f as the input and y as the output.

b. The spacecraft accelerates to a height of y_o and the maintains a constant speed v_o, still moving in the same vertical (Y) direction. Determine an expression for the rocket force that is needed for this constant-speed motion. Express your answer in terms of y_o, v_o, time t, and system parameters m, g, R, r, and k. Show that this force decreases as the spacecraft ascends.

c. Linearize the input-output model (Part (a)) about the steady operating condition (part (b)), for small variations $\hat{y}$ and $\dot{\hat{y}}$ in the position and speed of the spacecraft, due to a force disturbance $\hat{f}(t)$.

d. Treating y and $\dot{y}$ as state variables and y as the output, derive a complete (nonlinear) state-space model for the vertical dynamics of the spacecraft.

e. Linearize the state-space model in (d) about the steady conditions in (b) for small variations $\hat{y}$ and $\dot{\hat{y}}$ in the position and speed of the spacecraft, due to force disturbance $\hat{f}(t)$.

f. From the linear state model (Part (e)) derive the linear input-output model and show that the result is identical to what you obtained in Part (c).

SOLUTION

a. Newton's second law in the Y-direction:

$$m\ddot{y} = -mg\left(\frac{R}{R+y}\right)^2 - k\,|\dot{y}|\,\dot{y}e^{-y/r} + f(t) \tag{i}$$

b. At constant speed v_o we have

$$\dot{y} = v_o \tag{ii}$$

$$\ddot{y} = \frac{d}{dt}v_o = 0 \tag{iii}$$

Integrate Equation ii and use the initial condition $y = y_o$ at $t = 0$. Position under steady conditions:

$$\bar{y} = v_o t + y_o \tag{iv}$$

Substitute Equation ii and Equation iii in Equation i:

$$0 = -\frac{mg}{\left(1 + \dfrac{v_o t + y_o}{R}\right)^2} - k\,|v_o|\,v_o e^{-(v_o t + y_o)/r} + f_s(t)$$

where $f_s(t)$ is the force of rocket at constant speed v_o.
Since v_o is positive, we have

$$f_s(t) = \frac{mg}{\left(1 + \dfrac{v_o t + y_o}{R}\right)^2} + kv_o^2 e^{-(v_o t + y_o)/r}$$

Note that this expression decreases as t increases, reaching zero in the limit.

c. Derivatives needed for the linearization (O(1) Taylor series terms):

$$\frac{d}{dy}\frac{1}{\left(1 + \dfrac{y}{R}\right)^2} = -\frac{2}{R}\frac{1}{\left(1 + \dfrac{y}{R}\right)^3}$$

$$\frac{d}{dy}|\dot{y}|\,\dot{y}e^{-y/r} = -\frac{|\dot{y}|\,\dot{y}}{r}e^{-y/r}$$

$$\frac{d}{d\dot{y}}|\dot{y}|\,\dot{y}e^{-y/r} = 2\,|\dot{y}|\,e^{-y/r}$$

The linearized input-output equation becomes

$$m\ddot{\hat{y}} = \frac{2mg}{R} \frac{1}{\left(1+\dfrac{\bar{y}}{R}\right)^3} \hat{y} + \frac{k}{r} |\dot{\bar{y}}| \, \dot{\bar{y}} e^{-\bar{y}/r} \hat{y} - 2k \, |\dot{\bar{y}}| \, e^{-\bar{y}/r} \dot{\hat{y}} + \hat{f}(t) \qquad \text{(v)}$$

where $\bar{y}$ is as given by Equation iv.
Since, under steady conditions, $\dot{y} = \dot{\bar{y}} = v_o > 0$ we have

$$m\ddot{\hat{y}} = \frac{2mg}{R} \frac{1}{\left(1+\dfrac{\bar{y}}{R}\right)^3} \hat{y} + \frac{k}{r} v_o^2 e^{-\bar{y}/r} \hat{y} - 2kv_o e^{-\bar{y}/r} \dot{\hat{y}} + \hat{f}(t) \qquad \text{(v)}$$

NOTE Unstable system.

d. State vector

$$x = [x_1 \quad x_2]^T = [y \quad \dot{y}]^T$$

Then from Equation i, the state equations are

$$\dot{x}_1 = x_2 \qquad \text{(vi)}$$

$$\dot{x}_2 = -\frac{g}{\left(1+\dfrac{x_1}{R}\right)^2} - \frac{k}{m} |x_2| \, x_2 e^{-x_1/r} + \frac{1}{m} f(t) \qquad \text{(vii)}$$

The output equation is

$$y_1 = x_1$$

e. To linearize, we use the derivatives (local slopes) as before:

$$\frac{d}{dx_1} \frac{1}{\left(1+\dfrac{x_1}{R}\right)^2} = -\frac{2}{R} \frac{1}{\left(1+\dfrac{x_1}{R}\right)^3}$$

$$\frac{d}{dx_1} |x_2| \, x_2 e^{-x_1/r} = -\frac{|x_2| \, x_2}{r} e^{-x_1/r}$$

$$\frac{d}{dx_2} |x_2| \, x_2 e^{-x_1/r} = 2 \, |x_2| \, e^{-x_1/r}$$

at the steady operating (constant speed) conditions:

$$\bar{x}_1 = v_o t + y_o \quad \text{and} \quad \bar{x}_2 = v_o > 0$$

Accordingly, the linearized state-space model is

$$\dot{\hat{x}}_1 = \hat{x}_2 \tag{viii}$$

$$\dot{\hat{x}}_2 = \frac{2g}{R\left(1+\dfrac{\bar{x}_1}{R}\right)^3}\hat{x}_1 + \frac{k}{mr}v_o^2 e^{-(\bar{x}_1)/r}\,\hat{x}_1 - 2\frac{k}{m}v_o e^{-\bar{x}_1/r}\,\hat{x}_2 + \frac{1}{m}\hat{f}(t) \tag{ix}$$

with the output equation

$$\hat{y}_1 = \hat{x}_1 \tag{x}$$

f. Substitute Equation viii in Equation ix. We get

$$\ddot{\hat{x}}_1 = \frac{2mg}{R\left(1+\dfrac{\bar{x}_1}{R}\right)^3}\hat{x}_1 + \frac{k}{mr}v_o^2 e^{-(\bar{x}_1)/r}\,\hat{x}_1 - 2\frac{k}{m}v_o e^{-\bar{x}_1/r}\,\dot{\hat{x}}_1 + \frac{1}{m}\hat{f}(t)$$

which is identical to Equation v, since $\hat{x}_1 = \hat{y}$.

2.5 Linear Graphs

Lumped-parameter dynamic systems can be represented by *linear graphs*, which use interconnected line segments (called branches) to represent a dynamic model. The term "linear graph" stems from the use of line segments, and does not mean that the system itself is linear. In particular, linear graphs are a convenient tool with which to develop a state-space model of a system.

2.5.1 Through Variables and Across Variables

Each branch in the linear graph model has one *through variable* and one *across variable* associated with it. Some related concepts have been summarized in Table 2.1. For instance, in a hydraulic or pneumatic system, a pressure "across" an element causes some change of flow "through" the element. The across variable is pressure, the through variable is flow. The product of a through variable and an across variable is a power. Table 2.3 lists the through and across pairs for hydraulic/pneumatic, electrical, thermal and mechanical

TABLE 2.3

Through and Across Variables of Several Types of Systems

System type	Through variable	Across variable
Hydraulic/pneumatic	Flow rate	Pressure
Electrical	Current	Voltage
Mechanical	Force	Velocity
Thermal	Heat transfer	Temperature

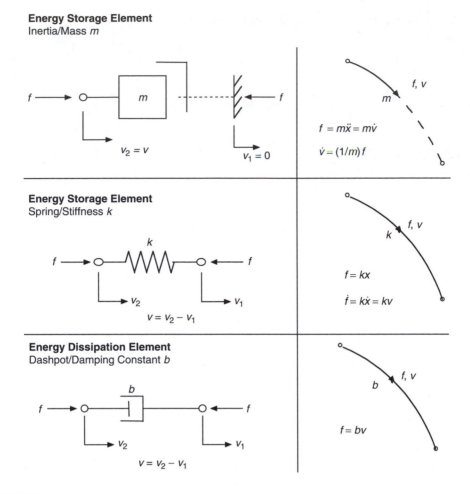

FIGURE 2.14
Mechanical system elements and their linear-graph representations.

systems. Figure 2.14 shows the lumped-parameter mechanical-system elements (mass, spring, dashpot/damper) and their linear-graph representations. Although translatory mechanical elements are presented in Figure 2.14, corresponding rotary elements are easy to visualize—f denotes an applied torque and v the relative angular velocity in the same direction. Analogous electrical elements may be represented in a similar manner. Note that the linear graph of an inertia element has a broken line segment. This is because the force does not physically travel from one end of this linear graph to the other end through the inertia, but rather "felt" at the two ends. This will be further discussed using an example. Linearity of the elements is not a requirement in order to represent them by linear graph segments.

2.5.2   Sign Convention

One end of any branch is considered the *point of reference* and the other end the *point of action*. The choice is somewhat arbitrary, and can reflect the physics of the actual system. An *oriented* branch is one to which a direction is assigned, using an arrowhead, which is picked to denote the positive direction of power flow at each end. By convention, the

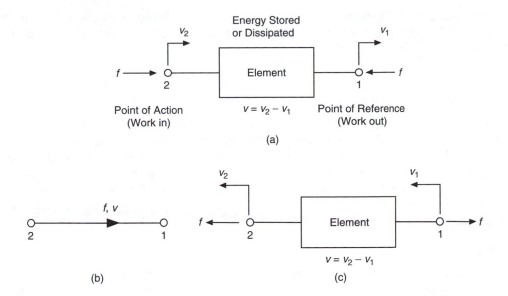

FIGURE 2.15
Sign convention for a linear graph.

positive direction of power is taken as "into" the element at the point of action, and "out of" the element at the point of reference. According to this convention, the arrowhead of a branch is always pointed toward the point of reference. In this manner the reference point and the action point are easily identified.

The across variable is always given relative to the point of reference. It is also convenient to give the through variable and the across variable as an ordered pair (f, v) on each branch. This nomenclature is illustrated in Figure 2.15(a) and Figure 2.15(b). It should be noted that the direction of a branch does not represent the positive direction of f or v. For example, when the positive directions of both f and v are changed, as in Figure 2.15(c), the linear graph given in Figure 5.2(b) remains unchanged because the positive direction of power flow is the same. In a given problem, the positive direction of any one of the two variables f and v should be pre-established for each branch. Then the corresponding positive direction of the other variable is automatically determined by the convention used to orient linear graphs. It is less confusing to assign the same positive direction for v as for power flow at the point of action (i.e., the convention shown in Figure 2.15(a) is preferred over that in Figure 2.15(c)).

Note that a force is transmitted through the element with no change; it is the "through" variable. Velocity is the "across" variable; it changes across the element, as it is measured relative to one end. In summary,

1. The through variable is the same at both the input and the output of the element; the across variable differs.
2. The across variable requires a reference point; the through variable does not.

According to the sign convention shown in Figure 2.15, the work done on the element at the point of action (by an external device) is positive, and work done by the element at the point of reference (on an external load) is positive. The amount of work done on the element that exceeds the amount of work done by a system is either stored as energy (kinetic and potential), which has the capacity to do additional work, or dissipated (damping) through various mechanisms manifested as heat transfer, noise, and other phenomena.

2.5.3 Single-Port Elements

Single-port (or, *single energy port)* elements are those which can be represented by a single branch (line segment). These elements possess only one power (or energy) variable; hence the nomenclature. They have two terminals. The general form of these elements is shown in Figure 2.15. In modeling mechanical systems we require three passive single-port elements, as shown at the right in Figure 2.14. The analogous three elements are needed for electrical systems.

2.5.3.1 *Use of Linear Graphs*

Linear-graph representation is particularly useful in understanding rates of energy transfer (power) associated with various phenomena, and dynamic interactions in a mechanical system can be interpreted in terms of power transfer. As mentioned, power is the product of a generalized force variable and the corresponding generalized velocity variable. The total work done on a mechanical system is, in part, used as stored energy (kinetic and potential); the remainder is dissipated. Stored energy can be completely recovered when the system is brought back to its original state (i.e., when the cycle is completed). Such a process is *reversible.* On the other hand, dissipation corresponds to irreversible energy transfer that cannot be recovered by returning the system to its initial state. (A fraction of the mechanical energy lost in this manner could be recovered, in principle, by operating a heat engine, but we shall not go into these details.) Energy dissipation may appear in many forms including temperature rise (a molecular phenomenon), noise (an acoustic phenomenon), or work in wear mechanisms.

These energy transfer characteristics are distributed phenomena, in general. Consider, for example, a coil spring oscillating under an external force (Figure 2.16(a)). The coil has a distributed mass and hence the capacity to store *kinetic energy* by acquiring velocity. Stored kinetic energy can be recovered as work done through a process of deceleration. Furthermore, the flexibility of the coil is distributed as well, and each small element in the coil has the capacity to store *elastic potential energy* through reversible (elastic) deflection. If the coil was moving in the vertical direction, there would be changes in *gravitational potential energy,* but we can disregard this in dynamic response studies if the deflections are measured from the static equilibrium position of the system. The coil will undoubtedly get warmer, make creaking noises, and wear out at the joints, clear evidence of its capacity to dissipate energy. A further indication of damping is provided by the fact that when the coil is pressed and released, it will eventually come to rest; the work done by pressing the coil is completely dissipated. For most purposes, a lumped-parameter model (such as the one shown in Figure 2.16(b)) is adequate. In this model, the three effects are considered separately. Even though these effects are distributed in the actual system, the discrete model is usually sufficient to predict the system response to a forcing function.

Further approximations are possible under certain circumstances. For instance, if the maximum kinetic energy is small in comparison with the maximum elastic potential energy in general (particularly true for light stiff coils, and at low frequencies of oscillation), and if in addition the rate of energy dissipation is relatively small (determined with respect to the time span of interest), the coil can be modeled by a discrete stiffness (spring)

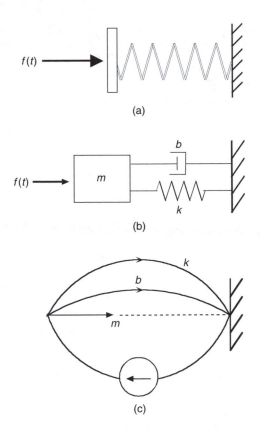

FIGURE 2.16
Coil spring oscillating against an external force: (a) System; (b) Lumped-parameter model; (c) Linear-graph representation of the model.

element. These are modeling decisions. Following this decision, it is relatively easy to represent the lumped-parameter model by its linear graph, as shown in Figure 2.16(c). In this figure a force source (f) has been introduced, which is discussed next.

2.5.3.2 Force and Velocity Sources

In linear-graph models, system *inputs* are represented by *source elements*. An ideal force source (a through-variable source) is able to supply a force input that is not affected by interactions with the rest of the system. The corresponding relative velocity across the force source, however, is determined by the overall system. A forcing function $f(t)$ applied at a point (Figure 2.17(a))—a force source—can be represented by the linear graph in Figure 2.17(b). The arrowhead indicates the direction of the applied force when $f(t)$ is positive. Note that it also determines the positive direction of power. It should be clear that the direction of $f(t)$ as shown in Figure 2.17(a) is the applied force. The reaction on the source would be in the opposite direction.

It is possible for velocity to be a source (input), as represented in Figure 2.17(c) and Figure 2.17(d). The + sign is placed to indicate the point of action, and − for the point of reference. An ideal velocity source (across-variable source) supplies a velocity input independent of the system to which it is applied. The corresponding force is, of course, determined by the system.

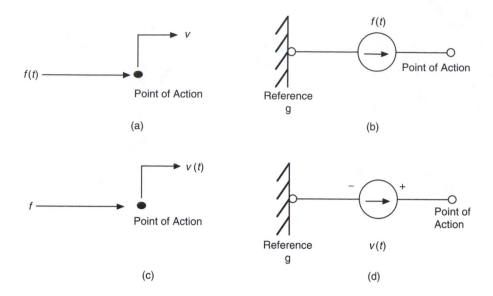

FIGURE 2.17
(a) Force source (force input); (b) Linear graph representation of a force source; (c) Velocity source; (d) Linear graph representation of a velocity source.

Source elements can serve as means of inhibiting interactions between systems. Specifically, it follows from the definition of an ideal source that dynamic behavior of a system is not affected by connecting a new system in series with an existing force source or in parallel with an existing velocity source. Furthermore, the original system is not affected in each case, by separating it into two new systems. These two cases are shown in Figure 2.18. In general, linking (networking) a subsystem will change the order of the overall system (because new dynamic interactions are introduced) although the two situations in Figure 2.18 are examples where this does not happen. Another way to interpret these situations is to consider the original system and the new system as two uncoupled subsystems driven by the same input source. In this sense, the order of the overall system is the sum of the order of these individual subsystems.

2.5.4 Two-Port Elements

There are two basic types of two-port elements that interest us in modeling mechanical systems—the (mechanical) *transformer* and the *gyrator*. Examples of mechanical transformers are a lever and pulley for translatory motions and a meshed pair of gear wheels for rotation (Figure 2.19). A gyrator is typically an element that displays gyroscopic properties (Figure 2.20). These elements can be interpreted as a pair of single-port elements whose net power is zero. In this respect the two basic elements are related. The linear graph of a two-port element (Figure 2.19(c)) has two coupled branches. We shall consider only the linear case, that is, ideal transformers and ideal gyrators only. The extension to the nonlinear case should be clear.

2.5.4.1 Mechanical Transformer

As for a single-port passive element, the arrows on each branch (line segment) of the linear graph in Figure 2.19 give the direction when the product of force and velocity variables for that segment is positive. Note that v_i and f_i are the velocity and force at the input port;

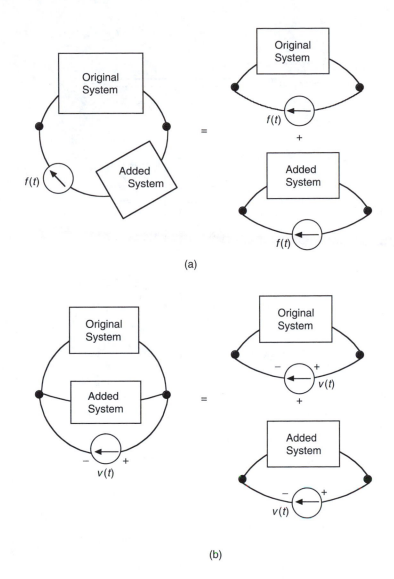

(a)

(b)

FIGURE 2.18
(a) Two systems connected in series to a force source; (b) Two systems connected in parallel to a velocity source.

v_o and f_o are the velocity and force at the output port. The (linear) transformation ratio r is given by

$$v_o = r\,v_i \tag{2.66}$$

Due to the conservation of power:

$$f_i v_i + f_o v_o = 0 \tag{2.67}$$

This gives

$$f_o = -\frac{1}{r} f_i \tag{2.68}$$

Note that r is a non-dimensional parameter.

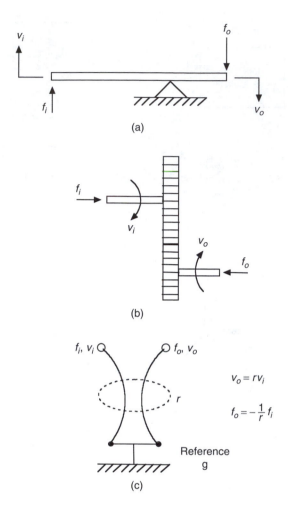

FIGURE 2.19
Mechanical Transformer—a two-port element: (a) Lever and pulley; (b) Meshed gear wheels; (c) Linear graph representation.

2.5.4.2 *Gyrator*

An ideal gyroscope is an example of a mechanical *gyrator* (Figure 2.20). It is simply a spinning top that rotates about its own axis at a high angular speed ω (positive in x direction), assumed to remain unaffected by other small motions that may be present. If the moment of inertia about this axis of rotation (x in the shown configuration) is J, the corresponding angular momentum is $J\omega$. If a velocity v_i is given to the free end in the y direction (which would result in a force f_i at that point, whose positive direction is also taken as y) the corresponding rate of change of angular momentum would be $J\omega v_i/L$ about the positive y-axis. Note that v_i/L is the angular velocity due to v_i. By Newton's second law, to sustain this rate of change of angular momentum, it would require a torque equal to $J\omega v_i/L$ in the same direction. If the corresponding force at the free end is denoted by f_o in the z-direction, the corresponding torque is $f_o L$ acting in the negative y-direction. It follows that

$$-f_o L = J\omega v_i/L$$

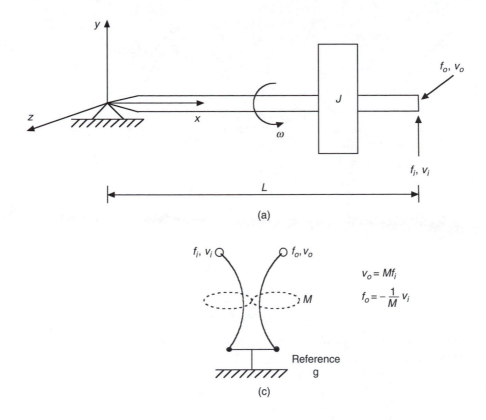

FIGURE 2.20
(a) Gyrator (gyroscope or spinning top)—a two-port element; (b) linear-graph representation.

in which L is the length of the gyroscope. This may be expressed as,

$$f_o = -\frac{1}{M} v_i \qquad (2.69)$$

in which

$$M = \frac{L^2}{J\omega} \qquad (2.70)$$

By the conservation of power (Equation 2.67) it follows that

$$v_o = Mf_i \qquad (2.71)$$

Note that M is a *mobility* parameter (velocity/force).

2.5.5 Loop and Node Equations

Figure 2.21 shows a mass-spring-damper system and its linear graph. Each element in the linear graph has two nodes, forming a branch. As noted before, an inertia element is connected to the reference (ground) point by a dotted line because the mass is not

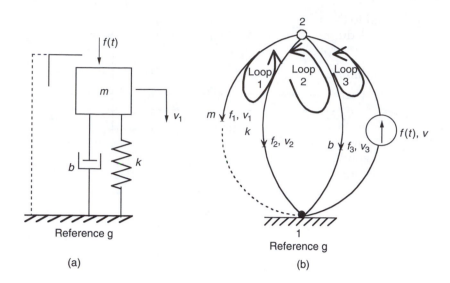

FIGURE 2.21
(a) Mass-spring-damper system; (b) Linear graph having 2 nodes and 3 primary loops.

physically connected to ground, but all measurements must be referenced to the ground reference point. This reference point "feels" the inertia force of the mass. To understand this further, suppose that we push a mass upwards by our hands. An equal force is transmitted to the ground though our feet. Note that in this case the mass is not directly connected to the ground, yet the force applied to the mass and the force "felt" at the ground are equal. Hence the force "appears" to travel directly through the mass element to the ground. Similarly, in Figure 2.21, the input force from the "force source" also travels to ("felt at") the reference point.

2.5.5.1 *Number of Loops*

A loop is a closed path formed by two or more branches of the linear graph. The number of "primary" loops is an important consideration in a linear graph. There are three primary loops in Figure 2.21. Note that loops closed by broken-line (inertia) branches are included in counting primary loops. The primary loop set can be chosen as (*b-k*, *m-b*, and *m-f*), or as (*b-k*, *m-b*, and *f-k*), or any three closed paths. Once one has selected a primary set of loops (3 loops in this example), any other loop will depend on this primary set. For example, an *m-k* loop can be obtained by algebraically adding the *m-b* loop and *b-k* loop. Similarly, the *f-m* loop is obtained by adding the *f-b* and *b-m* loops. Thus the primary loop set becomes an "independent" set that is the minimum number of loops required to obtain all the independent loop equations.

2.5.5.2 *Compatibility (Loop) Equations*

A loop equation (or, compatibility equation) is simply the sum of the across variables in a loop equated to zero. The arrow in each branch is important—but we need not (and indeed cannot) always go in the direction of the arrows in the branches when forming a loop. If we do go in the direction of the arrow in a branch, the associated across variable is considered positive; when we go opposite to the arrow, the associated across variable is considered negative. Note that, physically, a loop equation dictates that the across variables are the same (i.e., unique) at any given point in the loop, for example, a mass and spring connected to the same point must have the same velocity. This guarantees that the joint does not break during operation—that the system is compatible (hence the name).

2.5.5.3 Node (*Continuity*) Equations

A node is the point where two or more branches meet. A node equation (or, continuity equation) is created by equating to zero the sum of all the through variables at a node in effect saying, "what goes in must come out." A node equation dictates the continuity of through variables at a node. For example, for a mechanical system, the continuity equation is a force balance or equilibrium equation. For this reason one must use proper signs for the variables when writing either node equations or loop equations.

2.5.5.4 Series and Parallel Connections

Let us consider two systems with a spring (*k*) and a damper (*b*). In Figure 2.22(a) they are connected in parallel, and in Figure 2.22(b) they are connected in series. Note their linear graphs as shown in the figure. The linear graph in (a) has two primary loops (two elements in parallel), whereas in (b) it has only one loop, corresponding to all elements in series with the force. In Table 2.4 we note the differences in their node and loop equations. These observations should be intuitively clear, without even writing loop or node equations.

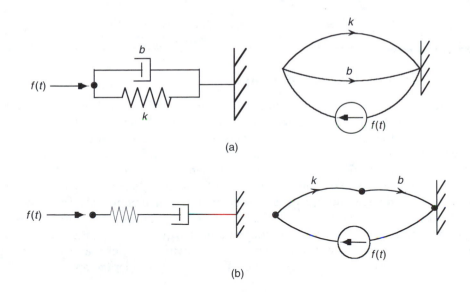

FIGURE 2.22
Spring-damper systems and their linear graphs: (a) elements in parallel; (b) elements in series.

TABLE 2.4

Series-Connected Systems and Parallel-Connected Systems

Series system	Parallel system
Through variables are the same	Across variables are the same
Across variables are not the same	Through variables are not the same

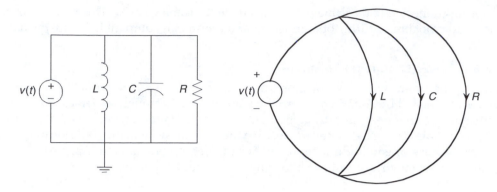

FIGURE 2.23
An *L-C-R* circuit and its linear graph.

Let us next consider the *L-C-R* circuit and its linear graph, as shown in Figure 2.23. This electrical system is analogous to the mechanical system of Figure 2.21. We may select any three loops as primary loops, for example, (*v–L, L–C, C–R*) or (*v–C, L–C, C–R*) or (*v–L, v–C, v–R*), etc. No matter what set we choose, we will get the same "equivalent" loop equations. In particular, the across variables for all four branches of this linear graph are the same.

2.6   State Models From Linear Graphs

We can obtain a state model of a system from its linear graph. In particular, each branch in a linear graph is a "model" of an actual system element. In mechanical systems it is common to use forces and velocities as state variables; specifically, the velocities of independent inertia elements in the system, and the forces associated with independent stiffness elements, because the state of an inertia element (mass element) can be represented by its velocity, and the state of a stiffness element (spring element) can be represented by its force. Masses and springs are independent storage elements, defined as elements that can possess or be assigned energy values, independently, at any given instant. The *system order* depends on (and can be defined as) the number of independent energy-storage elements in the system. The total number of masses and springs in a system can be greater than the system order because some of these energy storage elements might not be independent.

2.6.1   Sign Convention

The important first step of developing a state-space model using linear graphs is indeed to draw a linear graph for the considered system. A sign convention should be established, and as discussed before, a useful convention is given below:

1. Power flows into the action point and out of the reference point of a system element. This direction is shown by an arrow (unless the element is a source).
2. Through variable (*f*) and the across variable (*v*) are taken to be positive in the same direction at the action point as given by the linear graph arrow. At the reference point *v* is taken to be positive in the same direction as the linear-graph arrow, but *f* is taken positive in the opposite direction.

3. Flow into a node is taken as positive (in writing node equations).

4. Loop direction is taken to be counterclockwise (in writing loop equations). Potential "drop" is taken as positive, which direction is the same as the linear-graph (branch) arrow, except for a source element (For a source, potential increases in the arrow direction).

Note that these are conventions which need to be established first. Then, the actual values of the variables could be positive or negative depending on their actual direction.

2.6.2 Steps in Obtaining a State Model

The following four steps will create a set of state equations (a state-space model) from a linear graph:

1. Choose the state variables (forces and velocities for a mechanical system; currents and voltages would be chosen for an electrical system).

2. Write the constitutive equations (characteristic relationships) for the independent inertia and stiffness elements in the system (independent capacitors and inductors in an electrical system).

3. Do the same for the remaining elements (dependent stiffness and inertia elements, and damping elements; dependent inductors and capacitors, and resistors).

4. Develop the state equation (i.e., retain the state and input variables) by eliminating all other variables, using continuity (node) equations and compatibility (loop) equations.

If a linear graph has s sources (forcing functions) and a number of branches equal to b, then the total number of unknown variables in the system is

$$\text{Number of unknowns} = 2b - s \tag{2.72}$$

This is true because each passive branch contributes 2 unknowns (f, v), and each source $f(t)$ contributes one known variable.

Example 2.8

Let us develop a state-space model for the system shown in Figure 2.21, using its linear graph. There are 4 branches and one source. Thus $2b - s = 7$; we will need 7 equations to solve for unknowns. Note in Figure 2.21 that there are 3 primary loops. In particular, in this example we have:

Number of line branches $b = 4$

Number of nodes $n = 2$

Number of sources $s = 1$

Number of primary loops $l = 3$

Number of unknowns $= v_1, f_1, v_2, f_2, v_3, f_3, v = 7$

(Note: $f(t)$, the input variable, is known)

Number of constitutive equations (one each for m, k, b) $= b - s = 3$

Number of node equations $= n - 1 = 1$

Number of loop equations $= 3$ (because there are three primary loops)

Also,

unknowns − constitutive equations − node equations = 7 − 3 − 1 = 3 = loops.

Hence the system is solvable (7 unknowns and 7 equations).

Step 1. **Select state variables:** Since this is a second-order system (two independent energy-storage elements—mass m, spring k), we select v_1 and f_2 (the across variable of m and the through variable of k) as our state variables, and let:

$$x_1 = v_1$$
$$x_2 = f_2$$

The input variable is the applied forcing function (force source) $f(t)$.

Step 2. **Constitutive equations for *m* and k:**
From Newton's second law

$$f_1 = m\dot{v}_1$$

or,
$$\dot{v}_1 = (1/m)f_1 \tag{i}$$

Hooke's law, for spring

$$\dot{f}_2 = kv_2 \tag{ii}$$

The Equation i and Equation ii are our skeleton state equations (for v_1, and f_2).

Step 3. **Remaining constitutive equation** (for damper)

$$f_3 = bv_3 \tag{iii}$$

Step 4. **Node equation** (for node 2):

$$f - f_1 - f_2 - f_3 = 0 \tag{iv}$$

We see that the signs in Figure 2.21 give this result directly. Note that the arrow of the source branch is opposite to the other three branches.

The loop equation for loop 1: $v_1 - v_2 = 0$ $\qquad\qquad$ (v)

The loop equation for loop 2: $v_2 - v_3 = 0$ $\qquad\qquad$ (vi)

The loop equation for loop 3: $v_3 - v = 0$ $\qquad\qquad$ (vii)

The velocity (v) of the force source is not a state variable, and we need not use Equation vii. But, for the sake of completeness, let us comment about the sign of v. According to the present convention, v_3 and v have the same sign (see Equation vii). When v_3 and $f(t)$ are positive, for example, power from the source flows into it at node 2. The positive direction of the source variable $f(t)$ indicates that for that branch, node 2 should be the action point for measuring the across variable (v) of the force source. Hence, v = velocity at node 2 minus the velocity at node 1:

$$v = v_3 \tag{viii}$$

To obtain the state model, we wish to retain v_1 and f_2, by eliminating the auxiliary variables f_1 and v_2 in Equation i and Equation ii, by using Equation iii and Equation vi. The result is our two state equations:

$$\dot{v}_1 = -\frac{b}{m}v_1 - \frac{1}{m}f_2 + \frac{1}{m}f$$

$$\dot{f}_2 = kv_1$$

In the standard notation, with the state vector $x = [x_1 \quad x_2]^T = [v_1 \quad f_2]^T$ and the input vector $u = f(t)$ we have the system matrix:

$$A = \begin{bmatrix} -b/m & -1/m \\ k & 0 \end{bmatrix}$$

and the input distribution (gain) matrix:

$$B = \begin{bmatrix} 1/m \\ 0 \end{bmatrix}$$

Note that this is a second-order system, as clear from the fact that the state vector x is a second-order vector and, further, from the fact that the system matrix A is a 2×2 matrix. Also, note that in this system, the input vector u has only one element, $f(t)$. Hence it is actually a scalar variable, not a vector.

Example 2.9

A dynamic absorber is a passive vibration-control device that is mounted on a rotating system. By properly tuning (selecting the parameters of) the absorber, it is possible to absorb most of the power supplied by an unwanted excitation (e.g., support motion) in sustaining the absorber motion such that, in steady operation, the vibratory motions of the main system are inhibited. In practice, there should be some damping present in the absorber to dissipate the supplied energy, without generating excessive motions in the absorber mass. In the example shown in Figure 2.24(a), the main system and the absorber are modeled as simple oscillators with parameters (m_2, k_2, b_2) and (m_1, k_1, b_1), respectively.

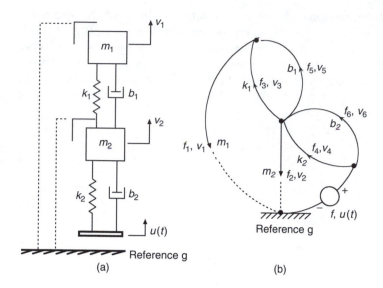

FIGURE 2.24
(a) Shock-absorber system; (b) Linear graph of the system.

The linear graph of this system is shown in Figure 2.24(b). The external excitation is the velocity $u(t)$ of the support. We note the following:

Number of branches $= b = 7$
Number of nodes $= n = 4$
Number of sources $= s = 1$
Number of independent loops $= l = 4$
Number of unknowns $= 2b - s = 13$
Number of constitutive equations $= b - s = 6$
Number of node equations $= n - 1 = 3$
Number of loop equations $= 4$

The four loop equations will be provided by the four independent loops.

CHECK   Number of unknowns $= 2b - s = 13$
        Number of equations $= (b - s) + (n - 1) + l = 6 + 3 + 4 = 13$

Step 1. Since the system has four independent energy storage elements (m_1, m_2, k_1, k_2) it is a fourth-order system. The state variables are chosen as the across variables of the two masses (velocities v_1 and v_2) and the through variables of the two springs (forces f_1 and f_2). Hence,

$$x = [x_1, \quad x_2 \quad x_3 \quad x_4]^T = [v_1, \quad v_2 \quad f_3 \quad f_4]^T$$

The input variable is $u(t)$.

Step 2. The skeleton state equations (model shell) are:

Newton's second law for mass m_1: $\quad\quad \dot{v}_1 = \dfrac{1}{m_1} f_1$

Newton's second law for mass m_2: $\quad\quad \dot{v}_2 = \dfrac{1}{m_2} f_2$

Hooke's law for spring k_1: $\quad\quad\quad\quad\quad \dot{f}_3 = k_1 v_3$

Hooke's law for spring k_2: $\quad\quad\quad\quad\quad \dot{f}_4 = k_2 v_4$

Step 3. The remaining constitutive equations:

For damper b_1: $\quad\quad\quad\quad\quad\quad\quad f_5 = b_1 v_5$

For damper b_2: $\quad\quad\quad\quad\quad\quad\quad f_6 = b_2 v_6$

Step 4. The node equations:

$$-f_1 + f_3 + f_5 = 0$$
$$-f_3 - f_5 - f_2 + f_4 + f_6 = 0$$
$$-f_4 - f_6 + f = 0$$

The loop equations:

$$v_1 - v_2 + v_3 = 0$$
$$v_2 - u + v_4 = 0$$
$$-v_4 + v_6 = 0$$
$$-v_3 + v_5 = 0$$

By eliminating the auxiliary variables, the following state equations are obtained:

$$\dot{v}_1 = -(b_1/m_1)v_1 + (b_1/m_1)v_2 + (1/m_1)f_3$$
$$\dot{v}_2 = (b_1/m_2)v_1 - [(b_1 + b_2)/m_2]v_2 - (1/m_2)f_3 + (1/m_2)f_4 + (b_2/m_2)u(t)$$
$$\dot{f}_3 = -k_1 v_1 + k_1 v_2$$
$$\dot{f}_4 = -k_2 v_2 + k_2 u(t)$$

This corresponds to the system matrix is

$$
A = \begin{bmatrix}
-b_1/m_1 & b_1/m_1 & 1/m_1 & 0 \\
b_1/m_2 & -(b_1+b_2)/m_2 & -1/m_2 & 1/m_2 \\
-k_1 & k_1 & 0 & 0 \\
0 & -k_2 & 0 & 0
\end{bmatrix}
$$

The input distribution matrix is

$$
B = \begin{bmatrix}
0 \\
b_2/m_2 \\
0 \\
k_2
\end{bmatrix}
$$

Example 2.10

Commercial motion controllers are digitally controlled (microprocessor-controlled) high-torque devices capable of applying a prescribed motion to a system. Such controlled actuators can be considered as velocity sources. Consider an application where a rotary motion controller is used to position an object, which is coupled through a gear box. The system is modeled as in Figure 2.25. We will develop a state-space model for this system.

Step 1. Note that the two inertial elements m_1 and m_2 are not independent, and together comprise one storage element. Thus, along with the stiffness element, there are only two independent energy storage elements. Hence the system is second order. Let us choose as state variables, v_1 and f_2 – the across variable of one of the inertias (because the other inertia will be "dependent") and the through variable of the spring.
We let

$$
x_1 = v_1 \text{ and } x_2 = f_2; \text{ hence } [x_1 \quad x_2]^T = [v_1 \quad f_2]^T
$$

Step 2. The constitutive equations for m_1 and k:

$$
\dot{v}_1 = \frac{1}{m_1} f_1 \qquad \dot{f}_2 = k v_2
$$

Step 3. The remaining constitutive equations:
For damper:

$$
f_3 = b v_3
$$

For the "dependent" inertia m_2:

$$
\dot{v}_4 = \frac{1}{m_2} f_4
$$

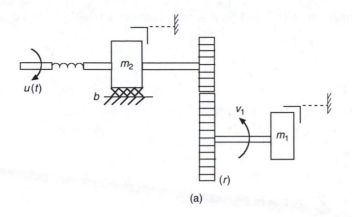

(a)

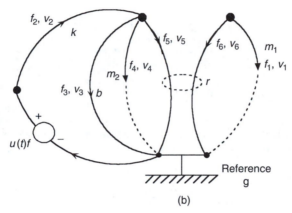

(b)

FIGURE 2.25
(a) Rotary-motion system with a gear box; (b) Linear graph of the system.

For the transformer (pair of meshed gear wheels):

$$v_6 = r v_5 \qquad f_6 = -\frac{1}{r} f_5$$

Step 4. The node equations:

$$-f_6 - f_1 = 0$$

$$f - f_2 = 0$$

$$f_2 - f_3 - f_4 - f_5 = 0$$

The loop equations:

$$v_6 - v_1 = 0$$

$$v_3 - v_4 = 0$$

$$v_4 - v_5 = 0$$

$$-v_2 + u(t) - v_3 = 0$$

Using equations from Step 3 and Step 4, the auxiliary variable f_1 can be expressed as

$$f_1 = \frac{1}{r}\left[f_2 - \frac{b}{r}v_1 - \frac{m_2}{r}\dot{v}_1 \right]$$

The auxiliary variable v_2 can be expressed as

$$v_2 = -\frac{1}{r}v_1 + u(t)$$

This results in the following two state equations:

$$\dot{v}_1 = -\left[\frac{b}{(m_1r^2 + m_2)} \right]v_1 + \left[\frac{r}{(m_1r^2 + m_2)} \right]f_2$$

$$\dot{f}_2 = -\frac{k}{r}v_1 + ku(t)$$

Note that the system is second order; only two state equations are obtained. The corresponding system matrix and the input-gain matrix (input distribution matrix) are:

$$A = \begin{bmatrix} -b/m & r/m \\ -k/m & 0 \end{bmatrix}, \quad B = \begin{bmatrix} 0 \\ k \end{bmatrix}$$

where $m = m_1r^2 + m_2$, which is the *equivalent inertia* of m_1 and m_2 when determined at the location of inertia m_2.

Example 2.11

a. List several advantages of using linear graphs in developing a state-space model of a dynamic system.
b. Electrodynamic shakers are commonly used in the dynamic testing of products. One possible configuration of a shaker/test-object system is shown in Figure 2.26(a). A simple, linear, lumped-parameter model of the mechanical system is shown in Figure 2.26(b).

Note that the driving motor is represented by a torque source T_m. Also, the following parameters are indicated:

J_m = equivalent moment of inertia of motor rotor, shaft, coupling, gears, and shaker platform
r_1 = pitch circle radius of the gear wheel attached to the motor shaft
r_1 = pitch circle radius of the gear wheel rocking the shaker platform
l = lever arm from the rocking gear center to the support location of the test object
m_L = equivalent mass of the test object and support fixture

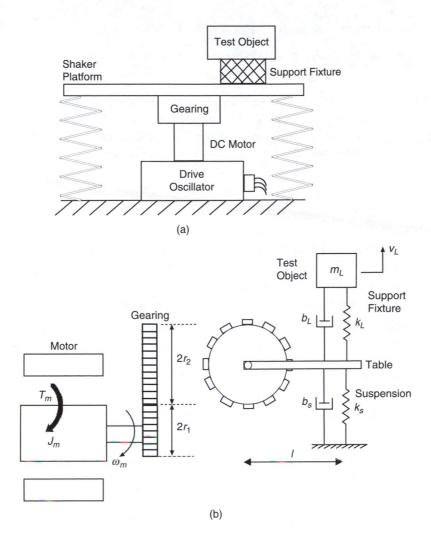

FIGURE 2.26
(a) A dynamic-testing system; (b) A model of the dynamic testing system.

k_L = stiffness of the support fixture
b_L = equivalent viscous damping constant of the support fixture
k_s = stiffness of the suspension system of the shaker table
b_s = equivalent viscous damping constant of the suspension system

Note that, since the inertia effects are lumped into equivalent elements it may be assumed that the shafts, gearing, platform and the support fixtures are light. The following variables are of interest:

ω_m = angular speed of the drive motor
v_L = vertical speed of motion of the test object
f_L = equivalent dynamic force of the support fixture (in spring k_L)
f_s = equivalent dynamic force of the suspension system (in spring k_s)

i. Obtain an expression for the motion ratio

$$r = \frac{\text{vertical movement of the shaker table at the test object support location}}{\text{angular movement of the drive motor shaft}}$$

ii. Draw a linear graph to represent the dynamic model.

iii. Using $x = [\omega_m \quad f_s \quad f_L \quad v_L]^T$ as the state vector, $u = [T_m]$ as the input, and $y = [v_L \quad f_L]^T$ as the output vector, obtain a complete state-space model for the system. You must use the linear graph drawn in Part (ii).

SOLUTION

a. Linear graphs

   • use physical variables as states.

   • provide a generalized approach for mechanical, electrical, fluid, and thermal systems

   • show the directions of power flows

   • provide a graphical representation of the system model

   • provide a systematic approach to automatically (computer) generate state equations

b. i. Let θ_m = rotation of the motor (drive gear).

   Hence, rotation of the output gear = $\frac{r_1}{r_2}\theta_m$

   Hence, displacement of the table at the test-object support point = $l\frac{r_1}{r_2}\theta_m$

   Hence, $r = l\frac{r_1}{r_2}$

   ii. The linear graph is shown in Figure 2.27.

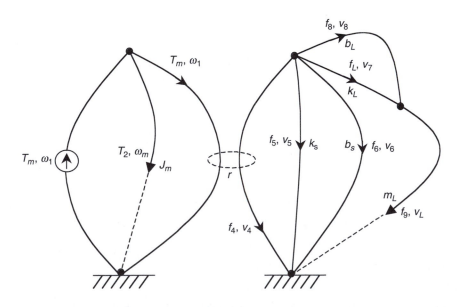

FIGURE 2.27
Linear graph of the shaker system.

iii. Constitutive equations

State space shell:

$$J_m \frac{d\omega_m}{dt} = T_2$$

$$\frac{df_s}{dt} = k_s v_5$$

$$\frac{df_L}{dt} = k_L v_L$$

$$m_L \frac{dv_L}{dt} = f_9$$

Others:

$$v_4 = r\omega \qquad \text{where } r = \frac{r_1}{r_2} l$$

$$f_4 = -\frac{1}{r} T_3$$

$$f_6 = b_s v_6$$

$$f_8 = b_L v_8$$

Continuity (node) equations

$$T_m - T_2 - T_3 = 0$$

$$-f_4 - f_s - f_6 - f_L - f_8 = 0$$

$$f_L + f_8 - f_9 = 0$$

Compatibility (loop) equations

$$-\omega_m + \omega_3 = 0$$

$$-v_4 + v_5 = 0$$

$$-v_5 + v_6 = 0$$

$$-v_6 + v_7 + v_L = 0$$

$$-v_7 + v_8 = 0$$

Elimination/substitution results in the following:

$$J_m \frac{d\omega_m}{dt} = T_2 = T_m - T_3 = T_m + rf_4 = T_m + r(-f_s - f_6 - f_L - f_8)$$

$$= T_m - r(f_s + b_s v_6 + f_L + b_L v_8)$$

$$= T_m - r(f_s + f_L + b_s v_6 + b_L v_8)$$

$$v_6 = v_5 = v_4 = r\omega_3 = r\omega_m$$

$$v_8 = v_7 = v_6 - v_L = v_5 - v_L = v_4 - v_L = r\omega_3 - v_L = r\omega_m - v_L$$

Hence,

$$J_m \frac{d\omega_m}{dt} = T_m - r(f_s + f_L) - r^2 b_s \omega_m - rb_L(r\omega_m - v_L) \qquad \text{(i)}$$

$$\frac{df_s}{dt} = k_s v_5 = k_s v_4 = k_s r\omega_3 = k_s r\omega_m \qquad \text{(ii)}$$

$$\frac{df_L}{dt} = k_L v_7 = k_L(v_6 - v_L) = k_L(v_4 - v_L) = k_L(r\omega_3 - v_L) = k_L(r\omega_m - v_L) \qquad \text{(iii)}$$

$$m_L \frac{dv_L}{dt} = f_9 = f_L + f_8 = f_L + b_L v_8 = f_L + b_L(r\omega_m - v_L) \qquad \text{(iv)}$$

In summary, we have the following state equations:

$$J_m \frac{d\omega_m}{dt} = T_m - rf_s - rf_L - r^2(b_s + b_L)\omega_m + rb_L v_L$$

$$\frac{df_s}{dt} = rk_s \omega_m$$

$$\frac{df_L}{dt} = rk_L \omega_m - k_L v_L$$

$$m_L \frac{dv_L}{dt} = f_L + rb_L \omega_m - b_L v_L$$

with v_L and f_L as the outputs.

Or, $\dot{x} = Ax + Bu$ and $y = Cx + Du$

where

$$x = [\omega_m \quad f_s \quad f_L \quad v_L]^T, \quad u = [T_m], \quad y = [v_L \quad f_L]^T$$

$$A = \begin{bmatrix} -\dfrac{r^2}{J_m}(b_s + b_L) & -\dfrac{r}{J_m} & -\dfrac{r}{J_m} & -\dfrac{rb_L}{J_m} \\ rk_s & 0 & 0 & 0 \\ rk_L & 0 & 0 & -k_L \\ \dfrac{rb_L}{m_L} & 0 & \dfrac{1}{m_L} & -\dfrac{b_L}{m_L} \end{bmatrix}, \quad B = \begin{bmatrix} 1/J_m \\ 0 \\ 0 \\ 0 \end{bmatrix}$$

$$C = \begin{bmatrix} 0 & 0 & 0 & 1 \\ 0 & 0 & 1 & 0 \end{bmatrix}, \qquad D = 0$$

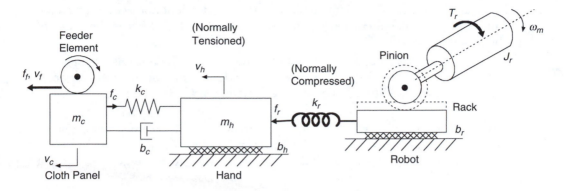

FIGURE 2.28
A robotic sewing system.

Example 2.12

A robotic sewing system consists of a conventional sewing head. During operation, a panel of garment is fed by a robotic hand into the sewing head. The sensing and control system of the robotic hand ensures that the seam is accurate and the cloth tension is correct in order to guarantee the quality of the stitch. The sewing head has a frictional feed mechanism, which pulls the fabric in a cyclic manner away from the robotic hand, using a toothed feeding element. When there is slip between the feeding element and the garment, the feeder functions as a *force source* and the applied force is assumed cyclic with a constant amplitude. When there is no slip, however, the feeder functions as a *velocity source*, which is the case during normal operation. The robot hand has inertia. There is some flexibility at the mounting location of the hand on the robot. The links of the robot are assumed rigid and some of its joints can be locked to reduce the number of degrees of freedom when desired.

Consider the simplified case of single-degree-of-freedom robot. The corresponding robotic sewing system is modeled as in Figure 2.28. Note that the robot is modeled as a single moment of inertia J_r that is linked to the hand with a light rack-and-pinion device of speed transmission parameter given by,

$$\frac{\text{rack tanslatory movement}}{\text{pinion rotatory movement}} = r$$

The drive torque of the robot is T_r and the associated rotatory speed is ω_r. Under conditions of slip the feeder input to the cloth panel is the force f_f, and with no slip the input is the velocity v_f. Various energy dissipation mechanisms are modeled as linear viscous damping of damping constant b (with appropriate subscripts). The flexibility of various system elements is modeled by linear springs with stiffness k. The inertia effects of the cloth panel and the robotic hand are denoted by the lumped masses m_c and m_h, respectively, having velocities v_c and v_h, as shown in Figure 2.28. Note that the cloth panel is normally in tension with tensile force f_c. In order to push the panel, the robotic wrist is normally in compression with compressive force f_r.

First consider the case of the feeding element with slip.

a. Draw a linear graph for the model shown in Figure 2.28, orient the graph, and mark all the element parameters, through variables and across variables on the graph.

b. Write all the constitutive (element) equations, independent node equations (continuity), and independent loop equations (compatibility). What is the order of the model?

c. Develop a complete state-space model for the system. The outputs are taken as the cloth tension f_c, and the robot speed ω_r, which represent the two variables that have to be measured to control the system. Obtain the system matrices A, B, C, and D.

d. Now consider the case where there is no slip at the feeder element. What is the order of the system now? Modify the linear graph of the model for this situation. Then modify the state-space model obtained earlier to represent the present situation and give the new matrices A, B, C, and D.

e. Generally comment on the validity of the assumptions made in obtaining the model shown in Figure 2.28 for a robotic sewing system.

SOLUTION

a. Linear graph of the system is given in Figure 2.29.

b. In the present operation f_f is an input. This case corresponds to a fifth-order model.

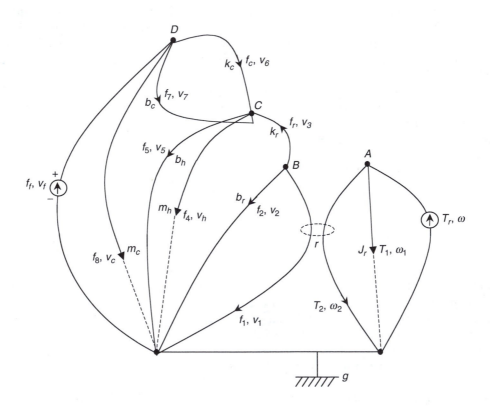

FIGURE 2.29
Linear graph of the robotic sewing system.

Constitutive equations:

$$J_r \frac{d\omega_r}{dt} = T_1$$

$$\frac{df_r}{dt} = k_r v_3$$

$$m_h \frac{dv_h}{dt} = f_4 \quad \Big\} \text{State-space shell}$$

$$\frac{df_c}{dt} = k_c v_6$$

$$m_c \frac{dv_c}{dt} = f_8$$

$$v_1 = r\omega_2$$

$$f_1 = -\frac{1}{r} T_2$$

$$f_2 = -b_r v_2$$

$$f_5 = -b_h v_5$$

$$f_7 = -b_c v_7$$

Continuity equations (node equations):

Node *A*: $-T_r + T_1 + T_2 = 0$

Node *B*: $f_1 + f_2 + f_r = 0$

Node *C*: $-f_r - f_c - f_7 + f_5 + f_4 = 0$

Node *D*: $f_c - f_f + f_8 + f_7 = 0$

Compatibility equations (loop equations):

$$-\omega + \omega_r = 0$$

$$-\omega_r + \omega_2 = 0$$

$$-v_1 + v_2 = 0$$

$$-v_1 + v_3 + v_h = 0$$

$$-v_h + v_5 = 0$$

$$-v_6 + v_7 = 0$$

$$-v_h - v_7 + v_c = 0$$

$$-v_c + v_f = 0$$

c. Eliminate unwanted variables as follows:

$$T_1 = T_r - T_2 = T_r + rf_1 = T_r + r(-f_2 - f_r)$$
$$= T_r - rb_r v_2 - rf_r = T_r - rb_r v_1 - rf_r$$
$$= T_r - rb_r r\omega_2 - rf_r$$
$$= T_r - r^2 b_r \omega_2 - rf_r$$

$$v_3 = v_1 - v_h = r\omega_2 - v_h = r\omega_r - v_h$$
$$f_4 = f_r + f_c + f_7 - f_5 = f_r + f_c + b_c v_7 - b_h v_5$$
$$= f_r + f_c + b_c(v_c - v_h) - b_h v_h$$
$$v_6 = v_7 = v_c - v_h$$
$$f_8 = f_f - f_c - f_7 = f_f - f_c - b_c v_7 = f_f - f_c - b_c(v_c - v_h)$$

State space model:

$$J_r \frac{d\omega_r}{dt} = -r^2 b_r \omega_r - rf_r + T_r$$

$$\frac{df_r}{dt} = k_r(r\omega_r - v_h)$$

$$m_h \frac{dv_h}{dt} = f_r - (b_c + b_h)v_h + f_c + b_c v_c$$

$$\frac{df_c}{dt} = k_c(-v_h + v_c)$$

$$m_c \frac{dv_c}{dt} = b_c v_h - f_c - b_c v_c + f_f$$

with $x = [\omega_r \quad f_r \quad v_h \quad f_c \quad v_c]^T$, $u = [T_r \quad f_f]^T$, $y = [f_c \quad \omega_r]^T$

$$A = \begin{bmatrix} -r^2 b_r/J_r & -r/J_r & 0 & 0 & 0 \\ rk_r & 0 & -k_r & 0 & 0 \\ 0 & 1/m_h & -(b_c + b_h)/m_h & 1/m_h & b_c/m_h \\ 0 & 0 & -k_c & 0 & k_c \\ 0 & 0 & b_c/m_c & -1/m_c & -b_c/m_c \end{bmatrix}, \quad B = \begin{bmatrix} 1/J_r & 0 \\ 0 & 0 \\ 0 & 0 \\ 0 & 0 \\ 0 & 1/m_c \end{bmatrix}$$

$$C = \begin{bmatrix} 0 & 0 & 0 & 1 & 0 \\ 1 & 0 & 0 & 0 & 0 \end{bmatrix}, \quad D = 0$$

d. In this case, v_f is an input. Then, the inertia element m_c ceases to influence the dynamics of the overall system because, $v_c = v_f$ in this case and is completely specified. Hence, we have a fourth- order model. Now

$$x = [\omega_r \quad f_r \quad v_h \quad f_c]^T \quad u = [T_r \quad f_f]^T$$

State model:

$$J_r \frac{d\omega_r}{dt} = -r^2 b_r \omega_r - rf_r + T_r$$

$$\frac{df_r}{dt} = k_r(r\omega_r - v_h)$$

$$m_h \frac{dv_h}{dt} = f_r - (b_c + b_h)v_h + f_c + b_c v_f$$

$$\frac{df_c}{dt} = k_c(-v_h + v_f)$$

The corresponding matrices are:

$$A = \begin{bmatrix} -r^2 b_r/J_r & -r/J_r & 0 & 0 \\ rk_r & 0 & -k_r & 0 \\ 0 & 1/m_h & -(b_c + b_h)/m_h & 1/m_h \\ 0 & 0 & -k_c & 0 \end{bmatrix}, \quad B = \begin{bmatrix} 1/J_r & 0 \\ 0 & 0 \\ 0 & b_c/m_h \\ 0 & k_c \end{bmatrix}$$

$$C = \begin{bmatrix} 0 & 0 & 0 & 1 \\ 1 & 0 & 0 & 0 \end{bmatrix}, \quad D = 0$$

2.7 Electrical Systems

Thus far we have primarily considered the modeling of mechanical systems—systems with inertia, flexibility, and mechanical energy dissipation. In view of the analogies that exist between mechanical, electrical, fluid, and thermal components and associated variables, there is an "analytical" similarity between these four types of physical systems. Accordingly, once we have developed procedures for modeling and analysis of one type of systems (say, mechanical systems) the same procedures may be extended (in an "analogous" manner) to the other three types of systems. First we will make use of these analogies to model electrical systems, by making use of the same procedures that have been used for mechanical systems. Next we will specifically consider fluid systems and thermal systems. These procedures can be extended to mixed systems—systems that use a combination of two or more types of physical components (mechanical, electrical, fluid, and thermal) in an integrated manner. Since the general procedures have been given in the previous sections, we will mainly employ illustrative examples to show how the procedures are used in specific types of systems. Futher considerations of electrical systems are found in Chapter 4, Chapter 5, Chapter 9 and Chaper 10.

Table 2.1 gives the constitutive equations for the three passive electrical elements: capacitor (an A-type of element with the *across variable* voltage as the state variable); inductor (a T-type element with the *through variable* current as the state variable); and resistor (D-type element representing energy dissipation, and no state variable is associated with it). The two types of energy present in an electrical system are:

- Electrostatic energy (in the stored charge of a capacitor)
- Electromagnetic energy (in the magnetic field of an inductor).

2.7.1 Capacitor

Electrical charge (q) is a function of the voltage (v) across the capacitor:

$$q = q(v) \tag{2.73}$$

For the linear case we have

$$q = Cv \tag{2.74}$$

where C is the *capacitance*. Then the current (i), which is $\dfrac{dq}{dt}$, is given by differentiating 2.74:

$$i = C\frac{dv}{dt} + v\frac{dC}{dt} \tag{2.75}$$

where we have allowed for time-varying capacitance. But, if C is constant, we have the familiar linear constitutive equation

$$i = C\frac{dv}{dt} \tag{2.76}$$

2.7.2 Inductor

Magnetic flux linkage (λ) of an inductor is a function of the current (i) through the inductor:

$$\lambda = \lambda(i) \tag{2.77}$$

For the linear case we have

$$\lambda = Li \tag{2.78}$$

where L is the *inductance*. The voltage induced in an inductor is equal to the rate of change of the flux linkage. Hence, by differentiating Equation 2.78 we get

$$v = L\frac{di}{dt} + i\frac{dL}{dt} \tag{2.79}$$

Assuming that the inductance is constant, we have the familiar linear constitutive equation

$$v = L \frac{di}{dt} \tag{2.80}$$

2.7.3   Resistor

In general the voltage across a (nonlinear) resistor is a function of the current through the resistor:

$$v = v(i) \tag{2.81}$$

In the linear case

$$v = R\, i \tag{2.82}$$

where R is the *resistance*, which can be time-varying in general. In most cases, however, we assume R to be constant.

Circuit representations of these three (passive) elements are shown in Figure 2.30. Also shown are two other useful elements: the transformer and the operational amplifier.

2.7.4   Transformer

A transformer has a primary coil, which is energized by an ac voltage (v_p), a secondary coil in which an ac voltage (v_s) is induced, and a common core, which helps the linkage of magnetic flux between the two coils. Note that a transformer converts v_p to v_s without

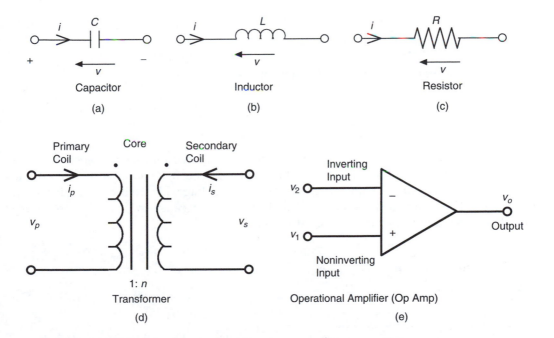

FIGURE 2.30
Basic electrical circuit elements: (a) Capacitor; (b) Inductor; (c) Resistor; (d) Transformer; (e) Operational amplifier.

making use of an external power source. Hence it is a *passive* device, just like a capacitor, inductor, or resistor. The turn ratio of the transformer

$$n = \frac{\text{number of turns in the secondary coil}}{\text{number of turns in the primary coil}}$$

In Figure 2.30(c), the two dots on the top side of the two coils indicate that the two coils are wound in the same direction.

In a "pure" and "ideal" transformer, there will be full flux linkage without any dissipation. Then, the flux linkage will be proportional to the number of turns. Hence

$$\lambda_s = n\lambda_p \tag{2.83}$$

where λ denotes the flux linkage in each coil. Differentiation of Equation 2.83 noting that the induced voltage in coil is given by the rate of charge of flux, gives

$$v_s = nv_p \tag{2.84}$$

For an *ideal transformer*, there is no energy dissipation and also the signals will be in phase. Hence, the output power will be equal to the input power; thus,

$$v_s i_s = v_p i_p \tag{2.85}$$

Hence, the current relation becomes

$$i_s = \frac{1}{n} i_p \tag{2.86}$$

2.7.5 Source Elements

An electrical system has two types of source elements:

- Voltage source
- Current source.

A voltage source is able to provide a specified voltage without being affected by the current (loading). Hence it has a low output impedance. A current source is able to provide a specified current without being affected by the load voltage. Hence it has a high output impedance. These are idealizations of actual elements, because in practice, the source output changes due to loading (See Chapter 4).

2.7.6 Circuit Equations

As usual we write

1. Node equations for currents:
   The sum of currents into a circuit node is zero. This is the well-known Kirchhoff's current law.

2. Loop equations for voltages:
   The sum of voltages around a circuit loop is zero. This is the celebrated Kirchhoff's voltage law.

Finally, we eliminate the unwanted (auxiliary) variables from the three types of equations (constitutive, node, loop) to obtain the analytical model (say, state equations). Linear graphs can be used for this purpose as usual.

2.7.7 Operational Amplifier

This is an active device (needs an external power source for operation) that can be very useful in practical circuits. With respect to the circuit element shown in Figure 2.30(e), the input output equation is

$$v_2 - v_1 = -\frac{v_o}{k_a} \tag{2.87}$$

where k_a = open-loop gain of the op amp. Since k_a is very high (10^5–10^9) for a practical op amp, the inverting input voltage and the noninverting input voltage are nearly equal: $v_2 = v_1$.

The current through the input leads is given by

$$i = \frac{v_2 - v_1}{Z_i} \tag{2.88}$$

where Z_i is the input impedance of the op amp. For a practical op amp Z_i is very high (1 MΩ or more). Hence, the input current is also almost zero. The output impedance Z_o of an op amp is quite low. The impedance conversion property (with high Z_i and low Z_o) is a practical advantage of an op amp in instrumentation applications. Hence, an op amp is an impedance transformer. Since the open-loop gain k_a is quite variable and not precisely known (even though very high), an op amp is not practically used as an open-loop device. A feedback loop is completed from the output side to an input terminal of the op amp, in order to make it stable and practically useful. Details are found in Chapter 4.

2.7.8 DC Motor

The dc motor is a commonly used electrical actuator. It converts direct current (dc) electrical energy into mechanical energy. The principle of operation is that when a conductor carrying current is placed in a magnetic field, a force is generated. Details are found in Chapter 9.

Example 2.13

A classic problem in robotics is the case of a robotic hand gripping and turning a doorknob to open a door. The mechanism is schematically shown in Figure 2.31(a). Suppose that the actuator of the robotic hand is an armature-controlled dc motor. The associated circuit is shown in Figure 2.31(b). Note that the field circuit provides a constant magnetic field to the motor, and is not important in the present problem. The armature (motor rotor winding) circuit has a back e.m.f. v_b, a leakage inductance L_a, and a resistance R_a. The input signal to the robotic hand is the armature voltage $v_a(t)$ as shown. The rotation of the motor (at an

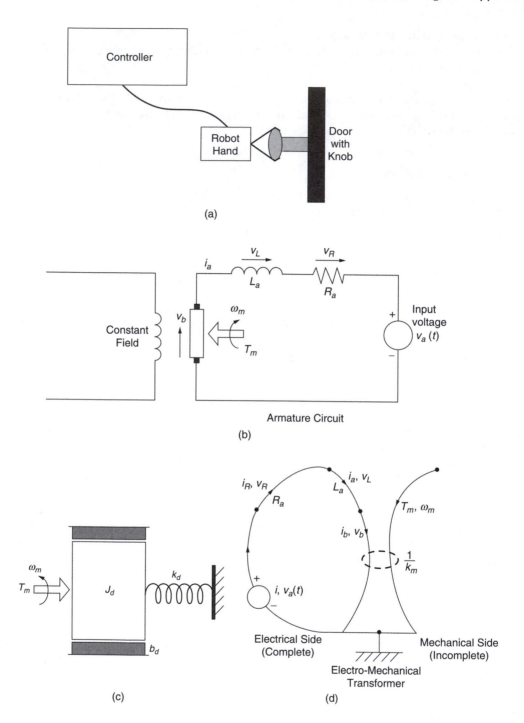

FIGURE 2.31
(a) Robotic hand turning a doorknob; (b) Armature-controlled dc motor of the robotic hand; (c) Mechanical model of the hand/doorknob system; (d) Incomplete linear graph.

angular speed ω_m) in the two systems of magnetic field generates a torque T_m (which is negative as marked in Figure 2.31(b) during normal operation). This torque (magnetic torque) is available to turn the doorknob, and is resisted by the inertia force (moment of inertia J_d), the friction (modeled as linear viscous damping of damping constant b_d) and the spring (of stiffness k_d) of the hand-knob-lock combination. A mechanical model is shown in Figure 2.31(c). The dc motor may be considered as an ideal electromechanical transducer which is represented by a linear-graph transformer. The associated equations are

$$\omega_m = \frac{1}{k_m} v_b \tag{2.89}$$

$$T_m = -k_m i_b \tag{2.90}$$

Note that the negative sign in Equation 2.90 arises due to the specific sign convention. The linear graph may be easily drawn, as shown in Figure 2.31(d), for the electrical side of the system.

Answer the following questions:

a. Complete the linear graph by including the mechanical side of the system.
b. Give the number of branches (b), nodes (n), and the independent loops (l) in the completed linear graph. Verify your answer.
c. Take current through the inductor (i_a), speed of rotation of the door knob (ω_d), and the resisting torque of the spring within the door lock (T_k) as the state variables, the armature voltage $v_a(t)$ as the input variable, and ω_d and T_k as the output variables. Write the independent node equations, independent loop equations, and the constitutive equations for the completed linear graph. Clearly show the state-space shell. Also verify that the number of unknown variables is equal to the number of equations obtained in this manner.
d. Eliminate the auxiliary variables and obtain a complete state-space model for the system, using the equations written in Part (c) above.

SOLUTION

a. The complete linear graph is shown in Figure 2.32.

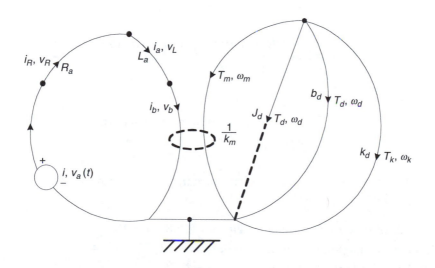

FIGURE 2.32
The complete linear graph of the system.

b. $b = 8$, $n = 5$, $l = 4$ for this linear graph. It satisfies the topological relationship $l = b - n + 1$

c. *Independent node equations:*

$$i - i_R = 0$$

$$i_R - i_a = 0$$

$$i_a - i_b = 0$$

$$-T_m - T_d - T_b - T_k = 0$$

Independent loop equations:

$$-v_a(t) + v_R + v_L + v_b = 0$$

$$-\omega_m + \omega_d = 0$$

$$-\omega_d + \omega_b = 0$$

$$-\omega_b + \omega_k = 0$$

Constitutive equations:

$$\left. \begin{array}{l} L_a \dfrac{di_a}{dt} = v_L \\[2mm] J_d \dfrac{d\omega_d}{dt} = T_d \\[2mm] \dfrac{dT_k}{dt} = k_d \omega_k \end{array} \right\} \text{State-space shell}$$

$$\left. \begin{array}{l} v_R = R_a i_R \\[2mm] T_b = b_d \omega_b \end{array} \right\} \text{Auxiliary constitutive equations}$$

$$\left. \begin{array}{l} \omega_m = \dfrac{1}{k_m} v_b \\[2mm] T_m = -k_m i_b \end{array} \right\} \text{Electromechanical transformer}$$

Note that there are 15 unknown variables (i, i_R, i_a, i_b, T_m, T_d, T_b, T_k, v_R, v_L, v_b, ω_m, ω_d, ω_b, ω_k) and 15 equations.

Number of unknown variables = $2b - s = 2 \times 8 - 1 = 15$
Number of independent node equations = $n - 1 = 5 - 1 = 4$
Number of independent loop equations = $l = 4$
Number of constitutive equations = $b - s = 8 - 1 = 7$

d. Eliminate the auxiliary variables from the state-space shell, by substitution:

$$v_L = v_a(t) - v_R - v_b = v_a(t) - R_a i_a - k_m \omega_m$$

$$= v_a(t) - R_a i_a - k_m \omega_d$$

$$T_d = -T_k - T_m - T_b = -T_k + k_m i_b - b_d \omega_b$$

$$= -T_k + k_m i_a - b_d \omega_d$$

$$\omega_k = \omega_b = \omega_d$$

Hence, we have the state-space equations:

$$L_a \frac{di_a}{dt} = -R_a i_a - k_m \omega_d + v_a(t)$$

$$J_d \frac{d\omega_d}{dt} = k_m i_a - b_d \omega_d - T_k$$

$$\frac{dT_k}{dt} = k_d \omega_d$$

Now with $x = [i_a \quad \omega_d \quad T_k]^T$, $u = [v_a(t)]$, and $y = [\omega_d \quad T_k]^T$ we have

$$\dot{x} = Ax + Bu$$

$$y = Cx + Du$$

where

$$A = \begin{bmatrix} -R_a/L_a & -k_m/L_a & 0 \\ -k_m/J_d & -b_d/J_d & -1/J_d \\ 0 & k_d & 0 \end{bmatrix}, \qquad B = \begin{bmatrix} 1/L_a \\ 0 \\ 0 \end{bmatrix}$$

$$C = \begin{bmatrix} 0 & 1 & 0 \\ 0 & 0 & 1 \end{bmatrix} \qquad D = 0$$

2.8 Fluid Systems

Pressure (P) is the across variable and the volume flow rate (Q) is the through variable in a fluid component. The three basic fluid elements are discussed below. Note the following:

1. The elements are usually distributed, but lumped-parameter approximations are used here.

2. The elements are usually nonlinear (particularly, the fluid resistor), but linear models are used here.

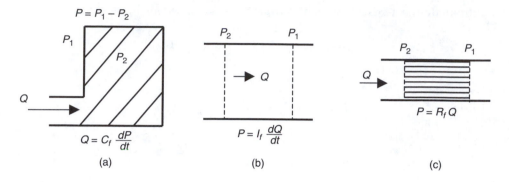

FIGURE 2.33
Basic fluid elements: (a) Capacitor; (b) Inertor; (c) Resistor.

2.8.1   Fluid Capacitor or Accumulator (A-type element)

Consider a rigid container with a single inlet through which fluid is pumped in at the volume rate Q, as shown in Figure 2.33(a). The pressure inside the container with respect to the outside is P. Then, we can write the linear constitutive equation

$$Q = C_f \frac{dP}{dt} \tag{2.91}$$

where C_f = fluid capacitance (capacity). Several special cases of fluid capacitance will be discussed later.

A fluid capacitor stores potential energy, given by $\frac{1}{2}C_f P^2$. Hence, this element is like a fluid spring. The appropriate state variable is the pressure difference (across variable) P. Contrast here that the mechanical spring is a T-type element.

2.8.2   Fluid Inertor (T-type element)

Consider a conduit carrying an accelerating flow of fluid, as shown in Figure 2.33(b). The associated linear constitutive equation may be written as

$$P = I_f \frac{dQ}{dt} \tag{2.92}$$

where I_f in the fluid inertance (inertia).

A fluid inertor stores kinetic energy, given by $\frac{1}{2}I_f Q^2$. Hence, this element is a fluid inertia. The appropriate state variable is the volume flow rate (through variable) Q. Contrast here that the mechanical inertia is an A-type element. Energy exchange between a fluid capacitor and a fluid inertor leads to oscillations (e.g., water hammer) in fluid systems, analogous to mechanical and electrical systems.

2.8.3   Fluid Resistor (D-type element)

Consider the flow of fluid through a narrow element such as a thin pipe, orifice, or valve. The associated flow will result in energy dissipation due to fluid friction. The linear constitutive equation is (see Figure 2.33(c)).

$$P = R_f Q \tag{2.93}$$

2.8.4 Fluid Source Element

The input elements in a fluid system are

- Pressure source (e.g., large reservoir or accumulator)
- Flow source (e.g., regulated pump).

These are idealizations of actual source devices where the source variable will be somewhat affected due to system loading.

2.8.5 System Equations

In addition to the constitutive equations, we need to write:

- Node equations (sum of flow into a junction is zero)
- Loop equations (sum of pressure drop around a closed path is zero).

The unwanted variables are eliminated from these equations to arrive at the analytical model (e.g., a state model or an input-output model).

2.8.6 Derivation of Constitutive Equations

We now indicate the derivation of the constitutive equations for fluid elements.

2.8.6.1 *Fluid Capacitor*

The capacitance in a fluid element may originate from

1. Bulk modulus effects of liquids
2. Compressibility effects of gases
3. Flexibility of the fluid container itself
4. Gravity head of a fluid column

Derivation of the associated constitutive equations is indicated below.

2.8.6.1.1 *Bulk Modulus Effect of Liquids*

Consider a rigid container. A liquid is pumped in at the volume rate of Q. An increase in the pressure in the container will result in compression of the liquid volume and thereby letting in more liquid (see Figure 2.34(a)). From calculus we can write

$$\frac{dP}{dt} = \frac{\partial P}{\partial V}\frac{dV}{dt}$$

where V is the control volume of liquid. Now, volume flow rate (into the container) $Q = -\frac{dV}{dt}$. By definition, bulk modulus of liquid:

$$\beta = -V\frac{\partial P}{\partial V} \tag{2.94}$$

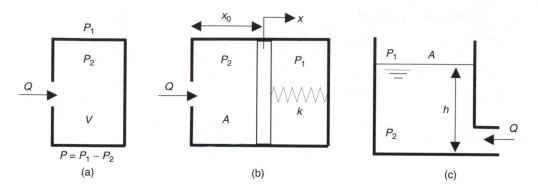

FIGURE 2.34
Three types of fluid capacitance: (a) Bulk modulus or compressibility; (b) Flexibility of container; (c) Gravity head of fluid column.

Hence,

$$\frac{dP}{dt} = \frac{\beta}{V}Q \quad \text{or} \quad Q = \frac{V}{\beta}\frac{dP}{dt} \tag{2.95}$$

and the associated capacitance is

$$C_{bulk} = \frac{V}{\beta} \tag{2.96}$$

2.8.6.1.2   Compression of Gases

Consider a perfect (ideal) gas, which is governed by the gas law

$$PV = mRT \tag{2.97}$$

where
P = pressure (units are pascals: $1\,\text{Pa} = 1\,\text{N/m}^2$)
V = volume (units are m^3)
T = absolute temperature (units are K or degrees Kelvin)
m = mass (units are kg)
R = specific gas constant (units: kJ/kg/K where $1\,\text{J} = 1\,\text{joule} = 1\,\text{N}\cdot\text{m}$; $1\,\text{kJ} = 1000\,\text{J}$).

Isothermal Case
Consider a slow flow of gas into a rigid container (see Figure 2.34(a)) so that the heat transfer is allowed to maintain the temperature constant (isothermal). Differentiate Equation 2.97 with T constant (i.e., RHS is constant)

$$P\frac{dV}{dt} + V\frac{dP}{dt} = 0$$

Noting that $Q = -\frac{dV}{dt}$ and substituting the above equation and Equation 2.97 we get

$$Q = \frac{V}{P}\frac{dP}{dt} = \frac{mRT}{P^2}\frac{dP}{dt} \tag{2.98}$$

Hence, the corresponding capacitance is

$$C_{comp} = \frac{V}{P} = \frac{mRT}{P^2} \tag{2.99}$$

Adiabatic Case
Consider a fast flow of gas (see Figure 2.34(a)) into a rigid container so that there is no time for heat transfer (adiabatic $\Rightarrow$ zero heat transfer). The associated gas law is known to be

$$PV^k = C \quad \text{with} \quad k = C_P/C_V \tag{2.100}$$

where
  C_p = specific heat when the pressure is maintained constant
  C_v = specific heat when the volume is maintained constant
  C = constant
  k = specific heat ratio

Differentiate Equation 2.100:

$$PkV^{k-1}\frac{dV}{dt} + V^k\frac{dP}{dt} = 0$$

Divide by V^k:

$$\frac{Pk}{V}\frac{dV}{dt} + \frac{dP}{dt} = 0$$

Now use $Q = -\frac{dV}{dt}$ as usual, and also substitute Equation 2.97:

$$Q = \frac{V}{kP}\frac{dP}{dt} = \frac{mRT}{kP^2}\frac{dP}{dt} \tag{2.101}$$

The corresponding capacitance is

$$C_{comp} = \frac{V}{kP} = \frac{mRT}{kP^2} \tag{2.102}$$

2.8.6.1.3 *Effect of Flexible Container*

Without loss of generality, consider a cylinder of cross-sectional area A with a spring-loaded wall (stiffness k) as shown in Figure 2.34(b). As a fluid (assumed incompressible) is pumped into the cylinder, the flexible wall will move through x.

Conservation of flow:
$$Q = \frac{d(A(x_0 + x))}{dt} = A\frac{dx}{dt} \qquad \text{(i)}$$

Equilibrium of spring:
$$A(P_2 - P_1) = kx \quad \text{or} \quad x = \frac{A}{k}P \qquad \text{(ii)}$$

Substitute Equation ii in Equation i. We get

$$Q = \frac{A^2}{k}\frac{dP}{dt} \qquad (2.103)$$

The corresponding capacitance

$$C_{\text{elastic}} = \frac{A^2}{k} \qquad (2.104)$$

NOTE   For an elastic container and a fluid having bulk modulus, the combined capacitance will be additive:

$$C_{eq} = C_{\text{bulk}} + C_{\text{elastic}}$$

Similar result holds for a compressible gas and an elastic container.

2.8.6.1.4   *Gravity Head of a Fluid Column*

Consider a liquid column (tank) having area of across section A, height h, and density ρ, as shown in Figure 2.34(c). The liquid is pumped into the tank at the volume rate Q, and as a result, the liquid level rises.

Relative pressure at the foot of the column $P = P_2 - P_1 = \rho g h$

Flow rate
$$Q = \frac{d(Ah)}{dt} = A\frac{dh}{dt}$$

Direct substitution gives

$$Q = \frac{A}{\rho g}\frac{dP}{dt} \qquad (2.105)$$

The corresponding capacitance is

$$C_{\text{grav}} = \frac{A}{\rho g} \qquad (2.106)$$

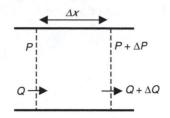

FIGURE 2.35
A fluid flow element.

2.8.6.2 *Fluid Inertor*

First assume a fluid flow in a conduit, with a uniform velocity distribution across it. Along a small element of length Δx of fluid, as shown in Figure 2.35, the pressure will change from P to $P + \Delta P$, and the volume flow rate will change from Q to $Q + \Delta Q$.

Mass of the fluid element $= \rho A \Delta x$

Net force in the direction of flow $= -\Delta PA$

Velocity of flow $= Q/A$

where
ρ = mass density of the fluid
A = area of cross section

Assuming A to be constant,

$$\text{Fluid acceleration} = \frac{1}{A}\frac{dQ}{dt}$$

Hence, Newton's second law gives

$$-\Delta PA = (\rho A \Delta x)\frac{1}{A}\frac{dQ}{dt}$$

or,

$$-\Delta P = \frac{\rho \Delta x}{A}\frac{dQ}{dt} \tag{2.107}$$

Hence,

$$\text{Fluid inertance } I_f = \frac{\rho \Delta x}{A} \tag{2.108a}$$

For a nonuniform cross-section, $A = A(x)$. Then for a length L

$$I_f = \int_0^L \frac{\rho}{A(x)}\,dx \tag{2.108b}$$

For a circular cross-section and a parabolic velocity profile, we have

$$I_f = \frac{2\rho\Delta x}{A} \tag{2.108c}$$

or, in general,

$$I_f = \alpha\frac{\rho\Delta x}{A} \tag{2.109}$$

where, α is a suitable correction factor.

2.8.6.3   *Fluid Resistor*

For the ideal case of viscous, laminar flow,

$$P = R_f Q \tag{2.110}$$

with

$$R_f = 128\frac{\mu L}{\pi d^4} \quad \text{for a circular pipe of diameter } d.$$

$$R_f = 12\frac{\mu L}{wb^3} \quad \begin{array}{l}\text{for a pipe of rectangular cross section (width } w \text{ and height } b)\\ \text{with } b \ll w\end{array}$$

where
   $L =$ length of pipe segment
   $\mu =$ absolute viscosity of fluid (dynamic viscosity).

NOTE   Fluid stress $= \mu\frac{du}{dy}$, where $\frac{du}{dy}$ is the velocity gradient across the pipe.

$v = \dfrac{\mu}{\rho} =$ kinematic viscosity.

Reynold's number $R_e = \dfrac{uL}{v} = \dfrac{\rho uL}{\mu}$

$u =$ fluid velocity along the pipe.
For turbulent flow, the resistance equation will be nonlinear:

$$P = K_R Q^n \tag{2.111}$$

Example 2.14

Consider two water tanks joined by a horizontal pipe with an on-off valve. With the valve closed, the water levels in the two tanks were initially maintained unequal. When the valve was suddenly opened, some oscillations were observed in the water levels of the tanks. Suppose that the system is modeled as two gravity-type capacitors linked by a fluid resistor. Would this model exhibit oscillations in the water levels when subjected to an initial-condition excitation? Clearly explain your answer.

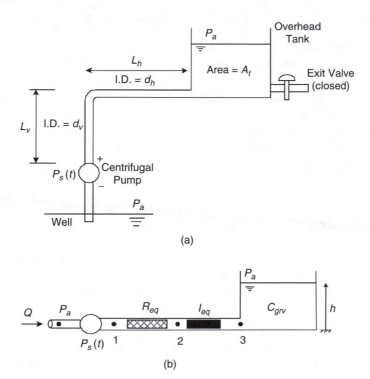

FIGURE 2.36
(a) A system for pumping water from a well into an overhead tank, (b) A lumped parameter model of the fluid system.

A centrifugal pump is used to pump water from a well into an overhead tank. This fluid system is schematically shown in Figure 2.36(a). The pump is considered as a pressure source $P_s(t)$ and the water level h in the overhead tank is the system output. The ambient pressure is denoted by P_a. The following parameters are given:

L_v, d_v = length and internal diameter of the vertical segment of pipe
L_h, d_h = length and internal diameter of the horizontal segment of pipe
A_t = area of cross section of overhead tank (uniform)
ρ = mass density of water
μ = dynamic viscosity of water
g = acceleration due to gravity

Suppose that this fluid system is approximated by the lumped parameter model shown in Figure 2.36(b).

a. Give expressions for the equivalent linear fluid resistance of the overall pipe (i.e., combined vertical and horizontal segments) R_{eq}, the equivalent fluid inertance within the overall pipe I_{eq}, and the gravitational fluid capacitance of the overhead tank C_{grv}, in terms of the system parameters defined above.

b. Treating $x = [P_{3a} \quad Q]^T$ as the state vector,

where

P_{3a} = pressure head of the overhead tank

Q = volume flow rate through the pipe

develop a complete state-space model for the system. Specifically, obtain the matrices **A**, **B**, **C**, and **D**.

c. Obtain the input-output differential equation of the system. What is the characteristic equation of this system?

d. Using the following numerical values for the system parameters:

L_v = 10.0 m, L_h = 4.0 m, d_v = 0.025 m, d_h = 0.02 m

ρ = 1000.0 kg/m³, μ = 1.0 × 10⁻³ N · s/m², and tank diameter = 0.5 m

compute the undamped natural frequency ω_n and the damping ratio ζ of the system. Will this system provide an oscillatory natural response? If so what is the corresponding frequency? If not, explain the reasons.

SOLUTION

Since inertia effects are neglected in the model, and only two capacitors are used as the energy storage elements, this model cannot provide an oscillatory response to an initial condition excitation. But, the actual physical system has fluid inertia, and hence the system can exhibit an oscillatory response.

a. Assuming a parabolic velocity profile, the fluid inertance in a pipe of uniform cross-section A and length L, is given by

$$I = \frac{2\rho L}{A}$$

Since the same volume flow rate Q is present in both segment of piping (continuity) we have, for series connection,

$$I_{eq} = \frac{2\rho L_v}{\frac{\pi}{4} d_v^2} + \frac{2\rho L_h}{\frac{\pi}{4} d_h^2} = \frac{8\rho}{\pi} \left[\frac{L_v}{d_v^2} + \frac{L_h}{d_h^2} \right]$$

The linear fluid resistance in a circular pipe is

$$R = \frac{128\mu L}{\pi d^4}$$

where d is the internal diameter.

Again, since the same Q exists in both segments of the series-connected pipe,

$$R_{eq} = \frac{128\mu}{\pi} \left[\frac{L_v}{d_v^4} + \frac{L_h}{d_h^4} \right]$$

Also

$$C_{grv} = \frac{A_t}{\rho g}$$

b. *State-Space Shell*:

$$C_{grv} \frac{dP_{3a}}{dt} = Q$$

$$I_{eq} \frac{dQ}{dt} = P_{23}$$

Remaining Constitutive Equation:

$$P_{12} = R_{eq} Q$$

NOTE Constitutive (node) equations are already satisfied.

Compatibility (loop) equations:

$$P_{1a} = P_{12} + P_{23} + P_{3a} \quad \text{with} \quad P_{1a} = P_s(t) \text{ and } P_{3a} = \rho g h$$

Now eliminate the auxiliary variable P_{23} in the state-space shell, using the remaining equations; thus

$$P_{23} = P_{1a} - P_{12} - P_{3a}$$
$$= P_s(t) - R_{eq} Q - P_{3a}$$

Hence, the state-space model is

$$\frac{dP_{3a}}{dt} = \frac{1}{C_{grv}} Q \tag{i}$$

$$\frac{dQ}{dt} = \frac{1}{I_{eq}} \left[P_s(t) - P_{3a} - R_{eq} Q \right] \tag{ii}$$

Output
$$h = \frac{1}{\rho g} P_{3a} \tag{iii}$$

or,

$$A = \begin{bmatrix} 0 & 1/C_{grv} \\ -1/I_{eq} & -R_{eq}/I_{eq} \end{bmatrix}, \quad B = \begin{bmatrix} 0 \\ 1/I_{eq} \end{bmatrix}$$

$$C = \begin{bmatrix} \dfrac{1}{\rho g} & 0 \end{bmatrix}, \quad D = 0$$

c. Substitute Equation i in Equation ii:

$$I_{eq}C_{grv}\frac{d^2P_{3a}}{dt^2} = P_s(t) - P_{3a} - R_{eq}C_{grv}\frac{dP_{3a}}{dt}$$

Now substitute Equation iii for P_{3a}:

$$I_{eq}C_{grv}\frac{d^2h}{dt^2} + R_{eq}C_{grv}\frac{dh}{dt} + h = \frac{1}{\rho g}P_s(t)$$

Characteristic equation of this system is

$$I_{eq}C_{grv}s^2 + R_{eq}C_{grv}s + 1 = 0$$

d. Substitute numerical values

$$I_{eq}C_{grv} = 2\left[\frac{10}{0.025^2} + \frac{4}{0.02^2}\right]\frac{0.5^2}{9.81} = 1.325 \times 10^3$$

$$R_{eq}C_{grv} = 32 \times 10^{-3}\left[\frac{10}{0.025^4} + \frac{4}{0.02^4}\right]\frac{0.5^2}{9.81 \times 1000} = 41.25$$

Undamped natural frequency $\omega_n = \dfrac{1}{\sqrt{I_{eq}C_{grv}}} = \dfrac{1}{\sqrt{1.325 \times 10^3}} = 2.75 \times 10^{-2}$ rad/s

Damping ratio $\zeta = \dfrac{1}{2\omega_n}\dfrac{R_{eq}}{I_{eq}} = \dfrac{41.25}{2 \times 2.75 \times 10^{-2}} = 0.564$

Damped natural frequency $\omega_d = \sqrt{1-\zeta^2}\,\omega_n = 2.27 \times 10^{-2}$ rad/s

Since $\zeta < 1$, the system will exhibit oscillatory natural response at $\omega_d = 2.27 \times 10^{-2}$ rad/s.

2.9 Thermal Systems

Thermal systems have temperature (T) as the across variable, as it is always measured with respect to some reference (or as a temperature difference across an element), and heat transfer (flow) rate (Q) as the through variable. Heat source and temperature source are the two types of source elements. The former is more common. The latter may correspond to a large reservoir whose temperature is hardly affected by heat transfer into or out of it. There is only one type of energy (thermal energy) in a thermal system. Hence there is only one type (A-type) energy storage element with the associated state variable,

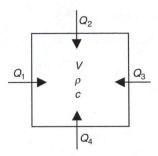

FIGURE 2.37
A control volume of a thermal system.

temperature. There is no T-type element in a thermal system. As a direct result of the absence of two different types of energy storage elements (unlike the case of mechanical, electrical, and fluid systems) a pure thermal system cannot exhibit natural oscillations. It can exhibit "forced" oscillations, however, when excited by an oscillatory input source.

2.9.1 Constitutive Equations

The constitutive equations in a thermal system are for thermal capacitances (A-type element) and thermal resistances (D-type elements). There are no T-type elements. There are three types of thermal resistance—conduction, convection, and radiation.

2.9.2 Thermal Capacitance

Consider a control volume of an object, with various heat transfer processes Q_i taking place at the boundary of the object (see Figure 2.37). The level of thermal energy in the object $= \rho V c T$, where

T = temperature of the object (assumed uniform)

V = volume of the object

ρ = mass density of the object

c = specific heat of the object

Since the net heat inflow is equal to the rate of change (increase) of thermal energy, the associated constitutive relation is

$$\sum Q_i = \rho V c \frac{dT}{dt} \tag{2.112}$$

where $\rho V c$ is assumed constant. We write this as

$$Q = C_h \frac{dT}{dt} \tag{2.113}$$

where, $C_h = \rho V c = mc =$ thermal capacitance.

Here $m = \rho V$ is the mass of the element. Note that thermal capacitance means the "capacity" to store thermal energy in a body. There are no thermal inductors.

2.9.3 Thermal Resistance

These elements provide resistance to heat transfer in a body or a medium. The three general types are:

- Conduction
- Convection
- Radiation

We will now give constitutive relations for each of these three types of thermal resistance elements.

2.9.3.1 *Conduction*

The heat transfer in a medium takes place by conduction when the molecules of the medium itself do not move to transfer the heat. Heat transfer takes place from a point of higher temperature to one of lower temperature. Specifically, heat conduction rate is proportional to the negative temperature gradient, and is given by the *Fourier equation*:

$$Q = -kA\frac{\partial T}{\partial x} \tag{2.114}$$

where
x = direction of heat transfer
A = area of cross section of the element along which heat transfer takes place
k = thermal conductivity

The above (Fourier) equation is a "local" equation. If we consider a finite object of length Δx and cross section A, with temperatures T_2 and T_1 at the two ends, as shown in Figure 2.38, the one-dimensional heat transfer rate Q can be written according Equation 2.114 as

$$Q = kA\frac{(T_2 - T_1)}{\Delta x} \tag{2.115}$$

or,

$$Q = \frac{1}{R_k}(T_2 - T_1) \tag{2.116a}$$

where

$$R_k = \frac{\Delta x}{kA} = \text{conductive thermal resistance} \tag{2.117}$$

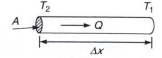

FIGURE 2.38
An element of 1-D heat conduction.

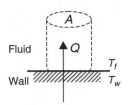

FIGURE 2.39
A control volume for heat transfer by convection.

2.9.3.2 Convection

In convection, the heat transfer takes place by the physical movement of the heat-carrying molecules in the medium. An example will be the case of fluid flowing against a wall, as shown in Figure 2.39. Then,

$$Q = h_c A(T_w - T_f) \tag{2.118}$$

where
T_w = wall temperature
T_f = fluid temperature at the wall interface
A = area of cross-section of the fluid control volume across which heat transfer Q takes place
h_c = convection heat transfer coefficient.

In practice h_c may depend on the temperature itself, and hence Equation 2.118 is non-linear in general. But, by approximating to a linear constitutive equation we have

$$Q = \frac{1}{R_c}(T_w - T_f) \tag{2.119a}$$

where

$$R_c = \frac{1}{h_c A} = \text{convective thermal resistance} \tag{2.120}$$

In natural convention, the particles in the heat transfer medium move naturally. In forced convection, they are moved by an actuator such as a fan or pump.

2.9.3.3 Radiation

In radiation, the heat transfer takes place from a higher temperature object (source) to a lower temperature object (receiver) through energy radiation, without needing a physical medium between the two objects (unlike in conduction and convection), as shown in Figure 2.40. The associated constitutive equation is the *Stefan-Boltzman law*:

$$Q = \sigma c_e c_r A(T_1^4 - T_2^4) \tag{2.121}$$

where
A = effective (normal) area of the receiver
c_e = effective emmissivity of the source
c_r = shape factor of the receiver
σ = Stefan-Boltzman constant ($= 5.7 \times 10^{-8} \text{ W/m}^2/\text{K}^4$)

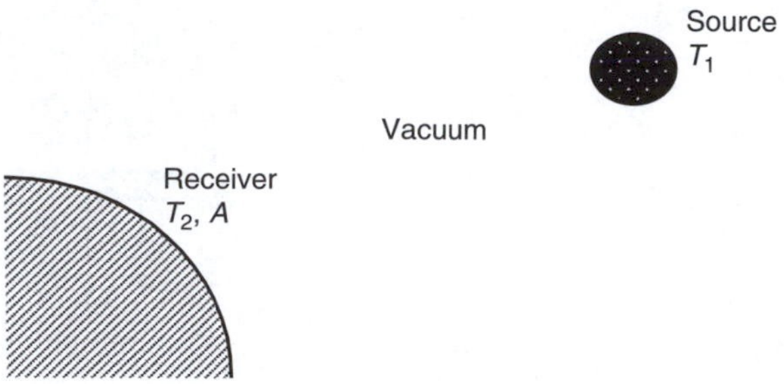

Vacuum

Receiver
T_2, A

FIGURE 2.40
Heat transfer by radiation.

Heat transfer rate is measured in watts (W), the area in square meters (m²), and the temperature in degrees Kelvin (K). The relation in Equation 2.121 is nonlinear, which may be linearized as

$$Q = \frac{1}{R_r}(T_1 - T_2) \qquad (2.122a)$$

where R_r = radiation thermal resistance.

Since the slope $\frac{\partial Q}{\partial T}$ at an operating point may be given by $4\sigma c_e c_r A \overline{T}^3$, where $\overline{T}$ is the representative temperature (which is variable) at the operating point, we have

$$R_r = \frac{1}{4\sigma c_e c_r A \overline{T}^3} \qquad (2.123)$$

Alternatively, since $T_1^4 - T_2^4 = (T_1^2 + T_2^2)(T_1 + T_2)(T_1 - T_2)$, we may use the approximate expression:

$$R_r = \frac{1}{\sigma c_e c_r A(\overline{T}_1^2 + \overline{T}_2^2)(\overline{T}_1 + \overline{T}_2)} \qquad (2.124a)$$

where, the over-bar denotes a representative (operating point) temperature.

2.9.4 Three-Dimensional Conduction

Conduction heat transfer in a continuous 3-D medium is represented by a distributed-parameter model. In this case the Fourier Equation 2.114 is applicable in each of the three orthogonal directions (x, y, z). In addition, to obtain a model for the thermal capacitance Equation 2.112 has to be applied.

Consider the small 3-D model element of sides dx, dy, and dz, in a conduction medium as shown in Figure 2.41. First consider heat transfer into the bottom ($dx \times dy$) surface in the z direction, which according to Equation 2.114 is $-k\,dx\,dy\,\frac{\partial T}{\partial z}$.

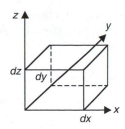

FIGURE 2.41
A 3-D heat conduction element.

Since the temperature gradient at the top ($dx \times dy$) surface is $\frac{\partial T}{\partial z} + \frac{\partial^2 T}{\partial z^2} dx$ (from calculus), the heat transfer out of this surface is $k\, dx\, dy(\frac{\partial T}{\partial z} + \frac{\partial^2 T}{\partial z^2} dz)$. Hence, the net heat transfer into the element in the z direction is $k\, dx\, dy \frac{\partial^2 T}{\partial z^2} dz$ or $k\, dx\, dy\, dz \frac{\partial^2 T}{\partial z^2}$. Similarly, the net heat transfer in the x and y directions are $k\, dx\, dy\, dz \frac{\partial^2 T}{\partial x^2}$ and $k\, dx\, dy\, dz \frac{\partial^2 T}{\partial y^2}$, respectively.

The thermal energy of the element is $\rho\, dx\, dy\, dz\, c_p T$ where $\rho\, dx\, dy\, dz$ is the mass of the element and c_p is the specific heat (at constant pressure). Hence, the capacitance Equation 2.112 gives

$$k\, dx\, dy\, dz\left(\frac{\partial^2 T}{\partial x^2} + \frac{\partial^2 T}{\partial y^2} + \frac{\partial^2 T}{\partial z^2}\right) = \rho\, dx\, dy\, dz\, c_p \frac{\partial T}{\partial t}$$

or,

$$\frac{\partial^2 T}{\partial x^2} + \frac{\partial^2 T}{\partial y^2} + \frac{\partial^2 T}{\partial z^2} = \frac{1}{\alpha}\frac{\partial T}{\partial t} \qquad (2.125)$$

where

$$\alpha = \frac{k}{\rho c_p} = \text{thermal diffusivity.}$$

Equation 2.125 is called the **Laplace equation**. Note that partial derivatives are used because T is a function of many variables; and derivatives with respect to x, y, z, and t would be needed. Hence, in general, distributed-parameter models have spatial variables (x, y, z) as well as the temporal variable (t) as independent variables, and are represented by partial differential equations.

2.9.5 Biot Number

This is a nondimesional parameter giving the ratio: conductive resistance/convective resistance. Hence from Equation 2.117 and Equation 2.120 we have

$$\text{Biot number} = \frac{R_k}{R_c} = \frac{\Delta x h_c A}{kA} = \frac{h_c \Delta x}{k} \qquad (2.126)$$

This parameter may be used as the basis for approximating the distributed-parameter model Equation 2.125 by a lumped parameter one. Specifically, divide the conduction medium into slabs of thickness Δx. If the corresponding Biot number ≤ 0.1, a lumped-parameter model may be used for each slab.

2.9.6   Model Equations

In developing the model equations for a thermal system, the usual procedure is followed. Specifically we write

1. Constitutive equations (for thermal resistance and capacitance elements)
2. Node equations (the sum of heat transfer rate at a node is zero)
3. Loop equations (the sum of the temperature drop around a closed thermal path is zero)

Finally, we eliminate the auxiliary variables that are not needed. The linear graph approach may be used.

Example 2.15

The pudding called "watalappam" is a delicacy, which is quite popular in Sri Lanka. Traditionally, it is made by blending roughly equal portions by volume of treacle (a palm honey similar to maple syrup), coconut milk, and eggs, spiced with cloves and cardamoms, and baking in a special oven for about one hour. The traditional oven uses a charcoal fire in an earthen pit that is well insulated, as the heat source. An aluminum container half filled with water is placed on fire. A smaller aluminum pot containing the dessert mixture is placed inside the water bath and covered fully with an aluminum lid. Both the water and the dessert mixture are well stirred and assumed to have uniform temperatures.

A simplified model of the watalappam oven is shown in Figure 2.42(a).

Assume that the thermal capacitances of the aluminum water container, dessert pot, and the lid are negligible. Also, the following equivalent (linear) parameters and variables are defined:

C_r = thermal capacitance of the water bath

C_d = thermal capacitance of the dessert mixture

R_r = thermal resistance between the water bath and the ambient air

R_d = thermal resistance between the water bath and the dessert mixture

R_c = thermal resistance between the dessert mixture and the ambient air, through the covering lid

T_r = temperature of the water bath

T_d = temperature of the dessert mixture

T_S = ambient temperature

Q = input heat flow rate from the charcoal fire into the water bath

a. Assuming that T_d is the output of the system, develop a complete state-space model for the system. What are the system inputs?

b. In part (a) suppose that the thermal capacitance of the dessert pot is not negligible, and is given by C_p. Also, as shown in Figure 2.42(b), thermal resistances R_{p1} and R_{p2} are defined for the two interfaces of the pot. Assuming that the pot temperature is maintained uniform at T_p show how the state-space model of part

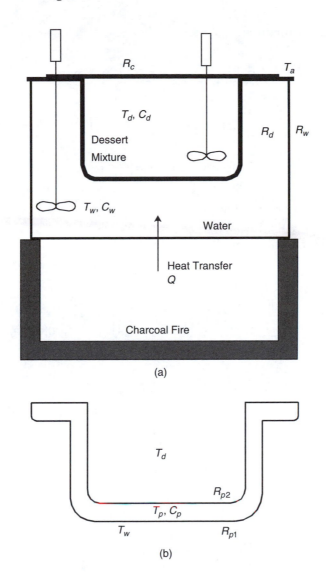

FIGURE 2.42
(a) A simplified model of a Sri Lankan dessert oven; (b) An improved model of the dessert pot.

   (a) should be modified to include this improvement. What parameters do R_{p1} and R_{p2} depend on?

c. Draw the linear graphs for the systems in (a) and (b). Indicate in the graph only the system parameters, input variables, and state variables.

SOLUTION

a. For the water bath:

$$C_w \frac{dT_w}{dt} = Q - \frac{1}{R_w}(T_w - T_a) - \frac{1}{R_d}(T_w - T_d) \qquad \text{(i)}$$

For the dessert mixture:

$$C_d \frac{dT_d}{dt} = \frac{1}{R_d}(T_w - T_d) - \frac{1}{R_c}(T_d - T_a) \qquad \text{(ii)}$$

State vector $x = [T_w \quad T_d]^T$

Input vector $u = [Q \quad T_a]^T$

Output vector $y = [T_d]^T$

We have the state-space model matrices:

$$A = \begin{bmatrix} -\dfrac{1}{C_w}\left(\dfrac{1}{R_w} + \dfrac{1}{R_d}\right) & \dfrac{1}{C_w R_d} \\[3mm] \dfrac{1}{C_d R_d} & -\dfrac{1}{C_d}\left(\dfrac{1}{R_d} + \dfrac{1}{R_c}\right) \end{bmatrix}, \quad B = \begin{bmatrix} \dfrac{1}{C_w} & \dfrac{1}{C_w R_w} \\[3mm] 0 & \dfrac{1}{C_d R_c} \end{bmatrix}$$

$$C = [0 \quad 1], \qquad D = [0 \quad 0]$$

b. For the dessert pot:

$$C_p \frac{dT_p}{dt} = \frac{1}{R_{p1}}(T_w - T_p) - \frac{1}{R_{p2}}(T_p - T_d) \qquad \text{(iii)}$$

Equation i and Equation ii have to be modified as

$$C_w \frac{dT_w}{dt} = Q - \frac{1}{R_w}(T_w - T_a) - \frac{1}{R_{p1}}(T_w - T_p) \qquad \text{(i)*}$$

$$C_d \frac{dT_d}{dt} = \frac{1}{R_{p2}}(T_w - T_d) - \frac{1}{R_c}(T_d - T_a) \qquad \text{(ii)*}$$

The system has become third order now, with

$$x = [T_w \quad T_d \quad T_p]^T$$

But *u* and *y* remain the same as before. Matrices *A*, *B*, and *C* have to be modified accordingly.

The resistance R_{pi} depends on the heat-transfer area A_i and the heat transfer coefficient h_i. Specifically,

$$R_{pi} = \frac{1}{h_i A_i}$$

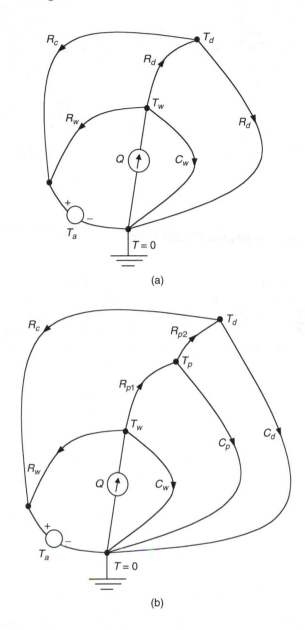

FIGURE 2.43
Linear graph of the: (a) simplified model; (b) improved model.

   c. The linear graph for Case (a) is shown in Figure 2.43(a). The linear graph for Case (b) is shown in Figure 2.43(b).

2.10 Bond Graphs

Bond graphs, like linear graphs, are graphical representations of lumped-parameter models of dynamic systems. Figure 2.44 shows a typical line segment of bond graph. Each line or branch is called a *bond* because it connects two elements in the model, analogous to a chemical bond in chemistry, which links two atoms.

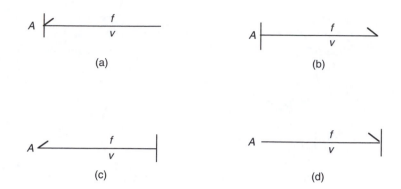

FIGURE 2.44
A bond graph showing various possibilities of causality and power flow.

First, let us comment on the similarities of bond graphs to linear graphs: (1) both represent dynamic models, (2) both are convenient and systematic graphic tools to obtain state equations, (3) both use lines to represent a model, and (4) both characterize a line with two variables representing cause/effect or input/output. However, one should not rush to establish a one-to-one correspondence between linear graphs and bond graphs because that does not exist. Some of the major differences between linear graphs and bond graphs are:

1. In linear graphs, a line segment represents an element (such as mass, stiffness, damper, source, etc.), whereas in bond graphs a line segment (a bond) emerges from an element. Hence, a bond connects two elements (like in chemical bonds—hence the name).

2. An arrow in a linear graph can represent the direction of power transmitted through an element. A half arrow in a bond graph will represent the direction of power flow (whether into or out of the element).

3. Linear graphs do not explicitly represent "causality" of variables associated with an element (i.e., which one is the input and which one the output). Bond graphs use "causality strokes" to explicitly indicate causality of the two variables associated with each bond.

4. Linear graphs do not explicitly indicate the correct "order" of a dynamic system. Bond graphs immediately show the system order. Specifically:

$$\text{system order} = \text{number of energy-storage elements} - \text{number of causality conflicts}$$

5. Linear graphs use "through variables" and "across variables" whereas bond graphs use "flow variables" and "effort variables."

6. Continuity equations are node equations for a linear graph. Continuity equations are explicitly represented by common-velocity-junction elements in bond graphs.

7. Compatibility equations are loop equations in a linear graph. Compatibility equations are explicitly represented by common-force-junction elements in bond graphs.

In our study of bond graphs we will continue to use the variables we used in linear graphs: the velocity of an inertia element and the force in a stiffness element, to develop state models. Alternatively, (1) momentum of inertia elements and (2) displacement of stiffness elements may be used. Note that momentum is directly related to velocity in an inertia element, and displacement is directly related to force in a stiffness element. In bond graphs, "effort" and "flow" denote force and velocity in mechanical systems, voltage and current in electrical systems, pressure and flow rate in fluid systems, and temperature and heat transfer rate in thermal systems. In linear graphs, force and current are through variables, and velocity and voltage are across variables. In this sense, force-voltage analogy is used in bond graphs whereas force-current analogy is used in linear graphs. In the present study of bond graphs we will present the principles primarily using mechanical elements and systems even though the techniques may be similarly applied to electrical, fluid, and thermal systems. We will present examples to illustrate the application of bond graphs to non-mechanical systems.

2.10.1 Single-Port Elements

Figure 2.44 shows four possible configurations for a single bond. The "half arrow" indicates the direction of power flow when the flow and the effort variables are positive according to some convention. In Figure 2.44(a), for example, when the effort (force) variable (f) and the flow (velocity) variable (v) are positive according to a preestablished sign convention, the power would flow into the element (denoted by A). Hence, once the positive direction of v is assigned, the half arrow will also determine the positive direction of f (if the signs of two of the three variables: effort, flow, and power, are known, the sign of the third is known because power = effort × flow).

The short (vertical) stroke across a bond indicates the causality of each port. In other words, the causality stroke determines which of the two variables (f and v) in a bond (or port) is the input variable and which is the output variable. If the stroke is at the near end of element A (as in Figure 2.44(a)), the input variable to A is f and the output variable of A is v. Similarly, if the "causality stroke" is at the far end of the bond to A, then v is the input to A and f is the output of A. This notation is further explained in Figure 2.45. Note that the causality is a completely independent consideration from the direction of power flow, as is clear from the four possible combinations of causality and power-flow directions shown in Figure 2.44. Also by convention, the effort variable (force) is marked above or to the left of each bond.

2.10.2 Source Elements

In modeling lumped-parameter mechanical systems we may use two ideal source elements: a *force source* and a *velocity source*. Their bond graph representations are shown in Table 2.5. From the practical point of view, the positive direction of power is always taken

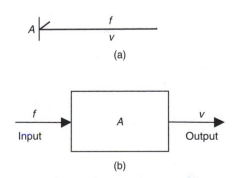

FIGURE 2.45
(a) Bond graph representation of element A with input f and output v with power flow into A; (b) Block diagram representation.

to be out of the source. However, it would be acceptable to use the reverse direction as positive because it is simply a sign convention. In a source element only one variable is specified (output variable) and the value of the other variable depends on the remainder of the system to which the source is connected.

The causality of a source is unique by its definition — the output of a force source is force and the output of a velocity source is velocity. It is clear that the causality stroke is completely determined by the source type, and hence is redundant. For this reason, causalities in source elements are known as necessary causalities. Causality strokes are often omitted from the final bond graph because the strokes are useful primarily in establishing that there are no causality conflicts and, hence, that no dependent energy-storage elements exist in the model. Causality conflict is an indication that the system order is less than the number of energy-storage elements in the model.

2.10.3 Energy Storage and Dissipation Elements

The two energy-storage elements in mechanical models are the mass (inertia) element and the stiffness (tension/compression spring, torsional spring, or flexural spring) element. In electrical models, these are inductors and capacitors. Energy dissipation is represented by a damping element for mechanical systems and by a resistance for electrical systems. In linear mechanical models, viscous damping is assumed. These single-port mechanical elements and their bond-graph representations are summarized in Table 2.5.

TABLE 2.5

Basic Single-Port Bond Graph Elements for Mechanical Systems

2.10.4 Derivative Causality and Integral Causality

When the constitutive (characterizing) relations are written in the derivative form, derivative causality is said to apply. For example, for the mass element; if we use Newton's law in the form: $f = m\frac{dv}{dt}$ this denotes that f is the output variable and v is therefore the input variable. This corresponds to derivative causality. On the other hand, if we use the same Newton's law in the form: $v = \frac{1}{m}\int f\,dt$ this implies that v is the output variable (and f is therefore the input variable). This is an example of integral causality. It is natural to use integral causality for mass and stiffness elements because, as discussed before, velocity (v) for mass and force (f) for stiffness as the natural state variables (which are a particular type of output variables). Once this decision is made, their causalities become fixed (i.e., velocity is the output for all mass elements and force is the output for all stiffness elements). For damper elements the causality is algebraic and hence arbitrary. Both forms of causality are shown in Table 2.5. Note that the integral causality is shown by the bottom figures for the bond graphs of the inertia and stiffness elements.

2.10.4.1 Causality Conflicts and System Order

Consider a system model having several energy-storage elements. First we draw its bond graph and indicate causality using the integral-causality convention. If it is not possible to assign causalities to all bonds without violating the assumed (integral) causality, then there exists a conflict in causality. This indicates that the energy-storage elements are not independent and hence the system order is less than the total number of energy-storage elements. It is seen that bond graphs are particularly useful in identifying the correct order of a system (model).

2.10.5 Two-Port Elements

When considering linear graphs, we examined systems with transformer and gyrator. These are two-port elements. According to the sign convention for linear graphs, power at the output port as well as the input port is taken to be positive into the element. On the contrary, in bond-graph notation it is customary to take the positive direction of power as into the element at the input port, and out of the element at the output port. The bond-graph representations for an ideal transformer and gyrator are given in Table 2.6. In each,

TABLE 2.6

Basic Two-Port Mechanical Bond Graph Elements

Element	Conventional Representation	Bond Graph	Constitutive Relation
Transformer			$v_o = rv_i$ $f_i = rf_o$ $f_o = \frac{1}{r}f_i$ $v_i = \frac{1}{r}v_o$
Gyrator			$v_o = Mf_i$ $v_i = Mf_o$ $f_o = \frac{1}{m}v_i$ $f_i = \frac{1}{m}v_o$

there are two choices for causality. For the transformer it is possible to choose either v_o and f_i or v_i and f_o as outputs. For the gyrator, either v_o and v_i or f_o and f_i may be chosen as outputs. Both choices are shown in Table 2.6 in the bond-graph column, along with the corresponding constitutive relations.

2.10.6 Multiport Junction Elements

Junction elements are used to represent continuity (conservation) relations and compatibility relations in bond-graph models (Table 2.7). Multiport junctions consisting of more than three ports can be represented by a combination of three-port junctions; it follows that only three-port junctions need be considered. For convenience and conciseness of representation, however, it is acceptable to use multiport junctions having more than three ports.

2.10.6.1 *Common-Force Junction*

The three port, common-force junction is shown in Figure 2.46(a) and Table 2.7. This junction has the property that the force variables at the port are identical, and the velocity variables add up to zero. This element, therefore, represents a compatibility condition (or a loop

TABLE 2.7

Three-Port Junction Elements.

Element	Significance	Bond Graph Representation	Constitutive Relation
Common-force junction	Compatibility (Sum of velocities in a loop = 0)	(f)	$f_1 = f_2 = f_3 = f$ $v_1 + v_2 + v_3 = 0$
Common-Velocity junction	Continuity (Sum of forces at a node = 0)	(v)	$v_1 = v_2 = v_3 = v$ $f_1 + f_2 + f_3 = 0$

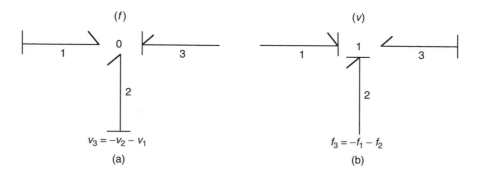

(a) $v_3 = -v_2 - v_1$

(b) $f_3 = -f_1 - f_2$

FIGURE 2.46

(a) Common-force junction; (b) Common-velocity junction.

equation for linear graphs). Since the force variable is common, the positive direction of power flow (half arrow) in each bond can be used to represent the positive direction of velocity at that port. In other words, if the direction of half arrow at a port is reversed, the corresponding positive direction of velocity, as well as the positive direction of power flow, is reversed. Note that causality is not indicated in Table 2.7. The approach to assigning causalities to a common-force junction is simple: Any two of the three velocity variables are selected as inputs to the element; the third velocity variable is necessarily the output. This, therefore, is a case of restricted causality. The particular choice, however, should be compatible with the rest of the bond graph. In Figure 2.46(a), for example, v_3 is considered output.

2.10.6.2 *Common-Velocity Junction*

A three-port, common-velocity junction is shown in Figure 2.46(b) and in Table 2.7. This element represents a continuity (or conservation, or force-balance) condition (a node equation for linear graphs) in a model. Since the velocity variable is common at the three ports, the positive direction of power flow (as given by half arrow) is also used to indicate the positive direction of force. The summing relation dictates the causality; any two force variables can be chosen as inputs and the third force variable is necessarily the output. This is also a case of restricted causality. The causality stroke for the case when f_3 is taken as the output is shown in Figure 2.46(b).

2.10.7 State-Models From Bond Graphs

The main steps of obtaining a state-space model from a bond graph are as follows:

1. Draw the bond-graph structure for the lumped-parameter model.
2. Augment the bond graph with causality (input-output) strokes and positive power-flow-direction half arrows. Use integral causality.
3. If no conflict of causality exists (as evidenced by being able to complete all causality strokes correctly), the order of the system (model) is equal to the number of energy-storage elements. For mechanical systems pick velocities of inertia elements and forces in stiffness elements as state variables. For electrical systems pick currents of inductors and voltages of capacitors as state variables.
4. Write constitutive relations for independent energy-storage elements to obtain the state-model skeleton.
5. Write constitutive relations for the remaining elements.
6. Eliminate auxiliary variables using the relations in Step (5).

Two circumstances deserve special attention here:

1. **Arbitrary Causality:** In order to complete the causality assignment in a bond graph, it may be required to assign causality to one or more dissipation (mechanical damping or electrical resistance) elements, arbitrarily. In this case the bond graph itself is not unique (i.e., more than one bond graph exists for the system).
2. **Conflicts in Causality:** In some bond-graph models a causality conflict can exist when integral causality is used for energy-storage elements. This means that, even though we started by assuming integral causality for the energy-storage elements, it becomes imperative to use derivative causality for one or more of these elements. This will imply that these energy-storage elements depend,

algebraically, on the remaining energy-storage elements. These algebraic relations can be used to eliminate the corresponding redundant state variables. The
order of the model, in this case, is given by:

$$\text{Model order} = n_1 - n_2$$

where n_1 is the number of energy-storage elements, and n_2 is the storage elements
with derivative causality.

When causality conflicts occur it might be necessary to include derivatives of input
variables in the system equations. These correspond to feedforward paths. To eliminate
these derivative terms from the state equations, it would be necessary to redefine the state
variables, resulting in an output equation that depends on the inputs as well as the states.
From a practical point of view, causality conflicts usually mean modeling errors and
modeling redundancies. Next we will consider several illustrative examples.

Example 2.16

Consider a mechanical system subjected to a support-motion excitation, modeled as in
Figure 2.47. Gravitational forces at inertia elements may be incorporated into the model
as constant force sources. Alternatively, if we apply a constant external force to support
gravity, and for stiffness elements if we use as state variables the changes in forces from
the static equilibrium position, the gravity forces do not enter into the state equations.
This is similar to the situation where system motion is in a horizontal plane. Let us make
this assumption.

Steps of developing a state-space model for this system using bond graphs are given below.

Step 1: The bond graph of the given model is drawn as shown in Figure 2.48. There are
four energy-storage elements (m_1, m_2, k_1, k_2). Prior to making causality assignments, we
should not state that the system (model) is fourth order because, should there be causality
conflicts, the order would be less than four.

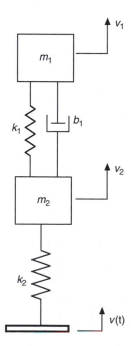

FIGURE 2.47
A system subject to support-motion excitation.

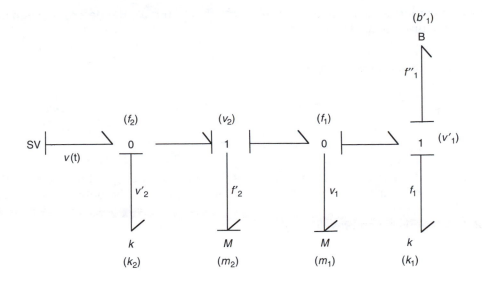

FIGURE 2.48
Bond graph of system in Figure 2.47.

Step 2: The half-arrows, indicating the direction of positive power, are assigned using the standard convention (see Table 2.5 and Table 2.7) of "out of the sources" and "into the *m*, *k*, and *b* elements." Causality strokes are assigned to each port using integral causality for *m* and *k* elements; necessary causality (i.e., velocity is the output) for the velocity source; and restricted causality for the common-force junctions (i.e., one velocity has to be the output) and common-velocity junctions (i.e., one force has to be the output). It follows that the causality of the damping element (*B*) is uniquely determined by these assignments; no arbitrary causality is needed. Hence the bond graph is unique. Furthermore, all *m* and *k* elements retain their assumed integral causality.

Step 3: There are no causality conflicts. Hence the system is fourth order. The state variables are chosen as v_1, v_2, f_1, and f_2—the velocities of m_1 and m_2 and the "changes" from static equilibrium configuration of the forces in k_1 and k_2.

Step 4: The skeleton state model is obtained by writing the constitutive equations for m_1, m_2, k_1, and k_2:

$$v_1 = \frac{1}{m_1}\int f_1' dt \quad \text{or} \quad \dot{v}_1 = \frac{1}{m_1}f_1'$$

$$v_2 = \frac{1}{m_2}\int f_2' dt \quad \text{or} \quad \dot{v}_2 = \frac{1}{m_2}f_2'$$

$$f_1 = k_1\int v_1' dt \quad \text{or} \quad \dot{f}_1 = k_1 v_1'$$

$$f_2 = k_2\int v_2' dt \quad \text{or} \quad \dot{f}_2 = k_2 v_2'$$

Step 5: To eliminate the auxiliary variables f_1', f_2', v_2' and v_2', the descriptive equations for the remaining elements are used:

For 0-junctions (common force):

$$-v_2' = -v(t) + v_2 \quad \text{or} \quad v_2' = v(t) - v_2$$

$$-v_1' = -v_2 + v_1 \quad \text{or} \quad v_1' = v_2 - v_1$$

Note that the signs are assigned to velocity variables according to the direction of the half arrows, and equations are written according to causality (the output variable is on the left-hand side.)

For 1 -junctions (common velocity):

$$-f_2' = -f_2 + f_1' \quad \text{or} \quad f_2' = f_2 - f_1'$$

$$f_1' = f_1 + f_1''$$

(Again, signs are assigned to the force variables according to the direction of half arrows, and equations are written according to causality.)

For B-element:

$$f_1'' = b_1 v_1'$$

Step 6: The elimination of the auxiliary variables is a straightforward algebraic exercise. Note that five constitutive equations are necessary because an additional auxiliary variable f_1'' is introduced in the process and has to be eliminated as well. The final state equations are:

$$\dot{v}_1 = \frac{1}{m_1}[f_1 + b_1(v_2 - v_1)]$$

$$\dot{v}_2 = \frac{1}{m_2}[f_2 + f_1 - b_1(v_2 - v_1)]$$

$$\dot{f}_1 = k_1(v_2 - v_1)$$

$$\dot{f}_2 = k_2[v(t) - v_2]$$

The corresponding *system matrix* and the *input-gain matrix* (input distribution matrix) are:

$$A = \begin{bmatrix} -b_1/m_1 & b_1/m_1 & 1/m_1 & 0 \\ b_1/m_2 & -b_1/m_2 & -1/m_2 & 1/m_2 \\ -k_1 & k_1 & 0 & 0 \\ 0 & -k_2 & 0 & 0 \end{bmatrix}, \quad B = \begin{bmatrix} 0 \\ 0 \\ 0 \\ k_2 \end{bmatrix}$$

with the state vector $x = [v_1 \quad v_2 \quad f_1 \quad f_2]^T$ and the input vector (which is a scalar for this model) $u = v(t)$.

It should be recognized that there is a hidden force (not shown in Figure 2.47) equal in magnitude to the force in k_2 that has to be applied to the support in order to generate the input velocity $v(t)$. This force (f_2) is a dependent variable and is considered as the input to the velocity source to generate its output $v(t)$. It is clear that a velocity source has an associated force as much as a force source has an associated velocity.

Example 2.17

An interesting exercise on identifying the order of a system is provided by the lumped-parameter model shown in Figure 2.49. Note that the velocity input $v(t)$ is applied directly to mass m_2 and the forcing input $f(t)$ is applied to mass m_1.

The bond graph of this model is shown in Figure 2.50. There is a conflict in causality, which has surfaced as derivative causality for mass m_2. (Note the causality stroke of the bond connected to m_2. It is at the far end meaning force is the output. But according to

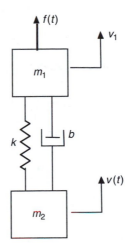

FIGURE 2.49
A system with velocity and force inputs.

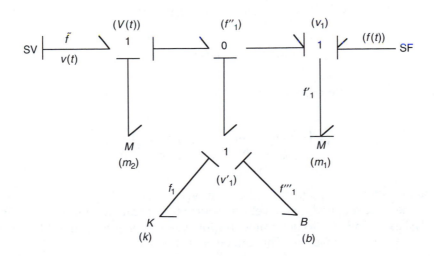

FIGURE 2.50
Bond graph of system in Figure 2.49.

the integral causality, velocity of m_2 should be the output – hence the conflict.) Thus, the order of the system is 2:

$$\text{order} = \#\text{ energy-storage elements} - \#\text{ causality conflicts}$$

$$= 3 - 1 = 2$$

On closer examination it appears that the velocity of m_2 is completely specified by the velocity source—this velocity is directly applied to the common ends of k and b. Consequently, the presence of m_2 is not felt by the rest of the system. This corresponds to a second-order system.

The state equations are obtained in terms of the state variables v_1 and f_1. The skeleton state model is

$$v_1 = \frac{1}{m_1}\int f_1' \, dt \quad \text{or} \quad \dot{v}_1 = \frac{1}{m_1} f_1'$$

$$f_1 = k\int v_1' dt \quad \text{or} \quad \dot{f}_1 = k v_1'$$

The remaining constitutive equations are:

$$-v_1' = -v(t) + v_1 \quad \text{(for the common-force junction } f_1'')$$

$$f_1'' = f_1 + f_1''' \quad \text{(for the common-velocity junction } v_1')$$

$$f_1' = -f'' - f(t) \quad \text{(for the common-velocity junction } v_1)$$

$$f_1''' = b v_1'$$

The final state equations are obtained by eliminating the auxiliary variables. We get

$$\dot{v}_1 = \frac{1}{m_1}[f(t) + f_1 + b(v(t) - v_1)]$$

$$\dot{f}_1 = k[v(t) - v_2]$$

The corresponding state-model matrices are:

$$A = \begin{bmatrix} -b/m_1 & 1/m_1 \\ -k & 0 \end{bmatrix}, \quad B = \begin{bmatrix} b/m_1 & 1/m_1 \\ k & 0 \end{bmatrix}$$

with the state vector $x = [v_1 \quad f_1]^T$ and the input vector $u = [v(t) \quad f(t)]^T$

Note that the force required by the velocity source to generate its velocity $v(t)$ is indicated in Figure 2.50 as $\tilde{f}$. In order to determine this force, however, it is required to consider the constitutive equation for mass m_2. This can be expressed as:

$$\tilde{f} - f_1'' = m_2 \dot{v}(t)$$

The value of f_1'' is known from a previous relation. It can be shown that:

$$\tilde{f} = m_2 \dot{v}(t) + f_1 + b[v(t) - v_1]$$

This expression contains the first derivative of the input variable $v(t)$, as expected.

Example 2.18

As another example, let us consider an ideal rack-and-pinion arrangement (with no back-lash and friction) shown in Figure 2.51(a). A torque τ_i is applied to the pinion causing it to rotate at angular velocity ω_i. The corresponding translational velocity of the rack is v_o. The load resisting this motion is indicated as a force f_o. The radius of the pinion is r.

If we neglect the inertia of both rack and pinion, the system corresponds to an ideal transformer. Its bond graph is shown in Figure 2.51(b). The constitutive relations written according to the causality indicated in the figure are:

$$f_o = \frac{1}{r}\tau_i$$

$$\omega_i = \frac{1}{r}v_o$$

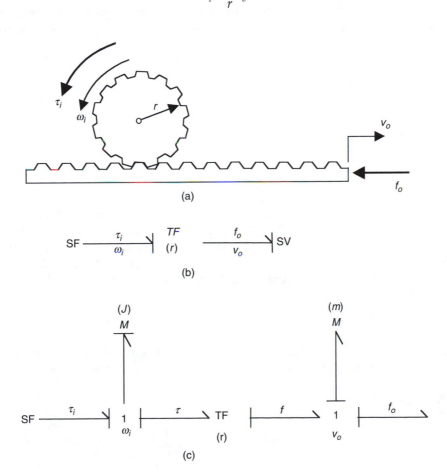

FIGURE 2.51
(a) A rack-and-pinion system; (b) Bond graph when inertia is neglected; (c) Bond graph with inertia included.

Now let us include the polar moment of inertia J of the pinion, and the mass m of the rack. The corresponding bond graph is shown in Figure 2.51(c). It is observed that there is a conflict in causality, which is indicated as derivative causality for mass m. Furthermore, it is required to make v_o an output variable for TF in Figure 2.51(c), whereas in Figure 2.51(b) it is an input variable. The system (model) is obviously first order.

The skeleton state model is

$$\omega_i = \frac{1}{J}\int(\tau_i - \tau)dt \quad \text{or} \quad \dot{\omega}_i = \frac{1}{J}(\tau_i - \tau)$$

The remaining constitutive equations are written according to the indicated causalities and power flow directions (half arrows):

$$f - f_o = m\dot{v}_o$$

$$v_o = r\omega_i$$

$$\tau = rf$$

In particular, note that the power flow of the force source f_o is "into" the force source because the positive direction of f_o is opposite to that of v_o (Figure 2.51(a)). But the power flow of torque source τ_i is "out of" the source because τ_i and ω_i are in the same direction.

By eliminating the auxiliary variable τ using these relations, the final state equation is obtained as

$$\dot{\omega}_i = \frac{1}{(J + mr^2)}(\tau_i - rf_o)$$

This is analogous to a simple mass driven by a force and resisted by a force (such as friction).

2.10.8   Bond Graphs of Electrical Systems

The concepts used in mechanical systems may be directly extended to electrical systems using the force-voltage analogy. Specifically, force and voltage are *effort variables* and velocity and current are *flow variables*. Then a spring is analogous to a capacitor and an inertia is analogous to an inductor. Also, we have voltage sources (SV) analogous to force sources (effort sources) and current sources (SI) analogous to velocity sources (flow sources).

Example 2.19

Consider the circuit shown in Figure 2.52(a). Its bond graph is shown in Figure 2.52(b). A state-space model is obtained by following the same procedure as for a mechanical system.

The state-space shell:

$$i = \frac{1}{L}\int v'dt \quad \text{or} \quad \frac{di}{dt} = \frac{1}{L}v'$$

$$v_o = \frac{1}{C}\int i_1 dt \quad \text{or} \quad \frac{dv_o}{dt} = \frac{1}{C}i_1$$

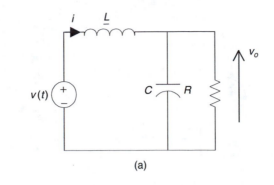

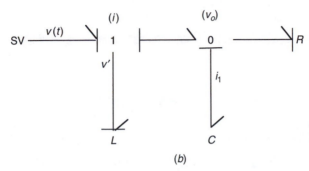

FIGURE 2.52
(a) An electrical circuit; (b) Bond graph of the circuit.

The remaining constitutive equations:

Resistor: $v_o = Ri_2$

Common-flow (1) junction: $\quad -v' = -v(t) + v_o \quad$ or $\quad v' = v(t) - v_o$

Common-effort (0) junction: $\quad -i_1 = -i + i_2 \quad$ or $\quad i_1 = i - i_2$

By eliminating the auxiliary variables, we get the following state-space model.

State vector: $\quad x = [i \quad v_o]^T$

Input vector: $\quad u = [v(t)]$

State equations:

$$\frac{di}{dt} = -\frac{1}{L}v_o + \frac{1}{L}v(t)$$

$$\frac{dv_o}{dt} = \frac{1}{C}i - \frac{1}{RC}v_o$$

There are ways to represent electronic circuits containing active devices such as operational amplifiers and nonlinear elements using bond graphs. Such topics are beyond the scope of this introductory section.

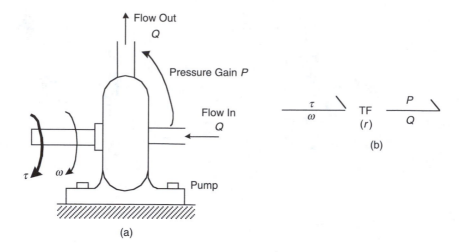

FIGURE 2.53
(a) Mechanical-fluid transformer (pump); (b) Bond graph representation.

2.10.9   Fluid and Thermal Systems

In fluid systems the effort variable is pressure and the flow variable is the fluid flow rate. Accordingly, the bond graph concepts can be extended to these systems. The model elements will include fluid capacitor, fluid inertor, fluid resistor, pressure source (effort source), and fluid flow source. In addition, a mechanical-fluid transformer is useful. This would represent a fluid pump, as shown in Figure 2.53. The reverse operation, which corresponds to a fluid motor or a hydraulic actuator is also important.

In thermal systems, the effort variable is temperature and the flow variable is heat transfer rate. The system elements are thermal capacitor, thermal resistor, temperature (effort) source, and heat (flow) source. As noted before, there is no thermal inertia element.

Example 2.20

A fluid of mass m and specific heat c is maintained at a uniform temperature T using heat source of rate Q_s. The container, which provides a thermal resistance R, loses heat to the environment (temperature T_a) at the rate Q_r. This thermal system is shown in Figure 2.54(a). A bond graph model for the system is shown in Figure 2.54(b).

Constitutive equations are written using the bond graph, as follows:

Capacitor: $mc\dfrac{dT}{dt} = Q_c$ (This is the state-space shell. The system is first order.)

Common-T (or, 0) junction: $-Q_c = -Q_s + Q_r$ or $Q_c = Q_s - Q_r$

Common-Q (or, 1) junction: $-T_r = -T + T_a$ or $T_r = T - T_a$

Resistor: $T_r = RQ_r$

NOTE

Q_c = heat transfer rate to capacitor

T_r = temperature across the thermal resistor.

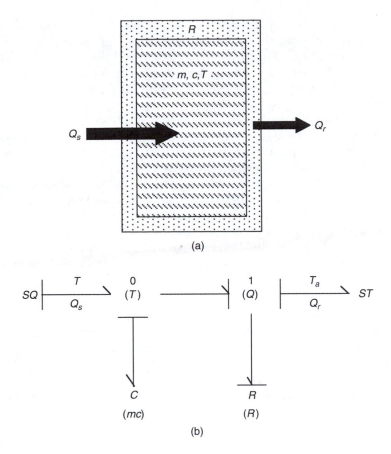

FIGURE 2.54
(a) A thermal system; (b) Bond graph model.

By substitution we get

$$mc\frac{dT}{dt} = Q_s - Q_r = Q_s - \frac{T_r}{R} = Q_s - \frac{T - T_a}{R}$$

Hence, the final state model is

$$mc\frac{dT}{dt} = -\frac{T}{R} + \frac{T_a}{R} + Q_s$$

2.11 Transfer-Function Models

Transfer-function models are based on the Laplace transform, and are versatile means of representing linear systems with constant (time-invariant) parameters. For a system with just one excitation input, the response characteristics at a given location (more correctly, in a given degree of freedom) can be determined using a single frequency-domain transfer function. For systems with multiple excitations, a transfer-function matrix of the appropriate order is necessary to determine the response at various degrees of freedom. Transfer-function models are said to be frequency-domain models because they provide the response of a system to a (sinusoidal) input at any frequency over an entire range of time; time (t) is integrated out. The resultant transform is in the "s domain." Since $s = a + j\omega$, where ω is the frequency of excitation, this is also the general frequency domain representation, as discussed in another section. Appendix A presents useful information on both Laplace transform and Fourier transform.

Transfer-function models were widely used in early studies of dynamic systems because a substantial amount of information regarding the dynamic behavior of a system can be obtained with minimal computational effort. This is the primary reason for the popularity enjoyed by transfer-function methods prior to the advent of the digital computer. Many thought that the abundance of high-speed, low-cost, digital computers would lead to the dominance of time-domain (direct differential equation solving) methods, over frequency-domain transfer-function methods. But there is evidence to the contrary in many areas, particularly in mechatronics and process control, due to analytical simplicity of transfer-function techniques. Only a minimal knowledge of Laplace-transform theory is needed to use transfer-function methods in system analysis, as we shall see here.

2.11.1   Transfer Function

The transfer function $G(s)$ of a linear, time-invariant, single-input single-output (SISO) system is given by the ratio of the Laplace-transformed output to the Laplace-transformed input, assuming zero initial conditions. This is a unique function that represents the system (a model); it does not depend on the input, the output, or the initial conditions. A physically realizable linear, constant-parameter system possesses a unique transfer function even if the Laplace transforms of a particular input and the corresponding output do not exist. In particular, suppose that the Laplace transform of a particular input $u(t)$ is infinite. Then the Laplace transform of the corresponding output $y(t)$ will also be infinite. But the transfer function itself will be finite.

Consider the nth-order linear, constant-parameter system given by

$$a_n \frac{d^n y}{dt^n} + a_{n-1} \frac{d^{n-1} y}{dt^{n-1}} + \cdots + a_0 y = b_0 u + b_1 \frac{du}{dt} + \cdots + b_m \frac{d^m u}{dt^m} \tag{2.127}$$

For systems that possess dynamic delay (i.e., systems whose response does not tend to feel the excitation either instantly or ahead of time, or systems whose excitation or its derivatives are not directly fed forward to the output, we will have $m < n$. These are the systems that concern us most in process control and mechanical dynamics. In fact, when $m \leq n$, the corresponding systems are said to be *physically realizable*.

Note from the Laplace-transform table in Appendix A that when zero initial conditions are assumed,

$$\mathcal{L} \frac{d^k f(t)}{dt^k} = s^k F(s) \tag{2.128}$$

where, "$\mathcal{L}$" is the Laplace transform operator. Accordingly, Equation 2.127 can be Laplace transformed to obtain the transfer function:

$$\frac{Y(s)}{U(s)} = G(s) = \frac{b_0 + b_1 s + \cdots + b_m s^m}{a_0 + a_1 s + \cdots + a_n s^n} \tag{2.129}$$

Note from Equation 2.128 that the Laplace variable s can be interpreted as the *derivative operator* in the context of Laplace transfer functions. Consequently, the transfer function corresponding to a system differential equation can be written simply by inspection, without requiring any knowledge of Laplace-transform theory. Conversely, once the transfer function is given, the corresponding time-domain (differential) equation should be immediately obvious since s corresponds to d/dt, s^2 corresponds to d^2/dt^2, and so on. Also from the Laplace-transform table (Appendix A) it is seen that, for a signal starting at $t = 0$, we have:

$$L \int_{-\infty}^{t} y(t)\, dt = \frac{Y(s)}{s} \tag{2.130}$$

It follows that $1/s$ can be interpreted as the *integration operator*, in the context of a dynamic system.

Transfer functions are simple algebraic expressions. Differential equations are transformed into simple algebraic relations through the Laplace transform. This is a major advantage of the transfer-function approach. Once the analysis is performed using transfer functions, the inverse Laplace transform can convert the results into the corresponding time-domain results. This can be accomplished simply by using Laplace transform tables.

2.11.1.1 Transfer-Function Matrix

Consider the state variable representation of a linear, time-invariant system given by the state equations:

$$\dot{x} = Ax + Bu \tag{2.131}$$

$$y = Cx + Du \tag{2.132}$$

This is a multi-input multi-output (MIMO) system. The corresponding transfer function model relates the mth-order response (output) vector y to the nth-order excitation (input) vector u. It follows that we will need $m \times n$ transfer functions, or a transfer-function matrix, to represent this MIMO system. To obtain an expression for this matrix, let us Laplace transform the Equation 2.131 and Equation 2.132 and use zero initial conditions for x. We get

$$sX(s) = AX(s) + BU(s) \tag{2.133a}$$

$$Y(s) = CX(s) + DU(s) \tag{2.133b}$$

From Equation 2.133a it follows that:

$$X(s) = (sI - A)^{-1} BU(s) \tag{2.134}$$

in which I is the nth-order identity matrix (note that $x(t)$ is the nth-order state vector). By substituting Equation 2.134 into Equation 2.133b we get the transfer function relation:

$$Y(s) = [C(sI - A)^{-1} B + D]U(s) \tag{2.135a}$$

or

$$Y(s) = G(s)U(s) \tag{2.136}$$

The transfer function matrix $G(s)$ is an $m \times n$ matrix given by

$$G(s) = C(sI - A)^{-1}B + D \tag{2.137a}$$

In practical systems with dynamic delay, the excitation $u(t)$ is not naturally fed forward to the response y; consequently, $D = 0$. For such systems,

$$G(s) = C(sI - A)^{-1}B \tag{2.138}$$

Several examples are presented now to illustrate some approaches of obtaining transfer-function models when the time-domain (differential-equation) models are given.

Example 2.21

Consider the simple oscillator (mass-spring-damper) equation given by

$$m\ddot{y} + b\dot{y} + ky = ku(t) \tag{i}$$

where $u(t)$ can be considered a displacement input (e.g., support motion); alternatively, $ku(t)$ can be considered an input force applied to a mass. Take the Laplace transform of the system equation with zero initial conditions:

$$(ms^2 + bs + k)Y(s) = kU(s)$$

The corresponding transfer function is

$$G(s) = \frac{Y(s)}{U(s)} = \frac{k}{(ms^2 + bs + k)} \tag{ii}$$

or, in terms of the undamped natural frequency (ω_n) and the damping ratio (ζ), where

$$\omega_n^2 = k/m \quad \text{and} \quad 2\zeta\omega_n = b/m \tag{iii}$$

the transfer function is given by

$$G(s) = \frac{\omega_n^2}{\left(s^2 + 2\zeta\omega_n s + \omega_n^2\right)} \tag{iv}$$

This is the transfer function corresponding to the displacement output. It follows that the output velocity transfer function (i.e., the transfer function if the output is taken to be the velocity) is

$$\frac{sY(s)}{U(s)} = sG(s) = \frac{s\omega_n^2}{(s^2 + 2\zeta\omega_n s + \omega_n^2)} \tag{v}$$

Note that velocity $\dot{y}(t)$ has the Laplace transform $sY(s)$, assuming zero initial conditions. Similarly, the output acceleration transfer function is

$$\frac{s^2Y(s)}{U(s)} = s^2G(s) = \frac{s^2\omega_n^2}{\left(s^2 + 2\zeta\omega_n s + \omega_n^2\right)} \tag{vi}$$

Here we used the fact that the Laplace transform of acceleration $\ddot{y}(t)$ is $s^2 Y(s)$ with zero initial conditions.

In the output acceleration transfer function, $m = n = 2$. That is, the order of the numerator equals that of the denominator. This means that the input (applied force) is instantly felt by the acceleration of the mass, which may be verified by sensing using an accelerometer. This corresponds to a feedforward of the input, or zero dynamic delay. For example, this is the primary mechanism through which road disturbances are felt inside a vehicle having hard suspensions.

If we apply a unit step input to the system given by Equation iv starting from rest (i.e., zero initial coordinates) the corresponding response is given by

$$y = 1 - \frac{e^{-\zeta \omega_n t}}{\sqrt{1 - \zeta^2}} \sin(\omega_d t + \phi) \qquad \text{(vii)}$$

where

$$\omega_d = \sqrt{1 - \zeta^2}\, \omega_n \qquad \text{(viii)}$$

and

$$\cos \phi = \zeta \qquad \text{(ix)}$$

Note that ω_d is the natural frequency of this damped (i.e., decaying response) system; it is the *damped natural frequency*. The damping ratio ζ is define as

$$\zeta = \frac{b}{b_c} \qquad \text{(x)}$$

where b is the damping constant (see Equation i) of the system and b_c is the critical damping constant (i.e., the value of b when the system just ceases to be oscillatory).

If $b < b_c$ the system will be naturally oscillatory ($\zeta < 1$). It is an *underdamped* system.
If $b > b_c$ the system will be naturally nonoscillatory ($\zeta > 1$). It is an *overdamped* system.

When $\zeta = 0$, we have an undamped system and then $\omega_d = \omega_n$. Hence ω_n is called the undamped natural frequency of the system.

From Equation vii it can be shown that (Also see Chapter 5)

$$\text{Peak time} \quad T_p = \frac{\pi}{\omega_d} \qquad \text{(xi)}$$

$$\text{Rise time} \quad T_r = \frac{\pi - \phi}{\omega_d} \qquad \text{(xii)}$$

$$\text{Peak value} \quad M_p = e^{-\pi \zeta / \sqrt{1 - \zeta^2}} \qquad \text{(xiii)}$$

and

$$\text{Percentage overshoot (PO)} = 100(M_p - 1) \qquad \text{(xiv)}$$

where peak time is the time when the first peak occurs in the response, M_p is the corresponding response value, and rise time is the time at which the response reaches the steady-state (i.e., 1.0) for the first time. These considerations will be revisited in Chapter 5.

Example 2.22

Let us again consider the simple oscillator differential equation:

$$\ddot{y} + 2\zeta\omega_n\dot{y} + \omega_n^2 y = \omega_n^2 u(t) \tag{i}$$

By defining the state variables as:

$$x = [x_1 \quad x_2]^T = [y \quad \dot{y}]^T \tag{ii}$$

where y is the position and $\dot{y}$ is the velocity, a state model for this system can be expressed as

$$\dot{x} = \begin{bmatrix} 0 & 1 \\ -\omega_n^2 & -2\zeta\omega_n \end{bmatrix} x + \begin{bmatrix} 0 \\ \omega_n^2 \end{bmatrix} u(t) \tag{iii}$$

If we consider both displacement and velocity as outputs, we have

$$y = x \tag{iv}$$

Note that the output gain matrix (measurement matrix) C is the identity matrix in this case. From Equation 2.136 and Equation 2.138 it follows that

$$Y(s) = \begin{bmatrix} s & -1 \\ \omega_n^2 & s+2\zeta\omega_n \end{bmatrix}^{-1} \begin{bmatrix} 0 \\ \omega_n^2 \end{bmatrix} U(s)$$

$$= \frac{1}{\left(s^2 + 2\zeta\omega_n s + \omega_n^2\right)} \begin{bmatrix} s+2\zeta\omega_n & 1 \\ -\omega_n^2 & s \end{bmatrix} \begin{bmatrix} 0 \\ \omega_n^2 \end{bmatrix} U(s) \tag{v}$$

$$= \frac{1}{\left(s^2 + 2\zeta\omega_n s + \omega_n^2\right)} \begin{bmatrix} \omega_n^2 \\ s\omega_n^2 \end{bmatrix} U(s)$$

We observe that the transfer function matrix is

$$G(s) = \begin{bmatrix} \omega_n^2/\Delta(s) \\ s\omega_n^2/\Delta(s) \end{bmatrix}$$

in which $\Delta(s) = s^2 + 2\zeta\omega_n s + \omega_n^2$. This function $\Delta(s)$ is termed the *characteristic polynomial* of the system.

Also, the characteristic equation is given by

$$s^2 + 2\zeta\omega_n s + \omega_n^2 = 0 \tag{vii}$$

whose roots are

$$s = -\zeta\omega_n \pm \sqrt{\zeta^2 - 1}\,\omega_n \tag{viii}$$

These characteristic roots are called *poles* or *eigenvalues* of the system and determine the nature of the natural response. In particular if at least one pole of a system has a positive real part, the natural response will grow exponentially and the system is said to be *unstable* (See Chapter 12). Note from Equation viii that three cases of system poles can be identified:

$$\text{Underdamped system } (\zeta < 1): \quad s = -\zeta \omega_n \pm j\omega_d \qquad \text{(ix)}$$

$$\text{Overdamped system } (\zeta > 1): \quad s = -\zeta \omega_n \pm \sqrt{\zeta^2 - 1}\, \omega_n \qquad \text{(x)}$$

$$\text{Critically damped system } (\zeta = 1): \quad s = -\zeta \omega_n \qquad \text{(xi)}$$

In the first case, the system is oscillatory, in the second case the system is non-oscillatory, and in the third case the system has two identical real poles and the system is marginally nonoscillatory. In the present example, the system is stable because the real parts of the poles are negative. In the undamped case ($\zeta = 0$) however, system will oscillate at frequency ω_n without growth or decay, and then the system is said to be *marginally stable*.

The first element in the only column in $G(s)$ is the displacement-output transfer function, and the second element is the velocity-output transfer function. These results agree with the expressions obtained in the previous example.

Now, let us consider the acceleration $\ddot{y}$ as an output, and denote it by y_3. It is clear from the system Equation i that:

$$y_3 = \ddot{y} = -2\zeta\omega_n\dot{y} - \omega_n^2 y + \omega_n^2 u(t)$$

or, in terms of the state variables:

$$y_3 = -2\zeta\omega_n x_2 - \omega_n^2 x_1 + \omega_n^2 u(t)$$

Note that this output explicitly contains the input variable. This feedforward situation implies that the matrix D becomes nonzero when acceleration $\ddot{y}$ is chosen as an output. In this case,

$$Y_3(s) = -2\zeta\omega_n X_2(s) - \omega_n^2 X_1(s) + \omega_n^2 U(s)$$

$$= -2\zeta\omega_n \frac{s\omega_n^2}{\Delta(s)} U(s) - \omega_n^2 \frac{\omega_n^2}{\Delta(s)} U(s) + \omega_n^2 U(s)$$

which simplifies to

$$Y_3(s) = \frac{s^2\omega_n^2}{\Delta(s)} U(s)$$

This confirms the result for the acceleration-output transfer function obtained in the previous example.

Example 2.23

Consider the simplified model of a vehicle shown in Figure 2.55, which can be used to study the heave (vertical up and down) and pitch (front-back rotation) motions due to the road profile and disturbances. For our purposes, let us assume that the road disturbances

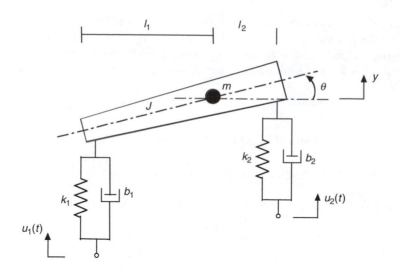

FIGURE 2.55
A model for vehicle suspension system.

exciting the front and back suspensions are independent. The equations of motion for heave (y) and pitch (θ) are written about the static equilibrium configuration of the vehicle model (hence, gravity does not enter into the equations) for small motions:

$$m\ddot{y} = k_1(u_1 - y + l_1\theta) + k_2(u_2 - y + l_2\theta) + b_1(\dot{u}_1 - \dot{y} + l_1\dot{\theta}) + b_2(\dot{u}_2 - \dot{y} + l_2\dot{\theta})$$

$$J\ddot{\theta} = -l_1[k_1(u_1 - y + l_1\theta) + b_1(\dot{u}_1 - \dot{y} + l_1\dot{\theta})] + l_2[k_2(u_2 - y + l_2\theta) + b_2(\dot{u}_2 - \dot{y} + l_2\dot{\theta})]$$

Take the Laplace transform of these two equations with zero initial conditions (i.e., substitute s^2Y for $\ddot{y}$, sY for $\dot{y}$, etc.):

$$[ms^2 + (b_1 + b_2)s + (k_1 + k_2)]Y(s) + [(b_2l_2 - b_1l_1)s + (k_2l_2 - k_1l_1)]\theta(s)$$
$$= (b_1s + k_1)U_1(s) + (b_2s + k_2)U_2(s)$$

$$[(b_2l_2 - b_1l_1)s + (k_2l_2 - k_1l_1)]Y(s) + \left[Js^2 + \left(b_1l_1^2 + b_2l_2^2\right)s + \left(k_1l_1^2 + k_2l_2^2\right)\right]\theta(s)$$
$$= -l_1(b_1s + k_1)U_1(s) + l_2(b_2s + k_2)U_2(s)$$

Let the coefficients be expressed as

$$C_1 = m \qquad\qquad C_7 = b_2s + k_2$$

$$C_2 = b_1 + b_2 \qquad\qquad C_8 = J$$

$$C_3 = k_1 + k_2 \qquad\qquad C_9 = b_1l_1^2 + b_2l_2^2$$

$$C_4 = b_2l_2 - b_1l_1 \qquad\qquad C_{10} = k_1l_1^2 + k_2l_2^2$$

$$C_5 = k_2l_2 - k_1l_1 \qquad\qquad C_{11} = -l_1(b_1s + k_1)$$

$$C_6 = b_1s + k_1 \qquad\qquad C_{12} = l_2(b_2s + k_2)$$

Then

$$[C_1s^2 + C_2s + C_3]Y(s) + [C_4s + C_5]\theta(s) = C_6U_1(s) + C_7U_2(s)$$

$$[C_4s + C_5]Y(s) + [C_8s^2 + C_9s + C_{10}]\theta(s) = C_{11}U_1(s) + C_{12}U_2(s)$$

In matrix form:

$$\begin{bmatrix} C_1s^2 + C_2s + C_3 & C_4s + C_5 \\ C_4s + C_5 & C_8s^2 + C_9s + C_{10} \end{bmatrix} \begin{bmatrix} Y(s) \\ \theta(s) \end{bmatrix} = \begin{bmatrix} C_6 \\ C_{11} \end{bmatrix} U_1(s) + \begin{bmatrix} C_7 \\ C_{12} \end{bmatrix} U_2(s)$$

Now, by taking the inverse of the left hand side matrix we get:

$$\begin{bmatrix} Y(s) \\ \theta(s) \end{bmatrix} = \frac{1}{\Delta(s)} \begin{bmatrix} P(s) & Q(s) \\ Q(s) & R(s) \end{bmatrix} \begin{bmatrix} 1 & 1 \\ -l_1 & l_2 \end{bmatrix} \begin{bmatrix} C_6U_1(s) \\ C_7U_2(s) \end{bmatrix}$$

in which

$$P(s) = Js^2 + C_9s + C_{10}$$

$$Q(s) = -C_4s - C_5s$$

$$R(s) = C_1s^2 + C_2s + C_3$$

and $\Delta(s)$ is the characteristic polynomial of the system as given by the determinant of the transformed system matrix:

$$\Delta(s) = \det \begin{bmatrix} P(s) & -Q(s) \\ -Q(s) & R(s) \end{bmatrix}$$

The transfer-function matrix is given by

$$G(s) = \frac{1}{\Delta(s)} \begin{bmatrix} P(s) & Q(s) \\ Q(s) & R(s) \end{bmatrix} \begin{bmatrix} C_6 & C_7 \\ C_{11} & C_{12} \end{bmatrix}$$

The individual transfer functions are given by the elements of $G(s)$:

$$\frac{Y(s)}{U_1(s)} = \frac{[P(s) - l_1Q(s)]}{\Delta(s)} C_6, \qquad \frac{\theta(s)}{U_1(s)} = \frac{[Q(s) - l_1R(s)]}{\Delta(s)} C_6$$

$$\frac{Y(s)}{U_2(s)} = \frac{[P(s) + l_2Q(s)]}{\Delta(s)} C_7, \qquad \frac{\theta(s)}{U_2(s)} = \frac{[Q(s) + l_2R(s)]}{\Delta(s)} C_7$$

2.11.2 Block Diagrams and State-Space Models

The transfer-function models $G(s)$ for a single-input single-output (SISO) system can be represented by the block diagram shown in Figure 2.56(a). For a multi-input multi-output (MIMO) system the inputs and outputs are vectors *u* and *y*. The corresponding information

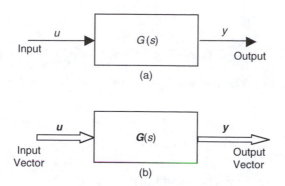

FIGURE 2.56

Block-diagram representation of a transfer-function model: (a) Single-input single-output (SISO) system; (b) Multi-input-multi-output (MIMO) system.

(signal) lines are drawn thicker as in Figure 2.56(b) to indicate that they represent vectors. One disadvantage of the transfer-function representation is obvious from Figure 2.56. No information regarding how various elements are connected within the system can be uniquely determined from the transfer function. It contains only a unique input-output description. For this reason the same transfer function can correspond to different state-space models. We identify the transfer function of a dynamic model by its inputs and outputs, not by its state variables, which are internal variables. However, the internal structure of a dynamic system can be indicated by a more elaborate block diagram. Such detailed diagrams are often used to uniquely indicate the state variables used in a particular model.

For example, consider the state-space model 2.131 and 2.132. A block diagram that uniquely possesses this model is shown in Figure 2.57. Note the *feedforward* path corresponding to **D**. The feedback paths (corresponding to **A**) do not necessarily represent a feedback control system; such paths are termed *natural feedback* paths. Strictly speaking thicker signal lines should be used in this diagram since we are dealing with vector variables. Two or more blocks in cascade can be replaced by a single block having the product of individual transfer functions. The circle in Figure 2.57 is a *summing junction*. A negative sign at the arrow-head of an incoming signal corresponds to subtraction of that signal. As mentioned earlier, $1/s$ can be interpreted as integration, and s as differentiation.

The equivalence of Figure 2.57 and the relations in Equation 2.131 and Equation 2.132 should be obvious. Alternatively, the rules for block diagram reduction (given in Table 2.8) can be used to show that the system transfer function is given by:

$$\frac{Y(s)}{U(s)} = G(s) = \frac{CB}{(s-A)} + D \tag{2.139}$$

This is the scalar version of Equation 2.137.

Using the same input-output differential equation, we now illustrate several methods of obtaining a state-space model through a special type of block diagrams called simulation block diagrams. In these block diagrams each block contains either an integrator ($1/s$) or a constant gain term. The name originates from classical analog computer applications in which hardware modules of summing amplifiers and integrators (along with other units such as potentiometers and resistors) are interconnected to simulate dynamic systems.

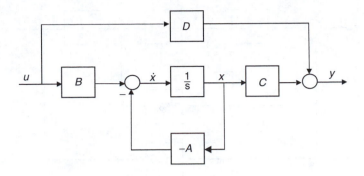

FIGURE 2.57

A block-diagram representation of a state-space model.

TABLE 2.8

Basic Relations for Block-Diagram Reduction

Description	Equivalent Representation	
Summing junction		$x_3 = x_1 + x_2$
Cascade (series) connection		
Parallel connection		
Shifting signal-pickoff point		
Shifting signal-application point		
Reduction of feedback loop		

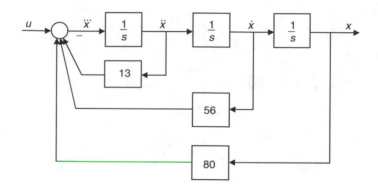

FIGURE 2.58
The simulation diagram of system $\dddot{x} + 13\ddot{x} + 56\dot{x} + 80x = u$.

Example 2.24: Superposition Method

Consider the time domain model given by the input-output differential equation:

$$\dddot{y} + 13\ddot{y} + 56\dot{y} + 80y = \dddot{u} + 6\ddot{u} + 11\dot{u} + 6u \qquad\qquad (i)$$

For a linear system, the *principle of superposition* applies. In particular if, with zero initial conditions, x is the response to an input u, then $d^r x/dt^r$ is the response to the input $d^r u/dt^r$ and, consequently, by the principle of superposition, $\alpha_1 x + \alpha_2 \, d^r x/dt^r$ is the response to the input $\alpha_1 u + \alpha_2 d^r u/dt^r$. To use this concept in the present method, consider the differential equation:

$$\dddot{x} + 13\ddot{x} + 56\dot{x} + 80x = u \qquad\qquad (ii)$$

This defines the "parent" (or, auxiliary) system. The simulation diagram for Equation ii is shown in Figure 2.58. Steps of obtaining this diagram are as follows: start with the highest-order derivative of the response variable (i.e., $\dddot{x}$); successively integrate it until the variable itself (x) is obtained; feed the resulting derivatives of different orders to a summing junction (along with the input variable) to produce the highest-order derivative of the response variable such that the original differential Equation ii is satisfied.

By the principle of superposition, it follows from Equation i and Equation ii that:

$$y = \dddot{x} + 6\ddot{x} + 11\dot{x} + 6x \qquad\qquad (iii)$$

Hence, the simulation diagram for the original system (Equation i) is obtained from Figure 2.58, as shown in Figure 2.59. In particular, note the resulting feedforward paths. The corresponding state model employs x and its derivatives as state variables:

$$[x_1 \quad x_2 \quad x_3]^T = [x \quad \dot{x} \quad \ddot{x}]^T$$

Note that these are outputs of the integrators in Figure 2.59. The state equations are written by considering the signal that goes into each integration block, to form the first derivative of the corresponding state variable. Specifically we have

$$\dot{x}_1 = x_2$$

$$\dot{x}_2 = x_3 \qquad\qquad (iv)$$

$$\dot{x}_3 = -80x_1 - 56x_2 - 13x_3 + u$$

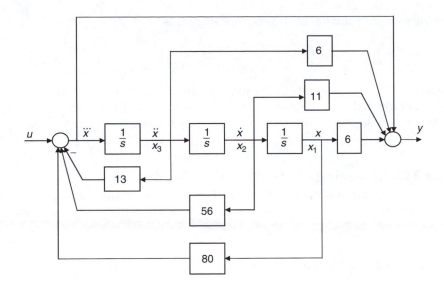

FIGURE 2.59
The simulation diagram of $\dddot{y}+13\ddot{y}+56\dot{y}+80y=\dddot{u}+6\ddot{u}+11\dot{u}+6u$.

The algebraic output equation is obtained by writing the signal summation equation for the summing junction (far right), which generates y. Specifically

$$y = 6x_1 + 11x_2 + 6x_3 + (-80x_1 - 56x_2 - 13x_3 + u)$$

or,

$$y = -74x_1 - 45x_2 - 7x_3 + u \tag{v}$$

The corresponding model matrices are:

$$A = \begin{bmatrix} 0 & 1 & 0 \\ 0 & 0 & 1 \\ -80 & -56 & -13 \end{bmatrix}, \qquad B = \begin{bmatrix} 0 \\ 0 \\ 1 \end{bmatrix}$$

$$C = [-74 \quad -45 \quad -7], \qquad D = 1,$$

The system matrix pair (A, B) is said to be in the *companion form* in this state model. Note that the system model is third order. Hence the simulation diagram needs three integrators, and the system matrix A is 3×3.

Note further that the "parent" (or, auxiliary) transfer function (that of (Equation ii) is given by

$$\frac{X}{U} = \frac{1}{s^3 + 13s^2 + 56s + 80}$$

From Equation iii, the output of the original system is given by

$$Y = s^3 X + 6s^2 X + 11sX + 6X$$

$$= (s^3 + 6s^2 + 11s + 6)X$$

Hence the transfer function of the original system is

$$G(s) = \frac{Y}{U} = \frac{s^3 + 6s^2 + 11s + 6}{s^3 + 13s^2 + 56s + 80}$$

which agrees with the original differential Equation i. Furthermore, in $G(s)$, since the numerator polynomial is of the same order (third order) as the denominator polynomial (characteristic polynomial), a nonzero feedforward gain matrix D is generated in the state model.

Example 2.25: Grouping Like-Derivatives Method

Consider the same input-output differential Equation i as in the previous example. By grouping derivatives of the same order, it can be written in the following form:

$$\dddot{y} = \dddot{u} + (6\ddot{u} - 13\ddot{y}) + (11\dot{u} - 56\dot{y}) + (6u - 80y)$$

By successively integrating this equation three times, we obtain:

$$y = u + \int \left[6u - 13y + \int \left\{ 11u - 56y + \int (6u - 80y) d\tau \right\} d\tau' \right] d\tau'' \tag{i*}$$

Note the three integrations on the right-hand side of this equation. Now draw the simulation diagram as follows: Assume that y is available. Form the integrand of the innermost integration in Equation i* through feedforward of the necessary u term and feedback of the necessary y term. Perform the innermost integration. The result will form a part of the integrand of the next integration. Complete the integrand through feedforwarding the necessary u term and feedback of the necessary y term. Perform this second integration. The result will form a part of the integrand of the next (outermost) integration. Proceed as before to complete the integrand and perform the outermost integration. Feedforward the necessary u term to generate y, which was assumed to be known in the beginning. The result is shown in Figure 2.60. Note that the "innermost" integration in Equation i* forms the "outermost" feedback loop in the block diagram.

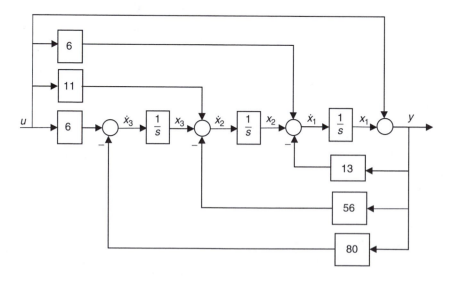

FIGURE 2.60
Simulation diagram obtained by grouping like-derivatives.

As in the previous example, the state variables are defined as the outputs of the integrators. The state equations are written by considering the signals that enter each integration block, to form the first derivative of the corresponding state variable. We get

$$\dot{x}_1 = 6u + x_2 - 13(x_1 + u) = -13x_1 + x_2 - 7u$$

$$\dot{x}_2 = 11u + x_3 - 56(x_1 + u) = -56x_1 + x_3 - 45u \qquad \text{(iv)}$$

$$\dot{x}_3 = 6u - 80(x_1 + u) = -80x_1 - 74u$$

The algebraic output equation is obtained by writing the equation for the summing junction (far right), which generates y. We get

$$y = x_1 + u$$

This corresponds to

$$A = \begin{bmatrix} -13 & 1 & 0 \\ -56 & 0 & 1 \\ -80 & 0 & 0 \end{bmatrix} \qquad B = \begin{bmatrix} -7 \\ -45 \\ -74 \end{bmatrix}$$

$$C = [1 \quad 0 \quad 0], \qquad D = 1$$

This state model is the *dual* of the state model obtained in the previous example.

Example 2.26: Factored-Transfer-Function Method

The method illustrated in this example is appropriate when the system transfer function is available in the factorized form, with first-order terms of the form:

$$G_1(s) = \frac{(s+b)}{(s+a)}$$

Since the block diagram of the transfer function $1/(s + a)$ is given by Figure 2.61(a), it follows from the superposition method that the block diagram for $(s+b)/(s + a)$ is as in Figure 2.61(b). This is one form of the basic block-diagram module, which is used in this method.

An alternative form of block diagram for this basic transfer function module is obtained by noting the equivalence shown in Figure 2.62(a). In other words, when it is necessary to

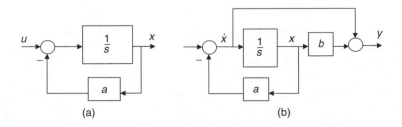

(a)   (b)

FIGURE 2.61
The simulation diagrams of: (a) $1/(s+a)$; (b) $(s+b)/(s+a)$.

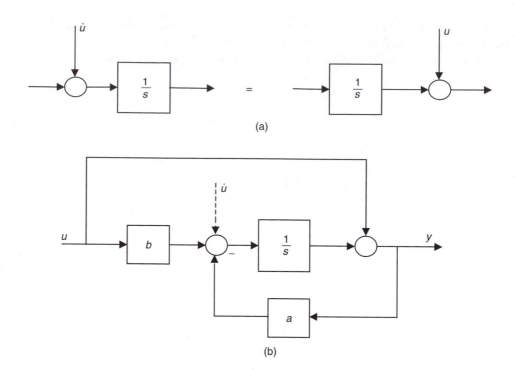

FIGURE 2.62
(a) Two equivalent ways of providing an input derivative ($\dot{u}$); (b) Equivalent simulation diagram for $(s+b)/(s+a)$.

supply a derivative signal $\dot{u}$ at the input to an integrator, the signal u itself can be supplied at the output of the integrator instead. Now, note that the first-order transfer function unit $(s+b)/(s+a)$ has the terms $\dot{u}+bu$ on the input side. The term bu is generated by cascading a block with simple gain b, as in Figure 2.62(b). To provide $\dot{u}$, instead of using the dotted input path in Figure 2.62(b), that would require differentiating the input signal, the signal u itself is applied at the output of the integrator. It follows that the block diagram in Figure 2.61(b) is equivalent to that in Figure 2.62(b).

Now, returning to our common Example (i), the transfer function is written as:

$$G(s) = \frac{s^3 + 6s^2 + 11s + 6}{s^3 + 13s^2 + 56s + 80}$$

This can be factored into the form:

$$G(s) = \frac{(s+1)}{(s+4)} \times \frac{(s+2)}{(s+4)} \times \frac{(s+3)}{(s+5)} \tag{i}**$$

Note that there are two common factors (corresponding to "repeated poles" or "repeated eigenvalues") in the characteristic polynomial (denominator). This has no special implications in the present method. The two versions of block diagram for this transfer function, in the present methods, are shown in Figure 2.63 and Figure 2.64. Here we have used the fact that the product of two transfer functions corresponds to cascading the corresponding simulation block diagrams. As before, the state variables are chosen as outputs of the integrators, and the state equations are written for the input terms of the integrator blocks. The output equation comes from the summation block at the far right, which generates the output.

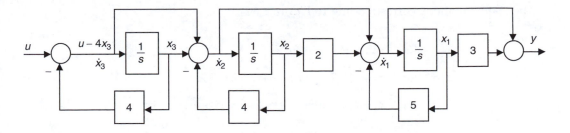

FIGURE 2.63
Simulation block diagram obtained by factorizing the transfer function.

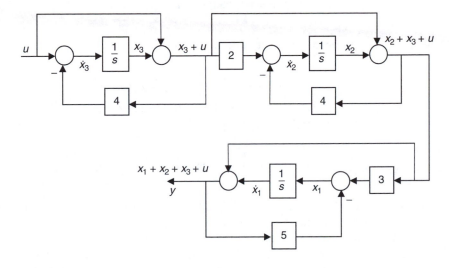

FIGURE 2.64
An alternative simulation diagram obtained by factorizing the transfer function.

From Figure 2.63, the state equations are obtained as

$$\dot{x}_3 = u - 4x_3$$

$$\dot{x}_2 = x_3 - 4x_2 + (u - 4x_3) = -4x_2 - 3x_3 + u \qquad \text{(iv)**}$$

$$\dot{x}_1 = 2x_2 - 5x_1 + (-4x_2 - 3x_3 + u) = -5x_1 - 2x_2 - 3x_3 + u$$

The algebraic output equation is

$$y = 3x_1 + (-5x_1 - 2x_2 - 3x_3 + u)$$

$$= -2x_1 - 2x_2 - 3x_3 + u \qquad \text{(v)**}$$

These correspond to the state model matrices:

$$A = \begin{bmatrix} -5 & -2 & -3 \\ 0 & -4 & -3 \\ 0 & 0 & -4 \end{bmatrix}, \qquad B = \begin{bmatrix} 1 \\ 1 \\ 1 \end{bmatrix}$$

$$C = [-2 \quad -2 \quad -3], \qquad D = 1$$

The state equations corresponding to Figure 2.64 are:

$$\dot{x}_3 = -4(x_3 + u) + u = -4x_3 - 3u$$

$$\dot{x}_2 = -4(x_2 + x_3 + u) + 2(x_3 + u) = -4x_2 - 2x_3 - 2u \qquad \text{(iv')**}$$

$$\dot{x}_1 = -5(x_1 + x_2 + x_3 + u) + 3(x_2 + x_3 + u) = -5x_1 - 2x_2 - 3x_3 - 2u$$

The algebraic output equation is

$$y = x_1 + x_2 + x_3 + u \qquad \text{(v')**}$$

These equations correspond to the state model matrices:

$$A = \begin{bmatrix} -5 & -2 & -2 \\ 0 & -4 & -2 \\ 0 & 0 & -4 \end{bmatrix}, \qquad B = \begin{bmatrix} -2 \\ -2 \\ -3 \end{bmatrix}$$

$$C = \begin{bmatrix} 1 & 1 & 1 \end{bmatrix}, \qquad D = 1$$

Both system matrices are upper-diagonal (i.e., all the elements below the main diagonal are zero), and the main diagonal consists of the poles (eigenvalues) of the system. These are the roots of the characteristic equation. We should note the duality in these two state models (Equation iv** and Equation iv'**). Note also that, if we group the original transfer function into different factor terms, we get different state models. In particular, the state equations will be interchanged.

Example 2.27: Partial-Fraction Method

The partial fractions of the transfer function Equation i** considered in the previous example can be written in the form:

$$G(s) = \frac{s^3 + 6s^2 + 11s + 6}{s^3 + 13s^2 + 56s + 80}$$

$$= 1 - \frac{a}{(s+4)} - \frac{b}{(s+4)^2} - \frac{c}{(s+5)}$$

By equating the like terms on the two sides of this identity, or by using the fact that:

$$c = -(s+5)G(s)\big|_{s=-5}$$

$$b = -(s+4)^2 G(s)\big|_{s=-4}$$

$$a = -\left[\frac{d}{ds}(s+4)^2 G(s)\right]_{s=-4}$$

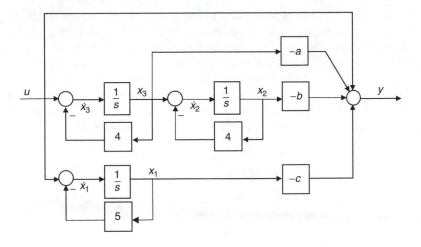

FIGURE 2.65
Simulation block diagram obtained by the partial fraction method.

we can determine the unknown coefficients; thus

$$a = -17, b = 6, c = 24$$

The simulation block diagram corresponding to the partial-fraction representation of the transfer function is shown in Figure 2.65. We have used the fact that the sum of two transfer functions corresponds to combining their block diagrams in parallel. Again the state variables are chosen as the outputs of the integrators. By following the same procedure as before, the corresponding state equations are obtained as

$$\dot{x}_1 = -5x_1 + u$$
$$\dot{x}_2 = -4x_2 + x_3$$
$$\dot{x}_3 = -4x_3 + u$$

The algebraic output equation is

$$y = -cx_1 - bx_2 - ax_3 + u$$

This corresponds to the state-model matrices:

$$A = \begin{bmatrix} -5 & 0 & 0 \\ 0 & -4 & 0 \\ 0 & 0 & -4 \end{bmatrix}, \quad B = \begin{bmatrix} 1 \\ 0 \\ 1 \end{bmatrix}$$

$$C = \begin{bmatrix} -c & -b & -a \end{bmatrix}, \quad D = 1$$

In this case, the system matrix is said to be in the *Jordan canonical form*. If the eigenvalues are distinct (unequal), the matrix A, when expressed in the Jordan form, will be diagonal, and the diagonal elements will be the eigenvalues. When repeated eigenvalues are present, as in the present example, the matrix A will consist of diagonal blocks (or Jordan blocks) consisting of upper-diagonal submatrices with the repeated eigenvalues lying on the main diagonal, elements of unity at locations immediately above the main diagonal, and zero elements elsewhere. More than one Jordan block can exist for the same repeated eigenvalue. These considerations are beyond the scope of the present study.

2.11.3 Causality and Physical Realizability

Consider a dynamic system that is represented by the single input-output differential Equation 2.127. The causality (cause-effect) of this system should dictate that u is the input and y is the output. Its transfer function is given by Equation 2.129. Here, n is the order of the system, $\Delta(s)$ is the characteristic polynomial, and $N(s)$ is the numerator polynomial of the system.

Suppose that $m > n$. Then, if we integrate Equation 2.127 n times, we will have y and its integrals on the LHS but the RHS will contain at least one derivative of u. Since the derivative of a step function is an impulse, this implies that a finite change in input will result in an infinite change in the response. Such a scenario will require infinite power, and is not physically realizable. It follows that a physically realizable system cannot have a numerator order greater than the denominator order, in its transfer function. If in fact $m > n$, then, what it means physically is that y should be the system input and u should be the system output. In other words, the causality should be reversed in this case. For a physically realizable system, a simulation block diagram can be established using integrals ($1/s$) alone, without the need of derivatives (s). Note that pure derivatives are physically not realizable. If $m > n$, the simulation block diagram will need at least one derivative for linking u to y. That will not be physically realizable, again, because it would imply the possibility of producing an infinite response by a finite input. In other words, the simulation block diagram of a physical realizable system will not require feedforward paths with pure derivatives.

Example 2.28

A manufacturer of rubber parts uses a conventional process of steam-cured molding of latex. The molded rubber parts are first cooled and buffed (polished) and then sent for inspection and packing. A simple version of a rubber buffing machine is shown in Figure 2.66(a). It consists of a large hexagonal drum whose inside surfaces are all coated with a layer of bonded emery. The drum is supported horizontally along its axis on two heavy duty, self-aligning bearings at the two ends and is rotated using a three-phase induction motor. The drive shaft of the drum is connected to the motor shaft through a flexible coupling. The buffing process consists of filling the drum with rubber parts, steadily rotating the drum for a specified period of time, and finally vacuum cleaning the drum and its contents. Dynamics of the machine affects loading on various parts such as motor, coupling, bearings, shafts and support structure.

In order to study the dynamic behavior, particularly at the startup stage and under disturbances during steady-state operation, an engineer develops a simplified model of the buffing machine. This model is shown in Figure 2.66(b). The motor is modeled as a torque source T_m, which is applied on the rotor having moment of inertia J_m and resisted by a viscous damping torque of damping constant b_m. The connecting shafts and the

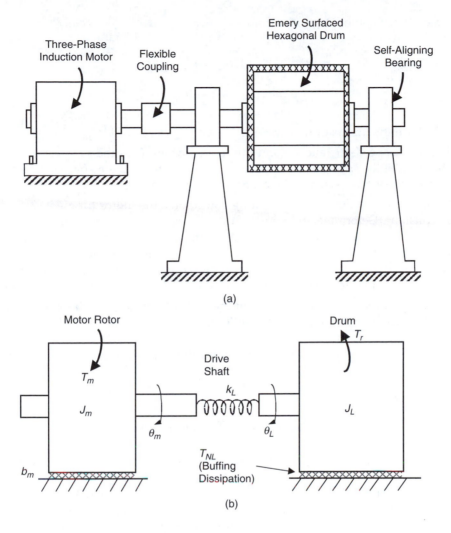

FIGURE 2.66
A rubber buffing machine: (a) Schematic diagram; (b) Dynamic model.

coupling unit are represented by a torsional spring of stiffness k_L. The drum and its contents are represented by an equivalent constant moment of inertia J_L. There is a resisting torque on the drum, even at steady operating speed, due to the eccentricity of the contents of the drum. This is represented by a constant torque T_r. Furthermore, energy dissipation due to the buffing action (between the rubber parts and the emery surfaces of the drum) is represented by a nonlinear damping torque T_{NL}, which may be approximated as

$$T_{NL} = c|\dot{\theta}_L|\dot{\theta}_L \text{ with } c > 0$$

Note that θ_m and θ_L are the angles of rotation of the motor rotor and the drum, respectively, and these are measured from inertial reference lines that correspond to a relaxed configuration of spring k_L.

a. Comment on the assumptions made in the modeling process of this problem and briefly discuss the validity (or accuracy) of the model.

b. Show that the model equations are

$$J_m \ddot{\theta}_m = T_m - k_L(\theta_m - \theta_L) - b_m \dot{\theta}_m$$

$$J_L \ddot{\theta}_L = k_L(\theta_m - \theta_L) - c|\dot{\theta}_L|\dot{\theta}_L - T_r$$

What are the inputs of this system?

c. Using the speeds $\dot{\theta}_m$ and $\dot{\theta}_L$, and the spring torque T_k as the state variables, and the twist of the spring as the output, obtain a complete state-space model for his nonlinear system. What is the order of the state model?

d. Suppose that under steady operating conditions, the motor torque is $\overline{T}_m$, which is constant. Determine an expression for the constant speed $\overline{\omega}$ of the drum in terms of $\overline{T}_m$, T_r and appropriate system parameters under these conditions. Show that, as intuitively clear, we must have $\overline{T}_m > T_r$ for this steady operation to be feasible. Also obtain an expression for the spring twist at steady state, in terms of $\overline{\omega}$, T_r and system parameters.

e. Linearize the system equations about the steady operation condition and express the two equations in terms of the following "incremental" variables:

$$q_1 = \text{variation of } \theta_m \text{ about the steady value}$$

$$q_2 = \text{variation of } \theta_L \text{ about the steady value}$$

$$u = \text{disturbance increment of } T_m \text{ from the steady value } \overline{T}_m.$$

f. For the linearized system obtain the input-output differential equation, first considering q_1 as the output and next considering q_2 as the output. Comment about and justify the nature of the homogeneous (characteristic) parts of the two equations. Discuss, by examining the physical nature of the system, why only the derivatives of q_1 and q_2 and not the variables themselves are present in these input-output equations.

Explain why the derivation of the input-output differential equations will become considerably more difficult if a damper is present between the two inertia elements J_m and J_L.

g. Consider the input-output differential equation for q_1. By introducing an auxiliary variable draw a simulation block diagram for this system. (Use integrators, summers, and coefficient blocks only). Show how this block diagram can be easily modified to represent the following cases:

   i. q_2 is the output
   ii. $\dot{q}_2$ is the output
   iii. $\dot{q}_1$ is the output.

What is the order the system (or the number of free integrators needed) in each of the four cases of output considered here?

h. Considering the spring twist $(q_1 - q_2)$ as the output draw a simulation block diagram for the system. What is the order of the system in this case?

HINT For this purpose you may use the two linearized second order differential equations obtained in part (e).

(i) Comment on why the "system order" is not the same for the five cases of output considered in parts (g) and (h).

SOLUTION

a. The assumptions are satisfactory for a preliminary model, particularly because very accurate control is not required in this process. Some sources of error and concern are as follows.

   i. Since rubber parts are moving inside the drum, J_L is not constant and the inertia contribution does not represent a rigid system.

   ii. Inertia of the shafts and coupling is either neglected or lumped with J_m and J_L.

   iii. Coulomb and other nonlinear types of damping in the motor and bearings have been approximated by viscous damping.

   iv. The torque source model (T_m) is only an approximation to a real induction motor.

   v. The resisting torque of the rubber parts (T_r) is not constant during rotation.

   vi. Dissipation due to relative movements between rubber parts and the inside surfaces of the drum may take a different form from what is given (a quadratic damping model).

b. For J_m, Newton's second law gives (see Figure 2.67(a))

$$J_m \frac{d^2\theta_m}{dt^2} = T_m - b_m \dot{\theta}_m - T_k \tag{i}$$

For spring k_L, Hooke's law gives (see Figure 2.67(b))

$$T_k = k_L(\theta_m - \theta_L) \tag{ii}$$

For J_L, Newtons' second law gives (see Figure 2.67(c))

$$J_L \frac{d^2\theta_L}{dt^2} = T_k - T_{NL} - T_r \tag{iii}$$

with

$$T_{NL} = c|\dot{\theta}_L|\dot{\theta}_L \tag{iv}$$

Substitute Equation ii into Equation i:

$$J_m \ddot{\theta}_m = T_m - b_m \dot{\theta}_m - k_L(\theta_m - \theta_L) \tag{v}$$

Substitute Equation ii and Equation iv into Equation iii:

$$J_L \ddot{\theta}_L = k_L(\theta_m - \theta_L) - c|\dot{\theta}_L|\dot{\theta}_L - T_r \tag{vi}$$

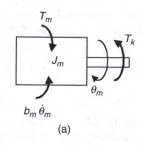

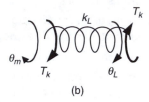

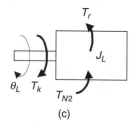

FIGURE 2.67
(a) Motor inertia; (b) Drive shaft; (c) Drum inertia.

Clearly, T_m and T_r are the inputs to the system (see Equation v and Equation vi).

c. Let $\dot{\theta}_m = \omega_m$ and $\dot{\theta}_L = \omega_L$.

From Equation i: $\dfrac{d\omega_m}{dt} = -\dfrac{b_m}{J_m}\omega_m - \dfrac{1}{J_m}T_k + \dfrac{1}{J_m}T_m$

From Equation iii and iv: $\dfrac{d\omega_L}{dt} = -\dfrac{c}{J_L}|\omega_L|\omega_L + \dfrac{1}{J_L}T_k - \dfrac{1}{J_L}T_r$

Differentiate Equation ii: $\dfrac{dT_k}{dt} = k_L\omega_m - k_L\omega_L$

The above three equations are the state equations. Now, output y = spring twist = $\theta_m - \theta_L$. Hence, from Equation ii we have

$$y = \frac{1}{k_L}T$$

which is the output equation. The system is third order (three state equations).

d. Under steady conditions,

$$\omega_m = \omega_L = \overline{\omega}, \quad \dot{\omega}_m = 0 = \dot{\omega}_L, \quad T_m = \overline{T}_m, \quad \text{and} \quad \theta_m - \theta_L = \Delta\overline{\theta}$$

T_r remains a constant. Then from Equation v and Equation vi

$$\overline{T}_m - b_m\overline{\omega} - k_L\Delta\overline{\theta} = 0 \tag{vii}$$

and

$$k_L\Delta\overline{\theta} - c\overline{\omega}^2 - T_r = 0 \tag{viii}$$

Note that, without loss of generality, $\overline{\omega}$ is assumed to be positive. Add Equation vii and Equation viii to eliminate $k_L\Delta\overline{\theta}$; thus,

$$\overline{T}_m - b_m\overline{\omega} - c\overline{\omega}^2 - T_r = 0$$

or,

$$c\overline{\omega}^2 + b_m\overline{\omega} - (\overline{T}_m - T_r) = 0$$

Hence

$$\overline{\omega} = -\frac{b_m}{2c} \pm \sqrt{\left(\frac{b_m}{2c}\right)^2 + \frac{(\overline{T}_m - T_r)}{c}}$$

The proper solution is

$$\overline{\omega} = \sqrt{\left(\frac{b_m}{2c}\right)^2 + \frac{(\overline{T}_m - T_r)}{c}} - \frac{b_m}{2c}$$

and for this to be positive, we must have $\overline{T}_m > T_r$.

Next, from Equation viii, the steady-state twist of the spring:

$$\Delta\overline{\theta} = \frac{(c\overline{\omega}^2 + T_r)}{k_L}$$

e. Taylor series expansion up to the first-order term gives

For Equation v: $\quad J_m\dot{\overline{\omega}} + J_m\ddot{q}_1 = \overline{T}_m + u - b_m\overline{\omega} - b_m\dot{q}_1 - k_L\Delta\overline{\theta} - k_L(q_1 - q_2)$

For Equation vi: $\quad J_L\dot{\overline{\omega}} + J_L\ddot{q}_2 = k_L\Delta\overline{\theta} + k_L(q_1 - q_2) - c\overline{\omega}^2 - 2c\overline{\omega}\dot{q}_2 - T_r$

The steady-state terms cancel out (also, $\dot{\overline{\omega}} = 0$). Hence, we have the following linearized equations:

$$J_m\ddot{q}_1 = u - b_m\dot{q}_1 - k_L(q_1 - q_2) \tag{ix}$$

$$J_L\ddot{q}_2 = k_L(q_1 - q_2) - 2c\overline{\omega}\dot{q}_2 \tag{x}$$

These two equations represent the linear model.

f. From Equation ix:

$$q_2 = \left[q_1 + \frac{b_m}{k_L} \dot{q}_1 + \frac{J_m}{k_L} \ddot{q}_1 - \frac{u}{k_L} \right] \quad \text{(xi)}$$

From Equation x:

$$q_1 = \left[q_2 + \frac{2c\overline{\omega}}{k_L} \dot{q}_2 + \frac{J_L}{k_L} \ddot{q}_2 \right] \quad \text{(xii)}$$

Substitute Equation xi into Equation xii for q_2:

$$q_1 = \left[q_1 + \frac{b_m}{k_L} \dot{q}_1 + \frac{J_m}{k_L} \ddot{q}_1 - \frac{u}{k_L} \right] + \frac{2c\overline{\omega}}{k_L} \left[\dot{q}_1 + \frac{b_m}{k_L} \ddot{q}_1 + \frac{J_m}{k_L} \dddot{q}_1 - \frac{\dot{u}}{k_L} \right]$$

$$+ \frac{J_L}{k_L} \left[\ddot{q}_1 + \frac{b_m}{k_L} \dddot{q}_1 + \frac{J_m}{k_L} \ddddot{q}_1 - \frac{\ddot{u}}{k_L} \right]$$

which gives

$$\frac{J_m J_L}{k_L^2} \frac{d^4 q_1}{dt^4} + \left(\frac{b_m J_L}{k_L^2} + \frac{2c\overline{\omega} J_m}{k_L^2} \right) \frac{d^3 q_1}{dt^3} + \left(\frac{J_m}{k_L} + \frac{2c\overline{\omega} b_m}{k_L^2} + \frac{J_L}{k_L} \right) \frac{d^2 q_1}{dt^2}$$

$$+ \left(\frac{2c\overline{\omega}}{k_L} + \frac{b_m}{k_L} \right) \frac{dq_1}{dt} = \frac{1}{k_L} u + \frac{2c\overline{\omega}}{k_L^2} \frac{du}{dt} + \frac{J_L}{k_L^2} \frac{d^2 u}{dt^2} \quad \text{(xiii)}$$

Next, substitute Equation xii into Equation xi for q_1. Here we get

$$\frac{J_m J_L}{k_L^2} \frac{d^4 q_2}{dt^4} + \left(\frac{b_m J_L}{k_L^2} + \frac{2c\overline{\omega} J_m}{k_L^2} \right) \frac{d^3 q_2}{dt^3} + \left(\frac{J_L}{k_L} + \frac{2c\overline{\omega} b_m}{k_L^2} + \frac{J_m}{k_L} \right) \frac{d^2 q_2}{dt^2}$$

$$+ \left(\frac{2c\overline{\omega}}{k_L} + \frac{b_m}{k_L} \right) \frac{dq_2}{dt} = \frac{1}{k_L} u \quad \text{(xiv)}$$

Observe that the left hand sides (homogenous or characteristic parts) of these two input-output differential equations are identical. This represents the "natural" dynamics of the system and should be common and independent of the input (u). Hence the result is justified. Furthermore, derivatives of u are present only in the q_1 equation. This is justified because motion q_1 is closer than q_2 to the input u. Also, only the derivatives of q_1 and q_2 are present in the two equations. This is a property of a mechanical system that is not anchored (by a spring) to ground. Here the reference value for q_1 or q_2 could be chosen arbitrarily, regardless of the relaxed position of the inter-component spring (k_L) and should not depend on u either. Hence the absolute displacements q_1 and q_2 themselves should not appear in the input-output equations, as clear from Equation xiii and Equation xiv. Such systems are said to possess *rigid body modes*. Even though the differential equations are fourth order, they can be directly integrated once, and the system is actually third order. The position itself can be defined by an arbitrary reference and should not be used as a state in order to avoid this ambiguity. However, if position (q_1 or q_2 and not the twist $q_1 - q_2$) is chosen as an output, the system has to be treated as fourth order. Compare this to the simple problem of a single mass subjected to an external force, and without any anchoring springs.

If there is a damper between J_m and J_L we cannot write simple expressions for q_2 in terms of q_1, and q_1 in terms of q_2, as in Equation xi and Equation xii. Here, the derivative operator $D = \frac{d}{dt}$ has to be introduced for the elimination process, and the solution of one variable by eliminating the other one becomes much more complicated.

g. Use the auxiliary equation

$$a_4 \frac{d^4x}{dt^4} + a_3 \frac{d^3x}{dt^3} + a_2 \frac{d^2x}{dt^2} + a_1 \frac{dx}{dt} = u$$

where

$$a_4 = \frac{J_m J_L}{k_L}, \qquad a_3 = \left(\frac{b_m J_L}{k_L} + \frac{2c\overline{\omega} J_m}{k_L} \right), \qquad a_2 = \left(J_m + \frac{2c\overline{\omega} b_m}{k_L} + J_L \right), \qquad a_1 = b_m + 2c\overline{\omega}$$

It follows from Equation xiv that

$$q_2 = x$$

and from Equation xii that

$$q_1 = x + b_1 \dot{x} + b_2 \ddot{x}$$

where $$b_1 = \frac{2c\overline{\omega}}{k_L}, \quad \text{and} \quad b_2 = \frac{J_L}{k_L}.$$

Hence, we have the block diagram shown in Figure 2.68(a) for the $u \rightarrow q_1$ relationship.

Note that four integrators are needed. Hence this is a fourth order system.

i. In this case the simulation block diagram is as shown in Figure 2.68(b). This also needs four integrators (a fourth-order system).

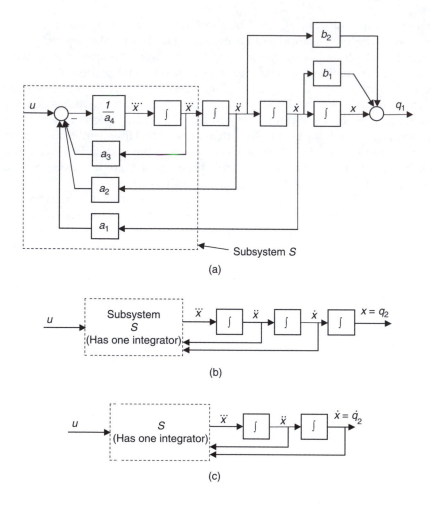

FIGURE 2.68
Simulation block diagram: (a) when q_1 is the output; (b) when q_2 is the output; (c) when $\dot{q}_2$ is the output; (d) when $\dot{q}_1$ is the output; (e) when the spring twist q_1–q_2 is the output.

    ii. In this case the simulation block diagram is as shown in Figure 2.68(c). This only needs three integrators (a third-order system).

    iii. By differentiating the expression for q_1, we have $\dot{q}_1 = \dot{x} + b_1\ddot{x} + b_2\dddot{x}$. Hence the block diagram in this case is as shown in Figure 2.68(d). This needs three integrators (a third-order system).

  h. Using Equation ix and Equation x we get

$$J_m\ddot{q}_1 = u - b_m\dot{q}_1 - k_L(q_1 - q_2)$$

$$J_L\ddot{q}_2 = k_L(q_1 - q_2) - 2c\overline{\omega}\dot{q}.$$

Accordingly, we can draw the block diagram shown in Figure 2.68(e). There are three integrators in this case. The system is third order.

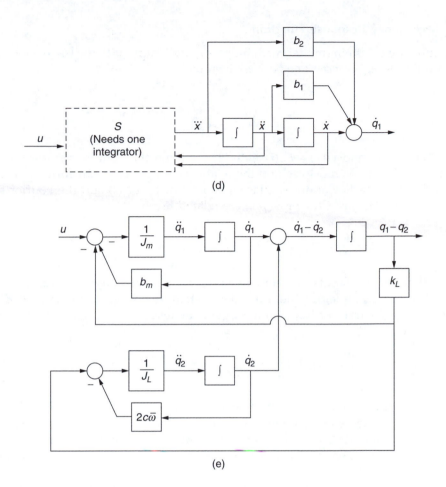

(d)

(e)

FIGURE 2.68
(Continued)

    i. When q_1 and q_2 are used as outputs, the system order increases to four. But, as discussed in Part (f), q_1 and q_2 are not realistic state variables for the present problem.

2.12 Frequency Domain Models

Any transfer function is defined as the ratio of output to input. If the output and input are expressed in the frequency domain, the frequency transfer function is given by the ratio of the Fourier transforms of the output to the input. The Laplace "transfer function" as discussed in the previous section can be easily converted into the "frequency transfer function," as we will learn. Frequency-domain representations are particularly useful in the analysis, design, control, and testing of mechatronic systems. Sinusoidal excitation is often used in testing of equipment and components. The signal waveforms encountered in a mechatronic system can be interpreted and represented as a series of sinusoidal components. Indeed, any waveform can be so represented. It is usually easier to obtain frequency-domain models than the associated time-domain models by testing.

2.12.1   Frequency Response Function

By definition, for a system (model) with input $u(t)$ and output $y(t)$, the frequency response function (or, *frequency transfer function*) is given by

$$G(f) = \frac{Y(f)}{U(f)} \tag{2.140}$$

where $Y(f) = \mathcal{F} y(t)$ and $U(f) = \mathcal{F} u(t)$ with $\mathcal{F}$ denoting the Fourier transform operator. The concepts of Fourier transform and the relation with Laplace transform are presented in Appendix A. In particular, the Laplace transfer function $G(s)$ and the Fourier transfer function $G(f)$ (or $G(jf)$ or $G(\omega)$ or $G(j\omega)$) are related through:

$$G(f) = G(s)\big|_{s=j2\pi f} \tag{2.141}$$

It follows that $G(f)$ constitutes a complete model for a linear, constant-parameter system, as does $G(s)$. For example, for the nth-order system given by the differential Equation 2.127, the frequency transfer function (or, frequency response function) is given by

$$G(f) = \frac{b_0 + b_1(j2\pi f) + \cdots + b_m(j2\pi f)^m}{a_0 + a_1(j2\pi f) + \cdots + a_n(j2\pi f)^n} \tag{2.142}$$

Compare this with Equation 2.129 which gives the Laplace transfer function. It should be clear that, even though $G(f)$ is defined in terms of $U(f)$ and $Y(f)$, it is a system model and is independent of the input (and hence the output): For a physically realizable linear constant-parameter system, $G(f)$ exists even if $U(f)$ and $Y(f)$ do not exist for a particular input.

The frequency transfer function $G(f)$ is, in general, a complex function of frequency f (which is a real variable), having *magnitude* denoted by $|G(f)|$ and *phase angle* denoted by $\angle G(f)$. If a harmonic (i.e., sinusoidal) excitation of frequency f is applied to a stable (i.e., finite natural response), linear, constant-parameter system, its steady-state response will be harmonic with the same frequency f, but the amplitude will be magnified by the factor $|G(f)|$ and the phase will lead by the angle $\angle G(f)$. This appears to be a convenient method of experimental determination of a system model. This approach of "experimental modeling" is termed *model identification*. Either a *sine-sweep* or a *sine-dwell* excitation may be used with these tests. Specifically, a sinusoidal excitation is applied (i.e., input) to the system and the amplification factor and the phase-lead angle of the resulting response are determined at steady state. The frequency of excitation is varied continuously for a sine sweep, and in steps for a sine dwell. Sweep rate should be sufficiently slow, or dwell times should be sufficiently long, to guarantee achieving steady-state response in these methods. The results are usually presented as either a pair of curves of $|G(f)|$ and $\angle G(f)$ versus f, or on the complex $G(f)$ plane with the real part plotted on the horizontal axis and the imaginary part on the vertical axis. The former pair of plots is termed *Bode plot* or *Bode diagram*; the latter is termed *Nyquist diagram* or *argand plot* or *polar plot* (Also see Chapter 12). The shape of these plots for a simple oscillator is shown in Figure 2.69.

In a Bode diagram the frequency is shown explicitly on one axis, whereas in a Nyquist plot the frequency is a parameter on the curve, and is not explicitly shown unless the curve itself is calibrated. In Bode diagrams, it is customary and convenient to give the magnitude in decibels ($20\log_0 |G(f)|$) and scale the frequency axis in logarithmic units

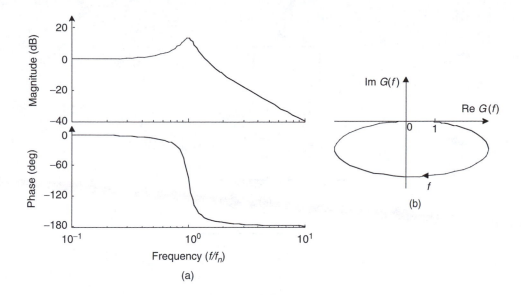

FIGURE 2.69
Frequency domain model of a simple oscillator: (a) Bode diagram; (b) Nyquist plot.

(typically factors of 10 or decades). Since the argument of a logarithm should necessarily be a dimensionless quantity, $Y(f)$ and $U(f)$ should have the same units, or the ratio, of $G(f)$ with respect to some base value such as $G(0)$ should be used.

The arrow on the Nyquist curve indicates the direction of increasing frequency. Only the part corresponding to positive frequencies is actually shown. The frequency response function corresponding to negative frequencies is obtained by replacing f by $-f$ or, equivalently, $j2\pi f$ by $-j2\pi f$. The result is clearly the complex conjugate of $G(f)$, and is denoted $G^*(f)$:

$$G^*(f) = |G(S)|_{s=-j2\pi f} \qquad (2.143)$$

Since, in complex conjugation, the magnitude does not change and the phase angle changes sign, it follows that the Nyquist plot for $G^*(f)$ is the mirror image of that for $G(f)$ about the real axis. In other words, the Nyquist plot for the entire frequency range $f(-\infty, +\infty)$ is symmetric about the real axis.

2.12.2 Significance of Frequency Transfer Function

The significance of frequency transfer function as a dynamic model can be explained by considering the simple oscillator (i.e., a single degree-of-freedom mass-spring-damper system, as shown in Figure 2.70(a)). Its force-displacement transfer function can be written as

$$G(\omega) = \frac{1}{ms^2 + bs + k} \quad \text{with } s = j\omega$$

in which m, b, and k denote mass, damping constant, and stiffness, respectively. Now when the excitation frequency ω is small in comparison to the system natural frequency

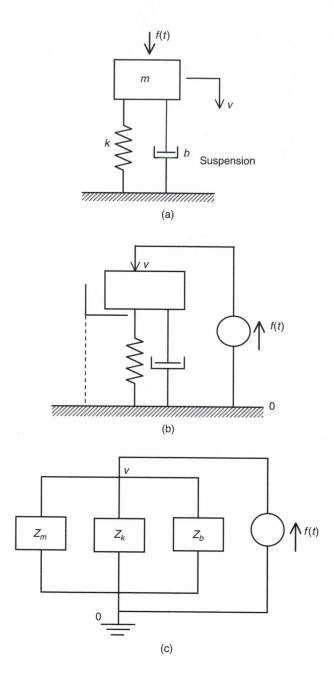

FIGURE 2.70
(a) Ground-based mechanical oscillator; (b) Schematic mechanical circuit; (c) Impedance circuit.

$\sqrt{k/m}$, the terms ms^2 and bs can be neglected with respect to k; and the system behaves as a simple spring. When the excitation frequency ω is much larger than the system natural frequency, the terms bs and k can be neglected in comparison to ms^2. In this case the system behaves like a simple mass element. When the excitation frequency ω is very close to the natural frequency (i.e., $s = j\omega \approx j\sqrt{k/m}$), it is seen that the term $ms^2 + k$ in the denominator of the transfer function (i.e., the characteristic polynomial) becomes almost zero, and can be neglected. Then the transfer function can be approximated by $G(\omega) = 1/(bs)$ with $s = j\omega$.

TABLE 2.9

Mechanical Impedance and Mobility Relations

Transfer function	Symbol	Definition	Combination Rule	
			Series	Parallel
Impedance	Z	$\dfrac{\text{Output force spectrum}}{\text{Input velocity spectrum}}$	$\dfrac{1}{Z} = \dfrac{1}{Z_1} + \dfrac{1}{Z_2}$	$Z = Z_1 + Z_2$
Mobility	M	$\dfrac{\text{Output velocity spectrum}}{\text{Input force spectrum}}$	$M = M_1 + M_2$	$\dfrac{1}{M} = \dfrac{1}{M_1} + \dfrac{1}{M_2}$

It follows that in the neighborhood of a resonance (i.e., for intermediate excitation frequencies), system damping becomes the most important parameter; whereas at low excitation frequencies it is the stiffness; and at high excitation frequencies it is the mass. These considerations use the physical parameters, mass, stiffness, and damping, as the system parameters. Instead we could use natural frequency $\omega_n = \sqrt{k/m}$ and the damping ratio $\zeta = b/(2\sqrt{mk})$ as the system parameters. In that case the number of system parameters reduces to two, which is an advantage in parametric or sensitivity studies.

2.12.3   Mechanical Impedance and Mobility

Any type of force or motion variable may be used as input and output variables in defining a system transfer function. In studies of mechanical system, for example, three types of frequency transfer functions are useful: *Impedance Functions, Mobility Functions*, and *Transmissibility Functions*. These are described now.

In the case of impedance function, velocity is considered the input variable and the force is the output variable, whereas in the case of mobility function the converse applies. These definitions are described in Table 2.9. It is clear that mobility is the inverse of impedance. Either transfer function may be used in a given problem.

Also given in Table 2.9 are the combination relations for interconnected elements. In parallel connection, the across variable (velocity) is common and the through variable (force) is additive. Accordingly, the impedance will be additive. In other words, the inverse of mobility will be additive. In series connection, the through-variable (force) will be common and the across-variable (velocity) will be additive. Accordingly, the mobility or the inverse of impedance will be additive. It follows that it is more convenient to use impedance for parallel combination, and mobility for series combination.

In the earlier sections of the chapter, the linear constitutive relations for the mass, spring and the damper elements have been presented as time-domain relations. The corresponding transfer relations are obtained by replacing the derivative operator d/dt by the Laplace operator s. The frequency transfer functions are obtained by substituting $j\omega$ or $j2\pi f$ for s. The results are summarized in Table 2.10

One can define several other versions of frequency transfer functions that might be useful in modeling and analysis of mechanical systems. Some relatively common ones are given in Table 2.11.

Note that in the frequency domain:

Acceleration $= (j\omega)(\text{velocity})$

Displacement $= \text{velocity}/j\omega$

TABLE 2.10

Mechanical Impedance and Mobility of Discrete Mechanical Elements

Element	Mechanical Circuit Element	Frequency Transfer Function (Set $s = j\omega = j2\pi f$)	
		Impedance	Mobility
Mass m	f $\rightarrow$ v_2 $\quad$ Z_m or M_m $\quad$ v_1 $\quad$ $v = v_2 - v_1$	$Z_m = ms$	$M_m = \dfrac{1}{ms}$
Spring k	f $\rightarrow$ v_2 $\quad$ Z_k or M_k $\quad$ v_1 $\quad$ $v = v_2 - v_1$	$Z_k = \dfrac{k}{s}$	$M_k = \dfrac{s}{k}$
Damper b	f $\rightarrow$ v_2 $\quad$ Z_b or M_b $\quad$ v_1 $\quad$ $v = v_2 - v_1$	$Z_b = b$	$M_b = \dfrac{1}{b}$

TABLE 2.11

Definitions of Useful Mechanical Transfer Functions

Transfer Function	Definition (in frequency domain)
Dynamic stiffness	Force/displacement
Receptance, dynamic flexibility, or compliance	Displacement/Force
Impedance (Z)	Force/velocity
Mobility (M)	Velocity/force
Dynamic inertia	Force/acceleration
Accelerance	Acceleration/force
Force transmissibility (T_f)	Transmitted force/applied force
Motion transmissibility (T_m)	Transmitted velocity/applied velocity

In view of these relations, many of the alternative types of transfer functions as defined in Table 2.11 are related to mechanical impedance and mobility through a factor of $j\omega$; specifically,

Dynamic Inertia = Force/Acceleration = Impedance/($j\omega$)

Accelerance = Acceleration/Force = Mobility $\times j\omega$

Dynamic Stiffness = Force/Displacement = Impedance $\times j\omega$

Dynamic Flexibility = Displacement/Force or Compliance = Mobility/($j\omega$)

In these definitions the variables force, acceleration and displacement should be interpreted as the corresponding Fourier spectra. Three examples are given next to demonstrate the use of impedance and mobility methods in frequency-domain models.

Example 2.29: Simple Oscillator

Consider the simple oscillator shown in Figure 2.70(a). Its mechanical circuit representation is given in Figure 2.70(b). If the input is the force $f(t)$, the source element is a force source. The corresponding response is the velocity v, and in this situation the transfer function $V(f)/F(f)$ is a mobility function. On the other hand, if the input is the velocity $v(t)$, the source element is a velocity source. Then, f is the output, and the transfer function $F(f)/V(f)$ is an impedance function.

Suppose that using a force source, a known forcing function is applied to this system (with zero initial conditions) and the velocity is measured. Now if we were to move the mass exactly at this predetermined velocity (using a velocity source), the force generated at the source will be identical to the originally applied force. In other words, mobility is the reciprocal (inverse) of impedance, as noted earlier. This reciprocity should be intuitively clear because we are dealing with the same system and same initial conditions. Due to this property, we may use either the impedance representation or the mobility representation, depending on whether the elements are connected in parallel or in series, irrespective of whether the input is a force or a velocity. Once the transfer function is determined in one form, its reciprocal gives the other form.

In the present example, the three elements are connected in parallel. Hence, as is clear from the impedance circuit shown in Figure 2.70(c), impedance representation is appropriate. The overall impedance function of the system is

$$Z(f) = \frac{F(f)}{V(f)} = Z_m + Z_k + Z_b = ms + \frac{k}{s} + b \Big|_{s=j2\pi f}$$

$$= \frac{ms^2 + bs + k}{s} \Big|_{s=j2\pi f}$$

The mobility function is the inverse of $Z(f)$:

$$M(f) = \frac{V(f)}{F(f)} = \frac{s}{ms^2 + bs + k} \Big|_{s=j2\pi f}$$

Note that if the input is in fact the force, the mobility function governs the system behavior. In this case, the characteristic polynomial of the system is $s^2 + bs + k$, which corresponds to a simple oscillator and, accordingly, the (dependent) velocity response of the system would be governed by this characteristic polynomial. If, on the other hand, the input is the velocity, the impedance function governs the system behavior. The characteristic polynomial of the system, in this case, is s—which corresponds to a simple integrator. The (dependent) force response of the system would be governed by an integrator type behavior. To explore this behavior further, suppose the velocity source has a constant value. The inertia force will be zero. The damping force will be constant. The spring force will increase linearly. Hence, the net force will have an integration (linearly increasing) effect. If the velocity source provides a linearly increasing velocity (constant acceleration), the inertia force will be constant, the damping force will increase linearly, and the spring force will increase quadratically.

Example 2.30: A Degenerate Case

Consider an intuitively degenerate example of a system as shown in Figure 2.71(a). Note that the support motion is not associated with an external force. The mass m has an external force f and velocity v. At this point we shall not specify which of these variables is the input. It should be clear, however, that v_1 cannot be logically considered an input because

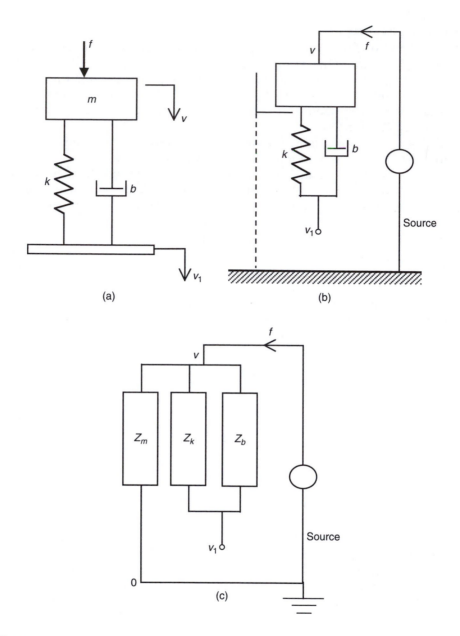

FIGURE 2.71
(a) A mechanical oscillator with support motion; (b) Schematic mechanical circuit; (c) Impedance circuit.

the application of any arbitrary velocity to the support will generate a force at that location and this is not allowed for in the given system. However, since $v_1 = v$ it follows from the mechanical circuit representation shown in Figure 2.71(b), and its impedance circuit shown in Figure 2.71(c), that it is acceptable to indirectly consider v_1 also as the input to the system when v is the input.

When v is the input to the system, the source element in Figure 2.71(b) becomes a velocity source. This corresponds to the impedance function

$$\frac{F(f)}{V(f)} = Z_m = ms\big|_{s=j2\pi f}$$

If, on the other hand, f is the input and v is the output, the mobility function is valid, as given by

$$\frac{V(f)}{F(f)} = M_m = \frac{1}{ms}\bigg|_{s=j2\pi f}$$

Furthermore, since $v_1 = v$, an alternative impedance function

$$\frac{F(f)}{V_1(f)} = ms\bigg|_{s=j2\pi f}$$

and a mobility function

$$\frac{V_1(f)}{F(f)} = \frac{1}{ms}\bigg|_{s=j2\pi f}$$

could be defined.

Example 2.31: Oscillator with Support Motion

To show an interesting reciprocity property, consider the system shown in Figure 2.72(a). In this example the motion of the mass m is not associated with an external force. The support motion, however, is associated with the force f. A schematic mechanical circuit for the system is shown in Figure 2.72(b) and the corresponding impedance circuit is shown in Figure 2.72(c). They clearly indicate that the spring and the damper are connected in parallel, and the mass is connected in series with this pair. By impedance addition for parallel elements, and mobility addition for series elements, it follows that the overall mobility function of the system is

$$\frac{V(f)}{F(f)} = M_m + \frac{1}{(Z_k + Z_b)} = \frac{1}{ms} + \frac{1}{(k/s + b)}\bigg|_{s=j2\pi f}$$

$$= \frac{ms^2 + bs + k}{ms(bs + k)}\bigg|_{s=j2\pi f}$$

It follows that when force is the input (force source) and the support velocity is the output, the system characteristic polynomial is $ms(bs + k)$, which is known to be inherently unstable due to the presence of a free integrator, and has a nonoscillatory transient response.

The impedance function that corresponds to support velocity input (velocity source) is the reciprocal of the previous mobility function, and is given by

$$\frac{F(f)}{V(f)} = \frac{ms(bs + k)}{ms^2 + bs + k}\bigg|_{s=j2\pi f}$$

Furthermore,

$$\frac{V_1(f)}{F(f)} = \frac{1}{ms}\bigg|_{s=j2\pi f}$$

The impedance function $F(f)/V_1(f)$ is not admissible and is physically unrealizable because V_1 cannot be an input (as in Example 2.30) for there is no associated force. This is

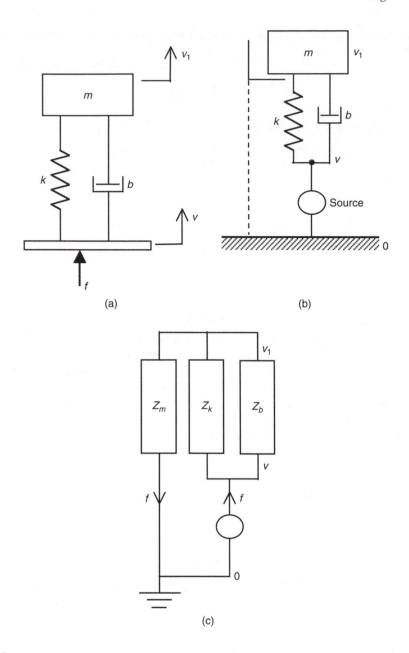

FIGURE 2.72
(a) A mechanical oscillator with support motion; (b) Schematic mechanical circuit; (c) Impedance circuit.

confirmed by the fact that the corresponding transfer function is a differentiator. The mobility function $V_1(f)/F(f)$ corresponds to a simple integrator. Physically, when a force f is applied to the support it transmits to the mass, unchanged, through the parallel spring-damper unit. Accordingly, when f is constant, a constant acceleration is produced at the mass, causing its velocity to increase linearly (an "integration" behavior).

Maxwell's principle of reciprocity is demonstrated by noting that in Example 2.30 and Example 2.31 the mobility functions $V_1(f)/F(f)$ are identical. What this means is that the support motion produced by applying a forcing excitation to the mass (system in Figure 2.71(a)) is equal to the motion of the mass when the same forcing excitation is applied to the support (system in Figure 2.72(a)), with the same initial conditions.

This reciprocity property is valid for linear, constant-parameter systems in general, and is particularly useful in testing of multi-degree-of-freedom mechanical systems; for example, to determine a transfer function that is difficult to measure, by measuring its symmetrical counterpart in the transfer function matrix.

2.12.4 Transmissibility Function

Transmissibility functions are transfer functions that are particularly useful in the design and analysis of fixtures, mounts, and support structures for machinery and other dynamic systems. In particular they are used in the studies of vibration isolation. Two types of transmissibility functions—force transmissibility and motion transmissibility—can be defined. Due to a reciprocity characteristic in linear systems, it can be shown that these two transfer functions are equal and, consequently, it is sufficient to consider only one of them. Let us first consider both types and show their equivalence.

2.12.4.1 Force Transmissibility

Consider a mechanical system supported on a rigid foundation through a suspension system. If a forcing excitation is applied to the system it is not directly transmitted to the foundation. The suspension system acts as an "isolation" device. Force transmissibility determines the fraction of the forcing excitation that is transmitted to the foundation through the suspension system at different frequencies, and is defined as

$$\text{Force Transmissibility } T_f = \frac{\text{Suspension Force } F_s}{\text{Applied Force } F}$$

Note that this function is defined in the frequency domain, and accordingly F_s and F should be interpreted as the Fourier spectra of the corresponding forces.

A schematic diagram of a force transmissibility mechanism is shown in Figure 2.73. The reason for the suspension force f_s not being equal to the applied force f is attributed to the inertia paths (broken line in Figure 2.73) that are present in the mechanical system.

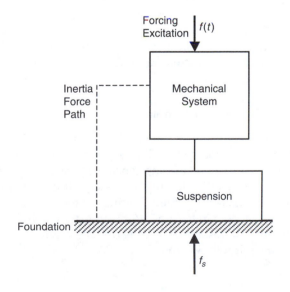

FIGURE 2.73
Force transmissibility mechanism.

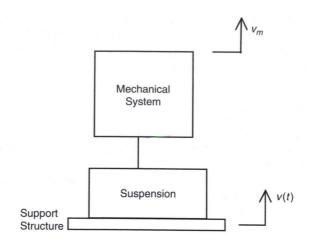

FIGURE 2.74
Motion transmissibility mechanism.

2.12.4.2   *Motion Transmissibility*

Consider a mechanical system supported through a suspension mechanism on a structure, which may be subjected to undesirable motions (e.g., seismic disturbances, road disturbances, machinery disturbances). Motion transmissibility determines the fraction of the support motion which is transmitted to the system through its suspension at different frequencies. It is defined as

$$\text{Motion Transmissibility } T_m = \frac{\text{System Motion } V_m}{\text{Support Motion } V}$$

The velocities V_m and V are expressed in the frequency domain, as Fourier spectra.

A schematic representation of the motion transmissibility mechanism is shown in Figure 2.74. Typically, the motion of the system is taken as the velocity of one of its critical masses. Different transmissibility functions are obtained when different mass points (or degrees of freedom) of the system are considered.

Next, two examples are given to show the reciprocity property, which makes the force transmissibility and the motion transmissibility functions identical.

2.12.5   Case of Single Degree of Freedom

Consider the single-degree-of-freedom systems shown in Figure 2.75. In this example the system is represented by a point mass m, and the suspension system is modeled as a spring of stiffness k and a viscous damper of damping constant b. The model shown in Figure 2.75(a) is used to study force transmissibility. Its impedance circuit is shown in Figure 2.76(a). The model shown in Figure 2.75(b) is used in determining the motion transmissibility. Its impedance (or, mobility) circuit is shown in Figure 2.76(b). Note that mobility elements are suitable for motion transmissibility studies. Since force is divided among parallel branches in proportion to their impedances it follows from Figure 2.76(a) that:

$$T_f = \frac{F_s}{F} = \frac{Z_s}{Z_m + Z_s} \tag{2.144}$$

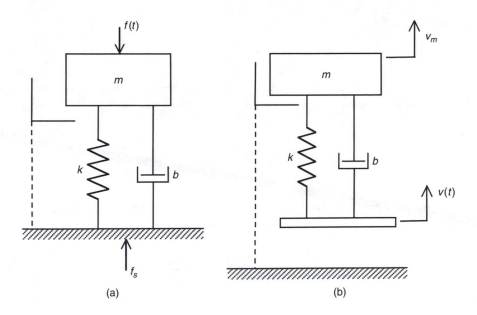

FIGURE 2.75
Single-degree-of-freedom systems: (a) Fixed on ground; (b) With support motion.

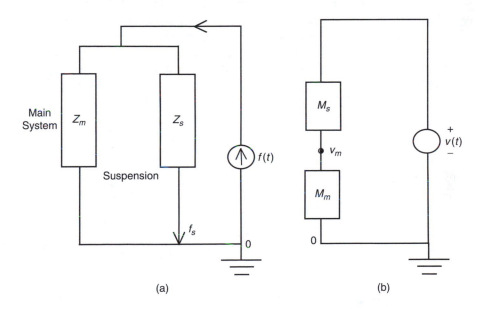

FIGURE 2.76
Impedance circuits of: (a) System in Figure 2.75(a); (b) System in Figure 2.75(b).

Since velocity is divided among series elements in proportion to their mobilities, it is clear from Figure 2.76(b) that:

$$T_m = \frac{V_m}{V} = \frac{M_m}{M_m + M_s}$$

$$= \frac{1/Z_m}{1/Z_m + 1/Z_s} = \frac{Z_s}{Z_s + Z_m}$$

(2.145)

Consequently,

$$T_f = T_m$$

and a distinction between the two types of transmissibility is not necessary. Let us denote them by a common *transmissibility function T*. Since, $Z_m = ms$ and $Z_s = k/s + b$, it follows that

$$T = \left[\frac{bs + k}{ms^2 + bs + k} \right]_{s = j2\pi f} \tag{2.146}$$

It is customary to consider only the magnitude of this complex transmissibility function. This, termed *magnitude transmissibility*, is given by

$$T = \left[\frac{4\pi^2 f^2 b^2 + k^2}{4\pi^2 f^2 b^2 + (k - 4\pi^2 f^2 m^2)^2} \right]^{\frac{1}{2}} \tag{2.147}$$

2.12.6 Case of Two Degrees of Freedom

Consider the two-degree-of-freedom systems shown in Figure 2.77. The main system is represented by two masses linked through a spring and a damper. Mass m_1 is considered the critical mass (It is equally acceptable to consider mass m_2 as the critical mass). To determine the force transmissibility, using Figure 2.78(a), note that the applied force is divided in the ratio of the impedances among the two parallel branches. The mobility of the main right-hand side branch is

$$M = \frac{1}{Z_{s1}} + \frac{1}{Z_{m2} + Z_s} \tag{i}$$

and the force through that branch is

$$F' = \left[\frac{\frac{1}{M}}{Z_{m1} + \frac{1}{M}} \right] F = \left[\frac{1}{MZ_{m1} + 1} \right] F$$

The force F_s through Z_s is given by

$$F_s = \left[\frac{Z_s}{Z_{m2} + Z_s} \right] F'$$

Consequently, the force transmissibility

$$T_f = \frac{F_s}{F} = \left[\frac{1}{MZ_{m1} + 1} \right] \left[\frac{Z_s}{Z_{m2} + Z_s} \right] \tag{ii}$$

where M is as given in Equation i.

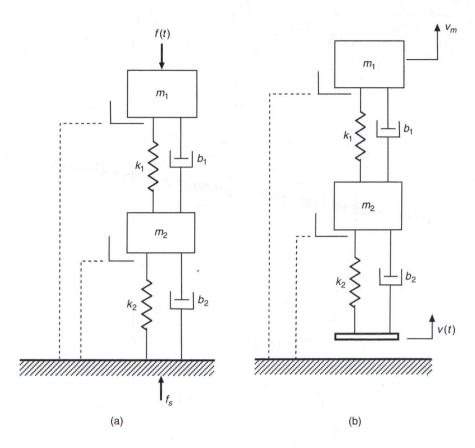

FIGURE 2.77
Systems with two degrees of freedom: (a) Fixed on ground; (b) With support motion.

To determine the motion transmissibility, using Figure 2.77(b) and the associated Figure 2.78(b), note that the velocity is distributed in proportion to the mobilities among the series elements. The impedance of the second composite series unit is

$$Z = \frac{1}{M_{m2}} + \frac{1}{M_{s1} + M_{m1}}$$

and the velocity across this unit is

$$V' = \left[\frac{\frac{1}{Z}}{M_s + \frac{1}{Z}} \right] V = \left[\frac{1}{M_s Z + 1} \right] V$$

The velocity V_m of mass m_1 is given by

$$V_m = \left[\frac{M_{m1}}{M_{s1} + M_{m1}} \right] V'$$

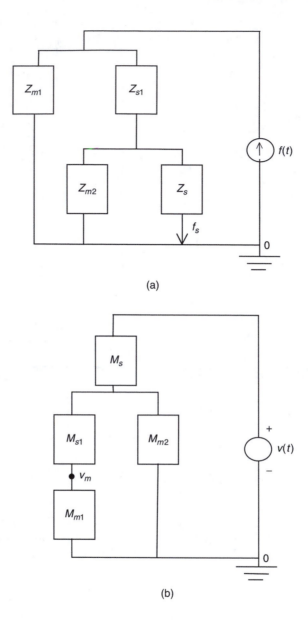

FIGURE 2.78
Impedance circuits of: (a) System in Figure 2.77(a); (b) System in Figure 2.77(b).

As a result, the motion transmissibility can be expressed as

$$T_m = \frac{V_m}{V} = \left[\frac{1}{M_s Z + 1} \right] \left[\frac{M_{m1}}{M_{s1} + M_{m1}} \right] \tag{iii}$$

It remains to show that $T_m = T_f$. To this end, let us examine the expression for T_m. Since $Z_s = 1/M_s$, T_m can be written as

$$T_m = \left[\frac{Z_s}{Z + Z_s} \right] \left[\frac{M_{m1}}{M_{s1} + M_{m1}} \right]$$

Note that

$$Z = \frac{1}{M_{s1} + M_{m1}} + Z_{m2}$$

Hence,

$$T_m = \left[\frac{Z_s}{\dfrac{1}{M_{s1} + M_{m1}} + Z_{m2} + Z_s} \right] \left[\frac{M_{m1}}{M_{s1} + M_{m1}} \right] = \left[\frac{M_{m1}}{\dfrac{1}{Z_{m2} + Z_s} + M_{s1} + M_{m1}} \right] \left[\frac{Z_s}{Z_{m2} + Z_s} \right]$$

$$= \left[\frac{1}{\dfrac{1}{M_{m1}} \left[\dfrac{1}{Z_{m2} + Z_s} + M_{s1} \right] + 1} \right] \left[\frac{Z_s}{Z_{m2} + Z_s} \right] = \left[\frac{1}{Z_{m1} \left[\dfrac{1}{Z_{m2} + Z_s} + \dfrac{1}{Z_{s1}} \right] + 1} \right] \left[\frac{Z_s}{Z_{m2} + Z_s} \right]$$

which is clearly identical to T_f as given in Equation ii, in view of Equation i.

The equivalence of T_f and T_m can be shown in a similar straightforward manner for higher degree-of-freedom systems as well.

2.13 Response Analysis and Simulation

An analytical model, which is a set of differential equations, has many uses. In particular, it can provide information regarding how the system responds when a specific excitation (input) is applied. Such a study may be carried out by

1. Solution of the differential equations (analytical)
2. Computer simulation (numerical)

In this section we will address these two approaches. A response analysis carried out using either approach, is valuable in many applications such as design, control, testing, validation, and qualification of mechatronic systems. For large-scale and complex systems, a purely analytical study may not be feasible, and we will have to increasingly rely on numerical approaches and computer simulation.

2.13.1 Analytical Solution

The response of a dynamic system may be obtained analytically by solving the associated differential equations, subject to the initial conditions. This may be done by

1. Direct solution (in the time domain)
2. Solution using Laplace transform

Consider a linear time-invariant model given by the input-output differential equation

$$a_n \frac{d^n y}{dt^n} + a_{n-1} \frac{d^{n-1} y}{dt^{n-1}} + \cdots + a_0 y = u \tag{2.148}$$

At the outset, note that it is not necessary to specifically include derivative terms on the RHS; for example, $b_0 u + b_1 \frac{du}{dt} + \cdots + b_m \frac{d^m u}{dt^m}$ because, once we have the solution (say, y_s) for Equation 2.148 we can use the *principle of superposition* to obtain the solution for the general case, and is given by: $b_0 y_s + b_1 \frac{dy_s}{dt} + \cdots + b_m \frac{d^m y_s}{dt^m}$. Hence, we will consider only the case of Equation 2.148.

2.13.1.1 Homogeneous Solution

The natural characteristics of a dynamic system do not depend on the input to the system. Hence, the natural behavior (or free response) of Equation 2.148 is determined by the homogeneous equation (i.e., the input = 0):

$$a_n \frac{d^n y}{dt^n} + a_{n-1} \frac{d^{n-1} y}{dt^{n-1}} + \cdots + a_0 y = 0 \tag{2.149}$$

Its solution is denoted by y_h and it depends on the system initial conditions. For a linear system the natural response is known to take an exponential form given by

$$y_h = ce^{\lambda t} \tag{2.150}$$

where c is an arbitrary constant and, in general, λ can be complex. Substitute Equation 2.149 in Equation 2.150 with the knowledge that

$$\frac{d}{dt} e^{\lambda t} = \lambda e^{\lambda t} \tag{2.151}$$

and cancel the common term $ce^{\lambda t}$, since u cannot be zero at all times. Then we have

$$a_n \lambda^n + a_{n-1} \lambda^{n-1} + \cdots + a_0 = 0 \tag{2.152}$$

This is called the *characteristic equation* of the system.

NOTE the LHS polynomial of Equation 2.152 is the *characteristic polynomial*. Equation 2.152 has n roots $\lambda_1, \lambda_2, \ldots, \lambda_n$. These are called *poles* or *eigenvalues* of the system. Assuming that they are distinct (i.e., unequal), the overall solution to Equation 2.149 becomes

$$y_h = c_1 e^{\lambda_1 t} + c_2 e^{\lambda_2 t} + \cdots + c_n e^{\lambda_n} \tag{2.153}$$

The unknown constants $c_1, c_2, \ldots, c_n$ are determined using the necessary n initial conditions $y(0), \dot{y}(0), \ldots, \frac{d^{n-1} y(0)}{dt^{n-1}}$.

2.13.1.1.1 Repeated Poles

Suppose that at least two eigenvalues are equal. Without loss of generality suppose in Equation 2.153 that $\lambda_1 = \lambda_2$. Then the first two terms in Equation 2.153 can be combined into the single unknown $(c_1 + c_2)$. Consequently there are only $n - 1$ unknowns in Equation 2.153 but there are n initial conditions. It follows that another unknown needs to be introduced for obtaining a complete solution. Since a repeated pole is equivalent to a double integration, the logical (and correct) solution for Equation 2.152 in the case $\lambda_1 = \lambda_2$ is

$$y_h = (c_1 + c_2 t)e^{\lambda_1 t} + c_3 e^{\lambda_3 t} + \cdots + c_n e^{\lambda_n} \tag{2.154}$$

2.13.1.2 Particular Solution

The homogeneous solution corresponds to the "free" or "unforced" response of a system, and it does not take into account the input function. The effect of the input is incorporated into the particular solution, which is defined as one possible function for y that satisfies Equation 2.148. We denote this by y_p. Several important input functions and the corresponding form of y_p which satisfies Equation 2.148 are given in Table 2.12.

The parameters A, B, A_1, A_2, B_1, B_2, and D are determined by substituting the pair $u(t)$ and y_p into Equation 2.148 and then equating the like terms. This approach is called the *method of undetermined coefficients*.

The total response is given by

$$y = y_h + y_p \tag{2.155}$$

The unknown constants $c_1, c_2, \ldots, c_n$ in this result are determined by substituting the initial conditions of the system into Equation 2.155. Note that it is incorrect to first determine $c_1, c_2, \ldots, c_n$ by substituting the ICs into y_h and then adding y_p to the resulting y_h. Furthermore, when $u = 0$, the homogeneous solution is same as the free response, initial condition response, or zero-input response. When an input is present, however, the homogeneous solution is not identical to the other three types of response. These ideas are summarized in Table 2.13

TABLE 2.12

Particular Solutions for Useful Input Functions

Input $u(t)$	Particular Solution y_p
c	A
ct	$B_1 t + B_2$
$\sin ct$	$A_1 \sin ct + A_2 \cos ct$
$\cos ct$	$B_1 \sin ct + B_2 \cos ct$
e^{ct}	De^{ct}

TABLE 2.13

Some Concepts of System Response

Total response (T)	= homogeneous solution + particular integral
	(H) (P)
	= free response + forced response
	(X) (F)
	= initial-condition response + zero-initial-condition response
	(X) (F)
	= zero-input response + zero-state response
	(X) (F)

Note: In general, $H \neq X$ and $P \neq F$

With no input (no forcing excitation), by definition, $H \equiv X$

At steady state, F becomes equal to P.

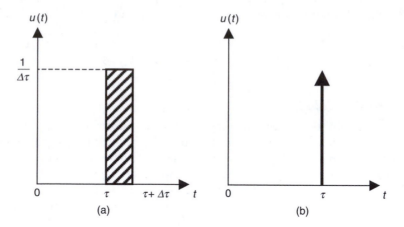

FIGURE 2.79
Illustration of: (a) Unit pulse; (b) Unit impulse.

2.13.1.3 Impulse Response Function

Consider a linear dynamic system. The principle of superposition holds. More specifically, if y_1 is the system response to excitation $u_1(t)$, and y_2 is the response to excitation $u_2(t)$, then $\alpha y_1 + \beta y_2$ is the system response to input $\alpha u_1(t) + \beta u_2(t)$ for any constants α and β and any excitation functions $u_1(t)$ and $u_2(t)$. This is true for both time-variant-parameter linear systems and constant-parameter linear systems.

A unit pulse of width $\Delta \tau$ starting at time $t = \tau$ is shown in Figure 2.79(a). Its area is unity. A unit impulse is the limiting case of a unit pulse for $\Delta \tau \to 0$. A unit impulse acting at time $t = \tau$ is denoted by $\delta(t - \tau)$ and is graphically represented as in Figure 2.79(b). In mathematical analysis, this is known as the *Dirac delta function*, and is defined by the two conditions:

$$\delta(t-\tau) = 0 \quad \text{for} \quad t \neq \tau \tag{2.156}$$

$$\to \infty \quad \text{at} \quad t = \tau$$

and

$$\int_{-\infty}^{\infty} \delta(t-\tau)\,dt = 1 \tag{2.157}$$

The Dirac delta function has the following well-known and useful properties:

$$\int_{-\infty}^{\infty} f(t)\delta(t-\tau)\,dt = f(\tau) \tag{2.158}$$

and

$$\int_{-\infty}^{\infty} \frac{d^n f(t)}{dt^n}\delta(t-\tau)\,dt = \frac{d^n f(t)}{dt^n}\bigg|_{t=\tau} \tag{2.159}$$

for any well-behaved time function $f(t)$. The system response (output) to a unit-impulse excitation (input) acted at time $t = 0$, is known as the *impulse-response function* and is denoted by $h(t)$.

2.13.1.4   Convolution Integral

The system output in response to an arbitrary input may be expressed in terms of its impulse-response function. This is the essence of the impulse-response approach to determining the forced response of a dynamic system. Without loss of generality we shall assume that the system input $u(t)$ starts at $t = 0$; that is,

$$u(t) = 0 \quad \text{for} \quad t < 0 \tag{2.160}$$

For physically realizable systems, the response does not depend on the future values of the input. Consequently,

$$y(t) = 0 \quad \text{for} \quad t < 0 \tag{2.161}$$

and

$$h(t) = 0 \quad \text{for} \quad t < 0 \tag{2.162}$$

where $y(t)$ is the response of the system, to any general excitation $u(t)$.

Furthermore, if the system is a constant-parameter system, then the response does not depend on the time origin used for the input. Mathematically, this is stated as follows: if the response to input $u(t)$ satisfying Equation 2.160 is $y(t)$, which in turn satisfies Equation 2.161, then the response to input $u(t - \tau)$, which satisfies,

$$u(t - \tau) = 0 \quad \text{for} \quad t < \tau \tag{2.163}$$

is $y(t - \tau)$, and it satisfies

$$y(t - \tau) = 0 \quad \text{for} \quad t < \tau \tag{2.164}$$

This situation is illustrated in Figure 2.80. It follows that the delayed-impulse input $\delta(t - \tau)$, having time delay τ, produces the delayed response $h(t - \tau)$.

A given input $u(t)$ can be divided approximately into a series of pulses of width $\Delta\tau$ and magnitude $u(\tau) \cdot \Delta\tau$. In Figure 2.81, for $\Delta\tau \to 0$, the pulse shown by the shaded area becomes an impulse acting at $t = \tau$, having the magnitude $u(\tau) \cdot d\tau$. This impulse is given by $\delta(t - \tau)u(\tau)d\tau$. In a linear, constant-parameter system, it produces the response $h(t - \tau)u(\tau)d\tau$. By integrating over the entire time duration of the input $u(t)$, the overall response $y(t)$ is obtained as

$$y(t) = \int_0^\infty h(t - \tau)u(\tau)d\tau$$

$$= \int_0^\infty h(\tau)u(t - \tau)d\tau \tag{2.165}$$

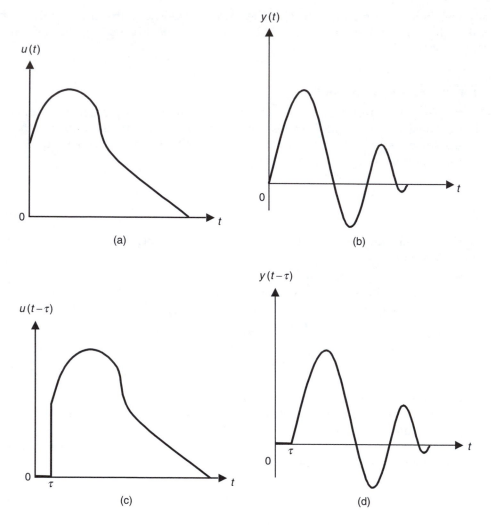

FIGURE 2.80
Response to a delayed input.

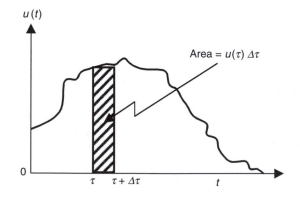

FIGURE 2.81
General input treated as a continuous series of impulses.

Equation 2.150 is known as the *convolution integral*. This is in fact the forced response, under zero initial conditions.

2.13.2 Stability

Many definitions are available for stability of a system. For example, a stable system may be defined as one whose natural response (i.e., free, initial-condition response) decays to zero. This is in fact the well-known *asymptotic stability*. If the initial-condition response oscillates within finite bounds we say the system is *marginally stable*. For a linear, time-invariant system of the type Equation 2.148, the free response is of the form Equation 2.153. Hence, if none of the eigenvalues λ_i have positive real parts, the system is considered stable, because in that case, the response Equation 2.153 does not grow unbounded. In particular, if the system has a single eigenvalue that is zero, or if the eigenvalues are purely imaginary, the system is marginally stable. If the system has two or more poles that are zero, we will have terms of the form $c_1 + ct$ as in Equation 2.154 and hence it will grow polynomially (not exponentially). Then the system will be *unstable*. Also note that, since physical systems have real parameters, their eigenvalues must occur as conjugate pairs, if complex. Since stability is governed by the sign of the real part of the eigenvalues, it can be represented on the eigenvalue plane (or the pole plane or root plane). This is illustrated in Figure 2.82.

2.13.3 First Order Systems

Consider the first order dynamic system with time constant τ, input u, and output y, as given by

$$\tau \dot{y} + y = u(t) \tag{2.166}$$

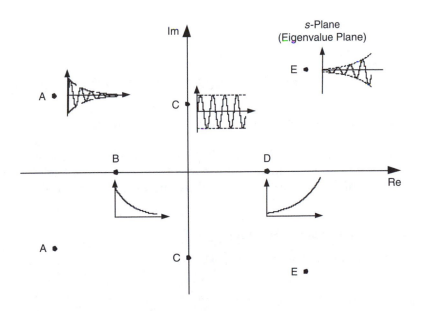

FIGURE 2.82
Dependence of stability on the pole location (**A** and **B** are stable; **C** is marginally stable; **D** and **E** are unstable).

Suppose that the system is starting from $y(0) = y_0$ and a step input of magnitude A is applied. The homogeneous solution is

$$y_h = ce^{-t/\tau}$$

The particular solution (see Table 2.12) is given by $y_p = A$. Hence, the total solution is

$$y = y_h + y_p = ce^{-t/\tau} + A$$

Substitute the IC: $y(0) = y_0$. We get $c + A = y_0$. Hence

$$y_{step} = \underbrace{(y_0 - A)e^{-t/\tau}}_{\substack{\text{Homogeneous} \\ y_h}} + \underbrace{A}_{\substack{\text{Particular} \\ y_p}} = \underbrace{y_0 e^{-t/\tau}}_{\substack{\text{Free Response} \\ y_x}} + \underbrace{A(1 - e^{-t/\tau})}_{\substack{\text{Forced Response} \\ y_f}} \qquad (2.167)$$

The steady-state value is given by $t \to \infty$. Hence

$$y_{ss} = A \qquad (2.168)$$

It is seen from Equation 2.167 that the forced response to a unit step input (i.e., $A = 1$) is $(1 - e^{-t/\tau})$. Due to linearity, the forced response to a unit impulse input is $\frac{d}{dt}(1 - e^{-t/\tau}) = \frac{1}{\tau}e^{-t/\tau}$. Hence, the total response to an impulse input of magnitude P is

$$y_{impulse} = y_0 e^{-t/\tau} + \frac{P}{\tau}e^{-t/\tau} \qquad (2.169)$$

This result follows from the fact that

$$\frac{d}{dt}(\text{Step Function}) = \text{Impulse Function}$$

and, due to linearity, when the input is differentiated, the output is correspondingly differentiated.

Note from Equation 2.167 and Equation 2.169 that if we know the response of a first order system to a step input, or to an impulse input, the system itself can be determined. This is known as *model identification*. We will illustrate this by an example.

2.13.4   Model Identification Example

Consider the first order system (model)

$$\tau \dot{y} + y = ku \qquad (i)$$

Note the gain parameter k. The initial condition is $y(0) = y_0$.

Due to linearity, using Equation 2.167 we can derive the response of the system to a step input of magnitude A:

$$y_{step} = y_0 e^{-t/\tau} + Ak(1 - e^{-t/\tau})$$

(ii)

Now suppose that the unit step response of a first order system with zero ICs, was found to be (say, by curve fitting of experimental data)

$$y_{step} = 2.25(1 - e^{-5.2t})$$

Then, it is clear from Equation ii that

$$k = 2.25 \text{ and } \tau = 1/5.2 = 0.192$$

2.13.5   Second Order Systems

A general high-order system can be represented by a suitable combination of first-order and second-order models, using the principles of modal analysis. Hence, it is useful to study the response behavior of second-order systems as well. Examples of second-order systems include mass-spring-damper systems and capacitor-inductor-resistor circuits, which we have studied in previous sections. These are called simple oscillators because they exhibit oscillations in the natural response (free response) when the level of damping is sufficiently low. We will study both free response and forced response.

2.13.5.1   *Free Response of an Undamped Oscillator*

We note that the equation of free (i.e., no excitation force) motion of an undamped simple oscillator is of the general form

$$\ddot{x} + \omega_n^2 x = 0$$

(2.170)

For a mechanical system of mass m and stiffness k, we have

$$\omega_n = \sqrt{\frac{k}{m}}$$

(2.171)

For an electrical circuit with capacitance C and inductance L we have

$$\omega_n = \sqrt{\frac{1}{LC}}$$

(2.172)

To determine the time response x of this system, we use the trial solution:

$$x = A \sin(\omega_n t + \phi)$$

(2.173)

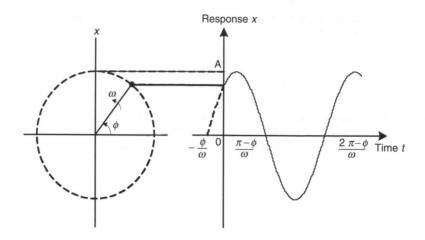

FIGURE 2.83
Free response of an undamped simple oscillator.

in which A and ϕ are unknown constants, to be determined by the initial conditions (for x and $\dot{x}$); say,

$$x(0) = x_o, \quad \dot{x}(0) = v_o \tag{2.174}$$

Substitute the trial solution into Equation (2.170). We get

$$\left(-A\omega_n^2 + A\omega_n^2\right)\sin(\omega_n t + \phi) = 0$$

This equation is identically satisfied for all t. Hence, the general solution of Equation 2.170 is indeed Equation 2.173, which is periodic and sinusoidal.

This response is sketched in Figure 2.83 (the subscript in ω_n is dropped for convenience). Note that this sinusoidal, oscillatory motion has a *frequency* of oscillation of ω (radians/s). Hence, a system that provides this type of natural motion is called a *simple oscillator*. In other words, the system response exactly repeats itself in time periods of T or at a *cyclic frequency* $f = \frac{1}{T}$ (Hz). The frequency ω is in fact the *angular frequency* given by $\omega = 2\pi f$. Also, the response has an *amplitude A*, which is the peak value of the sinusoidal response. Now, suppose that we shift this response curve to the right through ϕ/ω. Consider the resulting curve to be the reference signal (with signal value = 0 at $t = 0$, and increasing). It should be clear that the response shown in Figure 2.83 leads the reference signal by a time period of ϕ/ω. This may be verified from the fact that the value of the reference signal at time t is the same as that of the signal in Figure 2.83 at time $t - \phi/\omega$. Hence ϕ is termed the *phase angle* of the response, and it is a *phase lead*.

The left-hand-side portion of Figure 2.83 is the *phasor representation* of a sinusoidal response. In this representation, an arm of length A rotates in the counterclockwise direction at angular speed ω. This is the phasor. The arm starts at an angular position ϕ from the horizontal axis, at time $t = 0$. The projection of the arm onto the vertical (x) axis is the time response. In this manner, the phasor representation can conveniently indicate the amplitude, frequency, phase angle, and the actual time response (at any time t) of a sinusoidal motion.

2.13.5.2   *Free Response of a Damped Oscillator*

Energy dissipation may be added to a mechanical oscillator by using a damping element. For an electrical circuit, a resistor may be added to achieve this. In either case, the equation motion of the damped simple oscillator without an input, may be expressed as

$$\ddot{x} + 2\zeta\omega_n\dot{x} + \omega_n^2 x = 0 \tag{2.175}$$

Note that ζ is called the *damping ratio*.

Assume an exponential solution:

$$x = Ce^{\lambda t} \tag{2.176}$$

This is justified by the fact that linear systems have exponential or oscillatory (i.e., complex exponential) free responses. A more detailed justification will be provided later.

Substitute, Equation 2.176 into Equation 2.175. We get

$$\left[\lambda^2 + 2\zeta\omega_n\lambda + \omega_n^2\right] Ce^{\lambda t} = 0$$

Note that $Ce^{\lambda t}$ is not zero in general. It follows that, when λ satisfies the equation:

$$\lambda^2 + 2\zeta\omega_n\lambda + \omega_n^2 = 0 \tag{2.177}$$

then, Equation 2.176 will represent a solution of Equation 2.175. As noted before, Equation 2.177 is the *characteristic equation* of the system. This equation depends on the natural dynamics of the system, not the forcing excitation or the initial conditions. Solution of Equation 2.177 gives the two roots:

$$\lambda = -\zeta\omega_n \pm \sqrt{\zeta^2 - 1}\ \omega_n$$
$$= \lambda_1 \text{ and } \lambda_2 \tag{2.178}$$

These are the *eigenvalues* or *poles* of the system. When $\lambda_1 \neq \lambda_2$, the general solution is

$$x = C_1 e^{\lambda_1 t} + C_2 e^{\lambda_2 t} \tag{2.179}$$

The two unknown constants C_1 and C_2 are related to the integration constants, and can be determined by two initial conditions which should be known.

If $\lambda_1 = \lambda_2 = \lambda$; we have the case of repeated roots. In this case, the general solution Equation 2.179 does not hold because C_1 and C_2 would no longer be independent constants, to be determined by two initial conditions. The repetition of the roots suggests that one term of the homogenous solution should have the multiplier t (a result of the double integration of zero). Then the general solution is,

$$x = C_1 e^{\lambda t} + C_2 t e^{\lambda t} \tag{2.180}$$

We can identify three ranges of damping, as discussed below, and the nature of the response will depend on the particular range of damping.

Case 1: Underdamped Motion ($\zeta < 1$)

In this case it follows from Equation 2.178 that the roots of the characteristic equation are

$$\lambda = -\zeta\omega_n \pm j\sqrt{1-\zeta^2}\,\omega_n = -\zeta\omega_n \pm j\omega_d = \lambda_1 \text{ and } \lambda_2 \qquad (2.181)$$

where the *damped natural frequency* is given by

$$\omega_d = \sqrt{1-\zeta^2}\,\omega_n \qquad (2.182)$$

Note that λ_1 and λ_2 are complex conjugates, as required. The response (Equation 2.179), in this case, may be expressed as

$$x = e^{-\zeta\omega_n t}\left[C_1 e^{j\omega_d t} + C_2 e^{-j\omega_d t}\right] \qquad (2.183)$$

The term within the square brackets of Equation 2.183 has to be real, because it represents the time response of a real physical system. It follows that C_1 and C_2 as well, have to be complex conjugates.

NOTE
$$e^{j\omega_d t} = \cos\omega_d t + j\sin\omega_d t$$

$$e^{-j\omega_d t} = \cos\omega_d t - j\sin\omega_d t$$

So, an alternative form of the general solution would be

$$x = e^{-\zeta\omega_n t}[A_1 \cos\omega_d t + A_2 \sin\omega_d t] \qquad (2.184)$$

Here A_1 and A_2 are the two unknown constants. By equating the coefficients it can be shown that

$$A_1 = C_1 + C_2$$
$$A_2 = j(C_1 - C_2) \qquad (2.185)$$

Hence

$$C_1 = \frac{1}{2}(A_1 - jA_2)$$
$$C_2 = \frac{1}{2}(A_1 + jA_2) \qquad (2.186)$$

Initial Conditions:
Let
$x(0) = x_o$, $\dot{x}(0) = v_o$ as before. Then,

$$x_o = A_1 \quad \text{and} \quad v_o = -\zeta\omega_n A_1 + \omega_d A_2 \tag{2.187}$$

or,

$$A_2 = \frac{v_o}{\omega_d} + \frac{\zeta\omega_n x_o}{\omega_d} \tag{2.188}$$

Yet, another form of the solution would be:

$$x = A e^{-\zeta\omega_n t} \sin(\omega_d t + \phi) \tag{2.189}$$

Here A and ϕ are the unknown constants with

$$A = \sqrt{A_1^2 + A_2^2} \quad \text{and} \quad \sin\phi = \frac{A_1}{\sqrt{A_1^2 + A_2^2}}. \tag{2.190}$$

Also

$$\cos\phi = \frac{A_2}{\sqrt{A_1^2 + A_2^2}} \quad \text{and} \quad \tan\phi = \frac{A_1}{A_2} \tag{2.191}$$

Note that the response $x \to 0$ as $t \to \infty$. This means the system is *asymptotically stable*.

Case 2: Overdamped Motion ($\zeta > 1$)

In this case, roots λ_1 and λ_2 of the characteristic Equation 2.177 are real and negative. Specifically, we have

$$\lambda_1 = -\zeta\omega_n + \sqrt{\zeta^2 - 1} \quad \omega_n < 0 \tag{2.192}$$

$$\lambda_2 = -\zeta\omega_n - \sqrt{\zeta^2 - 1} \quad \omega_n < 0 \tag{2.193}$$

and the response Equation 2.179 is nonoscillatory. Also, since both λ_1 and λ_2 are negative, $x \to 0$ as $t \to \infty$. This means the system is asymptotically stable.

From the initial conditions $x(0) = x_o$, $\dot{x}(0) = v_o$ we get

$$x_o = C_1 + C_2 \tag{i}$$

and

$$v_o = \lambda_1 C_1 + \lambda_2 C_2 \tag{ii}$$

Multiply the first IC Equation i by λ_1: $\qquad \lambda_1 x_o = \lambda_1 C_1 + \lambda_1 C_2 \tag{iii}$

Subtract Equation iii from Equation ii: $\qquad v_o - \lambda_1 x_o = C_2(\lambda_2 - \lambda_1)$

We get:

$$C_2 = \frac{v_o - \lambda_1 x_o}{\lambda_2 - \lambda_1} \tag{2.194}$$

Similarly, multiply the first IC Equation i by λ_2 and subtract from Equation ii. We get

$$v_o - \lambda_2 x_o = C_1(\lambda_1 - \lambda_2)$$

Hence

$$C_1 = \frac{v_o - \lambda_2 x_o}{\lambda_1 - \lambda_2} \tag{2.195}$$

Case 3: Critically Damped Motion ($\zeta = 1$)

Here, we have repeated roots, given by

$$\lambda_1 = \lambda_2 = -\omega_n \tag{2.196}$$

The response, for this case is given by (see Equation 2.180)

$$x = C_1 e^{-\omega_n t} + C_2 t e^{-\omega_n t} \tag{2.197}$$

Since the term $e^{-\omega_n t}$ goes to zero faster than t goes to infinity, we have

$$t e^{-\omega_n t} \to 0 \text{ as } t \to \infty.$$

Hence the system is asymptotically stable.

Now use the initial conditions $x(0) = x_o$, $\dot{x}(0) = v_o$. We get,

$$x_o = C_1$$
$$v_o = -\omega_n C_1 + C_2$$

Hence

$$C_1 = x_o \tag{2.198}$$

$$C_2 = v_o + \omega_n x_o \tag{2.199}$$

NOTE When $\zeta = 1$ we have the critically damped response because below this value, the response is oscillatory (underdamped) and above this value, the response is nonoscillatory

TABLE 2.14

Free (natural) Response of a Damped Simple Oscillator

System Equation:

$$\ddot{x} + 2\zeta\omega_n\dot{x} + \omega_n^2 x = 0$$

Undamped natural frequency $\quad \omega_n = \sqrt{\dfrac{k}{m}}$

Damping ratio $\zeta = \dfrac{b}{2\sqrt{km}}$

Characteristic Equation: $\quad \lambda^2 + 2\zeta\omega_n\lambda + \omega_n^2 = 0$

Roots (eigenvalues or poles): $\quad \lambda_1$ and $\lambda_2 = -\zeta\omega_n \pm \sqrt{\zeta^2 - 1}\,\omega_n$

Response: $\quad x = C_1 e^{\lambda_1 t} + C_2 e^{\lambda_2 t}$ for unequal roots $(\lambda_1 \neq \lambda_2)$

$\qquad\qquad x = (C_1 + C_2 t)e^{\lambda t}$ for equal roots $(\lambda_1 = \lambda_2 = \lambda)$

Initial Conditions: $\quad x(0) = x_0$ and $\dot{x}(0) = v_0$

Case 1: Underdamped ($\zeta < 1$)

Poles are complex conjugates: $\quad -\zeta\omega_n \pm j\omega_d$

Damped natural frequency $\quad \omega_d = \sqrt{1 - \zeta^2}\,\omega_n$

$$x = e^{-\zeta\omega_n t}\left[C_1 e^{j\omega_d t} + C_2 e^{-j\omega_d t}\right]$$
$$= e^{-\zeta\omega_n t}\left[A_1 \cos\omega_d t + A_2 \sin\omega_d t\right]$$
$$= A e^{-\zeta\omega_n t}\sin(\omega_d t + \phi)$$

$A_1 = C_1 + C_2$ and $A_2 = j(C_1 - C_2)$

$C_1 = \frac{1}{2}(A_1 - jA_2)$ and $C_2 = \frac{1}{2}(A_1 + jA_2)$

$A = \sqrt{A_1^2 + A_2^2}$ and $\tan\phi = \dfrac{A_1}{A_2}$

ICs give: $\quad A_1 = x_0$ and $A_2 = \dfrac{v_0 + \zeta\omega_n x_0}{\omega_d}$

Logarithmic Decrement per Radian: $\quad \alpha = \dfrac{1}{2\pi n}\ln r = \dfrac{\zeta}{\sqrt{1 - \zeta^2}}$

where $r = \dfrac{x(t)}{x(t + nT)}$ = decay ratio over n complete cycles. For small ζ: $\zeta \cong \alpha$

Case 2: Overdamped ($\zeta > 1$)

Poles are real and negative: $\quad \lambda_1, \lambda_2 = -\zeta\omega_n \pm \sqrt{\zeta^2 - 1}\,\omega_n$

$$x = C_1 e^{\lambda_1 t} + C_2 e^{\lambda_2 t}$$

$$C_1 = \frac{v_0 - \lambda_2 x_0}{\lambda_1 - \lambda_2} \quad \text{and} \quad C_2 = \frac{v_0 - \lambda_1 x_0}{\lambda_2 - \lambda_1}$$

Case 3: Critically Damped ($\zeta = 1$)

Two identical poles: $\quad \lambda_1 = \lambda_2 = \lambda = -\omega_n$

$$x = (C_1 + C_2 t)e^{-\omega_n t} \quad \text{with} \quad C_1 = x_0 \quad \text{and} \quad C_2 = v_0 + \omega_n x_0$$

(overdamped). It follows that we may define the damping ratio as

$$\zeta = \text{damping ratio} = \frac{\text{damping constant}}{\text{damping constant for critically damped conditions}}$$

The main results for free (natural) response of a damped oscillator are given in Table 2.14. The response of a damped simple oscillator is shown in Figure 2.84.

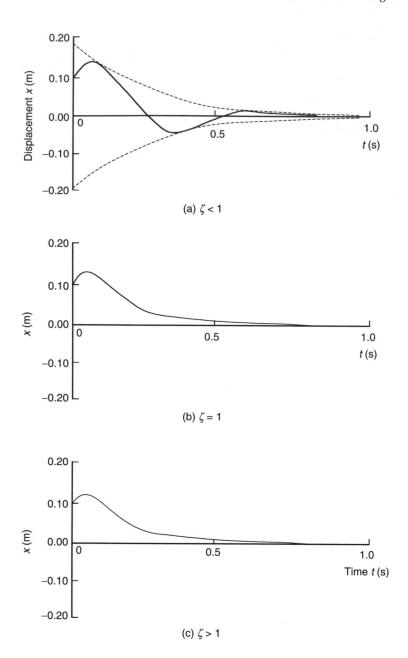

FIGURE 2.84
Free response of a damped oscillator: (a) Underdamped; (b) Critically Damped; (c) Overdamped.

2.13.5.3 Forced Response of a Damped Oscillator

The forced response depends on both the natural characteristics of the system (free response) and the nature of the input. Mathematically, as noted before, the total response is the sum of the homogeneous solution and the particular solution. Consider a damped simple oscillator, with input $u(t)$ scaled such that it has the same units as the response y; thus

$$\ddot{y} + 2\zeta\omega_n\dot{y} + \omega_n^2 y = \omega_n^2 u(t) \tag{2.200}$$

We will consider the response of this system to three types of inputs:

1. Impulse input
2. Step input
3. Harmonic (sinusoidal) input

Impulse Response: Many important characteristics of a system can be studied by analyzing the system response to a baseline excitation such as an impulse, a step, or a sinusoidal (harmonic) input. Characteristics which may be studied in this manner may include system stability, speed of response, time constants, damping properties, and natural frequencies. As well, an insight into the system response to an arbitrary excitation can be gained. Responses to such test inputs can also serve as the basis for system comparison. For example, it is possible to determine the degree of nonlinearity in a system by exciting it with two input intensity levels, separately, and checking whether the proportionality is retained at the output; or when the excitation is harmonic, whether limit cycles are encountered by the response.

The response of the system (Equation 2.200) to a unit impulse $u(t) = \delta(t)$ may be conveniently determined by the Laplace transform approach (See Appendix A). However, in the present section we will use a time-domain approach, instead. First integrate Equation 2.200, over the almost zero interval from $t = 0^-$ to $t = 0^+$. We get

$$\dot{y}(0^+) = \dot{y}(0^-) - 2\zeta\omega_n[y(0^+) - y(0^-)] - \omega_n^2 \int_{0^-}^{0^+} y \, dt + \omega_n^2 \int_{0^-}^{0^+} u(t) \, dt \qquad (2.201)$$

Suppose that the system starts from rest. Hence, $y(0^-) = 0$ and $\dot{y}(0^-) = 0$. Also, when an impulse is applied over an infinitesimally short time period $[0^-, 0^+]$ the system will not be able to move through a finite distance during that time. Hence, $y(0^+) = 0$ as well, and furthermore, the integral of y on the RHS of Equation 2.201 also will be zero. Now by definition of a unit impulse, the integral of u on the RHS of Equation 2.201 will be unity. Hence, we have $\dot{y}(0^+) = \omega_n^2$. It follows that as soon as a unit impulse is applied to the system (Equation 2.200) the initial conditions will become

$$y(0^+) = 0 \quad \text{and} \quad \dot{y}(0^+) = \omega_n^2 \qquad (2.202)$$

Also, beyond $t = 0^+$ the excitation $u(t) = 0$, according to the definition of an impulse. Hence, the impulse response of the system (Equation 2.200) is obtained by its homogeneous solution (as carried out before, under free response), but with the initial conditions given by Equation 2.202. The three cases of damping ratio ($\zeta < 1$, $\zeta > 1$, and $\zeta = 1$) should be considered separately. Then, we can conveniently obtain the following results:

$$y_{\text{impulse}}(t) = h(t) = \frac{\omega_n}{\sqrt{1-\zeta^2}} \exp(-\zeta\omega_n t)\sin\omega_d t \quad \text{for } \zeta < 1 \qquad (2.203a)$$

$$y_{\text{impulse}}(t) = h(t) = \frac{\omega_n}{2\sqrt{\zeta^2-1}}[\exp\lambda_1 t - \exp\lambda_2 t] \quad \text{for } \zeta > 1 \qquad (2.203b)$$

$$y_{\text{impulse}}(t) = h(t) = \omega_n^2 t \exp(-\omega_n t) \quad \text{for } \zeta = 1 \qquad (2.203c)$$

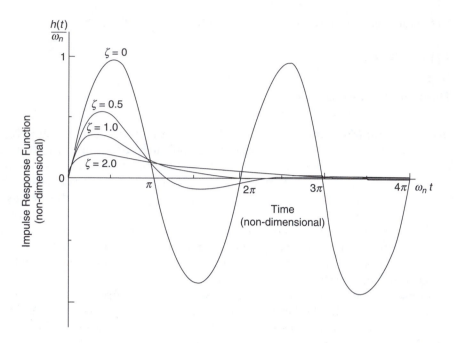

FIGURE 2.85
Impulse-response function of a damped oscillator.

An explanation concerning the dimensions of $h(t)$ is appropriate at this juncture. Note that $y(t)$ has the same dimensions as $u(t)$. Since $h(t)$ is the response to a unit impulse $\delta(t)$, it follows that these two have the same dimensions. The magnitude of $\delta(t)$ is represented by a unit area in the $u(t)$ versus t plane. Consequently, $\delta(t)$ has the dimensions of (1/time) or (frequency). It follows that $h(t)$ also has the dimensions of (1/time) or (frequency).

The impulse-response functions given by Equation 2.203 are plotted in Figure 2.85 for some representative values of damping ratio. It should be noted that, for $0 < \zeta < 1$, the angular frequency of damped vibrations is ω_d, which is smaller than the undamped natural frequency ω_n.

The Riddle of Zero Initial Conditions: For a second-order system, zero initial conditions correspond to $y(0) = 0$ and $\dot{y}(0) = 0$. It is clear from Equations 2.203 that $h(0) = 0$, but $\dot{h}(0) \neq 0$, which appears to violate the zero-initial-conditions assumption. This situation is characteristic in a system response to an impulse and its higher derivatives. This may be explained as follows. When an impulse is applied to a system at rest (zero initial state), the highest derivative of the system differential equation momentarily becomes infinity. As a result, the next lower derivative becomes finite (nonzero) at $t = 0^+$. The remaining lower derivatives maintain their zero values at that instant. When an impulse is applied to the mechanical system given by Equation 2.200 for example, the acceleration $\ddot{y}(t)$ becomes infinity, and the velocity $\dot{y}(t)$ takes a nonzero (finite) value shortly after its application ($t = 0^+$). The displacement $y(t)$, however, would not have sufficient time to change at $t = 0^+$. The impulse input is therefore equivalent to a velocity initial condition in this case. This initial condition is determined by using the integrated version (Equation 2.201) of the system Equation 2.200, as has been done.

Step Response: A unit step excitation is defined by

$$\mathcal{U}(t) = 1 \quad \text{for } t > 0$$

$$= 0 \quad \text{for } t \leq 0$$

(2.204)

Unit impulse excitation $\delta(t)$ may be interpreted as the time derivative of $\mathcal{U}(t)$:

$$\delta(t) = \frac{d\mathcal{U}(t)}{dt} \tag{2.205}$$

Note that Equation 2.205 re-establishes the fact that for nondimensional $\mathcal{U}(t)$, the dimension of $\delta(t)$ is (time)$^{-1}$. Since a unit step is the integral of a unit impulse, the step response can be obtained directly as the integral of the impulse response; thus

$$y_{\text{step}}(t) = \int_0^t h(\tau)d\tau \tag{2.206}$$

This result also follows from the convolution integral (2.165) because, for a delayed unit step, we have

$$\mathcal{U}(t - \tau) = 1 \quad \text{for} \quad \tau < t$$
$$= 0 \quad \text{for} \quad \tau \geq t \tag{2.207}$$

Thus, by integrating Equations 2.203 with zero initial conditions the following results are obtained for step response:

$$y_{\text{step}}(t) = 1 - \frac{1}{\sqrt{1-\zeta^2}} \exp(-\zeta\omega_n t)\sin(\omega_d t + \phi) \quad \text{for } \zeta < 1 \tag{2.208a}$$

$$y_{\text{step}} = 1 - \frac{1}{2\sqrt{1-\zeta^2}\,\omega_n}[\lambda_1 \exp\lambda_2 t - \lambda_2 \exp\lambda_1 t] \quad \text{for } \zeta > 1 \tag{2.208b}$$

$$y_{\text{step}} = 1 - (\omega_n t + 1)\exp(-\omega_n t) \quad \text{for } \zeta = 1 \tag{2.208c}$$

with

$$\cos\phi = \zeta \tag{2.195}$$

The step responses given by Equations 2.208 are plotted in Figure 2.86, for several values of damping ratio.

Note that, since a step input does not cause the highest derivative of the system equation to approach infinity at $t = 0^+$, the initial conditions which are required to solve the system equation remain unchanged at $t = 0^+$, provided that there are no derivative terms on the input side of the system equation. If there are derivative terms in the input, then, a step will be converted into an impulse (due to differentiation), and the situation can change.

It should be emphasized that the response given by the convolution integral is based on the assumption that the initial state is zero. Hence, it is known as the *zero-state response*. In particular, the impulse response assumes a zero initial state. As we have stated, the

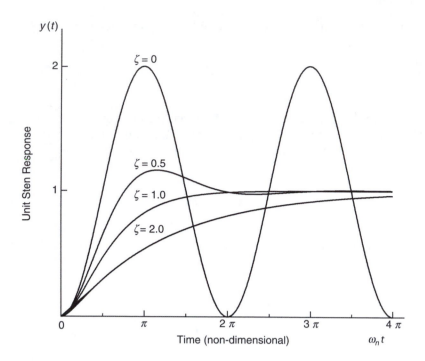

FIGURE 2.86
Unit step response of a damped oscillator.

zero-state response is not necessarily equal to the "particular solution" in mathematical analysis. Also, as t increases $(t \to \infty)$, this solution approaches the *steady-state response* denoted by y_{ss}, which is typically the particular solution. The impulse response of a system is the inverse Laplace transform of the transfer function. Hence, it can be determined using Laplace transform techniques (See Appendix A). Some useful concepts of forced response are summarized in Table 2.15.

2.13.5.4   *Response to Harmonic Excitation*

In many engineering problems the primary excitation typically has a repetitive periodic nature and in some cases this periodic input function may even be purely sinusoidal. Examples are excitations due to mass eccentricity and misalignments in rotational components, tooth meshing in gears, and electromagnetic devices excited by ac or periodic electrical signals. In basic terms, the frequency response of a dynamic system is the response to a pure sinusoidal excitation. As the amplitude and the frequency of the excitation are changed, the response also changes. In this manner the response of the system over a range of excitation frequencies can be determined, and this set of input-output data represents the frequency response. In this case frequency (ω) is the independent variable and hence we are dealing with the *frequency domain*.

Consider the damped oscillator with a harmonic input, as given by

$$\ddot{x} + 2\zeta\omega_n\dot{x} + \omega_n^2 x = a\cos\omega t = u(t) \tag{2.210}$$

TABLE 2.15

Useful Concepts of Forced Response

Convolution Integral: Response $y = \int_0^t h(t-\tau)u(\tau)d\tau = \int_0^t h(\tau)u(t-\tau)d\tau$

where u = excitation (input) and h = impulse response function (response to a unit impulse input).

Damped Simple Oscillator: $\ddot{y} + 2\zeta\omega_n\dot{y} + \omega_n^2 y = \omega_n^2 u(t)$

Poles (eigenvalues) $\lambda_1, \lambda_2 = -\zeta\omega_n \pm \sqrt{\zeta^2 - 1}\,\omega_n \quad$ for $\zeta \geq 1$

$$= -\zeta\omega_n \pm j\omega_d \text{ for } \zeta < 1$$

ω_n = undamped natural frequency, ω_d = damped natural frequency

ζ = damping ratio.

Note: $\omega_d = \sqrt{1-\zeta^2}\,\omega_n$

Impulse Response Function:
(zero ICs)

$h(t) = \dfrac{\omega_n}{\sqrt{1-\zeta^2}}\exp(-\zeta\omega_n t)\sin\omega_d t \quad$ for $\zeta < 1$

$$= \frac{\omega_n}{2\sqrt{\zeta^2-1}}[\exp\lambda_1 t - \exp\lambda_2 t] \quad \text{for } \zeta > 1$$

$$= \omega_n^2 t\exp(-\omega_n t) \quad \text{for } \zeta = 1$$

Unit Step Response:
(zero ICs)

$y(t)_{step} = 1 - \dfrac{1}{\sqrt{1-\zeta^2}}\exp(-\zeta\omega_n t)\sin(\omega_d t + \phi) \quad$ for $\zeta < 1$

$$= 1 - \frac{1}{2\sqrt{\zeta^2-1}\,\omega_n}[\lambda_1\exp\lambda_2 t - \lambda_2\exp\lambda_1 t] \quad \text{for } \zeta > 1$$

$$= 1 - (\omega_n t + 1)\exp(-\omega_n t) \quad \text{for } \zeta = 1$$

$$\cos\phi = \zeta$$

Note: Impulse Response $= \dfrac{d}{dt}$ (Step Response).

The particular solution x_p that satisfies Equation 2.210 is of the form (see Table 2.12)

$$x_p = a_1\cos\omega t + a_2\sin\omega t \quad \{\text{Except for the case: } \zeta = 0 \text{ and } \omega = \omega_n\} \tag{2.211}$$

where the constants a_1 and a_2 are determined by substituting Equation 2.211 into the system Equation 2.210 and equating the like coefficient; the method of undertermined coefficients. We will consider several important cases.

1. *Undamped Oscillator with Excitation Frequency ≠ Natural Frequency:*
We have

$$\ddot{x} + \omega_n^2 x = a\cos\omega t \quad \text{with} \quad \omega \neq \omega_n \tag{2.212}$$

Homogeneous solution:

$$x_h = A_1\cos\omega_n t + A_2\sin\omega_n t \tag{2.213}$$

Particular solution:

$$x_p = \frac{a}{\left(\omega_n^2 - \omega^2\right)} \cos \omega t \qquad (2.214)$$

NOTE   It can be easily verified that x_p given by Equation 2.214 satisfies the forced system Equation 2.210, with $\zeta = 0$. Hence it is a particular solution.

Complete solution:

$$x = \underbrace{A_1 \cos \omega_n t + A_2 \sin \omega_n t}_{H} \qquad \underbrace{+ \frac{a}{\left(\omega_n^2 - \omega^2\right)} \cos \omega t}_{P} \qquad (2.215)$$

Satisfies the homogeneous   Satisfies the equation with input.
equation.

Now A_1 and A_2 are determined using the initial conditions (ICs):

$$x(0) = x_o \quad \text{and} \quad \dot{x}(0) = v_o \qquad (2.216)$$

Specifically, we obtain

$$x_o = A_1 + \frac{a}{\omega_n^2 - \omega^2} \qquad (2.217a)$$

$$v_o = A_2 \omega_n \qquad (2.217b)$$

Hence, the complete response is

$$x = \underbrace{\left[x_o - \frac{a}{\left(\omega_n^2 - \omega^2\right)} \right] \cos \omega_n t + \frac{v_o}{\omega_n} \sin \omega_n t}_{H} \qquad \underbrace{+ \frac{a}{\omega_n^2 - \omega^2} \cos \omega t}_{P} \qquad (2.218a)$$

Homogeneous solution.   Particular solution.

$$= \underbrace{x_o \cos \omega_n t + \frac{v_o}{\omega_n} \sin \omega_n t}_{X} \; + \underbrace{\frac{a}{\left(\omega_n^2 - \omega^2\right)} \underbrace{\left[\cos \omega t - \cos \omega_n t\right]}_{2 \sin \frac{(\omega_n + \omega)}{2} t \sin \frac{(\omega_n - \omega)}{2} t}}_{F} \qquad (2.218b)$$

Free response	*Forced response (depends on input)
(Depends only on ICs)	Comes from both x_h and x_p.
Comes from x_h; Sinusodal at ω_n	*Will exhibit a beat phenomenon for

small $\omega_n - \omega$; i.e., $\dfrac{(\omega_n + \omega)}{2}$ wave

"modulated" by $\dfrac{(\omega_n - \omega)}{2}$ wave.

This is a "stable" response in the sense of bounded-input bounded-output (BIBO) stability, as it is bounded and does not increase steadily.

NOTE If there is no forcing excitation, the homogeneous solution H and the free response X will be identical. With a forcing input, the natural response (the homogeneous solution) will be influenced by it in general, as clear from Equation 2.218a.

2. *Undamped Oscillator with* $\omega = \omega_n$ *(Resonant Condition)*:
This is the degenerate case. In this case the x_p that was used before is no longer valid because, otherwise the particular solution could not be distinguished from the homogeneous solution and the former would be completely absorbed into the latter. Instead, in view of the "double-integration" nature of the forced system equation when $\omega = \omega_n$, we use the particular solution (P):

$$x_p = \frac{at}{2\omega}\sin\omega t \tag{2.219}$$

This choice of particular solution is strictly justified by the fact that it satisfies the forced system equation.
   Complete solution:

$$x = A_1\cos\omega t + A_2\sin\omega t + \frac{at}{2\omega}\sin\omega t \tag{2.220}$$

ICs:

$$x(0) = x_o \quad \text{and} \quad \dot{x}(0) = v_o.$$

By substitution we get

$$x_o = A_1 \tag{2.221}$$

$$v_o = \omega A_2 \tag{2.222}$$

The total response:

$$\underbrace{x = x_o\cos\omega t + \frac{v_o}{\omega}\sin\omega t}_{X} \qquad \underbrace{+\frac{at}{2\omega}\sin\omega t}_{F} \tag{2.223}$$

Free response (Depends on ICs)   Forced response (Depends on Input)
*Sinusoidal with frequency ω.   *Amplitude increases linearly.

Since the forced response increases steadily, this is an unstable response in the bounded-input-bounded-output (BIBO) sense. Furthermore, the homogeneous solution H and the free response X are identical, and the particular solution P is identical to the forced response F in this case.
   Note that the same system (undamped oscillator) gives a bounded response for some excitations while producing an unstable (steady linear increase) response when the excitation frequency is equal to its natural frequency. Hence, the system is not quite

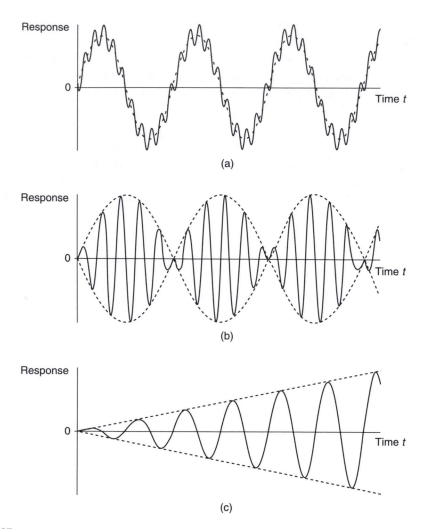

FIGURE 2.87
Forced response of a harmonic-excited undamped simple oscillator: (a) For a large frequency difference; (b) For a small frequency difference (beat phenomenon); (c) Response at resonance.

unstable, but is not quite stable either. In fact, the undamped oscillator is said to be marginally stable. When the excitation frequency is equal to the natural frequency it is reasonable for the system to respond in a complementary and steadily increasing manner because this corresponds to the most "receptive" excitation. Specifically, in this case, the excitation complements and reinforces the natural response of the system. In other words, the system is "in resonance" with the excitation, and the condition is called a *resonance*. Later on we will address this aspect for the more general case of a damped oscillator.

Figure 2.87 shows typical forced responses of an undamped oscillator for a large difference in excitation and natural frequencies (Case 1); for a small difference in excitation and natural frequencies (also Case 1), where a beat-phenomenon is clearly manifested; and for the resonant case (Case 2).

3. Damped Oscillator:

The equation of forced motion is

$$\ddot{x} + 2\zeta\omega_n\dot{x} + \omega_n^2 x = a\cos\omega t \tag{2.224}$$

Particular Solution: Since derivatives of both odd order and even order are present in this equation, the particular solution should have terms corresponding to odd and even derivatives of the forcing function (i.e., $\sin\omega t$ and $\cos\omega t$). Hence, the appropriate particular solution will be of the form:

$$x_p = a_1\cos\omega t + a_2\sin\omega t \tag{2.225}$$

Substitute Equation 2.225 into Equation 2.224. We get

$$-\omega^2 a_1\cos\omega t - \omega^2 a_2\sin\omega t + 2\zeta\omega_n[-\omega a_1\sin\omega t + \omega a_2\cos\omega t]$$

$$+\omega_n^2[a_1\cos\omega t + a_2\sin\omega t] = a\cos\omega t$$

Equate like coefficients:

$$-\omega^2 a_1 + 2\zeta\omega_n\omega a_2 + \omega_n^2 a_1 = a$$

$$-\omega^2 a_2 - 2\zeta\omega_n\omega a_1 + \omega_n^2 a_2 = 0$$

Hence, we have

$$\left(\omega_n^2 - \omega^2\right)a_1 + 2\zeta\omega_n\omega a_2 = a \tag{2.226a}$$

$$-2\zeta\omega_n\omega a_1 + \left(\omega_n^2 - \omega^2\right)a_2 = 0 \tag{2.226b}$$

This can be written in the vector-matrix form:

$$\begin{bmatrix} (\omega_n^2 - \omega^2) & 2\zeta\omega_n\omega \\ -2\zeta\omega_n\omega & (\omega_n^2 - \omega^2) \end{bmatrix} \begin{bmatrix} a_1 \\ a_2 \end{bmatrix} = \begin{bmatrix} a \\ 0 \end{bmatrix} \tag{2.226c}$$

Solution is

$$\begin{bmatrix} a_1 \\ a_2 \end{bmatrix} = \frac{1}{D} \begin{bmatrix} (\omega_n^2 - \omega^2) & -2\zeta\omega_n\omega \\ 2\zeta\omega_n\omega & (\omega_n^2 - \omega^2) \end{bmatrix} \begin{bmatrix} a \\ 0 \end{bmatrix} \tag{2.227}$$

with the determinant

$$D = \left(\omega_n^2 - \omega^2\right)^2 + (2\zeta\omega_n\omega)^2 \tag{2.228}$$

TABLE 2.16

Harmonic Response of a Simple Oscillator

Undamped Oscillator: $\ddot{x} + \omega_n^2 x = a\cos \omega t$ $;$ $x(0) = x_o$, $\dot{x}(0) = v_o$

For $\omega \neq \omega_n$:

$$\underbrace{x = x_0 \cos \omega_n t + \frac{v_0}{\omega_n} \sin \omega_n t}_{X} + \underbrace{\frac{a}{\omega_n^2 - \omega^2}[\cos \omega t - \cos \omega_n t]}_{F}$$

For $\omega = \omega_n$ **(resonance):** $x = $ Same $X + \dfrac{at}{2\omega} \sin \omega t$

Damped Oscillator: $\ddot{x} + 2\zeta\omega_n \dot{x} + \omega_n^2 x = a\cos \omega t$

$$x = H + \underbrace{\frac{a}{\left|\omega_n^2 - \omega^2 + 2j\zeta\omega_n\omega\right|} \cos(\omega t - \phi)}_{P}$$

where, $\tan \phi = \dfrac{2\zeta\omega_n\omega}{\omega_n^2 - \omega^2}$; $\phi = $ phase lag.

Particular solution P is also the steady-state response.

Homogeneous solution $H = A_1 e^{\lambda_1 t} + A_2 e^{\lambda_2 t}$

where, λ_1 and λ_2 are roots of $\lambda^2 + 2\zeta\omega_n\lambda + \omega_n^2 = 0$ (characteristic equation)

A_1 and A_2 are determined from ICs: $x(0) = x_0$, $\dot{x}(0) = v_0$

Resonant Frequency: $\omega_r = \sqrt{1 - 2\zeta^2}\,\omega_n$

The magnitude of P will peak at resonance.

Damping Ratio: $\zeta = \dfrac{\Delta\omega}{2\omega_n} = \dfrac{\omega_2 - \omega_1}{\omega_2 + \omega_1}$ for low damping

where, $\Delta\omega = $ half-power bandwidth $= \omega_2 - \omega_1$

Note: Q-factor $= \dfrac{\omega_n}{\Delta\omega} = \dfrac{1}{2\zeta}$ for low damping

---

On simplification, we get

$$a_1 = \frac{\left(\omega_n^2 - \omega^2\right)}{D} a \tag{2.229a}$$

$$a_2 = \frac{2\zeta\omega_n\omega}{D} a \tag{2.229b}$$

This is the method of "undetermined coefficients."

Some useful results on the frequency response of a simple oscillator are summarized in Table 2.16.

2.13.6   Response Using Laplace Transform

Transfer function concepts were discussed in previous sections, and transform techniques are outlined in Appendix A. Once a transfer function model of a system is available, its response can be determined using the Laplace transform approach. The steps are:

1. Using Laplace transform table (Appendix A) determine the Laplace transform ($U(s)$) of the input.
2. Multiply by the transfer function ($G(s)$) to obtain the Laplace transform of the output: $Y(s) = G(s)U(s)$
3. Convert the expression in Step 2 into a convenient form (e.g., by partial fractions).
4. Using Laplace transform table, obtain the inverse Laplace transform of $Y(s)$, which gives the response $y(t)$.

Let us illustrate this approach by determining again the step response of a simple oscillator.

2.13.6.1 *Step Response Using Laplace Transforms*

Consider the oscillator system given by Equation 2.200. Since $\mathcal{L}\mathcal{U}(t) = 1/s$, the unit step response of the dynamic system (Equation 2.200) can be obtained by taking the inverse Laplace transform of

$$Y_{step}(s) = \frac{1}{s}\frac{\omega_n^2}{\left(s^2 + 2\zeta\omega_n s + \omega_n^2\right)} \tag{2.230a}$$

To facilitate using the Laplace transform table, partial fractions of Equation 2.230 are determined in the form

$$\frac{a_1}{s} + \frac{a_2 + a_3 s}{\left(s^2 + 2\zeta\omega_n s + \omega_n^2\right)}$$

in which, the constants a_1, a_2, and a_3 are determined by comparing the numerator polynomial; thus,

$$\omega_n^2 = a_1\left(s^2 + 2\zeta\omega_n s + \omega_n^2\right) + s(a_2 + a_3 s)$$

Then, $a_1 = 1$, $a_2 = -2\zeta\omega_n$, and $a_3 = -1$.
   Hence,

$$Y_{Step}(s) = \frac{1}{s} + \frac{-s - \zeta\omega_n}{\left(s^2 + 2\zeta\omega_n s + \omega_n^2\right)} \tag{2.230b}$$

Next, using Laplace transform tables, the inverse transform of Equation 2.230b is obtained, and verified to be identical to Equation 2.208.

2.13.7 Computer Simulation

Simulation of the response of a dynamic system by using a digital computer is perhaps the most convenient and popular approach to response analysis. An important advantage is that any complex, nonlinear, and time variant system may be analyzed in this manner.

The main disadvantage is that the solution is not analytic and valid only for a specific excitation. Of course, symbolic approaches of obtaining analytical solutions using a digital computer are available as well. We will consider here numerical simulation only.

The digital simulation typically involves integration of a differential equation of the form

$$\dot{y} = f(y, u, t) \tag{2.231}$$

The most straightforward approach to digital integration of this equation is by using *trapezoidal rule*, which is Euler's method, as given by

$$y_{n+1} = y_n + f(y_n, u_n, t_n)\Delta t \quad n = 0, 1, \ldots \tag{2.232}$$

Here t_n is the nth time instant, $u_n = u(t_n)$, $y_n = y(t_n)$; and Δt is the integration time step $(\Delta t = t_{n+1} - t_n)$. This approach is generally robust. But depending on the nature of the function f, the integration can be ill behaved. Also, Δt has to be chosen sufficiently small.

For complex nonlinearities, a better approach of digital integration is the Runge-Kutta method. In this approach, in each time step, first the following four quantities are computed:

$$g_1 = f(y_n, u_n, t_n) \tag{2.233a}$$

$$g_2 = f\left[\left(y_n + g_1 \frac{\Delta t}{2}\right), u_{n+\frac{1}{2}}, \left(t_n + \frac{\Delta t}{2}\right)\right] \tag{2.233b}$$

$$g_3 = f\left[\left(y_n + g_2 \frac{\Delta t}{2}\right), u_{n+\frac{1}{2}}, \left(t_n + \frac{\Delta t}{2}\right)\right] \tag{2.233c}$$

$$g_4 = f[(y_n + g_3 \Delta t), u_{n+1}, t_{n+1}] \tag{2.233d}$$

Then, the integration step is carried out according to

$$y_{n+1} = y_n + (g_1 + 2g_2 + 2g_3 + g_4)\frac{\Delta t}{6} \tag{2.234}$$

Note that $u_{n+\frac{1}{2}} = u\left(t_n + \frac{\Delta t}{2}\right)$.

Other sophisticated approaches of digital simulation are available as well. Perhaps the most convenient computer-based approach to simulation of a dynamic model is by using a graphic environment that uses block diagrams. Several such environments are commercially available. One that is widely used is SIMULINK, which is an extension to MATLAB (See Appendix B).

2.14 Problems

2.1 What is a "dynamic" system, a special case of any general system?
A typical input variable is identified for each of the following examples of dynamic systems. Give at least one output variable for each system.

a. Human body: neuroelectric pulses

b. Company: information

c. Power plant: fuel rate

d. Automobile: steering wheel movement

e. Robot: voltage to joint motor

f. Highway bridge: vehicle force

2.2 Real systems are nonlinear. Under what conditions a linear model is sufficient in studying a real systems?

Consider the following system equations:

a. $\ddot{y} + (2\sin\omega t + 3)\dot{y} + 5y = u(t)$

b. $3\ddot{y} - 2y = u(t)$

c. $3\ddot{y} + 2\dot{y}^3 + y = u(t)$

d. $5\ddot{y} + 2\dot{y} + 3y = 5u(t)$

    i. Which ones of these are linear?

    ii. Which ones are nonlinear?

    iii. Which ones are time-variant?

2.3 Give four categories of uses of dynamic modeling.
List advantages and disadvantages of experimental modeling over analytical modeling.

2.4 What are the basic lumped elements of

    i. a mechanical system

    ii. an electrical system?

Indicate whether a distributed-parameter method is needed or a lumped-parameter model is adequate in the study of following dynamic systems:

a. vehicle suspension system (motion)

b. elevated vehicle guideway (transverse motion)

c. oscillator circuit (electrical signals)

d. environment (weather) system (temperature)

e. aircraft (motion and stresses)

f. large transmission cable (capacitance and inductance).

NOTE: Variables/parameters of interest are given in parentheses.

2.5 Write down the order of each of the systems shown in Figure P2.5.

2.6

a. Give logical steps of the analytical modeling process for a general physical system.

b. Once a dynamic model is derived, what other information would be needed for analyzing its time response (or for computer simulation)?

c. A system is divided into two subsystems, and models are developed for these subsystems. What other information would be needed to obtain a model for the overall system?

2.7 Various possibilities of model development for a physical system are shown in Figure P2.7. Give advantages and disadvantages of the SM approach of developing an approximate model in comparison to a combined DM+MR approach.

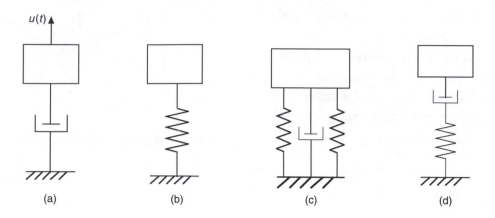

FIGURE P2.5
Models of four mechanical systems

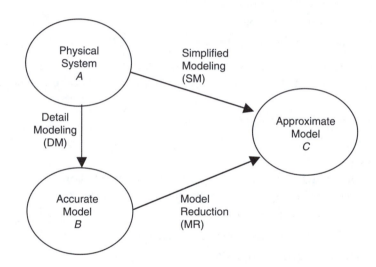

FIGURE P2.7
Approaches of model development

2.8 Describe two approaches of determining the parameters of a lumped-parameter model that is (approximately) equivalent to a distributed-parameter (i.e., continuous) dynamic system.

One end of a heavy spring of mass m_s and stiffness k_s is attached to a lumped mass m. The other end is attached to a support that is free to move, as shown in Figure P2.8.

Using the method of natural frequency equivalence, determine an equivalent lumped-parameter model for the spring where the equivalent lumped mass is located at the free end (support end) of the system. The natural frequencies of a heavy spring with one end fixed and the other end free is given by

$$\omega_n = \frac{\pi}{2}(2n-1)\sqrt{\frac{k_s}{m_s}}$$

where n is the mode number.

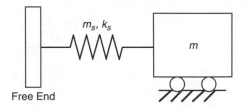

FIGURE P2.8
A mechanical system with a heavy spring and attached mass

2.9

a. Why are analogies important in modeling of dynamic systems?

b. In the force-current analogy, what mechanical element corresponds to an electrical capacitor?

c. In the velocity-pressure analogy, is the fluid inertia element analogous to the mechanical inertia element?

2.10

a. What are through variables in mechanical, electrical, fluid, and thermal systems?

b. What are across variables in mechanical, electrical, fluid, and thermal systems?

c. Can the velocity of a mass change instantaneously?

d. Can the voltage across a capacitor change instantaneously?

e. Can the force in a spring change instantaneously?

f. Can the current in an inductor change instantaneously?

g. Can purely thermal systems oscillate?

2.11 Answer the following questions true or false:

a. A state-space model is unique.

b. The number of state variables in a state vector is equal to the order of the system.

c. The outputs of a system are always identical to the state variables.

d. Outputs can be expressed in terms of state variables.

e. State model is a time domain model.

2.12 Consider a system given by the state equations

$$\dot{x}_1 = x_1 + 2x_2$$
$$\dot{x}_2 = -x_1 + 2u$$

in which x_1 and x_2 are the state variables and u is the input variable. Suppose that the output y is given by

$$y = 2x_1 - x_2.$$

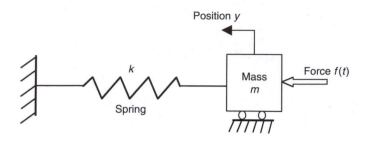

FIGURE P2.13
A mechanical system.

a. Write this state space model in the vector-matrix form:

$$\dot{x} = Ax + Bu$$

$$y = Cx$$

and identify the elements of the matrices A, B, and C.

b. What is the order of the system?

2.13 Consider the mass-spring system shown in Figure P2.13.

The mass m is supported by a spring of stiffness k and is excited by a dynamic force $f(t)$.

a. Taking $f(t)$ as the input, and position and speed of the mass as the two outputs, obtain a state-space model for the system.

b. What is the order of the system?

c. Repeat the problem, this time taking the compression force in the spring as the only output.

d. How many initial conditions are needed to determine the complete response of the system?

2.14 What precautions may be taken in developing and operating a mechanical system, in order to reduce system nonlinearities?
Read about the following nonlinear phenomena:

i. saturation

ii. hysteresis

iii. jump phenomena

iv. frequency creation

v. limit cycle

vi. deadband.

Two types of nonlinearities are shown in Figure P2.14
In each case indicate the difficulties of developing an analytical model for operation near:

i. point O

ii. point A.

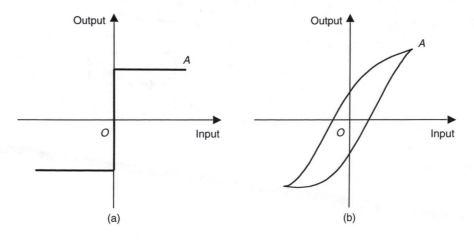

FIGURE P2.14
Two types of nonlinearities: (a) Ideal saturation; (b) Hysteresis.

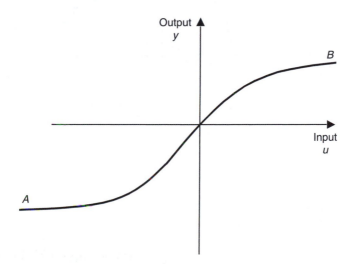

FIGURE P2.17
The characteristic curve of a nonlinear device.

2.15 An excitation was applied to a system and its response was observed. Then the excitation was doubled. It was found that the response also doubled. Is the system linear?

2.16

a. Determine the derivative $\dfrac{d}{dx} x\,|x|$.

b. Linearize the following terms about the operating point $\bar{x} = 2$:

    (i) $3x^3$ (ii) $|x|$ (iii) $\dot{x}^2$

2.17 A nonlinear device obeys the relationship $y = y(u)$ and has an operating curve as shown in Figure P2.17.

    i. Is this device a dynamic system?

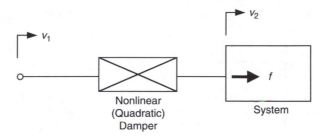

FIGURE P2.18
A nonlinear mechanical system.

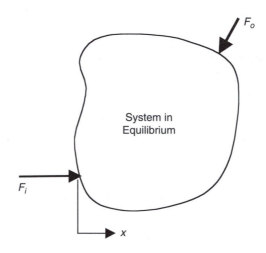

FIGURE P2.19
Virtual displacement of a system in equilibrium.

A linear model of the form $y = ku$ is to be determined for operation of the device:

ii.   in a small neighborhood of point B

iii. over the entire range from A to B.

Suggest a suitable value for k in each case.

2.18 A nonlinear damper is connected to a mechanical system as shown in Figure P2.18. The force f, which is exerted by the damper on the system is $c(v_2 - v_1)^2$ where c is a constant parameter.

i. Give an analytical expression for f in terms of v_1, v_2, and c, which would be generally valid.

ii. Give an appropriate linear model.

iii. If the operating velocities v_1 and v_2 are equal, what will be the linear model about this operating point?

2.19 Suppose that a system is in equilibrium under the forces F_i and F_o as shown in Figure P2.19. If the point of application of F_i is given a small "virtual" displacement x in the same direction, suppose that the location of F_o moves through $y = k\,x$ in the opposite direction to F_o.

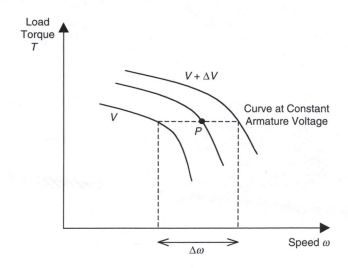

FIGURE P2.20
Characteristic curves of an armature-controlled dc motor.

i. Determine F_o in terms of F_i (This is a result of the "principle of virtual work").
ii. What is the relationship between the small changes $\hat{F}_i$ and $\hat{F}_o$, about operating conditions $\overline{F}_i$ and $\overline{F}_o$, assuming equilibrium?

2.20 Characteristic curves of an armature-controlled dc motor are as shown in Figure P2.20. These are torque versus speed curves, measured at a constant armature voltage. For the neighborhood of point P, a linear model of the form

$$\hat{\omega} = k_1\hat{v} + k_2\hat{T}$$

needs to be determined, for use in motor control. The following information is given:

The slope of the curve at $P = -a$

Voltage changes for the two adjacent curves of point $P = \Delta V$

Corresponding speed change for constant load torque through $P = \Delta\omega$.

Estimate the parameters k_1 and k_2.

2.21 An air circulation fan system of a building is shown in Figure P2.21(a), and a simplified model of the system may be developed, as represented in Figure P2.21(b).

The induction motor is represented as a torque source $\tau(t)$. The speed ω of the fan, which determines the volume flow rate of air, is of interest. The moment of inertia of the fan impeller is J. The energy dissipation in the fan is modeled by a linear viscous component (of damping constant b) and a quadratic aerodynamic component (of coefficient d).

a. Show that the system equation may be given by

$$J\dot{\omega} + b\omega + d\,|\omega|\,\omega = \tau(t)$$

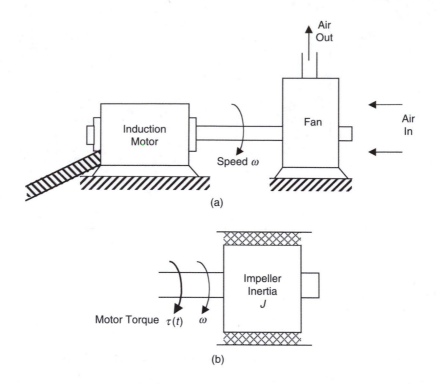

FIGURE P2.21
(a) A motor/fan combination of a building ventilation system; (b) A simplified model of the ventilation fan.

b. Suppose that the motor torque is given by

$$\tau(t) = \bar{\tau} + \hat{\tau}_a \sin \Omega t$$

in which $\bar{\tau}$ is the steady torque and $\hat{\tau}_a$ is a very small amplitude (compared to $\bar{\tau}$) of the torque fluctuations at frequency Ω. Determine the steady-state operating speed $\bar{\omega}$, which is assumed positive, of the fan.

(c) Linearize the model about the steady-state operating conditions and express it in terms of the speed fluctuations $\hat{\omega}$. From this, estimate the amplitude of the speed fluctuations.

2.22

a. Linearized models of nonlinear systems are commonly used in model-based control of processes. What is the main assumption that is made in using a linearized model to represent a nonlinear system?

b. A three-phase induction motor is used to drive a centrifugal pump for incompressible fluids. To reduce misalignment and associated problems such as vibration, noise, and wear, a flexible coupling is used for connecting the motor shaft to the pump shaft. A schematic representation of the system is shown in Figure P2.22.

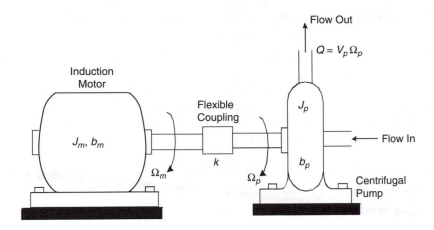

FIGURE P2.22
A centrifugal pump driven by an inductor motor.

Assume that the motor is a "torque source" of torque T_m, which is being applied to the motor of inertia J_m. Also, the following variables and parameters are defined:

J_p = moment of inertia of the pump impeller assembly
Ω_m = angular speed of the motor rotor/shaft
Ω_p = angular speed of the pump impeller/shaft
k = torsional stiffness of the flexible coupling
T_f = torque transmitted through the flexible coupling
Q = volume flow rate of the pump
b_m = equivalent viscous damping constant of the motor rotor.

Also, assume that the net torque required at the pump shaft, to pump fluid steadily at a volume flow rate Q, is given by $b_p\Omega_p$, where

$$Q = V_p\Omega_p$$

and V_p = volumetric parameter of the pump (assumed constant).

Using T_m as the input and Q as the output of the system, develop a complete state-space model for the system. Identify the matrices A, B, C, and D in the usual notation, in this model. What is the order of the system?

c. In Part (a) suppose that the motor torque is given by

$$T_m = \frac{aSV_f^2}{\left[1+(S/S_b)^2\right]}$$

where motor slip S is defined as

$$S = 1 - \frac{\Omega_m}{\Omega_s}$$

Note that a and S_b are constant parameters of the motor. Also,

Ω_s = no-load (i.e., synchronous) speed of the motor

V_f = amplitude of voltage applied to each phase winding (field) of the motor

In *voltage control* V_f is used as the input, and in *frequency control* Ω_s is used as the input. For combined voltage and frequency control, derive a linearized state-space model, using the incremental variables $\hat{V}_f$ and $\hat{\Omega}_s$, about the operating values $\overline{V}_f$ and $\overline{\Omega}_s$, as the inputs to the system, and the incremental flow $\hat{Q}$ as the output.

2.23 Select the correct answer for each of the following multiple-choice questions.

   i. A through variable is characterized by

      a. Being the same at both ends of the element

      b. Being listed first in the pair representation of a linear graph

      c. Requiring no reference value

      d. All the above

   ii. An across variable is characterized by

      a. Having different values across the element

      b. Being listed second in the pair representation

      c. Requiring a reference point

      d. All the above

   iii. Which of the following could be a through variable?

      a. Pressure

      b. Voltage

      c. Force

      d. All the above

   iv. Which of the following could be an across variable?

      a. Motion (velocity)

      b. Fluid flow

      c. current

      d. All the above

   v. If angular velocity is selected as an element's across variable, the accompanying through variable is

      a. Force

      b. Flow

      c. Torque

      d. Distance

   vi. The equation written for through variables at a node is called a

      a. Continuity equation

      b. Constitutive equation

      c. Compatibility equation

      d. All the above

vii. The functional relation between a through variable and its across variable is called a
   a. Continuity equation
   b. Constitutive equation
   c. Compatibility equation
   d. Node equation

viii. The equation that equates the sum of across variables in a loop to zero is known as
   a. Continuity equation
   b. Constitutive equation
   c. Compatibility equation
   d. Node equation

ix. A node equation is also known as
   a. An equilibrium equation
   b. A continuity equation
   c. The balance of through variables at the node
   d. All the above

x. A loop equation is
   a. A balance of across variables
   b. A balance of through variables
   c. A constitutive relationship
   d. All the above

2.24 A linear graph has 10 branches, two sources, and six nodes:
   i. How many unknown variables are there?
   ii. What is the number of independent loops?
   iii. How many inputs are present in the system?
   iv. How many constitutive equations could be written?
   v. How many independent continuity equations could be written?
   vi. How many independent compatibility equations could be written?
   vii. Do a quick check on your answers.

2.25 The circuit shown in Figure P2.25 has an inductor L, a capacitor C, a resistor R and a voltage source $v(t)$. Considering that L can be analogous to a spring, and C to be analogous

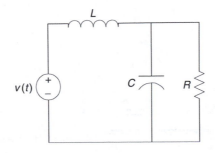

FIGURE P2.25
An electrical circuit.

to an inertia, follow the four steps outlined in the text to obtain the state equations. First sketch the linear graph denoting the currents through and the voltages across the elements L, C, and R by (f_1, v_1), (f_2, v_2) and (f_3, v_3), respectively, and then proceed in the usual manner.

  i. What is the system matrix and what is the input distribution matrix for your choice of state variables?

  ii. What is the order of the system?

  iii. Briefly explain what happens if the voltage source $v(t)$ is replaced by a current source $i(t)$.

2.26 Consider an automobile traveling at a constant speed on a rough road, as shown in Figure P2.26(a). The disturbance input due to road irregularities can be considered as a velocity source $u(t)$ at the tires in the vertical direction. An approximate one-dimensional model shown in Figure P2.26(b) may be used to study the "heave" (up and down) motion of the automobile. Note that v_1 and v_2 are the velocities of the lumped masses m_1 and m_2, respectively.

  a. Briefly state what physical components of the automobile are represented by the model parameters k_1, m_1, k_2, m_2, and b_2. Also, discuss the validity of the assumptions that are made in arriving at this model.

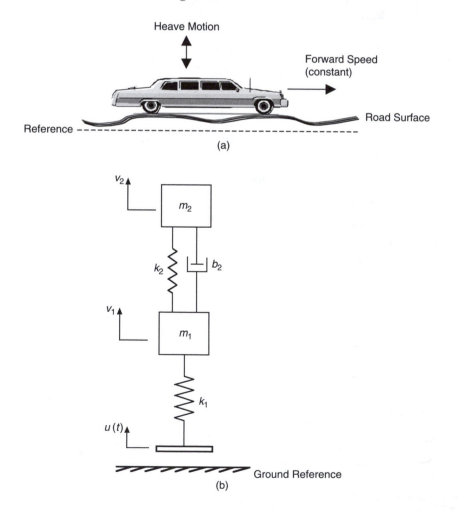

FIGURE P2.26
(a) An automobile traveling at constant speed; (b) A crude model of an automobile for the heave motion analysis.

b. Draw a linear graph for this model, orient it (i.e., mark the directions of the branches), and completely indicate the system variables and parameters.

c. By following the step-by-step- procedure of writing constitutive equations, node equations and loop equations, develop a state-space model for this system. The outputs are v_1 and v_2. What is the order of the system?

d. If instead of the velocity source $u(t)$, a force source $f(t)$ which is applied at the same location, is considered as the system input, draw a linear graph for this modified model. Also, obtain the state equations for this model. What is the order of the system now?

NOTE   In this problem you may assume that gravitational effects are completely balanced by the initial compression of the springs with reference to which all motions are defined.

2.27 Suppose that a linear graph has the following characteristics:

n = number of nodes

b = number of branches (segments)

s = number of sources

l = number of independent loops

Carefully explaining the underlying reasoning, answer the following questions regarding this linear graph:

a. From the topology of linear graph show that $l = b - n + 1$

b. What is the number of continuity equations required (in terms of n)?

c. What is the number of lumped elements including source elements in the model (expressed in terms of b and s)?

d. What is the number of unknown variables, both state and auxiliary, (expressed in terms of b and s)? Verify that this is equal to the number of available equations, and hence the problem is solvable.

2.28 An approximate model for a motor-compressor combination used in a process control application is shown Figure P2.28.

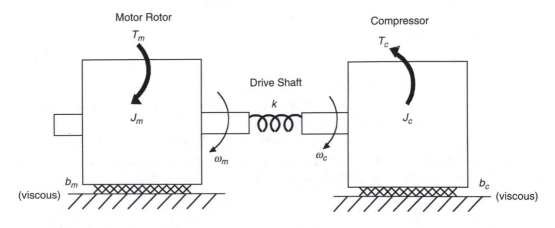

FIGURE P2.28
A model of a motor-compressor unit.

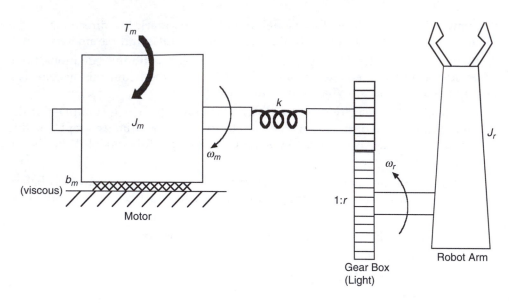

FIGURE P2.29
A model of a single-degree-of-freedom robot.

Note that T, J, k, b, and ω denote torque, moment of inertia, torsional stiffness, angular viscous damping constant, and angular speed, respectively, and the subscripts m and c denote the motor rotor and compressor impeller, respectively.
  a. Sketch a translatory mechanical model that is analogous to this rotary mechanical model.
  b. Draw a linear graph for the given model, orient it, and indicate all necessary variables and parameters on the graph.
  c. By following a systematic procedure and using the linear graph, obtain a state-space representation of the given model. The outputs of the system are compressor speed ω_c and the torque T transmitted through the drive shaft.

2.29 A model for a single joint of a robotic manipulator is shown in Figure P2.29. The usual notation is used. The gear inertia is neglected and the gear reduction ratio is taken s 1:r (note: $r < 1$). Joint flexibility is included in the model.
  a. Draw a linear graph for the model, assuming that no external (load) torque is present at the robot arm.
  b. Using the linear graph derive a state model for this system. The input is the motor magnetic torque T_m and the output is the angular speed ω_r of the robot arm. What is the order of the system?
  c. Discuss the validity of various assumptions made in arriving at this simplified model for a commercial robotic manipulator.

2.30 Consider the rotatory feedback control system shown schematically by Figure P2.30. The load has inertia J, stiffness K and equivalent viscous damping B as shown. The armature circuit for the dc fixed field motor is shown in Figure P2.30(b).

$$\text{The back e.m.f.} \qquad v_B = K_V \omega$$

$$\text{The motor torque} \qquad T_m = K_T i$$

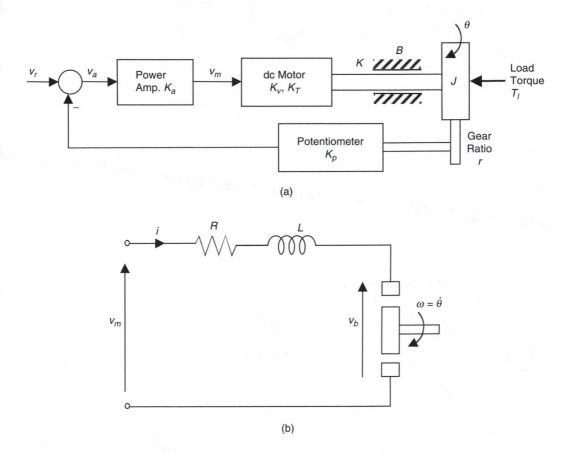

(a)

(b)

FIGURE P2.30
(a) A rotatory electromechanical system; (b) The armature circuit.

a. Identify the system inputs.

b. Write the linear system equations.

2.31

a. What are A-type elements and T-type elements?

Classify mechanical inertia, mechanical spring, fluid inertia and fluid capacitor into these two types. Explain a possible conflict that could arise due to this classification.

b. A system that is used to pump an incompressible fluid from a reservoir into an open overhead tank is schematically shown in Figure P2.31. The tank has a uniform across section of area A.

The pump is considered as a pressure source of pressure difference $P(t)$. A valve of constant k_v is placed near the pump in the long pipe line, which leads to the overhead tank. The valve equation is $Q = k_v \sqrt{P_1 - P_2}$ in which Q is the volume flow rate of the fluid. The resistance to the fluid flow in the pipe may be modeled as $Q = k_p \sqrt{P_2 - P_3}$ in which k_p is a pipe flow constant. The linear effect of the accelerating fluid is represented by the equation $I \frac{dQ}{dt} = P_3 - P_4$ in which I denotes the fluids inertance. Pressures $P_1, P_2, P_3,$ and P_4 are as marked along the pipe length, in Figure P2.31. Also P_0 denotes the ambient pressure.

    i. Using Q and P_{40} as the state variables, the pump pressure $P(t)$ as the input variable, and the fluid level H in the tank as the output variable, obtain a

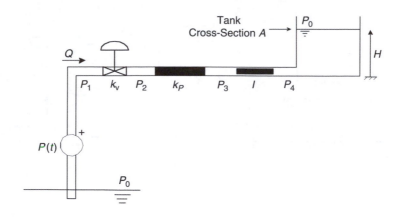

FIGURE P2.31
A pumping system for an overhead tank

complete (nonlinear) state-space model for the system. Note that $P_{40} = P_4 - P_0$. The density of the fluid is ρ.

ii. Lienarize the state equations about an operating point of flow rate $\overline{Q}$. Give the matrices A, B, C, and D for the linear model.

iii. What is the combined linear resistance of the valve and piping?

2.32

a. Briefly explain why a purely thermal system typically does not have an oscillatory response whereas a fluid system can.

b. Figure P2.32 shows a pressure-regulated system that can provide a high-speed jet of liquid. The system consists of a pump, a spring-loaded accumulator, and a fairly long section of piping which ends with a nozzle. The pump is considered as a flow source of value Q_s. The following parameters are important:

A = area of cross section (uniform) of the accumulator cylinder

k = spring stiffness of the accumulator piston

L = length of the section of piping from the accumulator to the nozzle

A_p = area of cross section (uniform, circular) of the piping

A_o = discharge area of the nozzle

C_d = discharge coefficient of the nozzle

Q = mass density of the liquid.

Assume that the liquid is incompressible. The following variables are important:

$P_{1r} = P_1 - P_r$ = pressure at the inlet of the accumulator with respect to the ambient reference r

Q = volume flow rate through the nozzle

h = height of the liquid column in the accumulator

Note that the piston (wall) of the accumulator can move against the spring, thereby varying h.

i. Considering the effects of the movement of the spring loaded wall and also the gravity head of the liquid, obtain an expression for the equivalent fluid capacitance C_a of the accumulator in terms of k, A, ρ, and g. Are the two capacitances

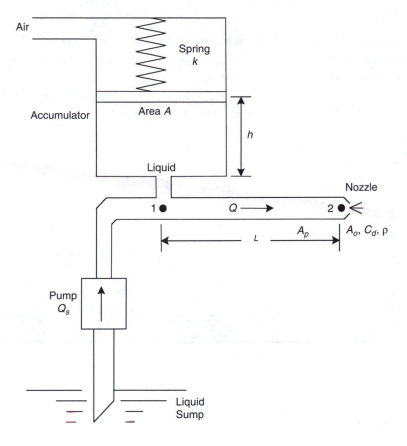

FIGURE P2.32
Pressure regulated liquid jet system.

which contribute to C_a (i.e., wall stretching and gravity) connected in parallel or in series?

NOTE   Neglect the effect of bulk modulus of the liquid.

ii. Considering the capacitance C_a, the inertance I of the fluid volume in the piping (length L and cross section area A_p), and the resistance of the nozzle only, develop a nonlinear state-space model for the system. The state vector $x = [P_{1r} \quad Q]^T$, and the input $u = [Q_s]$.

For flow in the (circular) pipe with a parabolic velocity profile, the inertance $I = \frac{2\rho L}{A_p}$ and for the discharge through the nozzle

$$Q = A_o c_d \sqrt{\frac{2P_{2r}}{\rho}}$$

in which

P_{2r} = pressure inside the nozzle with respect to the outside reference (r).

c_d = discharge coefficient.

2.33

a. What is the main physical reason for oscillatory behavior in a purely fluid system? Why do purely fluid systems with large tanks connected by small-diameter pipes rarely exhibit an oscillatory response?

b. Two large tanks connected by a thin horizontal pipe at the bottom level are shown in Figure P2.33(a). Tank 1 receives an inflow of liquid at the volume

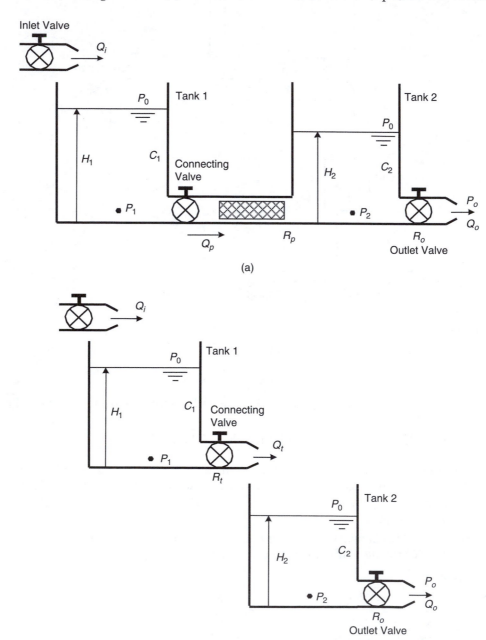

(a)

(b)

FIGURE P2.33
(a) An interacting two-tank fluid system; (b) A non-interacting two-tank fluid system.

rate Q_i when the inlet valve is open. Tank 2 has an outlet valve, which has a fluid flow resistance of R_o and a flow rate of Q_o when opened. The connecting pipe also has a valve, and when opened, the combined fluid flow resistance of the valve and the thin pipe is R_p. The following parameters and variables are defined:

C_1, C_2 = fluid (gravity head) capacitances of tanks 1 and 2

ρ = mass density of the fluid

g = acceleration due to gravity

P_1, P_2 = pressure at the bottom of tanks 1 and 2

P_0 = ambient pressure.

Using $P_{10} = P_1 - P_0$ and $P_{20} = P_2 - P_0$ as the state variables and the liquid levels H_1 and H_2 in the two tanks as the output variables, derive a complete, linear, state-space model for the system.

Defining the time constants $\tau_1 = C_1 R_p$ and $\tau_2 = C_2 R_o$, and the gain parameter $k = R_o/R_p$ express the characteristic equation of the system in terms of these three parameters.

Show that the poles of the system are real and negative but the system is coupled (interacting).

(c) Suppose that the two tanks are as in Figure P2.33(b). Here Tank 1 has an outlet valve at its bottom whose resistance is R_t and the volume flow rate is Q_t when open. This flow directly enters Tank 2, without a connecting pipe. The remaining characteristics of the tanks are the same as in Part (b).

Derive a state-space model for the modified system in terms of the same variables as in Part (b). With $\tau_1 = C_1 R_t$, $\tau_2 = C_2 R_o$, and $k = R_o/R_t$ obtain the characteristic equation of this system. What are the poles of the system? Show that the modified system is non-interacting.

2.34 A model for the automatic gage control (AGC) system of a steel rolling mill is shown in Figure P2.34. The rolls are pressed using a single acting hydraulic actuator with a valve displacement of u. The rolls are displaced through y, thereby pressing the steel that is being rolled. The rolling force F is completely known from the steel parameters for a given y.

    i. Identify the inputs and the controlled variable in this control system.

    ii. In terms of the variables and system parameters indicated in Figure P2.34, write dynamic equations for the system, including valve nonlinearities.

    iii. What is the order of the system? Identify the response variables.

    iv. What variables would you measure (and feed back through suitable controllers) in order to improve the performance of the control system?

2.35 A simplified model of a hotwater heating system is shown in Figure P2.35.

Q_s = rate of heat supplied by the furnace to the water heater (1000 kW)

T_a = ambient temperature (°C)

T_h = temperature of water in the water heater – assumed uniform (°C)

T_o = temperature of the water leaving the radiator (°C)

Q_r = rate of heat transfer from the radiator to the ambience (kW)

M = mass of water in the water heater (500 kg)

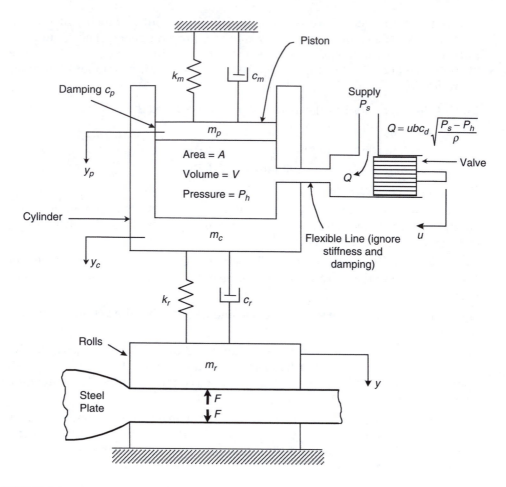

FIGURE P2.34
Automatic gage control (AGC) system of a steel rolling mill.

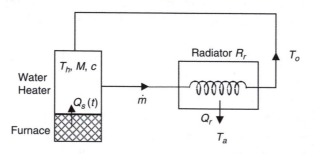

FIGURE P2.35
A household heating system.

$\dot{m}$ = mass rate of water flow through the radiator (25 kg/min)

c = specific heat of water (4200 J/kg/°C).

The radiator satisfies the equation

$$T_h - T_a = R_r Q_r$$

where

R_r = thermal resistance of the radiator $(2 \times 10^{-3} \ °C/kW)$

   a. What are the inputs to the system?

   b. Using T_h as a state variable, develop a state-space model for the system.

   c. Give the output equations for Q_r and T_o.

2.36 Consider a hollow cylinder of length l, inside diameter d_i, and the outside diameter d_o. If the conductivity of the material is k, what the conductive thermal resistance of the cylinder in the radial direction?

2.37 When two dissimilar metal wires are jointed at the two ends, to form a loop, and one junction is maintained at a different temperature from the other, a voltage is generated between the two junctions. A temperature sensor, which makes use of this property is the thermocouple. The cold junction is maintained at a known temperature (say, by dipping into an ice-water bath). The hot junction is then used to measure the temperature at some location. The temperature of the hot junction (T) does not instantaneously reach that of the sensed location (T_f), in view of the thermal capacitance of the junction. Derive an expression for the thermal time constant of a thermocouple in terms of the following parameters of the hot junction:

m = mass of the junction

c = specific heat of the junction

h = heat transfer coefficient of the junction

A = surface area of the junction

2.38

1. In the electro-thermal analogy of thermal systems, where voltage is analogous to temperature and current is analogous to heat transfer rate, explain why there exists a thermal capacitor but not a thermal inductor. What is a direct consequence of this fact with regard to the natural (free or unforced) response of a purely thermal system?

2. A package of semiconductor material consisting primarily of wafers of crystalline silicon substrate with minute amounts of silicon dioxide is heat treated at high temperature as an intermediate step in the production of transistor elements. An approximate model of the heating process is shown in Figure P2.38.

The package is placed inside a heating chamber whose walls are uniformly heated by a distributed heating element. The associated heat transfer rate into the wall is Q_i. The interior of the chamber contains a gas of mass m_c and specific heat c_c, and is maintained at a uniform temperature T_c. The temperature of silicon is T_s and that of the wall is T_w. The outside environment is maintained at temperature T_o. The specific heats of the silicon package and the wall are denoted by c_s and c_w, respectively, and the corresponding masses are denoted by m_s and m_w as shown. The convective heat transfer coefficient at the interface of silicon and gas inside the chamber is h_s, and the effective surface area is A_s. Similarly, h_i and h_o denote the convective heat transfer coefficients at the inside and outside surfaces of the chamber wall, and the corresponding surface areas are A_i and A_o, respectively.

   a. Using T_s, T_c, and Tw as state variables, write three state equations for the process.

   b. Express these equations in terms of the parameters $C_{hs} = m_s c_s$, $C_{hc} = m_c c_c$, $C_{hw} = m_w c_w$, $R_f = \frac{1}{h_s A_s}$, $R_f = \frac{1}{h_i A_i}$, and $R_o = \frac{1}{h_o A_o}$. Explain the electrical analogy and physical significance of these parameters.

   c. What are the inputs to the process? If T_s is the output of importance, obtain the matrices A, B, C, and D of the state-space model.

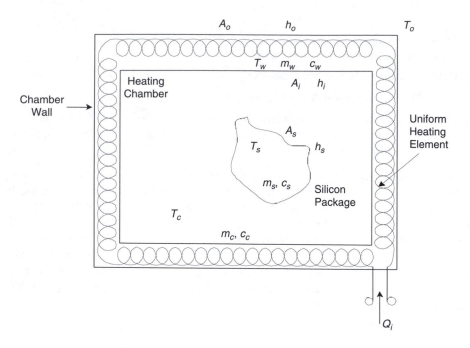

FIGURE P2.38
A model of the heat treatment of a package of silicon.

    d. Comment on the accuracy of the model in the context of the actual physical process of producing semiconductor elements.

2.39 State whether true (T) or false (F):

    a. A bond represents a power flow link between two elements.

    b. The half arrow represents the positive direction of power flow at an element.

    c. The causality stroke indicates which variable is input and which is output for an element.

    d. There is a direct relationship between the assignment of the half arrow and the assignment of the causality stroke for a bond.

    e. Force, voltage, pressure, and temperature are effort variables.

    f. Velocity, current, fluid flow rate, and heat transfer rate are flow variables.

    g. Force and current are through variables.

    h. Velocity and voltage are across variables.

2.40 Suppose that velocity is considered an input to a mechanical element.

    a. The associated force is automatically considered an output of the element.

    b. The associated force is also an input to the element.

    c. We cannot make a definitive statement about the causality of associated force.

    d. The causality of the associated force depends on the direction of power flow.

2.41 Write velocity causality equations for the common-force junctions shown in Figure P2.41.

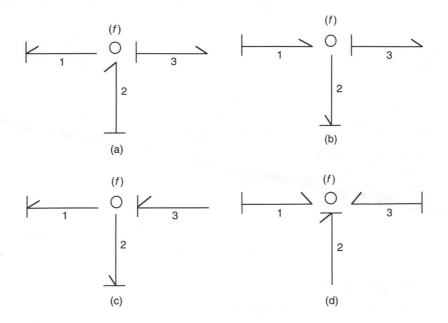

FIGURE P2.41
Four examples of common-force (effort) junctions.

2.42 State whether true (T) or false (F):

a. If one causality stroke of a common-effort junction is indicated, the remaining two strokes are uniquely determined.

b. If two causality strokes of a common-effort junction are indicated, the remaining stroke is uniquely determined.

c. If one of the causality strokes of a common-effort junction is marked at the near end of a bond, then the other two strokes are at the far ends of the other two bonds.

d. If one causality stroke of a common-flow junction is marked at the far end of a bond, then the other two strokes are at the near ends of the other two bonds.

2.43 Write constitutive relations using integral causality for

(a) an inductor L (b) a capacitor C

Let v = voltage across element, and i = current through element.

2.44 We can use the analogy between force and voltage (effort variables) and velocity and current (flow variables) to extend the bond graph concepts to electrical systems. Draw bond graph representations for a voltage source and a current source.

2.45 Consider a system consisting of five energy-storage elements and three energy dissipation elements. Suppose that we draw the bond graph of the system and mark causality strokes, assuming integral causality for the energy-storage elements. As a result of causalities that exist due to other elements in the system, two of the five energy-storage elements were found to have derivative causality. What is the order of the system?

2.46 Consider the damped simple oscillator shown in Figure P2.46.

a. Draw a bond graph for this system model and mark the power-flow half arrows and causality strokes.

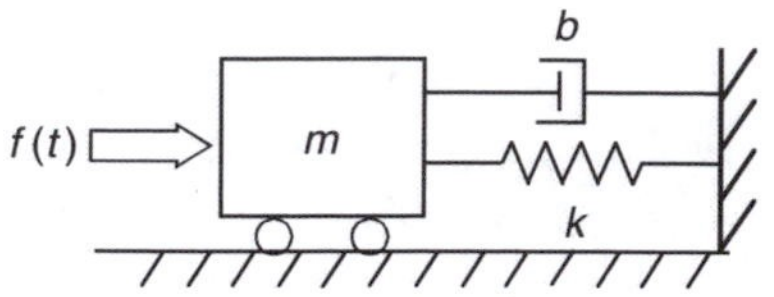

FIGURE P2.46
A simple oscillator

b. Is there a need for arbitrary causality in the damping element b?

c. Is there any conflict in causality?

d. What is the order of the system model?

e. Introducing any auxiliary variables that might be needed, write the skeletal state equations.

f. Write the characteristic (constitutive) equations for the remaining elements.

g. By eliminating the auxiliary equations, obtain the state equations for the system model.

2.47 State whether true (T) or false (F).

a. The output of a system will depend on the input.

b. The output of a system will depend on the system transfer function.

c. The transfer function of a system will depend on the input signal.

d. If the Laplace transform of the input signal does not exist (say, infinite), then the transfer function itself does not exist.

e. If the Laplace transform of the output signal does not exist, then the transfer function itself does not exist.

2.48 State whether true (T) or false (F).

a. A transfer function provides an algebraic expression for a system.

b. The Laplace variable s can be interpreted as time derivative operator d/dt, assuming zero initial conditions.

c. The variable $1/s$ may be interpreted as the integration of a signal starting at $t = 0$.

d. The numerator of a transfer function is characteristic polynomial.

e. A single-input single-output, linear, time-invariant system has a unique (one and only one) transfer function.

2.49 Consider the system given by the differential equation:

$$\ddot{y} + 4\dot{y} + 3y = 2u + \dot{u}$$

a. What is the order of the system?

b. What is the system transfer function?

c. Do we need Laplace tables to obtain the transfer function?

d. What are the poles?

e. What is the characteristic equation?

f. Consider the parent system: $\ddot{x} + 4\dot{x} + 3x = u$

Express y in terms of x, using the principle of superposition.

g. Using $[x_1 \quad x_2]^T = [x \quad \dot{x}]^T$ as the state variables, obtain a state-space model for the given original system (not the parent system).

h. Using the superposition approach, draw a simulation block diagram for the system.

i. Express the system differential equation in a form suitable for drawing a simulation diagram by the "grouping like-derivatives" method.

j. From Part (i) draw the simulation block diagram.

k. Express the transfer function $(s + 2)/(s + 3)$ in two forms of simulation block diagrams.

l. Using one of the two forms obtained in Part (k), draw the simulation block diagram for the original second order system.

m. What are the partial fractions of the original transfer function?

n. Using the partial-fraction method, draw a simulation block diagram for the system. What is the corresponding state-space model?

o. Obtain a state-space model for the system using Part (j).

p. Obtain at least one state model for the system using the block diagram obtained in Part (l).

q. What can you say about the diagonal elements of the system matrix A in Part (n) and in Part (p)?

2.50

a. List several characteristics of a physically realizable system. How would you recognize the physically realizability of a system by drawing a simulation block diagram, which uses integrators, summing junctions, and gain blocks?

b. Consider the system given by the following input/output differential equation:

$$\dddot{y} + a_2\ddot{y} + a_1\dot{y} + a_0 y = b_2\ddot{u} + b_1\dot{u} + b_0 u$$

in which u is the input and y is the output.

Is this system physically realizable?

Draw a simulation block diagram for this system using integrators, gains, and summing junctions only.

2.51 For the control system of Problem 2.30 draw a simulation block diagram.

2.52 It is required to study the dynamic behavior of an automobile during the very brief period of a sudden start from rest. Specifically, the vehicle acceleration a in the direction of primary motion, as shown in Figure P2.52(a), is of interest and should be considered as the system output. The equivalent force $f(t)$ of the engine, applied in the direction of primary motion, is considered as the system input. A simple dynamic model that may be used for the study is shown in Figure P2.52(b).

Note that k is the equivalent stiffness, primarily due to tire flexibility, and b is the equivalent viscous damping constant, primarily due to dissipations at the tires and other moving parts of the vehicle, taken in the direction of a. Also, m is the mass of the vehicle.

a. Discuss advantages and limitations of the proposed model for the particular purpose.

b. Using force f_k of the spring (stiffness k) and velocity v of the vehicle as the state variables, engine force $f(t)$ as the input and the vehicle acceleration a as the output, develop a complete state-space model for the system.

(Note: You must derive the matrices A, B, C, and D for the model).

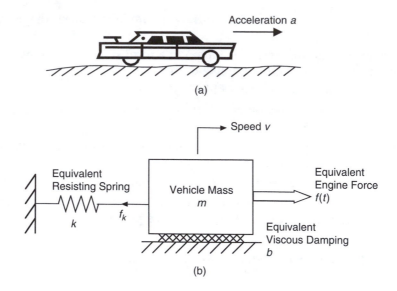

FIGURE P2.52

(a) Vehicle suddenly accelerating from rest, (b) A simplified model of the accelerating vehicle.

  c. Draw a simulation block diagram for the model, employing integration and gain blocks, and summation junctions only.

  d. Obtain the input/output differential equation of the system. From this, derive the transfer function (a/f in the Laplace domain).

  e. Discuss the characteristics of this model by observing the nature of matrix **D**, feed-forwardness of the block diagram, input and output orders of the I/O differential equation, and the numerator and denominator orders of the system transfer function.

2.53 Consider a dynamic system, which is represented by the transfer function (output/input):

$$G(s) = \frac{3s^3 + 2s^2 + 2s + 1}{s^3 + 4s^2 + s + 3}$$

System output = y; system input = u.

  a. What is the input/output differential equation of the system?

    What is the order of the system?

    Is this system physically realizable?

  b. Based on the "superposition method" draw a simulation block diagram for the system, using integrators, constant gain blocks, and summing junctions only.

    Obtain a state-space model using this simulation block diagram, clearly giving the matrices **A**, **B**, **C**, and **D**.

  c. Based on the "grouping like-derivatives method" draw a simulation block diagram, which should be different from what was drawn in Part (b), again using integrators, constant gain blocks, and summing junctions only.

    Give a state-space model for the system, now using this simulation block diagram. This state space model should be different from that in Part (b), which further illustrates that the state-space representation is not unique.

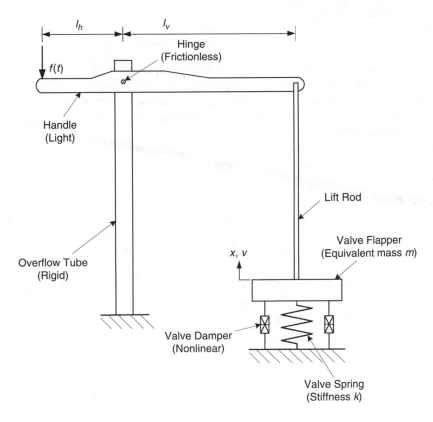

FIGURE P2.54
Simplified model of a toilet-flushing mechanism.

2.54 Give reasons for the common experience that in the flushing tank of a household toilet, some effort is needed to move the handle for the flushing action but virtually no effort is needed to release the handle at the end of the flush.

A simple model for the valve movement mechanism of a household flushing tank is shown in Figure P2.54. The overflow tube on which the handle lever is hinged, is assumed rigid. Also, the handle rocker is assumed light, and the rocker hinge is assumed frictionless. The following parameters are indicated in the figure:

$r = \frac{l_v}{l_h}$ = the lever arm ratio of the handle rocker

m = equivalent lumped mass of the value flapper and the lift rod

k = stiffness of spring action on the value flapper.

The damping force f_{NLD} on the valve is assumed quadratic and is given by

$$f_{NLD} = a|v_{VLD}|v_{VLD}$$

where, the positive parameter

$$a = a_u \text{ for upward motion of the flapper } (v_{NLD} \geq 0)$$
$$= a_d \text{ for downward motion of the flapper } (v_{NLD} < 0)$$

with

$$a_u \gg a_d$$

The force applied at the handle is $f(t)$, as shown.

We are interested in studying the dynamic response of the flapper valve. Specially, the valve displacement x and the valve speed v are considered outputs, as shown in Figure P2.54. Note that x is measured from the static equilibrium point of the spring where the weight mg is balanced by the spring force.

a. By defining appropriate through variables and across variables, draw a linear graph for the system shown in Figure P2.54, clearly indicating the power flow arrows.

b. Using valve speed and the spring force as the state variables, develop a (non-linear) state-space model for the system, with the aid of the linear graph. Start with all the constitutive, continuity, and compatibility equations, and eliminate the auxiliary variables systematically, in obtaining the state-space model.

c. Linearize the state-space model about an operating point where the valve speed is $\bar{v}$. For the linearized model, obtain the model matrices A, B, C, and D, in the usual notation. Note that the incremental variables $\hat{x}$ and $\hat{v}$ are the outputs in the linear model, and the incremental variable $\hat{f}(t)$ is the input.

d. From the linearized state-space model, derive the input-output model (differential equation) relating $\hat{f}(t)$ and $\hat{x}$.

e. Give expressions for the undamped natural frequency and the damping ratio of the linear model, in terms of the parameters a, $\bar{v}$, m, and k. Show that the damping ratio increases with the operating speed.

2.55 The electrical circuit shown in Figure P2.55 has two resistor R_1 and R_2, an inductor L, a capacitor C, and a voltage source $u(t)$. The voltage across the capacitor is considered the output y of the circuit.

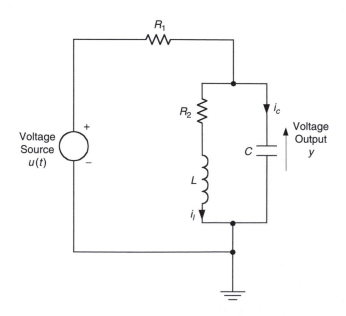

FIGURE P2.55
An *RLC* circuit driven by a voltage source.

a. What is the order of the system and why?

b. Show that the input-output equation of the circuit is given by

$$a_2 \frac{d^2y}{dt^2} + a_1 \frac{dy}{dt} + a_0 y = b_1 \frac{du}{dt} + b_0 u$$

Express the coefficients a_0, a_1, a_2, b_0 and b_1 in terms of the circuit parameters R_1, R_2, L, and C.

c. Starting with the auxiliary differential equation

$$a_2 \ddot{x} + a_1 \dot{x} + a_0 x = u$$

and using $x = [x \quad \dot{x}]^T$ as the state vector, obtain a complete state-space model for the system in Figure P2.55. Note that this is the "superposition method" of developing a state model.

d. Clearly explain why, for the system in Figure P2.55, neither the current i_c through the capacitor nor the time derivative of the output ($\hat{v}$) can be chosen as a state variable.

2.56 Consider an nth order, linear, time-invariant dynamic system with input $u(t)$ and output y. When a step input was applied to this system it was observed that the output jumped instantaneously in the very beginning. Which of the following statements are true for this system?

a. Any simulation block diagram of this system (consisting only of integrators, constant-gain blocks, and summation junctions) will have at least one feedforward path.

b. In its state-space model:

$$\dot{x} = Ax + Bu$$

$$y = Cx + Du$$

the D matrix does not vanish (i.e., $D \neq 0$).

c. This is not a physically realizable system.

d. The number of zeros in the system is equal to n.

e. The number of poles in the system is equal to n.

In each case briefly justify your answer.

2.57 In relation to a dynamic system, briefly explain your interpretation of the terms

a. Causality

b. Physical Realizability.

Using integrator blocks, summing junctions, and coefficient blocks only, unless it is absolutely necessary to use other types of blocks, draw simulation block diagrams for the following three input-output differential equations:

i. $a_1 \dfrac{dy}{dt} + a_0 y = u$

ii. $a_1 \dfrac{dy}{dt} + a_0 y = u + b_1 \dfrac{du}{dt}$

iii. $a_1 \dfrac{dy}{dt} + a_0 y = u + b_1 \dfrac{du}{dt} + b_2 \dfrac{d^2u}{dt^2}$

Note that u denotes the input and y denotes the output. Comment about causality and physical realizability of these three systems.

2.58 The Fourier transform of a position measurement $y(t)$ is $Y(j\omega)$.

  i. The Fourier transform of the corresponding velocity signal is:

  a. $Y(j\omega)$

  b. $j\omega\, Y(j\omega)$

  c. $Y(j\omega)/(j\omega)$

  d. $\omega\, Y(j\omega)$

  ii. The Fourier transform of the acceleration signal is:

  a. $Y(j\omega)$

  b. $j\omega\, Y(j\omega)$

  c. $-\omega^2\, Y(j\omega)$

  d. $Y(j\omega)/(j\omega)$

2.59 Answer true (T) or false (F):

  i. Mechanical impedances are additive for two elements connected in parallel.

  ii. Mobilities are additive for two elements connected in series.

2.60 The movable arm with read/write head of a disk drive unit is modeled as a simple oscillator. The unit has an equivalent bending stiffness $k = 10$ dyne.cm/rad and damping constant b. An equivalent rotation $u(t)$ radians is imparted at the read/write head. This in turn produces a (bending) moment to the read/write arm, which has an equivalent moment of inertia $J = 1 \times 10^{-3}$ gm.cm^2, and bends the unit at an equivalent angle θ about the centroid.

  a. Write the input-output differential equation of motion for the read/write arm unit.

  b. What is the undamped natural frequency of the unit in rad/s?

  c. Determine the value of b for 5% critical damping.

  d. Write the frequency transfer function of the model.

2.61 A rotating machine of mass M is placed on a rigid concrete floor. There is an isolation pad made of elastomeric material between the machine and the floor, and is modeled as a viscous damper of damping constant b. In steady operation there is a predominant harmonic force component $f(t)$, which is acting on the machine in the vertical direction at a frequency equal to the speed of rotation (n rev/s) of the machine. To control the vibrations produced by this force, a dynamic absorber of mass m and stiffness k is mounted on the machine. A model of the system is shown in Figure P2.61.

  a. Determine the frequency transfer function of the system, with force $f(t)$ as the input and the vertical velocity v of mass M as the output.

  b. What is the mass of the dynamic absorber that should be used in order to virtually eliminate the machine vibration (a tuned absorber)?

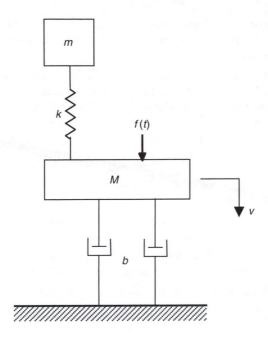

FIGURE P2.61
A mounted machine wit a dynamic absorber.

2.62 The frequency transfer function for a simple oscillator is given by

$$G(\omega) = \frac{\omega_n^2}{[\omega_n^2 - \omega^2 + 2j\zeta\omega_n\omega]}$$

a. If a harmonic excitation $u(t) = a\cos\omega_n t$ is applied to this system what is the steady-state response?

b. What is the magnitude of the resonant peak?

c. Using your answers to parts (a) and (b) suggest a method to measure damping in a mechanical system.

d. At what excitation frequency is the response amplitude maximum under steady state conditions?

e. Determine an approximate expression for the half-power (3 dB) bandwidth at low damping. Using this result, suggest an alternative method for the damping measurement.

2.63

a. An approximate frequency transfer function of a system was determined by Fourier analysis of measured excitation-response data and fitting into an appropriate analytical expression (by curve fitting using the least squares method). This was found to be

$$G(f) = \frac{5}{10 + j2\pi f}$$

What is its magnitude, phase angle, real part, and imaginary part at $f = 2$ Hz? If the reference frequency is taken as 1 Hz, what is the transfer function magnitude at 2 Hz expressed in dB?

b. A dynamic test on a structure using a portable shaker revealed the following: The accelerance between two locations (shaker location and accelerometer location) measured at a frequency ratio of 10 was 35 dB. Determine the corresponding mobility and mechanical impedance at this frequency ratio.

2.64 Answer true (T) or false (F):

a. Electrical impedances are additive for two elements connected in parallel.

b. Impedance, both mechanical and electrical, is given by the ratio of effort/flow, in the frequency domain.

c. Impedance, both mechanical and electrical, is given by the ratio of across variable/through variable, in the frequency domain.

d. Mechanical impedance is analogous to electrical impedance when determining the equivalent impedance of several interconnected impedances.

e. Mobility is analogous to electrical admittance (Current/Voltage in the frequency domain) when determining the equivalent value of several interconnected elements.

2.65 The unit step response of a system, with zero initial conditions, was found to be $1.5(1 - e^{-10t})$. What is the input-output differential equation of the system? What is the transfer function?

2.66 Discuss why the convolution integrals given below (where u is the input, y is the output, and h is the impulse response function) are all identical

$$y(t) = \int_0^\infty h(\tau)u(t-\tau)\,d\tau$$

$$y(t) = \int_{-\infty}^\infty h(t-\tau)u(\tau)\,d\tau$$

$$y(t) = \int_{-\infty}^\infty h(\tau)u(t-\tau)\,d\tau$$

$$y(t) = \int_{-\infty}^t h(t-\tau)u(\tau)\,d\tau$$

$$y(t) = \int_{-\infty}^t h(\tau)u(t-\tau)\,d\tau$$

$$y(t) = \int_0^t h(t-\tau)u(\tau)\,d\tau$$

$$y(t) = \int_0^t h(\tau)u(t-\tau)\,d\tau$$

2.67 A system at rest is subjected to a unit step input $\mathcal{U}(t)$. Its response is given by

$$y = 2e^{-t}(\cos t - \sin t)\mathcal{U}(t)$$

a. Write the input-output differential equation for the system
b. What is its transfer function?
c. Determine the damped natural frequency, undamped natural frequency, and the damped ratio.
d. Write the response of the system to a unit impulse and sketch it.

2.68 Consider the dynamic system given by the transfer function

$$\frac{Y(s)}{U(s)} = \frac{(s+4)}{(s^2 + 3s + 2)}$$

a. Plot the poles and zeros of the systems on the s-plane.
b. Indicate the correct statement among the following:
   i. The system is stable
   ii. The system is unstable
   iii. The system stability depends on the input
   iv. None of the above.
c. Obtain the system differential equation.
d. Using the Laplace transform technique determine the system response $y(t)$ to a unit step input, with zero initial conditions.

2.69 A dynamic system is represented by the transfer function

$$\frac{Y(s)}{U(s)} = G(s) = \frac{\omega_n^2}{s^2 + 2\zeta\omega_n s + \omega_n^2}$$

a. Is the system stable?
b. If the system is given an impulse input, at what frequency will it oscillate?
c. If the system is given a unit step input, what is the frequency of the resulting output oscillations? What is its steady state value?
d. The system is given the sinusoidal input

$$u(t) = a \sin \omega t$$

Determine an expression for the output $y(t)$ at steady state in terms of a, ω, ω_n, and ζ. At what value of ω will the output $y(t)$ be maximum at steady state?

2.70 A system at rest is subjected to a unit step input $\mathcal{U}(t)$ Its response is given by:

$$y = [2e^{-t} \sin t]\mathcal{U}(t)$$

a. Write the input-output differential equation for the system
b. What is its transfer function?

    c. Determine the damped natural frequency, undamped natural frequency, and the damping ratio.

    d. Write the response of the system to a unit impulse and find $y(0^+)$.

    e. What is the steady state response for a unit step input?

2.71

  a. Define the following terms with reference to the response of a dynamic system:

    i. Homogeneous solution.

    ii. Particular solution.

    iii. Zero-input (or free) response.

    iv. Zero-state (or forced) response.

    v. Steady-state response.

  b. Consider the first order system

$$\tau \frac{dy}{dt} + y = u(t)$$

in which u is the input, y is the output, and τ is a system constant.

    i. Suppose that the system is initially at rest with $u = 0$ and $y = 0$, and suddenly a unit step input is applied. Obtain an expression for the ensuing response of the system. Into which of the above five categories does this response fall? What is the corresponding steady-state response?

    ii. If the step input in Part (i) above is of magnitude A what is the corresponding response?

    iii. If the input in Part (i) above was an impulse of magnitude P what would be the response?

2.72 An "iron butcher" is a head-cutting machine which is commonly used in the fish processing industry. Millions of dollars worth salmon, is wasted annually due to inaccurate head cutting using these somewhat outdated machines. The main cause of wastage is the "over-feed problem." This occurs when a salmon is inaccurately positioned with respect to the cutter blade so that the cutting location is beyond the collar bone and into the body of a salmon. An effort has been made to correct this situation by sensing the position of the collar bone and automatically positioning the cutter blade accordingly.

A schematic representation of an electromechanical positioning system of a salmon-head cutter is shown in Figure P2.72(a). Positioning of the cutter is achieved through a lead screw and nut arrangement, which is driven by a brushless dc motor. The cutter carriage is integral with the nut of the lead screw and the ac motor which drives the cutter blade, and has an overall mass of m (kg). The carriage slides along a lubricated guideway and provides an equivalent viscous damping force of damping constant b (N/m/s). The overall moment of inertia of the motor rotor and the lead screw is J (N·m²) about the axis of rotation. The motor is driven by a drive system, which provides a voltage v to the stator field windings of the motor. Note that the motor has a permanent magnet rotor. The interaction between the field circuit and the motor rotor is represented by Figure P2.72(b).

The magnetic torque T_m generated by the motor is given by

$$T_m = k_m i_f$$

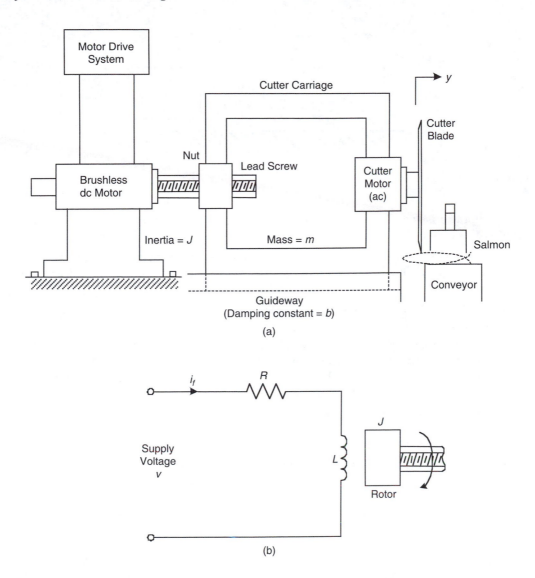

FIGURE P2.72
(a) A positioning system for an automated fish cutting machine, (b) The field circuit of the permanent-magnet rotor dc motor.

and the force F_L exerted by the lead screw in the y-direction of the cutter carriage is given by $F_L = \frac{e}{h} T_L$, in which,

$$h = \frac{\text{Translatory Motion of the Nut}}{\text{Rotatory Motion of Lead Screw}}$$

and e is the mechanical efficiency of the lead screw-nut unit.

Other parameters and variables, as indicated in Figure P2.72 should be self-explanatory.

a. Write the necessary equations to study the displacement y of the cutter in response to an applied voltage v to the motor. What is the order of the system? Obtain the input-output differential equation for the system and from that determine the

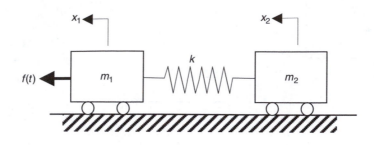

FIGURE P2.73
A two-car train.

characteristic equation. What are the roots (poles or eigenvalues) of the characteristic equation?

b. Using summation junctions, integration blocks, and constant gain blocks only, draw a complete block diagram of the system, with v as the input and y as the output.

c. Obtain a state-space model for the system, using v as the input and y as the output.

d. Assume that L/R ratio is very small and can be neglected. Obtain an expression for the response y of the system to a step input with zero initial conditions. Show from this expression that the behavior of the system is unstable in the present form (i.e., without feedback control).

2.73 Consider the two-mass system shown in Figure P2.73.

a. What is the transfer function x_1/f?

b. For a harmonic excitation $f(t)$, at what frequency will m_1 be motionless?

3

Mechanical Components and Robotic Manipulators

The field of mechatronics deals with the integration of mechanics and electronics. In a mechatronic product, mechanics plays a vital role, which may include structural support or load bearing, mobility, transmission of motion and power or energy, and actuation and manipulation. The mechanical system has to be designed (integral with electronics, controls, etc.) to satisfy such desirable characteristics as light weight, high strength, high speed, low noise and vibration, long design life, fewer moving parts, high reliability, low-cost production and distribution, and infrequent and low-cost maintenance. Clearly, the requirements can be conflicting and there is a need for design optimization.

Even in an integrated electromechanical system, there are good reasons why a distinction has to be made between the mechanical components and the electronic and computer (hardware and software) components. One relates to the energy (or power) conversion. The types of energy that are involved will differ in these different types of components (or functions). The level of energy (or power) can differ greatly as well. For example, digital electronic circuits and computer hardware typically use low levels of power and voltage. Analog devices such as amplifiers and power supplies can accommodate high voltages and power. Motors and other actuators (e.g., ac motors and hydraulic actuators in particular) can receive high levels of electrical power and generate similar high levels of mechanical power. Analog to digital conversion (ADC) and digital to analog conversion (DAC) involve relatively low levels of power. But, drive (power) amplifiers of electrical motors, pumps and compressors of hydraulic and pneumatic systems typically deal with much higher levels of power. It follows that the level of power needed for a task and the nature of energy conversion that is involved can separate mechanical components from others in a mechatronic system.

Another important consideration that separates a mechanical component from electronic components and computing components (hardware/software) is the bandwidth (speed, time constant, etc.). Typically, mechanical components have lower time constants than electronic components. Accordingly their speeds of operation will differ and furthermore, the bandwidth (useful frequency content) of the associated signals will differ as well. For example, process plants can have time constants as large as minutes and robotic devices and machine tools have time constant in the ms range. The time constants of analog electrical circuitry can be quite low (μs range). Software-based computer devices can conveniently generate digital actions in the kHz rate (i.e., ms time scale). If faster speeds are needed, one will have to go for faster processors, efficient computing algorithms, and computers with faster operation cycles. In order to carry out digital control and other digital actions at much faster speeds (MHz speed, μs cycle time) one will have to rely on hardware (not software) solutions using dedicated analog and digital electronics.

It should be clear from the foregoing discussion that even though a mechatronic system is designed using an integrated approach with respect to its functions and components, it will still be necessary to make a distinction between its mechanical components and nonmechanical components. In this chapter some important types of mechanical components will be discussed. Only typical cases will be studied in detail, where modeling and analysis will be presented. It is expected that these techniques and knowledge may be extended to other types of mechanical components and devices. Particular attention will be given to robotic manipulators, which are examples of mechatronic systems. Hydraulic and pneumatic components, which belong to the general class of mechanical components will not be studied in the present chapter, as they will be treated separately in Chapter 9.

3.1 Mechanical Components

Common mechanical components in a mechatronic system may be classified into some useful groups, as follows:

1. Load bearing/structural components (strength and surface properties)
2. Fasteners (strength)
3. Dynamic isolation components (transmissibility)
4. Transmission components (motion conversion)
5. Mechanical actuators (generated force/torque)
6. Mechanical controllers (controlled energy dissipation)

In each category we have indicated within parentheses the main property or attribute that is characteristic of the function of that category.

In load bearing or structural components the main function is to provide structural support. In this context, mechanical strength and surface properties (e.g., hardness, wear resistance, friction) of the component are crucial. The component may be rigid or flexible and stationary or moving. Examples of load bearing and structural components include bearings, springs, shafts, beams, columns, flanges, and similar load-bearing structures.

Fasteners are closely related to load bearing/structural components. The purpose of a fastener is to join two mechanical components. Here as well, the primary property of importance is the mechanical strength. Examples are bolts and nuts, locks and keys, screws, rivets, and spring retainers. Welding, bracing, and soldering are processes of fastening and will fall into the same category.

Dynamic-isolation components perform the main task of isolating a system from another system (or environment) with respect to motion and forces. These involve the "filtering" of motions and forces/torques. Hence motion transmissibility and force transmissibility are the key considerations in these components. Springs, dampers, and inertia elements may form the isolation element. Shock and vibration mounts for machinery, inertia blocks, and the suspension systems of vehicles are examples of isolation dynamic components.

Transmission components may be related to isolation components in principle, but their functions are rather different. The main purpose of a transmission component is the conversion of motion (in magnitude and from). In the process the force/torque of the input member is also converted in magnitude and form. In fact in some applications the modification of the force/torque may be the primary requirement of the transmission component. Examples of transmission components are gears, harmonic drives, lead screws and nuts (or power screws), racks and pinions, cams and followers, chains and sprockets, belts and pulleys (or drums), differentials, kinematic linkages, flexible couplings, and fluid transmissions.

Mechanical actuators are used to generate forces (and torques) for various applications. The common actuators are electromagnetic in form (i.e., electric motors) and not purely mechanical. Since the magnetic forces are "mechanical" forces which generate mechanical torques, electric motors may be considered as electromechanical devices. Other types of actuators that use fluids for generating the required effort may be considered in the category of mechanical actuators. Examples are hydraulic pistons and cylinders (rams), hydraulic motors, their pneumatic counterparts, and thermal power units (prime movers) such as steam/gas turbines. Of particular interest in mechatronic systems are the electro-mechanical actuators and hydraulic and pneumatic actuators.

Mechanical controllers perform the task of modifying dynamic response (motion and force/torque) in a desired manner. Purely mechanical controllers carry out this task by controlled dissipation of energy. These are not as common as electrical/electronic control-lers and hydraulic/pneumatic controllers. In fact hydraulic/pneumatic servo valves may be treated in the category of purely mechanical controllers. Furthermore, mechanical controllers are closely related to transmission components and mechanical actuators. Examples of mechanical controllers are clutches and brakes.

In selecting a mechanical component for a mechatronic application, many engineering aspects have to be considered. The foremost are the capability and performance of the component with respect to the design requirements (or specifications) of the system. For example, motion and torque specifications, flexibility and deflection limits, strength char-acteristics including stress-strain behavior, failure modes and limits and fatigue life, sur-face and material properties (e.g., friction, nonmagnetic, noncorrosive), operating range, and design life will be important. Other factors such as size, shape, cost, and commercial availability can be quite crucial.

The foregoing classification of mechanical components is summarized in Figure 3.1. It is not within the scope of the present chapter to study all the types of mechanical components that are summarized here. Rather, we select for further analysis a few important mechan-ical components that are particularly useful in mechatronic systems.

3.2 Transmission Components

Transmission devices are indispensable in mechatronic applications. We will undertake to discuss a few representative transmission devices here. It should be cautioned that in the present treatment, a transmission is isolated and treated as a separate unit. In an actual application, however, a transmission device works as an integral unit with other compo-nents, particularly the actuator, the electronic drive unit, and the load of the system. Hence a transmission design or selection should involve an integrated treatment of all interacting components. This should be clear in the subsequent chapters.

Perhaps the most common transmission device is a gearbox. In its simplest form, a gearbox consists of two gear wheels, which contain teeth of identical pitch (tooth separa-tion) and of unequal wheel diameter. The two wheels are meshed (i.e., the teeth are engaged) at one location. This device changes the rotational speed by a specific ratio (gear ratio) as dictated by the ratio of the diameters (or radii) of the two gear wheels. In particular, by stepping down the speed (in which case the diameter of the output gear is larger than that of the input gear), the output torque can be increased. Larger gear ratios can be realized by employing more than one pair of meshed gear wheels. Gear transmissions

Mechanical Components

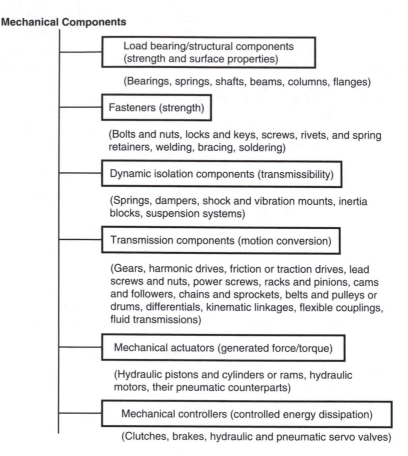

FIGURE 3.1
Classification of mechanical components.

are used in a variety of applications including automotive, industrial-drive and robotics. Specific gear designs range from conventional spur gears to harmonic drives, as discussed later in the present section.

Gear drives have several disadvantages. In particular, they exhibit "backlash" because the tooth width is smaller than the tooth space of the mating gear. Some degree of backlash is necessary for proper meshing. Otherwise jamming will occur. Unfortunately, backlash is a nonlinearity, which can cause irregular and noisy operation with brief intervals of zero torque transmission. It can lead to rapid wear and tear and even instability. The degree of backlash can be reduced by using proper profiles (shapes) for the gear teeth. Backlash can be eliminated through the use of spring-loaded gears. Sophisticated feedback control may be used as well to reduce the effects of gear backlash.

Conventional gear transmissions, such as those used in automobiles with standard gearboxes, contain several gear stages. The gear ratio can be changed by disengaging the drive-gear wheel (pinion) from a driven wheel of one gear stage, and engaging it with another wheel of a different number of teeth (different diameter) of another gear stage, while the power source (input) is disconnected by means of a clutch. Such a gearbox provides only a few fixed gear ratios. The advantages of a standard gearbox include relative simplicity of design and the ease with which it can be adapted to operate over a reasonably wide range of speed ratios, albeit in a few discrete increments of large steps. There are many disadvantages: Since each gear ratio is provided by a separate gear stage, the size, weight, and complexity (and associated cost, wear, and unreliability) of the transmission increases directly with the number of gear ratios provided. Also, since the drive source has to be disconnected by a clutch during the shifting of gears, the speed transitions are generally not smooth, and operation is noisy. There is also dissipation of power during the transmission steps, and wear and damage can be caused by inexperienced operators. These shortcomings can be reduced or eliminated if the transmission is able to vary the speed ratio continuously rather than in a stepped manner. Further, the output speed and corresponding torque can be matched to the load requirements closely and continuously for a fixed input power. This results in more efficient and smooth operation, and many other related advantages. A continuously-variable transmission, which has these desirable characteristics, will be discussed later in this section. First we will discuss a power screw, which is a converter of angular motion into rectilinear motion.

3.2.1 Lead Screw and Nut

A lead-screw drive is a transmission component, which converts rotatory motion into rectilinear motion. Lead screws, power screws, and ball screws are rather synonymous. Lead screw and nut units are used in numerous applications including positioning tables, machine tools, gantry and bridge systems, automated manipulators, and valve actuators. Figure 3.2 shows the main components of a lead-screw unit. The screw is rotated by a

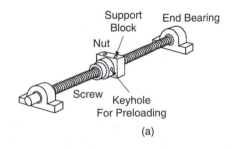

(a)

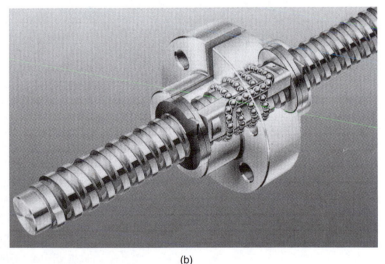

(b)

FIGURE 3.2
(a) A lead screw and nut unit, (b) A commercial ball screw unit (Deutsche Star GmbH, Scheweinfurt, Germany. With permission.)

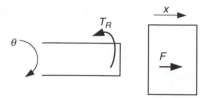

FIGURE 3.3
Effort and motion transmission at the screw and nut interface.

motor, and as a result the nut assembly moves along the axis of the screw. The support block, which is attached to the nut, provides means for supporting the device that has to be moved using the lead-screw drive. The screw holes that are drilled on the support block may be used for this purpose. Since there can be backlash between the screw and the nut as a result of the assembly clearance and/or wear and tear, a keyhole is provided in the nut to apply a preload through some form of a clamping arrangement that is designed into the nut. The end bearings support the moving load. Typically these are ball bearings that can carry axial loads as well, by means of an angular-contact thrust bearing design.

The basic equation for operation of a lead-screw drive is obtained now. As shown in Figure 3.3, suppose that a torque T_R is provided by the screw at (and reacted by) the nut. Note that this is the net torque after deducting the inertia torque (due to inertia of the motor rotor and the lead screw) and the frictional torque of the bearings, from the motor (magnetic) torque. Torque T_R is not completely available to move the load that is supported on the nut. The reason is the energy dissipation (friction) at the screw and nut interface. Suppose that the net force available from the nut to drive the load in the axial direction is F. Denote the screw rotation by θ and the rectilinear motion of the nut by x.

When the screw is rotated (by a motor) through $\delta\theta$, the nut (which is restrained from rotating due to the guides along which the support block moves) will move through δx along the axial direction. The work done by the screw is $T_R.\delta\theta$ and the work done in moving the nut (with its load) is $F.\delta x$. The lead screw efficiency e is given by

$$e = \frac{F.\delta x}{T_R.\delta\theta} \tag{i}$$

Now, $r\delta\theta = \delta x$, where the transmission parameter of the lead screw is r (axial distance moved per one radian of screw rotation). The "lead" l of the lead screw is the axial distance moved by the nut in one revolution of the screw, and it satisfies

$$l = 2\pi r \tag{3.1}$$

In general, the lead is not the same as the "pitch" p of the screw, which is the axial distance between two adjacent threads. For a screw with n threads,

$$l = np \tag{3.2}$$

Substituting r in Equation (i) we have

$$F = \frac{e}{r}T_R = \frac{2\pi e}{l}T_R \tag{3.3}$$

This result is the representative equation of a lead screw, and may be used in the design and selection of components in a lead-screw drive system.

For a screw of mean diameter d, the helix angle α is given by

$$\tan\alpha = \frac{l}{\pi d} = \frac{2r}{d} \tag{3.4}$$

Assuming square threads, we obtain a simplified equation for the screw efficiency in terms of the coefficient of friction μ. First, for a screw of 100% efficiency ($e = 1$), from Equation 3.3, a torque T_R at the nut can support an axial force (load) of T_R/r. The corresponding frictional force F_f is $\mu T_R/r$. The torque required to overcome this frictional force is $T_f = F_f d/2$. Hence, the frictional torque is given by

$$T_f = \frac{\mu d}{2r} T_R \tag{3.5}$$

The screw efficiency is

$$e = \frac{T_R - T_f}{T_R} = 1 - \frac{\mu d}{2r} = 1 - \frac{\mu}{\tan\alpha} \tag{3.6}$$

For threads that are not square (e.g., for slanted threads such as Acme threads, Buttress threads, modified square threads), Equation 3.6 has to be appropriately modified.

It is clear from Equation 3.6 that the efficiency of a lead-screw unit can be increased by decreasing the friction and increasing the helix angle. Of course, there are limits. For example, typically the efficiency will not increase by increasing the helix angle beyond 30°. In fact, a helix angle of 50° or more will cause the efficiency to drop significantly. The friction can be decreased by proper choice of material for screw and nut and through surface treatments, particularly lubrication. Typical values for the coefficient of friction (for identical mating material) are given in Table 3.1. Note that the static (starting) friction will be higher (as much as 30%) than the dynamic (operating) friction. An ingenious way to reduce friction is by using a nut with a helical track of balls instead of threads. In this case the mating between the screw and the nut is not through threads but through ball bearings. Such a lead-screw unit is termed a *ball screw*. A screw efficiency of 90% or greater is possible with a ball screw unit.

In the driving mode of a lead screw, the frictional torque acts in the opposite direction to (and has to be overcome by) the driving torque. In the "free" mode where the load is not driven by an external torque from the screw, it is expected that the load will try to "back-drive" the screw (say, due to gravitational load). Then, however, the frictional torque

TABLE 3.1

Some Useful Values for Coefficient of Friction

Material	Coefficient of Friction
Steel (dry)	0.2
Steel (lubricated)	0.15
Bronze	0.10
Plastic	0.10

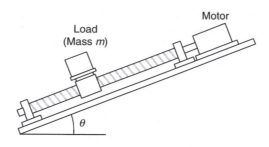

FIGURE 3.4
A lead-screw unit driving an inclined load.

will change direction and the back motion has to overcome it. If the back-driving torque is less than the frictional torque, motion will not be possible and the screw is said to be self-locking.

Example 3.1

A lead-screw unit is used to drive a load of mass up an incline of angle θ, as shown in Figure 3.4. Under quasi-static conditions (i.e., neglecting inertial loads) determine the drive torque needed by the motor to operate the device. The total mass of the moving unit (load, nut, and fixtures) is m. The efficiency of the lead screw is e and the lead is l. Assume that the axial load (thrust) due to gravity is taken up entirely by the nut (In practice, a significant part of the axial load is supported by the end bearings, which have the thrust-bearing capability.).

SOLUTION
The effective load that has to be acted upon by the net torque (after allowing for friction) in this example is

$$F = mg \sin \theta$$

Substitute into Equation 3.3. The required torque at the nut is

$$T_R = \frac{mgr}{e} \sin \theta = \frac{mgl}{2\pi e} \sin \theta \tag{3.7}$$

3.2.2 Harmonic Drives

Usually, motors run efficiently at high speeds. Yet in many practical applications, low speeds and high torques are needed. A straightforward way to reduce the effective speed and increase the output torque of a motor is to employ a gear system with high gear reduction. Gear transmission has several disadvantages, however. For example, backlash in gears would be unacceptable in high-precision applications. Frictional loss of torque, wear problems, and the need for lubrication must also be considered. Furthermore, the mass of the gear system consumes energy from the actuator (motor), and reduces the overall torque/mass ratio and the useful bandwidth of the actuator.

A harmonic drive is a special type of transmission device that provides very large speed reductions (e.g., 200: 1) without backlash problems. Also, a harmonic drive is comparatively

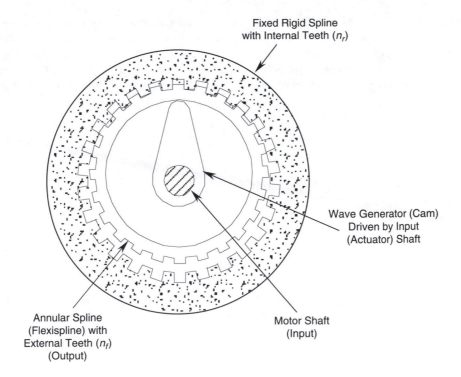

FIGURE 3.5

The principle of operation of a harmonic drive.

much lighter than a standard gearbox. The harmonic drive is often integrated with conventional motors to provide very high torques, particularly in direct-drive and servo applications. The principle of operation of a harmonic drive is shown in Figure 3.5. The rigid circular spline of the drive is the outer gear and it has internal teeth. An annular flexispline has external teeth that can mesh with the internal teeth of the rigid spline in a limited region when pressed in the radial direction. The external radius of the flexispline is slightly smaller than the internal radius of the rigid spline. As its name implies, the flexispline undergoes some elastic deformation during the meshing process. This results in a tight mesh without any clearance between meshed teeth, and hence the motion is backlash free.

In the design shown in Figure 3.5, the rigid spline is fixed and may also serve as the housing of the harmonic drive. The rotation of the flexispline is the output of the drive; hence, it is connected to the driven load. The input shaft (motor shaft) drives the wave generator (represented by a cam in Figure 3.5). The wave generator motion brings about controlled backlash-free meshing between the rigid spline and the flexispline.

Suppose that

n_r = number of teeth (internal) in the rigid spline
n_f = number of teeth (external) in the flexispline

It follows that

$$\text{tooth pitch of the rigid spline} = \frac{2\pi}{n_r} \text{ radians}$$

$$\text{tooth pitch of the flexispline} = \frac{2\pi}{n_f} \text{ radians.}$$

Further, suppose that n_r is slightly smaller than n_f. Then, during a single tooth engagement, the flexispline rotates through $(2\pi/n_r - 2\pi/n_f)$ radians in the direction of rotation of the wave generator. During one full rotation of the wave generator, there will be a total of n_r tooth engagements in the rigid spline (which is stationary in this design). Hence, the rotation of the flexispline during one rotation of the wave generator (around the rigid spline) is

$$n_r\left(\frac{2\pi}{n_r} - \frac{2\pi}{n_f}\right) = \frac{2\pi}{n_f}(n_f - n_r)$$

It follows that the gear reduction ratio ($r : 1$) representing the ratio: input speed/output speed, is given by

$$r = \frac{n_f}{n_f - n_r} \tag{3.8a}$$

We can see that by making n_r very close to n_f, very high gear reductions can be obtained. Furthermore, since the efficiency of a harmonic drive is given by

$$\text{efficiency } e = \frac{\text{output power}}{\text{input power}} \tag{3.9}$$

we have

$$\text{output torque} = \frac{e n_f}{(n_f - n_r)} \times \text{input torque} \tag{3.10}$$

This result illustrates the torque amplification capability of a harmonic drive.

An inherent shortcoming of the harmonic drive sketched in Figure 3.5 is that the motion of the output device (flexispline) is eccentric (or epicyclic). This problem is not serious when the eccentricity is small (which is the case for typical harmonic drives) and is further reduced because of the flexibility of the flexispline. For improved performance, however, this epicyclic rotation has to be reconverted into a concentric rotation. This may be accomplished by various means, including flexible coupling and pin-slot transmissions. The output device of a pin-slot transmission is a flange that has pins arranged on the circumference of a circle centered at the axis of the output shaft. The input to the pin-slot transmission is the flexispline motion, which is transmitted through a set of holes on the flexispline. The pin diameter is smaller than the hole diameter, the associated clearance being adequate to take up the eccentricity in the flexispline motion. This principle is shown schematically in Figure 3.6. Alternatively, pins could be attached to the flexispline and the slots on the output flange. The eccentricity problem can be eliminated altogether by using a double-ended cam in place of the single-ended cam wave generator shown in Figure 3.5. With this new arrangement, meshing takes place at two diametrical ends simultaneously, and the flexispline is deformed elliptically in doing this. The center of rotation of the flexispline now coincides with the center of the input shaft. This double-mesh design is more robust and is quite common in industrial harmonic drives.

Other designs of harmonic drive are possible. For example, if $n_f < n_r$ then r in Equation 3.8 will be negative and the flexipline will rotate in the opposite direction to the wave generator

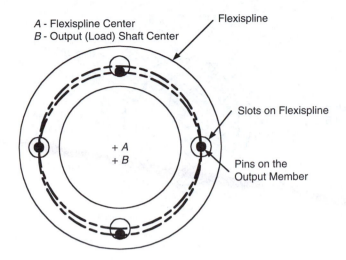

A - Flexispline Center
B - Output (Load) Shaft Center

Flexispline

Slots on Flexispline

+ A
+ B

Pins on the
Output Member

FIGURE 3.6
The principle of a pin-slot transmission.

(input shaft). Also, as indicated in the example below, the flexipline may be fixed and the rigid spline may serve as the output (rotating) member.

Traction drives (or friction drives) employ frictional coupling to eliminate backlash and overloading problems. These are not harmonic drives. In a traction drive, the drive member (input roller) is frictionally engaged with the driven member (output roller). The disadvantages of traction drives include indeterminacy of the speed ratio under slipping (overload) conditions and large size and weight for a specified speed ratio.

Example 3.2

An alternative design of a harmonic drive is sketched in Figure 3.7(a). In this design the flexipline is fixed. It loosely fits inside the rigid spline and is pressed against the internal teeth of the rigid spline at diametrically opposite locations. Tooth meshing occurs at these two locations only. The rigid spline is the output member of the harmonic drive (see Figure 3.7b).

1. Show that the speed reduction ratio is given by

$$r = \frac{\omega_i}{\omega_f} = \frac{n_r}{(n_r - n_f)} \qquad (3.8b)$$

   Note that if $n_f > n_r$ the output shaft will rotate in the opposite direction to the input shaft.

2. Now consider the free-body diagram shown in Figure 3.7(c). The axial moment of inertia of the rigid spline is J. Neglecting the inertia of the wave generator, write approximate equations for the system. The variables shown in Figure 3.7(c) are defined as:

   T_i = torque applied on the harmonic drive by the input shaft
   T_o = torque transmitted to the driven load by the output shaft (rigid spline)
   T_f = torque transmitted by the flexispline to the rigid spline
   T_r = reaction torque on the flexispline at the fixture
   T_w = torque transmitted by the wave generator

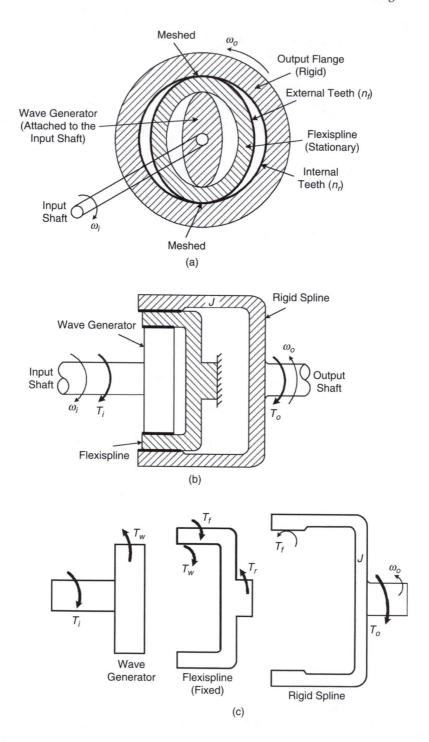

FIGURE 3.7

(a) An alternative design of harmonic drive, (b) Torque and speed transmission of the harmonic drive, (c) Free-body diagrams.

SOLUTION

Part 1:

Suppose that n_r is slightly larger than n_f. Then, during a single tooth engagement, the rigid spline rotates through $(2\pi/n_f - 2\pi/n_r)$ radians in the direction of rotation of the wave generator. During one full rotation of the wave generator, there will be a total of n_f tooth engagements in the flexispline (which is stationary in the present design). Hence, the rotation of the rigid spline during one rotation of the wave generator (around the flexispline) is

$$n_f \left(\frac{2\pi}{n_f} - \frac{2\pi}{n_r} \right) = \frac{2\pi}{n_r} (n_r - n_f)$$

It follows that the gear reduction ratio ($r : 1$) representing the ratio: input speed/output speed, is given by

$$r = \frac{n_r}{n_r - n_f} \tag{3.8c}$$

It should be clear that if $n_f > n_r$, the output shaft will rotate in the opposite direction to the input shaft.

Part 2:

Equations of motion for the three components are as follows:

1. Wave Generator

     Here, since inertia is neglected, we have

$$T_i - T_w = 0 \tag{3.11a}$$

2. Flexispline

     Here, since the component is fixed, the equilibrium condition is

$$T_w + T_f - T_r = 0 \tag{3.11b}$$

3. Rigid spline

     Newton's second law gives,

$$T_f - T_o = J \frac{d\omega_o}{dt} \tag{3.11c}$$

3.2.3 Continuously-Variable Transmission

A continuously-variable transmission (CVT) is one whose gear ratio (speed ratio) can be changed continuously—that is, infinitesimal increments or infinitesimal resolution—over its design range. Because of perceived practical advantages of a CVT over a conventional fixed-gear-ratio transmission, there has been significant interest in the development of a

CVT that can be particularly competitive in automotive applications. For example, in the *Van Doorne belt*, a belt-and-pulley arrangement is used and the speed ratio is varied by adjusting the effective diameter of the pulleys in a continuous manner. The mechanism that changes the pulley diameter is not straightforward. Further, belt life and geometry are practical limitations.

An early automotive application of a CVT used the *friction-drive* principle. This used a pair of friction disks, with one rolling on the face of the other. By changing the relative position of the disks, the output speed can be changed for a constant input speed. All friction drives have the advantage of overload protection, but the performance will depend on the frictional properties of the disks, and will deteriorate with age. Thermal problems, power loss, and component wear can be significant. Also, the range of speed ratios will depend on the disk dimension, which can be a limiting factor in applications with geometric constraints.

The *infinitely-variable transmission* (IVT), developed by Epilogies Inc. (Los Gatos, CA), is different in principle to the other types of CVTs mentioned. The IVT achieves the variation of speed ratio by first converting the input rotation to a reciprocating motion using a planetary assembly of several components (a planetary plate, four epicyclic shafts with crank arms, an overrunning clutch called a mechanical diode, etc.), then adjusting the effective output speed by varying the offset of an index plate with respect to the input shaft, recovering the effective rotation of the output shaft through a differential-gear assembly. One obvious disadvantage of this design is the large number of components and moving parts that are needed.

Now we will describe an innovative design of a continuously-variable transmission that has many advantages over existing CVTs. In particular, this CVT uses simple and conventional components such as racks and a pinion, and is easy to manufacture and operate. It has few moving parts and, as a result, has high mechanical efficiency and needs less maintenance than conventional designs.

3.2.3.1   *Principle of Operation*

Consider the rack-and-pinion arrangement shown in Figure 3.8(a). The pinion (radius r) rotates at an angular speed (ω) about a fixed axis (P). If the rack is not constrained in some manner, its kinematics will be indeterminate. For example, as in a conventional drive arrangement, if the direction of the rack is fixed, it will move at a rectilinear speed of ωr with zero angular speed. Instead, suppose that the rack is placed in a housing and is only allowed a rectilinear (sliding) lateral movement relative to the housing, and that the housing itself is "free" to rotate about an axis parallel to the pinion axis, at O. Let the offset between the two axes (OP) be denoted by e.

It should be clear that if the pinion is turned, the housing (along with the rack) will also turn. Suppose that the resulting angular speed of the housing (and the rack) is Ω. Let us determine an expression for Ω in terms of ω. The rack must move at rectilinear speed v relative to the housing. The operation of the CVT is governed by the kinematic arrangement of Figure 3.8, with ω as the input speed, Ω as the output speed, and offset e as the parameter that is varied to achieve the variable speed ratio. Note that perfect meshing between the rack and the pinion is assumed and backlash is neglected. Dimensions such as r are given with regard to the pitch line of the rack and the pitch circle of the pinion.

Suppose that Figure 3.8(a) represents the reference configuration of the kinematic system. Now consider a general configuration as shown in Figure 3.8(b). Here the output shaft has rotated through angle θ from the reference configuration. Note that this rotation is equal to the rotation of the housing (with which the racks rotate). Hence the angle θ

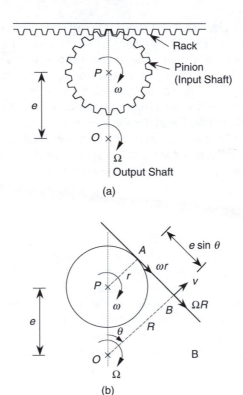

FIGURE 3.8
The kinematic configuration of the pinion and a meshed rack. (a) Reference configuration, (b) A general configuration.

can also be represented by the rotation of the line drawn perpendicular to a rack from the center of rotation O of the output shaft, as shown in Figure 3.8(b). This line intersects the rack at point B, which is the middle point of the rack. Point A is a general point of meshing. Note that A and B coincide in the reference configuration (Figure 3.8(a)). The velocity of point B has two components—the component perpendicular to AB and the component along AB. Since the rack (with its housing) rotates about O at angular speed Ω, the component of velocity of B along AB is ΩR. This component has to be equal to the velocity of A along AB, because the rack (AB) is rigid and does not stretch. The latter velocity is given by ωr. It follows that:

$$\omega r = \Omega R \tag{3.12}$$

From geometry (see Figure 3.8(b)),

$$R = r + e\cos\theta \tag{3.13}$$

By substituting Equation 3.13 in Equation 3.12, we get the speed ratio (p) of the transmission as

$$p = \frac{\omega}{\Omega} = 1 + \frac{e}{r}\cos\theta \tag{3.14}$$

From Equation 3.14 it is clear that the kinematic arrangement shown in Figure 3.8 can serve as a gear transmission. It is also obvious, however, that if only one rack is made to continuously mesh around the pinion, the speed ratio p will simply vary sinusoidally about an average value of unity. This, then, will not be a very useful arrangement for a CVT. If, instead, the angle of mesh is limited to a fraction of the cycle, say from $\theta = -\pi/4$ to $+\pi/4$, and at the end of this duration another rack is engaged with the pinion to repeat the same motion while the first rack is moved around a cam without meshing with the pinion, then the speed reduction p can be maintained at an average value greater than unity. Furthermore, with such a system the speed ratio can be continuously changed by varying the offset parameter e. This is the basis of the two-slider CVT.

3.2.3.2 Two-Slider CVT

A graphic representation of a CVT that operates according to the kinematic principles described above is shown in Figure 3.9, a two-slider arrangement (U.S. Patent No. 4,800,768). Specifically, each slider unit consists of two parallel racks. The spacing of the racks (w) is greater than the diameter of the pinion. The meshing of a rack with the pinion is maintained by means of a suitably profiled cam, as shown. The two slider units are placed orthogonally. It follows that each rack engages with the pinion at $\theta = -\pi/4$ and disengages at $\theta = +\pi/4$, according to the nomenclature given in Figure 3.8.

We note from Equation 3.14 that the speed ratio fluctuates periodically over periods of $\pi/2$ of the output-shaft rotation. For example, Figure 3.10 shows the variation of the output

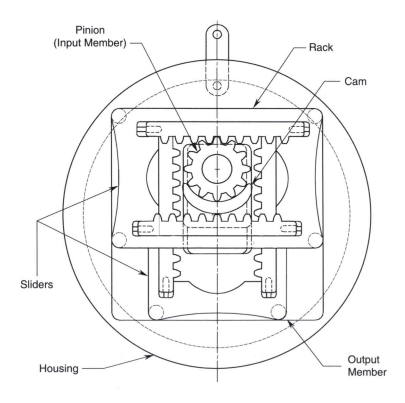

FIGURE 3.9
A drawing of a two-slider CVT.

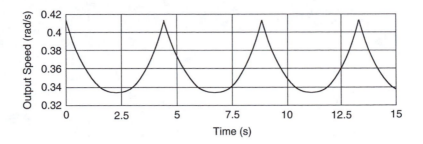

FIGURE 3.10
The response of the two-slider CVT for an input speed of 1.0 rad/s. Offset ratio $e/r = 2.0$.

speed of the transmission for a constant input speed of 1.0 rad/s and an offset ratio of $e/r = 2.0$. It can be easily verified that the average speed ratio p is given by:

$$\bar{p} = 1 + \frac{2\sqrt{2}}{\pi}\frac{e}{r} \tag{3.15}$$

Note that $2\sqrt{2}/\pi \approx 9$. Also, the maximum value of speed ratio p occurs at $\theta = 0$ and the minimum value of p occurs at $\theta = \pm\,\pi/4$.

In summary we can make the following observations regarding the present design of the CVT:

1. Speed ratio p (Input shaft speed/Output shaft speed) is not constant and changes with the shaft rotation.
2. The minimum speed ratio (p_{min}) occurs at the engaging and disengaging instants of a rack. The maximum speed ratio (p_{max}) occurs at halfway between these two points.
3. The maximum deviation from the average speed ratio is approximately 0.2 e/r and this occurs at the engaging and disengaging points.
4. Speed ratio increases linearly with e/r and hence the speed ratio of the transmission can be adjusted by changing the shaft-to-shaft offset e.
5. The larger the speed ratio the larger the deviation from the average value (see items 3 and 4 above).

It has been indicated that the speed ratio of the transmission depends linearly on the offset ratio (the offset between the output shaft and the input pinion/pinion radius). Figure 3.11 shows the variation of the average speed ratio p with the offset ratio. Note that a continuous variation of the speed reduction in a range of more than 1– 7 can be achieved by continuously varying the offset ratio e/r from 0–7.

3.2.3.3 Three-Slider CVT

A three-slider, continuously-variable transmission has been designed by us with the objective of reducing the fluctuations in the output speed and torque (Figure 3.12). The three-slider system consists of three rectangular pairs of racks (instead of two pairs), which slide along their slotted guideways, similar to the two-slider system. The main difference in the three-slider system is that each rack engages with the pinion for only 60° in a cycle

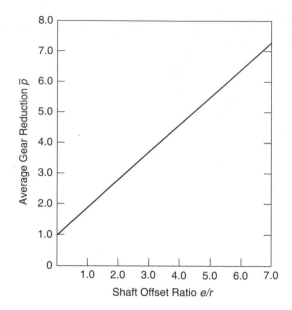

FIGURE 3.11
Average gear reduction curve for the two-slider CVT.

FIGURE 3.12
A three-slider CVT.

of 360°. Hence, the fluctuating (sinusoidal) component of the speed ratio varies over an angle of 60°, in comparison to a 90° angle in the two-slider CVT. As a result, the fluctuations of the speed ratio will be less in the three-slider CVT. The six racks will engage and disengage sequentially during transmission. The cam profile of the three-slider system will be different from that of the two-slider system as well.

The speed reduction ratio of the three-slider CVT (for θ between $-\pi/6$ and $\pi/6$) is given by:

$$p = \frac{\omega}{\Omega} = 1 + \frac{e}{r}\cos\theta \tag{3.16}$$

If we neglect inertia, elastic effects, and power dissipation (friction), the torque ratio of the transmission is given by the same equation. An advantage of the CVT is its ability to continuously change the torque ratio according to output torque requirements and input torque (source) conditions. An obvious disadvantage in high-precision applications is the fluctuation in speed and torque ratios. This is not crucial in moderate-to-low-precision applications such as bicycles, golf carts, snowmobiles, hydraulic cement mixers and generators. As a comparison, the percentage speed fluctuation of the two-slider CVT at an offset ratio of 6.0 (average speed ratio of approximately 6.5) is 18%, whereas for the three-slider CVT it is less than 8%.

3.3   Robotic Manipulators

A robot is a mechanical manipulator which can be programmed to perform various physical tasks. Robots have been demonstrated to play soccer, operate switches, turn doorknobs, and climb stairs, in addition to performing such industrial tasks as assembly of machine parts, welding and spray-painting of automobile bodies, and inspection of products. A properly designed robot is truly a mechatronic system. Programmability and the associated flexibility of carrying out tasks are necessary characteristics for a robot, according to this commonly used definition. Furthermore, a robotic task might be complex to the extent that some degree of intelligence would be required for satisfactory performance of the task. There is an increasing awareness of this and there have been calls to include intelligence, which would encompass abilities to perceive, reason, learn and infer from incomplete information, as a requirement in characterizing a robot.

Productivity and product quality of an automated manufacturing process rely on the accuracy of the individual manufacturing tasks such as parts transfer, assembly, welding, and inspection. In modern manufacturing workcells many of these tasks are carried out by robotic manipulators. The performance of a robotic manipulator depends considerably on the way the manipulator is controlled, and this has a direct impact on the overall performance of the manufacturing system. In this context, a robot can be interpreted as a control system. Its basic functional components are the structural skeleton of the robot; the actuator system which drives the robot; the sensor system which measures signals for performance monitoring, task learning and playback, and for control; the signal modification system for functions such as signal conversion, filtering, amplification, modulation, and demodulation; and the direct digital controller which generates drive signals for the actuator system so as to reduce response error. Higher level tasks such as path planning, activity coordination and supervisory control have to be treated as well within the overall control system.

3.3.1   Robot Classification

The physical structure of a robot may have anthropomorphic features, but this is a rather narrow perception. There is a particularly useful classification of industrial robots that is based on their kinematic structure. For example, consider the classification shown in Figure 3.13. Six degrees of freedom are required for a robot to arbitrarily position and

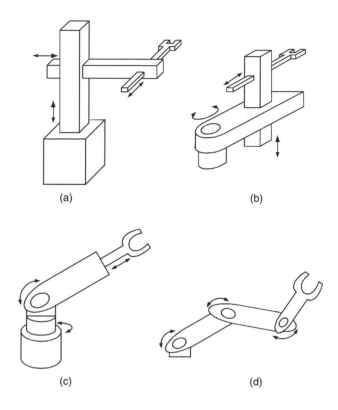

FIGURE 3.13
A kinematic classification for robotic manipulators. (a) Rectangular or Cartesian, (b) Cylindrical polar (R-P-P), (c) Spherical polar (R-R-P), (d) Jointed spherical or articulated (3R).

orient an object in the three-dimensional space. It is customary to assign three of these degrees of freedom to the wrist that manipulates the end effector (hand), and the remaining three to the arm of the robot. Since kinematic decoupling is desired for analytical simplicity, spherical wrists having three revolute (R) degrees of freedom with axes of motion coinciding at a single point (at the wrist) are commonly employed. Having decided on this configuration for the wrist, the kinematic structure of the arm can then be used as a basis for robot classification.

Specifically, the sequence of rotatory or revolute (R) joints and rectilinear or prismatic (P) joints employed in the arm structure will classify a robot. Four common classifications are shown in Figure 3.13: rectangular or Cartesian (3P), cylindrical (R-P-P or P-R-P), spherical or polar (R-R-P), and articulated or jointed spherical (3R). Furthermore, selective compliant assembly robot arm (SCARA) configurations, where at least the first two of the three degrees of arm freedom do not face gravity (i.e., they employ vertical revolute axes or horizontal prismatic axes), are desired so that the actuators of the most demanding joints are not subjected to gravity loads.

Other classifications are possible as well. For example, robots may be classified according to the actuator type (e.g., hydraulic, dc servo, ac servo, stepper motor), by the transmission type (e.g., geared, direct-drive, harmonic-drive, timing-belt, chain and sprocket, tendoned, and traction-drive or friction-drive), by capacity and accuracy (e.g., heavy-duty industrial robots and microminiaturized finger robots), and by mobility (e.g., mobile robots and AGVs or automated guided vehicles).

Robotic tasks can be grouped broadly into (1) gross manipulation tasks and (2) fine manipulation tasks. Control of the motion trajectory of the robot end effector is directly applicable to tasks in the first category. Examples of such tasks are seam tracking in arc welding, spray painting, contour cutting (e.g., laser and water jet) and joining (e.g., gluing, sewing, ultrasonic and laser merging), and contour inspection (e.g., ultrasonic, electromagnetic, and optical). Force and tactile considerations are generally crucial to tasks in the second category. Part assembly, robotic surgery, machining, forging, and engraving are examples of fine manipulation tasks. It is intuitively clear that gross manipulation can be accomplished through motion control. But force control (including compliance control) also would be needed for accurate fine manipulation, particularly because small motion errors can result in excessive and damaging forces in this class of tasks.

For predefined gross-manipulation tasks, a robot is usually taught the desired trajectory either by using a mechanical input device such as a teaching pendant or joystick, or by off-line programming. Precise path planning and continuous path generation are essential in trajectory tracking applications. For tasks such as pick-and-place operations where the end positions (and orientations) are of primary interest, point-to-point interpolation may be employed. Trajectory segmentation and segmental interpolation also are commonly used in continuous trajectory control. Once an end effector trajectory is specified, the desired joint trajectories may be determined by direct measurement using joint sensors during the teaching (learning) mode of operation, or alternatively by offline computation using kinematic relations for the particular robot. During the task-repeat (playback) mode of operation, the desired joint trajectories are compared with the measured joint trajectories, and the associated joint error values are used by the manipulator servos for compensation.

Fine manipulation control, which incorporates force and tactile information, is generally more complex. Dexterity comes into play quite prominently and conventional control techniques have to be augmented by more sophisticated control approaches such as hybrid force/position control, active compliance control and impedance control. Except in academic and research environments, a robot user is normally buffered from the intricate and complex programming activities that are needed to implement various control strategies. A typical user would program a robot through an appropriate high-level programming language, using simple English-like commands.

3.3.2 Robot Kinematics

It is important to know the position and orientation (geometric configuration) of a robot, along with velocities and accelerations of the robot components (links) in order to monitor and properly control the robot. Determination of these geometric configuration parameters and their derivatives is the kinematics problem of a robot. Coordinate transformation plays an important role in this problem. Now we will address proper determination and representation of robot kinematics.

Each degree of freedom of a robotic manipulator has an associated joint coordinate q_i. The robot is actuated by driving its joints, but a robotic task is normally specified in terms of end effector motions. The end effector of a robot can be represented by a Cartesian coordinate frame fixed to it (a body frame). The frame can be represented as a coordinate transformation with respect to some inertial frame (the *world coordinate frame*), typically a frame fixed to the stationary base of the robot (the *base frame*). The basic kinematics problem in modeling a robot is the expression of this coordinate transformation in terms of the joint coordinates q_i.

3.3.2.1 Homogeneous Transformation

Consider the Cartesian frame (x_0, y_0, z_0) shown in Figure 3.14. If this frame is rotated about the z_0 axis through an angle θ_1, we get the Cartesian frame (x_1, y_1, z_1) as shown. The coordinate transformation associated with this frame rotation may be represented by the

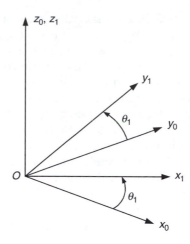

FIGURE 3.14
A coordinate transformation.

transformation matrix:

$$R_1 = \begin{bmatrix} \cos\theta_1 & -\sin\theta_1 & 0 \\ \sin\theta_1 & \cos\theta_1 & 0 \\ 0 & 0 & 1 \end{bmatrix} \tag{3.17}$$

Note that the positive direction of rotation is determined by the right-handed corkscrew rule.

We can make several important observations concerning R_1. The columns of this matrix give the *direction cosines* of the axes of the new frame, expressed in the old frame. Consider any arbitrary vector r whose components are expressed in the new frame. Then, if we premultiply r by R_1, we get the components of the same vector expressed in the old frame.

Note that the same observations hold if the frame rotation were to be made about any arbitrary axis through the origin, not just z_0. In summary, we can make the following general statements:

1. A coordinate transformation R represents a rotation of a coordinate frame to a new position.

2. The columns of R give the direction cosines of the new frame axes expressed in the old frame.

3. Premultiplication of a vector r by R is equivalent to fixing this vector r in the old frame and rotating the entire unit to the new frame position.

4. The product Rr gives the components of the rotated vector r expressed in the old frame, r itself giving the components of the rotated vector in the new frame.

Suppose that the new frame (x_1, y_1, z_1) is next rotated to another position represented by the Cartesian frame (x_2, y_2, z_2). The columns of the corresponding transformation matrix R_2 give the direction cosines of the axes of (x_2, y_2, z_2) when expressed in the frame (x_1, y_1, z_1). It follows that the matrix product $R_1 R_2$ gives the direction cosines of the axes of (x_2, y_2, z_2) expressed in the original frame (x_0, y_0, z_0). If the second transformation R_2 represents a rotation expressed in the frame (x_0, y_0, z_0), not frame (x_1, y_1, z_1), then the direction cosines of the resulting frame expressed in the original frame (x_0, y_0, z_0) are given by the columns of the product $R_2 R_1$. These ideas can be extended to a product of more than two transformation matrices.

The foregoing discussion considered rotation about an axis through the origin of a coordinate frame. Now let us consider pure *translations* (i.e., displacements without any rotations). Consider a vector r that has three components expressed in a Cartesian frame as in Figure 3.14. Let us augment this vector with a unity element, to form the fourth-order column vector r_a, given by:

$$r_a = \begin{bmatrix} r \\ 1 \end{bmatrix} \tag{3.18}$$

Now consider a 4×4 matrix T given by:

$$T = \begin{bmatrix} 1 & p \\ 0 & 1 \end{bmatrix} \tag{3.19}$$

in which 1 denotes the 3×3 *identity matrix* and p is a vector expressed in the original Cartesian frame, representing a pure translation. It is easy to verify that the product Tr_a is given by:

$$Tr_a = \begin{bmatrix} r + p \\ 1 \end{bmatrix} \tag{3.20}$$

It follows that the matrix T can be considered as a transformation matrix which represents a pure translation. Since this is a 4×4 matrix, in order to combine rotations and translations into a single transformation, we must first convert the 3×3 rotation matrix R into an equivalent 4×4 matrix. Since vectors are augmented by a unity element in this approach, it is easily seen that the corresponding 4×4 rotation matrix is:

$$R_a = \begin{bmatrix} R & 0 \\ 0 & 1 \end{bmatrix} \tag{3.21}$$

where 0 denotes a null column or row of compatible order. Now suppose that we translate a frame through vector p and then rotate the resulting frame about an axis through the origin of this new frame according to R. The overall transformation A is given by:

$$A = TR_a \tag{3.22}$$

By direct matrix multiplication we get:

$$A = \begin{bmatrix} R & p \\ 0 & 1 \end{bmatrix} \tag{3.23}$$

There is no rotation from the original frame to the intermediate frame. Hence, the direction cosines of the axes of the final frame, expressed in either the intermediate frame or the original frame, are given by the columns of R. It follows that the transformation matrix

A contains all the information about the final frame, expressed in terms of the original frame. Specifically, *p* gives the position of the frame origin and *R* gives the orientation of the frame. Matrix *A* is a unified or "homogenized" representation of translations and rotations of a coordinate frame. For that reason *A* is known as a 4 × 4 *"homogeneous transformation matrix."*

As a matter of interest, suppose that we first rotate the frame and then translate the resulting frame through *p* (of course, expressed in the intermediate frame with respect to which the translation is made). Then the overall homogeneous transformation matrix becomes:

$$R_a T = \begin{bmatrix} R & Rp \\ 0 & 1 \end{bmatrix}$$

Indeed, this result is compatible with Equation 3.23 because *Rp* is the translation expressed in the original coordinate frame.

3.3.2.2 Denavit-Hartenberg Notation

To formulate robot kinematics, we wish to present a homogeneous transformation matrix representing a general coordinate transformation from one link of a robot to an adjacent link. For this purpose, body frames, fixed to links of the robot, are chosen according to the Denavit-Hartenberg notation. This is explained in Figure 3.15. Note that joint *i* joins link *i*–1 with link *i*. Frame *i*, which is the body frame of link *i*, has its z axis located at joint *i* + 1. If the joint is *revolute*, then the joint rotation is about the z axis. If the joint is *prismatic*, the joint translation is along the z axis. It is seen from Figure 3.15 that frame *i* can be obtained by transforming frame *i* – 1 as follows:

1. Rotate frame *i* – 1 through θ_i about the z axis.
2. Translate the new frame through d_i along the z axis.
3. Translate the new frame through a_i along the new x axis.
4. Rotate the new frame through α_i about the current x axis.

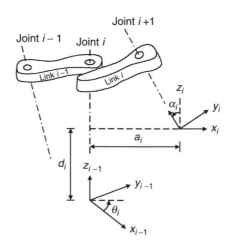

FIGURE 3.15
The Denavit-Hartenberg notation.

Note that all these movements are carried out in the positive sense of a right-handed Cartesian frame. The homogeneous transformation matrix A_i which represents the overall link-to-link transformation is obtained by the product of the four 4×4 transformation matrices corresponding to the above four steps, when taken in the proper order (i.e., $1 \times 2 \times 3 \times 4$). It can be easily verified that this matrix is given by:

$$A_i = \begin{bmatrix} \cos\theta_i & -\sin\theta_i\cos\alpha_i & \sin\theta_i\sin\alpha_i & a_i\cos\theta_i \\ \sin\theta_i & \cos\theta_i\sin\alpha_i & -\cos\theta_i\sin\alpha_i & a_i\sin\theta_i \\ 0 & \sin\alpha_i & \cos\alpha_i & d_i \\ 0 & 0 & 0 & 1 \end{bmatrix} \tag{3.25}$$

For a revolute joint, the joint coordinate would be:

$$q_i = \theta_i \tag{3.26}$$

and, for a prismatic joint, the joint coordinate would be:

$$q_i = a_i \tag{3.27}$$

with the remaining parameters in A_i kept constant. Hence, the only variable in A_i is q_i.

The base frame, frame 0, is assumed fixed. This is taken as the inertial frame with respect to which a robotic task is specified. For an n degree-of-freedom robot, the body frame of the end effector is frame n, and this frame moves with the end effector. It follows that the position and orientation of the end effector frame, expressed in the base frame, is given by the columns of the overall homogeneous transformation matrix T

$$T = A_1(q_1)A_2(q_2)\cdots A_n(q_n) \tag{3.28}$$

Equation 3.28 represents the kinematic formulation for a robotic manipulator.

Example 3.3

Consider the two-degree-of-freedom, revolute manipulator sketched in Figure P3.28. Suppose that a body frame for the end effector may be defined using the following four transformations, starting from the base frame (x, y, z) that is shown in the figure:

Step 1. Rotate the base frame about the z axis through an angle q_1.

Step 2. Move the new frame along the new x axis through a distance l_1.

Step 3. Rotate the resulting frame about the z axis through an angle q_2.

Step 4. Move the latest frame along the latest x axis through a distance l_2.

a. Give the 4×4 homogeneous transformations corresponding to each of the steps 1 through 4 above.

b. Multiply the transformations in Part (a) in the proper order to describe the kinematics of the manipulator (i.e., to express the end effector frame with respect to the base frame).

c. From Part (b) obtain the coordinates of the origin of the end effector frame, with respect to, and expressed in, the base frame.

SOLUTION

a. It is seen that the transformation matrices are:

$$A_1 = \begin{bmatrix} \cos q_1 & -\sin q_1 & 0 & 0 \\ \sin q_1 & \cos q_1 & 0 & 0 \\ 0 & 0 & 1 & 0 \\ 0 & 0 & 0 & 1 \end{bmatrix}$$

$$A_2 = \begin{bmatrix} 1 & 0 & 0 & l_1 \\ 0 & 1 & 0 & 0 \\ 0 & 0 & 1 & 0 \\ 0 & 0 & 0 & 1 \end{bmatrix}$$

$$A_3 = \begin{bmatrix} \cos q_2 & -\sin q_2 & 0 & 0 \\ \sin q_2 & \cos q_2 & 0 & 0 \\ 0 & 0 & 1 & 0 \\ 0 & 0 & 0 & 1 \end{bmatrix}$$

$$A_4 = \begin{bmatrix} 1 & 0 & 0 & l_2 \\ 0 & 1 & 0 & 0 \\ 0 & 0 & 1 & 0 \\ 0 & 0 & 0 & 1 \end{bmatrix}$$

b. The overall transformation matrix is

$$A = A_1 A_2 A_3 A_4$$

On multiplication and simplification we get:

$$A = \begin{bmatrix} \cos(q_1+q_2) & -\sin(q_1+q_2) & 0 & l_2\cos(q_1+q_2)+l_1\cos q_1 \\ \sin(q_1+q_2) & \cos(q_1+q_2) & 0 & l_2\sin(q_1+q_2)+l_1\sin q_1 \\ 0 & 0 & 1 & 0 \\ 0 & 0 & 0 & 1 \end{bmatrix}$$

c. The first three elements of the last (4th) column of A, as obtained in Part (c), give the origin of the end effector frame.

NOTE We have used the following trigonometric identities:

$$\cos(q_1 + q_2) = \cos q_1 \cos q_2 - \sin q_1 \sin q_2$$

$$\sin(q_1 + q_2) = \sin q_1 \cos q_2 + \cos q_1 \sin q_2$$

3.3.2.3 Inverse Kinematics

Typically, a robotic task is specified in terms of the *T* matrix in Equation 3.28. Since the drive variables are the joint variables, for the purposes of actuating and controlling a robot, it is necessary to solve Equation 3.28 and determine the joint motion vector *q* corresponding to a specified *T*. This is the *inverse-kinematics* problem associated with a robot.

Since six coordinates are needed to specify a rigid body (or a body frame) in the three-dimensional space, *T* is specified using six independent quantities (typically three position coordinates and three angles of rotation). It follows that Equation 3.28, in general, represents a set of six algebraic equations. These equations contain highly nonlinear trigonometric functions (of coordinate transformations) and are coupled. Hence a simple and unique solution for the joint coordinate vector *q* might not exist even in the absence of redundant kinematics (Note: If $n = 6$, the robot does not have redundant kinematics in the 3-D space). Some simplification is possible by proper design of robot geometry. For example, by using a spherical wrist so that three of the six degrees of freedom are provided by three revolute joints whose axes coincide at the wrist of the end effector, it is possible to decouple the six equations in Equation 3.28 into two sets of three simpler equations. In general, however, one must resort to numerical approaches to obtain the inverse-kinematics solution. In the presence of redundant kinematics ($n > 6$), an infinite set of solutions would be possible for the inverse-kinematics problem. In this case, it is necessary to employ a useful set of constraints for joint motions, or minimize a suitable cost function, in order to obtain a unique solution.

3.3.2.4 Differential Kinematics

The Jacobian matrix *J* of a robot is given by the relation:

$$\delta r = J \delta q \tag{3.29}$$

where, in 3-D space, *r* is a sixth-order vector, of which the first three elements represent the end effector position (distance coordinates), and the remaining three elements represent the end effector orientation (angles). It is important to determine *J* and its inverse in the computation of joint velocities and accelerations. This is the basic problem in *differential kinematics* for a robotic manipulator.

Example 3.4

Consider the two-link manipulator that carries a point load (weight *W*) at the end effector, as shown in Figure P3.28. The link lengths are l_1 and l_2, and the corresponding joint angles are q_1 and q_2 as indicated.

a. Express the position coordinates *x* and *y* at the end effector in terms of l_1, l_2, q_1, and q_2.

b. The Jacobian matrix *J* of this manipulator is given by the expression:

$$\begin{bmatrix} \delta x \\ \delta y \end{bmatrix} = J \begin{bmatrix} \delta q_1 \\ \delta q_2 \end{bmatrix}$$

where

$$J = \begin{bmatrix} \dfrac{\partial x}{\partial q_1} & \dfrac{\partial x}{\partial q_2} \\ \dfrac{\partial y}{\partial q_1} & \dfrac{\partial y}{\partial q_2} \end{bmatrix}$$ (3.30)

Obtain an expression for J in terms of l_1, l_2, q_1 and q_2.

c. Express the end effector velocity vector v in terms of the joint velocity vector $\dot{q}$.

SOLUTION

   a. From the geometry it is seen that

$$x = l_1 \cos q_1 + l_2 \cos(q_1 + q_2)$$

$$y = l_1 \sin q_1 + l_2 \sin(q_1 + q_2)$$

   b. Differentiate each of the above expressions separately with respect to q_1 and q_2. This gives the elements of J as follows:

$$J = \begin{bmatrix} -[l_1 \sin q_1 + l\sin(q_1 + q_2)] & -l_2 \sin(q_1 + q_2) \\ l_1 \cos q_1 + l_2 \cos(q_1 + q_2) & l_2 \cos(q_1 + q_2) \end{bmatrix}$$

   c. $V = J\dot{q}$ with J as given above.

Consider Figure 3.16, which uses the Denavit–Hartenberg notation. In particular k_{i-1} is a unit vector representing the axis of motion (rotation or translation) of joint i, expressed in the base frame. This is the z axis of the local frame (frame $i-1$). Vector r_{i-1} is the position

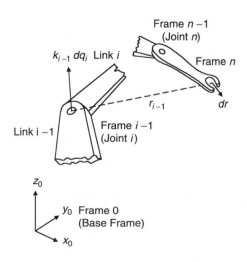

FIGURE 3.16
Representation of differential kinematics.

vector from joint i to the end effector frame, expressed in the base frame. Also, δr is an incremental motion at the end effector, again expressed in the base frame, caused by an incremental joint motion δq_i. It is easy to see that, if joint i is prismatic:

$$\delta r = \begin{bmatrix} k_{i-1} \\ 0 \end{bmatrix} \delta q_i \qquad (3.31)$$

and if the joint is revolute:

$$\delta r = \begin{bmatrix} k_{i-1} \times r_{i-1} \\ k_{i-1} \end{bmatrix} \delta q_i \qquad (3.32)$$

In view of Equation 3.29, the right-hand-side vectors of Equation 3.31 and Equation 3.32 give the ith column of the Jacobian matrix J, depending on whether the joint is prismatic or revolute. In this manner, the Jacobian matrix can be constructed for a robotic manipulator.

Example 3.5
Consider a vector

$$p = \begin{bmatrix} p_1 \\ p_2 \\ p_3 \end{bmatrix}$$

that is fixed to a coordinate frame (body frame). If the coordinate frame (and hence the vector p) is rotated through a small angle $\delta\theta$ about a unit vector

$$k = \begin{bmatrix} k_1/k \\ k_2/k \\ k_3/k \end{bmatrix}$$

what is the corresponding movement δp of the vector p expressed in the original coordinate frame?

SOLUTION
Perform the cross-product operation to obtain

$$\delta p = k \times p \delta\theta = \begin{bmatrix} k_2 p_3 - k_3 p_2 \\ k_3 p_1 - k_1 p_3 \\ k_1 p_2 - k_2 p_1 \end{bmatrix} \frac{\delta\theta}{k}$$

where

$$k = \sqrt{k_1^2 + k_2^2 + k_3^2}$$

If the manipulator does not contain redundant kinematics ($n = 6$), J would be a square matrix. Then the Jacobian can be inverted, provided that it is not *singular* at the particular orientation of the robot. In the case of a *redundant manipulator* ($n > 6$), however, additional constraints have to be introduced to joint motions in order to obtain a generalized inverse for J.

3.3.3   Robot Dynamics

Formulation of the equations of motion, or dynamics (or kinetics), of a robot is essential in the analysis, design, and control of a robot. In this section, the energy-based Lagrangian approach and the direct, Newton-Euler, vector mechanics approach are outlined for expressing the dynamics of a robot. The first approach is relatively more convenient to formulate and implement. But the tradeoff is that physical insight and part of the useful information (e.g., reaction forces at joints which are useful in computing friction and backlash) are lost in the process.

3.3.3.1   *Lagrangian Approach*

In the Lagrangian approach to the inverse-dynamics problem of a robot, first *kinetic energy* T and *potential energy* U are expressed in terms of joint motion variables q_i and $\dot{q}_i$. Next, the Lagrangian:

$$L = T - U \tag{3.33}$$

is formed and the *Lagrange's equations* of motion are written according to:

$$\frac{d}{dt}\frac{\partial L}{\partial \dot{q}_i} - \frac{\partial L}{\partial q_i} = f_i \quad i = 1, 2, \ldots, n \tag{3.34}$$

where f_i are the input forces/torques at the joints, the *generalized forces* in the Lagrange formulation. By adopting this approach, we can obtain the following set of equations for f_i:

$$f_i = \sum_{j=i}^{n}\left[\sum_{k=1}^{j}\left(tr\left(\frac{\partial T_j}{\partial q_i}J_j\frac{\partial T_j^T}{\partial q_k}\right)\ddot{q}_k + \sum_{p=1}^{j}tr\left(\frac{\partial T_j}{\partial q_i}J_j\frac{\partial^2 T_j^T}{\partial q_k\partial q_p}\right)\dot{q}_k\dot{q}_p\right) - m_j g^T\frac{\partial T_j}{\partial q_i}r_j\right] \quad i = 1, 2, \ldots, n \tag{3.35}$$

where T_j is the homogeneous transformation which gives the position and orientation of frame j, when expressed in the base frame; thus:

$$T_j = A_1(q_1)A_2(q_2) \ldots A_j(q_j) \tag{3.36}$$

as given before J_j is the moment of inertia matrix of link j expressed in the body frame j of the link, m_j is the mass of the link, r_j is the position vector of the centroid of the link j

expressed relative to frame j, and g is the gravity vector expressed in the base frame. Also, *tr* denotes the *trace* of a square matrix, the sum of the diagonal elements.

Note that Equation 3.35 represents a set of nonlinear and coupled differential equations which can be put into the form:

$$M(q)\ddot{q} = n(q, \dot{q}) + f(t) \tag{3.37}$$

where M is the inertia matrix of the robot and f is the vector of drive forces or torques. The vector n represents the nonlinear terms contributed by Coriolis and centrifugal accelerations and gravity.

Nonlinearities are present in terms of both q and $\dot{q}$. Nonlinearities in q are caused by the coordinate transformations that are used in the dynamic formulation. These are *trigonometric nonlinearities*, and they appear in the potential energy (gravity) term as well as in the kinetic energy (inertia) terms. Nonlinearities in $\dot{q}$ are quadratic functions, which arise from centrifugal and Coriolis acceleration components. These are *dynamic nonlinearities*. Also, note that each computation of f_i involves three summations over a range of up to n, and that there are n such computations. It follows that the direct computation of the joint force vector f represents an $O(n^4)$ computation. This high order in the joint force/torque computation is not acceptable in real-time control situations. A more efficient algorithm is needed.

3.3.3.2 Newton–Euler Formulation

For the sake of clarity, the manipulator is assumed to be an open-link chain, having revolute joints. The development can be extended to other types of manipulators in a straightforward manner. Consider the *i*th link of an *n*-link manipulator, as shown in Figure 3.17. The Newton-Euler equations for this link consist of the force-momentum equations:

$$f_{i-1} - f_i + mg = \frac{d}{dt}(m_i v_{ci}) \tag{3.38}$$

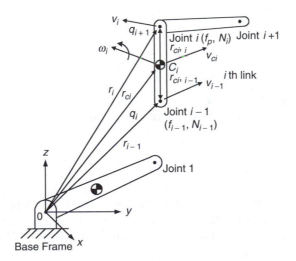

FIGURE 3.17
Link nomenclature for the Newton-Euler formulation.

and the moment-angular momentum equations about the centroid C_i of the link:

$$N_{i-1} - N_i + r_{ci,i-1} \times f_{i-1} - r_{ci,i} \times f_i = \frac{d}{dt}(I_{ci}\omega_i) \tag{3.39}$$

The following notation has been used:

f_{i-1} = force vector at the i–1th joint of the ith link.
N_{i-1} = moment (torque) vector at the i–1th joint of the ith link.
g = vector representing acceleration due to gravity.
m_i = mass of the ith link.
I_{ci} = moment of inertia matrix of the ith link about C_i.
v_{ci} = velocity of the centroid of the ith link.
ω_i = angular velocity vector of the ith link.
$r_{a,b}$ = position vector from point a to point b.

All vectors are expressed in the base frame. The forces and the moments at the ith joint of the ith link are $-f_i$ and $-N_i$, respectively, as dictated by Newton's third law (action is equal and opposite to reaction). This fact has been used in obtaining Equation 3.38 and Equation 3.39.

Next, by substituting:

$$r_{ci,i-1} = r_{i-1} - r_{ci}$$

$$r_{ci,i} = r_i - r_{ci}$$

in Equation 3.38, one obtains:

$$N_{i-1} - N_i + r_{i-1} \times f_{i-1} - r_i \times f_i - r_{ci} \times (f_{i-1} - f_i) = \frac{d}{dt}(I_{ci}\omega_i)$$

which, in view of Equation 3.37, becomes:

$$N_{i-1} - N_i + r_{i-1} \times f_{i-1} - r_i \times f_i - r_{ci} \times \left(\frac{d}{dt}(m_i v_{ci}) - m_i g \right) = \frac{d}{dt}(I_{ci}\omega_i) \tag{3.40}$$

It is clear that the inertia matrix I_{ci} is constant with respect to a body frame fixed to the ith link. Accordingly, since this body frame has an angular velocity ω_i, it is clear that:

$$\frac{d}{dt}(I_{ci}\omega_i) = I_{ci}\dot{\omega}_i + \omega_i \times (I_{ci}\omega_i) \tag{3.41}$$

The first term in Equation 3.41 represents the derivative with respect to the body frame, and the second term is a consequence of the fact that the body frame itself rotates at ω_i with respect to the inertial base frame. By substituting Equation 3.41 in Equation 3.40 one gets:

$$N_{i-1} - N_i + r_{i-1} \times f_{i-1} - R_i \times f_i - m_i r_{ci} \times (\dot{v}_{ci} - g) = I_{ci}\dot{\omega}_i + \omega_i \times (I_{ci}\omega_i) \tag{3.42}$$

For the n links of the manipulator there are n equations in Equation 3.38 and n equations in Equation 3.42. An equation for the base reaction force is obtained by summing the n equations given by Equation 3.38; thus:

$$f_o = \sum_{i=1}^{n} m_i (\dot{v}_{ci} - g) \tag{3.43}$$

Similarly, by summing the n Equations in Equation 3.42 and using the fact that $r_o = 0$ (see Figure 3.17), the equation for the base reaction moment is obtained; thus:

$$N_o = \sum_{i=1}^{n} [I_{ci} \dot{\omega}_i + \omega_i \times (I_{ci} \omega_i) + m_i r_{ci} \times (\dot{v}_{ci} - g)] \tag{3.44}$$

Equation 3.43 and Equation 3.44 can be expressed in terms of joint trajectories. Specifically, ω_i, $\dot{\omega}_i$, and $\dot{v}_{ci}$ can be expressed in terms of q and $\dot{q}$. It is easy to verify that the associated kinematic relations are the following:

Angular Velocities:

$$\omega_1 = \dot{q}_1$$
$$\omega_i = \omega_{i-1} + \dot{q}_i \tag{3.45}$$

Angular Accelerations:

$$\dot{\omega}_1 = \ddot{q}_1$$
$$\dot{\omega}_i = \dot{\omega}_{i-1} + \ddot{q}_i + \omega_{i-1} \times \dot{q}_i \tag{3.46}$$

Note that the scalars $\dot{q}_i$ and $\ddot{q}_i$ are joint angular velocities and joint angular accelerations, respectively, which are "relative" variables. They should be expressed as vectors $\dot{q}$ and $\ddot{q}$ in the base frame, in using Equation 3.45 and Equation 3.46.

Rectilinear Velocities:

$$v_o = 0$$
$$v_{ci} = v_{i-1} + \omega_i \times r_{i-1,ci} \tag{3.47}$$
$$v_i = v_{i-1} + \omega_i \times r_{i-1,i}$$

Rectilinear Accelerations:

$$a_o = 0$$
$$\dot{v}_{ci} = a_{i-1} + \dot{\omega}_i \times r_{i-1,ci} + \omega_i \times (\omega_i \times r_{i-1,ci}) \tag{3.48}$$
$$a_i = a_{i-1} + \dot{\omega}_i \times r_{i-1,i} + \omega_i \times (\omega_i \times r_{i-1,i})$$

Equation 3.43 through Equation 3.48 express the base reactions in terms of q, $\dot{q}$ and $\ddot{q}$. Note that q does not explicitly appear in these equations, but is present in the vectors r_{ci}, $r_{i-1,ci}$ and $r_{i-1,1}$ through the coordinate transformations that are necessary to express these vectors in the base frame.

3.3.4 Space-Station Robotics

Many mechanical tasks in the microgravity environment on space vehicles and space stations can be efficiently carried out by robotic manipulators. Not only can the objective of "minimal intervention by crew members" be satisfied in this manner, it is also possible to meet various task specifications in the dynamic environment of a space application more effectively (particularly in terms of time, precision, and reliability) by employing robots. Tasks of interest include delicate experiments as well as production and mainte-nance operations in space. High load-capacity/mass ratio, autonomous operation, high accuracy and repeatability, high stiffness, and high dexterity are some of the generally preferred characteristics for robotic manipulators used in space applications.

Gear transmission at joints is known to introduce undesirable backlash resulting in low stiffness, degraded accuracy and repeatability, and high friction with associated high levels of power dissipation, and thermal and wear problems. Direct-drive manipulation appears to reduce these problems, but in this case, manipulator joints tend to be rather massive. The traction-drive principle developed by NASA promises improvements in this direction, while providing gearless transmission.

The base reactions of a space manipulator are directly transmitted to the supporting structure, which is generally a part of the space vehicle or space station. These dynamic forces (and torques) are in fact disturbances on the supporting structure, as well as on other equipment and operations in the robot's environment. Furthermore, since the base reactions represent dynamic coupling between the robot and the space structure, not only will the environment be affected by these disturbances, but also the performance of the robot itself.

It is not trivial to take into account this coupling in the control schemes for a space structure and for a space robot. Ideally, one would desire zero base reactions, but in practice, minimization of an appropriate cost function would be acceptable. This latter approach has been taken by us. Specifically, the redundant degrees of freedom in a redundant robot were employed to dynamically minimize a quadratic cost function of base reactions. A four-degree-of-freedom robot having two traction-drive joints has been studied by us using this approach, providing encouraging results. Another aspect that requires attention is the handling of disturbance-sensitive specimens in space. The approach taken by us was to design the end effector trajectory of a robotic task in space such that acceleration and jerk are constrained while meeting the desired time and position objectives of the particular task. Specifically, cycloidal trajectories were employed.

There are several research and project-specific issues that have to be addressed under space-station robotics. Some are dynamic analysis and design issues pertaining to space robotics and some others are associated control issues. Several of these issues are as follows:

1. Effects of unplanned influences, payload variations and disturbances (e.g., obsta-cles and collisions) on operating conditions and ways to minimize the adverse effects.

2. Ways to include, in analysis and design of a space-robotic task, the effects of dynamic coupling between a robot and its supporting structure, and ways to minimize these effects.

3. Accounting of the effect of the initial configuration of a robot on the performance of a given task in an optimal manner.

4. What improvements in trajectory design for base reaction minimization could be achieved by using alternative cost functions and optimization schemes?

5. Could an algorithmic control approach such as adaptive control or nonlinear feedback control effectively solve the problems of base reaction minimization and disturbance (acceleration, jerk) limitation on payload?

6. How can more intelligent control approaches, knowledge-based control in particular, be employed to meet the performance objectives of a space robot?

7. How can the performance of a space robot be improved through the use of traction-drive joints?

In fact these issues may not be limited to space robotics and can have implications in other applications such as industrial robotics.

3.3.5 Robot Control Architecture

Most commercial robots have "closed" controllers and cannot be programmed at the low, direct-control level. The programming is done at a high, task level where the control strategy itself is not transparent. The low-level direct controllers in commercial robots are typically motion servos, which utilize strategies such as proportional-integral-derivative (PID) control, velocity and current feedback, and lead-lag compensation. To implement a control scheme such as "computed-torque" or "linearizing feedback" in a robot, it is necessary that the robot controller be programmable by the user, which requires an "open" architecture with direct access to sensory signals and actuator drivers of the robot. In addition to direct control, some monitoring and supervisory control may be needed as well at a higher, task level of robot. A hierarchical control architecture would be useful in implementing such multi-layered control.

The hardware components of a PUMA 560, a popular commercial robot, are shown in Figure 3.18. The overall system consists of a six-degree-of-freedom robot arm (having six revolute joints and associated dc motors, drive amplifiers, sensory encoders and

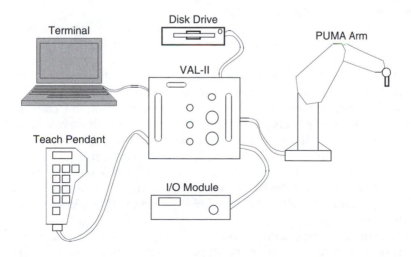

FIGURE 3.18
Components of PUMA 560 robot system.

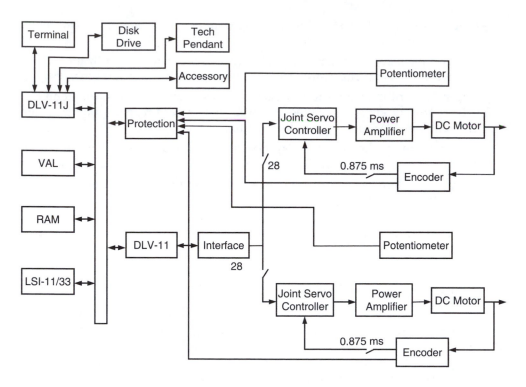

FIGURE 3.19
Hardware architecture of the PUMA 560 controller.

potentiometers), a controller using an LSI-11/33 microprocessor supported by a multi-tasking operating system, and peripherals such as a terminal, disk drive, and teach pendant. The system uses a high-level controller and operating system termed VAL II. This well-designed system uses a complete control language for conveniently programming motion tasks of the robot arm. Note that, in programming with VAL II, the user considers task kinematics only, without explicitly incorporating kinetics (force-motion dynamics) of the robot and its environment. Other functions, for example, for file editing and manipulation, are available as well with this controller.

The hardware architecture of a PUMA 560 robot is shown in Figure 3.19. Each joint is controlled by a dc servo. The position and velocity feedback signals for this purpose are provided by optical encoders, which generate digital pulses corresponding to motion increments at the joints. Potentiometers provide absolute position signals. Once the robot is programmed using VAL II, the control system computes the necessary joint motions, based on robot kinematics, that would achieve the desired motion trajectory, and provides them as the reference commands to the joint servos. Protection devices and brakes are available for discontinuing the operation under abnormal conditions.

The controller of a commercial robot such as PUMA 560 cannot be directly programmed according an advanced control scheme such as computed-torque or linearizing feedback. A commercial robot may be retrofitted, however, with an "open" controller by incorporating, for example, a powerful PC with a digital signal processor (DSP) board and an encoder board into the control loop. Typically, the drive system has to be developed as well, with new amplifiers, current sensing means, and analog-digital conversion hardware. The architecture of an open control system of this type is shown in Figure 3.20. The retrofitted controller may be programmed using a language like C, according an appropriate low-level control algorithm.

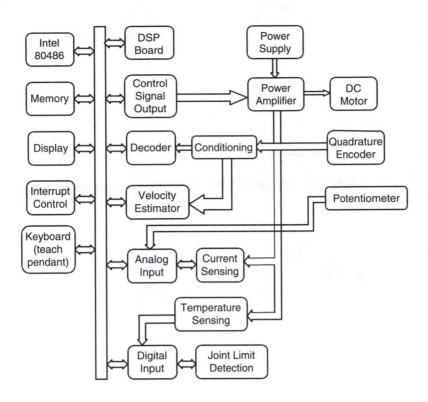

FIGURE 3.20
An open-architecture control system for a commercial robot.

In an industrial environment it is quite unlikely that a robot will function independently purely as a stand-alone device. For example, a flexible manufacturing cell (FMC) intended for the production of small batches of various parts might consist of one or more robots, several machine tools (milling machines, drills, forging machines, grinders, etc.), programmable fixtures (e.g., positioning tables, flexible jigs), parts transfer mechanisms (e.g., conveyors, gantry mechanisms), and inspection and gauging stations (e.g., vision systems, laser-based gauging devices), all coordinated and managed by a cell host computer through direct communication links. Several cells linked via a local area network (LAN) will form a flexible manufacturing system (FMS). A complex system of this nature is usually designed and operated in a distributed and hierarchical architecture. A typical three-level hierarchy will employ a top-level supervisory computer to handle general managerial functions, task scheduling, coordination of machines and material flow, and fault management; it will employ an intermediate-level host computer to generate desired trajectories, tool speeds, feed rates and other reference signals for low-level controllers; and it will employ a set of low-level computers or hardware controllers for direct digital control. Each hierarchical level may also contain laterally distributed structures.

A hierarchical structure of the form shown in Figure 3.21 is known to be particularly suitable for applications where operations of different time scales are involved. In the layered architecture that is shown, direct processing devices of the robot, for example, cutters, holding mechanisms, positioning platforms, conveyors and other object transfer devices with their sensors, actuators and direct controllers occupy the lowest level, with the highest bandwidth or speed of operation. Since the associated schemes and algorithms are direct, there may not be a great need to incorporate intelligence into this layer. But, sensor generalization and fusion, with associated preprocessing and interpretation, may

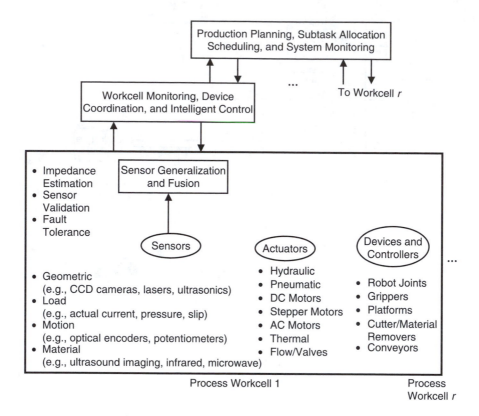

FIGURE 3.21
A hierarchical control structure for a robotic system.

be involved. Specifically, it may be desirable to use the same sensory signal to extract multiple items of information. This would be the case, for example, in camera images which can provide geometric, textural, color, and even weight information. Also, multiple sensory signals may be "fused" to provide more accurate and reliable information and, furthermore, processed jointly to validate various sensory data. The associated decision making may make use of a knowledge base at an upper level, and may be termed "intelligent fusion." Upper layers of the structure shown in Figure 3.21 may be responsible for tasks such as process monitoring, intelligent tuning, supervisory control, and device coordination as in traditional, workcell-type system architectures. Still higher layers may handle system tasks such as production planning, subtask identification and allocation, scheduling, and system restructuring.

3.3.6 Friction and Backlash

Modeling and computing the joint forces are often necessary for control of a manipulator. In these situations, bearing friction and gear friction are usually represented by equivalent viscous friction models, and usually backlash is completely neglected. These are not realistic assumptions, except for a few special types of manipulator; for instance, backlash is negligible in direct-drive arms. An accurate computation of the inverse dynamics would necessitate more realistic models.

A realistic friction model for robotic manipulator joints is given in Figure 3.22. The coefficient of friction is defined such that its product with an equivalent joint reaction

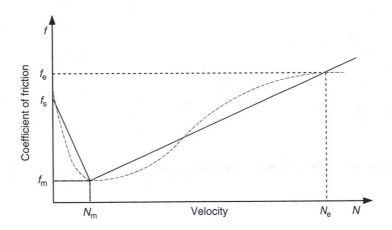

FIGURE 3.22
A friction model for a joint of a robot.

gives the frictional force (torque) in the direction of $-\dot{q}_j$. The coefficient of friction is known to vary with the relative speed as shown by a broken line in the figure. This relationship can be approximated by two straight-line segments, as indicated. Then frictional terms can be included in the inverse dynamics by modifying the Newton-Euler recursive formulation using the following computational steps:

1. Compute joint velocities and joint forces/torques (including reactions using Newton-Euler recursions and neglecting joint friction).
2. Obtain the corresponding coefficient of friction for each joint using the data in Figure 3.22 (say, by table look-up or by using a programmed analytical relationship).
3. Compute the frictional force (torque) associated with q_j, and modify the drive force (torque) accordingly.

Strictly speaking, the reactions themselves (in Step 1) would change due to the presence of friction, and hence further cycles of computation would be needed until the values converge. But, in practice, a single cycle is known to provide accurate results.

Backlash is another effect that can significantly affect the drive forces and torques of a robotic manipulator. If the backlash frequency is sufficiently higher than the control bandwidth, then backlash may be treated as an unknown high-frequency disturbance. On the other hand, if the backlash frequency is low enough, the following steps may be included in the Newton-Euler recursive formulation, to account for backlash.

1. Include gear stages in the dynamic formulation and compute the drive forces (torques) at every gear stage using the Newton-Euler recursive formulation.
2. If the drive torque at a gear stage changes sign, then there is backlash at that stage. Thus, disengage that stage, and assume zero transmitted torque there.
3. Apply the Newton-Euler recursion to the last disengaged manipulator segment that includes the end effector, and compute the drive forces (torques) for the specified end effector trajectory.
4. Compute motion of the remaining manipulator segments using drive forces (torques) computed in Step 1, and then use this information to check whether the segments will remain disengaged at the end of the present control cycle.

3.3.7 Robotic Sensors

Sensors play an important role in the operation of a robotic system, both within the robot itself and in its interactions with other components, parts, and environment. Analog devices that are available for motion sensing include resolvers, potentiometers, linear-variable differential transformers (LVDT), tachometers, accelerometers, Hall-effect sensors, and eddy current sensors. Pulse-generating (or digital) motion transducers such as optical encoders (both absolute and incremental) and binary (limit) switches are also commonly used. Force, torque, and tactile (distributed touch) sensors are also quite useful in robotic tasks. They may employ piezoresistive (including strain-gauge), piezoelectric, and optical principles. Cameras (linear or matrix, charge-coupled-device or CCD) and optical detectors with structured lighting such as lasers that can generate either single or multiple light stripes may be used in tasks such as object detection, recognition, and sensing of geometric features. See Chapter 6 and Chapter 7 for futher details.

3.4 Robotic Grippers

The end effector or mechanical hand plays an important role in robotic manipulation. Consequently, the control problem of multifingered mechanical hands has received much attention. Control of a robotic hand is facilitated through proper understanding and modeling of the associated system. Here, contact analysis between a robotic finger and an object is of interest. Characteristics and phenomena such as contact friction, flexibility of finger and the object, material properties and nonelastic behavior have been studied in this context. An innovative robotic gripper has been designed, developed, and tested by us. In the present section, the key features of the gripper are outlined. Next, an analysis of contact mechanics and kinematics of the gripper is presented. This will form an analytical model, which has been used in computer simulation and also in design development of the gripper. What is presented here may be used as a typical example in modeling, analysis, and design of robotic grippers.

3.4.1 Gripper Features

In theory, a gripper may contain any number of fingers, and each finger may consist of any number of links. In the present design, each finger, not each link joint, is driven by a single actuator. Actuation begins with the link that is directly coupled to the particular motor. When this link makes contact with the object, subsequent actuation will result in overloading of the corresponding joint. An innovative mechanical switch causes the next joint in that finger to be actuated. This actuation sequence will continue for all the joints of the finger, being driven by a single motor, until the mechanically-preset load thresholds of the joints are reached. Mechanical switching uses friction between two rotating members, one being in internal contact with the other. The level of frictional force/torque is set by adjusting the normal reaction force. When the transmitted torque is less than the frictional torque threshold, the two members rotate as one integral unit. When the torque to be transmitted exceeds this limit, that is, when the joint is overloaded, a relative motion between the two members will result. This motion will actuate the next joint in the finger, the torque at the current joint being decided by the limiting friction. A picture of the gripper prototype is shown in Figure 3.23.

The particular gripper design has several advantages. Notably, it uses fewer number of actuators than it has degrees of freedom, thereby providing quantifiable savings in weight,

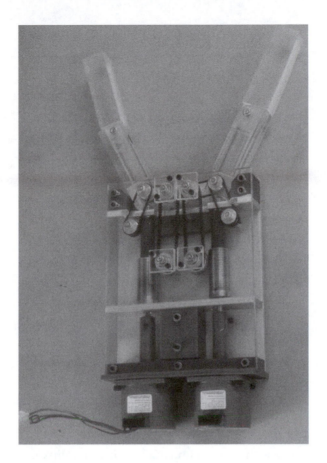

FIGURE 3.23
A view of the gripper.

size, complexity, and cost. Also, it is capable of conforming to different shapes and sizes of object through autonomous, sequential switching of the actuator drives between links, that are driven by the same actuator.

3.4.2 Analytical Model

Now, an analysis of motion and contact in a planar gripping process, using the present gripper, is formulated. The contact analysis is presented for the case of single-link contact. The analysis may be extended in a straightforward manner for multiple-link contacts. The analysis is based on the following three assumptions:

1. Object and links are rigid structures. This assumption does not cause significant errors in the determination of the final outcome of the grasp (i.e., whether or not a grasp is successful, and the final position of the object).

2. Motors are motion sources. For stepper motors, this assumption is valid during steady state motion if no motor steps are missed, as in the present gripper.

3. Contact bounce is neglected. This assumption is also valid under the conditions that prevail during the grasping process. In particular, link angular speeds are small, and rubber pads on the link contact surfaces, as well as friction between the object and the supporting surface, will tend to dampen the impact response characteristics.

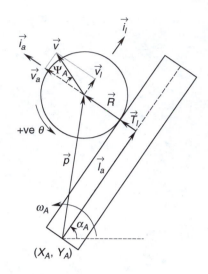

FIGURE 3.24
Position and velocity vectors (single contact, lower link).

The main implication of these assumptions is that the object will exhibit sudden changes in velocity between different phases of the grasp, when the number of contacted links changes.

3.4.2.1 Contact with the Lower Link

Consider the nomenclature shown in Figure 3.24, where the link comes into contact with a circular object.

$$\vec{p} = \vec{l}_A + \vec{T}_1 + \vec{R} = l_A \vec{i}_l + (R + T_1)\vec{i}_\alpha \tag{3.49}$$

$$\dot{\vec{p}} = v = [\dot{l}_A - \omega_A(R + T_1)]\vec{i}_l + [\omega_A l_A]\vec{i}_\alpha \tag{3.50}$$

$$\ddot{\vec{p}} = a = \left[\ddot{l}_A - \omega_A^2 l_A\right]\vec{i} + \left[2\omega_A \dot{l}_A - (R + T_1)\omega_A^2\right]\vec{i}_\alpha \tag{3.51}$$

From Equation 3.50

$$\psi_A = \arctan \frac{\dot{l}_A - \omega_A(R + T_1)}{\omega_A l_A} \tag{3.52}$$

Case 1: No Slip at Object-Link Contact Point: Transverse and radial equations of motion are:

$$F_{NA} - F_{fk} \cos \psi_A = 2m\omega_A \dot{l}_A - m\omega_A^2(R + T_1) \tag{3.53}$$

$$m\omega_A^2 l_A - F_{jk} \sin \psi_A = m\ddot{l}_A - \frac{mR\ddot{\theta}}{2} + \frac{T_{jk}}{R} \tag{3.54}$$

The following points should be noted with respect to Equation 3.54: The negative sign before the $\ddot{\theta}$ term is due to the direction of rotation of the object. The inertia torque $(-mR^2\ddot{\theta}/2)$ and the frictional torque (T_{fk}), resisting rotation of the object, manifest themselves as forces of $-mR^2\ddot{\theta}/2$ and T_{fh}/R, respectively, acting in the positive $\vec{i}_l$ direction at the point of contact between the link and the object. These are equivalent to forces of $-mR^2\ddot{\theta}/2$ and T_{fh}/R resisting motion at the object centroid.

The relationship between l_A and θ is derived by considering the link and object movements in two steps (these steps actually occur simultaneously). In Step 1, the object rolls a distance Δl_A up the link, and rotates through an angle $\theta'(=\Delta l_A/R)$. In Step 2 the link rotates through an angle $\Delta\alpha_A(=\omega_A\Delta_t)$ with no relative movement between the object and the link. The net change in angular position of the object relative to the flat surface is given by $-\theta = \theta' - \Delta_{\alpha A} = \frac{\Delta l_A}{R} - \omega_A\Delta t$ or $-R\theta = \Delta l_A - R\omega_A\Delta t$. Hence, the distance l_A in time t_A is given by

$$l_A(t_A) = l_A(0) + R(\omega_A t_A - \theta) \tag{3.55}$$

Thus

$$\dot{l}_A = R(\omega_A - \dot{\theta}) \tag{3.56}$$

and

$$\ddot{l}_A = -R\ddot{\theta} \tag{3.57}$$

By substituting Equation 3.57 into Equation 3.54 the radial equation becomes

$$m\omega_A^2 l_A - F_{fk}\sin\psi_A = \frac{T_{fk}}{R} - \frac{3mR\ddot{\theta}}{2} \tag{3.58}$$

In the X-Y reference frame, object position and velocity coordinates are given by:

$$x = x_A + l_A\cos\alpha_A - (R+T_1)\sin\alpha_A \tag{3.59}$$

$$y = y_A + l_A\sin\alpha_A + (R+T_1)\cos\alpha_A \tag{3.60}$$

$$\dot{x} = [\dot{l} - (R+T_1)\omega_A]\cos\alpha_A - \omega_A l_A\sin\alpha_A \tag{3.61}$$

$$\dot{y} = [\dot{l}_A - (R+T_1)\omega_A]\sin\alpha_A + \omega_A l_A\cos\alpha_A \tag{3.62}$$

3.4.2.2 Object Initial Velocities

The point of contact between link and object has a velocity V_p. Thus for the no-slip condition, the object is assumed to have the following initial angular and linear velocities at the instant of first contact:

$$\theta(0) = \frac{-V_R}{R} = \frac{-V_p\sin\alpha_{off}}{R} \tag{3.63}$$

$$\vec{v} = [V_R]\vec{i}_l + [V_N]\vec{i}_\alpha = [V_p\sin\alpha_{off}]\vec{i}_l + [V_p\cos\alpha_{off}]\vec{i}_\alpha \tag{3.64}$$

The variables on the right hand side of Equation 3.63 and Equation 3.64 are given by:

$$V_p = \omega_A L \tag{3.65}$$

$$L = \sqrt{l_A^2 + T_1^2} \tag{3.66}$$

$$\alpha_{\text{off}} = \arctan \frac{T_1}{l_A} \tag{3.67}$$

Case 2: Slip at Object-Link Contact Point: This situation occurs if

$$\frac{T_{fk}}{R} - \frac{mR\ddot{\theta}}{2} \succ \mu_L F_{NA} \tag{3.68}$$

When there is a slip at the link, l_A and θ become independent of each other, and the tangential contact force F_{CA} becomes a function of F_{NA} only, that is,

$$F_{CA} = \mu_{Lk} F_{NA} \tag{3.69}$$

Equation 3.53 is still valid, however, now the radial and angular equations of motion are:

$$m\omega_A^2 l_A - F_{jk} \sin \psi_A = m\ddot{l}_A + \mu_{Lk} F_{NA} \tag{3.70}$$

$$\mu_{Lk} F_{NA} = -\frac{mR\ddot{\theta}}{2} + \frac{T_{jk}}{R} \tag{3.71}$$

Equation 3.59 through 3.62 for the centroid position and velocity are still valid in this case; however, the initial angular velocity of the object is zero.

3.4.3   Contact with the Upper Link

Consider the nomenclature shown in Figure 3.25.

$$\vec{p} = \vec{L}_A + \vec{l}_C + \vec{T}_1 + \vec{R} = [L_A \cos \alpha_C + l_C] \vec{i}_l + [R + T_1 - L_A \sin \alpha_C] \vec{i}_\alpha \tag{3.72}$$

$$\dot{\vec{p}} = [\dot{l}_C + \omega_A (L_A \sin \alpha_C - R - T_1)] \vec{i}_l + [\omega_A (l_C + L_A \cos \alpha_C)] \vec{i}_\alpha \tag{3.73}$$

$$\ddot{\vec{p}} = [\ddot{l}_C - \omega_A^2 (l_C + L_A \cos \alpha_C)] \vec{i}_l + [2\omega_A \dot{l}_C + \omega_A^2 (L_A \sin \alpha_C - R - T_1)] \vec{i}_a \tag{3.74}$$

From Equation 3.73

$$\psi_C = \arctan \frac{[\dot{l}_C + \omega_A (L_A \sin \alpha_C - R - T_1)]}{[\omega_A (l_C + L_A \cos \alpha_C)]} \tag{3.75}$$

Case 1: No Slip at Object-Link Contact Point: Equation 3.76 and Equation 3.77 below are derived as for the single contact with the lower link. Transverse and radial equations

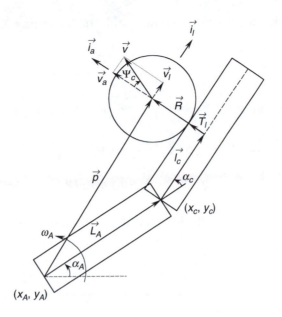

FIGURE 3.25
Position and velocity vectors (single contact, upper link).

are given by

$$F_{NC} - F_{fk}\cos\psi_C = 2m\omega_A \dot{l}_C + m\omega_A^2(L_A\sin\alpha_C - R - T_1) \qquad (3.76)$$

$$m_A^2(l_C + L_A\cos\alpha_C) - F_{fk}\sin\psi_C = T_{fk}/R - 3mR\ddot{\theta}/2 \qquad (3.77)$$

In the X-Y reference frame, object position and velocity coordinates are given by:

$$x = x_C + l_C\cos(\alpha_A + \alpha_C) - (R + T_1)\sin(\alpha_A + \alpha_C) \qquad (3.78)$$

$$y = y_C + l_C\sin(\alpha_A + \alpha_C) + (R + T_1)\cos(\alpha_A + \alpha_C) \qquad (3.79)$$

$$\dot{x} = [\dot{l}_C + \omega_A(L_A\sin\alpha_C - R - T_1)]\cos(\alpha_A + \alpha_C) - [\omega_A(l_C + L_A\cos\alpha_C)]\sin(\alpha_A + \alpha_C) \qquad (3.80)$$

$$\dot{y} = [\dot{l}_C + \omega_A(L_A\sin\alpha_C - R - T_1)]\sin(\alpha_A + \alpha_C) + [\omega_A(l_C + L_A\cos\alpha_C)]\cos(\alpha_A + \alpha_C) \qquad (3.81)$$

Object initial velocities are:

$$\dot{\theta}(0) = -V_R/R = -V_p\sin\gamma_{\text{off}}/R \qquad (3.82)$$

$$\vec{v}(0) = [V_p\sin\gamma_{\text{off}}]\vec{i}_l + [V_p\cos\gamma_{\text{off}}]\vec{i}_\alpha \qquad (3.83)$$

where $V_p = \omega_A L$ and γ_{off} is the direction of the contact point velocity with respect to the normal.

Case 2: Slip at Object-Link Contact Point: This situation will occur if

$$T_{fk}/R - mR\ddot{\theta}/2 > \mu_L F_{NC} \qquad (3.84)$$

which is similar to that for the single contact with the lower link (Case 2). The tangential contact force is

$$F_{CC} = \mu_{Lk}F_{NC} \tag{3.85}$$

Equation 3.76 is still valid here; however, now the radial and angular equations of motion are

$$m\omega_A^2(l_C + L_A\cos\alpha_C) - F_{fk}\sin\psi_C = m\ddot{l}_C + \mu_{Lk}F_{NC} \tag{3.86}$$

$$\mu_{Lk}F_{NC} = -mR\ddot{\theta}/2 + T_{fk}/R \tag{3.87}$$

These equations are useful in analysis, computer simulation, and design development of the gripper.

3.5 Problems

3.1 In a lead-screw unit, the coefficient of friction μ was found to be greater than $\tan\alpha$, where α is the helix angle. Discuss the implications of this condition.

3.2 The nut of a lead-screw unit may have means of preloading, which can eliminate backlash. What are disadvantages of preloading?

3.3 A load is moved in a vertical direction using a lead-screw drive, as shown in Figure P3.3. The following variables and parameters are given:

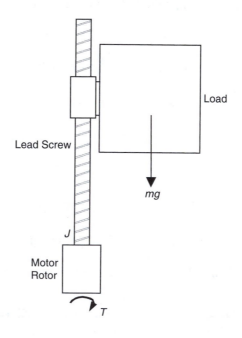

FIGURE P3.3
Moving a vertical load using a lead-screw drive.

T = motor torque
J = overall moment of inertia of the motor rotor and the lead screw
m = overall mass of the load and the nut
a = upward acceleration of the load
r = transmission ratio (rectilinear motion/angular motion) of the lead screw
e = fractional efficiency of the lead screw

Show that

$$T = \left(J + \frac{mr^2}{e}\right)\frac{a}{r} + \frac{r}{e}mg$$

In a particular application the system parameters are: $m = 500$ kg, $J = 0.25$ kg$\cdot$m², and the screw lead is 5.0 mm. In view of the static friction, the starting efficiency is 50% and the operating efficiency is 65%. Determine the torque required to start the load and then move it upwards at an acceleration of 3.0 m/s². What is the torque required to move the load downwards at the same acceleration? Show that in either case much of the torque is used in accelerating the rotor (J). Note that, in view of this observation it is advisable to pick a motor rotor and a lead screw with least moment of inertia.

3.4 Consider the planetary gear unit shown in Figure P3.4. The pinion (pitch-circle radius r_p) is the input gear and it rotates at angular velocity ω_i. If the outer gear is fixed, determine the angular velocities of the planetary gear (pitch-circle radius r_g) and the connecting arm. Note that the pitch-circle radius of the outer gear is $r_p + 2r_g$.

3.5 List some advantages and shortcomings of conventional gear drives in speed transmission applications. Indicate ways to overcome or reduce some of the shortcomings.

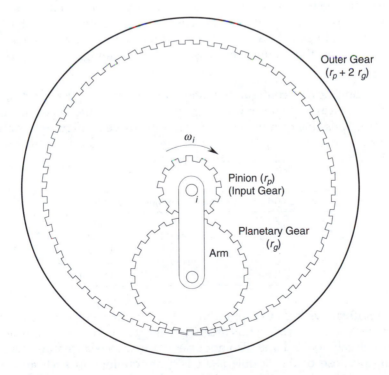

FIGURE P3.4
A planetary gear unit.

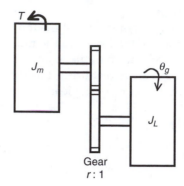

FIGURE P3.6
An inertial load driven by a motor
through a gear transmission.

3.6 A motor of torque T and moment of inertia J_m is used to drive an inertial load of moment of inertia J_L through an ideal (loss free) gear of motor-to-load speed ratio $r:1$, as shown in Figure P3.6. Obtain an expression for the angular acceleration $\ddot{\theta}_g$ of the load. Neglect the flexibility of the connecting shaft. Note that the gear inertia may be incorporated into the terms J_m and J_L.

3.7 In drive units of mechatronic systems, it is necessary to minimize backlash. Discuss the reasons for this. Conventional techniques for reducing backlash in gear drives include preloading (or, spring loading), the use of bronze bearings that automatically compensate for tooth wear, and the use of high-strength steel and other alloys that can be machined accurately to obtain tooth profiles of low backlash and that have minimal wear problems. Discuss the shortcomings of some of the conventional methods of backlash reduction. Discuss the operation of a drive unit that has virtually no backlash problems.

3.8 In some types of (indirect-drive) robotic manipulators, joint motors are located away from the joints and torques are transmitted to the joints through transmission devices such as gears, chains, cables, and timing belts. In some other types of (direct-drive) manipulators, joint motors are located at the joints themselves, the rotor being on one link and the stator being on the joining link. Discuss the advantages and disadvantages of these two designs.

3.9 In the harmonic drive configuration shown in Figure 3.5, the outer rigid spline is *fixed* (stationary), the wave generator is the input member, and the flexispline is the *output* member. Five other possible combinations of harmonic drive configurations are tabulated below. In each case, obtain an expression for the gear ratio in terms of the standard ratio (for Figure 3.5) and comment on the drive operation.

Case	Rigid Spline	Wave Generator	Flexispline
1	Fixed	Output	Input
2	Output	Input	Fixed
3	Input	Output	Fixed
4	Output	Fixed	Input
5	Input	Fixed	Output

3.10 Figure P3.10 shows a picture of an induction motor connected to a flexible shaft through a flexible coupling. Using this arrangement, the motor may be used to drive a load that is not closely located and also not oriented in a coaxial manner with respect to the motor. The purpose of the flexible shaft is quite obvious in such an arrangement. Indicate the purpose of the flexible coupling. Could a flexible coupling be used with a rigid shaft instead of a flexible shaft?

FIGURE P3.10
An induction motor linked to flexible shaft through a flexible coupling.

3.11 Backlash is a nonlinearity, which is often displayed by robots having gear transmissions. Indicate why it is difficult to compensate for backlash by using sensing and feedback control. What are preferred ways to eliminate backlash in robots?

3.12 Friction drives (traction drives), which use rollers that make frictional contact have been used as transmission devices. One possible application is for joint drives in robotic manipulators that typically use gear transmissions. An advantage of friction roller drives is the absence of backlash. Another advantage is finer motion resolution in comparison to gear drives.

    a. Give two other possible advantages and several disadvantages of friction roller drives.

    b. A schematic representation of the NASA traction drive joint is shown in Figure P3.12. Write dynamic equations for this model for evaluating its behavior.

3.13 A single-degree-of-freedom robot arm (inverted pendulum) moving in a vertical plane is shown in Figure P3.13. The centroid of the arm is at a distance l from the driven joint. The mass of the arm is m and the moment of inertia about the drive axis is I. A direct-drive motor (without gears) with torque τ is used to drive the arm. Angle of rotation of the arm is θ, as measured from a horizontal axis. The dissipation at the joint is represented by a linear viscous damping coefficient b and a Coulomb friction constant c. Obtain an expression for the drive torque τ (which may be used in control).

3.14 Consider a single joint of a robot driven by a motor through gear transmission, as shown in Figure P3.14. The joint inertia is represented by an axial load of inertia J_l whose angular rotation is θ_l. The motor rotation is θ_m and the inertia of the motor rotor is J_m. The equivalent viscous damping constant at the load is b_l, and that at the motor rotor is b_m. The gear reduction ratio is r (i.e., $\theta_m : \theta_l = r : 1$). The fractional efficiency of the gear transmission is e (Note: $0 < e < 1$). Derive an expression for the drive motor torque τ, assuming that the motor speed $\dot{\theta}_m$ and the acceleration $\ddot{\theta}_m$ are known. What are the overall moment of inertia and the overall viscous damping constant of the system, as seen from the motor end?

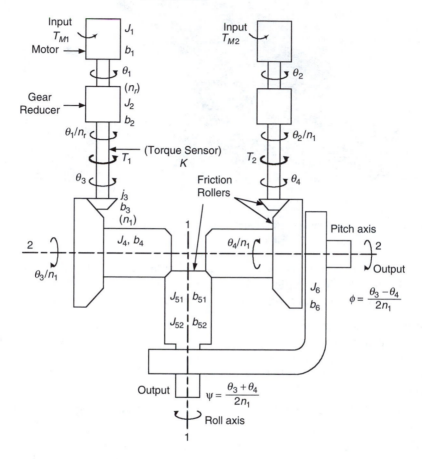

J = Moment of inertia
b = Damping Constant
K = Stiffness
T = Torque
n = Gear Ratio

FIGURE P3.12
A traction-drive joint.

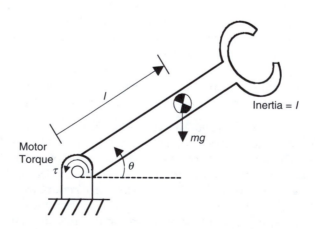

FIGURE P3.13
A single-link robot (inverted pendulum).

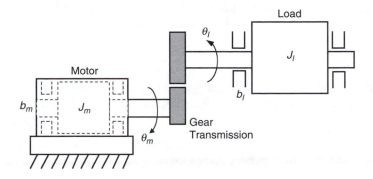

FIGURE P3.14
A geared robot joint.

3.15 a. What is a redundant manipulator?

b. Consider a six-degree-of-freedom robot that is being used to position a point object in 3-D space. Is this a redundant robot for the task? If so, how many redundant degrees of freedom are there?

3.16 In each of the following cases indicate the minimum number of degrees of freedom needed to carry out the task:

a. Positioning a point object on a two-dimensional (2-D) surface (i.e., a plane).

b. Positioning and orienting (i.e., direction) of a solid object on a plane.

c. Positioning a point object in a 3-D space.

d. Positioning and orienting a solid object in a 3-D space.

3.17 a. A robotic task calls for moving the end effector from point A to point B in a specified time. List several reasons that could make this movement infeasible.

b. If, in addition, the end effector is required to follow a specified path from A to B, what further problems could arise?

3.18 In each of the following two cases determine the 3×3 coordinate transformation matrix R.

a. A positive rotation of θ about the x axis.

b. A negative rotation of θ about the x axis.

c. A positive rotation of θ about the y axis.

d. A negative rotation of θ about the y axis.

What are the direction cosines of the new axes, expressed in the old axis frame?

3.19 In each of the four cases of Problem 3.18, show that the "inverse" of the transformation matrix is the same as the "transpose." What does this general result tell us?

3.20 Consider a vector $r = \begin{bmatrix} 1 \\ 1 \\ 0 \end{bmatrix}$ in a coordinate frame (x_1, y_1, z_1). Suppose that this vector is fixed to the frame and the entire coordinate frame is rotated (with the vector) through the angle θ about the z_1 axis, to a new frame (x_2, y_2, z_2).

a. Express the new (rotated) vector r in the old coordinate frame (x_1, y_1, z_1).

b. Express the new vector r in the new coordinate frame (x_2, y_2, z_2).

3.21 In Problem 3.20 suppose that the vector r is kept at its old position and only the coordinate frame is rotated. Express the vector in the:

a. New coordinate frame.

b. Old coordinate frame.

3.22 A coordinate frame is moved without rotation (i.e., translated) so that its origin is at the point (a, b, c). What is the 4×4 homogeneous transformation matrix corresponding to this movement?

3.23 A coordinate frame is rotated through angle θ about the z axis. What is the 4×4 homogeneous transformation matrix corresponding to this movement?

3.24 (a) If a frame is translated as in Problem 3.22, and then rotated with respect to the new frame as in Problem 3.23, what is the corresponding homogeneous transformation matrix? (b) If the two movements in (a) are carried out in the reverse order, what is the corresponding homogeneous transformation?

3.25 Consider the two vectors $r = \begin{bmatrix} r_1 \\ r_2 \\ r_3 \end{bmatrix}$ and $a = \begin{bmatrix} a_1 \\ a_2 \\ a_3 \end{bmatrix}$ expressed in the same coordinate frame.

a. What is the cross product $r \times a$?

b. What is the cross product $a \times r$?

c. What is the magnitude r of vector r?

d. What is the normalized version of vector r such that its magnitude is unity?

3.26 Consider a two-degree-of-freedom robot with two revolute joints, as shown in Figure P3.26. The link lengths are l_1 and l_2. A Cartesian $(x–y)$ coordinate frame is used to represent the motion of this planar robot. The origin 0 of the coordinate frame is also the location of the base joint of the robot. Show that the end effector of the robot will not be able to reach point P unless $l_1 \leq 1/4\, l_2$. Point P has coordinates $x = 3l_1$ and $y = l_2$.

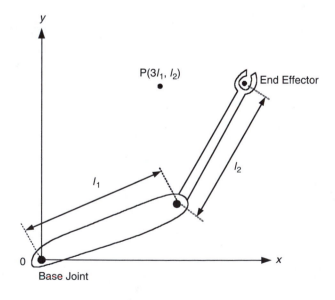

FIGURE P3.26
A planar revolute robot having two degrees of freedom.

3.27 Consider the kinematic relation for velocity of a robot:

$$\dot{r} = J\dot{q}$$

where

$\dot{r}$ = end effector velocity vector.
$\dot{q}$ = joint velocity vector.
J = Jacobian matrix of robot

Suppose that J is an $m \times n$ matrix with $m < n$. Here a unique inverse J^{-1} does not exist. But suppose that it is also required to minimize the cost function:

$$C = \dot{q}^T W \dot{q}$$

where W is a positive-definite weighting matrix.
Then, show that a *pseudo-inverse* relation can be written as

$$\dot{q} = W^{-1}J^T(JW^{-1}J^T)^{-1}\dot{r}$$

When the weighting matrix W is the identity matrix, what is the pseudo-inverse relation?

3.28 For the manipulator shown in Figure P3.28, compute the Jacobian J at the following configurations: (a) $q_1 = 90°$, $q_2 = 90°$. (b) $q_1 = 90°$, $q_2 = 90°$.
Comment on your results.

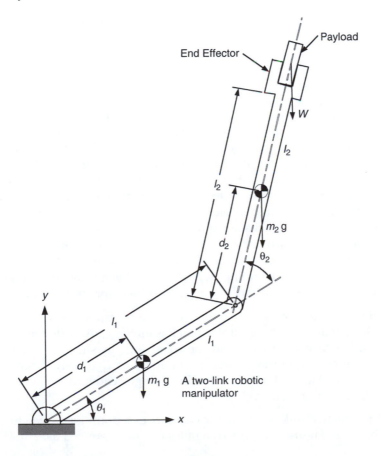

FIGURE P3.28
Parameters of a two-link planar robot with revolute joints.

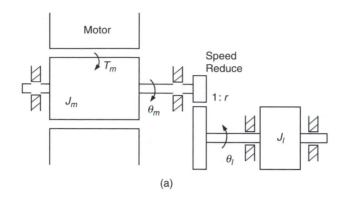

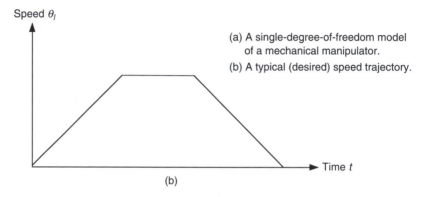

(a) A single-degree-of-freedom model
of a mechanical manipulator.

(b) A typical (desired) speed trajectory.

FIGURE P3.29
(a) A single-degree-of-freedom drive unit, (b) A desired speed trajectory.

3.29 A single-degree-of-freedom drive unit is shown in Figure P3.29(a). The joint motor has rotor inertia J_m. It drives an inertial load that has moment of inertia J_l through a speed reducer of gear ratio $1 : r$ (Note: $r <1$). The control scheme used in this system is the so-called feedforward control (strictly, *computed-torque control*) method. Specifically, the motor torque T_m that is required to accelerate or decelerate the load is computed using a suitable dynamic model and a desired motion trajectory for the load, and the motor windings are excited so as to generate that torque. A typical trajectory would consist of a constant angular acceleration segment followed by a constant angular velocity segment, and finally a constant deceleration segment, as shown in Figure P3.29(b). Neglecting friction (particularly bearing friction) and inertia of the speed reducer, obtain a dynamic model for torque computation during accelerating and decelerating segments of the motion trajectory. Specifically, obtain an expression for T_m in terms of J_m, J_l, r, and the angular acceleration $\ddot{\theta}_l$.

3.30 Consider a single-degree-of-freedom revolute manipulator of link inertia (load) J_l that is driven by a motor with a rigid rotor that has inertia J_m. A torsion member of stiffness K_s is connected between the motor and the robot link, as shown in Figure P3.30(a), in order to measure the torque transmitted to the robot. A free-body diagram of the robot system is shown in Figure P3.30(b). Obtain a single equation to represent the twist $\theta = \theta_m - \theta_l$ of the torsion member.

3.31 Consider the two-link manipulator that carries a point load (weight W) at the end effector, as shown in Figure P3.28. Its dynamics can be expressed as:

$$I\ddot{q} + b = \tau$$

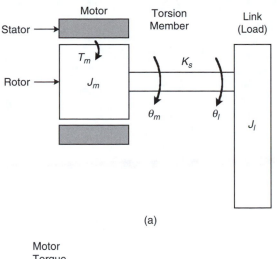

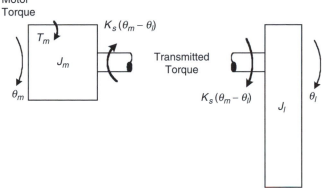

FIGURE P3.30
(a) A single-degree-of-freedom revolute manipulator, (b) Free-body diagram of the system.

where:
q = vector of (relative) joint rotations q_1 and q_2.
τ = vector of drive torques τ_1 and τ_2 at the two joints, corresponding to the coordinates q_1 and q_2.
I = second-order inertia matrix

$$= \begin{bmatrix} I_{11} & I_{12} \\ I_{21} & I_{22} \end{bmatrix}$$

b = vector of joint-friction, gravitational, centrifugal, and Coriolis torques (components are b_1 and b_2).

Assume zero payload ($W = 0$), and neglect friction. Obtain expressions for the system parameters I_{11}, I_{12}, I_{21}, I_{22}, b_1, and b_2 in terms of:

I_1, I_2 = moments of inertia of the links about their centroids.
m_1, m_2 = masses of the links.

and the geometric parameters l_1, l_2, d_1, and d_2 which are as defined in the figure.

4

Component Interconnection and Signal Conditioning

Mechatronic systems are mixed systems, which consist of more than one type of components properly interconnected and integrated. In particular, mechanical, electrical, electronic, and computer hardware are integrated to form a mechatronic system. It follows that component interconnection is an important topic in the field of Mechatronics. When two components are interconnected, signals flow through them. The nature and type of the signals that are present at the interface of two components will depend on the nature and type of the components. For example, when a motor is coupled with a load through a gear (transmission) unit, mechanical power flows at the interfaces of these components. Then, we are particularly interested in such signals as angular velocity and torque. In particular, these signals would be modified or "conditioned" as they are transmitted through the gear transmission. Similarly, when a motor is connected to its electronic drive system, command signals of motor control, typically available as voltages, would be converted into appropriate currents for energizing the motor windings so as to generate the necessary torque. Again, signal conditioning or conversion is important here. In general, then, signal conditioning is important in the context of component interconnection and integration, and becomes an important subject in the study of Mechatronics.

This chapter addresses interconnection of components such as sensors, signal conditioning circuitry, actuators, and power transmission devices in a mechatronic system. Desirable impedance characteristics for such components are discussed. Signal modification plays a crucial role in component interconnection or interfacing. When two devices are interfaced, it is essential to guarantee that a signal leaving one device and entering the other will do so at proper signal levels (the values of voltage, current, speed, force, power, etc.), in the proper form (electrical, mechanical, analog, digital, modulated, demodulated, etc.), and without distortion (where loading problems, nonlinearities, and noise have to be eliminated, and where impedance considerations become important). Particularly for transmission, a signal should be properly modified (by amplification, modulation, digitizing, etc.) so that the signal/noise ratio of the transmitted signal is sufficiently large at the receiver. The significance of signal modification is clear from these observations.

The tasks of signal-modification may include *signal conditioning* (e.g., amplification, and analog and digital filtering), *signal conversion* (e.g., analog-to-digital conversion, digital-to-analog conversion, voltage-to-frequency conversion, and frequency-to-voltage conversion), *modulation* (e.g., amplitude modulation, frequency modulation, phase modulation, pulse-width modulation, pulse-frequency modulation, and pulse-code modulation), and *demodulation* (the reverse process of modulation). In addition, many other types of useful signal modification operations can be identified. For example, *sample and hold circuits* are used in digital data acquisition systems. Devices such as *analog and digital multiplexers* and *comparators* are needed in many applications of data acquisition and processing. Phase shifting, curve shaping, offsetting, and linearization can also be classified as signal modification.

This chapter describes signal conditioning and modification operations that are useful in mechatronic applications. The operational amplifier is introduced as a basic element in signal conditioning and impedance matching circuitry for electronic systems. Various types of signal conditioning and modification devices such as amplifiers, filters, modulators, demodulators, bridge circuits, analog-to-digital converters and digital-to-analog converters are discussed.

4.1   Component Interconnection

A mechatronic system can consist a wide variety of components, which are interconnected to perform the intended functions. When two or more components are interconnected, the behavior of the individual components in the integrated system can deviate significantly from their behavior when each component operates independently. Matching of components in a multicomponent system, particularly with respect to their impedance characteristics, should be done carefully in order to improve the system performance and accuracy. In this chapter, first we shall study basic concepts of impedance and component matching. The concepts presented here are applicable to many types of components in a general mechatronic system. Discussions and developments given here can be quite general. Nevertheless, specific hardware components and designs are considered particularly in relation to component interfacing and signal conditioning.

4.2   Impedance Characteristics

When components such as sensors and transducers, control boards, process (plant) equipment, and signal-conditioning hardware are interconnected, it is necessary to *match* impedances properly at each interface in order to realize their rated performance level. One adverse effect of improper impedance matching is the *loading effect*. For example, in a measuring system, the measuring instrument can distort the signal that is being measured. The resulting error can far exceed other types of measurement error. Both electrical and mechanical loading are possible. Electrical loading errors result from connecting an output unit such as a measuring device that has a low input impedance to an input device such as a signal source. Mechanical loading errors can result in an input device due to inertia, friction, and other resistive forces generated by an interconnected output component.

Impedance can be interpreted either in the traditional electrical sense or in the mechanical sense, depending on the type of signals that are involved. For example, a heavy accelerometer can introduce an additional dynamic load, which will modify the actual acceleration at the monitoring location. Similarly, a voltmeter can modify the currents (and voltages) in a circuit, and a thermocouple junction can modify the temperature that is being measured as a result of the heat transfer into the junction. In mechanical and electrical systems, loading errors can appear as phase distortions as well. Digital hardware also can produce loading errors. For example, an analog-to-digital conversion (ADC) board can load the amplifier output from a strain gage bridge circuit, thereby affecting digitized data.

Another adverse effect of improper impedance consideration is inadequate output signal levels, which can make the output functions such as signal processing and transmission, component driving, and actuation of a final control element or plant very difficult. In the context of sensor-transducer technology it should be noted here that many types of transducers (e.g., piezoelectric accelerometers, impedance heads, and microphones) have high output impedances on the order of a thousand megohms (1 megohm or $1\ \text{M}\Omega = 1 \times 10^6\ \Omega$).

These devices generate low output signals, and they would require conditioning to step up the signal level. *Impedance-matching amplifiers*, which have high input impedances and low output impedances (a few ohms), are used for this purpose (e.g., charge amplifiers are used in conjunction with piezoelectric sensors). A device with a high input impedance has the further advantage that it usually consumes less power (v^2/R is low) for a given input voltage. The fact that a low input impedance device extracts a high level of power from the preceding output device may be interpreted as the reason for loading error.

4.2.1 Cascade Connection of Devices

Consider a standard two-port electrical device. The *output impedance Z_o* of such a device is defined as the ratio of the open-circuit (i.e., no-load) voltage at the output port to the short-circuit current at the output port.

Open-circuit voltage at output is the output voltage present when there is no current flowing at the output port. This is the case if the output port is not connected to a load (impedance). As soon as a load is connected at the output of the device, a current will flow through it, and the output voltage will drop to a value less than that of the open-circuit voltage. To measure the open-circuit voltage, the rated input voltage is applied at the input port and maintained constant, and the output voltage is measured using a voltmeter that has a very high (input) impedance. To measure the short-circuit current, a very low-impedance ammeter is connected at the output port.

The *input impedance Z_i* is defined as the ratio of the rated input voltage to the corresponding current through the input terminals while the output terminals are maintained as an open circuit.

Note that these definitions are associated with electrical devices. A generalization is possible by interpreting voltage and velocity as *across variables*, and current and force as *through variables*. Then mechanical *mobility* should be used in place of electrical impedance, in the associated analysis.

Using these definitions, input impedance Z_i and output impedance Z_o can be represented schematically as in Figure 4.1(a). Note that v_o is the open-circuit output voltage. When a

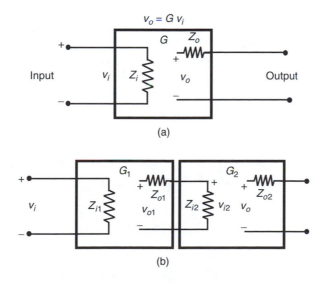

FIGURE 4.1

(a) Schematic representation of input impedance and output impedance, (b) Cascade connection of two two-port devices.

load is connected at the output port, the voltage across the load will be different from v_o. This is caused by the presence of a current through Z_o. In the frequency domain, v_i and v_o are represented by their respective *Fourier spectra*. The corresponding transfer relation can be expressed in terms of the complex frequency response (transfer) function $G(j\omega)$ under open-circuit (no-load) conditions:

$$v_o = Gv_i \tag{4.1}$$

Now consider two devices connected in cascade, as shown in Figure 4.1b. It can be easily verified that the following relations apply:

$$v_{o1} = G_1 v_i \tag{4.2}$$

$$v_{i2} = \frac{Z_{i2}}{Z_{o1} + Z_{i2}} v_{o1} \tag{4.3}$$

$$v_o = G_2 v_{i2} \tag{4.4}$$

These relations can be combined to give the overall input/output relation:

$$v_o = \frac{Z_{i2}}{Z_{o1} + Z_{i2}} G_2 G_1 v_i \tag{4.5}$$

We see from Equation 4.5 that the overall frequency transfer function differs from the ideally expected product (G_2G_1) by the factor

$$\frac{Z_{i2}}{Z_{o1} + Z_{i2}} = \frac{1}{Z_{o1}/Z_{i2} + 1} \tag{4.6}$$

Note that cascading has "distorted" the frequency response characteristics of the two devices. If $Z_{o1}/Z_{i2} \ll 1$, this deviation becomes insignificant. From this observation, it can be concluded that when frequency response characteristics (i.e., dynamic characteristics) are important in a cascaded device, cascading should be done such that the output impedance of the first device is much smaller than the input impedance of the second device.

Example 4.1

A lag network used as the compensation element of a control system is shown in Figure 4.2(a). Show that its transfer function is given by

$$\frac{v_o}{v_i} = \frac{Z_2}{R_1 + Z_2}$$

where

$$Z_2 = R_2 + \frac{1}{Cs}$$

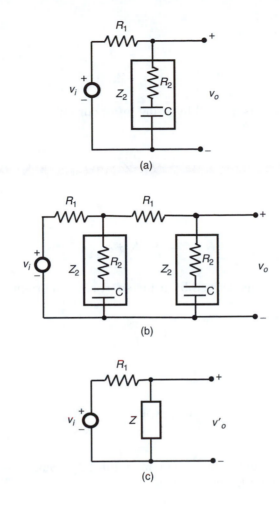

FIGURE 4.2
(a) A single circuit module, (b) Cascade connection of two modules, (c) An equivalent circuit for (b).

What is the input impedance and what is the output impedance for this circuit? Also, if two such lag circuits are cascaded as shown in Figure 4.2(b), what is the overall transfer function? How would you make this transfer function become close to the ideal result:

$$\left\{ \frac{Z_2}{R_1 + Z_2} \right\}^2$$

SOLUTION
To solve this problem, first note that in Figure 4.2(a), voltage drop across the element $R_2 + 1/(Cs)$ is

$$v_o = \left(R_2 + \frac{1}{Cs} \right) \Big/ \left\{ R_1 + R_2 + \frac{1}{Cs} \right\} v_i$$

Hence,

$$\frac{v_o}{v_i} = \frac{Z_2}{R_1 + Z_2}$$

Now, input impedance Z_i is derived by using input current

$$i = \frac{v_i}{R_1 + Z_2}$$

as

$$Z_i = \frac{v_i}{i} = R_1 + Z_2$$

and output impedance Z_o, is derived by using short-circuit current

$$i_{sc} = \frac{v_i}{R_1}$$

as

$$Z_o = \frac{v_o}{i_{sc}} = \frac{Z_2/(R_1 + Z_2)v_i}{v_i/R_1} = \frac{R_1 Z_2}{R_1 + Z_2} \qquad \text{(i)}$$

Next, consider the equivalent circuit shown in Figure 4.2(c). Since Z is formed by connecting Z_2 and $(R_1 + Z_2)$ in parallel, we have

$$\frac{1}{Z} = \frac{1}{Z_2} + \frac{1}{R_1 + Z_2} \qquad \text{(ii)}$$

Voltage drop across Z is

$$v_o' = \frac{Z}{R_1 + Z} v_i \qquad \text{(iii)}$$

Now apply the single-circuit module result Equation i to the second circuit stage in Figure 4.2(b); thus,

$$v_o = \frac{Z_2}{R_1 + Z_2} v_o'$$

Substituting Equation iii, we get

$$v_o = \frac{Z_2}{(R_1 + Z_2)} \frac{Z}{(R_1 + Z)} v_i$$

The overall transfer function for the cascaded circuit is

$$G = \frac{v_o}{v_i} = \frac{Z_2}{(R_1 + Z_2)} \frac{Z}{(R_1 + Z)} = \frac{Z_2}{(R_1 + Z_2)} \frac{1}{(R_1/Z + 1)}$$

Now substituting Equation ii we get

$$G = \left[\frac{Z_2}{R_1 + Z_2} \right]^2 \frac{1}{1 + R_1 Z_2 / (R_1 + Z_2)^2}$$

We observe that the ideal transfer function is approached by making $R_1 Z_2 / (R_1 + Z_2)^2$ small compared to unity.

4.2.2 Impedance Matching

When two electrical components are interconnected, current (and energy) will flow between the two components. This will change the original (unconnected) conditions. This is known as the (electrical) loading effect, and it has to be minimized. At the same time, adequate power and current would be needed for signal communication, conditioning, display, etc. Both situations can be accommodated through proper matching of impedances when the two components are connected. Usually an impedance matching amplifier (impedance transformer) would be needed between the two components.

From the analysis given in the preceding section, it is clear that the signal-conditioning circuitry should have a considerably large input impedance in comparison to the output impedance of the sensor-transducer unit in order to reduce loading errors. The problem is quite serious in measuring devices such as piezoelectric sensors, which have very high output impedances. In such cases, the input impedance of the signal-conditioning unit might be inadequate to reduce loading effects; also, the output signal level of these high-impedance sensors is quite low for signal transmission, processing, actuation, and control. The solution for this problem is to introduce several stages of amplifier circuitry between the output of the first hardware unit (e.g., sensor) and the input of the second hardware unit (e.g., data acquisition unit). The first stage of such an interfacing device is typically an *impedance-matching amplifier* that has very high input impedance, very low output impedance, and almost unity gain. The last stage is typically a stable high-gain amplifier stage to step up the signal level. Impedance-matching amplifiers are, in fact, *operational amplifiers* with feedback.

When connecting a device to a signal source, loading problems can be reduced by making sure that the device has a high input impedance. Unfortunately, this will also reduce the level (amplitude, power) of the signal received by the device. In fact, a high-impedance device may reflect back some harmonics of the source signal. A termination resistance may be connected in parallel with the device in order to reduce this problem.

In many data acquisition systems, output impedance of the output amplifier is made equal to the transmission line impedance. When maximum power amplification is desired, *conjugate matching* is recommended. In this case, input impedance and output impedance of the matching amplifier are made equal to the complex conjugates of the source impedance and the load impedance, respectively.

Example 4.2

Consider a dc power supply of voltage v_s and output impedance (resistance) R_s. It is used to power a load of resistance R_l, as shown in Figure 4.3. What should be the relationship between R_s and R_l if the objective is to maximize the power absorbed by the load?

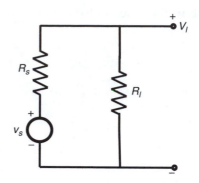

FIGURE 4.3
A load powered by a dc power supply.

SOLUTION
Current through the circuit is

$$i_l = \frac{v_s}{R_l + R_s}$$

Accordingly, the voltage across the load is

$$v_l = i_l R_l = \frac{v_s R_l}{R_l + R_s}$$

The power absorbed by the load is

$$p_l = i_l v_l = \frac{v_s^2 R_l}{[R_l + R_s]^2} \tag{i}$$

For maximum power, we need

$$\frac{dp_l}{dR_l} = 0 \tag{ii}$$

We differentiate the RHS expression of Equation i with respect to R_l in order to satisfy Equation ii. This gives the requirement for maximum power as

$$R_l = R_s$$

4.2.3 Impedance Matching in Mechanical Systems

The concepts of impedance matching can be extended to mechanical systems and to mixed and mechatronic systems in a straightforward manner. The procedure follows from the familiar electro-mechanical analogies. As a specific application, consider a mechanical load driven by a motor. Often, direct driving is not practical due to the limitations of the speed-torque characteristics of the available motors. By including a suitable gear transmission between the motor and the load, it is possible to modify the speed-torque characteristics of the drive system as felt by the load. This is a process of impedance matching.

Example 4.3

Consider the mechanical system where a torque source (motor) of torque T and moment of inertia J_m is used to drive a purely inertial load of moment of inertia J_L as shown in Figure 4.4(a). What is the resulting angular acceleration $\ddot{\theta}$ of the system? Neglect the flexibility of the connecting shaft. Now suppose that the load is connected to the same torque source through an ideal (loss free) gear of motor-to-load speed ratio $r : 1$, as shown in Figure 4.4(b). What is the resulting acceleration $\ddot{\theta}_g$ of the load?

Obtain an expression for the normalized load acceleration $a = \ddot{\theta}_g/\ddot{\theta}$ in terms of r and $p = J_L/J_m$. Sketch a versus r for $p = 0.1$, 1.0, and 10.0. Determine the value of r in terms of p that will maximize the load acceleration a.

Comment on the results obtained in this problem.

SOLUTION

For the unit without the gear transmission: Newton's second law gives

$$(J_m + J_L)\ddot{\theta} = T$$

Hence

$$\ddot{\theta} = \frac{T}{J_m + J_L} \tag{i}$$

For the unit with the gear transmission: See the free-body diagram shown in Figure 4.5, in the case of a loss-free (i.e., 100% efficient) gear transmission.

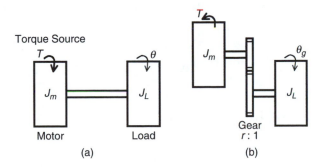

Torque Source

Motor   Load   Gear
(a)   (b) $r : 1$

FIGURE 4.4

An inertial load driven by a motor: (a) Without gear transmission, (b) With a gear transmission.

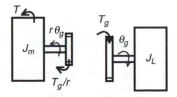

FIGURE 4.5
Free-body diagram.

Newton's second law gives

$$J_m r \ddot{\theta}_g = T - \frac{T_g}{r} \tag{ii}$$

and

$$J_L \ddot{\theta}_g = T_g \tag{iii}$$

where T_g = gear torque on the load inertia. Eliminate T_g in Equation ii and Equation iii. We get

$$\ddot{\theta}_g = \frac{rT}{\left(r^2 J_m + J_L\right)} \tag{iv}$$

Divide Equation iv by Equation i.

$$\frac{\ddot{\theta}_g}{\ddot{\theta}} = a = \frac{r(J_m + J_L)}{\left(r^2 J_m + J_L\right)} = \frac{r(1 + J_L/J_m)}{\left(r^2 + J_L/J_m\right)}$$

or,

$$a = \frac{r(1+p)}{(r^2 + p)} \tag{v}$$

where, $p = J_L/J_m$.

From Equation v note that for $r = 0$, $a = 0$ and for $r \to \infty$, $a \to 0$. Peak value of a is obtained through differentiation:

$$\frac{\partial a}{\partial r} = \frac{(1+p)[(r^2 + p) - r \times 2r]}{(r^2 + p)^2} = 0$$

We get, by taking the positive root,

$$r_p = \sqrt{p} \tag{vi}$$

where r_p is the value of r corresponding to peak a. The peak value of a is obtained by substituting Equation vi in Equation v; thus,

$$a_p = \frac{1+p}{2\sqrt{p}} \tag{vii}$$

Also, note from Equation v that when $r = 1$ we have $a = r = 1$. Hence, all curves given by Equation v should pass through the point (1,1).

The relation Equation v is sketched in Figure 4.6 for $p = 0.1$, 1.0, and 10.0. The peak values are tabulated below.

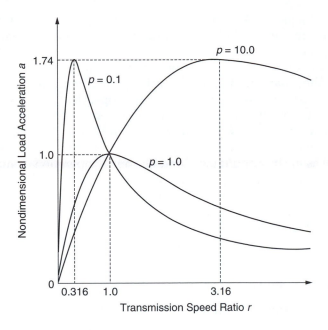

FIGURE 4.6
Normalized acceleration versus speed ratio.

p	r_p	a_p
0.1	0.316	1.74
1.0	1.0	1.0
10.0	3.16	1.74

Note from Figure 4.6 that the transmission speed ratio can be chosen, depending on the inertia ratio, to maximize the load acceleration. In particular, we can state the following:

1. When $J_L = J_m$, pick a direct-drive system (no gear transmission; i.e., $r = 1$).
2. When $J_L < J_m$, pick a speed-up gear at the peak value of r ($= \sqrt{J_L/J_m}$).

When $J_L > J_m$, pick a speed-down gear at the peak value of r.

4.3 Amplifiers

The level of an electrical signal can be represented by variables such as voltage, current, and power. Analogous across variables, through variables, and power variables can be defined for other types of signals (e.g., mechanical) as well. Signal levels at various interface locations of components in a mechatronic system have to be properly adjusted for satisfactory performance of these components and of the overall system. For example, input to an actuator should possess adequate power to drive the actuator. A signal should maintain its signal level above some threshold during transmission so that errors due to signal weakening would not be excessive. Signals applied to digital devices must remain within the specified logic levels. Many types of sensors produce weak signals that have to be upgraded before they could be fed into a monitoring system, data processor, controller, or data logger.

Signal amplification concerns proper adjustment of the signal level for performing a specific task. Amplifiers are used to accomplish signal amplification. An amplifier is an active device that needs an external power source to operate. Even though various active circuits, amplifiers in particular, are commonly produced in the monolithic form using an original integrated-circuit (IC) layout so as to accomplish a particular amplification task, it is convenient to study their performance using discrete circuit models with the *operational amplifier* (op-amp) as the basic building block. Of course, operational amplifiers are themselves available as monolithic IC packages. They are widely used as the basic building blocks in producing other types of amplifiers, and in turn for modeling and analyzing these various kinds of amplifiers. For these reasons, our discussion on amplifiers will evolve from the operational amplifier.

4.3.1 Operational Amplifier

The origin of the operational amplifier dates back to the 1940s when the vacuum tube operational amplifier was introduced. Operational amplifier or *op-amp* got its name due to the fact that originally it was used almost exclusively to perform mathematical operations; for example, in analog computers. Subsequently, in the 1950s the transistorized op-amp was developed. It used discrete elements such as *bipolar junction transistors* and *resistors*. Still it was too large in size, consumed too much power, and was too expensive for widespread use in general applications. This situation changed in the late 1960s when *integrated-circuit* (IC) op-amp was developed in the monolithic form, as a single IC chip. Today, the IC op-amp, which consists of a large number of circuit elements on a *substrate* of typically a single *silicon crystal* (the monolithic form), is a valuable component in almost any signal modification device. Bipolar-CMOS (complementary metal oxide semiconductor) op-amps in various plastic packages and pin configurations are commonly available.

An op-amp could be manufactured in the discrete-element form using, say, ten bipolar junction transistors and as many discrete resistors or alternatively (and preferably) in the modern monolithic form as an IC chip that may be equivalent to over 100 discrete elements. In any form, the device has an *input impedance* Z_i, an *output impedance* Z_o and a gain K. Hence, a schematic model for an op-amp can be given as in Figure 4.7(a). Op-amp packages are available in several forms. Very common is the 8-pin dual in-line package (DIP) or V package, as shown in Figure 4.7(b). The assignment of the pins (pin configuration or pinout) is as shown in the figure, which should be compared with Figure 4.7(a). Note the counter-clockwise numbering sequence starting with the top left pin next to the semicircular notch (or, dot). This convention of numbering is standard for any type of IC package, not just op-amp packages. Other packages include 8-pin metal-can package or T package, which has a circular shape instead of the rectangular shape of the previous package, and the 14-pin rectangular "Quad" package which contains four op-amps (with a total of eight input pins, four output pins, and two power supply pins). The conventional symbol of an op-amp is shown in Figure 4.7(c). Typically, there are five terminals (pins or lead connections) to an op-amp. Specifically, there are two input leads (a positive or noninverting lead with voltage v_{ip} and a negative or inverting lead with voltage v_{in}), an output lead (voltage v_o), and two bipolar power supply leads (+ v_s or v_{CC} or collector supply and $-v_s$ or v_{EE} or emitter supply). The typical supply voltage is ±22 V. Some of the pins may not be normally connected; for example, pins 1, 5, and 8 in Figure 4.7(b).

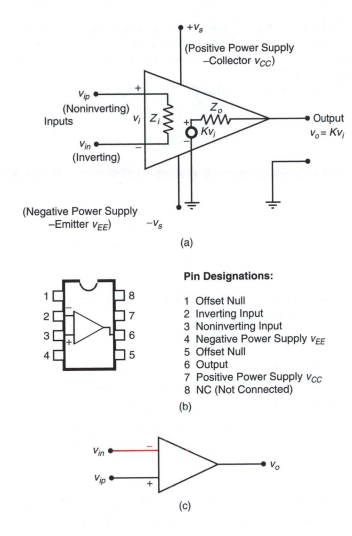

Pin Designations:

1 Offset Null
2 Inverting Input
3 Noninverting Input
4 Negative Power Supply v_{EE}
5 Offset Null
6 Output
7 Positive Power Supply v_{CC}
8 NC (Not Connected)

(b)

(c)

FIGURE 4.7
Operational amplifier: (a) A schematic model, (b) Eight-pin dual in-line package (DIP), (c) Conventional circuit symbol.

Note from Figure 4.7(a) that under open-loop (no feedback) conditions

$$v_o = Kv_i \tag{4.7}$$

in which the input voltage v_i is the differential input voltage defined as the algebraic difference between the voltages at the positive and negative lead; thus

$$v_i = v_{ip} - v_{in} \tag{4.8}$$

The open loop voltage gain K is very high ($10^5 - 10^9$) for a typical op-amp. Furthermore, the input impedance Z_i could be as high as 10 MΩ (typical is 2 MΩ) and the output impedance is low, of the order of 10 Ω and may reach about 75 Ω for some op-amps. Since v_o is typically 1–15 V, from Equation 4.7 it follows that $v_i \cong 0$ since K is very large. Hence, from Equation 4.8 we have $v_{ip} \cong v_{in}$. In other words, the voltages at the two input leads are nearly equal. Now if we apply a large voltage differential v_i (say, 10 V) at the input, then according to Equation 4.7, the output voltage should be extremely high. This never happens in practice, however, since the device saturates quickly beyond moderate output voltages (of the order of 15 V).

From Equation 4.7 and Equation 4.8 it is clear that if the negative input lead is grounded (i.e., $v_{in} = 0$) then

$$v_o = Kv_{ip} \tag{4.9}$$

and if the positive input lead is grounded (i.e., $v_{ip} = 0$)

$$v_o = -Kv_{in} \tag{4.10}$$

This is the reason why v_{ip} is termed *noninverting input* and v_{in} is termed *inverting input*.

Example 4.4

Consider an op-amp having an open loop gain of 1×10^5. If the saturation voltage is 15 V, determine the output voltage in the following cases:

a. 5 µV at the positive lead and 2 µV at the negative lead
b. –5 µV at the positive lead and 2 µV at the negative lead
c. 5 µV at the positive lead and –2 µV at the negative lead
d. –5 µV at the positive lead and –2 µV at the negative lead
e. 1 V at the positive lead, and the negative lead is grounded
f. 1 V at the negative lead, and the positive lead is grounded

SOLUTION

This problem can be solved using Equation 4.7 and Equation 4.8. The results are given in Table 4.1. Note that in the last two cases the output will saturate and Equation 4.7 will no longer hold.

Field effect transistors (FET), for example, metal oxide semiconductor field effect transistors (MOSFET), are commonly used in the IC form of an op-amp. The MOSFET type has advantages over many other types; for example, higher input impedance and more stable output (almost equal to the power supply voltage) at saturation, making the MOSFET op-amps preferable over bipolar junction transistor op-amps in applications.

TABLE 4.1

Solution to Example 4.4

v_{ip}	v_{in}	v_i	v_o
5 µV	2 µV	3 µV	0.3 V
– 5 µV	2 µV	– 7 µV	– 0.7 V
5 µV	– 2 µV	7 µV	0.7 V
– 5µV	– 2 µV	– 3 µV	– 0.3 V
1 V	0	1 V	15 V
0	1 V	– 1 V	– 15 V

In analyzing operational amplifier circuits under unsaturated conditions, we use the following two characteristics of an op-amp:

1. Voltages of the two input leads should be (almost) equal
2. Currents through each of the two input leads should be (almost) zero

As explained earlier, the first property is credited to high open-loop gain, and the second property to high input impedance in an operational amplifier. We shall repeatedly use these two properties, to obtain input-output equations for amplifier systems.

4.3.1.1   Use of Feedback in Op-Amps

Operational amplifier is a very versatile device, primarily due to its very high input impedance, low output impedance, and very high gain. But, it cannot be used without modification as an amplifier because it is not very stable in the form shown in Figure 4.7. The two main factors which contribute to this problem are:

1. Frequency response
2. Drift.

Stated in another way, op-amp gain K does not remain constant; it can vary with frequency of the input signal (i.e., frequency response function is not flat in the operating range); and, also it can vary with time (i.e., drift). The frequency response problem arises due to circuit dynamics of an operational amplifier. This problem is usually not severe unless the device is operated at very high frequencies. The drift problem arises due to the sensitivity of gain K to environmental factors such as temperature, light, humidity, and vibration, and also as a result of the variation of K due to aging. Drift in an op-amp can be significant and steps should be taken to eliminate that problem.

It is virtually impossible to avoid the drift in gain and the frequency response error in an operational amplifier. But an ingenious way has been found to remove the effect of these two problems at the amplifier output. Since gain K is very large, by using feedback we can virtually eliminate its effect at the amplifier output. This *closed loop* form of an op-amp has the advantage that the characteristics and the accuracy of the output of the overall circuit depends on the passive components (e.g., resistors and capacitors) in it, which can be provided at high precision, and not the parameters of the op amp itself. The closed loop form is preferred in almost every application; in particular, *voltage follower* and *charge amplifier* are devices that use the properties of high Z_i, low Z_o, and high K of an op-amp along with feedback through a high-precision resistor, to eliminate errors due to nonconstant K. In summary, operational amplifier is not very useful in its open-loop form, particularly because gain K is not steady. But since K is very large, the problem can be removed by using feedback. It is this closed-loop form that is commonly used in practical applications of an op-amp.

In addition to the unsteady nature of gain, there are other sources of error that contribute to less-than-ideal performance of an operational amplifier circuit. Noteworthy are:

i. The *offset current* present at the input leads due to bias currents that are needed to operate the solid-state circuitry.
ii. The *offset voltage* that might be present at the output even when the input leads are open.
iii. The unequal gains corresponding to the two input leads (i.e., the *inverting gain* not equal to the *noninverting gain*).

Such problems can produce nonlinear behavior in op-amp circuits, and they can be reduced by proper circuit design and through the use of compensating circuit elements.

4.3.2   Voltage, Current, and Power Amplifiers

Any type of amplifier can be constructed from scratch in the monolithic form as an IC chip, or in the discrete form as a circuit containing several discrete elements such as discrete bipolar junction transistors or discrete field effect transistors, discrete diodes, and discrete resistors. But, almost all types of amplifiers can also be built using operational amplifier as the basic building block. Since we are already familiar with op-amps and since op-amps are extensively used in general amplifier circuitry, we prefer to use the latter approach, which uses discrete op-amps for building general amplifiers. Furthermore, modeling, analysis, and design of a general amplifier may be performed on this basis.

If an electronic amplifier performs a voltage amplification function, it is termed a *voltage amplifier*. These amplifiers are so common that, the term "amplifier" is often used to denote a voltage amplifier. A voltage amplifier can be modeled as

$$v_o = K_v v_i \qquad (4.11)$$

in which
   v_o = output voltage
   v_i = input voltage
   K_v = voltage gain

Voltage amplifiers are used to achieve voltage compatibility (or level shifting) in circuits.

Current amplifiers are used to achieve current compatibility in electronic circuits. A *current amplifier* may be modeled by

$$i_o = K_i i_i \qquad (4.12)$$

in which
   i_o = output current
   i_i = input current
   K_i = current gain

A voltage follower has $K_v = 1$ and, hence, it may be considered as a current amplifier. Besides, it provides impedance compatibility and acts as a buffer between a low-current (high-impedance) output device (signal source or the device that provides the signal) and a high-current (low-impedance) input device (signal receiver or the device that receives the signal) that are interconnected. Hence, the name *buffer amplifier* or *impedance transformer* is sometimes used for a current amplifier with unity voltage gain.

If the objective of signal amplification is to upgrade the associated power level, then a *power amplifier* should be used for that purpose. A simple model for a power amplifier is

$$p_o = K_p p_i \qquad (4.13)$$

in which
   p_o = output power
   p_i = input power
   K_p = power gain

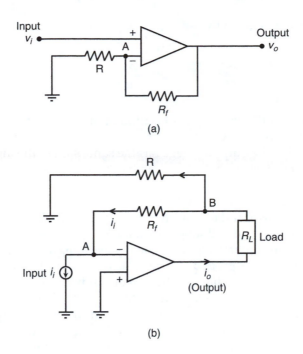

FIGURE 4.8
(a) A voltage amplifier, (b) A current amplifier.

It is easy to see from Equation 4.11, Equation 4.12, and Equation 4.13 that

$$K_p = K_v K_i \tag{4.14}$$

Note that all three types of amplification could be achieved simultaneously from the same amplifier. Furthermore, a current amplifier with unity voltage gain (e.g., a voltage follower) is a power amplifier as well. Usually voltage amplifiers and current amplifiers are used in the first stages of a signal path (e.g., sensing, data acquisition and signal generation) where signal levels and power levels are relatively low, while power amplifiers are typically used in the final stages (e.g., final control, actuation, recording, display) where high signal levels and power levels are usually required.

Figure 4.8(a) gives an op-amp circuit for a voltage amplifier. Note the feedback resistor R_f, which serves the purposes of stabilizing the op-amp and providing an accurate voltage gain. The negative lead is grounded through an accurately-known resistor R. To determine the voltage gain, recall that the voltages at the two input leads of an op-amp should be equal (in the ideal case). The input voltage v_i is applied to the positive lead of the op-amp. Then the voltage at point A should also be equal to v_i. Next, recall that the current through the input lead of an op-amp is ideally zero. Hence, by writing the current balance equation for the node point A we have,

$$\frac{v_o - v_i}{R_f} = \frac{v_i}{R}$$

This gives the amplifier equation

$$v_o = \left(1 + \frac{R_f}{R}\right) v_i \tag{4.15}$$

Hence, the voltage gain is given by

$$K_v = 1 + \frac{R_f}{R} \qquad (4.16)$$

Note the K_v depends on R and R_f and not on the op-amp gain. Hence, the voltage gain can be accurately determined by selecting the two passive elements (resistors) R and R_f precisely. Also, note that the output voltage has the same sign as the input voltage. Hence, this is a *noninverting amplifier*. If the voltages are of the opposite sign, we have an *inverting amplifier*.

A current amplifier is shown in Figure 4.8(b). The input current i_i is applied to the negative lead of the op-amp as shown, and the positive lead is grounded. There is a feedback resistor R_f connected to the negative lead through the load R_L. The resistor R_f provides a path for the input current since the op-amp takes in virtually zero current. There is a second resistor R through which the output is grounded. This resistor is needed for current amplification. To analyze the amplifier, use the fact that the voltage at point A (i.e., at the negative lead) should be zero because the positive lead of the op-amp is grounded (zero voltage). Furthermore, the entire input current i_i passes through the resistor R_f as shown. Hence, the voltage at point B is $R_f i_i$. Consequently, current through the resistor R is $R_f i_i / R$, which is positive in the direction shown. It follows that the output current i_o is given by

$$i_o = i_i + \frac{R_f}{R} i_i$$

or

$$i_o = \left(1 + \frac{R_f}{R} \right) i_i \qquad (4.17)$$

The current gain of the amplifier is

$$K_i = 1 + \frac{R_f}{R} \qquad (4.18)$$

As before, the amplifier gain can be accurately set using the high-precision resistors R and R_f.

4.3.3   Instrumentation Amplifiers

An instrumentation amplifier is typically a special-purpose voltage amplifier dedicated to instrumentation applications. Examples include amplifiers used for producing the output from a bridge circuit (bridge amplifier) and amplifiers used with various sensors and transducers. An important characteristic of an instrumentation amplifier is the adjustable-gain capability. The gain value can be adjusted manually in most instrumentation amplifiers. In more sophisticated instrumentation amplifiers the gain is *programmable* and can be set by means of digital logic. Instrumentation amplifiers are normally used with low-voltage signals.

4.3.3.1 Differential Amplifier

Usually, an instrumentation amplifier is also a *differential amplifier* (sometimes termed *difference amplifier*). Note that in a differential amplifier both input leads are used for signal input, whereas in a single-ended amplifier one of the leads is grounded and only one lead is used for signal input. Ground-loop noise can be a serious problem in single-ended amplifiers. Ground-loop noise can be effectively eliminated using a differential amplifier because noise loops are formed with both inputs of the amplifier and, hence, these noise signals are subtracted at the amplifier output. Since the noise level is almost the same for both inputs, it is canceled out. Any other noise (e.g., 60 Hz line noise) that might enter both inputs with the same intensity will also be canceled out at the output of a differential amplifier.

A basic differential amplifier that uses a single op-amp is shown in Figure 4.9(a). The input-output equation for this amplifier can be obtained in the usual manner. For instance, since current through an op-amp is negligible, the current balance at point B gives

$$\frac{v_{i2} - v_B}{R} = \frac{v_B}{R_f} \tag{i}$$

in which v_B is the voltage at B. Similarly, current balance at point A gives

$$\frac{v_o - v_A}{R_f} = \frac{v_A - v_{i1}}{R} \tag{ii}$$

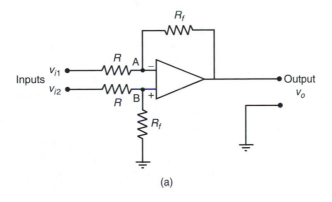

(a)

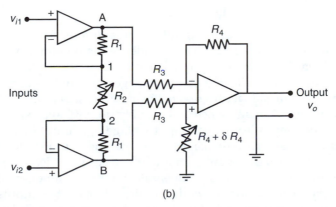

(b)

FIGURE 4.9
(a) A basic differential amplifier, (b) A basic instrumentation amplifier.

Now we use the property

$$v_A = v_B \tag{iii}$$

for an operational amplifier, to eliminate v_A and v_B from Equation i and Equation ii. This gives

$$\frac{v_{i2}}{(1+R/R_f)} = \frac{(v_o R/R_f + v_{i1})}{(1+R/R_f)}$$

or

$$v_o = \frac{R_f}{R}(v_{i2} - v_{i1}) \tag{4.19}$$

Two things are clear from Equation 4.19. First, the amplifier output is proportional to the "difference" and not the absolute value of the two inputs v_{i1} and v_{i2}. Second, voltage gain of the amplifier is R_f/R. This is known as the *differential gain*. It is clear that the differential gain can be accurately set by using high-precision resistors R and R_f.

The basic differential amplifier, shown in Figure 4.9(a) and discussed above, is an important component of an instrumentation amplifier. In addition, an instrumentation amplifier should possess the capability of adjustable gain. Furthermore, it is desirable to have a very high input impedance and very low output impedance at each input lead. It is desirable for an instrumentation amplifier to possess a higher and more stable gain, and also a higher input impedance than a basic differential amplifier. An instrumentation amplifier that possesses these basic requirements may be fabricated in the monolithic IC form as a single package. Alternatively, it may be built using three differential amplifiers and high precision resistors, as shown in Figure 4.9(b). The amplifier gain can be adjusted using the fine-tunable resistor R_2. Impedance requirements are provided by two voltage-follower type amplifiers, one for each input, as shown. The variable resistance δR_4 is necessary to compensate for errors due to unequal common-mode gain. Let us first consider this aspect and then obtain an equation for the instrumentation amplifier.

4.3.3.2   Common Mode

The voltage that is "common" to both input leads of a differential amplifier is known as the *common-mode voltage*. This is equal to the smaller of the two input voltages. If the two inputs are equal, then the common-mode voltage is obviously equal to each one of the two inputs. When $v_{i1} = v_{i2}$, ideally, the output voltage v_o should be zero. In other words, ideally, any common-mode signals are rejected by a differential amplifier. But, since commercial op-amps are not ideal and since they usually do not have exactly identical gains with respect to the two input leads, the output voltage v_o will not be zero when the two inputs are identical. This *common-mode error* can be compensated for by providing a variable resistor with fine resolution at one of the two input leads of the differential amplifier. Hence, in Figure 4.9(b), to compensate for the common-mode error (i.e., to achieve a satisfactory level of common-mode rejection), first the two inputs are made equal and then δR_4 is varied carefully until the output voltage level is sufficiently small (minimum). Usually, δR_4 that is required to achieve this compensation is small compared to the nominal feedback resistance R_4.

Since ideally $\delta R_4 = 0$ we can neglect δR_4 in the derivation of the instrumentation amplifier equation. Now, note from a basic property of an op-amp with no saturation (specifically, the voltages at the two input leads have to be almost identical) that in Figure 4.9(b), the voltage at point 2 should be v_{i2} and the voltage at point 1 should be v_{i1}. Next we use the property that the current through each input lead of an op-amp is negligible. Accordingly, current through the circuit path $B \rightarrow 2 \rightarrow 1 \rightarrow A$ has to be the same. This gives the current continuity equations

$$\frac{v_B - v_{i2}}{R_1} = \frac{v_{i2} - v_{i1}}{R_2} = \frac{v_{i1} - v_A}{R_1}$$

in which v_A and v_B are the voltages at points A and B, respectively. Hence, we get the following two equations:

$$v_B = v_{i2} + \frac{R_1}{R_2}(v_{i2} - v_{i1})$$

$$v_A = v_{i1} - \frac{R_1}{R_2}(v_{i2} - v_{i1})$$

Now, by subtracting the second equation from the first, we have the equation for the first stage of the amplifier; thus,

$$v_B - v_A = \left(1 + \frac{2R_1}{R_2}\right)(v_{i2} - v_{i1}) \tag{i}$$

Next from the previous result (see Equation 4.19) for a differential amplifier, we have (with $\delta R_4 = 0$)

$$v_o = \frac{R_4}{R_3}(v_B - v_A) \tag{ii}$$

Note that only the resistor R_2 is varied to adjust the gain (differential gain) of the amplifier. In Figure 4.9(b), the two input op-amps (the voltage-follower op-amps) do not have to be identical as long as the resistors R_1 and R_2 are chosen to be accurate. This is so because the op-amp parameters such as open-loop gain and input impedance do not enter into the amplifier equations provided that their values are sufficiently high, as noted earlier.

4.3.4 Amplifier Performance Ratings

Main factors that affect the performance of an amplifier are

1. Stability
2. Speed of response (bandwidth, slew rate)
3. Unmodeled signals

We have already discussed the significance of some of these factors.

The level of stability of an amplifier, in the conventional sense, is governed by the dynamics of the amplifier circuitry, and may be represented by a *time constant*. But more important consideration for an amplifier is the "parameter variation" due to aging, temperature, and other environmental factors. Parameter variation is also classified as a stability issue, in the context of devices such as amplifiers, because it pertains to the steadiness of the response when the input is maintained steady. Of particular importance is the *temperature drift*. This may be specified as a drift in the output signal per unity change in temperature (e.g., $\mu V/°C$).

The speed of response of an amplifier dictates the ability of the amplifier to faithfully respond to transient inputs. Conventional time-domain parameters such as *rise time* may be used to represent this. Alternatively, in the frequency domain, speed of response may be represented by a *bandwidth* parameter. For example, the frequency range over which the frequency response function is considered constant (flat) may be taken as a measure of bandwidth. Since there is some nonlinearity in any amplifier, bandwidth can depend on the signal level itself. Specifically, *small-signal bandwidth* refers to the bandwidth that is determined using small input signal amplitudes.

Another measure of the speed of response is the *slew rate*, which is defined as the largest possible rate of change of the amplifier output for a particular frequency of operation. Since for a given input amplitude, the output amplitude depends on the amplifier gain, slew rate is usually defined for unity gain.

Ideally, for a linear device, the frequency response function (transfer function) does not depend on the output amplitude (i.e., the product of the dc gain and the input amplitude). But for a device that has a limited slew rate, the bandwidth (or the maximum operating frequency at which output distortions may be neglected) will depend on the output amplitude. The larger the output amplitude, the smaller the bandwidth for a given slew rate limit. A bandwidth parameter that is usually specified for a commercial op-amp is the *gain-bandwidth product* (GBP). This is the product of the open-loop gain and the bandwidth of the op-amp. For example, for an op-amp with GBP = 15 MHz and an open-loop gain of 100 dB (i.e., 10^5), the bandwidth = $15 \times 10^6/10^5$ Hz = 150 Hz. Clearly, this bandwidth value is rather low. Since, the gain of an op-amp with feedback is significantly lower than 100 dB, its effective bandwidth is much higher than that of an open-loop op-amp.

Example 4.5

Obtain a relationship between the slew rate and the bandwidth for a slew rate-limited device. An amplifier has a slew rate of 1 V/μs. Determine the bandwidth of this amplifier when operating at an output amplitude of 5 V.

SOLUTION

Clearly, the amplitude of the rate of change signal divided by the amplitude of the output signal gives an estimate of the output frequency. Consider a sinusoidal output voltage given by

$$v_o = a \sin 2\pi f t \tag{4.20}$$

The rate of change of output is

$$\frac{dv_o}{dt} = 2\pi f a \cos 2\pi f t$$

Hence, the maximum rate of change of output is $2\pi fa$. Since this corresponds to the slew rate when f is the maximum allowable frequency, we have

$$s = 2\pi f_b a \tag{4.21}$$

in which

   s = slew rate
   f_b = bandwidth
   a = output amplitude

Now, with $s = 1$ V/µs and $a = 5$ V we get

$$f_b = \frac{1}{2\pi} \times \frac{1}{1 \times 10^{-8}} \times \frac{1}{5} \text{Hz}$$

$$= 31.8 \text{ kHz}$$

We have noted that stability problems and frequency response errors are prevalent in the open loop form of an operational amplifier. These problems can be eliminated using feedback because the effect of the open loop transfer function on the closed loop transfer function is negligible if the open loop gain is very large, which is the case for an operational amplifier.

Unmodeled signals can be a major source of amplifier error, and these signals include

1. Bias currents
2. Offset signals
3. Common mode output voltage
4. Internal noise

In analyzing operational amplifiers we assume that the current through the input leads is zero. This is not strictly true because bias currents for the transistors within the amplifier circuit have to flow through these leads. As a result, the output signal of the amplifier will deviate slightly from the ideal value.

Another assumption that we make in analyzing op-amps is that the voltage is equal at the two input leads. In practice, however, offset currents and voltages are present at the input leads, due to minute discrepancies inherent to the internal circuits within an op-amp.

4.3.4.1 Common-Mode Rejection Ratio (CMRR)

Common-mode error in a differential amplifier was discussed earlier. We note that ideally the common mode input voltage (the voltage common to both input leads) should have no effect on the output voltage of a differential amplifier. But, since any practical amplifier has some unbalances in the internal circuitry (e.g., gain with respect to one input lead is not equal to the gain with respect to the other input lead and, furthermore, bias signals are needed for operation of the internal circuitry), there will be an error voltage at the output, which depends on the common-mode input. Common-mode rejection ratio of a differential amplifier is defined as

$$\text{CMRR} = \frac{K v_{cm}}{v_{ocm}} \tag{4.22}$$

in which

K = gain of the differential amplifier (i.e., differential gain)

v_{cm} = common-mode voltage (i.e., voltage common to both input leads)

v_{ocm} = common-mode output voltage (i.e., output voltage due to common-mode input voltage)

Note that ideally $v_{ocm} = 0$ and CMRR should be infinity. It follows that the larger the CMRR the better the differential amplifier performance.

The three types of unmodeled signals mentioned above can be considered as noise. In addition, there are other types of noise signals that degrade the performance of an amplifier. For example, ground-loop noise can enter the output signal. Furthermore, stray capacitances and other types of unmodeled circuit effects can generate internal noise. Usually in amplifier analysis, unmodeled signals (including noise) can be represented by a noise voltage source at one of the input leads. Effects of unmodeled signals can be reduced by using suitably connected compensating circuitry including variable resistors that can be adjusted to eliminate the effect of unmodeled signals at the amplifier output (e.g., see δR_4 in Figure 4.9(b)). Some useful information about operational amplifiers is summarized in Box 4.1.

4.3.4.2   AC-Coupled Amplifiers

The dc component of a signal can be blocked off by connecting the signal through a capacitor (Note that the impedance of a capacitor is $1/(j\omega C)$ and hence, at zero frequency there will be an infinite impedance). If the input lead of a device has a series capacitor, we say that the input is ac-coupled and if the output lead has a series capacitor, then the output is ac-coupled. Typically, an ac-coupled amplifier has a series capacitor both at the input lead and the output lead. Hence, its frequency response function will have a high-pass characteristic; in particular, the dc components will be filtered out. Errors due to bias currents and offset signals are negligible for an ac-coupled amplifier. Furthermore, in an ac-coupled amplifier, stability problems are not very serious.

4.3.5   Ground Loop Noise

In instruments that handle low-level signals (e.g., sensors such as accelerometers; signal conditioning circuitry such as strain gage bridges; and sophisticated and delicate electronic components such as computer disk drives and automobile control modules) electrical noise can cause excessive error unless proper corrective actions are taken. One form of noise is caused by fluctuating magnetic fields due to nearby ac power lines or electric machinery. This is commonly known as *electromagnetic interference* (EMI). This problem can be avoided by removing the source of EMI so that fluctuating external magnetic fields and currents are not present near the affected instrument. Another solution would be to use *fiber optic* (optically coupled) signal transmission so that there is no noise conduction along with the transmitted signal from the source to the subject instrument. In the case of hard-wired transmission, if the two signal leads (positive and negative or hot and neutral) are twisted or if shielded cables are used, the induced noise voltages become equal in the two leads, which cancel each other.

Proper grounding practices are important to mitigate unnecessary electrical noise problems and more importantly, to avoid electrical safety hazards. A standard single-phase ac outlet (120 V, 60 Hz) has three terminals, one carrying power (hot), the second being neutral, and the third connected to earth ground (which is maintained at zero potential rather uniformly from point to point in the power network). Correspondingly, the power

---

BOX 4.1

Operational Amplifiers.

Ideal Op-Amp Properties:
- Infinite open-loop differential gain
- Infinite input impedance
- Zero output impedance
- Infinite bandwidth
- Zero output for zero differential input

Ideal Analysis Assumptions:
- Voltages at the two input leads are equal
- Current through either input lead is zero

Definitions:

$$\text{Open-loop gain} = \left| \frac{\text{Output voltage}}{\text{Voltage difference at input leads}} \right| \text{ with no feedback.}$$

$$\text{Input impedance} = \frac{\text{Voltage between an input lead and ground}}{\text{Current through that lead}}$$

(with other input lead grounded and the output in open circuit)

$$\text{Output impedance} = \frac{\text{Voltage between output lead and ground in open circuit}}{\text{Closed-circuit current}}$$

(with normal input conditions)
- Bandwidth = frequency range in which the frequency response is flat (gain is constant).
- Gain bandwidth product (GBP) = Openloop gain × Bandwidth at that gain
- Input bias current = average (dc) current through one input lead
- Input offset current = difference in the two input bias currents
- Differential input voltage = voltage at one input lead with the other grounded when the output voltage is zero.

$$\text{Common-mode gain} = \frac{\text{Output voltage when input leads are at the same voltage}}{\text{Common input voltage}}$$

$$\text{Common-mode rejection ratio (CMRR)} = \frac{\text{Open loop differential gain}}{\text{Common-mode gain}}$$

- Slew rate = rate of change of output of a unity-gain op-amp, for a step input

---

plug of an instrument should have three prongs. The shorter flat prong is connected to a black wire (hot) and the longer flat prong is connected to a white wire (neutral). The round prong is connected to a green wire (ground), which at the other end is connected to the chassis (or, casing) of the instrument (chassis ground). In view of grounding the chassis in this manner, the instrument housing is maintained at zero potential even in the presence of a fault in the power circuit (e.g., a leakage or a short). The power circuitry of an instrument also has a local ground (signal ground), with reference to which its power signal is measured. This is a sufficiently thick conductor within the instrument and it provides a common and uniform reference of 0 V. Consider the sensor signal conditioning example shown in Figure 4.10. The dc power supply can provide both positive (+) and negative (−) outputs. Its zero voltage reference is denoted by COM, and it is the common ground (signal ground) of the device. It should be noted that COM of the dc power supply is not connected to the chassis ground, the latter being connected to the earth ground through the round prong of the power plug of the power supply. This is necessary to avoid the danger of an electric shock. Note that COM of the power supply is connected

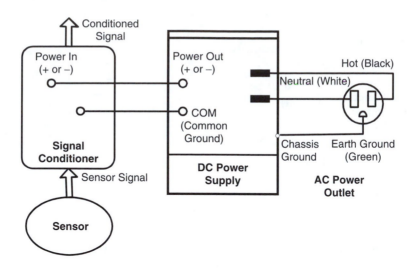

FIGURE 4.10
An example of grounding of instruments.

to the signal ground of the signal conditioning module. In this manner a common 0 V reference is provided for the dc voltage that is supplied to the signal conditioning module.

A main cause of electrical noise is the ground loops, which are created due to improper grounding of instruments. If two interconnected instruments are grounded at two separate locations that are far apart (multiple grounding), ground loop noise can enter the signal leads because of the possible potential difference between the two ground points. The reason is that ground itself is not generally a uniform-potential medium, and a nonzero (and finite) impedance may exist from point to point within this medium. This is, in fact, the case with a typical ground medium such as a common ground wire. An example is shown schematically in Figure 4.11(a). In this example, the two leads of a sensor are directly connected to a signal-conditioning device such as an amplifier, one of its input leads (+) being grounded (at point B). The 0 V reference lead of the sensor is grounded through its housing to the earth ground (at point A). Because of nonuniform ground potentials, the two ground points A and B are subjected to a potential difference v_g. This will create a ground loop with the common reference lead, which interconnects the two devices. The solution to this problem is to isolate (i.e., provide an infinite impedance to) either one of the two devices. Figure 4.11b shows internal isolation of the sensor. External isolation, by insulating the housing of the sensor, will also remove the ground loop. Floating off the common ground (COM) of a power supply (see Figure 4.10) is another approach to eliminating ground loops. Specifically, COM is not connected to earth ground.

4.4   Analog Filters

A filter is a device that allows through only the desirable part of a signal, rejecting the unwanted part. Unwanted signals can seriously degrade the performance of a mechatronic system. External disturbances, error components in excitations, and noise generated internally within system components and instrumentation are such spurious signals, which may be removed by a filter. As well, a filter is capable of shaping a signal in a desired manner.

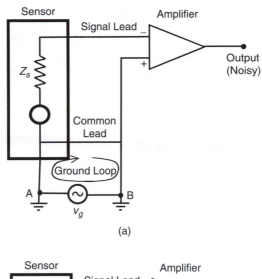

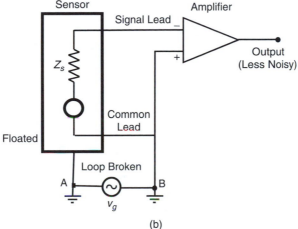

FIGURE 4.11
(a) Illustration of a ground loop, (b) Device isolation to eliminate ground loops (an example of internal isolation).

In typical applications of acquisition and processing of signals in a mechatronic system, the filtering task would involve the removal of signal components in a specific frequency range. In this context we can identify the following four broad categories of filters:

1. Low-pass filters
2. High-pass filters
3. Band-pass filters
4. Band-reject (or notch) filters

The ideal frequency-response characteristic of each of these four types of filters is shown in Figure 4.12. Note that only the magnitude of the frequency response function (magnitude of the frequency transfer function) is shown. It is understood, however, that the phase distortion of the input signal also should be small within the *pass band* (the allowed frequency range). Practical filters are less than ideal. Their frequency response functions do not exhibit sharp cutoffs as in Figure 4.12 and, furthermore, some phase distortion will be unavoidable.

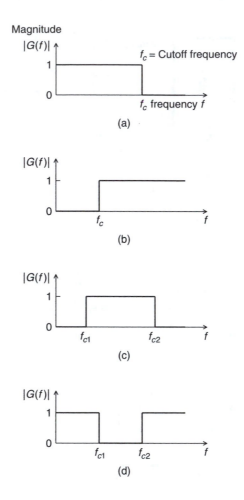

FIGURE 4.12
Ideal filter characteristics: (a) Low-pass filter, (b) High-pass filter, (c) Band-pass filter, (d) Band-reject (notch) filter.

A special type of band-pass filter that is widely used in acquisition and monitoring of response signals (e.g., in product dynamic testing) is *tracking filter*. This is simply a band-pass filter with a narrow pass band that is frequency-tunable. The center frequency (mid value) of the pass band is variable, usually by coupling it to the frequency of a carrier signal (e.g., drive signal). In this manner, signals whose frequency varies with some basic variable in the system (e.g., rotor speed, frequency of a harmonic excitation signal, frequency of a sweep oscillator) can be accurately tracked in the presence of noise. The inputs to a tracking filter are the signal that is being tracked and the variable *tracking frequency* (*carrier input*). A typical tracking filter that can simultaneously track two signals is schematically shown in Figure 4.13.

Filtering can be achieved by *digital filters* as well as *analog filters*. Before digital signal processing became efficient and economical, analog filters were exclusively used for signal filtering, and are still widely used. An analog filter is typically an *active filter* containing active components such as transistors or op-amps. In an analog filter, the input signal is passed through an analog circuit. Dynamics of the circuit will determine which (desired) signal components would be allowed through and which (unwanted) signal components would be rejected. Earlier versions of analog filters employed discrete circuit elements such as discrete transistors, capacitors, resistors and even discrete inductors. Since inductors

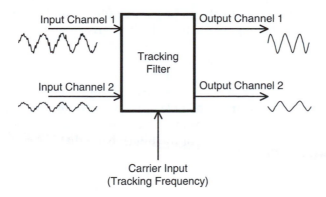

FIGURE 4.13
Schematic representation of a two-channel tracking filter.

have several shortcomings such as susceptibility to electromagnetic noise, unknown resistance effects, and large size, today they are rarely used in filter circuits. Furthermore, due to well-known advantages of integrated circuit (IC) devices, today analog filters in the form of monolithic IC chips are extensively used in mechatronic applications and are preferred over discrete-element filters. Digital filters, which employ digital signal processing to achieve filtering, are also widely used today.

4.4.1 Passive Filters and Active Filters

Passive analog filters employ analog circuits containing passive elements such as resistors and capacitors (and sometimes inductors) only. An external power source is not needed in a passive filter. Active analog filters employ active elements and components such as transistors and operational amplifiers in addition to passive elements. Since external power is needed for the operation of the active elements and components, an active filter is characterized by the need of an external power supply. Active filters are widely available in a monolithic integrated-circuit (IC) package and are usually preferred over passive filters.
   Advantages of active filters include the following:

1. Loading effects and interaction with other components are negligible because active filters can provide a very high input impedance and a very low output impedance.
2. They can be used with low signal levels because both signal amplification and filtering can be provided by the same active circuit.
3. They are widely available in a low-cost and compact integrated-circuit form.
4. They can be easily integrated with digital devices.
5. They are less susceptible to noise from electromagnetic interference.

Commonly mentioned disadvantages of active filters are the following:

1. They need an external power supply.
2. They are susceptible to "saturation" type nonlinearity at high signal levels.
3. They can introduce many types of internal noise and unmodeled signal errors (offset, bias signals, etc.).

Note that advantages and disadvantages of passive filters can be directly inferred from the disadvantages and advantages of active filters as given above.

4.4.1.1   Number of Poles

Analog filters are dynamic systems and they can be represented by transfer functions, assuming linear dynamics. Number of poles of a filter is the number of poles in the associated transfer function. This is also equal to the order of the characteristic polynomial of the filter transfer function (i.e., *order* of the filter). Note that poles (or, eigenvalues) are the roots of the characteristic equation.

In our discussion we will show simplified versions of filters, typically consisting of a single filter stage. Performance of such a basic filter can be improved at the expense of circuit complexity (and increased pole count). Only simple discrete-element circuits are shown for passive filters. Basic operational-amplifier circuits are given for active filters. Even here, much more complex devices are commercially available, but our purpose is to illustrate the underlying principles rather than to provide complete descriptions and data sheets for commercial filters.

4.4.2   Low-Pass Filters

The purpose of a low-pass filter is to allow through all signal components below a certain (cutoff) frequency and block off all signal components above that cutoff. Analog low-pass fitters are widely used as *antialiasing filters* in digital signal processing. An error known as aliasing will enter the digitally processed results of a signal if the original signal has frequency components above half the *sampling frequency* (half the sampling frequency is called the *Nyquist frequency*). Hence, aliasing distortion can be eliminated if the signal is filtered using a low-pass filter with its cutoff set at Nyquist frequency, prior to sampling and digital processing (See Chapter 5). This is one of numerous applications of analog low-pass filters. Another typical application would be to eliminate high-frequency noise in a measured system response.

A single-pole passive low-pass filter circuit is shown in Figure 4.14(a). An active filter corresponding to the same low-pass filter is shown in Figure 4.14(b). It can be shown that the two circuits have identical transfer functions. Hence, it might seem that the op-amp in Figure 4.14(b) is redundant. This is not true, however. If two passive filter stages, each similar to Figure 4.14(a) are connected together, the overall transfer function is not equal to the product of the transfer functions of the individual stages. The reason for this apparent ambiguity is the circuit loading (interaction) that arises due to the fact that the input impedance of the second stage is not sufficiently larger than the output impedance of the first stage. But, if two active filter stages similar to Figure 4.14(b) are connected together, such loading errors will be negligible because the op-amp with feedback (i.e., a voltage follower) introduces a very high input impedance and very low output impedance, while maintaining the voltage gain at unity. With similar reasoning it can be concluded that an active filter has the desirable property of very low interaction with any other connected component.

To obtain the filter equation for Figure 4.14(a) note that since the output is in open circuit (zero load current), the current through capacitor C is equal to the current through resistor R. Hence,

$$C\frac{dv_o}{dt} = \frac{v_i - v_o}{R}$$

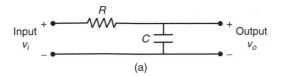

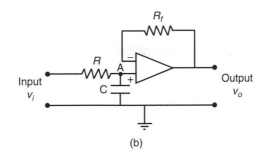

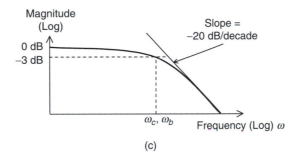

FIGURE 4.14
A single-pole low-pass filter: (a) A passive filter stage, (b) An active filter stage, (c) The frequency response characteristic.

or

$$\tau \frac{dv_o}{dt} + v_o = v_i \tag{4.23}$$

where the filter time constant is

$$\tau = RC \tag{4.24}$$

Now from Equation 4.23 it follows that the filter transfer function is

$$\frac{v_o}{v_i} = G(s) = \frac{1}{(\tau s + 1)} \tag{4.25}$$

From this transfer function it is clear that an analog low-pass filter is essentially a *lag circuit* (i.e., it provides a phase lag).

It can be shown that the active filter stage in Figure 4.14(b) has the same input/output equation. First, since current through an op-amp lead is almost zero, it follows from the previous analysis of the passive circuit stage that

$$\frac{v_A}{v_i} = \frac{1}{(\tau s + 1)} \tag{i}$$

in which v_A is the voltage at the node point A. Now, since the op-amp with feedback resistor is in fact a voltage follower, we have

$$\frac{v_o}{v_A} = 1 \tag{ii}$$

Next, by combining Equation i and Equation ii we get Equation 4.25 as required. Repeating, a main advantage of the active filter version is that the resulting loading error is negligible.

The frequency response function corresponding to Equation 4.25 is obtained by setting $s = j\omega$; thus

$$G(j\omega) = \frac{1}{(\tau j\omega + 1)} \tag{2.26}$$

This gives the response of the filter when a sinusoidal signal of frequency ω is applied. The magnitude $|G(j\omega)|$ of the frequency transfer function gives the signal amplification and phase angle $\angle G(j\omega)$ gives the phase lead of the output signal with respect to the input. The magnitude curve (*Bode magnitude curve*) is shown in Figure 4.14(c). Note from Equation 4.26 that for small frequencies (i.e., $\omega \ll 1/\tau$) the magnitude is approximately unity. Hence, $1/\tau$ can be considered the cutoff frequency ω_c:

$$\omega_c = \frac{1}{\tau} \tag{4.27}$$

Example 4.6
Show that the cutoff frequency given by Equation 4.27 is also the *half-power bandwidth* for the low-pass filter. Show that for frequencies much larger than this, the filter transfer function on the Bode magnitude plane (i.e., log magnitude versus log frequency) can be approximated by a straight line with slope −20 dB/decade. This slope is known as the *roll-off rate*.

SOLUTION
The frequency corresponding to half power (or $1/\sqrt{2}$ magnitude) is given by

$$\frac{1}{|\tau j\omega + 1|} = \frac{1}{\sqrt{2}}$$

By cross multiplying, squaring, and simplifying the equation we get

$$\tau^2 \omega^2 = 1$$

Hence, the half-power bandwidth is

$$\omega_b = \frac{1}{\tau} \tag{2.28}$$

This is identical to the cutoff frequency give by Equation 4.27.

Now for $\omega \gg 1/\tau$ (i.e., $\tau\omega \gg 1/\tau$) Equation 4.26 can be approximated by

$$G(j\omega) = \frac{1}{\tau j\omega}$$

This has the magnitude

$$|G(j\omega)| = \frac{1}{\tau\omega}$$

Converting to the log scale,

$$\log_{10}|G(j\omega)| = -\log_{10}\omega - \log_{10}\tau$$

It follows that the $\log_{10}$ (magnitude) versus $\log_{10}$ (frequency) curve is a straight line with slope −1. In other words when frequency increases by a factor of 10 (i.e., a decade) the $\log_{10}$ magnitude decreases by unity (i.e., by 20 dB). Hence, the roll-off rate is −20 dB/decade. These observations are shown in Figure 4.14(c). Note that an amplitude change by a factor of $\sqrt{2}$ (or power by a factor of 2) corresponds to 3 dB. Hence, when the dc (zero-frequency magnitude) value is unity (0 dB), the half power magnitude is −3 dB.

Cutoff frequency and the roll-off rate are the two main design specifications for a low-pass filter. Ideally, we would like a low-pass filter magnitude curve to be flat up to the required pass-band limit (cutoff frequency) and then roll off very rapidly. The low-pass filter shown in Figure 4.14 only approximately meets these requirements. In particular, the roll-off rate is not large enough. We would prefer a roll-off rate of at least −40 dB/decade and even −60 dB/decade in practical filters. This can be realized by using a high order filter (i.e., a filter having many poles). Low-pass Butterworth filter is of this type and is widely used.

4.4.2.1 Low-Pass Butterworth Filter

A low-pass Butterworth filter having 2 poles can provide a roll-off rate of −40 dB/decade and, one having 3 poles can provide a roll-off rate of −60 dB/decade. Furthermore, the steeper the roll-off slope, the flatter the filter magnitude curve within the pass band.

A two-pole, low-pass Butterworth filter is shown in Figure 4.15. We could construct a two-pole filter simply by connecting together two single-pole stages of the type shown in

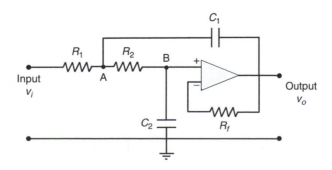

FIGURE 4.15
A two-pole low-pass Butterworth filter.

Figure 4.14(b). Then, we would require two op-amps, whereas the circuit shown in Figure 4.15 achieves the same objective by using only one op-amp (i.e., at a lower cost).

Example 4.7
Show that the op-amp circuit in Figure 4.15 is a low-pass filter having two poles. What is the transfer function of the filter? Estimate the cutoff frequency under suitable conditions. Show that the roll-off rate is –40 dB/decade.

SOLUTION
To obtain the filter equation, we write the current balance equations first. Specifically, the sum of the currents through R_1 and C_1 passes through R_2. The same current has to pass through C_2 because the current through the op-amp lead is zero (a property of an op-amp). Hence,

$$\frac{v_i - v_A}{R_1} + C_1 \frac{d}{dt}(v_o - v_A) = \frac{v_A - v_B}{R_2} = C_2 \frac{dv_B}{dt} \tag{i}$$

Also, since the op-amp with a feedback resistor R_f is a voltage follower (with unity gain), we have

$$v_B = v_o \tag{ii}$$

From Equation i and Equation ii we get

$$\frac{v_i - v_A}{R_1} + C_1 \frac{dv_o}{dt} - C_1 \frac{dv_A}{dt} = C_2 \frac{dv_o}{dt} \tag{iii}$$

$$\frac{v_A - v_o}{R_2} = C_2 \frac{dv_o}{dt} \tag{iv}$$

Now defining the constants

$$\tau_1 = R_1 C_1 \tag{4.29}$$

$$\tau_2 = R_2 C_2 \tag{4.30}$$

$$\tau_3 = R_1 C_2 \tag{4.31}$$

and introducing the Laplace variable s, we can eliminate v_A by substituting Equation iv into Equation iii; thus,

$$\frac{v_o}{v_i} = \frac{1}{[\tau_1\tau_2 s^2 + (\tau_2 + \tau_3)s + 1]} = \frac{\omega_n^2}{[s^2 + 2\zeta\omega_n s + \omega_n^2]} \tag{4.32}$$

This second order transfer function becomes oscillatory if $(\tau_2 + \tau_3)^2 < 4\tau_1\tau_2$. Ideally, we would like to have a zero resonant frequency, which corresponds to a damping ratio value

$\zeta = 1/\sqrt{2}$. Since the undamped natural frequency is

$$\omega_n = \frac{1}{\sqrt{\tau_1 \tau_2}} \tag{4.33}$$

the damping ratio is

$$\zeta = \frac{\tau_2 + \tau_3}{\sqrt{4\tau_1 \tau_2}} \tag{4.34}$$

and the resonant frequency is

$$\omega_r = \sqrt{1 - 2\zeta^2} \, \omega_n \tag{4.35}$$

we have, under ideal conditions (i.e., for $\omega_r = 0$),

$$(\tau_2 + \tau_3)^2 = 2\tau_1 \tau_2 \tag{4.36}$$

The frequency response function of the filter is (see Equation 4.32)

$$G(j\omega) = \frac{\omega_n^2}{\left[\omega_n^2 - \omega^2 + 2j\zeta\omega_n\omega\right]} \tag{4.37}$$

Now for $\omega \ll \omega_n$, , the filter frequency response is flat with a unity gain. For $\omega \gg \omega_n$, the filter frequency response can be approximated by

$$G(j\omega) = -\frac{\omega_n^2}{\omega^2}$$

In a log (magnitude) versus log (frequency) scale, this function is a straight line with slope = –2. Hence, when the frequency increases by a factor of 10 (i.e., one decade), the $\log_{10}$ (magnitude) drops by 2 units (i.e., 40 dB). In other words, the roll-off rate is – 40 dB/decade. Also, ω_n can be taken as the filter cutoff frequency. Hence,

$$\omega_c = \frac{1}{\sqrt{\tau_1 \tau_2}} \tag{4.38}$$

It can be easily verified that when $\zeta = 1/\sqrt{2}$, this frequency is identical to the half-power bandwidth (i.e, the frequency at which the transfer function magnitude becomes $1/\sqrt{2}$).

Note that if two single-pole stages (of the type shown in Figure 4.14(b)) are cascaded, the resulting two-pole filter has an overdamped (nonoscillatory) transfer function, and it is not possible to achieve $\zeta = 1/\sqrt{2}$ as in the present case. Also, note that a three-pole low-pass Butterworth filter can be obtained by cascading the two-pole unit shown in Figure 4.15 with a single-pole unit shown in Figure 4.14(b). Higher order low-pass Butterworth filters can be obtained in a similar manner by cascading an appropriate selection of basic units.

4.4.3   High-Pass Filters

Ideally, a high-pass filter allows through it all signal components above a certain (cutoff) frequency and blocks off all signal components below that frequency. A single-pole high-pass filter is shown in Figure 4.16. As for the low-pass filter that was discussed earlier, the passive filter stage (Figure 4.16(a)) and the active filter stage (Figure 4.16(b)) have identical transfer functions. The active filter is desired, however, because of its many advantages, including negligible loading error due to high input impedance and low output impedance of the op-amp voltage follower that is present in this circuit.

Filter equation is obtained by considering current balance in Figure 4.16(a), noting that the output is in open circuit (zero load current). Accordingly,

$$C\frac{d}{dt}(v_1 - v_o) = \frac{v_o}{R}$$

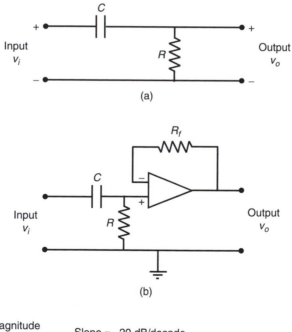

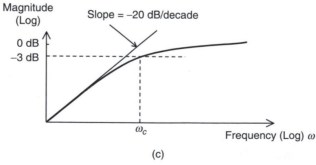

FIGURE 4.16
A single-pole high-pass filter: (a) A passive filter stage, (b) An active filter stage, (c) Frequency response characteristic.

or

$$\tau \frac{dv_o}{dt} + v_o = \tau \frac{dv_i}{dt} \tag{4.39}$$

in which the filter time constant

$$\tau = RC \tag{4.40}$$

Introducing the Laplace variable s, the filter transfer function is obtained as

$$\frac{v_o}{v_i} = G(s) = \frac{\tau s}{(\tau s + 1)} \tag{4.41}$$

Note that this corresponds to a "lead circuit" (i.e., an overall phase lead is provided by this transfer function), as discussed in Chapter 12. The frequency response function is

$$G(j\omega) = \frac{\tau j\omega}{(\tau j\omega + 1)} \tag{4.42}$$

Since its magnitude is zero for $\omega \ll 1/\tau$ and it is unity for $\omega \gg 1/\tau$ we have the cutoff frequency

$$\omega_c = \frac{1}{\tau} \tag{4.43}$$

Signals above this cutoff frequency should be allowed undistorted, by an ideal high-pass filter, and signals below the cutoff should be completely blocked off. The actual behavior of the basic high-pass filter discussed above is not that perfect, as observed from the frequency response characteristic shown in Figure 4.16(c). It can be easily verified that the half-power bandwidth of the basic high-pass filter is equal to the cutoff frequency given by Equation 4.43, as in the case of the basic low-pass filter. The roll-up slope of the single-pole high-pass filter is 20 dB/decade. Steeper slopes are desirable. Multiple-pole, high-pass, Butterworth filters can be constructed to give steeper roll-up slopes and reasonably flat pass-band magnitude characteristics.

4.4.4 Band-Pass Filters

An ideal band-pass filter passes all signal components within a finite frequency band and blocks off all signal components outside that band. The lower frequency limit of the pass band is called the *lower cutoff frequency* (ω_{c1}) and the upper frequency limit of the band is called the *upper cutoff frequency* (ω_{c2}). The most straightforward way to form a band-pass filter is to cascade a high-pass filter of cutoff frequency ω_{c1} with a low-pass filter of cutoff frequency ω_{c2}. Such an arrangement is shown in Figure 4.17. The passive circuit shown in Figure 4.17(a) is obtained by connecting together the circuits shown in Figure 4.14(a) and Figure 4.16(a). The active circuit shown in Figure 4.17(b) is obtained by connecting a voltage follower op-amp circuit to the original passive circuit. Passive and active filters have the same transfer function, assuming that loading problems (component interaction) are not present in the passive filter. Since loading errors and interactions can be serious in practice, however, the active version is preferred.

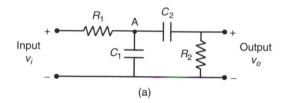

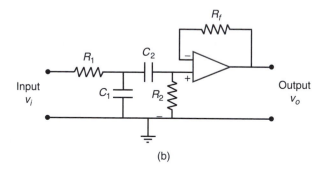

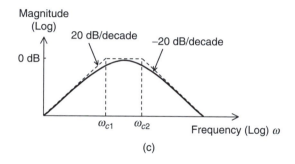

FIGURE 4.17
Band-pass filter: (a) A basic passive filter stage, (b) A basic active filter stage, (c) Frequency response characteristic.

To obtain the filter equation, first consider the high-pass portion of the circuit shown in Figure 4.17(a). Since the output is in open-circuit (zero current) we have from Equation 4.41

$$\frac{v_o}{v_A} = \frac{\tau_2 s}{(\tau_2 s + 1)} \tag{i}$$

in which

$$\tau_2 = R_2 C_2 \tag{4.44}$$

Next, on writing the current balance at node A of the circuit we have

$$\frac{v_i - v_A}{R_1} = C_1 \frac{dv_A}{dt} + C_2 \frac{d}{dt}(v_A - v_o) \tag{ii}$$

Introducing the Laplace variable s (see Appendix A) we get

$$v_i = (\tau_1 s + \tau_3 s + 1)v_A - \tau_3 s v_o \tag{iii}$$

in which

$$\tau_1 = R_1 C_1 \tag{4.45}$$

and

$$\tau_3 = R_1 C_2 \tag{4.46}$$

Now on eliminating v_A by substituting Equation i in Equation iii we get the band-pass filter transfer function:

$$\frac{v_o}{v_i} = G(s) = \frac{\tau_2 s}{\left[\tau_1 \tau_2 s^2 + (\tau_1 + \tau_2 + \tau_3)s + 1\right]} \tag{4.47}$$

We can show that the roots of the characteristic equation

$$\tau_1 \tau_2 s^2 + (\tau_1 + \tau_2 + \tau_3)s + 1 = 0 \tag{4.48}$$

are real and negatives. The two roots are denoted by $-\omega_{c1}$ and $-\omega_{c2}$ and they provide the two cutoff frequencies shown in Figure 4.17(c). It can be verified that, for this basic band-pass filter, the roll-up slope is +20 dB/decade and the roll-down slope is –20 dB/decade. These slope magnitudes are not sufficient in many applications. Furthermore, the flatness of the frequency response within the pass band of the basic filter is not adequate as well. More complex (higher order) band-pass filters with sharper cutoffs and flatter pass bands are commercially available.

4.4.4.1 Resonance-Type Band-Pass Filters

There are many applications where a filter with a very narrow pass band is required. The tracking filter mentioned in the beginning of the section on analog filters is one such application. A filter circuit with a sharp resonance can serve as a narrow-band filter. Note that the cascaded RC circuit shown in Figure 4.17 does not provide an oscillatory response (filter poles are all real) and, hence, it does not form a resonance-type filter. A slight modification to this circuit using an additional resistor R_1 as shown in Figure 4.18(a) will produce the desired effect.

To obtain the filter equation, note that for the voltage follower unit

$$v_A = v_o \tag{i}$$

Next, since current through an op-amp lead is zero, for the high-pass circuit unit (see Equation 4.41), we have

$$\frac{v_A}{v_B} = \frac{\tau_2 s}{(\tau_2 s + 1)} \tag{ii}$$

in which

$$\tau_2 = R_2 C_2$$

Finally, the current balance at node B gives

$$\frac{v_i - v_B}{R_1} = C_1 \frac{dv_B}{dt} + C_2 \frac{d}{dt}(v_B - v_A) + \frac{v_B - v_o}{R_1}$$

or, by using the Laplace variable, we get

$$v_i = (\tau_1 s + \tau_3 s + 2) v_B - \tau_3 s v_A - v_o \qquad \text{(iii)}$$

Now, by eliminating v_A and v_B in Equations i through Equation iii we get the filter transfer function

$$\frac{v_o}{v_i} = G(s) = \frac{\tau_2 s}{\left[\tau_1 \tau_2 s^2 + (\tau_1 + \tau_2 + \tau_3)s + 2\right]} \qquad \text{(4.49)}$$

It can be shown that, unlike Equation 4.48, the present characteristic equation

$$\tau_1 \tau_2 s^2 + (\tau_1 + \tau_2 + \tau_3)s + 2 = 0 \qquad \text{(4.50)}$$

can possess complex roots.

Example 4.8
Verify that the band-pass filter shown in Figure 4.18(a) can have a frequency response with a resonant peak as shown in Figure 4.18(b). Verify that the half-power bandwidth $\Delta\omega$ of the filter is given by $2\zeta\omega_r$ at low damping values. (Note: ζ = damping ratio and ω_r = resonant frequency).

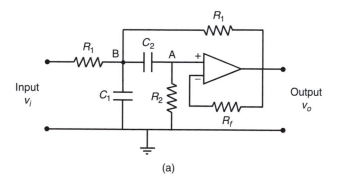

(a)

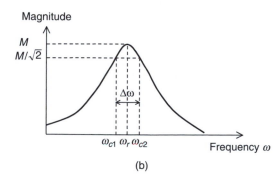

(b)

FIGURE 4.18
A resonance-type narrow band-pass filter: (a) An active filter stage, (b) Frequency response characteristic.

SOLUTION
We may verify that the transfer function given by Equation 4.49 can have a resonant peak by showing that the characteristic Equation 4.50 can have complex roots. For example, if we use parameter values $C_1 = 2$, $C_2 = 1$, $R_1 = 1$, and $R_2 = 2$ we have $\tau_1 = 2$, $\tau_2 = 2$, and $\tau_3 = 1$. The corresponding characteristic equation is

$$4s^2 + 5s + 2 = 0$$

which has the roots

$$-\frac{5}{8} \pm j\frac{\sqrt{7}}{8}$$

and are obviously complex.

To obtain an expression for the half-power bandwidth of the filter, note that the filter transfer function may be written as

$$G(s) = \frac{ks}{\left(s^2 + 2\zeta\omega_n s + \omega_n^2\right)} \tag{4.51}$$

in which
ω_n = undamped natural frequency
ζ = damping ratio
k = a gain parameter

The frequency response function is given by

$$G(j\omega) = \frac{kj\omega}{\left[\omega_n^2 - \omega^2 + 2j\zeta\omega_n\omega\right]} \tag{4.52}$$

For low damping, resonant frequency $\omega_r \cong \omega_n$. The corresponding peak magnitude M is obtained by substituting $\omega = \omega_n$ in Equation 4.52 and taking the transfer function magnitude; thus,

$$M = \frac{k}{2\zeta\omega_n} \tag{4.53}$$

At half-power frequencies we have

$$|G(j\omega)| = \frac{M}{\sqrt{2}}$$

or

$$\frac{k\omega}{\sqrt{\left(\omega_n^2 - \omega^2\right)^2 + 4\zeta^2\omega_n^2\omega^2}} = \frac{k}{2\sqrt{2}\zeta\omega_n}$$

This gives

$$\left(\omega_n^2 - \omega^2\right)^2 = 4\zeta^2\omega_n^2\omega^2 \tag{4.54}$$

the positive roots of which provide the pass band frequencies ω_{c1} and ω_{c2}. Note that the roots are given by

$$\omega_n^2 - \omega^2 = \pm 2\zeta\omega_n\omega$$

Hence, the two roots ω_{c1} and ω_{c2} satisfy the following two equations:

$$\omega_{c1}^2 + 2\zeta\omega_n\omega_{c1} - \omega_n^2 = 0$$

$$\omega_{c2}^2 - 2\zeta\omega_n\omega_{c2} - \omega_n^2 = 0$$

Accordingly, by solving these two quadratic equations and selecting the appropriate sign, we get

$$\omega_{c1} = -\zeta\omega_n + \sqrt{\omega_n^2 + \zeta^2\omega_n^2} \tag{4.55}$$

$$\omega_{c2} = \zeta\omega_n + \sqrt{\omega_n^2 + \zeta^2\omega_n^2} \tag{4.56}$$

The half-power bandwidth is

$$\Delta\omega = \omega_{c2} - \omega_{c1} = 2\zeta\omega_n \tag{4.57}$$

Now, since $\omega_n \cong \omega_r$ for low ζ we have

$$\Delta\omega = 2\zeta\omega_r \tag{4.58}$$

This result is identical to what was reported in Chapter 2.

A notable shortcoming of a resonance-type filter is that the frequency response within the bandwidth (pass band) is not flat. Hence, quite nonuniform signal attenuation takes place inside the pass band.

4.4.5 Band-Reject Filters

Band-reject filters or *notch filters* are commonly used to filter out a narrow band of noise components from a signal. For example, 60 Hz line noise in a signal can be eliminated by using a notch filter with a notch frequency of 60 Hz.

An active circuit that could serve as a notch-filter is shown in Figure 4.19(a). This is known as the Twin T circuit because its geometric configuration resembles two T-shaped circuits connected together. To obtain the filter equation, note that the voltage at point P is v_o because of unity gain of the voltage follower. Now, we write the current balance at nodes A and B; thus,

$$\frac{v_i - v_B}{R} = 2C\frac{dv_B}{dt} + \frac{v_B - v_o}{R}$$

$$C\frac{d}{dt}(v_i - v_A) = \frac{v_A}{R/2} + C\frac{d}{dt}(v_A - v_o)$$

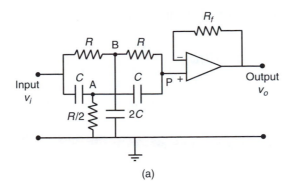

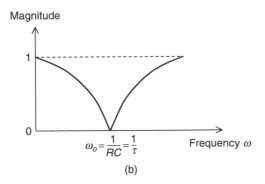

FIGURE 4.19
A notch filter: (a) An active twin T filter circuit, (b) Frequency response characteristic.

Next, since the current through the + lead of the op-amp (voltage follower) is zero, we have the current through point P as

$$\frac{v_B - v_o}{R} = C\frac{d}{dt}(v_o - v_A)$$

These three equations are written in the Laplace form:

$$v_i = 2(\tau s + 1)v_B - v_o \qquad \text{(i)}$$

$$\tau s v_i = 2(\tau s + 1)v_A - \tau s v_o \qquad \text{(ii)}$$

$$v_B = (\tau s + 1)v_o - \tau s v_A \qquad \text{(iii)}$$

in which

$$\tau = RC \qquad (4.59)$$

Finally, eliminating v_A and v_B in Equations i through iii we get

$$\frac{v_o}{v_i} = G(s) = \frac{(\tau^2 s^2 + 1)}{(\tau^2 s^2 + 4\tau s + 1)} \qquad (4.60)$$

The frequency response function of the filter (with $s = j\omega$) is

$$G(j\omega) = \frac{(1 - \tau^2\omega^2)}{(1 - \tau^2\omega^2 + 4j\tau\omega)} \tag{4.61}$$

Note that the magnitude of this function becomes zero at frequency

$$\omega_o = \frac{1}{\tau} \tag{4.62}$$

This is known as the *notch frequency*. The magnitude of the frequency response function of the notch filter is sketched in Figure 4.19(b). It is noticed that any signal component at frequency ω_o will be completely eliminated by the notch filter. Sharp roll-down and roll-up are needed to allow the other (desirable) signal components through without too much attenuation.

Whereas the previous three types of filters achieve their frequency response characteristics through the poles of the filter transfer function, a notch filter achieves its frequency response characteristic through its zeros (roots of the numerator polynomial equation). Some useful information about filters is summarized in Box 4.2.

BOX 4.2

Filters

Active Filters (Need External Power)

Advantages:

- Smaller loading errors and interaction (have high input impedance and low output impedance, and hence don't affect the input circuit conditions, output signals and other components).
- Better accuracy

Passive Filters (No External Power, Use Passive Elements)

Advantages:

- Useable at very high frequencies (e.g., radio frequency)
- No need of a power supply
- Lower cost

Filter Types

- Low Pass: Allows frequency components up to cutoff and rejects the higher frequency components.
- High Pass: Rejects frequency components up to cutoff and allows the higher frequency components.
- Band Pass: Allows frequency components within an interval and rejects the rest.
- Notch (or, Band Reject): Rejects frequency components within an interval (usually, a narrow band) and allows the rest.

Definitions

- Filter Order: Number of poles in the filter circuit or transfer function
- Anti-aliasing Filter: Low-pass filter with cutoff at less than half the sampling rate (i.e., at less than Nyquist frequency), for digital processing.
- Butterworth Filter: A high-order filter with a quite flat pass band.
- Chebyshev Filter: An optimal filter with uniform ripples in the pass band
- Sallen-Key Filter: An active filter whose output is in phase with input.

4.5 Modulators and Demodulators

Sometimes signals are deliberately modified to maintain the accuracy during their transmission, conditioning, and processing. In signal modulation, the data signal, known as the *modulating signal*, is used to vary a property (such as amplitude or frequency) of a *carrier signal*. In this manner the carrier signal is "modulated" by the data signal. After transmitting or conditioning the modulated signal, typically the data signal has to be recovered by removing the carrier signal. This is known as *demodulation* or *discrimination*.

A variety of modulation techniques exist, and several other types of signal modification (e.g., digitizing) could be classified as signal modulation even though they might not be commonly termed as such. Four types of modulation are illustrated in Figure 4.20. In *amplitude modulation* (AM) the amplitude of a periodic carrier signal is varied according to the amplitude of the data signal (modulating signal), frequency of the carrier signal (*carrier frequency*) being kept constant. Suppose that the transient signal shown in Figure 4.20(a) is the modulating signal, and a high-frequency sinusoidal signal is used as the carrier signal. The resulting amplitude-modulated signal is shown in Figure 4.20(b). Amplitude modulation is used in telecommunication, transmission of radio and TV signals, instrumentation, and signal conditioning. The underlying principle is particularly useful in applications such as sensing and control instrumentation of mechatronic systems, and fault detection and diagnosis in rotating machinery.

In *frequency modulation* (FM), the frequency of the carrier signal is varied in proportion to the amplitude of the data signal (modulating signal), while keeping the amplitude of the carrier signal constant. Suppose that the data signal shown in Figure 4.20(a) is used to frequency-modulate a sinusoidal carrier signal. The modulated result will appear as in Figure 4.20(c). Since information is carried as frequency rather than amplitude, any noise that might alter the signal amplitude will have virtually no effect on the transmitted data. Hence, FM is less susceptible to noise than AM. Furthermore, since in FM the carrier amplitude is kept constant, signal weakening and noise effects that are unavoidable in long-distance data communication will have less effect than in the case of AM, particularly if the data signal level is low in the beginning. But more sophisticated techniques and hardware are needed for signal recovery (demodulation) in FM transmission, because FM demodulation involves frequency discrimination rather than amplitude detection. Frequency modulation is also widely used in radio transmission and in data recording and replay.

In *pulse-width modulation* (PWM) the carrier signal is a pulse sequence. The pulse width is changed in proportion to the amplitude of the data signal, while keeping the pulse spacing constant. This is illustrated in Figure 4.20(d). Suppose that the high level of the PWM signal corresponds to the "on" condition of a circuit and the low level corresponds to the "off" condition. Then, as shown in Figure 4.21, the pulse width is equal to the on time ΔT of the circuit within each signal cycle period T. The duty cycle of the PWM is defined as the percentage on time in a pulse period, and is given by

$$\text{Duty cycle} = \frac{\Delta T}{T} \times 100\% \qquad (4.63)$$

Pulse-width modulated signals are extensively used in mechatronic systems, for controlling electric motors and other mechanical devices such as valves (hydraulic, pneumatic) and machine tools. Note that in a given (short) time interval, the average value of the pulse-width modulated signal is an estimate of the average value of the data signal in

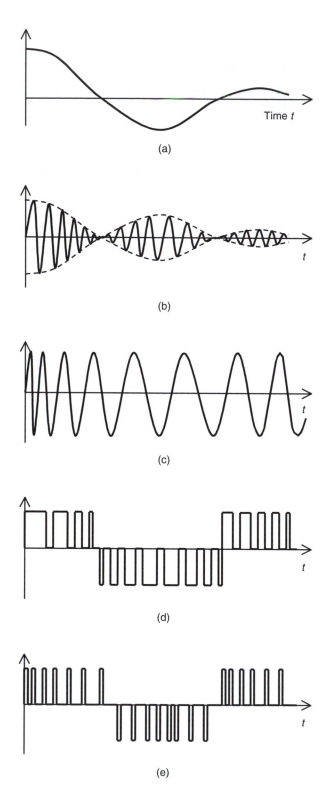

FIGURE 4.20
(a) Modulating signal (data signal), (b) Amplitude-modulated (AM) signal, (c) Frequency-modulated (FM) signal, (d) Pulse-width-modulated (PWM) signal, (e) Pulse-frequency-modulated (PFM) signal.

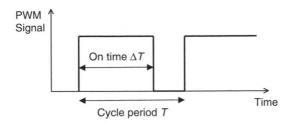

FIGURE 4.21
Duty cycle of a PWM signal.

that period. Hence, PWM signals can be used directly in controlling a process, without having to demodulate it. Advantages of pulse-width modulation include better energy efficiency (less dissipation) and better performance with nonlinear devices. For example, a device may stick at low speeds, due to Coulomb friction. This can be avoided by using a PWM signal having an amplitude that is sufficient to overcome friction, while maintaining the required average control signal, which might be very small.

In *pulse-frequency modulation* (PFM), as well, the carrier signal is a pulse sequence. In this method, the frequency of the pulses is changed in proportion to the value of the data signal, while keeping the pulse width constant. Pulse-frequency modulation has the advantages of ordinary frequency modulation. Additional advantages result due to the fact that electronic circuits (digital circuits in particular) can handle pulses very efficiently. Furthermore, pulse detection is not susceptible to noise because it involves distinguishing between the presence and absence of a pulse rather than accurate determination of the pulse amplitude (or width). Pulse-frequency modulation may be used in place of pulse-width modulation in most applications, with better results.

Another type of modulation is *phase modulation* (PM). In this method, the phase angle of the carrier signal is varied in proportion to the amplitude of the data signal.

Conversion of discrete (sampled) data into the digital (binary) form is also considered a form of modulation. In fact, this is termed *pulse-code modulation* (PCM). In this case each discrete data sample is represented by a binary number containing a fixed number of binary digits (bits). Since each digit in the binary number can take only two values, 0 or 1, it can be represented by the absence or presence of a voltage pulse. Hence, each data sample can be transmitted using a set of pulses. This is known as *encoding*. At the receiver, the pulses have to be interpreted (or decoded) in order to determine the data value. As with any other pulse technique, PCM is quite immune to noise because decoding involves detection of the presence or absence of a pulse rather than determination of the exact magnitude of the pulse signal level. Also, since pulse amplitude is constant, long distance signal transmission (of this digital data) can be accomplished without the danger of signal weakening and associated distortion. Of course, there will be some error introduced by the digitization process itself, which is governed by the finite word size (or dynamic range) of the binary data element. This is known as the *quantization* error and is unavoidable in signal digitization.

In any type of signal modulation it is essential to preserve the algebraic sign of the modulating signal (data). Different types of modulators handle this in different ways. For example, in pulse-code modulation (PCM) an extra *sign bit* is added to represent the sign of the transmitted data sample. In amplitude modulation and frequency modulation, a *phase-sensitive demodulator* is used to extract the original (modulating) signal with the correct algebraic sign. Note that in these two modulation techniques a sign change in the modulating signal can be represented by a 180° phase change in the modulated signal. This is not quite noticeable in Figure 4.20(b) and Figure 4.20(c). In pulse width modulation and pulse frequency modulation, a sign change in the modulating signal can be represented by changing the sign of the pulses, as shown in Figure 4.20(d) and Figure 4.20(e). In phase modulation a positive range of phase angles (say 0–π) could be assigned for the positive values of the data signal and a negative range of phase angles (say $-\pi$–0) could be assigned for the negative values of the signal.

4.5.1   Amplitude Modulation

Amplitude modulation can naturally enter into many physical phenomena. More important perhaps is the deliberate (artificial) use of amplitude modulation to facilitate data transmission and signal conditioning. Let us first examine the related mathematics.

Amplitude modulation is achieved by multiplying the data signal (modulating signal) $x(t)$ by a high frequency (periodic) carrier signal $x_c(t)$. Hence, amplitude-modulated signal $x_a(t)$ is given by

$$x_a(t) = x(t)x_c(t) \tag{4.64}$$

Note that the carrier could be any periodic signal such as harmonic (sinusoidal), square wave, or triangular. The main requirement is that the fundamental frequency of the carrier signal (carrier frequency) f_c be significantly large (say, by a factor of 5 or 10) than the highest frequency of interest (bandwidth) of the data signal. Analysis can be simplified by assuming a sinusoidal carrier frequency; thus,

$$x_c(t) = a_c \cos 2\pi f_c t \tag{4.65}$$

4.5.1.1   *Modulation Theorem*

This is also known as the *frequency-shifting theorem* and relates the fact that if a signal is multiplied by a sinusoidal signal, the Fourier spectrum of the product signal is simply the Fourier spectrum of the original signal shifted through the frequency of the sinusoidal signal. In other words, the Fourier spectrum $X_a(f)$ of the amplitude-modulated signal $x_a(t)$ can be obtained from the Fourier spectrum $X(f)$ of the original data signal $x(t)$ simply by shifting it through the carrier frequency f_c.

To mathematically explain the modulation theorem, we use the definition of the Fourier integral transform to get

$$X_a(f) = a_c \int_{-\infty}^{\infty} x(t) \cos 2\pi f_c t \exp(-j2\pi f t) \, dt$$

Next, since

$$\cos 2\pi f_c t = \frac{1}{2}[\exp(j2\pi f_c t) + \exp(-j2\pi f_c t)]$$

we have

$$X_a(f) = \frac{1}{2} a_c \int_{-\infty}^{\infty} x(t) \exp[-j2\pi(f - f_c)t] \, dt + \frac{1}{2} a_c \int_{-\infty}^{\infty} x(t) \exp[-j2\pi(f + f_c)t] \, dt$$

or,

$$X_a(f) = \frac{1}{2} a_c [X(f - f_c) + X(f + f_c)] \qquad (4.66)$$

Equation 4.66 is the mathematical statement of the modulation theorem. It is illustrated by an example in Figure 4.22. Consider a transient signal $x(t)$ with a (continuous) Fourier spectrum $X(f)$ whose magnitude $|X(f)|$ is as shown in Figure 4.22(a). If this signal is used

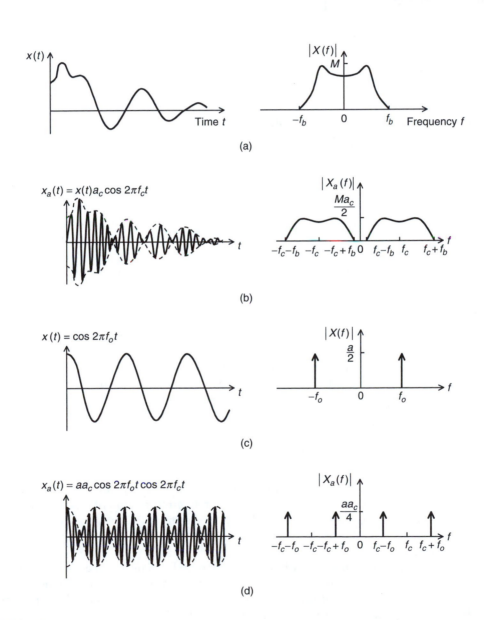

FIGURE 4.22
Illustration of the modulation theorem: (a) A transient data signal and its Fourier spectrum magnitude, (b) Amplitude-modulated signal and its Fourier spectrum magnitude, (c) A sinusoidal data signal, (d) Amplitude modulation by a sinusoidal signal.

to amplitude modulate a high-frequency sinusoidal signal, the resulting modulated signal $x_a(t)$ and the magnitude of its Fourier spectrum are as shown in Figure 4.22(b). It should be kept in mind that the magnitude has been multiplied by $a_c/2$. Furthermore, the data signal is assumed to be *band limited*, with bandwidth f_b. Of course, the theorem is not limited to band-limited signals, but for practical reasons, we need to have some upper limit on the useful frequency of the data signal. Also, for practical reasons (not for the theorem itself), the carrier frequency f_c should be several times larger than f_o so that there is a reasonably wide frequency band from 0 to $(f_c - f_b)$ within which the magnitude of the modulated signal is virtually zero. The significance of this should be clear when we discuss applications of amplitude modulation.

Figure 4.22 shows only the magnitude of the frequency spectra. It should be remembered, however, that every Fourier spectrum has a phase angle spectrum as well. This is not shown for the sake of conciseness. But, clearly the phase-angle spectrum is also similarly affected (frequency shifted) by amplitude modulation.

4.5.1.2   Side Frequencies and Side Bands

The modulation theorem as described above, assumed transient data signals with associated continuous Fourier spectra. The same ideas are applicable as well to periodic signals (with discrete spectra). Periodic signals represent merely a special case of what was discussed above. This case can be analyzed by directly using the Fourier integral transform. In that case, however, we will have to cope with impulsive spectral lines. Alternatively, Fourier series expansion may be employed thereby avoiding the introduction of impulsive discrete spectra into the analysis. As shown in Figure 4.22(c) and Figure 4.22(d), however, no analysis is actually needed for the case of periodic signals because the final answer can be deduced from the results for a transient signal. Specifically, in the Fourier series expansion of the data signal, each frequency component f_o with amplitude $a/2$ will be shifted by $\pm f_o$ to the two new frequency locations $f_c + f_o$ and $-f_c + f_o$ with an associated amplitude $aa_c/4$. The negative frequency component $-f_o$ should also be considered in the same way, as illustrated in Figure 4.22(d). Note that the modulated signal does not have a spectral component at the carrier frequency f_c but rather, on each side of it, at $f_c \pm f_o$. Hence, these spectral components are termed side frequencies. When a band of *side frequencies* is present, it is termed a *side band*. Side frequencies are very useful in fault detection and diagnosis of rotating machinery.

4.5.2   Application of Amplitude Modulation

The main hardware component of an amplitude modulator is an *analog multiplier*. It is commercially available in the monolithic IC form. Alternatively, one can be assembled using integrated-circuit op-amps and various discrete circuit elements. Schematic representation of an amplitude modulator is shown in Figure 4.23. In practice, in order to achieve satisfactory modulation, other components such as signal preamplifiers and filters would be needed.

There are many applications of amplitude modulation. In some applications, modulation is performed intentionally. In others, modulation occurs naturally as a consequence of the physical process, and the resulting signal is used to meet a practical objective. Typical applications of amplitude modulation include the following:

1. Conditioning of general signals (including dc, transient, and low-frequency) by exploiting the advantages of ac signal conditioning hardware.

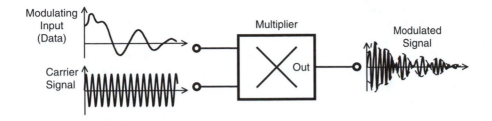

FIGURE 4.23
Representation of an amplitude modulator.

2. Improvement of the immunity of low-frequency signals to low-frequency noise.

3. Transmission of general signals (dc, low-frequency, etc.) by exploiting the advantages of ac signals.

4. Transmission of low-level signals under noisy conditions.

5. Transmission of several signals simultaneously through the same medium (e.g., same telephone line, same transmission antenna, etc.)

6. Fault detection and diagnosis of rotating machinery.

The role of amplitude modulation in many of these applications should be obvious if one understands the frequency-shifting property of amplitude modulation. Several other types of applications are also feasible due to the fact that power of the carrier signal can be increased somewhat arbitrarily, irrespective of the power level of the data (modulating) signal. Let us discuss, one by one, the six categories of applications mentioned above.

AC signal conditioning devices such as ac amplifiers are known to be more "stable" than their dc counterparts. In particular, *drift* problems are not as severe and nonlinearity effects are lower in ac signal conditioning devices. Hence, instead of conditioning a dc signal using dc hardware, we can first use the signal to modulate a high-frequency carrier signal. Then, the resulting high-frequency modulated signal may be conditioned more effectively using ac hardware.

The frequency-shifting property of amplitude modulation can be exploited in making low-frequency signals immune to low-frequency noise. Note from Figure 4.22 that using amplitude modulation, low-frequency spectrum of the modulating signal can be shifted out into a very high frequency region, by choosing a carrier frequency f_c that is sufficiently large. Then, any low-frequency noise (within the band 0 to $f_c - f_b$) would not distort the spectrum of the modulated signal. Hence, this noise could be removed by a high-pass filter (with cutoff at $f_c - f_b$) so that it would not affect the data. Finally, the original data signal can be recovered using demodulation. Since the frequency of a noise component can very well be within the bandwidth f_b of the data signal, if amplitude modulation was not employed, noise could directly distort the data signal.

Transmission of ac signals is more efficient than that of dc signals. Advantages of ac transmission include lower energy dissipation problems. As a result, a modulated signal can be transmitted over long distances more effectively than could the original data signal alone. Furthermore, the transmission of low-frequency (large wave-length) signals requires large antennas. Hence, when amplitude modulation is employed (with an associated reduction in signal wave length), the size of broadcast antenna can be effectively reduced.

Transmission of weak signals over long distances is not desirable because further signal weakening and corruption by noise could produce disastrous results. By increasing the power of the carrier signal to a sufficiently high level, the strength of the modulated signal can be elevated to an adequate level for long-distance transmission.

It is impossible to transmit two or more signals in the same frequency range simultaneously using a single telephone line. This problem can be resolved by using carrier signals with significantly different carrier frequencies to amplitude modulate the data signals. By picking the carrier frequencies sufficiently farther apart, the spectra of the modulated signals can be made nonoverlapping, thereby making simultaneous transmission possible. Similarly, with amplitude modulation, simultaneous broadcasting by several radio (AM) broadcast stations in the same broadcast area has become possible.

4.5.2.1   Fault Detection and Diagnosis

A use of the principle of amplitude modulation that is particularly important in the practice of electromechanical systems, is in the fault detection and diagnosis of rotating machinery. In this method, modulation is not deliberately introduced, but rather results from the dynamics of the machine. Flaws and faults in a rotating machine are known to produce periodic forcing signals at frequencies higher than, and typically at an integer multiple of, the rotating speed of the machine. For example, backlash in a gear pair will generate forces at the tooth-meshing frequency (equal to the product: number of teeth × gear rotating speed). Flaws in roller bearings can generate forcing signals at frequencies proportional to the rotating speed times the number of rollers in the bearing race. Similarly, blade passing in turbines and compressors, and eccentricity and unbalance in the rotor can generate forcing components at frequencies that are integer multiples of the rotating speed. The resulting system response is clearly an amplitude-modulated signal, where the rotating response of the machine modulates the high frequency forcing response. This can be confirmed experimentally through Fourier analysis (fast Fourier transform or FFT) of the resulting response signals. For a gearbox, for example, it will be noticed that, instead of getting a spectral peak at the gear tooth-meshing frequency, two side bands are produced around that frequency. Faults can be detected by monitoring the evolution of these side bands. Furthermore, since side bands are the result of modulation of a specific forcing phenomenon (e.g., gear-tooth meshing, bearing-roller hammer, turbine-blade passing, unbalance, eccentricity, misalignment, etc.), one can trace the source of a particular fault (i.e., diagnose the fault) by studying the Fourier spectrum of the measured response.

Amplitude modulation is an integral part of many types of sensors. In these sensors a high-frequency carrier signal (typically the ac excitation in a primary winding) is modulated by the motion. Actual motion can be detected by demodulation of the output. Examples of sensors that generate modulated outputs are differential transformers (LVDT, RVDT), magnetic-induction proximity sensors, eddy-current proximity sensors, ac tachometers, and strain-gage devices that use ac bridge circuits (See Chapter 6). Signal conditioning and transmission would be facilitated by amplitude modulation in these cases. The signal has to be demodulated at the end, for most practical purposes such as analysis and recording.

4.5.3   Demodulation

Demodulation or discrimination, or detection is the process of extracting the original data signal from a modulated signal. In general, demodulation has to be phase sensitive in the sense that, algebraic sign of the data signal should be preserved and determined by the demodulation process. In *full-wave demodulation* an output is generated continuously. In *half-wave demodulation* no output is generated for every alternate half-period of the carrier signal.

A simple and straightforward method of demodulation is by detection of the envelope of the modulated signal. For this method to be feasible, the carrier signal must be quite powerful (i.e., signal level has to be high) and the carrier frequency also should be very high. An alternative method of demodulation, which generally provides more reliable results involves a further step of modulation performed on the already-modulated signal followed by low-pass filtering. This method can be explained by referring to Figure 4.22.

Consider the amplitude-modulated signal $x_a(t)$ shown in Figure 4.22(b). If this signal is multiplied by the sinusoidal carrier signal $2/a_c \cos 2\pi f_c t$, we get

$$\tilde{x}(t) = \frac{2}{a_c} x_a(t) \cos 2\pi f_c t \tag{4.67}$$

Now, by applying the modulation theorem (Equation 4.66) to Equation 4.67 we get the Fourier spectrum of $\tilde{x}(t)$ as

$$\tilde{X}(f) = \frac{2}{a_c} \left[\frac{1}{2} a_c \{ X(f - 2f_c) + X(f) \} + \frac{1}{2} a_c \{ X(f) + X(f + 2f_c) \} \right]$$

or

$$\tilde{X}(f) = X(f) + \frac{1}{2} X(f - 2f_c) + \frac{1}{2} X(f + 2f_c) \tag{4.68}$$

The magnitude of this spectrum is shown in Figure 4.24(a). Observe that we have recovered the spectrum $X(f)$ of the original data signal, except for the two side bands that are present at locations far removed (centered at $\pm 2f_c$) from the bandwidth of the

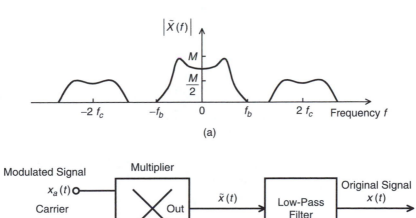

(a)

(b)

FIGURE 4.24
Amplitude demodulation: (a) Spectrum of the signal after the second modulation, (b) Demodulation schematic diagram (modulation + filtering).

original signal. We can conveniently low-pass filter the signal $\tilde{x}(t)$ using a filter with cutoff at f_b to recover the original data signal. A schematic representation of this method of amplitude demodulation is shown in Figure 4.24(b).

4.6   Analog-Digital Conversion

Mechatronic systems use digital data acquisition for a variety of purposes such as process condition monitoring and performance evaluation, fault detection and diagnosis, product quality assessment, dynamic testing, system identification (i.e., experimental modeling), and feedback control. Consider the feedback control system shown in Figure 4.25. Typically, the measured responses (outputs) of a physical system (process, plant) are available in the analog form, as a continuous signal (function of continuous time). Furthermore, typically, the excitation signals (or control inputs) for a physical system have to be provided in the analog form. A digital computer is an integral component of a modern control system; particularly a mechatronic system, and is commonly incorporated in the form of microprocessors and single-board computers together with such components as digital signal processors (DSP). In a mechatronic system, a digital computer will perform tasks such as signal processing, data analysis and reduction, parameter estimation and model identification, decision-making, and control.

Inputs to a digital device (typically, a digital computer) and outputs from a digital device are necessarily present in the digital form. Hence, when a digital device is interfaced with an analog device, the interface hardware and associated driver software have to perform several important functions. Two of the most important interface functions are *digital to analog conversion* (DAC) and *analog to digital conversion*. A digital output from a digital device has to be converted into the analog form for feeding into an analog device such as actuator or analog recording or display unit. Also, an analog signal has to be converted into the

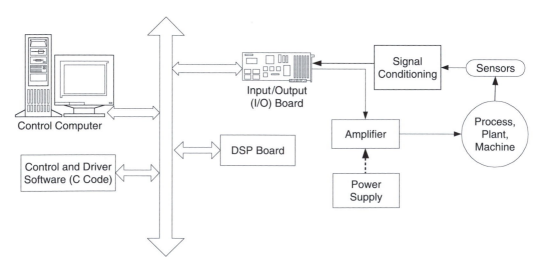

FIGURE 4.25
Components of a data acquisition and control loop.

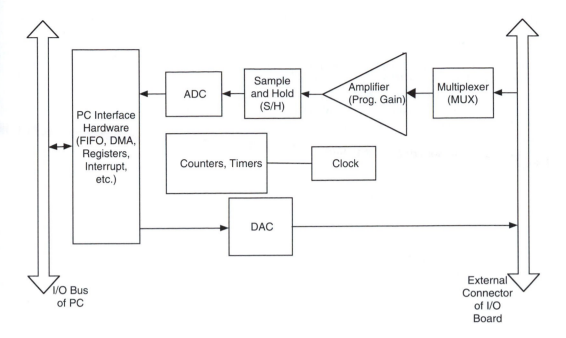

FIGURE 4.26
Main components of an I/O board of a PC.

digital form, according to an appropriate code, before being read by a digital processor or computer.

Both ADC and DAC are elements are components in a typical input/output board (or, I/O board, or, data acquisition and control card, or, DAC or DAQ). Complete I/O boards for mechatronic applications are available from such companies as National Instruments, Servo to Go, Inc., Precision MicroDynamics, Inc., and Keithly Instruments (Metrabyte), Inc. An I/O board can be plugged into a slot of a personal computer (PC) and automatically linked with the bus of the PC. The main components of an I/O board are shown in Figure 4.26. The multiplexer selects the appropriate input channel. The signal is amplified by a programmable amplifier prior to analog-to-digital conversion. As discussed in a later section, the sample-and-hold element (S/H) samples the analog signal and maintains its value at the sampled level until conversion by the ADC. The first-in-first-out (FIFO) element stores the ADC output until it is accessed by the PC for digital processing. The I/O board can provide an analog output through the DAC. Furthermore, a typical I/O board can provide digital outputs as well. An encoder (i.e., a pulse-generating position sensors) can be directly interfaced to I/O boards that are intended for use in motion control applications. Specifications of a typical I/O board are given in Box 4.3. Many of the indicated parameters are discussed in the present chapter. Others are either self-explanatory or discussed elsewhere in the book. Particular note should be made about the sampling rate. This is the rate at which an analog input signal is sampled by the ADC. The Nyquist frequency (or the bandwidth limit) of the sampled data would be half this number (50 kHz for the I/O board specified in Box 4.3). When multiplexing is used (i.e., several input channels are read at the same time), the effective sampling rate for each channel will be reduced by a factor equal to the number of channels. For the I/O board specified in Box 4.3, when 16 channels are sampled simultaneously, the effective sampling rate will be 100 kHz/16 = 6.25 kHz, giving a Nyquist frequency of 3.125 kHz.

BOX 4.3

Typical specifications of a plug-in input/output (I/O) board for a PC

Number of analog input channels = 16 single ended or 8 differential
Analog input ranges = ± 5 V; 0–10 V; ± 10 V; 0–20 V
Input gain ranges (programmable) = 1, 2, 5, 10, 20, 50, 100
Sampling rate for A/D conversion = 100, 000 samples/s (100 kHz)
Word size (resolution) of ADC = 12 bits
Number of D/A output channels = 4
Word size (resolution) of DAC = 12 bits
Ranges of analog output = 0 – 10 V (unipolar mode); ± 10 V (bipolar mode)
Number of digital input lines = 12
Low voltage of input logic = 0.8 V (maximum)
High voltage of input logic = 2.0 V (minimum)
Number of digital output lines = 12
Low voltage of output logic = 0.45 V (maximum)
High voltage of output logic = 2.4 V (minimum)
Number of counters/timers = 3
Resolution of a counter/timer = 16 bits

Since DAC and ADC play important functions in a mechatronic system, they are discussed now. Digital to analog converters are simpler and lower in cost than analog to digital converters. Furthermore, some types of analog to digital converters employ a digital to analog converter to perform their function. For these reasons, we will first discuss DAC.

4.6.1 Digital to Analog Conversion

The function of a digital to analog converter is to convert a sequence of digital words stored in a *data register* (called DAC register), typically in the straight binary form, into an analog signal. The data in the DAC register may be arriving from a data bus of a computer. Each binary digit (bit) of information in the register may be present as a state of a bistable (two-stage) logic device, which can generate a voltage pulse or a voltage level to represent that bit. For example, the "off state" of a bistable logic element or "absence" of a voltage pulse or "low level" of a voltage signal or "no change" in a voltage level can represent binary 0. Conversely, the "on state" of a bistable device or "presence" of a voltage pulse or "high level" of a voltage signal or "change" in a voltage level will represent binary 1. The combination of these bits forming the digital word in the DAC register, will correspond to some numerical value for the analog output signal. Then, the purpose of the DAC is to generate an output voltage (signal level) that has this numerical value, and maintain the value until the next digital word is converted into the analog form. Since a voltage output cannot be arbitrarily large or small for practical reasons, some form of scaling would have to be employed in the DAC process. This scale will depend on the reference voltage v_{ref} used in the particular DAC circuit.

A typical DAC unit is an active circuit in the integrated circuit form and may consist of a data register (digital circuits), solid-state switching circuits, resistors, and operational amplifiers powered by an external power supply, which can provide the reference voltage for the DAC. The reference voltage will determine the maximum value of the output (full-scale voltage). As noted before, the integrated circuit (IC) chip that represents the DAC is usually one of many components mounted on a printed circuit (PC) board, which is the input/output (I/O) board (or, I/O card; or, interface board; or, data acquisition and control board). This board is plugged into a slot of the data acquisition and control PC (see Figure 4.25 and Figure 4.26)

There are many types and forms of DAC circuits. The form will depend mainly on the manufacturer, and requirements of the user or of the particular application. Most types of DAC are variations of two basic types: the weighted type (or summer type or adder type) and the ladder type. The latter type of DAC is more desirable even though the former type could be somewhat simpler and less expensive.

4.6.1.1 Weighted Resistor DAC

A schematic representation of a weighted-resistor DAC (or *summer DAC* or *adder DAC*) is shown in Figure 4.27. Note that this is a general n-bit DAC and n is the number of bits in the output register. The binary word in the register is

$$w = [b_{n-1}b_{n-1}b_{n-3}\ldots b_1b_0] \tag{4.69}$$

in which b_i is the bit in the ith position and it can take the value 0 or 1 depending on the value of the digital output. The decimal value of this binary word is given by (See Chapter 10)

$$D = 2^{n-1}b_{n-1} + 2^{n-2}b_{n-2} + \cdots + 2^0 b_0 \tag{4.70}$$

Note that the least significant bit (LSB) is b_0 and the most significant bit (MSB) is b_{n-1}. The analog output voltage v of the DAC has to be proportional to D.

Each bit b_i in the digital word w will activate a solid-state microswitch in the switching circuit, typically by sending a switching voltage pulse. If $b_i = 1$ the circuit lead will be connected to the $-v_{\text{ref}}$ supply, providing an input voltage $v_i = -v_{\text{ref}}$ to the corresponding weighting resistor $2^{n-i-1}R$. If, on the other hand $b_i = 0$, then the circuit lead will be connected to ground, thereby providing an input voltage $v_i = 0$ to the same resistor. Note that the

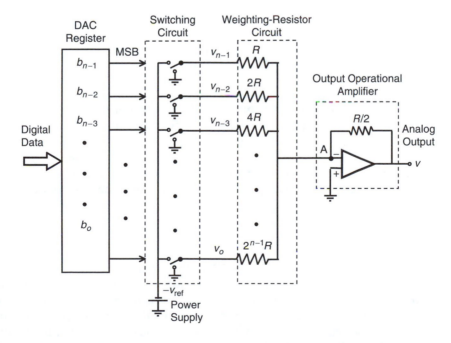

FIGURE 4.27
Weighted-resistor (adder) DAC.

MSB is connected to the smallest resistor (R) and the LSB is connected to the largest resistor ($2^{n-1}R$). By writing the summation of currents at node A of the output op-amp, we get

$$\frac{v_{n-1}}{R} + \frac{v_{n-2}}{2R} + \cdots + \frac{v_0}{2^{n-1}R} + \frac{v}{R/2} = 0$$

In writing this equation, we have used the two principal facts for an op-amp: the voltage is the same at both input leads, and the current through each lead is zero. Note that the + lead is grounded and hence node A should have zero voltage. Now since $v_i = -b_i v_{ref}$ where $b_i = 0$ or 1 depending on the bit value (state of the corresponding switch), we have

$$v = \left[b_{n-1} + \frac{b_{n-2}}{2} + \cdots + \frac{b_0}{2^{n-1}} \right] \frac{v_{ref}}{2} \qquad (4.71)$$

Clearly, as required, the output voltage v is proportional to the value D of the digital word w.

The *full-scale value* (*FSV*) of the analog output occurs when all b_i are equal to 1. Hence,

$$FSV = \left[1 + \frac{1}{2} + \cdots + \frac{1}{2^{n-1}} \right] \frac{v_{ref}}{2}$$

Using the commonly known formula for the sum of a geometric series

$$1 + r + r^2 + \cdots + r^{n-1} = \frac{(1 - r^n)}{(1 - r)} \qquad (4.72)$$

we get

$$FSV = \left(1 - \frac{1}{2^n} \right) v_{ref} \qquad (4.73)$$

Note that this value is slightly smaller than the reference voltage v_{ref}.

A major drawback of the weighted-resistor DAC is that the range of the resistance value in the weighting circuit is very wide. This presents a practical difficulty, particularly when the size (number of bits n) of the DAC is large. Use of resistors having widely different magnitudes in the same circuit can create accuracy problems. For example, since the MSB corresponds to the smallest weighting resistor, it follows that the resistors must have a very high precision.

4.6.1.2 Ladder DAC

A digital to analog converter that uses an R-2R ladder circuit is known as a ladder DAC. This circuit uses only two types of resistors, one having resistance R and the other having $2R$. Hence, the precision of the resistors is not as stringent as what is needed for the weighted-resistor DAC. Schematic representation of an R-2R ladder DAC is shown in Figure 4.28. In this case the switching circuit can operate just like in the previous case of weighted-resistor DAC. To obtain the input-output equation for the ladder DAC, suppose that, as before, the voltage output from the solid-state switch associated with b_i of the

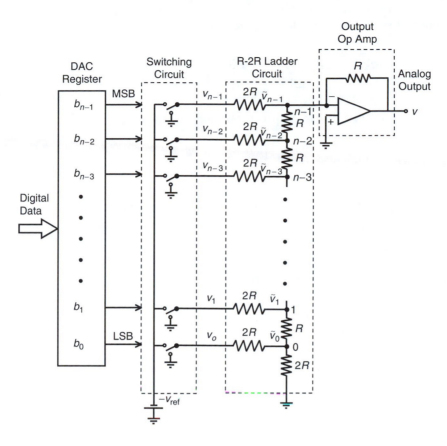

FIGURE 4.28
Ladder DAC.

digital word is v_i. Furthermore, suppose that $\tilde{v}_i$ is the voltage at node i of the ladder circuit, as shown in Figure 4.28. Now, writing the current summation at node i we get,

$$\frac{v_i - \tilde{v}_i}{2R} + \frac{\tilde{v}_{i+1} - \tilde{v}_i}{R} + \frac{\tilde{v}_{i-1} - \tilde{v}_i}{R} = 0$$

or

$$\frac{1}{2}v_i = \frac{5}{2}\tilde{v}_i - \tilde{v}_{i-1} - \tilde{v}_{i+1} \quad \text{for } i = 1, 2, \ldots, n-2 \tag{i}$$

Note that Equation i is valid for all nodes except node 0 and node n–1. It is seen that the current summation for node 0 gives

$$\frac{v_0 - \tilde{v}_0}{2R} + \frac{\tilde{v}_1 - \tilde{v}_0}{R} + \frac{0 - \tilde{v}_0}{2R} = 0$$

or

$$\frac{1}{2}v_0 = 2\tilde{v}_0 - \tilde{v}_1 \tag{ii}$$

and the current summation for node $n-1$ gives

$$\frac{v_{n-1} - \tilde{v}_{n-1}}{2R} + \frac{v - \tilde{v}_{n-1}}{R} + \frac{\tilde{v}_{n-2} - \tilde{v}_{n-1}}{R} = 0$$

Now, since the + lead of the op-amp is grounded, we have $\tilde{v}_{n-1} = 0$. Hence,

$$\frac{1}{2} v_{n-1} = -\tilde{v}_{n-2} - v \qquad \text{(iii)}$$

Next, by using Equation i through iii along with the fact that $\tilde{v}_{n-1} = 0$, we can write the following series of equations:

$$\frac{1}{2} v_{n-1} = -\tilde{v}_{n-2} - v$$

$$\frac{1}{2^2} v_{n-2} = \frac{1}{2}\frac{5}{2}\tilde{v}_{n-2} - \frac{1}{2}\tilde{v}_{n-3}$$

$$\frac{1}{2^3} v_{n-3} = \frac{1}{2^2}\frac{5}{2}\tilde{v}_{n-3} - \frac{1}{2^2}\tilde{v}_{n-4} - \frac{1}{2^2}\tilde{v}_{n-2}$$

$$\vdots \qquad\qquad \text{(iv)}$$

$$\frac{1}{2^{n-1}} v_1 = \frac{1}{2^{n-2}}\frac{5}{2}\tilde{v}_1 - \frac{1}{2^{n-2}}\tilde{v}_0 - \frac{1}{2^{n-2}}\tilde{v}_2$$

$$\frac{1}{2^n} v_0 = \frac{1}{2^{n-1}} 2\tilde{v}_0 - \frac{1}{2^{n-1}}\tilde{v}_1$$

If we sum these n equations, first denoting

$$S = \frac{1}{2^2}\tilde{v}_{n-2} + \frac{1}{2^3}\tilde{v}_{n-3} + \cdots + \frac{1}{2^{n-1}}\tilde{v}_1$$

we get

$$\frac{1}{2} v_{n-1} + \frac{1}{2^2} v_{n-2} + \cdots + \frac{1}{2^n} v_0 = 5S - 4S - S + \frac{1}{2^{n-1}} 2\tilde{v}_0 - \frac{1}{2^{n-2}}\tilde{v}_0 - v = -v$$

Finally, since $v_i = -b_i v_{\text{ref}}$ we have the analog output as

$$v = \left[\frac{1}{2} b_{n-1} + \frac{1}{2^2} b_{n-2} + \cdots + \frac{1}{2^n} b_0 \right] v_{\text{ref}} \qquad (4.74)$$

This result is identical to Equation 4.71 which we obtained for the weighted-resistor DAC. Hence, as before, the analog output is proportional to the value D of the digital word and, furthermore, the full-scale value of the ladder DAC as well is given by the previous Equation 4.73.

4.6.1.3 DAC Error Sources

For a given digital word, the analog output voltage from a DAC would not be exactly equal to what is given by the analytical formulas (e.g., Equation 4.71) that were derived earlier. The difference between the actual output and the ideal output is the error. The DAC error could be normalized with respect to the full-scale value.

There are many causes of DAC error. Typical error sources include parametric uncertainties and variations, circuit time constants, switching errors, and variations and noise in the reference voltage. Several types of error sources and representations are discussed below.

1. *Code Ambiguity.* In many digital codes (e.g., in the straight binary code), incrementing a number by an LSB will involve more than one bit switching. If the speed of switching from 0–1 is different from that for 1–0, and if switching pulses are not applied to the switching circuit simultaneously, the switching of the bits will not take place simultaneously. For example, in a 4-bit DAC, incrementing from decimal 2 to decimal 4 will involve changing the digital word from 0011–0100. This requires two bit-switchings from 1–0 and one bit switching from 0–1. If 1–0 switching is faster than the 0–1 switching, then an intermediate value given by 0000 (decimal zero) will be generated, with a corresponding analog output. Hence, there will be a momentary code ambiguity and associated error in the DAC signal. This problem can be reduced (and eliminated in the case of single bit increments) if a *gray code* is used to represent the digital data (See Chapter 7). Improving the switching circuitry will also help reduce this error.

2. *Settling Time.* The circuit hardware in a DAC unit will have some dynamics, with associated time constants and perhaps oscillations (underdamped response). Hence, the output voltage cannot instantaneously settle to its ideal value upon switching. The time required for the analog output to settle within a certain band (say ±2% of the final value or $\pm\frac{1}{2}$ resolution), following the application of the digital data, is termed settling time. Naturally, settling time should be smaller for better (faster and more accurate) performance. As a rule of thumb, the settling time should be approximately half the data arrival time. Note that the data arrival time is the time interval between the arrival of two successive data values, and is given by the inverse of the data arrival rate.

3. *Glitches.* Switching of a circuit will involve sudden changes in magnetic flux due to current changes. This will induce the voltages that produce unwanted signal components. In a DAC circuit, these induced voltages due to rapid switching can cause signal spikes, which will appear at the output. The error due to these noise signals is not significant at low conversion rates.

4. *Parametric Errors.* As discussed before, resistor elements in a DAC might not be very precise, particularly when resistors within a wide range of magnitudes are employed, as in the case of weighted-resistor DAC. These errors appear at the analog output. Furthermore, aging and environmental changes (primarily, change in temperature) will change the values of circuit parameters, resistance in particular. This also will result in DAC error. These types of errors due to imprecision of circuit parameters and variations of parameter values are termed parametric errors. Effects of such errors can be reduced by several ways including the use of compensation hardware (and perhaps software), and directly by using precise and robust circuit components and employing good manufacturing practices.

5. *Reference Voltage Variations.* Since the analog output of a DAC is proportional to the reference voltage v_{ref}, any variations in the voltage supply will directly appear

as an error. This problem can be overcome by using stabilized voltage sources with sufficiently low output impedance.

6. *Monotonicity.* Clearly, the output of a DAC should change by its resolution ($\delta y = v_{ref}/2^n$) for each step of one LSB (Least-significant bit) increment in the digital value. This ideal behavior might not exist in some practical DACs due to such errors as those mentioned above. At least the analog output should not decrease as the value of the digital input increases. This is known as the monotonicity requirement, and it should be met by a practical digital-to-analog converter.

7. *Nonlinearity.* Suppose that the digital input to a DAC is varied from [0 0 ... 0] to [1 1 ... 1] in steps of one LSB. As mentioned above, ideally the analog output should increase in constant jumps of $\delta y = v_{ref}/2^n$ giving a staircase-shaped analog output. If we draw the best linear fit for this ideally montonic staircase response, it will have a slope equal to the resolution/step. This slope is known as the ideal scale factor. Nonlinearity of a DAC is measured by the largest deviation of the DAC output from this best linear fit. Note that in the ideal case, the nonlinearity is limited to half the resolution ($\frac{1}{2}\delta y$).

One cause of nonlinearity is clearly the faulty bit-transitions. Another cause is circuit nonlinearity in the conventional sense. Specifically, due to nonlinearities in circuit elements such as op-amps and resistors, the analog output will not be proportional to the value of the digital word dictated by the bit switchings (faulty or not). This latter type of nonlinearity can be accounted for by using calibration.

4.6.2 Analog to Digital Conversion

Analog signals, which are continuously defined with respect to time, have to be sampled at discrete time points and the sample values have to be represented in the digital form (according to a suitable code) to be read into a digital system such as a microcomputer. An ADC is used to accomplish this. For example, since response measurements of a mechatronic systems are usually available as analog signals, these signals have to be converted into the digital form before passing on to a digital computer for analysis and possibly generating a control command. Hence, the computer interface for the measurement channels should contain one or more ADCs (see Figure 4.25).

DACs and ADCs are usually situated on the same digital interface board (see Figure 4.26). The analog to digital conversion process is more complex and time consuming than the digital to analog conversion process. Furthermore, many types of ADCs use DACs to accomplish the analog to digital conversation. Hence, ADCs are usually more costly, and their conversion rate is usually slower in comparison to DACs. Several types of analog to digital converters are commercially available. The principle of operation may vary depending on the type. A few commonly known types are discussed here.

4.6.2.1 *Successive Approximation ADC*

This type of analog to digital converter is very fast, and is suitable for high-speed applications. The speed of conversion depends on the number of bits in the output register of ADC but is virtually independent of the nature of the analog input signal. A schematic diagram for a successive approximation ADC is shown in Figure 4.29. Note that a DAC is an integral component of this ADC. The sampled analog signal (from a *sample and hold circuit*) is applied to a *comparator* (typically a *differential amplifier*). Simultaneously, a "start conversion" (SC) control pulse is sent into the *control logic unit* by the external device

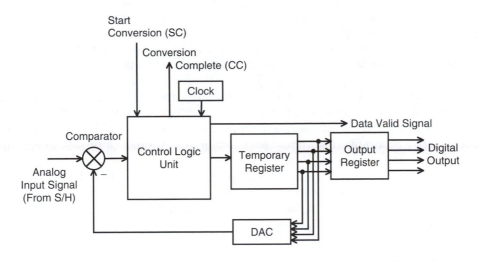

FIGURE 4.29
Successive approximation ADC.

(perhaps a microcomputer) that controls the operation of the ADC. Then, no new data will be accepted by the ADC until a "conversion complete" (CC) pulse is sent out by the control logic unit. Initially, the registers are cleared so that they contain all zero bits. Now, the ADC is ready for its first conversion approximation.

The first approximation begins with a clock pulse. Then, the control logic unit will set the most significant bit (MSB) of the temporary register (DAC control register) to 1, all the remaining bits in that register being zero. This digital word in the temporary register is supplied to the DAC. Note that the analog output of the DAC is now equal to half the full-scale value. This analog signal is subtracted from the analog input by the comparator. If the output of the comparator is positive, the control logic unit will keep the MSB of the temporary register at binary 1 and will proceed to the next approximation. If the comparator output is negative, the control logic unit will change the MSB to binary 0 before proceeding to the next approximation.

The second approximation will start at another clock pulse. This approximation will consider the second most significant bit of the temporary register. As before, this bit is set to 1 and the comparison is made. If the comparator output is positive, this bit is left at value 1 and the third most significant bit is considered. If the comparator output is negative, the bit value will be changed to 0 before proceeding to the third most significant bit.

In this manner, all bits in the temporary register are set successively starting from the MSB and ending with the LSB. The contents of the temporary register are then transferred to the output register, and a "data valid" signal is sent by the control logic unit, signaling the interfaced device (computer) to read the contents of the output register. The interfaced device will not read the register if a data valid signal is not present. Next, a "conversion complete" (CC) pulse is sent out by the control logic unit, and the temporary register is cleared. The ADC is now ready to accept another data sample for digital conversion. Note that the conversion process is essentially the same for every bit in the temporary register. Hence, the total conversion time is approximately n times the conversion time for one bit. Typically, one bit conversion can be completed within one clock period.

It should be clear that if the maximum value of an analog input signal exceeds the full-scale value of a DAC, then the excess signal value cannot be converted by the ADC. The excess value will directly contribute to error in the digital output of the ADC. Hence, this situation should be avoided either by properly scaling the analog input or by properly selecting the reference voltage for the internal DAC unit.

In the foregoing discussion we have assumed that the value of the analog input signal is always positive. Otherwise, the sign of the signal has to be accounted for by some means. For example, the sign of the signal can be detected from the sign of the comparator output initially, when all bits are zero. If the sign is negative, then the same A/D conversion process as for a positive signal is carried out after switching the polarity of the comparator. Finally, the sign is correctly represented in the digital output (e.g., by the two's complement representation for negative quantities. See Chapter 10). Another approach to account for signed (bipolar) input signals is to offset the signal by a sufficiently large constant voltage such that the analog input is always positive. After the conversion, the digital number corresponding to this offset is subtracted from the converted data in the output register in order to obtain the correct digital output. In what follows, we shall assume that the analog input signal is positive.

4.6.2.2 Dual Slope ADC

This analog to digital converter uses an RC integrating circuit. Hence, it is also known as an *integrating ADC*. This ADC is simple and inexpensive. In particular, an internal DAC is not utilized and hence, DAC errors as mentioned previously will not enter the ADC output. Furthermore, the parameters R and C in the integrating circuit do not enter the ADC output. As a result, the device is self-compensating with regard to circuit-parameter variations due to temperature, aging, etc. A shortcoming of this ADC is its slow conversion rate because, for accurate results, the signal integration has to proceed for a longer time in comparison to the conversion time for a successive approximation ADC.

Analog-to-digital conversion in a dual slope ADC is based on timing (i.e., counting the number of clock pulses during) a capacitor-charging process. The principle of operation can be explained with reference to the integrating circuit shown in Figure 4.30(a). Note that v_i is a constant input voltage to the circuit and v is the output voltage. Since the "+" lead of the op-amp is grounded, the "−" lead (and node A) also will have zero voltage. Also, the currents through the op-amp leads are negligible. Hence, the current balance at node A gives

$$\frac{v_i}{R} + C\frac{dv}{dt} = 0$$

Integrating this equation for constant v_i we have

$$v(t) = v(0) - \frac{v_i t}{RC} \tag{4.75}$$

Equation 4.75 will be utilized in obtaining a principal result for the dual slope ADC.

A schematic diagram for a dual slope ADC is shown in Figure 4.30(b). Initially, the capacitor C in the integrating circuit is discharged (zero voltage). Then, the analog signal v_s is supplied to the switching element and held constant by the sample and hold circuit (S/H). Simultaneously, a "conversion start" (CS) control signal is sent to the control logic unit. This will clear the timer and the output register (i.e., all bits are set to zero) and will send a pulse to the switching element to connect the input v_s to the integrating circuit. Also, a signal is sent to the timer to initiate timing (counting). The capacitor C will begin to charge. Equation 4.75 is now applicable with input $v_i = v_s$ and the initial state $v = (0) = 0$. Suppose that the integrator output v becomes $-v_c$ at time $t = t_1$. Hence, from Equation 4.75

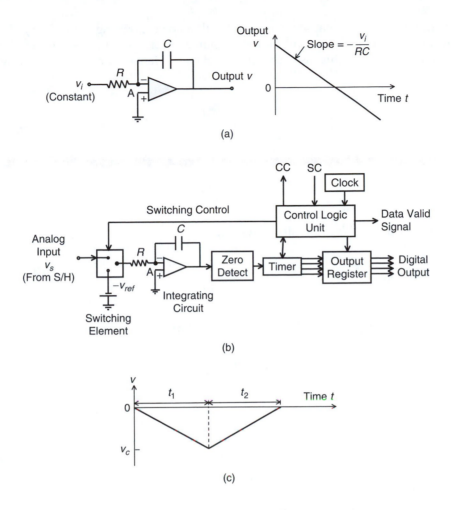

FIGURE 4.30
(a) RC integrating circuit, (b) Dual slope ADC, (c) Dual slope charging-discharging curve.

we have

$$v_c = \frac{v_s t_1}{RC} \tag{i}$$

The timer will keep track of the capacitor charging time (as a clock pulse count n) and will inform the control logic unit when the elapsed time is t_1 (i.e., when the count is n_1). Note that t_1 and n_1 are fixed (and known) parameters but voltage v_c depends on the value of v_s, and is unknown.

At this point the control logic unit will send a signal to the switching unit, which will connect the input lead of the integrator to a negative supply voltage $-v_{\text{ref}}$. Simultaneously, a signal is sent to the timer to clear its contents and start timing (counting) again. Now the capacitor begins to discharge. The output of the integrating circuit is monitored by the "zero-detect" unit. When this output becomes zero the zero-detect unit sends a signal to the timer to stop counting. The zero-detect unit could be a comparator (differential amplifier) having one of the two input leads set at zero potential.

Now suppose that the elapsed time is t_2 (with a corresponding count of n_2). It should be clear that Equation 4.75 is valid for the capacitor discharging process as well. Note that $v_i = -v_{\text{ref}}$ and $v(0) = -v_c$ in this case. Also, $v(t) = 0$ at $t = t_2$. Hence, from Equation 4.75 we have

$$0 = -v_c + \frac{v_{\text{ref}} t_2}{RC}$$

or

$$v_c = \frac{v_{\text{ref}} t_2}{RC} \tag{ii}$$

On dividing Equation i by Equation ii we get

$$v_s = v_{\text{ref}} \frac{t_2}{t_1}$$

But, the timer pulse count is proportional to the elapsed time. Hence,

$$\frac{t_2}{t_1} = \frac{n_2}{n_1}$$

Substituting, we have

$$v_s = \frac{v_{\text{ref}}}{n_1} n_2 \tag{4.76}$$

Since v_{ref} and n_1 are fixed quantities, v_{ref}/n_1 can be interpreted as a scaling factor for the analog input. Then, it follows from Equation 4.76 that the second count n_2 is proportional to the analog signal sample v_s. Note that the timer output is available in the digital form. Accordingly, the count n_2 is used as the digital output of the ADC.

At the end of the capacitor discharge period, the count n_2 in the timer is transferred to the output register of the ADC, and the "data valid" signal is set. The contents of the output register are now ready to be read by the interfaced digital system, and the ADC is ready to convert a new sample.

The charging-discharging curve for the capacitor during the conversion process is shown in Figure 4.30(c). The slope of the curve during charging is $-\frac{v_s}{RC}$ and the slope during discharging is $+\frac{v_{\text{ref}}}{RC}$. The reason for the use of the term "dual slope" to denote this ADC is therefore clear.

As mentioned before, any variations in R and C do not affect the accuracy of the output. But, it should be clear from the foregoing discussion that the conversion time depends on the capacitor discharging time t_2 (note that t_1 is fixed), which in turn depends on v_c and hence on the input signal value v_s (see Equation i). It follows that, unlike the successive approximation ADC, the dual slope ADC has a conversion time that directly depends on the magnitude of the input data sample. This is a disadvantage in a way because in many applications we prefer to have a constant conversion rate.

The above discussion assumed that the input signal is positive. For a negative signal, the polarity of the supply voltage v_{ref} has to be changed. Furthermore, the sign has to be properly represented in the contents of the output register as, for example, in the case of successive approximation ADC.

4.6.2.3 Counter ADC

The counter-type ADC has several aspects in common with the successive approximation ADC. Both are comparison-type (or closed-loop) ADCs. Both use a DAC unit internally to compare the input signal with the converted signal. The main difference is that in a counter ADC the comparison starts with the LSB and proceeds down. It follows that, in a counter ADC, the conversion time depends on the signal level, because the counting (comparison) stops when a match is made, resulting in shorter conversion times for smaller signal values.

A schematic diagram for a counter ADC is shown in Figure 4.31. Note that this is quite similar to Figure 4.29. Initially, all registers are cleared (i.e., all bits and counts are set to zero). As an analog data signal (from the sample and hold circuit) arrives at the comparator, a "start conversion" (SC) pulse is sent to the control logic unit. When the ADC is ready for conversion (i.e., when "data valid" signal is on) the control logic unit initiates the counter. Now, the counter sets its count to 1, and the LSB of the DAC register is set to 1 as well. The resulting DAC output is subtracted from the analog input, by means of the comparator. If the comparator output is positive, the count is incremented by one and this causes the binary number in the DAC register to be incremented by one LSB. The new (increased) output of the DAC is now compared with the input signal. This cycle of count incrementing and comparison is repeated until the comparator output becomes less than or equal to zero. At that point the control logic unit sends out a "conversion complete" (CC) signal and transfers the contents of the counter to the output register. Finally, the "data valid" signal is turned on, indicating that the ADC is ready for a new conversion cycle, and the contents of the output register (the digital output) is available to be read by the interfaced digital system.

The count of the counter is available in the binary form, which is compatible with the output register as well as the DAC register. Hence, the count can be transferred directly to these registers. The count when the analog signal is equal to (or slightly less than) the output of the DAC, is proportional to the analog signal value. Hence, this count represents the digital output. Again, sign of the input signal has to be properly accounted for in the bipolar operation.

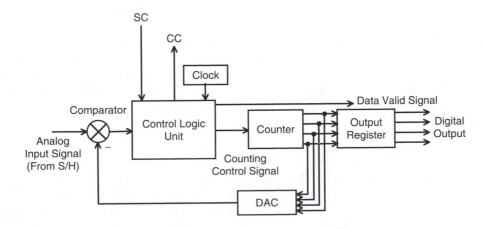

FIGURE 4.31
Counter ADC.

4.6.2.4   ADC Performance Characteristics

For ADCs that use a DAC internally, the same error sources that were discussed previously for DACs will apply. Code ambiguity at the output register will not be a problem because the converted digital quantity is transferred instantaneously to the output register. Code ambiguity in the DAC register can still cause error in ADCs that use a DAC. Conversion time is a major factor, this being much larger for an ADC. In addition to resolution and dynamic range, quantization error will be applicable to an ADC. These considerations, which govern the performance of an ADC are discussed next.

4.6.2.4.1   Resolution and Quantization Error

The number of bits n in an ADC register determines the resolution and dynamic range of an ADC. For an n-bit ADC the size of the output register is n bits. Hence, the smallest possible increment of the digital output is one LSB. The change in the analog input that results in a change of one LSB at the output is the resolution of the ADC. For the unipolar (unsigned) case, the available range of the digital outputs is from 0 to $2^n - 1$. This represents the dynamic range. It follows that, as for a DAC, the dynamic range of an n-bit ADC is given by the ratio:

$$DR = 2^n - 1 \tag{4.77}$$

or, in decibels:

$$DR = 20\log_{10}(2^n - 1)\,\mathrm{dB} \tag{4.78}$$

The *full-scale value* of an ADC is the value of the analog input that corresponds to the maximum digital output.

  Suppose that an analog signal within the dynamic range of a particular ADC is converted by that ADC. Since the analog input (sample value) has an infinitesimal resolution and the digital representation has a finite resolution (one LSB), an error is introduced in the process of analog to digital conversion. This is known as the *quantization error*. A digital number undergoes successive increments in constant steps of 1 LSB. If an analog value falls at an intermediate point within a step of single LSB, a quantization error is caused as a result. Rounding off of the digital output can be accomplished as follows: The magnitude of the error when quantized up, is compared with that when quantized down; say, using two hold elements and a differential amplifier. Then, we retain the digital value corresponding to the lower error magnitude. If the analog value is below the 1/2 LSB mark, then the corresponding digital value is represented by the value at the beginning of the step. If the analog value is above the 1/2 LSB mark, then the corresponding digital value is the value at the end of the step. It follows that with this type of rounding off, the quantization error does not exceed 1/2 LSB.

4.6.2.4.2   Monotonicity, Nonlinearity, and Offset Error

Considerations of monotonicity and nonlinearity are important for an ADC as well as for a DAC. In the case of an ADC, the input is an analog signal and the output is digital. Disregarding quantization error, the digital output of an ADC will increase in constant steps in the shape of an ideal staircase function, when the analog input is increased from 0 in steps of the device resolution (δy). This is the ideally monotonic case. The best straight-line fit to this curve has a slope equal to $1/\delta y$(LSB/Volts). This is the *ideal gain* or *ideal scale factor.* Still there will be an *offset error* of 1/2 LSB because the best linear fit will not pass through the origin. Adjustments can be made for this offset error.

Incorrect bit-transitions can take place in an ADC, due to various errors that might be present and also possibly due to circuit malfunctions. The best linear fit under such faulty conditions will have a slope different from the ideal gain. The difference is the gain error. Nonlinearity is the maximum deviation of the output from the best linear fit. It is clear that with perfect bit transitions, in the ideal case, a nonlinearity of 1/2 LSB would be present. Nonlinearities larger than this would result due to incorrect bit transitions. As in the case of a DAC, another source of nonlinearity in an ADC is circuit nonlinearities, which would deform the analog input signal before being converted into the digital form.

4.6.2.4.3 ADC Conversion Rate

It is clear that analog to digital conversion is much more time consuming than digital to analog conversion. The conversion time is a very important factor because the rate at which conversion can take place governs many aspects of data acquisition, particularly in real-time applications. For example, the data sampling rate has to synchronize with the ADC conversion rate. This, in turn, will determine the *Nyquist Frequency* (half the sampling rate), which corresponds to the bandwidth of the sampled signal, and is the maximum value of useful frequency that is retained as a result of sampling (See Chapter 5). Furthermore, the sampling rate will dictate the requirements of storage and memory. Another important consideration related to the conversion rate of an ADC is the fact that a signal sample has to be maintained at the same value during the entire process of conversion into the digital form. This would require a *hold circuit*, and this circuit should be able to perform accurately at the largest possible conversion time for the particular ADC unit.

The time needed for a sampled analog input to be converted into the digital form, will depend on the type of ADC. Usually in a comparison type ADC (which uses an internal DAC) each bit transition will take place in one clock period Δt. Also, in an integrating (dual slope) ADC each clock count will need a time of Δt. On this basis, for the three types of ADC that we have discussed, the following figures can be given for their conversion times:

1. Successive-approximation ADC. In this case, for an *n*-bit ADC, *n* comparisons are needed. Hence, the conversion time is given by

$$t_c = n \cdot \Delta t \tag{4.79}$$

   in which Δt is the clock period. Note that for this ADC, t_c does not depend on the signal level (analog input).

2. Dual-slope (integrating) ADC. In this case, the conversion time is the total time needed to generate the two counts n_1 and n_2 (see Figure 4.30(c)). Hence,

$$t_c = (n_1 + n_2) \Delta t \tag{4.80}$$

   Note that n_1 is a fixed count. But n_2 is a variable count, which represents the digital output, and is proportional to the analog input (signal level). Hence in this type of ADC, conversion time depends on the analog input level. The largest output for an *n*-bit converter is $2^n - 1$. Hence, the largest conversion time may be given by

$$t_{c\max} = (n_1 + 2^n - 1) \Delta t \tag{4.81}$$

3. **Counter ADC.** For a counter ADC, the conversion time is proportional to the number of bit transitions (1 LSB per step) from zero to the digital output n_o. Hence, the conversion time is given by

$$t_c = n_o \Delta t \qquad (4.82)$$

in which n_o is the digital output value (in decimal). Note that for this ADC as well, t_c depends on the magnitude of the input data sample. For an n-bit ADC, since the maximum value of n_o is $2^n - 1$, we have the maximum conversion time

$$t_{c\max} = (2^n - 1)\Delta t \qquad (4.83)$$

By comparing Equation 4.79, Equation 4.81, and Equation 4.83 it can be concluded that the successive-approximation ADC is the fastest of the three types discussed.

The total time taken to convert an analog signal will depend on other factors besides the time taken for the conversion of sampled data into digital form. For example, in multiple-channel data acquisition (multiplexing), the time taken to select the channels has to be counted in. Furthermore, time needed to sample the data and time needed to transfer the converted digital data into the output register have to be included. In fact, the *conversion rate* for an ADC is the inverse of this overall time needed for a conversion cycle. Typically, however, the conversion rate depends primarily on the bit conversion time in the case of a comparison-type ADC, and on the integration time in the case of an integration-type ADC. A typical time period for a comparison step or counting step in an ADC is $\Delta t = 5$ μs. Hence, for an 8-bit successive approximation ADC the conversion time is 40 μs. The corresponding sampling rate would be of the order of (less than) $1/(40 \times 10^{-6}) = 25 \times 10^3$ samples/s (or 25 kHz). The maximum conversion rate for an 8-bit counter ADC would be about $5 \times (2^8 - 1) = 1275$ μs. The corresponding sampling rate would be of the order of 780 samples/s. Note that this is considerably slow. The maximum conversion time for a dual slope ADC would likely be larger (i.e., slower rate).

4.7 Sample-and-Hold (S/H) Circuitry

Typical applications of data acquisition use analog to digital conversion. The analog input to an ADC can be very transient, and furthermore, the process of analog to digital conversion itself is not instantaneous (ADC time can be much larger than the digital to analog conversion time). Specifically, the incoming analog signal might be changing at a rate higher than the ADC conversion rate. Then, the input signal value will vary during the conversion period and there will be an ambiguity as to what analog input value corresponds to a particular digital output value. Hence it is necessary to sample the analog input signal and maintain the input to the ADC at this sampled value until the analog to digital conversion is completed. In other words, since we are typically dealing with analog signals that can vary at a high speed, it would be necessary to sample and hold (S/H) the input signal during each analog to digital conversion cycle. Each data sample must be generated and captured by the S/H circuit on the issue of the "start conversion" (SC) control signal, and the captured voltage level has to be maintained constant until a "conversion complete" (CC) control signal is issued by the ADC unit.

The main element in an S/H circuit is the holding capacitor. A schematic diagram of a sample and hold circuit is shown in Figure 4.32. The analog input signal is supplied

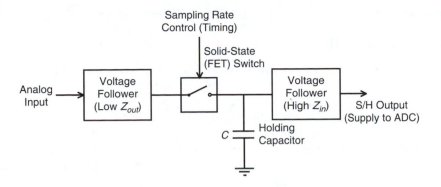

FIGURE 4.32
A sample and hold circuit.

through a voltage follower to a solid-state switch. The switch typically uses a field-effect transistor (FET), such as the metal-oxide semiconductor field effect transistor (MOSFET). The switch is closed in response to a "sample pulse" and is opened in response to a "hold pulse." Both control pulses are generated by the control logic unit of the ADC. During the time interval between these two pulses, the holding capacitor is charged to the voltage of the sampled input. This capacitor voltage is then supplied to the ADC through a second voltage follower.

The functions of the two voltage followers are explained now. When the FET switch is closed in response to a sample command (pulse), the capacitor has to be charged as quickly as possible. The associated time constant (charging time constant) τ_c is given by

$$\tau_c = R_s C \tag{4.84}$$

in which
  R_s = source resistance
  C = capacitance of the holding capacitor

Since τ_c has to be very small for fast charging, and since C is fixed by the holding requirements (typically C is of the order of 100 pF where 1 pF = 1×10^{-12} F), we need a very small source resistance. The requirement is met by the input voltage follower (which is known to have a very low output impedance), thereby providing a very small R_s. Furthermore, since a voltage follower has a unity gain, the voltage at the output of this input voltage follower would be equal to the voltage of the analog input signal, as required.

Next, once the FET switch is opened in response to a hold command (pulse), the capacitor should not discharge. This requirement is met due to the presence of the output voltage follower. Since the input impedance of a voltage follower is very high, the current through its leads would be almost zero. Because of this, the holding capacitor will have a virtually zero discharge rate under "hold" conditions. Furthermore, we like the output of this second voltage follower to be equal to the voltage of the capacitor. This condition is also satisfied due to the fact that a voltage follower has a unity gain. Hence, the sampling would be almost instantaneous and the output of the S/H circuit would be maintained (almost) constant during the holding period, due to the presence of the two voltage followers. Note that the practical S/H circuits are *zero-order-hold* devices, by definition.

4.8 Multiplexers (MUX)

A multiplexer (also known as a *scanner*) is used to select one channel at a time from a bank of signal channels and connect it to a common hardware unit. In this manner a costly and complex hardware unit can be time-shared among several signal channels. Typically, channel selection is done in a sequential manner at a fixed channel-select rate.

There are two types of multiplexers: analog multiplexers and digital multiplexers. An analog multiplexer is used to scan a group of analog channels. Alternatively, a digital multiplexer is used to read one data word at a time sequentially from a set of digital data words.

The process of distributing a single channel of data among several output channels is known as *demultiplexing*. A demultiplexer (or data distributor) performs the reverse function of a multiplexer (or scanner). A demultiplexer may be used, for example, when the same (processed) signal from a digital computer is needed for several purposes (e.g., digital display, analog reading, digital plotting, and control).

Multiplexing used in short-distance signal transmission applications (e.g., data logging and process control) is usually *time-division multiplexing*. In this method, channel selection is made with respect to time. Hence, only one input channel is connected to the output channel of the multiplexer. This is the method described here. Another method of multiplexing, used particularly in long-distance transmission of several data signals, is known as *frequency-division multiplexing*. In this method, the input signals are modulated (e.g., by amplitude modulation, as discussed previously) with carrier signals having different frequencies and are transmitted simultaneously through the same data channel. The signals are separated by demodulation at the receiving end.

4.8.1 Analog Multiplexers

Monitoring of a mechatronic system often requires the measurement of several process responses. These signals have to be conditioned (e.g., amplification and filtering) and modified in some manner (e.g., analog to digital conversion) before being supplied to a common-purpose system such as a digital computer or data logger. Usually, data modification devices are costly. In particular, we have noted that ADCs are more expensive than DACs. An expensive option for interfacing several analog signals with a common-purpose system such as a digital computer would be to provide separate data modification hardware for each signal channel. This method has the advantage of high speed. An alternative, low-cost method is to use an analog multiplexer (analog scanner) to select one signal channel at a time sequentially and connect it to a common signal-modification hardware unit (consisting of amplifiers, filters, S/H, ADC, etc.). In this way, by time-sharing expensive hardware among many data channels, the data acquisition speed is traded off to some extent for significant cost savings. Because very high speeds of channel selection are possible with solid-state switching (e.g., solid-state speeds of the order of 10 MHz), the speed reduction due to multiplexing is not a significant drawback in most applications. On the other hand, since the cost of hardware components such as ADC is declining due to rapid advances in solid-state technologies, cost reductions attainable through the use of multiplexing might not be substantial in some applications. Hence, some economic evaluation and engineering judgment would be needed when deciding on the use of signal multiplexing for a particular data acquisition and control application.

A schematic diagram of an analog multiplexer is shown in Figure 4.33. The figure represents the general case of N input channels and one output channel. This is called an $N \times 1$ analog multiplexer. Each input channel is connected to the output through a

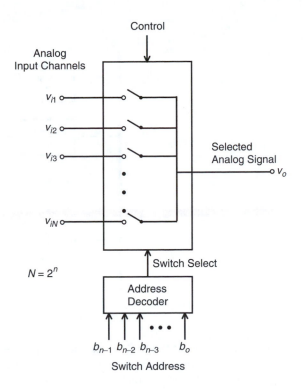

FIGURE 4.33
An N-channel analog multiplexer (analog scanner).

solid-state switch, typically a field-effect transistor (FET) switch. One switch is closed (turned on) at a time. A switch is selected by a digital word, which contains the corresponding *channel address*. Note that an n-bit address can assume 2^n digital values in the range of $0-2^n-1$. Hence, a multiplexer (MUX) with an n-bit address can handle $N = 2^n$ channels. Channel selection can be done by an external microprocessor, which places the address of the channel on the address bus and simultaneously sends a control signal to the MUX to enable the MUX. The address decoder decodes the address and activates the corresponding FET switch. In this manner, channel selection can be done in an arbitrary order and with arbitrary timing, controlled by the microprocessor. In simple versions of multiplexers, the channel selection is made in a fixed order at a fixed speed, however.

Typically, the output of an analog MUX is connected to an S/H circuit and an ADC. Voltage followers can be provided both at the input and the output in order to reduce loading problems. A differential amplifier (or instrumentation amplifier) could be used at the output to reduce noise problems, particularly to reject common-mode interference, as discussed earlier in this chapter. Note that the channel-select speed has to be synchronized with the sampling and ADC speeds for each signal channel. The multiplexer speed is not a major limitation because very high speeds (solid-state speeds of 10 MHz or more) are available with solid-state switching.

4.8.2 Digital Multiplexers

Sometimes it is required to select one data word at a time from a set of digital data words, to be fed into a common device. For example, the set of data may be the outputs from a bank of digital transducers (e.g., shaft encoders, which measure angular motions.

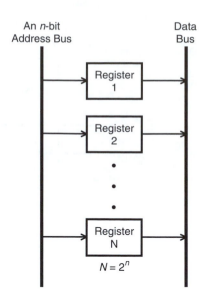

FIGURE 4.34
An N × 1 digital multiplexer.

See Chapter 7) or outputs from a set of ADCs that are connected to a series of analog signal channels. Then the selection of the particular digital output (data word) can be made using techniques of *addressing* and *data-bus transfer*, which are commonly used in digital systems.

A digital multiplexing (or logic multiplexing) configuration is shown in Figure 4.34. The N registers of the multiplexer hold a set of N data words. The contents of each register may correspond to a response measurement, and, hence, will change regularly. The registers may represent separate hardware devices (e.g., output registers of a bank of ADCs) or may represent locations in a computer memory to which data are being transferred (read in) regularly. Each register has a unique binary address. As in the case of analog MUX, an n-bit address can select (address) 2^n registers. Hence, the number of registers will be given by $N = 2^n$, as before. When the address of the register to be selected is placed on the address bus, it enables the particular register. This causes the contents of that register to be placed on the data bus. Now the data bus is read by the device (e.g., computer), which is time-shared among the N data registers. Placing a different address on the address bus will result in selecting another register and reading the contents of that register, as before.

Digital multiplexing is usually faster than analog multiplexing, and has the usual advantages of digital devices; for example, high accuracy, better noise immunity, robustness (no drift and errors due to parameter variations), long-distance data transmission capability without associated errors due to signal weakening, and capability to handle very large numbers of data channels. Furthermore, a digital multiplexer can be modified using software, usually without the need for hardware changes. If, however, instead of using an analog multiplexer followed by a single ADC, a separate ADC is used for each analog signal channel and then digital multiplexing is used, it would be quite possible for the digital multiplexing approach to be more costly. If, on the other hand, the measurements are already available in the digital form (for instance, as encoder outputs of displacement measurement), then digital multiplexing tends to be very cost effective and most desirable.

Transfer of a digital word from a single data source (e.g., a data bus) into several data registers, which are to be accessed independently, may be interpreted as *digital demultiplexing*. This is also a straightforward process of digital data transfer and reading.

4.9 Digital Filters

A filter is a device that eliminates undesirable frequency components in a signal and passes only the desirable frequency components through it. In *analog filtering*, the filter is a physical dynamic system; typically an electric circuit. The signal to be filtered is applied (input) to this dynamic system. The output of the dynamic system is the filtered signal. It follows that any physical dynamic system can be interpreted as an analog filter.

An analog filter can be represented by a differential equation with respect to time. It takes an analog input signal $u(t)$, which is defined continuously in time t, and generates an analog output $y(t)$. A digital filter is a device that accepts a sequence of discrete input values (say, sampled from an analog signal at sampling period Δt), represented by

$$\{u_k\} = \{u_0, u_1, u_2, \ldots\} \tag{4.85}$$

and generates a sequence of discrete output values:

$$\{y_k\} = \{y_0, y_1, y_2, \ldots\} \tag{4.86}$$

It follows that a digital filter is a discrete-time system and it can be represented by a *difference equation* (See Chapter 12).

An nth order linear difference equation can be written in the form

$$a_0 y_k + a_1 y_{k-1} + \cdots + a_n y_{k-n} = b_0 u_k + b_1 u_{k-1} + \cdots + b_m u_{k-m} \tag{4.87}$$

This is a *recursive algorithm* in the sense that it generates one value of the output sequence using previous values of the output sequence and all values of the input sequence up to the present time point. Digital filters represented in this manner are termed *recursive digital filters*. There are filters that employ digital processing where a block (a collection of samples) of the input sequence is converted by a one-shot computation into a block of the output sequence. They are not recursive filters. Nonrecursive filters usually employ digital Fourier analysis, the *fast Fourier transform* (FFT) algorithm, in particular. We restrict our discussion below to recursive digital filters. Our intention in the present section is to give a brief (and nonexhaustive) introduction to the subject of digital filtering.

4.9.1 Software Implementation and Hardware Implementation

In digital filters, signal filtering is accomplished through digital processing of the input signal. The sequence of input data (usually obtained by sampling and digitizing the corresponding analog signal) is processed according to the recursive algorithm of the particular digital filter. This generates the output sequence. The resulting digital output can be converted into an analog signal using a DAC if so desired.

A recursive digital filter is an implementation of a recursive algorithm that governs the particular filtering scheme (e.g., low-pass, high-pass, band-pass, and band-reject). The filter algorithm can be implemented either by software or by hardware. In software implementation, the filter algorithm is programmed into a digital computer. The processor (e.g., microprocessor or digital signal processor or DSP) of the computer can process an input data sequence according to the run-time filter program stored in the memory (in machine code), to generate the filtered output sequence.

Digital processing of data is accomplished by means of logic circuitry that can perform basic arithmetic operations such as addition. In the software approach, the processor of a digital computer makes use of these basic logic circuits to perform digital processing according to the instructions of a software program stored in the computer memory. Alternatively, a hardware digital processor can be built to perform a somewhat complex, yet fixed, processing operation. In this approach the program of computation is said to be in hardware. The hardware processor is then available as an IC chip whose processing operation is fixed and cannot be modified. The logic circuitry in the IC chip is designed to accomplish the required processing function. Digital filters implemented by this hardware approach are termed *hardware digital filters.*

The software implementation of digital filters has the advantage of flexibility; specifically, the filter algorithm can be easily modified by changing the software program that is stored in the computer. If, on the other hand, a large number of filters of a particular (fixed) structure is commercially needed then it would be economical to design the filter as an IC chip and replicate the chip in mass production. In this manner, very low-cost digital filters can be produced. A hardware filter can operate at a much faster speed in comparison to a software filter because in the former case, processing takes place automatically through logic circuitry in the filter chip without having to access by the processor, a software program and various data items stored in the memory. The main disadvantage of a hardware filter is that its algorithm and parameter values cannot be modified, and the filter is dedicated to perform a fixed function.

4.10 Bridge Circuits

A full bridge is a circuit having four arms connected in a lattice form. Four nodes are formed in this manner. Two opposite nodes are used for excitation (by a voltage or current supply) of the bridge and the remaining two opposite nodes provide the bridge output.

A bridge circuit is used to make some form of measurement. Typical measurements include change in resistance, change in inductance, change in capacitance, oscillating frequency, or some variable (stimulus) that causes these changes. There are two basic methods of making the measurement; namely,

1. Bridge balance method
2. Imbalance output method

A bridge is said to be balanced when its output voltage is zero.

In the bridge-balance method, we start with a balanced bridge. When making a measurement, the balance of the bridge will be upset due to the associated variation. As a result a nonzero output voltage will be produced. The bridge is balanced again by varying one of the arms of the bridge (assuming, of course, that some means is provided for fine adjustments that may be required). The change that is required to restore the balance is in fact the measurement. The bridge can be balanced precisely using a servo device, in this method.

In the imbalance output method as well, we usually start with a balanced bridge. As before, the balance of the bridge will be upset as a result of the change in the variable that is being measured. Now, instead of balancing the bridge again, the output voltage of the bridge due to the resulted imbalance is measured and used as the bridge measurement.

There are many types of bridge circuits. If the supply to the bridge is dc, then we have a *dc bridge.* Similarly, an *ac bridge* has an ac excitation. A resistance bridge has only resistance elements in its four arms, and it is typically a dc bridge. An impedance bridge has impedance elements consisting of resistors, capacitors, and inductors in one or more of its arms. This is necessarily an ac bridge. If the bridge excitation is a constant voltage supply, we have a constant-voltage bridge. If the bridge supply is a constant current source, we get a constant-current bridge.

4.10.1 Wheatstone Bridge

This is a resistance bridge with a constant dc voltage supply (i.e., it is a constant-voltage resistance bridge). A Wheatstone bridge is particularly useful in strain-gage measurements, and consequently in force, torque and tactile sensors that employ strain-gage techniques (See Chapter 6). Since a Wheatstone bridge is used primarily in the measurement of small changes in resistance, it could be used in other types of sensing applications as well. For example, in resistance temperature detectors (RTD) the change in resistance in a metallic (e.g., platinum) element, as caused by a change in temperature, is measured using a bridge circuit. Note that the temperature coefficient of resistance is positive for a typical metal (i.e., the resistance increases with temperature). For platinum, this value (change in resistance per unit resistance per unit change in temperature) is about 0.00385/°C.

Consider the Wheatstone bridge circuit shown in Figure 4.35(a). Assuming that the bridge output is in open-circuit (i.e., very high load resistance), the output v_o may be expressed as

$$v_o = v_A - v_B = \frac{R_1 v_{ref}}{(R_1 + R_2)} - \frac{R_3 v_{ref}}{(R_3 + R_4)} = \frac{(R_1 R_4 - R_2 R_3)}{(R_1 + R_2)(R_3 + R_4)} v_{ref} \qquad (4.88)$$

For a balanced bridge, the numerator of the RHS expression of 4.88 must vanish. Hence, the condition for bridge balance is

$$\frac{R_1}{R_2} = \frac{R_3}{R_4} \qquad (4.89)$$

Suppose that at first $R_1 = R_2 = R_3 = R_4 = R$. Then, according to Equation 4.89, the bridge is balanced. Now increase R_1 by δR. For example, R_1 may represent the only active strain gage while the remaining three elements in the bridge are identical dummy elements. In view of Equation 4.88, the change in the bridge output due to the change δR is given by

$$\delta v_o = \frac{[(R + \delta R)R - R^2]}{(R + \delta R + R)(R + R)} v_{ref} - 0$$

or

$$\frac{\delta v_o}{v_{ref}} = \frac{\delta R/R}{(4 + 2\delta R/R)} \qquad (4.90a)$$

Note that the output is nonlinear in $\delta R/R$. If, however, $\delta R/R$ is assumed small in comparison to 2, we have the linearized relationship.

$$\frac{\delta v_o}{v_{ref}} = \frac{\delta R}{4R} \qquad (4.91)$$

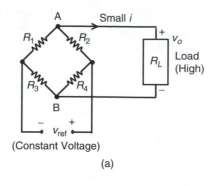

(a)

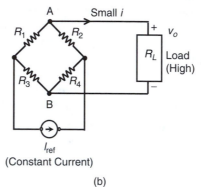

(b)

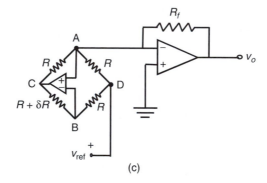

(c)

FIGURE 4.35
(a) Wheatstone bridge (Constant-voltage resistance bridge), (b) Constant-current resistance bridge, (c) A linearized bridge.

The factor $\frac{1}{4}$ on the RHS of Equation 4.91 represents the *sensitivity* in the bridge, as it gives the change in the bridge output for a given change in the active resistance, while the other parameters are kept fixed. Strictly speaking the bridge sensitivity is given by $\delta v_o / \delta R$, which is equal to $v_{ref}/(4R)$.

The error due to linearization, which is a measure of nonlinearity, may be given as the percentage,

$$N_p = 100 \left(1 - \frac{\text{linearized output}}{\text{actual output}} \right)\%$$

(4.92)

Hence, from Equation 4.90a and Equation 4.91 we have

$$N_P = 50 \frac{\delta R}{R} \%$$ (4.93)

Example 4.9

Suppose that in Figure 4.35(a), at first $R_1 = R_2 = R_3 = R_4 = R$. Now increase R_1 by δR decrease R_2 by δR. This will represent two active elements that act in reverse, as in the case of two strain gage elements mounted on the top and bottom surfaces of a beam in bending. Show that the bridge output is linear in δR in this case.

SOLUTION
From Equation 4.88 we get

$$\delta v_o = \frac{[(R + \delta R)R - (R - \delta R)R]}{(R + \delta R + R - \delta R)(R + R)} v_{\text{ref}} - 0$$

This simplifies to

$$\frac{\delta v_o}{v_{\text{ref}}} = \frac{\delta R}{2R}$$

which is linear. Similarly, it can be shown using Equation 4.88 that the pair of changes: $R_3 \rightarrow R + \delta R$ and $R_4 \rightarrow R - \delta R$ will result in a linear relation for the bridge output.

4.10.2 Constant-Current Bridge

When large resistance variations δR are required for a measurement, the Wheatstone bridge may not be satisfactory due to its nonlinearity, as indicated by Equation 4.90. The constant-current bridge is less nonlinear and is preferred in such applications. However, it needs a current-regulated power supply, which is typically more costly than a voltage-regulated power supply.

As shown in Figure 4.35(b), the constant-current bridge uses a constant-current excitation i_{ref} instead of a constant-voltage supply. The output equation for a constant-current bridge can be determined from Equation 4.88 simply by knowing the voltage at the current source. Suppose that this voltage is v_{ref} with the polarity as shown in Figure 4.35(a). Now, since the load current is assumed small (high-impedance load), the current through R_2 is equal to the current through R_1 and is given by $\frac{v_{\text{ref}}}{(R_1 + R_2)}$. Similarly, current through R_4 and R_3 is given by $\frac{v_{\text{ref}}}{(R_3 + R_4)}$. Accordingly, by current summation we get

$$i_{\text{ref}} = \frac{v_{\text{ref}}}{(R_1 + R_2)} + \frac{v_{\text{ref}}}{(R_3 + R_4)}$$

or

$$v_{\text{ref}} = \frac{(R_1 + R_2)(R_3 + R_4)}{(R_1 + R_2 + R_3 + R_4)} i_{\text{ref}}$$ (4.94)

This result may be directly obtained from the equivalent resistance of the bridge, as seen by the current source. Substituting Equation 4.94 in 4.88 we have the output equation for the constant-current bridge; thus,

$$v_o = \frac{(R_1 R_4 - R_2 R_3)}{(R_1 + R_2 + R_3 + R_4)} i_{ref} \tag{4.95}$$

Note from Equation 4.95 that the bridge-balance requirement (i.e., $v_o = 0$) is again given by Equation 4.89.

To estimate the nonlinearity of a constant-current bridge, we start with the balanced condition: $R_1 = R_2 = R_3 = R_4 = R$ and change R_1 by δR while keeping the remaining resistors inactive. Again, R_1 will represent the active element (sensing element) of the bridge, and may correspond to an active strain gage. The change in output δv_o is given by

$$\delta v_o = \frac{[(R+\delta R)R - R^2]}{(R+\delta R + R + R + R)} i_{ref} - 0$$

or

$$\frac{\delta v_o}{R i_{ref}} = \frac{\delta R/R}{(4 + \delta R/R)} \tag{4.96a}$$

By comparing the denominator on the RHS of this equation with Equation 4.90a, we observe that the constant-current bridge is less nonlinear. Specifically, using the definition given by Equation 4.92, the percentage nonlinearity may be expressed as

$$N_p = 25 \frac{\delta R}{R} \% \tag{4.97}$$

It is noted that the nonlinearity is halved by using a constant-current excitation instead of a constant-voltage excitation.

Example 4.10

Suppose that in the constant-current bridge circuit shown in Figure 4.35(b), at first $R_1 = R_2 = R_3 = R_4 = R$. Assume that R_1 and R_4 represent strain gages mounted on the same side of a rod in tension. Due to the tension, R_1 increases by δR and R_4 also increases by δR. Derive an expression for the bridge output (normalized) in this case, and show that it is linear. What would be the result if R_2 and R_3 represent the active tensile strain gages in this example?

SOLUTION
From Equation 4.95 we get

$$\delta v_o = \frac{[(R+\delta R)(R+\delta R) - R^2]}{(R+\delta R + R + R + R + \delta R)} i_{ref} - 0$$

By simplifying and canceling the common term in the numerator and the denominator, we get the linear relation:

$$\frac{\delta v_o}{R i_{ref}} = \frac{\delta R/R}{2} \tag{4.96b}$$

If R_2 and R_3 are the active elements, it is clear from Equation 4.95 that we get the same linear result, except for a sign change; specifically,

$$\frac{\delta v_o}{Ri_{\text{ref}}} = -\frac{\delta R/R}{2} \tag{4.96c}$$

4.10.3 Hardware Linearization of Bridge Outputs

From the foregoing developments and as illustrated in the examples, it should be clear that the output of a resistance bridge is not linear in general, with respect to the change in resistance of the active elements. Particular arrangements of the active elements can result in a linear output. It is seen from Equation 4.88 and Equation 4.95 that, when there is only one active element the bridge output is nonlinear. Such a nonlinear bridge can be linearized using hardware; particularly op-amp elements. To illustrate this approach, consider a constant-voltage resistance bridge. We modify it by connecting two op-amp elements, as shown in Figure 4.35 (c). The output amplifier has a feedback resistor R_f. The output equation for this circuit can be obtained by using the properties of an op-amp, in the usual manner. In particular, the potentials at the two input leads must be equal and the current through these leads must be zero. From the first property it follows that the potentials at the nodes A and B are both zero. Let the potential at node C be denoted by v. Now use the second property, and write current summations at nodes A and B.

Node A:
$$\frac{v}{R} + \frac{v_{\text{ref}}}{R} + \frac{v_o}{R_f} = 0 \tag{i}$$

Node B:
$$\frac{v_{\text{ref}}}{R} + \frac{v}{R + \delta R} = 0 \tag{ii}$$

Substitute Equation ii in Equation i to eliminate v, and simplify to get the linear result:

$$\frac{\delta v_o}{v_{\text{ref}}} = \frac{R_f}{R}\frac{\delta R}{R} \tag{4.90b}$$

Compare this result with Equation 4.90a for the original bridge with a single active element. Note that, when $\delta R = 0$, from Equation ii we get, $v = v_{\text{ref}}$, and from Equation i we get $v_o = 0$. Hence, v_o and δv_o are identical, as used in Equation 4.90b.

4.10.4 Bridge Amplifiers

The output signal from a resistance bridge is usually very small in comparison to the reference signal, and it has to be amplified in order to increase its voltage level to a useful value (e.g., for use in system monitoring, data logging, or control). A bridge amplifier is used for this purpose. This is typically an instrumentation amplifier, which is essentially a sophisticated differential amplifier. The bridge amplifier is modeled as a simple gain K_a, which multiplies the bridge output.

4.10.5 Half-Bridge Circuits

A half bridge may be used in some applications that require a bridge circuit. A half bridge has only two arms and the output is tapped from the mid-point of these two arms. The ends of the two arms are excited by two voltages, one of which is positive and the other negative.

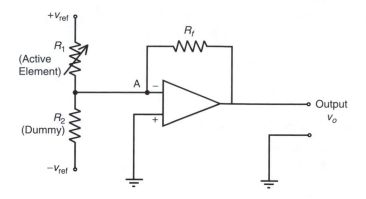

FIGURE 4.36
A half bridge with an output amplifier.

Initially, the two arms have equal resistances so that nominally the bridge output is zero. One of the arms has the active element. Its change in resistance results in a nonzero output voltage. It is noted that the half-bridge circuit is somewhat similar to a potentiometer circuit (a voltage divider).

A half-bridge amplifier consisting of a resistance half-bridge and an output amplifier is shown in Figure 4.36. The two bridge arms have resistances R_1 and R_2, and the output amplifier uses a feedback resistance R_f. To get the output equation we use the two basic facts for an unsaturated op-amp; the voltages at the two input leads are equal (due to high gain) and the current in either lead is zero (due to high input impedance). Hence, voltage at node A is zero and the current balance equation at node A is given by

$$\frac{v_{\text{ref}}}{R_1} + \frac{(-v_{\text{ref}})}{R_2} + \frac{v_o}{R_f} = 0$$

This gives

$$v_o = R_f \left(\frac{1}{R_2} - \frac{1}{R_1} \right) v_{\text{ref}} \tag{4.98}$$

Now, suppose that initially $R_1 = R_2 = R$ and the active element R_1 changes by δR. The corresponding change in output is

$$\delta v_o = R_f \left(\frac{1}{R} - \frac{1}{R + \delta R} \right) v_{\text{ref}} - 0$$

or

$$\frac{\delta v_o}{v_{\text{ref}}} = \frac{R_f}{R} \frac{\delta R/R}{(1 + \delta R/R)} \tag{4.99}$$

Note that R_f/R is the amplifier gain. Now in view of Equation 4.92, the percentage nonlinearity of the half-bridge circuit is

$$N_p = 100 \frac{\delta R}{R} \% \tag{4.100}$$

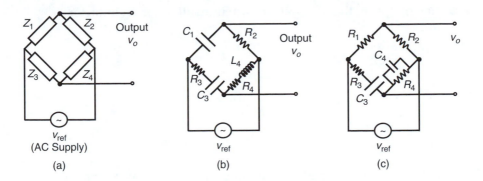

FIGURE 4.37
(a) General impedance bridge, (b) Owen bridge, (c) Wien-bridge oscillator.

It follows that the nonlinearity of a half-bridge circuit is worse than that for the Wheatstone bridge.

4.10.6 Impedance Bridges

An impedance bridge is an ac bridge. It contains general impedance elements Z_1, Z_2 Z_3, and Z_4 in its four arms, as shown in Figure 4.37(a). The bridge is excited by an ac (supply) voltage v_{ref}. Note that v_{ref} would represent a carrier signal and the output voltage v_o has to be demodulated if a transient signal representative of the variation in one of the bridge elements is needed. Impedance bridges could be used, for example, to measure capacitances in *capacitive sensors* and changes of inductance in *variable-inductance sensors* and *eddy-current sensors* (See Chapter 6). Also, impedance bridges can be used as *oscillator circuits*. An oscillator circuit could serve as a constant-frequency source of a signal generator (e.g., in product dynamic testing) or it could be used to determine an unknown circuit parameter by measuring the oscillating frequency.

Analyzing by using frequency-domain concepts it is seen that the frequency spectrum of the impedance-bridge output is given by

$$v_o(\omega) = \frac{(Z_1 Z_4 - Z_2 Z_3)}{(Z_1 + Z_2)(Z_3 + Z_4)} v_{ref}(\omega) \tag{4.101}$$

This reduces to Equation 4.88 in the dc case of a Wheatstone bridge. The balanced condition is given by

$$\frac{Z_1}{Z_2} = \frac{Z_3}{Z_4} \tag{4.102}$$

This equation is used to measure an unknown circuit parameter in the bridge. Let us consider two particular impedance bridges.

4.10.6.1 Owen Bridge

The Owen bridge is shown in Figure 4.37(b). It may be used, for example, to measure both inductance L_4 and capacitance C_3, by the bridge-balance method. To derive the necessary equation, note that the voltage-current relation for an inductor is

$$v = L\frac{di}{dt} \tag{4.103}$$

and for a capacitor it is

$$i = C\frac{dv}{dt} \tag{4.104}$$

It follows that the voltage/current transfer function (in the Laplace domain) for an inductor is

$$\frac{v(s)}{i(s)} = Ls \tag{4.105}$$

and, that for a capacitor is

$$\frac{v(s)}{i(s)} = \frac{1}{Cs} \tag{4.106}$$

Accordingly, the impedance of an inductor element at frequency ω is

$$Z_L = j\omega L \tag{4.107}$$

and the impedance of a capacitor element at frequency ω is

$$Z_c = \frac{1}{j\omega C} \tag{4.108}$$

Applying these results for the Owen bridge we have

$$Z_1 = \frac{1}{j\omega C_1}$$

$$Z_2 = R_2$$

$$Z_3 = R_3 + \frac{1}{j\omega C_3}$$

$$Z_4 = R_4 + j\omega L_4$$

in which ω is the excitation frequency. Now, from Equation 4.102 we have

$$\frac{1}{j\omega C_1}(R_4 + j\omega L_4) = R_2\left(R_3 + \frac{1}{j\omega C_3}\right)$$

By equating the real parts and the imaginary parts of this equation, we get the two equations

$$\frac{L_4}{C_1} = R_2 R_3$$

and

$$\frac{R_4}{C_1} = \frac{R_2}{C_3}$$

Hence, we have

$$L_4 = C_1 R_2 R_3 \qquad (4.109)$$

and

$$C_3 = C_1 \frac{R_2}{R_4} \qquad (4.110)$$

It follows that L_4 and C_3 can be determined with the knowledge of C_1, R_1, R_3, and R_4 under balanced conditions. For example, with fixed C_1 and R_2, an adjustable R_3 could be used to measure the variable L_4, and an adjustable R_4 could be used to measure the variable C_3.

4.10.6.2   Wien-Bridge Oscillator

Now consider the Wien-bridge oscillator shown in Figure 4.37(c). For this circuit we have

$$Z_1 = R_1$$

$$Z_2 = R_2$$

$$Z_3 = R_3 + \frac{1}{j\omega C_3}$$

$$\frac{1}{Z_4} = \frac{1}{R_4} + j\omega C_4$$

Hence, from Equation 4.102, the bridge-balance requirement is

$$\frac{R_1}{R_2} = \left(R_3 + \frac{1}{j\omega C_3} \right)\left(\frac{1}{R_4} + j\omega C_4 \right)$$

Equating the real parts we get

$$\frac{R_1}{R_2} = \frac{R_3}{R_4} + \frac{C_4}{C_3} \qquad (4.111)$$

and by equating the imaginary parts we get

$$0 = \omega C_4 R_3 - \frac{1}{\omega C_3 R_4}$$

Hence

$$\omega = \frac{1}{\sqrt{C_3 C_4 R_3 R_4}} \qquad (4.112)$$

Equation 4.112 tells us that the circuit is an oscillator whose natural frequency is given by this equation, under balanced conditions. If the frequency of the supply is equal to the natural frequency of the circuit, large-amplitude oscillations will take place. The circuit can be used to measure an unknown resistance (e.g., in strain gage devices) by first measuring the frequency of the bridge signals at resonance (natural frequency). Alternatively, an oscillator that is excited at its natural frequency can be used as an accurate source of periodic signals (signal generator).

4.11   Linearizing Devices

Nonlinearity is present in any physical device, to varying levels. If the level of nonlinearity in a system (component, device, or equipment) can be neglected without exceeding the error tolerance, then the system can be assumed linear.

In general, a linear system is one that can be expressed as one or more linear differential equations. Note that the *principle of superposition* holds for linear systems. Specifically, if the system response to an input u_1 is y_1 and the response to another input u_2 is y_2, then the response to $\alpha_1 u_1 + \alpha_2 u_2$ would be $\alpha_1 y_1 + \alpha_2 y_2$.

Nonlinearities in a system can appear in two forms:

1. Dynamic manifestation of nonlinearities
2. Static manifestation of nonlinearities

In many applications, the useful operating region of a system can exceed the frequency range where the frequency response function is flat. The operating response of such a system is said to be dynamic. Examples include a typical mechatronic system (e.g., automobile, aircraft, milling machine, and robot), actuator (e.g., hydraulic motor), and controller (e.g., proportional-integral-derivative or PID control circuitry). Nonlinearities of such systems can manifest themselves in a dynamic form such as the *jump phenomenon* (also known as the *fold catastrophe*), *limit cycles*, and *frequency creation* (See Chapter 5). Design changes, extensive adjustments, or reduction of the operating signal levels and bandwidths would be necessary in general, to reduce or eliminate these dynamic manifestations of nonlinearity. In many instances such changes would not be practical, and we may have to somehow cope with the presence of these nonlinearities under dynamic conditions. Design changes might involve replacing conventional gear drives by devices such as harmonic drives in order to reduce backlash, replacing nonlinear actuators by linear actuators, and using components that have negligible Coulomb friction and that make small motion excursions.

A wide majority of sensors, transducers, and signal modification devices is expected to operate in the flat region of their frequency response function. The input/output relation of these types of devices, in the operating range, is expressed (modeled) as a *static curve* rather than a differential equation. Nonlinearities in these devices will manifest themselves in the static operating curve in many forms. These manifestations include *saturation*, *hysteresis*, and *offset*.

In the first category of systems (e.g., plants, actuators, and compensators) if a nonlinearity is exhibited in the dynamic form, proper modeling and control practices should be employed in order to avoid unsatisfactory degradation of the system performance. In the second category of systems (e.g., sensors, transducers and signal modification devices) if nonlinearities are exhibited in the "static" operating curve, again the overall performance of the system will be degraded. Hence it is important to "linearize" the output of such devices. Note that in dynamic manifestations it is not possible to realistically linearize the output because the response is generated in the dynamic form. The solution in that case is either to minimize nonlinearities within the system by design modifications and adjustments, so that a linear approximation to the system would be valid, or alternatively to take the nonlinearities into account in system modeling and control. In the present section we are not concerned with this aspect; that is, dynamic nonlinearities. Instead, we are interested in the "linearization" of devices in the second category whose operating characteristics can be expressed by static input-output curves.

Linearization of a static device can be attempted as well by making design changes and adjustments, as in the case of dynamic devices. But, since the response is "static," and since we normally deal with an available device (fixed design) whose internal hardware cannot be modified, we should consider ways of linearizing the input-output characteristic by modifying the output itself.

Static linearization of a device can be made in three ways:

1. Linearization using digital software
2. Linearization using digital (logic) hardware
3. Linearization using analog circuitry

In the software approach to linearization, the output of the device is read into a digital processor with software-programmable memory, and the output is modified according to the program instructions. In the hardware approach, the output is read by a device having fixed logic circuitry for processing (modifying) the data. In the analog approach, a linearizing circuit is directly connected at the output of the device so that the output of the linearizing circuit is proportional to the input to the original device. An example of this type of (analog) linearization was given in Section 4.7.2.1. We shall discuss these three approaches in the rest of the present section, heavily emphasizing the analog-circuit approach.

Hysteresis type static nonlinearity characteristics have the property that the input-output curve is not one-to-one. In other words, one input value may correspond to more than one (static) output value, and one output value may correspond to more than one input value. If we disregard these types of nonlinearities, our main concern would be with the linearization of a device having a single-valued static response curve that is not a straight line. An example of a typical nonlinear input-output characteristic is shown in Figure 4.38(a). Strictly speaking, a straight-line characteristic with a simple offset, as shown in Figure 4.38(b), is also a nonlinearity. In particular, note that superposition does not hold for an input-output characteristic of this type, given by

$$y = ku + c \qquad (4.113)$$

It is very easy, however, to linearize such a device because a simple addition of a dc component will convert the characteristic into the linear form given by

$$y = ku \qquad (4.114)$$

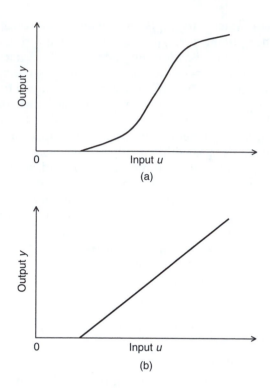

FIGURE 4.38
(a) A general static nonlinear characteristic, (b) An offset nonlinearity.

This method of linearization is known as *offsetting*. Linearization is more difficult in the general case where the characteristic curve could be much more complex.

4.11.1   Linearization by Software

If the nonlinear relationship between the input and the output of a nonlinear device is known, the input can be "computed" for a known value of the output. In the software approach of linearization, a processor and memory that can be programmed using software (i.e., a digital computer) is used to compute the input using output data. Two approaches can be used. They are

1. Equation inversion
2. Table lookup

In the first method, the nonlinear characteristic of the device is known in the analytic (equation) form:

$$y = f(u) \tag{4.115}$$

in which
   u = device input
   y = device output.

Assuming that this is a one-to-one relationship, a unique inverse given by the equation

$$u = f^{-1}(y) \tag{4.116}$$

can be determined. This equation is programmed as a computation *algorithm*, into the read-and-write memory (RAM) of the computer. When the output values y are supplied to the computer, the processor will compute the corresponding input values u using the instructions (executable program) stored in the RAM (See Chapter 11).

In the table lookup method, a sufficiently large number of pairs of values (y, u) are stored in the memory of the computer in the form of a table of ordered pairs. These values should cover the entire operating range of the device. Then when a value for y is entered into the computer, the processor scans the stored data to check whether that value is present. If so, the corresponding value of u will be read, and this is the linearized output. If the value of y is not present in the data table, then the processor will *interpolate* the data in the vicinity of the value and will compute the corresponding output. In the linear interpolation method, the neighborhood of the data table where the y value falls is fitted with a straight line and the corresponding u value is computed using this straight line. Higher order interpolations use nonlinear interpolation curves such as quadratic and cubic polynomial equations (splines).

Note that the equation inversion method is usually more accurate than the table lookup method and the former does not need excessive memory for data storage. But it is relatively slow because data are transferred and processed within the computer using program instructions, which are stored in the memory and which typically have to be accessed in a sequential manner. The table lookup method is fast. Since accuracy depends on the amount of stored data values, this is a memory-intensive method. For better accuracy more data should be stored. But, since the entire data table has to be scanned to check for a given data value, this increase in accuracy is derived at the expanse of speed as well as memory requirements.

4.11.2 Linearization by Hardware Logic

The software approach of linearization is "flexible" because the linearization algorithm can be modified (e.g., improved, changed) simply by modifying the program stored in the RAM. Furthermore, highly complex nonlinearities can be handled by the software method. As mentioned before, the method is relatively slow, however.

In the hardware logic method of linearization, the linearization algorithm is permanently implemented in the integrated-circuit (IC) form using appropriate *digital logic circuitry* for data processing, and memory elements (e.g., *flip-flops*). Note that the algorithm and numerical values of parameters (except input values) cannot be modified without redesigning the IC chip, because a hardware device typically does not have programmable memory. Furthermore, it will be difficult to implement very complex linearization algorithms by this method, and unless the chips are mass produced for an extensive commercial market, the initial chip development cost will make the production of linearizing chips economically infeasible. In bulk production, however, the per-unit cost will be very small. Furthermore, since both the access of stored program instructions and extensive data manipulation are not involved, the hardware method of linearization can be substantially faster than the software method.

A digital linearizing unit having a processor and a *read-only memory* (ROM) whose program cannot be modified, also lacks the flexibility of a programmable software device. Hence, such a ROM-based device also falls into the category of hardware logic devices.

4.11.3   Analog Linearizing Circuitry

Three types of analog linearizing circuitry can be identified:

1. Offsetting circuitry
2. Circuitry that provides a proportional output
3. Curve shapers

We will describe each of these categories now.

An offset is a nonlinearity that can be easily removed using an analog device. This is accomplished by simply adding a dc offset of equal value to the response, in the opposite direction. Deliberate addition of an offset in this manner is known as *offsetting*. The associated removal of original offset is known as *offset compensation*. There are many applications of offsetting. Unwanted offsets such as those present in the results of ADC and DAC can be removed by analog offsetting. Constant (dc) error components such as steady-state errors in dynamic systems due to load changes, gain changes and other disturbances, can be eliminated by offsetting. Common-mode error signals in amplifiers and other analog devices can also be removed by offsetting. In measurement circuitry such as potentiometer (ballast) circuits, where the actual measurement signal is a small "change" δv_o of a steady output signal v_o, the measurement can be completely masked by noise. To reduce this problem, first the output should be offset by $-v_o$ so that the net output is δv_o and not $v_o + \delta v_o$. Subsequently, this output can be conditioned through filtering and amplification. Another application of offsetting is the additive change of the scale of a measurement, for example, from a relative scale to an absolute scale (e.g., in the case of velocity). In summary, some applications of offsetting are:

1. Removal of unwanted offsets and dc components in signals (e.g., in ADC, DAC, signal integration).
2. Removal of steady-state error components in dynamic system responses (e.g., due to load changes and gain changes in Type 0 systems. Note: Type 0 systems are open-loop systems having no free integrators. See Chapter 12).
3. Rejection of common-mode levels (e.g., in amplifiers and filters).
4. Error reduction when a measurement is an increment of a large steady output level (e.g., in ballast circuits for strain-gage and RTD sensors. See Chapter 6).
5. Scale changes in an additive manner (e.g., conversion from relative to absolute units or from absolute to relative units).

We can remove unwanted offsets in the simple manner as discussed above. Let us now consider more complex nonlinear responses that are nonlinear in the sense that the input-output curve is not a straight line. Analog circuitry can be used to linearize this type of responses as well. The linearizing circuit used will generally depend on the particular device and the nature of its nonlinearity. Hence, often linearizing circuits of this type have to be discussed with respect to a particular application. For example, such linearization circuits are useful in a transverse-displacement capacitive sensor. Several useful circuits are described below.

Consider the type of linearization that is known as *curve shaping*. A curve shaper is a linear device whose gain (output/input) can be adjusted so that response curves with different slopes can be obtained. Suppose that a nonlinear device having an irregular (nonlinear) input-output characteristic is to be linearized. First, we apply the operating input simultaneously to both the device and the curve shaper, and the gain of the curve shaper is adjusted such that it closely matches that of the device in a small range of operation. Now the output of the curve shaper can be utilized for any task that requires the device output. The advantage here is that linear assumptions are valid with the curve shaper, which is not the case for the actual device. When the operating range changes, the curve shaper has to be adjusted to the new range. Comparison (calibration) of the curve shaper and the nonlinear device can be done off line and, once a set of gain values corresponding to a set of operating ranges is determined in this manner for the curve shaper, it is possible to completely replace the nonlinear device by the curve shaper. Then the gain of the curve shaper can be adjusted depending on the actual operating range during system operation. This is known as *gain scheduling*. Note that we can replace a nonlinear device by a linear device (curve shaper) within a multi-component system in this manner without greatly sacrificing the accuracy of the overall system.

4.11.4 Offsetting Circuitry

Common-mode outputs and offsets in amplifiers and other analog devices can be minimized by including a *compensating resistor*, which will provide fine adjustments at one of the input leads. Furthermore, the larger the magnitude of the feedback signal in a control system, the smaller the steady-state error (See Chapter 12). Hence, steady-state offsets can be reduced by reducing the *feedback resistance* (thereby increasing the feedback signal). Furthermore, since a ballast (potentiometer) circuit provides an output of $v_o + \delta v_o$ and a bridge circuit provides an output of δv_o, the use of a *bridge circuit* can be interpreted as an offset compensation method.

The most straightforward way of offsetting a nonlinear device is by using a *differential amplifier* (or a summing amplifier) to subtract (or add) a dc voltage to the output of the device. The dc level has to be variable so that various levels of offset can be provided with the same circuit. This is accomplished by using an adjustable resistance at the dc input lead of the amplifier.

An operational-amplifier circuit that can be used for offsetting is shown in Figure 4.39. Since the input v_i is connected to the "−" lead of the op-amp, we have an inverting amplifier, and the input signal will appear in the output v_o with its sign reversed. This is also a summing amplifier because two signals can be added together by this circuit. If the input v_i is connected to the "+" lead of the op-amp, we will have a noninverting amplifier.

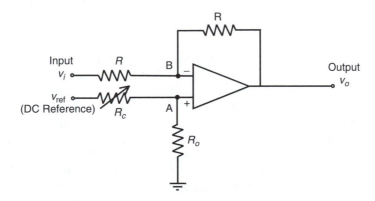

FIGURE 4.39
An inverting amplifier circuit for offset compensation.

The dc voltage v_{ref} provides the offsetting voltage. The resistor R_c (compensating resistor) is variable so that different values of offset can be compensated using the same circuit. To obtain the circuit equation, we write the current balance equation for node A, using the usual assumption that the current through an input lead is zero for an op-amp (because of very high input impedance); thus

$$\frac{v_{ref} - v_A}{R_c} = \frac{v_A}{R_o}$$

or

$$v_A = \frac{R_o}{(R_o + R_c)} v_{ref}$$

Similarly, the current balance at node B gives

$$\frac{v_i - v_B}{R} + \frac{v_o - v_B}{R} = 0$$

or

$$v_o = -v_i + 2v_B \tag{ii}$$

Since $v_A = v_B$ for the op-amp (because of very high open-loop gain), we can substitute Equation i in Equation ii. Then,

$$v_o = -v_i + \frac{2R_o}{(R_o + R_c)} v_{ref} \tag{4.117}$$

Note the sign reversal of v_i at the output (because this is an inverting amplifier). This is not a problem because the polarity can be reversed at input or output in connecting this circuit to other circuitry, thereby recovering the original sign. The important result here is the presence of a constant offset term on the RHS of Equation 4.117. This term can be adjusted by picking the proper value for R_c so as to compensate for a given offset in v_i.

4.11.5 Proportional-Output Circuitry

An operational-amplifier circuit may be employed to linearize the output of a capacitive transverse-displacement sensor. We have noted that in constant-voltage and constant-current resistance bridges and in a constant-voltage half bridge, the relation between the bridge output δv_o and the measurand (change in resistance in the active element) is nonlinear in general. The nonlinearity is least for the constant-current bridge and it is highest for the half bridge. Since δR is small compared to R, however, the nonlinear relations can be linearized without introducing large errors. But the linear relations are inexact, and are not suitable if δR cannot be neglected in comparison to R. Under these circumstances the use of a linearizing circuit would be appropriate.

One way to obtain a proportional output from a Wheatstone bridge is to feedback a suitable factor of the bridge output into the bridge supply v_{ref}. This approach was illustrated previously (see Figure 4.35(c)). Another way is to use the op-amp circuit shown in

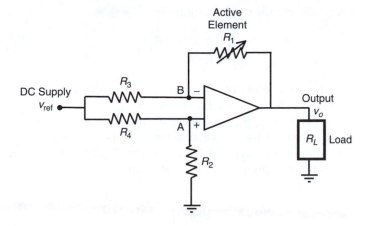

FIGURE 4.40
A proportional-output circuit for an active resistance element (strain gage).

Figure 4.40. This should be compared with the Wheatstone bridge shown in Figure 4.35(a). Note that R_1 represents the only active element (e.g., an active strain gage).

First let us show that the output equation for the circuit in Figure 4.40 is quite similar to Equation 4.88. Using the fact that the current through an input lead of an unsaturated op-amp can be neglected, we have the following current balance equations for nodes A and B:

$$\frac{v_{ref} - v_A}{R_4} = \frac{v_A}{R_2}$$

$$\frac{v_{ref} - v_B}{R_3} + \frac{v_o - v_B}{R_1} = 0$$

Hence,

$$v_A = \frac{R_2}{(R_2 + R_4)} v_{ref}$$

and

$$v_B = \frac{R_1 v_{ref} + R_3 v_o}{(R_1 + R_3)}$$

Now using the fact $v_A = v_B$ for an op-amp, we get

$$\frac{R_1 v_{ref} + R_3 v_o}{(R_1 + R_3)} = \frac{R_2}{(R_2 + R_4)} v_{ref}$$

Accordingly, we have the circuit output equation

$$v_o = \frac{(R_2 R_3 - R_1 R_4)}{R_3 (R_2 + R_4)} v_{ref} \tag{4.118}$$

Note that this relation is quite similar to the Wheatstone bridge Equation 4.88. The balance condition (i.e., $v_o = 0$) is again given by Equation 4.89.

Suppose that $R_1 = R_2 = R_3 = R_4 = R$ in the beginning (hence, the circuit is balanced), so that $v_o = 0$. Next suppose that the active resistance R_1 is changed by δR (say, due to a change in strain in the strain gage R_1). Then, using Equation 4.118 we can write an expression for the resulting change in the circuit output as

$$\delta v_o = \frac{[R^2 - R(R + \delta R)]}{R(R + R)} v_{ref} - 0$$

or

$$\frac{\delta v_o}{v_{ref}} = -\frac{1}{2} \frac{\delta R}{R} \tag{4.119}$$

By comparing this result with Equation 4.90a we observe that the circuit output δv_o is proportional to the measurand δR. Furthermore, note that the *sensitivity* (1/2) of the circuit in Figure 4.40 is double that of a Wheatstone bridge (1/4) with one active element, which is a further advantage of the proportional-output circuit. The sign reversal is not a drawback because it can be accounted for by reversing the load polarity.

4.11.6   Curve Shaping Circuitry

A curve shaper can be interpreted as an amplifier whose gain is adjustable. A typical arrangement for a curve shaping circuit is shown in Figure 4.41. The feedback resistance R_f is adjustable by some means. For example, a switching circuit with a bank of resistors (say, connected in parallel through solid-state switches as in the case of weighted-resistor DAC) can be used to switch the feedback resistance to the required value. Automatic switching can be realized by using Zener diodes, which will start conducting at certain voltage levels. In both cases (external switching by switching pulses or automatic switching using Zener diodes), amplifier gain is variable in discrete steps.

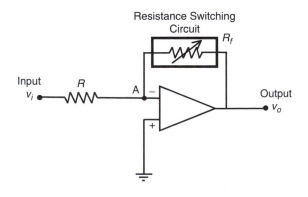

FIGURE 4.41
A curve-shaping circuit.

Alternatively, a potentiometer may be used as R_f so that the gain can be continuously adjusted (manually or automatically).

The output equation for the curve-shaping circuit shown in Figure 4.41 is obtained by writing the current balance at node A, noting that $v_A = 0$; thus,

$$\frac{v_i}{R} + \frac{v_o}{R_f} = 0$$

or

$$v_o = -\frac{R_f}{R} v_i \tag{4.120}$$

It follows that the gain (R_f/R) of the amplifier can be adjusted by changing R_f.

4.12 Miscellaneous Signal Modification Circuitry

In addition to the signal modification devices discussed so far in this chapter there are many other types of circuitry that are used for signal modification and related tasks. Examples are phase shifters, voltage-to-frequency converters, frequency-to-voltage converters, voltage-to-current converters, and peak-hold circuits. The objective of the present section is to briefly discuss several of such miscellaneous circuits and components that are useful in the instrumentation of mechatronic systems.

4.12.1 Phase Shifters

A sinusoidal signal given by

$$v = v_a \sin (\omega t + \phi) \tag{4.121}$$

has the following three representative parameters:

v_a = amplitude
ω = frequency
ϕ = phase angle.

Note that the phase angle represents the time reference (starting point) of the signal. The phase angle is an important consideration only when two or more signal components are compared. In particular, the Fourier spectrum of a signal is presented as its amplitude (magnitude) and the phase angle with respect to frequency.

Phase shifting circuits have many applications. When a signal passes through a system its phase angle changes due to dynamic characteristics of the system. Consequently, the phase change provides very useful information about the dynamic characteristics of the system. Specifically, for a linear constant-coefficient (time-invariant) system, this phase shift is equal to the phase angle of the *frequency-response function* (i.e., *frequency-transfer function*) of the system at that particular frequency. This phase shifting behavior is, of course, not limited to electrical systems and is equally exhibited by other types of systems including mechanical systems and mechatronic systems. The phase shift between two signals can be determined by converting the signals into the electrical form (using suitable transducers), and shifting the phase angle of one signal through known amounts using a phase-shifting circuit until the two signals are in phase.

Another application of phase shifters is in signal demodulation. For example, as noted earlier in the present chapter, one method of amplitude demodulation involves processing the modulated signal together with the carrier signal. This, however, requires the modulated signal and the carrier signal to be in phase. But, usually, since the modulated signal has already transmitted through electrical circuitry having impedance characteristics, its phase angle will have changed. Then, it is necessary to shift the phase angle of the carrier until the two signals are in phase, so that demodulation can be performed accurately. Hence phase shifters are used in demodulating, for example, in the outputs of LVDT (linear variable transformer) displacement sensors (See Chapter 6).

A phase shifter circuit, ideally, should not change the signal amplitude while changing the phase angle by a required amount. Practical phase shifters could introduce some degree of amplitude distortion (with respect to frequency) as well. A simple phase shifter circuit can be constructed using resistance (R) and capacitance (C) elements. A resistance or a capacitor of such an RC circuit is made fine-adjustable so as to obtain a variable phase shifter.

An op-amp-based phase shifter circuit is shown in Figure 4.42. We can show that this circuit provides a phase shift without distorting the signal amplitude. The circuit equation is obtained by writing the current balance equations at nodes A and B, as usual, noting that the current through the op-amp leads can be neglected due to high input impedance; thus,

$$\frac{v_i - v_A}{R_C} = C\frac{dv_A}{dt}$$

$$\frac{v_i - v_B}{R} + \frac{v_o - v_B}{R} = 0$$

On simplifying and introducing the Laplace variable s, we get

$$v_i = (\tau s + 1)v_A \qquad \text{(i)}$$

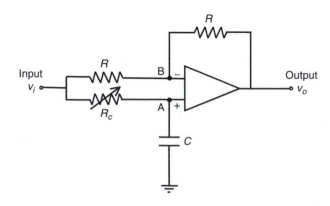

FIGURE 4.42
A phase shifter circuit.

and

$$v_B = \frac{1}{2}(v_i + v_o) \tag{ii}$$

in which, the circuit *time constant* τ is given by

$$\tau = R_c C$$

Since $v_A = v_B$, as a result of very high gain in the op-amp, we have by substituting Equation ii in Equation i,

$$v_i = \frac{1}{2}(\tau s + 1)(v_i + v_o)$$

It follows that the transfer function $G(s)$ of the circuit is given by

$$\frac{v_o}{v_i} = G(s) = \frac{(1 - \tau s)}{(1 + \tau s)} \tag{4.122}$$

It is seen that the magnitude of the frequency-response function $G(j\omega)$ is

$$|G(j\omega)| = \frac{\sqrt{1 + \tau^2 \omega^2}}{\sqrt{1 + \tau^2 \omega^2}}$$

or,

$$|G(j\omega)| = 1 \tag{4.123}$$

and the phase angle of $G(j\omega)$ is

$$\angle G(j\omega) = -\tan^{-1}\tau\omega - \tan^{-1}\tau\omega$$

or,

$$\angle G(j\omega) = -2\tan^{-1}\tau\omega = -2\tan^{-1}R_c C\omega \tag{4.124}$$

As needed, the transfer function magnitude is unity, indicating that the circuit does not distort the signal amplitude over the entire bandwidth. Equation 4.124 gives the *phase lead* of the output v_o with respect to the input v_i. Note that this angle is negative, indicating that actually a *phase lag* is introduced. The phase shift can be adjusted by varying the resistance R_c.

4.12.2 Voltage-to-Frequency Converters (VFC)

A voltage-to-frequency converter generates a periodic output signal whose frequency is proportional to the level of an input voltage. Since such an oscillator generates a periodic output according to the voltage excitation, it is also called a *voltage-controlled oscillator* (VCO).

A common type of VFC uses a capacitor. The time needed for the capacitor to be charged to a fixed voltage level depends on (inversely proportional to) the charging voltage. Suppose that this voltage is governed by the input voltage. Then if the capacitor is made to periodically charge and discharge, we have an output whose frequency (inverse of the charge-discharge period) is proportional to the charging voltage. The output amplitude will be given by the fixed voltage level to which capacitor is charged in each cycle. Consequently, we have a signal with a fixed amplitude, and a frequency that depends on the charging voltage (input).

A voltage-to-frequency converter (or voltage-controlled oscillator) circuit is shown in Figure 4.43(a). The voltage-sensitive switch closes when the voltage across it exceeds a reference level v_s and it will open again when the voltage across it falls below a lower limit $v_o(0)$. The *programmable unijunction transistor* (PUT) is such a switching device.

Note that the polarity of the input voltage v_i is reversed. Suppose that the switch is open. Then, current balance at node A of the op-amp circuit gives,

$$\frac{v_i}{R} = C\frac{dv_o}{dt}$$

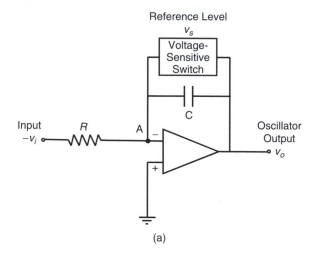

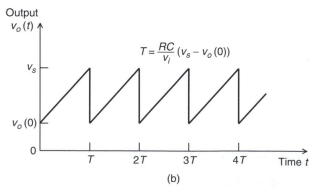

(b)

FIGURE 4.43
A voltage-to-frequency converter (voltage-controlled oscillator): (a) Circuit; (b) Output signal.

As usual, voltage at "−" lead v_A = voltage at "+" lead = 0 because the op-amp has a very high gain, and current through the op-amp leads = 0 because the op-amp has a very high input impedance. The capacitor charging equation can be integrated for a given value of v_i. This gives

$$v_o(t) = \frac{1}{RC} v_i t + v_o(0)$$

The switch will be closed when the voltage across the capacitor $v_o(t)$ equals the reference level v_s. Then the capacitor will be immediately discharged through the closed switch. Hence, the capacitor charging time T is given by

$$v_s = \frac{1}{RC} v_i T + v_o(0)$$

Accordingly,

$$T = \frac{RC}{v_i}(v_s - v_o(0)) \qquad (4.125)$$

The switch will be open again when the voltage across the capacitor drops to $v_o(0)$ and the capacitor will again begin to charge from $v_o(0)$ up to v_s. This cycle of charging and instantaneous discharge will repeat periodically. The corresponding output signal will be as shown in Figure 4.43(b). This is a periodic (*saw tooth*) wave with period T. The frequency of oscillation $(1/T)$ of the output is given by

$$f = \frac{v_i}{RC(v_s - v_o(0))} \qquad (4.126)$$

It is seen that the oscillator frequency is proportional to the input voltage v_i. The oscillator amplitude is v_s which is fixed.

Voltage-controlled oscillators have many applications. One application is in analog to digital conversion. In the VCO type analog-to-digital converters, the analog signal is converted into an oscillating signal using a VCO. Then the oscillator frequency is measured using a digital counter. This count, which is available in the digital form, is representative of the input analog signal level. Another application is in digital voltmeters. Here the same method as for ADC is used. Specifically, the voltage is converted into an oscillator signal and its frequency is measured using a digital counter. The count can be scaled and displayed to provide the voltage measurement. A direct application of VCO is apparent from the fact that VCO is actually a *frequency modulator* (FM), providing a signal whose frequency is proportional to the input (modulating) signal. Hence, VCO is useful in applications that require frequency modulation. Also, a VCO can be used as a signal (wave) generator for variable-frequency applications; for example, excitation inputs for shakers in product dynamic testing, excitations for frequency-controlled dc motors, and pulse signals for translator circuits of stepping motors (See Chapter 8 and Chapter 9).

4.12.3 Frequency-to-Voltage Converter (FVC)

A frequency-to-voltage converter generates an output voltage whose level is proportional to the frequency of its input signal. One way to obtain an FVC is to use a digital counter

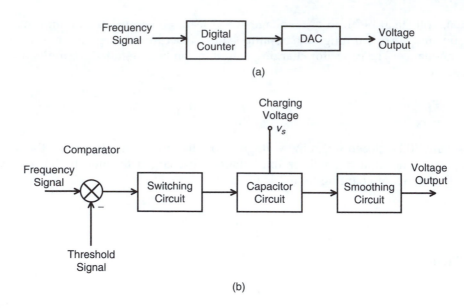

FIGURE 4.44
Frequency-to-voltage converters: (a) Digital counter method, (b) Capacitor charging method.

to count the signal frequency and then use a DAC to obtain a voltage proportional to the frequency. A schematic representation of this type of FVC is shown in Figure 4.44(a).

An alternative FVC circuit is schematically shown in Figure 4.44(b). In this method the frequency signal is supplied to a comparator along with a threshold voltage level. The sign of the comparator output will depend on whether the input signal level is larger or smaller than the threshold level. The first sign change (–ve to +ve) in the comparator output is used to trigger a switching circuit that will respond by connecting a capacitor to a fixed charging voltage. This will charge the capacitor. The next sign change (+ve to –ve) of the comparator output will cause the switching circuit to short the capacitor thereby instantaneously discharging it. This charging-discharging process will be repeated in response to the oscillator input. Note that the voltage level to which the capacitor is charged each time, will depend on the switching period (charging voltage is fixed), which is in turn governed by the frequency of the input signal. Hence, the output voltage of the capacitor circuit will be representative of the frequency of the input signal. Since the output is not steady due to the ramp-like charging curve and instantaneous discharge, a smoothing circuit is provided at the output to remove the resulting noise ripples.

Applications of FVC include demodulation of frequency-modulated signals, frequency measurement in mechatronic applications, and conversion of pulse outputs in some types of sensors and transducers into analog voltage signals (See Chapter 7).

4.12.4 Voltage-to-Current Converter (VCC)

Measurement and feedback signals are usually transmitted as current levels in the range of 4–20 mA rather than as voltage levels. This is particularly useful when the measurement site is not close to the monitoring room. Since the measurement itself is usually available as a voltage, it has to be converted into current by using a voltage-to-current converter. For example, pressure transmitters and temperature transmitters in operability testing systems provide current outputs that are proportional to the measured values of pressure and temperature.

There are many advantages to transmitting current rather than voltage. In particular, the voltage level will drop due to resistance in the transmitting path, but the current through a conductor will remain uncharged unless the conductor is branched. Hence current signals are less likely to acquire errors due to signal weakening. Another advantage of using current instead of voltage as the measurement signal is that the same signal can be used to operate several devices in series (e.g., a display, a plotter, and a signal processor simultaneously), again without causing errors due to signal weakening by the power lost at each device, because the same current is applied to all devices. A voltage-to-current converter (VCC) should provide a current proportional to an input voltage, without being affected by the load resistance to which the current is supplied.

An operational-amplifier-based voltage-to-current converter circuit is shown in Figure 4.45. Using the fact that the currents through the input leads of an unsaturated op-amp can be neglected (due to very high input impedance), we write the current summation equations for the two nodes A and B; thus,

$$\frac{v_A}{R} = \frac{v_p - v_A}{R}$$

and

$$\frac{v_i - v_B}{R} + \frac{v_p - v_B}{R} = i_o$$

Accordingly, we have

$$2v_A = v_P \tag{i}$$

and

$$v_i - 2v_B + v_P = R\, i_o \tag{ii}$$

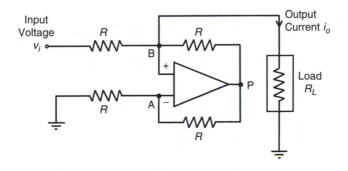

FIGURE 4.45
A voltage-to-current converter.

Now using the fact that $v_A = v_B$ for the op-amp (due to very high gain) we substitute Equation i in Equation ii. This gives

$$i_o = \frac{v_i}{R}$$

in which
   i_o = output current
   v_i = input voltage

It follows that the output current is proportional to the input voltage, irrespective of the value of the load resistance R_L, as required for a VCC.

4.12.5   Peak-Hold Circuits

Unlike a sample-and-hold circuit, which holds every sampled value of a signal, a peak-hold circuit holds only the largest value reached by the signal during the monitored period. Peak holding is useful in a variety of applications. In signal processing for shock and vibration studies of dynamic systems, what is known as *response spectra* (e.g., *shock response spectrum*) are determined by using a *response spectrum analyzer*, which exploits a peak holding scheme. Suppose that a signal is applied to a simple oscillator (a single-degree-of-freedom second-order system with no zeros) and the peak value of the response (output) is determined. A plot of the peak output as a function of the natural frequency of the oscillator, for a specified damping ratio, is known as the response spectrum of the signal for that damping ratio. Peak detection is also useful in machine monitoring and alarm systems. In short, when just one representative value of a signal is needed in a particular application, the peak value would be a leading contender.

Peak detection of a signal can be conveniently done using digital processing. For example the signal is sampled and the previous sample value is replaced by the present sample value if and only if the latter is larger than the former. In this manner, the peak value of the signal is retained by sampling and then holding one value. Note that, usually the time instant at which the peak occurs is not retained.

Peak detection can be done using analog circuitry as well. This is in fact the basis of analog spectrum analyzers. A peak-holding circuit is shown in Figure 4.46. The circuit

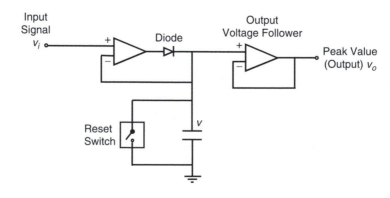

FIGURE 4.46
A peak-holding circuit.

consists of two voltage followers. The first voltage follower has a diode at its output that is forward-biased by the positive output of the voltage follower and reverse-biased by a low-leakage capacitor, as shown. The second voltage follower presents the peak voltage that is held by the capacitor to the circuit output at a low output impedance, without loading the previous circuit stage (capacitor and first voltage follower). To explain the operation of the circuit, suppose that the input voltage v_i is larger than the voltage to which capacitor is charged (v). Since the voltage at the "+" lead of the op-amp is v_i and the voltage at the "−" lead is v, the first op-amp will be saturated. Since the differential input to the op-amp is positive under these conditions, the op-amp output will be positive. The output will charge the capacitor until the capacitor voltage v equals the input voltage v_i. This voltage (call it v_o) is in turn supplied to the second voltage follower, which presents the same value to its output (Note: gain = 1 for a voltage follower), but at a very low impedance level. The op-amp output remains at the saturated value only for a very short time (the time taken by the capacitor to charge). Now suppose that v_i is smaller than v. Then the differential input of the op-amp will be negative, and the op-amp output will be saturated at the negative saturation level. This will reverse bias the diode. Hence, the output of the first op-amp will be in open-circuit, and as a result the voltage supplied to the output voltage follower will still be the capacitor voltage and not the output of the first op-amp. It follows that the voltage level of the capacitor (and hence the output of the second voltage follower) will always be the peak value of the input signal. The circuit can be reset by discharging the capacitor through a solid-state switch that is activated by an external pulse.

4.13 Signal Analyzers and Display Devices

Since signal analysis involves processing of a signal to generate useful information, it is appropriate to consider the topic within the present context of signal modification as well. Signal analysis may employ both analog and digital procedures. In the present section we will introduce digital signal analyzers. Signal display devices also make use of at least some basic types of signal processing. This may involve filtering and change of signal level and format. More sophisticated signal display devices, particularly digital oscilloscopes, can carry out more complex signal analysis functions such as those normally available with digital signal analyzers. Oscilloscopes, which are primarily instruments for signal display and monitoring, are introduced as well in the present section. They typically employ basic types of signal analysis, and may be treated under signal analysis and instrumentation.

Signal-recording equipment commonly employed in the mechatronic practice includes digital storage devices such as hard drives, floppy disks and CD ROMs, analog devices like tape recorders, strip-chart recorders and X-Y plotters, and digital printers. Tape recorders are used to record system response data (transducer outputs) that are subsequently reproduced for processing or examination. Often, tape recorded waveforms are also used to generate (by replay) signals that drive dynamic test exciters (shakers). Tape recorders use tapes made of a plastic material that has a thin coating of a specially treated ferromagnetic substance. During the recording process, magnetic flux proportional to the recorded signal is produced by the recording head (essentially an electromagnet), which magnetizes the tape surface in proportion to the signal variation. Reproduction is the reverse process, whereby an electrical signal is generated at the reproduction head by electromagnetic induction in accordance with the magnetic flux of the magnetized (recorded) tape. Several signal-conditioning circuitries are involved in the recording and reproducing stages. Recording by FM is very common in dynamic testing.

Strip-chart recorders are usually employed to plot time histories (i.e., quantities that vary with time), although they also may be used to plot such data as frequency-response functions and response spectra. In these recorders, a paper roll unwinds at a constant linear speed, and the writing head moves across the paper (perpendicular to the paper motion) proportionally to the signal level. There are many kinds of strip-chart recorders, which are grouped according to the type of writing head that is employed. Graphic-level recorders, which use ordinary paper, employ such heads as ink pens or brushes, fiber pens, and sapphire styli. Visicoders are simple oscilloscopes that are capable of producing permanent records; they employ light-sensitive paper for this. Several channels of input data can be incorporated with a visicoder. Obviously, graphic-level recorders are generally limited by the number of writing heads available (typically, one or two), but visicoders can have many more input channels (typically, twenty-four). Performance specifications of these devices include paper speed, frequency range of operation, dynamic range, and power requirements.

In electro-mechanical experimentation, X-Y plotters are generally employed to plot frequency data (e.g., power spectral densities or psd, frequency-response functions, response spectra, and transmissibility curves), although they also can be used to plot time-history data. Many types of X-Y plotters are available, most of them using ink pens on ordinary paper. There are also hard-copy units that use heat-sensitive paper in conjunction with a heating element as the writing head. The writing head of an X-Y plotter is moved in the X and Y directions on the paper by two input signals, which form the coordinates for the plot. In this manner, a trace is made on stationary plotting paper. Performance specifications of X-Y plotters are governed by such factors as paper size; writing speed (in/sec, cm/sec); dead band (expressed as a percentage of the full scale) which measures the resolution of the plotter head; linearity (expressed as a percentage of the full scale) which measures the accuracy of the plot or deviation from a reference straight line; minimum trace separation (in, cm) for multiple plots on the same axes; dynamic range; input impedance; and maximum input (mV/in, mV/cm).

Today, the most widespread signal recording device is in fact the digital computer (memory, storage) and printer combination. Digital computer and other (analog) devices used in signal recording and display, generally make use of some form of signal modification to accomplish their functions. But, we will not discuss these devices in the present section.

4.13.1 Signal Analyzers

Modern signal analyzers employ digital techniques of signal analysis to extract useful information that is carried by the signal. Digital Fourier analysis using fast Fourier transform (FFT) is perhaps the single common procedure that is used in the vast majority of signal analyzers. Fourier analysis produces the frequency spectrum of a time signal. It should be clear, therefore, why the terms "digital signal analyzer," "FFT analyzer," "frequency analyzer," "spectrum analyzer," and "digital Fourier analyzer" are synonymous to some extent, as used in the commercial instrumentation literature.

A signal analyzer typically has two (dual) or more (multiple) input signal channels. To generate results such as frequency response (transfer) functions, cross spectra, coherence functions, and cross-correlation functions we need at least two data signals and hence a dual-channel analyzer.

In hardware analyzers, digital circuitry rather than software is used to carry out the mathematical operations. Clearly they are very fast but less flexible (in terms of programmability and functional capability) for this reason. Digital signal analyzers, regardless of whether they use the hardware approach or the software approach, employ some basic operations. These operations, carried out in sequence, are

1. Anti-alias filtering (analog)
2. Analog to digital conversion (i.e., single sampling and digitization)
3. Truncation of a block of data and multiplication by a window function
4. FFT analysis of the block of data

The following facts are important in the present context of digital signal analysis. If the sampling period of the ADC is ΔT (i.e., the sampling frequency is $1/\Delta T$) then the Nyquist frequency $f_c = 1/2\Delta T$. This Nyquist frequency is the upper limit of the useful frequency content of the sampled signal. The cut-off frequency of the anti-aliasing filter should be set at f_c or less (See Chapter 5). If there are N data samples in the block of data that is used in the FFT analysis, the corresponding record length is $T = N \cdot \Delta T$. Then, the spectral lines in the FFT results are separated at a frequency spacing of $\Delta F = 1/T$. In view of the Nyquist frequency limit, however, there will be only $N/2$ useful spectral lines in the FFT result.

Strictly speaking, a real-time signal analyzer should analyze a signal instantaneously and continuously, as the signal is received by the analyzer. This is usually the case with an analog signal analyzer. But, in digital signal analyzers, which are usually based on digital Fourier analysis, a block of data (i.e., N samples of record length T) is analyzed together to produce $N/2$ useful spectral lines (at frequency spacing $1/T$). This then is not a truly real-time analysis. But for practical purposes, if the speed of analysis is sufficiently fast, the analyzer may be considered real time, which is usually the case with hardware analyzers and also modern, high-speed, software analyzers.

The bandwidth B of a digital signal analyzer is a measure of its speed of signal processing. Specifically, for an analyzer that uses N data samples in each block of signal analysis, the associated processing time may be given by

$$T_c = \frac{N}{B} \qquad (4.128)$$

Note that the larger the B, the smaller the T_c. Then, the analyzer is considered a real-time one if the analysis time (T_c) of the data record is less than the generation time ($T = N \cdot \Delta T$) of the data record. Hence, we need $T_c < T$ or $N/B < T$ or $N/B < N \cdot \Delta T$, which gives

$$\frac{1}{\Delta T} < B \qquad (4.129)$$

In other words, a real-time analyzer should have a bandwidth greater than its sampling rate.

A multichannel digital signal analyzer can analyze one or more signals simultaneously and generate (and display) results such as Fourier spectra, power spectral densities, cross spectral densities, frequency response functions, coherence functions, autocorrelations, and cross correlations. They are able to perform high-resolution analysis on a small segment of the frequency spectrum of a signal. This is termed *zoom analysis*. Essentially, in this case, the spectral line spacing ΔF is decreased while keeping unchanged the number of lines (N) and hence the number of time data samples. That means the record length ($T = 1/\Delta F$) has to be increased in proportion, for zoom analysis.

4.13.2 Oscilloscopes

An oscilloscope is used to display and observe one or two signals separately or simultaneously. Amplitude, frequency, and phase information of the signals can be obtained using an oscilloscope. In this sense it is a signal analysis/modification device as well as a measurement (monitoring) and display device. While both analog and digital oscilloscopes are commercially available, the latter is far more common. A typical application of an oscilloscope is to observe (monitor) experimental data such as response signals of machinery and processes, as obtained from sensors and transducers. They are also useful in observing and examining dynamic test results, such as frequency-response plots, psd curves, and response spectra. Typically, only temporary records are available on an analog oscilloscope screen. The main component of an analog oscilloscope is the cathode-ray tube (CRT), which consists of an electron gun (cathode), which deflects an electron ray according to the input-signal level. The oscilloscope screen has a coating of electron-sensitive material, so that the electron ray that impinges on the screen leaves a temporary trace on it. The electron ray sweeps across the screen horizontally, so that waveform traces can be recorded and observed. Typically, two input channels are available. Each input may be observed separately, or the variations in one input may be observed against those of the other. In this manner, signal phasing can be examined. Several sensitivity settings for the input-signal-amplitude scale (in the vertical direction) and sweep-speed selections are available on the oscilloscope panel.

4.13.2.1 Triggering

The voltage level of the input signal deflects the electron gun in proportion, in the vertical (y-axis) direction on the CRT screen of an oscilloscope. This alone will not show the time evolution of the signal. The true time variation of the signal is achieved by means of a saw-tooth signal, which is generated internally in the oscilloscope and used to move the electron gun in the horizontal (x-axis) direction. As the name implies, the saw-tooth signal increases linearly in amplitude up to a threshold value and then suddenly drops to zero, and repeats this cycle over and over again. In this manner, the observed signal is repetitively swept across the screen and a trace of it can be observed as a result of the temporary retention of the illumination of the electron gun on the fluorescent screen. The saw-tooth signal may be controlled (triggered) in several ways. For example, the *external trigger* mode uses an external signal form another channel (not the observed channel) to generate and synchronize the saw-tooth signal. In the *line trigger* mode, the saw-tooth signal is synchronized with the ac line supply (60 or 50 Hz). In the *internal trigger* mode, the observed signal (which is used to deflect the electron beam in the y direction) itself is used to generate (synchronize) the saw-tooth signal. Since the frequency and the phase of the observed signal and the trigger signal are perfectly synchronized in this case, the trace on the oscilloscope screen will appear stationary. Careful observation of a signal can be made in this manner.

4.13.2.2 Lissajous Patterns

Suppose that two signals x and y are provided to the two channels of an oscilloscope. If they are used to deflect the electron beam in the horizontal and the vertical directions, respectively, a pattern known as the *Lissajous pattern* will be observed on the oscilloscope screen. Useful information about the amplitude and phasing of the two signals may be observed by means of these patterns. Consider two sine waves x and y. Several special cases of Lissajous patterns are given below:

1. *Same frequency, same phase:*
   Here,

   $$x = x_o \sin(\omega t + \phi)$$

   $$y = y_o \sin(\omega t + \phi)$$

   Then we have,

   $$\frac{x}{x_o} = \frac{y}{y_o}$$

   which gives a straight-line trace with a positive slope, as shown in Figure 4.47(a).

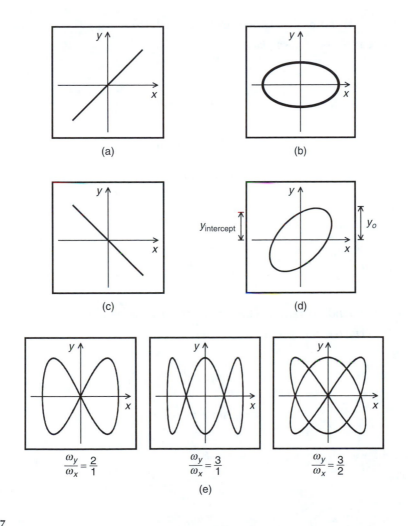

FIGURE 4.47
Some Lissajous patterns for: (a) Equal frequency and in-phase, (b) Equal frequency and 90° out-of-phase, (c) Equal frequency and 180° out-of-phase, (d) Equal frequency and θ out-of-phase, (e) Integral frequency ratio.

2. *Same frequency, 90° out-of-phase*
   Here,

$$x = x_o \sin(\omega t + \phi)$$

$$y = y_o \sin(\omega t + \phi + \pi/2)$$

$$= y_o \cos(\omega t + \phi)$$

Then we have

$$\left(\frac{x}{x_o}\right)^2 + \left(\frac{y}{y_o}\right)^2 = 1$$

which gives an ellipse, as shown in Figure 4.47(b).

3. *Same frequency, 180° out-of-phase*
   Here,

$$x = x_o \sin(\omega t + \phi)$$

$$y = y_o \sin(\omega t + \phi + \pi)$$

$$= -y_o \sin(\omega t + \phi)$$

Hence,

$$\frac{x}{x_o} + \frac{y}{y_o} = 0$$

which corresponds to a straight line with a negative slope, as shown in Figure 4.47(c).

4. *Same frequency, θ out-of-phase*

$$x = x_o \sin(\omega t + \phi)$$

$$y = y_o \sin(\omega t + \phi + \theta)$$

When $\omega t + \phi = 0$, $y = y_{intercept} = y_o \sin \theta$
Hence,

$$\sin \theta = \frac{y_{intercept}}{y_o} \qquad\qquad\qquad (i)$$

In this case, we get a tilted ellipse as shown in Figure 4.47(d). The phase difference θ is obtained from the Lissajous pattern using Equation (i).

5. *Integral frequency ratio*

$$\frac{\omega_y}{\omega_x} = \frac{\text{number of } y - \text{peaks}}{\text{number of } x - \text{peaks}}$$

Three examples are shown in Figure 4.47(e).

$$\frac{\omega_y}{\omega_x} = \frac{2}{1} \qquad \frac{\omega_y}{\omega_x} = \frac{3}{1} \qquad \frac{\omega_y}{\omega_x} = \frac{3}{2}$$

NOTE The above observations hold true as well for narrowband signals, which can be approximated as sinusoidal signals. Broadband random signals produce scattered (irregular) Lissajous patterns.

4.13.2.3 *Digital Oscilloscopes*

The basic uses of a digital oscilloscope are quite similar to those of a traditional analog oscilloscope. The main differences stem from the manner in which information is represented and processed "internally" within the oscilloscope. Specifically, a digital oscilloscope first samples a signal that arrives at one of its input channels and stores the resulting digital data within a memory segment. This is essentially a typical ADC operation. This digital data may be processed to extract and display the necessary information. The sampled data and the processed information may be stored on a floppy disk, if needed, for further processing using a digital computer. Also, some digital oscilloscopes have the communication capability so that the information may be displayed on a video monitor or printed to provide a hard copy.

A typical digital oscilloscope has four channels so that four different signals may be acquired (sampled) into the oscilloscope and displayed. Also, it has various triggering options so that the acquisition of a signal may be initiated and synchronized by means of either an internal trigger or an external trigger. Apart from the typical capabilities that were listed in the context of an analog oscilloscope, a digital oscilloscope can automatically provide other useful features such as the following:

1. Automatic scaling of the acquired signal.
2. Computation of signal features such as frequency, period, amplitude, mean, root-mean-square (rms) value, and rise time.
3. Zooming into regions of interest of a signal record.
4. Averaging of multiple signal records.
5. Enveloping of multiple signal records.
6. Fast Fourier transform (FFT) capability, with various window options and anti-aliasing.

These various functions are menu selectable. Typically, first a channel of the incoming data (signal) is selected and then an appropriate operation on the data is chosen from the menu (through menu buttons).

4.14 Problems

4.1 Define electrical impedance and mechanical impedance. Identify a defect in these definitions in relation to the force-current analogy. What improvements would you suggest? What roles do input impedance and output impedance play in relation to the accuracy of a measuring device?

4.2 What is meant by "loading error" in a signal measurement? Also, suppose that a piezoelectric sensor of output impedance Z_s is connected to a voltage-follower amplifier of input impedance Z_i as shown in Figure P4.2. The sensor signal is v_i volts and the amplifier output is v_o volts. The amplifier output is connected to a device with very high input impedance. Plot to scale the signal ratio v_o/v_i against the impedance ratio Z_i/Z_s for values of the impedance ratio in the range 0.1–10.

4.3 Thevenin's theorem states that with respect to the characteristics at an output port, an unknown subsystem consisting of linear passive elements and ideal source elements may be represented by a single across-variable (voltage) source v_{eq} connected in series with a single impedance Z_{eq}. This is illustrated in Figure P4.3(a) and (b). Note that v_{eq} is equal to the open-circuit across variable v_{oc} at the output port, because the current through Z_{eq} is zero. Consider the network shown in Figure P4.3(c). Determine the equivalent voltage source v_{eq} and the equivalent series impedance Z_{eq}, in the frequency domain, for this circuit.

4.4 Explain why a voltmeter should have a high resistance and an ammeter should have a very low resistance. What are some of the design implications of these general

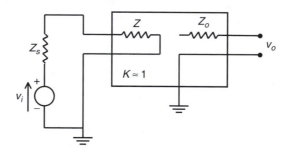

FIGURE P4.2
System with a piezoelectric sensor.

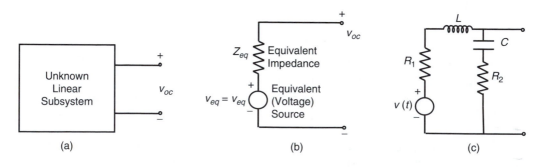

FIGURE P4.3
Illustration of Thevenin's theorem: (a) Unknown linear subsystem, (b) Equivalent representation, (c) Example.

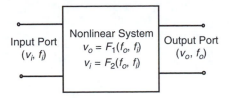

FIGURE P4.5
Impedance characteristics of a nonlinear system.

requirements for the two types of measuring instruments, particularly with respect to instrument sensitivity, speed of response, and robustness? Use a classical moving-coil meter as the model for your discussion.

4.5 A two-port nonlinear device is shown schematically in Figure P4.5. The transfer relations under static equilibrium conditions are given by

$$v_o = F_1(f_o, f_i)$$
$$v_i = F_2(f_o, f_i)$$

where v denotes an across variable, f denotes a through variable, and the subscripts o and i represent the output port and the input port, respectively. Obtain expressions for input impedance and output impedance of the system in the neighborhood of an operating point, under static conditions, in terms of partial derivatives of the functions F_1 and F_2. Explain how these impedances could be determined experimentally.

4.6 Define the terms
   a. Mechanical loading
   b. Electrical loading

in the context of motion sensing, and explain how these loading effects can be reduced.

The following table gives ideal values for some parameters of an operational amplifier. Give typical, practical values for these parameters (e.g., output impedance of 50 Ω).

Parameter	Ideal value	Typical value
Input impedance	Infinity	?
Output impedance	Zero	50Ω
Gain	Infinity	?
Bandwidth	Infinity	?

Also note that, under ideal conditions, inverting-lead voltage is equal to the noninverting-lead voltage (i.e., offset voltage is zero).

4.7 Linear variable differential transformer (LVDT) is a displacement sensor, which is commonly used in mechatronic control systems. Consider a digital control loop that uses an LVDT measurement, for position control of a machine. Typically, the LVDT is energized by a dc power supply. An oscillator provides an excitation signal in the kilohertz range to the primary windings of the LVDT. The secondary winding segments are connected in series opposition. An ac amplifier, demodulator, low-pass filter, amplifier, and ADC are used in the monitoring path. Figure P4.7 shows the various hardware components in the control loop. Indicate the functions of these components.

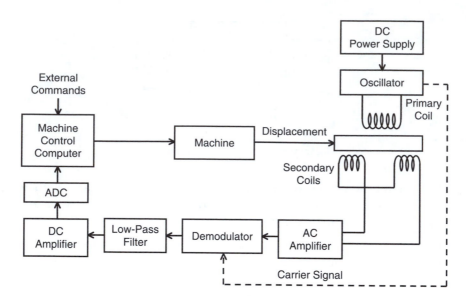

FIGURE P4.7
Components of an LVDT-based machine control loop.

At null position of the LVDT stroke, there was a residual voltage. A compensating resistor is used to eliminate this voltage. Indicate the connections for this compensating resistor.

4.8 Today, *machine vision* is used in many industrial tasks including process control and monitoring. In an industrial system based on machine vision, an imaging device such as a charge-coupled-device (CCD) camera is used as the sensing element. The camera provides to an image processor an image (picture) of a scene related to the industrial process (the measurement). The computed results from the image processor are used to determine the necessary information about the process (plant).

A CCD camera has an image plate consisting of a matrix of metal-oxide-semiconductor field-effect-transistor (MOSFET) elements. The electrical charge that is held by each MOSFET element is proportional to the intensity of light falling on the element. The output circuit of the camera has a charge-amplifier like device (capacitor-coupled) which is supplied by each MOSFET element. The MOSFET element that is to be connected to the output circuit at a given instant is determined by the control logic, which systematically scans the matrix of MOSFET elements. The capacitor circuit provides a voltage that is proportional to the charge in each MOSFET element.

   a. Draw a schematic diagram for a process monitoring system based on machine vision, which uses a CCD camera. Indicate the necessary signal modification operations at various stages in the monitoring loop, showing whether analog filters, amplifiers, ADC, and DAC are needed and if so, at which locations. An image may be divided into *pixels* (or picture elements) for representation and subsequent processing. A pixel has a well-defined coordinate location in the picture frame, relative to some reference coordinate frame. In a CCD camera, the number of pixels per image frame is equal to the number of CCD elements in the image plate. The information carried by a pixel (in addition to its location) is the photointensity (or *gray level*) at the image location. This number has to be expressed in the digital form (using a certain number of bits) for digital image processing. The need for very large data-handling rates is a serious limitation on a real-time controller that uses machine vision.

b. Consider a CCD image frame of the size 488×380 pixels. The refresh rate of the picture frame is 30 frames/s. If 8 bits are needed to represent the gray level of each pixel, what is the associated data (baud) rate (in bits/s)?

c. Discuss whether you prefer hardware processing or programmable-software-based processing in a process monitoring system based on machine vision.

4.9 Usually, an operational amplifier circuit is analyzed making use of the following two assumptions:

i. The potential at the "+" input lead is equal to the potential at the "–" input lead.

ii. The current through each of the two input leads is zero.

Explain why these assumptions are valid under unsaturated conditions of an op-amp.

An amateur electronics enthusiast connects to a circuit an op-amp without a feedback element. Even when there is no signal applied to the op-amp, the output was found to oscillate between +12 and –12 V once the power supply is turned on. Give a reason for this behavior.

An operational amplifier has an open-loop gain of 5×10^5 and a saturated output of ±14 V. If the noninverting input is –1 µV and the inverting input is +0.5 µV, what is the output? If the inverting input is 5 µV and the noninverting input is grounded, what is the output?

4.10 Define the following terms in connection with an operational amplifier:

a. Offset current

b. Offset voltage (at input and output)

c. Unequal gains

d. Slew rate

Give typical values for these parameters. The open-loop gain and the input impedance of an op-amp are known to vary with frequency and are known to drift (with time) as well. Still, the op-amp circuits are known to behave very accurately. What is the main reason for this?

4.11 What is a voltage follower? Discuss the validity of the following statements:

a. Voltage follower is a current amplifier

b. Voltage follower is a power amplifier

c. Voltage follower is an impedance transformer

Consider the amplifier circuit shown in Figure P4.11. Determine an expression for the voltage gain K_v of the amplifier in terms of the resistances R and R_f. Is this an inverting amplifier or a noninverting amplifier?

4.12 The speed of response of an amplifier may be represented using the three parameters: bandwidth, rise time, and slew rate. For an idealized linear model (transfer function) it can be verified that the rise time and the bandwidth are independent of the size of the input and the dc gain of the system. Since the size of the output (under steady conditions) may be expressed as the product of the input size and the dc gain it is seen that rise time and the bandwidth are independent of the amplitude of the output, for a linear model.

Discuss how slew rate is related to bandwidth and rise time of a practical amplifier. Usually, amplifiers have a limiting slew rate value. Show that bandwidth decreases with the output amplitude in this case.

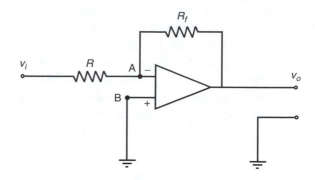

FIGURE P4.11
An amplifier circuit.

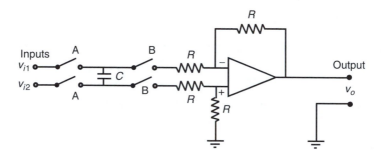

FIGURE P4.13
A differential amplifier with a flying capacitor for common-mode rejection.

A voltage follower has a slew rate of 0.5 V/μs. If a sinusoidal voltage of amplitude 2.5 V is applied to this amplifier, estimate the operating bandwidth. If, instead, a step input of magnitude 5 V is applied, estimate the time required for the output to reach 5 V.

4.13 Define the terms

    a. Common-mode voltage

    b. Common-mode gain

    c. Common-mode rejection ratio (CMRR)

What is a typical value for the CMRR of an op-amp? Figure P4.13 shows a differential amplifier circuit with a flying capacitor. The switch pairs A and B are turned on and off alternately during operation. For example first the switches denoted by A are turned on (closed) with the switches B off (open). Next, the switches A are opened and the switches B are closed. Explain why this arrangement provides good common-mode rejection characteristics.

4.14 Compare the conventional (textbook) meaning of system stability and the practical interpretation of instrument stability. An amplifier is known to have a temperature drift of 1 mV/°C and a long-term drift of 25 μV/month. Define the terms temperature drift and long-term drift. Suggest ways to reduce drift in an instrument.

4.15 Electrical isolation of a device (or circuit) from another device (or circuit) is very useful in the mechatronic practice. An isolation amplifier may be used to achieve this.

It provides a transmission link, which is almost one way, and avoids loading problems. In this manner, damage in one component due to increase in signal levels in the other components (perhaps due to short-circuits, malfunctions, noise, high common-mode signals, etc.) could be reduced. An isolation amplifier can be constructed from a transformer and a demodulator with other auxiliary components such as filters and amplifiers. Draw a suitable schematic diagram for an isolation amplifier, and explain the operation of this device.

4.16 What are passive filters? List several advantages and disadvantages of passive (analog) filters in comparison to active filters.

A simple way to construct an active filter is to start with a passive filter of the same type and add a voltage follower to the output. What is the purpose of such a voltage follower?

4.17 Give one application each for the following types of analog filters:

a. Low-pass filter

b. High-pass filter

c. Band-pass filter

d. Notch filter

Suppose that several single-pole active filter stages are cascaded. Is it possible for the overall (cascaded) filter to possess a resonant peak? Explain.

4.18 Butterworth filter is said to have a "maximally flat magnitude." Explain what is meant by this. Give another characteristic that is desired from a practical filter.

4.19 An active filter circuit is given in Figure P4.19.

a. Obtain the input-output differential equation for the circuit.

b. What is the filter transfer function?

c. What is the order of the filter?

d. Sketch the magnitude of the frequency transfer function and state what type of filter it represents.

e. Estimate the cutoff frequency and the roll-off slope.

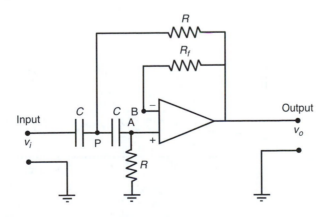

FIGURE P4.19
An active filter circuit.

4.20 What is meant by each of the following terms: modulation, modulating signal, carrier signal, modulated signal, and demodulation? Explain the following types of signal modulation giving an application for each case:

 a. Amplitude modulation
 b. Frequency modulation
 c. Phase modulation
 d. Pulse-width modulation
 e. Pulse-frequency modulation
 f. Pulse-code modulation.

How could the sign of the modulating signal be accounted for during demodulation in each of these types of modulation?

4.21 Give two situations where amplitude modulation is intentionally introduced and in each situation explain how amplitude modulation is beneficial. Also, describe two devices where amplitude modulation might be naturally present. Could the fact that amplitude modulation is present be exploited to our advantage in these two natural situations as well? Explain.

4.22 A monitoring system for a ball bearing of a rotating machine is schematically shown in Figure P4.22(a). It consists of an accelerometer to measure the bearing vibration and an FFT analyzer to compute the Fourier spectrum of the response signal. This spectrum is

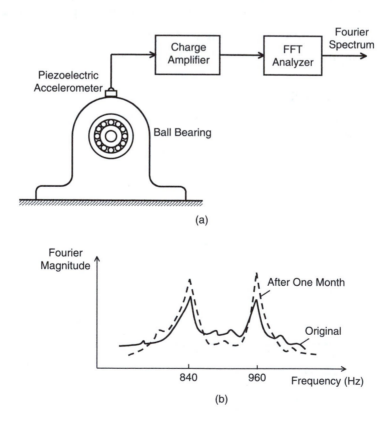

FIGURE P4.22
(a) A monitoring system for a ball bearing, (b) A zoomed Fourier spectrum.

examined over a period of one month after installation of the rotating machine in order to detect any degradation in the bearing performance. An interested segment of the Fourier spectrum can be examined with high resolution by using the "zoom analysis" capability of the FFT analyzer. The magnitude of the original spectrum and that of the spectrum determined one month later, in the same zoom region, are shown in Figure P4.22(b).

   a. Estimate the operating speed of the rotating machine and the number of balls in the bearing.

   b. Do you suspect any bearing problems?

4.23 Explain the following terms:

   a. Phase sensitive demodulation

   b. Half-wave demodulation

   c. Full-wave demodulation

When vibrations in rotating machinery such as gear boxes, bearings, turbines, and compressors are monitored, it is observed that a peak of the spectral magnitude curve does not usually occur at the frequency corresponding to the forcing function (e.g., tooth meshing, ball or roller hammer, blade passing). But, instead, two peaks occur on the two sides of this frequency. Explain the reason for this fact.

4.24 Define the following terms in relation to an analog-to-digital converter:

   a. Resolution

   b. Dynamic range

   c. Full-scale value

   d. Quantization error.

4.25 Single-chip amplifiers with built-in compensation and filtering circuits are becoming popular for signal conditioning tasks in mechatronic systems, particularly those associated with data acquisition, machine monitoring, and control. Signal processing such as integration that would be needed to convert, say, an accelerometer into a velocity sensor, can also be accomplished in the analog form using an IC chip. What are advantages of such signal modification chips in comparison to the conventional analog signal conditioning hardware that employ discrete circuit elements and separate components to accomplish various signal conditioning tasks?

4.26 Compare the three types of bridge circuits: constant-voltage bridge; constant-current bridge; and half-bridge, in terms of nonlinearity, effect of change in temperature, and cost.

Obtain an expression for the percentage error in a half-bridge circuit output due to an error δv_{ref} in the voltage supply v_{ref}. Compute the percentage error in the output if voltage supply has a 1% error.

4.27 Suppose that in the constant-voltage bridge circuit shown in Figure 4.35(a), at first, $R_1 = R_2 = R_3 = R_4 = R$. Assume that R_1 represents a strain gage mounted on the tensile side of a bending beam element and that R_3 represents another strain gage mounted on the compressive side of the bending beam. Due to bending, R_1 increases by δR and R_3 decreases by δR. Derive an expression for the bridge output in this case, and show that it is nonlinear. What would be the result if R_2 represents the tensile strain gage and R_4 represents the compressive strain gage, instead?

4.28 Suppose that in the constant-current bridge circuit shown in Figure 4.35(b), at first, $R_1 = R_2 = R_3 = R_4 = R$. Assume that R_1 and R_2 represent strain gages mounted on a rotating shaft, at right angles and symmetrically about the axis of rotation. Also, in this configuration and in a particular direction of rotation of the shaft, suppose that R_1 increases by δR and R_2 decreases by δR. Derive an expression for the bridge output (normalized) in this case, and show that it is linear. What would be the result if R_4 and R_3 were to represent the active strain gages in this example, the former element being in tension and the latter in compression?

4.29 Consider the constant-voltage bridge shown in Figure 4.35(a). The output Equation 4.88 can be expressed as:

$$v_o = \frac{(R_1/R_2 - R_3/R_4)}{(R_1/R_2 + 1)(R_3/R_4 + 1)} v_{ref}$$

Now suppose that the bridge is balanced, with the resistors set according to:

$$\frac{R_1}{R_2} = \frac{R_3}{R_4} = p$$

Then, if the active element R_1 increases by δR_1, show that the resulting output of the bridge is given by

$$\delta v_o = \frac{p \delta r}{[p(1 + \delta r) + 1](p + 1)} v_{ref}$$

where $\delta r = \delta R_1/R_1$, which is the fractional change in resistance in the active element.

For a given δr, it should be clear that δv_o represents the sensitivity of the bridge. For what value of the resistance ratio p, would the bridge sensitivity be a maximum? Show that this ratio is almost equal to 1.

4.30 The Maxwell bridge circuit is shown in Figure P4.30. Obtain the conditions for a balanced Maxwell bridge in terms of the circuit parameters $R_1, R_2, R_3, R_4, C_1,$ and L_4. Explain how this circuit could be used to measure variations in both C_1 or L_4.

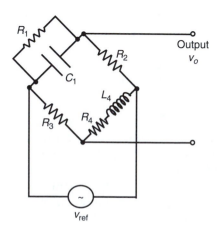

FIGURE P4.30
The Maxwell bridge.

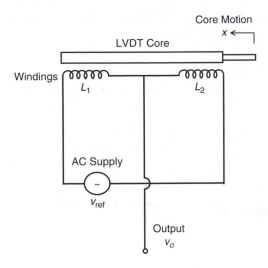

FIGURE P4.31
A half-bridge circuit for an LVDT.

4.31 The standard LVDT (linear variable differential transducer or transformer) arrangement has a primary coil and two secondary coil segments connected in series opposition. Alternatively, some LVDTs use a bridge circuit to produce their output. An example of a half-bridge circuit for an LVDT is shown in Figure P4.31. Explain the operation of this arrangement. Extend this idea to a full impedance bridge, for LVDT measurement.

4.32 The output of a Wheatstone bridge is nonlinear with respect to the variations in a bridge resistance. This nonlinearity is negligible for small changes in resistance. For large variations in resistance, however, some method of calibration or linearization should be employed. One way to linearize the bridge output is to positive feedback the output voltage signal into the bridge supply using a feed-back operational amplifier. Consider the Wheatstone bridge circuit shown in Figure 4.35(a). Initially, the bridge is balanced with $R_1 = R_2 = R_3 = R_4 = R$. Then, the resistor R_1 is varied to $R + \delta R$. Suppose that the bridge output δv_o is fed back (positive) with a gain of 2 into the bridge supply v_{ref}. Show that this will linearize the bridge equation.

4.33 A furnace used in a chemical process is controlled in the following manner. The furnace is turned on in the beginning of the process. When the temperature within the furnace reaches a certain threshold value T_o the (temperature) × (time) product is measured in the units of Celsius minutes. When this product reaches a specified value, the furnace is turned off. The available hardware includes a resistance temperature detector (RTD), a differential amplifier, a diode circuit which does not conduct when the input voltage is negative and conducts with a current proportional to the input voltage when the input is positive, a current-to-voltage converter circuit, a VFC, a counter, and an on/off control unit. Draw a block diagram for this control system and explain its operation. Clearly identify the signal-modification operations in this control system, indicating the purpose of each operation.

4.34 Typically, when a digital transducer is employed to generate the feedback signal for an analog controller, a DAC would be needed to convert the digital output from the transducer into a continuous (analog) signal. Similarly, when a digital controller is used to drive an analog process, a DAC has to be used to convert the digital output from the controller into the analog drive signal. There exist ways, however, to eliminate the need for a DAC in these two types of situations.

    a. Show how a shaft encoder and a frequency-to-voltage converter can replace an analog tachometer in an analog speed-control loop.

    b. Show how a digital controller with pulse width modulation (PWM) can be employed to drive a dc motor without the use of a DAC.

4.35 The noise in an electrical circuit can depend on the nature of the coupling mechanism. In particular, the following types of coupling are available:

    a. Conductive coupling

    b. Inductive coupling

    c. Capacitive coupling

    d. Optical coupling.

Compare these four types of coupling with respect to the nature and level of noise that is fed through or eliminated in each case. Discuss ways to reduce noise that is fed through in each type of coupling.

    The noise due to variations in ambient light can be a major problem in optically coupled systems. Briefly discuss a method that could be used in an optically-coupled device in order to make the device immune to variations in the ambient light level.

4.36 What are the advantages of using optical coupling in electrical circuits? For optical coupling, diodes that emit infrared radiation are often preferred over light emitting diodes (LEDs) which emit visible light. What are the reasons behind this? Discuss why pulse-modulated light (or pulse-modulated radiation) is used in many types of optical systems. List several advantages and disadvantages of laser-based optical systems.

    The Young's modulus of a material with known density can be determined by measuring the frequency of the fundamental mode of transverse vibration of a uniform cantilever beam specimen of the material. A photosensor and a timer can be used for this measurement. Describe an experimental setup for this method of determining the modulus of elasticity.

5

Performance Specification and Analysis

A mechatronic system consists of an integration of several components such as sensors, transducers, signal conditioning/modification devices, controllers, and a variety of other electronic and digital hardware. In the design, selection, and prescription of these components their performance requirements have to be specified or established within the functional needs of the overall mechatronic system. Engineering parameters for performance specification, particularly for mechatronic-system components, may be defined either in the time domain or in the frequency domain. Instrument ratings of commercial products are often developed on the basis of these engineering parameters. This chapter will address these and related issues of performance specification.

A sensor detects (feels) the quantity that is being measured (the measurand). The transducer converts the detected measurand into a convenient form for subsequent use (recording, control, actuation, etc.). The transducer signal may be filtered, amplified, and suitably modified prior to this. Transfer function models, in the frequency domain, are quite useful in representing, analyzing, designing and evaluating sensors, transducers, controllers, actuators, and interface devices (including signal conditioning/modification devices). Bandwidth plays an important role in specifying and characterizing any component of a mechatronic system. In particular, useful frequency range, operating bandwidth, and control bandwidth are important considerations in mechatronic systems. In this chapter we will study several important issues related to system bandwidth as well.

In any multi-component systems, the overall error depends on the component error. Component error degrades the performance of a mechatronic system. This is particularly true for sensors and transducers as their error is directly manifested within the system as incorrectly-known system variables and parameters. Since error may be separated into a systematic (or deterministic) part and a random (or stochastic) part, statistical considerations are important in error analysis. This chapter also deals with such considerations of error analysis.

5.1 Parameters for Performance Specification

All devices that assist in the functions of a mechatronic system can be interpreted as components of the system. Selection of available components for a particular application, or design of new components, should rely heavily on performance specifications for these components. A great majority of instrument ratings provided by manufacturers are in the form of static parameters. In mechatronic applications, however, dynamic performance specifications are also very important. In this section, we will study instrument ratings and parameters for performance specification, pertaining to both static and dynamic characteristics of instruments.

5.1.1   Perfect Measurement Device

Consider a measuring device of a mechatronic system, for example. A *perfect measuring device* can be defined as one that possesses the following characteristics:

1. Output of the measuring device instantly reaches the measured value (fast response).
2. Transducer output is sufficiently large (high gain, low output impedance, high sensitivity).
3. Device output remains at the measured value (without drifting or being affected by environmental effects and other undesirable disturbances and noise) unless the measurand (i.e., what is measured) itself changes (stability and robustness).
4. The output signal level of the transducer varies in proportion to the signal level of the measurand (static linearity).
5. Connection of a measuring device does not distort the measurand itself (loading effects are absent and impedances are matched; see Chapter 4).
6. Power consumption is small (high input impedance; see Chapter 4).

All of these properties are based on dynamic characteristics and therefore can be explained in terms of dynamic behavior of the measuring device. In particular, items 1 through 4 can be specified in terms of the device response, either in the *time domain* or in the *frequency domain*. Items 2, 5, and 6 can be specified using the *impedance* characteristics of the device. First, we shall discuss response characteristics that are important in performance specification of a component of a mechatronic system.

5.2   Time Domain Specifications

Figure 5.1 shows a typical step response in the dominant mode of a device. Note that the curve is normalized with respect to the steady-state value. We have identified several parameters that are useful for the time domain performance specification of the device. Definitions of these parameters are as follows:

5.2.1   Rise Time T_r

This is the time taken to pass the steady-state value of the response for the first time. In overdamped systems, the response is nonoscillatory; consequently, there is no overshoot. So that the definition is valid for all systems, rise time is often defined as the time taken to pass 90% of the steady-state value. Rise time is often measured from 10% of the steady-state value in order to leave out start-up irregularities (e.g., non-minimum phase behavior) and time lags that might be present in a system. A modified rise time (T_{rd}) may be defined in this manner (see Figure 5.1). An alternative definition of rise time, particularly suitable for nonoscillatory responses, is the reciprocal slope of the step response curve at 50% of the steady-state value, multiplied by the steady-state value. In process control terminology, this is in fact the *cycle time*. Note that no matter what definition is used, rise time represents the speed of response of a device—a small rise time indicates a fast response.

5.2.2   Delay Time T_d

This is usually defined as the time taken to reach 50% of the steady-state value for the first time. This parameter is also a measure of speed of response.

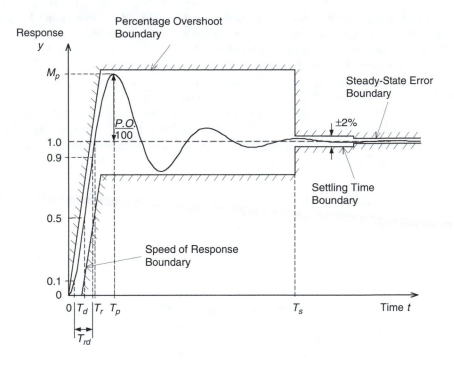

FIGURE 5.1
Response parameters for time-domain specification of performance.

5.2.3 Peak Time T_p

This is the time at the first peak of the device response. This parameter also represents the speed of response of the device.

5.2.4 Settling Time T_s

This is the time taken for the device response to settle down within a certain percentage (typically ±2 %) of the steady-state value. This parameter is related to the degree of damping present in the device as well as the degree of stability.

5.2.5 Percentage Overshoot (P.O.)

This is defined as

$$\text{P.O.} = 100\,(M_p - 1)\% \tag{5.1}$$

using the normalized-to-unity step response curve, where M_p is the peak value. Percentage overshoot is a measure of damping or relative stability in the device.

5.2.6 Steady-State Error

This is the deviation of the actual steady-state value of the device response from the desired value. Steady-state error may be expressed as a percentage with respect to the (desired) steady-state value. In a device output, the steady-state error manifests itself as an offset. This is a systematic (deterministic) error that normally can be corrected by recalibration.

In servo-controlled devices, steady-state error can be reduced by increasing loop gain or by introducing lag compensation. Steady-state error can be completely eliminated using the integral control (*reset*) action (See Chapter 12).

For the best performance of an output device (e.g., sensor/transducer), we wish to have the values of all the foregoing parameters as small as possible. In actual practice, however, it might be difficult to meet all the specifications, particularly for conflicting requirements. For instance, T_r can be decreased by increasing the dominant natural frequency ω_n of the device. This, however, increases the P.O. and sometimes the T_s. On the other hand, the P.O. and T_s can be decreased by increasing the device damping, but it has the undesirable effect of increasing T_r.

5.2.7 Simple Oscillator Model

The simple oscillator is a versatile model that can represent the performance of a variety of devices. Depending on the level of damping that is present, both oscillatory and non-oscillatory behavior can be represented by this model. The model can be expressed as (See Chapter 2)

$$\ddot{y} + 2\zeta\omega_n\dot{y} + \omega_n^2 y = \omega_n^2 u(t) \tag{5.2}$$

where u is the excitation (normalized), y is the response, ω_n = undamped natural frequency, and ζ = damping ratio. The damped natural frequency is given by

$$\omega_d = \sqrt{1-\zeta^2}\,\omega_n \tag{5.3}$$

The actual (damped) system executes free (natural) oscillations at this frequency. The response of the system to a unit step excitation, with zero initial conditions, is known to be

$$y = 1 - \frac{1}{\sqrt{1-\zeta^2}}\, e^{-\zeta\omega_n t}\, \sin(\omega_d t + \phi) \tag{5.4}$$

in which

$$\cos\phi = \zeta \tag{5.5}$$

As derived in Chapter 2, some important parameters for performance specification in the time domain, using the simple oscillator model, are given in Table 5.1.

With respect to time domain specifications of a mechatronic system component such as a transducer, it is desirable to have a very small rise time, and very small settling time in comparison to the time constants of the system whose response is being measured, and low percentage overshoot. These conflicting requirements guarantee fast and steady response.

Example 5.1

An automobile weighs 1000 kg. The equivalent stiffness at each wheel, including the suspension system, is approximately 60.0×10^3 N/m. If the suspension is designed for a percentage overshoot of 1%, estimate the damping constant that is needed at each wheel.

SOLUTION
For a quick estimate use a simple oscillator model, which is of the form

$$m\ddot{y} + b\dot{y} + ky = ku(t) \tag{i}$$

TABLE 5.1

Time-Domain Performance Parameters Using the Simple Oscillator Model

Performance Parameter	Expression
Rise time	$T_r = \dfrac{\pi - \phi}{\omega_d}$ with $\cos\phi = \zeta$
Peak time	$T_p = \dfrac{\pi}{\omega_d}$
Peak value	$M_p = 1 - e^{-\frac{\pi\zeta}{\sqrt{1-\zeta^2}}}$
Percentage overshoot (P.O.)	$\text{P.O.} = 100\, e^{-\frac{\pi\zeta}{\sqrt{1-\zeta^2}}}$
Time constant	$\tau = \dfrac{1}{\zeta\omega_n}$
Settling time (2%)	$T_s = -\dfrac{\ln\left[0.02\sqrt{1-\zeta^2}\right]}{\zeta\omega_n} \approx 4\tau = \dfrac{4}{\zeta\omega_n}$

in which,

m = equivalent mass = 250 kg
b = equivalent damping constant (to be determined)
k = equivalent stiffness = 60.0×10^3 N/m
u = displacement excitation at the wheel.

By comparing Equation i with Equation 5.2 we get

$$\zeta = \frac{b}{2\sqrt{km}} \tag{ii}$$

NOTE The equivalent mass at each wheel is taken as 1/4th the total mass.

For a P.O. of 1%, from Table 5.1, we have

$$1 = 100\exp\left(-\frac{\pi\zeta}{\sqrt{1-\zeta^2}}\right)$$

which gives $\zeta = 0.83$. Substitute the values in Equation ii. We get

$$0.83 = \frac{b}{2\sqrt{60 \times 10^3 \times 250.0}}$$

or

$$b = 6.43 \times 10^3 \text{ N/m/s}.$$

5.2.8   Stability and Speed of Response

The free response of a mechatronic device can provide valuable information concerning the natural characteristics of the device. The free (unforced) excitation may be obtained, for example, by giving an initial-condition excitation to the device and then allowing it to respond freely. Two important characteristics that can be determined in this manner are:

1. Stability
2. Speed of response

The stability of a dynamic system implies that the response will not grow without bounds when the excitation force itself is finite. Speed of response of a system indicates how fast the system responds to an excitation force. It is also a measure of how fast the free response (1) rises or falls if the system is oscillatory (i.e., underdamped); or (2) decays, if the system is non-oscillatory (i.e., overdamped). It follows that the two characteristics, stability and speed of response, are not completely independent. In particular, for non-oscillatory systems these two properties are very closely related.

   The level of stability of a linear dynamic system depends on the real parts of the eigenvalues (or poles), which are the roots of the characteristic equation. Specifically, if all the roots have real parts that are negative, then the system is stable. Also, the more negative the real part of a pole, the faster the decay of the free response component corresponding to that pole. The inverse of the negative real part is the *time constant*. Hence, the smaller the time constant, the faster the decay of the corresponding free response, and hence, the higher the level of stability associated with that pole. We can summarize these observations as follows:

   Level of Stability: Depends on decay rate of free response (and hence on time constants or real parts of poles).

   Speed of Response: Depends on natural frequency and damping for oscillatory systems and decay rate for non-oscillatory systems.

   Time Constant: Determines system stability and decay rate of free response (and speed of response in non-oscillatory systems).

Example 5.2

Consider an underdamped system and an overdamped system with the same undamped natural frequency, but damping ratios ζ_u and ζ_o, respectively. Show that the underdamped system is more stable and faster than the overdamped system if and only if:

$$\zeta_o > \frac{\zeta_u^2 + 1}{2\zeta_u}$$

where $\zeta_0 > 1 > \zeta_u > 0$ by definition.

SOLUTION

Use the simple oscillator model 5.2. The characteristic equation is

$$\lambda^2 + 2\zeta\omega_n\lambda + \omega_n^2 = 0 \tag{5.6}$$

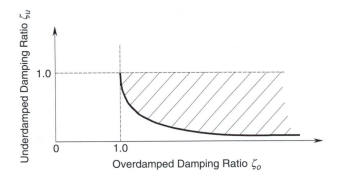

FIGURE 5.2
Region where underdamped system is faster and more stable than the corresponding overdamped system.

The eigenvalues (poles) are

$$\lambda = -\zeta\omega_n \pm \sqrt{\zeta^2 - 1}\,\omega_n \qquad (5.7)$$

To be more stable, we should have the underdamped pole located farther away from the origin than the dominant overdamped pole; thus

$$\zeta_u\omega_n > \zeta_o\omega_n - \sqrt{\zeta_o^2 - 1}\,\omega_n$$

This gives

$$\zeta_o > \frac{\zeta_u^2 + 1}{2\zeta_u} \qquad (5.8)$$

The corresponding region is shown as the shaded area in Figure 5.2

To explain this result further, consider an undamped ($\zeta = 0$) simple oscillator of natural frequency ω_n. Now let us add damping and increase ζ from 0 to 1. Then the complex conjugates poles $-\zeta\omega_n \pm j\omega_d$ will move away from the imaginary axis as ζ increases (because $\zeta\omega_n$ increases) and hence the level of stability will increase. When ζ reaches the value 1 (critical damping) we get two identical and real poles at $-\omega_n$. When ζ is increased beyond 1, the poles will be real and unequal, with one pole having a magnitude smaller than ω_n and the other having a magnitude larger than ω_n. The former (closer to the "origin" of zero) is the dominant pole, and will determine both stability and the speed of response of the resulting overdamped system. It follows that as ζ increases beyond 1, the two poles will branch out from the location $-\omega_n$, one moving towards the origin (becoming less stable) and the other moving away from the origin. It is now clear that as ζ is increased beyond the point of critical damping, the system becomes less stable. Specifically, for a given value of $\zeta_u < 1.0$, there is a value of $\zeta_o > 1$, governed by Equation 5.8, above which the overdamped system is less stable and slower than the underdamped system.

5.3 Frequency Domain Specifications

Figure 5.3 shows a representative *frequency transfer function* (often termed frequency response function) of a device. This constitutes the plots of *gain* and *phase angle*, using frequency as the independent variable. This pair of plots is commonly known as the *Bode diagram*,

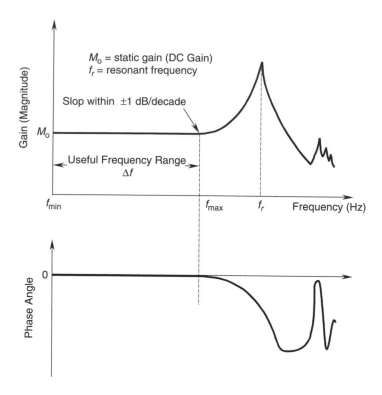

FIGURE 5.3
Response parameters for frequency-domain specification of performance.

particularly when the magnitude axis is calibrated in *decibels* (dB) and the frequency axis in a log scale such as *octaves* or *decades*. Experimental determination of these curves can be accomplished either by applying a harmonic excitation and noting the *amplitude gain* and the *phase lead* in the response signal at steady state or by Fourier analysis of the excitation and response signals for either transient or random excitations. Experimental determination of transfer functions is known as *system identification in the frequency domain*. Note that transfer functions provide complete information concerning the system response to a sinusoidal excitation. Since any time signal can be decomposed into sinusoidal components through Fourier transformation, it is clear that the response of a system to an arbitrary input excitation also can be determined using the transfer-function information for that system. In this sense, transfer functions are frequency domain models, which can completely describe a linear system. For this reason, one could argue that it is redundant to use both time domain specifications and frequency domain specifications, as they carry the same information. Often, however, both specifications are used simultaneously, because this can provide a better picture of the system performance. Frequency domain parameters are more suitable in representing some characteristics of a system under some types of excitation.

Some useful parameters for performance specification of a device, in the frequency domain, are:

- Useful Frequency Range (operating interval)
- Bandwidth (speed of response)
- Static Gain (steady-state performance)

- Resonant Frequency (speed and critical frequency region)
- Magnitude at Resonance (stability)
- Input Impedance (loading, efficiency, interconnectability)
- Output Impedance (loading, efficiency, interconnectability)
- Gain Margin (stability)
- Phase Margin (stability)

The first three items will be discussed in detail in this chapter, and is also indicated in Figure 5.3. Resonant frequency corresponds to a frequency where the response magnitude peaks. The dominant resonant frequency typically is the lowest resonant frequency, which usually also has the largest peak magnitude. It is shown as f_r in Figure 5.3. The term "Magnitude at Resonance" is self explanatory, and is the peak magnitude mentioned above and shown in Figure 5.3. Resonant frequency is a measure of speed of response and bandwidth, and is also a frequency region that should be avoided during normal operation and whenever possible. This is particularly true for devices that have poor stability (e.g., low damping). Specifically, a high magnitude at resonance is an indication of poor stability. Input impedance and output impedance are discussed in Chapter 4.

5.3.1 Gain Margin and Phase Margin

Gain and phase margins are measures of stability of a device. To define these two parameters consider the feedback system of Figure 5.4(a). The forward transfer function of the system is $G(s)$ and the feedback transfer function is $H(s)$. Note that these transfer functions are frequency-domain representations of the overall system which may include a variety

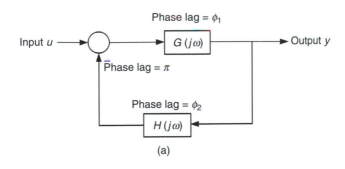

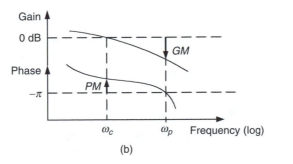

FIGURE 5.4
Illustration of gain and phase margins: (a) A feedback system; (b) Bode diagram.

of components such as the plant, sensors, transducers, actuators, controllers and interfacing and signal-modification devices.

The Bode diagram of the system constitutes the magnitude and phase "lead" plots of the loop transfer function $G(j\omega)H(j\omega)$ as a function of frequency (See Chapter 12). This is sketched in Figure 5.4(b).

Suppose that, at a particular frequency ω the forward transfer function $G(j\omega)$ provides a phase "lag" of ϕ_1, and the feedback transfer function $H(j\omega)$ provides a phase lag of ϕ_2. Now, in view of the negative feedback, which corresponds to a phase lag of π,

$$\text{Total phase lag in the loop} = \phi + \pi$$

where

$$\text{Phase lag of } GH = \phi_1 + \phi_2 = \phi$$

It follows that, when the overall phase lag of the *loop transfer function* $GH(j\omega)$ is equal to π, the loop phase lag becomes 2π, which means as a signal of frequency ω travels through the system loop it will not experience a net phase lag. Also, if at this particular frequency the loop gain $|GH(j\omega)|$ is unity, a sinusoidal signal having this frequency will be able to repeatedly travel through the loop without ever changing its phase or altering its magnitude, even in the absence of any external excitation input. This corresponds to a *marginally stable* condition.

If, on the other hand, the loop gain $|GH(j\omega)| > 1$ at this frequency where the loop phase gain phase lag is π, the signal magnitude will monotonically grow as the signal travels through the loop. This is an unstable situation.

In summary,

1. If $|GH(j\omega)| = 1$ when $\angle GH(j\omega) = -\pi$
   the system is marginally stable.
2. If $|GH(j\omega)| > 1$ when $\angle GH(j\omega) = -\pi$
   the system is unstable
3. If $|GH(j\omega)| < 1$ when $\angle GH(j\omega) = -\pi$
   the system is stable.

It follows that, the smallness of $|GH(j\omega)|$ compared to 1 at the frequency ω where $\angle GH(j\omega) = -\pi$, provides a measure of stability, and is termed *gain margin*. (see Figure 5.4(b)). Similarly, at the frequency ω where $|GH(j\omega)| = 1$, the amount of phase lag that can be added to the system so as to make the loop phase lag equal to π, is a measure of stability. This amount is termed *phase margin* (see Figure 5.4(b)).

In terms of frequency domain specifications, a mechatronic device such as a transducer should have a wide useful frequency range. For this it must have a high fundamental natural frequency (about 5 to 10 times the maximum frequency of the operating range) and a low damping ratio (slightly less than 1).

5.3.2   Simple Oscillator Model

As discussed in Chapter 2, the transfer function for the simple oscillator is given by

$$\frac{Y(s)}{U(s)} = G(s) = \left[\frac{\omega_n^2}{s^2 + 2\zeta\omega_n s + \omega_n^2} \right] \tag{5.9}$$

The frequency transfer function $G(j\omega)$ is defined as $G(s)\big|_{s=j\omega}$ where ω is the excitation frequency. Note that $G(j\omega)$ is a complex function in ω.

$$\text{Gain} = |G(j\omega)| = \text{magnitude of } G(j\omega)$$

$$\text{Phase lead} = \angle G(j\omega) = \text{phase angle of } G(j\omega)$$

These represent amplitude gain and phase lead of the output (response) when a sine input signal (excitation) of frequency ω is applied to the system.

Resonant frequency ω_r corresponds to the excitation frequency when the amplitude gain is maximum, and is given by

$$\omega_r = \sqrt{1 - 2\zeta^2}\; \omega_n \tag{5.10}$$

This expression is valid for $\zeta \le 1/\sqrt{2}$. It can be shown that

$$\text{Gain} = \frac{1}{2\zeta} \quad \text{and} \quad \text{Phase lead} = -\frac{\pi}{2} \quad \text{when } \omega = \omega_n \tag{5.11}$$

This concept is used in measure damping in simple systems, in addition to specifying the performance in the frequency domain. Frequency-domain concepts will be further discussed under bandwidth considerations (Section 5.4) in the present chapter.

5.4 Linearity

A device is considered linear if it can be modeled by linear differential equations, with time t as the independent variable. Nonlinear devices are often analyzed using linear techniques by considering small excursions about an operating point. This linearization is accomplished by introducing incremental variables for inputs and outputs. If one increment can cover the entire operating range of a device with sufficient accuracy, it is an indication that the device is linear. If the input/output relations are nonlinear algebraic equations, it represents a *static nonlinearity*. Such a situation can be handled simply by using nonlinear calibration curves, which linearize the device without introducing nonlinearity errors. If, on the other hand, the input/output relations are nonlinear differential equations, analysis usually becomes quite complex. This situation represents a *dynamic nonlinearity*. The subject of linearization is treated in Chapter 2.

Transfer-function representation of an instrument implicitly assumes linearity. According to industrial terminology, a linear measuring instrument provides a measured value that varies linearly with the value of the measurand—the variable that is measured. This is consistent with the definition of static linearity. All physical devices are *nonlinear* to some degree. This stems from deviation from the ideal behavior, due to causes such as saturation, deviation from Hooke's law in elastic elements, Coulomb friction, creep at joints, aerodynamic damping, backlash in gears and other loose components, and component wearout.

Nonlinearities in devices are often manifested as some peculiar characteristics. In particular, the following properties are important in detecting nonlinear behavior in dynamic systems:

5.4.1 Saturation

Nonlinear devices may exhibit saturation (see Figure 5.5(a)). This may result from such causes as magnetic saturation, which is common in magnetic-induction devices and transformer-like devices such as differential transformers, plastic mechanical components, and nonlinear springs.

5.4.2 Dead Zone

A dead zone is a region in which a device would not respond to an excitation. Stiction in mechanical devices with Coulomb friction is a good example. Due to stiction, a component would not move until the applied force reaches a certain minimum value. Once the motion is initiated, subsequent behavior can be either linear or nonlinear. A dead zone with subsequent linear behavior is shown in Figure 5.5(b).

5.4.3 Hysteresis

Nonlinear devices may produce hysteresis. In hysteresis, the input/output curve changes depending on the direction of motion (as indicated in Figure 5.5(c)), resulting in a hysteresis loop. This behavior is common in loose components such as gears, which have backlash; in components with nonlinear damping, such as Coulomb friction; and in magnetic devices with ferromagnetic media and various dissipative mechanisms (e.g., eddy current dissipation). For example, consider a coil wrapped around a ferromagnetic core. If a dc current is passed through the coil, a magnetic field is generated. As the current is

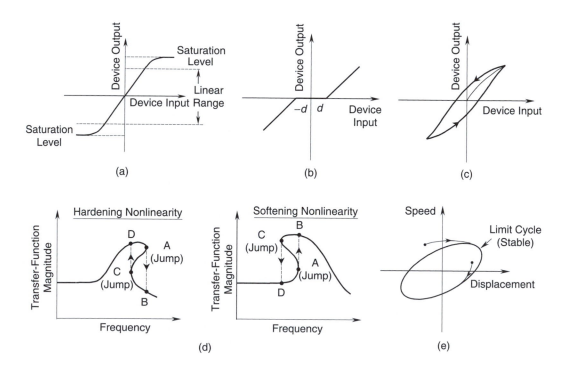

FIGURE 5.5
Common manifestations of nonlinearity in mechatronic devices: (a) Saturation; (b) Dead zone; (c) Hysteresis; (d) The jump phenomenon; (e) Limit cycle response.

increased from zero, the field strength will also increase. Now, if the current is decreased back to zero, the field strength will not return to zero because of residual magnetism in the ferromagnetic core. A negative current has to be applied to demagnetize the core. It follows that the field strength versus current curve looks somewhat like Figure 5.5(c). This is magnetic hysteresis. Note that linear viscous damping also exhibits a hysteresis loop in its force-displacement curve. This is a property of any mechanical component that dissipates energy. (Area within the hysteresis loop gives the energy dissipated in one cycle of motion). In general, if force depends on both displacement (as in the case of a spring) and velocity (as in the case of a damping element), the value of force at a given value of displacement will change with velocity. In particular, the force when the component is moving in one direction (say positive velocity) will be different from the force at the same location when the component is moving in the opposite direction (negative velocity), thereby giving a hysteresis loop in the force-displacement plane. If the relationship of displacement and velocity to force is linear (as in viscous damping), the hysteresis effect is linear. If on the other hand the relationship is nonlinear (as in Coulomb damping and aerodynamic damping), the resulting hysteresis is nonlinear.

5.4.4 The Jump Phenomenon

Some nonlinear devices exhibit an instability known as the jump phenomenon (or *fold catastrophe*) in the frequency response (transfer) function curve. This is shown in Figure 5.5(d) for both *hardening* devices and *softening* devices. With increasing frequency, jump occurs from A to B; and with decreasing frequency, it occurs from C to D. Furthermore, the transfer function itself may change with the level of input excitation in the case of nonlinear devices.

5.4.5 Limit Cycles

Nonlinear devices may produce limit cycles. An example is given in Figure 5.5(e) on the phase plane of velocity versus displacement. A limit cycle is a closed trajectory in the state space that corresponds to sustained oscillations at a specific frequency and amplitude, without decay or growth. Amplitude of these oscillations is independent of the initial location from which the response started. Also, an external input is not needed to sustain a limit-cycle oscillation. In the case of a stable limit cycle, the response will move onto the limit cycle irrespective of the location in the neighborhood of the limit cycle from which the response was initiated (see Figure 5.5(e)). In the case of an unstable limit cycle, the response will move away from it with the slightest disturbance.

5.4.6 Frequency Creation

At steady state, nonlinear devices can create frequencies that are not present in the excitation signals. These frequencies might be harmonics (integer multiples of the excitation frequency), subharmonics (integer fractions of the excitation frequency), or nonharmonics (usually rational fractions of the excitation frequency).

Example 5.3

Consider a nonlinear device modeled by the differential equation

$$\left\{ \frac{dy}{dt} \right\}^{1/2} = u(t)$$

in which $u(t)$ is the input and y is the output. Show that this device creates frequency components that are different from the excitation frequencies.

SOLUTION
First, note that the steady-state response is given by

$$y = \int_0^t u^2(t)\,dt + y(0)$$

Now, for an input given by

$$u(t) = a_1 \sin \omega_1 t + a_2 \sin \omega_2 t$$

straightforward integration using properties of trigonometric functions gives the following response:

$$y = \left(a_1^2 + a_2^2\right)\frac{t}{2} - \frac{a_1^2}{4\omega_1}\sin 2\omega_1 t - \frac{a_2^2}{4\omega_2}\sin 2\omega_2 t + \frac{a_1 a_2}{(\omega_1 - \omega_2)}\sin(\omega_1 - \omega_2)t$$

$$- \frac{a_1 a_2}{(\omega_1 + \omega_2)}\sin(\omega_1 + \omega_2)t - y(0)$$

Note that the discrete frequency components $2\omega_1$, $2\omega_2$, $(\omega_1 - \omega_2)$ and $(\omega_1 + \omega_2)$ are created. Also, there is a continuous spectrum that is contributed by the linear function of t present in the response.

———

Nonlinear systems can be analyzed using the *describing function* approach. When a harmonic input (at a specific frequency) is applied to a nonlinear device, the resulting output at steady state will have a component at this fundamental frequency and also components at other frequencies (due to frequency creation by the nonlinear device), typically harmonics. The response may be represented by a Fourier series, which has frequency components that are multiples of the input frequency. The describing function approach neglects all the higher harmonics in the response and retains only the fundamental component. This output component, when divided by the input, produces the describing function of the device. This is similar to the transfer function of a linear device, but unlike for a linear device, the gain and the phase shift will be dependent on the input amplitude. Details of the describing function approach can be found in textbooks on nonlinear control theory.

Several methods are available to reduce or eliminate nonlinear behavior in systems. They include calibration (in the static case), use of linearizing elements such as resistors and amplifiers to neutralize the nonlinear effects, and the use of nonlinear feedback. It is also a good practice to take the following precautions:

1. Avoid operating the device over a wide range of signal levels.
2. Avoid operation over a wide frequency band.
3. Use devices that do not generate large mechanical motions.
4. Minimize Coulomb friction and stiction (e.g., using proper lubrication).
5. Avoid loose joints and gear coupling (i.e., use *direct-drive* mechanisms).

5.5 Instrument Ratings

Instrument manufacturers do not usually provide complete dynamic information for their products. In most cases, it is unrealistic to expect complete dynamic models (in the time domain or the frequency domain) and associated parameter values for complex instruments in mechatronic systems. Performance characteristics provided by manufacturers and vendors are primarily static parameters. Known as instrument ratings, these are available as parameter values, tables, charts, calibration curves, and empirical equations. Dynamic characteristics such as transfer functions (e.g., transmissibility curves expressed with respect to excitation frequency) might also be provided for more sophisticated instruments, but the available dynamic information is never complete. Furthermore, definitions of rating parameters used by manufacturers and vendors of instruments are in some cases not the same as analytical definitions used in textbooks. This is particularly true in relation to the term *linearity*. Nevertheless, instrument ratings provided by manufacturers and vendors are very useful in the selection, installation, operation, and maintenance of components in a mechatronic system. Now, we shall examine some of these performance parameters.

5.5.1 Rating Parameters

Typical rating parameters supplied by instrument manufacturers are

1. Sensitivity
2. Dynamic range
3. Resolution
4. Linearity
5. Zero drift and full-scale drift
6. Useful frequency range
7. Bandwidth
8. Input and output impedances

We have already discussed the meaning and significance of some of these terms. In this section, we shall look at the conventional definitions given by instrument manufacturers and vendors.

Sensitivity of a device (e.g., transducer) is measured by the magnitude (peak, rms value, etc.) of the output signal corresponding to a unit input (e.g., measurand). This may be expressed as the ratio of (incremental output)/(incremental input) or, analytically, as the corresponding partial derivative. In the case of vectorial or tensorial signals (e.g., displacement, velocity, acceleration, strain, force), the direction of sensitivity should be specified.

Cross-sensitivity is the sensitivity along directions that are orthogonal to the primary direction of sensitivity. It is normally expressed as a percentage of direct sensitivity. High sensitivity and low cross-sensitivity are desirable for any input/output device (e.g., measuring instrument). Sensitivity to parameter changes and noise has to be small in any device, however, and this is an indication of its *robustness*. On the other hand, in *adaptive control* and *self-tuning control*, the sensitivity of the system to control parameters has to be sufficiently high. Often, sensitivity and robustness are conflicting requirements.

Dynamic range of an instrument is determined by the allowed lower and upper limits of its input or output (response) so as to maintain a required level of output accuracy.

This range is usually expressed as a ratio, in *decibels*. In many situations, the lower limit of dynamic range is equal to the resolution of the device. Hence, the dynamic range ratio is usually expressed as (range of operation)/(resolution).

Resolution of an input/output instrument is the smallest change in a signal (input) that can be detected and accurately indicated (output) by a transducer, a display unit, or any pertinent instrument. It is usually expressed as a percentage of the maximum range of the instrument or as the inverse of the dynamic range ratio. It follows that dynamic range and resolution are very closely related.

Example 5.4

The meaning of dynamic range (and resolution) can easily be extended to cover digital instruments. For example, consider an instrument that has a 12-bit analog-to-digital converter (ADC). Estimate the dynamic range of the instrument.

SOLUTION

In this example, dynamic range is determined (primarily) by the word size of the ADC. Each bit can take the binary value 0 or 1. Since the resolution is given by the smallest possible increment, that is, a change by the least significant bit (LSB), it is clear that digital resolution = 1. The largest value represented by a 12-bit word corresponds to the case when all 12 bits are unity. This value is decimal $2^{12} - 1$. The smallest value (when all twelve bits are zero) is zero. Now, use the definition

$$\text{Dynamic range} = 20 \log_{10}\left[\frac{\text{Range of operation}}{\text{Resolution}}\right] \tag{5.12}$$

The dynamic range of the instrument is given by

$$20 \log_{10}\left[\frac{2^{12} - 1}{1}\right] = 72 \text{ dB}$$

Another (perhaps more correct) way of looking at this problem is to consider the resolution to be some value δy, rather than unity, depending on the particular application. For example δy may represent an output signal increment of 0.0025 V. Next, we note that a 12-bit word can represent a combination of 2^{12} values (i.e., 4,096 values), the smallest being some y_{min} and the largest value being

$$y_{max} = y_{min} + (2^{12} - 1)\delta y$$

Note that y_{min} can be zero, positive, or negative. The smallest increment between values is δy, which is by definition, the resolution. There are 2^{12} values with y_{min} and y_{max} (the two end values) inclusive. Then

$$\text{Dynamic range} = \frac{y_{max} - y_{min}}{\delta y} = \frac{(2^{12} - 1)\delta y}{\delta y} = 12^{12} - 1 = 4,095 = 72 \text{ dB}$$

So we end up with the same result for dynamic range, but the interpretation of resolution is somewhat different.

Linearity is determined by the calibration curve of an instrument. The curve of output amplitude (peak or rms value) versus input amplitude under static conditions within the dynamic range of an instrument is known as the *static calibration curve*. Its closeness to a straight line measures the degree of linearity. Manufacturers provide this information either as the maximum deviation of the calibration curve from the least squares straight-line fit of the calibration curve or from some other reference straight line. If the least-squares fit is used as the reference straight line, the maximum deviation is called *independent linearity* (more correctly, independent nonlinearity, because the larger the deviation, the greater the nonlinearity). Nonlinearity may be expressed as a percentage of either the actual reading at an operating point or the full-scale reading.

Zero drift is defined as the drift from the null reading of the instrument when the input is maintained steady for a long period. Note that in this case, the input is kept at zero or any other level that corresponds to the null reading of the instrument. Similarly, *full-scale drift* is defined with respect to the full-scale reading (the input is maintained at the full-scale value). Usual causes of drift include instrument instability (e.g., instability in amplifiers), ambient changes (e.g., changes in temperature, pressure, humidity, and vibration level), changes in power supply (e.g., changes in reference dc voltage or ac line voltage), and parameter changes in an instrument (due to aging, wear and tear, nonlinearities, etc.). Drift due to parameter changes that are caused by instrument nonlinearities is known as *parametric drift, sensitivity drift*, or *scale-factor drift*. For example, a change in spring stiffness or electrical resistance due to changes in ambient temperature results in a parametric drift. Note that parametric drift depends on the input level. Zero drift, however, is assumed to be the same at any input level if the other conditions are kept constant. For example, a change in reading caused by thermal expansion of the readout mechanism due to changes in ambient temperature is considered a zero drift. Drift in electronic devices can be reduced by using alternating current (ac) circuitry rather than direct current (dc) circuitry. For example, ac-coupled amplifiers have fewer drift problems than dc amplifiers. Intermittent checking for instrument response level with zero input is a popular way to calibrate for zero drift. In digital devices, for example, this can be done automatically from time to time between sample points, when the input signal can be bypassed without affecting the system operation.

Useful frequency range corresponds to a flat gain curve and a zero phase curve in the frequency response characteristics of an instrument. The maximum frequency in this band is typically less than half (say, one-fifth) of the dominant resonant frequency of the instrument. This is a measure of the instrument bandwidth.

Bandwidth of an instrument determines the maximum speed or frequency at which the instrument is capable of operating. High bandwidth implies faster speed of response. Bandwidth is determined by the dominant natural frequency ω_n or the dominant resonant frequency ω_r of the device. (Note: For low damping, ω_r is approximately equal to ω_n). It is inversely proportional to rise time and the dominant time constant. Half-power bandwidth is also a useful parameter. Instrument bandwidth has to be several times greater than the maximum frequency of interest in the input signals. For example, bandwidth of a measuring device is important particularly when measuring transient signals. Note that bandwidth is directly related to the useful frequency range.

5.6   Bandwidth Design

Bandwidth plays an important role in specifying and characterizing the components of a mechatronic system. In particular, useful frequency range, operating bandwidth, and control bandwidth are important considerations. In this section we will study several important issues related to these topics.

5.6.1 Bandwidth

Bandwidth takes different meanings depending on the particular context and application. For example, when studying the response of a dynamic system, the bandwidth relates to the fundamental resonant frequency and correspondingly to the speed of response for a given excitation. In band-pass filters, the bandwidth refers to the frequency band within which the frequency components of the signal are allowed through the filter, the frequency components outside the band being rejected by it. With respect to measuring instruments, bandwidth refers to the range frequencies within which the instrument measures a signal accurately. In digital communication networks (e.g., the Internet), the bandwidth denotes the "capacity" of the network in terms of information rate (bits/s). Note that these various interpretations of bandwidth are somewhat related. As a particular note, if a signal passes through a band-pass filter we know that its frequency content is within the bandwidth of the filter, but we cannot determine the actual frequency content of the signal on the basis of that observation. In this context, the bandwidth appears to represent a frequency uncertainty in the observation (i.e., the larger the bandwidth of the filter, less certain you are about the actual frequency content of a signal that passes through the filter).

5.6.1.1 Transmission Level of a Band-Pass Filter

Practical filters can be interpreted as dynamic systems. In fact all physical, dynamic systems (e.g., mechatronic systems) are analog filters. It follows that the filter characteristic can be represented by the frequency transfer function $G(f)$ of the filter. A magnitude squared plot of such a filter transfer function is shown in Figure 5.6. In a logarithmic plot the magnitude-squared curve is obtained by simply doubling the corresponding magnitude curve (in the Bode plot). Note that the actual filter transfer function (Figure 5.6(b)) is not flat like the ideal filter shown in Figure 5.6(a). The reference level G_r is the average value of the transfer function magnitude in the neighborhood of its peak.

5.6.1.2 Effective Noise Bandwidth

Effective noise bandwidth of a filter is equal to the bandwidth of an ideal filter that has the same reference level and that transmits the same amount of power from a white noise source. Note that white noise has a constant (flat) power spectral density (psd).

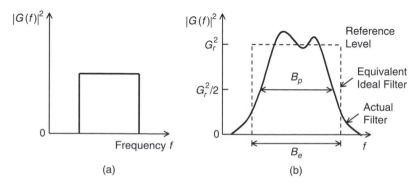

FIGURE 5.6
Characteristics of: (a) An ideal band-pass filter; (b) A practical band-pass filter.

Hence, for a noise source of unity power spectral density (psd) the power transmitted by the practical filter is given by

$$\int_0^\infty |G(f)|^2 \, df$$

which, by definition, is equal to the power $G_r^2 B_e$ transmitted by the equivalent ideal filter. Here, the effective noise bandwidth B_e is given by,

$$B_e = \int_0^\infty |G(f)|^2 \, df / G_r^2 \qquad (5.13)$$

5.6.1.3 Half-Power (or 3 dB) Bandwidth

Half of the power from a unity-psd noise source as transmitted by an ideal filter, is $G_r^2 B_e / 2$. Hence, $G_r / \sqrt{2}$ is referred to as the *half-power level*. This is also known as a 3 dB level because $20 \log_{10} \sqrt{2} = 10 \log_{10} 2 = 3$ dB (Note: 3 dB refers to a power ratio of 2 or an amplitude ratio of $\sqrt{2}$. Furthermore, 20 dB corresponds to an amplitude ratio of 10 or a power ratio of 100). The 3 dB (or half-power) bandwidth corresponds to the width of the filter transfer function at the half-power level. This is denoted by B_p in Figure 5.6(b). Note that B_e and B_p are different in general. In an ideal case where the magnitude-squared filter characteristic has linear rise and fall-off segments, however, these two bandwidths are equal (see Figure 5.7).

5.6.1.4 Fourier Analysis Bandwidth

In Fourier analysis, bandwidth is interpreted as the *frequency uncertainty* in the spectral results. In analytical Fourier integral transform (FIT) results, which assume that the entire signal is available for analysis, the spectrum is continuously defined over the entire frequency range $[-\infty, \infty]$ and the frequency increment df is infinitesimally small $(df \to 0)$. There is no frequency uncertainty in this case, and the analysis bandwidth is infinitesimally narrow. In digital Fourier transform, the discrete spectral lines are generated at frequency intervals of ΔF. This finite frequency increment ΔF, which is the frequency uncertainty, is therefore, the analysis bandwidth B for this analysis. It is known that $\Delta F = 1/T$ where T is the record length of the signal (or window length when a rectangular window is used to select the signal segment for analysis). It follows also that the minimum frequency that

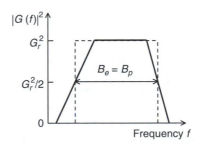

FIGURE 5.7
An idealized filter with linear segments.

has a meaningful accuracy is the analysis bandwidth. This interpretation for analysis bandwidth is confirmed by noting the fact that harmonic components of frequency less than ΔF (or period greater than T) cannot be studied by observing a signal record of length less than T. Analysis bandwidth carries information regarding distinguishable minimum frequency separation in computed results. In this sense bandwidth is directly related to the frequency resolution of analyzed results. The accuracy of analysis increases by increasing the record length T (or decreasing the analysis bandwidth B).

When a time window other than the rectangular window is used to truncate a signal, then reshaping of the signal segment (data) occurs according to the shape of the window. This reshaping suppresses the sidelobes of the Fourier spectrum of the window and hence reduces the frequency leakage that arises from truncation of the signal. At the same time, however, an error is introduced due to the information loss as a result of data reshaping. This error is proportional to the bandwidth of the window itself. The effective noise bandwidth of a rectangular window is only slightly less than $1/T$, because the main lobe of its Fourier spectrum is nearly rectangular. Hence, for all practical purposes, the effective noise bandwidth can be taken as the analysis bandwidth. Note that data truncation (multiplication in the time domain) is equivalent to convolution of the Fourier spectrum (in the frequency domain). The main lobe of the spectrum uniformly affects all spectral lines in the discrete spectrum of the data signal. It follows that a window main lobe having a broader effective-noise bandwidth introduces a larger error into the spectral results. Hence, in digital Fourier analysis, bandwidth is taken as the effective noise bandwidth of the time window that is employed.

5.6.1.5 Useful Frequency Range

This corresponds to the flat region (static region) in the gain curve and the zero-phase-lead region in the phase curve of a device (with respect to frequency). It is determined by the dominant (i.e., the lowest) resonant frequency f_r of the device. The upper frequency limit f_{max} in the useful frequency range is several times smaller than f_r for a typical input/output device (e.g., $f_{max} = 0.25\,f_r$). Useful frequency range may also be determined by specifying the flatness of the static portion of the frequency response curve. For example, since a single pole or a single zero introduces a slope on the order of ∓ 20 dB/decade to the Bode magnitude curve of the device, a slope within 5% of this value (i.e., ± 1 dB/decade) may be considered flat for most practical purposes. For a measuring instrument, for example, operation in the useful frequency range implies that the significant frequency content of the measured signal is limited to this band. In that case, faithful measurement and fast response are guaranteed, because dynamics of the measuring device will not corrupt the measurement.

5.6.1.6 Instrument Bandwidth

This is a measure of the useful frequency range of an instrument. Furthermore, the larger the bandwidth of the device, the faster will be the speed of response. Unfortunately, the larger the bandwidth, the more susceptible the instrument will be to high-frequency noise as well as stability problems. Filtering will be needed to eliminate unwanted noise. Stability can be improved by dynamic compensation. Common definitions of bandwidth include the frequency range over which the transfer-function magnitude is flat; the resonant frequency; and the frequency at which the transfer-function magnitude drops to $1/\sqrt{2}$ (or 70.7%) of the zero-frequency (or static) level. The last definition corresponds to the *half-power bandwidth*, because a reduction of amplitude level by a factor of $\sqrt{2}$ corresponds to a power drop by a factor of 2.

5.6.1.7 Control Bandwidth

This is used to specify the maximum possible speed of control. It is an important specification in both analog control and digital control. In digital control, the data sampling rate (in samples/second) has to be several times higher than the control bandwidth (in hertz) so that sufficient data would be available to compute the control action. Also, from *Shannon's sampling theorem*, control bandwidth is given by half the rate at which the control action is computed (See under the topic of aliasing distortion). The control bandwidth provides the frequency range within which a system can be controlled (assuming that all the devices in the system can operate within this bandwidth).

5.6.2 Static Gain

This is the gain (i.e., transfer function magnitude) of a measuring instrument within the useful (flat) range (or at very low frequencies) of the instrument. It is also termed *dc gain*. A high value for static gain results in a high-sensitivity measuring device, which is a desirable characteristic.

Example 5.5

A mechanical device for measuring angular velocity is shown in Figure 5.8. The main element of the tachometer is a rotary viscous damper (damping constant b) consisting of two cylinders. The outer cylinder carries a viscous fluid within which the inner cylinder rotates. The inner cylinder is connected to the shaft whose speed ω_i is to be measured. The outer cylinder is resisted by a linear torsional spring of stiffness k. The rotation θ_o of the outer cylinder is indicated by a pointer on a suitably calibrated scale. Neglecting the inertia of moving parts, perform a bandwidth analysis on this device.

SOLUTION
The damping torque is proportional to the relative velocity of the two cylinders and is resisted by the spring torque. The equation of motion is given by

$$b(\omega_i - \dot{\theta}_o) = k\theta_o$$

or

$$b\dot{\theta}_o + k\theta_o = b\omega_i \tag{i}$$

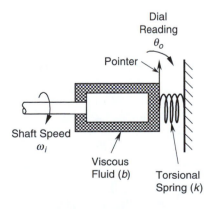

FIGURE 5.8
A mechanical tachometer.

The transfer function is determined by first replacing the time derivative by the Laplace operator *s*; thus,

$$\frac{\theta_o}{\omega_i} = \frac{b}{[bs + k]} = \frac{b/k}{[(b/k)s + 1]} = \frac{k_g}{[\tau s + 1]} \tag{ii}$$

Note that the static gain or dc gain (transfer-function magnitude at *s* = 0) is

$$k_g = \frac{b}{k} \tag{iii}$$

and the time constant is

$$\tau = \frac{b}{k} \tag{iv}$$

We face conflicting design requirements in this case. On the one hand, we like to have a large static gain so that a sufficiently large reading is available. On the other hand, the time constant must be small in order to obtain a quick reading that faithfully follows the measured speed. A compromise must be reached here, depending on the specific design requirements. Alternatively, a signal-conditioning device could be employed to amplify the sensor output.

Now, let us examine the half-power bandwidth of the device. The frequency transfer function is

$$G(j\omega) = \frac{k_g}{\tau j\omega + 1} \tag{v}$$

Since the maximum magnitude of this transfer function is k_g, which occursat $\omega = 0$, by definition, the half-power bandwidth ω_b is given by

$$\frac{k_g}{|\tau j\omega_b + 1|} = \frac{k_g}{\sqrt{2}}$$

Hence

$$(\tau\omega_b)^2 + 1 = 2$$

Since both τ and ω_b are positive we have

$$\tau\omega_b = 1$$

or

$$\omega_b = \frac{1}{\tau} \tag{vi}$$

Note that the bandwidth is inversely proportional to the time constant. This confirms our earlier statement that bandwidth is a measure of the speed of response.

Example 5.6
Part 1

i. Briefly discuss any conflicts that can arise in specifying parameters that can be used to predominantly represent the speed of response and the degree of stability of a process (plant).

ii. Consider a measuring device that is connected to a plant for feedback control. Explain the significance of

   a. Bandwidth

   b. Resolution

   c. Linearity

   d. Input impedance

   e. Output impedance

   of the measuring device, in the performance of the feedback control system.

Part 2
Consider the speed control system schematically shown in Figure 5.9. Suppose that the plant and the controller together are approximated by the transfer function

$$G_p(s) = \frac{k}{(\tau_p s + 1)}$$

where τ_p is the plant time constant.

   a. Give an expression for the bandwidth ω_p of the plant, in the absence of feedback.

   b. If the feedback tachometer is ideal and is represented by a unity (negative) feedback, what is the bandwidth ω_c of the feedback control system?

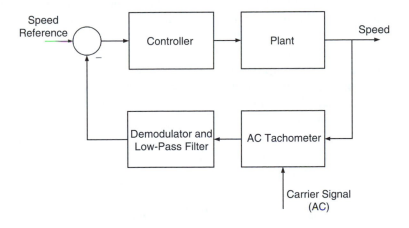

FIGURE 5.9
A speed control system.

c. If the feedback tachometer can be represented by the transfer function

$$G_s(s) = \frac{1}{(\tau_s s + 1)}$$

where τ_s is the sensor time constant, explain why the bandwidth ω_{cs} of the feedback control system is given by the smaller quantity of $1/\tau_s$ and $(k+1)/(\tau_p + \tau_s)$. Assume that both τ_p and τ_s are sufficiently small.

Next suppose that approximately $\tau_p = 0.016$ sec. Estimate a sufficient bandwidth in Hz for the tachometer. Also, if $k = 1$, estimate the overall bandwidth of the feedback control system. If $k = 49$, what would be the representative bandwidth of the feedback control system? For the particular ac tachometer (with the bandwidth value as chosen in the present numerical example), what should be the frequency of the carrier signal? Also, what should be the cutoff frequency of the low-pass filter that is used with its demodulator circuit?

SOLUTION
Part 1

   i. Usually speeding up a system has a destabilizing effect. For example, if gain is increased to speed up a system, the percent overshoot can increase.
  ii. a. Measuring device bandwidth should cover the entire bandwidth of possible operation of the system. (Typically make it several times larger than the required bandwidth). Otherwise useful frequency components in the measured could be distorted.
      b. Resolution of the measuring device should be less than half the error tolerance of the control system. Otherwise the sensor tolerance alone can provide an unacceptable error level in the control system (even when the control itself is satisfactory).
      c. If the measuring device has a static nonlinearity, an accurate calibration curve will be needed. Otherwise the operating range has to be limited. Dynamic nonlinearity can cause undesirable effects such as limit cycles, hysteresis, frequency creation, jump phenomenon, saturation and related errors.
      d. Input impedance of the measuring device has to be significantly higher than the output impedance of the process. Otherwise the signal will be subjected to loading error and distortion.
      e. The output impedance of the measuring device has to be small. Otherwise, the devices connected to that end should have a very high impedance. Also the output level of a high-output-impedance device will be low in general (not satisfactory). Then, additional, expensive hardware will be necessary to condition the measured signal.

Part 2

   a.

$$G_p(s) = \frac{k}{(\tau_p s + 1)}$$

Bandwidth $\omega_p = 1/\tau_p$

b. With unity feedback, closed-loop transfer function is (See Chapter 2 and Chapter 12)

$$G_c(s) = \frac{k/(\tau_p s + 1)}{1 + k/(\tau_p s + 1)}, \text{ which simplifies to}$$

$$G_c(s) = \frac{k}{(\tau_p s + 1 + k)}$$

Hence

$$\text{Bandwidth} \quad \omega_c = \frac{1+k}{\tau_p}$$

NOTE   The bandwidth has increased.

c. With feedback sensor of $G_s(s) = \frac{1}{(\tau_s s + 1)}$ the closed-loop transfer function is

$$G_{cs}(s) = \frac{k/(\tau_p s + 1)}{1 + k/\{(\tau_p s + 1)(\tau_s s + 1)\}}$$

$$= \frac{k(\tau_s s + 1)}{\tau_p \tau_s s^2 + (\tau_p + \tau_s)s + 1 + k}$$

$$\cong \frac{k(\tau_s s + 1)}{(\tau_p + \tau_s)s + 1 + k} \qquad \{\text{Neglecting } \tau_p \tau_s\}$$

Hence to avoid the dynamic effect of the sensor (which has introduced a zero at $s = 1/\tau_s$ in $G_{cs}(s)$) we should limit the bandwidth to $1/\tau_s$.

Also, from the denominator of G_{cs}, it is seen that the closed-loop bandwidth is given by $\frac{1+k}{(\tau_p + \tau_s)}$. Hence, for satisfactory performance, the bandwidth has to be limited to

$$\text{Bandwidth} \quad \min \left[\frac{1}{\tau_s}, \frac{1+k}{(\tau_p + \tau_s)} \right].$$

With $\tau_p = 0.016$ sec

$$\omega_p = \frac{1}{0.016} = 62.5 \text{ rad/s} = 10.0 \text{ Hz}$$

Hence, pick a sensor bandwidth of 10 times this value.

$$\Rightarrow \quad \omega_s = 100.0 \text{ Hz} = 625.0 \text{ rad/s}$$

Then

$$\tau_s = \frac{1}{\omega_s} = 0.0016 \text{ sec} .$$

With $k = 1$:

$$\frac{1+k}{(\tau_p + \tau_s)} = \frac{(1+1)}{(0.016 + 0.0016)} \text{ rad/s} = 18.0 \text{ Hz}$$

Hence

$$\omega_{cs} = \min [10 \text{ Hz}, 18.0 \text{ Hz}]$$

or,

$$\text{Bandwidth} \qquad \omega_{cs} \cong 18.0 \text{ Hz}.$$

With $k = 49$:

$$\frac{1+k}{(\tau_p + \tau_s)} = \frac{1+49}{(0.016 + 0.0016)} \text{ rad/s} = 450.0 \text{ Hz}$$

Then

$$\omega_{cs} = \min [100 \text{ Hz}, 450 \text{ Hz}] \simeq 100 \text{ Hz}$$

Hence, now bandwidth ω_{cs} is 100.0 Hz or less.

For a sensor with 100 Hz bandwidth:

$$\text{Carrier frequency} \cong 10 \times 100 \text{ Hz} = 1000.00 \text{ Hz}$$

$$2 \times \text{carrier frequency} = 2000 \text{ HZ}$$

$$\text{Low-pass filter cutoff} = \frac{1}{10} \times 2000 \text{ Hz} = 200.0 \text{ Hz}.$$

5.7 Aliasing Distortion Due to Signal Sampling

Aliasing distortion is an important consideration when dealing with sampled data from a continuous signal. This error may enter into computation in both the time domain and the frequency doman, depending on the domain in which the data are sampled.

5.7.1 Sampling Theorem

If a time signal $x(t)$ is sampled at equal steps of ΔT, no information regarding its frequency spectrum $X(f)$ is obtained for frequencies higher than $f_c = 1/(2\Delta T)$. This fact is known as *Shannon's sampling theorem*, and the limiting (cut-off) frequency is called the *Nyquist frequency*.

It can be shown that the aliasing error is caused by "folding" of the high-frequency segment of the frequency spectrum beyond the Nyquist frequency into the low-frequency segment.

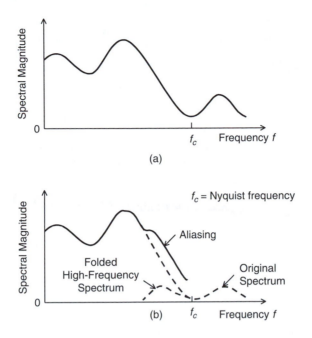

FIGURE 5.10
Aliasing distortion of a frequency spectrum: (a) Original spectrum (b) Distorted spectrum due to aliasing.

This is illustrated in Figure 5.10. The aliasing error becomes more and more prominent for frequencies of the spectrum closer to the Nyquist frequency. In signal analysis, a sufficiently small sample step ΔT should be chosen in order to reduce aliasing distortion in the frequency domain, depending on the highest frequency of interest in the analyzed signal. This however, increases the signal processing time and the computer storage requirements, which is undesirable particularly in real-time analysis. It also can result in stability problems in numerical computations. The Nyquist sampling criterion requires that the sampling rate $(1/\Delta T)$ for a signal should be at least twice the highest frequency of interest. Instead of making the sampling rate very high, a moderate value that satisfies the Nyquist sampling criterion is used in practice, together with an *anti-aliasing filter* to remove the distorted frequency components.

5.7.2   Anti-Aliasing Filter

It should be clear from Figure 5.10 that, if the original signal is low-pass filtered at a cut-off frequency equal to the Nyquist frequency, then the aliasing distortion due to sampling would not occur. A filter of this type is called an anit-aliasing filter. Analog hardware filters may be used for this purpose. In practice, it is not possible to achieve perfect filtering. Hence, some aliasing could remain even after using an anti-aliasing filter. Such residual errors may be reduced by using a filter cut-off frequency that is somewhat less than the Nyquist frequency. Then the resulting spectrum would only be valid up to this filter cut-off frequency (and not up to the theoretical limit of Nyquist frequency). Aliasing reduces the valid frequency range in digital Fourier results. Typically, the useful frequency limit is $f_c/1.28$ so that the last 20% of the spectral points near the Nyquist frequency should be neglected. Note that sometimes $f_c/1.28 (\cong 0.8f_c)$ is used as the filter cutoff frequency. In this case the computed spectrum is accurate up to $0.8\,f_c$ and not up to f_c.

Example 5.7

Consider 1024 data points from a signal, sampled at 1 millisecond (ms) intervals.

$$\text{Sample rate } f_s = 1/0.001 \text{ samples/s} = 1000 \text{ Hz} = 1 \text{ kHz}$$

$$\text{Nyquist frequency} = 1000/2 \text{ Hz} = 500 \text{ Hz}$$

Due to aliasing, approximately 20% of the spectrum (i.e., spectrum beyond 400 Hz) will be distorted. Here we may use an anti-aliasing filter.

Suppose that a digital Fourier transform computation provides 1024 frequency points of data up to 1000 Hz. Half of this number is beyond the Nyquist frequency, and will not give any new information about the signal.

$$\text{Spectral line separation} = 1000/1024 \text{ Hz} = 1 \text{ Hz (approx.)}$$

Keep only the first 400 spectral lines as the useful spectrum.

NOTE   Almost 500 spectral lines may be retained if an accurate anti-aliasing filter is used.

Example 5.8

a. If a sensor signal is sampled at f_s Hz, suggest a suitable cutoff frequency for an antialiasing filter to be used in this application.

b. Suppose that a sinusoidal signal of frequency f_1 Hz is sampled at the rate of f_s samples/s. Another sinusoidal signal of the same amplitude, but of a higher frequency f_2 Hz was found to yield the same data when sampled at f_s. What is the likely analytical relationship between f_1, f_2, and f_s?

c. Consider a plant of transfer function

$$G(s) = \frac{k}{(1 + \tau s)}$$

What is the static gain of this plant? Show that the magnitude of the transfer function reaches $1/\sqrt{2}$ of the static gain when the excitation frequency is $1/\tau$ rad/s. Note that the frequency, $\omega_b = 1/\tau$ rad/s, may be taken as the operating bandwidth of the plant.

d. Consider a chip refiner that is used in the pulp and paper industry. The machine is used for mechanical pulping of wood chips. It has a fixed plate and a rotating plate, driven by an induction motor. The gap between the plates is sensed and is adjustable as well. As the plate rotates, the chips are ground into a pulp within the gap. A block diagram of the plate-positioning control system is shown in Figure 5.11.

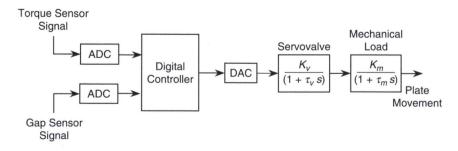

FIGURE 5.11
Block diagram of the plate positioning control system for a chip refiner.

Suppose that the torque sensor signal and the gap sensor signal are sampled at 100 Hz and 200 Hz, respectively, into the digital controller, which takes 0.05 sec to compute each positioning command for the servovalve. The time constant of the servovalve is $\frac{0.05}{2\pi}$ sec and that of the mechanical load is $\frac{0.2}{2\pi}$ sec. Estimate the control bandwidth and the operating bandwidth of the positioning system.

SOLUTION

a. In theory, the cutoff frequency of the antialiasing filter has to be $\frac{1}{2}f_s$, which is the Nyquist frequency. In practice, however, $0.4 f_s$ would be desirable, providing a useful spectrum of only up to $0.4 f_s$.

b. It is likely that f_1 and f_2 are symmetrically located on either side of $f_c (= \frac{1}{2}f_s)$. Hence,

$$f_2 - f_c = f_c - f_1$$

or

$$f_2 = f_s - f_1 \tag{5.14}$$

c.

$$G(j\omega) = \frac{k}{(1+\tau j\omega)} = \text{frequency transfer function where, } \omega \text{ is in rad/s.}$$

Static gain is the transfer function magnitude at steady state (i.e., at zero frequency). Hence,

$$\text{Static gain} = G(0) = k$$

$$\text{When } \omega = \frac{1}{\tau}$$

$$G(j\omega) = \frac{k}{(1+j)}$$

Hence,

$$|G(j\omega)| = \frac{k}{\sqrt{2}} \text{ at this frequency.}$$

This corresponds to the half-power bandwidth.

d. Due to sampling, the torque signal has a bandwidth of $1/2 \times 100$ Hz = 50 Hz, and the gap sensor signal has a bandwidth of $1/2 \times 200$ Hz = 100 Hz. Control cycle time = 0.05 sec, which provides control signals at a rate of $1/0.05$ Hz = 20 Hz. Since

$$20 \text{ Hz} < \min (50 \text{ Hz}, 100 \text{ Hz})$$

we have adequate bandwidth from the sampled sensor signals to compute the control signal. The control bandwidth from the digital controller

$$= 1/2 \times 20 \text{ Hz (From Shannon's sampling theorem)}$$

$$= 10 \text{ Hz}$$

But, the servovalve is also part of the controller. Its bandwidth

$$\frac{1}{\tau_v} \text{rad/s} = \frac{1}{2\pi\tau_m} \text{Hz}$$

$$= \frac{2\pi}{2\pi \times 0.2} \text{Hz} = 5 \text{ Hz}$$

Hence,

$$\text{Control bandwidth} = \min (10 \text{ Hz}, 20 \text{ Hz}) = 10 \text{ Hz}.$$

Bandwidth of the mechanical load

$$\frac{1}{\tau_v} \text{rad/s} = \frac{1}{2\pi\tau_m} \text{Hz} = \frac{2\pi}{2\pi \times 0.2} \text{Hz} = 5 \text{ Hz}$$

Hence,

Operating bandwidth of the system $= \min (10 \text{ Hz}, 5 \text{ Hz}) = 5 \text{ Hz}.$

5.7.3 Another Illustration of Aliasing

A simple illustration of aliasing is given in Figure 5.12. Here, two sinusoidal signals of frequency $f_1 = 0.2$ Hz and $f_2 = 0.8$ Hz are shown (Figure 5.12(a)). Suppose that the two signals are sampled at the rate of $f_s = 1$ sample/s. The corresponding Nyquist frequency is $f_c = 0.5$ Hz. It is seen that, at this sampling rate, the data samples from the two signals are identical. In other words, the high-frequency signal cannot be distinguished from the low-frequency signal. Hence, a high-frequency signal component of frequency 0.8 Hz will appear as a low-frequency signal component of frequency 0.2 Hz. This is aliasing, as clear from the signal spectrum shown in Figure 5.12(b). Specifically, the spectral segment of the signal beyond the Nyquist frequency (f_c) cannot be recovered.

Example 5.9

Suppose that the frequency range of interest in a particular signal is 0–200 Hz. We are interested in determining the sampling rate (digitization speed) and the cutoff frequency for the antialiasing (low pass) filter.

The Nyquist frequency f_c is given by $f_c/1.28 = 200$.
Hence, $f_c = 256$ Hz.

The sampling rate (or digitization speed) for the time signal that is needed to achieve this range of analysis is $F = 2f_c = 512$ Hz. With this sampling frequency, the cutoff frequency for the antialiasing filter could be set at a value between 200 and 256 Hz.

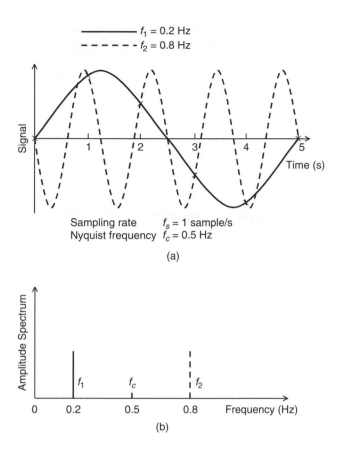

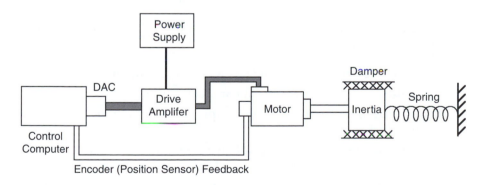

FIGURE 5.12
A simple illustration of aliasing: (a) Two harmonic signals with identical sampled data. (b) Frequency spectra of the two harmonic signals.

FIGURE 5.13
Digital control system for a mechanical positioning application.

Example 5.10

Consider the digital control system for a mechanical positioning application, as schematically shown in Figure 5.13. The control computer generates a control signal according to an algorithm, on the basis of the desired position and actual position, as measured by an optical encoder. This digital signal is converted into the analog form using a digital-to-analog converter (DAC) and is supplied to the drive amplifier. Accordingly, the current signals needed to energize the motor windings are generated by the amplifier. The inertial element, which has to be positioned is directly (and rigidly) linked to the motor rotor and is resisted by a spring and a damper, as shown.

Suppose that the combined transfer function of the drive amplifier and the electromagnetic circuit (torque generator) of the motor is given by

$$\frac{k_e}{\left(s^2 + 2\zeta_e \omega_e s + \omega_e^2\right)}$$

and the transfer function of the mechanical system including the inertia of the motor rotor is given by

$$\frac{k_m}{\left(s^2 + 2\zeta_m \omega_m s + \omega_m^2\right)}$$

Here

$$k = \text{equivalent gain}$$

$$\zeta = \text{damping ratio}$$

$$\omega = \text{natural frequency}$$

with the subscripts *e* and *m* denoting the electrical and mechanical components, respectively. Also,

$$\Delta T_c = \text{time taken to compute each control action}$$

$$\Delta T_p = \text{pulse period of the position sensing encoder.}$$

The following numerical values are given:

$$\omega_e = 1000\pi \text{ rad/s}, \quad \zeta_e = 0.5, \quad \omega_m = 100\pi \text{ rad/s, and } \zeta_m = 0.3$$

For the purpose of this example, you may neglect loading effects and coupling effects due to component cascading and signal feedback.

   i. Explain why the control bandwidth of this system cannot be much larger than 50 Hz.

   ii. If $\Delta T_c = 0.02$ sec, estimate the control bandwidth of the system.

   iii Explain the significance of ΔT_p in this application. Why, typically, ΔT_p should not be greater than $0.5\Delta T_c$?

   iv. Estimate the operating bandwidth of the positioning system, assuming that significant plant dynamics are to be avoided.

   v. If $\omega_m = 500\pi$ rad/s and $\Delta T_c = 0.02$ sec, with the remaining parameters kept as specified above, estimate the operating bandwidth of the system, again in order not to excite significant plant dynamics.

SOLUTION

   i. The drive system has a resonant frequency less than 500 Hz. Hence the flat region of the spectrum of the drive system would be about 1/10th of this; i.e., 50 Hz. This would limit the maximum spectral component of the drive signal to about 50 Hz, so that the drive system dynamics are avoided. Hence the control bandwidth would be limited by this value.

ii. Rate at which the digital control signal is generated $= \frac{1}{0.02}$ Hz $= 50$ Hz. By Shannon's sampling theorem, the effective (useful) spectrum of the control signal is limited to $\frac{1}{2} \times 50$ Hz $= 25$ Hz. Even though the drive system can accommodate a bandwidth of about 50 Hz, the control bandwidth would be limited to 25 Hz, due to digital control, in this case.

iii. Note that ΔT_p corresponds to the sampling period of the measurement signal (for feedback). Hence its useful spectrum would be limited to $1/2\Delta T_p$, by Shannon's sampling theorem. Consequently, the feedback signal will not be able to provide any useful information of the process beyond the frequency $1/2\Delta T_p$. To generate a control signal at the rate of $1/\Delta T_c$ samples/s, the process information has to be provided at least up to $1/\Delta T_c$ Hz. To provide this information we must have:

$$\frac{1}{2\Delta T_p} \geq \frac{1}{\Delta T_c} \quad \text{or} \quad \Delta T_p \leq 0.5 \, \Delta T_c. \tag{5.15}$$

Note that this guarantees that at least two points of sampled data from the sensor are available for computing each control action.

iv. The resonant frequency of the plant (positioning system) is approximately (less than) $\frac{100\pi}{2\pi}$ Hz $\doteq 50$ Hz. At frequencies near this, the resonance will interfere with control, and should be avoided if possible, unless the resonances (or modes) of the plant themselves need to be modified through control. At frequencies much larger than this, the process will not significantly respond to the control action, and will not be of much use (the plant will be felt like a rigid wall). Hence, the operating bandwidth has to be sufficiently smaller than 50 Hz, say 25 Hz, in order to avoid plant dynamics.

   NOTE   This is a matter of design judgment, based on the nature of the application (e.g., excavator, disk drive). Typically, however, one needs to control the plant dynamics. In that case it is necessary to use the entire control bandwidth (i.e., maximum possible control speed) as the operating bandwidth. In the present case, even if the entire control BW (i.e., 25 Hz) is used as the operating BW, it still avoids the plant resonance.

v. The plant resonance in this case is about $\frac{500\pi}{2\pi}$ Hz $\doteq 250$ Hz. This limits the operating bandwidth to about $\frac{250\pi}{2}$ Hz $\doteq 125$ Hz, so as to avoid plant dynamics. But, the control bandwidth is about 25 Hz because $\Delta T_c = 0.02$ sec, as obtained in Part ii. The operating bandwidth cannot be greater than this value, and would be ≈ 25 Hz.

5.8   Bandwidth Design of a Mechatronic System

Based on the foregoing concepts, it is now possible to give a set of simple steps for designing a mechatronic control system on the basis of bandwidth considerations.

   Step 1: Decide on the maximum frequency of operation (BW_o)of the system based on the requirements of the particular application.

   Step 2: Select the process components (electro-mechanical) that have the capacity to operate at BW_o and to perform the required tasks at this bandwidth.

   Step 3: Select feedback sensors with a flat frequency spectrum (operating frequency range) greater than $4 \times BW_o$.

Step 4: Develop a digital controller with a sampling rate greater than $4 \times BW_o$ for the sensor feedback signals (keeping within the flat spectrum of the sensors) and a direct-digital control cycle time (period) of $1/(2 \times BW_o)$. Note that the digital control actions are generated at a rate of $2 \times BW_o$.

Step 5: Select the control drive system (interface analog hardware, filters, amplifiers, actuators, etc.) that have a flat frequency spectrum of at least BW_o.

Step 6: Integrate the system and test the performance. If the performance specifications are not satisfied, make necessary adjustments and test again.

5.8.1 Comment About Control Cycle Time

In the engineering literature it is often used that $\Delta T_c = \Delta T_p$, where ΔT_c = control cycle time (period at which the digital control actions are generated) and ΔT_p = period at which the feedback sensor signals are sampled (See Figure 5.14(a)). This is acceptable in systems

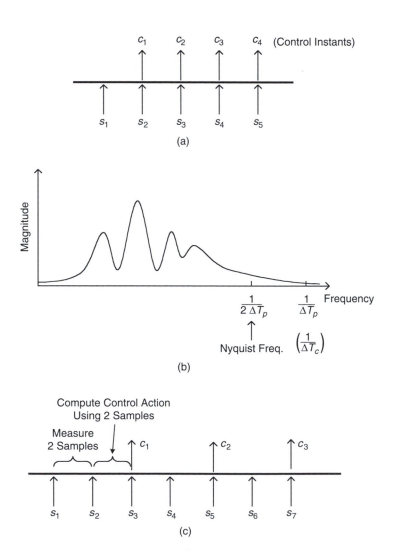

FIGURE 5.14
(a) Conventional sampling of feedback sensor signals for direct digital control; (b) Acceptable frequency characteristic of a plant for case (a); (c) Improved sampling criterion for feedback signals in direct digital control.

where the important frequency range of the plant is sufficiently smaller than $1/\Delta T_p$ (and $1/\Delta T_c$). In that case the sampling rate $1/\Delta T_p$ of the feedback measurements (and the Nyquist frequency $0.5/\Delta T_p$) will still be sufficiently larger than the operating frequency range of the plant (See Figure 5.14(b)), and hence the control system will function satisfactorily. But, the bandwidth criterion presented in this section satisfies $\Delta T_p \leq \Delta T_c$. This is a more desirable option. For example, in Figure 5.14(c), two measurement samples are used in computing each control action. Here, the Nyquist frequency of the sampled feedback signals is double that of the previous case, and it will cover a larger (double) frequency range of the plant.

5.9 Instrument Error Analysis

Analysis of error in an instrument or a multi-component mechatronic system, is a very challenging task. Difficulties arise for many reasons, particularly the following:

1. True value is usually unknown.
2. The instrument reading may contain random error that cannot be determined exactly.
3. The error may be a complex (i.e., not simple) function of many variables (input variables and state variables or response variables).
4. The system/instrument may be made up of many components that have complex interrelations (dynamic coupling, multiple degree-of-freedom responses, nonlinearities, etc.), and each component may contribute to the overall error.

The first item is a philosophical issue that would lead to an argument similar to the chicken-and-egg controversy. For instance, if the true value is known, there is no need to measure it; and if the true value is unknown, it is impossible to determine exactly how inaccurate a particular reading is. In fact, this situation can be addressed to some extent by using statistical representations of error, which takes us to the second item listed. The third and fourth items may be addressed by error combination in multivariable systems and by error propagation in complex multicomponent systems. It is not feasible here to provide a full treatment of all these topics. Only an introduction to simple analytical techniques will be given, using illustrative examples. The concepts discussed here are useful not only in statistical error analysis but also in the field of *statistical process control* (SPC)—the use of statistical signals to improve performance of a process. Performing statistical analysis of a response signal and drawing its *control chart*, along with an *upper control line* and a *lower control line*, are key procedures in statistical process control.

5.9.1 Statistical Representation

In general, error is a random variable. It is defined as:

$$\text{Error} = (\text{instrument reading}) - (\text{true value}).$$

Randomness associated with a measurand can be interpreted in two ways. First, since the true value of the measurand is a fixed quantity, randomness can be interpreted as the randomness in error that is usually originating from the random factors in instrument response. Second, looking at the issue in a more practical manner, error analysis can be interpreted as an "estimation problem" in which the objective is to estimate the true value of a measurand from a known set of readings. In this latter point of view, the "estimated" true value itself becomes a random variable. No matter what approach is used, however, the same statistical concepts may be used in representing error.

5.9.2   Accuracy and Precision

The instrument ratings as mentioned before, affect the overall *accuracy* of an instrument. Accuracy can be assigned either to a particular reading or to an instrument. Note that instrument accuracy depends not only on the physical hardware of the instrument but also on the operating conditions (e.g., design conditions that are the normal, steady operating conditions or extreme transient conditions, such as emergency start-up and shutdown). *Measurement accuracy* determines the closeness of the measured value to the true value. *Instrument accuracy* is related to the worst accuracy obtainable within the dynamic range of the instrument in a specific operating environment. *Measurement error* is defined as

$$\text{error} = (\text{measured value}) - (\text{true value}) \tag{5.16}$$

Correction, which is the negative of error, is defined as

$$\text{correction} = (\text{true value}) - (\text{measured value}) \tag{5.17}$$

Each of these can also be expressed as a percentage of the true values. Accuracy of an instrument may be determined by measuring a parameter whose true value is known, near the extremes of the dynamic range of the instrument, under certain operating conditions. For this purpose, standard parameters or signals that can be generated at very high levels of accuracy would be needed. The National Institute for Standards and Testing (NIST) is usually responsible for the generation of these standards. Nevertheless, accuracy and error values cannot be determined to 100 percent exactness in typical applications, because the true value is not known to begin with. In a given situation, we can only make estimates for accuracy, by using ratings provided by the instrument manufacturer or by analyzing data from previous measurements and models.

Causes of error include instrument instability, external noise (disturbances), poor calibration, inaccurate information (e.g., poor analytical models, inaccurate control laws and digital control algorithms), parameter changes (e.g., due to environmental changes, aging, and wearout), unknown nonlinearities, and improper use of instrument.

Errors can be classified as *deterministic* (or *systematic*) and *random* (or *stochastic*). Deterministic errors are those caused by well-defined factors, including nonlinearities and offsets in readings. These usually can be accounted for by proper calibration and analysis practices. Error ratings and calibration charts are used to remove systematic errors from instrument readings. Random errors are caused by uncertain factors entering into instrument response. These include device noise, line noise, and effects of unknown random variations in the operating environment. A statistical analysis using sufficiently large amounts of data is necessary to estimate random errors. The results are usually expressed as a mean error, which is the systematic part of random error, and a standard deviation or confidence interval for instrument response. *Precision* is not synonymous with accuracy. Reproducibility (or repeatability) of an instrument reading determines the precision of an instrument. An instrument that has a high offset error might be able to generate a response at high precision, even though this output is clearly inaccurate. For example, consider a timing device (clock) that very accurately indicates time increments (say, up to the nearest nanosecond). If the reference time (starting time) is set incorrectly, the time readings will be in error, even though the clock has a very high precision.

Instrument error may be represented by a random variable that has a mean value μ_e and a standard deviation σ_e. If the standard deviation is zero, the variable is considered deterministic. In that case, the error is said to be deterministic or repeatable. Otherwise, the error is said to be random. The precision of an instrument is determined by the standard deviation of error in the instrument response. Readings of an instrument may have a large mean value of error (e.g., large offset), but if the standard deviation is small, the instrument has high precision. Hence, a quantitative definition for precision would be:

$$\text{Precision} = (\text{measurement range}) / \sigma_e \qquad (5.18)$$

Lack of precision originates from random causes and poor construction practices. It cannot be compensated for by recalibration, just as precision of a clock cannot be improved by resetting the time. On the other hand, accuracy can be improved by recalibration. Repeatable (deterministic) accuracy is inversely proportional to the magnitude of the mean error μ_e.

Matching instrument ratings with specifications is very important in selecting instruments for a mechatronic application. Several additional considerations should be looked into as well. These include geometric limitations (size, shape, etc.), environmental conditions (e.g., chemical reactions including corrosion, extreme temperatures, light, dirt accumulation, electromagnetic fields, radioactive environments, shock and vibration), power requirements, operational simplicity, availability, past record and reputation of the manufacturer and of the particular instrument, and cost-related economic aspects (initial cost, maintenance cost, cost of supplementary components such as signal-conditioning and processing devices, design life and associated frequency of replacement, and cost of disposal and replacement). Often, these considerations become the ultimate deciding factors in the selection process.

5.9.3 Error Combination

Error in a response variable of a device or in an estimated parameter of a system would depend on errors present in measured variables and parameter values that are used to determine the unknown variable or parameter. Knowing how component errors are propagated within a multicomponent system and how individual errors in system variables and parameters contribute toward the overall error in a particular response variable or parameter would be important in estimating error limits in complex mechatronic systems. For example, if the output power of a rotational manipulator is computed by measuring torque and speed at the output shaft, error margins in the two measured "response variables" (torque and speed) would be directly combined into the error in the power computation. Similarly, if the natural frequency of a simple suspension system is determined by measuring mass and spring stiffness "parameters" of the suspension, the natural frequency estimate would be directly affected by possible errors in mass and stiffness measurements. Extending this idea further, the overall error in a mechatronic system depends on individual error levels in various components (sensors, actuators, controller hardware, filters, amplifiers, etc.) of the system and on the manner in which these components are physically interconnected and physically interrelated. For example, in a robotic manipulator, the accuracy of the actual trajectory of the end effector will depend on the accuracy of sensors and actuators at the manipulator joints and on the accuracy of the robot controller. Note that we are dealing with a generalized idea of error propagation that considers errors in system variables (e.g., input and output signals, such as velocities, forces, voltages, currents, temperatures, heat transfer rates, pressures and fluid flow rates) system parameters (e.g., mass, stiffness, damping, capacitance, inductance, resistance, thermal conductivity, and viscosity), and system components (e.g., sensors, actuators, filters, amplifiers, control circuits, thermal conductors, and valves).

For the analytical development of a basic result in error combination, we will start with a functional relationship of the form

$$y = f(x_1, x_2, \ldots, x_r) \tag{5.19}$$

Here, x_i are the independent system variables (or parameter values) whose error is propagated into a dependent variable (or parameter value) y. Determination of this functional relationship is not always simple, and the relationship itself may be in error. Since our intention is to make a reasonable estimate for possible error in y due to the combined effect of errors from x_i, an approximate functional relationship would be adequate in most cases. Let us denote error in a variable by the differential of that variable. Taking the differential of Equation 5.19, we get

$$\delta y = \frac{\partial f}{\partial x_1} \delta x_1 + \frac{\partial f}{\partial x_2} \delta x_2 + \cdots + \frac{\partial f}{\partial x_r} \delta x_r \tag{5.20}$$

for small errors. For those who are not familiar with differential calculus, Equation 5.20 may be interpreted as the first-order terms in a *Taylor series expansion* of Equation 5.19. Now, rewriting Equation 5.20 in the fractional form, we get

$$\frac{\delta y}{y} = \sum_{i=1}^{r} \left[\frac{x_i}{y} \frac{\partial f}{\partial x_i} \frac{\delta x_i}{x_i} \right] \tag{5.21}$$

Here, $\delta y / y$ represents the overall error and $\delta x_i / x_i$ represents the component error, expressed as fractions. We shall consider two types of estimates for overall error.

5.9.3.1   Absolute Error

Since error δx_i could be either positive or negative, an upper bound for the overall error is obtained by summing the absolute value of each right-hand-side term in Equation 5.21. This estimate e_{ABS}, which is termed *absolute error*, is given by

$$e_{ABS} = \sum_{i=1}^{r} \left| \frac{x_i}{y} \frac{\partial f}{\partial x_i} \right| e_i \tag{5.22}$$

Note that component error e_i and absolute error e_{ABS} in Equation 5.22 are always positive quantities; when specifying error, however, both positive and negative limits should be indicated or implied (e.g., $\pm e_{ABS}, \pm e_i$).

5.9.3.2   SRSS Error

Equation 5.22 provides a conservative (upper bound) estimate for overall error. Since the estimate itself is not precise, it is often wasteful to introduce such a high conservatism. A nonconservative error estimate that is frequently used in practice is the *square root of sum of squares* (SRSS) error. As the name implies, this is given by

$$e_{SRSS} = \left[\sum_{i=1}^{r} \left(\frac{x_i}{y} \frac{\partial f}{\partial x_i} e_i \right)^2 \right]^{1/2} \tag{5.23}$$

This is not an upper bound estimate for error. In particular, $e_{SRSS} < e_{ABS}$ when more than one nonzero error contribution is present. The SRSS error relation is particularly suitable when component error is represented by the standard deviation of the associated variable or parameter value and when the corresponding error sources are independent. Now we will present several examples of error combination.

Example 5.11

Using the absolute value method for error combination, determine the fractional error in each item x_i so that the contribution from each item to the overall error e_{ABS} is the same.

SOLUTION

For equal contribution, we must have

$$\left| \frac{x_1}{y} \frac{\partial f}{\partial x_1} \right| e_1 = \left| \frac{x_2}{y} \frac{\partial f}{\partial x_2} \right| e_2 = \cdots = \left| \frac{x_r}{y} \frac{\partial f}{\partial x_r} \right| e_r$$

Hence,

$$r \left| \frac{x_i}{y} \frac{\partial f}{\partial x_i} \right| e_i = e_{ABS}$$

Thus,

$$e_i = e_{ABS} \bigg/ \left(r \left| \frac{x_i}{y} \frac{\partial f}{\partial x_i} \right| \right) \tag{5.24}$$

Example 5.12

The result obtained in the previous example is useful in the design of multicomponent systems and in the cost-effective selection of instrumentation for a particular application. Using Equation 5.24, arrange the items x_i in their order of significance.

SOLUTION

Note that Equation 5.24 may be written as

$$e_i = K \bigg/ \left| x_i \frac{\partial f}{\partial x_i} \right| \tag{5.25}$$

where K is a quantity that does not vary with x_i. It follows that for equal error contribution from all items, error in x_i should be inversely proportional to $|x_i(\partial f/\partial x_i)|$. In particular, the item with the largest $|x(\partial f/\partial x)|$ should be made most accurate. In this manner, allowable relative accuracy for various components can be estimated. Since, in general, the most accurate device is also the most costly one, instrumentation cost can be optimized if components are selected according to the required overall accuracy, using a criterion such as that implied by Equation 5.25.

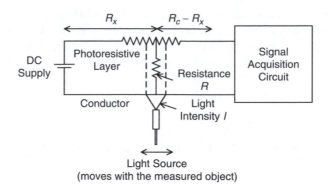

FIGURE 5.15
An optical displacement sensor.

Example 5.13

Figure 5.15 schematically shows an optical device for measuring displacement. This sensor is essentially an optical potentiometer. The potentiometer element is uniform and has a resistance R_c. A photoresistive layer is sandwiched between this element and a perfect conductor of electricity. A light source, which moves with the object whose displacement is being measured, directs a beam of light whose intensity is I, on to a narrow rectangular region of the photoresistive layer. As a result, this region becomes resistive with resistance R that bridges the potentiometer element and the conductor element, as shown. An empirical relation between R and I was found to be

$$\ln\left(\frac{R}{R_0}\right) = \left(\frac{I_0}{I}\right)^{1/4}$$

in which the resistance R is in kΩ and the light intensity I is expressed in watts per square metre (W/m^2). The parameters R_0 and I_0 are empirical constants having the same units as R and I, respectively. These two parameters generally have some experimental error.

a. Sketch the curve of R versus I and explain the significance of the parameters R_0 and I_0.

b. Using the absolute error method, show that the combined fractional error e_R in the bridging resistance R can be expressed as

$$e_R = e_{R0} + \frac{1}{4}\left(\frac{I_0}{I}\right)^{1/4}[e_I + e_{I0}]$$

in which e_{R0}, e_I, and e_{I0} are the fractional errors in R_0, I, and I_0 respectively.

c. Suppose that the empirical error in the sensor model can be expressed as $e_{R0} = \pm 0.01$ and $e_{I0} = \pm 0.01$, and due to variations in the supply to the light source and in ambient lighting conditions, the fractional error in I is also ± 0.01. If the error E_R is to be maintained within ± 0.02, at what light intensity level (I) should the light source operate? Assume that the empirical value of I_0 is 2.0 W/m^2.

d. Discuss advantages and disadvantages this device has as a dynamic displacement sensor.

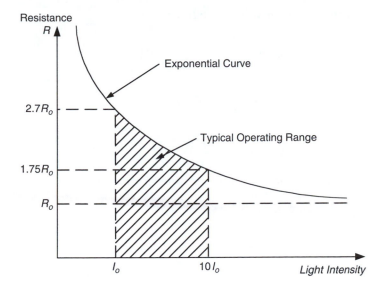

FIGURE 5.16
Characteristic curve of the sensor.

SOLUTION

a. $\ln \dfrac{R}{R_0} = \left(\dfrac{I_0}{I}\right)^{1/4}$

R_0 represents the minimum resistance provided by the photoresistive bridge (i.e., at very high light intensity levels). When $I = I_0$ the bridge resistance R is about 2.7 R_0 and hence I_0 represents a lower bound for the intensity for proper operation of the sensor. For intensity levels lower than this, the effects of noise (e.g., ambient light) and power supply can be unacceptable. A suitable upper bound for the intensity would be 10 I_0, for satisfactory operation. For intensity levels larger than this, the sensor sensitivity may be inadequate and the power consumption of the light source can be excessive. At this value, $R = 1.75 R_0$ as shown in Figure 5.16.

(b) $\ln R - \ln R_0 = \left(\dfrac{I_0}{I}\right)^{1/4}$

Differentiate,

$$\frac{\delta R}{R} - \frac{\delta R_0}{R_0} = \frac{1}{4}\left(\frac{I_0}{I}\right)^{-3/4}\left[\frac{\delta I_0}{I} - \frac{I_0}{I^2}\delta I\right]$$

$$= \frac{1}{4}\left(\frac{I_0}{I}\right)^{1/4}\left[\frac{\delta I_0}{I_0} - \frac{\delta I}{I}\right]$$

Hence, with the absolute method of error combination,

$$e_R = e_{R_0} + \frac{1}{4}\left(\frac{I_0}{I}\right)^{1/4}[e_{I_0} + e_I]$$

c. With the given numerical values, we have

$$0.02 = 0.01 + \frac{1}{4}\left(\frac{I_0}{I}\right)^{1/4} [0.01 + 0.01]$$

$$\Rightarrow \left(\frac{I_0}{I}\right)^{1/4} = 2$$

or, $I = \frac{1}{16} I_0 = \frac{2.0}{16} \text{ W/m}^2 = 0.125 \text{ W/m}^2$

NOTE For larger values of I the absolute error in R_o would be smaller. For example, for $I = 10\, I_0$ we have,

$$e_R = 0.01 + \frac{1}{4}\left(\frac{1}{10}\right)^{1/4} [0.01 + 0.01] = 0.013.$$

d. *Advantages*

- Noncontacting
- Small moving mass (low inertial loading)
- All advantages of a potentiometer.

Disadvantages

- Nonlinear and exponential variation of R
- Effect of ambient lighting
- Possible nonlinear behavior of the device (input-output relation)
- Effect of variations in the supply to the light source
- Effect of aging of the light source.

Example 5.14

a. You are required to select a sensor for a position control application. List several important considerations that you have to take into account in this selection. Briefly indicate why each of them is important.

b. A schematic diagram of a chip refiner that is used in the pulp and paper industry is shown in Figure 5.17. This machine is used for mechanical pulping of wood chips. The refiner has one fixed disc and one rotating disc (typical diameter = 2 m). The plate is rotated by an ac induction motor. The plate separation (typical gap = 0.5 mm) is controlled using a hydraulic actuator (piston-cylinder unit with servovalve). Wood chips are supplied to the eye of the refiner by a screw conveyor and are diluted with water. As the refiner plate rotates the chips are ground into a pulp within the internal grooves of the plates. This is accompanied by the generation of steam due to energy dissipation. The pulp is drawn and further processed for making paper.

An empirical formula relating the plate gap (h) and the motor torque (T) is given by

$$T = \frac{ah}{(1 + bh^2)}$$

with the model parameters a and b are known to be positive.

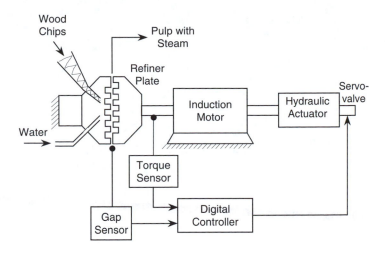

FIGURE 5.17
A single-disc chip refiner.

i. Sketch the curve T versus h. Express the maximum torque $T_{\max}$ and the plate gap (h_0) at this torque in terms of a and b only.

ii. Suppose that the motor torque is measured and the plate gap is adjusted by the hydraulic actuator according to the formula given above. Show that the fractional error in h may be expressed as

$$e_h = \left[e_T + e_a + \frac{bh^2}{(1+bh^2)} e_b \right] \frac{(1+bh^2)}{|1-bh^2|}$$

where e_T, e_a, and e_b are the fractional errors in T, a and b, respectively, the latter two being representative of model error.

iii. The normal operating region of the refiner corresponds to $h > h_0$. The interval $0 < h < h_0$ is known as the "pad collapse region" and should be avoided. If the operating value of the plate gap is $h = 2/\sqrt{b}$ and if the error values are given as $e_T = \pm 0.05$, $e_a = \pm 0.02$, and $e_b = \pm 0.025$, compute the corresponding error in the plate gap estimate.

iv. Discuss why operation at $h = 1/\sqrt{b}$ is not desirable.

SOLUTION

a. **Bandwidth:** Determines the useful (or flat) frequency range of operation, and also the speed of response.

Accuracy: Low random error or high precision. Low systematic error or reduced need for recalibration.

Dynamic Range: Range of operating amplitude.

Resolution: Determines the smallest signal change that could be correctly measured.

Input Impedance: High value means low loading error or distortion of the measurement.

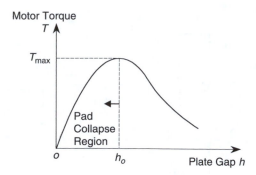

FIGURE 5.18
Characteristic curve of the chip refiner.

Output Impedance: Low value means high output level and low distortion due to subsequent signal conditioning.

Linearity: Proportionality of input and output. Ease of recalibration.

Sensitivity: High value means high output for a given input.

Drift: Specifies output stability under steady conditions.

Size: Smaller size typically means higher (finer) resolution and bandwidth, lower mechanical loading, and reduced space requirements.

Cost: The cheaper the better.

b. i. See the sketch in Figure 5.18.

$$T = \frac{ah}{1+bh^2} \tag{i}$$

$$\frac{\partial T}{\partial h} = \frac{(1+bh^2)a - ah(2bh)}{(1+bh^2)^2} = 0 \text{ at maximum } T.$$

Hence,

$$1 - bh^2 = 0$$

or,

$$h_o = 1/\sqrt{b}$$

Substitute in Equation i:

$$T_{max} = \frac{a}{2\sqrt{b}}$$

ii. The differential relation of Equation i is

$$\delta T = \frac{h}{(1+bh^2)}\delta a + \frac{\partial T}{\partial h}\delta h - \frac{ah}{(1+bh^2)^2}\cdot h^2\delta b$$

Substitute for $\frac{\partial T}{\partial h}$ from Part (i)

$$\delta T = \frac{h}{(1+bh^2)}\delta a + a\frac{(1-bh^2)}{(1+bh^2)^2}\delta h - \frac{ah^3}{(1+bh^2)^2}\delta b$$

Divide throughout by Equation i

$$\frac{\delta T}{T} = \frac{\delta a}{a} + \left[\frac{1-bh^2}{1+bh^2}\right]\frac{\delta h}{h} - \frac{bh^2}{(1+bh^2)}\frac{\delta b}{b}$$

or,

$$\frac{\delta h}{h} = \left[\frac{\delta T}{T} - \frac{\delta a}{a} + \frac{bh^2}{(1+bh^2)}\frac{\delta b}{b}\right]\left[\frac{1+bh^2}{1-bh^2}\right]$$

Now representing the fractional errors by fractional deviations (differentials), and using the absolute value method of error combination, we have

$$e_h = \left[e_T + e_a + \frac{bh^2}{(1+bh^2)}e_b\right]\frac{(1+bh^2)}{|1-bh^2|} \qquad \text{(ii)}$$

iii. With $h = \frac{2}{\sqrt{b}}$ we have $bh^2 = 4$

Substitute the given numerical values for fractional error, in Equation ii.

$$e_h = \left[0.05 + 0.02 + \frac{4}{5}\times 0.025\right]\frac{(1+4)}{|1-4|}$$

$$= \pm 0.15$$

iv. When $h = \frac{1}{\sqrt{b}}$ we see from Equation ii that $e_h \to \infty$. Also, from the curve in Part (i), Figure 5.18, we see that at this point the motor torque is not sensitive to changes in the plate gap. Hence operation at this point is not appropriate.

5.10 Statistical Process Control

In statistical process control (SPC), statistical analysis of process responses is used to generate control actions. This method of control is applicable in many situations of process control, including manufacturing quality control, control of chemical process plants, computerized office management systems, inventory control systems, and urban transit control systems. A major step in statistical process control is to compute control limits (or action lines) on the basis of measured data from the process.

5.10.1 Control Limits or Action Lines

Since a very high percentage of readings from an instrument should lie within $\pm 3\sigma$ about the mean value, according to the normal distribution, these boundaries (-3σ and $+3\sigma$) drawn about the mean value may be considered *control limits* or *action lines* in statistical process control. Here σ denotes the standard deviation. If any measurements fall outside the action lines, corrective measures such as recalibration, controller adjustment, or redesign should be carried out.

5.10.2   Steps of SPC

The main steps of statistical process control are as follows:

1. Collect measurements of appropriate response variables of the process.
2. Compute the mean value and the standard deviation of the data, the upper control limit, and the lower control limit.
3. Plot the measured data and draw the two control limits on a control chart.
4. If measurements fall outside the control limits, take corrective action and repeat the control cycle (go to step 1).

If the measurements always fall within the control limits, the process is said to be in statistical control.

Example 5.15

Error in a satellite tracking system was monitored on-line for a period of one hour to determine whether recalibration or gain adjustment of the tracking controller would be necessary. Four measurements of the tracking deviation were taken in a period of five minutes, and twelve such data groups were acquired during the one-hour period. Sample means and sample variance values of the twelve groups of data were computed. The results are tabulated as follows:

Period i	1	2	3	4	5	6	7	8	9	10	11	12
Sample mean $\overline{X}_i$	1.34	1.10	1.20	1.15	1.30	1.12	1.26	1.10	1.15	1.32	1.35	1.18
Sample variance S_i^2	0.11	0.02	0.08	0.10	0.09	0.02	0.06	0.05	0.08	0.12	0.03	0.07

Draw a control chart for the error process, with control limits (action lines) at $\overline{X} \pm 3\sigma$. Establish whether the tracking controller is in statistical control or needs adjustment.

SOLUTION

The overall mean tracking deviation, $\overline{X} = \frac{1}{12}\sum_{i=1}^{12} \overline{X}_i$ is computed to be $\overline{X} = 1.214$. The average sample variance, $\overline{S}^2 = \frac{1}{11}\sum_{I=1}^{12} S_i^2$ is computed to be $\overline{S}^2 = 0.075$. Since there are four readings within each period, the standard deviation σ of group mean $\overline{X}_i$ can be estimated as

$$S = \frac{\overline{S}}{\sqrt{3}} = \frac{\sqrt{0.075}}{\sqrt{3}} = 0.137$$

The upper control limit (action line) is at (approximately)

$$x = \overline{X} + 3S = 1.214 + 3 \times 0.137 = 1.625$$

The lower control limit (action line) is at

$$x = \overline{X} - 3S = 0.803$$

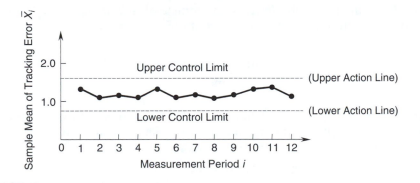

FIGURE 5.19
Control chart for the satellite tracking error example.

These two lines are shown on the control chart in Figure 5.19. Since the sample means (See Table) lie within the two action lines, the process is considered to be in statistical control, and controller adjustments would not be necessary. Note that if better resolution is required in making this decision, individual readings, rather than group means, should be plotted in Figure 5.19.

5.11 Problems

5.1 What do you consider a perfect measuring device? Suppose that you are asked to develop an analog device for measuring angular position in an application related to control of a kinematic linkage system (a robotic manipulator, for example). What instrument ratings (or specifications) would you consider crucial in this application? Discuss their significance.

5.2 List and explain some time-domain parameters and frequency-domain parameters that can be used to predominantly represent

a. speed of response
b. degree of stability

of a mechatronic system. Also, briefly discuss any conflicts that can arise in specifying these parameters.

5.3 A tactile (distributed touch) sensor of the gripper of a robotic manipulator consists of a matrix of piezoelectric sensor elements placed at 2 mm apart. Each element generates an electric charge when it is strained by an external load. Sensor elements are multiplexed at very high speed in order to avoid charge leakage and to read all data channels using a single high-performance charge amplifier. Load distribution on the surface of the tactile sensor is determined from the charge amplifier readings, since the multiplexing sequence is known. Each sensor element can read a maximum load of 50 N and can detect load changes on the order of 0.01 N.

a. What is the spatial resolution of the tactile sensor?
b. What is the load resolution (in N/m^2) of the tactile sensor?
c. What is the dynamic range?

5.4 A useful rating parameter for a mechatronic tool is *dexterity*. Though not complete, an appropriate analytical definition for dexterity of a device is

$$\text{dexterity} = \frac{\text{number of degrees of freedom}}{\text{motion resolution}}$$

where the number of degrees of freedom is equal to the number of independent variables that is required to completely define an arbitrary position increment of the tool (i.e., for an arbitrary change in its kinematic configuration).

a. Explain the physical significance of dexterity and give an example of a mechatronic device for which the specification of dexterity would be very important.

b. The power rating of a tool may be defined as the product of maximum force that can be applied by it in a controlled manner and the corresponding maximum speed. Discuss why the power rating of a manipulating device is usually related to the dexterity of the device. Sketch a typical curve of power versus dexterity.

5.5 Resolution of a feedback sensor (or resolution of a response measurement used in feedback) has a direct effect on the accuracy that is achievable in a control system. This is true because the controller cannot correct a deviation of the response from the desired value (set point) unless the response sensor can detect that change. It follows that the resolution of a feedback sensor will govern the minimum (best) possible deviation band (about the desired value) of the system response, under feedback control. An angular position servo uses a resolver as its feedback sensor. If peak-to-peak oscillations of the servo load (plant) under steady-state conditions have to be limited to no more than two degrees, what is the worst tolerable resolution of the resolver? Note that, in practice, the feedback sensor should have a resolution better (smaller) than this worst value.

5.6 Consider a simple mechatronic device (single degree of freedom) having low damping. An approximate design relationship between the two performance parameters T_r and f_b may be given as

$$T_r f_b = k$$

where
T_r = rise time in nanoseconds (ns)
f_b = bandwidth in megahertz (MHz)

Estimate a suitable value for k.

5.7 List several response characteristics of nonlinear mechatronic systems that are not exhibited by linear mechatronic systems in general. Also, determine the response y of the nonlinear system

$$\left[\frac{dy}{dt}\right]^{1/3} = u(t)$$

when excited by the input $u(t) = a_1 \sin \omega_1 t + a_2 \sin \omega_2 t$. What characteristic of a nonlinear system does this result illustrate?

5.8 Consider a mechanical component whose response x is governed by the relationship

$$f = f(x, \dot{x})$$

where f denotes applied (input) force and $\dot{x}$ denotes velocity. Three special cases are

a. Linear spring:

$$f = kx$$

b. Linear spring with a viscous (linear) damper:

$$f = kx + b\dot{x}$$

c. Linear spring with Coulomb friction:

$$f = kx + f_c \text{sgn}(\dot{x})$$

Suppose that a harmonic excitation of the form $f = f_o \sin \omega t$ is applied in each case. Sketch the force-displacement curves for the three cases at steady state. Which components exhibit hysteresis? Which components are nonlinear? Discuss your answers.

5.9 Discuss how the accuracy of a digital controller may be affected by

a. Stability and bandwidth of amplifier circuitry

b. Load impedance of the analog-to-digital conversion circuitry.

Also, what methods do you suggest to minimize problems associated with these parameters?

5.10 a. Sketch (not to scale) the magnitude versus frequency curves of the following two transfer functions.

(i) $G_i(s) = \dfrac{1}{\tau_i s + 1}$

(ii) $G_d(s) = \dfrac{1}{1 + \frac{1}{\tau_d s}}$

Explain why these two transfer fractions may be used as an integrator; a low-pass filter; a differentiator; and a high-pass filter. In your magnitude versus frequency curves indicate in which frequency bands these four respective realizations are feasible. You may make appropriate assumptions for the time-constant parameters τ_i and τ_d.

b. Active vibration isolators, known as electronic mounts, have been considered for sophisticated automobile engines. The purpose is to actively filter out the cyclic excitation forces generated by the internal-combustion engines before they would adversely vibrate the components such as seats, floor, and steering column, which come into contact with the vehicle occupants. Consider a four-stroke, four-cylinder engine. It is known that the excitation frequency on the engine mounts is twice the crank-shaft speed, as a result of the firing cycles of the cylinders. A schematic representation of an active engine mount is shown in Figure P5.10(a). The crank-shaft speed is measured and supplied to the controller of a valve actuator. The servo valve of a hydraulic cylinder is operated on the basis of this measurement. The hydraulic cylinder functions as an active suspension with a variable (active) spring and a damper. A simplified model of the mechanical interactions is shown in Figure P5.10(b).

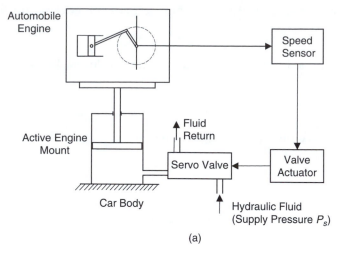

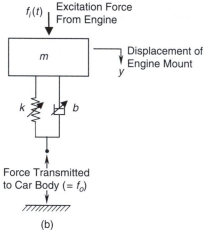

FIGURE P5.10
An active engine mount for an automobile: (a) Schematic diagram, (b) Approximate model.

    i. Neglecting gravity forces (which cancel out due to the static spring force) show that a linear model for system dynamics may be expressed as

$$m\ddot{y} + b\dot{y} + ky = f_i$$

$$b\dot{y} + ky - f_o = 0$$

in which,

f_i = excitation force from the engine
f_o = force transmitted to the passenger compartment
y = displacement of the engine mount with respect to a frame fixed to the passsenger compartment
m = mass of the engine unit
k = equivalent stiffness of the active mount
b = equivalent viscous damping constant of the active mount.

ii. Determine the transfer function (with the Laplace variable s) f_o/f_i for the system.

iii. Sketch the magnitude versus frequency curve of the transfer function obtained in Part (ii) and show a suitable operating range for the active mount.

iv. For a damping ratio $\zeta = 0.2$ what is the magnitude of the transfer function when the excitation frequency ω is 5 times the natural frequency ω_n of the suspension (engine mount) system?

v. Suppose that the magnitude estimated in Part (iv) is satisfactory for the purpose of vibration isolation. If the engine speed varies from 600 rpm to 1200 rpm, what is the range in which the spring stiffness k (N/m) should be varied by the control system in order to maintain this level of vibration isolation? Assume that the engine mass $m = 100$ kg and the damping ratio is approximately constant at $\zeta = 0.2$.

5.11 Consider the mechanical tachometer shown in Figure 5.8. Write expressions for sensitivity and bandwidth for the device. Using the example, show that the two performance ratings, sensitivity and bandwidth, generally conflict. Discuss ways to improve the sensitivity of this mechanical tachometer.

5.12 a. What is an antialiasing filter? In a particular application, the sensor signal is sampled at f_s Hz. Suggest a suitable cutoff frequency for an antialiasing filter to be used in this application.

5.13 a. Consider a multi-degree-of-freedom robotic arm with flexible joints and links. The purpose of the manipulator is to accurately place a payload. Suppose that the second natural frequency (i.e., the natural frequency of the 2nd flexible mode) of bending of the robot, in the plane of its motion, is more than four times the first natural frequency.

Discuss pertinent issues of sensing and control (e.g., types and locations of the sensors, types of control, operating bandwidth, control bandwidth, sampling rate of sensing information) if the primary frequency of the payload motion is:

i. One-tenth of the first natural frequency of the robot.

ii. Very close to the first natural frequency of the robot.

iii. Twice the first natural frequency of the robot.

b. A single-link space robot is shown in Figure P5.13. The link is assumed to be uniform with length 10 m and mass 400 kg. The total mass of the end effector and the payload is also 400 kg. The robot link is assumed to be flexible while

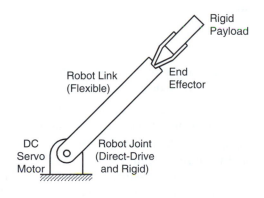

FIGURE P5.13
A single-link robotic manipulator.

the other components are rigid. The modulus of rigidity of bending deflection of the link in the plane of robot motion is known to be $EI = 8.25 \times 10^9$ N.m². The primary natural frequency of bending motion of a uniform cantilever beam with an end mass is given by

$$\omega_1 = \lambda_1^2 \sqrt{\frac{EI}{m}}$$

where
    m = mass per unit length
    λ_1 = mode shape parameter for mode 1

For (beam mass/end mass) = 1.0, it is known that $\lambda_1 l = 1.875$ where l = beam length. Give a suitable operating bandwidth for the robot manipulator. Estimate a suitable sampling rate for response measurements, to be used in feedback control. What is the corresponding control bandwidth, assuming that the actuator and the signal conditioning hardware can accommodate this bandwidth?

5.14 a. Define the following terms:
  - Sensor
  - Transducer
  - Actuator
  - Controller
  - Control system
  - Operating bandwidth of a control system
  - Control bandwidth
  - Nyquist frequency

b. Choose three practical dynamic systems each of which has at least one sensor, one actuator and a feedback controller.
  i. Briefly describe the purpose and operation of each dynamic system.
  ii. For each system give a suitable value for the operating bandwidth, control bandwidth, operating frequency range of the sensor, and sampling rate for sensor signal for feedback control. Clearly justify the values that you have given.

5.15 Discuss and contrast the following terms:
  a. Measurement accuracy
  b. Instrument accuracy
  c. Measurement error
  d. Precision

Also, for an analog sensor-transducer unit of your choice, identify and discuss various sources of error and ways to minimize or account for their influence.

5.16 a. Explain why mechanical loading error due to tachometer inertia can be significantly higher when measuring transient speeds than when measuring constant speeds.

b. A dc tachometer has an equivalent resistance $R_a = 20 \ \Omega$ in its rotor windings. In a position plus velocity servo system, the tachometer signal is connected to a feedback control circuit with equivalent resistance 2 kΩ. Estimate the percentage error due to electrical loading of the tachometer at steady state.

c. If the conditions were not steady, how would the electrical loading be affected in this application?

5.17 Briefly explain what is meant by the terms *systematic error* and *random error* of a measuring device. What statistical parameters may be used to quantify these two types of error? State, giving an example, how *precision* is related to error.

5.18 Four sets of measurements were taken on the same response variable of a process using four different sensors. The true value of the response was known to be constant. Suppose that the four sets of data are as shown in Figure P5.18(a-d). Classify these data sets, and hence the corresponding sensors, with respect to precision and deterministic (repeatable) accuracy.

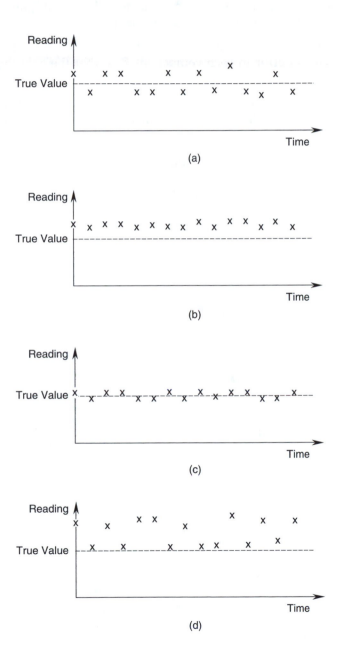

FIGURE P5.18
Four sets of measurements on the same response variable using different sensors.

5.19 The damping constant b of the mounting structure of a machine is determined experimentally. First, the spring stiffness k is determined by applying a static load and measuring the resulting displacement. Next, mass m of the structure is directly measured. Finally, damping ratio ζ is determined using the logarithmic decrement method, by conducting an impact test and measuring the free response of the structure. A model for the structure is shown in Figure P5.19. Show that the damping constant is given by

$$b = 2\zeta\sqrt{km}$$

If the allowable levels of error in the measurements of k, m, and ζ are ±2%, ±1%, and ±6% respectively, estimate a percentage absolute error limit for b.

5.20 Using the square root of sum of squares (SRRSS) method for error combination, determine the fractional error in each component x_i so that the contribution from each component to the overall error e_{SRSS} is the same.

5.21 A single-degree-of-freedom model of a robotic manipulator is shown in Figure P5.21(a). The joint motor has rotor inertia J_m. It drives an inertial load that has moment of inertia J_l

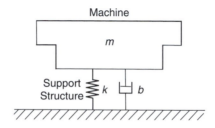

FIGURE P5.19
A model for the mounting structure of a machine.

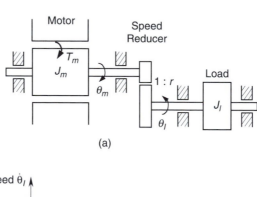

(a)

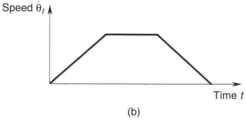

(b)

FIGURE P5.21
(a) A single-degree-of-freedom model of a robotic manipulator, (b) A typical reference (desired) speed trajectory for computed-torque control.

through a speed reducer of gear ratio 1 : r (Note: $r < 1$). The control scheme used in this system is the so-called feedforward control (strictly, *computed-torque control*) method. Specifically, the motor torque T_m that is required to accelerate or decelerate the load is computed using a suitable dynamic model and a desired motion trajectory for the manipulator, and the motor windings are excited so as to generate that torque. A typical trajectory would consist of a constant angular acceleration segment followed by a constant angular velocity segment, and finally a constant deceleration segment, as shown in Figure P5.21(b).

a. Neglecting friction (particularly bearing friction) and inertia of the speed reducer, show that a dynamic model for torque computation during accelerating and decelerating segments of the motion trajectory would be

$$T_m = (J_m + r^2 J_l)\ddot{\theta}_l / r$$

where $\ddot{\theta}_l$ is the angular acceleration of the load, hereafter denoted by α_l. Show that the overall system can be modeled as a single inertia rotating at the motor speed. Using this result, discuss the effect of gearing on a mechanical drive.

b. Given that $r = 0.1$, $J_m = 0.1$ kg $\cdot$ m², $J_l = 1.0$ kg $\cdot$ m², and $\alpha_l = 5.0$ rad/s², estimate the allowable error for these four quantities so that the combined error in the computed torque is limited to $\pm 4\%$ and so that each of the four quantities contributes equally toward this error in the computed T_m. Use the absolute value method for error combination.

c. Arrange the four quantities r, J_m, J_l, and α_l in the descending order of required accuracy for the numerical values given in the problem.

d. Suppose that $J_m = r^2 J_l$. Discuss the effect of error in r on the error in T_m.

5.22 An actuator (e.g., electric motor, hydraulic piston-cylinder) is used to drive a terminal device (e.g., gripper, hand, wrist with active remote center compliance) of a robotic manipulator. The terminal device functions as a force generator. A schematic diagram for the system is shown in Figure P5.22. Show that the displacement error e_x is related to the force error e_f through

$$e_f = \frac{x}{f}\frac{df}{dx}e_x$$

The actuator is known to be 100% accurate for practical purposes, but there is an initial position error δx_o (at $x = x_o$). Obtain a suitable transfer relation $f(x)$ for the terminal device so that the force error e_f remains constant throughout the dynamic range of the device.

5.23 a. Clearly explain why the "Square-Root of Sum of Squares" (SRSS) method of error combination is preferred to the "Absolute" method when the error parameters are assumed Gaussian and independent.

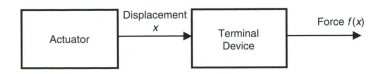

FIGURE P5.22
Block diagram for a terminal device of a robotic manipulator.

b. Hydraulic pulse generators (HPG) may be used in a variety of applications such as rock blasting, projectile driving, and seismic signal generation. In a typical HPG, water at very high pressure is supplied intermittently from an accumulator into the discharge gun, through a high-speed control valve. The pulsating water jet is discharged through a shock tube and may be used, for example, for blasting granite. A model for an HPG was found to be

$$E = aV\left(b + \frac{c}{V^{1/3}}\right)$$

in which

    E = hydraulic pulse energy (kJ)

    V = volume of blast burden (m³)

and, a, b, and c are model parameters that may be determined experimentally. Suppose that this model is used to estimate the blast volume of material (V) for a specific amount of pulse energy (E).

  i. Assuming that the estimation error values in the model parameters a, b, and c are independent and may be represented by appropriate standard deviations, obtain an equation relating these fractional errors e_a, e_b, and e_c, to the fractional error e_v of the estimated blast volume.

  ii. Assuming that $a = 2175.0$, $b = 0.3$, and $c = 0.07$ with consistent units, show that a pulse energy of $E = 219.0$ kJ can blast a material volume of approximately 0.6^3 m³. If $e_a = e_b = e_c = \pm 0.1$, estimate the fractional error e_v of this predicted volume.

5.24 The absolute method of error combination is suitable when the error contributions are additive (same sign). Under what circumstances would the square root of sum of squares (SRSS) method be more appropriate than the absolute method?

A simplified block diagram of a dc motor speed control system is shown in Figure P5.24. Show that in the Laplace domain, the fractional error e_y in the motor speed y is given by

$$e_y = -\frac{\tau s}{(\tau s + 1 + k)}e_\tau + \frac{(\tau s + 1)}{(\tau s + 1 + k)}e_k$$

in which

    e_τ = fractional error in the time constant τ

    e_k = fractional error in the open-loop gain k.

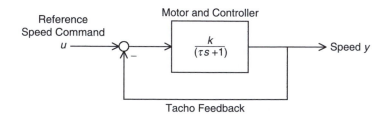

FIGURE P5.24
A dc motor speed control system.

The reference speed command u is assumed error free. Express the absolute error combination relation for this system in the frequency domain ($s = j\omega$). Using it show that

a. At low frequencies the contribution from the error in k will dominate and the error can be reduced by increasing the gain.

b. At high frequencies, k and τ will make equal contributions toward the speed error and the error cannot be reduced by increasing the gain.

5.25 a. Compare and contrast the "Absolute Error Method" with/against the "Square Root of Sum of Squares Method" in analyzing error combination of multicomponent systems. Indicate situations where one method is preferred over the other.

b. Figure P5.25 shows a schematic diagram of a machine that is used to produce steel billets. The molten steel in the vessel (called "tundish") is poured into the copper mould having a rectangular cross section. The mould has a steel jacket with channels to carry cooling water upwards around the copper mould. The mould, which is properly lubricated, is oscillated using a shaker (electromechanical or hydraulic) in order to facilitate stripping of the solidified steel inside it. A set of power-driven friction rollers is used to provide the withdrawal force for delivering the solidified steel strand to the cutting station. A billet cutter (torch or shear type) is used to cut the strand into billets of appropriate length.

The quality of the steel billets produced by this machine is determined on the basis of several factors, which include various types of cracks, deformation

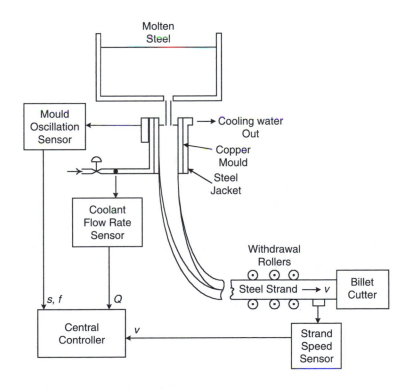

FIGURE P5.25
A steel-billet casting machine.

problems such as rhomboidity, and oscillation marks. It is known that the quality can be improved through proper control of the following variables:

Q = coolant (water) flow rate
v = speed of the steel strand (withdrawal speed)
s = stroke of the mould oscillations
f = cyclic frequency of the mould oscillations.

Specifically, these variables are measured and transmitted to the central controller of the billet-casting machine, which in turn generates proper control commands for the coolant-valve controller, the drive controller of the withdrawal rollers, and the shaker controller.

A nondimensional quality index q has been expressed in terms of the measured variables, as

$$q = \left[1 + \frac{s}{s_0} \sin \frac{\pi}{2}\left(\frac{f}{f_0 + f}\right)\right] \bigg/ (1 + \beta v/Q)$$

in which s_0, f_0, and β are operating parameters of the control system and are exactly known. Under normal operating conditions, the following conditions are (approximately) satisfied:

$$Q \approx \beta v$$

$$f \approx f_0$$

$$s \approx s_0$$

Note that if the sensor readings are incorrect, the control system will not function properly, and the quality of the billets will deteriorate. It is proposed to use the "Absolute Error Method" to determine the influence of the sensor errors on the billet quality.

i. Obtain an expression for the quality deterioration δq in terms of the fractional errors $\delta v/v$, $\delta Q/Q$, $\delta s/s$, and $\delta f/f$ of the sensor readings.

ii. If the sensor of the strand speed is known to have an error of 1% determine the allowable error percentages for the other three sensors so that there is equal contribution of error to the quality index from all four sensors, under normal operating conditions.

5.26 Consider the servo control system that is modeled as in Figure P5.24. Note that k is the equivalent gain and τ is the overall time constant of the motor and its controller.

a. Obtain an expression for the closed-loop transfer function $\frac{y}{u}$.

b. In the frequency domain, show that for equal contribution of parameter error towards the system response, we should have

$$\frac{e_k}{e_\tau} = \frac{\tau \omega}{\sqrt{\tau^2 \omega^2 + 1}}$$

where, fractional errors (or variations) are: for the gain, $e_k = \left|\frac{\delta k}{k}\right|$; and for the time constant, $e_\tau = \left|\frac{\delta \tau}{\tau}\right|$.

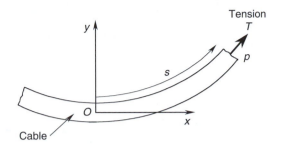

FIGURE P5.27
Cable tension example of error combination.

Using this relation explain why, at low frequencies the control system has a larger tolerance to error in τ than to that in k. Also show that, at very high frequencies the two error tolerance levels are almost equal.

5.27 Tension T at point P in a cable can be computed with the knowledge of the cable sag y, cable length s, cable weight w per unit length, and the minimum tension T_o at point O (see Figure P5.27). The applicable relationship is

$$1 + \frac{w}{T_o} y = \sqrt{1 + \frac{w^2}{T^2} s^2}$$

For a particular arrangement, it is given that $T_o = 100$ lbf. The following parameter values were measured:

$$w = 1 \text{ lb/ft}, \qquad s = 10 \text{ ft}, \qquad y = 0.412 \text{ ft}$$

Calculate the tension T.

In addition, if the measurements y and s each have 1% error and the measurement w has 2% error in this example, estimate the percentage error in T. Now suppose that equal contributions to error in T are made by y, s, and w. What are the corresponding percentage error values for y, s, and w so that the overall error in T is equal to the value computed in the previous part of the problem? Which of the three quantities y, s, and w should be measured most accurately, according to the equal contribution criterion?

5.28 In Problem 5.27, suppose that the percentage error values specified are in fact standard deviations in the measurements of y, s, and w. Estimate the standard deviation in the estimated value of tension T.

5.29 The quality control system in a steel rolling mill uses a proximity sensor to measure the thickness of rolled steel (steel gage) at every two feet along the sheet, and the mill controller adjustments are made on the basis of the last twenty measurements. Specifically, the controller is adjusted unless the probability that the mean thickness lies within ±1% of the sample mean, exceeds 0.99. A typical set of twenty measurements in millimeters is as follows:

5.10	5.05	4.94	4.98	5.10	5.12	5.07	4.96	4.99	4.95
4.99	4.97	5.00	5.08	5.10	5.11	4.99	4.96	4.90	4.10

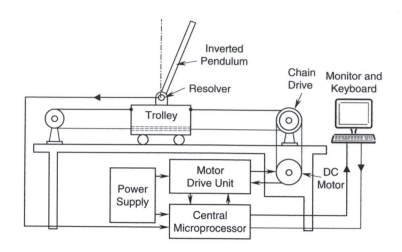

FIGURE P5.30
A microprocessor-controlled inverted pendulum—an application of statistical process control.

Check whether adjustments would be made in the gage controller on the basis of these measurements.

5.30 Dynamics and control of inherently unstable systems, such as rockets, can be studied experimentally using simple scaled-down physical models of the prototype systems. One such study is the classic inverted pendulum problem. An experimental setup for the inverted pendulum is shown in Figure P5.30. The inverted pendulum is supported on a trolley that is driven on a tabletop along a straight line, using a chain-and-sprocket transmission operated by a dc motor. The motor is turned by commands from a microprocessor that is interfaced with the drive system of the motor. The angular position of the pendulum rod is measured using a resolver and is transmitted (fed back) to the microprocessor. A strategy of statistical process control is used to balance the pendulum rod. Specifically, control limits are established from an initial set of measurement samples of the pendulum angle. Subsequently, if the angle exceeds one control limit, the trolley is accelerated in the opposite direction, using an automatic command to the motor drive unit. The control limits are also updated regularly. Suppose that the following twenty readings of the pendulum angle were measured (in degrees) after the system had operated for a few minutes:

| 0.5 | −0.5 | 0.4 | −0.3 | 0.3 | 0.1 | −0.3 | 0.3 | 4.0 | 0.0 |
| 0.4 | −0.4 | 0.5 | −0.5 | −5.0 | 0.4 | −0.4 | 0.3 | −0.3 | −0.1 |

Establish whether the system was in statistical control during the period in which the readings were taken. Comment on this method of control.

6

Analog Sensors and Transducers

Proper selection and integration of sensors and transducers are crucial in instrumenting a mechatronic system. Sensors may be used in a mechatronic system for a variety of purposes. In particular, output signals are measured for feedback control; input signals are measured for feedforward control; output signals are measured in system monitoring, tuning and supervisory control; and input-output signal pairs are measured for experimental modeling and evaluation of a plant. Ideal characteristics of sensors and transducers are indicated in Chapter 5. Even though real sensors and transducers can behave quite differently in practice, when developing a mechatronic system we should use the ideal behavior as a reference for the design specifications. In this chapter the significance of sensors and transducers in a mechatronic system is indicated; important criteria in selecting sensors and transducer for mechatronic applications are presented; and several representative sensors and transducers and their operating principles, characteristics, and applications are described.

6.1 Terminology

Potentiometers, differential transformers, resolvers, synchros, strain gages, tachometers, piezoelectric devices, bellows, diaphragms, flow meters, thermocouples, thermistors, and resistance temperature detectors (RTDs) are examples of sensors used in mechatronic systems. In a mechatronic control system, sensors are used to measure the system response, and it enables the controller to take corrective actions if the system does not operate properly. A mechatronic system may have unknown excitations and disturbances, which can make the associated tasks (performance monitoring, experimental modeling, control, etc.) particularly difficult. Removing such excitations at the source level is desirable, through proper design or system isolation. But, in the context of control, if these disturbances can be measured, or if some information about them is available, then they can be compensated for within the controller itself. This is in fact the approach of feedforward control. In summary, sensors may be used in a mechatronic system in several ways:

1. To measure the system outputs for feedback control.
2. To measure system inputs (desirable inputs, unknown inputs, and disturbances) for feedforward control.
3. To measure output signals for system monitoring, diagnosis, evaluation, parameter adjustment, and supervisory control.
4. To measure input and output signals for system testing and experimental modeling (i.e., for system identification).

The variable that is being measured is termed the *measurand*. Examples are acceleration and velocity of a vehicle, torque into robotic joint, temperature and pressure of a process plant, and current through an electric circuit. A measuring device passes through two stages while measuring a signal. First, the measurand is *felt* or *sensed*. Then, the measured signal is *transduced* (or converted) into the form of the device output. In fact the sensor, which "senses" the response automatically converts (i.e., transduce) this "measurement" into the sensor output—the response of the sensor element. For example, a piezoelectric accelerometer senses acceleration and converts it into an electric charge; an electromagnetic tachometer senses velocity and converts it into a voltage; and a shaft encoder senses a rotation and converts it into a sequence of voltage pulses. Hence, the terms sensor and transducer are used interchangeably to denote a sensor-transducer unit. Sensor and transducer stages are functional stages, and sometimes it is not easy or even feasible to separately identify physical elements associated with them. Furthermore, this separation is not very important in using existing devices. Proper separation of sensor and transducer stages (physically as well as functionally) can be crucial, however, when designing new measuring devices.

Typically, the measured signal is *transduced* (or converted) into a form that is particularly suitable for transmitting, recording, conditioning, processing, activating a controller, or driving an actuator. For this reason, output of a transducer is often an electrical signal. The measurand is usually an analog signal, because it represents the output of a dynamic system. For example, the charge signal from a piezoelectric accelerometer has to be converted into a voltage signal of appropriate level using a charge amplifier. For use in a digital controller it has to be digitized using an analog-to-digital converter (ADC). In digital transducers the transducer output is discrete. This facilitates the direct interface of a transducer with a digital processor.

A complex measuring device can have more than one sensing stage. Often, the measurand goes through several transducer stages before it is available for control and actuating purposes. Furthermore, filtering may be needed to remove measurement noise. Hence signal conditioning is usually needed between the sensor and the controller as well as the controller and the actuator. Charge amplifiers, lock-in amplifiers, power amplifiers, switching amplifiers, linear amplifiers, tracking filters, low-pass filters, high-pass filters, and notch filters are some of the signal-conditioning devices used in mechatronic systems. The subject of signal conditioning is studied in Chapter 5. In some literature, signal-conditioning devices such as electronic amplifiers are also classified as transducers. Since we are treating signal-conditioning and modification devices separately from measuring devices, this unified classification is avoided whenever possible, and the term *transducer* is used primarily in relation to measuring instruments. Note that it is somewhat redundant to consider electrical-to-electrical transducers as measuring devices, because electrical signals need conditioning only before they are used to carry out a useful task. In this sense, electrical-to-electrical transduction should be considered a "conditioning" function rather than a "measuring" function. Additional components, such as power supplies and surge-protection units, are often needed in mechatronic systems, but they are only indirectly related to control functions. Relays and other switching devices and modulators and demodulators may also be included.

Pure transducers depend on nondissipative coupling in the transduction stage. *Passive transducers* (sometimes called *self-generating transducers*) depend on their power transfer characteristics for operation, and do not need an external power source. It follows that pure transducers are essentially passive devices. Some examples are *electromagnetic, thermoelectric, radioactive, piezoelectric,* and *photovoltaic* transducers. External power is required to operate active sensors/transducers, and they do not depend on power conversion characteristics for their operation. A good example is a *resistive* transducer, such as a potentiometer, which depends on its power dissipation through a resistor to generate the output signal. Note that an active transducer requires a separate power source (power supply) for operation, whereas a passive transducer draws its power from a measured signal (measurand). Since passive transducers derive their energy almost entirely from the measurand, they generally tend to distort (or load) the measured signal to a greater extent than an active transducer would. Precautions can be taken to reduce such loading effects. On the other hand, passive transducers are generally simple in design, more reliable, and less costly. In the present classification of transducers, we are dealing with power in the immediate transducer stage associated with the measurand, not the power used in subsequent signal conditioning. For example, a piezoelectric charge generation is a passive process. But, a charge amplifier, which uses an auxiliary power source, would be needed in order to condition the generated charge.

Next, we will study several analog sensor-transducer devices that are commonly used in mechatronic system instrumentation. We will not attempt to present an exhaustive discussion of all types of sensors; rather, we will consider a representative selection. Such an approach is reasonable in view of the fact that even though the scientific principles behind various sensors may differ, many other aspects (e.g., performance parameters, signal conditioning, interfacing, and modeling procedures) can be common to a large extent.

6.1.1 Motion Transducers

By motion, we mean the four kinematic variables:

- Displacement (including position, distance, proximity, and size or gage)
- Velocity
- Acceleration
- Jerk

Note that each variable is the time derivative of the preceding one. Motion measurements are extremely useful in controlling mechanical responses and interactions in mechatronic systems. Numerous examples can be cited: The rotating speed of a work piece and the feed rate of a tool are measured in controlling machining operations. Displacements and speeds (both angular and translatory) at joints (revolute and prismatic) of robotic manipulators or kinematic linkages are used in controlling manipulator trajectory. In high-speed ground transit vehicles, acceleration and jerk measurements can be used for active suspension control to obtain improved ride quality. Angular speed is a crucial measurement that is used in the control of rotating machinery, such as turbines, pumps, compressors, motors, and generators in power-generating plants. Proximity sensors (to measure displacement) and accelerometers (to measure acceleration) are the two most common types of measuring devices used in machine protection systems for condition monitoring, fault detection, diagnostic, and on-line (often real-time) control of large and complex machinery. The accelerometer is often the only measuring device used in controlling dynamic test rigs. Displacement measurements are used for valve control in process applications. Plate thickness (or gage) is continuously monitored by the automatic gage control (AGC) system in steel rolling mills.

A one-to-one relationship may not always exist between a measuring device and a measured variable. For example, although strain gages are devices that measure strains (and, hence, stresses and forces), they can be adapted to measure displacements by using a suitable *front-end auxiliary sensor element*, such as a cantilever (or spring). Furthermore, the same measuring device may be used to measure different variables through appropriate data interpretation techniques. For example, piezoelectric accelerometers with built-in microelectronic integrated circuitry are marketed as piezoelectric velocity transducers. Resolver signals, which provide angular displacements, are differentiated to get angular velocities. Pulse-generating (or digital) transducers, such as optical encoders and digital tachometers, can serve as both displacement transducers and velocity transducers, depending on whether the absolute number of pulses are counted or the pulse rate is measured. Note that pulse rate can be measured either by counting the number of pulses during a unit interval of time or by gating a high-frequency clock signal through the pulse width. Furthermore, in principle, any force sensor can be used as an acceleration sensor, velocity sensor, or displacement sensor, depending on whether:

1. An inertia element (converting acceleration into force)
2. A damping element (converting velocity into force), or
3. A spring element (converting displacement into force)

respectively, is used as the front-end auxiliary sensor.

We might question the need for separate transducers to measure the four kinematic variables—displacement, velocity, acceleration, and jerk—because any one variable is related to any other through simple integration or differentiation. It should be possible, in theory, to measure only one of these four variables and use either analog processing (through analog circuit hardware) or digital processing (through a dedicated processor) to obtain any of the remaining motion variables. The feasibility of this approach is highly limited, however, and it depends crucially on several factors, including the following:

1. The nature of the measured signal (e.g., steady, highly transient, periodic, narrow-band, broad-band)
2. The required frequency content of the processed signal (or the frequency range of interest)
3. The signal-to-noise ratio (SNR) of the measurement
4. Available processing capabilities (e.g., analog or digital processing, limitations of the digital processor, and interface, such as the speed of processing, sampling rate, and buffer size)
5. Controller requirements and the nature of the plant (e.g., time constants, delays, complexity, hardware limitations)
6. Required accuracy in the end objective (on which processing requirements and hardware costs will depend)

For instance, differentiation of a signal (in the time domain) is often unacceptable for noisy and high-frequency narrow-band signals. In any event, costly signal-conditioning hardware might be needed for preprocessing prior to differentiating a signal. As a rule of thumb, in low-frequency applications (on the order of 1 Hz), displacement measurements generally provide good accuracies. In intermediate-frequency applications (less than 1 kHz), velocity measurement is usually favored. In measuring high-frequency motions with high noise levels, acceleration measurement is preferred. Jerk is particularly useful in ground transit (ride quality), manufacturing (forging, rolling, and similar impact-type operations), and shock isolation applications (for delicate and sensitive equipment).

6.2 Potentiometer

The potentiometer, or *pot*, is a displacement transducer. This active transducer consists of a uniform coil of wire or a film of high-resistive material—such as carbon, platinum, or conductive plastic—whose resistance is proportional to its length. A constant voltage v_{ref} is applied across the coil (or film) using an external dc voltage supply. The transducer output signal v_o is the dc voltage between the movable contact (wiper arm) sliding on the coil and one terminal of the coil, as shown schematically in Figure 6.1(a). Slider displacement x is proportional to the output voltage:

$$v_o = kx \tag{6.1}$$

This relationship assumes that the output terminals are in open-circuit; that is, a load of infinite impedance (or resistance in the present dc case) is present at the output terminal, so that the output current is zero. In actual practice, however, the load (the circuitry into which the pot signal is fed—e.g., conditioning, interfacing, processing, or control circuitry) has a finite impedance. Consequently, the output current (the current through the load) is nonzero, as shown in Figure 6.1(b). The output voltage thus drops to $\tilde{v}_o$, even if the reference voltage v_{ref} is assumed to remain constant under load variations (i.e., output impedance of the voltage source is zero); this consequence is known as the *loading effect* of the transducer. Under these conditions, the linear relationship given by Equation 6.1 would no longer be valid, causing an error in the displacement reading. Loading can affect the transducer reading in two ways: by changing the reference voltage (i.e., loading the voltage source) and by loading the transducer. To reduce these effects, a voltage source that is not seriously affected by load variations (e.g., a regulated or stabilized power supply that has a low output impedance), and data acquisition circuitry (including signal-conditioning circuitry) that has a high input impedance should be used.

The resistance of a potentiometer should be chosen with care. On the one hand, an element with high resistance is preferred because this results in reduced power dissipation for a given voltage, which has the added benefit of reduced thermal effects. On the other hand, increased resistance increases the output impedance of the potentiometer and results in loading nonlinearity error unless the load resistance is also increased proportionately. Low-resistance pots have resistances less than 10 Ω. High-resistance pots can have resistances on the order of 100 kΩ. Conductive plastics can provide high resistances—typically about 100 Ω/mm—and are increasingly used in potentiometers. Reduced friction (low mechanical loading), reduced wear, reduced weight, and increased resolution are advantages of using conductive plastics in potentiometers.

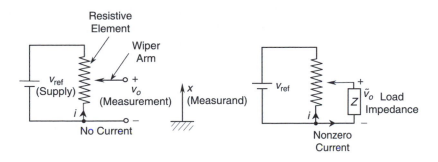

FIGURE 6.1

(a) Schematic diagram of a potentiometer, (b) Potentiometer loading.

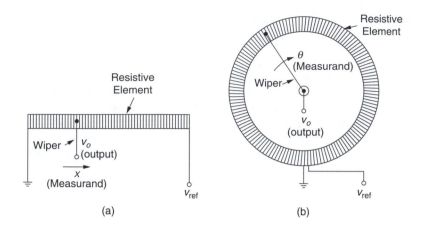

FIGURE 6.2
Practical potentiometer configurations for measuring: (a) Rectilinear motions; (b) Angular motions.

6.2.1 Rotary Potentiometers

Potentiometers that measure angular (rotatory) displacements are more common and convenient, because in conventional designs of rectilinear (translatory) potentiometers, the length of the resistive element has to be increased in proportion to the measurement range or stroke. Figure 6.2 presents schematic representations of translatory and rotary potentiometers. Helix-type rotary potentiometers are available for measuring absolute angles exceeding 360°. The same function may be accomplished with a standard single-cycle rotary pot simply by including a counter to record full 360° rotations.

Note that angular displacement transducers, such as rotary potentiometers, can be used to measure large rectilinear displacements on the order of 3 m. A cable extension mechanism may be employed to accomplish this. A light cable wrapped around a spool, which moves with the rotary element of the transducer, is the cable extension mechanism. The free end of the cable is attached to the moving object, and the potentiometer housing is mounted on a stationary structure. The device is properly calibrated so that as the object moves, the rotation count and fractional rotation measure will directly provide the rectilinear displacement. A spring-loaded recoil device, such as a spring motor, will wind the cable back when the object moves toward the transducer.

6.2.1.1 Loading Nonlinearity

Consider the rotatory potentiometer shown in Figure 6.3. Let us now discuss the significance of the "loading nonlinearity" error caused by a purely resistive load connected to the pot. For a general position θ of the pot slider arm, suppose that the resistance in the output (pick-off) segment of the coil is R_θ. Note that, assuming a uniform coil,

$$R_\theta = \frac{\theta}{\theta_{max}} R_c \tag{6.2}$$

where R_c is the total resistance of the potentiometer coil. The current balance at the sliding contact (node) point gives

$$\frac{v_{ref} - v_o}{R_c - R_\theta} = \frac{v_o}{R_\theta} + \frac{v_o}{R_L} \tag{i}$$

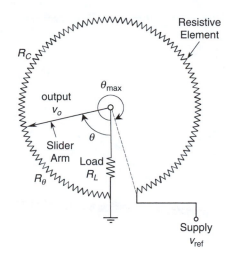

FIGURE 6.3
A rotatory potentiometer with a resistive load.

where R_L is the load resistance. Multiply throughout Equation i by R_c and use Equation 6.2; thus,

$$\frac{v_{\text{ref}} - v_o}{1 - \theta/\theta_{\text{max}}} = \frac{v_o}{\theta/\theta_{\text{max}}} + \frac{v_o}{R_L/R_c}$$

By using straightforward algebra, we have

$$\frac{v_o}{v_{\text{ref}}} = \left[\frac{(\theta/\theta_{\text{max}})(R_L/R_C)}{(R_L/R_C + (1 - R_L/R_C)(\theta/\theta_{\text{max}}) - (\theta/\theta_{\text{max}})^2)} \right] \tag{6.3}$$

Equation 6.3 is plotted in Figure 6.4. Loading error appears to be high for low values of the R_L/R_C ratio. Good accuracy is possible for $R_L/R_C > 10$, particularly for small values of $\theta/\theta_{\text{max}}$. It should be clear that the following actions can be taken to reduce loading error in pots:

1. Increase R_L/R_C (increase load impedance, reduce coil impedance).
2. Use pots only to measure small values of $\theta/\theta_{\text{max}}$ (or calibrate only a small segment of the element for linear reading).

The loading-nonlinearity error is defined by

$$e = \frac{(v_o/v_{\text{ref}} - \theta/\theta_{\text{max}})}{\theta/\theta_{\text{max}}} 100\% \tag{6.4}$$

The error at $\theta/\theta_{\text{max}} = 0.5$ is tabulated in Table 6.1. Note that this error is always negative. Using only a segment of the resistance element as the range of the potentiometer, is similar

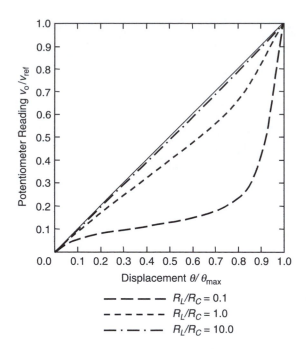

FIGURE 6.4
Loading nonlinearity in a potentiometer.

TABLE 6.1

Loading Nonlinearity Error in a Potentiometer

Load Resistance Ratio R_L/R_C	Loading Nonlinearity Error (e) at $\theta/\theta_{max} = 0.5$.
0.1	–71.4%
1.0	–20%
10.0	–2.4%

to adding two end resistors to the elements. It is known that this tends to linearize the pot. If the load resistance is known to be small, a voltage follower may be used at the potentiometer output to virtually eliminate loading error, since this arrangement provides a high load impedance to the pot and a low impedance at the output of the amplifier.

6.2.2 Performance Considerations

The potentiometer is a *resistively coupled transducer*. The force required to move the slider arm comes from the motion source, and the resulting energy is dissipated through friction. This energy conversion, unlike pure mechanical-to-electrical conversions, involves relatively high forces, and the energy is wasted rather than being converted into the output signal of the transducer. Furthermore, the electrical energy from the reference source is also dissipated through the resistor element (coil or film), resulting in an undesirable temperature rise and coil degradation. These are two obvious disadvantages of a potentiometer. In coil-type pots there is another disadvantage, which is the finite *resolution*.

A coil, instead of a straight wire, is used to increase the resistance per unit travel of the slider arm. But the slider contact jumps from one turn to the next in this case. Accordingly, the resolution of a coil-type potentiometer is determined by the number of turns in the coil. For a coil that has N turns, the resolution r, expressed as a percentage of the output range, is given by

$$r = \frac{100}{N}\%$$ (6.5)

Resolutions better (smaller) than 0.1% (i.e., 1000 turns) are available with coil potentiometers. Virtually infinitesimal (incorrectly termed infinite) resolutions are now possible with high-quality resistive film potentiometers, which use conductive plastics. In this case, the resolution is limited by other factors, such as mechanical limitations and signal-to-noise ratio. Nevertheless, resolutions on the order of 0.01 mm are possible with good rectilinear potentiometers.

Selection of a potentiometer involves many considerations. A primary factor is the required resolution for the specific application. Power consumption, loading, and size are also important factors. The following design example highlights some of these considerations.

Example 6.1

A high-precision mobile robot uses a potentiometer attached to the drive wheel to record its travel during autonomous navigation. The required resolution for robot motion is 1 mm, and the diameter of the drive wheel of the robot is 20 cm. Examine the design considerations for a standard (single-coil) rotary potentiometer to be used in this application.

SOLUTION
Assume that the potentiometer is directly connected (without gears) to the drive wheel. The required resolution for the pot is

$$\frac{0.1}{\pi \times 20} \times 100\% = 0.16\%$$

This resolution is feasible with a coil-type rotatory pot. From Equation 6.5 the number of turns in the coil = 100/0.16 = 625 turns. Assuming an average pot diameter of 10 cm and denoting the wire diameter by d, we have:

$$\text{Potentiometer circumference} = \pi \times 10 = 625 \times d$$

or

$$d = 0.5 \text{ mm}.$$

Now, taking the resistance of the potentiometer to be 5 Ω and the resistivity of the wire to be 4$\mu\Omega$ cm, the diameter D of the core of the coil is given by

$$\frac{4 \times 10^{-6} \times \pi D \times 625}{\pi(0.05/2)^2} = 5\,\Omega$$

NOTE   Resistivity = (resistance) × (cross-section area)/(length).

Hence,

$$D = 1.25 \text{ cm.}$$

___

The *sensitivity* of a potentiometer represents the change (Δv_o) in the output signal associated with a given small change ($\Delta \theta$) in the measurand (the displacement). This is usually nondimensionalized, using the actual value of the output signal (v_o) and the actual value of the displacement (θ). For a rotatory potentiometer in particular, the sensitivity S is given by

$$S = \frac{\Delta v_o}{\Delta \theta} \tag{6.6}$$

or, in the limit:

$$S = \frac{\partial v_o}{\partial \theta} \tag{6.7}$$

These relations may be nondimensionalized by multiplying by θ/v_o. An expression for S may be obtained by simply substituting Equation 6.3 into Equation 6.7.

Some limitations and disadvantages of the potentiometer as a displacement measuring device are given below:

1. The force needed to move the slider (against friction and arm inertia) is provided by the displacement source. This mechanical loading distorts the measured signal itself.
2. High-frequency (or highly transient) measurements are not feasible because of such factors as slider bounce, friction and inertia resistance, and induced voltages in the wiper arm and primary coil.
3. Variations in the supply voltage cause error.
4. Electrical loading error can be significant when the load resistance is low.
5. Resolution is limited by the number of turns in the coil and by the coil uniformity. This will limit small-displacement measurements.
6. Wearout and heating up (with associated oxidation) in the coil or film, and slider contact cause accelerated degradation.

There are several advantages associated with potentiometer devices, however, including the following:

1. They are relatively inexpensive.
2. Potentiometers provide high-voltage (low-impedance) output signals, requiring no amplification in most applications. Transducer impedance can be varied simply by changing the coil resistance and supply voltage.

Example 6.2

A rectilinear potentiometer was tested with its slider arm moving horizontally. It was found that at a speed of 1 cm/s, a driving force of 3×10^{-4} N was necessary to maintain the speed. At 10 cm/s, a force of 3×10^{-3} N was necessary. The slider weighs 5 gm, and the potentiometer stroke is ± 8 cm. If this potentiometer is used to measure the damped natural frequency of a simple mechanical oscillator of mass 10 kg, stiffness 10 N/m, and damping constant 2 N/m/s, estimate the percentage error due to mechanical loading. Justify this procedure for the estimation of damping.

SOLUTION
Suppose that the mass, stiffness, and damping constant of the simple oscillator are denoted by M, K, and B, respectively. The equation of free motion of the simple oscillator is given by

$$M\ddot{y} + B\dot{y} + Ky = 0 \tag{i}$$

where y denotes the displacement of the mass from the static equilibrium position. This equation is of the form

$$\ddot{y} + 2\zeta\omega_n\dot{y} + \omega_n^2 y = 0 \tag{ii}$$

where ω_n is the undamped natural frequency of the oscillator and ζ is the damping ratio. By direct comparison, it is seen that

$$\omega_n = \sqrt{\frac{K}{M}} \quad \text{and} \quad \zeta = \frac{B}{2\sqrt{MK}} \tag{ii}$$

The damped natural frequency is

$$\omega_d = \sqrt{1-\zeta^2}\,\omega_n \quad \text{for } 0 < \zeta < 1 \tag{iii}$$

Hence,

$$\omega_d = \sqrt{\left(1-\frac{B^2}{4MK}\right)\frac{K}{M}} \tag{iv}$$

Now, if the wiper arm mass and the damping constant of the potentiometer are denoted by m and b, respectively, the measured damped natural frequency (using the potentiometer) is given by

$$\tilde{\omega}_d = \sqrt{\left[1-\frac{(B+b)^2}{4(M+m)K}\right]\frac{K}{(M+m)}} \tag{v}$$

Assuming linear viscous friction (which is not quite realistic), the damping constant b of the potentiometer may be estimated as

$$b = \text{damping force/steady state velocity of the wiper}$$

For the present example, in the two speeds tested, we have

$$b_1 = 7\times10^{-4}/1\times10^{-2} \text{ N/m/s} = 7\times10^{-2} \text{ N/m/s at 1 cm/s}$$

$$b_2 = 3\times10^{-3}/10\times10^{-2} \text{ N/m/s} = 3\times10^{-2} \text{ N/m/s at 10 cm/s}$$

We should use some form of interpolation to estimate b for the actual measuring conditions. Let us estimate the average velocity of the wiper. The natural frequency of the oscillator is

$$\omega_n = \sqrt{\frac{10}{10}} = 1\,\text{rad/s} = \frac{1}{2\pi}\,\text{Hz}$$

The wiper travels a maximum distance of 4×8 cm $= 32$ cm in one cycle. Hence, the average operating speed of the wiper may be estimated as $32/(2\pi)$ cm/s, which is approximately equal to 5 cm/s. Therefore, the operating damping constant may be estimated as the average of b_1 and b_2:

$$b = 5 \times 10^{-2}\,\text{N/m/s}$$

With the foregoing numerical values.

$$\omega_d = \sqrt{\left(1 - \frac{2^2}{4 \times 10 \times 10}\right)\frac{10}{10}} = 0.99499\,\text{rad/s}$$

$$\tilde{\omega}_d = \sqrt{\left(1 - \frac{2.05^2}{4 \times 10.005 \times 10}\right)\frac{10}{10.005}} = 0.99449\,\text{rad/s}$$

$$\text{Percentage error} = \left[\frac{\tilde{\omega}_d - \omega_d}{\omega_d}\right] \times 100\% = -0.05\%$$

Although pots are primarily used as displacement transducers, they can be adapted to measure other types of signals, such as pressure and force, using appropriate auxiliary sensor (front-end) elements. For instance, a bourdon tube or bellows may be used to convert pressure into displacement, and a cantilever element may be used to convert force or moment into displacement.

6.2.3 Optical Potentiometer

The optical potentiometer, shown schematically in Figure 6.5(a), is a displacement sensor. A layer of photoresistive material is sandwiched between a layer of ordinary resistive material and a layer of conductive material. The layer of resistive material has a total resistance of R_c, and it is uniform (i.e., it has a constant resistance per unit length). This corresponds to the coil resistance of a conventional potentiometer. The photoresistive layer is practically an electrical insulator when no light is projected on it. The displacement of the moving object (whose displacement is being measured) causes a moving light beam to be projected on a small rectangular area of the photoresistive layer. This light-activated area attains a resistance of R_p, which links the resistive layer that is above the photoresistive layer and the conductive layer that is below the photoresistive layer. The supply voltage to the potentiometer is v_{ref}, and the length of the resistive layer is L. The light spot is projected at a distance x from one end of the resistive element, as shown in the figure.

An equivalent circuit for the optical potentiometer is shown in Figure 6.5(b). Here it is assumed that a load of resistance R_L is present at the output of the potentiometer, voltage across which being v_o. Current through the load is v_o/R_L. Hence, the voltage drop across

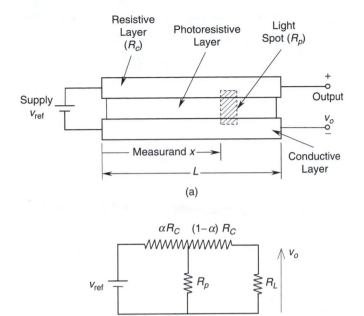

FIGURE 6.5
(a) An optical potentiometer, (b) Equivalent circuit ($\alpha = x/L$).

$(1 - \alpha)R_c + R_L$ which is also the voltage across R_p, is given by $[(1 - \alpha)R_c + R_L]v_o/R_L$. Note that $\alpha = x/L$, is the fractional position of the light spot. The current balance at the junction of the resistors in Figure 6.5(b) is

$$\frac{v_{\text{ref}} - [(1-\alpha)R_c + R_L]v_o/R_L}{\alpha R_c} = \frac{v_o}{R_L} + \frac{[(1-\alpha)R_c + R_L]v_o/R_L}{R_p}$$

which can be written as

$$\frac{v_o}{v_{\text{ref}}}\left\{\frac{R_c}{R_L} + 1 + \frac{x}{L}\frac{R_c}{R_p}\left[\left(1 - \frac{x}{L}\right)\frac{R_c}{R_L} + 1\right]\right\} = 1 \tag{6.8}$$

When the load resistance R_L is quite large in comparison to the element resistance R_c we have $R_c/R_L \simeq 0$. Hence, Equation 6.8 becomes

$$\frac{v_o}{v_{\text{ref}}} = \frac{1}{\left[\dfrac{x}{L}\dfrac{R_c}{R_p} + 1\right]} \tag{6.9}$$

This relationship is still nonlinear in v_o/v_{ref} vs. x/L. The nonlinearity decreases, however, with decreasing R_c/R_p. This is also seen from Figure 6.6 where Equation 6.9 is plotted for several values of R_c/R_p. Then, for the case of $R_c/R_p = 0.1$, the original Equation 6.8 is plotted in Figure 6.7, for several values of load resistance ratio. As expected, the behavior of the optical potentiometer becomes more linear for higher values of load resistance. This should also be clear from the Taylor series expansion of the right hand side of Equation 6.9.

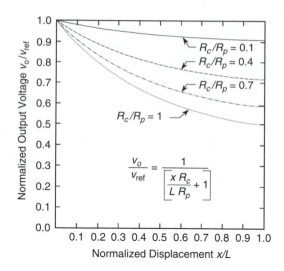

FIGURE 6.6
Behavior of the optical potentiometer at high load resistance.

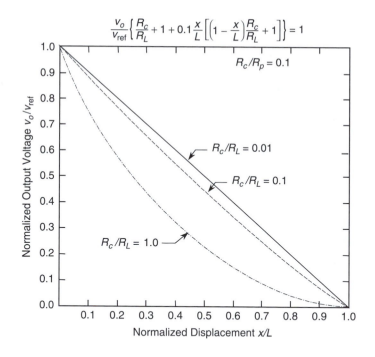

FIGURE 6.7
Behavior of the optical potentiometer for $R_c/R_p = 0.1$.

The potentiometer has disadvantages such as loading problems (both mechanical and electrical), limited speed of operation, considerable time constants, wear, noise, and thermal effects. Many of these problems arise from the fact that it is a "contact" device where its slider has to be in intimate contact with the resistance element of the pot, and also has to be an integral part of the moving object whose displacements need to be measured. Next we will consider several noncontact motion sensors.

6.3 Variable-Inductance Transducers

Motion transducers that employ the principle of electromagnetic induction are termed variable-inductance transducers. When the flux linkage (defined as magnetic flux density times the number of turns in the conductor) through an electrical conductor changes, a voltage is induced in the conductor. This, in turn, generates a magnetic field, which opposes the primary field. Hence, a mechanical force is necessary to sustain the change of flux linkage. If the change in flux linkage is brought about by a relative motion, the associated mechanical energy is directly converted (induced) into electrical energy. This is the basis of *electromagnetic induction,* and it is the principle of operation of electrical generators and variable-inductance transducers. Note that in these devices, the change of flux linkage is caused by a mechanical motion, and mechanical-to-electrical energy transfer takes place under near-ideal conditions. The induced voltage or change in inductance may be used as a measure of the motion. Variable-inductance transducers are generally electromechanical devices coupled by a magnetic field.

There are many different types of *variable-inductance transducers.* Three primary types can be identified:

1. Mutual-induction transducers
2. Self-induction transducers
3. Permanent-magnet transducers

Those variable-inductance transducers that use a nonmagnetized ferromagnetic medium to alter the reluctance (magnetic resistance) of the flux path are known as *variable-reluctance transducers.* Some of the mutual-induction transducers and most of the self-induction transducers are of this type. Permanent-magnet transducers are not considered variable-reluctance transducers.

6.3.1 Mutual-Induction Transducers

The basic arrangement of a mutual-induction transducer constitutes two coils, the *primary winding* and the *secondary winding.* One of the coils (primary winding) carries an alternating-current (ac) excitation, which induces a steady ac voltage in the other coil (secondary winding). The level (amplitude, rms value, etc.) of the induced voltage depends on the flux linkage between the coils. None of these transducers employ contact sliders or slip-rings and brushes as do resistively coupled transducers (potentiometer). Consequently, they will have an increased design life and low mechanical loading. In mutual-induction transducers, a change in the flux linkage is effected by one of two common techniques. One technique is to move an object made of ferromagnetic material within the flux path. This changes the reluctance of the flux path, with an associated change of the flux linkage in the secondary coil. This is the operating principle of the linear-variable differential transformer (LVDT), the rotatory-variable differential transformer (RVDT), and the mutual-induction proximity probe. All of these are, in fact, variable-reluctance transducers. The other common way to change the flux linkage is to move one coil with respect to the other. This is the operating principle of the resolver, the synchro-transformer, and some types of ac tachometer. These are not variable-reluctance transducers, however.

The motion can be measured by using the secondary signal in several ways. For example, the ac signal in the secondary coil may be "demodulated" by removing the *carrier signal* (i.e., the signal component at the excitation frequency) and directly measuring the resulting signal, which represents the motion. This method is particularly suitable for measuring transient motions. Alternatively, the amplitude or the rms (root-mean-square) value of the secondary (induced) voltage may be measured. Another method is to measure the change of inductance (or, reactance, which is equal to $Lj\omega$, since $v = L\frac{di}{dt}$) in the secondary circuit directly, by using a device such as an inductance bridge circuit see Chapter 4).

6.3.2   Linear-Variable Differential Transformer (LVDT)

Differential transformer is a noncontact displacement sensor, which does not possess many of the shortcomings of the potentiometer. It is a variable-inductance transducer, and is also a variable-reluctance transducer and a mutual-induction transducer. Furthermore, unlike the potentiometer, the differential transformer is a passive device. First we will discuss the linear-variable differential transformer, which is used for measuring rectilinear (or translatory) displacements. Next we will describe the rotatory-variable differential transformer (RVDT), which is used for measuring angular (or rotatory) displacements.

The LVDT is considered a passive transducer because the measured displacement provides energy for "changing" the induced voltage, even though an external power supply is used to energize the primary coil, which in turn induces a steady voltage at the carrier frequency in the secondary coil. In its simplest form (see Figure 6.8), the LVDT consists of an insulating, nonmagnetic "form" (a cylindrical structure on which a coil is wound, and is integral with the housing), which has a primary coil in the mid-segment and a secondary coil symmetrically wound in the two end segments, as depicted schematically in Figure 6.8(b). The housing is made of magnetized stainless steel in order to shield the sensor from outside fields. The primary coil is energized by an ac supply of voltage v_{ref}. This will generate, by mutual induction, an ac of the same frequency in the secondary coil. A core made of ferromagnetic material is inserted coaxially through the cylindrical form without actually touching it, as shown. As the core moves, the reluctance of the flux path changes. The degree of flux linkage depends on the axial position of the core. Since the two secondary coils are connected in series opposition (as shown in Figure 6.9), so that the potentials induced in the two secondary coil segments oppose each other, it is seen that the net induced voltage is zero when the core is centered between the two secondary winding segments. This is known as the *null position*. When the core is displaced from this position, a nonzero induced voltage will be generated. At steady state, the amplitude v_o of this induced voltage is proportional to the core displacement x in the linear (operating) region (see Figure 6.8(c)). Consequently, v_o may be used as a measure of the displacement. Note that because of opposed secondary windings, the LVDT provides the direction as well as the magnitude of displacement. If the output signal is not demodulated, the direction is determined by the phase angle between the primary (reference) voltage and the secondary (output) voltage, which includes the carrier signal.

For an LVDT to measure transient motions accurately, the frequency of the reference voltage (the carrier frequency) has to be at least ten times larger than the largest significant frequency component in the measured motion, and typically can be as high as 20 kHz. For quasi-dynamic displacements and slow transients on the order of a few hertz, a standard ac supply (at 60 Hz line frequency) is adequate. The performance (particularly sensitivity and accuracy) is known to improve with the excitation frequency, however. Since the amplitude of the output signal is proportional to the amplitude of the primary signal, the reference voltage should be regulated to get accurate results. In particular, the power source should have a low output impedance.

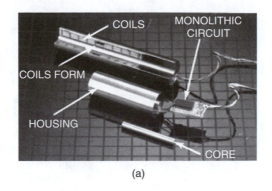

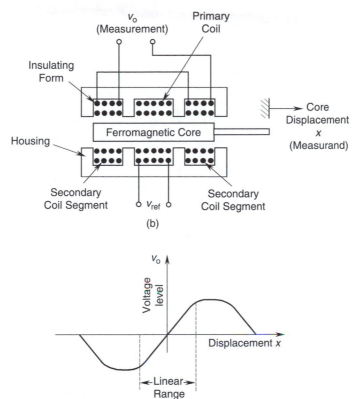

FIGURE 6.8
LVDT: (a) A commercial unit (Scheavitz Sensors, Measurement Specialties, Inc., Hampton, VA. With permission); (b) Schematic diagram; (c) A typical operating curve.

6.3.2.1 Phase Shift and Null Voltage

An error known as *null voltage* is present in some differential transformers. This manifests itself as a nonzero reading at the null position (i.e., at zero displacement). This is usually 90° out of phase from the main output signal and, hence, is known as *quadrature error*. Nonuniformities in the windings (unequal impedances in the two segments of the secondary winding) are a major reason for this error. The null voltage may also result from harmonic noise components in the primary signal and nonlinearities in the device. Null voltage is usually negligible (typically about 0.1% of the full scale). This error can be

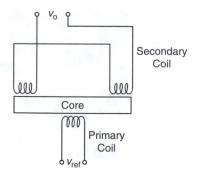

FIGURE 6.9
Series opposition connection of
secondary windings.

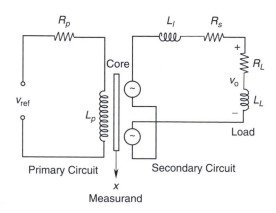

FIGURE 6.10
Equivalent circuit for a differential transformer.

eliminated from the measurements by employing appropriate signal-conditioning and calibration practices.

The output signal from a differential transformer is normally not in phase with the reference voltage. Inductance in the primary coil and the leakage inductance in the secondary coil are mainly responsible for this phase shift. Since demodulation involves extraction of the modulating signal by rejecting the carrier frequency component from the secondary signal, it is important to understand the size of this phase shift. An equivalent circuit for a differential transformer is shown in Figure 6.10. The resistance in the primary coil is denoted by R_p and the corresponding inductance is denoted by L_p. The total resistance of the secondary coil is R_s. The net leakage inductance, due to magnetic flux leakage, in the two segments is denoted by L_l. The load resistance is R_L and the load inductance is L_L. First, let us derive an expression for the phase shift in the output signal.

The magnetizing voltage in the primary coil is given by

$$v_p = v_{\text{ref}}\left[\frac{j\omega L_p}{R_p + j\omega L_p}\right] \tag{6.10}$$

in the frequency domain. Now suppose that the core, length L, is moved through a distance x from the null position. The induced voltage in one segment (a) of the secondary coil would be

$$v_a = v_p k_a (L/2 + x) \qquad (6.11)$$

and the induced voltage in the other segment (b) would be

$$v_b = v_p k_b (L/2 - x) \qquad (6.12)$$

Here, k_a and k_b are nonlinear functions of the position of the core, and also are complex functions of the frequency variable ω. Furthermore, each function will depend on the mutual-induction properties between the primary coil and the corresponding secondary-coil segment, through the core element. Due to series opposition connection of the two secondary segments, the net secondary voltage induced would be

$$v_s = v_a - v_b = v_p [k_a (L/2 + x) - k_b (L/2 - x)] \qquad (6.13)$$

In the ideal case, the two functions $k_a(.)$ and $k_b(.)$ would be identical. Then, at $x = 0$ we have $v_s = 0$. Hence, the null voltage would be zero in the ideal case. Suppose that, at $x = 0$, the magnitudes of $k_a(.)$ and $k_b(.)$ are equal, but there is a slight phase difference. Then, $k_a (L/2) - k_b (L/2)$ will have a small magnitude value, but its phase will be almost 90° with respect to both k_a and k_b. This is the *quadrature error*.

For small x, the Taylor series expansion of Equation 6.13 gives

$$v_s = v_p \left[k_a (L/2) + \frac{\partial k_a}{\partial x} (L/2) x - k_b (L/2) + \frac{\partial k_b}{\partial x} (L/2) x \right]$$

Then, assuming that $k_a(.) = k_b(.)$ and be denoted by $k_o(.)$ we have

$$v_s = 2 v_p \frac{\partial k_o}{\partial x} (L/2) x$$

or,

$$v_s = v_p k x \qquad (6.14)$$

where

$$k = 2 \frac{\partial k_o}{\partial x} (L/2) \qquad (6.15)$$

In this case, the net induced voltage is proportional to x and is given by

$$v_s = v_{ref} \left[\frac{j \omega L_p}{R_p + j \omega L_p} \right] k x \qquad (6.16)$$

It follows that the output voltage v_o at the load is given by

$$v_o = \left[\frac{j\omega L_p}{R_p + j\omega L_p} \right] \left[\frac{R_L + j\omega L_L}{(R_L + R_s) + j\omega(L_L + L_l)} \right] kx \tag{6.17}$$

Hence, for small displacements, the amplitude of the net output voltage of the LVDT is proportional to the displacement x. The phase *lead* at the output is, given by

$$\phi = 90° - \tan^{-1} \frac{\omega L_p}{R_p} + \tan^{-1} \frac{\omega L_L}{R_L} - \tan^{-1} \frac{\omega(L_L + L_l)}{R_L + R_s} \tag{6.18}$$

Note that the level of dependence of the phase shift on the load (including the secondary circuit) can be reduced by increasing the load impedance.

6.3.2.2 Signal Conditioning

Signal conditioning associated with differential transformers includes filtering and amplification. Filtering is needed to improve the signal-to-noise ratio of the output signal. Amplification is necessary to increase the signal strength for data acquisition and processing. Since the reference frequency (carrier frequency) is induced into (and embedded in) the output signal, it is also necessary to interpret the output signal properly, particularly for transient motions.

The secondary (output) signal of an LVDT is an amplitude-modulated signal where the signal component at the carrier frequency is modulated by the lower-frequency transient signal produced as a result of the core motion (x). Two methods are commonly used to interpret the crude output signal from a differential transformer: rectification and demodulation. Block diagram representations of these two procedures are given in Figure 6.11. In the first method (*rectification*) the ac output from the differential transformer is rectified to obtain a dc signal. This signal is amplified and then low-pass filtered to eliminate any high-frequency noise components. The amplitude of the resulting signal provides the transducer reading. In this method, phase shift in the LVDT output has to be checked separately to determine the direction of motion. In the second method (*demodulation*) the carrier frequency component is rejected from the output signal by comparing it with a phase-shifted and amplitude-adjusted version of the primary (reference) signal. Note that phase shifting is necessary because, as discussed before, the output signal is not in phase with the reference signal. The result is the modulating signal (proportional to x), which is subsequently amplified and filtered.

As a result of advances in miniature integrated circuit technology, differential transformers with built-in microelectronics for signal conditioning are commonly available today. A dc differential transformer uses a dc power supply (typically, ±15 V) to activate it. A built-in oscillator circuit generates the carrier signal. The rest of the device is identical to an ac differential transformer. The amplified full-scale output voltage can be as high as ±10 V. Let us illustrate the demodulation approach of signal conditioning for an LVDT, using an example.

Example 6.3

Figure 6.12 shows a schematic diagram of a simplified signal conditioning system for an LVDT. The system variables and parameters are as indicated in the figure.

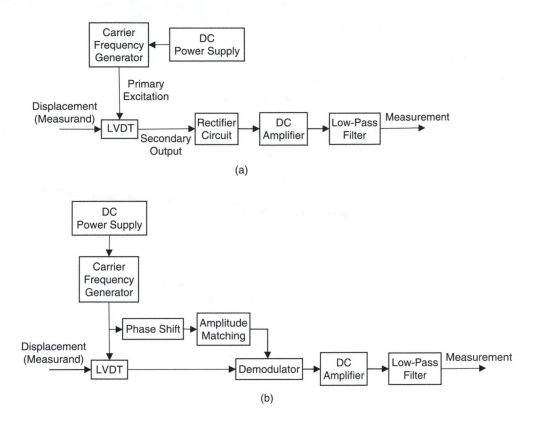

FIGURE 6.11
Signal-conditioning methods for a differential transformer: (a) Rectification; (b) Demodulation.

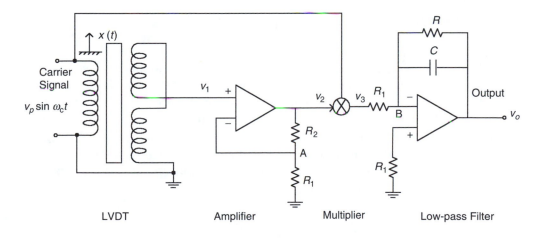

FIGURE 6.12
Signal conditioning system for an LVDT.

In particular,

> $x(t)$ = displacement of the LVDT core (measurand, to be measured)
> ω_c = frequency of the carrier voltage
> v_o = output signal of the system (measurement)

The resistances R_1, R_2, and R, and the capacitance C are as marked. In addition, we may introduce a transformer parameter r for the LVDT, as required.

i.   Explain the functions of the various components of the system shown in Figure 6.12.

ii. Write equations for the amplifier and filter circuits and, using them, give expressions for the voltage signals v_1, v_2, v_3, and v_o marked in Figure 6.12. Note that the excitation in the primary coil is $v_p \sin \omega_c t$.

iii. Suppose that the carrier frequency is ω_c = 500 rad/s and the filter resistance R = 100 kΩ. If no more than 5% of the carrier component should pass through the filter, estimate the required value of the filter capacitance C. Also, what is the useful frequency range (measurement bandwidth) of the system in rad/s, with these parameter values?

iv. If the displacement $x(t)$ is linearly increasing (i.e., speed is constant), sketch the signals $u(t)$, v_1, v_2, v_3, and v_o as functions of time.

SOLUTION
i.   The LVDT has a primary coil, which is excited by an ac voltage of $v_p \sin \omega_c t$. The ferromagnetic core is attached to the moving object whose displacement $x(t)$ is to be measured. The two secondary coils are connected in series opposition so that the LVDT output is zero at the null position, and that the direction of motion can be detected as well. The amplifier is a noninverting type. It amplifies the output of the LVDT which is an ac (carrier) signal of frequency ω_c that is modulated by the core displacement $x(t)$.

   The multiplier circuit produces the product of the primary (carrier) signal and the secondary (LVDT output) signal. This is an important step in demodulating the LVDT output.

   The product signal from the multiplier has a high-frequency ($2\omega_c$) carrier component, added to the modulating component ($x(t)$). The low-pass filter removes this unnecessary high-frequency component, to obtain the demodulated signal, which is proportional to the core displacement $x(t)$.

ii. **Non-Inverting Amplifier:** Note that the potentials at the + and − terminals of the opamp are nearly equal. Also, currents through these leads are nearly zero. (These are the two common assumptions used for an opamp; see Chapter 4). Then, the current balance at node A gives,

$$\frac{v_2 - v_1}{R_2} = \frac{v_1}{R_1}$$

Hence,

$$v_2 = kv_1 \tag{i}$$

with

$$k = \frac{R_1 + R_2}{R_1} = \text{amplifier gain} \qquad \text{(ii)}$$

Low-Pass Filter: Since the + lead of the opamp has approximately zero potential (ground), the voltage at point B is also approximately zero. The current balance for node B gives

$$\frac{v_3}{R_1} + \frac{v_o}{R} + C\dot{v}_o = 0$$

Hence,

$$\tau \frac{dv_o}{dt} + v_o = -\frac{R}{R_1} v_3 \qquad \text{(iii)}$$

where

$$\tau = RC = \text{filter time constant} \qquad \text{(iv)}$$

The transfer function of the filter is

$$\frac{v_o}{v_3} = -\frac{k_o}{(1 + \tau s)} \qquad \text{(v)}$$

with the filter gain

$$k_o = R/R_1 \qquad \text{(vi)}$$

In the frequency domain,

$$\frac{v_o}{v_3} = -\frac{k_o}{(1 + \tau j \omega)} \qquad \text{(vii)}$$

Finally, neglecting the phase shift in the LVDT, we have

$$v_1 = v_p r \, x(t) \sin \omega_c t$$

$$v_2 = v_p r k \, x(t) \sin \omega_c t$$

$$v_3 = v_p^2 r k \, x(t) \sin^2 \omega_c t$$

or

$$v_3 = \frac{v_p^2 r k}{2} x(t) [1 - \cos 2\omega_c t] \qquad \text{(viii)}$$

The carrier signal will be filtered out by the low-pass filter with an appropriate cut-off frequency. Then,

$$v_o = \frac{v_p^2 r k k_o}{2} x(t) \qquad \text{(ix)}$$

iii. $$\text{Filter magnitude} = \frac{k_o}{\sqrt{1+\tau^2\omega^2}} \qquad (6.19)$$

For no more than 5% of the carrier ($2\omega_c$) component to pass through, we must have

$$\frac{k_o}{\sqrt{1+\tau^2(2\omega_c)^2}} \leq \frac{5}{100} k_o$$

or,

$$\tau\omega_c \geq 10 \text{ (approximately)} \qquad (6.20)$$

$$\text{Pick } \tau\omega_c = 10.$$

With $R = 100$ kΩ, $\omega_c = 500$ rad/s we have

$$C \times 100 \times 10^3 \times 500 = 10$$

Hence,

$$C = 0.2 \text{ } \mu\text{F}$$

According to the carrier frequency (500 rad/s) we should be able to measure displacements $x(t)$ up to about 50 rad/s. But the flat region of the filter is up to about $\omega\tau = 0.1$, which with the present value of $\tau = 0.02$ sec, gives a bandwidth of only 5 rad/s.

iv. See Figure 6.13 for a sketch of various signals in the LVDT measurement system.

Advantages of the LVDT include the following:

1. It is essentially a noncontacting device with no frictional resistance. Near-ideal electromechanical energy conversion and light-weight core will result in very small resistive forces. Hysteresis (both magnetic hysteresis and mechanical backlash) is negligible.
2. It has low output impedance, typically on the order of 100 Ω. (Signal amplification is usually not needed beyond what is provided by the conditioning circuit.)
3. Directional measurements (positive/negative) are obtained.
4. It is available in small sizes (e.g., 1 cm long with maximum travel of 2 mm).
5. It has a simple and robust construction (inexpensive and durable).
6. Fine resolutions are possible (theoretically, infinitesimal resolution; practically, much better than that of a coil potentiometer).

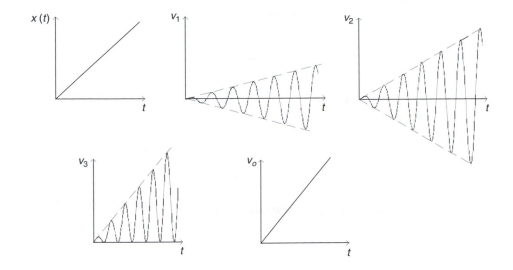

FIGURE 6.13
Nature of the signals at various locations in an LVDT measurement circuit.

6.3.3 Rotatory-Variable Differential Transformer (RVDT)

The RVDT operates using the same principle as the LVDT, except that in an RVDT, a rotating ferromagnetic core is used. The RVDT is used for measuring angular displacements. A schematic diagram of the device is shown in Figure 6.14(a), and a typical operating curve is shown in Figure 6.14(b). The rotating core is shaped such that a reasonably wide linear operating region is obtained. Advantages of the RVDT are essentially the same as those cited for the LVDT. Since the RVDT measures angular motions directly, without requiring nonlinear transformations (which is the case in resolvers, as will be discussed subsequently), its use is convenient in angular position servos. The linear range is typically ±40° with a nonlinearity error less than ±0.5% of full scale.

In variable-inductance devices, the induced voltage is generated through the rate of change of the magnetic flux linkage. Therefore, displacement readings are distorted by velocity; similarly, velocity readings are affected by acceleration. For the same displacement value, the transducer reading will depend on the velocity at that displacement. This error known as the *rate error*, increases with the ratio (cyclic velocity of the core)/(carrier frequency). Hence, the rate error can be reduced by increasing carrier frequency. The reason for this is as follows:

At high frequencies, the induced voltage due to the transformer effect (having frequency of the primary signal) is greater than the induced voltage due to the rate (velocity) effect of the moving member. Hence the error will be small. To estimate a lower limit for the carrier frequency in order to reduce rate effects, we may proceed as follows:

1. For an LVDT: Let

$$\frac{\text{Maximum speed of operation}}{\text{Stroke of LVDT}} = \omega_o \qquad (6.21)$$

   The excitation frequency of the primary coil should be chosen $5\omega_o$ or more.
2. For an RVDT: For ω_o above, use the maximum angular frequency of operation (of the rotor).

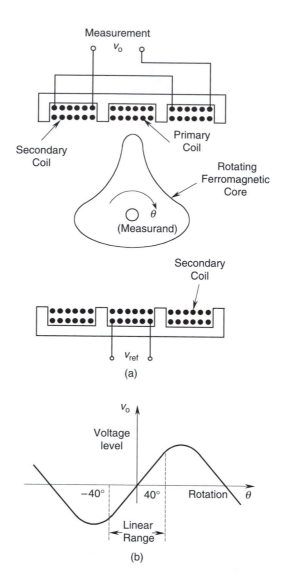

FIGURE 6.14
(a) Schematic diagram of an RVDT, (b) Operating curve.

6.3.4 Mutual-Induction Proximity Sensor

This displacement transducer operates on the mutual-induction principle. A simplified schematic diagram of such a device is shown in Figure 6.15(a). The insulating "E core" carries the primary winding in its middle limb. The two end limbs carry secondary windings, which are connected in series. Unlike the LVDT and the RVDT, the two voltages induced in the secondary winding segments are additive in this case. The region of the moving surface (target object) that faces the coils has to be made of ferromagnetic material so that as the object moves, the magnetic reluctance and the flux linkage will change. This, in turn, will change the induced voltage in the secondary coil, and this change is a measure of the displacement.

Note that, unlike the LVDT, which has an "axial" displacement configuration, the proximity probe has a "transverse" (or, lateral) displacement configuration. Hence, it is

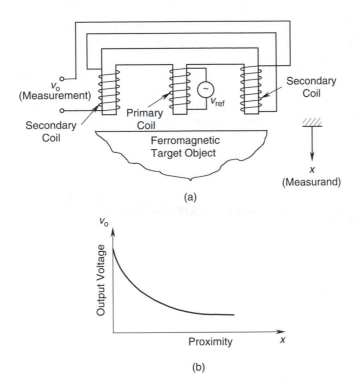

(a)

(b)

FIGURE 6.15
(a) Schematic diagram of a mutual-induction proximity sensor, (b) Operating curve.

particularly suitable for measuring transverse displacements or proximities of moving objects (e.g., transverse motion of a beam or whirling shaft). We can see from the operating curve shown in Figure 6.15(b) that the displacement-voltage relation of a proximity probe is nonlinear. Hence, these proximity sensors should be used only for measuring small displacements (e.g., a linear range of 5.0 mm or 0.2 in.), unless accurate nonlinear calibration curves are available. Since the proximity sensor is a noncontacting device, mechanical loading is small and the product life is high. Because a ferromagnetic object is used to alter the reluctance of the flux path, the mutual-induction proximity sensor is a variable-reluctance device. The operating frequency limit is about 1/10th the excitation frequency of the primary coil (carrier frequency). As for an LVDT, demodulation of the induced voltage (secondary voltage) is required to obtain direct (dc) output readings.

Proximity sensors are used in a wide variety of applications pertaining to noncontacting displacement sensing and dimensional gauging. Some typical applications are:

1. Measurement and control of the gap between a robotic welding torch head and the work surface
2. Gauging the thickness of metal plates in manufacturing operations (e.g., rolling and forming)
3. Detecting surface irregularities in machined parts
4. Angular speed measurement at steady state, by counting the number of rotations per unit time
5. Measurement of vibration in rotating machinery, gears, bearings, etc.

6. Level detection (e.g., in the filling and chemical process industries)
7. Monitoring of bearing assembly processes

Some mutual-induction displacement transducers use the relative motion between the primary coil and the secondary coil to produce a change in flux linkage. Two such devices are the resolver and the synchro-transformer. These are not variable-reluctance transducers because they do not employ a ferromagnetic moving element.

6.3.5 Resolver

This mutual-induction transducer is widely used for measuring angular displacements. A simplified schematic diagram of the resolver is shown in Figure 6.16. The *rotor* contains the primary coil. It consists of a single two-pole winding element energized by an ac supply voltage v_{ref}. The rotor is directly attached to the object whose rotation is being measured. The *stator* consists of two sets of windings placed 90° apart. If the angular position of the rotor with respect to one pair of stator windings is denoted by θ, the induced voltage in this pair of windings is given by

$$v_{o1} = av_{ref} \cos\theta \tag{6.22}$$

The induced voltage in the other pair of windings is given by

$$v_{o2} = av_{ref} \sin\theta \tag{6.23}$$

Note that these are amplitude-modulated signals—the carrier signal v_{ref} is modulated by the motion θ. The constant parameter a depends primarily on geometric and material characteristics of the device, for example, the ratio of the number of turns in the rotor and stator windings.

Either of the two output signals v_{o1} and v_{o2} may be used to determine the angular position in the first quadrant ($0 \leq \theta \leq 90°$). Both signals are needed, however, to determine the displacement (direction as well as magnitude) in all four quadrants ($0 \leq \theta \leq 360°$) without causing any ambiguity. For instance, the same sine value is obtained for both $90° + \theta$ and

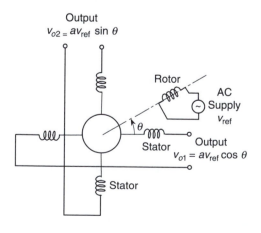

FIGURE 6.16
Schematic diagram of a resolver.

$90° - \theta$ (i.e., a positive rotation and a negative rotation from the 90° position), but the corresponding cosine values have opposite signs, thus providing the proper direction.

6.3.5.1 Demodulation

As for differential transformers (i.e., LVDT and RVDT) transient displacement signals of a resolver can be extracted by demodulating its modulated outputs. This is accomplished by filtering out the carrier signal, thereby extracting the modulating signal. The two output signals v_{o1} and v_{o2} of a resolver are termed quadrature signals. Suppose that the carrier (primary) signal is

$$v_{ref} = v_a \sin \omega t \tag{6.24}$$

The induced quadrate signals are:

$$v_{o1} = a v_a \cos \theta \sin \omega t \tag{6.25}$$

$$v_{o2} = a v_a \sin \theta \sin \omega t \tag{6.26}$$

Multiply each quadrature signal by v_{ref} to get

$$v_{m1} = v_{o1} v_{ref} = a v_a^2 \cos \theta \sin^2 \omega t = \frac{1}{2} a v_a^2 \cos \theta [1 - \cos 2\omega t] \tag{6.27}$$

$$v_{m2} = v_{o2} v_{ref} = a v_a^2 \sin \theta \sin^2 \omega t = \frac{1}{2} a v_a^2 \sin \theta [1 - \cos 2\omega t] \tag{6.28}$$

Since the carrier frequency ω is about 10 times the maximum frequency content in the angular displacement θ, one can use a low-pass filter with a cut-off set at $\omega/20$ in order to remove the carrier components in v_{m1} and v_{m2}. This gives the demodulated outputs

$$v_{f1} = \frac{1}{2} a v_a^2 \cos \theta \tag{6.29}$$

$$v_{f2} = \frac{1}{2} a v_a^2 \sin \theta \tag{6.30}$$

Note that Equation 6.29 and Equation 6.30 provide both $\cos \theta$ and $\sin \theta$, and hence magnitude and sign of θ.

6.3.5.2 Resolver with Rotor Output

An alternative form of resolver uses two ac voltages 90° out of phase, generated from a digital signal-generator board, to power the two coils of the stator. The rotor is the secondary winding in this case. The phase shift of the induced voltage determines the angular position of the rotor. An advantage of this arrangement is that it does not require slip rings and brushes to energize the windings, as needed in the previous arrangement where the rotor has the primary winding. But it will need some mechanism to pick off the output signal from the rotor. To illustrate this alternative design, suppose that the excitation signals in the two stator coils are

$$v_1 = v_a \sin \omega t \qquad\qquad (6.31)$$

$$v_2 = v_a \cos \omega t \qquad\qquad (6.32)$$

When the rotor coil is oriented at angular position θ with respect to the stator-coil pair 2, it will be at an angular position $\pi/2 - \theta$ from the stator-coil pair 1 (assuming that the rotor coil is in the first quadrant: $0 \leq \theta \leq \pi/2$). Hence the voltage induced by stator coil 1 in the rotor coil would be $v_a \sin \omega t \sin \theta$, and the voltage induced by the stator coil 2 in the rotor coil would be $v_a \cos \omega t \cos \theta$. It follows that the total induced voltage in the rotor coil is given by

$$v_r = v_a \sin \omega t \sin \theta + v_a \cos \omega t \cos \theta$$

or,

$$v_r = v_a \cos(\omega t - \theta) \qquad\qquad (6.33)$$

It is seen that the phase angle of the rotor output signal with respect to the stator excitation signals v_1 and v_2 will provide both magnitude and sign of the rotor position θ.

The output signals of a resolver are nonlinear (trigonometric) functions of the angle of rotation. (Historically, resolvers were used to compute trigonometric functions or to "resolve" a vector into orthogonal components). In robot control applications, this is sometimes viewed as a blessing. For computed torque control of robotic manipulators, for example, trigonometric functions of the joint angles are needed in order to compute the required input signals (joint torques). Consequently, when resolvers are used to measure joint angles in manipulators, there is an associated reduction in processing time because the trigonometric functions are available as direct measurements.

The primary advantages of the resolver include:

1. Fine resolution and high accuracy
2. Low output impedance (high signal levels)
3. Small size (e.g., 10 mm diameter)

Its main limitations are:

1. Nonlinear output signals (an advantage in some applications where trigonometric functions of the rotations are needed)
2. Bandwidth limited by supply frequency
3. Slip rings and brushes would be needed if complete and multiple rotations have to be measured (which adds mechanical loading and also creates component wear, oxidation, and thermal and noise problems)

6.3.6   Synchro Transformer

The synchro is somewhat similar in operation to the resolver. The main differences are that the synchro employs two identical rotor stator pairs, and each stator has three sets of windings, which are placed 120° apart around the rotor shaft. A schematic diagram for

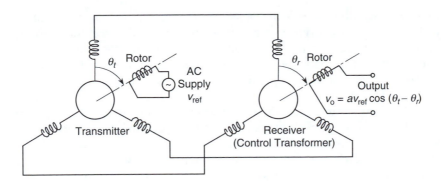

FIGURE 6.17
Schematic diagram of a synchro-transformer.

this arrangement is shown in Figure 6.17. Both rotors have single-phase windings and, contrary to popular belief, the synchro is essentially a single-phase device. One of the rotors is energized with an ac supply voltage v_{ref}. This induces voltages in the three winding segments of the corresponding stator. These voltages have different amplitudes, which depend on the angular position of the rotor but are in phase. This drive rotor-stator pair is known as the *transmitter*. The other rotor-stator pair is known as the *receiver* or the *control transformer*. Windings of the transmitter stator are connected correspondingly to the windings of the receiving stator, as shown in Figure 6.17. This induces a voltage v_o in the rotor of the receiver. Suppose that the angle between the drive rotor and one set of windings in its stator is denoted by θ_t. The resultant magnetic field on the receiver stator will make the same angle with the corresponding winding of that stator. If the receiver rotor is aligned with this direction (i.e., $\theta_r = \theta_t$), then the induced voltage v_o will be maximum. If the receiver rotor is placed at 90° to this resultant magnetic field, then $v_o = 0$. Therefore, an appropriate expression for the synchro output is

$$v_o = av_{ref}\cos(\theta_t - \theta_r) \qquad (6.34)$$

Synchros are operated near $\theta_r = \theta_t + 90°$, where the output voltage is zero. Hence, we define a new angle θ such that

$$\theta_r = \theta_t + 90° - \theta \qquad (6.35)$$

As a result, Equation 6.34 becomes

$$v_o = av_{ref}\sin\theta \qquad (6.36)$$

Synchro-transformers can be used to measure relative displacements between two rotating objects. For measuring absolute displacements, one of the rotors is attached to the rotating member (e.g., the shaft), while the other rotor is fixed to a stationary member (e.g., the bearing). As is clear from the previous discussion, a zero reading corresponds to the case where the two rotors are 90° apart.

Synchros have been used extensively in position servos, particularly for the position control of rotating objects. Typically, the input command is applied to the transmitter rotor. The receiver rotor is attached to the object that is being controlled. The initial physical orientations of the two rotors should ensure that for a given command, the desired position of the object corresponds to zero output voltage v_o, that is, when the two rotors are 90° apart.

In this manner, v_o can be used as the position error signal, which is fed into the control circuitry that generates a drive signal so as to compensate for the error (e.g., using proportional plus derivative control). For small angles θ, the output voltage may be assumed proportional to the angle. For large angles, inverse sine should be taken. Note that ambiguities arise when the angle θ exceeds 90°. Hence, synchro readings should be limited to ±90°. In this range, the synchro provides directional measurements. As for a resolver or LVDT, demodulation is required to extract transient measurements from the output signal. This is accomplished, as usual, by suppressing the carrier from the modulated signal, as demonstrated for the resolver.

The advantages and disadvantages of the synchro are essentially the same as those of the resolver. In particular, quadrature error (at null voltage) may be present because of impedance nonuniformities in the winding segments. Furthermore, velocity error (i.e., velocity-dependent displacement readings) is also a possibility. This may be reduced by increasing the carrier frequency, as in the case of a differential transformer and a resolver.

6.3.7 Self-Induction Transducers

These transducers are based on the principle of self-induction. Unlike mutual-induction transducers, only a single coil is employed. This coil is activated by an ac supply voltage v_{ref} of sufficiently high frequency. The current produces a magnetic flux, which is linked back with the coil. The level of flux linkage (or self-inductance) can be varied by moving a ferromagnetic object within the magnetic field. This movement changes the reluctance of the flux path and the inductance in the coil. The change in self-inductance, which can be measured using an inductance-measuring circuit (e.g., an inductance bridge; see Chapter 4), represents the measurand (displacement of the object). Note that self-induction transducers are usually *variable-reluctance* devices.

A typical self-induction transducer is a *self-induction proximity sensor*. A schematic diagram of this device is shown in Figure 6.18. This device can be used as a displacement sensor for transverse displacements. For instance, the distance between the sensor tip and ferromagnetic surface of a moving object, such as a beam or shaft, can be measured. Other applications include those mentioned for mutual-induction proximity sensors. High-speed displacement measurements can give rise to velocity error (rate error) when

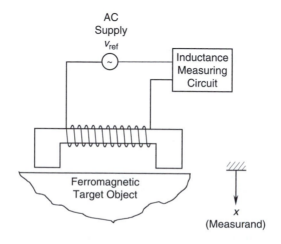

FIGURE 6.18
Schematic diagram of a self-induction proximity sensor.

variable-inductance displacement sensors (including self-induction transducers) are used. This effect may be reduced, as in other ac-activated variable-inductance sensors, by increasing the carrier frequency.

6.4 Permanent-Magnet Transducers

In discussing this third category of variable-inductance transducer, we will present several types of velocity transducers termed tachometers. A distinctive feature of permanent-magnet transducers is that they have a permanent magnet to generate a uniform and steady magnetic field. A relative motion between the magnetic field and an electrical conductor induces a voltage, which is proportional to the speed at which the conductor crosses the magnetic field (i.e., the rate of change of flux linkage). In some designs, a unidirectional magnetic field generated by a dc supply (i.e., an electromagnet) is used in place of a permanent magnet. Nevertheless, they are generally termed permanent-magnet transducers.

6.4.1 DC Tachometer

This is a permanent-magnet dc velocity sensor in which the principle of electromagnetic induction between a permanent magnet and a conducting coil is used. Depending on the configuration, either rectilinear speeds or angular speeds can be measured. Schematic diagrams of the two configurations are shown in Figure 6.19. Note that these are passive transducers, because the energy for the output signal v_o is derived from the motion

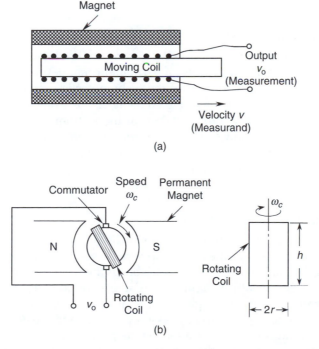

FIGURE 6.19
Permanent-magnet dc transducers: (a) Rectilinear velocity transducer; (b) DC tachometer.

(measured signal) itself. The entire device is usually enclosed in a steel casing to shield (isolate) it from ambient magnetic fields.

In the rectilinear velocity transducer (Figure 6.19(a)), the conductor coil is wound on a core and placed centrally between two magnetic poles, which produce a cross-magnetic field. The core is attached to the moving object whose velocity v must be measured. This velocity is proportional to the induced voltage v_o. Alternatively, a moving magnet and a fixed coil may be used as a dc tachometer. This arrangement is perhaps more desirable since it eliminates the need for any sliding contacts (slip rings and brushes) for the output leads, thereby reducing mechanical loading error, wear, and related problems.

The dc tachometer (or, tachogenerator) is a common transducer for measuring angular velocities. Its principle of operation is the same as that for a dc generator (or, back-driving of a dc motor). This principle of operation is illustrated in Figure 6.19(b). The rotor is directly connected to the rotating object. The output signal that is induced in the rotating coil is picked up as dc voltage v_o using a suitable *commutator* device—typically consisting of a pair of low-resistance carbon brushes—that is stationary but makes contact with the rotating coil through split slip rings so as to maintain the direction of the induced voltage the same throughout each revolution (see commutation in dc motors—Chapter 9). According to Faraday's law, the induced voltage is proportional to the rate of change of magnetic flux linkage. For a coil of height h and width $2r$ that has n turns, moving at an angular speed ω_c in a uniform magnetic field of flux density β, this is given by

$$v_o = (2nhr\beta)\omega_c = K\omega_c \qquad (6.37)$$

This proportionality between v_o and ω_c is used to measure the angular speed ω_c. The proportionality constant K is known as the *back-emf constant* or the *voltage constant*.

6.4.1.1 Electronic Commutation

Slip rings and brushes and associated drawbacks can be eliminated in a dc tachometer by using electronic commutation. In this case a permanent-magnet rotor together with a set of stator windings are used. The output of the tachometer is drawn from the stationary (stator) coil. It has to be converted to a dc signal using an electronic switching mechanism, which has to be synchronized with the rotation of the tachometer (see Chapter 9, under brushless dc motors). As a result of switching and associated changes in the magnetic field of the output signal, induced voltages known as *switching transients* will result, This is a drawback in electronic commutation.

6.4.1.2 Modeling and Design Example

A dc tachometer is shown schematically in Figure 6.20(a). The field windings are powered by dc voltage v_f. The across variable at the input port is the measured angular speed ω_i. The corresponding torque T_i is the through variable at the input port (See Chapter 2). The output voltage v_o of the armature circuit is the across variable at the output port. The corresponding current i_o is the through variable at the output port. Obtain a transfer-function model for this device. Discuss the assumptions needed to "decouple" this result into a practical input-output model for a tachometer. What are the corresponding design implications? In particular discuss the significance of the mechanical time constant and the electrical time constant of the tachometer.

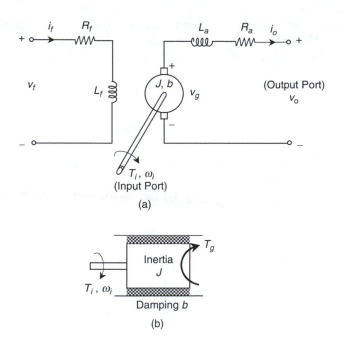

FIGURE 6.20
A dc tachometer example: (a) Equivalent circuit; (b) Armature free-body diagram.

SOLUTION
The generated voltage v_g at the armature (rotor) is proportional to the magnetic field strength of field windings (which, in turn, is proportional to the filed current i_f) and the speed of the armature ω_i. Hence, $v_g = K' i_f \omega_i$. Now assuming a constant field current, we have

$$v_g = K\omega_i \tag{i}$$

The rotor magnetic torque T_g, which resists that applied torque T_i, is proportional to the magnetic field strengths of the field windings and armature windings. Consequently, $T_g = K' i_f i_o$. Since i_f is assumed constant, we get

$$T_g = K i_o \tag{ii}$$

Note that the same constant K is used in both Equation i and Equation ii. This is valid when the same units are used to measure mechanical power and electrical power and when the internal energy-dissipation mechanisms are not significant in the associated internal coupling. The equation for the armature circuit is

$$v_o = v_g - R_a i_o - L_a \frac{di_o}{dt} \tag{iii}$$

where R_a is the armature resistance and L_a is the *leakage inductance* in the armature circuit. With reference to Figure 6.20(b), Newton's second law for a tachometer armature having inertia J and damping constant b is expressed as

$$J \frac{d\omega_i}{dt} = T_i - T_g - b\omega_i \tag{iv}$$

Now Equation i is substituted into Equation iii in order to eliminate v_g. Similarly, Equation ii is substituted into Equation iv in order to eliminate T_g. Next, the time derivatives are replaced by the Laplace variable s. This results in the two algebraic relations:

$$v_o = K\omega_i - (R_a + sL_a)i_o \tag{v}$$

$$(b + sJ)\omega_i = T_i - Ki_o \tag{vi}$$

Note that the variables v_i, i_o, ω_i, and T_i in Equation v and Equation vi are actually Laplace transforms (functions of s), not functions of t, as in Equations i through iv. Finally, i_o in Equation v is eliminated using Equation vi. This gives the matrix transfer function relation

$$\begin{bmatrix} v_o \\ i_o \end{bmatrix} = \begin{bmatrix} K + (R_a + sL_a)(b + sJ)/K & -(R_a + sL_a)/K \\ -(b + sJ)/K & 1/K \end{bmatrix} \begin{pmatrix} \omega_i \\ T_i \end{pmatrix} \tag{vii}$$

The corresponding frequency domain relations are obtained by replacing s with $j\omega$, where ω represents the angular frequency (radians/second) in the frequency spectrum of a signal.

Even though transducers are more accurately modeled as two-port elements, which have two variables associated with each port, it is useful and often essential, for practical reasons, to relate just one input variable (measurand) and one output variable (measurement) so that only one (scalar) transfer function relating these two variables need be specified. This assumes some form of decoupling in the true model. If this assumption does not hold in the range of operation of the transducer, a measurement error would result. In particular, for a tachometer, we like to express the output voltage v_o in terms of the measured speed ω_i. In this case, the off-diagonal term $-(R_a + sL_a)/K$ in Equation vii has to be neglected. This is valid when the tachometer gain parameter K is large and the armature resistance R_a is negligible, since the leakage inductance L_a is negligible in any case for most practical purposes. Note from Equation i and Equation ii that the tachometer gain K can be increased by increasing the field current i_f. This will not be feasible if the field windings are already saturated, however. Furthermore, K (or K') depends on parameters such as number of turns and dimensions of the stator windings and magnetic properties of the stator core. Since there is a limitation on the physical size of the tachometer and the types of materials used in the construction, it is clear that K cannot be increased arbitrarily. The instrument designer should take such factors into consideration in developing a design that is optimal in many respects. In practical transducers, the operating range is specified in order to minimize the effect of coupling terms, and the residual errors are accounted for by using correction curves. This approach is more convenient than using a coupled model, which introduces three more (scalar) transfer functions (in general) into the model.

Another desirable feature for practical transducers is to have a *static* (algebraic, non-dynamic) input/output relationship so that the output instantly reaches the input value (or the measured variable). In this case, the transducer transfer function is a pure gain. This happens when the transducer time constants are small (i.e., the transducer bandwidth is high). In the present tachometer example, it is clear from Equation vii that the transfer-function relations become static (frequency-independent) when both electrical time constant:

$$\tau_e = \frac{L_a}{R_a} \tag{6.38}$$

and mechanical time constant:

$$\tau_m = \frac{J}{b} \tag{6.39}$$

are negligibly small. The electrical time constant is usually an order of magnitude smaller than the mechanical time constant. Hence, one must first concentrate on the mechanical time constant. Note from Equation 6.39 that τ_m can be reduced by decreasing rotor inertia and increasing rotor damping. Unfortunately, rotor inertia depends on rotor dimensions, and this determines the gain parameter K, as we saw earlier. Hence, we face some constraint in reducing K. Furthermore, when the rotor size is reduced (in order to reduce J), the number of turns in the windings has to be reduced as well. Then, the air gap between the rotor and the stator will be less uniform, which will create a voltage ripple in the induced voltage (tachometer output). The resulting measurement error can be significant. Next turning to damping, it is intuitively clear that if we increase b, it will require a larger torque T_i to drive the tachometer, and this will load the measured object, which generates the measurand ω_i, possibly affecting the measurand itself. Furthermore, increased damping will result in increased thermal problems. Hence, increasing b also has to be done cautiously. Now, going back to Equation vii, we note that the dynamic terms in the transfer function between ω_i and v_o decrease as K is increased. So we notice that increasing K has two benefits: reduction of coupling and reduction of dynamic effects (i.e., increasing the useful frequency range and bandwidth or speed of response).

6.4.1.3 Loading Considerations

The torque required to drive a tachometer is proportional to the current generated (in the dc output). The associated proportionality constant is the *torque constant*. With consistent units, in the case of ideal energy conversion, this constant is equal to the voltage constant. Since the tachometer torque acts on the moving object whose speed is measured, high torque corresponds to high mechanical loading, which is not desirable. Hence, it is needed to reduce the tachometer current as much as possible. This can be realized by making the input impedance of the signal-acquisition device (i.e., hardware for voltage reading and interface) for the tachometer as large as possible. Furthermore, distortion in the tachometer output signal (voltage) can result because of the reactive (inductive and capacitive) loading of the tachometer. When dc tachometers are used to measure transient velocities, some error will result from the rate (acceleration) effect. This error generally increases with the maximum significant frequency that must be retained in the transient velocity signal, which in turn depends on the maximum speed that has to be measured. All these types of error can be reduced by increasing the load impedance.

For illustration, consider the equivalent circuit of a tachometer with an impedance load Z_L connected to the output port of the armature circuit in Figure 6.20(a). The induced voltage $K\omega_i$ is represented by a voltage source. Note that the constant K depends on the coil geometry, the number of turns, and the magnetic flux density (see Equation 6.37). Coil resistance is denoted by R_a, and leakage inductance is denoted by L_a. The load impedance is Z_L. From straightforward circuit analysis in the frequency domain, the output voltage at the load is given by

$$v_o = \left[\frac{Z_L}{R_a + j\omega L_a + Z_L} \right] k\omega_i \tag{6.40}$$

It can be seen that because of the leakage inductance, the output signal attenuates more at higher frequencies ω of the velocity transient. In addition, a loading error is present.

If Z_L is much larger than the coil impedance, however, the ideal proportionality, as given by $v_o = K\omega_i$ is achieved.

A *digital tachometer* is a velocity transducer, which is governed by somewhat different principles. It generates voltage pulses at a frequency proportional to the angular speed. Hence, it is considered a digital transducer, as discussed in Chapter 7.

6.4.2 Permanent-Magnet AC Tachometer

This device has a permanent magnet rotor and two separate sets of stator windings as schematically shown in Figure 6.21(a). One set of windings is energized using an ac reference (carrier) voltage. Induced voltage in the other set of windings is the tachometer output. When the rotor is stationary or moving in a quasi-static manner, the output voltage is a constant-amplitude signal much like the reference voltage. As the rotor moves at a finite speed, an additional induced voltage, which is proportional to the rotor speed, is generated in the secondary winding. This is due to the rate of change of flux linkage into the secondary coil from the rotating magnet. The overall output from the secondary coil is an amplitude-modulated signal whose amplitude is proportional to the rotor speed. For transient velocities, it will be necessary to demodulate this signal in order to extract the transient velocity signal (i.e., the modulating signal) from the overall (modulated) output. The direction of velocity is determined from the phase angle of the modulated signal with respect to the carrier signal. Note that in an LVDT, the amplitude of the ac magnetic flux (linkage) is altered by the position of the ferromagnetic core. But in an ac permanent-magnet tachometer, a dc magnetic flux is generated by the magnetic rotor, and when the rotor is stationary it does not induce a voltage in the coils. The flux linked with the stator windings changes due to the rotation of the rotor, and the rate of change of linked flux is proportional to the speed of the rotor.

For low-frequency applications (5 Hz or less), a standard ac supply at line frequency (60 Hz) may be adequate to power an ac tachometer. For moderate-frequency applications, a 400 Hz supply may be used. For high-frequency (high-bandwidth) applications a high-frequency signal generator (oscillator) may be used as the primary signal. In high-bandwidth applications, carrier frequencies as high as 1.5 kHz are commonly used. Typical sensitivity of an ac permanent-magnet tachometer is on the order of 50–100 mV/rad/s.

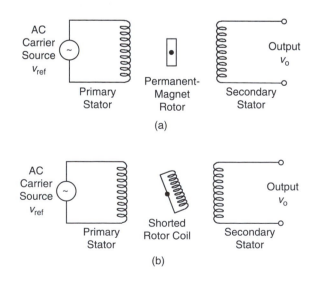

FIGURE 6.21
(a) An ac permanent-magnet tachometer, (b) An ac induction tachometer.

6.4.3 AC Induction Tachometer

This tachometer is similar in construction to a two-phase induction motor (see Chapter 9). The stator arrangement is identical to that of the ac permanent-magnet tachometer, as presented before. The rotor has windings, which are shorted and not energized by an external source, as shown in Figure 6.21(b). One of the stator windings is powered by an ac supply. This induces a voltage in the rotor windings, and it is a modulated signal. The high-frequency (carrier) component of this induced signal is due to the direct transformer action of the primary ac. The other (modulating) component is induced by the speed of rotation of the rotor, and its magnitude is proportional to the speed of rotation. The nonenergized stator (secondary) winding provides the output of the tachometer. This voltage output is a result of both the stator (primary) windings and the rotor windings. As a result, the tachometer output has a carrier ac component whose frequency is the same as the primary signal frequency, and a modulating component, which is proportional to the speed of rotation. Demodulation would be needed to extract the component that is proportional to the angular speed of the rotor.

The main advantage of ac tachometers over their conventional dc counterparts is the absence of slip-ring and brush devices, since the output is obtained from the stator. In particular, the signal from a dc tachometer usually has a voltage ripple, known as the *commutator ripple* or *brush noise*, which are generated as the split ends of the slip ring pass over the brushes, and as a result of contact bounce, etc. The frequency of the commutator ripple is proportional to the speed of operation; consequently, filtering it out using a notch filter is difficult (a speed-tracking notch filter would be needed). Also, there are problems with frictional loading and contact bounce in dc tachometers, and these problems are absent in ac tachometers. Note, however, that a dc tachometer with electronic commutation does not use slip rings and brushes. But they produce switching transients, which are undesirable.

As for any sensor, the noise components will dominate at low levels of output signal. In particular, since the output of a tachometer is proportional to the measured speed, at low speeds, the level of noise, as a fraction of the output signal, can be large. Hence, removal of noise takes an increased importance at low speeds.

It is known that at high speeds the output from an ac tachometer is somewhat nonlinear (primarily due to the saturation effect). Furthermore, signal demodulation is necessary, particularly for measuring transient speeds. Another disadvantage of ac tachometers is that the output signal level depends on the supply voltage; hence, a stabilized voltage source, which has a very small output-impedance is necessary for accurate measurements.

6.4.4 Eddy Current Transducers

If a conducting (i.e., low-resistivity) medium is subjected to a fluctuating magnetic field, eddy currents are generated in the medium. The strength of eddy currents increases with the strength of the magnetic field and the frequency of the magnetic flux. This principle is used in eddy current proximity sensors. Eddy current sensors may be used as either dimensional gauging devices or displacement sensors.

A schematic diagram of an eddy current proximity sensor is shown in Figure 6.22(a). Unlike variable-inductance proximity sensors, the target object of the eddy current sensor does not have to be made of a ferromagnetic material. A conducting target object is needed, but a thin film of conducting material—such as household aluminum foil glued onto a nonconducting target object—would be adequate. The probe head has two identical coils, which will form two arms of an impedance bridge. The coil closer to the probe face is the *active coil*. The other coil is the *compensating coil*. It compensates for ambient changes, particularly thermal effects. The remaining two arms of the bridge will consist of purely resistive

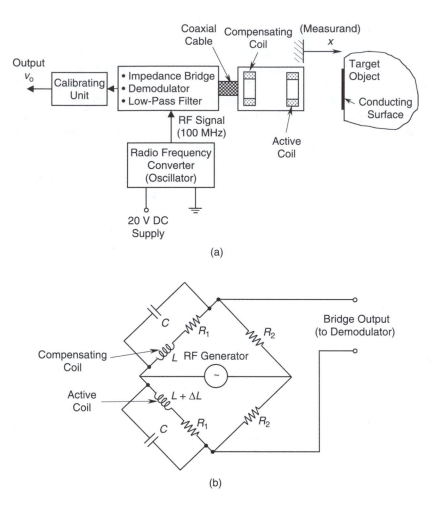

FIGURE 6.22
Eddy current proximity sensor: (a) Schematic diagram; (b) Impedance bridge.

elements (see Figure 6.22(b)). The bridge is excited by a radio-frequency voltage supply. The frequency may range from 1–100 MHz. This signal is generated from a radio-frequency converter (an oscillator) that is typically powered by a 20 V DC supply. When the target (sensed) object is absence, the output of the impedance bridge is zero, which corresponds to the balanced condition. When the target object is moved close to the sensor, eddy currents are generated in the conducting medium because of the radio-frequency magnetic flux from the active coil. The magnetic field of the eddy currents opposes the primary field, which generates these currents. Hence, the inductance of the active coil increases, creating an imbalance in the bridge. The resulting output from the bridge is an amplitude-modulated signal containing the radio-frequency carrier. This signal can be demodulated by removing the carrier. The resulting signal (modulating signal) measures transient displacement of the target object. Low-pass filtering is used to remove high-frequency leftover noise in the output signal once the carrier is removed. For large displacements, the output is not linearly related to the displacement. Furthermore, the sensitivity of an eddy current probe depends nonlinearly on the nature of the conducting medium, particularly the resistivity. For example, for low resistivities, sensitivity increases with resistivity; for high resistivities, sensitivity decreases with resistivity. A calibrating unit is usually available with commercial eddy current sensors to accommodate various target objects and nonlinearities. The gage factor is usually expressed in volts/millimeter. Note that eddy current probes can also be used to measure resistivity and surface hardness (which affects resistivity) in metals.

The facial area of the conducting medium on the target object has to be slightly larger than the frontal area of the eddy current probe head. If the target object has a curved surface, its radius of curvature has to be at least four times the diameter of the probe. These are not serious restrictions, because the typical diameter of a probe head is about 2 mm. Eddy current sensors are medium-impedance devices; 1000 Ω output impedance is typical. Sensitivity is on the order of 5 V/mm. Since the carrier frequency is very high, eddy current devices are suitable for highly transient displacement measurements—for example, bandwidths up to 100 kHz. Another advantage of the eddy current sensor is that it is a noncontacting device; hence, there is no mechanical loading on the moving (target) object.

6.5 Variable-Capacitance Transducers

Variable-inductance devices and variable capacitance devices are *variable-reactance* devices. (Note that the *reactance* of an inductance L is given by $j\omega L$ and that of a capacitance C is given by $1/(j\omega C)$, since $v = L\frac{di}{dt}$ and $i = C\frac{dv}{dt}$). For this reason, capacitive transducers fall into the category of *reactive* transducers. They are typically high impedance sensors, particularly at low frequencies, as clear from the impedance (reactance) expression for a capacitor. Also, capacitive sensors are noncontacting devices in the common usage. They require specific signal conditioning hardware. In addition to analog capacitive sensors, digital (pulse-generating) capacitive transducers such as digital tachometers are also available.

A capacitor is formed by two plates, which can store an electric charge. The charge generates a potential difference, which may be maintained using an external voltage. The capacitance C of a two-plate capacitor is given by

$$C = \frac{kA}{x} \tag{6.41}$$

where A is the common (overlapping) area of the two plates, x is the gap width between the two plates, and k is the dielectric constant (or, permittivity $k = \varepsilon = \varepsilon_r \varepsilon_o$; ε_r = relative permittivity, ε_o = permittivity in vacuum) which depends on dielectric properties of the medium between the two plates. A change in any one of the three parameters in Equation 6.41 may be used in the sensing process, for example, to measure: small transverse displacements, large rotations, and fluid levels. Schematic diagrams for measuring devices that use this feature are shown in Figure 6.23. In Figure 6.23(a), angular displacement of one of the plates causes a change in A. In Figure 6.23(b), a transverse displacement of one of the plates changes x. Finally, in Figure 6.23(c), a change in k is produced as the fluid level between the capacitor plates changes. In all cases, the associated change in capacitance is measured directly or indirectly and is used to estimate the measurand. A popular method is to use a capacitance bridge circuit to measure the change in capacitance, in a manner similar to how an inductance bridge (see Chapter 4) is used to measure changes in inductance. Other methods include measuring a change in such quantities as charge (using a charge amplifier), voltage (using a high input-impedance device in parallel), and current (using a very low impedance device in series) that will result from the change in capacitance in a suitable circuit. An alternative method is to make the capacitor a part of an

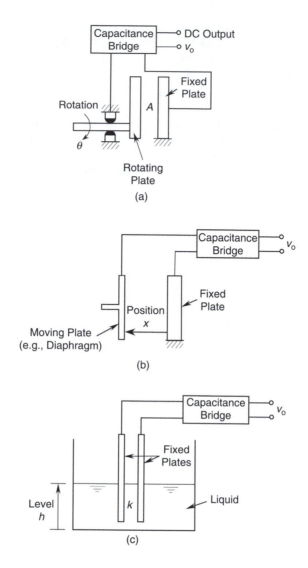

FIGURE 6.23
Schematic diagrams of capacitive sensors: (a) Capacitive rotation sensor; (b) Capacitive displacement sensor; (c) Capacitive liquid level sensor.

inductance-capacitance (*L-C*) oscillator circuit—the natural frequency of the oscillator ($1/\sqrt{LC}$) measures the capacitance. (Incidentally, this method may also be used to measure inductance.)

6.5.1 Capacitive Rotation Sensor

In the arrangement shown in Figure 6.23(a), one plate of the capacitor rotates with a rotating object (shaft) and the other plate is kept stationary. Since the common area *A* is proportional to the angle of rotation θ, Equation 6.41 may be written as

$$C = K\theta \tag{6.42}$$

where *K* is a sensor constant. This is a linear relationship between *C* and θ. The capacitance may be measured by any convenient method. The sensor is linearly calibrated to give the angle of rotation.

The sensitivity of this angular displacement sensor is

$$S = \frac{\partial C}{\partial \theta} = K \tag{6.43}$$

which is constant throughout the measurement. This is expected because the sensor relationship is linear. Note that in the nondimensional form, the sensitivity of the sensor is unity, implying "direct" sensitivity.

6.5.2 Capacitive Displacement Sensor

The arrangement shown in Figure 6.23(b) provides a sensor for measuring transverse displacements and proximities. One of the capacitor plates is attached to the moving object and the other plate is kept stationary. The sensor relationship is

$$C = \frac{K}{x} \tag{6.44}$$

The constant K has a different meaning here. The corresponding sensitivity is given by

$$S = \frac{\partial C}{\partial x} = -\frac{K}{x^2} \tag{6.45}$$

Again, the sensitivity is unity (negative) in the nondimensional form, which indicates direct sensitivity of the sensor (in the reciprocal sense).

Note that Equation 6.44 is a nonlinear relationship. A simple way to linearize this transverse displacement sensor is to use an inverting amplifier, as shown in Figure 6.24. Note that C_{ref} is a fixed, reference capacitance whose value is accurately known. Since the gain of the operational amplifier is very high, the voltage at the negative lead (point A) is zero for most practical purposes (because the positive lead is grounded). Furthermore, since the input impedance of the opamp is also very high, the current through the input leads is negligible. These are the two common assumptions used in opamp analysis (see Chapter 4). Accordingly, the charge balance equation for node point A is: $v_{ref} C_{ref} + v_o C = 0$. Now, in view of Equation 6.44, we get the following linear relationship for the output voltage v_o in terms of the displacement x:

$$v_o = -\frac{v_{ref} C_{ref}}{K} x \tag{6.46}$$

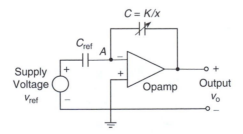

FIGURE 6.24
Linearizing amplifier circuit for a capacitive transverse displacement sensor.

Hence, measurement of v_o gives the displacement through a linear relationship. The sensitivity of the device can be increased by increasing v_{ref} and C_{ref}. The reference voltage may be either dc or ac with frequency as high as 25 kHz (for high-bandwidth measurements). With an ac reference voltage, the output voltage is a modulated signal, which has to be demodulated to measure transient displacements, as discussed before in the context of variable-inductance sensors.

Example 6.4
Consider the circuit shown in Figure 6.25. Examine how this arrangement could be used to measure displacements.

SOLUTION
Assuming that a very high impedance device is used to measure the output voltage v_o, the current through the capacitor is the same as that through the resistor. Thus,

$$i = \frac{d}{dt}(Cv_o) = \frac{v_{ref} - v_o}{R} \tag{6.47}$$

If a transverse displacement capacitor is considered, for example, from Equation 6.44 we have

$$x = RKv_o \Big/ \int_0^t (v_{ref} - v_o)\,dt \tag{6.48}$$

This is a nonlinear differential relationship. To measure x, we need to measure the output voltage and perform an integration by either analog or digital means. That reduces the operating speed (frequency range, bandwidth). Furthermore, since $v_o = v_{ref}$ and $v_o = 0$ at steady state, it follows that this approach cannot be used to make steady-state (or quasi-static) measurements. This situation can be corrected by using an ac source as the supply. If the supply frequency is ω, the frequency-domain transfer function between the supply and the output is given by

$$\frac{v_o}{v_{ref}} = \frac{1}{[1 + RKj\omega/x]} \tag{6.49}$$

Now the displacement x may be determined by measuring either the signal amplification (i.e., amplitude ratio or magnitude) M at the output or the phase "lag" ϕ of the output signal. The corresponding relations are

$$x = \frac{RK\omega}{\sqrt{1/M^2 - 1}} \tag{6.50}$$

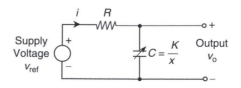

FIGURE 6.25
Capacitive displacement sensor example.

and

$$x = RK\omega/\tan\phi \tag{6.51}$$

Note that the differential equation of the circuit is not linear or, at least time-invariant unless x is constant. The foregoing transfer-function relations do not strictly hold if the displacement is transient. Nevertheless, reasonably accurate results are obtained when the measured displacement is slowly varying.

The arrangement shown in Figure 6.23(c) can be used as well for displacement sensing. In this case a solid dielectric element, which is free to move in the longitudinal direction of the capacitor plates, is attached to the moving object whose displacement is to be measured. The dielectric constant of the capacitor changes as the common area between the dielectric element and the capacitor plates varies due to the motion. The same arrangement may be used as a liquid level sensor, in which case the dielectric medium is the measured liquid, as shown in Figure 6.23(c).

6.5.3 Capacitive Angular Velocity Sensor

The schematic diagram for an angular velocity sensor that uses a rotating-plate capacitor is shown in Figure 6.26. Since the current sensor has negligible resistance, the voltage across the capacitor is almost equal to the supply voltage v_{ref}, which is kept constant. It follows that the current in the circuit is given by

$$i = \frac{d}{dt}(Cv_{ref}) = v_{ref}\frac{dC}{dt}$$

which, in view of Equation 6.42, may be expressed as

$$\frac{d\theta}{dt} = \frac{i}{Kv_{ref}} \tag{6.52}$$

This is a linear relationship for angular velocity in terms of the measured current i. Care must be exercised to ensure that the current-measuring device does not interfere with (e.g., does not load) the basic circuit.

An advantage of capacitance transducers is that because they are noncontacting devices, mechanical loading effects are negligible. There is some loading due to inertial forces of the moving plate and frictional resistance in associated sliding mechanisms, bearings, etc. Such influences can be eliminated by using the moving object itself as the moving plate. Variations in the dielectric properties due to humidity, temperature, pressure, and impurities introduce errors. A capacitance bridge circuit can compensate for these effects.

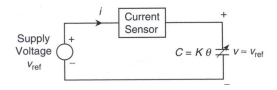

FIGURE 6.26
Rotating-plate capacitive angular velocity sensor.

Extraneous capacitances, such as cable capacitance, can produce erroneous readings in capacitive sensors. This problem can be overcome by using proper conditioning circuitry such as a charge amplifier for the sensor signal. Another drawback of capacitance displacement sensors is low sensitivity. For a transverse displacement transducer, the sensitivity is typically less than one picofarad (pF) per millimeter (1 pF = 10^{-12} F). This problem is not serious, because high supply voltages and amplifier circuitry can be used to increase the sensor sensitivity.

6.5.4   Capacitance Bridge Circuit

Sensors that are based on the change in capacitance (reactance) will require some means of measuring that change. Furthermore, changing capacitance that is not caused by a change in measurand, for example, due to change in humidity, temperature, etc., will cause errors and should be compensated for. Both these goals are accomplished using a capacitance bridge circuit. An example is shown in Figure 6.27.

In this circuit,

$Z_2 = \frac{1}{j\omega C_2}$ = reactance (i.e., capacitive impedance) of the capacitive sensor (of capacitance C_2)

$Z_1 = \frac{1}{j\omega C_1}$ = reactance of the compensating capacitor C_1

Z_4, Z_3 = bridge completing impedances (typically, reactances)

$v_{ref} = v_a \sin \omega t$ = excitation ac voltage

ϕ = phase lag of the output with respect to the excitation.

Using the two assumptions for an opamp (potentials at the negative and positive leads are equal and the current through these leads is zero; see Chapter 4) we can write the current balance equations:

$$\frac{v_{ref} - v}{Z_1} + \frac{v_o - v}{Z_2} = 0 \tag{i}$$

$$\frac{v_{ref} - v}{Z_3} + \frac{0 - v}{Z_4} = 0 \tag{ii}$$

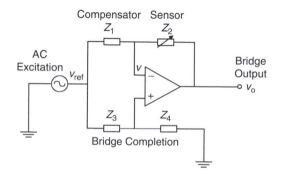

FIGURE 6.27
A bridge circuit for capacitive sensors.

where, v is the common voltage at the opamp leads. Next, eliminate v in Equation i and Equation ii to obtain

$$v_o = \frac{(Z_4/Z_3 - Z_2/Z_1)}{1 + Z_4/Z_3} v_{\text{ref}} \qquad (6.53)$$

It is noted that when

$$\frac{Z_2}{Z_1} = \frac{Z_4}{Z_3} \qquad (6.54)$$

the bridge output $v_o = 0$, and the bridge is said to be balanced. Since all capacitors in the bridge are similarly affected by ambient changes, a balanced bridge will maintain that condition even under ambient changes, unless the sensor reactance Z_2 is changed due to the measurand itself. It follows that the ambient effects are compensated (at least up to the first order) by a bridge circuit. From Equation 6.53 it is clear that the bridge output due to a sensor change of δZ, starting from a balanced state, is given by

$$\delta v_o = -\frac{v_{\text{ref}}}{Z_1(1 + Z_4/Z_3)} \delta Z \qquad (6.55)$$

The amplitude and phase angle of δv_o with respect to v_{ref} will determine δZ, assuming that Z_1 and Z_4/Z_3 are known.

6.5.5 Differential (Push-Pull) Displacement Sensor

Consider the capacitor shown in Figure 6.28 where the two end plates are fixed and the middle plate is attached to a moving object whose displacement (δx) needs to be measured. Suppose that the capacitor plates are connected to the bridge circuit of Figure 6.27 as shown, forming the reactances Z_3 and Z_4.

If initially the middle plate is placed at an equal separation of x from the end plates, and then the middle plate is moved by δx, we have

$$Z_3 = \frac{1}{j\omega C_3} = \frac{x - \delta x}{j\omega K} \qquad \text{(i)}$$

$$Z_4 = \frac{1}{j\omega C_4} = \frac{x + \delta x}{j\omega K} \qquad \text{(ii)}$$

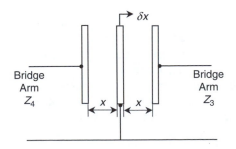

FIGURE 6.28
A linear push-pull displacement sensor.

where K is a capacitor constant as in Equation 6.44. Also assume that Z_1 and Z_2 are bridge-completion impedances and that they are equal. Then Equation 6.53 becomes

$$v_o = \left[\frac{Z_4 - Z_3}{Z_4 + Z_3}\right] v_{\text{ref}} \tag{6.56}$$

This, in view of the results Equation i and Equation ii above, becomes:

$$v_o = \frac{\delta x}{x} v_{\text{ref}} \tag{6.57}$$

This is a convenient linear relation for the displacement.

6.6 Piezoelectric Sensors

Some substances, such as barium titanate, single-crystal quartz, and lead zirconate-titanate (PZT) can generate an electrical charge and an associated potential difference when they are subjected to mechanical stress or strain. This piezoelectric effect is used in piezoelectric transducers. Direct application of the piezoelectric effect is found in pressure and strain measuring devices, touch screens of computer monitors, and a variety of microsensors. Many indirect applications also exist. They include piezoelectric accelerometers and velocity sensors and piezoelectric torque sensors and force sensors. It is also interesting to note that piezoelectric materials deform when subjected to a potential difference (or charge or electric field). Some delicate test equipment (e.g., in vibration testing) use piezoelectric actuating elements (reverse piezoelectric action) to create fine motions. Also, piezoelectric valves (e.g., flapper valves), with direct actuation using voltage signals, are used in pneumatic and hydraulic control applications and in ink-jet printers. Miniature stepper motors based on the reverse piezoelectric action are available. Microactuators based on the piezoelectric effect are found in a number of applications including hard-disk drives (HDD) and micro-electromechanical systems (MEMS). Modern piezoelectric material include lanthanum modified PZT (or, PLZT) and piezoelectric polymeric polyvinylidene fluoride (PVDF).

The piezoelectric effect arises as a result of charge polarization in an anisotropic material (having nonsymmetric molecular structure), as a result of an applied strain. This is a reversible effect. In particular, when an electric field is applied to the material so as to change the ionic polarization, the material will regain its original shape. Natural piezoelectric materials are by and large crystalline whereas synthetic piezoelectric materials tend to be ceramics. When the direction of the electric field and the direction of strain (or stress) are the same, we have direct sensitivity. Other cross-sensitivities can be defined in a 6×6 matrix with reference to three orthogonal direct axes and three rotations about these axes.

Consider a piezoelectric crystal in the form of a disc with two electrodes plated on the two opposite faces. Since the crystal is a dielectric medium, this device is essentially a capacitor, which may be modeled by a capacitance C, as in Equation 6.41. Accordingly, a piezoelectric sensor may be represented as a *charge source* with a capacitive impedance in parallel (Figure 6.29). An equivalent circuit (Thevenin equivalent representation) can be given as well, where the capacitor is in series with an equivalent voltage source. The impedance from the capacitor is given by

$$Z = \frac{1}{j\omega C} \tag{6.58}$$

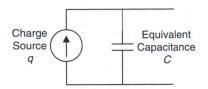

FIGURE 6.29
Equivalent circuit representation of a
piezoelectric sensor.

As is clear from Equation 6.58, the output impedance of piezoelectric sensors is very high, particularly at low frequencies. For example, a quartz crystal may present an impedance of several megohms at 100 Hz, increasing hyperbolically with decreasing frequencies. This is one reason why piezoelectric sensors have a limitation on the useful lower frequency. The other reason is the charge leakage.

6.6.1 Sensitivity

The sensitivity of a piezoelectric crystal may be represented either by its *charge sensitivity* or by its *voltage sensitivity*. Charge sensitivity is defined as

$$S_q = \frac{\partial q}{\partial F} \tag{6.59}$$

where, q denotes the generated charge and F denotes the applied force. For a crystal with surface area A, Equation 6.59 may be expressed as

$$S_q = \frac{1}{A}\frac{\partial q}{\partial p} \tag{6.60}$$

where, p is the stress (normal or shear) or pressure applied to the crystal surface. Voltage sensitivity S_v is given by the change in voltage due to a unit increment in pressure (or stress) per unit thickness of the crystal. Thus, in the limit, we have

$$S_v = \frac{1}{d}\frac{\partial v}{\partial p} \tag{6.61}$$

where d denotes the crystal thickness. Now, since

$$\delta q = C\delta v \tag{6.62}$$

by using Equation 6.41 for a capacitor element, the following relationship between charge sensitivity and voltage sensitivity is obtained:

$$S_q = kS_v \tag{6.63}$$

Note that k is the dielectric constant (permittivity) of the crystal capacitor, as defined in Equation 6.41. The overall sensitivity of a piezoelectric device can be increased through the use of properly designed multielement structures (bimorphs).

Example 6.5

A barium titanate crystal has a charge sensitivity of 150.0 picocoulombs per newton (pC/N). (*Note*: 1 pC = 1 × 10⁻¹² coulombs; coulombs = farads × volts). The dielectric constant for the crystal is 1.25×10^{-8} farads per meter (F/m). From Equation 6.63, the voltage sensitivity of the crystal is given by

$$S_v = \frac{150.0 \text{ pC/N}}{1.25 \times 10^{-8} \text{ F/m}} = \frac{150.0 \times 10^{-12} \text{ C/N}}{1.25 \times 10^{-8} \text{ F/m}} = 12.0 \times 10^{-3} \text{ V} \cdot \text{m/N} = 12.0 \text{ mV} \cdot \text{m/N}$$

The sensitivity of a piezoelectric element is dependent on the direction of loading. This is because the sensitivity depends on the molecular structure (e.g., crystal axis). Direct sensitivities of several piezoelectric material along their most sensitive crystal axis are listed in Table 6.2.

6.6.2 Types of Accelerometers

It is known from Newton's second law that a force (f) is necessary to accelerate a mass (or inertia element), and its magnitude is given by the product of mass (M) and acceleration (a). This product (Ma) is commonly termed *inertia force*. The rationale for this terminology is that if a force of magnitude Ma were applied to the accelerating mass in the direction opposing the acceleration, then the system could be analyzed using static equilibrium considerations. This is known as *d'Alembert's principle* (Figure 6.30). The force that causes acceleration is itself a measure of the acceleration (mass is kept constant). Accordingly, mass can serve as a "front-end" element to convert acceleration into a force. This is the principle of operation of common accelerometers. There are many different types of accelerometers, ranging from strain gage devices to those that use electromagnetic induction. For example, the force which causes acceleration may be converted into a proportional displacement using a spring element, and this displacement may be measured using

TABLE 6.2

Sensitivities of Several Piezoelectric Material

Material	Charge Sensitivity S_q (pC/N)	Voltage Sensitivity S_v (mV · m/N)
Lead zirconate titanate (PZT)	110	10
Barium titanate	140	6
Quartz	2.5	50
Rochelle salt	275	90

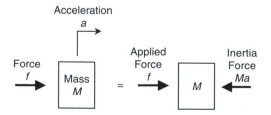

FIGURE 6.30
Illustration of d'Alembert's principle.

a convenient displacement sensor. Examples of this type are differential-transformer accelerometers, potentiometer accelerometers, and variable-capacitance accelerometers. Alternatively, the strain at a suitable location of a member that was deflected due to inertia force may be determined using a strain gage. This method is used in strain gage accelerometers. Vibrating-wire accelerometers use the accelerating force to tension a wire. The force is measured by detecting the natural frequency of vibration of the wire (which is proportional to the square root of tension). In servo force-balance (or null-balance) accelerometers, the inertia element is restrained from accelerating by detecting its motion and feeding back a force (or torque) to exactly cancel out the accelerating force (torque). This feedback force is determined, for instance, by knowing the motor current, and it is a measure of the acceleration.

6.6.3   Piezoelectric Accelerometer

The piezoelectric accelerometer (or, *crystal accelerometer*) is an acceleration sensor, which uses a piezoelectric element to measure the inertia force caused by acceleration. A piezoelectric velocity transducer is simply a piezoelectric accelerometer with a built-in integrating amplifier in the form of a miniature integrated circuit (IC).

The advantages of piezoelectric accelerometers over other types of accelerometers are their light weight and high-frequency response (up to about 1 MHz). However, piezoelectric transducers are inherently high-output impedance devices, which generate small voltages (on the order of 1 mV). For this reason, special impedance-transforming amplifiers (e.g., charge amplifiers) have to be employed to condition the output signal and to reduce loading error.

A schematic diagram for a compression-type piezoelectric accelerometer is shown in Figure 6.31. The crystal and the inertia mass are restrained by a spring of very high stiffness. Consequently, the fundamental natural frequency or resonant frequency of the device becomes high (typically 20 kHz). This gives a reasonably wide useful range (typically up to 5 kHz). The lower limit of the useful range (typically 1 Hz) is set by factors such as the limitations of the signal-conditioning system, the mounting methods, the charge leakage in the piezoelectric element, the time constant of the charge-generating dynamics, and the signal-to-noise ratio. A typical frequency response curve of a piezoelectric accelerometer is shown in Figure 6.32.

In a compression-type crystal accelerometer, the inertia force is sensed as a compressive normal stress in the piezoelectric element. There are also piezoelectric accelerometers

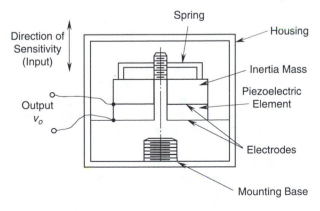

FIGURE 6.31
A compression-type piezoelectric accelerometer.

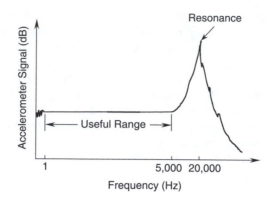

FIGURE 6.32
A typical frequency response curve for a piezoelectric accelerometer.

where the inertia force is applied to the piezoelectric element as a shear strain or as a tensile strain.

For an accelerometer, acceleration is the signal that is being measured (the measurand). Hence, accelerometer sensitivity is commonly expressed in terms of electrical charge per unit acceleration or voltage per unit acceleration (compare this with Equation 6.60 and Equation 6.61). Acceleration is measured in units of acceleration due to gravity (g), and charge is measured in picocoulombs (pC), which are units of 10^{-12} coulombs (C). Typical accelerometer sensitivities are 10 pC/g and 5 mV/g. Sensitivity depends on the piezoelectric properties, the way in which the inertia force is applied to the piezoelectric element (e.g., compressive, tensile, shear), and the mass of the inertia element. If a large mass is used, the reaction inertia force on the crystal will be large for a given acceleration, thus generating a relatively large output signal. Large accelerometer mass results in several disadvantages, however. In particular:

1. The accelerometer mass distorts the measured motion variable (mechanical loading effect).
2. A heavy accelerometer has a lower resonant frequency and hence a lower useful frequency range (Figure 6.32).

For a given accelerometer size, improved sensitivity can be obtained by using the shear-strain configuration. In this configuration, several shear layers can be used (e.g., in a *delta arrangement*) within the accelerometer housing, thereby increasing the effective shear area and hence the sensitivity in proportion to the shear area. Another factor that should be considered in selecting an accelerometer is its *cross-sensitivity* or transverse sensitivity. Cross-sensitivity is present because a piezoelectric element can generate a charge in response to forces and moments (or, torques) in orthogonal directions as well. The problem can be aggravated due to manufacturing irregularities of the piezoelectric element, including material unevenness and incorrect orientation of the sensing element, and due to poor design. Cross-sensitivity should be less than the maximum error (percentage) that is allowed for the device (typically 1%).

The technique employed to mount the accelerometer on an object can significantly affect the useful frequency range of the accelerometer. Some common mounting techniques are:

1. Screw-in base
2. Glue, cement, or wax

3. Magnetic base
4. Spring-base mount
5. Hand-held probe

Drilling holes in the object can be avoided by using the second through fifth methods, but the useful range can decrease significantly when spring-base mounts or hand-held probes are used (typical upper limit of 500 Hz). The first two methods usually maintain the full useful range (e.g., 5 kHz), whereas the magnetic attachment method reduces the upper frequency limit to some extent (typically 3 kHz).

6.6.4 Charge Amplifier

Piezoelectric signals cannot be read using low-impedance devices. The two primary reasons for this are:

1. High output impedance in the sensor results in small output signal levels and large loading errors.
2. The charge can quickly leak out through the load.

A charge amplifier is commonly used as the signal-conditioning device for piezoelectric sensors, in order to overcome these problems to a great extent (see Chapter 4). Because of impedance transformation, the impedance at the output of the charge amplifier becomes much smaller than the output impedance of the piezoelectric sensor. This virtually eliminates loading error and provides a low-impedance output for purposes such as signal communication, acquisition, recording, processing, and control. Also, by using a charge amplifier circuit with a relatively large time constant, speed of charge leakage can be decreased. For example, consider a piezoelectric sensor and charge amplifier combination, as represented by the circuit in Figure 6.33. Let us examine how the rate of charge leakage is reduced by using this arrangement. Sensor capacitance, feedback capacitance of the charge amplifier, and feedback resistance of the charge amplifier are denoted by C, C_f, and R_f, respectively. The capacitance of the cable, which connects the sensor to the charge amplifier, is denoted by C_c.

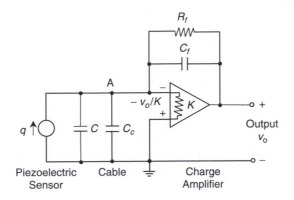

FIGURE 6.33
A piezoelectric sensor and charge amplifier combination.

For an unsaturated opamp of gain K, the voltage at its inverting (negative) input is $-v_o/K$, where v_o is the voltage at the amplifier output. Note that the noninverting (positive) input of the opamp is grounded (i.e., maintained at zero potential). Due to very high input impedance of the opamp, the currents through its input leads will be negligible. Current balance at point A gives:

$$\dot{q} + C\frac{\dot{v}_o}{K} + C_c\frac{\dot{v}_o}{K} + C_f\left(\dot{v}_o + \frac{\dot{v}_o}{K}\right) + \frac{v_o + v_o/K}{R_f} = 0 \tag{6.64}$$

Since gain K is very large (typically 10^5–10^9) compared to unity, this differential equation may be approximated as

$$R_f C_f \frac{dv_o}{dt} + v_o = -R_f \frac{dq}{dt} \tag{6.65}$$

Alternatively, instead of using Equation 6.64, it is possible to directly obtain Equation 6.65 from the two common assumptions (equal inverting and noninverting lead potentials and zero lead currents; see Chapter 4) for an opamp. Then the potential at the negative (inverting) lead would be zero, as the positive lead is grounded. Also, as a result, the voltage across C_c would be zero. Hence, the current balance at point A gives:

$$\dot{q} + \frac{v_o}{R_f} + C_f \dot{v}_o = 0$$

which is identical to Equation 6.65. The corresponding transfer function is

$$\frac{v_o(s)}{q(s)} = -\frac{R_f s}{[R_f C_f s + 1]} \tag{6.65a}$$

where s is the Laplace variable. Now, in the frequency domain ($s = j\omega$), we have

$$\frac{v_o(j\omega)}{q(j\omega)} = -\frac{R_f j\omega}{[R_f C_f j\omega + 1]} \tag{6.65b}$$

Note that the output is zero at zero frequency ($\omega = 0$). Hence, a piezoelectric sensor cannot be used for measuring constant (dc) signals. At very high frequencies, on the other hand, the transfer function approaches the constant value $-1/C_f$. which is the calibration constant for the device.

From Equation 6.65 or Equation 6.65a which represent a first-order system, it is clear that the time constant τ_c of the sensor-amplifier unit is

$$\tau_c = R_f C_f \tag{6.66}$$

Suppose that the charge amplifier is properly calibrated (by the factor $-1/C_f$) so that the frequency transfer function (Equation 6.65b) can be written as

$$G(j\omega) = \frac{j\tau_c\omega}{[j\tau_c\omega + 1]} \tag{6.67}$$

Magnitude M of this transfer function is given by

$$M = \frac{\tau_c \omega}{\sqrt{\tau_c^2 \omega^2 + 1}} \tag{6.68}$$

As $\omega \to \infty$, note that $M \to 1$. Hence, at infinite frequency there is no sensor error. Measurement accuracy depends on the closeness of M to 1. Suppose that we want the accuracy to be better than a specified value M_o. Accordingly, we must have

$$\frac{\tau_c \omega}{\sqrt{\tau_c^2 \omega^2 + 1}} > M_0 \tag{6.69a}$$

or

$$\tau_c \omega > \frac{M_o}{\sqrt{1 - M_0^2}} \tag{6.69}$$

If the required lower frequency limit is ω_{min}, the time constant requirement is

$$\tau_c > \frac{M_o}{\omega_{min} \sqrt{1 - M_0^2}} \tag{6.70}$$

or

$$R_f C_f > \frac{M_o}{\omega_{min} \sqrt{1 - M_0^2}} \tag{6.70a}$$

It follows that, for a specified level of accuracy, a specified lower limit on frequency of operation may be achieved by increasing the time constant (i.e., by increasing R_f, C_f, or both). The feasible lower limit on the frequency of operation (ω_{min}) can be set by adjusting the time constant.

Example 6.6

For a piezoelectric accelerometer with a charge amplifier, an accuracy level better than 99% is obtained if $\frac{\tau_c \omega}{\sqrt{\tau_c^2 \omega^2 + 1}} > 0.99$ or: $\tau_c \omega > 7.0$. The minimum frequency of a transient signal, which can tolerate this level of accuracy is

$$\omega_{min} = \frac{7.0}{\tau_c}$$

In theory, it is possible to measure velocity by first converting velocity into a force using a viscous damping element and measuring the resulting force using a piezoelectric sensor. This principle may be used to develop a piezoelectric velocity transducer. The practical implementation of an ideal velocity-force transducer is quite difficult, however. Hence, commercial piezoelectric velocity transducers use a piezoelectric accelerometer and a built-in

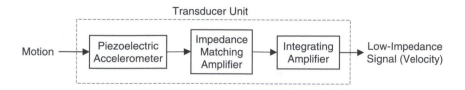

FIGURE 6.34
Schematic diagram of a piezoelectric velocity transducer:

(miniature) integrating amplifier. A schematic diagram of the arrangement of a piezoelectric velocity transducer is shown in Figure 6.34. The overall size of such a unit can be as small as 1 cm. With double integration hardware, a piezoelectric displacement transducer is obtained. Alternatively, an ideal spring element (or, cantilever), which converts displacement into a force (or, bending moment or strain), may be employed to stress the piezoelectric element, resulting in a displacement transducer. Such devices are usually not practical for low-frequency (few hertz) applications because of the poor low-frequency characteristics of piezoelectric elements.

6.7 Effort Sensors

The response of a mechanical system depends on "effort" excitations (forces and torques) applied to the system (See chapter 2). Many applications exist in which process performance is specified in terms of forces and torques. Examples include machine-tool operations, such as grinding, cutting, forging, extrusion, and rolling; manipulator tasks, such as parts handling, assembly, engraving, and robotic fine manipulation; and actuation tasks, such as locomotion. The forces and torques present in a dynamic system are generally functions of time. Performance monitoring and evaluation, failure detection and diagnosis, testing, and control of mechatronic systems can depend heavily on accurate measurement of associated forces and torques. One example in which force (and torque) sensing can be very useful is a drilling robot. The drill bit is held at the end effector by the gripper of the robot, and the workpiece is rigidly fixed to a support structure by clamps. Although a displacement sensor (such as a potentiometer or a differential transformer) can be used to measure drill motion in the axial direction, this alone does not determine the drill performance. Depending on the material properties of the workpiece (e.g., hardness) and the nature of the drill bit (e.g., degree of wear), a small misalignment or slight deviation in feed (axial movement) or speed (rotational speed of the drill) can create large normal (axial) and lateral forces and resistance torques. This can create problems such as excessive vibrations, uneven drilling, excessive tool wear, and poor product quality and eventually may lead to a major mechanical failure. By sensing the axial force or motor torque, for example, and using the information to adjust process variables (speed, feed rate, etc.), or even to provide warning signals and eventually stop the process, can significantly improve the system performance. Another example in which force sensing is useful is in *nonlinear feedback control* (or, feedback linearization technique or FLT) of mechanical systems such as robotic manipulators.

Since both force and torque are *effort variables*, the term *force* may be used to represent both these variables. This generalization is adopted here except when discrimination might be necessary,—for example, when discussing specific applications.

6.7.1 Force Causality Issues

One important application of force (and torque) sensing is in the area of control. Since forces are variables in a mechanical system, their measurement can lead to effective control. There are applications in which force control is invaluable. This is particularly evident in situations where a small error in motion can lead to the generation of large forces, which is the case, for example, in parts assembly operations. In assembly, a slight misalignment (or position error) can cause jamming and generation of damaging forces. As another example, consider-high precision machining of a hard workpiece. A slight error in motion could generate large cutting forces, which might lead to unacceptable product quality or even to rapid degradation of the machine tool. In such situations, measurement and control of forces seem an effective way to improve the system performance. First, we shall address the force control problem from a generalized and unified point of view. The concepts introduced here will be illustrated further by examples.

6.7.1.1 Force-Motion Causality

The response of a mechanical control system does not necessarily have to be a motion variable. When the objective of a control system is to produce a desired motion, the response variables (outputs) are the associated motion variables. A good example is the response of a spray-painting robot whose end effector is expected to follow a specific trajectory of motion. On the other hand, when the objective is to exert a desired set of forces (or torques)—which is the case in some tasks of machining, forging, gripping, engraving, and assembly—the outputs are the associated force variables. The choice of inputs and outputs cannot be done arbitrarily, however, for a physical system. The conditions of "physical realizability" have to be satisfied, as we will illustrate.

A lumped-parameter mechanical system can be treated as a set of basic mechanical elements (springs, inertia elements, dampers, levers, gyros, force sources, and velocity sources, etc.) that are interconnected through ports (or *bonds*) through which power (or energy) flows (see Chapter 2 for bond graphs). Each port actually consists of two terminals. There is an effort variable (force) f and a flow variable (motion, such as velocity) v associated with each port. In particular, consider a port that connects two subsystems A and B, as shown in Figure 6.35(a). The two subsystems interact, and power flows through the port. Hence, for example, if we assume that A pushes B with a force f, then f is considered input to B. Now, B responds with motion v—the output of B—which, in turn causes A to respond with force f. Consequently, the input to A is v and the output of A is f. This cause-effect relationship (*causality*) is clearly shown by the block diagram representation in Figure 6.35(b). A similar argument would lead to the opposite causality if we had started with the assumption that B is pushing A with force f. It follows that one has to make some causal decisions for a dynamic system, based on system requirements and physical realizability, and the rest of the causalities will be decided automatically. It can be shown that conflicts in causality indicate that the energy storage elements (e.g., springs, masses) in the particular system model are not independent (see Chapter 2, under bond graphs). Note that causality is not governed by the positive direction assigned to the variables f and v. This is clear from the free-body diagram representation in Figure 6.35(c). In particular, if the force f on A (the action) is taken as positive in one direction, then according to Newton's third law, the force f on B (the reaction) is positive in the opposite direction. Similarly, if the velocity v of B relative to A is positive in one direction, then the velocity v of A relative to B is positive in the opposite direction. These directions are unrelated to the input/output (causality) directions. The above discussion shows that forces can be considered outputs (responses) as well as inputs (excitations), depending on the functional requirements of a particular system. In particular, when force is considered

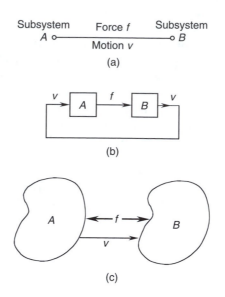

FIGURE 6.35
Force-motion causality: (a) Bond or port representation; (b) Block diagram representation; (c) Free-body diagram representation.

an output, we should imagine that there is an interacting subsystem, which receives this force as an input. In these considerations, however, the requirements of physical realizability have to be satisfied.

6.7.1.2 Physical Realizability

A primary requirement for physical realizability of a linear system is that the its transfer function

$$G(s) = \frac{N(s)}{\Delta(s)} \tag{6.71}$$

must satisfy:

$$O(N) \le O(\Delta) \tag{6.72}$$

where N and Δ are the numerator polynomial and the denominator (characteristic) polynomial, respectively, of the transfer function, and $O(N)$ and $O(\Delta)$ denote their orders. It should be intuitively clear why this condition must be satisfied. In particular, if the condition 6.72 were violated, then we would encounter the following unrealistic characteristics in the system, particularly at very high frequencies:

1. A finite input will produce an infinite output.
2. Infinite levels of power will be needed to drive the system.
3. Saturation will not be possible. The magnitude (gain) can become infinite as the excitation frequency approaches infinity.
4. Future inputs will affect the present outputs.

Example 6.7

In the linear system given by

$$\frac{dy}{dt} + a_0 y = b_0 u + b_1 \frac{du}{dt} + b_2 \frac{d^2 u}{dt^2}$$

suppose that u denotes the input and y denotes the output. Show that the physical realizability is violated by this system. Using the same reasoning show that if u and y were switched the physical realizability would be satisfied.

SOLUTION

Integrate the system equation. We get

$$y = -a_0 \int y\, dt + b_0 \int u\, dt + b_1 u + b_2 \frac{du}{dt}$$

If u is a step input, the derivative term on the RHS becomes infinite at $t = 0$. That means, the output will become infinite instantaneously, which is not feasible. Hence we conclude that the system is not physically realizable. Also note that the system transfer function is

$$G(s) = \frac{b_2 s^2 + b_1 s + b_0}{s + a_0}$$

where $O(N) = 2$ and $O(\Delta) = 1$. Hence the condition 6.72 is not satisfied.

Now if we switch u and y in the system equation, and integrate twice to express the output, there will not be any derivatives of the input. Specifically:

$$b_2 y = -b_1 \int y\, dt - b_0 \iint y\, dt + a_1 \int u\, dt + a_0 \iint u\, dt$$

Hence, in this case, a finite input will not produce an infinite output, because all the integrals will be finite. Furthermore, now the system transfer function is

$$G(s) = \frac{s + a_0}{b_2 s^2 + b_1 s + b_0}$$

which satisfies 6.72.

Example 6.8

A mechanical system is modeled as in Figure 6.36. The system parameters are the masses m_1 and m_2 damping constants b_1 and b_2, and stiffness values k_1 and k_2. Forces f_1 and f_2 exist at masses m_1 and m_2, respectively. The displacements of the corresponding masses are y_1 and y_2. In general, a displacement y_i may be considered either as an input or as an output. Similarly, a force f_i may be considered either as an input or as an output. In particular, when a displacement is treated as an input, we may assume it to be generated by an "internal" actuator of adequate capability without producing an external torque that acts through the environment. When a force is treated as an output, we may consider it to be

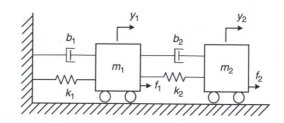

FIGURE 6.36
A model of a mechanical system.

applied on a system that is external to the original system (i.e., on the environment of the original system).

It is easy to verify, by applying Newton's second law to each mass that the equations of motion of the given system are:

$$m_1 \ddot{y}_1 = -b_1 \dot{y}_1 - k_1 y_1 + b_2 (\dot{y}_2 - \dot{y}_1) + k_2 (y_2 - y_1) + f_1$$

$$m_2 \ddot{y}_2 = -b_2 (\dot{y}_2 - \dot{y}_1) - k_2 (y_2 - y_1) + f_2$$

Using the Laplace variable s, these equations may be expressed as the transfer-function relations:

$$[m_1 s^2 + (b_1 + b_2)s + (k_1 + k_2)]y_1 - [b_2 s + k_2]y_2 = f_1$$

$$-[b_2 s + k_2]y_1 + [m_2 s^2 + b_2 s + k_2]y_2 = f_2$$

   i. Express in the Laplace domain, y_1 in terms of f_1 and f_2, and also y_2 in terms of f_1 and f_2 together with the system parameters.

   ii. Indicate, giving reasons, whether f_1 and f_2 can be treated as inputs and y_1 and y_2 as outputs.

   iii. Also indicate, giving reasons, whether y_1 and y_2 can be treated as inputs f_1 and f_2 as outputs.

SOLUTION

   i. The system transfer equations may be expressed as:

$$\begin{bmatrix} a_{11} & a_{12} \\ a_{21} & a_{22} \end{bmatrix} \begin{bmatrix} y_1 \\ y_2 \end{bmatrix} = \begin{bmatrix} f_1 \\ f_2 \end{bmatrix} \qquad \text{(i)}$$

   where
$$a_{11} = m_1 s^2 + (b_1 + b_2)s + k_1 + k_2$$
$$a_{22} = m_2 s^2 + b_2 s + k_2$$
$$a_{12} = a_{21} = -(b_2 s + k_2)$$

Then, by matrix inversion, we get

$$\begin{bmatrix} y_1 \\ y_2 \end{bmatrix} = \frac{1}{(a_{11}a_{22} - a_{12}a_{21})} \begin{bmatrix} a_{22} & -a_{12} \\ -a_{21} & a_{11} \end{bmatrix} \begin{bmatrix} f_1 \\ f_2 \end{bmatrix} \qquad (ii)$$

or,

$$y_1 = \frac{a_{22}}{\Delta} f_1 - \frac{a_{12}}{\Delta} f_2$$

$$y_2 = -\frac{a_{21}}{\Delta} f_1 + \frac{a_{11}}{\Delta} f_2 \qquad (iii)$$

where

$$\Delta = (a_{11}a_{22} - a_{12}a_{21})$$

ii. From Equations iii note that it is feasible to treat f_1 and f_2 as inputs and y_1 and y_2 as the resulting outputs. The reason is, the associated transfer functions $\frac{a_{22}}{\Delta}$, $-\frac{a_{12}}{\Delta}$, $-\frac{a_{21}}{\Delta}$, and $\frac{a_{11}}{\Delta}$ are all physically realizable. This is true because the order of the characteristic polynomial Δ (i.e., the denominator of the transfer functions) is 4, while the order of the numerator polynomials are: for a_{22} it is 2; for a_{12} it is 1, for a_{21} it is 1, and for a_{11} it is 2. Accordingly, for each transfer function we have,

Numerator Order < Denominator Order

which is a requirement for physical realizability.

iii. From Equations i

$$f_1 = a_{11}y_1 + a_{12}y_2$$

$$f_2 = a_{21}y_1 + a_{22}y_2$$

Then, for y_1 and y_2 to be inputs and f_1 and f_2 to be the resulting outputs, the associated transfer functions a_{11}, a_{12}, a_{21} and a_{22} all must be physically realizable. But, this is not the case because the denominator order = 0 and the numerator order is ≥ 1.

6.7.2 Force Control Problems

Some forces in a control system are actuating or excitation (input) forces, and some others are response (output) forces. For example, torques (or forces) driving the joints of a robotic manipulator are considered inputs to the robot. (From the control point of view, however, joint input is the voltage applied to the motor drive amplifier, which produces a field current that generates the motor torque that is resisted by the load torque at the joint.) Gripping or tactile (touch) forces at the end effector of a manipulator, tool tip forces in a milling machine, and forces at the die in a forging machine can be considered output forces, which are "exerted" on objects in the outside environment of the system. Output forces should be completely determined by the inputs (motions as well as forces) to the system, however. Unknown input forces and output forces can be measured using appropriate force sensors, and force control may be implemented using these measurements.

6.7.2.1   Force Feedback Control

Consider the system shown in Figure 6.37, which is connected to its environment through two ports. Force variable f_A and motion variable v_A (both in the same direction) are associated with port A, and force variable f_B and motion variable v_B (in the same direction) are associated with port B. Suppose that f_A is an input to the system and f_B is an output. It follows that v_B is also an input to the system. In other words, the motion at B is constrained. For example, the point might be completely restrained in the direction of f_B, resulting in the constraint $v_B = 0$. Ideally, if we knew the dynamic behavior of the system—and assuming that there are no extraneous inputs (disturbances and noise)—we would be able to analytically determine the input f_A that will generate a desired f_B for a specified v_B. Then we could control the output force f_B simply by supplying the predetermined f_A and by subjecting B to the specified motion v_B. The corresponding open-loop configuration is shown in Figure 6.38(a). Since it is practically impossible to achieve accurate system

FIGURE 6.37
A two-port system.

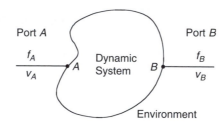

(a)

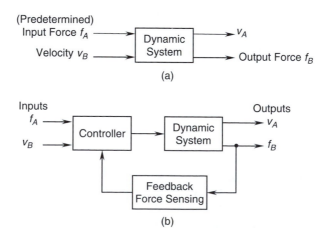

(b)

FIGURE 6.38
(a) An open-loop (feedforward control) system, (b) A force feedback control system.

performance with this open-loop control arrangement (inappropriately known as feedforward control), except in a few simple situations, the feedback control loop shown in Figure 6.38(b) has to be added. In this case, the response force f_B is measured and fed back into the controller, which will modify the control input signals according to a suitable control law, so as to correct any deviations in f_B from the desired value.

6.7.2.2 Feedforward Force Control

Once again, consider the system shown in Figure 6.37, which has two ports, *A* and *B*. Here again, suppose that f_A is an actuating (input) force that can be generated according to a given specification (a "known" input). But suppose that f_B is an unknown input force. It could be a disturbance force, such as that resulting from a collision, or a useful force, such as a gripping force whose value is not known. This configuration is shown in Figure 6.39(a). Since f_B is unknown, it might not be possible to accurately control the system response (v_A and v_B). One solution is to measure the unknown force f_B, using a suitable force sensor, and feed it forward into the controller (Figure 6.39(b)). The controller can use this additional information to compensate for the influence of f_B on the system and produce the desired response. This is an example of feedforward control. Sometimes, if an input force to a system (e.g., a joint force or torque of a robotic manipulator) is computed using an analytical model and is supplied to the actuator, the associated control is inappropriately termed feedforward control. It should be termed *computed force/torque control* or *computed input control*, to be exact.

Example 6.9

A schematic representation of a single joint of a direct-drive robotic manipulator is shown in Figure 6.40(a). Note that direct-drive joints have no speed transmission devices such as gears; the stator of the drive motor is rigidly attached to one link, and the rotor is rigidly attached to the next link. Motor torque is T_m, and the joint torque that is transmitted to

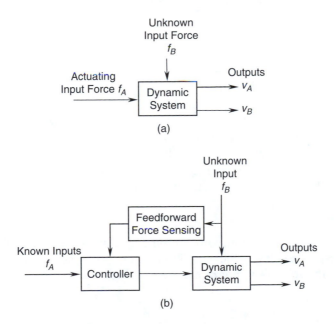

FIGURE 6.39
(a) A system with an unknown input force, (b) Feedforward force control.

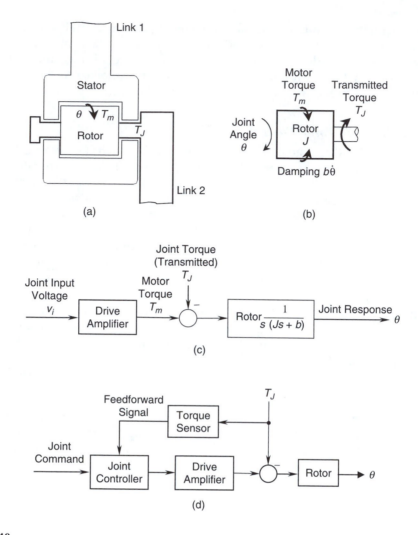

FIGURE 6.40
(a) A single joint of a direct-drive arm, (b) Free-body diagram of the motor rotor, (c) Block diagram of the joint,
(d) Feedforward control using the joint torque.

the driven link (link 2) is T_J. Draw a block diagram for the joint, and show that T_J is
represented as an input to the joint control system. If T_J is measured directly, using a
semiconductor strain gage torque sensor, what type of control would you recommend for
improving the manipulator performance? Extend the discussion to the case in which the
joint actuator is a hydraulic piston-cylinder mechanism.

SOLUTION
Let us assume linear viscous friction, with damping constant b. If θ denotes the relative rotation
of Link 2 with respect to Link 1, the equation of motion for the motor rotor (inertia J) may
be written using the free-body diagram shown in Figure 6.40(b); thus,

$$J\ddot{\theta} = T_m - b\dot{\theta} - T_J$$

Here, for simplicity we have assumed that Link 1 is at rest (or moving with constant velocity).

The motor torque T_m is generated from the field current provided by the drive amplifier, in response to a command voltage v_i to the joint. Figure 6.40(c) shows a block diagram for the joint. Note that T_J enters as an input. Hence, an appropriate control method would be to measure the joint torque T_J directly and feed it forward so as to correct any deviations in the joint response θ. This feedforward control structure is shown by the block diagram in Figure 6.40(d).

The same concepts are true in the case of a hydraulic actuator. Note that the control input is the voltage signal to the hydraulic valve actuator. The pressure in the hydraulic fluid is analogous to the motor torque. Joint torque/force is provided by the force exerted on the driven link by the piston. This force should be measured for feed forward control.

6.7.3 Impedance Control

Consider a mechanical operation where we push against a spring that has constant stiffness. Here, the value of the force completely determines the displacement; similarly the value of the displacement completely determines the force. It follows that, in this example, we are unable to control the force and the displacement independently at the same time. Also, it is not possible, in this example, to apply a command force that has an arbitrarily specified relationship with displacement. In other words, *stiffness control* is not possible. Now suppose that we push against a complex dynamic system, not a simple spring element. In this case, we should be able to command a pushing force in response to the displacement of the dynamic system so that the ratio of force to displacement varies in a desired manner. This is a stiffness control (or *compliance control*) action. Dynamic stiffness is defined as the ratio (output force)/(input displacement), expressed in the frequency domain (see Chapter 2). Mechanical impedance is defined as the ratio (output force)/(input velocity), in the frequency domain. Note that stiffness and impedance both relate force and motion variables in a mechanical system. The objective of impedance control is to make the impedance function equal to some specified function (without separately controlling or independently constraining the associated force variable and velocity variable). Force control and motion control can be considered extreme cases of impedance control (and stiffness control). Specifically, since the objective of force control is to keep the force variable from deviating from a desired level, in the presence of independent variations of the associated motion variable (an input), force control can be considered zero-impedance control if velocity is chosen as the motion variable (or zero-stiffness control if displacement is chosen as the motion variable). Similarly, displacement control can be considered infinite-stiffness control and velocity control can be considered infinite-impedance control.

Impedance control has to be accomplished through active means, generally by generating forces as specified functions of associated motions. Impedance control is particularly useful in mechanical manipulation against physical constraints, which is the case in assembly and machining tasks. In particular, very high impedance is naturally present in the direction of a motion constraint, and very low impedance is naturally present in the direction of a free motion. Problems that arise by using motion control in applications where small motion errors would create large forces, can be avoided if stiffness control or impedance control is used. Furthermore, the stability of the overall system can be guaranteed and the robustness of the system improved by properly bounding the values of impedance parameters.

Impedance control is useful as well in tasks of fine/flexible manipulation, for example, in the processing of flexible and inhomogeneous natural material such as meat. In this case, the mechanical impedance of the task interface (i.e., in the region where the mechanical processor interacts with the processed object) will provide valuable characteristics of the process, which can be used in fine control of the processing task. Since impedance relates an input velocity to an output force, it is a transfer function. The concepts of impedance control can be applied to situations where the input is not a velocity and the output is not a force. Still, the term "impedance control" is used, even though the corresponding transfer function is, strictly speaking, not an impedance.

Example 6.10

The control of processes such as machine tools and robotic manipulators may be addressed from the point of view of impedance control. For example, consider a milling machine that performs a straight cut on a workpiece, as shown in Figure 6.41(a). The tool position is stationary, and the machine table, which holds the workpiece, moves along a horizontal axis at speed v—the feed rate. The cutting force in the direction of feed is f. Suppose that the machine table is driven using the speed error, according to the law

$$F = Z_d(V_{ref} - V) \tag{i}$$

where Z_d denotes the drive impedance of the table and V_{ref} is the reference (command) feed rate. (The uppercase letters are used to represent frequency domain variables of the system). Cutting impedance Z_w, of the workpiece satisfies the relation

$$F = Z_w V \tag{ii}$$

Note that Z_w depends on system properties, and we usually don't have direct control over it. The overall system is represented by the block diagram in Figure 6.41(b). An impedance

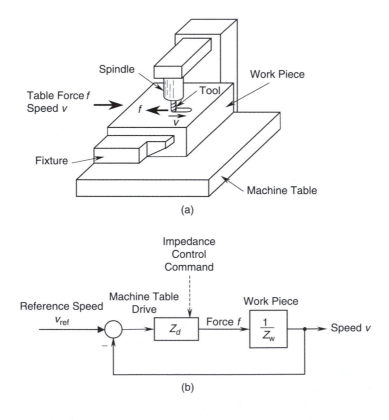

(a)

(b)

FIGURE 6.41
(a) A straight-cut milling operation, (b) Impedance block representation.

control problem would be to adjust (or "adapt") the drive impedance Z_d so as to maintain the feed rate near V_{ref} and the cutting force near F_{ref}. We wish to determine an adaptive control law for Z_d.

SOLUTION
The control objective is satisfied by minimizing the objective function

$$J = \frac{1}{2}\left[\frac{F - F_{ref}}{f_0}\right] + \frac{1}{2}\left[\frac{V - V_{ref}}{v_o}\right]^2 \tag{iii}$$

where f_o denotes the *force tolerance* and v_o denotes the *speed tolerance*. For example, if we desire stringent control of the feed rate, we need to choose a small value for v_o, which corresponds to a heavy weighting on the feed rate term in J. Hence, these two parameters are weighting parameters as well, in the cost function.

The optimal solution is given by

$$\frac{\partial J}{\partial Z_d} = 0 = \frac{(F - F_{ref})}{f_0^2}\frac{\partial F}{\partial Z_d} + \frac{(V - V_{ref})}{v_0^2}\frac{\partial V}{\partial Z_d} \tag{iv}$$

Now, from Equation i and Equation ii, we obtain

$$V = \left[\frac{Z_d}{Z_d + Z_w}\right]V_{ref} \tag{v}$$

$$F = \left[\frac{Z_d Z_w}{Z_d + Z_w}\right]V_{ref} \tag{vi}$$

On differentiating Equation v and Equation vi, we get

$$\frac{\partial V}{\partial Z_d} = \frac{Z_w}{(Z_d + Z_w)^2}V_{ref} \tag{vii}$$

and

$$\frac{\partial F}{\partial Z_d} = \frac{Z_w^2}{(Z_d + Z_w)^2}V_{ref} \tag{viii}$$

Next, we substitute Equation vii and Equation viii in Equation iv and divide by the common term; thus,

$$\frac{(F - F_{ref})}{f_0^2}Z_w + \frac{(V - V_{ref})}{v_0^2} = 0 \tag{ix}$$

Equation ix is expanded after substituting Equation v and Equation vi in order to get the required expression for Z_d:

$$Z_d = \left[\frac{Z_0^2 + Z_w Z_{ref}}{Z_w - Z_{ref}}\right] \tag{x}$$

where

$$Z_o = \frac{f_o}{v_o} \qquad\qquad\qquad \text{(xi)}$$

and

$$Z_{ref} = \frac{F_{ref}}{V_{ref}} \qquad\qquad\qquad \text{(xii)}$$

Equation x is the impedance control law for the table drive. Specifically, since Z_w—which depends on workpiece characteristics, tool bit characteristics, and the rotating speed of the tool bit—may be known through a suitable model or may be experimentally determined (identified) by monitoring v and f, and since Z_d and Z_{ref} are specified, we are able to determine the necessary drive impedance Z_d using Equation x. Parameters of the table drive controller—particularly gain—can be adjusted to match this optimal impedance. Unfortunately, exact matching is virtually impossible, because Z_d is generally a function of frequency. If the component bandwidths are high, we may assume that the impedance functions are independent of frequency, and this somewhat simplifies the impedance control task.

Note from Equation ii and Equation xii that for the ideal case of $V = V_{ref}$ and $F = F_{ref}$, we have $Z_w = Z_{ref}$. Then, from Equation x, it follows that a drive impedance of infinite magnitude is needed for exact control. This is impossible to achieve in practice, however. Of course, an upper limit for the drive impedance should be set in any practical scheme of impedance control.

6.7.4   Force Sensor Location

In force feedback control, the location of the force sensor with respect to the location of actuation can have a crucial effect on system performance, stability in particular. For example, in robotic manipulator applications, it has been experienced that with some locations and configurations of a force-sensing wrist at the robot end effector, dynamic instabilities were present in the manipulator response for some (large) values of control gains in the force feedback loop. These instabilities were found to be limit-cycle-type motions in most cases. Generally, it is known that when the force sensors are more remotely located with respect to the drive actuators of a mechanical system, the system is more likely to exhibit instabilities under force feedback control. Hence, it is desirable to make force measurements very close to the actuator locations when force feedback is used.

Consider a mechanical processing task. The tool actuator generates the processing force, which is applied to the workpiece. The force transmitted to the workpiece by the tool is measured by a force sensor and is used by a feedback controller to generate the correct actuator force. The machine tool is a dynamic system, which consists of a tool subsystem (dynamic) and a tool actuator (dynamic). The workpiece is also a dynamic system.

Relative location of the tool actuator with respect to the force sensor (at the tool-workpiece interface) can affect stability of the feedback control system. In general, the closer the actuator to the sensor the more stable the feedback control system. Two scenarios are shown in Figure 6.42, which can determine the stability of the overall control system. In both cases the processing force at the interface between the tool and the workpiece is measured using a force sensor, and is used by the feedback controller to generate the actuator drive signal. In Figure 6.42(a) the tool actuator, which generates the drive signal of the actuator, is located right next to the force sensor. In Figure 6.42(b) the tool actuator is separated from the force sensor by a dynamic system of the processing machine. It is known that the arrangement (b) in the Figure is less stable than arrangement in Figure (a).

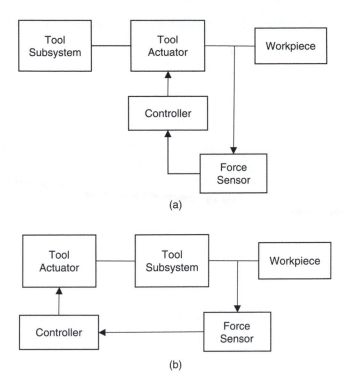

FIGURE 6.42
(a) Force sensor located next to the actuator, (b) Force sensor separated from the actuator by a dynamic subsystem.

The reason is simple. Arrangement in Figure 6.42 (b) introduces more dynamic delay into the feedback control loop. It is well known that the time delay has a destabilizing effect on feedback control systems, particularly at high control gains. The root locus method (See Chapter 12) is useful in stability studies of this type.

6.8   Strain Gages

Many types of force/torque sensors are based on strain gage measurements. Although strain gages measure strain, the measurements can be directly related to stress and force. Hence, it is appropriate to discuss strain gages under force/torque sensors. Note, however, that strain gages may be used in a somewhat indirect manner (using auxiliary front-end elements) to measure other types of variables, including displacement, acceleration, pressure, and temperature. Two common types of resistance strain gages will be discussed next. Specific types of force/torque sensors will be studied in the subsequent sections.

6.8.1   Equations for Strain Gage Measurements

The change of electrical resistance in material when mechanically deformed, is the property used in resistance-type strain gages. The resistance R of a conductor that has length ℓ and area of cross-section A is given by

$$R = \rho \frac{\ell}{A} \qquad (6.73)$$

where ρ denotes the *resistivity* of the material. Taking the logarithm of Equation 6.73, we have

$$\log R = \log \rho + \log(\ell/A)$$

Now, taking the differential, we obtain

$$\frac{dR}{R} = \frac{d\rho}{\rho} + \frac{d(\ell/A)}{\ell/A} \tag{6.74}$$

The first term on the right-hand side of Equation 6.74 depends on the change in resistivity, and the second term represents deformation. It follows that the change in resistance comes from the change in shape as well as from the change in resistivity of the material. For linear deformations, the two terms on the right-hand side of Equation 6.74 are linear functions of strain ε; the proportionality constant of the second term, in particular, depends on Poisson's ratio of the material. Hence, the following relationship can be written for a strain gage element:

$$\frac{\delta R}{R} = S_s \varepsilon \tag{6.75}$$

The constant S_s is known as the *gage factor* or *sensitivity* of the strain gage element. The numerical value of this constant ranges from 2–6 for most *metallic strain gage* elements and from 40–200 for *semiconductor strain gages*. These two types of strain gages will be discussed later.

The change in resistance of a strain gage element, which determines the associated strain (Equation 6.75), is measured using a suitable electrical circuit. Many variables—including displacement, acceleration, pressure, temperature, liquid level, stress, force, and torque—can be determined using strain measurements. Some variables (e.g., stress, force, and torque) can be determined by measuring the strain of the dynamic object itself at suitable locations. In other situations, an auxiliary front-end device may be required to convert the measurand into a proportional strain. For instance, pressure or displacement may be measured by converting them to a measurable strain using a diaphragm, bellows, or bending element. Acceleration may be measured by first converting it into an inertia force of a suitable mass (seismic) element, then subjecting a cantilever (strain member) to that inertia force and, finally, measuring the strain at a high-sensitivity location of the cantilever element (see Figure 6.43). Temperature may be measured by measuring the thermal expansion or deformation in a

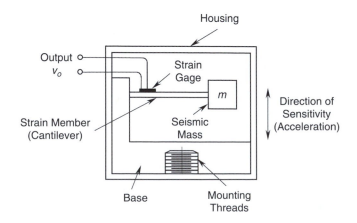

FIGURE 6.43
A strain gage accelerometer.

bimetallic element. *Thermistors* are temperature sensors made of semiconductor material whose resistance changes with temperature. *Resistance temperature detectors* (RTDs) operate by the same principle, except that they are made of metals, not of semiconductor material. These temperature sensors, and the piezoelectric sensors discussed previously, should not be confused with strain gages. Resistance strain gages are based on resistance change due to strain, or the *piezoresistive* property of materials.

Early strain gages were fine metal filaments. Modern strain gages are manufactured primarily as metallic foil (e.g., using the copper-nickel alloy known as *constantan*) or semiconductor elements (e.g., silicon with trace impurity boron). They are manufactured by first forming a thin film (foil) of metal or a single crystal of semiconductor material and then cutting it into a suitable grid pattern, either mechanically or by using photoetching (opto-chemical) techniques. This process is much more economical and is more precise than making strain gages with metal filaments. The strain gage element is formed on a backing film of electrically insulated material (e.g., polymide plastic). This element is cemented or bonded using epoxy, onto the member whose strain is to be measured. Alternatively, a thin film of insulating ceramic substrate is melted onto the measurement surface, on which the strain gage is mounted directly. The direction of sensitivity is the major direction of elongation of the strain gage element (Figure 6.44(a)). To measure strains in more than one direction, multiple strain gages (e.g., various rosette configurations) are available as single units. These units have more than one direction of sensitivity. Principal strains in a given plane (the surface of the object on which the strain gage is mounted) can be determined by using these multiple strain gage units. Typical foil-type gages are shown in Figure 6.44(b), and a semiconductor strain gage is shown in Figure 6.44(c).

A direct way to obtain strain gage measurement is to apply a constant dc voltage across a series-connected pair of strain gage element (of resistance R) and a suitable resistor R_c, and to measure the output voltage v_o across the strain gage under open-circuit conditions (using a voltmeter with high input impedance). It is known as a *potentiometer circuit* or *ballast circuit*. This arrangement has several weaknesses. Any ambient temperature variation will directly introduce some error because of associated change in the strain gage resistance and the resistance of the connecting circuitry. Also, measurement accuracy will be affected by possible variations in the supply voltage v_{ref}. Furthermore, the electrical loading error will be significant unless the load impedance is very high. Perhaps the most serious disadvantage of this circuit is that the change in the signal due to strain is usually a very small percentage of the total signal level in the circuit output. This problem can be reduced to some extent by decreasing v_o, which may be accomplished by increasing the resistance R_c. This, however, reduces the sensitivity of the circuit. Any changes in the strain gage resistance due to ambient changes will directly enter the strain gage reading unless R and R_c have identical coefficients with respect to ambient changes.

A more favorable circuit for use in strain gage measurements is the *Wheatstone bridge*, as discussed in Chapter 4. One or more of the four resistors R_1, R_2, R_3, and R_4 in the bridge (Figure 6.45) may represent strain gages. The output relationship for the Wheatstone bridge circuit is given by (see Chapter 4)

$$v_o = \frac{R_1 v_{ref}}{(R_1 + R_2)} - \frac{R_3 v_{ref}}{(R_3 + R_4)} = \frac{(R_1 R_4 - R_2 R_3)}{(R_1 + R_2)(R_3 + R_4)} v_{ref} \tag{6.76}$$

When this output voltage is zero, the bridge is "balanced." It follows from Equation 6.76 that for a balanced bridge,

$$\frac{R_1}{R_2} = \frac{R_3}{R_4} \tag{6.77}$$

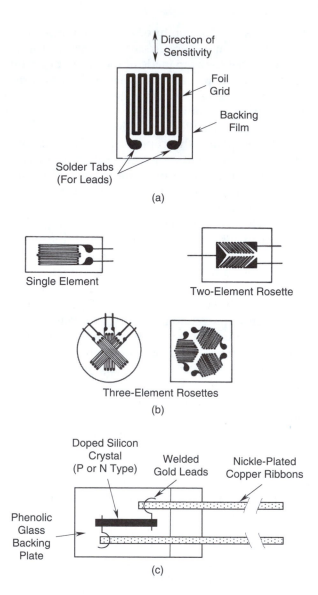

FIGURE 6.44
(a) Strain gage nomenclature; (b) Typical foil-type strain gages; (c) A semiconductor strain gage.

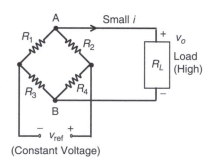

FIGURE 6.45
Wheatstone bridge circuit.

Note that Equation 6.77 is valid for any value of R_L, not just for large R_L, because when the bridge is balanced, current i through the load will be zero, even for small R_L.

6.8.1.1 Bridge Sensitivity

Strain gage measurements are calibrated with respect to a balanced bridge. When the strain gages in the bridge deform, the balance is upset. If one of the arms of the bridge has a variable resistor, it can be changed to restore balance. The amount of this change measures the amount by which the resistance of the strain gages changed, thereby measuring the applied strain. This is known as the *null-balance method* of strain measurement. This method is inherently slow because of the time required to balance the bridge each time a reading is taken. A more common method, which is particularly suitable for making dynamic readings from a strain gage bridge, is to measure the output voltage resulting from the imbalance caused by the deformation of active strain gages in the bridge. To determine the *calibration constant* of a strain gage bridge, the sensitivity of the bridge output to changes in the four resistors in the bridge should be known. For small changes in resistance, using straightforward calculus, this may be determined as

$$\frac{\delta v_o}{v_{ref}} = \frac{(R_2 \delta R_1 - R_1 \delta R_2)}{(R_1 + R_2)^2} - \frac{(R_4 \delta R_3 - R_3 \delta R_4)}{(R_3 + R_4)^2} \tag{6.78}$$

This result is subject to Equation 6.77, because changes are measured from the balanced condition. Note from Equation 6.78 that if all four resistors are identical (in value and material), resistance changes due to ambient effects cancel out among the first-order terms $(\delta R_1, \delta R_2, \delta R_3, \delta R_4)$, producing no net effect on the output voltage from the bridge. Closer examination of Equation 6.78 will reveal that only the adjacent pairs of resistors (e.g., R_1 with R_2 and R_3 with R_4) have to be identical in order to achieve this environmental compensation. Even this requirement can be relaxed. In fact, compensation is achieved if R_1 and R_2 have the same temperature coefficient and if R_3 and R_4 have the same temperature coefficient.

Example 6.11

Suppose that R_1 represents the only active strain gage and R_2 represents an identical "dummy" gage in Figure 6.45. The other two elements of the bridge are *bridge-completion resistors*, which do not have to be identical to the strain gages. For a balanced bridge, we must have $R_3 = R_4$, but not necessarily equal to the resistance of the strain gage. Let us determine the output of the bridge.

In this example, only R_1 changes. Hence, from Equation 6.78, we have

$$\frac{\delta v_o}{v_{ref}} = \frac{\delta R}{4R} \tag{6.79a}$$

where R denotes the strain gage resistance.

6.8.1.2 The Bridge Constant

Equation 6.79a assumes that only one resistance (strain gage) in the Wheatstone bridge (Figure 6.45) is active. Numerous other activating combinations are possible, however, for example, tension in R_1 and compression in R_2, as in the case of two strain gages mounted symmetrically at 45° about the axis of a shaft in torsion. In this manner, the overall sensitivity of a strain gage bridge can be increased. It is clear from Equation 6.78 that if all four resistors in the bridge are active, the best sensitivity is obtained if, for example, R_1 and R_4 are in tension and R_2 and R_3 are in compression, so that all four differential terms have the same sign. If more than one strain gage is active, the bridge output may be expressed as

$$\frac{\delta v_o}{v_{ref}} = k\frac{\delta R}{4R}$$

(6.79)

where

$$k = \frac{\text{bridge output in the general case}}{\text{bridge output if only one strain gage is active}}$$

This constant is known as the *bridge constant*. The larger the bridge constant, the better the sensitivity of the bridge.

Example 6.12

A strain gage load cell (force sensor) consists of four identical strain gages, forming a Wheatstone bridge, that are mounted on a rod that has square cross-section. One opposite pair of strain gages is mounted axially and the other pair is mounted in the transverse direction, as shown in Figure 6.46(a). To maximize the bridge sensitivity, the strain gages

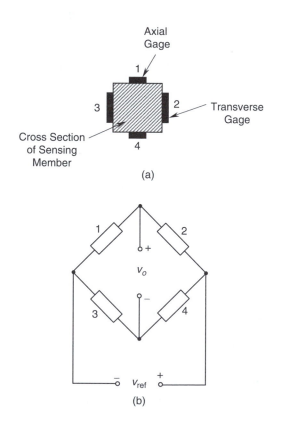

FIGURE 6.46
An example of four active strain gages: (a) mounting configuration on the load cell; (b) bridge circuit.

are connected to the bridge as shown in Figure 6.46(b). Determine the bridge constant k in terms of *Poisson's ratio v* of the rod material.

SOLUTION
Suppose that $\delta R_1 = \delta R$. Then, for the given configuration, we have

$$\delta R_2 = -v\delta R$$
$$\delta R_3 = -v\delta R$$
$$\delta R_4 = \delta R$$

Note that from the definition of Poisson's ratio:

$$\text{Transverse strain} = (-v) \times \text{longitudinal strain}$$

Now, it follows from Equation 6.78 that

$$\frac{\delta v_o}{v_{ref}} = 2(1+v)\frac{\delta R}{4R}$$

according to which the bridge constant is given by

$$k = 2(1 + v)$$

6.8.1.3 The Calibration Constant

The calibration constant C of a strain gage bridge relates the strain that is measured to the output of the bridge. Specifically,

$$\frac{\delta v_o}{v_{ref}} = C\varepsilon \qquad (6.80)$$

Now, in view of Equation 6.75 and Equation 6.79, the calibration constant may be expressed as

$$C = \frac{k}{4}S_s \qquad (6.81)$$

where k is the *bridge constant* and S_s is the *sensitivity* or *gage factor* of the strain gage. Ideally, the calibration constant should remain constant over the measurement range of the bridge (i.e., independent of strain ε and time t) and should be stable with respect to ambient conditions. In particular, there should not be any creep, nonlinearities such as hysteresis, or thermal effects.

Example 6.13

A schematic diagram of a strain gage accelerometer is shown in Figure 6.47(a). A point mass of weight W is used as the acceleration sensing element, and a light cantilever with rectangular cross-section, mounted inside the accelerometer casing, converts the inertia force of the mass into a strain. The maximum bending strain at the root of the cantilever is measured using four identical active semiconductor strain gages. Two of the strain gages (A and B) are mounted axially on the top surface of the cantilever, and the remaining two (C and D) are mounted on the bottom surface, as shown in Figure 6.47(b). In order to

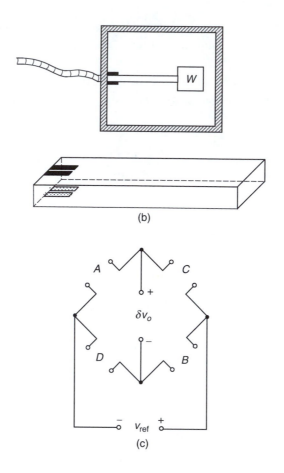

(b)

(c)

FIGURE 6.47
A miniature accelerometer using strain gages: (a) Schematic diagram; (b) Mounting configuration of the strain gages; (c) Bridge connection.

maximize the sensitivity of the accelerometer, indicate the manner in which the four strain gages—A, B, C, and D—should be connected to a Wheatstone bridge circuit. What is the bridge constant of the resulting circuit?

Obtain an expression relating the applied acceleration a (in units of g, which denotes the acceleration due to gravity) to the bridge output δv_o (measured using a bridge balanced at zero acceleration) in terms of the following parameters:

$W = Mg$ = weight of the seismic mass at the free end of the cantilever element
E = Young's modulus of the cantilever
ℓ = length of the cantilever
b = cross-section width of the cantilever
h = cross-section height of the cantilever
S_s = gage factor (sensitivity) of each strain gage
v_{ref} = supply voltage to the bridge

If $M = 5$ gm, $E = 5 \times 10^{10}$ N/m², $\ell = 1$ cm, $b = 1$ mm, $h = 0.5$ mm, $S_s = 200$, and $v_{ref} = 20$ V, determine the sensitivity of the accelerometer in mV/g.

If the yield strength of the cantilever element is 5×10^7 N/m², what is the maximum acceleration that could be measured using the accelerometer? If the ADC which reads the strain signal into a process computer has the range 0–10 V, how much amplification (bridge amplifier gain) would be needed at the bridge output so that this maximum acceleration corresponds to the upper limit of the ADC (10 V)?

Is the cross-sensitivity (i.e., the sensitivity in the two directions orthogonal to the direction of sensitivity shown in Figure 6.47(a)) small with your arrangement of the strain gage bridge? Explain.

HINT For a cantilever subjected to force F at the free end, the maximum stress at the root is given by

$$\sigma = \frac{6F\ell}{bh^2}$$

with the present notation.

NOTE Microelectromechanical (MEMS) accelerometers where the cantilever member, inertia element, and the strain gage are all integrated into a single semiconductor (silicon) unit are available in commercial applications such as air bag activation sensors for automobiles.

SOLUTION
Clearly, the bridge sensitivity is maximized by connecting the strain gages A, B, C, and D to the bridge as shown in Figure 6.47(c). This follows from Equation 6.78, noting that the contributions from the four strain gages are positive when δR_1 and δR_4 are positive, δR_2 and δR_3 and are negative. The bridge constant for the resulting arrangement is $k = 4$. Hence, from Equation 6.79,

$$\frac{\delta v_o}{v_{\text{ref}}} = \frac{\delta R}{R}$$

or, from Equation 6.80 and Equation 6.81,

$$\frac{\delta v_o}{v_{\text{ref}}} = S_s \varepsilon$$

Also,

$$\varepsilon = \frac{\sigma}{E} = \frac{6F\ell}{Ebh^2}$$

where F denotes the inertia force;

$$F = \frac{W}{g}\ddot{x} = Wa$$

Note that $\ddot{x}$ is the acceleration in the direction of sensitivity and $\ddot{x}/g = a$ is the acceleration in units of g.
    Thus,

$$\varepsilon = \frac{6W\ell}{Ebh^2}a$$

or

$$\delta v_o = \frac{6W\ell}{Ebh^2} S_s v_{ref} \, a$$

Now, with the given numerical values,

$$\frac{\delta v_o}{a} = \frac{6 \times 5 \times 10^{-3} \times 9.81 \times 1 \times 10^{-2} \times 200 \times 20}{5 \times 10^{10} \times 1 \times 10^{-3} \times (0.5 \times 10^{-3})^2} \text{ V/g}$$

$$= 0.94 \text{ V/g}$$

$$\frac{\varepsilon}{a} = \frac{1}{S_s v_{ref}} \frac{\delta v_o}{a} = \frac{0.94}{200 \times 20} \text{ strain/g}$$

$$= 2.35 \times 10^{-4} \ e/g = 235.0 \ \mu\varepsilon/g$$

$$\text{Yield strain} = \frac{\text{Yield strength}}{E} = \frac{5 \times 10^7}{5 \times 10^{10}} = 1 \times 10^{-3} \text{ strain}$$

Hence,

$$\text{Number of } g\text{'s to yield point} = \frac{1 \times 10^{-3}}{2.35 \times 10^{-4}} \ g = 4.26 \ g$$

$$\text{Corresponding voltage} = 0.94 \times 4.26 \text{ V} = 4.0 \text{ V}$$

Hence, the amplifier gain = 10.0/4.0 = 2.25

Cross-sensitivity comes from accelerations in the two directions y and z, which are orthogonal to the direction of sensitivity (x). In the lateral (y) direction, the inertia force causes lateral bending. This will produce equal tensile (or compressive) strains in B and D and equal compressive (or tensile) strains in A and C. According to the bridge circuit, we see that these contributions cancel each other. In the axial (z) direction, the inertia force causes equal tensile (or compressive) stresses in all four strain gages. These also will cancel out, as is clear from the relationship 6.78 for the bridge, which gives

$$\frac{\delta v_o}{v_{ref}} = \frac{(\delta R_A - \delta R_C - \delta R_D + \delta R_B)}{4R}$$

It follows that this arrangement compensates for cross-sensitivity problems.

6.8.1.4   *Data Acquisition*

For measuring dynamic strains, either the servo null-balance method or the imbalance output method should be employed (see Chapter 4). A schematic diagram for the imbalance output method is shown in Figure 6.48. In this method, the output from the active bridge is directly measured as a voltage signal and calibrated to provide the measured strain. Figure 6.48 shows the use of an ac bridge. In this case, the bridge is powered by an ac voltage. The supply frequency should be about 10 times the maximum frequency of interest in the dynamic strain signal (bandwidth). A supply frequency on the order of 1 kHz is typical. This signal is generated by an oscillator and is fed into the bridge. The transient component of the output from the bridge is very small (typically less than 1 mV and possibly a few microvolts). This signal has to be amplified, demodulated (especially

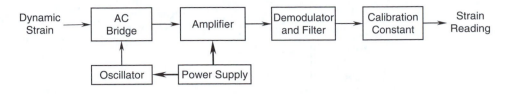

FIGURE 6.48
Measurement of dynamic strains using an ac bridge.

if the signals are transient), and filtered to provide the strain reading. The calibration constant of the bridge should be known in order to convert the output voltage to strain.

Strain gage bridges powered by dc voltages are common. They have the advantages of simplicity with regard to necessary circuitry and portability. The advantages of ac bridges include improved stability (reduced drift) and accuracy, and reduced power consumption.

6.8.1.5 Accuracy Considerations

Foil gages are available with resistances as low as 50 Ω and as high as several kilohms. The power consumption of a bridge circuit decreases with increased resistance. This has the added advantage of decreased heat generation. Bridges with a high range of measurement (e.g., a maximum strain of 0.01 m/m) are available. The accuracy depends on the linearity of the bridge, environmental effects (particularly temperature), and mounting techniques. For example, zero shift, due to strains produced when the cement or epoxy that is used to mount the strain gage dries, will result in calibration error. Creep will introduce errors during static and low-frequency measurements. Flexibility and hysteresis of the bonding cement (or epoxy) will bring about errors during high-frequency strain measurements. Resolutions on the order of 1 μm (i.e., one *microstrain*) are common.

As noted earlier, the cross-sensitivity of a strain gage is the sensitivity to strains that are orthogonal to the measured strain. This cross-sensitivity should be small (say, less than 1% of the direct sensitivity). Manufacturers usually provide cross-sensitivity factors for their strain gages. This factor, when multiplied by the cross strain present in a given application, gives the error in the strain reading due to cross-sensitivity.

Often, strains in moving members are sensed for control purposes. Examples include real-time monitoring and failure detection in machine tools, measurement of power, measurement of force and torque for feedforward and feedback control in dynamic systems, biomechanical devices, and tactile sensing using instrumented hands in industrial robots. If the motion is small or the device has a limited stroke, strain gages mounted on the moving member can be connected to the signal-conditioning circuitry and power source, using coiled flexible cables. For large motions, particularly in rotating shafts, some form of commutating arrangement has to be used. Slip rings and brushes are commonly used for this purpose. When ac bridges are used, a mutual-induction device (rotary transformer) may be used, with one coil located on the moving member and the other coil stationary. To accommodate and compensate for errors (e.g., losses and glitches in the output signal) caused by commutation, it is desirable to place all four arms of the bridge, rather than just the active arms, on the moving member.

6.8.2 Semiconductor Strain Gages

In some low-strain applications (e.g., dynamic torque measurement), the sensitivity of foil gages is not adequate to produce an acceptable strain gage signal. Semiconductor (SC) strain gages are particularly useful in such situations. The strain element of an SC strain

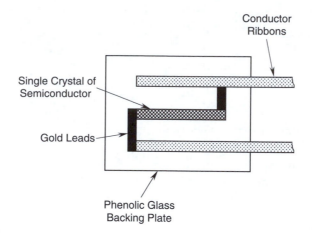

FIGURE 6.49
Component details of a semiconductor strain gage.

TABLE 6.3

Properties of Common Strain Gage Material

Material	Composition	Gage Factor (Sensitivity)	Temperature Coefficient of Resistance (10^{-6}/°C)
Constantan	45% Ni, 55% Cu	2.0	15
Isoelastic	36% Ni, 52% Fe, 8% Cr, 4% (Mn, Si, Mo)	3.5	200
Karma	74% Ni, 20% Cr, 3% Fe, 3% Al	2.3	20
Monel	67% Ni, 33% Cu	1.9	2000
Silicon	*p*-type	100–170	70–700
Silicon	*n*-type	−140 to −100	70–700

gage is made of a single crystal of *piezoresistive* material such as silicon, doped with a trace impurity such as boron. A typical construction is shown in Figure 6.49. The gage factor (sensitivity) of an SC strain gage is about two orders of magnitude higher than that of a metallic foil gage (typically, 40–200), as seen for Silicon, from the data given in Table 6.3. The resistivity is also higher, providing reduced power consumption and heat generation. Another advantage of SC strain gages is that they deform elastically to fracture. In particular, mechanical hysteresis is negligible. Furthermore, they are smaller and lighter, providing less cross-sensitivity, reduced distribution error (i.e., improved spatial resolution), and negligible error due to mechanical loading. The maximum strain that is measurable using a semiconductor strain gage is typically 0.003 m/m (i.e., 3000 $\mu\varepsilon$). Strain gage resistance can be several hundred ohms (typically, 120 Ω or 350 Ω).

There are several disadvantages associated with semiconductor strain gages, however, which can be interpreted as advantages of foil gages. Undesirable characteristics of SC gages include the following:

1. The strain-resistance relationship is more nonlinear.
2. They are brittle and difficult to mount on curved surfaces.
3. The maximum strain that can be measured is an order of magnitude smaller (typically, less than 0.01 m/m).

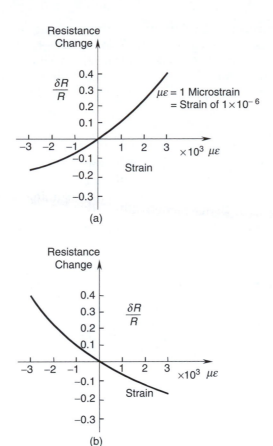

FIGURE 6.50
Nonlinear behavior of a semiconductor (silicon/boron) strain gage: (a) A *p*-type gage; (b) An *n*-type gage.

   4. They are more costly.

   5. They have a much larger temperature sensitivity.

The first disadvantage is illustrated in Figure 6.50. There are two types of semiconductor strain gages: the *p*-type, which are made of a semiconductor (e.g., silicon) doped with an acceptor impurity (e.g., boron), and the *n*-type, which are made of a semiconductor doped with a donor impurity (e.g., arsenic). In *p*-type strain gages, the direction of sensitivity is along the (1, 1, 1) crystal axis, and the element produces a "positive" (*p*) change in resistance in response to a positive strain. In *n*-type strain gages, the direction of sensitivity is along the (1, 0, 0) crystal axis, and the element responds with a "negative" (*n*) change in resistance to a positive strain. In both types, the response is nonlinear and can be approximated by the quadratic relationship

$$\frac{\delta R}{R} = S_1\varepsilon + S_2\varepsilon^2 \tag{6.82}$$

The parameter S_1 represents the *linear gage factor* (*linear sensitivity*), which is positive for *p*-type gages and negative for *n*-type gages. Its magnitude is usually somewhat larger for *p*-type gages, corresponding to better sensitivity. The parameter S_2 represents the degree of nonlinearity, which is usually positive for both types of gages. Its magnitude, however, is typically a little smaller for p-type gages. It follows that p-type gages are less nonlinear and have higher strain sensitivities. The nonlinear relationship given by Equation 6.82 or the nonlinear characteristic curve (Figure 6.50) should be used when measuring moderate to large strains with semiconductor strain gages. Otherwise, the nonlinearity error would be excessive.

Example 6.14

For a semiconductor strain gage characterized by the quadratic strain-resistance relationship, Equation 6.82, obtain an expression for the equivalent gage factor (sensitivity) S_s, using least squares error linear approximation and assuming that strains in the range $\pm\varepsilon_{max}$, have to be measured. Derive an expression for the percentage nonlinearity.

Taking $S_1 = 117$, $S_2 = 3600$, and $\varepsilon_{max} = 1 \times 10^{-2}$, calculate S_s and the percentage nonlinearity.

SOLUTION

The linear approximation of Equation 6.82 may be expressed as

$$\left[\frac{\delta R}{R}\right]_L = S_s \varepsilon$$

The error is given by

$$e = \frac{\delta R}{R} - \left[\frac{\delta R}{R}\right]_L = S_1 \varepsilon + S_2 \varepsilon^2 - S_s \varepsilon = (S_1 - S_s)\varepsilon + S_2 \varepsilon^2 \tag{i}$$

The quadratic integral error is

$$J = \int_{-\varepsilon_{max}}^{\varepsilon_{max}} e^2 d\varepsilon = \int_{-\varepsilon_{max}}^{\varepsilon_{max}} \left[(S_1 - S_s)\varepsilon + S_2 \varepsilon^2\right]^2 d\varepsilon \tag{ii}$$

We have to determine S_s that will result in minimum J. Hence, we use $\frac{\partial J}{\partial S_s} = 0$. Thus, from Equation ii:

$$\int_{-\varepsilon_{max}}^{\varepsilon_{max}} (-2\varepsilon)\left[(S_1 - S_s)\varepsilon + S_2 \varepsilon^2\right]^2 d\varepsilon = 0$$

On performing the integration, we get

$$S_s = S_1 \tag{6.83}$$

The quadratic curve and the linear approximation are shown in Figure 6.51. Note that the maximum error is at $\varepsilon = \pm\varepsilon_{max}$. The maximum error value is obtained from Equation i, with $S_s = S_1$ and $\varepsilon = \pm\varepsilon_{max}$, as

$$e_{max} = S_2 \varepsilon_{max}^2$$

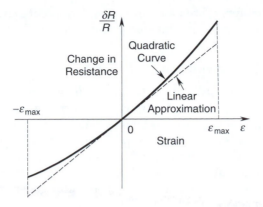

FIGURE 6.51
Least squares linear approximation for a semiconductor strain gage.

The true change in resistance (nondimensional) from to $-\varepsilon_{max}$ to $+\varepsilon_{max}$ is obtained using Equation 6.82; thus,

$$\frac{\Delta R}{R} = \left(S_1\varepsilon_{max} + S_2\varepsilon_{max}^2\right) - \left(-S_1\varepsilon_{max} + S_2\varepsilon_{max}^2\right)$$

$$= 2S_1\varepsilon_{max}$$

Hence, the percentage nonlinearity is given by

$$N_p = \frac{\max\,\text{error}}{\text{range}} \times 100\% = \frac{S_2\varepsilon_{max}^2}{2S_1\varepsilon_{max}} \times 100\%$$

or

$$N_p = 50S_2\varepsilon_{max}/S_1\%　　　　　　　　　(6.84)$$

Now, with the given numerical values, we have

$$S_s = 117$$

and

$$N_p = 50 \times 3,600 \times 1 \times 10^{-2}/117\% = 15.4\%$$

Note that we obtained this high value for nonlinearity because the given strain limits were high. Usually, the linear approximation is adequate for strains up to $\pm 1 \times 10^{-3}$.

The higher temperature sensitivity listed as a disadvantage of semiconductor strain gages may be considered an advantage in some situations. For instance, it is this property of high temperature sensitivity that is used in piezoresistive temperature sensors. Furthermore, using the fact that the temperature sensitivity of a semiconductor strain gage can be determined very accurately, accurate methods can be employed for temperature compensation in strain gage circuitry, and temperature calibration can also be done accurately. In particular, a passive SC strain gage may be used as an accurate temperature sensor for compensation purposes.

6.8.3 Automatic (Self) Compensation for Temperature

In foil gages the change in resistance due to temperature is typically small. Then the linear (first-order) approximation for the contribution from each arm of the bridge to the output signal, as given by Equation 6.78, would be adequate. These contributions cancel out if we pick strain gage elements and bridge completion resistors properly—for example, R_1 identical to R_2 and R_3 identical R_4. If this is the case, the only remaining effect of temperature change on the bridge output signal will be due to changes in the parameter values k and S_s (see Equation 6.80 and Equation 6.81). For foil gages, such changes are also typically negligible. Hence, for small to moderate temperature changes, additional compensation will not be required when foil gage bridge circuits are employed.

In semiconductor gages, not only the change in resistance with temperature (and with strain) is larger; the change in S_s with temperature is also larger compared to the corresponding values for foil gages. Hence, the linear approximation given by Equation 6.78 might not be accurate under variable temperature conditions; furthermore, the bridge sensitivity could change significantly with temperature. Under such conditions, temperature compensation will be necessary.

Directly measuring temperature and correcting strain gage readings accordingly, using calibration data, is a straightforward way to account for temperature changes. Another method of temperature compensation is described now. This method assumes that the linear approximation given by Equation 6.78 is valid; hence, Equation 6.80 is applicable.

The resistance R and strain sensitivity (or gage factor) S_s of a semiconductor strain gage are highly dependent on the concentration of the trace impurity, in a nonlinear manner. The typical behavior of the temperature coefficients of these two parameters for a p-type semiconductor strain gage is shown in Figure 6.52. The *temperature coefficient of resistance* α and the *temperature coefficient of sensitivity* β are defined by

$$R = R_o(1 + \alpha \cdot \Delta T) \tag{6.85}$$

$$S_s = S_{so}(1 + \beta \cdot \Delta T) \tag{6.86}$$

where ΔT denotes the temperature increase. Note from Figure 6.52 that β is a negative quantity and that for some dope concentrations, its magnitude is less than the value of the temperature coefficient of resistance (α). This property can be used in self-compensation with regard to temperature.

Consider a constant-voltage bridge circuit with a compensating resistor R_c connected to the supply lead, as shown in Figure 6.53(a). It can be shown that self-compensation can result if R_c is set to a value predetermined on the basis of the temperature coefficients of the strain gages. Consider the case where load impedance is very high and the bridge has four identical SC strain gages, which have resistance R. In this case, the bridge can be represented by the circuit shown in Figure 6.53(b). Since series impedances are additive and parallel admittances (inverse of impedance) are additive, the equivalent resistance of the bridge is R. Hence, the voltage supplied to the bridge, allowing for the voltage drop across R_c, is not v_{ref} but v_i, as given by

$$v_i = \frac{R}{(R + R_c)} v_{ref} \tag{6.87}$$

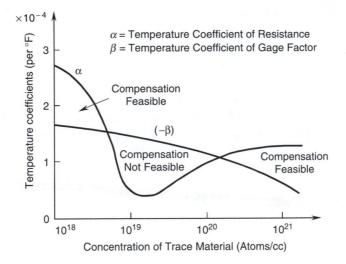

FIGURE 6.52
Temperature coefficients of resistance and gage factor for a *p*-type semiconductor (silicon) strain gage.

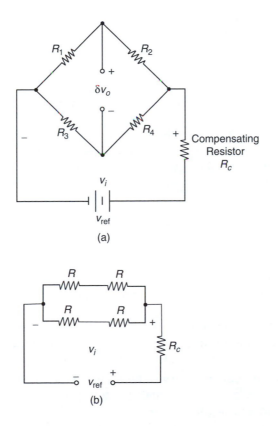

FIGURE 6.53
A strain gage bridge with a compensating resistor: (a) Constant voltage dc bridge; (b) Equivalent circuit with high load impedance.

Now, from Equation 6.80, we have

$$\frac{\delta v_o}{v_{ref}} = \frac{R}{(R+R_c)}\frac{kS_s}{4}\varepsilon \tag{6.88}$$

We will assume that the bridge constant k does not change with temperature. Otherwise, the following procedure will still hold, provided that the calibration constant C is used in place of the gage factor S_s (see Equation 6.81). For self-compensation, we must have the same output after the temperature changes through ΔT. Hence, from Equation 6.88, we have

$$\frac{R_o}{(R_o + R_c)}S_{so} = \frac{R_o(1 + \alpha \cdot \Delta T)}{[R_o(1 + \alpha \cdot \Delta T) + R_c]}S_{so}(1 + \beta \cdot \Delta T)$$

where the subscript o denotes values before the temperature change. Cancellation of the common terms and cross-multiplication gives

$$R_o\beta + R_c(\alpha + \beta) = (R_o + R_c)\alpha\beta\Delta T$$

Now, since both $\alpha \cdot \Delta T$ and $\beta \cdot \Delta T$ are usually much smaller than unity, we may neglect the right-hand-side (second-order) term in the preceding equation. This gives the following expression for the compensating resistance:

$$R_c = -\left[\frac{\beta}{\alpha + \beta}\right]R_o \tag{6.89}$$

Note that compensation is possible because the temperature coefficient of the strain gage sensitivity (β) is negative. The feasible ranges of operation, which correspond to positive R_c are indicated in Figure 6.52. This method requires that R_c be maintained constant at the chosen value under changing temperature conditions. One way to accomplish this is by selecting a material with negligible temperature coefficient of resistance for R_c. Another way is to locate R_c in a separate, temperature-controlled environment.

6.9   Torque Sensors

Sensing of torque and force is useful in many applications, including the following:

1. In robotic tactile and manufacturing applications—such as gripping, surface gauging, and material forming—where exerting an adequate load on an object is the primary purpose of the task.

2. In the control of fine motions (e.g., fine manipulation and micromanipulation) and in assembly tasks, where a small motion error can cause large damaging forces or performance degradation.

3. In control systems that are not fast enough when motion feedback alone is employed, where force feedback and feedforward force control can be used to improve accuracy and bandwidth.

4. In process testing, monitoring, and diagnostic applications, where torque sensing can detect, predict, and identify abnormal operation, malfunction, component failure, or excessive wear (e.g., in monitoring machine tools such as milling machines and drills).

5. In the measurement of power transmitted through a rotating device, where power is given by the product of torque and angular velocity in the same direction.

6. In controlling complex nonlinear mechanical systems, where measurement of force and acceleration can be used to estimate unknown nonlinear terms and an appropriate nonlinear feedback can linearize or simplify the system (nonlinear feedback control).

In most applications, sensing is done by detecting an effect of torque or the cause of torque. As well, there are methods for measuring torque directly. Common methods of torque sensing include the following:

1. Measuring strain in a sensing member between the drive element and the driven load, using a strain gage bridge.

2. Measuring displacement in a sensing member (as in the first method)—either directly, using a displacement sensor, or indirectly, by measuring a variable, such as magnetic inductance or capacitance, that varies with displacement.

3. Measuring reaction in support structure or housing (by measuring a force) and the associated lever arm length.

4. In electric motors, measuring the field or armature current that produces motor torque; in hydraulic or pneumatic actuators, measuring actuator pressure.

5. Measuring torque directly, using piezoelectric sensors, for example.

6. Employing a servo method—balancing the unknown torque with a feedback torque generated by an active device (say, a servomotor) whose torque characteristics are precisely known.

7. Measuring the angular acceleration caused by the unknown torque in a known inertia element.

The remainder of this section will be devoted to a discussion of torque measurement using some of these methods. Note that force sensing may be accomplished by essentially the same techniques. For the sake of brevity, however, we will limit our treatment primarily to torque sensing. The extension of torque-sensing techniques to force sensing is challenging.

6.9.1 Strain Gage Torque Sensors

The most straightforward method of torque sensing is to connect a torsion member between the drive unit and the load in series, as shown in Figure 6.54, and to measure the torque in the torsion member. If a circular shaft (solid or hollow) is used as the torsion member, the torque-strain relationship becomes relatively simple, and is given by:

$$\varepsilon = \frac{r}{2GJ}T \qquad (6.90)$$

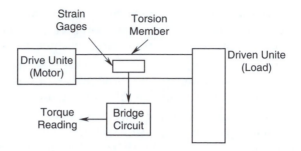

FIGURE 6.54
Torque sensing using a torsion member.

where
 T = torque transmitted through the member
 ε = principal strain (45° to axis) at radius r of the member
 J = polar moment of area of cross-section of the member
 G = shear modulus of the material

Also, the shear stress τ at a radius r of the shaft is given by

$$\tau = \frac{Tr}{J} \tag{6.91}$$

It follows from Equation 6.90 that torque T can determined by measuring direct strain ε on the shaft surface along a principal stress direction (i.e., at 45° to the shaft axis). This is the basis of torque sensing using strain measurements. Using the general bridge Equation 6.80 along with Equation 6.81 in Equation 6.90, we can obtain torque T from bridge output δv_o:

$$T = \frac{8GJ}{kS_s r} \frac{\delta v_o}{v_{\text{ref}}} \tag{6.92}$$

where S_s is the gage factor (or sensitivity) of the strain gages. The bridge constant k depends on the number of active strain gages used. Strain gages are assumed to be mounted along a principal direction. Three possible configurations are shown in Figure 6.55. In configurations (a) and (b) only two strain gages are used, and the bridge constant $k = 2$. Note that both axial and bending loads are compensated with the given configurations because resistance in both gages will be changed by the same amount (same sign and same magnitude), which cancels out up to first order, for the bridge circuit connection shown in Figure 6.55. Configuration (c) has two pairs of gages, mounted on the two opposite surfaces of the shaft. The bridge constant is doubled in this configuration, and here again, the sensor self-compensates for axial and bending loads up to first order $[O(\delta R)]$.

Configuration	(a)	(b)	(c)
Bridge constant (k):	2	2	4
Axial loads compensated:	Yes	Yes	Yes
Bending loads compensated:	Yes	Yes	Yes

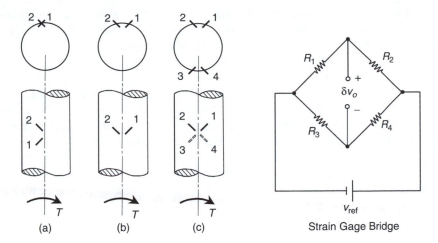

FIGURE 6.55
Strain gage configurations for a circular shaft torque sensor.

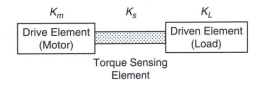

FIGURE 6.56
Stiffness degradation due to flexibility of the torque-sensing element.

6.9.2 Design Considerations

Two conflicting requirements in the design of a torsion element for torque sensing are sensitivity and bandwidth. The element has to be sufficiently flexible in order to get an acceptable level of sensor sensitivity (i.e., a sufficiently large output signal). According to Equation 6.90, this requires a small torsional rigidity GJ, so as to produce a large strain for a given torque. Unfortunately, since the torsion-sensing element is connected in series between a drive element and a driven element, an increase in flexibility of the torsion element will result in reduction of the overall stiffness of the system. Specifically, with reference to Figure 6.56, the overall stiffness K_{old} prior to connecting the torsion element is given by

$$\frac{1}{K_{old}} = \frac{1}{K_m} + \frac{1}{K_L} \tag{6.93}$$

and the stiffness K_{new} after connecting the torsion member is given by

$$\frac{1}{K_{new}} = \frac{1}{K_m} + \frac{1}{K_L} + \frac{1}{K_s} \tag{6.94}$$

where K_m is the equivalent stiffness of the drive unit (motor), K_L is the equivalent stiffness of the load, and K_s is the stiffness of the torque-sensing element. It is clear from Equation 6.93 and Equation 6.94 that $1/K_{new} > 1/K_{old}$. Hence, $K_{new} < K_{old}$. This reduction in stiffness is associated with a reduction in natural frequency and bandwidth, resulting in slower response to control commands in the overall system. Furthermore, a reduction in stiffness causes a reduction in the loop gain. As a result, the steady-state error in some motion variables can increase, which will demand more effort from the controller to achieve a required level of accuracy. One aspect in the design of the torsion element is to guarantee that the element stiffness is small enough to provide adequate sensitivity but large enough to maintain adequate bandwidth and system gain. In situations where K_s cannot be increased adequately without seriously jeopardizing the sensor sensitivity, the system bandwidth can be improved by decreasing either the load inertia or the drive unit (motor) inertia.

Example 6.15

Consider a rigid load, which has a polar moment of inertia J_L, and driven by a motor with a rigid rotor, which has inertia J_m. A torsion member of stiffness K_s is connected between the rotor and the load, as shown in Figure 6.57(a), in order to measure the torque transmitted to the load. Determine the transfer function between the motor torque T_m and the twist angle θ of the torsion member. What is the torsional natural frequency ω_n of the system? Discuss why the system bandwidth depends on ω_n. Show that the bandwidth can be improved by increasing K_s, by decreasing J_m, or by decreasing J_L. Mention some advantages and disadvantages of introducing a gearbox at the motor output.

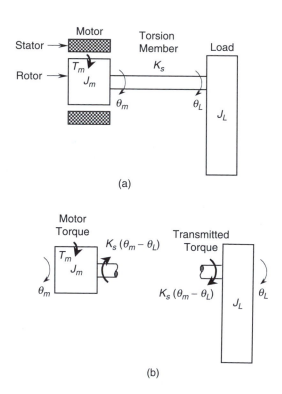

(a)

(b)

FIGURE 6.57
An example of bandwidth analysis of a system with a torque sensor: (a) System model; (b) Free-body diagram.

SOLUTION
From the free-body diagram shown in Figure 6.57(b), the equations of motion can be written:

$$\text{For motor:} \quad J_m \ddot{\theta}_m = T_m - K_s(\theta_m - \theta_L) \tag{i}$$

$$\text{For load:} \quad J_L \ddot{\theta}_L = K_s(\theta_m - \theta_L) \tag{ii}$$

Note that θ_m is the motor rotation and θ_L is the load rotation. Divide Equation i by J_m, divide Equation ii by J_L, and subtract the second equation from the first; thus,

$$\ddot{\theta}_m - \ddot{\theta}_L = \frac{T_m}{J_m} - \frac{K_s}{J_m}(\theta_m - \theta_L) - \frac{K_s}{J_L}(\theta_m - \theta_L)$$

This equation can be expressed in terms of the twist angle:

$$\theta = \theta_m - \theta_L \tag{iii}$$

$$\ddot{\theta} + K_s\left(\frac{1}{J_m} + \frac{1}{J_L}\right)\theta = \frac{T_m}{J_m} \tag{iv}$$

Hence, the transfer function $G(s)$ between input T_m and output θ is obtained by introducing the Laplace variable s in place of the time derivative d/dt. Specifically, we have:

$$G(s) = \frac{1/J_m}{s^2 + K_s(1/J_m + 1/J_L)} \tag{v}$$

The characteristic equation of the twisting system is

$$s^2 + K_s\left(\frac{1}{J_m} + \frac{1}{J_L}\right) = 0 \tag{vi}$$

It follows that the torsional (twisting) natural frequency ω_n is given by

$$\omega_n = \sqrt{K_s\left(\frac{1}{J_m} + \frac{1}{J_L}\right)} \tag{vii}$$

In addition to this natural frequency, there is a zero natural frequency in the overall system, which corresponds to rotation of the entire system as a rigid body without any twisting in the torsion member (the *rigid-body mode*). Note that both natural frequencies will be obtained if the output is taken as either θ_m or θ_L, not the twist angle θ. When the output is taken as the twist angle θ, the response is measured relative to the rigid-body mode; hence, the zero-frequency term disappears from the characteristic equation.

The transfer function given by Equation v may be written as

$$G(s) = \frac{1/J_m}{s^2 + \omega_n^2} \tag{viii}$$

In the frequency domain, $s = j\omega$, and the resulting frequency transfer function is

$$G(j\omega) = \frac{1/J_m}{\omega_n^2 - \omega^2} \qquad (ix)$$

It follows that if ω is small in comparison to ω_n, the transfer function can be approximated by

$$G(j\omega) = \frac{1/J_m}{\omega_n^2} \qquad (x)$$

which is a static relationship, implying an instantaneous response without any dynamic delay. Since system bandwidth represents the excitation frequency range ω within which the system responds sufficiently fast, it follows that system bandwidth improves when ω_n is increased. Hence, ω_n is a measure of system bandwidth.

Now, observe from Equation vii that ω_n (and the system bandwidth) increases when K_s is increased, when J_m is decreased, or when J_L is decreased. If a gearbox is added to the system, the equivalent inertia increases and the equivalent stiffness decreases. This reduces the system bandwidth, resulting in a slower response. Another disadvantage of a gearbox is the backlash and friction that enter the system. The main advantage, however, is that torque transmitted to the load is amplified through speed reduction between motor and load. However, high torques and low speeds can be achieved by using torque motors without employing any speed reducers or by using backlash-free transmissions such as harmonic drives and traction (friction) drives (See Chapter 9 and Chapter 3).

_____

The design of a torsion element for torque sensing can be viewed as the selection of the polar moment of area J of the element to meet the following four requirements:

1. The strain capacity limit specified by the strain gage manufacturer is not exceeded.
2. A specified upper limit on nonlinearity for the strain gage is not exceeded for linear operation.
3. Sensor sensitivity is acceptable in terms of the output signal level of the differential amplifier (see Chapter 4) in the bridge circuit.
4. The overall stiffness (bandwidth, steady-state error, etc.) of the system is acceptable.

Now we will develop design criteria for each of these requirements.

6.9.2.1   Strain Capacity of the Gage

The maximum strain handled by a strain gage element is limited by factors such as strength, creep problems associated with the bonding material (epoxy), and hysteresis. This limit ε_{max} is specified by the strain gage manufacturer. For a typical semiconductor gage, the maximum strain limit is on the order of 3000 $\mu\varepsilon$. If the maximum torque that needs to be handled by the sensor is T_{max}, we have, from Equation 6.90:

$$\frac{r}{2GJ} T_{max} < \varepsilon_{max}$$

which gives

$$J > \frac{r}{2G} \frac{T_{max}}{\varepsilon_{max}} \qquad (6.95)$$

where ε_{max} and T_{max} are specified.

6.9.2.2 Strain Gage Nonlinearity Limit

For large strains, the characteristic equation of a strain gage will become increasingly nonlinear. This is particularly true for semiconductor gages. If we assume the quadratic Equation 6.82, the percentage nonlinearity N_p is given by Equation 6.84. For a specified nonlinearity, an upper limit for strain can be determined using this result; thus,

$$\frac{r}{2GJ} T_{max} = \varepsilon_{max} \leq \frac{N_p S_1}{50 S_2} \tag{6.96}$$

The corresponding J is given by,

$$J \geq \frac{25 S_2}{G S_1} \frac{T_{max}}{N_p} \tag{6.97}$$

where N_p is specified.

6.9.2.3 Sensitivity Requirement

The output signal from the strain gage bridge is provided by a differential amplifier (see Chapter 4), which detects the voltages at the two output nodes of the bridge (A and B in Figure 6.45), takes the difference, and amplifies it by a gain K_a. This output signal is supplied to an analog-to-digital converter (ADC), which provides a digital signal to the computer for performing further processing and control. The signal level of the amplifier output has to be sufficiently high so that the signal-to-noise ratio (SNR) is adequate. Otherwise, serious noise problems will result. Typically, a maximum voltage on the order of ±10 V is desired.

Amplifier output v is given by

$$v = K_a \delta v_o \tag{6.98}$$

where δv_o is the bridge output before amplification. It follows that the desired signal level can be obtained by simply increasing the amplifier gain. There are limits to this approach, however. In particular, a large gain will increase the susceptibility of the amplifier to saturation and instability problems, such as drift and errors due to parameter changes. Hence, sensitivity has to be improved as much as possible through mechanical considerations.

By substituting Equation 6.92 into Equation 6.98, we get the signal level requirement as:

$$v_o \leq \frac{K_a k S_s r v_{ref}}{8GJ} T_{max}$$

where v_o is the specified lower limit on the output signal from the bridge amplifier, and T_{max} is also specified. Then the limiting design value for J is given by:

$$J \leq \frac{K_a k S_s r v_{ref}}{8G} \frac{T_{max}}{v_o} \tag{6.99}$$

where v_o and T_{max} are specified.

6.9.2.4 Stiffness Requirement

The lower limit of the overall stiffness of the system is constrained by factors such as speed of response (represented by system bandwidth) and steady-state error (represented by system gain). The polar moment of area J should be chosen such that the stiffness of the torsion element does not fall below a specified limit K. First, we will obtain an expression for the torsional stiffness of a circular shaft. For a shaft of length L and radius r, a twist angle of θ corresponds to a shear strain of

$$\gamma = \frac{r\theta}{L} \tag{6.100}$$

on the outer surface. Accordingly, shear stress is given by

$$\tau = \frac{Gr\theta}{L} \tag{6.101}$$

Now in view of Equation 6.91, the torsional stiffness of the shaft is given by

$$K_s = \frac{T}{\theta} = \frac{GJ}{L} \tag{6.102}$$

Note that the stiffness can be increased by increasing GJ. However, this decreases the sensor sensitivity because, in view of Equation 6.90, measured direct strain ε decreases for a given torque when GJ is increased. Note that there are two other parameters—outer radius r and length L of the torsion element—which we can manipulate. Although for a solid shaft J increases (to the fourth power) with r, for hollow shafts it is possible to manipulate J and r independently, with practical limitations. For this reason, hollow members are commonly used as torque-sensing elements. With these design freedoms, for a given value of GJ, we can increase r to increase the sensitivity of the strain gage bridge without changing the system stiffness, and we can decrease L to increase the system stiffness without affecting the bridge sensitivity.

Assuming that the shortest possible length L is used in the sensor, for a specified stiffness limit K we should have $\frac{GJ}{L} \geq K$. The limiting design value for J is given by

$$J \geq \frac{L}{G} K \tag{6.103}$$

where K is specified.

The governing formulas for the polar moment of area J of the torque sensor, based on the four criteria discussed earlier, are summarized in Table 6.4.

TABLE 6.4

Design Criteria for a Strain Gage Torque-Sensing Element

Criterion	Specification	Governing formula for Polar Moment of Area (J)
Strain capacity of strain gage element	ε_{max} and T_{max}	$> \dfrac{r}{2G} \cdot \dfrac{T_{max}}{\varepsilon_{max}}$
Strain gage nonlinearity	N_p and T_{max}	$> \dfrac{25 r S_2}{G S_1} \cdot \dfrac{T_{max}}{N_p}$
Sensor sensitivity	v_o and T_{max}	$\leq \dfrac{K_a k S_s r v_{ref}}{8G} \cdot \dfrac{T_{max}}{v_o}$
Sensor stiffness (system bandwidth and gain)	K	$\geq \dfrac{L}{G} \cdot K$

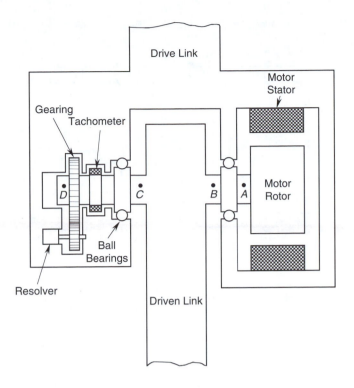

FIGURE 6.58
A joint of a direct-drive robotic arm.

Example 6.16

A joint of a direct-drive robotic arm is sketched in Figure 6.58. Note that the rotor of the drive motor is an integral part of the driven link, without the use of gears or any other speed reducers. Also, the motor stator is an integral part of the drive link. A tachometer measures the joint speed (relative), and a resolver measures the joint rotation (relative). Gearing is used to improve the performance of the resolver. Neglecting mechanical loading from sensors and gearing, but including bearing friction, sketch the torque distribution along the joint axis. Suggest a location (or locations) for measuring the net torque transmitted to the driven link using a strain gage torque sensor.

SOLUTION
For simplicity, assume point torques. Denoting the motor torque by T_m; the total rotor inertia torque and frictional torque in the motor by T_i; and the frictional torques at the two bearings by T_{f1}, and T_{f2}; the torque distribution can be sketched as shown in Figure 6.59. The net torque transmitted to the driven link is T_L. Locations available to install strain gages include, A, B, C, and D. Note that T_L is given by the difference between the torques at B and C. Hence, strain gage torque sensors should be mounted at B and C and the difference of the readings should be taken for accurate measurement of T_L. Since bearing friction is small for most practical purposes, a single torque sensor located at B will provide reasonably accurate results. The motor torque T_m is also approximately equal to the transmitted torque when bearing friction and motor loading effects (inertia and friction) are negligible. This is the reason behind using motor current (field or armature) to measure joint torque in some mechatronics applications.

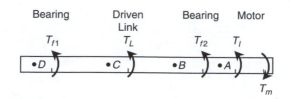

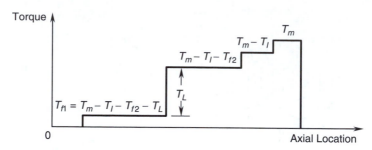

FIGURE 6.59
Torque distribution along the axis of a direct-drive manipulator joint.

Example 6.17

Consider the design of a tubular torsion element. Using the notation of Table 6.4, the following design specifications are given: $\varepsilon_{max} = 3000\ \mu\varepsilon$; $N_p = 5\%$; $v_o = 10$ V; and for a system bandwidth of 50 Hz, $K = 2.5 \times 10^3$ N · m/rad. A bridge with four active strain gages is used to measure torque in the torsion element. The following parameter values are provided:

1. For strain gages:

$$S_s = S_1 = 115,\ S_2 = 3500$$

2. For the torsion element:

$$\text{outer radius } r = 2 \text{ cm}$$

$$\text{shear modulus } G = 3 \times 10^{10} \text{ N/m}^2$$

$$\text{length } L = 2 \text{ cm}$$

3. For the bridge circuitry:

$$v_{ref} = 20 \text{ V} \quad \text{and} \quad K_a = 100$$

The maximum torque that is expected is

$$T_{max} = 10 \text{ N} \cdot \text{m}.$$

Using these values, design a torsion element for the sensor. Compute the operating parameter limits for the designed sensor.

SOLUTION

Let us assume a safety factor of 1 (i.e., use the limiting values of the design formulas). We can compute the polar moment of area J using each of the four criteria given in Table 6.4:

1. For $\varepsilon_{max} = 3000\ \mu\varepsilon$

$$J = \frac{0.02 \times 10}{2 \times 3 \times 10^{10} \times 3 \times 10^{-3}}\ \text{m}^4 = 1.11 \times 10^{-9}\ \text{m}^4$$

2. For $N_p = 5$:

$$J = \frac{25 \times 0.02 \times 3500 \times 10}{3 \times 10^{10} \times 115 \times 5}\ \text{m}^4 = 1.01 \times 10^{-9}\ \text{m}^4$$

3. For $v_o = 10$ V:

$$J = \frac{100 \times 4 \times 115 \times 0.02 \times 20 \times 10}{8 \times 3 \times 10^{10} \times 10}\ \text{m}^4 = 7.67 \times 10^{-8}\ \text{m}^4$$

4. For $K = 2.5 \times 10^3$ N · m/rad:

$$J = \frac{0.02 \times 2.5 \times 10^3}{3 \times 10^{10}}\ \text{m}^4 = 1.67 \times 10^{-9}\ \text{m}^4$$

It follows that for an acceptable sensor, we should satisfy

$$J \geq (1.11 \times 10^{-9})\ \text{and}\ (1.01 \times 10^{-9})\ \text{and}\ (1.67 \times 10^{-9})\ \text{and}\ J \leq 7.67 \times 10^{-8}\ \text{m}^4$$

We pick $J = 7.67 \times 10^{-8}$ m⁴ so that the tube thickness is sufficiently large to transmit load without buckling or yielding. Since, for a tubular shaft,

$$J = \frac{\pi}{2}\left(r_o^4 - r_i^4\right)$$

where r_o is the outer radius and r_i is the inner radius, we have

$$7.67 \times 10^{-8} = \frac{\pi}{2}\left(0.02^4 - r_i^4\right)$$

or

$$r_i = 1.8\ \text{cm}.$$

Now, with the chosen value for J:

$$\varepsilon_{max} = \frac{7.67 \times 10^{-8}}{1.11 \times 10^{-9}} \times 3000\ \mu\varepsilon = 2.07 \times 10^5\ \mu\varepsilon$$

$$N_p = \frac{1.01 \times 10^{-9}}{7.67 \times 10^{-8}} \times 5\% = 0.07\%$$

$$v_o = 10\ \text{V}$$

$$K = \frac{7.67 \times 10^{-8}}{1.67 \times 10^{-9}} \times 2.5 \times 10^3 = 1.15 \times 10^5\ \text{N} \cdot \text{m/rad}$$

Since natural frequency is proportional to the square root of stiffness, for a given inertia, we note that a bandwidth of

$$50\sqrt{\frac{1.15\times10^5}{2.5\times10^3}} = 339 \text{ Hz}$$

is possible with this design.

Although the manner in which strain gages are configured on a torque sensor can be exploited to compensate for cross-sensitivity effects arising from factors such as tensile and bending loads, it is advisable to use a torque-sensing element that inherently possesses low sensitivity to these factors, which cause error in a torque measurement. The tubular torsion element discussed in this section is convenient for analytical purposes because of the simplicity of the associated expressions for design parameters. Its mechanical design and integration into a practical system are convenient as well. Unfortunately, this member is not optimal with respect to rigidity (stiffness) for both bending and tensile loads. Alternative shapes and structural arrangements have to be considered when inherent rigidity (insensitivity) to cross-loads is needed. Furthermore, a tubular element has the same principal strain at all locations on the element surface. This does not give us a choice with respect to mounting locations of strain gages in order to maximize the torque sensor sensitivity. Another disadvantage of the basic tubular torsion member is that, due to curved surface, much care is needed in mounting fragile semiconductor gages, which could be easily damaged even with slight bending. Hence, a sensor element that has flat surfaces to mount the strain gages would be desirable.

A torque-sensing element that has the foregoing desirable characteristics (i.e., inherent insensitivity to cross-loading, nonuniform strain distribution on the surface, and availability of flat surfaces to mount strain gages) is shown in Figure 6.60. Notice that two sensing elements are connected radially between the drive unit and the driven member. The sensing elements undergo bending to transmit a torque between the driver and the driven member. Bending strains are measured at locations of high sensitivity and are taken to be proportional to the transmitted torque. Analytical determination of the calibration constant is not easy for such complex sensing elements, but experimental determination is straightforward. Finite element analysis may be used as well for this purpose. Note that the strain gage torque sensors measure the direction as well as the magnitude of the torque transmitted through it.

6.9.3 Deflection Torque Sensors

Instead of measuring strain in the sensor element, the actual deflection (twisting or bending) can be measured and used to determine torque, through a suitable calibration constant. For a circular shaft (solid or hollow) torsion element, the governing relationship is given by Equation 6.102, which can be written in the form

$$T = \frac{GJ}{L}\theta \tag{6.104}$$

The calibration constant GJ/L has to be small in order to achieve high sensitivity. This means that the element stiffness should be low. This limits the bandwidth (which measures the speed of response) and the gain (which determines the steady-state error) of the overall system. The twist angle θ is very small (e.g., a fraction of a degree) in systems

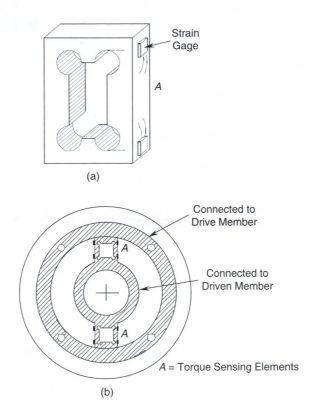

FIGURE 6.60
A bending element for torque sensing: (a) Shape of the sensing element; (b) Element location.

with high bandwidth. This requires very accurate measurement of θ for accurate determination of torque T. Two types of displacement torque sensors will be described next. One sensor directly measures the angle of twist, and the other sensor uses the change in magnetic induction associated with sensor deformation.

6.9.3.1 Direct-Deflection Torque Sensor

Direct measurement of the twist angle between two axial locations in a torsion member, using an angular displacement sensor, can be used to determine torque. The difficulty in this case is that under dynamic conditions, relative deflection has to be measured while the torsion element is rotating. One type of displacement sensor that could be used here is a synchro transformer. The two rotors of the synchro are mounted at the two ends of the torsion member. The synchro output gives the relative angle of rotation of the two rotors. Another type of displacement sensor that could be used is shown in Figure 6.61(a). Two ferromagnetic gear wheels are splined at two axial locations of the torsion element. Two stationary proximity probes of the magnetic induction type (self-induction or mutual induction) are placed radially, facing the gear teeth, at the two locations. As the shaft rotates, the gear teeth cause a change in flux linkage with the proximity sensor coils. The resulting output signals of the two probes are pulse sequences, shaped somewhat like sine waves. The phase shift of one signal with respect to the other determines the relative angular deflection of one gear wheel with respect to the other, assuming that the two probes are synchronized under no-torque conditions. Both the magnitude and the direction of the transmitted torque are determined using this method. A 360° phase shift corresponds

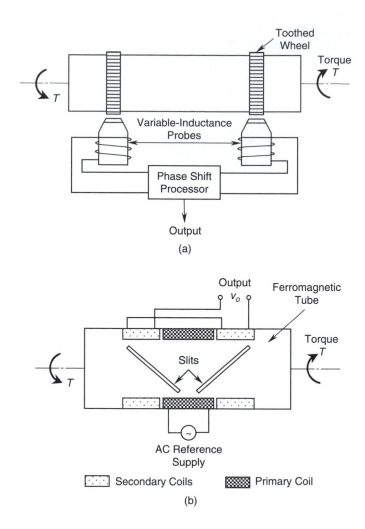

FIGURE 6.61
Deflection torque sensors: (a) A direct-deflection torque sensor; (b) A variable-reluctance torque sensor.

to a relative deflection by an integer multiple of the gear pitch. It follows that deflections less than half the pitch can be measured without ambiguity. Assuming that the output signals of the two probes are sine waves (narrow-band filtering can be used to achieve this), the phase shift ϕ will be proportional to the angular twist θ. If the gear wheel has n teeth, a primary phase shift of 2π corresponds to a twist angle of $2\pi/n$ radians. Hence, $\theta = \phi/n$ and from Equation 6.104, we get

$$T = \frac{GJ\phi}{Ln} \qquad (6.105)$$

where
   G = shear modulus of the torsion element
   J = polar moment of area of the torsion element
   ϕ = phase shift between the two proximity probe signals
   L = axial separation of the proximity probes
   n = number of teeth in each gear wheel

Note that the proximity probes are non-contact devices, unlike dc strain gage sensors. Also, note that eddy current proximity probes and Hall effect proximity probes (see Chapter 7) could be used instead of magnetic induction probes in this method of torque sensing.

6.9.3.2   Variable Reluctance Torque Sensor

A torque sensor that is based on sensor element deformation and that does not require a contacting commutator is shown in Figure 6.61(b). This is a variable-reluctance device, which operates like a differential transformer (RVDT or LVDT). The torque-sensing element is a ferromagnetic tube, which has two sets of slits, typically oriented along the two principal stress directions of the tube (45°) under torsion. When a torque is applied to the torsion member, one set of gaps closes and the other set opens as a result of the principal stresses normal to the slit axes. Primary and secondary coils are placed around the slitted tube, and they remain stationary. One segment of the secondary coil is placed around one set of slits, and the second segment is placed around the other (perpendicular) set. The primary coil is excited by an ac supply, and the induced voltage v_o in the secondary coil is measured. As the tube deforms, it changes the magnetic reluctance in the flux linkage path, thus changing the induced voltage. To obtain the best sensitivity, the two segments of the secondary coil, as shown in Figure 6.61(b), should be connected so that the induced voltages are absolutely additive (algebraically subtractive), because one voltage increases and the other decreases. The output signal should be demodulated (by removing the carrier frequency component) to effectively measure transient torques. Note that the direction of torque is given by the sign of the demodulated signal.

6.9.4   Reaction Torque Sensors

The foregoing methods of torque sensing use a sensing element, which is connected between the drive member and the driven member. A major drawback of such an arrangement is that the sensing element modifies the original system in an undesirable manner, particularly by decreasing the system stiffness and adding inertia. Not only will the overall bandwidth of the system decrease, but the original torque will also be changed (mechanical loading) because of the inclusion of an auxiliary sensing element. Furthermore, under dynamic conditions, the sensing element will be in motion, thereby making torque measurement more difficult. The reaction method of torque sensing eliminates these problems to a large degree. This method can be used to measure torque in a rotating machine. The supporting structure (or housing) of the rotating machine (e.g., motor, pump, compressor, turbine, generator) is cradled by releasing the fixtures, and the effort necessary to keep the structure from moving is measured. A schematic representation of the method is shown in Figure 6.62(a). Ideally, a lever arm is mounted on the cradled housing, and the force required to maintain the housing stationary is measured using a force sensor (load cell). The reaction torque on the housing is given by

$$T_R = F_R \cdot L \tag{6.106}$$

where
  F_R = reaction force measured using load cell
  L = lever arm length

Alternatively, strain gages or other types of force sensors may be mounted directly at the fixture locations (e.g., on the mounting bolts) of the housing, to measure the reaction forces

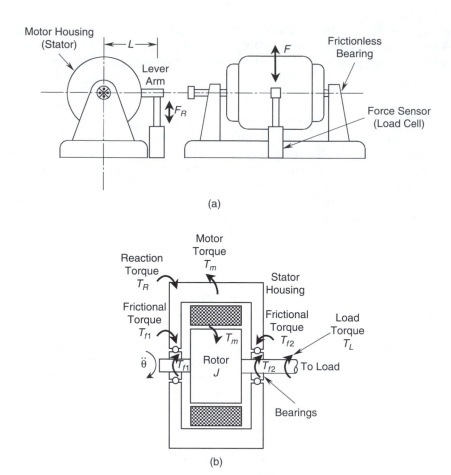

FIGURE 6.62
(a) Schematic representation of a reaction torque sensor setup (reaction dynamometer), (b) The relationship between reaction torque and load torque.

without cradling the housing. Then the reaction torque is determined with knowledge of the distance of the fixture locations from the shaft axis.

The reaction-torque method of torque sensing is widely used in dynamometers (reaction dynamometers), which determine the transmitted power in rotating machinery through the measurement of torque and shaft speed. A drawback of the reaction-type torque sensors can be explained using Figure 6.62(b). A motor of rotor inertia J, which rotates at angular acceleration $\ddot{\theta}$, is shown. By Newton's third law (action = reaction), the electromagnetic torque generated at the rotor of the motor T_m will be reacted back onto the stator and housing. In the figure, T_{f1} and T_{f2} denote the frictional torques at the two bearings and T_L is the torque transmitted to the driven load.

When applying Newton's second law to the entire system, note that the friction torques and the motor (magnetic) torque all cancel out, giving $J\ddot{\theta} = T_R - T_L$, or

$$T_L = T_R - J\ddot{\theta} \qquad (6.107)$$

Note that T_L is what must be measured. Under accelerating or decelerating conditions, the reaction torque T_R, which is measured, is not equal to the actual torque T_L that is transmitted. One method of compensating for this error is to measure shaft acceleration, compute the inertia torque, and adjust the measured reaction torque using this inertia torque. Note that the frictional torque in the bearings does not enter the final equation, which is an advantage of this method.

6.9.5 Motor Current Torque Sensors

Torque in an electric motor is generated as a result of the electromagnetic interaction between the armature windings and the field windings of the motor (see Chapter 9 for details). For a dc motor, the armature windings are located on the rotor and the field windings are on the stator. The (magnetic) torque T_m in a dc motor is given by

$$T_m = k i_f i_a \tag{6.108}$$

where
i_f = field current
i_a = armature current
k = torque constant.

It is seen from Equation 6.108 that the motor torque can be determined by measuring either i_a or i_f while the other is kept constant at a known value. In particular, note that i_f is assumed constant in armature control and i_a is assumed constant in field control.

As noted before (e.g., see Figure 6.62(b)), the magnetic torque of a motor is not quite equal to the transmitted torque, which is what needs to be sensed in most applications. It follows that the motor current provides only an approximation for the needed torque. The actual torque that is transmitted through the motor shaft (the load torque) is different from the motor torque generated at the stator-rotor interface of the motor. This difference is necessary for overcoming the inertia torque of the moving parts of the motor unit (particularly rotor inertia) and frictional torque (particularly bearing friction). Methods are available to adjust (compensate for) the readings of magnetic toque, so as to estimate the transmitted torque at reasonable accuracy. One approach is to incorporate a suitable dynamic model for the electromechanical system of the motor and the load into a *Kalman filter* whose input is the measured current and the estimated output is the transmitted load. A detailed presentation of this approach is beyond the present scope. The current can be measured by sensing the voltage across a known resistor (of low resistance) placed in series with the current circuit.

In the past, dc motors were predominantly used in complex control applications. Although ac synchronous motors were limited mainly to constant-speed applications in the past, they are finding numerous uses in variable-speed applications (e.g., robotic manipulators) and servo systems, because of rapid advances in solid-state drives. Today, ac motor drive systems incorporate both frequency control and voltage control using advanced semiconductor technologies (see Chapter 9). Torque in an ac motor may also be determined by sensing the motor current. For example, consider the three-phase synchronous motor shown schematically in Figure 6.63.

The armature windings of a conventional synchronous motor are carried by the stator (in contrast to the case of a dc motor). Suppose that the currents in the three phases (armature currents) are denoted by i_1, i_2, and i_3. The dc field current in the rotor windings

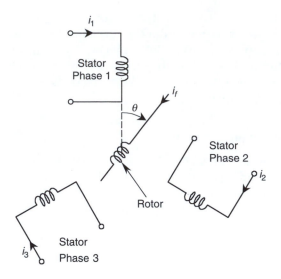

FIGURE 6.63
Schematic representation of a three-phase synchronous motor.

is denoted by i_f. The motor torque T_m can be expressed as

$$T_m = ki_f \left[i_1 \sin \theta + i_2 \sin\left(\theta - \frac{2\pi}{3}\right) + i_3 \sin\left(\theta - \frac{4\pi}{3}\right) \right] \tag{6.109}$$

where θ denotes the angular rotation of the rotor and k is the torque constant of the synchronous motor. Note that since i_f is assumed fixed, the motor torque can be determined by measuring the phase currents. For the special case of a "balanced" three-phase supply, we have

$$i_1 = i_a \sin \omega t \tag{6.110}$$

$$i_2 = i_a \sin\left(\omega t - \frac{2\pi}{3}\right) \tag{6.111}$$

$$i_3 = i_a \sin\left(\omega t - \frac{4\pi}{3}\right) \tag{6.112}$$

where ω denotes the line frequency (frequency of the current in each supply phase) and i_a is the amplitude of the phase current. Substituting Equations 6.110 through Equation 6.112 into Equation 6.109, and simplifying using well-known trigonometric identities, we get

$$T_m = 1.5ki_f ia \cos\left(\theta - \omega t\right) \tag{6.113}$$

We know that the angular speed of a three-phase synchronous motor with one pole pair per phase is equal to the line frequency ω (see Chapter 9). Accordingly, we have

$$\theta = \theta_0 + \omega t \tag{6.114}$$

where θ_0 denotes the angular position of the rotor at $t = 0$. It follows that with a balanced three-phase supply, the torque of a synchronous motor is given by

$$T_m = 1.5ki_f i_a \cos\theta_0 \qquad (6.115)$$

This expression is quite similar to the one for a dc motor, as given by Equation 6.108.

6.9.6 Force Sensors

Force sensors are useful in numerous applications. For example, cutting forces generated by a machine tool may be monitored to detect tool wear and an impending failure and to diagnose the causes of failure; to control the machine tool, with feedback; and to evaluate the product quality. In vehicle testing, force sensors are used to monitor impact forces on the vehicles and crash-test dummies. Robotic handling and assembly tasks are controlled by measuring the forces generated at the end effector. Measurement of excitation forces and corresponding responses is employed in experimental modeling (model identification) of mechanical systems. Direct measurement of forces is useful in nonlinear feedback control of mechanical systems.

Force sensors that employ strain gage elements or piezoelectric (quartz) crystals with built-in microelectronics are common. For example, thin-film and foil sensors that employ the strain gage principle for measuring forces and pressures are commercially available. A sketch of an industrial load cell, which uses the strain-gage method is shown in Figure 6.64. Both impulsive forces and slowly varying forces can be monitored using this sensor. Some types of force sensors are based on measuring a deflection caused by the force. Relatively high deflections (fraction of a mm) would be necessary for this technique to be feasible. Commercially available sensors range from sensitive devices, which can detect forces on the order of thousandth of a newton to heavy-duty load cells, which can handle very large forces (e.g., 10,000 N). Since the techniques of torque sensing can be extended in a straightforward manner to force sensing, further discussion of the topic is not undertaken here. Typical rating parameters for several types of sensors are given in Table 6.5. The optical encoder is discussed in Chapter 7.

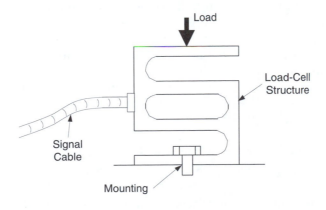

FIGURE 6.64
An industrial force sensor (load cell).

TABLE 6.5

Rating Parameters of Several Sensors and Transducers

Transducer	Measurand	Measurand Frequency max/min	Output Impedanc	Typical Resolution	Accuracy	Sensitivity
Potentiometer	Displacement	5 Hz/DC	Low	0.1 mm	0.1%	200 mV/mm
LVDT	Displacement	2500 Hz/DC	Moderate	0.001 mm or less	0.3%	50 mV/mm
Resolver	Angular displacement	500 Hz/DC (limited by excitation frequency)	Low	2 min.	0.2%	10 mV/deg
Tachometer	Velocity	700 Hz/DC	Moderate (50 Ω)	0.2 mm/s	0.5%	5 mV/mm/s 75 mV/rad/s
Eddy current proximity sensor	Displacement	100 kHz/DC	Moderate	0.001 mm 0.05% full scale	0.5%	5 V/mm
Piezoelectric accelerometer	Acceleration (and velocity, etc.)	25 kHz/1Hz	High	1 mm/s^2	1%	0.5 mV/m/s^2
Semiconductor strain gage	Strain (displacement, acceleration, etc.)	1 kHz/DC (limited by fatigue)	200 Ω	$1 - 10\ \mu\varepsilon$ ($1\mu\varepsilon=10^{-6}$ unity strain)	1%	$1\ V/\varepsilon$, 2000 $\mu\varepsilon$ max
Loadcell	Force (10–1000 *N*)	500 Hz/DC	Moderate	0.01 N	0.05%	1 mV/N
Laser	Displacement/ Shape	1 kHz/DC	100 Ω	1.0 μm	0.5%	1 V/mm
Optical encoder	Motion	100 kHz/DC	500 Ω	10 bit	±1/2 bit	10^4/rev.

6.10 Tactile Sensing

Tactile sensing is usually interpreted as *touch sensing*, but tactile sensing is different from a simple "clamping" where very few discrete force measurements are made. In tactile sensing, a force "distribution" is measured, using a closely spaced array of force sensors and usually exploiting the skin-like properties of the sensor array.

Tactile sensing is particularly important in two types of operations: (1) grasping and (2) object identification. In grasping, the object has to be held in a stable manner without being allowed to slip and without being damaged. Object identification includes recognizing or determining the shape, location, and orientation of an object as well as detecting or identifying surface properties (e.g., density, hardness, texture, flexibility), and defects. Ideally, these tasks would require two types of sensing:

1. Continuous spatial sensing of time-variable contact forces
2. Sensing of surface deformation profiles (time-variable)

These two types of data are generally related through the constitutive relations (e.g., stress-strain relations) of the touch surface of the tactile sensor or of the object that is being grasped. As a result, either the almost-continuous-spatial sensing of tactile forces or the sensing of a tactile deflection profile, separately, is often termed tactile sensing. Note that the learning experience is also an important part of tactile sensing. For example, picking up a fragile object such as an egg and picking up an object such as a lemon that has the same shape but is made of a flexible material, are not identical processes; they require some learning through touch, particularly when vision capability is not available.

6.10.1 Tactile Sensor Requirements

Significant advances in tactile sensing have taken place in the robotics area. Applications, which are very general and numerous, include: automated inspection of surface profiles and joints (e.g., welded or glued parts) for defects; material handling or parts transfer (e.g., pick and place); parts assembly (e.g., parts mating); parts identification and gaging in manufacturing applications (e.g., determining the size and shape of a turbine blade picked from a bin); and fine-manipulation tasks (e.g., production of arts and craft, robotic engraving, and robotic microsurgery). Note that some of these applications might need only simple touch (force-torque) sensing if the parts being grasped are properly oriented and if adequate information about the process and the objects is already available.

Naturally, the frequently expressed design objective for tactile sensing devices has been to mimic the capabilities of human fingers. Specifically, the tactile sensor should have a compliant covering with skin-like properties, along with enough degrees of freedom for flexibility and dexterity, adequate sensitivity and resolution for information acquisition, adequate robustness and stability to accomplish various tasks, and some local intelligence for identification and learning purposes. Although the spatial resolution of a human fingertip is about 2 mm, still finer spatial resolutions (less than 1 mm) can be realized if information through other senses (e.g., vision), prior experience, and intelligence are used simultaneously during the touch. The force resolution (or sensitivity) of a human fingertip is on the order of 1 gm. Also, human fingers can predict "impending slip" during grasping, so that corrective actions can be taken before the object actually slips. At an elementary level, this requires the knowledge of shear stress distribution and friction properties at the common surface between the object and the hand. Additional information and an "intelligent" processing capability are also needed to predict slip accurately and to take corrective actions to prevent slipping. These are, of course, ideal goals for a tactile sensor, but they are not unrealistic in the long run. Typical specifications for an industrial tactile sensor are as follows:

1. Spatial resolution of about 2 mm
2. Force resolution (sensitivity) of about 2 gm
3. Force capacity (maximum touch force) of about 1 kg
4. Response time of 5 ms or less
5. Low hysteresis (low energy dissipation)
6. Durability under harsh working conditions
7. Robustness and insensitivity to change in environmental conditions (temperature, dust, humidity, vibration, etc.)
8. Capability to detect and even predict slip

Although the technology of tactile sensing has not peaked yet, and the wide-spread use of tactile sensors in industrial applications is still to come, several types of tactile sensors that meet and even exceed the foregoing specifications are commercially available. In future developments of these sensors, two separate groups of issues need to be addressed:

1. Ways to improve the mechanical characteristics and design of a tactile sensor so that accurate data with high resolution can be acquired quickly using the sensor
2. Ways to improve signal analysis and processing capabilities so that useful information can be extracted accurately and quickly from the data acquired through tactile sensing

Under the second category, we also have to consider techniques for using tactile information in the feedback control of dynamic processes. In this context, the development of control algorithms, rules, and inference techniques for intelligent controllers that use tactile information, has to be addressed.

6.10.2 Construction and Operation of Tactile Sensors

The touch surface of a tactile sensor is usually made of an elastomeric pad or flexible membrane. Starting from this common basis, the principle of operation of a tactile sensor differs primarily depending on whether the distributed force is sensed or the deflection of the tactile surface is measured. The common methods of tactile sensing include the following:

1. Use a closely spaced set of strain gages or other types of force sensors to sense the distributed force.
2. Use a conductive elastomer as the tactile surface. The change in its resistance as it deforms, will determine the distributed force.
3. Use a closely spaced array of deflection sensors or proximity sensors (e.g., optical sensors) to determine the deflection profile of the tactile surface.

Note that since force and deflection are related through a constitutive law for the tactile sensor (touch pad), only one type of measurement, not both force and deflection, is needed in tactile sensing. A force distribution profile or a deflection profile obtained in this manner may be treated as a 2-D array or an "image" and may be processed (filtered, function-fitted, etc.) and displayed as a tactile image, or used in applications (object identification, manipulation control, etc.).

The contact force distribution in a tactile sensor is commonly measured using an array of force sensors located under the flexible membrane. Arrays of piezoelectric sensors and metallic or semiconductor strain gages (piezoresistive sensors) in sufficient density (number of elements per unit area) may be used for the measurement of the tactile force distribution. In particular, semiconductor elements are poor in mechanical strength but have good sensitivity. Alternatively, the skin-like membrane itself can be made from a conductive elastomer (e.g., graphite-leaded neoprene rubber) whose resistance changes can be sensed and used in determining the force and deflection distribution. In particular, as the tactile pressure increases, the resistance of the particular elastomer segment decreases and the current conducted through it (due to an applied constant voltage) will increase. Conductors can be etched underneath the elastomeric pad to detect the current distribution in the pad, through proper signal acquisition circuitry. Common problems with conductive elastomers are electrical noise, nonlinearity, hysteresis, low sensitivity, drift, low bandwidth, and poor material strength.

The deflection profile of a tactile surface may be determined using a matrix of proximity sensors or deflection sensors. Electromagnetic and capacitive sensors may be used in obtaining this information. The principles of operation of these types of sensors have been discussed previously, in this chapter. Optical tactile sensors use light-sensitive elements (photosensors) to sense the intensity of light (or laser beams) reflected from the tactile surface.

In one approach (extrinsic), the proximity of a light-reflecting surface, which is attached to the back of a transparent tactile pad, is measured. Since the light intensity depends on the distance from the light-reflecting surface to the photosensor, the deflection profile can be determined. In another approach (intrinsic), the deformation of the tactile pad alters the light transmission characteristics of the pad. As a result, the intensity distribution of the transmitted light, as detected by an array of photosensors, determines the deflection profile. Optical methods have the advantages of being free from electromagnetic noise and safe in explosive environments, but they can have errors due to stray light reaching the sensor, variation in intensity of the light source, and changes in environmental conditions (e.g., dirt, humidity, and smoke).

Example 6.18

A tactile sensor pad consists of a matrix of conductive elastomer elements. The resistance R_t in each tactile element is given by

$$R_t = \frac{a}{F_t}$$

where F_t is the tactile force applied to the element and a is a constant. The circuit shown in Figure 6.65 is used to acquire the tactile sensor signal v_o which measures the local tactile force F_t. The entire matrix of tactile elements may be scanned by addressing the corresponding elements through an appropriate switching arrangement.

For the signal acquisition circuit shown in Figure 6.65 obtain a relationship for the output voltage v_o in terms of the parameters a, R_o and others if necessary, and the variable F_t.

Show that $v_o = 0$ when the tactile element is not addressed (i.e., when the circuit is switched to the reference voltage 2.5 V).

SOLUTION
Define,

v_i = input to the circuit (2.5 V or 0.0 V)

v_{o1} = output of the first opamp

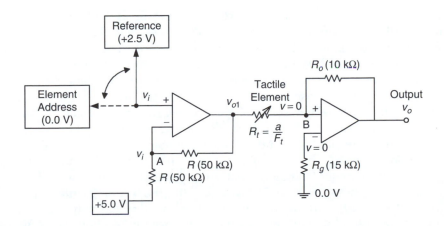

FIGURE 6.65
A signal acquisition circuit for a conductive-elastomer tactile sensor.

According to the properties of an opamp (see Chapter 4):

1. Voltages at the two input leads are equal
2. Currents through the two input leads are zero

Hence, note the same v_i at both input leads of the first opamp (and at node A); and the same zero voltage at both input leads of the second opamp (and at node B), because one of the leads is grounded.

Current balance at A:

$$\frac{5.0 - v_i}{R} = \frac{v_i - v_{o1}}{R} \quad \Rightarrow \quad v_{o1} = 2v_i - 5.0 \tag{i}$$

Current balance at B:

$$\frac{v_{o1} - 0}{R_t} = \frac{0 - v_o}{R_o} \quad \Rightarrow \quad v_o = -v_{o1}\frac{R_o}{R_t} \tag{ii}$$

Substitute Equation i into Equation ii and also substitute the given expression for R_t. We get,

$$v_o = \frac{R_o}{a} F_t (5.0 - 2v_i) \tag{6.116}$$

Substitute the two switching values for v_i. We have

$$v_o = \frac{5R_o}{a} F_t \quad \text{when addressed} \tag{6.116a}$$

$$= 0 \quad \text{when reference.}$$

6.10.3   Optical Tactile Sensors

A schematic representation of an optical tactile sensor (built at the Man-Machine Systems Laboratory at Massachusetts Institute of Technology—MIT) is shown in Figure 6.66. If a beam of light (or laser) is projected onto a reflecting surface, the intensity of light reflected back and received by a light receiver depends on the distance (proximity) of the reflecting surface. For example, in Figure 6.66(a), more light is received by the light receiver when the reflecting surface is at Position 2 than when it is at Position 1. But if the reflecting surface actually touches the light source, light will be completely blocked off, and no light will reach the receiver. Hence, in general, the proximity-intensity curve for an optical proximity sensor will be nonlinear and will have the shape shown in Figure 6.66(a). Using this (calibration) curve, we can determine the position (x) once the intensity of the light received at the photosensor is known. This is the principle of operation of many optical tactile sensors. In the system shown in Figure 6.66(b), the flexible tactile element consists of a thin, light-reflecting surface embedded within an outer layer (touch pad) of high-strength rubber and an inner layer of transparent rubber. Optical fibers are uniformly and rigidly mounted across this inner layer of rubber so that light can be projected directly onto the reflecting surface.

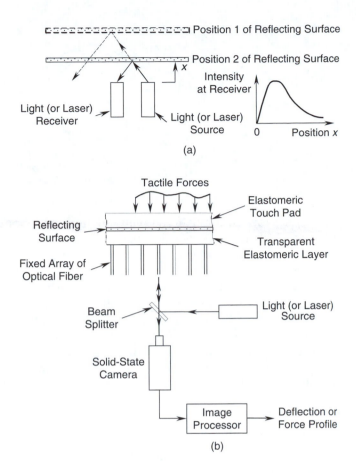

FIGURE 6.66
(a) The principle of an optical proximity sensor, (b) Schematic representation of a fiber optic tactile sensor.

The light source, the beam splitter, and the solid-state (charge-coupled device, or CCD) camera form an integral unit, which can be moved laterally in known steps to scan the entire array of optical fiber if a single image frame of the camera does not cover the entire array. The splitter plate reflects part of the light from the light source onto a bundle of optical fiber. This light is reflected by the reflecting surface and is received by the solid-state camera. Since the intensity of the light received by the camera depends on the proximity of the reflecting surface, the gray-scale intensity image detected by the camera will determine the deflection profile of the tactile surface. Using appropriate constitutive relations for the tactile sensor pad, the tactile force distribution can be determined as well. The image processor carries out conditioning of (filtering, segmenting, etc.) the successive image frames received by the frame grabber, and computes the deflection profile and the associated tactile force distribution in this manner. The image resolution will depend on the pixel (picture-element) size of each image frame (e.g., 512×512 pixels, 1024×1024 pixels, etc.) as well as the spacing of the fiber optic matrix. Note that the force resolution or sensitivity of the tactile sensor can be improved at the expense of the thickness of the elastomeric layer, which determines the robustness of the sensor.

In the described fiber optic tactile sensor (Figure 6.66), the optical fibers serve as the medium through which light or laser rays are transmitted to the tactile site. This is an

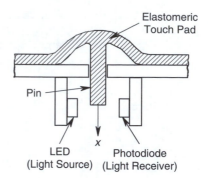

FIGURE 6.67
An optical tactile sensor with
localized light sources and
photosensors.

"extrinsic" use of fiber-optics for sensing. Alternatively, an "intrinsic" application can be developed where an optical fiber serves as the sensing element itself. Specifically, the tactile pressure is directly applied to a mesh of optical fibers. Since the amount of light transmitted through a fiber will decrease due to deformation caused by the tactile pressure, the light intensity at a receiver can be used to determine the tactile pressure distribution.

Yet another alternative of an optical tactile sensor is available. In this design, the light source and the receiver are located at the tactile site itself; optical fibers are not used. The principle of operation of this type of tactile sensor is shown in Figure 6.67. When the elastomeric touch pad is pressed at a particular location, a pin attached to the pad at that point moves (in the x direction), thereby obstructing the light received by the photodiode from the light-emitting diode (LED). The output signal of the photodiode measures the pin movement.

6.10.4   Piezoresistive Tactile Sensors

One type of piezoresistive tactile sensor uses an array of semiconductor strain gages mounted under the touch pad on a rigid base. In this manner, the force distribution on the touch pad is measured directly.

Example 6.19

When is tactile sensing preferred over sensing of a few point forces. A piezoelectric tactile sensor has 25 force-sensing elements per square centimeter. Each sensor element in the sensor can withstand a maximum load of 40 N and can detect load changes on the order of 0.01 N. What is the force resolution of the tactile sensor? What is the spatial resolution of the sensor? What is the dynamic range of the sensor in decibels?

SOLUTION

Tactile sensing is preferred when it is not a simple-touch application. Shape, surface characteristics, flexibility characteristics of a manipulated (handled or grasped) object can be determined using tactile sensing.

$$\text{Force resolution} = 0.01\,\text{N}$$

$$\text{Spatial resolution} = \frac{\sqrt{1}}{\sqrt{25}}\ \text{cm} = 2\ \text{mm}$$

$$\text{Dynamic range} = 20\log_{10}\left(\frac{40}{0.01}\right) = 72\ \text{dB}$$

6.10.5 Dexterity

Dexterity is an important consideration in sophisticated manipulators and robotic hands, which employ tactile sensing. The dexterity of a device is conventionally defined as the ratio: (number of degrees of freedom in the device)/(motion resolution of the device). We will call this *motion dexterity.*

We can define another type of dexterity called *force dexterity,* as follows:

$$\text{Force dexterity} = \frac{\text{number of degrees of freedom}}{\text{force resolution}}$$

Both types of dexterity are useful in mechanical manipulation where tactile sensing is used.

6.10.6 Strain Gage Tactile Sensor

A strain gage tactile sensor has been developed by the Eaton Corporation in Troy, Michigan. The concept behind it can be used to determine the size and location of a point-contact force, which is useful, for example, in parts-mating applications. A square plate of length a is simply supported by frictionless hinges at its four corners on strain gage load cells, as shown in Figure 6.68(a). The magnitude, direction, and location of a point force P applied normally to the plate can be determined using the readings of the four (strain-gage) load cells. To illustrate this principle, consider the free-body diagram shown in Figure 6.68(b). The location of force P is given by the coordinates (x, y) in the Cartesian

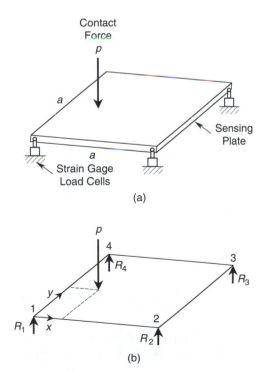

(a)

(b)

FIGURE 6.68

(a) Schematic representation of a strain gage point-contact sensor, (b) Free-body diagram.

coordinate system (x, y, z), with origin located at 1, as shown. The load cell reading at location i is denoted by R_i. Equilibrium in the z direction gives the force balance

$$P = R_1 + R_2 + R_3 + R_4 \tag{6.117}$$

Equilibrium about the y-axis gives the moment balance

$$Px = R_2 a + R_3 a \tag{i}$$

or,

$$x = \frac{a}{P}(R_2 + R_3) \tag{6.118}$$

Similarly, equilibrium about the x-axis gives

$$y = \frac{a}{P}(R_3 + R_4) \tag{6.119}$$

It follows from Equations 6.117 through 6.119 that the force P (direction as well as magnitude) and its location (x, y) are completely determined by the load cell readings. Note that typical values for the plate length a and the maximum force P are 5 cm and 10 kg, respectively.

Example 6.20

In a particular parts-mating process using the principle of strain-gage tactile sensor described above, suppose that the tolerance on the measurement error of the force location is limited to δr. Determine the tolerance δF on the load-cell error.

SOLUTION

Take the differentials of Equation 6.117 and Equation ii:

$$\delta P = \delta R_1 + \delta R_2 + \delta R_3 + \delta R_4$$

$$P\,\delta x + x\,\delta P = a\,\delta R_2 + a\,\delta R_3$$

Direct substitution gives

$$\delta x = \frac{a}{P}(\delta R_2 + \delta R_3) - \frac{x}{P}(\delta R_1 + \delta R_2 + \delta R_3 + \delta R_4)$$

Note that x lies between 0 and a, and each δR_i can vary up to $\pm \delta F$. Hence, the largest error in x is given by $(2a/P)\delta F$. This is limited to δr. Hence, we have

$$\delta r = \frac{2a}{P}\delta F$$

or

$$\delta F = \frac{P}{2a}\delta r$$

which gives the tolerance on the force error. The same result is obtained by considering y instead of x.

6.10.7 Other Types of Tactile Sensors

Ultrasonic tactile sensors are based, for example, on pulse-echo ranging. In this method, the tactile surface is two membranes separated by an air gap. The time taken for an ultrasonic pulse to travel through the gap and be reflected back onto a receiver depends, in particular on the thickness of the air gap. Since this time interval changes with deformation of the tactile surface, it can serve as a measure of the deformation of the tactile surface at a given location. Other possibilities for tactile sensors include the use of chemical effects that might be present when an object is touched and the influence of grasping on the natural frequencies of an array of sensing elements.

Sensor density or resolution, dynamic range, response time or bandwidth, strength and physical robustness, size, stability (dynamic robustness), linearity, flexibility, and localized intelligence (including data processing and learning) are important factors, which require consideration in the analysis, design, or selection of a tactile sensor. The specifications chosen will depend on the particular application. Typical values are 100 sensor elements spaced at 1 mm, a dynamic range of 60 dB, and a bandwidth of over 100 Hz (a response time of 10 ms or less). Because of the large number of sensor elements, signal conditioning and processing for tactile sensors present enormous difficulties. For instance, in piezoelectric tactile sensors, it is usually impractical to use a separate charge amplifier (or a voltage amplifier) for each piezoelectric element, even when built-in microelectronic amplifiers are available. Instead, signal multiplexing could be employed, along with a few high-performance signal amplifiers. The sensor signals could then be serially transmitted for conditioning and digital processing. The obvious disadvantages here are the increase in data acquisition time and the resulting reduction in sensor bandwidth.

6.10.8 Passive Compliance

Tactile sensing is used for active control of processes. This is particularly useful in robotic applications that call for fine manipulation (e.g., microsurgery, assembly of delicate instruments, and robotic artistry). Heavy-duty industrial manipulators are often not suitable for fine manipulation because of errors that arise from such factors as backlash, friction, drift, errors in control hardware and algorithms, and generally poor dexterity. This situation can be improved to a great extent by using a heavy-duty robot for gross manipulations and using a miniature robot (or hand or gripper or finger manipulator) to serve as an end effector for fine-manipulation purposes. The end effector will use tactile sensing and localized control to realize required levels of accuracy in fine manipulation. This approach to accurate manipulation is expensive, primarily because of the sophisticated instrumentation and local processing needed at the end effector. A more cost-effective approach is to use passive remote-center compliance (RCC) at the end effector. With this method, passive devices (linear or nonlinear springs) are incorporated in the end-effector design (typically, at the wrist) so that some compliance (flexibility) is present. Errors in manipulation (e.g., jamming during parts mating) will generate forces and moments, which will deflect the end effector so as to self-correct the situation. Active compliance, which employs local sensors and control to adaptively change the end-effector compliance, is used in operations where passive compliance alone is not adequate. Impedance control is useful here.

6.11 Gyroscopic Sensors

Gyroscopic sensors are used for measuring angular orientations and angular speeds of aircraft, ships, vehicles, and various mechanical devices. These sensors are commonly used in control systems for stabilizing vehicle systems. Since a spinning body (a gyroscope) requires an external torque to turn (precess) its axis of spin, it is clear that if this gyro is mounted on a rigid vehicle with frictionless joints, so that there are a sufficient number of degrees of freedom (at most three) between the gyro and the vehicle, the spin axis will remain unchanged in space, regardless of the motion of the vehicle. Hence the axis of spin of the gyro will provide a reference with respect to which the vehicle orientation (e.g., azimuth or yaw, pitch, and roll angles) and angular speed can be measured. The orientation can be measured by using angular sensors at the pivots of the structure, which mounts the gyro on the vehicle. The angular speed about an orthogonal axis can be determined; for example, by measuring the precession torque (which is proportional to the angular speed) using a strain-gage sensor; or by measuring using a resolver, the deflection of a torsional spring that restrains the precession. The angular deflection in the latter case is proportional to the precession torque and hence the angular speed.

Consider a rigid disk spinning about an axis at angular speed ω. If the moment of inertia of the disk about that axis (polar moment of inertia) is J, the angular momentum H about the same axis is given by

$$H = J\omega \tag{6.120}$$

Newton's second law (torque = rate of change of angular momentum) tells us that to rotate (precess) the spinning axis slightly, a torque has to be applied, because precession causes a change in the spinning angular momentum vector (the magnitude remains constant but the direction changes), as shown in Figure 6.69(a). This explains the principle of operation of a gyroscope.

In the gyroscope shown in Figure 6.69(b), the disk is spun about frictionless bearings using a constant-speed motor. Since the gimbal (the framework on which the disk is supported) is free to turn about frictionless bearings in the vertical axis, it will remain fixed with respect to an inertial frame, even if the bearing housing (the main structure in which the gyroscope is located) rotates about the same vertical axis. Hence, the relative angle between the gimbal and the bearing housing (angle θ in the figure) can be measured (using a resolver, RVDT, encoder, etc.) and this gives the angle of rotation of the main structure. Note that bearing friction introduces an error, which has to be compensated for, perhaps by recalibration before a reading is taken or by active feedback using a motor (*torquer*).

Figure 6.69(b) illustrates the case of a single-axis gyro sensor. The idea can be extended to the three-axis (3 dof) case, by providing two further frames, which are mounted on gimbals with their axes orthogonal to each other. The angular displacement sensors are mounted at all three gimbal bearings. For small rotations, these three angles can be considered uncoupled and will provide the orientation of the body on which the gyro

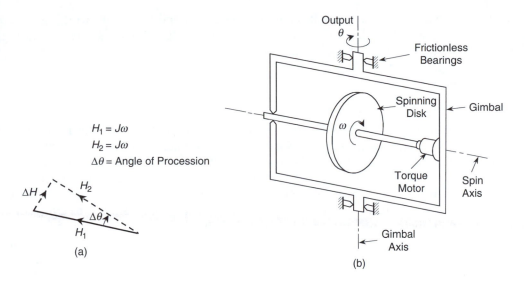

FIGURE 6.69
(a) Illustration of the gyroscopic torque needed to change the direction of an angular momentum vector, (b) A simple single-axis gyroscope for sensing angular displacement.

unit is mounted. For large rotations, proper coordinate transformation (see Chapter 3 on robot kinematics) has to be applied in converting the sensor readings to the orientation of the moving body (vehicle).

6.11.1 Rate Gyro

A rate gyro is used to measure angular speeds. The same arrangement shown in Figure 6.69(b), with a slight modification, can be used. In this case, the gimbal is not free; and may be restrained by a torsional spring. A viscous damper is provided to suppress any oscillations. By analyzing this gyro, we will note that the relative angle of rotation θ gives the angular speed Ω of the structure (vehicle) about the axis that is orthogonal to both gimbal axis and spin axis.

For a simplified analysis assume that the angles of rotation are small and that the moment of inertia of the gimbal frame is negligible (compared to J). Newton's second law for motion of the unit about the gimbal axis gives:

$$J\omega\Omega = K\theta + B\dot{\theta}$$

where
   K = torsional stiffness of the spring restraint at the gimbal bearings
   B = damping constant of rotational motion about the gimbal axis.

We get,

$$\Omega = \frac{K\theta + B\dot{\theta}}{J\omega} \tag{6.121}$$

In particular, when B is very small, angular rotation at the gimbal bearings (e.g., as measured by a resolver) will be proportional to the angular speed of the system including the main body (vehicle) about the axis orthogonal to both gimbal axis and spin axis.

6.11.2   Coriolis Force Devices

Consider a mass m moving at velocity v relative to a rigid frame. If the frame itself rotates at an angular velocity ω it is known that the acceleration of m has term given by $2\omega \times v$ This is known as the *Coriolis acceleration*. The associated force $2m\omega \times v$ is the *Coriolis force*. This force can be sensed either directly using a force sensor or by measuring a resulting deflection in a flexible element, and may be used to determine the variables (ω or v) in the Coriolis force. Note that Coriolis force is somewhat similar to gyroscopic force even though the concepts are different. For this reason, devices based on the Coriolis effect are also commonly termed gyroscopes. Coriolis concepts are gaining popularity in MEMS-based sensors, which use microelectromechanical systems (MEMS) technologies.

6.12   Optical Sensors and Lasers

The laser (*l*ight *a*mplification by *s*timulated *e*mission of *r*adiation) produces electromagnetic radiation in the ultraviolet, visible, or infrared bands of the spectrum (see Chapter 4). A laser can provide a single-frequency (*monochromatic*) light source. Furthermore, the electromagnetic radiation in a laser is *coherent* in the sense that all waves generated have constant phase angles. The laser uses oscillations of atoms or molecules of various elements. The laser is useful in fiber optics, and it can also be used directly in sensing and gauging applications. The helium-neon (HeNe) laser and the semiconductor laser are commonly used in optical sensor applications.

The characteristic component in a fiber-optic sensor is a bundle of glass fibers (typically a few hundred) that can carry light. Each optical fiber may have a diameter on the order of a few μm to about 0.01 mm. There are two basic types of fiber-optic sensors. In one type—the "indirect" or the *extrinsic* type—the optical fiber acts only as the medium in which the sensor light is transmitted. In this type, the sensing element itself does not consist of optical fibers. In the second type—the "direct" or the *intrinsic* type—the optical fiber itself acts as the sensing element. When the conditions of the sensed medium change, the light-propagation properties of the optical fibers change (e.g., due to microbending of a straight fiber as a result of an applied force), providing a measurement of the change in conditions. Examples of the first (extrinsic) type of sensor include fiber-optic position sensors, proximity sensors, and tactile sensors. The second (intrinsic) type of sensor is found, for example, in fiber-optic gyroscopes, fiber-optic hydrophones, and some types of micro-displacement or force sensors in MEMS devices).

6.12.1   Fiber-Optic Position Sensor

A schematic representation of a fiber-optic position sensor (or proximity sensor or displacement sensor) is shown in Figure 6.70. The optical fiber bundle is divided into two groups: transmitting fibers and receiving fibers. Light from the light source is

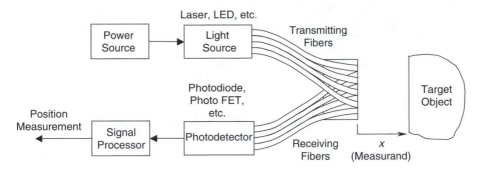

FIGURE 6.70
A fiber-optic position sensor.

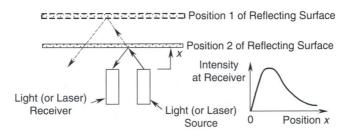

FIGURE 6.71
The principle of a fiber-optic proximity sensor.

transmitted along the first bundle of fibers to the target object whose position is being measured. Light reflected (or, diffused) onto the receiving fibers by the surface of the target object is carried to a photodetector. The intensity of the light received by the photodetector will depend on position x of the target object. In particular, if $x = 0$, the transmitting bundle will be completely blocked off and the light intensity at the receiver will be zero. As x is increased, intensity of the received light will increase, because more and more light will be reflected onto the tip of the receiving bundle. This will reach a peak at some value of x. When x is increased beyond that value, more and more light will be reflected outside the receiving bundle; hence, the intensity of the received light will drop. In general then, the proximity-intensity curve for an optical proximity sensor will be nonlinear and will have the shape shown in Figure 6.71. Using this (calibration) curve, we can determine the position (x) once the intensity of the light received at the photosensor is known. The light source could be a laser (structured light), infrared light-source, or some other type, such as a light-emitting diode (LED). The light sensor (photodetector) could be some device such as a photodiode or a photo field effect transistor (photo FET). This type of fiber-optic sensors can be used, with a suitable front-end device (such as bellows, springs, etc.) to measure pressure, force, etc. as well.

6.12.2   Laser Interferometer

This sensor is useful in the accurate measurement of small displacements. In this fiber-optic position sensor, the same bundle of fibers is used for sending and receiving a monochromatic beam of light (typically, laser). Alternatively, monomode fibers, which

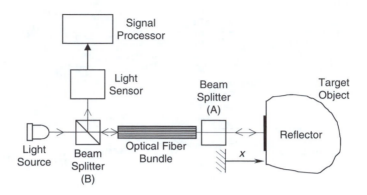

FIGURE 6.72
Laser interferometer position sensor.

transmit only monochromatic light (of a specific wavelength) may be used for this purpose. In either case, as shown in Figure 6.72, a beam splitter (A) is used so that part of the light is directly reflected back to the bundle tip and the other part reaches the target object (as in Figure 6.70) and reflected back from it (using a reflector mounted on the object) on to the bundle tip. In this manner, part of the light returning through the bundle had not traveled beyond the beam splitter while the other part had traveled between the beam splitter (A) and the object (through an extra distance equal to twice the separation between the beam splitter and the object). As a result, the two components of light will have a phase difference ϕ, which is given by

$$\phi = \frac{2x}{\lambda} \times 2\pi \tag{6.122}$$

where
  x = distance of the target object from the beam splitter
  λ = wavelength of monochromatic light

The returning light is directed to a light sensor using a beam splitter (B). The sensed signal is processed using principles of interferometry to determine ϕ, and from Equation 6.122, the distance x. Very fine resolutions better than a fraction of a micrometer (μm) can be obtained using this type of fiber-optic position sensors.

   The advantages of fiber optics include insensitivity to electrical and magnetic noise (in view of optical coupling), safe operation in explosive, high-temperature, corrosive, and hazardous environments, and high sensitivity. Furthermore, mechanical loading and wear problems do not exist because fiber-optic position sensors are noncontact devices with no moving parts. The disadvantages include direct sensitivity to variations in the intensity of the light source and dependence on ambient conditions (temperature, dirt, moisture, smoke, etc.). Compensation can be made, however, with respect to temperature. An *optical encoder* is a digital (or pulse-generating) motion transducer (see Chapter 7). Here, a light beam is intercepted by a moving disk that has a pattern of transparent windows. The light that passes through, as detected by a photosensor, provides the transducer output. These sensors may also be considered in the extrinsic category.

As an *intrinsic* application of fiber optics in sensing, consider a straight optical fiber element that is supported at the two ends. In this configuration almost 100% of the light at the source end will transmit through the optical fiber and reach the detector (receiver) end. Now, suppose that a slight load is applied to the optical fiber segment at its mid span. It will deflect slightly due to the load, and as a result the amount of light received at the detector can drop significantly. For example, a microdeflection of just 50 µm can result in a drop in intensity at the detector by a factor of 25. Such an arrangement may be used in deflection, force, and tactile sensing. Another intrinsic application is the fiber-optic gyroscope, as described next.

6.12.3   Fiber-Optic Gyroscope

This is an angular speed sensor that uses fiber optics. Contrary to the implication of its name, however, it is not a gyroscope in the conventional sense. Two loops of optical fibers wrapped around a cylinder are used in this sensor, and they rotate with the cylinder, at the same angular speed, which needs to be sensed. One loop carries a monochromatic light (or laser) beam in the clockwise direction; the other loop carries a beam from the same light (laser) source in the counterclockwise direction (see Figure 6.73). Since the laser beam traveling in the direction of rotation of the cylinder attains a higher frequency than that of the other beam, the difference in frequencies (known as the *Sagnac effect*) of the two laser beams received at a common location will measure the angular speed of the cylinder. This may be accomplished through interferometry, because the combined signal is a sine beat. As a result, light and dark patterns (*fringes*) will be present in the detected light, and they will measure the frequency difference and hence the rotating speed of the optical fibers.

In a laser (ring) gyroscope, it is not necessary to have a circular path for the laser. Triangular and square paths are commonly used as well. In general the beat frequency $\Delta\omega$ of the combined light from the two laser beams traveling in opposite directions is given by

$$\Delta\omega = \frac{4A}{p\lambda}\Omega \qquad (6.123)$$

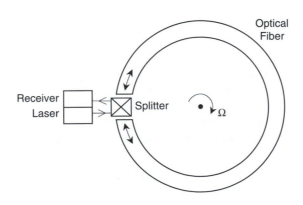

FIGURE 6.73
A fiber-optic, laser gyroscope.

where
  A = area enclosed by the travel path (πr^2 for a cylinder of radius r)
  p = length (perimeter) of the traveled path ($2\pi r$ for a cylinder)
  λ = wavelength of the laser
  Ω = angular speed of the object (or, optical fiber).

The length of the optical fiber wound around the rotating object can exceed 100 m and can be as large as 1 km. Angular displacements can be measured with a laser gyro simply by counting the number of cycles and clocking fractions of cycles. Acceleration can be determined by digitally determining the rate of change of speed. In a laser gyro, there is an alternative to use two separate loops of optical fiber, wound in opposite directions. The same loop can be used to transmit light from the same laser from the opposite ends of the fiber. A beam splitter has to be used in this case, as shown in Figure 6.73.

6.12.4   Laser Doppler Interferometer

The laser Doppler interferometer is used for accurate measurement of speed. To understand the operation of this device, we should explain two phenomena: the Doppler effect and light wave interference. The latter phenomenon is used in the laser interferometer position sensor, which was discussed before. Consider a wave source (e.g., a light source or sound source) that is moving with respect to a receiver (observer). If the source moves toward the receiver, the frequency of the received wave appears to have increased; if the source moves away from the receiver, the frequency of the received wave appears to have decreased. The change in frequency is proportional to the velocity of the source relative to the receiver. This phenomenon is known as the *Doppler effect*. Now consider a monochromatic (single-frequency) light wave of frequency f (say, 5×10^{14} Hz) emitted by a laser source. If this ray is reflected by a target object and received by a light detector, the frequency of the received wave would be

$$f_2 = f + \Delta f \tag{6.124}$$

The frequency increase Δf will be proportional to the velocity v of the target object, which is assumed positive when moving toward the light source. Specifically,

$$\Delta f = \frac{2f}{c} v = kv \tag{6.125}$$

where c is the speed of light in the particular medium (typically, air). Now by comparing the frequency f_2 of the reflected wave, with the frequency $f_1 = f$ of the original wave, we can determine Δf and, hence, the velocity v of the target object.

The change in frequency Δf due to the Doppler effect can be determined by observing the *fringe pattern* due to light wave interference. To understand this, consider the two waves

$$v_1 = a \sin 2\pi f_1 t \tag{6.126}$$

and

$$v_2 = a \sin 2\pi f_2 t \tag{6.127}$$

If we add these two waves, the resulting wave would be

$$v = v_1 + v_2 = a(\sin 2\pi f_1 t + \sin 2\pi f_2 t) \tag{6.128}$$

which can be expressed as

$$v = 2a \sin \pi (f_2 + f_1) t \cos \pi (f_2 - f_1) t \tag{6.129}$$

It follows that the combined signal beats at the beat frequency $\Delta f/2$. Since f_2 is very close to f_1 (because Δf is small compared to f), these beats will appear as dark and light lines (fringes) in the resulting light wave. This is known as *wave interference*. Note that Δf can be determined by two methods:

1. By measuring the spacing of the fringes
2. By counting the beats in a given time interval or by timing successive beats using a high-frequency clock signal

The velocity of the target object is determined in this manner. Displacement can be obtained simply by digital integration (or by accumulating the count). A schematic diagram for the laser Doppler interferometer is shown in Figure 6.74. Industrial interferometers usually employ a helium-neon laser, which has waves of two frequencies close together. In that case, the arrangement shown in Figure 6.74 has to be modified to take into account the two frequency components.

Note that the laser interferometer discussed before (Figure 6.72) directly measures displacement rather than speed. It is based on measuring the phase difference between the direct and returning laser beams, not the Doppler effect (frequency difference).

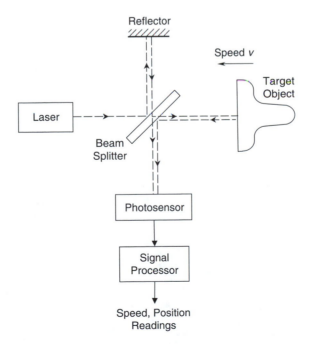

FIGURE 6.74
A laser-Doppler interferometer for measuring velocity and displacement.

6.13 Ultrasonic Sensors

Audible sound waves have frequencies in the range of 20 Hz to 20 kHz. Ultrasound waves are pressure waves, just like sound waves, but their frequencies are higher ("ultra") than the audible frequencies. Ultrasonic sensors are used in many applications, including medical imaging, ranging systems for cameras with autofocusing capability, level sensing, and speed sensing. For example, in medical applications, ultrasound probes of frequencies 40 kHz, 75 kHz, 7.5 MHz, and 10 MHz are commonly used. Ultrasound can be generated according to several principles. For example, high-frequency (gigahertz) oscillations in a piezoelectric crystal subjected to an electrical potential, is used to generate very high-frequency ultrasound. Another method is to use the *magnetostrictive* property of ferromagnetic material. Ferromagnetic materials deform when subjected to magnetic fields. Respondent oscillations generated by this principle can produce ultrasonic waves. Another method of generating ultrasound is to apply a high-frequency voltage to a metal-film capacitor. A microphone can serve as an ultrasound detector (receiver).

Analogous to fiber-optic sensing, there are two common ways of employing ultrasound in a sensor. In one approach—the *intrinsic* method—the ultrasound signal undergoes changes as it passes through an object, due to acoustic impedance and absorption characteristics of the object. The resulting signal (image) may be interpreted to determine properties of the object, such as texture, firmness, and deformation. This approach has been utilized, for example, in an innovative firmness sensor for herring roe. It is also the principle used in medical ultrasonic imaging. In the other approach—the *extrinsic* method—the time of flight of an ultrasound burst from its source to an object and then back to a receiver is measured. This approach is used in distance and position measurement and in dimensional gauging. For example, an ultrasound sensor of this category has been used in thickness measurement of fish. This is also the method used in camera autofocusing.

In distance (range, proximity, displacement) measurement using ultrasound, a burst of ultrasound is projected at the target object, and the time taken for the echo to be received is clocked. A signal processor computes the position of the target object, possibly compensating for environmental conditions. This configuration is shown in Figure 6.75. The applicable relation is

$$x = \frac{ct}{2} \tag{6.130}$$

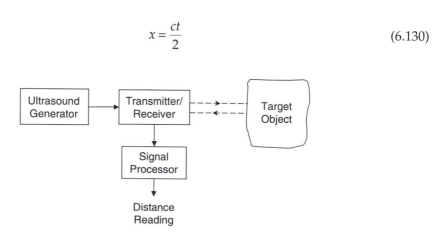

FIGURE 6.75
An ultrasonic position sensor.

where

t = time of flight of the ultrasound pulse (from generator to receiver)
x = distance between the ultrasound generator/receiver and the target object
c = speed of sound in the medium (typically, air).

Distances as small as a few cm to several meters may be accurately measured by this approach, with fine resolution (e.g., a millimeter or less). Since the speed of ultrasonic wave propagation depends on the medium and the temperature of the medium (typically air), errors will enter into the ultrasonic readings unless the sensor is compensated (calibrated) for the variations in the medium; particularly for temperature.

Alternatively, the velocity of the target object can be measured, using the Doppler effect, by measuring (clocking) the change in frequency between the transmitted wave and the received wave. The "beat" phenomenon is employed here. The applicable relation is Equation 6.125, except, now f is the frequency of the ultrasound signal and c is the speed of sound.

6.13.1 Magnetostrictive Displacement Sensors

The ultrasound-based time of flight method is used somewhat differently in a magnetostrictive displacement sensor (e.g., the sensor manufactured by Temposonics). The principle behind this method is illustrated in Figure 6.76. The sensor head generates an interrogation current pulse, which travels along the magnetostrictive wire. This pulse interacts with the magnetic field of the permanent magnet and generates an ultrasound pulse (by magnetostrictive action in the wire). This pulse is received (and timed) at the sensor head. The time of flight is proportional to the distance of the magnet from the sensor head. If the target object is attached to the magnet of the sensor, its position (x) can be determined using the time of flight as usual. Strokes (maximum displacements) ranging from a few cm to one or two meters, at resolutions better than 50 μm are possible with these sensors. With a 15 VDC power supply, the sensor can provide a dc output in the range ±5 V. Since the sensor uses a magnetostrictive medium with a protective nonferromagnetic tubing, some of the common sources of error in ultrasonic sensors that use air as the medium of propagation, can be avoided.

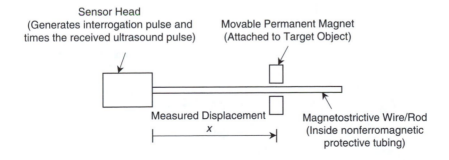

FIGURE 6.76
A magnetostrictive ultrasound displacement sensor.

6.14 Thermo-Fluid Sensors

Common thermo-fluid (mechanical engineering) sensors include those measuring pressure, fluid flow rate, temperature and heat transfer rate. Such sensors are useful in mechatronic applications as well in view of the fact that the plant (e.g., automobile, machine tool, aircraft) may involve these measurands. Several common types of sensors in this category are presented next.

6.14.1   Pressure Sensors

Common methods of pressure sensing are the following:

1. Balance the pressure with an opposing force (or head) and measure this force. Examples are liquid manometers and pistons.
2. Subject the pressure to a flexible front-end (auxiliary) member and measure the resulting deflection. Examples are Bourdon tube, bellows, and helical tube.
3. Subject the pressure to a front-end auxiliary member and measure the resulting strain (or stress). Examples are diaphragms and capsules.

Some of these devices are illustrated in Figure 6.77. In the manometer shown in Figure 6.77(a), the liquid column of height h and density ρ provides a counterbalancing pressure head to support the measured pressure p with respect to the reference (ambient) pressure p_{ref}. Accordingly, this device measures the gauge pressure given by

$$p - p_{ref} = \rho g h \qquad\qquad (6.131)$$

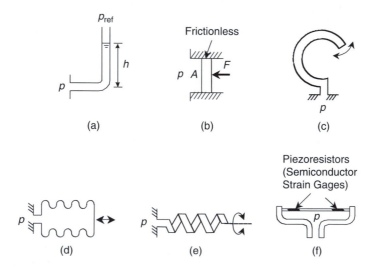

(a)   (b)   (c)

(d)   (e)   (f)

FIGURE 6.77
Typical pressure sensors: (a) Manometer; (b) Counterbalance piston; (c) Bourdon tube; (d) Bellows; (e) Helical tube; (f) Diaphragm.

where g is the acceleration due to gravity. In the pressure sensor shown in Figure 6.77(b), a frictionless piston of area A supports the pressure load with an external force F. The governing equation is

$$p = \frac{F}{A} \tag{6.132}$$

The pressure is determined by measuring F using a force sensor. The Bourdon tube shown in Figure 6.77(c) deflects with a straightening motion as a result of internal pressure. This deflection can be measured using a displacement sensor (typically, arotatory sensor) or indicated by a moving pointer. The bellows deflect with internal pressure, causing a linear motion, as shown in Figure 6.77(d). The deflection can be measured using a sensor such as LVDT or a capacitive sensor, and can be calibrated to indicate pressure. The helical tube shown in Figure 6.77(e) undergoes a twisting (rotational) motion when deflected by internal pressure. This deflection can be measured by an angular displacement sensor (RVDT, resolver, potentiometer, etc.), to provide pressure reading through proper calibration. Figure 6.77(f) illustrates the use of a diaphragm to measure pressure. The membrane (typically metal) will be strained due to pressure. The pressure can be measured by means of strain gauges (piezoresistive sensors) mounted on the diaphragm. MEMS pressure sensors that use this principle are available. In one such device, the diaphragm has a silicon wafer substrate integral with it. Through proper doping (using boron, phosphorous, etc.) a microminiature semiconductor strain gauge can be formed. In fact more than one piezoresistive sensor can be etched on the diaphragm, and used in a bridge circuit to provide the pressure reading, through proper calibration. The most sensitive locations for the piezoresitive sensors are closer to the edge of the diaphragm, where the strains reach the maximum.

6.14.2 Flow Sensors

The volume flow rate Q of a fluid is related to the mass flow rate Q_m through

$$Q_m = \rho Q \tag{6.133}$$

where ρ is the mass density of the fluid. Also, for a flow across an area A at average velocity v, we have

$$Q = Av \tag{6.134}$$

When the flow is not uniform, a suitable correction factor has to be included in Equation 6.134, if a local velocity of the maximum velocity is used.

According to *Bernoulli's equation* for incompressible, ideal flow (no energy dissipation) we have

$$p + \frac{1}{2}\rho v^2 = \text{constant} \tag{6.135}$$

This theorem may be interpreted as conservation of energy. Also, note that the pressure p due to fluid head of height h is given by (gravitational potential energy) ρgh. Using Equation 6.135

and allowing for dissipation (friction), the flow across a constriction (i.e., a fluid resistance element such as an orifice, nozzle, valve, etc.—see Chapter 2) of area A can be shown to obey the relation

$$Q = c_d A \sqrt{\frac{2\Delta p}{\rho}} \qquad (6.136)$$

where Δp is the pressure drop across the constriction and c_d is the *discharge coefficient* for the constriction.

Common methods of measuring fluid flow may be classified as follows:

1. Measure pressure across a known constriction or opening. Examples include nozzles, venturi meters, and orifice plates.
2. Measure the pressure head, which will bring the flow to static conditions. Pitot tube, liquid level sensing using floats, etc. are examples.
3. Measure the flow rate (volume or mass) directly. Turbine flow meter and angular-momentum flow meter are examples.
4. Measure the flow velocity. Coriolis meter, laser-Doppler velocimeter, and ultrasonic flow meter are examples.
5. Measure an effect of the flow and estimate the flow rate using that information. Hot-wire (or, hot-film) anemometer and magnetic induction flow meter are examples.

Several examples of flow meters are shown in Figure 6.78. For the orifice meter shown in Figure 6.78(a), Equation 6.136 is applied to measure the volume flow rate. The pressure drop is measured using the techniques outlined earlier. For the pitot tube shown in Figure 6.78(b), Bernoulli's Equation 6.135 is applicable, noting that the fluid velocity at the free surface of the tube is zero. This gives the flow velocity

$$v = \sqrt{2gh} \qquad (6.137)$$

Note that a correction factor is needed when determining the flow rate because the velocity is not uniform across the flow section. In the angular momentum method shown in Figure 6.78(c), the tube bundle through which the fluid flows, is rotated by a motor. The motor torque τ and the angular speed ω are measured. As the fluid mass passes through the tube bundle, it is imparted an angular momentum at a rate governed by the mass flow rate Q_m of the fluid. The motor torque provides the torque needed for this rate of change of angular momentum. Neglecting losses, the governing equation is

$$\tau = \omega r^2 Q_m \qquad (6.138)$$

where r is the radius of the centroid of the rotating fluid mass. In a turbine flow meter, the rotation of the turbine wheel located in the flow can be calibrated to directly give the flow rate. In the Coriolis method shown in Figure 6.78(d), the fluid is made to flow through a "U" segment, which is hinged to oscillate out of plane (at angular velocity ϖ) and restrained by springs (with known stiffness) in the lateral direction. If the fluid velocity is v, the resulting Coriolis force (due to Coriolis acceleration $2\varpi \times v$) is supported by the springs. The out-of-plane angular speed is measured by a motion sensor. Also, the spring

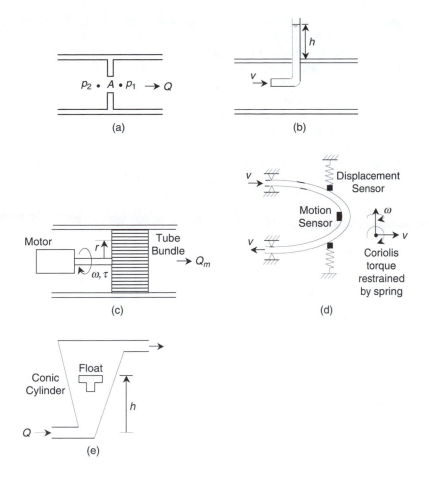

FIGURE 6.78
Several flow meters: (a) Orifice flow meter; (b) Pitot tube; (c) Angular-momentum flow meter; (d) Coriolis velocity meter; (e) Rotameter.

force is measured using a suitable sensor (e.g., displacement sensor). This information will determine the Coriolis acceleration of the fluid particles and hence their velocity.

In the laser-Doppler velocimeter, a laser beam is projected on the fluid flow (through a window) and its frequency shift due to the Doppler effect is measured (see under optical sensors, as described before). This is a measure of the speed of the fluid particles. As another method of sensing velocity of a fluid, an ultrasonic burst is sent in the direction of flow and the time of flight is measured. Increase in the speed of propagation is due to the fluid velocity, and may be determined as usual (see under ultrasonic sensors, as outlined earlier).

In the hot-wire anemometer, a conductor carrying current (i) is placed in the fluid flow. The temperatures of the wire (T) and the surrounding fluid (T_f) are measured along with the current. The coefficient of heat transfer (forced convection) at the boundary of the wire and the moving fluid is known to vary with $\sqrt{v}$, where v is the fluid velocity. Under steady conditions, the heat loss from the wire into the fluid is exactly balanced by the heat generated by the wire due to its resistance (R). The heat balance equation gives

$$i^2 R = c(a + \sqrt{v})(T - T_f) \tag{6.139}$$

This relation can be used to determine v. Instead of a wire, a metal film (e.g., platinum plated glass tube) may be used.

There are other indirect methods of measuring fluid flow rate. In one method, the drag force on an object suspended in the flow using a cantilever arm is measured (using a strain-gauge sensor at the clamped end of the cantilever). This force is known to vary quadratically with the fluid speed. A rotameter (see Figure 6.78(e)) is another device for measuring fluid flow. This device consists of a conic tube with uniformly increasing cross-sectional area, which is vertically oriented. A cylindrical object is floated in the conic tube, through which the fluid flows. The weight of the object is balanced by the pressure differential on the object. When the flow speed increases, the object rises within the conic tube, thereby allowing more area between the object and the tube for the fluid to pass. The pressure differential, however, still balances the weight of the object, and is constant. Equation 6.136 is used to measure fluid flow rate, since A increases quadratically with the height of the object. Consequently, the level of the object can be calibrated to give the flow rate.

6.14.3   Temperature Sensors

In most (if not all) temperature measuring devices, the temperature is sensed through "heat transfer" from the source to the measuring device. The physical (or chemical) change in the device caused by this heat transfer is the transducer stage. Several temperature sensors are outlined below.

6.14.3.1   *Thermocouple*

When the temperature at the junction formed by joining two unlike conductors, is changed, its electron configuration changes due to heat transfer. This electron reconfiguration produces a voltage (emf or electro-motive force), and is known as the *Seebeck effect*. Two junctions (or more) of a thermocouple are made by two unlike conductors such as iron and constant, copper and constantan, chrome and alumel, and so on. One junction is placed in a reference source (cold junction) and the other in the temperature source (hot junction), as shown in Figure 6.79. The voltage across the two junctions is measured to give the temperature of the hot junction with respect to the cold junction. Note that the presence of other junctions (such as the ones formed by the wires to the voltage sensor) do not affect the reading as long as these leads are maintained at the same temperature. Very low temperatures (e.g., –250°C) as well as very high temperatures (e.g., 3000°C) can be measured using a thermocouple. Since the temperature-voltage relationship is nonlinear, correction has to be made when

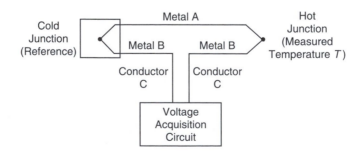

FIGURE 6.79
A thermocouple.

measuring changes in temperature; usually by making polynomial relations. Sensitivity is quite reasonable (e.g., 10 mV/°C), but signal conditioning may be needed in some applications. Fast measurements are possible with small thermocouples having low time constants (e.g., 1 ms).

6.14.3.2 Resistance Temperature Detector (RTD)

An RTD is simply a metal element (in a ceramic tube) whose resistance typically increases with temperature, according to a known function. A linear approximation, as given by Equation 6.140 is adequate when the temperature change is not too large. Temperature is measured using an RTD simply by measuring the change in resistance (say, using a bridge circuit). Metals used in RTDs include platinum, nickel, copper, and various alloys. The temperature coefficient of resistance (α) of several metals, which can be used in RTDs, are given in Table 6.6.

$$R = R_0(1 + \alpha T) \tag{6.140}$$

The useful temperature range of an RTD is about −200°C to +800°C. At high temperatures, an RTD may tend to be less accurate than a thermocouple. The speed of response can be lower as well (e.g., fraction of a second). Some commercial RTD units are shown in Figure 6.80.

TABLE 6.6

Temperature Coefficients of Resistance of Some RTD Metals

Metal	Temperature Coefficient of Resistance α (/°K)
Copper	0.0043
Nickel	0.0068
Platinum	0.0039

FIGURE 6.80
Commercial RTD units (RdF Corp. Hudson, NH. With permission).

6.14.3.3 Thermistor

Unlike an RTD, a thermistor is made of a semiconductor material (e.g., metal oxides such as those of chromium, cobalt, copper, iron, manganese, and nickel), which usually has a negative change in resistance with temperature (i.e., negative α). The resistance change is detected through a bridge circuit or a voltage circuit. Even though the accuracy provided by a thermistor is usually better than that of an RTD, the temperature-resistance relation is far more nonlinear, as given by

$$R = R_0 \exp \beta \left(\frac{1}{T} - \frac{1}{T_0} \right) \tag{6.141}$$

The characteristic temperature β (about 4000°K) itself is temperature dependent, thereby adding to the nonlinearity of the device. Hence, proper calibration is essential when operating in a wide temperature range (say, greater than 50°C). Thermistors are quite robust and they provide a fast response and high sensitivity (compared to RTDs).

6.14.3.4 Bi-Metal Strip Thermometer

Unequal thermal expansion of different materials is used in this device. If strips of the two materials (typically metals) are firmly bonded, thermal expansion causes this element to bend toward the material with lower expansion. This motion can be measured using a displacement sensor, or indicated using a needle and scale. Households thermostats commonly use this principle for temperature sensing and control (on-off).

6.15   Other Types of Sensors

There are many other types of sensors and transducers, which cannot be discussed here due to space limitation. But, the principles and techniques presented in this chapter may be extended to many of these devices. One area where a great variety of sensors are used is *factory automation*. Here, in applications of automated manufacturing and robotics it is important to use proper sensors for specific operations and needs. For example, mechanical and electronic switches (binary or two-state sensors), chemical sensors, camera-based vision systems, and ultrasonic motion detectors may be used for human safety requirements. Motion and force, power-line, debris, sound, vibration, temperature, pressure, flow and liquid-level sensing may be used in machine monitoring and diagnosis. Motion, force, torque, current, voltage, flow and pressure sensing are important in machine control. Vision, motion, proximity, tactile, force, torque, and pressure sensing and dimensional gauging are useful in task monitoring and control.

Several areas can be identified where new developments and innovations are being made in sensor technology:

1. Microminiature (MEMS and nano) sensors: (IC-based, with built-in signal processing)
2. Intelligent sensors: (Built-in reasoning or information preprocessing to provide high-level knowledge and decision-making capability)
3. Integrated and distributed sensors: (Sensors are integral with the components and agents, which communicate with each other in an overall multi-agent system)
4. Hierarchical sensory architectures: (Low-level sensory information is preprocessed to match higher level requirements)

These four areas of activity are also representative of future trends in sensor technology development.

6.16 Problems

6.1 In each of the following examples, indicate at least one (unknown) input, which should be measured and used for feedforward control to improve the accuracy of the control system.

   a. A servo system for positioning a mechanical load. The servo motor is a field-controlled dc motor, with position feedback using a potentiometer and velocity feedback using a tachometer.

   b. An electric heating system for a pipeline carrying a liquid. The exit temperature of the liquid is measured using a thermocouple and is used to adjust the power of the heater.

   c. A room heating system. Room temperature is measured and compared with the set point. If it is low, a valve of a steam radiator is opened; if it is high, the valve is shut.

   d. An assembly robot, which grips a delicate part to pick it up without damaging the part.

   e. A welding robot, which tracks the seam of a part to be welded.

6.2 A typical input variable is identified for each of the following examples of dynamic systems. Give at least one output variable for each system.

   a. Human body: neuroelectric pulses

   b. Company: information

   c. Power plant: fuel rate

   d. Automobile: steering wheel movement

   e. Robot: voltage to joint motor.

6.3 Measuring devices (sensors-transducers) are useful in measuring outputs of a process for feedback control.

   a. Give other situations in which signal measurement would be important.

   b. List at least five different sensors used in an automobile engine.

6.4 Give one situation (system) where output measurement is needed and give a second situation where input measurement is needed for proper control of the chosen system. In each case justify the need.

6.5 Giving examples, discuss situations in which measurement of more than one type of kinematic variables using the same measuring device is

   a. An advantage

   b. A disadvantage.

6.6 Giving examples for suitable auxiliary front-end elements, discuss the use of a force sensor to measure

   a. Displacement

   b. Velocity

   c. Acceleration.

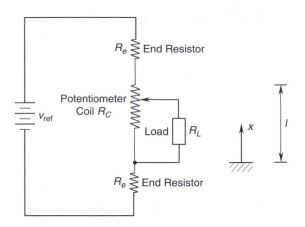

FIGURE P6.9
A potentiometer circuit with end resistors.

6.7 Write the expression for electrical-loading nonlinearity error (percentage) in a rotatory potentiometer in terms of the angular displacement, maximum displacement (stroke), potentiometer element resistance, and load resistance. Plot the percentage error as a function of the fractional displacement for the three cases $R_L/R_c = 0.1$, 1.0, and 10.0.

6.8 Determine the angular displacement of a rotatory potentiometer at which the loading nonlinearity error is the largest.

6.9 A potentiometer circuit with element resistance R_c and equal end resistors R_e is shown in Figure P6.9. Derive the necessary input/output relations. Show that the end resistors can produce a linearizing effect in the potentiometer. At half the maximum reading of the potentiometer shown in Figure P6.9, calculate the percentage loading error for the three values of the resistance ratio: $R_c/R_e = 0.1$, 1.0, and 10.0, assuming that the load resistance R_L is equal to the element resistance. Compare the results with the corresponding value for $R_e = 0$. Finally, choose a suitable value for R_c/R_e and plot the curve of percentage loading error versus fractional displacement x/x_{max}. From the graph, estimate the maximum loading error.

6.10 Derive an expression for the sensitivity (normalized) of a rotatory potentiometer as a function of displacement (normalized). Plot the corresponding curve in the nondimensional form for the three load values given by $R_L/R_c = 0.1$, 1.0, and 10.0. Where does the maximum sensitivity occur? Verify your observation using the analytical expression.

6.11 The range of a coil-type potentiometer is 10 cm. If the wire diameter is 0.1 mm, determine the resolution of the device.

6.12 The data acquisition system connected at the output of a differential transformer (say, an LVDT) has a very high resistive load. Obtain an expression for the phase lead of the output signal (at the load) of the differential transformer, with reference to the supply to the primary windings of the transformer, in terms of the impedance of the primary windings only.

6.13 At the null position, the impedances of the two secondary winding segments of an LVDT were found to be equal in magnitude but slightly unequal in phase. Show that the quadrature error (null voltage) is about 90° out of phase with reference to the predominant component of the output signal under open-circuit conditions.

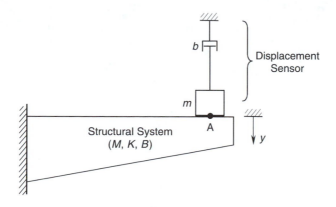

FIGURE P6.14
The use of a displacement sensor to measure the natural frequency and damping ratio of a structure.

HINT This may be proved either analytically or graphically by considering the difference between two rotating directed lines (phasors) that are separated by a very small angle.

6.14 A vibrating system has an effective mass M, an effective stiffness K, and an effective damping constant B in its primary mode of vibration at point A with respect to coordinate y. Write expressions for the undamped natural frequency, the damped natural frequency, and the damping ratio for this first mode of vibration of the system.

A displacement transducer is used to measure the fundamental undamped natural frequency and the damping ratio of the system by subjecting the system to an initial excitation and recording the displacement trace at a suitable location (point A along y in Figure P6.14) in the system. This trace will provide the period of damped oscillations and the logarithmic decrement of the exponential decay from which the required parameters can be computed using well-known relations. It was found, however, the mass m of the moving part of the displacement sensor and the associated equivalent viscous damping constant b are not negligible. Using the model shown in Figure P6.14, derive expressions for the measured undamped natural frequency and damping ratio. Suppose that $M = 10$ kg, $K = 10$ N/m, and $B = 2$ N/m/s. Consider an LVDT whose core weighs 5 gm and has negligible damping, and a potentiometer whose slider arm weighs 5 gm and has an equivalent viscous damping constant of 0.05 N/m/s. Estimate the percentage error of the results for the undamped natural frequency and damping ratio measured using each of these two displacement sensors.

6.15 Standard rectilinear displacement sensors such as the LVDT and the potentiometer are used to measure displacements up to 25 cm; within this limit, accuracies as high as ±0.2% can be obtained. For measuring large displacements on the order of 3 m, cable extension displacement sensors, which have an angular displacement sensor as the basic sensing unit, may be used. One type of rectilinear displacement sensor has a rotatory potentiometer and a light cable, which wraps around a spool that rotates with the wiper arm of the pot. In using this sensor, the free end of the cable is connected to the moving member whose displacement is to be measured. The sensor housing is mounted on a stationary platform, such as the support structure of the system being monitored. A spring motor winds the cable back as the cable retracts. Using suitable sketches, describe the operation of this displacement sensor. Discuss the shortcomings of this device.

6.16 It is known that some of the factors that should be considered in selecting an LVDT for a particular application are linearity, sensitivity, response time, size and weight of core,

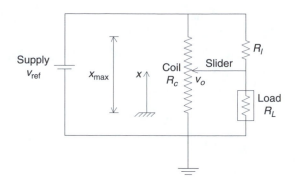

FIGURE P6.18
Potentiometer circuit with a linearizing resistor.

size of the housing, primary excitation frequency, output impedance, phase change between primary and secondary voltages, null voltage, stroke, and environmental effects (temperature compensation, magnetic shielding, etc.). Explain why and how each of these factors is an important consideration.

6.17 The signal-conditioning system for an LVDT has the following components: power supply, oscillator, synchronous demodulator, filter, and voltage amplifier. Using a block diagram, show how these components are connected to the LVDT. Describe the purpose of each component. A high-performance LVDT has a linearity rating of 0.01% in its output range of 0.1–1.0 VAC. The response time of the LVDT is known to be 10 ms. What should be the frequency of the primary excitation?

6.18 List merits and shortcomings of a potentiometer (pot) as a displacement sensing device, in comparison with a linear-variable differential transformer (LVDT). Give several ways to improve the measurement linearity of a potentiometer.

Suppose that a resistance R_l is added to the conventional potentiometer circuit as shown in Figure P6.18. With $R_l = R_L$ show that

$$\frac{v_o}{v_{ref}} = \frac{(R_L/R_c + 1 - x/x_{max})x/x_{max}}{\left[R_L/R_c + 2x/x_{max} - 2(x/x_{max})^2\right]}$$

where
    R_c = potentiometer coil resistance (total)
    R_L = load resistance
   v_{ref} = supply voltage to the coil
    v_o = output voltage
     x = slider displacement
 x_{max} = slider stroke (maximum displacement).

Explain why R_l will produce a linearizing effect.

6.19 Suppose that a sinusoidal carrier frequency is applied to the primary coil of an LVDT. Sketch the shape of the output voltage of the LVDT when the core is stationary at: **(a)** null position; **(b)** the left of null position; **(c)** the right of null position.

6.20 For directional sensing using an LVDT it is necessary to determine the phase angle of the induced signal. In other words, *phase-sensitive demodulation* would be needed.

a. First consider a linear core displacement starting from a positive value, moving to zero, and then returning to the same position in an equal time period. Sketch the output of the LVDT for this "triangular" core displacement.

b. Next sketch the output if the core continued to move to the negative side at the same speed.

By comparing the two outputs show that phase-sensitive demodulation would be needed to distinguish between the two cases of displacement.

6.21 Joint angles and angular speeds are the two basic measurements used in the direct (low-level) control of robotic manipulators. One type of robot arm uses resolvers to measure angles and differentiate these signals (digitally) to obtain angular speeds. A gear system is used to step up the measurement (typical gear ratio, 1:8). Since the gear wheels are ferromagnetic, an alternative measuring device would be a self-induction or mutual-induction proximity sensor located at a gear wheel. This arrangement, known as a pulse tachometer, generates a pulse (or near-sine) signal, which can be used to determine both angular displacement and angular speed. Discuss the advantages and disadvantages of these two arrangements (resolver and pulse tachometer) in this particular application.

6.22 Why is motion sensing important in trajectory-following control of robotic manipulators? Identify five types of motion sensors that could be used in robotic manipulators.

6.23 Compare and contrast the principles of operation of dc tachometer and ac tachometer (both permanent-magnet and induction types). What are the advantages and disadvantages of these two types of tachometers?

6.24 Describe three different types of proximity sensors. In some applications, it may be required to sense only two-state positions (e.g., presence or absence, go or no-go). Proximity sensors can be used in such applications, and in that context they are termed proximity switches (or, limit switches). For example, consider a parts-handling application in automated manufacturing in which a robot end effector grips a part and picks it up to move it from a conveyor to a machine tool. We can identify four separate steps in the gripping process. Explain how proximity switches can be used for sensing in each of these four tasks:

a. Make sure that the part is at the expected location on the conveyor.

b. Make sure that the gripper is open.

c. Make sure that the end effector has moved to the correct location so that the part is in between the gripper fingers.

d. Make sure that the part did not slip when the gripper was closed.

NOTE A similar use of limit switches is found in lumber mills, where tree logs are cut (bucked) into smaller logs; bark removed (de-barked); cut into a square/rectangular log using a chip'n saw operation; and sawed into smaller dimensions (e.g., two by four cross-sections) for marketing.

6.25 Discuss the relationships among displacement sensing, distance sensing, position sensing, and proximity sensing. Explain why the following characteristics are important in using some types of motion sensors:

a. Material of the moving (or target) object

b. Shape of the moving object

c. Size (including mass) of the moving object

d. Distance (large or small) of the target object

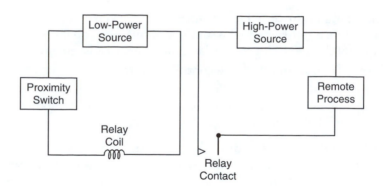

FIGURE P6.26
Proximity switch-operated relay circuit.

    e. Nature of motion (transient or not, what speed, etc.) of the moving object
    f. Environmental conditions (humidity, temperature, magnetic fields, dirt, lighting conditions, shock and vibration, etc.).

6.26 In some industrial processes, it is necessary to sense the condition of a system at one location and, depending on that condition, activate an operation at a location far from that location. For example, in a manufacturing environment, when the count of the finished parts exceeds some value, as sensed in the storage area, a milling machine could be shut down or started. A proximity switch could be used for sensing, and a networked (e.g., Ethernet-based) control system for process control. Since activation of the remote process usually requires a current that is larger than the rated load of a proximity switch, one would have to use a relay circuit, which is operated by the proximity switch. One such arrangement is shown in Figure P6.26. Note that the relay circuit can be used to operate a device such as a valve, a motor, a pump, or a heavy-duty switch. Discuss an application of the arrangement shown in Figure P6.26 in the food-packaging industry. A mutual-induction proximity sensor with the following ratings is used in this application:

    Sensor diameter = 1 cm

    Sensing distance (proximity) = 1 mm

    Supply to primary winding =110 AC at 60 Hz

    Load current rating (in secondary) = 200 mA.

Discuss the limitations of this proximity sensor.

6.27 Compression molding is used in making parts of complex shapes and varying sizes. Typically, the mold consists of two platens, the bottom platen fixtured to the press table and the top platen operated by a hydraulic press. Metal or plastic sheets—e.g., for the automotive industry—can be compression-molded in this manner. The main requirement in controlling the press is to position the top platen accurately with respect to the bottom platen (say, with a 0.001 in or 0.025 mm tolerance), and it has to be done quickly (say, in a few seconds). How many degrees of freedom have to be sensed (how many position sensors are needed) in controlling the mold? Suggest typical displacement measurements that would be made in this application and the types of sensors that could be employed. Indicate sources of error that cannot be perfectly compensated for in this application.

6.28 Seam tracking in robotic arc welding needs accurate position control under dynamic conditions. The welding seam has to be accurately followed (tracked) by the welding torch. Typically, the position error should not exceed 0.2 mm. A proximity sensor could be used for sensing the gap between the welding torch and the welded part. It is necessary to install the sensor on the robot end effector so that it tracks the seam at some distance (typically 1 in or 2.5 cm) ahead of the welding torch. Explain why this is important. If the speed of welding is not constant and the distance between the torch and the proximity sensor is fixed, what kind of compensation would be necessary in controlling the end effector position? Sensor sensitivity of several volts per millimeter is required in this position control application. What type of proximity sensor would you recommend?

6.29 An angular motion sensor, which operates somewhat like a conventional resolver has been developed at Wright State University. The rotor of this resolver is a permanent magnet. A 2 cm diameter Alnico-2 disk magnet, diametrically magnetized as a two-pole rotor, has been used. Instead of the two sets of stationary windings placed at 90° in a conventional resolver, two Hall-effect sensors (see Chapter 7) placed at 90° around the permanent-magnet rotor are used for detecting quadrature signals. Note that Hall-effect sensors can detect moving magnetic sources. Describe the operation of this modified resolver and explain how this device could be used to measure angular motions continuously. Compare this device with a conventional resolver, giving advantages and disadvantages.

6.30 Discuss factors that limit the lower frequency and upper frequency limits of the output from the following sensors:

   a. Potentiometer
   b. LVDT
   c. Resolver
   d Eddy current proximity sensor
   e. DC tachometer
   f. Piezoelectric transducer.

6.31 An active suspension system is proposed for a high-speed ground transit vehicle in order to achieve improved ride quality. The system senses jerk (rate of change of acceleration) due to road disturbances and adjusts system parameters accordingly.

   a. Draw a suitable schematic diagram for the proposed control system and describe appropriate measuring devices.
   b. Suggest a way to specify the "desired" ride quality for a given type of vehicle. (Would you specify one value of jerk, a jerk range, or a jerk curve with respect to time or frequency?)
   c. Discuss the drawbacks and limitations of the proposed control system with respect to such factors as reliability, cost, feasibility, and accuracy.

6.32 A design objective in most control system applications is to achieve small time constants. An exception is the time constant requirements for a piezoelectric sensor. Explain why a large time constant, on the order of 1 sec, is desirable for a piezoelectric sensor in combination with its signal conditioning system. An equivalent circuit for a piezoelectric accelerometer, which uses a quartz crystal as the sensing element, is shown in Figure P6.32. The charge generated is denoted by q, and the voltage output at the end of the accelerometer cable is v_o. The piezoelectric sensor capacitance is modeled by C_p, and the overall capacitance experienced at the sensor output, whose primary contribution is due to cable capacitance, is denoted by C_c. The resistance of the electric insulation in the accelerometer is denoted by R. Write a differential equation relating v_o to q. What is the corresponding transfer function? Using this result, show that the accuracy of accelerometer improves when the sensor time constant is large and when the frequency of the measured acceleration is high. For a quartz

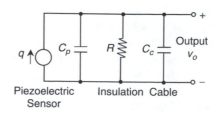

FIGURE P6.32
Equivalent circuit for a quartz crystal (piezoelectric) accelerometer.

crystal sensor with $R = 1 \times 10^{11}$ Ω and $C_p = 300$ pF, and a circuit with $C_c = 700$ pF compute the time constant.

6.33 Applications of accelerometers are found in the following areas:

a. Transit vehicles (automobiles—microsensors for airbag sensing in particular, aircraft, ships, etc.)

b. Power cable monitoring

c. Robotic manipulator control

d. Building structures

e. Shock and vibration testing

f. Position and velocity sensing.

Describe one direct use of acceleration measurement in each application area.

6.34 a A standard accelerometer that weighs 100 gm is mounted on a test object that has an equivalent mass of 3 kg. Estimate the accuracy in the first natural frequency of the object measured using this arrangement, considering mechanical loading due to accelerometer mass alone. If a miniature accelerometer that weighs 0.5 gm is used instead, what is the resulting accuracy?

b. A strain gage accelerometer uses a semiconductor strain gage mounted at the root of a cantilever element, with the seismic mass mounted at the free end of the cantilever. Suppose that the cantilever element has a square cross-section with dimension 1.5×1.5 mm². The equivalent length of the cantilever element is 25 mm, and the equivalent seismic mass is 0.2 gm. If the cantilever is made of an aluminum alloy with Young's modulus $E = 69 \times 10^9$ N/m², estimate the useful frequency range of the accelerometer in hertz. *Hint:* When force F is applied to the free end of a cantilever, the deflection y at that location may be approximated by the formula

$$y = \frac{Fl^3}{3EI}$$

where
l = cantilever length
I = second moment area of the cantilever cross-section about the bending axis = $bh^3/12$
b = cross-section width
h = cross-section height.

6.35 Applications of piezoelectric sensors are numerous; push-button devices and switches, airbag microelectromechanical (MEMS) sensors in vehicles, pressure and force sensing, robotic tactile sensing, accelerometers, glide testing of computer disk-drive heads, excitation sensing in dynamic testing, respiration sensing in medical diagnostics, and graphics input devices for computers. Discuss advantages and disadvantages of piezoelectric sensors.

What is cross-sensitivity of a sensor? Indicate how the anisotropy of piezoelectric crystals (i.e., charge sensitivity quite large along one particular crystal axis) is useful in reducing cross-sensitivity problems in a piezoelectric sensor.

6.36 As a result of advances in microelectronics, piezoelectric sensors (such as accelerometers and impedance heads) are now available in miniature form with built-in charge amplifiers in a single integral package. When such units are employed, additional signal conditioning is usually not necessary. An external power supply unit is needed, however, to provide power for the amplifier circuitry. Discuss the advantages and disadvantages of a piezoelectric sensor with built-in microelectronics for signal conditioning.

A piezoelectric accelerometer is connected to a charge amplifier. An equivalent circuit for this arrangement is shown in Figure 6.33.

a. Obtain a differential equation for the output v_o of the charge amplifier, with acceleration a as the input, in terms of the following parameters: S_a = charge sensitivity of the accelerometer (charge/acceleration); R_f = feedback resistance of the charge amplifier; τ_c = time constant of the system (charge amplifier).

b. If an acceleration pulse of magnitude a_o and duration T is applied to the accelerometer, sketch the time response of the amplifier output v_o. Show how this response varies with τ_c. Using this result, show that the larger the τ_c the more accurate the measurement.

6.37 Give typical values for the output impedance and the time constant of the following measuring devices:

a. Potentiometer

b. Differential transformer

c. Resolver

d. Piezoelectric accelerometer.

A resistance temperature detector (RTD) has an output impedance of 500 Ω. If the loading error has to be maintained near 5%, estimate a suitable value for the load impedance.

6.38 A signature verification pen has been developed by IBM Corporation. The purpose of the pen is to authenticate the person who provides the signature, by detecting whether the user is forging someone else's signature. The instrumented pen has analog sensors. Sensor signals are conditioned using microcircuitry built into the pen and sampled into a digital computer at the rate of 80 samples/second using an ADC. Typically about 1,000 data samples are collected per signature. Prior to the pen's use, authentic signatures are collected off-line and stored in a reference data base. When a signature and the corresponding identification code are supplied to the computer for verification, a program in the processor retrieves the authentic signature from the data base, by referring to the identification code, and then compares the two sets of data for authenticity. This process takes about 3 sec. Discuss the types of sensors that could be used in the pen. Estimate the total time required for a signal verification. What are the advantages and disadvantages of this method in comparison to having the user punch in an identification code alone or provide the signature without the identification code?

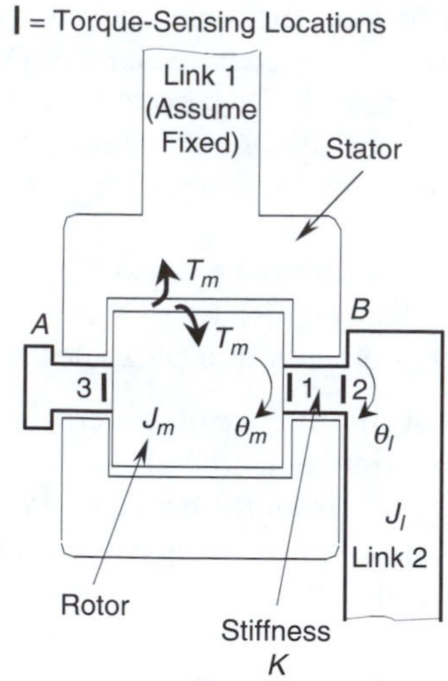

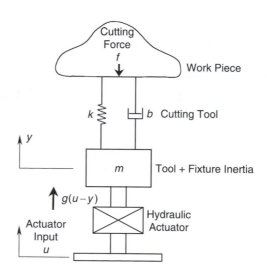

FIGURE P6.40
Torque sensing locations for a
manipulator joint.

FIGURE P6.41
A model for a machining operation.

6.39 Under what conditions can displacement control be treated as force control? Describe a situation in which this is not feasible.

6.40 Consider the joint of a robotic manipulator, shown schematically in Figure P6.40. Torque sensors are mounted at locations 1, 2, and 3. If the electromagnetic torque generated at the motor rotor is T_m write equations for the torque transmitted to link 2, the frictional torque at bearing A, the frictional torque at bearing B, and the reaction torque on link 1, in terms of the measured torques, the inertia torque of the rotor, and T_m.

6.41 A model for a machining operation is shown in Figure P6.41. The cutting force is denoted by f, and the cutting tool with its fixtures is modeled by a spring (stiffness k), a viscous damper (damping constant b), and a mass m. The actuator (hydraulic) with its controller is represented by an active stiffness g. Assuming linear g, obtain a transfer relation between the actuator input u and the cutting force f. Now determine an approximate expression for the gradient $(\partial g / \partial u)$. Discuss a control strategy for counteracting

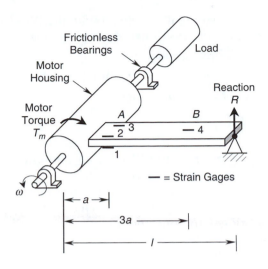

FIGURE P6.42
A strain gage sensor for measuring motor torque.

effects due to random variations in the cutting force. Note that this is important for controlling the product quality.

HINT You may use a reference-adaptive feedforward control strategy where a reference g and u are the inputs to the machine tool. The reference g is "adapted" using the gradient $(\partial g / \partial u)$, as u changes by Δu.

6.42 A strain gage sensor to measure the torque T_m generated by a motor is shown schematically in Figure P6.42. The motor is floated on frictionless bearings. A uniform rectangular lever arm is rigidly attached to the motor housing, and its projected end is restrained by a pin joint. Four identical strain gages are mounted on the lever arm, as shown. Three of the strain gages are at point A, which is located at a distance a from the motor shaft, and the fourth strain gage is at point B, which is located at a distance $3a$ from the motor shaft. The pin joint is at a distance l from the motor shaft. Strain gages 2, 3, and 4 are on the top surface of the lever arm, and gage 1 is on the bottom surface. Obtain an expression for T_m in terms of the bridge output δv_o and the following additional parameters:

S_s = gage factor (strain gage sensitivity)
v_{ref} = supply voltage to the bridge
b = width of the lever arm cross-section
h = height of the lever arm cross-section
E = Young's modulus of the lever arm

Verify that the bridge sensitivity does not depend on a and l. Describe means to improve the bridge sensitivity. Explain why the sensor reading is only an approximation to the torque transmitted to the load. Give a relation to determine the net normal reaction force at the bearings, using the bridge output.

6.43 The sensitivity S_s of a strain gage consists of two parts: the contribution from the change in resistivity of the material and the direct contribution due to the change in shape of the strain gage when deformed. Show that the second part may be approximated by $(1 + 2v)$, where v denotes the Poisson's ratio of the strain gage material.

6.44 Compare the potentiometer (ballast) circuit with the Wheatstone bridge circuit for strain gage measurements, with respect to the following considerations:

   a. Sensitivity to the measured strain
   b. Error due to ambient effects (e.g., temperature changes)
   c. Signal-to-noise ratio of the output voltage
   d. Circuit complexity and cost
   e. Linearity.

6.45 In the strain gage bridge shown in Figure 6.45, suppose that the load current i is not negligible. Derive an expression for the output voltage v_o in terms of R_1, R_2, R_3, R_4, R_L, and v_{ref}. Initially, the bridge was balanced, with the resistances in the four arms being equal. Then one of the resistances (say, R_1) was increased by 1%. Plot to scale the ratio (actual output from the bridge)/(output under open-circuit, or infinite-load-impedance, conditions) as a function of the nondimensionalized load resistance R_L/R in the range 0.1–10.0, where R denotes the initial resistance in each arm of the bridge.

6.46 What is meant by the term *bridge sensitivity* in strain gage measurements? Describe methods of increasing bridge sensitivity. Assuming the load resistance to be very high in comparison with the arm resistances in the strain gage bridge shown in Figure 6.45, obtain an expression for the power dissipation p in terms of the bridge resistances and the supply voltage. Discuss how the limitation on power dissipation can affect bridge sensitivity.

6.47 Consider the strain gage bridge shown in Figure 6.45. Initially, the bridge is balanced, with $R_1 = R_2 = R$. (Note: R_3 may not be equal to R_1.) Then R_1 is changed by δR. Assuming the load current to be negligible, derive an expression for the percentage error due to neglecting the second-order and higher-order terms in δR. If $\delta R/R = 0.05$, estimate this nonlinearity error.

6.48 Discuss the advantages and disadvantages of the following techniques in the context of measuring transient signals.

   a. DC bridge circuits versus ac bridge circuits
   b. Slip ring and brush commutators versus ac transformer commutators
   c. Strain gage torque sensors versus variable-inductance torque sensors
   d. Piezoelectric accelerometers versus strain gage accelerometers
   e. Tachometer velocity transducers versus piezoelectric velocity transducers.

6.49 For a semiconductor strain gage characterized by the quadratic strain-resistance relationship

$$\frac{\delta R}{R} = S_1 \varepsilon + S_2 \varepsilon^2$$

obtain an expression for the equivalent gage factor (sensitivity) S_s using the least squares error linear approximation. Assume that only positive strains up to ε_{max} are measured with the gage. Derive an expression for the percentage nonlinearity. Taking $S_1 = 117$, $S_2 = 3,600$, and $\varepsilon_{max} = 0.01$ strain, compute S_s and the percentage nonlinearity.

6.50 Briefly describe how strain gages may be used to measure

   a. Force
   b. Displacement

c. Acceleration

d. Pressure

e. Temperature.

Show that if a compensating resistance R_c is connected in series with the supply voltage v_{ref} to a strain gage bridge that has four identical members, each with resistance R, the output equation is given by

$$\frac{\delta v_o}{v_{ref}} = \frac{R}{(R+R_c)} \frac{kS_s}{4} \varepsilon$$

in the usual rotation.

A foil-gage load cell uses a simple (one-dimensional) tensile member to measure force. Suppose that k and S_s are insensitive to temperature change. If the temperature coefficient of R is α_1, that of the series compensating resistance R_c is α_2, and that of the Young's modulus of the tensile member is $(-\beta)$, determine an expression for R_c that would result in automatic (self) compensation for temperature effects. Under what conditions is this arrangement realizable?

6.51 Draw a block diagram for a single joint of a robot, identifying inputs and outputs. Using the diagram, explain the advantages of torque sensing in comparison to displacement and velocity sensing at the joint. What are the disadvantages of torque sensing?

6.52 Figure P6.52 shows a schematic diagram of a measuring device.

a. Identify the various components in this device.

b. Describe the operation of the device, explaining the function of each component and identifying the nature of the measurand and the output of the device.

c. List the advantages and disadvantages of the device

d. Describe a possible application of this device.

6.53 Discuss the advantages and disadvantages of torque sensing by the motor current method. Show that for a synchronous motor with a balanced three-phase supply, the electromagnetic torque generated at the rotor-stator interface is given by

$$T_m = ki_f i_a \cos(\theta - \omega t)$$

where

i_f = de current in the rotor (field) winding

i_a = amplitude of the supply current to each phase in the stator (armature)

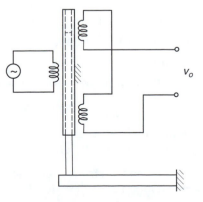

v_o

FIGURE P6.52

An analog sensor.

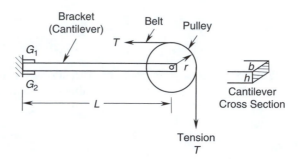

FIGURE P6.55
A strain gage tension sensor for a moving belt.

θ = angle of rotation
ω = frequency (angular) of the ac supply
t = time
k = motor torque constant

6.54 Discuss factors that limit the lower frequency and upper frequency limits of measurements obtained from the following devices:

a. Strain gage

b. Rotating shaft torque sensor

c. Reaction torque sensor.

6.55 Briefly describe a situation in which tension in a moving belt or cable has to be measured under transient conditions. What are some of the difficulties associated with measuring tension in a moving member? A strain gage tension sensor for a belt-drive system is shown in Figure P6.55. Two identical active strain gages, G_1 and G_2, are mounted at the root of a cantilever element with rectangular cross-section, as shown. A light, frictionless pulley is mounted at the free end of the cantilever element. The belt makes a 90° turn when passing over this idler pulley.

a. Using a circuit diagram, show the Wheatstone bridge connections necessary for the strain gages G_1 and G_2 so that strains due to axial forces in the cantilever member have no effect on the bridge output (i.e., effects of axial loads are compensated) and the sensitivity to bending loads is maximized.

b. Obtain an equation relating the belt tension T and the bridge output δv_o in terms of the following additional parameters:

S_s = gage factor (sensitivity) of each strain gage

E = Young's modulus of the cantilever element

L = length of the cantilever element

b = width of the cantilever cross-section

h = height of the cantilever cross-section

In particular, show that the radius of the pulley does not enter this equation. Give the main assumptions made in your derivation.

6.56 Consider a standard strain-gage bridge (Figure 6.45) with R_1 being the only active gage and $R_3 = R_4$. Obtain an expression for R_1 in terms of R_2, v_o and v_{ref}. Show that when

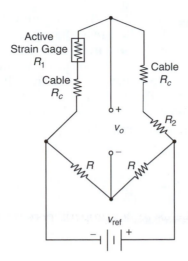

FIGURE P6.56
The influence of cable resistance on strain gage bridge measurements.

$R_1 = R_2$, we get $v_o = 0$—a balanced bridge—as required. Note that the equation for R_1, assuming that v_o is measured using a high-impedance voltmeter, can be used to detect large resistance changes in R_1. Now suppose that the active gage R_1 is connected to the bridge using a long, twisted wire pair, with each wire having a resistance of R_c. The bridge circuit has to be modified as in Figure P6.56 in this case. Using the expression obtained earlier for R_1, show that the equation of the modified bridge is given by

$$R_1 = R_2 \left[\frac{v_{ref} + 2v_o}{v_{ref} - 2v_o} \right] + 4R_c \frac{v_o}{[v_{ref} - 2v_o]}$$

Obtain an expression for the fractional error in the R_1 measurement due to cable resistance R_c. Show that this error can be decreased by increasing R_2 and v_{ref}.

6.57 The read-write head in a disk drive of a digital computer should float at a constant but small height (say, fraction of a µm) above the disk surface. Because of aerodynamics resulting from the surface roughness and the surface deformations of the disk, the head can be excited into vibrations that could cause head-disk contacts. These contacts, which are called head-disk interferences (HDIs), are clearly undesirable. They can occur at very high frequencies (say, 1 MHz). The purpose of a glide test is to detect HDIs and to determine the nature of these interferences. Glide testing can be used to determine the effect of parameters such as the flying height of the head and the speed of the disk, and to qualify (certify the quality of) disk drive units. Indicate the basic instrumentation needed in glide testing. In particular, suggest the types of sensors that could be used and their advantages and disadvantages.

6.58 What are the typical requirements for an industrial tactile sensor? Explain how a tactile sensor differs from a simple touch sensor. Define spatial resolution and force resolution (or sensitivity) of a tactile sensor.

The spatial resolution of your fingertip can be determined by a simple experiment using two pins and a helper. Close your eyes. Instruct the helper to apply one pin or both pins randomly to your fingertip so that you will feel the pressure of the tip of the pins.

You should respond by telling the helper whether you feel both pins or just one pin. If you feel both pins, the helper should decrease the spacing of the two pins in the next round of tests. The test should be repeated in this manner by successively decreasing the spacing between the pins until you feel only one pin when both pins are actually applied. Then measure the distance between the two pins in millimeters. The largest spacing between the two pins that will result in this incorrect sensation corresponds to the spatial resolution of your fingertip. Repeat this experiment for all your fingers, repeating the test several times on each finger. Compute the average and the standard deviation. Then perform the test on other subjects. Discuss your results. Do you notice large variations in the results?

6.59 Torque force, and tactile sensing can be very useful in many applications particularly in the manufacturing industry. For each of the following applications, indicate the types of sensors that would be useful for properly performing the task:

   a. Controlling the operation of inserting printed circuit boards in card cages using a robotic end effector

   b. Controlling a robotic end effector that screws a threaded part into a hole

   c. Failure prediction and diagnosis of a drilling operation

   d. Gripping a fragile, delicate, and somewhat flexible object by a robotic hand without damaging the object

   e. Gripping a metal part using a two-fingered gripper

   f. Quickly identifying and picking a complex part from a bin containing many different parts.

6.60 The *motion dexterity* of a device is defined as the ratio (Number of degrees of freedom in the device)/(Motion resolution of the device). The *force dexterity* may be defined as (Number of degrees of freedom in the device)/(Force resolution of the device). Giving a situation where both types of dexterity mean the same thing and a situation where the two terms mean different things. Outline how force dexterity of a device (say, an end effector) can be improved by using tactile sensors. Provide the dexterity requirements for the following tasks by indicating whether motion dexterity or force dexterity is preferred in each case:

   a. Gripping a hammer and driving a nail with it

   b. Threading a needle

   c. Seam tracking of a complex part in robotic arc welding

   d. Finishing the surface of a complex metal part using robotic grinding.

6.61 Describe four advantages and four disadvantages of a semiconductor strain gage weight sensor. A weight sensor is used in a robotic wrist. What would be the purpose of this sensor? How can the information obtained from the weight sensor be used in controlling the robotic manipulator?

6.62 Discuss whether there is any relationship between the dexterity and the stiffness of a manipulator hand. The stiffness of a robotic hand can be improved during gripping operations by temporarily decreasing the number of degrees of freedom of the hand using suitable fixtures. What purpose does this serve?

6.63 Using the usual equation for a dc strain-gage bridge (Figure 6.45) show that if the resistance elements R_1 and R_2 have the same temperature coefficient of resistance and if R_3 and R_4 have the same temperature coefficient of resistance, the temperature effects are compensated up to first order.

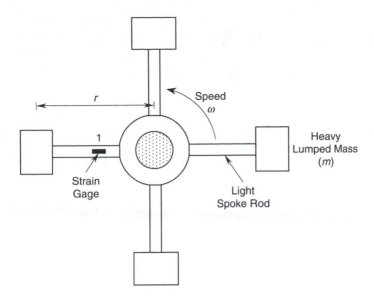

FIGURE P6.64
Strain-gage speed sensor for a fly wheel.

A microminiature (MEMS) strain-gage accelerometer uses two semiconductor strain gages, one integral with the cantilever element near the fixed end (root) and the other mounted at an unstrained location of the accelerometer. The entire unit including the cantilever and the strain gages, has a silicon integrated-circuit (IC) construction, and measures smaller than 1 mm in size. Outline the operation of the accelerometer. What is the purpose of the second strain gage?

6.64 a. List three advantages and three disadvantages of a semiconductor strain gage when compared with a foil strain gage.

b. A fly-wheel device is schematically shown in Figure P6.64. The wheel consists of four spokes which carry lumped masses at one end and are clamped to the rotating hub at the other end, as shown. Suppose that the inertia of the spokes can be neglected in comparison to that of the lumped masses. Four active strain gages are used in a bridge circuit for measuring speed.

    i. If the bridge can be calibrated to measure the tensile force F in each spoke, express the dynamic equation, which may be used to measure the rotating speed (ω). The following parameters may be used:

       m = mass of the lumped element at the end of a spoke
       r = radius of rotation of the center of mass of the lumped element

    ii. For good results with regard to high sensitivity of the bridge and also for compensation of secondary effects such as out-of-plane bending, indicate where the four strain gages (1,2,3, and 4) should be located on the spokes and in what configuration they should be connected in a dc bridge.

    iii. Compare this method of speed sensing to that using a tachometer and/or a potentiometer by giving three advantages and two disadvantages of the strain gage method.

6.65 a. Consider a simple mechanical manipulator. Explain why in some types of manipulation tasks, motion sensing alone might not be adequate for accurate control, and torque/force sensing might be needed as well.

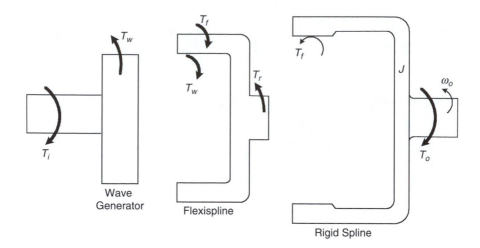

FIGURE P6.65
Free body diagram of a harmonic drive.

b. Discuss what factors should be considered when installing a torque sensor to measure the torque transmitted from an actuator to a rotating load.

c. A harmonic drive consists of the following three main components:

1. The input shaft with the elliptical wave generator (cam)
2. The circular flexispline with external teeth
3. The rigid circular spline with internal teeth

Consider the free-body diagrams shown in Figure P6.65. The following variables are defined:

ω_i = speed of the input shaft (wave generator)

ω_o = speed of the output shaft (rigid spline)

T_o = torque transmitted to the driven load by the output shaft (rigid spline)

T_i = torque applied on the harmonic drive by the input shaft

T_f = torque transmitted by the flexispline to the rigid spline

T_r = reaction torque on the flexispline at the fixture

T_w = torque transmitted by the wave generator

If strain gages are to be used to measure the output torque T_o, suggest suitable locations for mounting them and discuss how the torque measurement can be obtained in this manner. Using a block diagram for the system indicate whether you consider T_o to be an input to or an output of the harmonic drive. What are the implications of this consideration?

6.66 a. Describe three different principles of torque sensing. Discuss relative advantages and disadvantages of the three approaches.

b. A torque sensor is needed for measuring the drive torque that is transmitted to a link of a robot (i.e., joint torque). What characteristics and specifications of the sensor and the requirement of the system should be considered in selecting a suitable torque sensor for this application?

6.67 A simple rate gyro, which may be used to measure angular speeds, is shown in Figure P6.67. The angular speed of spin is ω, and is kept constant at a known value.

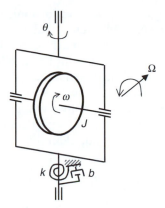

FIGURE P6.67
A rate gyro speed sensor.

The angle of rotation of the gyro about the gimbalaxis (or, the angle of twist of the torsional spring) is θ, and is measured using a displacement sensor. The angular speed of the gyro about the axis that is orthogonal to both gimbal axis and spin axis is Ω. This is the angular speed of the supporting structure (vehicle), which needs to be measured. Obtain a relationship between Ω and θ in terms of parameters such as the following:

J = moment of inertia of the spinning wheel
k = torsional stiffness of the spring restraint at the gimbal bearings
b = damping constant of rotational motion about the gimbal axis;

and the spinning speed ω. How would you improve the sensitivity of this device?

6.68 Level sensors are used in a wide variety of applications, including soft drink bottling, food packaging, monitoring of storage vessels, mixing tanks, and pipelines. Consider the following types of level sensors, and briefly explain the principle of operation of each type, in level sensing. Also, what are the limitations of each type?

   a. Capacitive sensors
   b. Inductive sensors
   c. Ultrasonic sensors
   d. Vibration sensors.

6.69 Consider the following types of position sensors: inductive, capacitive, eddy current, fiber-optic, and ultrasonic. For the following conditions, indicate which of these types are not suitable and explain why:

   a. Environment with variable humidity
   b. Target object made of aluminum
   c. Target object made of steel
   d. Target object made of plastic
   e. Target object several feet away from the sensor location
   f. Environment with significant temperature fluctuations
   g. Smoke-filled environment.

6.70 Discuss advantages and disadvantages of fiber-optic sensors. Consider the fiber-optic position sensor. In the curve of light intensity received versus x, in which region would you prefer to operate the sensor, and what are the corresponding limitations?

6.71 The manufacturer of an ultrasonic gage states that the device has applications in measuring cold roll steel thickness, determining parts positions in robotic assembly, lumber sorting, measurement of particle board and plywood thickness, ceramic tile dimensional inspection, sensing the fill level of food in a jar, pipe diameter gaging, rubber tire positioning during fabrication, gauging of fabricated automotive components, edge detection, location of flaws in products, and parts identification. Discuss whether the following types of sensors are also equally suitable for some or all of the foregoing applications. In each case where you think that a particular sensor in not suitable for a given application, give reasons to support your claim.

a. Fiber optic position sensors

b. Self-induction proximity sensors

c. Eddy current proximity sensors

d. Capacitive gages

e. Potentiometers

f. Differential transformers

6.72 a. Consider the motion control system that is shown by the block diagram in Figure P6.72.

   i. Giving examples of typical situations explain the meaning of the block represented as "Load" in this system.

   ii. Indicate advantages and shortcomings of moving the motion sensors from the motor shaft to the load response point, as indicated by the broken lines in the figure.

b. Indicate, giving reasons, what type of sensors you will recommend for the following applications:

   i. In a soft drink bottling line, for on-line detection of improperly fitted metal caps on glass bottles.

   ii. In a paper processing plant, to simultaneously measure both the diameter and eccentricity of rolls of newsprint.

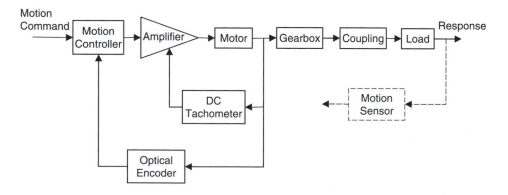

FIGURE P6.72
Block diagram of a motion control system.

iii To measure the dynamic force transmitted from a robot to its support structure, during operation.

iv. In a plywood manufacturing machine, for on-line measurement of the thickness of plywood.

v. In a food canning plant, to detect defective cans (with damage to flange, side seam, etc.,)

vi. To read codes on food packages.

7

Digital Transducers

A digital transducer is a measuring device that produces a digital output. A transducer whose output is a pulse signal may be considered in this category since the pulses can be counted and presented in the digital form using a counter. Similarly, a transducer whose output is a frequency falls into the same category since it can use a frequency counter to generate a digital output.

Sensors and transducers are useful in many industrial applications within the general area of mechatronics. Numerous examples are found in robotic manipulation, transit system, digital computers and accessories, process monitoring and control, material processing, fabrication, finishing, handling, inspection, testing, grading, and packaging. In Chapter 6 we discussed analog sensors and transducers. In this chapter we will study some useful types, concepts, operation, and utilization of digital transducers. Our discussion will primarily be limited to motion transducers. Note, however, that by using a suitable auxiliary front-end sensor, other *measurands*, such as force, torque, temperature, and pressure, may be converted into a motion and subsequently measured using a motion transducer. For example, altitude (or pressure) measurements in aircraft and aerospace applications are made using a pressure-sensing front end, such as a bellows or diaphragm device, in conjunction with an optical encoder (which is a digital transducer) to measure the resulting displacement. Similarly, a bimetallic element may be used to convert temperature into a displacement, which may be measured using a displacement sensor.

As we have done, it is acceptable to call an analog sensor an "analog transducer," since both the sensor stage and the transducer stage are analog in this case. The sensor stage of a digital transducer is typically analog as well. For example, motion, as manifested in physical systems, is continuous in time. Therefore, we cannot generally speak of digital motion sensors. Actually, it is the transducer stage that generates the discrete output signal in a digital motion-measuring device. Hence, we have chosen to term the present category of devices as digital transducers rather than digital sensors. Commercially available digital transducers are not as numerous as analog sensors, but what is available has found extensive application.

7.1 Advantages of Digital Transducers

Any measuring device that presents information as discrete samples and that does not introduce a *quantization error* when the reading is represented in the digital form may be classified as a digital transducer. According to this definition, for example, an analog sensor such as a thermocouple along with an analog-to-digital converter (ADC) is not a digital transducer. This is so because a quantization error is introduced by the ADC process (see Chapter 4).

A digital processor plays the role of controller in a digital control system. This facilitates complex processing of measured signals and other known quantities, thereby generating control signals for the plant of the control system. If the measured signals are available in analog form, an ADC stage is necessary prior to processing in a digital controller.

There are several advantages of digital signals (or, digital representation of information) in comparison to analog signals.

1. Digital signals are less susceptible to noise, disturbances, or parameter variation in instruments because data can be generated, represented, transmitted, and processed as binary words consisting of bits, which possess two identifiable states.

2. Complex signal processing with very high accuracy and speed are possible through digital means (Hardware implementation is faster than software implementation).

3. High reliability in a system can be achieved by minimizing analog hardware components.

4. Large amounts of data can be stored using compact, high-density, data storage methods.

5. Data can be stored or maintained for very long periods of time without any drift or being affected by adverse environmental conditions.

6. Fast data transmission is possible over long distances without introducing significant dynamic delays, as in analog systems.

7. Digital signals use low voltages (e.g., 0–12 V DC) and low power.

8. Digital devices typically have low overall cost.

These advantages help build a strong case in favor of digital measuring devices for mechatronic systems.

Digital measuring devices (or digital transducers, as they are commonly known) generate discrete output signals such as pulse trains or encoded data that can be directly read by a digital controller. Nevertheless, the sensor stage of a digital measuring device is usually quite similar to that of an analog counterpart. There are digital measuring devices that incorporate microprocessors to perform numerical manipulations and conditioning locally and provide output signals in either digital form or analog form. These measuring systems are particularly useful when the required variable is not directly measurable but could be computed using one or more measured outputs (e.g., power = force × speed). Although a microprocessor is an integral part of the measuring device in this case, it performs not a measuring task but, rather, a conditioning task. For our purposes, we shall consider the two tasks separately.

When the output of a digital transducer is a pulse signal, a common way of reading the signal is by using a counter, either to count the pulses (for high-frequency pulses) or to count clock cycles over one pulse duration (for low-frequency pulses). The count is placed as a digital word in a buffer, which can be accessed by the host (control) computer, typically at a constant frequency (sampling rate). On the other hand, if the output of a digital transducer is automatically available in a coded form (e.g., natural binary code or gray code), it can be directly read by a computer. In the latter case, the coded signal is normally generated by a parallel set of pulse signals; each pulse transition generates one bit of the digital word, and the numerical value of the word is determined by the pattern of the generated pulses. This is, for example, is the case with absolute encoders. Data acquisition from (i.e., computer interfacing) a digital transducer is commonly done using a general-purpose input/output

(I/O) card, for example, a motion control (servo) card, which may be able to accommodate multiple transducers (e.g., 8 channels of encoder inputs with 24-bit counters), or using a data acquisition card specific to the particular transducer.

7.2   Shaft Encoders

Any transducer that generates a "coded" (digital) reading of a measurement can be termed an encoder. Shaft encoders are digital transducers that are used for measuring "angular" displacements and "angular" velocities. Applications of these devices include motion measurement in performance monitoring and control of robotic manipulators, machine tools, industrial processes (e.g., food processing and packaging, pulp and paper), digital data storage devices, positioning tables, satellite mirror positioning systems, and rotating machinery such as motors, pumps, compressors, turbines, and generators. High resolution (depending on the word size of the encoder output and the number of pulses generated per revolution of the encoder), high accuracy (particularly due to noise immunity and reliability of digital signals and superior construction), and relative ease of adoption in digital control systems (because transducer output can be read as a digital word), with associated reduction in system cost and improvement of system reliability, are some of the relative advantages of digital transducers in general and shaft encoders in particular, in comparison to their analog counterparts.

7.2.1   Encoder Types

Shaft encoders can be classified into two categories, depending on the nature and the method of interpretation of the transducer output:

1. Incremental encoders
2. Absolute encoders

The output of an incremental encoder is a pulse signal, which is generated when the transducer disk rotates as a result of the motion that is being measured. By counting the pulses or by timing the pulse width using a clock signal, both angular displacement and angular velocity can be determined. With an incremental encoder, displacement is obtained with respect to some reference point. The reference point can be the home position of the moving component (say, determined by a limit switch); or a reference point on the encoder disk, as indicated by a reference pulse (*index pulse*) generated at that location on the disk. Furthermore, the index pulse count determines the number of full revolutions.

   An absolute encoder (or, whole-word encoder) has many pulse tracks on its transducer disk. When the disk of an absolute encoder rotates, several pulse trains—equal in number to the tracks on the disk—are generated simultaneously. At a given instant, the magnitude of each pulse signal will have one of two signal levels (i.e., a binary state), as determined by a level detector (or, edge detector). This signal level corresponds to a binary digit (0 or 1). Hence, the set of pulse trains gives an encoded binary number at any instant. The pulse windows on the tracks can be organized into some pattern (code) so that the generated binary number at a particular instant corresponds to the specific angular position of the encoder disk at that time. The pulse voltage can be made compatible with some digital interface logic (e.g., transistor-to-transistor logic, or TTL). Consequently, the direct digital readout of an angular position is possible with an absolute encoder, thereby expediting digital data acquisition and processing. Absolute encoders are commonly used to measure fractions of a revolution. However, complete revolutions can be measured using an additional track, which generates an index pulse, as in the case of incremental encoder.

The same signal generation (and pick-off) mechanism may be used in both types of transducers. Four techniques of transducer signal generation can be identified:

1. Optical (photosensor) method
2. Sliding contact (electrical conducting) method
3. Magnetic saturation (reluctance) method
4. Proximity sensor method

By far, the optical encoder is most popular and cost effective. The other three approaches may be used in special circumstances where an optical may not be suitable (e.g., under extreme temperatures) or may be redundant (e.g., where a code disk such as a toothed wheel is already available as an integral part of the moving member). For a given type of encoder (incremental or absolute), the method of signal interpretation is identical for all four types of signal generation listed above. We will briefly describe the principle of signal generation for all four techniques, and will consider only the optical encoder in the context of signal interpretation and processing.

The optical encoder uses an opaque disk (code disk) that has one or more circular tracks, with some arrangement of identical transparent windows (slits) in each track. A parallel beam of light (e.g., from a set of light-emitting diodes or LEDs) is projected to all tracks from one side of the disk. The transmitted light is picked off using a bank of photosensors on the other side of the disk, which typically has one sensor for each track. This arrangement is shown in Figure 7.1(a), which indicates just one track and one pick-off sensor. The light sensor could be a silicon photodiode or a phototransistor. Since the light from the source is interrupted by the opaque regions of the track, the output signal from the photosensor is a series of voltage pulses. This signal can be interpreted (e.g., through edge detection or level detection) to obtain the increments in the angular position and also angular velocity of the disk. Note that in the standard terminology, the sensor element of such a measuring device is the encoder disk, which is coupled to the rotating object (directly or through a gear mechanism). The transducer stage is the conversion of disk motion (analog) into the pulse signals (which can be coded into a digital word). The opaque background of transparent windows (the window pattern) on an encoder disk may be produced by contact printing techniques. The precision of this production procedure is a major factor that determines the accuracy of optical encoders. Note that a transparent disk with a track of opaque spots will work equally well as the encoder disk of an optical encoder. In either form, the track has a 50% duty cycle (i.e., length of the transparent region = length of the opaque region). A commercially available optical encoder is shown in Figure 7.1(b).

In a sliding contact encoder, the transducer disk is made of an electrically insulating material. Circular tracks on the disk are formed by implanting a pattern of conducting areas. These conducting regions correspond to the transparent windows on an optical encoder disk. All conducting areas are connected to a common slip ring on the encoder shaft. A constant voltage v_{ref} is applied to the slip ring using a brush mechanism. A sliding contact such as a brush touches each track, and as the disk rotates, a voltage pulse signal is picked off by it. The pulse pattern depends on the conducting-nonconducting pattern on each track as well as the nature of rotation of the disk. The signal interpretation is done as it is for optical encoders. The advantages of sliding contact encoders include high sensitivity (depending on the supply voltage) and simplicity of construction (low cost).

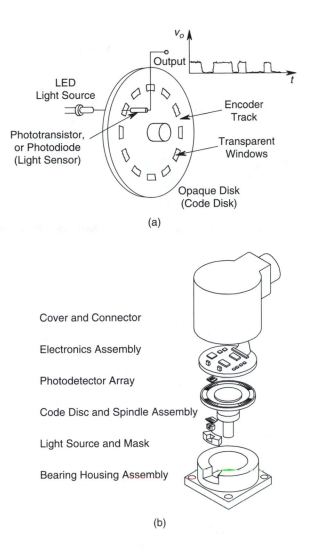

FIGURE 7.1

(a) Schematic representation of an (incremental) optical encoder, (b) Components of a commercial incremental encoder (BEI Electronics, Inc., Goleta, CA. With permission).

The disadvantages include the familiar drawbacks of contacting and commutating devices (e.g., friction, wear, brush bounce due to vibration, and signal glitches and metal oxidation due to electrical arcing). A transducer's accuracy is very much dependent upon the precision of the conducting patterns of the encoder disk. One method of generating the conducting pattern on the disk is electroplating.

Magnetic encoders have high-strength magnetic regions imprinted on the encoder disk using techniques such as etching, stamping, or recording (similar to magnetic data recording). These magnetic regions correspond to the transparent windows on an optical encoder disk. The signal pick-off device is a microtransformer, which has primary and secondary windings on a circular ferromagnetic core. This pick-off sensor resembles a core storage element in older generations of mainframe computers. The encoder arrangement is illustrated schematically in Figure 7.2. A high-frequency (typically 100 kHz) primary voltage induces a voltage in the secondary winding of the sensing element at the same frequency, operating as a transformer. A magnetic field of sufficient strength can saturate the core,

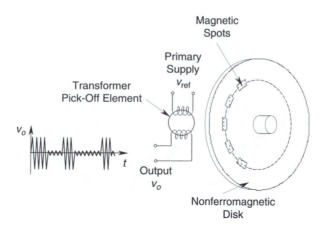

FIGURE 7.2
Schematic representation of a magnetic encoder.

however, thereby significantly increasing the reluctance and dropping the induced voltage. By demodulating the induced voltage, a pulse signal is obtained. This signal can be interpreted in the usual manner. Note that a pulse peak corresponds to a nonmagnetic area and a pulse valley corresponds to a magnetic area on each track. Magnetic encoders have noncontacting pick-off sensors, which is an advantage. They are more costly than the contacting devices, however, primarily because of the cost of transformer elements and demodulating circuitry for generating the output signal.

Proximity sensor encoders use a proximity sensor as the signal pick-off element. Any type of proximity sensor may be used—for example, a magnetic induction probe or an eddy current probe, as discussed in Chapter 6. In the magnetic induction probe, for example, the disk is made of ferromagnetic material. The encoder tracks have raised spots of the same material, serving a purpose analogous to that of the windows on an optical encoder disk. As a raised spot approaches the probe, the flux linkage increases as a result of the associated decrease in reluctance, thereby raising the induced voltage level. The output voltage is a pulse-modulated signal at the frequency of the supply (primary) voltage of the proximity sensor. This is then demodulated, and the resulting pulse signal is interpreted. Instead of a disk with a track of raised regions, a ferromagnetic toothed wheel may be used along with a proximity sensor placed in a radial orientation. In principle, this device operates like a conventional digital tachometer. If an eddy current probe is used, pulse areas in the track have to be plated with a conducting material.

Note that an incremental encoder disk requires only one primary track that has equally spaced and identical window (pick-off) regions. The window area is equal to the area of the inter-window gap (i.e., 50% duty cycle). Usually, a reference track that has just one window is also present in order to generate a pulse (known as the *index pulse*) to initiate pulse counting for angular position measurement and to detect complete revolutions. In contrast, absolute encoder disks have several rows of tracks, equal in number to the bit size of the output data word. Furthermore, the windows in a track are not equally spaced but are arranged in a specific pattern so as to obtain a binary code (or a gray code) for the output data from the transducer. It follows that absolute encoders need at least as many signal pick-off sensors as there are tracks, whereas incremental encoders need one pick-off sensor to detect the magnitude of rotation. As will be explained, it will also need a sensor at a quarter-pitch separation (pitch = center-to-center distance between adjacent windows) to generate a *quadrature signal*, which will identify the direction of rotation.

Some designs of incremental encoders have two identical tracks, one at a quarter-pitch offset from the other, and the two pick-off sensors are placed radially without offset. The two (quadrature) signals obtained with this arrangement will be similar to those with the previous arrangement. A pick-off sensor for receiving a reference pulse is also used in some designs of incremental encoders (three-track incremental encoders).

In many control applications, encoders are built into the plant itself, rather than being externally fitted onto a rotating shaft. For instance, in a robot arm, the encoder might be an integral part of the joint motor and may be located within its housing. This reduces coupling errors (e.g., errors due to backlash, shaft flexibility, and resonances added by the transducer and fixtures), installation errors (e.g., misalignment and eccentricity), and overall cost. Encoders are available in sizes as small as 2 cm and as large as 15 cm in diameter.

Since the techniques of signal interpretation are quite similar for the various types of encoders with different principles of signal generation, we shall limit further discussion to optical encoders. These encoders are in fact the most common types in practical applications. Signal interpretation differs depending on whether the particular optical encoder is an incremental device or an absolute device.

7.3   Incremental Optical Encoders

There are two possible configurations for an incremental encoder disk:

1. Offset sensor configuration
2. Offset track configuration

The first configuration is schematically shown in Figure 7.3. The disk has a single circular track with identical and equally spaced transparent windows. The area of the opaque region between adjacent windows is equal to the window area. Two photodiode sensors (pick-offs 1 and 2 in Figure 7.3) are positioned facing the track at a quarter-pitch (half the

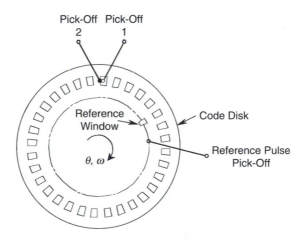

FIGURE 7.3
An incremental encoder disk (offset sensor configuration).

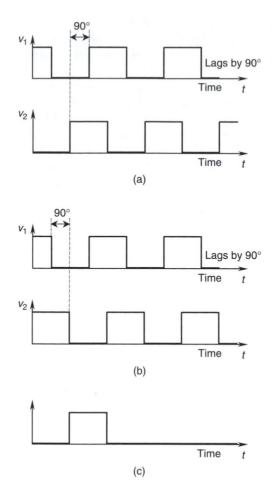

FIGURE 7.4
Shaped pulse signals from an incremental encoder: (a) For clockwise rotation; (b) For counterclockwise rotation; (c) Reference pulse signal.

window length) apart. The forms of their output signals (v_1 and v_2), after passing them through pulse-shaping circuitry (idealized), are shown in Figure 7.4(a) and Figure 7.4(b) for the two directions of rotation.

In the second configuration of incremental encoders, two identical tracks are used, one offset from the other by a quarter-pitch. Each track has its own pick-off sensor, oriented normally facing the track. The two pick-off sensors are positioned on a radial line facing the disk, without any circumferential offset unlike the previous configuration. The output signals from the two sensors are the same as before, however (Figure 7.4). Note that an output pulse signal is on half the time and off half the time, giving a 50% duty cycle.

In both configurations, an additional track with a lone window and associated sensor is also usually available. This track generates a reference pulse (index pulse) per revolution of the disk (see Figure 7.4(c)). This pulse is used to initiate the counting operation. Furthermore, the index pulse count gives the number of complete revolutions, which is required in measuring absolute angular rotations. Note that when the disk rotates at constant angular speed, the pulse width and pulse-to-pulse period (encoder cycle) are constant (with respect to time) in each sensor output. When the disk accelerates, the pulse width decreases continuously; when the disk decelerates, the pulse width increases.

7.3.1   Direction of Rotation

The quarter-pitch offset in sensor location (or, in track placement) is used to determine the direction of rotation of the disk. For example, Figure 7.4(a) shows the shaped (idealized) sensor outputs (v_1 and v_2) when the disk rotates in the clockwise (cw) direction; and Figure 7.4(b) shows the outputs when the disk rotates in the counterclockwise (ccw) direction. It is clear from these two figures that in cw rotation, v_1 lags v_2 by a quarter of a cycle (i.e., a phase lag of 90°); and in ccw rotation, v_1 leads v_2 by a quarter of a cycle. Hence, the direction of rotation is obtained by determining the phase difference of the two output signals, using phase-detecting circuitry.

One method for determining the phase difference is to time the pulses using a high-frequency clock signal. Suppose that the counting (timing) begins when the v_1 signal begins to rise (i.e., when a rising edge is detected). Let n_1 = number of clock cycles (time) up to the time when v_2 begins to rise; and n_2 = number of clock cycles up to the time when v_1 begins to rise again. Then, the following logic applies:

If $n_1 > n_2 - n_1 \Rightarrow$ cw rotation

If $n_1 < n_2 - n_1 \Rightarrow$ ccw rotation

This logic for direction detection should be clear from Figure 7.4(a) and Figure 7.4(b).

Another scheme can be given for direction. In this case, we first detect a high level (logic high or binary 1) in signal v_2, and then check whether the edge in signal v_1 rises or falls during this period. From Figure 7.4(a) and Figure 7.4(b), the following logic applies:

If rising edge in v_1 when v_2 is logic high $\Rightarrow$ cw rotation

If falling edge in v_1 when v_2 is logic high $\Rightarrow$ ccw rotation

7.3.2   Hardware Features

The actual hardware of commercial encoders is not as simple as what is suggested by Figure 7.3 (see Figure 7.1(b)). A more detailed schematic diagram of the signal generation mechanism of an optical incremental encoder is shown in Figure 7.5(a). The light generated by the light-emitting diode (LED) is collimated (forming parallel rays) using a lens. This pencil of parallel light passes through a window of the rotating code disk. The masking (grating) disk is stationary and has a track of windows identical to that in the code disk. Because of the presence of the masking disk, light from the LED will pass through more than one window of the code disk, thereby improving the intensity of light received by the photosensor but not introducing any error due to the diameter of the pencil of light being larger than the window length. When the windows of the code disk face the opaque areas of the masking disk, virtually no light is received by the photosensor. When the windows of the code disk face the transparent areas of the masking disk, maximum amount of light reaches the photosensor. Hence, as the code disk moves, a sequence of triangular (and positive) pulses of light is received by the photosensor. Pulse width in this case is a full cycle (i.e., it corresponds to the window pitch) and not a half cycle.

Fluctuation in the supply voltage to the encoder light source also directly influence the light level received by the photosensor. If the sensitivity of the photosensor is not high enough, a low light level might be interpreted as no light, which would result in measurement error. Such errors due to instabilities and changes in the supply voltage, can be eliminated by using two photosensors, one placed half a pitch away from the other along the window track, as shown in Figure 7.5(b). This arrangement is for contrast detection,

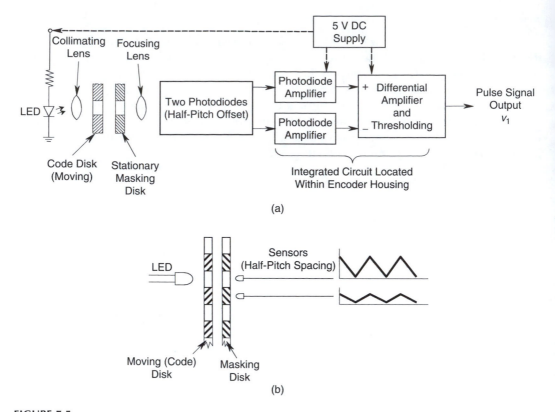

FIGURE 7.5
(a) Internal hardware of an optical incremental encoder, (b) Use of two sensors at 180° spacing to generate a shaped pulse.

and it should not be confused with the quarter-of-a-pitch offset arrangement that is required for direction detection. The sensor facing the opaque region of the masking disk will always read a low signal. The other sensor will read a triangular signal whose peak occurs when a moving widow overlaps with a window of the masking disk, and whose valley occurs when a moving window faces an opaque region of the masking disk. The two signals from these two sensors are amplified separately and fed into a differential amplifier. The result is a high-intensity triangular pulse signal. A shaped (or, binary) pulse signal can be generated by subtracting a threshold value from this signal and identifying the resulting positive (or, binary 1) and negative (or, binary 0) regions. This procedure will produce a more distinct (or, binary) pulse signal that is immune to noise.

The signal amplifiers are integrated circuit devices and are housed within the encoder itself. Additional pulse-shaping circuitry may also be present. The power supply has to be provided separately as an external component. The voltage level and pulse width of the output pulse signal are logic-compatible (e.g., transistor-to-transistor logic, or TTL) so that they may be read directly using a digital board. Note that if the output level v_1 is positive high, we have a logic high (or, binary 1) state (See Chapter 10). Otherwise, we have a logic low (or, binary 0) state. In this manner, a stable and accurate digital output can be obtained even under unstable voltage supply conditions. The schematic diagram in Figure 7.5 shows the generation of only one (v_1) of the two quadrature pulse signals. The other pulse signal (v_2) is generated using identical hardware but at a quarter of a pitch offset. The index pulse (reference pulse) signal is also generated in a similar manner. The cable of the encoder (usually a ribbon cable) has a multipin connector. Three of the pins provide the three output pulse signals. Another pin carries the dc supply voltage (typically 5 V) from the power supply into the encoder. Typically, a ground line (ground pin) is included as well. Note that the only moving part in the system shown in Figure 7.5 is the code disk.

7.3.3 Displacement Measurement

An incremental encoder measures displacement as a pulse count, and it measures velocity as a pulse frequency. A digital processor is able to express these readings in engineering units (radians, degrees, rad/s, etc.) using pertinent parameter values of the physical system. Suppose that the maximum count possible is M pulses and the range of the encoder is $\pm\theta_{max}$.

The angular position θ corresponding to a count of n pulses is computed as

$$\theta = \frac{n}{M}\theta_{max} \tag{7.1}$$

7.3.3.1 Digital Resolution

The resolution of an encoder represents the smallest change in measurement that can be measured realistically. Since an encoder can be used to measure both displacement and velocity, we can identify a resolution for each case. First we will consider *displacement resolution*, which is governed by the number of windows N in the code disk and the digital size (number of bits) of the buffer (counter output). First we will discuss digital resolution.

Suppose that the encoder count is stored as digital data of r bits. Allowing for a *sign bit*, we have

$$M = 2^{r-1} \tag{7.2}$$

The displacement resolution of an incremental encoder is given by the change in displacement corresponding to a unit change in the count (n). It follows from Equation 7.1 that the *displacement resolution* is given by

$$\Delta\theta = \frac{\theta_{max}}{M} \tag{7.3}$$

In particular, the *digital resolution* corresponds to a unit change in the bit value. By substituting Equation 7.2 into Equation 7.3, we have the digital resolution

$$\Delta\theta_d = \frac{\theta_{max}}{2^{r-1}} \tag{7.4a}$$

Typically,

$$\theta_{max} = \pm180° \quad \text{or} \quad 360°$$

Then,

$$\Delta\theta_d = \frac{180°}{2^{r-1}} = \frac{360°}{2^r} \tag{7.4}$$

Note that the minimum count corresponds to the case where all the bits are zero and the maximum count corresponds to the case where all the bits are unity. Suppose that these two readings represent the angular displacements θ_{min} and θ_{max}. We have,

$$\theta_{max} = \theta_{min} + (M-1)\Delta\theta \tag{7.5}$$

or, substituting Equation 7.2,

$$\theta_{max} = \theta_{min} + (2^{r-1} - 1)\Delta\theta_d \tag{7.5a}$$

Equation 7.5a leads to the conventional definition for digital resolution:

$$\Delta\theta_d = \frac{(\theta_{max} - \theta_{min})}{(2^{r-1} - 1)} \tag{7.6}$$

This result is exactly the same as what is given by Equation 7.4.

If θ_{max} is 2π and $\theta_{min} = 0$, then θ_{max} and θ_{min} will correspond to the same position of the code disk. To avoid this ambiguity, we use

$$\theta_{min} = \frac{\theta_{max}}{2^{r-1}} \tag{7.7}$$

Note that if we substitute Equation 7.7 into Equation 7.6 we get Equation 7.4a as required. Then, the digital resolution is given by $\frac{(360° - 360°/2^r)}{(2^r - 1)}$ which is identical to Equation 7.4.

7.3.3.2   *Physical Resolution*

The physical resolution of an encoder is governed by the number of windows N in the code disk. If only one pulse signal is used (i.e., no direction sensing), and if only the rising edges of the pulses are detected (i.e., full cycles of the encoder signal are counted), the physical resolution is given by the pitch angle of the track (i.e., angular separation between adjacent windows), which is $(360/N)°$. But if quadrature signals (i.e., two pulse signals, one out of phase with the other by 90° or quarter of a pitch angle) are available and the capability to detect both rising and falling edges of a pulse is also present, four counts can be made per encoder cycle, thereby improving the resolution by a factor of 4. Under these conditions, the physical resolution of an encoder is given by

$$\Delta\theta_p = \frac{360°}{4N} \tag{7.8}$$

To understand this, note in Figure 7.4(a) (or Figure 7.4(b)) that when the two signals v_1 and v_2 are added, the resulting signal has a transition at every quarter of the encoder cycle. This is illustrated in Figure 7.6. By detecting each transition (through edge detection or level detection), four pulses can be counted within every main cycle. It should be mentioned that each signal (v_1 or v_2) separately has a resolution of half a pitch, provided that all transitions (rising edges and falling edges) are detected and counted instead of pulses (or, high signal levels) being counted. Accordingly a disk with 10,000 windows has a resolution of 0.018° if only one pulse signal is used (and both transitions, rise and fall, are detected). When two signals (with a phase shift of a quarter of a cycle) are used, the

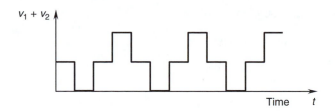

FIGURE 7.6
Quadrature signal addition to improve physical resolution.

resolution improves to 0.009°. This resolution is achieved directly from the mechanics of the transducer; no interpolation is involved. It assumes, however, that the pulses are nearly ideal and, in particular, that the transitions are perfect. In practice, this cannot be realized if the pulse signals are noisy. Then, pulse shaping will be necessary as mentioned before. The larger of the two resolutions given by Equation 7.4 and Equation 7.8 governs the displacement resolution of the encoder.

Example 7.1

For an ideal design of an incremental encoder, obtain an equation relating the parameters d, w, and r, where

  d = diameter of encoder disk
  w = number of windows per unit diameter of disk
  r = word size (bits) of the angle measurement

Assume that quadrature signals are available. If $r = 12$ and $w = 500/\text{cm}$, determine a suitable disk diameter.

SOLUTION

In this problem, we take the ideal design as the case where the physical resolution is equal to the digital resolution. The position resolution due to physical constraints (assuming that quadrature signals are available) is given by Equation 7.8. Hence,

$$\Delta\theta_p = \frac{1}{4}\left(\frac{360}{wd}\right)^{\circ}$$

The resolution limited by the digital word size of the buffer is given by Equation 7.4:

$$\Delta\theta_d = \left(\frac{360}{2^r}\right)^{\circ}$$

For an ideal design we need $\Delta\theta_p = \Delta\theta_d$, which gives

$$\frac{1}{4}\frac{360}{wd} = \frac{360}{2^r}$$

Simplifying, we have

$$wd = 2^{r-2} \qquad\qquad (7.9)$$

Substitute $r = 12$ and $w = 500/\text{cm}$ to obtain:

$$d = \left(\frac{2^{12-2}}{500}\right) \text{cm} = 2.05 \text{ cm}$$

7.3.3.3 Step-up Gearing

The physical resolution of an encoder can be improved by using step-up gearing so that one rotation of the moving object that is being monitored corresponds to several rotations of the code disk of the encoder. This improvement is directly proportional to the step-up gear ratio (p). Specifically, we have

$$\Delta\theta_p = \frac{360°}{4pN} \tag{7.10}$$

Backlash in the gearing introduces a new error, however. For best results, this backlash error should be several times smaller than the resolution with no backlash.

The digital resolution will not improve by gearing if the maximum angle of rotation of the moving object (say, 360°) still corresponds to the buffer/register size. Then the change in the least significant bit (LSB) of the buffer corresponds to the same change in the angle of rotation of the moving object. In fact, the overall displacement resolution can be harmed in this case if excessive backlash is present. But, if the buffer/register size corresponds to a full rotation of the code disk (i.e., a rotation of $360°/p$ in the object) and if the output register (or buffer) is cleared at the end of each such rotation and a separate count of full rotations of the code disk is kept, then the digital resolution will also improve by a factor of p. Specifically, from Equation 7.4 we get

$$\Delta\theta_d = \frac{180°}{p2^{r-1}} = \frac{360°}{p2^r} \tag{7.11}$$

Example 7.2

By using high-precision techniques to imprint window tracks on the code disk, it is possible to attain a window density of 500 windows/cm of diameter. Consider a 3000-window disk. Suppose that step-up gearing is used to improve resolution and the gear ratio is 10. If the word size of the output register is 16 bits, examine the displacement resolution of this device for the two cases where the register size corresponds to: (1) a full rotation of the object, (2) a full rotation of the code disk.

SOLUTION

First consider the case in which gearing is not present. With quadrature signals, the physical resolution is

$$\Delta\theta_p = \frac{360°}{4 \times 3000} = 0.03°$$

For a range of measurement given by ±180°, a 16-bit output provides a digital resolution of

$$\Delta\theta_d = \frac{180°}{2^{15}} = 0.005°$$

Hence, in the absence of gearing, the overall displacement resolution is 0.03°.

Next consider a geared encoder with gear ratio of 10, and neglect gear backlash. The physical resolution improves to 0.003°. But, in Case (1), the digital resolution remains unchanged at best. Hence, the overall displacement resolution improves to 0.005° as a result of gearing. In Case (2), the digital resolution improves to 0.0005°. Hence, the overall displacement resolution becomes 0.003°.

In summary, the displacement resolution of an incremental encoder depends on the following factors:

1. Number of windows on the code track (or disk diameter)
2. Gear ratio
3. Word size of the measurement buffer

Example 7.3

A positioning table uses a backlash-free high precision lead screw of lead 2 cm/rev, which is driven by a servo motor with a built-in optical encoder for feedback control. If the required positioning accuracy is ±10 μm determine the number of windows required in the encoder track. Also, what is the minimum bit size required for the digital data register/buffer of the encoder count?

SOLUTION
Required accuracy = ±10 μm
   To achieve this accuracy, the required resolution for a linear displacement sensor is ±5 μm.
   Lead of the lead screw = 2 cm/rev
   In order to achieve the required resolution, the number of pulses per encoder revolution is:

$$\frac{2 \times 10^{-2} \text{ m}}{5 \times 10^{-6} \text{ m}} = 4000 \text{ pulses}$$

Assuming that quadrature signals are available (with a resolution improvement of × 4), the required number of widows in the encoder track = 1000.

$$\text{Percentage physical resolution} = \frac{1}{4000} \times 100\% = 0.025\%$$

Consider a buffer size of r bits, including a sign bit. Then, we need

$$2^{r-1} = 4000$$

or,

$$r = 13 \text{ bits.}$$

7.3.3.4   Interpolation

The output resolution of an encoder can be further enhanced by interpolation. This is accomplished by adding equally spaced pulses in between every pair of pulses generated by the encoder circuit. These auxiliary pulses are not true measurements, and they can be interpreted as a linear interpolation scheme between true pulses. One method of accomplishing this interpolation is by using the two pick-off signals that are generated by the encoder (quadrature signals). These signals are nearly sinusoidal (or, triangular) prior to shaping (say, by level detection). They can be filtered to obtain two sine signals that are 90° out of phase (i.e., a sine signal and a cosine signal). By weighted combination of these two signals, a series of sine signals can be generated such that each signal lags the preceding signal by any integer fraction of 360°. By level detection or edge detection (rising and falling edges), these sine signals can be converted into square wave signals. Then, by logical combination of the square waves, an integer number of pulses can be generated within each encoder cycle. These are the interpolation pulses that are added to improve the encoder resolution. In practice, about twenty interpolation pulses can be added between a pair of adjacent main pulses.

7.3.4   Velocity Measurement

Two methods are available for determining velocities using an incremental encoder:

1. Pulse-counting method
2. Pulse-timing method

In the first method, the pulse count over a fixed time period (the successive time period at which the data buffer is read) is used to calculate the angular velocity. For a given period of data reading, there is a lower speed limit below which this method is not very accurate. To compute the angular velocity ω using this method, suppose that the count during a time period T is n pulses. Hence, the average time for one pulse is T/n. If there are N windows on the disk, assuming that qudrature signals are not used, the angle moved during one pulse is $2\pi/N$. Hence,

$$\text{Speed} \quad \omega = \frac{2\pi/N}{T/n} = \frac{2\pi n}{NT} \tag{7.12}$$

If quadrature signals are used, replace N by $4N$ in Equation 7.12.

In the second method, the time for one encoder cycle is measured using a high-frequency clock signal. This method is particularly suitable for accurately measuring low speeds. In this method, suppose that the clock frequency is f Hz. If m cycles of the clock signal are counted during an encoder period (interval between two adjacent windows, assuming that quadrature signals are not used), the time for that encoder cycle (i.e., the time to rotate through one encoder pitch) is given by m/f. With a total of N windows on the track, the angle of rotation during this period is $2\pi/N$ as before. Hence,

$$\text{Speed} \quad \omega = \frac{2\pi/N}{m/f} = \frac{2\pi f}{Nm} \tag{7.13}$$

If quadrature signals are used, replace N by $4N$ in Equation 7.13.

Note that a single incremental encoder can serve as both position sensor and speed sensor. Hence, a position loop and a speed loop in a control system can be closed using a single encoder, without having to use a conventional (analog) speed sensor such as a tachometer. The speed resolution of the encoder (depending on the method of speed computation—pulse counting or pulse timing) can be chosen to meet the accuracy requirements for the speed control loop. A further advantage of using an encoder rather than a conventional (analog) motion sensor is that an analog-to-digital converter (ADC) would be unnecessary. For example, the pulses generated by the encoder may be used as *interrupts* for the control computer. These interrupts are then directly counted (by an up/down counter or indexer) or timed (by a clock in the data acquisition computer) within the control computer, thereby providing position and velocity readings.

7.3.4.1   Velocity Resolution

The velocity resolution of an incremental encoder depends on the method that is employed to determine velocity. Since the pulse-counting method and the pulse-timing method are both based on counting, the velocity resolution is given by the change in angular velocity that corresponds to a change (increment or decrement) in the count by one.

For the pulse-counting method, it is clear from Equation 7.12 that a unity change in the count n corresponds to a speed change of

$$\Delta\omega_c = \frac{2\pi}{NT} \tag{7.14}$$

where N is the number of windows in the code track and T is the time period over which a pulse count is read. Equation 7.14 gives the velocity resolution by this method. Note that the engineering value (in rad/s) of this resolution is independent of the angular velocity itself, but when expressed as percentage of the speed, the resolution becomes better (smaller) at higher speeds. Note further from Equation 7.14 that the resolution improves with the number of windows and the count reading (sampling) period. Under transient conditions, the accuracy of a velocity reading decreases with increasing T (because, according to Shannon's sampling theorem—see Chapter 5—the sampling frequency has to be at least double the highest frequency of interest in the velocity signal). Hence, the sampling period should not be increased indiscriminately. As usual, if quadrature signals are used, N in Equation 7.14 has to be replaced by $4N$ (i.e., the resolution improves by ×4).

In the pulse-timing method, the velocity resolution is given by (see Equation 7.13)

$$\Delta\omega_t = \frac{2\pi f}{Nm} - \frac{2\pi f}{N(m+1)} = \frac{2\pi f}{Nm(m+1)} \tag{7.15a}$$

where f is the clock frequency. For large m, $(m + 1)$ can be approximated by m. Then, by substituting Equation 7.13 in Equation 7.15a, we get

$$\Delta\omega_t = \frac{2\pi f}{Nm^2} = \frac{N\omega^2}{2\pi f} \tag{7.15}$$

Note that in this case, the resolution degrades quadratically with speed. This resolution degrades with the speed even when it is considered as a fraction of the measured speed:

$$\frac{\Delta\omega_t}{\omega} = \frac{N\omega}{2\pi f} \tag{7.16}$$

This observation confirms the previous suggestion that the pulse-timing method is appropriate for low speeds. For a given speed and clock frequency, the resolution further degrades with increasing N. This is true because, when N is increased the pulse period shortens and hence the number of clock cycles per pulse period also decreases. The resolution can be improved, however, by increasing the clock frequency.

Example 7.4

An incremental encoder with 500 windows in its track is used for speed measurement. Suppose that:

    a. In the pulse-counting method, the count (buffer) is read at the rate of 10 Hz
    b. In the pulse-timing method, a clock of frequency 10 MHz is used

Determine the percentage resolution for each of these two methods when measuring a speed of:

    i. 1 rev/s
    ii. 100 rev/s

SOLUTION
    i. Speed = 1 rev/s
       With 500 windows, we have 500 pulses/s
       a. Pulse-counting method

$$\text{Counting Period} = \frac{1}{10 \text{ Hz}} = 0.1 \text{ s}$$

Pulse count (in 0.1 s) = $500 \times 0.1 = 50$

$$\text{Percentage resolution} = \frac{1}{50} \times 100\% = 2\%$$

       b. Pulse-timing method
         At 500 pulses/s,

$$\text{Pulse period} = \frac{1}{500} \text{ s} = 2 \times 10^{-3} \text{ s}$$

With a 10 MHz clock,

$$\text{Clock count} = 10 \times 10^{6} \times 2 \times 10^{-3} = 20 \times 10^{3}$$

$$\text{Percentage resolution} = \frac{1}{20 \times 10^{3}} \times 100\% = 0.005\%$$

    ii. Speed = 100 rev/s
       With 500 windows, we have 50,000 pulses/s

a. Pulse-counting method

Pulse count (in 0.1 s) = 50,000 × 0.1 = 5000

$$\text{Percentage resolution} = \frac{1}{5000} \times 100\% = 0.02\%$$

b. Pulse-timing method

At 50,000 pulses/s,

$$\text{Pulse period} = \frac{1}{50,000} \text{ s} = 20 \times 10^{-6} \text{ s}$$

With a 10 MHz clock,

$$\text{Clock count} = 10 \times 10^6 \times 20 \times 10^{-6} = 200$$

$$\text{Percentage resolution} = \frac{1}{200} \times 100\% = 0.5\%$$

The results are summarized in Table 7.1

Results given in Table 7.1 confirm that in the pulse-counting method the resolution improves with speed, and hence it is more suitable for measuring high speeds. Furthermore, in the pulse-timing method the resolution degrades with speed, and hence it is more suitable for measuring low speeds.

7.3.4.2 Step-Up Gearing

Consider an incremental encoder that has N windows per track, and connected to a rotating shaft through a gear unit with step-up gear ratio p. Formulas for computing angular velocity of the shaft by:

a. Pulse-counting method
b. Pulse-timing method

can be easily determined by using Equation 7.12 and Equation 7.13. Specifically, the angle of rotation of the shaft corresponding to window spacing (pitch) of the encoder disk now is $2\pi/(pN)$. Hence the corresponding formulas for speed can be obtained by replacing N

TABLE 7.1

Comparison of Speed Resolution from an Incremental Encoder

	Percentage Resolution	
Speed (rev/s)	Pulse Counting Method (%)	Pulse Timing Method (%)
1.0	2	0.005
100.0	0.02	0.5

by pN in Equation 7.12 and Equation 7.13. We have,

$$\text{For pulse count method:} \quad \omega = \frac{2\pi n}{pNT} \tag{7.17}$$

$$\text{For pulse time method:} \quad \omega = \frac{2\pi f}{pNm} \tag{7.18}$$

Note that these relations can also be obtained simply by dividing the encoder disk speed by the gear ratio, which gives the object speed.

As before, the speed resolution is given by the change in speed corresponding to a unity change in the court. Hence,

$$\text{For the pulse count method:} \quad \Delta\omega_c = \frac{2\pi(n+1)}{pNT} - \frac{2\pi n}{pNT} = \frac{2\pi}{pNT} \tag{7.19}$$

It follows that in the pulse count method, step-up gearing causes an improvement in the resolution.

For the pulse time method:

$$\Delta\omega_t = \frac{2\pi f}{pNm} - \frac{2\pi f}{pN(m+1)} = \frac{2\pi f}{pNm(m+1)} \cong \frac{pN}{2\pi f}\omega^2 \tag{7.20}$$

Note that in the pulse time approach, for a given speed, the resolution degrades with increasing p.

In summary, the speed resolution of an incremental encoder depends on the following factors:

1. Number of windows N
2. Count reading (sampling) period T
3. Clock frequency f
4. Speed ω
5. Gear ratio

In particular, gearing up has a detrimental effect on the speed resolution in the pulse-timing method, but it has a favorable effect in the pulse-counting method.

7.3.5   Data Acquisition Hardware

A method for interfacing an incremental encoder to a digital processor (digital controller) is shown schematically in Figure 7.7. In practice, a suitable interface card (e.g., servo card, encoder card, etc.) in the control computer will possess the necessary functional capabilities indicated in the figure. The pulse signals from the encoder are fed into an up/down counter, which has circuitry to detect pulses (e.g., by rising-edge detection, falling-edge detection, or by level detection) and logic circuitry to determine the direction of motion (i.e., sign of the reading) and to code the count. A pulse in one direction (say, clockwise will increment the count by one (an upcount), and a pulse in the opposite direction will

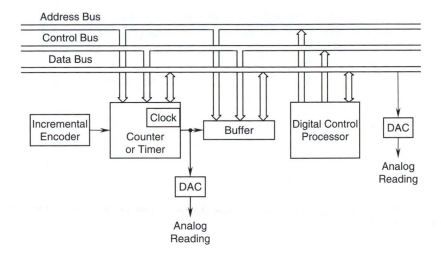

FIGURE 7.7
Computer interface for an incremental encoder.

decrement the count by one (a downcount). The coded count may be directly read by the host computer, through its input/output (I/O) board without the need for an analog to digital converter (ADC). The count is transferred to a latch buffer so that the measurement is read from the buffer rather than from the counter itself. This arrangement provides an efficient means of data acquisition because the counting process can continue without interruption while the computer reads the count from the latch buffer. The digital processor (computer) identifies various components in the measurement system using addresses, and this information is communicated to the individual components through the *address bus*. The start, end, and nature of an action (e.g., data read, clear the counter, clear the buffer) are communicated to various devices by the computer through its *control bus*. The computer can command an action to a component in one direction of the bus, and the component can respond with a message (e.g., job completed) in the opposite direction. The data (e.g., the count) are transmitted through the *data bus*. While the computer reads (samples) data from the buffer, the control signals guarantee that no data are transferred to that buffer from the counter. It is clear that the data acquisition consists of *handshake operations* between the main processor of the computer and auxiliary components. More than one encoder may be addressed, controlled, and read by the same three buses of the computer. The buses are conductors, for example, multicore cables carrying signals in parallel logic. Communication in serial logic is also common but is slower.

An incremental optical encoder generates two pulse signals one 1/4 of a pitch out of phase with the other. The internal electronics of the encoder may be powered by a 5 V dc supply. The two pulse signals determine the direction of rotation of the motor by one of various means (e.g., sign of the phase difference, timing of the consecutive rising edges). The encoder pulse count is stored in a buffer within the controller and is read at fixed intervals (say, 5 ms). The net count gives the joint position, and the difference in count at a fixed time increment gives joint speed.

While measuring a displacement (position) of an object using an incremental encoder, the counter may be continuously monitored as an analog signal through a digital-to-analog converter (DAC in Figure 7.7). On the other hand, the pulse count is read by the computer only at finite time intervals. Since a cumulative count is required in displacement measurement, the buffer is not cleared in this case once the count is read in by the computer.

In velocity measurement by the pulse-counting method, the buffer is read at fixed time intervals of T, which is also the counting-cycle time. The counter is cleared every time a count is transferred to the buffer, so that a new count can begin. With this method, a new reading is available at every sampling instant.

In the pulse-timing method of velocity computation, the counter is actually a timer. The encoder cycle is timed using a clock (internal or external), and the count is passed on to the buffer. The counter is then cleared and the next timing cycle is started. The buffer is periodically read by the computer. With this method, a new reading is available at every encoder cycle. Note that under transient velocities, the encoder-cycle time is variable and is not directly related to the data sampling period. In the pulse-timing method, it is desirable to make the sampling period slightly smaller than the encoder-cycle time, so that no count is missed by the processor.

More efficient use of the digital processor may be achieved by using an interrupt routine. With this method, the counter (or buffer) sends an interrupt request to the processor when a new count is ready. The processor then temporarily suspends the current operation and reads in the new data. Note that in this case the processor does not continuously wait for a reading.

7.4 Absolute Optical Encoders

An absolute encoder directly generates a coded digital word to represent each discrete angular position (sector) of its code disk. This is accomplished by producing a set of pulse signals (data channels) equal in number to the word size (number of bits) of the reading. Unlike with an incremental encoder, no pulse counting is involved. An absolute encoder may use various techniques (e.g., optical, sliding contact, magnetic saturation, proximity sensor) to generate the sensor signal, as for an incremental encoder. The optical method, which uses a code disk with transparent and opaque regions and pairs of light sources and photosensors, is the most common technique.

A simplified code pattern on the disk of an absolute encoder that utilizes the direct binary code, is shown in Figure 7.8(a). The number of tracks (n) in this case is 4. In practice n is on the order of 14, but may be as high as 22. The disk is divided into 2^n sectors. Each partitioned area of the matrix thus formed corresponds to a bit of data. For example, a transparent area will correspond to binary 1 and an opaque area to binary 0. Each track has a pick-off sensor similar to what is used in incremental encoders. The set of n pick-off sensors is arranged along a radial line and facing the tracks on one side of the disk. A light source (e.g., light-emitting diode—LED) illuminates the other side of the disk. As the disk rotates, the bank of pick-off sensors generates pulse signals, which are sent to n parallel data channels (or pins). At a given instant, the particular combination of signal levels in the data channels will provide a coded data word that uniquely determines the position of the disk at that time.

7.4.1 Gray Coding

There is a data interpretation problem associated with the straight binary code in absolute encoders. Notice in Table 7.2 that with the straight binary code, the transition from one sector to an adjacent sector may require more than one switching of bits in the binary data. For example, the transition from 0011–0100 or from 1011–1100 requires three bit

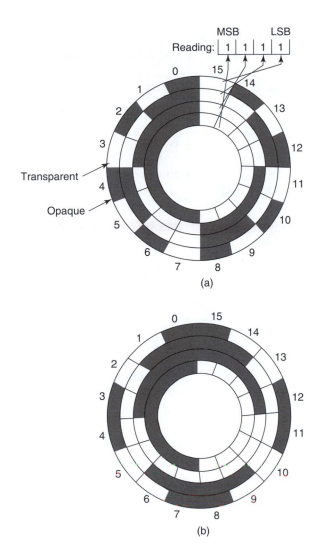

FIGURE 7.8
Illustration of the code pattern of an absolute encoder disk: (a) Binary code; (b) A gray code.

switchings, and the transition from 0111–1000 or from 1111–0000 requires four bit switchings. If the pick-off sensors are not properly aligned along a radius of the encoder disk, or if the manufacturing error tolerances for imprinting the code pattern on the disk were high, or if environmental effects have resulted in large irregularities in the sector matrix, or due to hardware (switch) delays, the bit switching from one reading to the next will not take place simultaneously. This results in ambiguous readings during the transition period. For example, in changing from 0011–0100, if the LSB switches first, the reading becomes 0010. In decimal form, this incorrectly indicates that the rotation was from angle 3 to angle 2, whereas, it was actually a rotation from angle 3 to angle 4. Such ambiguities in data interpretation can be avoided by using a gray code, as shown in Figure 7.8(b) for this example. The coded representation of the sectors is given in Table 7.2. Note that in the case of gray code, each adjacent transition involves only one bit switching.

For an absolute encoder, a gray code is not essential for removing the ambiguity in bit switchings of binary code. For example, for a given absolute reading, the two adjacent

TABLE 7.2

Sector Coding for a 4-Bit Absolute Encoder

Sector Number	Straight Binary Code (MSB → LSB)	A Gray Code (MSB → LSB)
0	0 0 0 0	0 0 0 0
1	0 0 0 1	0 0 0 1
2	0 0 1 0	0 0 1 1
3	0 0 1 1	0 0 1 0
4	0 1 0 0	0 1 1 0
5	0 1 0 1	0 1 1 1
6	0 1 1 0	0 1 0 1
7	0 1 1 1	0 1 0 0
8	1 0 0 0	1 1 0 0
9	1 0 0 1	1 1 0 1
10	1 0 1 0	1 1 1 1
11	1 0 1 1	1 1 1 0
12	1 1 0 0	1 0 1 0
13	1 1 0 1	1 0 1 1
14	1 1 1 0	1 0 0 1
15	1 1 1 1	1 0 0 0

absolute readings are automatically known. A reading can be checked against these two valid possibilities (or, a single possibility if the direction of rotation is known) to see whether the reading is correct. Another approach is to introduce a delay (e.g., Schmitt trigger) to reading the output. In this manner a reading will be taken only after all the bit switchings have taken place, thereby eliminating the possibility of an ambiguous reading.

7.4.1.1   Code Conversion Logic

A disadvantage of utilizing a gray code is that it requires additional logic to convert the gray-coded number to the corresponding binary number. This logic may be provided in hardware or software. In particular, an "Exclusive-Or" gate can implement the necessary logic (see Chapter 10), as given by

$$B_{n-1} = G_{n-1}$$
$$B_{k-1} = B_k \oplus G_{k-1} \qquad k = n-1, \ldots, 1 \tag{7.21}$$

This converts an n-bit gray-coded word $[G_{n-1} G_{n-2}, \ldots, G_0]$ into an n-bit binary coded word $[B_{n-1} B_{n-2}, \ldots, B_0]$ where the subscript $n-1$ denotes the MSB and 0 denotes the LSB. For small word sizes, the code may be given as a look-up table (see Table 7.2). Note that the gray code is not unique. Other gray codes, which provide single bit switching between adjacent numbers can be developed.

7.4.2   Resolution

The resolution of an absolute encoder is limited by the word size of the output data. Specifically, the displacement (position) resolution is given by the sector angle, which is also the angular separation between adjacent transparent and opaque regions on the outermost track of the code disk; thus

$$\Delta\theta = \frac{360°}{2^n} \tag{7.22}$$

In Figure 7.8(a), the word size of the data is 4 bits. This can represent decimal numbers from 0–15, as given by the sixteen sectors of the disk. In each sector, the outermost element is the least significant bit (LSB) and the innermost element is the most significant bit (MSB). The direct binary representation of the disk sectors (angular positions) is given in Table 7.2. The angular resolution for this simplified example is $(360/2^4)°$, or 22.5°. If $n = 14$, the angular resolution improves to $(360/2^{14})°$, or 0.022°. If $n = 22$, the resolution further improves to 0.000086°.

Step-up gear mechanisms can also be employed to improve encoder resolution. But, this has the same disadvantages as mentioned under incremental encoders (e.g., backlash, added weight and loading, increased cost). Furthermore, when a gear is included, the "absolute" nature of a reading will be limited to a fraction of rotation of the main shaft; specifically, 360°/gear ratio. If a count of the total rotations of the gear shaft (encoder disk) is maintained, this will not present a problem.

An ingenious method of improving the resolution of an absolute encoder is by generating auxiliary pulses in between the bit switchings of the coded word. This requires an auxiliary track (usually placed as the outermost track) with a sufficiently finer pitch than the LSB track, and some means of direction sensing (e.g., two pick-off sensors placed a quarter-pitch apart, to generate quadrature signals). This is equivalent to having an incremental encoder of finer resolution and an absolute encoder in a single integral unit. Knowing the reading of the absolute encoder (from its coded output, as usual) and the direction of motion (from the quadrature signal) it is possible to determine the angle corresponding to the successive incremental pulses (from the finer track) until the next absolute-word reading is reached. Of course, if a data failure occurs in between the absolute readings, the additional accuracy (and resolution) provided by the incremental pulses will be lost.

7.4.3   Velocity Measurement

An absolute encoder can be used for angular velocity measurement as well. For this, either the pulse-timing method or the angle-measurement method may be used. With the first method, the interval between two consecutive readings is strobed (or timed) using a high-frequency strobe (clock) signal, as in the case of an incremental encoder. Typical strobing frequency is 1 MHz. The start and stop of strobing are triggered by the coded data from the encoder. The clock cycles are counted by a counter, as in the case of an incremental encoder, and the count is reset (cleared) after each counting cycle. The angular speed can be computed using these data, as discussed earlier for an incremental encoder. With the second method, the change in angle is measured from one absolute angle reading to the next, and the angular speed is computed as the ratio (angle change)/(sampling period).

7.4.4   Advantages and Drawbacks

The main advantage of an absolute encoder is its ability to provide absolute angle readings (within a full 360° rotation). Hence, if a reading is missed, it will not affect the next reading. Specifically, the digital output uniquely corresponds to a physical rotation of the code disk, and hence a particular reading is not dependent on the accuracy of a previous reading. This provides immunity to data failure. A missed pulse (or, a data failure of some sort) in an incremental encoder would carry an error into the subsequent readings until the counter is cleared.

An incremental encoder has to be powered throughout operation of the device. Thus, a power failure can introduce an error unless the reading is reinitialized (or, calibrated). An absolute encoder has the advantage that it needs to be powered and monitored only when a reading is taken.

Because the code matrix on the disk is more complex in an absolute encoder, and because more light sensors are required, an absolute encoder can be nearly twice as expensive as an incremental encoder. Also, since the resolution depends on the number of tracks present, it is more costly to obtain finer resolutions. An absolute encoder does not require digital counters and buffers, however, unless resolution enhancement is done using an auxiliary track, or pulse-timing is used for velocity calculation.

7.5   Encoder Error

Errors in shaft encoder readings can come from several factors. The primary sources of these errors are as follows:

1. Quantization error (due to digital word size limitations)
2. Assembly error (eccentricity of rotation, etc.)
3. Coupling error (gear backlash, belt slippage, loose fit, etc.)
4. Structural limitations (disk deformation and shaft deformation due to loading)
5. Manufacturing tolerances (errors from inaccurately imprinted code patterns, inexact positioning of the pick-off sensors, limitations and irregularities in signal generation and sensing hardware, etc.)
6. Ambient effects (vibration, temperature, light noise, humidity, dirt, smoke, etc.)

These factors can result in inexact readings of displacement and velocity, and erroneous detection of the direction of motion.

One form of error in encoder readings is the hysteresis. For a given position of the moving object, if the encoder reading depends on the direction of motion, the measurement has a hysteresis error. In that case, if the object rotates from position A to position B and back to position A, for example, the initial and the final readings of the encoder will not match. The causes of hysteresis include backlash in gear couplings, loose fits, mechanical deformation in the code disk and shaft, delays in electronic circuitry and components (electrical time constants, nonlinearities, etc.), magnetic hystersis (in the case of a magnetic encoder), and noisy pulse signals that make the detection of pulses (say, by level detection or edge detection) less accurate.

The raw pulse signal from an optical encoder is somewhat irregular and does not consist of perfect pulses, primarily because of the variation (somewhat triangular) of the intensity of light received by the optical sensor as the code disk moves though a window, and because of noise in the signal generation circuitry, including the noise created by imperfect light sources and photosensors. Noisy pulses have imperfect edges. As a result, pulse detection through edge detection can result in errors such as multiple triggering for the same edge of a pulse. This can be avoided by including a Schmitt trigger (a logic circuit with electronic hysteresis) in the edge-detection circuit, so that slight irregularities in the pulse edges will not cause erroneous triggering, provided that the noise level is within the hysteresis band of the trigger. A disadvantage of this method, however, is that hysteresis will be present even when the encoder itself is perfect. Virtually noise-free pulses can be generated if two photosensors are used to detect adjacent transparent and opaque areas on a track simultaneously and a separate circuit (a comparator) is used to create a pulse that depends on the sign of the voltage difference of the two sensor signals. This method of pulse shaping has been described earlier, with reference to Figure 7.5.

7.5.1 Eccentricity Error

Eccentricity (denoted by e) of an encoder is defined as the distance between the center of rotation C of the code disk and the geometric center G of the circular code track. Nonzero eccentricity causes a measurement error known as the *eccentricity error*. The primary contributions to eccentricity are

1. Shaft eccentricity (e_s)
2. Assembly eccentricity (e_t)
3. Track eccentricity (e_1)
4. Radial play (e_p)

Shaft eccentricity results if the rotating shaft on which the code disk is mounted is imperfect, or due to shaft flexibility or whirling, so that its axis of rotation does not coincide with its geometric axis. Assembly eccentricity is caused if the code disk is improperly mounted on the shaft, so that the center of the code disk does not fall on the shaft axis. Track eccentricity comes from irregularities in the imprinting process of the code track, so that the center of the track circle does not coincide with the nominal geometric center of the disk. Radial play is caused by any looseness in the assembly in the radial direction. All four of these parameters are random variables. Let their mean values be μ_s, μ_a, μ_t, and μ_p, and the standard deviations be σ_s, σ_a, σ_t, and σ_p, respectively. A very conservative upper bound for the mean value of the overall eccentricity is given by the sum of the individual mean values, each value being considered positive (See chapter 5 on instrument error). A more reasonable estimate is provided by the *root-mean-square* (*rms*) value, as given by

$$\mu = \sqrt{\mu_s^2 + \mu_a^2 + \mu_t^2 + \mu_p^2} \qquad (7.23)$$

Furthermore, assuming that the individual eccentricities are independent random variables, the standard deviation of the overall eccentricity is given by

$$\sigma = \sqrt{\sigma_s^2 + \sigma_a^2 + \sigma_t^2 + \sigma_p^2} \qquad (7.24)$$

Knowing the mean value μ and the standard deviation σ of the overall eccentricity, it is possible to obtain a reasonable estimate for the maximum eccentricity that can occur. It is reasonable to assume that the eccentricity has a Gaussian (or normal) distribution, as shown in Figure 7.9. The probability that the eccentricity lies between two given values is obtained by the area under the probability density curve within these two values (points) on the x-axis. In particular, for the normal distribution, the probability that the eccentricity lies within $\mu - 2\sigma$ and $\mu + 2\sigma$ is 95.5%, and the probability that the eccentricity falls within $\mu - 3\sigma$ and $\mu + 3\sigma$ is 99.7%. We can say, for example, that at a confidence level of 99.7%, the net eccentricity will not exceed $\mu + 3\sigma$.

Example 7.5

The mean values and the standard deviations of the four primary contributions to eccentricity in a shaft encoder are as follows (in millimeters):

Shaft eccentricity = (0.1, 0.01)
Assembly eccentricity = (0.2, 0.05)
Track eccentricity = (0.05, 0.001)
Radial play = (0.1, 0.02)

Estimate the overall eccentricity at a confidence level of 96%.

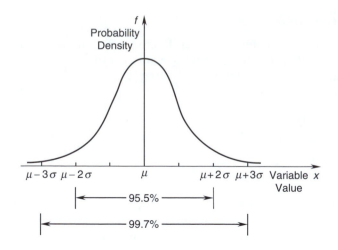

FIGURE 7.9
Gaussian (normal) probability density function.

SOLUTION
The mean value of the overall eccentricity may be estimated as the rms value of the individual means, as given by Equation 7.23; thus

$$\mu = \sqrt{0.1^2 + 0.2^2 + 0.05^2 + 0.1^2} = 0.25 \text{ mm}$$

Using Equation 7.24, the standard deviation of the overall eccentricity is estimated as

$$\sigma = \sqrt{0.01^2 + 0.05^2 + 0.001^2 + 0.02^2} = 0.055 \text{ mm}$$

Now, assuming a Gaussian distribution, an estimate for the overall eccentricity at a confidence level of 96% is given by

$$\hat{e} = 0.25 + 2 \times 0.055 = 0.36 \text{ mm}$$

Once the overall eccentricity is estimated in the foregoing manner, the corresponding measurement error can be determined. Suppose that the true angle of rotation is θ and the corresponding measurement is θ_m. The eccentricity error is given by

$$\Delta\theta = \theta_m - \theta \tag{7.25}$$

The maximum error can be shown to exist when the line of eccentricity (*CG*) is symmetrically located within the angle of rotation, as shown in Figure 7.10. For this configuration, the sine rule for triangles gives

$$\frac{\sin(\Delta\theta/2)}{e} = \frac{\sin(\theta/2)}{r}$$

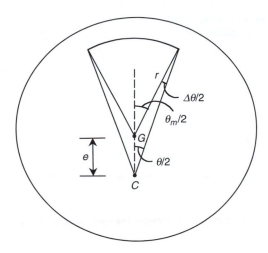

FIGURE 7.10
Nomenclature for eccentricity error (C = center of rotation, G = geometric center of the code track).

where r denotes the code track radius, which for most practical purposes can be taken as the disk radius. Hence, the eccentricity error is given by

$$\Delta\theta = 2\sin^{-1}\left(\frac{e}{r}\sin\frac{\theta}{2}\right) \quad (7.26)$$

It is intuitively clear that the eccentricity error should not enter measurements of complete revolutions, and this can be shown analytically by using Equation 7.26. In this case, $\theta = 2\pi$. Accordingly, $\Delta\theta = 0$. For multiple revolutions, the eccentricity error is periodic with period 2π.

For small angles, the sine of an angle is approximately equal to the angle itself, in radians. Hence, for small $\Delta\theta$, the eccentricity error may be expressed as

$$\Delta\theta = \frac{2e}{r}\sin\frac{\theta}{2} \quad (7.27)$$

Furthermore, for small angles of rotation, the fractional eccentricity error is given by

$$\frac{\Delta\theta}{\theta} = \frac{e}{r} \quad (7.28)$$

which is, in fact, the worst-case fractional error. As the angle of rotation increases, the fractional error decreases (as shown in Figure 7.11), reaching the zero value for a full revolution. From the point of view of gross error, the worst value occurs when $\theta = \pi$, which corresponds to half a revolution. From Equation 7.26, it is clear that the maximum gross error due to eccentricity is given by

$$\Delta\theta_{max} = 2\sin^{-1}\frac{e}{r} \quad (7.29)$$

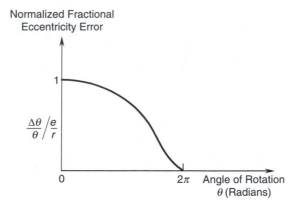

FIGURE 7.11
Fractional eccentricity error variation with the angle of rotation.

If this value is less than half the resolution of the encoder, the eccentricity error becomes inconsequential. For all practical purposes, since e is much less than r, we may use the following expression for the maximum eccentricity error:

$$\Delta\theta_{max} = \frac{2e}{r} \qquad (7.30)$$

Example 7.6

Suppose that in the previous example (7.5), the radius of the code disk is 5 cm. Estimate the maximum error due to eccentricity. If each track has 1000 windows, determine whether the eccentricity error is significant.

SOLUTION
With the given level of confidence, we have calculated the overall eccentricity to be 0.36 mm. Now, from Equation 7.29 or Equation 7.30, the maximum angular error is given by

$$\Delta\theta_{max} = \frac{2\times0.36}{50} = 0.014\,rad = 0.83°$$

Assuming that quadrature signals are used to improve the encoder resolution, we have

$$\text{Resolution} = \frac{360°}{4\times1000} = 0.09°$$

Note that the maximum error due to eccentricity is more than 10 times the encoder resolution. Hence, eccentricity will significantly affect the accuracy of the encoder.

———

Eccentricity of an incremental encoder also affects the phase angle between the quadrature signals, if a single track and two pick-off sensors (with circumferential offset) are used. This error can be reduced using the two-track arrangement, with the two sensors positioned along a radial line, so that eccentricity will equally affect the two outputs.

7.6 Miscellaneous Digital Transducers

Now several other types of digital transducers, which are useful in the mechatronics practice are described. In particular, digital rectilinear transducers are described. These are useful in many applications. Typical applications include *x-y* positioning tables, machine tools, valve actuators, read-write heads in disk drive systems, and robotic manipulators (e.g., at prismatic joints) and robot hands. The principles used in angular motion transducers described so far in this book can be used in measuring rectilinear motions as well. Techniques of signal acquisition, interpretation, conditioning, etc. may find similarities in the devices described below, with those presented thus far.

7.6.1 Digital Resolvers

Digital resolvers, or mutual induction encoders, operate somewhat like analog resolvers, using the principle of mutual induction. They are known commercially as *Inductosyns*. A digital resolver has two disks facing each other (but not in contact), one (the stator) stationary and the other (the rotor) coupled to the rotating object whose motion is measured. The rotor has a fine electric conductor foil imprinted on it, as schematically shown in Figure 7.12. The printed pattern is "pulse" shaped, closely spaced, and connected to a high-frequency ac supply (carrier) of voltage v_{ref}. The stator disk has two separate printed patterns that are identical to the rotor pattern, but one pattern on the stator is shifted by a quarter-pitch from the other pattern. The primary voltage in the rotor circuit induces voltages in the two secondary (stator) foils at the same frequency, that is, the rotor and the stator are *inductively coupled*. These induced voltages are "quadrature" signals. As the rotor turns, the level of the induced voltage changes, depending on the relative position of the foil patterns on the two disks. When the foil pulse patterns coincide, the induced voltage is a maximum (positive or negative), and when the rotor foil pattern has a half-pitch offset from the stator foil pattern, the induced voltages in the adjacent segments cancel each other, producing a zero output. The output (induced) voltages v_1 and v_2 in the

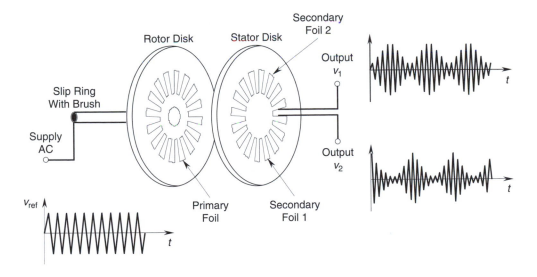

FIGURE 7.12
Schematic diagram of a digital resolver.

two foils of the stator have a carrier component at the supply frequency and a modulating component corresponding to the rotation of the disk. The latter (modulating component) can be extracted through demodulation (see Chapter 6), and converted into a proper pulse signal using pulse-shaping circuitry, as for an incremental encoder. When the rotating speed is constant, the two modulating components are periodic and nearly sinusoidal, with a phase shift of 90° (i.e., in quadrature). When the speed is not constant, the pulse width will vary with time.

As in the case of an incremental encoder, angular displacement is determined by counting the pulses. Furthermore, angular velocity is determined either by counting the pulses over a fixed time period (counter sampling period) or by timing the pulses. The direction of rotation is determined by the phase difference in the two modulating (output) signals. (In one direction, the phase shift is 90°; in the other direction, it is –90°.) Very fine resolutions (e.g., 0.0005°) may be obtained from a digital resolver, and it is usually not necessary to use step-up gearing or other techniques to improve the resolution. These transducers are usually more expensive than optical encoders. The use of a slip ring and brush to supply the carrier signal may be viewed as a disadvantage.

Consider the conventional resolver discussed in Chapter 6. Its outputs may be converted into digital form using appropriate hardware. Strictly speaking, such a device cannot be classified as a digital resolver.

7.6.2 Digital Tachometers

A pulse-generating transducer whose pulse train is synchronized with a mechanical motion may be treated as a digital transducer for motion measurement. In particular, pulse counting may be used for displacement measurement, and the pulse rate (or, pulse timing) may be used for velocity measurement. As studied in Chapter 6, tachometers are devices for measuring angular velocities. According to this terminology, a shaft encoder (particularly, an incremental optical encoder) may be considered as a digital tachometer. According to the popular terminology, however, a digital tachometer is a device that employs a toothed wheel to measure angular velocities.

A schematic diagram of a digital tachometer is shown in Figure 7.13. This is a magnetic induction, *pulse tachometer* of the variable-reluctance type. The teeth on the wheel are made of a ferromagnetic material. The two magnetic-induction (and variable-reluctance) proximity

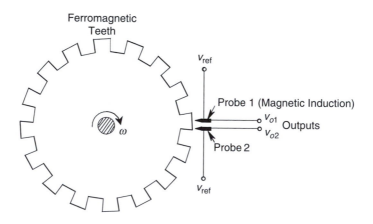

FIGURE 7.13
Schematic diagram of a pulse tachometer.

probes are placed radially facing the teeth, at quarter-pitch apart (pitch = tooth-to-tooth spacing). When the toothed wheel rotates, the two probes generate output signals that are 90° out of phase (i.e., quadrature signals). One signal leads the other in one direction of rotation and lags the other in the opposite direction of rotation. In this manner, a directional reading (i.e., velocity rather than speed) is obtained. The speed is computed either by counting the pulses over a sampling period or by timing the pulse width, as in the case of an incremental encoder.

Alternative types of digital tachometers use eddy current proximity probes or capacitive proximity probes (see Chapter 6). In the case of an eddy current tachometer, the teeth of the pulsing wheel are made of (or plated with) electricity-conducting material. The probe consists of an active coil connected to an ac bridge circuit excited by a radio-frequency (i.e., in the range 1–100 MHz) signal. The resulting magnetic field (at radio frequency is modulated by the tooth-passing action. The bridge output may be demodulated and shaped to generate the pulse signal. In the case of a capacitive tachometer, the toothed wheel forms one plate of the capacitor; the other plate is the probe and is kept stationary. As the wheel turns, the gap width of the capacitor fluctuates. If the capacitor is excited by an ac voltage of high frequency (typically 1 MHz), a nearly pulse-modulated signal at that carrier frequency is obtained. This can be detected through a bridge circuit as before (but using a capacitance bridge rather than an inductance bridge). By demodulating the output signal, the modulating signal can be extracted, which can be shaped to generate the pulse signal. The pulse signal generated in this manner is used in the angular velocity computation.

The advantages of digital (pulse) tachometers over optical encoders include simplicity, robustness, immunity to environmental effects and other common fouling mechanisms (except magnetic effects), and low cost. Both are noncontacting devices. The disadvantages of a pulse tachometer include poor resolution (determined by the number of teeth), size (bigger and heavier than optical encoders), mechanical errors due to loading, hysteresis (i.e., output is not symmetric depends on the direction of motion), and manufacturing irregularities. Note that mechanical loading will not be a factor if the toothed wheel already exists as an integral part of the original system that is sensed. The resolution (digital resolution) depends on the word size used for data acquisition).

7.6.3 Hall Effect Sensors

Consider a semiconductor element subject to a dc voltage v_{ref}. If a magnetic field is applied perpendicular to the direction of this voltage, a voltage v_o will be generated in the third orthogonal direction within the semiconductor element. This is known as the Hall effect (observed by E. H. Hall in 1879). A schematic representation of a Hall effect sensor is shown in Figure 7.14.

A Hall effect sensor may be used for motion sensing in many ways—for example, as an analog proximity sensor, a limit switch (digital), or a shaft encoder. Since the output voltage v_o increases as the distance from the magnetic source to the semiconductor element decreases, the output signal v_o can be used as a measure of proximity. This is the principle behind an analog proximity sensor. Alternatively, a certain threshold level of the output voltage v_o can be used to generate a binary output, which represents the presence/absence of an object. This principle is used in a digital limit switch. The use of a toothed ferromagnetic wheel (as for a digital tachometer) to alter the magnetic flux will result in a shaft encoder. The sensitivity of a practical sensor element is of the order of 10 V/tesla. For a Hall-effect device, the temperature coefficient of resistance is positive and the temperature coefficient of sensitivity is negative. In view of these properties, auto-compensation for temperature may be achieved (as for a semiconductor strain gage—see Chapter 6).

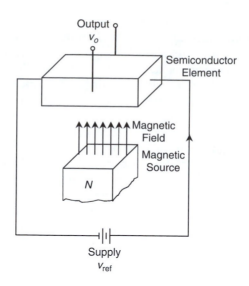

FIGURE 7.14
Schematic representation of a Hall effect sensor.

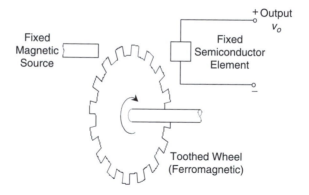

FIGURE 7.15
A Hall effect shaft encoder or digital tachometer.

The longitudinal arrangement of a proximity sensor, in which the moving element approaches head-on toward the sensor, is not suitable when there is a danger of over-shooting the target, since it will damage the sensor. A more desirable configuration is the lateral arrangement, in which the moving member slides by the sensing face of the sensor. The sensitivity will be lower, however, with this lateral arrangement. The relationship between the output voltage v_o and the distance x of a Hall effect sensor, measured from the moving member, is nonlinear. Linear Hall effect sensors use calibration to linearize their output.

A practical arrangement for a motion sensor based on the Hall effect would be to have the semiconductor element and the magnetic source fixed relative to one another in a single package. As a ferromagnetic member is moved into the air gap between the magnetic source and the semiconductor element, the flux linkage is varied. The output voltage v_o is changed accordingly. This arrangement is suitable for both an analog proximity sensor and a limit switch. By using a toothed ferromagnetic wheel as in Figure 7.15 to change v_o, and then by shaping the resulting signal, it is possible to generate a pulse train in proportion to the wheel rotation. This provides a shaft encoder or a digital tachometer. Apart from the familiar applications of motion sensing, Hall effect sensors are used for *electronic commutation* of brushless dc motors (see Chapter 9) where the field circuit of the motor is appropriately switched depending on the angular position of the rotor with respect to the stator.

Hall effect motion transducers are rugged devices and have many advantages. They are not affected by "rate effects" (specifically, the generated voltage is not affected by the rate of change of the magnetic field). Also, their performance is not severely affected by common environmental factors, except magnetic fields. They are noncontacting sensors with associated advantages as mentioned before. Some hysteresis will be present, but it is not a serious drawback in digital transducers. Miniature Hall-effect devices (mm scale) are available.

7.6.4 Linear Encoders

In rectilinear encoders (popularly called linear encoders, where "linear" does not imply linearity but refers to rectilinear motion) rectangular flat plates moving rectilinearly, instead of rotating disks, are used with the same types of signal generation and interpretation mechanisms as for shaft (rotary) encoders. A transparent plate with a series of opaque lines arranged in parallel in the transverse direction forms the stationary plate (grating plate or phase plate) of the transducer. This is called the *mask plate*. A second transparent plate, with an identical set of ruled lines, forms the moving plate (or the *code plate*). The lines on both plates are evenly spaced, and the line width is equal to the spacing between adjacent lines. A light source is placed on the moving plate side, and light transmitted through the common area of the two plates is detected on the other side using one or more photosensors. When the lines on the two plates coincide, the maximum amount of light will pass through the common area of the two plates. When the lines on one plate fall on the transparent spaces of the other plate, virtually no light will pass through the plates. Accordingly, as one plate moves relative to the other, a pulse train is generated by the photosensor, and it can be used to determine rectilinear displacement and velocity, as in the case of an incremental encoder.

A suitable arrangement is shown in Figure 7.16. The code plate is attached to the moving object whose rectilinear motion is to be measured. An LED light source and a photo-transistor light sensor are used to detect the motion pulses, which can be interpreted just like in the

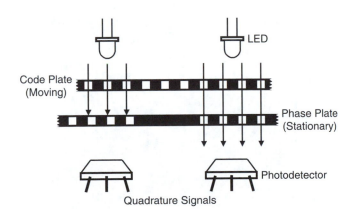

FIGURE 7.16
A rectilinear optical encoder.

case of a rotatory encoder. The phase plate is used, as with a shaft encoder, to enhance the intensity and the discrimination of the detected signal. Two tracks of windows in quadrature (i.e., 1/4 pitch offset) would be needed to determine the direction of motion, as shown in Figure 7.16. Another track of windows at ½ pitch offset with the main track (not shown in Figure 7.16) may be used as well on the phase plate, to further enhance the discrimination of the detected pulses. Specifically, when the sensor at the main track reads a high intensity (i.e., when the windows on the code plate and the phase plate are aligned) the sensor at the track that is ½ pitch away will read a low intensity (because the corresponding windows of the phase plate are blocked by the solid regions of the code plate).

7.6.5 Moiré Fringe Displacement Sensors

Suppose that a piece of transparent fabric is placed on another. If one piece is moved or deformed with respect to the other, we will notice various designs of light and dark patterns (lines) in motion. Dark lines of this type called moiré fringes. In fact the French term moiré refers to a silk-like fabric, which produces moiré fringe patterns. An example of a moiré fringe pattern is shown in Figure 7.17. Consider the rectilinear encoder, which was described above. When the window slits of one plate overlap with the window slits of the other plate, we get an alternating light and dark pattern. This is a special case of moiré fringes. A moiré device of this type may be used to measure rigid-body movements of one plate of the sensor with respect to the other.

The application of the moiré fringe technique is not limited to sensing rectilinear motions. This technology can be used to sense angular motions (rotations) and more generally, distributed deformations (e.g., elastic deformations) of one plate with respect to the other. Consider two plates with gratings (optical lines) of identical pitch (spacing) p. Suppose that initially the gratings of the two plates exactly coincide. Now if one plate is deformed in the direction of the grating lines, the transmission of light through the two plates will not be altered. But, if a plate is deformed in the perpendicular direction to the grating

FIGURE 7.17
A moiré fringe pattern.

lines, then window width of that plate will be deformed accordingly. In this case, depending on the nature of the plate deformation, some transparent lines of one plate will be completely covered by the opaque lines of the other plate, and some other transparent lines of the first plate will have coinciding transparent lines on the second plate. Thus, the observed image will have dark lines (moiré fringes) corresponding to the regions with clear/opaque overlaps of the two plates and bright lines corresponding to the regions with clear/clear overlaps of the two plates. The resulting moiré fringe pattern will provide the deformation pattern of one plate with respect to the other. Such two-dimensional fringe patterns can be detected and observed by arrays of optical sensors (e.g., using a charge-coupled device or CCD) and by photographic means. In particular, since the "presence" of a fringe is a binary information, binary optical sensing techniques (as for optical encoders) and digital imaging techniques may be used with these transducers. Accordingly, these devices may be classified as digital transducers. With the moiré fringe technique, very small resolutions (e.g., 0.002 mm) can be realized, because finer line spacing (in conjunction with wider light sensors) can be used.

To further understand and analyze the fundamentals of moiré fringe technology, consider two grating plates with identical line pitch (spacing between the windows) p. Suppose that one plate is kept stationary. This is the plate of master gratings (or reference gratings or main gratings). The other plate (which is termed the plate containing index gratings or model gratings) is placed over the fixed plated and rotated so that the index gratings form an angle α with the master gratings, as shown in Figure 7.18. The lines shown are in fact the opaque regions, which are identical in size and spacing to the windows in between the opaque regions. A uniform light source is placed on one side of the overlapping pair

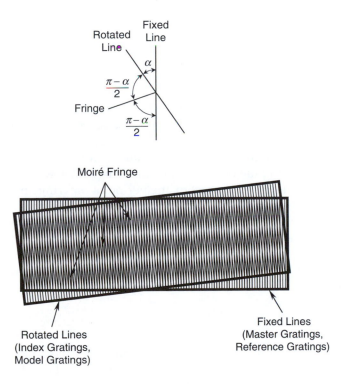

FIGURE 7.18
Formation of moiré fringes.

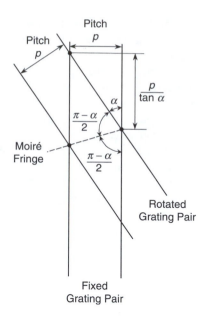

FIGURE 7.19
The orientation of moiré fringes.

of plates and the light transmitted through them is observed on the other side. Dark bands called moiré fringes are seen as a result, as in Figure 7.18.

A moiré fringe corresponds to the line joining a series of points of intersection of the opaque lines of the two plates, because no light can pass through such points. This is further shown in Figure 7.19. Note that in the present arrangement, the line pitch of the two plates is identical and equal to p. A fringe line formed is shown as the broken line in Figure 7.19. Since the line pattern in the two plates is identical, by symmetry of the arrangement, the fringe line should bisect the obtuse angle $(\pi - \alpha)$ formed by the intersecting opaque lines. In other words, a fringe line makes an angle of $\frac{\pi - \alpha}{2}$ with the fixed gratings. Furthermore, the vertical separation (or, the separation in the direction of the fixed gratings) of the moiré fringes is seen to be $p/\tan \alpha$.

In summary then, the rotation of the index plate with respect to the reference plate can be measured by sensing the orientation of the fringe lines with respect to the fixed (master or reference) gratings. Furthermore, the period of the fringe lines in the direction of the reference gratings is $p/\tan \alpha$ and when the index plate is moved rectilinearly by a distance of one grating pitch, the fringes also shift vertically by its period of $p/\tan \alpha$ (see Figure 7.19). It is clear then that the rectilinear displacement of the index plate can be measured by sensing the fringe spacing. In a two-dimensional pattern of moiré fringes, these facts can be used as local information in order to sense full-field motions and deformations.

Example 7.7

Suppose that each plate of a moiré fringe deformation sensor has a line pitch of 0.01 mm. A tensile load is applied to one plate in the direction perpendicular to the lines. Five moiré fringes are observed in 10 cm of the moiré image under tension. What is the tensile strain in the plate?

SOLUTION

There is one moiré fringe in every 10/5 = 2 cm of the plate. Hence, extension of a 2 cm portion of the plate = 0.01 mm, and

$$\text{tensile strain} = \frac{0.1 \text{ mm}}{2 \times 10 \text{ mm}} = 0.0005\varepsilon = 500 \text{ } \mu\varepsilon$$

In this example, we have assumed that the strain distribution (or deformation) of the plate is uniform. Under nonuniform strain distributions the observed moiré fringe pattern generally will not be parallel straight lines but rather complex shapes.

7.6.6 Cable Extension Sensors

In many applications, rectilinear motion is produced from a rotary motion (say, of a motor) through a suitable transmission device, such as rack and pinion or lead screw and nut. In these cases, rectilinear motion can be determined by measuring the associated rotary motion, assuming that errors due to backlash, flexibility, and so forth, in the transmission device can be neglected. Another way to measure rectilinear motions using a rotary sensor is to use a modified sensor that has the capability to convert a rectilinear motion into a rotary motion within the sensor itself. An example would be *the cable extension method* of sensing rectilinear motions. This method is particularly suitable for measuring motions that have large excursions. The cable extension method uses an angular motion sensor with a spool rigidly coupled to the rotating part of the sensor (e.g., the encoder disk) and a cable that wraps around the spool. The other end of the cable is attached to the object whose rectilinear motion is to be sensed. The housing of the rotary sensor is firmly mounted on a stationary platform, so that the cable can extend in the direction of motion. When the object moves, the cable extends, causing the spool to rotate. This angular motion is measured by the rotary sensor. With proper calibration, this device can give rectilinear measurements directly. As the object moves toward the sensor, the cable has to retract without slack. This is guaranteed by using a device such as a spring motor to wind the cable back. The disadvantages of the cable extension method include mechanical loading of the moving object, time delay in measurements, and errors caused by the cable, including irregularities, slack, and tensile deformation.

7.6.7 Binary Transducers

Digital binary transducers are two-state sensors. The information provided by such a device takes only two states (on/off, present/absent, go/no-go, etc.); it can be represented by one bit. For example, *limit switches* are sensors used in detecting whether an object has reached its limit (or, destination) of mechanical motion, and are useful in sensing presence/absence and in object counting. In this sense, a limit switch is considered a digital transducer. Additional logic is needed if the direction of contact is also needed. Limit switches are available for both rectilinear and angular motions. A limit of a movement can be detected by mechanical means using a simple contact mechanism to close a circuit or trigger a pulse. Although a purely mechanical device consisting of linkages, gears, ratchet wheels and pawls, and so forth, can serve as a limit switch, electronic and solid-state switches are usually preferred for such reasons as accuracy, durability, a low activating force (practically zero) requirement, low cost, and small size. Any proximity sensor could serve as the sensing element of a limit switch, to detect the presence of an object. The proximity sensor signal is then used in a desired manner—for example, to activate a counter, a mechanical switch, or a relay circuit, or simply as an input to a digital controller. A *microswitch* is a solid-state switch that can be used as a limit switch. Microswitches are commonly used in counting operations—for example, to keep a count of completed products in a factory warehouse.

There are many types of binary transducers that are applicable in detection and counting of objects. They include:

1. Electromechanical switches
2. Photoelectric devices
3. Magnetic (Hall-effect, eddy current) devices
4. Capacitive devices
5. Ultrasonic devices

An electromechanical switch is a mechanically activated electric switch. The contact with an arriving object turns on the switch, thereby completing a circuit and providing an electrical signal. This signal provides the "present" state of the object. When the object is removed, the contact is lost and the switch is turned off. This corresponds to the "absent" state.

In the other four types of binary transducers listed above, a signal (light beam, magnetic field, electric field, or ultrasonic wave) is generated by a source (emitter) and received by a receiver. A passing object interrupts the signal. This event can be detected by usual means, using the signal received at the receiver. In particular, the signal level, a rising edge, or a falling edge may be used to detect the event. The following three arrangements of the emitter-receiver pair are common:

1. Through (opposed) configuration
2. Reflective (reflex) configuration
3. Diffuse (proximity, interceptive) configuration

In the through configuration (Figure 7.20(a)), the receiver is placed directly facing the emitter. In the reflective configuration, the emitter-source pair is located in a single package. The emitted signal is reflected by a reflector, which is placed facing the emitter-receiver package (Figure 7.20(b)). In the diffuse configuration as well, the emitter-reflector pair is in a single package. In this case, a conventional proximity sensor can serve the purpose of detecting the presence of an object (Figure 7.20(c)) by using the signal diffused from the intercepting object. When the photoelectric method is used, a light-emitting diode (LED) may serve as the emitter and a phototransistor may serve as the receiver. Infrared LEDs are preferred emitters for phototransistors because their peak spectral responses match. Furthermore, they are not affected by ambient light. Many factors govern the performance of a digital transducer for object detection. They include:

1. Sensing range (operating distance between the sensor and the object)
2. Response time
3. Sensitivity
4. Linearity
5. Size and shape of the sensed object
6. Material of the object (e.g., color, reflectance, permeability, permittivity)

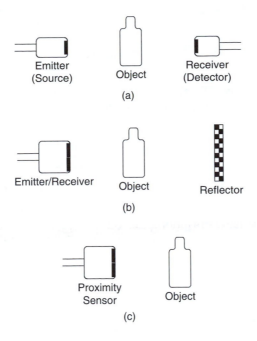

FIGURE 7.20
Two-state transducer configurations: (a) Through (opposed); (b) Reflective (reflex); (c) Interceptive (proximity).

7. Orientation and alignment (optical axis, reflector, object)
8. Ambient conditions (light, dust, moisture, magnetic field, etc.)
9. Signal conditioning considerations (modulation, demodulation, shaping, etc.)
10. Reliability, robustness, and design life.

Example 7.8

The response time of a binary transducer for object counting is the fastest (shortest) time the transducer needs to detect an absent-to-present condition or a present-to-absent condition and generate the counting signal (say, a pulse). Consider the counting process of packages on a conveyor. Suppose that, typically, packages of length 20 cm are placed on the conveyor at 15 cm spacing. A transducer of response time 10 ms is used for counting the packages. Estimate the allowable maximum operating speed of the conveyor.

SOLUTION
If the conveyor speed is v cm/ms, then

$$\text{Package-present time} = \frac{20.0}{v}\text{ ms}$$

$$\text{Package-absent time} = \frac{15.0}{v}\text{ ms}$$

We must have a transducer response time of at least $\frac{15.0}{v}$ ms. Hence,

$$10.0 \leq \frac{15.0}{v}$$

or,

$$v \leq 1.5 \, \text{cm/ms}$$

The maximum allowable operating speed is 1.5 cm/ms or 15.0 m/s. This corresponds to a counting rate of $\frac{1.5}{(20.0 + 15.0)}$ packages/ms or about 43 packages/s.

7.7 Problems

7.1 Identify active transducers among the following types of shaft encoders, and justify your claims. Also, discuss the relative merits and drawbacks of the four types of encoders.

a. Optical encoders

b. Sliding contact encoders

c. Magnetic encoders

d. Proximity sensor encoders.

7.2 Consider the two quadrature pulse signals (say, A and B) from an incremental encoder. Using sketches of these signals, show that in one direction of rotation, signal B is at a high level during the up-transition of signal A, and in the opposite direction of rotation, signal B is at a low level during the up-transition of signal A. Note that the direction of motion can be determined in this manner, by using level detection of one signal during the up-transition of the other signal.

7.3 Explain why the speed resolution of a shaft encoder depends on the speed itself. What are some of the other factors that affect speed resolution? The speed of a dc motor was increased from 50 rpm to 500 rpm. How would the speed resolution change if the speed were measured using an incremental encoder,

a. By the pulse-counting method?

b. By the pulse-timing method?

7.4 Describe methods of improving the displacement resolution and the velocity resolution in an encoder. An incremental encoder disk has 5000 windows. The word size of the output data is 12 bits. What is the angular (displacement) resolution of the device? Assume that quadrature signals are available but that no interpolation is used.

7.5 An incremental optical encoder that has N window per track is connected to a shaft through a gear system with gear ratio p. Derive formulas for calculating angular velocity of the shaft by the

a. Pulse-counting method

b. Pulse-timing method.

What is the speed resolution in each case? What effect does step-up gearing have on the speed resolution?

7.6 What is hysteresis in an optical encoder? List several causes of hysteresis and discuss ways to minimize hysteresis.

7.7 An optical encoder has n windows/cm diameter (in each track). What is the eccentricity tolerance e below which readings are not affected by eccentricity error?

7.8 Show that in the single-track, two-sensor design of an incremental encoder, the phase angle error (in quadrature signals) due to eccentricity is inversely proportional to the second power of the radius of the code disk for a given window density. Suggest a way to reduce this error.

7.9 Consider an encoder with 1000 windows in its track and capable of providing quadrature signals. What is the displacement resolution $\Delta\theta_r$ in radians? Obtain a value for the nondimensional eccentricity e/r below which the eccentricity error has no effect on the sensor reading. For this limiting value, what is $\frac{\Delta\theta_r}{e/r}$? Typically, the values for this parameter, as given by encoder manufacturers, range from 3 to 6. Note: e = track eccentricity, r = track radius.

7.10 What is the main advantage of using a gray code instead of straight binary code, in an encoder? Give a table corresponding to a gray code different from what is given in Table 7.2 for a 4-bit absolute encoder. What is the code pattern on the encoder disk in this case?

7.11 Discuss construction features and operation of an optical encoder for measuring *rectilinear* displacements and velocities.

7.12 A particular type of multiplexer can handle ninety-six sensors. Each sensor generates a pulse signal with variable pulse width. The multiplexer scans the incoming pulse sequences, one at a time, and passes the information onto a control computer.

a. What is the main objective of using a multiplexer?

b. What type of sensors could be used with this multiplexer?

7.13 A centrifuge is a device that is used to separate components in a mixture. In an industrial centrifugation process, the mixture to be separated is placed in the centrifuge and rotated at high speed. The centrifugal force on a particle depends on the mass, radial location, and the angular speed of the particle. This force is responsible for separating the particles in the mixture.

Angular motion and the temperature of the container are the two key variables that have to be controlled in a centrifuge. In particular, a specific centrifugation curve is used, which consists of an acceleration segment, a constant-speed segment, and a braking (deceleration) segment, and this corresponds to a trapezoidal speed profile. An optical encoder may be used as the sensor for microprocessor-based speed control in the centrifuge. Discuss whether an absolute encoder is preferred for this purpose. Give advantages and possible drawbacks of using an optical encoder in this application.

7.14 Suppose that a feedback control system (Figure P7.14) is expected to provide an accuracy within $\pm\Delta y$ for a response variable y. Explain why the sensor that measures y should have a resolution of $\pm(\Delta y/2)$ or better for this accuracy to be possible. An x-y table has a travel of 2 m. The feedback control system is expected to provide an accuracy of ± 1 mm.

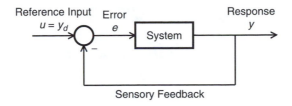

FIGURE P7.14
A feedback control loop.

An optical encoder is used to measure the position for feedback in each direction (x and y). What is the minimum bit size that is required for each encoder output buffer? If the motion sensor used is an absolute encoder, how many tracks and how many sectors should be present on the encoder disk?

7.15 Encoders that can provide 50,000 counts/turn with ± 1 count accuracy are commercially available. What is the resolution of such an encoder? Describe the physical construction of an encoder that has this resolution.

7.16 The pulses generated by the coding disk of an incremental optical encoder are approximately triangular (actually, upward shifted sinusoidal) in shape. Explain the reason for this. Describe a method for converting these triangular (or, shifted sinusoidal) pulses into sharp rectangular pulses.

7.17 Explain how resolution of a shaft encoder could be improved by pulse interpolation. Specifically, consider the arrangement shown in Figure P7.17. When the masking widows are completely covered by the opaque regions of the moving disk, no light is received by the photosensor. The peak level of light is received when the windows of the moving disk coincide with the windows of the masking disk. The variation of the light intensity from the minimum level to the peak level is approximately linear (generating a triangular pulse), but more accurately sinusoidal, and may be given by

$$v = v_{o}\left(1 - \cos\frac{2\pi\theta}{\Delta\theta}\right)$$

where θ denotes the angular position of the encoder window with respect to the masking window, as shown, and $\Delta\theta$ is the window pitch angle. Note that, in the sense of rectangular pulses, the pulse corresponds to the motion in the interval $\Delta\theta/4 \le \theta \le 3\Delta\theta/4$. By using this sinusoidal approximation for a pulse, as given above, show that one can improve the resolution of an encoder indefinitely simply by measuring the shape of each pulse at clock cycle intervals using a high-frequency clock signal.

7.18 A Schmitt trigger is a semiconductor device that can function as a level detector or a switching element, with hysteresis. The presence of hysteresis can be used, for example, to eliminate chattering during switching caused by noise in the switching signal. In an optical encoder, a noisy signal detected by the photosensor may be converted into a clean signal of rectangular pulses by this means. The input/output characteristic of a Schmitt trigger is shown in Figure P7.18(a). If the input signal is as shown in Figure P7.18(b), determine the output signal.

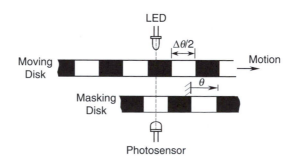

FIGURE P7.17
An encoder with a masking disk.

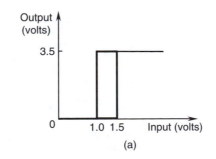

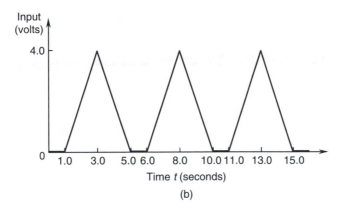

FIGURE P7.18
(a) The input/output characteristic of a Schmitt trigger, (b) A triangular input signal.

7.19 Displacement sensing and speed sensing are essential in a position servo. If a digital controller is employed to generate the servo signal, one option would be to use an analog displacement sensor and an analog speed sensor, along with analog-to-digital converters (ADCs) to produce the necessary digital feedback signals. Alternatively, an incremental encoder may be used to provide both displacement and speed feedbacks. In this case, ADCs are not needed. Encoder pulses will provide interrupts to the digital controller. Displacement is obtained by counting the interrupts. The speed is obtained by timing the interrupts. In some applications, analog speed signals are needed. Explain how an incremental encoder and a frequency-to-voltage converter (FVC) may be used to generate an analog speed signal.

7.20 Compare and contrast an optical incremental encoder against a potentiometer, by giving advantages and disadvantages, for an application involving the sensing of a rotatory motion.

A schematic diagram for the servo control loop of one joint of a robotic manipulator is given in Figure P7.20.

The motion command for each joint of the robot is generated by the robot controller, in accordance with the required trajectory. An optical incremental encoder is used for both position and velocity feedback in each servo loop. Note that for a six-degree-of-freedom robot there will be six such servo loops. Describe the function of each hardware component shown in the figure and explain the operation of the servo loop.

After several months of operation the motor of one joint of the robot was found to be faulty. An enthusiastic engineer quickly replaced the motor with an identical one without realizing that the encoder of the new motor was different. In particular, the original encoder generated 200 pulses/rev whereas the new encoder generated 720 pulses/rev. When the

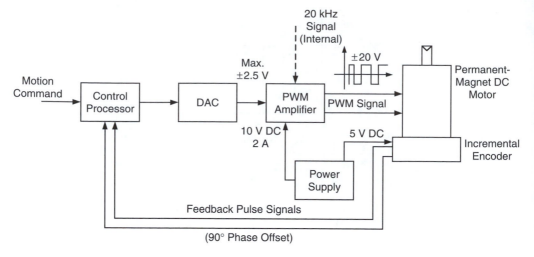

FIGURE P7.20
A servo loop of a robot.

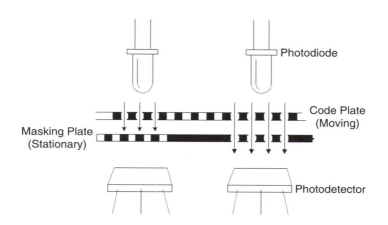

FIGURE P7.21
Photodiode-detector arrangement of a linear optical encoder.

robot was operated the engineer noticed an erratic and unstable behavior at the repaired joint. Discuss reasons for this malfunction and suggest a way to correct the situation.

   7.21 a. A position sensor is used in a microprocessor-based feedback control system for accurately moving the cutter blades of an automated meat-cutting machine. The machine is an integral part of the production line of a meat processing plant. What are primary considerations in selecting the position sensor for this application? Discuss advantages and disadvantages of using an optical encoder in comparison to a LVDT in this context.

   b. Figure P7.21 illustrates one arrangement of the optical components in a linear incremental encoder.

The moving code plate has uniformly spaced windows as usual, and the fixed masking plate has two groups of identical windows, one above each of the two photodetectors. These two groups of fixed windows are positioned in half-pitch out of phase so that when one detector receives light from its source directly through the aligned windows of the two plates, the other detector has the light from its source virtually obstructed by the masking plate.

Explain the purpose of the two sets of photodiode-detector units, giving a schematic diagram of the necessary electronics. Can the direction of motion be determined with the arrangement shown in Figure P7.21? If so, explain how this could be done. If not, describe a suitable arrangement for detecting the direction of motion.

7.22 a. What features and advantages of a digital transducer will distinguish it from a purely analog sensor?

b. Consider a "linear incremental encoder," which is used to measure rectilinear positions and speeds. The moving element is a nonmagnetic plate containing a series of identically magnetized areas uniformly distributed along its length. The pickoff transponder is a mutual-induction-type proximity sensor (i.e., a transformer) consisting of a toroidal core with a primary winding and a secondary winding. A schematic diagram of the encoder is shown in Figure P7.22. The primary excitation v_{ref} is a high-frequency sine wave.

Explain the operation of this position encoder, clearly indicating what types of signal conditioning would be needed to obtain a pure pulse signal. Also, sketch the output v_o of the proximity sensor as the code plate moves very slowly. Which position of the code plate does a high value of the pulse signal represent and which position does a low value represent?

c. Suppose that the "pulse period timing" method is used to measure speed (v) using this encoder. The pitch distance of the magnetic spots on the plate is p, as shown in Figure P7.22. If the clock frequency of the pulse period timer is f, give an expression for the speed v in terms of the clock cycle count m.

Show that the speed resolution Δv for this method may be approximated by

$$\Delta v = \frac{v^2}{pf}$$

It follows that the dynamic range $v/\Delta v = \frac{pf}{v}$.

If the clock frequency is 20 MHz, the code pitch is 0.1 mm, and the required dynamic range is 100 (i.e., 40 dB) what is the maximum speed in m/s that can be measured by this method?

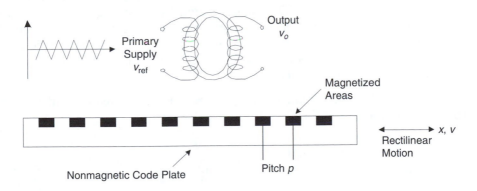

FIGURE P7.22
A linear incremental encoder of the magnetic induction type.

7.23 What is a Hall effect tachometer? Discuss the advantages and disadvantages of a Hall effect motion sensor in comparison to an optical motion sensor (e.g., an optical encoder).

7.24 Discuss the advantages of solid-state limit switches over mechanical limit switches. Solid-state limit switches are used in many applications, particularly in the aircraft and aerospace industries. One such application is in landing gear control, to detect up, down, and locked conditions of the landing gear. High reliability is of utmost importance in such applications. Mean time between failure (MTBF) of over 100,000 h is possible with solid-state limit switches. Using your engineering judgment, give an MTBF value for a mechanical limit switch.

7.25 Mechanical force switches are used in applications where only a force limit, rather than a continuous force signal, has to be detected. Examples include detecting closure force (torque) in valve closing, detecting fit in parts assembly, automated clamping devices, robotic grippers and hands, overload protection devices in process/machine monitoring, and product filling in containers by weight. Expensive and sophisticated force sensors are not needed in such applications because a continuous history of a force signal is not needed. Furthermore, they are robust and reliable, and can safely operate in hazardous environments. Using a sketch, describe the construction of a simple spring-loaded force switch.

7.26 Consider the following three types of photoelectric object counters (or object detectors or limit switches):

1. Through (opposed) type
2. Reflective (reflex) type
3. Diffuse (proximity, interceptive) type

Classify these devices into long-range (up to several meters), intermediate range (up to 1 m), and short-range (up to fraction of a meter) detection.

7.27 A brand of autofocusing camera uses a microprocessor-based feedback control system consisting of a charge-coupled device (CCD) imaging system, a microprocessor, a drive motor, and an optical encoder. The purpose of the control system is to focus the camera automatically based on the image of the subject as sensed by a matrix of CCDs (a set of metal oxide semiconductor field-effect transistors, or MOSFETs). The light rays from the subject that pass through the lens will fall onto the CCD matrix. This will generate a matrix (image frame) of charge signals, which are shifted one at a time, row by row, into an output buffer (or, frame grabber) and passed on to the microprocessor after conditioning the resulting video signal. The CCD image obtained by sampling the video signal is analyzed by the microprocessor to determine whether the camera is focused properly. If not, the lens is moved by the motor so as to achieve focusing. Draw a schematic diagram for the autofocusing control system and explain the function of each component in the control system, including the encoder.

7.28 Measuring devices with frequency outputs may be considered as digital transducers. Justify this statement.

8

Stepper Motors

The actuator is the device that mechanically drives a mechatronic system. Proper selection of actuators and their drive systems for a particular application is of utmost importance in the instrumentation and design of mechatronic systems. There is another perspective to the significance of actuators in the field of mechatronics. A typical actuator contains mechanical components like rotors, shafts, cylinders, coils, bearings, and seals while the control and drive systems are primarily electronic in nature. Integrated design, manufacture, and operation of these two categories of components are crucial to efficient operation of an actuator. This is essentially a mechatronic problem.

Stepper motors are a popular type of actuators. Unlike continuous-drive actuators (see Chapter 9), stepper motors are driven in fixed angular steps (increments). Each step of rotation is the response of the motor rotor to an input pulse (or a digital command). In this manner, the stepwise rotation of the rotor can be synchronized with pulses in a command-pulse train, assuming of course that no steps are missed, thereby making the motor respond faithfully to the input signal (pulse sequence) in an open-loop manner. From this perspective, it is reasonable to treat stepper motors as digital actuators. Nevertheless, like a conventional continuous-drive motor, a stepper motor is also an electromagnetic actuator, in that it converts electromagnetic energy into mechanical energy to perform mechanical work. The present chapter studies stepper motors, which are incremental-drive actuators. The next chapter will discuss continuous-drive actuators.

8.1   Principle of Operation

The terms *stepper motor, stepping motor,* and *step motor* are synonymous and are often used interchangeably. Actuators that can be classified as stepper motors have been in use for more than sixty years, but only after the incorporation of solid-state circuitry and logic devices in their drive systems have stepper motors emerged as cost-effective alternatives for dc servomotors in high-speed motion-control applications. Many kinds of actuators fall into the stepper motor category, but only those that are widely used in industry are discussed in this chapter. Note that even if the mechanism by which the incremental motion is generated differs from one type of stepper motor to the next, the same control techniques can be used in the associated control systems, making a general treatment of stepper motors possible, at least from the control point of view.

There are three basic types of stepper motors:

1. Variable-reluctance (VR) stepper motors, which have soft-iron rotors
2. Permanent-magnet (PM) stepper motors, which have magnetized rotors
3. Hybrid stepper motors, which have two stacks of rotor teeth forming the two poles of a permanent magnet located along the rotor axis

FIGURE 8.1
A commercial two-stack stepper motor. (Danaher Motion, Rockford, IL. With permission.)

The VR stepper motors and PM stepper motors operate in a somewhat similar manner. Hybrid motors possess characteristics of both VR steppers and PM steppers. A disadvantage of VR stepper motors is that since the rotor is not magnetized, the holding torque is zero when the stator windings are not energized (power off). Hence, there is no capability to hold a mechanical load at a given position under power-off conditions unless mechanical brakes are employed. A photograph of the internal components of a two-stack stepping motor is given in Figure 8.1.

8.1.1 Permanent Magnet Stepper Motor

To explain the operation of a permanent-magnet stepper motor, consider the simple schematic diagram shown in Figure 8.2. The stator has two sets of windings (i.e., two *phases*), placed at 90°. This arrangement has four *salient poles* in the stator, each pole being geometrically separated by a 90° angle from the adjacent one. The rotor is a two-pole permanent magnet. Each phase can take one of the three states 1, 0, and –1, which are defined as follows:

State 1: current in a specified direction
State –1: current in the opposite direction
State 0: no current

NOTE Since –1 is the complement state of 1, in some literature the notation 1' is used to denote the state –1.

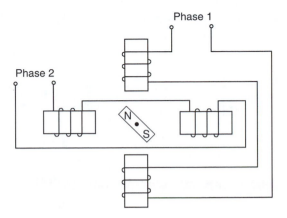

FIGURE 8.2
Schematic diagram of a two-phase permanent magnet stepper motor.

By switching the currents in the two phases in an appropriate sequence, either a clockwise (CW) rotation or a counterclockwise (CCW) rotation can be produced. The CW rotation sequence is shown in Figure 8.3. Note that ϕ_i denotes the state of the ith phase. The *step angle* for this motor is 45°. At the end of each step, the rotor assumes the *minimum reluctance position* that corresponds to the particular magnetic polarity pattern in the stator. (Reluctance measures the magnetic resistance in a flux path). This is a *stable equilibrium configuration* and is known as the *detent position* for that step. When the stator currents (phases) are switched for the next step, the minimum reluctance position changes (rotates by the step angle) and the rotor assumes the corresponding stable equilibrium position; the rotor turns through a single step (45° in this example). Table 8.1 gives the stepping sequences necessary for a complete clockwise rotation. Note that a separate pair of columns is not actually necessary to give the states for the CCW rotation; they are simply given by the CW rotation states themselves, but tracked in the opposite direction (bottom to top).

The switching sequence given in Table 8.1 corresponds to *half-stepping*, with a step angle of 45°. Full-stepping for the stator-rotor arrangement shown in Figure 8.2 corresponds to a step angle of 90°. In this case, only one phase is energized at a time. For half-stepping, both phases have to be energized simultaneously in alternate steps, as is clear from Table 8.1.

Typically, the phase activation (switching) sequence is triggered by the pulses of an input pulse sequence. The switching logic (which determines the states of the phases for a given step) may be digitally generated using appropriate logic circuitry (see Chapter 10) or by a simple table lookup procedure with just eight pairs of entries given in Table 8.1. The clockwise stepping sequence is generated by reading the table in the top-to-bottom direction, and the counterclockwise stepping sequence is generated by reading the same table in the opposite direction. A still more compact representation of switching cycles is also available. Note that in one complete rotation of the rotor, the state of each phase sweeps through one complete cycle of the switching sequence (shown in (Figure 8.4(a)) in the clockwise direction. For clockwise rotation of the motor, the state of phase 2 (ϕ_2) *lags* the state of phase 1 (ϕ_1) by two steps (Figure 8.4(b)). For counterclockwise rotation, ϕ_2 leads ϕ_1 by two steps (Figure 8.4(c)). Hence, instead of eight pairs of numbers, just eight numbers with a "delay" operation would suffice to generate the phase switching logic. Although the commands that generate the switching sequence for a phase winding could be supplied by a microprocessor or a personal computer (a software approach), it is

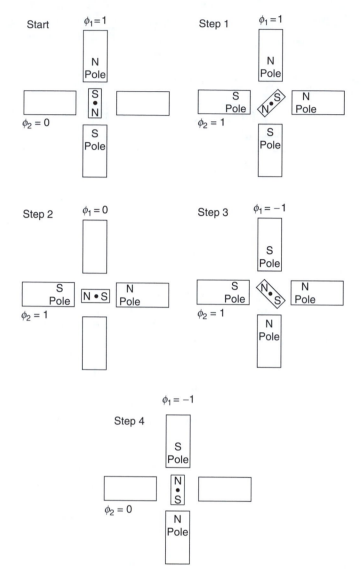

FIGURE 8.3
Stepping sequence (half-stepping) for a two-phase PM stepper motor for clockwise rotation.

customary to generate it through hardware logic in a device called a *translator* or an *indexer*. This approach is more effective because the switching logic for a stepper motor is fixed, as noted in the foregoing discussion. We shall say more about the translator in a later section, when dealing with motor drive electronics.

8.1.2 Variable Reluctance Stepper Motor

Now consider the variable-reluctance (VR) stepper motor shown schematically in Figure 8.5. The rotor is a nonmagnetized soft-iron bar. If only two phases are used in the stator, there will be ambiguity regarding the direction of rotation. At least three phases would be needed for this two-pole rotor geometry, as shown in Figure 8.5. The full-stepping sequence for clockwise rotation is shown in Figure 8.6. The step angle is 60°. Only one phase is energized

TABLE 8.1

Stepping Sequence (half-stepping) for a Two-Phase
PM Stepper Motor with Two Rotor Poles

	Clockwise Rotation	
Step Number	ϕ_1	ϕ_2
1	1	1
2	0	1
3	−1	1
4	−1	0
5	−1	−1
6	0	−1
7	1	−1
8	1	0

CCW
rotation

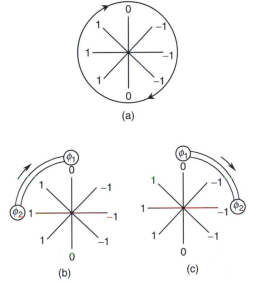

(a)

(b)   (c)

FIGURE 8.4
(a) Half-step switching states; (b) Switching logic for clockwise rotation; (c) Switching logic for counterclockwise rotation.

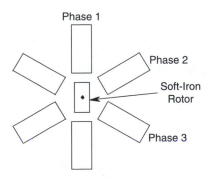

FIGURE 8.5
Schematic diagram of a three-phase variable-reluctance stepper motor.

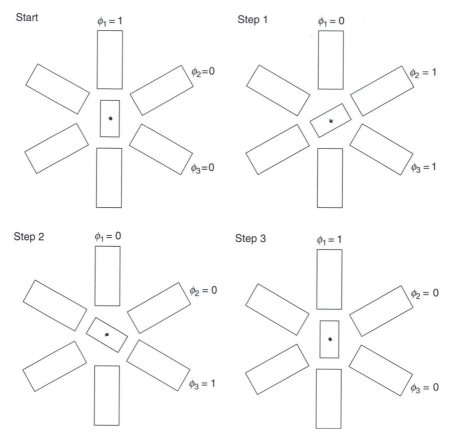

FIGURE 8.6
Full-stepping sequence for the three-phase VR stepper motor (step angle = 60°).

at a time in order to execute full-stepping. With VR stepping motors, the direction of the current (the polarity of a stator pole pair) is not reversed in the full-stepping sequence; only the states 1 and 0 (on and off) are used for each phase. In the case of half-stepping, however, two phases have to be energized simultaneously during some steps. Furthermore, current reversals are needed, thus requiring more elaborate switching circuitry. The advantage, however, is that the step angle has been halved to 30°, thereby providing improved motion resolution. When two phases are activated simultaneously, the minimum reluctance position is halfway between the corresponding pole pairs (i.e., 30° from the detent position that is obtained when only one of the two phases is energized), which enables half-stepping. It follows that, depending on the energizing sequence of the phases, either full-stepping or half-stepping would be possible. As will be discussed later, *microstepping* provides much smaller step angles. This is achieved by changing the phase currents by small increments (rather than on, off, and reversal) so that the detent (equilibrium) position of the rotor shifts in correspondingly small angular increments.

8.1.3 Polarity Reversal

One common feature in any stepper motor is that the stator of the motor contains several pairs of field windings that can be switched on to produce electromagnetic pole pairs (N and S). These pole pairs effectively pull the motor rotor in sequence so as to generate the torque for motor rotation. The polarities of a stator pole may have to be reversed in some types of stepper motors in order to carry out a stepping sequence. The polarity of a stator pole can be reversed in two ways:

1. There is only one set of windings for a group of stator poles (i.e., *unifilar* windings). Polarity of the poles is reversed by reversing the direction of current in the winding.
2. There are two sets of windings for a group of stator poles (i.e., *bifilar* windings). One set of windings, when energized, produces one polarity for this group of poles, and the other set of windings produces the opposite polarity.

Note that the drive circuitry for unifilar (i.e., single "file" or single-coil) windings is somewhat complex because current reversal (i.e., *bipolar*) circuitry is needed. Specifically, a bipolar drive system is needed for a motor with unifilar windings in order to reverse the polarities of the poles (if needed). With bifilar (i.e., double "file" or two- coil) windings, a relatively simpler on/off switching mechanism is adequate for reversing the polarity of a stator pole because one coil gives one polarity and the other coil gives the opposite polarity, and hence current reversal is not required. It follows that a *unipolar* drive system is adequate for a bifilar-wound motor. Of course, a more complex (and costly) bipolar drive system may be used with a bifilar motor as well (but not necessary). Bipolar winding simply means a winding that has the capability to reverse its polarity.

For a given torque rating, as twice the number of windings as in the unifilar-wound case would be required in bifilar-wound motors, at least half the windings being inactive at a given time. This increases the motor size for a given torque rating, and will increase the friction at the bearings thereby reducing the starting torque. Furthermore, since all the copper (windings) of a stator pole is utilized in the unifilar case, the motor torque tends to be higher. At high speeds, current reversal occurs at a higher frequency in unifilar windings. Consquently, the levels of induced voltages by self-induction and mutual induction (back emf) can be significant, resulting in a degradation of the available torque from the motor. For this reason, at high speeds (high stepping rates) the effective (dynamic) torque is typically larger for bifilar stepper motors than for their unifilar counterparts, for the same level of drive power (see Figure 8.7.) At very low stepping rates, however, dissipation (friction) effects will dominate induced-voltage effects, a drawback with bifilar-wound motors. Furthermore, all the copper in a stator pole is

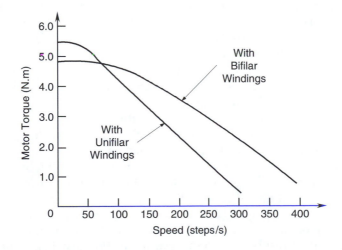

FIGURE 8.7
The effect of bifilar windings on motor torque.

utilized in a unifilar motor. As a result, unifilar windings provide better toque characteristics at low stepping rates, as shown in Figure 8.7.

The motor size can be reduced to some extent by decreasing the wire diameter, which
results in increased resistance for a given length of wire. This decreases the current level
(and torque) for a given voltage, which is a disadvantage. Increased resistance, however,
means decreased electrical time constant (L/R) of the motor, which results in an improved
(fast but less oscillatory) single-step response.

8.2 Stepper Motor Classification

Since any actuator that generates stepwise motion can be considered a stepper motor, it
is difficult to classify all such devices into a small number of useful categories. For example,
toothed devices such as harmonic drives (a class of flexible-gear drives—see Chapter 3)
and pawl-and-ratchet-wheel drives, which produce intermittent motions through purely
mechanical means are also classified as stepper motors. Such mechanical devices are in
fact motion transmision/conversion mechanisms (chapter 3) rather than actuators in the
present sensess. Of primary interest in today's mechatronic applications, however, are
actuators that generate stepwise motion directly by electromagnetic forces in response to
pulse (or digital) inputs. Even for these *electromagnetic incremental actuators*, however, no
standardized classification is available.

Most classifications of stepper motors are based on the nature of the motor rotor. One such
classification considers the magnetic character of the rotor. Specifically, a variable-reluctance
(VR) stepper motor has a soft-iron rotor while a permanent-magnet (PM) stepper motor has
a magnetized rotor. The two types of motors operate in a somewhat similar manner. Specifically the stator magnetic field (polarity) is stepped so as to change the minimum reluctance
(or detent) position of the rotor in increments. Hence both types of motors undergo similar
changes in reluctance (magnetic resistance) during operation. A disadvantage of VR stepper
motors is that since the rotor is not magnetized, the holding torque is zero when the stator
windings are not energized (power off). Hence, it has no capacity to hold the load at a given
position under power-off conditions unless mechanical brakes are employed. A hybrid stepper motor possesses characteristics of both VR steppers and PM steppers. The rotor of a
hybrid stepper motor consists of two rotor segments connected by a shaft. Each rotor segment
is a toothed wheel and is called a *stack*. The two rotor stacks form the two poles of a permanent
magnet located along the rotor axis. Hence an entire stack of rotor teeth is magnetized to be
a single pole (which is different from the case of a PM stepper where the rotor has multiple
poles). The rotor polarity of a hybrid stepper can be provided either by a permanent magnet,
or by an electromagnet using a coil activated by a unidirectional dc source and placed on the
stator to generate a magnetic field along the rotor axis.

Another practical classification that is used in this book is based on the number of
"stacks" of teeth (or rotor segments) present on the rotor shaft. In particular, a hybrid
stepper motor has two stacks of teeth. Further subclassifications are possible, depending
on the tooth pitch (angle between adjacent teeth) of the stator and tooth pitch of the rotor.
In a *single-stack stepper motor*, the rotor tooth pitch and the stator tooth pitch generally
have to be unequal so that not all teeth in the stator are ever aligned with the rotor teeth
at any instant. It is the misaligned teeth that exert the magnetic pull, generating the driving
torque. In each motion increment, the rotor turns to the minimum reluctance (stable
equilibrium) position corresponding to that particular polarity distribution of the stator.

In *multiple-stack stepper motors*, operation is possible even when the rotor tooth pitch is
equal to the stator tooth pitch, provided that at least one stack of rotor teeth is rotationally

shifted (misaligned) from the other stacks by a fraction of the rotor tooth pitch. In this design, it is this *inter-stack misalignment* that generates the drive torque for each motion step. It should be obvious that unequal-pitch multiple stack steppers are also a practical possibility. In this design, each rotor stack operates as a separate single-stack stepper motor. The stepper motor classifications described thus far are summarized in Figure 8.8.

Next we will describe some geometric, mechanical, design, and operational aspects of single-stack and multiple-stack stepper motors. One point to remember is that some form of geometric misalignment of teeth is necessary in both types of motors. A motion step is obtained by simply redistributing (i.e., switching) the polarities of the stator, thereby changing the minimum reluctance detent position of the motor. Once a stable equilibrium position is reached by the rotor, the stator polarities are switched again to produce a new detent position, and so on. In descriptive examples, it is more convenient to use variable-reluctance stepper motors. However, the principles can be extended in a straightforward manner to cover permanent-magnet stepper motors as well.

8.2.1 Single-Stack Stepper Motors

To establish somewhat general geometric relationships for a single-stack variable-reluctance stepper motor, consider Figure 8.9. The motor has three phases of winding ($p = 3$) in the

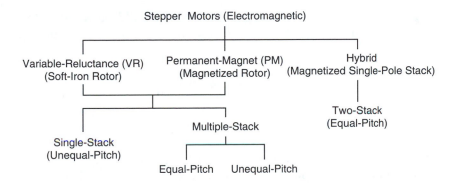

FIGURE 8.8
Classifications of stepper motors.

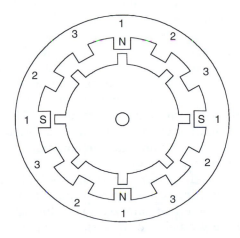

FIGURE 8.9
Three-phase single-stack VR stepper motor with twelve stator poles (teeth) and eight rotor teeth.

stator, and there are eight teeth in the soft-iron rotor ($n_r = 8$). The three phases are numbered 1, 2, and 3. Note that each phase represents a group of 4 stator poles wound together, and the total number of stator poles (n_s) is 12. When Phase 1 is energized, one pair of diametrically opposite poles becomes N (north) poles and the other pair in that phase (located at 90° from the first pair) becomes S (south) poles. Furthermore, a geometrically orthogonal set of four teeth on the rotor will align themselves perfectly with these four stator poles. This is the minimum reluctance, stable equilibrium configuration (detent position) for the rotor under the given activation state of the stator (i.e., Phase 1 is on and the other two phases are off). Observe, however, that there is a misalignment of 15° between the remaining rotor teeth and the nearest stator poles.

If the *pitch angle*, defined as the angle between two adjacent teeth, is denoted by θ (in degrees) and the number of teeth is denoted by n, we have

$$\text{Stator pitch} \quad \theta_s = \frac{360°}{n_s}$$

$$\text{Rotor pitch} \quad \theta_r = \frac{360°}{n_r}$$

For one-phase-on excitation, the step angle $\Delta\theta$, which should be equal to the smallest misalignment between a stator pole and an adjacent rotor tooth, in any stable equilibrium state, is given by

$$\Delta\theta = \theta_r - r\theta_s \quad (\text{for } \theta_r > \theta_s) \tag{8.1a}$$

$$\Delta\theta = \theta_s - r\theta_r \quad (\text{for } \theta_r < \theta_s) \tag{8.1b}$$

where r is the largest positive integer such that $\Delta\theta$ is positive (i.e., the largest feasible r such that a misalignment in rotor and stator teeth occurs). It is clear that for the arrangement shown in Figure 8.9, $\theta_r = 360°/8 = 45°$, $\theta_s = 360°/12 = 30°$, and hence, $\Delta\theta = 45° - 30° = 15°$, as stated earlier.

Now, if Phase 1 is turned off and Phase 2 is turned on, the rotor will turn 15° in the counterclockwise (CCW) direction to its new minimum reluctance position. If Phase 3 is energized instead of Phase 2, the rotor would turn 15° clockwise (CW). It should be clear that half this step size (7.5°) is also possible with the motor shown in Figure 8.9. Suppose, for example, that Phase 1 is on, as before. Next suppose that Phase 2 is energized while Phase 1 is on, so that two like poles are in adjacent locations. Since the equivalent field of the two adjoining like poles is halfway between the two poles, two rotor teeth will orient symmetrically about this pair of poles, which is the corresponding minimum reluctance position. It is clear that this corresponds to a rotation of 7.5° from the previous detent position, in the CCW direction. For executing the next half-step (in the CCW direction), Phase 1 is turned off while Phase 2 is on. Thus, in summary, the full-stepping sequence for CCW rotation is 1-2-3-1; for CW rotation, it is 1-3-2-1. The half-stepping sequence for CCW rotation is 1-12-2-23-3-31-1; for CW rotation, it is 1-31-3-23-2-12-1. Note that the switching sequence is simply reversed for reversing the direction of rotation.

Returning to full-stepping, note that since each switching of phases corresponds to a rotation of $\Delta\theta$ and there are p number of phases, the angle of rotation for a complete switching cycle of p switches is $p \cdot \Delta\theta$. In a switching cycle, the stator polarity distribution returns to the distribution that it had in the beginning. Hence, in one switching cycle (*p* switches), the rotor should assume a configuration exactly like what it had in the beginning of the cycle. That is, the rotor should turn through a complete pitch angle of θ_r. Hence, the following relationship holds for the one-phase-on case.

$$\theta_r = p \cdot \Delta\theta \tag{8.2}$$

Substituting this in Equation 8.1a, we have

$$\theta_r = r\theta_s + \frac{\theta_r}{p} \quad (\text{for } \theta_r > \theta_s) \tag{8.3a}$$

and similarly with Equation 8.1b we have,

$$\theta_s = r\theta_r + \frac{\theta_r}{p} \quad (\text{for } \theta_r < \theta_s) \tag{8.3b}$$

where

θ_r = rotor tooth pitch angle
θ_s = stator tooth pitch angle
p = number of phases in the stator
r = largest feasible positive integer

Now, by definition of the pitch angle, Equation 8.3a gives

$$\frac{360°}{n_r} = \frac{r \times 360°}{n_s} + \frac{360°}{p \cdot n_r}$$

or

$$n_s = rn_r + \frac{n_s}{p} \quad (\text{for } n_s > n_r) \tag{8.4a}$$

and similarly from Equation 8.3b

$$n_r = rn_s + \frac{n_s}{p} \quad (\text{for } n_s < n_r) \tag{8.4b}$$

where

n_r = number of rotor teeth
n_s = number of stator teeth
r = largest feasible positive integer

Finally, the number of steps per revolution is

$$n = \frac{360°}{\Delta\theta} \tag{8.5}$$

Example 8.1

Consider the stepper motor shown in Figure 8.9. The number of stator poles $n_s = 12$ and the number of phases $p = 3$. Assuming that $n_r < n_s$ (which is the case in Figure 8.9), substitute in Equation 8.4a:

$$12 = rn_r + \frac{12}{3}$$

or

$$rn_r = 8 \tag{i}$$

Now, $r = 1$ gives $n_r = 8$. This is the feasible case shown in Figure 8.9. Then the rotor pitch $\theta_r = 360°/8 = 45°$, and the step angle can be calculated from Equation 8.2 as

$$\Delta\theta = \frac{45°}{3} = 15°$$

Note that this is the full-step angle, as observed earlier. Furthermore, the stator pitch is $\theta_s = 360°/12 = 30°$, which further confirms that the step angle is $\theta_r - \theta_s = 45° - 30° = 15°$.

NOTE   In the result Equation i above, $r = 2$ and $r = 4$ are not feasible solutions because, then all the rotor teeth will be fully aligned with stator poles, and there will not be a misalignment to facilitate stepping.

Example 8.2

Consider the motor arrangement shown in Figure 8.5. Here $\theta_r = 180°$ and $\theta_s = 360°/6 = 60°$. Now from Equation 8.1a, $\Delta\theta = 180° - r \times 60°$. The largest feasible r in this case is 2, which gives $\Delta\theta = 180° - 2 \times 60° = 60°$. Furthermore, from Equation 8.2 we have $\Delta\theta = 180°/3 = 60°$.

Example 8.3

Consider a stepper motor with $n_r = 5$, $n_s = 2$, and $p = 2$. A schematic representation is given in Figure 8.10. In this case, $\theta = 360°/5 = 72°$ and $\theta_s = 360°/2 = 180°$. From Equation 8.1b,

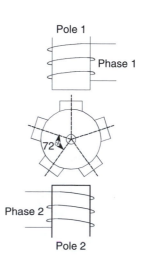

FIGURE 8.10
A two-phase two-pole stepper with a
five-tooth rotor.

$\Delta\theta = 180° - r \times 72°$. Here, the largest feasible value for r is 2, which corresponds to a step angle of $\Delta\theta = 180° - 2 \times 72° = 36°$. This is further confirmed by Equation 8.2, which gives $\Delta\theta = 72°/2 = 36°$.

Note that this particular arrangement is feasible for a PM stepper but not for a VR stepper. The reason is simple. In a detent position (equilibrium position), a rotor tooth will align itself with the stator pole (say, phase 1) that is on, and two other rotor teeth will orient symmetrically between the remaining stator pole (phase 2), which is off. Next, when phase 1 is turned off and phase 2 is turned on (for executing a full step), unless the two rotor teeth that are symmetric with phase 2 have opposite polarities, they will be exerted an equal force by this pole, trying to move the rotor in opposite directions, thereby forming an unstable equilibrium position (or, ambiguity in the direction of rotation).

8.2.2 Toothed-Pole Construction

The foregoing analysis indicates that the step angle can be reduced by increasing the number of poles in the stator and the number of teeth in the rotor. Obviously, there are practical limitations to the number of poles (windings) that can be incorporated in a stepper motor. A common solution to this problem is to use "toothed" poles in the stator, as shown in Figure 8.11(a). The toothed construction of the stator and the rotor not only improves the motion resolution (step angle) but also enhances the concentration of the magnetic field, which generates the motor torque. Also, the torque and motion characteristics become smother (smaller ripples and less jitter) as a result of the distributed tooth construction.

In the case shown in Figure 8.11(a), the stator teeth are equally spaced but their pitch (angular spacing) is not identical to the pitch of the rotor teeth. In the toothed-stator construction, n_s represents the number of teeth rather than the number of poles in the stator. The number of rotor teeth has to be increased in proportion. Note that in full-stepping (e.g., one-phase-on) after p number of switchings (steps), where p is the number of phases, the adjacent rotor tooth will take the previous position of a particular rotor tooth. It follows that the rotor rotates through θ_r (the tooth pitch of the rotor) in p steps. Thus, the relationship $\Delta\theta = \theta_r/p$ (Equation 8.2) still holds. But Equation 8.1 has to be modified to accommodate toothed poles. Toothed-stator construction can provide very small step angles—0.72° for example, or more commonly, 1.8°.

The equations for the step angle given so far in this chapter assume that the number of stator poles is identical to the number of stator teeth. In particular, Equation 8.1, Equation 8.3, and Equation 8.4 are obtained using this assumption. These equations have to be modified when there are many teeth on each stator pole. Generalization of the step angle equations for the case of toothed-pole construction can be made by referring to Figure 8.11(b). What is shown are the center tooth of each pole and the rotor tooth closest to that stator tooth. This is one possible geometry for a single-stack toothed construction. In this case, the rotor tooth pitch θ_r is not equal to the stator tooth pitch θ_s. Another possibility for a single-stack toothed construction (which is perhaps preferable from the practical point of view) will be described later. There, θ_r and θ_s are identical, but when all the stator teeth that are wound to one of the phases are aligned with the rotor teeth, all the stator teeth wound to another phase will have a "constant" misalignment with the rotor teeth in their immediate neighborhood. Unlike that case, in the construction used in the present case we have $\theta_r \neq \theta_s$; hence, only one tooth in a stator pole can be completely aligned with a rotor tooth.

Consider the case of $\theta_r > \theta_s$ (i.e., $n_r < n_s$). Since $(\theta_r - \theta_s)$ is the offset between the rotor tooth pitch and the stator tooth pitch, and since there are $n_s/(mp)$ rotor teeth in the sector

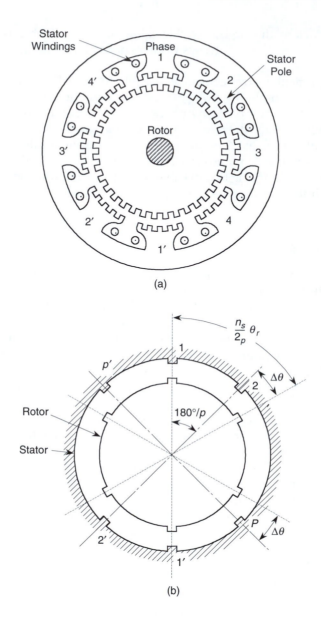

FIGURE 8.11
A possible toothed-pole construction for a stepper motor: (a) An eight-pole, four-phase motor; (b) Schematic diagram for generalizing the step angle equation.

made by two adjacent stator poles, we see that the total tooth offset (step angle) $\Delta\theta$ at the second pole is given by

$$\Delta\theta = \frac{n_s}{mp}(\theta_r - \theta_s) \quad \text{for} \quad \theta_r > \theta_s \tag{8.6}$$

where
   p = number of phases
   m = number of stator poles per phase

Now, noting that $\Delta\theta = \theta_r/p$ is true even for the toothed construction, and by substituting this in Equation 8.6 we get

$$n_s = n_r + m \quad \text{for} \quad n_r < n_s \tag{8.7}$$

or, in general (including the case $n_r > n_s$) we have

$$n_s = n_r \pm m \tag{8.8}$$

A further generalization is possible if $\theta_r > 2\theta_s$ or $\theta_r < 2\theta_s$, as in the non-toothed case, by introducing an integer r.

Also, we recall that when the stator teeth are interpreted as stator poles, $n_s = mp$, and then Equation 8.6 reduces to Equation 8.1a, as expected, for $r = 1$. For the toothed-pole construction, n_s is several times the value of mp. In this case, Equation 8.6 should be used in place of Equation 8.1a. Note that, in general, p has to be replaced by n_s/m in converting an equation for a nontoothed-pole construction to the corresponding equation for a toothed-pole construction. For example, then, Equation 8.4a becomes Equation 8.7, with $r = 1$.

Finally, we observe from Figure 8.11(b) that the switching sequence 1-2-3-,..., -p produces counterclockwise rotations, and the switching sequence 1-p-(p-1)-,..., -2 produces clockwise rotations.

Example 8.4

Consider a simple design example for a single-stack VR stepper. Suppose that the number of steps per revolution, which is a functional requirement, is specified as $n = 200$. This corresponds to a step angle of $\Delta\theta = 360°/200 = 1.8°$. Assume full-stepping. Design restrictions, such as size and the number of poles in the stator, govern t_s, the number of stator teeth per pole. Let us use the typical value of six teeth per pole. Also assume that there are two poles wound to the same stator phase. We are interested in designing a motor to meet these requirements.

SOLUTION

First, we will derive some useful relationships. Suppose that there are m poles per phase. Hence, there are mp poles in the stator. (Note: $n_s = mpt_s$). Then, assuming $n_r < n_s$, Equation 8.7: $n_s = n_r + m$ would apply. Dividing this equation by mp, we get

$$t_s = \frac{n_r}{mp} + \frac{1}{p} \tag{i}$$

Now

$$n_r = \frac{360°}{\theta_r} = \frac{360°}{p \cdot \Delta\theta} \quad \text{(from equation (8.2))}$$

or

$$n_r = \frac{n}{p} \tag{8.9}$$

Substituting this in Equation i we get

$$t_s = \frac{n}{mp^2} + \frac{1}{p} \tag{8.10a}$$

Now, since $1/p$ is less than 1 for a stepper motor and t_s is greater than 1 for the toothed-pole construction, an approximation for Equation 8.10a can be given by

$$t_s = \frac{n}{mp^2} \tag{8.10}$$

where
  t_s = number of teeth per stator pole
  m = number of stator poles per phase
  p = number of phases
  n = number of steps per revolution

In the present example, $t_s \approx 6$, $m = 2$, and $n = 200$. Hence, from Equation 8.10, we have

$$6 \sim \frac{200}{2 \times p^2}$$

which gives $p \approx 4$. Note that p has to be an integer. Now, using Equation 8.10a, we get two possible designs for $p = 4$. First, with the specified values $n = 200$ and $m = 2$, we get $t_s = 6.5$, which is slightly larger than the required value of 6. Alternatively, with the specified $t_s = 6$ and $m = 2$, we get $n = 184$, which is slightly smaller than the specified value of 200. Either of these two designs would be acceptable. The second design gives a slightly larger step angle. (Note that $\Delta\theta = 360°/n = 1.96°$ for the second design and $\Delta\theta = 360°/200 = 1.8°$ for the first design.) Summarizing the two designs, we have the following results:

For Design 1:

Number of phases $p = 4$
Number of stator poles = 8
Number of teeth per pole = 6.5
Number of steps per revolution = 200
Step angle = 1.8°
Number of rotor teeth = 50 (from Equation 8.9)
Number of stator teeth = 52

For Design 2:

Number of phases $p = 4$
Number of stator poles = 8
Number of teeth per pole = 6
Number of steps per revolution = 184
Step angle = 1.96°
Number of rotor teeth = 46 (from Equation 8.9)
Number of stator teeth = 48

NOTE The number of teeth per stator pole (t_s) does not have to be an integer (see Design 1). Since there are interpolar gaps around the stator, it is possible to construct a motor with an integer number of actual stator teeth, even when t_s and n_s are not integers.

8.2.3 Another Toothed Construction

In the single-stack toothed construction just presented, we have $\theta_r \neq \theta_s$. An alternative design possibility exists where $\theta_r = \theta_s$, but in this case the stator poles are located around the rotor such that when the stator teeth corresponding to one of the phases are fully aligned with the rotor teeth, the stator teeth in another phase will have a constant offset with the neighboring rotor teeth, thereby providing the misalignment that is needed for stepping. The torque magnitude of this construction is perhaps better because of this uniform tooth offset per phase, but torque ripples (jitter) would also be stronger (a disadvantage) because of sudden and more prominent changes in magnetic reluctance from pole to pole, during phase switching.

To obtain some relations that govern this construction, suppose that Figure 8.11(a) represents a stepper motor of this type. When the stator teeth in Pole 1 (and Pole 1') are perfectly aligned with the rotor teeth, the stator teeth in Pole 2 (and Pole 2') will have an offset of $\Delta\theta$ with the neighboring rotor teeth. This offset is in fact the step angle, in full stepping. This offset can be either in the counterclockwise direction (as in Figure 8.11(a)) or in the clockwise direction. Since the pole pitch is given by $360°/pm$, we must have

$$\frac{1}{\theta_r}\left[\frac{360°}{pm} \pm \Delta\theta\right] = r \qquad (8.11)$$

where r is the integer number of rotor teeth contained within the angular sector $360°/(pm) \pm \Delta\theta$. Also

$\Delta\theta$ = step angle (full-stepping)
θ_r = rotor tooth pitch
p = number of phases
m = number of stator poles per phase

It should be clear that within two consecutive poles wound to the same phase, there are n_r/m rotor teeth.

Since p switchings of magnitude $\Delta\theta$ each will result in a total rotation of θ_r (i.e., Equation 8.2, by substituting this along with $n_r = \frac{360°}{\theta_r}$ in Equation 8.11, and simplifying, we get

$$n_r \pm m = pmr \qquad (8.12)$$

where n_r is the number of rotor teeth.

Example 8.5

Consider the full-stepping operation of a single-stack, equal pitch stepper motor whose design is governed by Equation 8.12. Discuss the possibility of constructing a four-phase motor of this type that has fifty rotor teeth (i.e., step angle = 1.8°). Obtain a suitable design for a four-phase motor that uses eight stator poles. Specifically, determine the number of rotor teeth (n_r), the step angle $\Delta\theta$, the number of steps per revolution (n), and the number of teeth per stator pole (t_s).

SOLUTION
First, with $n_r = 50$ and $p = 4$, Equation 8.12 becomes

$$50 \pm m = 4mr$$

or

$$m = \frac{50}{(4r \mp 1)} \tag{i}$$

Note that m and r should be natural numbers (i.e., positive integers). Since the smallest such value for r is 1, we see from Equation i that the largest value for m is 16. It can be easily verified that only two solutions are valid in this range; namely, $r = 1$ and $m = 10$; $r = 6$ and $m = 2$, both corresponding to the + sign in the denominator of Equation i. The first solution is not very practical. Notably, in this case the number of poles $= 10 \times 4 = 40$, and hence the pole pitch is $360°/40 = 9°$. Since each stator pole will occupy nearly this angle, it cannot have more than one tooth of pitch 7.2° (the rotor tooth pitch $= 360°/50 = 7.2°$). The second solution is more practical. In this case, the number of poles $= 2 \times 4 = 8$ and the pole pitch is $360°/8 = 45°$. Each stator pole will occupy nearly this angle, and with a tooth of pitch 7.2°, a pole can have 6 full teeth.

Next, consider $p = 4$ and $m = 2$ (i.e., a four-phase motor with eight stator poles, as specified in the example). Then Equation 8.12 becomes

$$n_r = 8r \pm 2 \tag{ii}$$

Hence, one possible design that is close to the previously mentioned case of $n_r = 50$ is realized with $r = 6$ (giving $n_r = 50$, for the + sign in Equation ii) and $r = 7$ (giving $n_r = 54$, for the − sign in Equation ii). Consider the latter case, where $n_r = 54$. The corresponding tooth pitch (for both rotor and stator) is

$$\theta_r = \theta_s = \frac{360°}{54} = 6.67°$$

The step angle (for full-stepping) is

$$\Delta\theta = \frac{\theta_r}{p} = \frac{6.67°}{4} = 1.67°$$

The number of steps per revolution is

$$n = \frac{360°}{\Delta\theta} = pn_r = 4 \times 54 = 216$$

$$\text{Pole pitch} = \frac{360°}{mp} = \frac{360°}{8} = 45°$$

The maximum number of teeth that could be occupied within a pole pitch is

$$\frac{45°}{\theta_s} = \frac{45°}{6.67°} = 6.75$$

Hence, the maximum possible number of full teeth per pole is

$$t_s = 6$$

In a practical motor there can be an interpolar gap of nearly half the pole angle. In that case, a suitable number for t_s would be the integer value of

$$\frac{1}{\theta_s} \frac{360°}{(8+4)}$$

This gives

$$t_s = 4$$

Summarizing, we have the following design parameters:

Number of phases $p = 4$

Number of stator poles $mp = 8$

Number of teeth per pole, t_s = maximum 6 (typically 4)

Number of steps per revolution (full-stepping) $n = 216$

Step angle $\Delta\theta = 1.67°$

Number of rotor teeth $n_r = 54$

Tooth pitch (both rotor and stator) $\approx 6.67°$

8.2.4 Microstepping

We have seen how full-stepping or half-stepping can be achieved simply by using an appropriate switching scheme. For example, half-stepping occurs when phase switchings alternate between one-phase-on and two-phase-on states. Full-stepping occurs when either one-phase-on switching or two-phase-on switching is used exclusively for every step. In both these cases, the current level (or, state) of a phase is either 0 (off) or 1 (on). Rather than using two current levels (the binary case), it is possible to use several levels of phase current between these two extremes, thereby achieving much smaller step angles. This is the principle behind microstepping.

Microstepping is achieved by properly changing the phase currents in small steps instead of switching them on and off (as in the case of full-stepping and half-stepping). The principle behind this can be understood by considering two identical stator poles (wound with identical windings), as shown in Figure 8.12. When the currents through the windings are identical (in magnitude and direction), the resultant magnetic field will lie symmetrically between the two poles. If the current in one pole is decreased while the other current is kept unchanged, the resultant magnetic field will move closer to the pole with the larger current. Since the detent position (equilibrium position) depends on the position of the resultant magnetic field, it follows that very small step angles can be achieved simply by controlling (varying the relative magnitudes and directions of) the phase currents.

Step angles of 1/125 of a full step or smaller could be obtained through microstepping. For example, 10,000 steps/revolution may be achieved. Note that the step size in a sequence of microsteps is not identical. This is because stepping is done through microsteps of the phase current, which (and the magnetic field generated by it) has a nonlinear relation with the step angle.

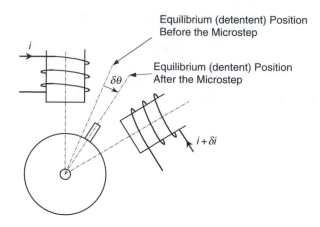

FIGURE 8.12
The principle of microstepping.

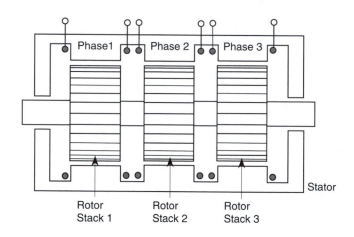

FIGURE 8.13
Longitudinal view of a three-stack (three-phase) stepper motor.

Motor drive units with the microstepping capability are more costly, but microstepping provides the advantages of accurate motion capabilities, including finer resolution, overshoot suppression, and smoother operation (reduced jitter and less noise) even in the neighborhood of a resonance in the motor-load combination. A disadvantage is that, usually there is a reduction in the motor torque as a result of microstepping.

8.2.5 Multiple-Stack Stepper Motors

For illustration purposes, consider the longitudinal view of a three-stack stepper motor shown schematically in Figure 8.13. In this example, there are three identical stacks of teeth mounted on the same rotor shaft. There is a separate stator segment surrounding each rotor stack. One straightforward approach to designing a multiple-stack stepper motor would be to treat it as a cascaded set of identical single-stack steppers with common phase windings for all the stator segments. Then the number of phases of the motor is fixed regardless of the number of stacks used. Such a design is simply a single-stack stepper with a longer rotor and a correspondingly longer stator, thereby generating a higher torque (proportional to the length of the motor, for a given winding density and a phase current). What is considered here is not such a trivial design, but somewhat complex designs where the phase windings of a stack can operate (on, off, reversal) independently of another stack.

Both equal-pitch construction $(\theta_r = \theta_s)$ and unequal-pitch construction $(\theta_r > \theta_s,$ or $\theta_r < \theta_s)$ are possible in multiple-stack steppers. An advantage of the unequal-pitch construction is that smaller step angles are possible than with an equal-pitch construction of the same size (i.e., same diameter and number of stacks). But the switching sequence is somewhat more complex for unequal-pitch, multiple-stack stepper motors. In particular, each stack has more than one phase and they can operate independently of the phases of another stack. First, we will examine the equal-pitch, multiple-stack construction. The operation of an unequal-pitch, multiple-stack motor should follow directly from the analysis of the single-stack case as given before. Subsequently, a hybrid stepper will be described. It has a two-stack rotor, but an entire stack is magnetized with a single polarity and the two stacks have opposite polarities.

8.2.5.1   Equal-Pitch Multiple-Stack Stepper

For each rotor stack, there is a toothed stator segment around it whose pitch angle is identical to that of the rotor $(\theta_s = \theta_r)$. A stator segment may appear to be similar to that of an equal-pitch single-stack stepper (discussed previously), but this is not the case. Each stator segment is wound to a single phase, thus being energized (polarized) or de-energized (depolarized) simultaneously. It follows that, in the equal pitch case,

$$p = s \tag{8.13}$$

where
p = number of phases and s = number of rotor stacks

The misalignment that is necessary to produce the motor torque may be introduced in one of two ways:

1. The teeth in the stator segments are perfectly aligned, but the teeth in the s rotor stacks are misaligned consecutively by $1/s \times$ pitch angle.
2. The teeth in the rotor stacks are perfectly aligned, but the teeth in the stator segments are misaligned consecutively by $1/s \times$ pitch angle.

Now consider the three-stack case. Suppose that phase 1 is energized. Then the teeth in the rotor stack 1 will align perfectly with the stator teeth in phase 1 (segment 1). But the teeth in the rotor stack 2 will be shifted from the stator teeth in phase 2 (segment 2) by a one-third-pitch angle in one direction, and the teeth in rotor stack 3 will be shifted from the stator teeth in phase 3 (segment 3) by a two-thirds-pitch angle in the same direction (or a one-third-pitch angle in the opposite direction). It follows that if phase 1 is now de-energized and phase 2 is energized, the rotor will turn through one-third pitch in one direction. If, instead, phase 3 is turned on after phase 1, the rotor will turn through one-third pitch in the opposite direction. Clearly, the step angle (for full-stepping) is a one-third-pitch angle for the three-stack, three-phase construction. The switching sequence 1-2-3-1 will turn the rotor in one direction, and the switching sequence 1-3-2-1 will turn the rotor in the opposite direction.

In general, for a stepper motor with s stacks of teeth on the rotor shaft, the full-stepping step angle is given by

$$\Delta\theta = \frac{\theta}{s} \qquad (8.14a)$$

where, $\theta = \theta_r = \theta_s =$ tooth pitch angle. In view of Equation 8.13 we have

$$\Delta\theta = \frac{\theta}{p} \qquad (8.14b)$$

Note that the step angle can be decreased by increasing the number of stacks of rotor teeth. Increased number of stacks also means more phase windings with associated increase in the magnetic field and the motor torque. However, the length of the motor shaft increases with the number of stacks, and can result in flexural (shaft bending) vibration problems (particularly whirling of the shaft), air gap contact problems, large bearing loads, wear and tear, and increased noise. As in the case of a single-stack stepper, half-stepping can be accomplished by energizing two phases at a time. Hence, in the three-stack stepper, for one direction, the half-stepping sequence is 1-12-2-23-3-31-1; in the opposite direction, it is 1-13-3-32-2-21-1.

8.2.5.2   Unequal-Pitch Multiple-Stack Stepper

Unequal-pitch, multiple-stack stepper motors are also of practical interest. Very fine angular resolutions (step angles) can be achieved by this design without compromising the length of the motor. In an unequal-pitch stepper motor, each stator segment has more than one phase (p number of phases), just like in a single-stack unequal-pitch stepper. Rather than a simple cascading, however, the phases of different stacks are not wound together and can be switched on and off independently. In this manner yet finer small angles are realized, together with the added benefit of increased torque provided by the multi-stack design.

For a single stack nontoothed-pole stepper, we have seen that the step angle is equal to $\theta_r - \theta_s$. In a multi-stack stepper, this misalignment is further subdivided into s equal steps using the interstack misalignment. Hence, the overall step angle for an unequal-pitch, multiple-stack stepper motor with nontoothed poles is given by

$$\Delta\theta = \frac{\theta_r - \theta_s}{s} \qquad \text{(for } \theta_r > \theta_s\text{)} \qquad (8.15)$$

For a toothed-pole multiple-stack stepper motor, we have

$$\Delta\theta = \frac{n_s(\theta_r - \theta_s)}{mps} \qquad (8.16)$$

where m is the number of stator poles per phase. Alternatively, using Equation 8.2, we have

$$\Delta\theta = \frac{\theta_r}{p \cdot s} \qquad (8.17)$$

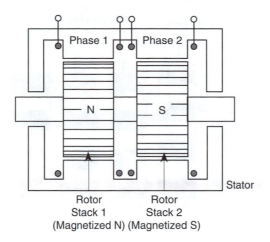

FIGURE 8.14
A hybrid stepper motor.

for both toothed-pole and nontoothed-pole motors, where

p = number of phases in each stator segment
s = number of rotor stacks

8.2.6 Hybrid Stepper Motor

Hybrid steppers are arguably the most common variety of stepping motors in engineering applications. A hybrid stepper motor has two stacks of rotor teeth on its shaft. The two rotor stacks are magnetized to have opposite polarities, as shown in Figure 8.14. There are two stator segments surrounding the two rotor stacks. Both rotor and stator have teeth and their pitch angles are equal. Each stator segment is wound to a single phase, and accordingly, the number of phases is two. It follows that a hybrid stepper is similar in mechanical design and stator winding to a multi-stack, equal-pitch, VR stepper. There are some dissimilarities, however. First, the rotor stacks are magnetized. Second, the inter-stack misalignment is 1/4 of a tooth pitch (see Figure 8.15).

A full cycle of the switching sequence for the two phases is given by [0 1], [−1 0], [0 −1], [1 0], [0 1] for one direction of rotation. In fact, this sequence produces a downward movement (CW rotation, looking from the left end) in the arrangement shown in Figure 8.15, starting from the state of [0 1] shown in the figure (phase 1 off and phase 2 on with N polarity). For the opposite direction, the sequence is simply reversed; thus, [0 1], [1 0], [0 −1], [−1 0], [0 1]. Clearly, the step angle is given by

$$\Delta\theta = \frac{\theta}{4} \tag{8.18}$$

where, $\theta = \theta_r = \theta_s$ = tooth pitch angle.

Just like in the case of a PM stepper motor, a hybrid stepper has the advantage providing a holding torque (detent torque) even under power-off conditions. Furthermore, a hybrid stepper can provide very small step angles, high stepping rates, and generally good torque-speed characteristics.

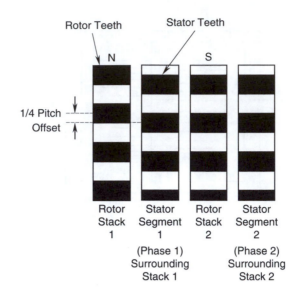

FIGURE 8.15
Rotor stack misalignment (1/4 pitch) in a hybrid stepper motor (Schematically shows the state where phase 1 is off and phase 2 is on with N polarity).

Example 8.6

The half-stepping sequence for the motor represented in Figure 8.14 and Figure 8.15 may be determined quite conveniently. Starting from the state [0, 1] as before, if phase 1 is turned on to state "−1" without turning off phase 2, then phase 1 will oppose the pull of phase 2, resulting in a detent position halfway between the full stepping detent position. Next if phase 2 is turned off while keeping phase 1 in "−1," the remaining half step of the original full step will be completed. In this manner, the half-stepping sequence for CW rotation is obtained as: [0, 1], [−1, 1], [−1, 0], [−1, −1], [0, −1], [1, −1], [1, 0], [1, 1], [0, 1]. For CCW rotation, this sequence is simply reversed. Note that, as expected, in half-stepping, both phases remain on during every other half step.

8.3 Driver and Controller

In principle, the stepper motor is an open-loop actuator. In its normal operating mode, the stepwise rotation of the motor is synchronized with the command pulse train. This justifies the term *digital synchronous motor,* which is sometimes used to denote the stepper motor. As a result of stepwise (incremental) synchronous operation, open-loop operation is adequate, at least in theory. An exception to this may result under highly transient conditions near rated torque and excessive external loading such as frictional torques when "pulse missing" can be a problem. We will address this situation in a later section.

A stepper needs a "control computer" or at least a hardware "indexer" to generate the pulse commands and a "driver" to interpret the commands and correspondingly generate the proper currents for the phase windings of the motor. This basic arrangement is shown in Figure 8.16(a). For feedback control, the response of the motor has to be sensed (say, using an optical encoder) and fed back into the controller (see the dotted line in

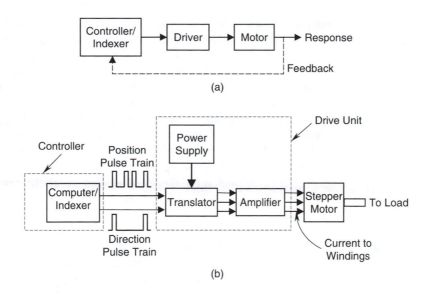

FIGURE 8.16
(a) The basic control system of a stepper motor; (b) The basic components of a driver.

Figure 8.16(a)) for taking the necessary corrective action to the pulse command, when an error is present. We will return to the subject of control later. The basic components of the driver for a stepper motor are identified in Figure 8.16(b). It consists of a logic circuit called "translator" to interpret the command pulses and switch the appropriate analog circuits to generate the phase currents. Since sufficiently high current levels are needed for the phase windings, depending on the motor capacity, the drive system includes amplifiers powered by a power supply.

The command pulses are generated either by a control computer (a desktop computer or a microprocessor)—the software approach, or by a variable-frequency oscillator (or, an indexer)—the basic hardware approach. For bidirectional motion, two pulse trains are necessary—the position-pulse train and the direction-pulse train, which are determined by the required motion trajectory. The position pulses identify the exact times at which angular steps should be initiated. The direction pulses identify the instants at which the direction of rotation should be reversed. Only a position pulse train is needed for unidirectional operation. Generation of the position pulse train for steady-state operation at a constant speed is a relatively simple task. In this case, a single command identifying the stepping rate (pulse rate), corresponding to the specified speed, would suffice. The logic circuitry within the translator will latch onto a constant-frequency oscillator, with the frequency determined by the required speed (stepping rate), and continuously cycle the switching sequence at this frequency. This is a hardware approach to open-loop control of a stepping motor. For steady-state operation, the stepping rate can be set by manually adjusting the knob of a potentiometer connected to the translator. For simple motions (e.g., starting from rest and stopping after reaching a certain angular position), the commands that generate the pulse train (commands to the oscillator) can be set manually. Under the more complex and transient operating conditions that are present when following intricate motion trajectories, however, a computer-based (or, microprocessor-based) generation of the pulse commands, using programmed logic, would be necessary. This is a software approach, which is usually slower than the hardware approach. Sophisticated feedback control schemes can be implemented as well through such a computer-based controller.

The *translator* module has logic circuitry to interpret a pulse train and "translate" it into the corresponding switching sequence for stator field windings (on/off/reverse state for each phase of the stator). The translator also has solid-state switching circuitry (using gates, latches, triggers, etc.) to direct the field currents to the appropriate phase windings according to the particular switching state. A "packaged" system typically includes both indexer (or, controller) functions and driver functions. As a minimum, it possesses the capability to generate command pulses at a steady rate, thus assuming the role of the pulse generator (or, indexer) as well as the translator and switching amplifier functions. The stepping rate or direction may be changed manually using knobs or through a user interface.

The translator may not have the capability to keep track of the number of steps taken by the motor (i.e., a step counter). A packaged device that has all these capabilities, including pulse generation, the standard translator functions, and drive amplifiers, is termed a *preset indexer*. It usually consists of an oscillator, digital microcircuitry (integrated-circuit chips) for counting and for various control functions, and a translator, and drive circuitry in a single package. The required angle of rotation, stepping rate, and direction are set manually, by turning the corresponding knobs. With a more sophisticated programmable preset indexer, these settings can be programmed through computer commands from a standard interface. An external pulse source is not needed in this case. A programmable indexer—consisting of a microprocessor and microelectronic circuitry for the control of position and speed and for other programmable functions, memory, a pulse source (an oscillator), a translator, drive amplifiers with switching circuitry, and a power supply—represents a "programmable" controller for a stepping motor. A programmable indexer can be programmed using a personal computer or a hand-held programmer (provided with the indexer) through a standard interface (e.g., RS232 serial interface). Control signals within the translator are on the order of 10 mA, whereas the phase windings of a stepper motor require large currents on the order of several amperes. Control signals from the translator have to be properly amplified and directed to the motor windings by means of "switching amplifiers" for activating the required phase sequence.

Power to operate the translator (for logic circuitry, switching circuitry, etc.) and to operate phase excitation amplifiers comes from a *dc power supply* (typically 24 V DC). A regulated (i.e., voltage level maintained constant irrespective of the load) power supply is preferred. A packaged unit that consists of the translator (or preset indexer), the switching amplifiers, and the power supply, is what is normally termed a *motor-drive system*. The leads of the output amplifiers of the drive system carry currents to the phase windings on the stator (and to the rotor magnetizing coils located on the stator in the case of an electromagnetic rotor) of the stepping motor. The *load* may be connected to the motor shaft directly or through some form of mechanical coupling device (e.g., harmonic drive, tooth-timing belt drive, hydraulic amplifier, rack and pinion; See Chapter 3).

8.3.1 Driver Hardware

The driver hardware consists of the following basic components:

1. Digital (logic) hardware to interpret the information carried by the stepping pulse signal and the direction pulse signal (i.e., step instants and the direction of motion) and provide appropriate signals to the switches (switching transistors) that actuate the phase windings. This is the "translator" component of the drive hardware.

2. The drive circuit for phase windings with switching transistors to actuate the phases (on, off, reverse, in the unifilar case; on, off, in the bifilar case).

3. Power supply to power the phase windings.

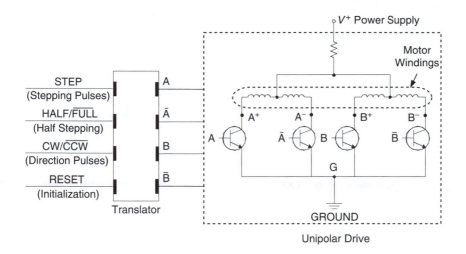

FIGURE 8.17
Basic drive hardware for a two-phase bifilar-wound stepper motor.

These three components are commercially available as a single package, to operate a corresponding class of stepper motors. Since there is considerable heat generation in a drive module, an integrated *heat sink* (or, some means of heat removal) is needed as well. Consider the drive hardware for a two-phase stepper motor. The phases are denoted by A and B. A schematic representation of the drive system, which is commercially available as a single package, is shown in Figure 8.17. What is indicated is a unipolar drive (no current reversal in a phase winding). As a result, a stepper motor with bifilar windings (two coil segments for each phase) has to be used. The motor has five leads, one of which is the "motor common" or ground (G) and the other four are the terminals of the two bifilar coil segments (A^+, A^-, B^+, B^-).

In the drive module there are several pins, some of which are connected to the motor controller/computer (driver inputs) and some are connected to the motor leads (driver outputs). There are other pins, which correspond to the dc power supply, common ground, various control signals, etc. The pin denoted by STEP (or PULSE) receives the stepping pulse signal (from the motor controller). This corresponds to the required stepping sequence of the motor. A transition from a low level to a high level (or, rising edge) of a pulse will cause the motor to move by one step. The direction in which the motor moves is determined by the state of the pin denoted by CW/$\overline{\text{CCW}}$. A logical high (see Chapter 10 for further details) state at this pin (or, open connection) will generate switching logic for the motor to move in the clockwise (CW) direction, and a logical low (or, logic common) will generate switching logic for the motor to move in the counterclockwise (CCW) direction. The pin denoted by HALF/$\overline{\text{FULL}}$ determines whether the half-stepping or full-stepping is carried out. Specifically, a logical low at this pin will generate the switching logic for full stepping, and a logical high will generate switching logic for half-stepping. The pin denoted by RESET receives the signal for initialization of a stepping sequence. There are several other pins, which are not necessary for the present discussion. The translator interprets the logical states at the STEP, HALF/$\overline{\text{FULL}}$ and CW/$\overline{\text{CCW}}$ pins and generates the proper logic to activate the switches in the unipolar drive. Specifically, four active logic signals are generated corresponding to A (Phase A on), $\overline{\text{A}}$ (Phase A reversed), B (Phase B on), and $\overline{\text{B}}$ (Phase B reversed). These logic signals activate the four switches in the bipolar drive, thereby sending current through the corresponding winding segments/leads (A^+, A^-, B^+, B^-) of the motor.

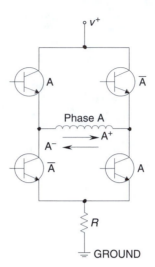

FIGURE 8.18
A bipolar drive for a single phase of a stepper motor (unifilar-wound).

The logic hardware is commonly available as compact chips in the monolithic form. If the motor is unifilar-wound (for a two-phase stepper there should be three leads—a ground wire and two power leads for the two phases), a bipolar drive will be necessary in order to change the direction of the current in a phase winding. A schematic representation of a bipolar drive for a single phase of a stepper is shown in Figure 8.18. Note that when the two transistors marked A are on, the current flows in one direction through the phase winding and when the two transistors marked $\overline{A}$ are on, the current flows in the opposite direction through the same phase winding. What is shown is an H-bridge circuit.

8.3.2   Motor Time Constant

Since the torque generated by a stepper motor is proportional to the phase current, it is desirable for a phase winding to reach its maximum current level as quickly as possible when it is switched on. Unfortunately, as a result of self-induction, the current in the energized phase does not build up instantaneously when switched on. As the stepping rate increases, the time period that is available for each step decreases. Consequently, a phase may be turned off before reaching its desired current level in order to turn on the next phase, thereby degrading the generated torque. This behavior is illustrated in Figure 8.19.

One way to increase the current level reached by a phase winding would be to simply increase the supply voltage as the stepping rate increases. Another approach would be to use a *chopper circuit* (a switching circuit) to switch on and off at high frequency, a supply voltage that is several times higher than the rated voltage of a phase winding. Specifically, a sensing element (typically, a resistor) in the drive circuit detects the current level and when the desired level is reached, the voltage supply is turned off. When the current level goes below the rated level, the supply is turned on again. The required switching rate (chopping rate) is governed by the electrical time constant of the motor.

The electrical time constant of a stepper motor is given by

$$\tau_e = \frac{L}{R} \tag{8.19}$$

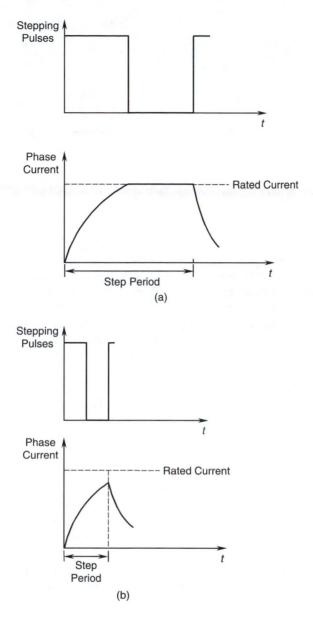

FIGURE 8.19
Torque degradation at higher stepping rates due to inductance: (a) Low stepping rate; (b) High stepping rate.

where L is the inductance of the energized phase winding and R is the resistance of the energized circuit, including winding resistance. It is well known that the current build-up is given by

$$i = \frac{v}{R}\exp(1 - t/\tau_e) \tag{8.20}$$

where v is the supply voltage (see Chapter 2). The larger the electrical time constant, the slower the current buildup. The driving torque of the motor decreases due to the lower phase current. Also, because of self-induction, the current does not die out instantaneously

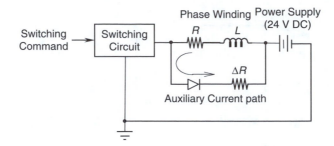

FIGURE 8.20
A diode circuit in a motor driver for decreasing the electrical time constant.

when the phase is switched off. The instantaneous voltages caused by self-induction can be high, and they can damage the translator and other circuitry. The torque characteristics of a stepper motor can be improved (particularly at high stepping rates) and the harmful effects of induced voltages can be reduced by decreasing the electrical time constant. A convenient way to accomplish this is by increasing the resistance R. But we want this increase in R to be effective only during the transient periods (switch-on and switch-off times). During the steady period, we like to have a smaller R, which will give a larger current (and magnetic field), producing a higher torque, and furthermore lower power dissipation and associated mitigation of thermal problems and reduction of efficiency. This can be accomplished by using a diode and a resistor ΔR, connected in parallel with the phase winding, as shown in Figure 8.20. In this case, the current will loop through R and ΔR, as shown, during the switch-on and switch-off periods, thereby decreasing the electrical time constant to

$$\tau_e = \frac{L}{R + \Delta R} \tag{8.21}$$

During steady conditions, however, no current flows through ΔR, as desired. Such circuits to improve the torque performance of stepping motors are commonly integrated into the motor drive hardware.

Note that the electrical time constant is typically much smaller than the mechanical time constant of a motor. Hence, increasing R is not a very effective way of increasing damping in a stepper motor.

8.4 Torque Motion Characteristics

It is useful to examine the response of a stepper motor to a single pulse input before studying the behavior under general stepping conditions. Ideally, when a single pulse is applied, the rotor should instantaneously turn through one step angle ($\Delta\theta$) and stop at that detent position (stable equilibrium position). Unfortunately, due to system dynamics, the actual single-pulse response is somewhat different from this ideal behavior. In particular, the rotor will oscillate for a while about the detent position before settling down. These oscillations result primarily from the interaction of motor load inertia (the combined inertia of rotor, load, etc.) with drive torque. This behavior can be explained using Figure 8.21.

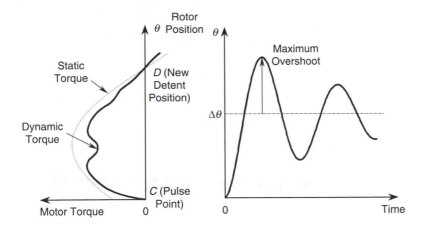

FIGURE 8.21
Single-pulse response and the corresponding single-phase torque.

Assume single-phase energization (i.e., only one phase is energized at a time). When a pulse is applied to the translator at C, the corresponding stator phase is energized. This generates a torque (due to magnetic attraction), causing the rotor to turn toward the corresponding minimum reluctance position (detent position D). The static torque curve (broken line in Figure 8.21) represents the torque applied on the rotor from the energized phase, as a function of the rotor position θ, under ideal conditions when dynamic effects are neglected. Under normal operating conditions, however, there will be induced voltages due to self-induction and mutual induction. Hence, a finite time is needed for the current to build up in the windings once the phases are switched on. Furthermore, there will be eddy currents generated in the rotor. These effects cause the magnetic field to deviate from the static conditions as the rotor moves at a finite speed, thereby making the dynamic torque curve different from the static torque curve, as shown in Figure 8.21. The true dynamic torque is somewhat unpredictable because of its dependence on many time-varying factors (rotor speed, rotor position, current level, etc.). The static torque curve is normally adequate to explain many characteristics of a stepper motor, including the oscillations in the single-pulse response.

It is important to note that the static torque is positive at the switching point, but is generally not maximum at that point. To explain this further, consider the three-phase VR stepper motor (with nontoothed poles) shown in Figure 8.5. The step angle $\Delta\theta$ for this arrangement is 60°, and the full-step switching sequence for clockwise rotation is 1-2-3-1. Suppose that Phase 1 is energized. The corresponding detent position is denoted by D in Figure 8.22(a). The static torque curve for this phase is shown in Figure 8.22(b), with the positive angle measured clockwise from the detent position D. Suppose that we turn the rotor counterclockwise from this stable equilibrium position, using an external rotating mechanism (e.g., by hand). At position C, which is the previous detent position where Phase 1 would have been energized under normal operation, there is a positive torque that tries to turn the rotor to its present detent position D. At position B, the static torque is zero, because the force from the N pole of Phase 1 exactly balances that from the S pole. This point, however, is an *unstable* equilibrium position; a slight push in either direction will move the rotor in that direction. Position A, which is located at a rotor tooth pitch ($\theta_r = 180°$) from position D, is also a "stable" equilibrium position. The maximum static torque occurs at position M, which is located approximately halfway between positions B and D (at an angle $\theta_r/4 = 45°$ from the detent position). This static maximum torque is also known

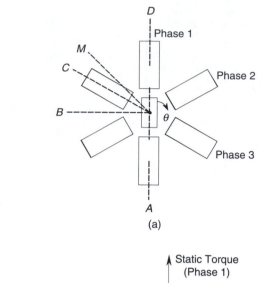

(a)

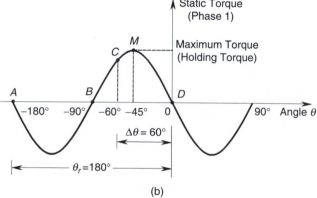

(b)

FIGURE 8.22
Static torque distribution in the VR stepper motor in Figure 8.5: (a) Schematic diagram; (b) Static torque curve for Phase 1.

as the "holding torque" because it is the maximum resisting torque an energized motor can exert if we try to turn the rotor away from the corresponding detent position. The torque at the normal switching position (C) is less than the maximum value.

For a simplified analysis, the static torque curve may be considered to be sinusoidal. In this case, with Phase 1 excited, and with the remaining phases inactive, the static torque distribution T_1 can be expressed as

$$T_1 = -T_{max} \sin n_r \theta \tag{8.22}$$

where θ is the angular position in radians, measured from the current detent position (with Phase 1 excited)

n_r = number of teeth on the rotor

T_{max} = maximum static torque (or, holding torque)

Equation 8.22 can be verified by referring to Figure 8.22 where $n_r = 2$. Note that Equation 8.22 is valid irrespective of whether the stator poles are toothed or not, even though the example considered in Figure 8.22 has nontoothed stator poles.

Returning to the single-pulse response shown in Figure 8.21, note that starting from rest at C, the rotor will have a positive velocity at the detent position D. Its kinetic energy (or momentum) will take it beyond the detent position. This is the first *overshoot*. Since the same phase is still on, the torque will be negative beyond the detent position; static torque always attracts the rotor to the detent position, which is a stable equilibrium position. The rotor will decelerate because of this negative torque and will attain zero velocity at the point of maximum overshoot. Then the rotor will be accelerated back toward the detent position and carried past this position by the kinetic energy, and so on. This oscillatory motion would continue forever with full amplitude ($\Delta\theta$) if there were no energy dissipation. In reality, however, there are numerous damping mechanisms—such as mechanical dissipation (frictional damping) and electrical dissipation (resistive damping through eddy currents and other induced voltages)—in the stepper motor, which will gradually slow down the rotor, as shown in Figure 8.21. Dissipated energy will appear primarily as thermal energy (temperature increase). For some stepper motors, the maximum overshoot could be as much as 80% of the step angle. Such high-amplitude oscillations with slow decay rate are clearly undesirable in most practical applications. Adequate damping should be provided by mechanical means (e.g., attaching mechanical dampers), electrical means (e.g., by further eddy current dissipation in the rotor or by using extra turns in the field windings), or by electronic means (electronic switching or multiple-phase energization) in order to suppress these oscillations. The first two techniques are wasteful, while the third approach requires switching control. The single-pulse response is often modeled using a simple oscillator transfer function.

Now we will examine the stepper motor response when a sequence of pulses is applied to the motor under normal operating conditions. If the pulses are sufficiently spaced—typically, more than the settling time T_s of the motor (Note: $T_s = 4 \times$ motor time constant)—then the rotor will come to rest at the end of each step before starting the next step. This is known as *single stepping*. In this case, the overall response is equivalent to a cascaded sequence of single-pulse responses; the motor will faithfully follow the command pulses in synchronism. In many practical applications, however, fast responses and reasonably continuous motor speeds (stepping rates) are desired. These objectives can be met, to some extent, by decreasing the motor settling time through increased dissipation (mechanical and electrical damping). This, beyond a certain optimal level of damping, could result in undesirable effects, such as excessive heat generation, reduced output torque, and sluggish response. Electronic damping, explained in a later section, can eliminate these problems.

Since there are practical limitations to achieving very small settling times, faster operation of a stepper motor would require switching before the rotor settles down in each step. Of particular interest under high-speed operating conditions is *slewing motion*. In this case, the motor operates at steady state in synchronism at a constant pulse rate called the *slew rate*. It is not necessary for the phase switching (i.e., pulse commands) to occur when the rotor is at the detent position of the old phase, but switchings (pulses) should occur in a uniform manner. Since the motor moves in harmony, practically at a constant speed, the torque required for slewing is smaller than that required for transient operation (accelerating and decelerating conditions). Specifically, at a constant speed there is no inertial torque, and as a result, a higher speed can be maintained for a given level of motor torque. But, since stepper motors generate heat in their windings, it is not desirable to operate them at high speeds for long periods.

A typical displacement time curve under slewing is shown in Figure 8.23. The slew rate is given by

$$R_s = \frac{1}{\Delta t} \quad \text{steps/second} \tag{8.23}$$

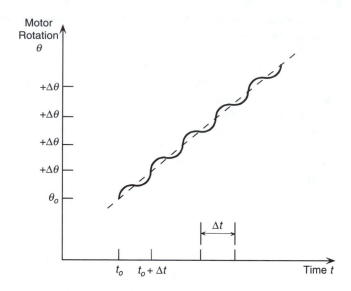

FIGURE 8.23
Typical slewing response of a stepping motor.

where Δt denotes the time between successive pulses under slewing conditions. Note that Δt could be significantly smaller than the motor settling time T_s. Some periodic oscillation (or *hunting*) is possible under slewing conditions, as seen in Figure 8.23. This is generally unavoidable, but its amplitude can be reduced by increasing damping. The slew rate depends as well on the external load connected to the motor. In particular, motor damping, bearing friction, and torque rating set an upper limit to the slew rate.

To attain slewing conditions, the stepper motor has to be accelerated from a low speed by *ramping.* This is accomplished by applying a sequence of pulses with a continuously increasing pulse rate $R(t)$. Strictly speaking, ramping represents a linear (straight-line) increase of the pulse rate, as given by

$$R(t) = R_o + \frac{(R_s - R_o)}{n \cdot \Delta t} \tag{8.24}$$

where
R_o = starting pulse rate (typically zero)
R_s = final pulse rate (slew rate)
n = total number of pulses

If exponential ramping is used, the pulse rate is given by

$$R(t) = R_s - (R_s - R_o)e^{-t/\tau} \tag{8.25}$$

If the time constant τ of the ramp is equal to $n \cdot \Delta t/4$, a pulse rate of $0.98R_s$ is reached in a total of n pulses. (Note: $e^{-4} = 0.02$.) In practice, the pulse rate is often increased beyond the slew rate, in a time interval shorter than what is specified for acceleration, and then decelerated to the slew rate by pulse subtraction at the end. In this manner, the slew rate is reached more quickly. In general, during *upramping* (acceleration), the rotor angle trails the pulse command, and during *downramping* (deceleration), the rotor angle leads the pulse command. These conditions are illustrated in Figure 8.24. The ramping rate cannot

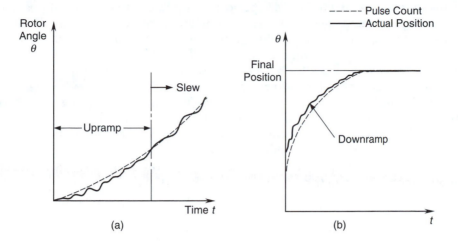

FIGURE 8.24
Ramping response: (a) Accelerating motion; (b) Decelerating motion.

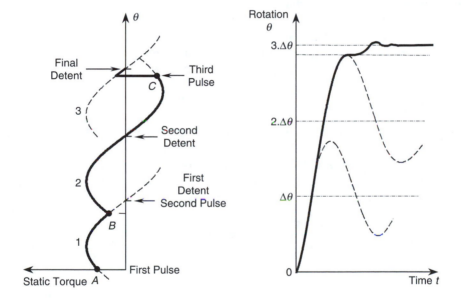

FIGURE 8.25
Torque-response diagram for a three-step drive sequence.

be chosen arbitrarily, and is limited by the torque-speed characteristics of the motor. If ramping rates beyond the capability of the particular motor are attempted, it is possible that the response will go completely out of synchronism and the motor will stall.

In transient operation of stepper motors, nonuniform stepping sequences might be necessary, depending on the complexity of the motion trajectory and the required accuracy. Consider, for example, the three-step drive sequence shown in Figure 8.25. The first pulse is applied at A when the motor is at rest. The resulting positive torque (curve 1) of the energized phase will accelerate the motor, causing an overshoot beyond the detent position (see broken line). The second pulse is applied at B, the point of intersection of the torque curves 1 and 2, which is before the detent position. This switches the torque to curve 2, which is the torque generated by the newly energized phase. Fast acceleration is possible in this manner because the torque is kept positive up to the second detent position. Note that the average torque is maximum when switching is done at the point of intersection of successive torque curves. The resulting torque produces a larger overshoot beyond the second detent position. As the rotor moves beyond the second detent position, the torque becomes negative, and the motor begins to slow down. The third pulse is applied at C when the rotor is closest to the required final position. The corresponding torque (curve 3) is relatively small, because the rotor is near its final (third) detent position. As a result of this and in view of the previous negative torque, the overshoot from the final detent position is relatively small, as desired. The rotor then quickly settles down to the final position, as there exists some damping/friction in the motor and its bearings.

Drive sequences can be designed in this manner to produce virtually any desired motion in a stepper motor. The motor controller is programmed to generate the appropriate pulse train in order to achieve the required phase switchings for a specified motion. Such drive sequences are useful also in compensating for missed pulses and in electronic damping. These two topics will be discussed later in the chapter.

Note that in order to simplify the discussion and illustration, we have used static torque curves in Figure 8.25. This assumes instant buildup of current in the energized phase and instant decay of current in the de-energized phase, thus neglecting all induced voltages and eddy currents (i.e., electrical dynamics are neglected). In reality, however, the switching torques will not be generated instantaneously. The horizontal lines with sharp ends to represent the switching torques, as shown in Figure 8.25, are simply approximations, and in practice the entire torque curve will be somewhat irregular. These dynamic torque curves should be used for accurate switching control in sophisticated practical applications.

8.4.1 Static Position Error

If a stepper motor does not support a static load (e.g., spring-like torsional element), the equilibrium position under power-on conditions would correspond to the zero-torque (detent) point of the energized phase. If there is a static load T_L, however, the equilibrium position would be shifted to $-\theta_e$, as shown in Figure 8.26. The offset angle θ_e is called the static position error.

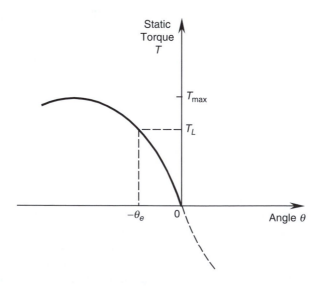

FIGURE 8.26
Representation of the static position error.

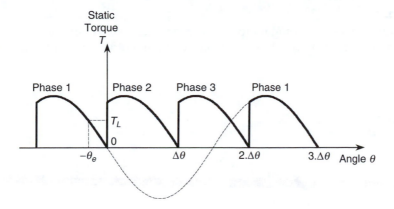

FIGURE 8.27
Periodicity of the single-phase static torque distribution (a three-phase example).

Assuming that the static torque curve is sinusoidal, we can obtain an expression for θ_e. First, note that the static torque curve for each phase is periodic with period $p \cdot \Delta\theta$ (equal to the rotor pitch θ_r), where p is the number of phases and $\Delta\theta$ is the step angle. For example, this relationship is shown for the three-phase case in Figure 8.27. Accordingly, the static torque curve may be expressed as

$$T = -T_{max} \sin\left(\frac{2\pi\theta}{p \cdot \Delta\theta}\right) \tag{8.26}$$

where T_{max} denotes the maximum torque. Equation 8.26 can be directly obtained by substituting Equation 8.2 in Equation 8.22. Note that under standard switching conditions, Equation 8.26 governs for $-\Delta\theta \le \theta \le 0$. With reference to Figure 8.26, the static position error is given by,

$$T_L = -T_{max} \sin\left[\frac{2\pi(-\theta_e)}{p \cdot \Delta\theta}\right]$$

or

$$\theta_e = \frac{p \cdot \Delta\theta}{2\pi} \sin^{-1}\left(\frac{T_L}{T_{max}}\right) \tag{8.27}$$

If n denotes the number of steps per revolution, Equation 8.27 may be expressed as

$$\theta_e = \frac{p}{n} \sin^{-1}\left(\frac{T_L}{T_{max}}\right) \tag{8.28}$$

It is intuitively clear that the static position error decreases with the number of steps per revolution.

Example 8.7

Consider a three-phase stepping motor with seventy-two steps per revolution. If the static load torque is 10% of the maximum static torque of the motor, determine the static position error.

SOLUTION

In this problem,

$$\frac{T_L}{T_{max}} = 0.1, \quad p = 3, \quad n = 72$$

Now, using Equation 8.28, we have

$$\theta_e = \frac{3}{72}\sin^{-1}0.1 = 0.0042 \text{ rad} = 0.24°$$

Note that this is less than 5% of the step angle.

8.5 Damping of Stepper Motors

Lightly damped oscillations in stepper motors are undesirable in applications that require single-step motions or accurate trajectory following under transient conditions. Also, in slewing motions (where the stepping rate is constant), high-amplitude oscillations (hunting) can result if the resonant frequency of the motor-shaft-load combination coincides with the stepping frequency. Damping has the advantages of suppressing overshoots, increasing the decay rate of oscillations (i.e., shorter settling time), and decreasing the amplitude of oscillations under resonant conditions. Unfortunately, heavy damping has drawbacks, such as sluggish response (longer rise time, peak time, or delay), large time constants, and reduction of the net output torque. On the average, however, the advantages of damping outweigh the disadvantages, in stepper motor applications.

Several techniques are employed to damp stepper motors. Most straightforward are the conventional techniques of damping, which use mechanical and electrical energy dissipation. Usually, mechanical damping is provided by a torsional damper attached to the motor shaft. Methods of electrical damping include eddy current dissipation in the rotor, the use of magnetic hysteresis and saturation effects, and increased resistive dissipation by adding extra windings to the motor stator. For example, solid-rotor construction has higher hysteresis losses due to magnetic saturation than laminated-rotor construction has. These direct techniques of damping have undesirable side effects, such as excessive heat generation, reduction of the net output torque of the motor, and decreased speed of response. Electronic damping methods have been developed to overcome such shortcomings. These methods are nondissipative, and are based on employing properly designed switching schemes for phase energization so as to inhibit overshoots in the final step of response. A general drawback of electronic damping is that the associated switching sequences are complex (irregular) and depend on the nature of a particular motion trajectory. A rather sophisticated controller may be necessary as a result. The level of damping achieved by electronic damping is highly sensitive to the time sequence of the switching

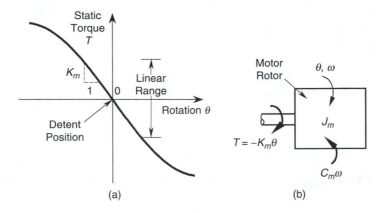

FIGURE 8.28
Model for single-step oscillations of a stepper motor: (a) Linear torque approximation; (b) Rotor free-body diagram.

scheme. Accordingly, a high level of intelligence concerning the actual response of the motor is required to effectively use electronic damping methods. Note, also, that in the design stage, damping in a stepper motor can be improved/optimized by judicious choice of values for motor parameters (e.g., resistance of the windings, rotor size, material properties of the rotor, air gap width).

8.5.1 Mechanical Damping

A convenient, practical method of damping a stepper motor is to connect an inertia element to the motor shaft through an energy dissipation medium, such as a viscous fluid (e.g., silicone) or a solid friction surface (e.g., brake lining). A common example for of the first type of torsional dampers is the *Houdaille damper* (or viscous *torsional damper*) and for the second type (which depends on Coulomb-type friction) it is the *Lanchester damper.*

The effectiveness of torsional dampers on stepper motors can be examined using a linear dynamic model for the single-step oscillations. From Figure 8.28(a) it is evident that in the neighborhood of the detent position, the static torque due to the energized phase is approximately linear, and this torque acts as an electromagnetic spring. In this region, the torque can be expressed by

$$T = -K_m\theta \tag{8.29}$$

where
   θ = angle of rotation measured from the detent position
   K_m = torque constant (or, *magnetic stiffness* or *torque gradient*) of the motor

Damping forces also come from such sources as bearing friction, resistive dissipation in windings, eddy current dissipation in the rotor, and magnetic hysteresis. If the combined contribution from these internal dissipation mechanisms is represented by a single damping constant C_m, the equation of motion for the rotor near its detent position (equilibrium position) can be written as

$$J_m\frac{d\omega}{dt} = -C_m\omega - K_m\theta \tag{8.30}$$

where
  J_m = overall inertia of the rotor
  $\omega = \frac{d\theta}{dt}$ = motor speed.

Note that for a motor with an external load, the *load inertia* has to be included in J_m. Equation 8.30 is expressed in terms of θ; thus,

$$J_m \ddot{\theta} + C_m \dot{\theta} + K_m \theta = 0 \qquad (8.31)$$

The solution of this second-order ordinary differential equation (See Chapter 2) is obtained using the maximum overshoot point as the initial state:

$$\dot{\theta}(0) = 0 \quad \text{and} \quad \theta(0) = \alpha \Delta \theta$$

The constant α represents the *fractional overshoot*. Its magnitude can be as high as 0.8. The undamped natural frequency of single-step oscillations is given by

$$\omega_n = \sqrt{\frac{K_m}{J_m}} \qquad (8.32)$$

and the damping ratio is given by

$$\zeta = \frac{C_m}{2\sqrt{K_m J_m}} \qquad (8.33)$$

With a Houdaille damper attached to the motor (see Figure 8.29), the equations of motion are

$$(J_m + J_h)\ddot{\theta} = -C_m \dot{\theta} - K_m \theta - C_d(\dot{\theta} - \dot{\theta}_d) \qquad (8.34)$$

$$J_d \ddot{\theta} = C_d(\dot{\theta} - \dot{\theta}_d) \qquad (8.35)$$

where
  θ_d = angle of rotation of the damper inertia
  J_d = moment of inertia of the damper
  J_h = moment of inertia of the damper housing

It is assumed that the damper housing is rigidly attached to the motor shaft.

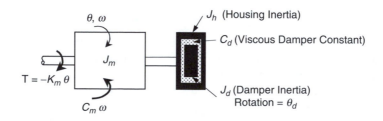

FIGURE 8.29
A stepper motor with a Houdaille damper.

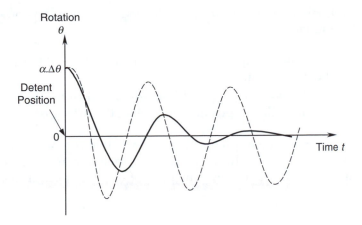

FIGURE 8.30
Typical single-step response of a stepper motor with a Houdaille damper.

In Figure 8.30, a typical response of a mechanically damped stepper motor is compared with the response when the external damper is disconnected. Observe the much faster decay when the external damper is present. One disadvantage of this method of damping, however, is that it always adds inertia to the motor (note the J_h term in Equation 8.34). This reduces the natural frequency of the motor (Equation 8.32) and, hence, decreases the speed of response (or bandwidth). Other disadvantages include reduction of the effective torque and increased heat generation, which may require a special cooling means.

A Lanchester damper is similar to a Houdaille damper except that the former depends on nonlinear (Coulomb) friction instead of viscous damping. Hence, a stepper motor with a Lanchester damper can be analyzed in a manner similar to what was presented before, but the equations of motion are nonlinear in this case, because the frictional torque is of Coulomb type. Coulomb frictional torque has a constant magnitude for a given reaction force but acts opposite to the direction of relative motion between the rotor (and damper housing) and the damper inertia element. The reaction force on the friction lining can be adjusted using spring-loaded bolts, thereby changing the frictional torque. There are two limiting states of operation: If the reaction force is very small, the motor is virtually uncoupled (disengaged) from the damper; if the reaction force is very large, the damper inertia will be rigidly attached to the damper housing, thus moving as a single unit. In either case, there is very little dissipation. Maximum energy dissipation takes place under some intermediate condition. For constant-speed operation, by adjusting the reaction force, the damper inertia element can be made to rotate at the same speed as the rotor, thereby eliminating dissipation and torque loss under these steady conditions in which damping is usually not needed. This is an advantage of friction dampers.

8.5.2 Electronic Damping

Damping of stepper motor response by electronic switching control is an attractive method of overshoot suppression for several reasons. For instance, it is not an energy dissipation method. In that sense, it is actually an electronic control technique rather than a damping technique. By properly timing the switching sequence, virtually a zero overshoot response can be realized. Another advantage is that the reduction in net output torque is insignificant in this case in comparison to the torque losses in direct (mechanical) damping methods.

A majority of electronic damping techniques depend on a two-step procedure that is straightforward in principle:

1. Decelerate the last-step response of the motor so as to avoid large overshoots from the final detent position.
2. Energize the final phase (i.e., apply the last pulse) when the motor response is very close to the final detent position (i.e., when the torque is very small).

It is possible to come up with many switching schemes that conform to these two steps. Generally, such schemes differ only in the manner in which response deceleration is brought about (in step 1). Three common methods of response deceleration are

1. The pulse turn-off method: Turn off the motor (all phases) for a short time.
2. The pulse reversal method: Apply a pulse in the opposite direction (i.e., energize the reverse phase) for a short time.
3. The pulse delay method: Maintain the present phase beyond its detent position for a short time.

These three types of switching schemes can be explained using the static torque response curves in Figures 8.31 through 8.33. In all three figures, the static torque curve corresponding to the last pulse (i.e., last energized phase) is denoted by 2. The static curve corresponding to the next-to-last pulse is denoted by 1.

In the pulse turn-off method (Figure 8.31), the last pulse is applied at *A*, as usual. This energizes Phase 2, turning off Phase 1. The rotor accelerates toward its final detent position because of the positive torque that is present. At point *B*, which is sufficiently close to the final detent position, Phase 2 is shut off. From *B* to *C*, all phases of the motor are inactive, and the static torque is zero. The motor decelerates during this interval, giving a peak response that is very close to, but below, the final detent position. At point *C*, the last phase (Phase 2) is energized again. Since the corresponding static torque is very small (in comparison to the maximum torque) but positive, the motor will accelerate slowly (assuming a pure inertial load) to the final detent position. By properly choosing the points *B*

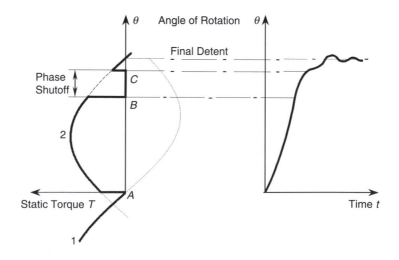

FIGURE 8.31
The pulse turn-off method of electronic damping.

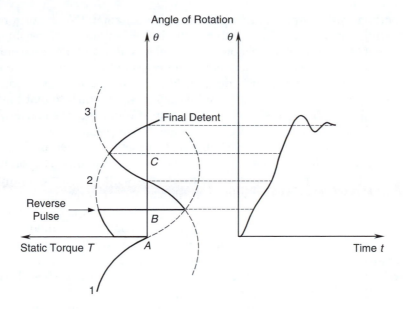

FIGURE 8.32
The pulse reversal method of electronic damping.

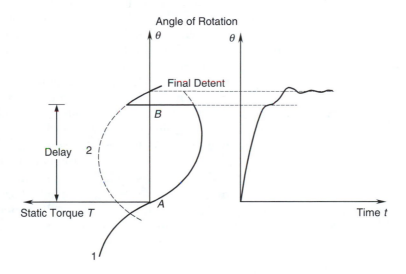

FIGURE 8.33
The pulse delay method of electronic damping. Damping of Stepper Motors.

and C, the overshoot can be made sufficiently small. This choice requires knowledge of the actual response of the motor. The amount of final overshoot can be very sensitive to the timing of the switching points B and C. Furthermore, the actual response θ will depend on mechanical damping and other load characteristics as well.

The pulse reversal method is illustrated in Figure 8.32. The static torque curve corresponding to the second pulse before last is denoted by 3. As usual, the last phase (Phase 2) is energized at A. The motor will accelerate toward the final detent position. At point B (located at less than half the step angle from A), Phase 2 is shut off and Phase 3 is turned on. (Note: The forward pulse sequence is 1-2-3-1, and the reverse pulse sequence is 1-3-2-1.)

The corresponding static torque is negative over some duration. (Note: For a three-phase stepper motor, this torque is usually negative up to the halfway point of the step angle and positive thereafter). Consequently, the motor will decelerate first and then accelerate (assuming a pure inertial load); the overall decelerating effect is not as strong as in the previous method. (Note: If faster deceleration is desired, Phase 1 should be energized, instead of Phase 3, at *B*.) At point *C*, the static torque of Phase 3 becomes equal to that of Phase 2. To avoid large overshoots, Phase 3 is turned off at point *C*, and the last phase (Phase 2) is energized again. This will drive the motor to its final detent position.

In the pulse delay method (Figure 8.33) the last phase is not energized at the detent position of the previous step (point *A*). Instead, Phase 1 is continued on beyond this point. The resulting negative torque will decelerate the response. If intentional damping is not employed, the overshoot beyond *A* could be as high as 80%. (Note: In the absence of any damping, 100% overshoot is possible). When the overshoot peak is reached at *B*, the last phase is energized. Since the static torque of Phase 2 is relatively small at this point and will reach zero at the final detent position, the acceleration of the motor is slow. Hence, the final overshoot is maintained within a very small value. It is interesting to note that if 100% overshoot is obtained with Phase 1 energized, the final overshoot becomes zero in this method, thus producing ideal results.

In all these techniques of electronic damping, the actual response depends on many factors-particularly the dynamic behavior of the load. Hence, the switching points cannot be exactly prespecified unless the true response is known ahead of time (through tests, simulations, etc.). In general, accurate switching may require measurement of the actual response, and use of this information in real time to apply the switching pulses.

In Figures 8.31 through 8.33 static torque curves are used to explain electronic damping. In practice, however, currents in the phase windings neither decay instantaneously nor build up instantaneously, following a pulse command. Induced voltages, eddy currents, and magnetic hysteresis effects are primarily responsible for this behavior. These factors, in addition to external loads, can complicate the nature of dynamic torque and, hence, the true response of a stepping motor. This can make accurate preplanning of switching points rather difficult in electronic damping. In the foregoing discussion, we have assumed that the mechanical damping (including bearing friction) of the motor is negligible and that the load connected to the motor is a pure inertia. In a practical situation the net torque available to drive the combined rotor-load inertia is smaller than the electromagnetic torque generated at the rotor. Hence, in practice, the accelerations obtained are not quite as high as what the Figures 8.31 through 8.33 suggest. Nevertheless, the general characteristics of motor response will be the same as those shown in these figures.

8.5.3 Multiple Phase Energization

A popular and relatively simple method that may be classified under electronic damping is multiple-phase energization. With this method, two phases are excited simultaneously (e.g., 13-21-32-13). One is the standard "stepping phase" and the other is the "damping phase." Note that the damping phase should result in a deceleration effect. Consequently, the damping phase should correspond to rotation in the reverse direction, but it is energized at a fraction of the stepping voltage (rated voltage), along with the stepping phase (which is energized with the full voltage). As noted earlier in the chapter, the step angle remains unchanged when more than one phase is energized simultaneously (as long as the number of phases activated at a time is the same, which is the case here). It has been observed that this switching sequence provides a better response (less overshoot) than the single-phase energization method (e.g., 1-2-3-1), particularly for single-stack stepper motors. The damping phase (which is the reverse phase) provides a negative torque, and it not only reduces the overshoot but also the speed of response. Increased magnetic hysteresis and saturation effects of the ferromagnetic materials in the motor as well as higher energy dissipation through eddy currents when two phases are energized simultaneously are other factors that enhance damping and reduce the speed of response, in simultaneous multiphase energization. Another factor is that multiple-phase excitation results in wider overlaps of magnetic flux between switchings, giving smoother torque transitions. Note, however, that there can be excessive heat generation with this method. This may be reduced, to some extent, by further reducing the voltage of the damping phase, typically to half the normal rated voltage.

8.6 Stepping Motor Models

In the preceding sections we have discussed variable-reluctance (VR) stepper motors, which have nonmagnetized soft-iron rotors, and permanent magnet (PM) stepper motors, which have magnetized rotors. As noted, hybrid stepper motors are a special type of PM stepping motors. Specifically, a hybrid motor has two rotor stacks that are magnetized to have opposite polarities (one rotor stack is the N pole and the other is the S pole). Also, there is a tooth misalignment between the two rotor stacks. As usual, stepping is achieved by switching the phase excitations.

In the analysis of stepper motors under steady operation at low speeds, we usually do not need to differentiate between VR motors and PM motors. But under transient conditions, the torque characteristics of the two types of motors can differ considerably. In particular, the torque in a PM motor varies somewhat linearly with the magnitude of the phase current (since rotor field is provided by permanent magnets), whereas the torque in a VR motor varies nearly quadratically with the phase current (since the stator field links with the rotor, which does not have its own field).

8.6.1 A Simplified Model

Under steady-state operation of a stepper motor at low speeds, the motor (magnetic) torque can be approximated by a sinusoidal function, as given by Equation 8.22 or Equation 8.26. Hence, the simplest model for any type of stepping motor (VR or PM) is the *torque source* given by

$$T = -T_{max} \sin n_r \theta \qquad (8.36)$$

or equivalently,

$$T = -T_{max} \sin\left(\frac{2\pi\theta}{p \cdot \Delta\theta}\right) \qquad (8.37)$$

where
T_{max} = maximum torque during a step (holding torque)
$\Delta\theta$ = step angle
n_r = number of rotor teeth
p = number of phases

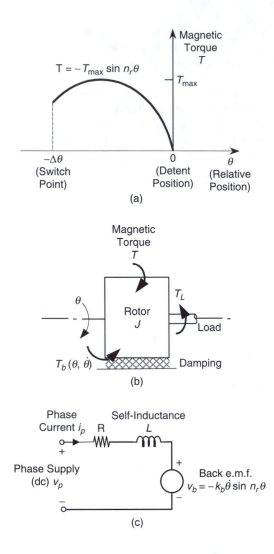

FIGURE 8.34
Stepper motor models: (a) Torque source model; (b) Mechanical model; (c) Equivalent circuit for an improved model.

Note that θ is the angular position of the rotor measured from the detent position of the presently excited phase, as indicated in Figure 8.34(a). Hence, $\theta = -\Delta\theta = -\theta_r/p$ at the previous detent position, where the present phase is switched on, and $\theta = 0$ at the approaching detent position. The coordinate frame of θ is then shifted again to a new origin $(+\Delta\theta)$ when the next phase is excited at the approaching detent position of the conventional method of switching. Hence, θ gives the relative position of the rotor during each step. The absolute position is obtained by adding θ to the absolute rotor angle at the approaching detent position.

The motor model becomes complete with the mechanical dynamic equation for the rotor. With reference to Figure 8.34(b), Newton's second law gives

$$T - T_L - T_b(\theta, \dot{\theta}) = J\ddot{\theta} \tag{8.38}$$

where

T_l = resisting torque (reaction) on the motor by the driven load

$T_b(\theta,\dot{\theta})$ = dissipative resisting torque (viscous damping torque, frictional torque, etc.) on the motor

J = motor rotor inertia

Note that T_L will depend on the nature of the external load. Furthermore, $T_b(\theta,\dot{\theta})$ will depend on the nature of damping. If viscous damping is assumed, T_b may be taken as proportional to $\dot{\theta}$. On the other hand, if Coulomb friction is assumed, the magnitude of T_b is taken to be constant, and the sign of T_b is the sign of $\dot{\theta}$. In the case of general dissipation (e.g., a combination of viscous, Coulomb, and structural damping), T_b is a nonlinear function of both θ and $\dot{\theta}$. Note that the torque source model may be used for both VR and PM types of stepping motors.

8.6.2 An Improved Model

Under high-speed and transient operation of a stepper motor, many of the quantities that were assumed constant in the torque source model will vary with time as well as rotor position. In particular, for a given supply voltage v_p to a phase winding, the associated phase current i_p will not be constant. Also, inductance L in the phase circuit will vary with the rotor position. Furthermore, a voltage v_b (a *back electromotive force*, or back e.m.f.) will be induced in the phase circuit because of the changes in magnetic flux resulting from the speed of rotation of the rotor (in both VR and PM motors). It follows that an improved dynamic model is needed to represent the behavior of a stepper motor under high-speed and transient conditions. Such a model that is described now. Instead of using rigorous derivations, motor equations are obtained from an equivalent circuit using qualitative considerations.

Since magnetic flux linkage of the phase windings changes as a result of variations in the phase current, a voltage is induced in the phase windings. Hence, a self-inductance (L) should be included in the circuit. Although a mutual inductance should also be included to account for voltages induced in a phase winding as a result of current variations in the other phase windings, this voltage is usually smaller than the self-induced voltage. Hence, in the present model we neglect mutual inductance. Furthermore, flux linkage of the phase windings changes as a result of the motion of the rotor. This induces a voltage v_b (termed a back e.m.f.) in the phase windings. This voltage is present irrespective of whether the rotor is a VR type or a PM type. Also, phase windings will have a finite resistance R. It follows that an approximate equivalent circuit (neglecting mutual induction, in particular) for one phase of a stepper motor may be represented as in Figure 8.34(c). The phase circuit equation is

$$v_p = Ri_p + L\frac{di_p}{dt} + v_b \tag{8.39}$$

where

v_p = phase supply voltage (dc)

i_p = phase current

v_b = back e.m.f., due to rotor motion

R = resistance in the phase winding

L = self-inductance of the phase winding

The back e.m.f. is proportional to the rotor speed $\dot{\theta}$, and it will also vary with the rotor position θ. The variation with position is periodic with period θ_r. Hence, using only the fundamental term in a Fourier series expansion, we have

$$v_b = -k_b \dot{\theta} \sin n_r \theta \tag{8.40}$$

where
$\dot{\theta}$ = rotor speed
θ = rotor position (as defined in Figure 8.34(a))
n_r = number of rotor teeth
k_b = back e.m.f. constant

Since θ is negative in a conventional step (from $\theta = -\Delta\theta$ to $\theta = 0$), we note that v_b is positive for positive $\dot{\theta}$.

Self-inductance L also varies with the rotor position θ. This variation is periodic with period θ_r. Now, retaining only the constant and the fundamental terms in a Fourier series expansion, we have

$$L = L_0 + L_a \cos n_r \theta \qquad (8.41)$$

where L_o and L_a are appropriate constants and angle θ is as defined in Figure 8.34(a).

Note that Equation 8.39 through Equation 8.41 are valid for both types of stepper motors (VR and PM). The torque equation will depend on the type of stepper motor, however.

8.6.2.1   Torque Equation for PM Motors

In a permanent-magnet stepper motor, magnetic flux is generated by both the phase current i_p and the magnetized rotor. The flux from the magnetic rotor is constant, but its linkage with the phase windings will be modulated by the rotor position θ. Hence, retaining only the fundamental term in a Fourier series expansion, we have

$$T = -k_m i_p \sin n_r \theta \qquad (8.42)$$

where i_p is the phase current and k_m is the *torque constant* for the PM motor.

8.6.2.2   Torque Equation for VR Motors

In a variable-reluctance stepper motor, the rotor is not magnetized; hence, there is no magnetic flux generation from the rotor. The flux generated by the phase current i_p is linked with the phase windings. The flux linkage is coupled with the motor rotor and as a result it is modulated by the motion of the VR rotor. Hence, retaining only the fundamental term in a Fourier series expansion, the torque equation for a VR stepper motor may be expressed as

$$T = -k_r i_p^2 \sin n_r \theta \qquad (8.43)$$

where k_r is the *torque constant* for the VR motor. Note that torque T depends on the phase current i_p in a quadratic manner in the VR stepper motor. This make a VR motor more nonlinear than a PM motor.

In summary, to compute the torque T at a given rotor position, we first have to solve the differential equation given by Equation 8.39 through Equation 8.41 for known values of the rotor position θ and the rotor speed $\dot{\theta}$, and for a given (constant) phase supply voltage v_p. Initially, as a phase is switched on, the phase current is zero. The model parameters R, L_o, L_a, and k_b are assumed to be known (either experimentally or from the manufacturer's data sheet). Then torque is computed using Equation 8.42 for a PM stepper motor or using Equation 8.43 for a VR stepper motor. Again, the torque constant (k_m or k_r) is assumed to be known. The simulation of the model then can be completed by using this torque in the mechanical dynamic Equation 8.38 to determine the rotor position θ and the rotor speed $\dot{\theta}$.

8.7 Control of Stepper Motors

Open-loop operation is adequate for many applications of stepper motors, particularly at low speeds and in steady-state operation. The main shortcoming of open-loop control is that the actual response of the motor is not measured; consequently, it is not known whether a significant error is present, for example due to missed pulses.

8.7.1 Pulse Missing

There are two main reasons for pulse missing:

1. Particularly under variable-speed conditions, if the successive pulses are received at a high frequency (high stepping rate), the phase translator might not respond to a received pulse, and the corresponding phase would not be energized before the next pulse arrives. This may occur, for instance, due to a malfunction in the translator or the drive circuit.

2. Because of a malfunction in the pulse source, a pulse might not actually be generated, even when the motor is operating at well below its rated capacity (low-torque, low-speed conditions). Extra (erroneous) pulses can be generated as well by a faulty pulse source or drive circuitry.

If a pulse is missed by the motor, the response has to catch up somehow (e.g., by a subsequent overshoot in motion), or else an erratic behavior may result, causing the rotor to oscillate and probably stall eventually. Under very favorable conditions, particularly with small step angles, if a single pulse is missed, the motor will decelerate so that a complete cycle of pulses is missed; then it will lock in again with the input pulse sequence. In this case, the motor will trail the correct trajectory by a rotor tooth pitch angle (θ_r). Here, pulses equal in number to the total phases (p) of the motor (Note: $\theta_r = p \cdot \Delta\theta$) are missed. In this manner it is also possible to lose accuracy by an integer multiple of θ_r because of a single missed pulse. Under adverse conditions, however, pulse missing can lead to a highly nonsynchronous response or even complete stalling of the motor.

In summary, the missing (or dropping) of a pulse can be interpreted in two ways. First, a pulse can be lost between the pulse generator (e.g., a command microcomputer or controller) and the translator. In this case, the logic sequence within the translator that energizes motor phases will remain intact. The next pulse to arrive at the translator will be interpreted as the lost pulse and will energize the phase corresponding to the lost pulse. As a result, a "time delay" is introduced to the command (pulse) sequence. The second interpretation of a missed pulse is that the pulse actually reached the translator, but the corresponding motor phase was not energized because of some hardware problem in the translator or other drive circuit. In this case, the next pulse reaching the translator will not energize the phase corresponding to the missed pulse but will energize the phase corresponding to the present pulse. This interpretation is termed *missed phase activation*.

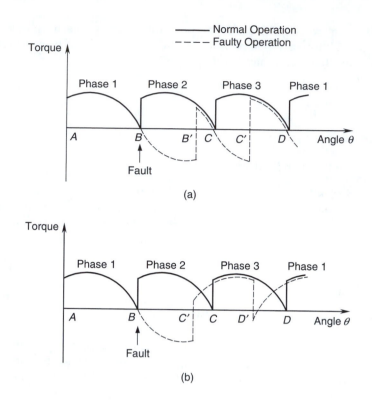

FIGURE 8.35
Motor deceleration due to pulse missing: (a) Case of a missed pulse; (b) Case of a missed phase activation.

In both interpretations of pulse missing, the motor will decelerate because of the negative torque from the phase that was not switched off. Depending on the timing of subsequent pulses, a negative torque can continue to exist in the motor, thereby eventually stalling the motor. Motor deceleration due to pulse missing can be explained using the static torque approximation, as shown in Figure 8.35. Consider a three-phase motor with one phase on excitation (i.e., only one phase is excited at a given time). Suppose that under normal operating conditions, the motor runs at a constant speed and phase activation is brought about at points A, B, C, D, etc. in Figure 8.35, using a pulse sequence sent into the translator. Note that these points are equally spaced (with the horizontal axis being the angle of rotation θ, not time t) because of constant speed operation. The torque generated by the motor under normal operation, without pulse missing, is shown as a solid line in Figure 8.35. Note that Phase 1 is excited at point A, Phase 2 is excited at point B, Phase 3 is excited at point C, and so on. Now let us examine the two cases of pulse missing.

In the first case (Figure 8.35(a)), a pulse is missed at B. Phase 1 continues to be active, providing a negative torque. This slows down the motor. The next pulse is received when the rotor is at position B' (not C) because of rotor deceleration (note that pulses are sent at equal time intervals for constant speed operation). At point B', Phase 2 (not Phase 3) is excited in this case, because the translator interprets the present pulse as the pulse that was lost. The next pulse is received at C', and so on. The resulting torque is shown by the broken line in Figure 8.35(a). Since this torque could be much less than the torque in the absence of missed pulses—depending on the locations of points B', C', and so on—the motor might decelerate continuously and finally stall.

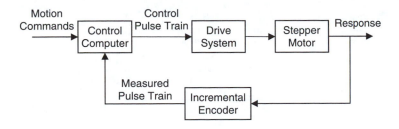

FIGURE 8.36
Feedback control of a stepper motor.

In the second case (Figure 8.35(b)), the pulse at *B* fails to energize Phase 2. This decelerates the motor because of the negative torque generated by the existing Phase 1. The next pulse is received at point *C'* (not C) because the motor has slowed. This pulse excites Phase 3 (not Phase 2, unlike in the previous case), because the translator assumes that Phase 2 has been excited by the previous pulse. The subsequent pulse arrives at point *D'* (not *D)* because of the slowed speed of the motor. The corresponding motor torque is shown by the broken line in Figure 8.35(b). In this case as well, the net torque can be much smaller than what is required to maintain the normal operating speed, and the motor may stall. To avoid this situation, pulse missing should be detected by response sensing (e.g., using an optical encoder), and proper corrective action taken by modifying the future switching sequence in order to accelerate the motor back into the desired trajectory. In other words, feedback control is required.

8.7.2 Feedback Control

Feedback control is used to compensate for motion errors in stepper motors. A block diagram for a typical closed-loop control system is shown in Figure 8.36. This should be compared with Figure 8.16(a). The noted improvement in the feedback control scheme is that the actual response of the stepper motor is sensed and compared with the desired response; if an error is detected, the pulse train to the drive system is modified appropriately to reduce the error. Typically, an *optical incremental encoder* (see Chapter 7) is employed as the motion transducer. This device provides two pulse trains that are in phase quadrature (or, a position pulse sequence and a direction change pulse), giving both the magnitude and the direction of rotation of the stepper motor. The encoder pitch angle should be made equal to the step angle of the motor for ease of comparison and error detection. Note that when feedback control is employed, the resulting closed-loop system can operate near the rated capacity (torque, speed, acceleration, etc.) of the stepper motor, perhaps exceeding these ratings at times but without introducing excessive error and stability problems (e.g., hunting).

A simple closed-loop device that does not utilize sophisticated control logic is the *feedback encoder-driven stepper motor*. In this case, the drive pulses, except for the very first pulse, are generated by a feedback encoder itself, which is mounted on the motor shaft. This mechanism is particularly useful for operations of steady acceleration and deceleration under possible overload conditions, when there is a likelihood of pulse missing. The principle of operation of a feedback encoder-driven stepper motor may be explained using Figure 8.37. The starting pulse is generated externally at the initial detent position *O*. This will energize Phase 1 and drive the rotor toward the corresponding detent position D_1. The encoder disk is positioned such that the first pulse from the encoder is generated at E_1. This pulse is automatically fed back as the second pulse input to the motor (translator).

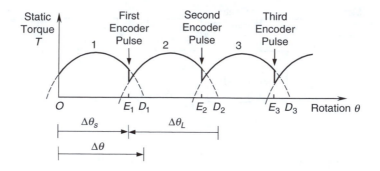

FIGURE 8.37
Operation of a feedback encoder-driven stepper motor.

This pulse will energize Phase 2 and drive the rotor toward the corresponding detent position D_2. During this step, the second pulse from the encoder is generated at E_2, which is automatically fed back as the third pulse input to the motor, energizing Phase 3 and driving the motor toward the detent position D_3, and so on. Note that phase switching occurs (because of an encoder pulse) every time the rotor has turned through a fixed angle $\Delta\theta_s$, from the previous detent position. This angle is termed the *switching angle*. The encoder pulse "leads" the corresponding detent position by an angle $\Delta\theta_L$. This angle is termed the *lead angle*. Note from Figure 8.37 that

$$\Delta\theta_s + \Delta\theta_L = 2\Delta\theta \qquad (8.44)$$

where $\Delta\theta$ denotes the step angle.

For the switching angle position (or lead angle position) shown in Figure 8.37, the static torque on the rotor is positive throughout the motion. As a result, the motor will accelerate steadily until the motor torque exactly balances damping torque, other speed-dependent resistive torques, and load torque. The resulting final steady-state condition corresponds to the maximum speed of operation for a feedback encoder-driven stepper motor. This maximum speed usually decreases as the switching angle is increased beyond the point of intersection of two adjacent torque curves. For example, if $\Delta\theta_s$ is increased beyond $\Delta\theta$, there is a negative static torque from the present phase (before switching), which tends to somewhat decelerate the motor. But the combined effect of the before-switching torque and the after-switching torque is to produce an overall increase in speed until the speed limit is reached. This is generally true, provided that the lead angle $\Delta\theta_L$ is positive (the positive direction, as indicated by the arrowhead in Figure 8.37). The lead angle may be adjusted either by physically moving the signal pick-off point on the encoder disk or by introducing a *time delay* into the feedback path of the encoder signal. The former method is less practical, however.

Steady decelerations can be achieved using feedback encoder-driven stepper motors if negative lead angles are employed. In this case, switching to a particular phase occurs when the rotor has actually passed the detent position for that phase. The resulting negative torque will steadily decelerate the rotor, eventually bringing it to a halt. Negative lead angles may be obtained by simply adding a time delay into the feedback path. Alternatively, the same effect (negative torque) can be generated by blanking out (using a blanking gate) the first two pulses generated by the encoder and using the third pulse to energize the phase that would be energized by the first pulse for accelerating operation.

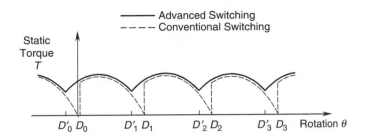

FIGURE 8.38
The effect of advancing the switching pulses.

The feedback encoder-driven stepper motor is a simple form of closed-loop control. Its application is normally limited to steadily accelerating (upramping), steadily decelerating (downramping), and steady-state (constant speed or slewing) operations. More sophisticated feedback control systems require point-by-point comparison of the encoder pulse train with the desired pulse train and injection of extra pulses or extraction (blanking out) of existing pulses at proper instants so as to reduce the error. A commercial version of such a feedback controller uses a *count-and-compare* card. More complex applications of closed-loop control include switching control for electronic damping (see Figures 8.31 through 8.33), transient drive sequencing (see Figure 8.25), and dynamic torque control.

8.7.3 Torque Control Through Switching

Under standard operating conditions for a stepper motor, phase switching (by a pulse) occurs at the present detent position. It is easy to see from the static torque diagram in Figure 8.38, however, that a higher average torque is possible by advancing the switching time to the point of intersection of the two adjacent torque curves (before and after switching). In the figure, the standard switching points are denoted as D_0, D_1, D_2, and so forth, and the advanced switching points as D_0', D_1', D_2', and so forth. In the case of advanced switching, the static torque always remains greater than the common torque value at the point of intersection. This confirms what is intuitively clear: Motor torque can be controlled by adjusting the switching point. The resulting actual magnitude of torque, however, will depend on the dynamic conditions that exist. For low speeds, the dynamic torque may be approximated by the static torque curve, making the analysis simpler. As the speed increases, the deviation from the static curve becomes more pronounced, for reasons that were mentioned earlier.

Example 8.8

Suppose that the switching point is advanced beyond the zero-torque point of the switched phase, as shown in Figure 8.39. The switching points are denoted by D_0', D_1', D_2', and so forth. Note that although the static torque curve takes negative values in some regions under this advanced switching sequence, the dynamic torque stays positive at all times. The main reason for this is that a finite time is needed for the current in the turned-off phase to decay completely because of induced voltages and eddy current effects.

8.7.4 Model-Based Feedback Control

The improved motor model, as presented before, is useful in computer simulation of stepper motors, for example, for performance evaluation. Such a model is also useful in model-based feedback control of stepper motors. In this latter case, the model provides

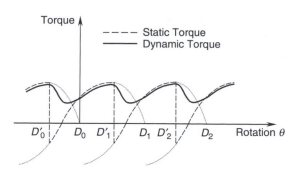

FIGURE 8.39
Dynamic torque at high speeds.

a relationship between the motor torque and the motion variables θ and $\dot\theta$. From the model then, we can determine the required phase-switching points in order to generate a desired motor torque (to drive the load). Actual values of θ and $\dot\theta$ (e.g., as measured using an incremental optical encoder) are used in model-based computations.

A simple feedback control strategy for a stepper motor is outlined now. Initially, when the motor is at rest, the phase current $i_p = 0$. Also, $\theta = -\Delta\theta$ and $\dot\theta = 0$. As a phase is switched on to drive the motor, the motor Equation 8.39 is integrated in real time, using a suitable integration algorithm and an appropriate time step. Simultaneously, the desired position is compared with the actual (measured) position of the load. If the two are sufficiently close, no phase-switching action is taken. But suppose that the actual position lags behind the desired position. Then we compute the present motor torque using the model: Equation 8.40, Equation 8.41, and Equation 8.42 or Equation 8.43, and repeat the computations, assuming (hypothetically) that the excitation is switched to one of the two adjoining phases. Since we need to accelerate the motor, we should actually switch to the phase that provides a torque larger than the present torque. If the actual position leads the desired position, however, we need to decelerate the motor. In this case, we switch to the phase that provides a torque smaller than the present torque or we turn off all the phases. The time taken by the phase current to build up to its full value is approximately equal to 4τ, where τ is the electrical time constant for each phase, as approximated by $\tau = L_o/R$. Hence, when a phase is hypothetically switched, numerical integration has to be performed for a time period of 4τ before the torques are compared. It follows that the performance of this control approach will depend on the operating speed of the motor and the computational efficiency of the integration algorithm. At high speeds, less time is available for control computations. Ironically, it is at high speeds that control problems are severe and sophisticated control techniques are needed; hence, hardware implementations of switching are desired. For better control, phase switching has to be based on the motor speed as well as the motor position.

8.8   Stepper Motor Selection and Applications

We have discussed design problem that addressed the selection of geometric parameters (number of stator poles, number of teeth per pole, number of rotor teeth, etc.) for a stepper motor. Selection of a stepper motor for a specific application cannot be made on the basis of geometric parameters alone, however. Torque and speed considerations are often more crucial in the selection process. For example, a faster speed of response is possible if a motor with a larger torque-to-inertia ratio is used. Furthermore, from the perspective of mechatronics, a motor has to be properly matched to its drive system as well as the driven load. In this context it is useful to review some terminology related to torque characteristics of a stepper motor.

8.8.1 Torque Characteristics and Terminology

The torque that can be provided to a load by a stepper motor depends on several factors. For example, the motor torque at constant speed is not the same as that when the motor "passes through" that speed (i.e., under acceleration, deceleration, or general transient conditions). In particular, at constant speed there is no inertia torque. Also, the torque losses due to magnetic induction are lower at constant stepping rates in comparison to variable stepping rates. It follows that the available torque is larger under steady (constant-speed) conditions. Another factor of influence is the magnitude of the speed. At low speeds (i.e., when the step period is considerably larger than the electrical time constant), the time taken for the phase current to build up or turned off is insignificant compared to the step time. Then the phase current waveform can be assumed rectangular. At high stepping rates, the induction effects dominate and as a result a phase may not reach its rated current over the duration of a step. As a result, the generated torque will be degraded. Furthermore, since the power provided by the power supply is limited, the torque × speed product of the motor is limited as well. Consequently, as the motor speed increases, the available torque must decrease in general. These two are the primary reasons for the characteristic shape of a speed-torque curve of a stepper motor where the peak torque occurs at a very low (typically zero) speed, and as the speed increases the available torque decreases. Eventually, at a particular limiting speed (known as the no-load speed) the available torque becomes zero.

The characteristic shape of the speed-torque curve of a stepper motor is shown in Figure 8.40. Some terminology is given as well. What is given may be interpreted as experimental data measured under steady operating conditions (and averaged and interpolated). The given torque is called the "pull-out torque" and the corresponding speed is the "pull-out speed." In industry, this curve is known as the "pull-out curve."

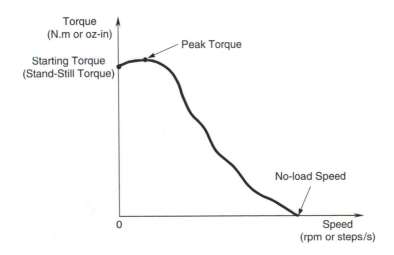

FIGURE 8.40
The speed-torque characteristics of a stepper motor.

Holding torque is the maximum static torque (see Equation 8.36, for instance), and is different from the maximum (pull-out) torque defined in Figure 8.40. In particular, the holding torque can be about 40% greater than the maximum pull-out torque, which is typically equal to the starting torque (or, stand-still torque). Furthermore, the static torque becomes higher if the motor has more than one stator pole per phase and if all these poles are excited at a time. The *residual torque* is the maximum static torque that is present when the motor phases are not energized (i.e., under power-off conditions). This torque is practically zero for a VR motor, but is not negligible for a PM motor. In some industrial literature, *detent torque* takes the same meaning as the residual torque. In this context, detent torque is defined as the torque ripple that is present under power-off conditions. A more appropriate definition for detent torque is the static torque at the present "detention" position (equilibrium position) of the motor, when the next phase is energized. According to this definition, detent toque is equal to $T_{max} \sin 2\pi/p$ where T_{max} is the holding torque, and p is the number of phases.

Some further definitions of speed-torque characteristics of a stepper motor are given in Figure 8.41. The pull-out curve or the *slew curve* here takes the same meaning as what is given in Figure 8.40. Another curve known as the *start-stop curve* or *pull-in curve* is given as well.

As noted before, the pull-out curve (or, slew curve) gives the speed at which motor can run under steady (constant-speed) conditions, under rated current and using appropriate drive circuitry. But, the motor is unable to steadily accelerate to the slew speed, starting from rest and applying a pulse sequence at constant rate corresponding to the slew speed. Instead, it should be accelerated first up to the pull-in speed by applying a pulse sequence corresponding to this speed. After reaching the start-stop region (pull-in region) in this manner, the motor can be accelerated to the pull-out speed (or to a speed lower than this, within the slew region). Similarly, when stopping the motor from a slew speed, it should be first decelerated (by down-ramping) to a speed in the start-stop region (pull-in region) and only when this region is reached satisfactorily, the stepping sequence should be turned off.

Since the drive system determines the current and the switching sequence of the motor phases and the rate at which the switching pulses are applied, it directly affects the speed-torque curve of a motor. Accordingly, what is given in a product data sheet should be interpreted as the speed-torque curve of the particular motor when used with a specified drive system and a matching power supply, and operating at rated values.

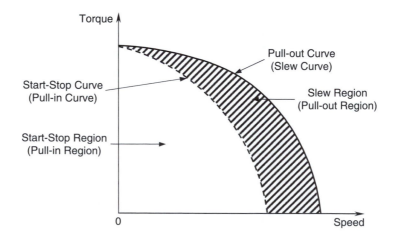

FIGURE 8.41
Further speed-torque characteristics and terminology.

8.8.2 Stepper Motor Selection

The effort required in selecting a stepper motor for a particular application can be reduced if the selection is done in a systematic manner. The following steps provide some guidelines for the selection process:

Step 1. List the main requirements for the particular application, according to the conditions and specifications for the application.

These include operational requirements such as speeds, accelerations, and required accuracy and resolution, and load characteristics, such as size, inertia, fundamental natural frequencies, and resistance torques.

Step 2. Compute the operating torque and stepping rate requirements for the particular application.

Newton's second law is the basic equation employed in this step. Specifically, the required torque rating is given by

$$T = T_R + J_{eq} \frac{\omega_{max}}{\Delta t} \tag{8.45}$$

where

T_R = net resistance torque
J_{eq} = equivalent moment of inertia (including rotor, load, gearing, dampers, etc.)
ω_{max} = maximum operating speed
Δt = time taken to accelerate the load to the maximum speed, starting from rest

Step 3. Using the torque versus stepping rate curves (pull-out curves) for a group of commercially available stepper motors, select a suitable stepper motor.

The torque and speed requirements determined in Step 2 and the accuracy and resolution requirements specified in Step 1 should be used in this step.

Step 4. If a stepper motor that meets the requirements is not available, modify the basic design.

This may be accomplished by changing the speed and torque requirements by adding devices such as gear systems (e.g., harmonic drive) and amplifiers (e.g., hydraulic amplifiers).

Step 5. Select a drive system that is compatible with the motor and that meets the operational requirements in Step 1.

Motors and appropriate drive systems are prescribed in product manuals and catalogs available from the vendors. For relatively simple applications, a manually controlled preset indexer or an open-loop system consisting of a pulse source (oscillator) and a translator could be used to generate the pulse signal to the translator in the drive unit. For more complex transient tasks, a software controller (a microprocessor or a personal computer), or a customized hardware controller may be used to generate the desired pulse command in open-loop operation. Further sophistication may be incorporated by using digital processor-based closed-loop control with encoder feedback, for tasks that require very high accuracy under transient conditions and for operation near the rated capacity of the motor.

The single most useful piece of information in selecting a stepper motor is the torque versus stepping rate curve (the pull-out curve). Other parameters that are valuable in the selection process include:

1. The step angle or the number of steps per revolution
2. The static holding torque (maximum static torque of motor when powered at rated voltage)

3. The maximum slew rate (maximum steady-state stepping rate possible at rated load)

4. The motor torque at the required slew rate (pull-out torque, available from the pull-out curve)

5. The maximum ramping slope (maximum acceleration and deceleration possible at rated load)

6. The motor time constants (no-load electrical time constant and mechanical time constant)

7. The motor natural frequency (without an external load and near detent position)

8. The motor size (dimensions of poles, stator and rotor teeth, air gap and housing, weight, rotor moment of inertia)

9. The power supply ratings (voltage, current, and power)

There are many parameters that determine the ratings of a stepper motor. For example, the static holding torque increases with the number of poles per phase that are energized, and decreases with the air gap width and tooth width, and increases with the rotor diameter and stack length. Furthermore, the minimum allowable air gap width, should exceed the combined maximum lateral (flexural) deflection of the rotor shaft caused by thermal deformations and the flexural loading, such as magnetic pull and static and dynamic mechanical loads. In this respect, the flexural stiffness of the shaft, the bearing characteristics, and the thermal expansion characteristics of the entire assembly become important. Field winding parameters (diameter, length, resistivity, etc.) are chosen by giving due consideration to the required torque, power, electrical time constant, heat generation rate, and motor dimensions. Note that a majority of these are design parameters that cannot be modified in a cost-effective manner during the motor selection stage.

8.8.2.1 Positioning (X-Y) Tables

A common application of stepper motors is in positioning tables (see Figure 8.42(a)). A two-axis (x-y) table requires two stepper motors of nearly equal capacity. The values of the following parameters are assumed to be known:

- Maximum positioning resolution (displacement/step)
- Maximum operating velocity, to be attained in less than a specified time
- Weight of the x-y table
- Maximum resistance force (primarily friction) against table motion

A schematic diagram of the mechanical arrangement for one of the two axes of the table is shown in Figure 8.42(b). A lead screw is used to convert the rotary motion of the motor into rectilinear motion. Free-body diagrams for the motor rotor and the table are shown in Figure 8.43.

Now we will derive a somewhat generalized relation for this type of application. The equations of motion (from Newton's second law) are

$$\text{For the rotor: } T - T_R = J\alpha \tag{8.46}$$

$$\text{For the table: } F - F_R = ma \tag{8.47}$$

where

T = motor torque

T_R = resistance torque from the lead screw

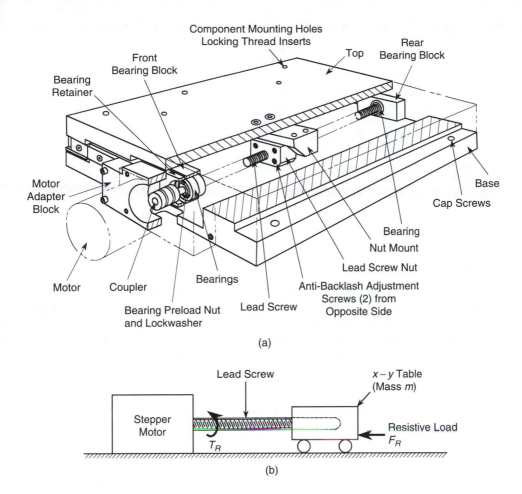

(a)

(b)

FIGURE 8.42
(a) A single axis of a positioning table; (b) An equivalent model.

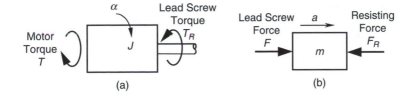

FIGURE 8.43
Free-body diagrams: (a) Motor rotor; (b) Table.

J = equivalent moment of inertia of the rotor
α = angular acceleration of the rotor
F = driving force from the lead screw
F_R = external resistance force on the table
M = equivalent mass of the table
a = acceleration of the table

Assuming a rigid lead screw without backlash, the compatibility condition is written as

$$\alpha = r\alpha \tag{8.48}$$

where r denotes the *transmission ratio* (rectilinear motion/angular motion) of the lead screw. The load transmission equation for the lead screw is

$$F = \frac{e}{r}T_R \tag{8.49}$$

where e denotes the *fractional efficiency* of the lead screw. Finally, Equation (8.46) through Equation 8.49 can be combined to give

$$T = \left(J + \frac{mr^2}{e}\right)\frac{a}{r} + \frac{r}{e}F_R \tag{8.50}$$

Example 8.9

A schematic diagram of an industrial conveyor unit is shown in Figure 8.44. In this application, the conveyor moves intermittently at a fixed rate, thereby indexing the objects on the conveyor through a fixed distance d in each time period T. A triangular speed profile is used for each motion interval, having an acceleration and a deceleration that are equal in magnitude (see Figure 8.45). The conveyor is driven by a stepper motor. A gear unit with step-down speed ratio p:1, where $p > 1$, may be used if necessary, as shown in Figure 8.44.

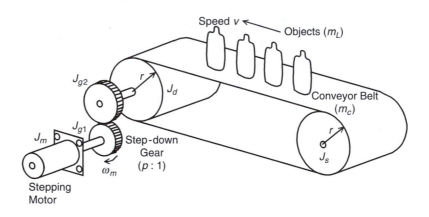

FIGURE 8.44
Conveyor unit with intermittent motion.

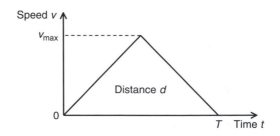

FIGURE 9.45
Speed profile for a motion period of the conveyor.

TABLE 8.2

Stepper Motor Data (Aerotech, Inc., pittsburg, PA. With Permission.)

		Stepping Motor Specifications			
Model		50SM	101SM	310SM	1010SM
Nema Motor Frame Size			23	34	42
Full Step Angle	degrees		1.8		
Accuracy	percent		±3 (noncumulative)		
Holding Torque	oz-in	38	90	370	1050
	N·m	0.27	0.64	2.61	7.42
Detent Torque	oz·in	6	18	25	20
	N·m	0.04	0.13	0.18	0.14
Rated Phase Current	Amps	1	5	6	8.6
Rotor-inertia	oz-in-sec²	1.66×10^3	5×10^{-3}	26.5×10^{-3}	114×10^{-3}
	kg·m²	11.8×10^6	35×10^{-6}	187×10^{-6}	805×10^{-6}
Maximum Radial Load	ib		15	35	40
	N		67	156	178
Maximum Thrust Load	ib		25	60	125
	N		111	267	556
Weight	ib	1.4	2.8	7.8	20
	kg	0.6	1.3	3.5	9.1
Operating Temperature	°C		−55 to +50		
Storage Temperature	°C		−55 to +130		

a. Explain why the equivalent moment of inertia J_e at the motor shaft, for the overall system, is given by

$$J_e = J_m + J_{g1} + \frac{1}{p^2}(J_{g2} + J_d + J_s) + \frac{r^2}{p^2}(m_c + m_L)$$

where J_m, J_{g1}, J_{g2}, J_d, and J_s are the moments of inertia of the motor rotor, drive gear, driven gear, drive cylinder of the conveyor, and the driven cylinder of the conveyor, respectively; m_c and m_L are the overall masses of the conveyor belt and the moved objects (load), respectively; and r is the radius of each of the two conveyor cylinders.

b. Four models of stepping motor are available for the application. Their specifications are given in Table 8.2 and the corresponding performance curves are given in Figure 8.46. The following values are known for the system:

$$d = 10 \text{ cm}, \quad T = 0.2 \text{ sec}, \quad r = 10 \text{ cm}, \quad m_c = 5 \text{ kg}, \quad m_L = 5 \text{ kg},$$

$$J_d = J_s = 2.0 \times 10^{-3} \text{ kg} \cdot \text{m}^2.$$

Also two gear units with $p = 2$ and 3 are available, and for each unit $J_{g1} = 50 \times 10^{-6} \text{ kg} \cdot \text{m}^2$ and $J_{g2} = 200 \times 10^{-6} \text{ kg} \cdot \text{m}^2$.

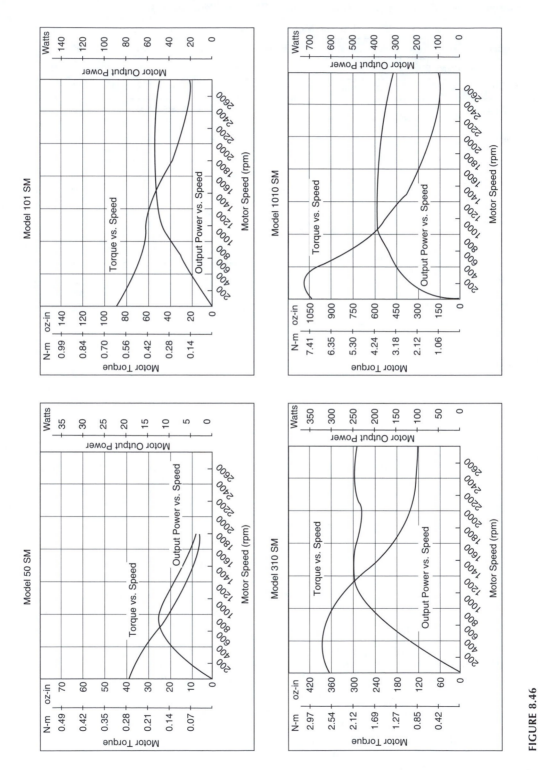

FIGURE 8.46
Stepper motor performance curves. (Aerotech, Inc., Pittsburgh, PA. With permission.)

Indicating all calculations and procedures, select a suitable motor unit for this application. You must not use a gear unit unless it is necessary to have one with the available motors.

What is the positioning resolution of the conveyor (rectilinear) for the final system?

NOTE Assume an overall system efficiency of 80% regardless of whether a gear unit is used.

SOLUTION

a. Angular speed of the motor and drive gear $= \omega_m$.

Angular speed of the driven gear and conveyor cylinders $= \frac{\omega_m}{p}$.

Rectilinear speed of the conveyor and objects $v = \frac{r\omega_m}{p}$.

Kinetic energy of the overall system

$$= \frac{1}{2}(J_m + J_{g1})\omega_m^2 + \frac{1}{2}(J_{g2} + J_d + J_s)\left(\frac{\omega_p}{p}\right)^2 + \frac{1}{2}(m_c + m_L)\left(\frac{r\omega_m}{p}\right)^2$$

$$= \frac{1}{2}[J_m + J_{g1} + \frac{1}{p^2}(J_{g2} + J_d + J_s) + \frac{r^2}{p^2}(m_c + m_L)]\omega_m^2$$

$$= \frac{1}{2}J_e\omega_m^2$$

Hence, the equivalent moment of inertia as felt at the motor rotor, is

$$J_e = J_m + J_{g1} + \frac{1}{p^2}(J_{g2} + J_d + J_s) + \frac{r^2}{p^2}(m_c + m_L)$$

b. From the triangular speed profile we have

$$d = \frac{1}{2}v_{max}T$$

Substitute numerical values:

$$0.1 = \frac{1}{2}v_{max}0.2$$

or

$$v_{max} = 1.0 \text{ m/s}$$

The acceleration/deceleration of the system

$$a = \frac{v_{max}}{T/2} = \frac{1.0}{0.2/2} \text{ m/s}^2 = 10.0 \text{ m/s}^2$$

Corresponding angular acceleration/deceleration of the motor

$$\alpha = \frac{pa}{r}$$

With an efficiency of η, the motor torque T_m that is needed to accelerate/decelerate the system is given by

$$\eta T_m = J_e \alpha = J_e \frac{pa}{r} = \left[J_m + J_{g1} + \frac{1}{p^2}(J_{g2} + J_d + J_s) + \frac{r^2}{p^2}(m_c + m_L) \right] \frac{pa}{r}$$

Maximum speed of the motor

$$\omega_{max} = \frac{pv_{max}}{r}$$

Without gears, we have

$$\eta T_m = [J_m + J_d + J_s + r^2(m_c + m_L)] \frac{a}{r}$$

and

$$\omega_{max} = \frac{v_{max}}{r}$$

Now we substitute numerical values.

Case 1: Without Gears
For an efficiency value $\eta = 0.8$ (i.e., 80% efficient), we have

$$0.8T_m = [J_m + 2 \times 10^{-3} + 2 \times 10^{-3} + 0.1^2(5+5)] \frac{10}{0.1} \ \text{N·m}$$

or

$$T_m = 125.0[J_m + 0.104] \ \text{N.m}$$

and

$$\omega_{max} = \frac{1.0}{0.1} \ \text{rad/s} = 10 \times \frac{60}{2\pi} \ \text{rpm} = 95.5 \ \text{rpm}$$

The operating speed range is 0–95.5 rpm. Note that the torque at 95.5 rpm is less than the starting torque for the first two motor models, and not so for the second two models (see the speed-torque curves in Figure 8.46). We must use the weakest point (i.e., lowest torque) in the operating speed range, in the motor selection process. Allowing for this requirement, Table 8.3 is formed for the four motor models.

It is seen that without a gear unit, the available motors cannot meet the system requirements.

Case 2: With Gears

TABLE 8.3

Data for Selecting a Motor Without a Gear Unit

Motor Model (SM)	Available Torque at ω_{max} (N·m)	Motor Rotor Inertia (kg·m²)	Required Torque (N·m)
50	0.26	11.8×10^{-6}	13.0
101	0.60	35.0×10^{-6}	13.0
310	2.58	187.0×10^{-6}	13.0
1010	7.41	805.0×10^{-6}	13.1

TABLE 8.4

Data for Selecting a Motor With a Gear Unit

Motor Model	Available Torque at ω_{max} (N·m)	Motor Rotor Inertia (kg·m²)	Required Torque (N·m)
50 SM	0.25	11.8×10^{-6}	6.53
101 SM	0.58	35.0×10^{-6}	6.53
310 SM	2.63	187.0×10^{-6}	6.57
1010 SM	7.41	805.0×10^{-6}	6.73

NOTE Usually the system efficiency drops when a gear unit is introduced. In the present exercise we use the same efficiency for reasons of simplicity.

With an efficiency of 80%, we have $\eta = 0.8$. Then,

$$0.8 T_m = \left[J_m + 50 \times 10^{-6} + \frac{1}{p^2}(200 \times 10^{-6} + 2 \times 10^{-3} + 2 \times 10^{-3}) + \frac{0.1^2}{p^2}(5+5) \right] p \times \frac{10}{0.1} \; \text{N·m}$$

and

$$\omega_{max} = \frac{1.0p}{0.1} \; \text{rad/s} = 10p \times \frac{60}{2\pi} \; \text{rpm}$$

or

$$T_m = 125.0 \left[J_m + 50 \times 10^{-6} + \frac{1}{p^2} \times 104.2 \times 10^{-3} \right] p \; \text{N·m}$$

and

$$\omega_{max} = 95.5p \; \text{rpm}$$

For the case of $p = 2$ we have $\omega_{max} = 191.0$ rpm. Table 8.4 is formed for the present case. It is seen that with a gear of speed ratio $p = 2$, motor model 1010 SM satisfies requirement. With full stepping, step angle of the rotor = 1.8°. Corresponding step in the conveyor motion is the positioning resolution.

With $p = 2$ and $r = 0.1$ m, the position resolution is $\frac{1.8°}{2} \times \frac{\pi}{180°} \times 0.1 = 1.57 \times 10^{-3}$ m.

8.8.3 Stepper Motor Applications

More than one type of actuator may be suitable for a given application. In the present discussion we indicate situations where stepper motor is a suitable choice as an actuator. It does not, however, rule out the use of other types of actuators for the same application.

Stepper motors are particularly suitable for positioning, ramping (constant acceleration and deceleration), and slewing (constant speed) applications at relatively low speeds. Typically they are suitable for short and repetitive motions at speeds lower than 2,000 rpm. They are not the best choice for servoing or trajectory following applications, because of jitter and step (pulse) missing problems (dc and ac servo motors are better for such applications). Encoder feedback will make the situation better, but at a higher cost and controller complexity. Generally, however, stepper motor provides a low-cost option in a variety of applications.

The stepper motor is a low-speed actuator that may be used in applications that require torques as high as 15 N·m (2121 oz.in). For heavy-duty applications, torque amplification may be necessary. One way to accomplish this is by using a hydraulic actuator in cascade with the motor. The hydraulic valve (typically a rectilinear spool valve as described in Chapter 9), which controls the hydraulic actuator (typically a piston-cylinder device), is driven by a stepper motor through suitable gearing for speed reduction as well as for rotary-rectilinear motion conversion. Torque amplification by an order of magnitude is possible with such an arrangement. Of course, the time constant will increase and operating bandwidth will decrease because of the sluggishness of hydraulic components. Also, a certain amount of backlash will be introduced by the gear system. Feedback control will be necessary to reduce the position error, which is usually present in open-loop hydraulic actuators.

Stepper motors are incremental actuators. As such, they are ideally suited for digital control applications. High-precision open-loop operation is possible as well, provided that the operating conditions are well within the motor capacity. Early applications of stepper motor were limited to low-speed, low-torque drives. With rapid developments in solid-state drives and microprocessor-based pulse generators and controllers, however, reasonably high-speed operation under transient conditions at high torques and closed-loop control have become feasible. Since brushes are not used in stepper motors, there is no danger in spark generation. Hence, they are suitable in hazardous environments. But, heat generation and associated thermal problems can be significant at high speeds. In particular, mainly because of heat generator, stepper motors are not suitable for applications that require operation over long periods without stopping.

There are numerous applications of stepper motors. For example, stepper motor is particularly suitable in printing applications (including graphic printers, plotters, and electronic typewriters) because the print characters are changed in steps and the printed lines (or paper feed) are also advanced in steps. Stepper motors are commonly used in x–y tables. In automated manufacturing applications, stepper motors are found as joint actuators and end effector (gripper) actuators of robotic manipulators and as drive units in programmable dies, parts-positioning tables, and tool holders of machine tools (milling machines, lathes, etc.). In automotive applications, pulse windshield wipers, power window drives, power seat mechanisms, automatic carburetor control, process control applications, valve actuators, and parts-handling systems use stepper motors. Other applications of stepper motors include source and object positioning in medical and metallurgical radiography, lens drives in auto-focus cameras, camera movement in computer vision systems, and paper feed mechanisms in photocopying machines.

The advantages of stepper motors include the following:

1. Position error is noncumulative. A high accuracy of motion is possible, even under open-loop control.

2. The cost is relatively low. Furthermore, considerable savings in sensor (measuring system) and controller costs are possible when the open-loop mode is used.

3. Because of the incremental nature of command and motion, stepper motors are easily adaptable to digital control applications.

4. No serious stability problems exist, even under open-loop control.

5. Torque capacity and power requirements can be optimized and the response can be controlled by electronic switching.

6. Brushless construction has obvious advantages (see Chapter 9).

The disadvantages of stepper motors include the following:

1. They are low speed actuators. The torque capacity is typically less than $15 \, \text{N} \cdot \text{m}$, which may be low compared to torque motors.

2. They have limited speed (limited by torque capacity and by pulse-missing problems due to faulty switching systems and drive circuits).

3. They have high vibration levels due to stepwise motion.

4. Large errors and oscillations can result when a pulse is missed under open-loop control.

5. Thermal problems can be significant when operating at high speeds over long periods.

In most applications, the merits of stepper motors outweigh the drawbacks.

8.9 Problems

8.1 Consider the two-phase PM stepper motor shown in Figure 8.2. Show that in full stepping, the sequence of states of the two phases is given by Table P8.1. What is the step angle in this case?

8.2 Consider the variable-reluctance (VR) stepper motor shown schematically in Figure 8.5. The rotor is a nonmagnetized soft-iron bar. The motor has a two-pole rotor geometry and a three-phase stator. Using a schematic diagram show the half-stepping sequence for a full clockwise rotation of this motor. What is the step angle? Indicate an advantage and a disadvantage of half-stepping over full-stepping.

8.3 Consider a stepper motor with three rotor teeth ($n_r = 3$), two rotor poles ($n_s = 2$), and two phases ($p = 2$). What is the step angle for this motor, in full stepping? Is this a VR motor or a PM motor? Explain.

TABLE P8.1

Stepping Sequence (full- stepping) for a Two-Phase PM Stepper Motor with Two Rotor Poles

State of ϕ_1	State of ϕ_2
1	0
0	↑ CW 1 ↓ CCW
−1	0
0	−1

8.4 Explain why a two-phase variable-reluctance stepper motor is not a physical reality. A single-stack variable-reluctance stepper motor with nontoothed poles has n_r teeth in the rotor, n_s poles in the stator, and p phases of winding. Show that

$$n_r = \left(r + \frac{1}{p} \right) n_s$$

where r is the largest positive integer (natural number) such that $n_r > rn_s$.

8.5 For a single-stack stepper motor that has toothed poles, for the case $\theta_s > \theta_r$, show that,

$$\Delta\theta = \frac{n_s}{mp}(\theta_s - r\theta_r)$$

$$\theta_s = r\theta_r + \frac{m\theta_r}{n_s}$$

$$n_r = rn_s + m$$

where
   $\Delta\theta$ = step angle
   θ_r = rotor tooth pitch
   θ_s = stator tooth pitch
   n_r = number of teeth in the rotor
   n_s = number of teeth in the stator
   p = number of phases
   m = number of stator poles per phase
   r = largest integer such that $\theta_s - r\theta_r > 0$

Assume that the stator teeth are uniformly distributed around the rotor. Derive the corresponding equations for the case $\theta_s < \theta_r$?

8.6 For a stepper motor with m stator poles per phase, show that the number of teeth in a stator pole is given by

$$t_s = \frac{n}{mp^2} - \frac{1}{p}$$

where n denotes the number of steps per revolution, for the case $n_r > n_s$. (Hint: This relation is the counterpart of Equation 8.10a). Pick suitable parameters for a four-phase, eight-pole motor, using this relation, if the step angle is required to be 1.8°. Can the same step be obtained using a three-phase stepper motor?

8.7 Consider the single-stack, three-phase VR stepper motor shown in Figure 8.9 ($n_r = 8$ and $n_s = 12$). For this arrangement, compare the following phase-switching sequences:
   i. 1-2-3-1
   ii. 1-12-2-23-3-31-1
   iii. 12-23-31-12
What is the step angle, and how would you reverse the direction of rotation in each case?

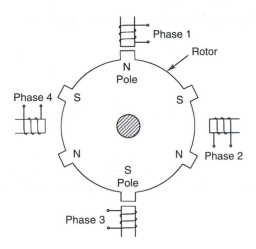

FIGURE P8.9
Schematic diagram of a stepper motor.

8.8 Describe the principle of operation of a single-stack VR stepper motor that has toothed poles in the stator. Assume that the stator teeth are uniformly distributed around the rotor. If the motor has five teeth per pole and two pole pairs per phase and provides 500 full steps per revolution, determine the number of phases in the stator. Also determine the number of stator poles, the step angle, and the number of teeth in the rotor.

8.9 Figure P8.9 shows a schematic diagram of a stepper motor. What type of stepper is this? Describe the operation of this motor. In particular, discuss whether 4 separate phases are needed or whether the phases of the opposite stator poles may be connected together, giving a 2-phase stepper. What is the step angle of the motor

a. In full-stepping?
b. In half-stepping?

8.10 So far, in the problems on toothed single-stack stepper motors, we have assumed that $\theta_r \neq \theta_s$. Now consider the case of $\theta_r = \theta_s$. In a single-stack stepper motor of this type, the stator-rotor tooth misalignment that is necessary to generate the driving torque is achieved by offsetting the entire group of teeth on a stator pole (not just the central tooth of the pole) by the step angle $\Delta\theta$ with respect to the teeth on the adjacent stator pole. The governing equations are Equation 8.11 and Equation 8.12. There are two possibilities, as given by the + sign and the − sign in these equations. The + sign governs the case in which the offset is generated by reducing the pole pitch. The − sign governs the case where the offset of $\Delta\theta$ is realized by increasing the pole pitch. Show that in this latter case it is possible to design a four-phase motor that has fifty rotor teeth. Obtain appropriate values for tooth pitch (θ_r and θ_s), full-stepping step angle $\Delta\theta$, number of steps per revolution (n), number of poles per phase (m), and number of stator teeth per pole (t_s) for this design.

8.11 The stepper motor shown in Figure 8.11(a) uses the *balanced pole arrangement*. Specifically, all the poles wound to the same phase are uniformly distributed around the rotor. In Figure 8.11, there are two poles per phase. Hence, the two poles connected to the same phase are placed at diametrically opposite locations. In general, in the case of m poles per phase, the poles connected to the same phase would be located at angular intervals of $360°/m$. What are the advantages of this balanced pole arrangement?

8.12 In connection with the phase windings of a stepper motor, explain the following terms:

   a. Unifilar (or monofilar) winding

   b. Bifilar winding

   c. Bipolar winding

Discuss why the torque characteristics of a bifilar-wound motor are better than those of a unifilar-wound motor at high stepping rates.

8.13 For a multiple-stack variable-reluctance stepper motor whose rotor tooth pitch angle is not equal to the stator tooth pitch angle (i.e., $\theta_r \neq \theta_s$), show that the step angle may be expressed by

$$\Delta\theta = \frac{\theta_r}{ps}$$

where

   p = number of phases in each stator segment
   s = number of stacks of rotor teeth on the shaft

8.14 Describe the principle of operation of a multiple-stack VR stepper motor that has toothed poles in each stator stack. Show that if $\theta_r < \theta_s$, the step angle of this type of motor is given by

$$\Delta\theta = \frac{n_s}{mps}(\theta_s - \theta_r)$$

where

   θ_r = rotor tooth pitch
   θ_s = stator tooth pitch
   n_s = number of teeth in the stator
   p = number of phases
   m = number of poles per phase
   s = number of stacks

Assume that the stator teeth are uniformly distributed around the rotor and that the phases of different stator segments are independent (can be activated independently).

8.15 The torque of a stepping motor can be increased by increasing its diameter, for a given coil density (the number of turns per unit area) of the stator poles, for a given current rating. Alternatively, the motor torque can be increased by introducing multiple stacks (i.e., a longer motor) for a given diameter, coil density, and current rating. Giving reasons indicate which design is generally preferred.

8.16 The principle of operation of a (hybrid) linear stepper motor is indicated in the schematic diagram of Figure P8.16. The toothed platen is a stationary member made of ferromagnetic material, which is not magnetized. The moving member is termed the "forcer," which has four groups of teeth (only one tooth per group is shown in the figure, for convenience). A permanent magnet has its N pole located at the first two groups of teeth and the S pole located at the next two groups of teeth, as shown. Accordingly, the first two groups are magnetized to take the N polarity and the next two groups take the

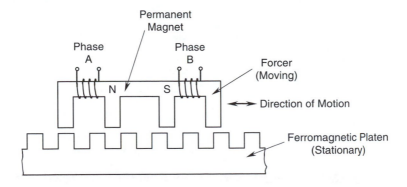

FIGURE P8.16
Schematic representation of a linear hybrid stepping motor.

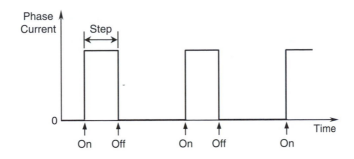

FIGURE P8.17
Ideal phase current waveform for a stepper motor.

S polarity. The motor has two phases, denoted by A and B. Phase A is wound between the first two groups of teeth and Phase B is wound between the second two groups of teeth of the forcer, as shown. In this manner, when Phase A is energized, it will create an electromagnet with opposite polarities located at the first two groups of teeth. Hence, one of this first two groups of teeth will have its magnetic polarity reinforced while the other group will have its polarity neutralized. Similarly, Phase B, when energized, will strengthen one of the next two groups of teeth while neutralizing the other group. The teeth in the four groups of the forcer have quadrature offsets as follows. Second group has an offset of 1/2 tooth pitch with respect to the first group. The third group of teeth has an offset of 1/4 tooth pitch with respect to the first group in one direction, and the fourth group has an offset of 1/4 pitch with respect to the first group in the opposite direction (Hence, the fourth group has an offset of 1/2 pitch with respect to the third group of teeth). The phase windings are bipolar (i.e., the current can be reversed) (a) Describe the full-stepping cycle of this motor, for motion to the right and for motion to the left; (b) Give the half-stepping cycle of this motor, for motion to the right and for motion to the left.

8.17 When a phase winding of a stepper motor is switched on, ideally the current in the winding should instantly reach the full value (hence providing the full magnetic field instantly). Similarly, when a phase is switched off, its current should become zero immediately. It follows that the ideal shape of phase current history is a rectangular pulse sequence, as shown in Figure P8.17. In actual motors, however, the current curves deviate from the ideal rectangular shape, primarily because of the magnetic induction in the phase

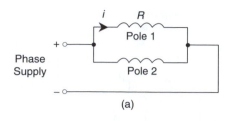

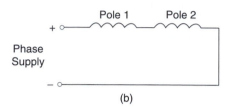

FIGURE P8.18
Pole windings in a phase of a stepper motor that has two poles per phase: (a) Parallel connection; (b) Series connection.

windings. Using sketches, indicate how the phase current waveform would deviate from this ideal shape under the following conditions:

a. Very slow stepping

b. Very fast stepping at a constant stepping rate

c. Very fast stepping at a variable (transient) stepping rate

A stepper motor has a phase inductance of 10 mH and a phase resistance of 5 Ω. What is the electrical time constant of each-phase in a stepper motor? Estimate the stepping rate below which magnetic induction effects can be neglected so that the phase current waveform is almost a rectangular pulse sequence.

8.18 Consider a stepper motor that has two poles per phase. The pole windings in each phase may be connected either in parallel or in series, as shown in Figure P8.18. In each case, determine the required ratings for phase power supply (rated current, rated voltage, rated power) in terms of current i and resistance R, indicated in Figure P8.18(a). Note that the power rating should be the same for both cases, as is intuitively clear.

8.19 Some industrial applications of stepper motors call for very high stepping rates under variable load (variable motor torque) conditions. Since motor torque depends directly on the current in the phase windings (typically 5 A per phase), one method of obtaining a variable-torque drive is to use an adjustable resistor in the drive circuit. An alternative method is to use a *chopper drive.* Switching transistors, diodes, or thyristors are used in a chopper circuit to periodically bypass (chop) the current through a phase winding. The chopped current passes through a free-wheeling diode back to the power supply. The chopping interval and chopping frequency are adjustable. Discuss the advantages of chopper drives compared to the resistance drive method.

8.20 Define and compare the following pairs of terms in the context of electromagnetic stepper motors:

a. Pulses and steps

b. Step angle and resolution

c. Residual torque and static holding torque

   d. Translator and drive system

   e. PM stepper motor and VR stepper motor

   f. Single-stack stepper and multiple-stack stepper

   g. Stator poles and stator phases

   h. Pulse rate and slew rate.

8.21 Compare the VR stepper motor with the PM stepper motor with respect to the following considerations:

   a. Torque capacity for a given motor size

   b. Holding torque

   c. Complexity of switching circuitry

   d. Step size

   e. Rotor inertia.

The hybrid stepper motor possesses characteristics of both the VR and the PM types of stepper motors. Consider a typical construction of a hybrid stepper motor, as shown schematically in Figure P8.21. The rotor has two stacks of teeth made of ferromagnetic material, joined together by a permanent magnet, that assigns opposite polarities to the two rotor stacks. The tooth pitch is the same for both stacks, but the two stacks have a tooth misalignment of half a tooth pitch ($\theta_r/2$). The stator may consist of a common tooth stack for both rotor stacks, or it may consist of two tooth stack segments that are in complete alignment, one for each rotor stack. The number of teeth in the stator is not equal to the number of teeth in each rotor stack. The stator is made up of several toothed poles that are equally spaced around the rotor. Half the poles are connected to one phase and the other half are connected to the second phase. The current in each phase may be turned on and off or reversed using switching amplifiers. The switching sequence for rotation in one direction (say, CW) would be A^+, B^+, A^-, B^-; for rotation in the opposite direction (CCW), it would be A^+, B^-, A^-, B^+, where A and B denote the two phases and the superscripts + and − denote the direction of current in each phase. This may also be denoted by [1, 0], [0, 1], [−1, 0], [0, −1] for CW rotation, and [1, 0], [0, −1], [−1, 0], [0, 1] for CCW rotation.

   Consider a motor that has 18 teeth in each rotor stack and eight poles in the stator, with two teeth per stator pole. The stator poles are wound to the two phases as follows: Two

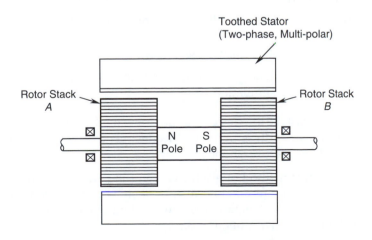

FIGURE P8.21
Schematic diagram of a hybrid stepper motor.

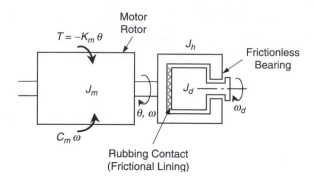

FIGURE P8.22
A Lanchester damper attached to a stepper motor.

radially opposite poles are wound to the same phase, with identical polarity. The two radially opposite poles that are at 90 deg. from this pair of poles is also wound to the same phase, but with the field in the opposite direction (i.e., opposite polarity) to the previous pair.

   a. Using suitable sketches of the rotor and stator configurations at the two stacks, describe the operation of this hybrid stepper motor.

   b. What is the step size of the motor?

8.22 A Lanchester damper is attached to a stepper motor. A sketch is shown in Figure P8.22. Write equations to describe the single-step response of the motor about the detent position. Assume that the flywheel of the damper is not locked onto its housing at any time. Let T_d denote the magnitude of the frictional torque of the damper. Give appropriate initial conditions. Using a computer simulation, plot the motor response, with and without the damper, for the following parameter values:

   Rotor + load inertia $J_m = 4.0 \times 10^{-3}$ N·m²

   Damper housing inertia $J_h = 0.2 \times 10^{-3}$ N·m²

   Damper flywheel inertia $J_d = 1.0 \times 10^{-3}$ N·m²

   Maximum overshoot $\theta(0) = 1.0°$

   Static torque constant (torque gradient) $K_m = 114.6$ N·m/rad

   Damping constant of the motor when the Lanchester damper is disconnected, $C_m = 0.08$ N·m/rad/s

   Magnitude of the frictional torque $T_d = 80.0$ N·m.

   Estimate the resonant frequency of the motor using the given parameter values, and verify it using the simulation results.

8.23 Compare and contrast the three electronic damping methods illustrated in Figures 8.31 through 8.33. In particular, address the issue of effectiveness in relation to the speed of response and the level of final overshoot.

8.24 In the pulse reversal method of electronic damping, suppose that Phase 1 is energized, instead of Phase 3, at point B in Figure 8.32. Sketch the corresponding static torque curve and the motor response. Compare this new method of electronic damping with the pulse reversal method illustrated in Figure 8.32.

8.25 A relatively convenient method of electronic damping uses simultaneous multiphase energization, where more than one phase are energized simultaneously and some of the simultaneous phases are excited with a fraction of the normal operating (rated) voltage. A simultaneous two-phase energization technique has been suggested for a three-phase, single-stack stepper motor. If the standard sequence of switching of the phases for forward motion is given by 1-2-3-1, what is the corresponding simultaneous two-phase energization sequence?

8.26 The torque versus speed curve of a stepper motor is approximated by a straight line, as shown in Figure P8.26. The following two parameters are given:

T_0 = torque at zero speed (starting torque or stand-still torque)

ω_0 = speed at zero torque (no-load speed)

Suppose that the load resistance is approximated by a rotary viscous damper with damping constant b. Assuming that the motor directly drives the load, without any speed reducers, determine the steady-state speed of the load and the corresponding drive torque of the stepper motor.

8.27 Speed-torque curve of a stepper motor is shown in Figure P8.27. Explain the shape, particularly the two dips, of this curve.

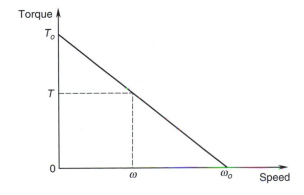

FIGURE P8.26
An approximate speed-torque curve for a stepper motor.

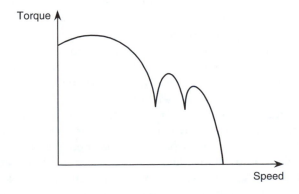

FIGURE P8.27
Typical speed-torque curve of a stepper motor.

Suppose that with one phase on, the torque of a stepper motor in the neighborhood of the detent position of the rotor is given by the linear relationship

$$T = -K_m \theta$$

where θ is the rotor displacement measured from the detent position and K_m is the motor torque constant (or, magnetic stiffness or torque gradient). The motor is directly coupled to an inertial load. The combined moment of inertia of the motor rotor and the inertial load is $J = 0.01$ kg · m². If $K_m = 628.3$ N · m/rad, at what stepping rates would you expect dips in the speed-torque curve of the motor-load combination?

8.28 The torque source model may be used to represent both VR and PM types of stepper motors at low speeds and under steady operating conditions. What assumptions are made in this model?

A stepper motor has an inertial load rigidly connected to its rotor. The equivalent moment of inertia of rotor and load is $J = 5.0 \times 10^{-3}$ kg · m². The equivalent viscous damping is $b = 0.5$ N · m/rad/sec. The number of phases $p = 4$, and the number of rotor teeth $n_r = 50$. Assume full-stepping (step angle $= 1.8°$). The mechanical model for the motor is

$$T = b\dot{\bar{\theta}} + J\ddot{\bar{\theta}}$$

where $\bar{\theta}$ is the absolute position of the rotor.

a. Assuming a torque source model with $T_{max} = 100$ N · m, simulate and plot the motor response $\bar{\theta}$ as a function of t for the first ten steps, starting from rest. Assume that in open-loop control, switching is always at the detent position of the present step. You should pay particular attention to the position coordinate, because $\bar{\theta}$ is the absolute position from the starting point and θ is the relative position measured from the approaching detent position of the current step. Plot the response on the phase plane (with speed $\dot{\bar{\theta}}$ as the vertical axis and position $\bar{\theta}$ as the horizontal axis).

b. Repeat part (a) for the first 150 steps of motion. Check whether a steady state (speed) is reached or whether there is an unstable response.

c. Consider the improved PM motor model with torque due to one excited phase being given by Equation 8.39 through Equation 8.42, with $R = 2.0$ Ω, $L_0 = 10.0$ mH, $L_0 = 2.0$ mH, $k_b = 0.05$ V/rad/sec, $v_p = 20.0$ V, and $k_m = 10.0$ N · m/A. Starting from rest and switching at each detent position, simulate the motor response for the first 10 steps. Plot $\bar{\theta}$ versus t to the same scale as in part (a). Also, plot the response on the phase plane to the same scale as in part (a). Note that at each switching point, the initial condition of the phase current i_p is zero. For example, simulation may be done by picking about 100 integration steps for each motor step. In each integration step, first for known θ and $\dot{\theta}$, integrate Equation 8.39 along with Equation 8.40 and Equation 8.41 to determine i_p. Substitute this in Equation 8.42 to compute torque T for the integration step. Then use this torque and integrate the mechanical equation to determine $\bar{\theta}$ and $\dot{\bar{\theta}}$. Repeat this for the subsequent integration steps. After the detent position is reached, repeat the integration steps for the new phase, with zero initial value for current, but using $\bar{\theta}$ and $\dot{\bar{\theta}}$, as computed before, as the initial values for position and speed. Note that $\bar{\theta} = \dot{\theta}$.

d. Repeat part (c) for the first 150 motor steps. Plot the curves to the same scale as in part (b).

e. Repeat parts (c) and (d), this time assuming a VR motor with torque given by Equation 8.43 and $k_r = 1.0 \text{ N} \cdot \text{m}/\text{A}^2$. The rest of the model is the same as for the PM motor.

f. Suppose that the fifth pulse did not reach the translator. Simulate the open-loop response of the three motor models during the first 10 steps of motion. Plot the response of all three motor models (torque source, improved PM, and improved VR) to the same scale as before. Give both the time history response and the phase plane trajectory for each model.

g. Suppose that the fifth pulse was generated and translated but the corresponding phase was not activated. Repeat part (f) under these conditions.

h. If the rotor position is measured, the motor can be accelerated back to the desired response by properly choosing the switching point. Note that the switching point for maximum average torque is the point of intersection of the two adjacent torque curves, not the detent point. Simulate the response under a feedback control scheme of this type to compensate for the missed pulse in parts (f) and (g). Plot the controlled responses to the same scale as for the earlier results. Each simulation should be done for all three motor models and the results should be presented as a time history as well as a phase plane trajectory. Also, both pulse losing and phase losing should be simulated in each case. Explain how the motor response would change if the mechanical dissipation were modeled by Coulomb friction rather than by viscous damping.

8.29 A stepper motor with rotor inertia J_m drives a free (no-load) gear train, as shown in Figure P8.29. The gear train has two meshed gear wheels. The gear wheel attached to the motor shaft has inertia J_1 and the other gear wheel has inertia J_2. The gear train steps down the motor speed by the ratio $1 : r$ ($r < 1$). One phase of the motor is energized, and once the steady state is reached, the gear system is turned (rotated) slightly from the corresponding detent position and released.

a. Explain why the system will oscillate about the detent position.

b. What is the natural frequency of oscillation (neglecting electrical and mechanical dissipations) in radians per second?

c. What is the significance of this frequency in a control system that uses a stepper as the actuator?

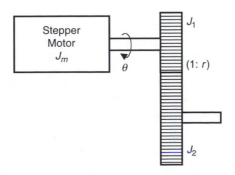

FIGURE P8.29
A stepper motor-driven gear system.

HINT   Static torque for the stepper motor may be taken as

$$T = -T_{max} \sin\left(\frac{2\pi\theta}{p \cdot \Delta\theta}\right)$$

with the usual notation.

8.30 Using the sinusoidal approximation for static torque in a three-phase variable-reluctance stepper motor, the torques T_1, T_2, and T_3 due to the three phases (1, 2, and 3) activated separately, may be expressed as

$$T_1 = -T_{max} \sin n_r\theta$$

$$T_2 = -T_{max} \sin\left(n_r\theta - \frac{2\pi}{3}\right)$$

$$T_3 = -T_{max} \sin\left(n_r\theta - \frac{4\pi}{3}\right)$$

where
   θ = angular position of the rotor measured from the detent position of Phase 1
   n_r = number of rotor teeth.

Using the trigonometric identity

$$\sin A + \sin B = 2\sin\left(\frac{A+B}{2}\right)\cos\left(\frac{A-B}{2}\right)$$

show that

$$T_1 + T_2 = -T_{max} \sin\left(n_r\theta - \frac{\pi}{3}\right)$$

$$T_2 + T_3 = -T_{max} \sin\left(n_r\theta - \pi\right)$$

$$T_3 + T_1 = -T_{max} \sin\left(n_r\theta - \frac{5\pi}{3}\right)$$

Using these expressions, show that the step angle for the switching sequence 1-2-3 is $\theta_r/3$ and the step angle for the switching sequence 1-12-2-23-3-31 is $\theta_r/6$. Determine the step angle for the two-phase-on switching sequence 12-23-31.

8.31 A stepper motor misses a pulse during slewing (high-speed stepping at a constant rate in steady state). Using a displacement versus time curve, explain how a logic controller may compensate for this error by injecting a special switching sequence.

8.32 Briefly discuss the operation of a microprocessor-controlled stepper motor. How would it differ from the standard setup in which a "preset indexer" is employed? Compare and contrast table lookup, programmed stepping, and hardware stepping methods for stepper motor translation.

8.33 Using a static torque diagram, indicate the locations of the first two encoder pulses for a feedback encoder-driven stepper motor for steady deceleration.

8.34 Suppose that the torque produced by a stepping motor when one of the phases is energized can be approximated by a sinusoidal function with amplitude T_{max}. Show that with the advanced switching sequence shown in Figure 8.38, for a three-phase stepper motor ($p = 3$), the average torque generated is approximately $0.827T_{max}$. What is the average torque generated with conventional switching?

8.35 A lectern (or podium) in an auditorium is designed to adjust its height automatically, depending on the height of the speaker. An ultrasonic gage measures the height of the speaker and sends a command to the logic hardware controller of a stepper motor, which adjusts the lectern vertically through a rack-and-pinion drive. The dead load of the moving parts is supported by a bellow device. A schematic diagram of this arrangement is shown in Figure P8.35. The following design requirements have been specified:

Time to adjust a maximum stroke of 1 m = 5 s

Mass of the lectern = 50 kg

Maximum resistance to vertical motion = 5 kg

Displacement resolution = 0.5 cm/step

Select a suitable stepper motor system for this application. You may use the ratings of the four commercial stepper motors as given in Table 8.2 and Figure 8.46.

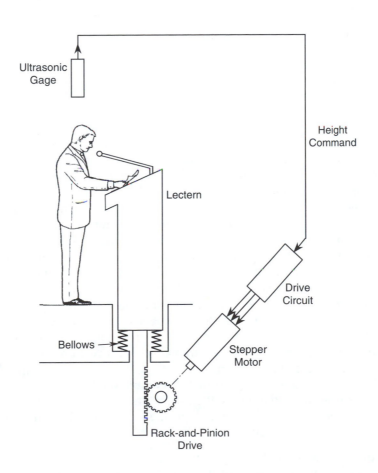

FIGURE P8.35
An automated lectern.

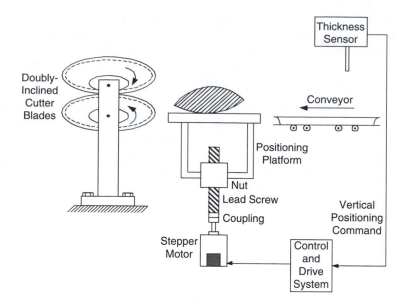

FIGURE P8.36(a)
An automated fish cutting system.

8.36 In connection with a stepper motor, explain the terms:

- Stand-still or stalling torque
- Residual torque
- Holding torque

Figure P8.36(a) shows an automated salmon heading system. The fish move horizontally along a conveyor towards a doubly inclined rotary cutter. The cutter mechanism generates a symmetric V-cut near the gill region of a fish thereby improving the overall product recovery. Before a fish enters the cutter it passes over a positioning platform. The vertical position of the platform is automatically adjusted using a lead screw and nut arrangement that is driven by a stepper motor in the open-loop mode. Specifically, the thickness of a fish is measured using an ultrasound sensor and is transmitted to the drive system of the stepper motor. The drive system commands the stepper motor to adjust the vertical position of the platform according to this measurement so that the fish will enter the cutter blade pair in a symmetrical orientation. The cutter blades are continuously driven by two ac motors. The positioning trajectory of the drive system of the stepper motor is always triangular, starting from rest, uniformly accelerating to a desired speed during the first half of the positioning time and then uniformly decelerating to rest during the second half.

The throughput of the machine is 2 fish/sec. Even though this would make available the full 500 ms for thickness sensing and transmission, it will only provide a fraction of the time for positioning of the platform. More specifically, the platform cannot be positioned for the next fish until the present fish completely leaves the platform, and the positioning has to be completed before the fish enters the cutter. For this reason, the time available for positioning of the platform is specified as 200 ms. Primary specifications for the positioning system are as follows:

- Positioning resolution of 0.1 mm/step
- Maximum positioning range of 2 cm with the positioning time not exceeding 200 ms, while following a triangular speed trajectory

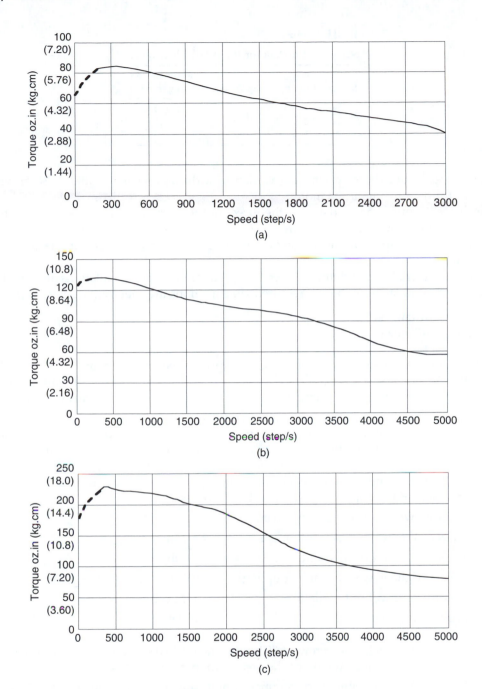

FIGURE P8.36(b)
Speed-torque characteristics of the motors: (a) Model 1; (b) Model 2, (c) Model 3.

- Equivalent mass of a fish, platform and the lead-screw nut = 10 kg
- The equivalent moment of inertia of the lead screw and coupling (excluding the motor rotor inertia) is given as 2.5 kg.cm²
- Lead screw efficiency may be taken as 80%

Suppose that three stepper motors (models 1,2, and 3) of a reputed manufacturer along

TABLE P8.36

Some Useful Data for the Available Motors

Stepper Motor	Step Angle	Rotor Inertia (kg·cm²)
Model 1	1.8°	0.23
Model 2	1.8°	0.67
Model 3	1.8°	1.23

with their respective drive systems are available for this application. Table P8.36 provides some useful data for the three motors.

The speed-torque characteristics of the three motors (when appropriate drive systems are incorporated) are shown in Figure P8.36(b).

Select the most appropriate motor out of the given three models, for the particular application.

Justify your choice by giving all necessary equations and calculations in detail. In particular you must show that all required specifications are met by the selected motor.

NOTE $g = 981$ cm/sec²

1 kg·cm = 13.9 oz.in

8.37 a. In theory, a stepper motor does not require a feedback sensor for its control. But, in practice, a feedback encoder is needed for accurate control, particularly under transient and dynamic loading conditions. Explain the reasons for this.

b. A material transfer unit in an automated factory is sketched in Figure P8.37. The unit consists of a conveyor, which moves objects on to a platform. When an object reaches the platform, the conveyor is stopped and the height of the object is measured using a laser triangulation unit. Then the stepper motor of the platform is activated to raise the object through a distance that is determined on the basis of the object height, for further processing of the object.

The following parameters are given:

Mass of the heaviest object that is raised	= 3.0 lb (1.36 kg)
Mass of the platform and nut	= 3.0 lb
Inertia of the lead screw and coupling	= 0.001 oz.in.s² (0.07 kg·m²)
Maximum travel of the platform	= 1.0 in (2.54 cm)
Positioning time	= 200 ms

Assume a 4-pitch lead screw of 80% efficiency. Also, neglect any external resistance to the vertical motion of the object, apart from gravity.

Out of the four choices of stepper motor that are given in Table 8.2 and Figure 8.46, which one would you pick to drive the platform? Justify your selection by giving all the computational details of the approach.

8.38 a. What parameters or features determine the step angle of a stepper motor? What is microstepping? Briefly explain how microstepping is achieved.

b. A Stepper-motor-driven positioning platform is schematically shown in Figure P8.38.

Suppose that the maximum travel of the platform is L and this is accomplished in a time period of Δt. A trapezoidal velocity profile is used with a region of

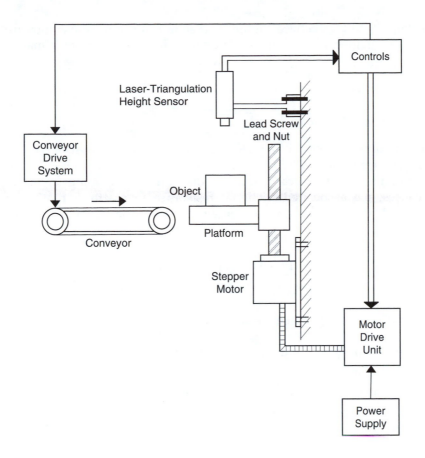

FIGURE P8.37
A material transfer unit in an automated factory.

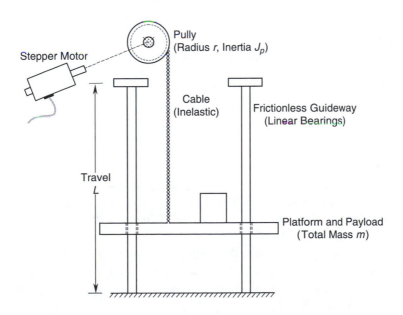

FIGURE P8.38
An automated positioning platform.

constant speed V in between an initial region of constant acceleration from rest and a final region of constant deceleration to rest, in a symmetric manner.

 i. Show that the acceleration is given by

$$a = \frac{V^2}{V \cdot \Delta t - L}$$

The platform is moved using a mechanism of light, inextensible cable and a pulley, which is directly (without gears) driven by a stepper motor. The platform moves on a pair of vertical guideways that use linear bearings and, for design purposes, the associated frictional resistance to platform motion may be neglected. The frictional torque at the bearings of the pulley is not negligible, however. Suppose that

$$\frac{\text{frictional touque of the pulley}}{\text{load torque on the pulley from the cable}} = e$$

Also, the following parameters are known:

J_p = moment of inertia of the pulley about the axis of rotation
r = radius of the pulley
m = equivalent mass of the platform and its payload

 ii. Show that the maximum operating torque that is required from the stepper motor is given by

$$T = [J_m + J_p + (1+e)mr^2]\frac{a}{r} + (1+e)rmg$$

in which
J_m = moment of inertia of the motor rotor.

iii. Suppose that
 $V = 8.0$ m/s
 $L = 1.0$ m
 $\Delta t = 1.0$ s
 $m = 1.0$ kg
 Four models of stepper motor are available, and their specifications given in Table 8.2 and Figure 8.46. Select the most appropriate motor (with the corresponding drive system) for this application. Clearly indicate all your computations and justify your choice.

 iv. What is the position resolution of the platform, as determined by the chosen motor?

8.39 a. Consider a stepper motor of moment of inertia J_m, which drives a purely inertial load of moment of inertia J_L, through a gearbox of speed reduction r:1, as shown in Figure P8.39(a).

Note that $\omega_L = \omega_m/r$

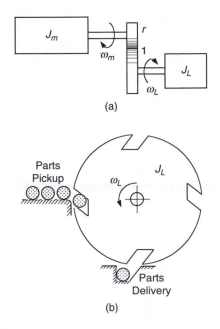

(a)

(b)

FIGURE P8.39

(a) Stepper motor driving an inertial load, (b) A parts transfer mechanism: an example of inertial load.

where

ω_m = motor speed
ω_L = load speed.

i. Show that the motor torque T_m may be expressed as

$$T_m = \left(rJ_m + \frac{J_L}{er} \right) \dot{\omega}_L$$

in which

e = gear efficiency

ii. For optimal conditions of load acceleration express the required gear ratio r in terms of J_L, J_m, and e.

b. An example of a rotary load that is driven by a stepper motor is shown in Figure P8.39(b). Here, in each quarter revolution of the load rotor, a part is transferred from the pickup position to the delivery position. The equivalent moment of inertia of the rotor, which carries a part, is denoted by J_L.

Suppose that $J_L = 12.0 \times 10^{-3}$ kg.m². The required rate of parts transfer is 7 parts/ sec. A stepper motor is used to drive the load. A gearbox may be employed as well. Four motor models are available, whose parameters are given in Table P8.39.

The speed-torque characteristics of the motors are given in Figure 8.46. Assume that the step angle of each motor is 1.8°. The gearbox efficiency may be taken as 0.8.

i. Prepare a table giving the optimal gear ratio, the operating speed of motor, the available torque, and the required torque, for each of the four models of motor, assuming that a gearbox with optimal gear ratio is employed in each case. On this basis, which motor would you choose for the present application?

TABLE P8.39

Motor Parameter Values

Motor model (SM)	Motor inertia (J_m kg·m²)
50	11.8×10^{-6}
101	35×10^{-6}
310	187×10^{-6}
1010	805×10^{-6}

ii. Now consider the motor chosen in Part (i). Suppose that three gearboxes of speed reduction 5, 8, and 10 respectively, may be available to you. Is a gearbox required in the present application, with the chosen motor? If so, which gearbox would you choose? Make your decision by computing the available torque and the required torque (with the motor chosen in Part (i)), for the four values of r given by 1, 5, 8, and 10.

iii 'What is the positioning resolution of the parts transfer system? What factors can affect this value?

8.40 Piezoelectric stepper motors are actuators that convert vibrations in a piezoelectric element (say, PZT) generated by an ac voltage (reverse piezoelectric effect, see Chapter 6) into rotary motion. Step angles on the order of $0.001°$ can be obtained by this method. In one design developed at University of Cambridge, England, as the piezoelectric PZT rings vibrate due to an applied ac voltage, radial bending vibrations are produced in a conical aluminum disk. These vibrations impart twisting (torsional) vibrations onto a beam element. The twisting motion is subsequently converted into a rotary motion of a frictional disk, which is frictionally coupled with the top surface of the beam. Essentially, because of the twisting motion, the two top edges of the beam push the frictional disk tangentially in a stepwise manner. This forms the output member of the piezoelectric stepper motor. List several advantages and disadvantages of this motor. Describe an application in which a miniature stepper motor of this type could be used.

9

Continuous-Drive Actuators

Actuator is the device that mechanically drives a mechatronic system. There are many classifications of actuators. Those that directly operate a process (load, plant) are termed *process actuators*. Joint motors in a robotic manipulator are good examples of process actuators. In process control applications in particular, actuators are often used to operate controller components (final control elements), such as servovalves, as well. Actuators in this category are termed *control actuators*. Actuators that automatically use response error signals from a process in feedback to correct the operation of the process (i.e., to drive the process according to a desired response) are termed *servoactuators*. In particular, the motors that use position, speed, and perhaps load torque measurements and armature current or field current in feedback, to drive a load according to a specified motion, are termed *servomotors*.

One broad classification of actuators separates them into the two types: *incremental-drive actuators* and *continuous-drive actuators*. Stepper motors, which are driven in fixed angular steps, represent the class of incremental-drive actuators. They can be considered as digital actuators, which are pulse-driven devices. Each pulse received at the driver of a digital actuator causes the actuator to move by a predetermined, fixed increment of displacement. Stepper motors were studied in Chapter 8. Most actuators used in mechatronic applications are continuous-drive devices. Examples are dc torque motors, induction motors, hydraulic and pneumatic motors, and piston-cylinder drives (rams). Microactuators are actuators that are able to generate very small (microscale) actuating forces/torques and motions. In general, they can be neither developed nor analyzed as scaled-down versions of regular actuators. Separate and more innovative procedures of design, construction, and analysis are necessary for *microactuators*. Micromachined, millimeter-size micromotors with submicron accuracy are useful in modern information storage systems. Distributed or multilayer actuators constructed using piezoelectric, electrostrictive, magnetostrictive, or photostrictive materials are used in advanced and complex applications such as adaptive structures and various microelectromechanical systems (MEMS).

In the early days of analog control, servo actuators were exclusively continuous-drive devices. Since the control signals in this early generation of (analog) control systems generally were not discrete pulses, the use of pulse-driven incremental actuators was not feasible in those systems. Direct current (dc) servomotors and servovalve-driven hydraulic and pneumatic actuators were the most widely used types of actuators in industrial control systems, particularly because digital control was not available. Furthermore, the control of alternating current (ac) actuators was a difficult task at that time. Today, multi-phase ac motors are also widely used as servomotors, employing modern methods of phase voltage control and frequency control through microelectronic drive systems and using field feedback compensation through digital signal processing (DSP) chips. It is interesting to note that actuator control using pulse signals is no longer limited to digital actuators. Pulse width-modulated (PWM) signals through PWM amplifiers (rather than linear amplifiers) are increasingly being used to drive continuous-drive actuators such as dc servomotors and hydraulic servos. Furthermore, it should be pointed out that electronic-switching commutation in dc motors is quite similar to the method of phase switching used in driving (actuating) stepper motors.

Requirements of size, torque/force, speed, power, stroke, motion resolution, repeatability, duty cycle, and operating bandwidth for an actuator can differ significantly, depending on the particular application and the specific function of the actuator within the mechatronic system. Furthermore, the capabilities of an actuator will be affected by its drive system. Although the cost of sensors and transducers is a deciding factor in low-power applications and in situations where precision, accuracy, and resolution are of primary importance, the cost of actuators can become crucial in moderate-to high power control applications. It follows that the proper design and selection of actuators can have a significant economical impact in many applications of industrial control. The applications of actuators are immense, spanning over industrial, medical, instrumentation, business and office automation, and household appliance fields. Most such applications are mechatronic in nature.

This chapter will discuss the principles of operation, mathematical modeling, analysis, characteristics, performance evaluation, methods of control, and sizing (or selection) of the more common types of continuous-drive actuators used in control applications. In particular, dc motors, ac induction motors, ac synchronous motors, and hydraulic and pneumatic actuators will be considered.

9.1   DC Motors

The dc motor converts direct current (dc) electrical energy into rotational mechanical energy. A major part of the torque generated in the rotor (armature) of the motor is available to drive an external load. The dc motor is probably the earliest form of electric motor. Because of features such as high torque, speed controllability over a wide range, portability, well-behaved speed-torque characteristics, easier and accurate modeling, and adaptability to various types of control methods, dc motors are still widely used in numerous mechatronic applications including robotic manipulators, transport mechanisms, disk drives, positioning tables, machine tools, and servovalve actuators.

The principle of operation of a dc motor is illustrated in Figure 9.1. Consider an electric conductor placed in a steady magnetic field at right angles to the direction of the field. Flux density B is assumed constant. If a dc current is passed through the conductor, the magnetic flux due to the current will loop around the conductor, as shown in the figure. Consider a plane through the conductor, parallel to the direction of flux of the magnet. On one side of this plane, the current flux and the field flux are additive; on the opposite side, the two magnetic fluxes oppose each other. As a result, an imbalance magnetic force F is generated on the conductor, normal to the plane. This force is given by (*Lorentz's law*)

$$F = Bil \qquad (9.1)$$

where
   B = flux density of the original field
   i = current through the conductor
   l = length of the conductor.

Note that if the field flux is not perpendicular to the length of the conductor, it can be resolved into a perpendicular component that generates the force and to a parallel component that has no effect. The active components of i, B, and F are mutually perpendicular

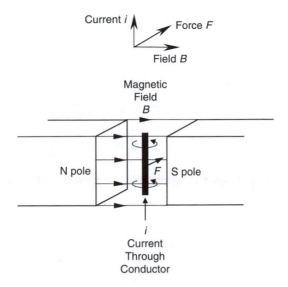

FIGURE 9.1
Operating principle of a dc motor.

and form a right-hand triad, as shown in Figure 9.1. Alternatively, in the vector representation of these three quantities, the vector F can be interpreted as the cross product of the vectors i and B. Specifically, $F = i \times B$.

If the conductor is free to move, the generated force will move it at some velocity v in the direction of the force. As a result of this motion in the magnetic field B, a voltage is induced in the conductor. This is known as the back electromotive force, or *back e.m.f.*, and is given by

$$v_b = Blv \qquad (9.2)$$

According to *Lenz's law*, the flux due to the back e.m.f. v_b will be opposing the flux due to the original current through the conductor, thereby trying to stop the motion. This is the cause of *electrical damping* in motors, which we will discuss later. Equation 9.1 determines the armature torque (motor torque), and Equation 9.2 governs the motor speed.

9.1.1 Rotor and Stator

A dc motor has a rotating element called rotor or *armature*. The rotor shaft is supported on two bearings in the motor housing. The rotor has many closely spaced slots on its periphery. These slots carry the rotor windings, as shown in Figure 9.2(a). Assuming the field flux is in the radial direction of the rotor, the force generated in each conductor will be in the tangential direction, thereby generating a torque (force × radius), which drives the rotor. The rotor is typically a laminated cylinder made from a ferromagnetic material. A ferromagnetic core helps concentrate the magnetic flux toward the rotor. The lamination reduces the problem of magnetic hysteresis and limits the generation of eddy currents and associated dissipation (energy loss by heat generation) within the ferromagnetic material. More advanced dc motors use powdered-iron-core rotors rather than the laminated-iron-core variety, thereby further restricting the generation and conduction/dissipation of eddy currents and reducing various nonlinearities such as hysteresis. The rotor windings (armature windings) are powered by the supply voltage v_a.

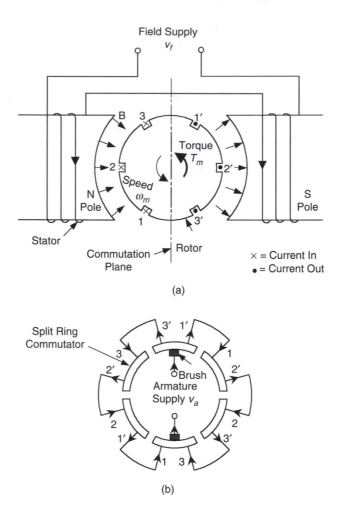

FIGURE 9.2
(a) Schematic diagram of a dc motor, (b) Commutator wiring.

The fixed magnetic field (which interacts with the rotor coil and generates the motor torque) is provided by a set of fixed magnetic poles around the rotor. These poles form the stator of the motor. The stator may consist of two opposing poles of a permanent magnet. In industrial dc motors, however, the field flux is usually generated not by a permanent magnet but electrically in the stator windings, by an electromagnet, as schematically shown in Figure 9.2(a). Stator poles are constructed from ferromagnetic sheets (i.e., a *laminated construction*). The stator windings are powered by supply voltage v_f, as shown in Figure 9.2(a). Furthermore, note that in Figure 9.2(a), the net stator magnetic field is perpendicular to the net rotor magnetic field, which is along the commutation plane. The resulting forces that attempt to pull the rotor field toward the stator field may be interpreted as the cause of the motor torque (which is maximum when the two fields are at right angles).

The rotor in a conventional dc motor is called the *armature* (voltage supply to the armature windings is denoted by v_a). This nomenclature is particularly suitable for electric generators because the windings within which the useful voltage is induced (generated) are termed armature windings. According to this nomenclature, armature windings of an ac machine are located in the stator, not in the rotor. Stator windings in a conventional dc motor are termed *field windings.* In an electric generator, the armature moves relative to the magnetic field of the field windings, generating the useful voltage output. In synchronous ac machines, the field windings are the rotor windings. Note that a dc motor may have more than two stator poles and far more conductor slots than what is shown in Figure 9.2(a). This enables the stator to provide a more uniform and radial magnetic field. For example, some rotors carry more than 100 conductor slots.

9.1.2 Commutation

A plane known as the "commutation plane" symmetrically divides two adjacent stator poles of opposite polarity. In the two-pole stator shown in Figure 9.2(a), the commutation plane is at right angles to the common axis of the two stator poles, which is the direction of the stator magnetic field. It is noted that on one side of the plane, the stator field is directed toward the rotor, while on the other side the stator field is directed away from the rotor. Accordingly, when a rotor conductor rotates from one side of the plane to the other side, the direction of the generated torque will be reversed. Such a scenario is not useful since the average torque will be zero in that case.

In order to maintain the direction of torque in each conductor group (one group is numbered 1, 2, 3, and the other group is numbered 1', 2', and 3', in Figure 9.2(a)), the direction of current in a conductor has to change as the conductor crosses the commutation plane. Physically, this may be accomplished by using a split ring and brush commutator, shown schematically in Figure 9.2(b). The armature voltage is applied to the rotor windings through a pair of stationary conducting blocks made of graphite (conducting soft carbon), which maintain a sliding contact with the split ring. These contact blocks are called "brushes" because historically, they were made of bristles of copper wire in the form of a brush. The graphite contacts are cheaper, more durable primarily due to reduced sparking (arcing) problems, and provide more contact area (less electrical contact resistance). Also the contact friction is lower. The split ring segments, equal in number to the conductor slots in the rotor, are electrically insulated from one another, but the adjacent segments are connected by the armature windings in each opposite pair of rotor slots, as shown in Figure 9.2(b). For the rotor position shown in Figure 9.2, note that when the split ring rotates in the counterclockwise direction through 30°, the current paths in conductors 1 and 1' reverse but the remaining current paths are unchanged, thus achieving the required commutation. Mechanically, this is possible because the split ring is rigidly mounted on the rotor shaft, as shown in Figure 9.3.

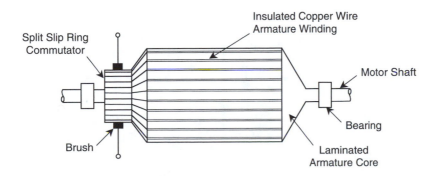

FIGURE 9.3
Physical construction of the rotor of a dc motor.

9.1.3 Static Torque Characteristics

Let us examine the nature of the static torque generated by a dc motor. Here we assume that the motor speed is low so that the dynamic effects need not be explicitly included into the discussion. Consider a two-pole permanent magnet stator and a planar coil that is free to rotate about the motor axis, as shown in Figure 9.4(a). The coil (rotor or armature) is energized by current i_a as shown. The flux density vector of the stator magnetic filed is B and the unit vector normal to the plane of the coil is n. The angle between B and n is δ, which is known as the torque angle. It should be clear from Figure 9.4(b) that the torque T generated in the rotor is given by

$$T = F \times 2r \sin \delta$$

which, in view of Equation 9.1 becomes

$$T = B i_a l \times 2r \sin \delta$$

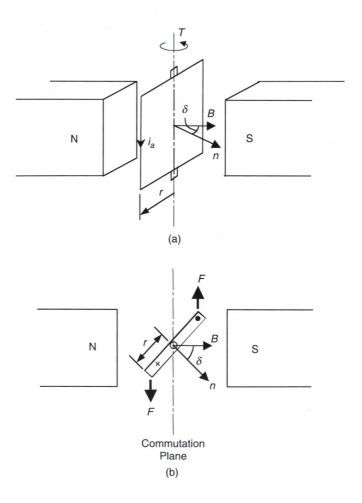

(a)

Commutation
Plane

(b)

FIGURE 9.4
(a) Torque generated in a planar rotor, (b) Nomenclature.

or

$$T = Ai_a B \sin \delta \qquad (9.3)$$

where
 l = length of the coil
 r = radius of the rotor
 A = face area of the planar rotor

Suppose that the rotor rotation starts by coinciding with the commutation plane (where $\delta = 0$ or π) and rotates through an angle of 2π. The resulting torque profile is shown in Figure 9.5(a). Next suppose that the rotor has three planar coil segments placed at 60° apart, and denoted by 1, 2, and 3, as in Figure 9.2. Note from Figure 9.2(b) that current switching occurs at every 60° rotation, Figure 9.5(b) shows the torque profile of each coil segment (with switching at 180°) and the overall torque profile due to the three-segment rotor of Figure 9.2. Note that the solid line in Figure 9.5(b) is intended to present the shape rather than the actual magnitude of the resultant torque. It is seen, however, that the torque profile has improved (i.e., larger torque magnitude and smaller variation) as a result of the multiple coil segments, with shorter commutation angles. The torque profile can be further improved by incorporating still more coil segments, with correspondingly shorter commutation angles, but the design of the split-ring and brush arrangement becomes more challenging then. Hence, there is a design

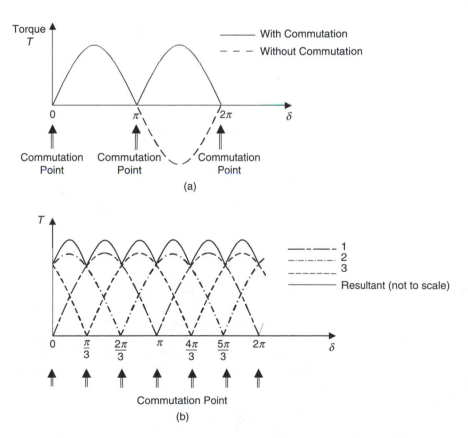

FIGURE 9.5
(a) Torque profile due to commutation, (b) Resultant torque from a rotor with three coil segments (not to scale).

limitation to achieving uniform torque profiles in a dc motor. It should be clear from Figure 9.2(a) that if the stator field can be made radial, then B is always perpendicular to n and hence sin δ becomes equal to 1. In that case, the torque profile is uniform, under ideal conditions.

9.1.4 Brushless DC Motors

There are several shortcomings of the slip ring and brush mechanisms, which are used for current transmission through moving members, even with the advances from the historical copper brushes to modern graphite contacts. The main disadvantages include rapid wearout, mechanical loading, and heating due to sliding friction, contact bounce, excessive noise, and electrical sparks (arcing) with the associated dangers in hazardous (e.g., chemical) environments, problems of oxidation, problems in applications that require wash-down (e.g., in food processing), and voltage ripples at switching points. Conventional remedies to these problems—such as the use of improved brush designs and modified brush positions to reduce arcing—are inadequate in sophisticated applications. Also, the required maintenance (to replace brushes and resurface the split-ring commutator) can be rather costly and time consuming, involving down time.

Brushless dc motors have permanent-magnet rotors. Since in this case the polarities of the rotor cannot be switched as the rotor crosses a commutation plane, commutation is accomplished by electronically switching the current in the stator winding segments. Note that this is the reverse of what is done in brushed commutation, where the stator polarities are fixed and the rotor polarities are switched when crossing a commutation plane. The stator windings of a brushless dc motor can be considered the armature windings whereas for a brushed dc motor, the rotor is the armature. In concept, brushless dc motors are somewhat similar to permanent-magnet stepper motors (see Chapter 8) and to some types of ac motors. By definition, a dc motor should use a dc supply to power the motor. The torque-speed characteristics of dc motor are different from those of stepper motor or ac motor. Furthermore, permanent-magnet motors are less nonlinear than the electro-magnet motors because the field strength generated by a permanent magnet is rather constant and independent of the current through a coil. This is true whether the permanent magnet is in the stator (i.e., a brushed motor) or in the rotor (i.e., a brushless dc motor or a PM stepper motor).

Figure 9.6 schematically shows a brushless dc motor and associated commutation circuitry. The rotor is a multiple-pole permanent magnet. Conventional ferrite magnets and alnico or ceramic magnets are economical but their "field-strength/mass" ratio is relatively low compared to more costly rare-earth magnets. Hence, for a given torque rating, the rotor inertia can be reduced by using rare-earth material for the rotor of a brushless dc motor. Examples of rare-earth magnetic material are samariam cobalt and neodymium-iron-boron, which have magnetic energy products that are more than 10 times those for ceramic-ferrite magnets, for a given mass. This is particularly desirable when a high torque is required, as in torque motors. The popular two-pole rotor design consists of a diametrically magnetized cylindrical magnet, as shown in Figure 9.6. The stator windings are distributed around the stator in segments of winding groups. Each winding segment has a separate supply lead. Figure 9.6 shows a four-segment stator. Two diametrically opposite segments are connected together so that they carry current simultaneously but in opposite directions. Commutation is accomplished by energizing each pair of diametrically opposite segments sequentially, at time instants determined by the rotor position. This commutation could be achieved through mechanical means, using a multiple contact switch driven by the motor itself. Such a mechanism would defeat the purpose, however, because it has most of the drawbacks of regular commutation using split rings and brushes. Modern brushless motors use microelectronic switching for commutation.

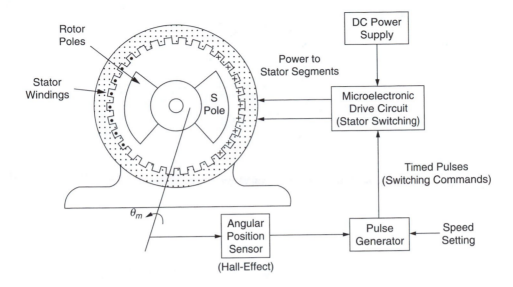

FIGURE 9.6
A brushless dc motor system.

9.1.4.1 Constant Speed Operation

For constant-speed operation, open-loop switching may be used. In this case, speed setting is provided as the input to a timing pulse generator. It generates a pulse sequence starting at zero pulse rate and increasing (ramping) to the final rate, which corresponds to the speed setting. Each pulse causes the driver circuit, which has proper switching circuitry, to energize a pair of stator segments. In this manner, the input pulse signal activates the stator segments sequentially, thereby generating a stator field, which rotates at a speed that is determined by the pulse rate. This rotating magnetic field would accelerate the rotor to its final speed. A separate command (or a separate pulse signal) is needed to reverse the direction of rotation, which is accomplished by reversing the switching sequence.

9.1.4.2 Transient Operation

Under transient motions of a brushless dc motor, it is necessary to know the actual position of the rotor for accurate switching of the stator field circuitry. An angular position sensor (e.g., a shaft encoder or more commonly a Hall effect sensor) is used for this purpose, as shown in Figure 9.6. By switching the stator segments at the proper instants, it is possible to maximize the motor torque. To explain this further, consider a brushless dc motor that has two rotor poles and four stator winding segments. Let us number the stator segments as in Figure 9.7(a) and also define the rotor angle θ_m as shown. The typical shape of the static torque curve of the motor when segment 1 is energized (segment 1' being automatically energized in the opposite direction) is shown in Figure 9.7(b), as a function of θ_m. When segment 2 is energized, the torque distribution would be identical, but it would be shifted to the right through 90°. Similarly, if segment 1' is energized in the positive direction (segment 1 being energized in the opposite direction), the corresponding torque distribution would be shifted to the right by a further 90°, and so on. The superposition of these individual torque curves is shown in Figure 9.7(c). It should be clear that to maximize the motor torque, switching has to be done at the points of intersection of the torque curves corresponding to the adjacent stator segments (as for stepper motors—see Chapter 8). These switching points are indicated as A, B, C, and D in Figure 9.7(c). Under transient

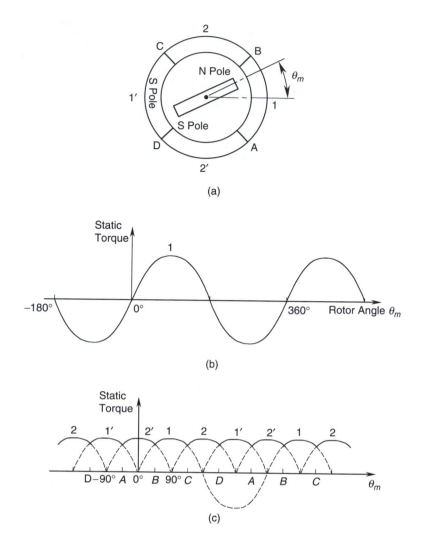

FIGURE 9.7
(a) A brushless dc motor, (b) Static torque curve with no switching (one stator segment energized), (c) Switching sequence for maximum average torque.

motions, position measurement would be required to determine these switching positions accurately. An effective solution is to mount Hall effect sensors at switching points (which are fixed) around the stator. The voltage pulse generated by each of these sensors when a rotor pole passes the sensor, is used to switch the appropriate field windings.

Note from Figure 9.7(c) that positive average torque is possible even if the switching positions are shifted from these ideal locations by less than 90° to either side. It follows that the motor torque can be controlled by adjusting the switching locations with respect to the actual position of the rotor. The smoothness and magnitude of the motor torque, the accuracy of operation, and motor controllability can be improved by increasing the number of winding segments in the stator. This however, increases the number of power lines and the complexity of the commutation circuitry. The commutation electronics for modern brushless dc motors is available as a single integrated-circuit (IC) instead of discrete circuits using transistor switches.

The foregoing discussion on electronics communication assumes that the curently energized winding segment is turned off as a new winding segment is turned on. This is a desirable scheme with regard to the efficiency of torque generation. If the currently energized winding segment is kept on while a new winding segment is energized, there will be two torque components generated by the two winding segments that are energized. It is clear that there is an angular region within which both torque components are positive, resulting in a higher torque. But since the torque profile is somewhat sinsoidal, one torque component can be quite small compared to other (even though the current in both winding segments is equal). For this reason, for the best torque efficiency, the currently energized coil segment should be turned off and the new coil segment should be turned on at the point of intersection of the torque profiles of the new coil segments, as assumed above. The foregoing discussion of electronic communication assumes that the previous winding segment is turned off when a new winding segment is turned on. This producer a higher efficiency with regard to the generated torque. It is possible, however, to maintain the current in the presently activated winding segment when a new winding segment is turned on. This will result in two torque components corresponding. Advantages of brushless dc motors primarily result from the disadvantages of using split-rings and brushes for commutation, as noted before. Primary among them are the high efficiency, low mass for a specified torque rating, low maintenance, lower mechanical loading, longer life, improved safety, and quieter operation. The drawbacks include the additional cost due to sensing and switching hardware. Two-state, on-off switching will result in torque ripples due to induction effects. This problem can be reduced by using transient (gradual) switching, simultaneous energization of multiple stator coil segments, or shaped (e.g., ramp, sinusoidal) drive signals. Independent switching of multiple coil segments, however, will require more complex switching hardware, with added cost. Brushless dc motors with neodymium iron boron rotors can generate high torques (over 30 N·m). Motors in the continuous-operating torque range of 0.5 N·m (75 oz.in) to 30 N·m (270 lb·in) are commercially available and used in general-purpose applications as well as in servo systems. For example, motors in the range 0.01–5 hp, operating at speeds up to 7200 rpm, are available. Fractional horse-power applications include optical scanners, computer disk drives, instrumentation applications, household tools, surgical drills, and other medical devices. Medium-to-high power applications include robots, positioning devices, power blowers, industrial refrigerators, HVAC systems, and positive displacement pump drives. Many of these applications are for constant speed operations, where ac motors are equally suitable. There are variable-speed servo applications (e.g., robotics and inspection devices) and high-acceleration applications (e.g., spinners and centrifuges) for which dc motors are preferred over ac motors.

9.1.5 Torque Motors

Conventionally, torque motors are high-torque dc motors with permanent magnet stators. These actuators characteristically possess a linear (straight-line) torque-speed relationship, primarily because of their high-strength permanent-magnet stators, which provide a fairly constant and uniform magnetic field. The magnet should have high flux density per unit volume of the magnet material, yielding a high torque/mass ratio for a torque motor. Furthermore, coercivity (resistance to demagnetization) should be high and the cost has to be moderate. Rare-earth materials (e.g., samarium cobalt, $SmCo_5$) possess most of these desirable characteristics, although their cost is typically high. Conventional and low-cost ferrite magnets and alnico (aluminum-nickel-cobalt) or ceramic magnets provide a relatively low torque/mass ratio. Hence, rare-earth magnets are widely used in torque motor and servomotor applications. As a comparison, a typical rare-earth motor may produce a peak torque of over 27 N·m, with a torque/mass ratio of over 6 N·m/kg, while an alnico motor of identical dimensions and mass may produce a peak torque of about half the value (less than 15 N·m, with a torque/mass ratio of about 3.4 N·m/kg).

When operating at high torques (e.g., a thousand or more $N \cdot m$; Note: $1\,N \cdot m = 0.74\,lbf \cdot ft$), the motor speeds have to be quite low for a given level of power. One straightforward way to increase the output torque of a motor (with a corresponding reduction in speed) is to employ a gear system (typically using worm gears) with high gear reduction. Gear drives introduce undesirable effects—such as backlash, additional inertia loading, higher friction, increased noise, lower efficiency, and increased maintenance. Backlash in gears would be unacceptable in high-precision applications. Frictional loss of torque, wear problems, and the need for lubrication must also be considered. Furthermore, the mass of the gear system reduces the overall torque/mass ratio and the useful bandwidth of the actuator. For these reasons, torque motors are particularly suitable for high-precision, direct-drive applications (e.g., direct-drive robot arms) that require high-torque drives without having to use speed reducers and gears. Torque motors are usually more expensive than the conventional types of dc motors. This is not a major drawback, however, because torque motors are often custom-made and are supplied as units that can be directly integrated with the process (load) within a common housing. For example, the stator might be integrated with one link of a robot arm and the rotor with the next link, thus forming a common joint in a direct-drive robot. Torque motors are widely used as valve actuators in hydraulic servovalves where large torques and very small displacements are required.

Brushless torque motors have permanent-magnet rotors and wound stators, with electronic commutation. Consider a brushless dc motor with electronic commutation. The output torque can be increased by increasing the number of magnetic poles. Since direct increase of the magnetic poles has serious physical limitations, a toothed construction, as in stepping motors, could be employed for this purpose. Torque motors of this type have toothed ferromagnetic stators with field windings on them. Their rotors are similar in construction to those of variable-reluctance (VR) stepping motors. A harmonic drive is a special type of gear reducer that provides very large speed reductions (e.g., 200: 1) without backlash problems. The harmonic drive is often integrated with conventional motors to provide very high torques, particularly in backlash-free servo applications. The principle of operation of a harmonic drive is discussed in Chapter 3.

9.2   DC Motor Equations

Consider a dc motor with separate windings in the stator and the rotor. Each coil has a resistance (R) and an inductance (L). When a voltage (v) is applied to the coil, a current (i) flows through the circuit, thereby generating a magnetic field. As discussed before, a force is produced in the rotor windings, and an associated torque (T_m), which turns the rotor. The rotor speed (ω_m) causes the flux linkage of the rotor coil with the stator field to change at a corresponding rate, thereby generating a voltage (back e.m.f) in the rotor coil.

Equivalent circuits for the stator and the rotor of a conventional dc motor are shown in Figure 9.8(a). Since the field flux is proportional to field current i_f, we can express the magnetic torque of the motor as

$$T_m = k i_f i_a \tag{9.4}$$

which directly follows from Equation 9.1. Next, in view of Equation 9.2, the back e.m.f generated in the armature of the motor is given by

$$v_b = k' i_f \omega_m \tag{9.5}$$

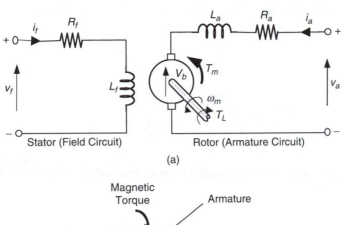

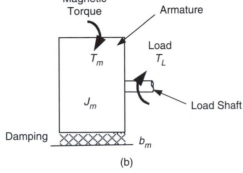

FIGURE 9.8
(a) The equivalent circuit of a conventional dc motor (separately excited), (b) Armature mechanical loading diagram.

The following notation has been used:
i_f = field current
i_a = armature current
ω_m = angular speed of the motor

and k and k' are motor constants, which depend on factors such as the rotor dimensions, the number of turns in the armature winding, and the permeability (inverse of reluctance—the magnetic resistance) of the magnetic medium. In the case of ideal electrical-to-mechanical energy conversion at the rotor (where the rotor coil links with the stator field), we have $T_m \omega_m = v_b i_a$ with consistent units (e.g., torque in Newton-meters, speed in radians per second, voltage in volts, and current in amperes). Then we observe that

$$k = k' \tag{9.6}$$

The field circuit equation is obtained by assuming that the stator magnetic field is not affected by the rotor magnetic field (i.e., the stator inductance is not affected by the rotor) and that there are no eddy current effects in the stator. Then, from Figure 9.8(a),

$$v_f = R_f i_f + L_f \frac{di_f}{dt} \tag{9.7}$$

where
v_f = supply voltage to the stator
R_f = resistance of the field winding
L_f = inductance of the field winding

The equation for the armature rotor circuit is written as (see Figure 9.8(a))

$$v_a = R_a i_a + L_a \frac{di_a}{dt} + v_b \tag{9.8}$$

where
  v_a = supply voltage to the armature
  R_a = resistance of the armature winding
  L_a = leakage inductance in the armature winding

It should be emphasized here that the primary inductance or *mutual inductance* in the armature winding is represented in the back e.m.f term v_b. The leakage inductance, which is usually neglected, represents the fraction of the armature flux that is not linked with the stator and is not used in the generation of useful torque. This includes self-inductance in the armature.

The mechanical equation of the motor is obtained by applying Newton's second law to the rotor. Assuming that the motor drives some load, which requires a load torque T_L to operate, and that the frictional resistance in the armature can be modeled by a linear viscous term, we have (see Figure 9.8(b))

$$J_m \frac{d\omega_m}{dt} = T_m - T_L - b_m \omega_m \tag{9.9}$$

where
  J_m = moment of inertia of the rotor
  b_m = equivalent (mechanical) damping constant for the rotor

Note that the load torque may be due, in part, to the inertia of the external load that is coupled to the motor shaft. If the coupling flexibility is neglected, the load inertia may be directly added to (i.e., lumped with) the rotor inertia after accounting for the possible existence of a speed reducer (gear, harmonic drive, etc.). In general, however, a separate set of equations is necessary to represent the dynamics of the external load.

Equations 9.4 through 9.9 form the dynamic model for a dc motor. In obtaining this model, we have made several assumptions and approximations. In particular, we have either approximated or neglected the following factors:

1. Coulomb friction and associated dead-band effects
2. Magnetic hysteresis (particularly in the stator core, but in the armature as well if not a brushless motor)
3. Magnetic saturation (in both stator and the armature)
4. Eddy current effects (laminated core will reduce this effect)
5. Nonlinear constitutive relations for magnetic induction (i.e., inductance L is not constant)
6. Brush contact resistance, finite-width contact of brushes, and other types of noise and nonlinearities in split ring commutators
7. The effect of the rotor magnetic flux (armature flux) on the stator magnetic flux (field flux).

9.2.1 Steady-State Characteristics

In selecting a motor for a given application, its steady-state characteristics are a major determining factor. In particular, steady-state torque-speed curves are employed for this purpose. The rationale is that, if the motor is able to meet the steady-state operating requirements, with some design conservatism, it should be able to tolerate small deviations under transient conditions of short duration. Furthermore, as will be seen in this chapter, it is possible to build a dynamic model for a motor using its steady-state characteristics as the "source" and then externally incorporating dynamic parameters such as inertia, damping, and flexibility into it. In the separately excited case shown in Figure 9.8(a), where the armature circuit and the field circuit are excited by separate and independent voltage sources, it can be shown that the steady-state torque-speed curve is a straight line. To verify this, we set the time derivative terms in Equation 9.7 and Equation 9.8 to zero; this corresponds to steady-state conditions. It follows that i_f is constant for a fixed voltage v_f. By substituting Equation 9.4 and Equation 9.5 in Equation 9.8, we get

$$v_a = \frac{R_a}{k i_f} T_m + k' i_f \omega_m$$

Under steady-state conditions in the field circuit, we have from Equation 9.7,

$$i_f = \frac{v_f}{R_f}$$

It follows that the steady-state torque-speed characteristics of a separately excited dc motor may be expressed as

$$\omega_m + \frac{R_a R_f^2}{k k' v_f^2} T_m = \frac{R_f v_a}{k' v_f} \tag{9.10a}$$

Now, since v_a and v_f are constant supply voltages from a regulated power supply, on defining the constant parameters T_s and ω_o, Equation 9.10 can be expressed as

$$\frac{\omega_m}{\omega_o} + \frac{T_m}{T_s} = 1 \tag{9.10b}$$

where
   ω_o = no-load speed (at steady state, assuming zero damping)
   T_s = stalling torque (or, starting torque) of the motor

It should be noted from Equation 9.9 that if there is no damping ($b_m = 0$), the steady-state magnetic torque (T_m) of the motor is equal to the load torque (T_L). In practice, however, there is mechanical damping on the rotor, and the load torque is smaller than the motor torque. In particular, the motor will stall at a load torque smaller than T_s. The idealized characteristic curve given by Equation 9.10 is shown in Figure 9.9.

9.2.1.1   *Bearing Friction*

The primary source of mechanical damping in a motor is the bearing friction. Ball bearings possess low friction and some degree of self-alignment. But, since the balls make point contacts on the bearing sleeve, they are prone to damage due to impact and wear problems. Roller bearings provide better contact capability (line contact), but can produce roller creep, alignment problems, and noisy operation. For ultraprecision and specialized applications, air bearings and magnetic bearings are suitable, which offer the capability of active control, automatic self-alignment, and very low friction. A linear viscous model is normally adequate to represent bearing damping. For more accurate analysis, sophisticated models (e.g., Stribeck model) may be incorporated.

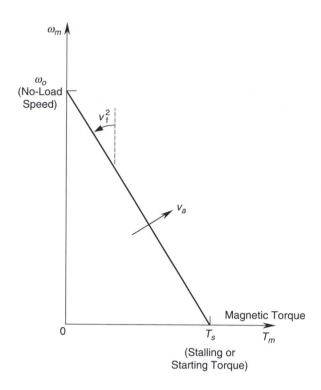

FIGURE 9.9
Steady-state speed-torque characteristics of a separately-wound dc motor.

Example 9.1

A load is driven at constant power under steady-state operating conditions, using a separately wound dc motor with constant supply voltages to the field and armature windings. Show that, in theory, two operating points are possible. Also show that one of the operating points is stable and the other one is unstable.

SOLUTION
As shown in Figure 9.9, the steady-state characteristic curve of a dc motor with windings that are separately excited by constant voltage supplies, is a straight line. The constant-power curve for the load is a hyperbola, because $T\omega_m$ = constant in this case. The two curves shown in Figure 9.10 intersect at points P and Q. At point P, if there is a slight decrease in the speed of operation, the motor (magnetic) torque will increase to T_{PM} and the load torque demand will increase to T_{PL}. But since $T_{PM} > T_{PL}$, the system will accelerate back to point P. It follows that point P is a stable operating point. Alternatively, at point Q, if the speed drops slightly, the magnetic torque of the motor will increase to T_{QM} and the load torque demand will increase to T_{QL}. However, in this case, $T_{QM} < T_{QL}$. As a result, the system will decelerate further, subsequently stalling the system. Therefore, it can be concluded that point Q is an unstable operating point.

9.2.1.2   Output Power

The output power of a motor is given by

$$p = T_m \omega_m \tag{9.11}$$

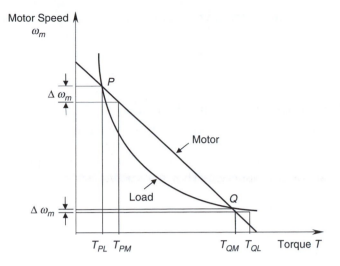

FIGURE 9.10
Operating points for a constant-power load driven by a dc motor.

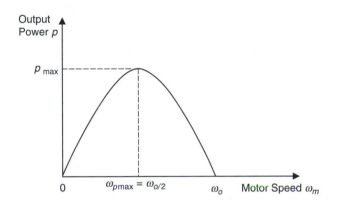

FIGURE 9.11
The output power curve of a dc motor at steady state.

Equation 9.9a applies for a dc motor excited by a regulated power supply, in steady state. Substitute this in Equation 9.11, for T_m. We get the output power

$$p = T_s \left(1 - \frac{\omega_m}{\omega_o} \right) \omega_m \qquad (9.12)$$

Equation 9.12 has a quadratic shape, as shown in Figure 9.11. The point of maximum power is obtained by differentiating Equation 9.12 and equating to zero; thus,

$$\frac{dp}{d\omega_m} = T_s \left(1 - \frac{\omega_m}{\omega_o} \right) - \frac{T_s}{\omega_o} \omega_m = T_s \left(1 - 2\frac{\omega_m}{\omega_o} \right) = 0$$

It follows that the speed at which the motor provides the maximum power is given by half the no-load speed:

$$\omega_{pmax} = \frac{\omega_o}{2} \qquad (9.13)$$

From Equation 9.12, the corresponding maximum power is

$$p_{max} = \frac{1}{4}T_s\omega_o \qquad (9.14)$$

9.2.1.3 Combined Excitation of Motor Windings

The shape of the steady-state speed-torque curve will be modified if a common voltage supply is used to excite both the field winding and the armature winding. Here, the two windings have to be connected together. There are three common ways the windings of the rotor and the stator are connected. They are known as:

1. Shunt-wound motor
2. Series-wound motor
3. Compound-wound motor

In a shunt-wound motor, the armature windings and the field windings are connected in parallel. In the series-wound motor, they are connected in series. In the compound-wound motor, part of the field windings is connected with the armature windings in series and the other part is connected in parallel. These three connection types of the rotor and the stator of a dc motor are shown in Figure 9.12. Note that in a shunt-wound motor at steady state, the back e.m.f. v_b depends directly on the supply voltage. Since the back e.m.f. is proportional to the speed, it follows that speed controllability is good with the shunt-wound configuration. In a series-wound motor, the relation between v_b and the supply voltage is coupled through both the armature windings and the field windings. Hence its speed controllability is relatively poor. But in this case, a relatively large current flows through both windings at low speeds of the motor, giving a higher starting torque. Also, the operation is approximately at constant power in this case. These properties are summarized in Table 9.1. Since both speed controllability and higher starting torque are desirable characteristics, compound-wound motors are used to obtain a compromise performance in between the two extremes.

9.2.1.4 Speed Regulation

Variation in the operating speed of a motor due to changes in the external load is measured by the percentage speed regulation. Specifically,

$$\text{Percentage speed regulation} = \frac{(\omega_o - \omega_f)}{\omega_f} \times 100\% \qquad (9.15)$$

where
ω_o = no-load speed
ω_f = full-load speed

This is a measure of the speed stability of a motor; the smaller the percentage speed regulation, the more stable the operating speed under varying load conditions (particularly in the presence of load disturbances). In the shunt-wound configuration, the back e.m.f., and hence the rotating speed, depend directly on the supply voltage. Consequently, the armature current and the related motor torque have virtually no effect on the speed.

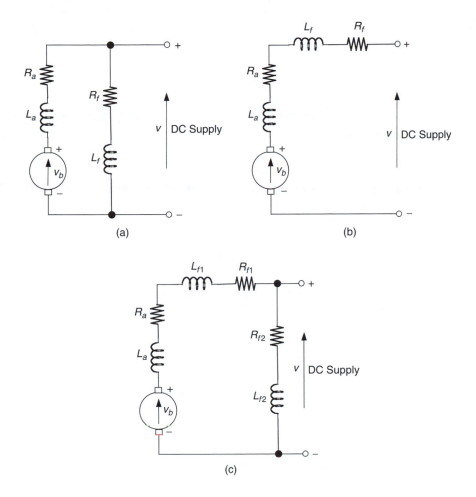

FIGURE 9.12
(a) A shunt-wound motor, (b) A series-wound motor, (c) A compound-wound motor.

TABLE 9.1

Influence of the Winding Configuration on the Steady-State Characteristics of a DC Motor

DC Motor Type	Field Coil Resistance	Speed Controllability	Starting Torque
Shunt-wound	High	Good	Average
Series-wound	Low	Poor	High
Compound-wound	Parallel high, series low	Average	Average

For this reason, the percentage speed regulation is relatively small for shunt-wound motors, resulting in improved speed stability.

Example 9.2

An automated guideway transit (AGT) vehicle uses a series-wound dc actuator in its magnetic suspension system. If the desired control bandwidth of the active suspension (in terms of the actuator force) is 40 Hz, what is the required minimum bandwidth for the input voltage signal?

SOLUTION
The actuating force is

$$F = ki_a i_f = ki^2 \tag{i}$$

where i denotes the common current through both windings of the actuator. Consider a harmonic component

$$v(\omega) = v_o \sin \omega t \tag{ii}$$

of the input voltage to the windings, where ω denotes the frequency of the chosen frequency component. The field current is given by

$$i(\omega) = i_o \sin(\omega t + \phi) \tag{iii}$$

at this frequency, where ϕ denotes the phase shift. Substitute Equation iii in Equation i to determine the corresponding actuating force:

$$F = ki_o^2 \sin^2(\omega t + \phi) = ki_o^2 [1 - \sin(2\omega t + 2\phi)] / 2$$

It follows that there is an inherent frequency doubling in the suspension system. As a result, the required minimum bandwidth for the input voltage signal is 20 Hz.

Example 9.3

Consider the three types of winding connections for dc motors, shown in Figure 9.12. Derive equations for the steady-state torque-speed characteristics in the three cases. Sketch the corresponding characteristic curves. Using these curves, discuss the behavior of the motor in each case.

SOLUTION
Shunt-Wound Motor
At steady state, the inductances are not present in the motor equivalent circuit. For the shunt-wound dc motor (Figure 9.12(a)), the field current is

$$i = v / R_f = \text{constant} \tag{i}$$

The armature current is

$$i_a = [v - v_b] / R_a \tag{ii}$$

The back e.m.f. for a motor speed of ω_m is given by (see Equation 9.5)

$$v_b = k' i_f \omega_m \tag{iii}$$

Substituting Equation i through iii in the motor magnetic torque equation (Equation 9.4)

$$T_m = k i_f i_a$$

we get

$$\omega_m + \left(\frac{R_a R_f^2}{kk'v^2}\right) T_m = \frac{R_f}{k'} \tag{9.16a}$$

Note that Equation 9.16a represents a straight line with a negative slope of magnitude:

$$\left(\frac{R_a R_f^2}{kk'v^2}\right)$$

Since this magnitude is typically small, it follows that good speed regulation (constant-speed operation and relatively low sensitivity of the speed to torque changes) can be obtained using a shunt-wound motor. The characteristic curve for the shunt-wound dc motor is shown in Figure 9.13(a). The starting torque T_s is obtained by setting $\omega_m = 0$ in Equation 9.16a. The no-load speed ω_o is obtained by setting $T_m = 0$ in the same equation. The corresponding expressions are tabulated in Table 9.2. Note that if the input voltage v is increased, the starting torque will increase but the no-load speed will remain unchanged, as sketched in Figure 9.13(a).

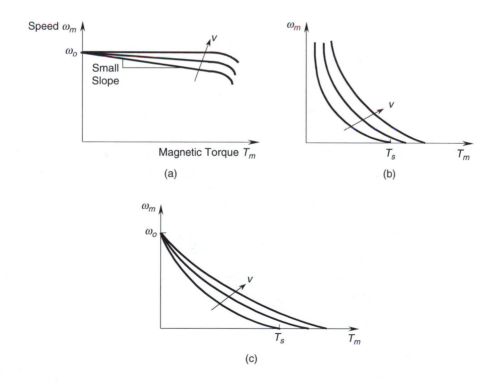

FIGURE 9.13
Torque-speed characteristic curves for dc motors: (a) Shunt-wound; (b) Series-wound; (c) Compound-wound; (d) General case.

TABLE 9.2

Comparison of dc Motor Winding Types

Winding Type	No-Load Speed ω_o	Starting Torque T_s
Shunt-wound	$\dfrac{R_f}{k'}$	$\dfrac{kv^2}{R_a R_f}$
Series-wound	∞	$\dfrac{kv^2}{(R_a + R_f)^2}$
Compound-wound	$\dfrac{R_{f2}}{k'}$	$\dfrac{kv^2}{R_a + R_{f1}}\left[\dfrac{1}{R_a + R_{f1}} + \dfrac{1}{R_{f2}}\right]$

Series-Wound Motor

At steady state, for the series-wound dc motor shown in Figure 9.12(b), the field current is equal to the armature current; thus,

$$i_a = i_f = \frac{v - v_b}{R_a + R_f} \tag{v}$$

The back e.m.f. is given by Equation iii, as before. The motor magnetic torque is given by

$$T_m = ki_f^2 \tag{vi}$$

From these relations, we get the following equation for the steady-state speed-torque relation of a series-wound motor:

$$\omega_m = \frac{v}{k'}\sqrt{\frac{k}{T_m}} - \frac{R_a + R_f}{k'} \tag{9.16b}$$

This equation is sketched in Figure 9.13(b). The starting torque, as given in Table 9.2, increases with the input voltage v. In the present case, the no-load speed is infinite. For this reason, the motor will coast at low loads. It follows that speed regulation in series-wound motors is poor. Starting torque and low-speed operation are satisfactory, however.

Compound-Wound Motor

Figure 9.12(c) gives the equivalent circuit for a compound-would dc motor. Note that part of the field coil is connected in series with the rotor windings and the other part is connected in parallel with the rotor coil. Under steady-state conditions, the currents in the two parallel branches of the circuit are given by

$$i_a = i_{f1} = \frac{v - v_b}{R_a + R_{f1}} \tag{vii}$$

$$i_{f2} = \frac{v}{R_{f2}} \tag{viii}$$

Note that the total field current that generates the stator field is

$$i_f = i_{f1} + i_{f2} \tag{ix}$$

which, in view of Equation vii, Equation viii, and Equation iii, becomes

$$i_f = v \left[\frac{1}{R_a + R_{f1}} + \frac{1}{R_{f2}} \right] - \frac{k' i_f \omega_m}{R_a + R_{f1}}$$

Consequently,

$$i_f = v \left[\frac{1}{R_a + R_{f1}} + \frac{1}{R_{f2}} \right] \Bigg/ \left[1 + \frac{k' \omega_m}{R_a + R_{f1}} \right] \tag{x}$$

The motor magnetic torque is given by

$$T_m = k i_f i_a = k i_f \frac{v - v_b}{R_a + R_{f1}} = k i_f \frac{v - k' i_f \omega_m}{R_a + R_{f1}} \tag{xi}$$

Finally, by substituting Equation x in Equation xi, we get the steady-state torque-speed relationship; thus,

$$T_m = \frac{k v^2 \left(\dfrac{1}{R_a + R_{f1}} + \dfrac{1}{R_{f2}} \right) \left[1 - k' \omega_m \left(\dfrac{1}{R_a + R_{f1}} + \dfrac{1}{R_{f2}} \right) \Big/ \left(1 + \dfrac{k' \omega_m}{R_a + R_{f1}} \right) \right]}{(R_a + R_{f1}) \left(1 + \dfrac{k' \omega_m}{R_a + R_{f1}} \right)}$$

$$= \frac{k v^2 \left(\dfrac{1}{R_a + R_{f1}} + \dfrac{1}{R_{f2}} \right) \left(1 - \dfrac{k' \omega_m}{R_{f2}} \right)}{(R_a + R_{f1}) \left(1 + \dfrac{k' \omega_m}{R_a + R_{f1}} \right)^2} \tag{9.16c}$$

This equation is sketched in Figure 9.13(c). The expressions for the starting torque and the no-load speed are given in Table 9.2.

By comparing the foregoing results, we can conclude that good speed regulation and high starting torques are available from a shunt-wound motor, and nearly constant-power operation is possible with a series-wound motor. The compound-wound motor provides a trade-off between these two.

9.2.2 Experimental Model

We have noticed that, in general, the speed-torque characteristic of a dc motor is nonlinear. A linearized dynamic model can be extracted from the speed-torque curves. One of the parameters of the model is the damping constant. First we will examine this.

9.2.2.1 Electrical Damping Constant

Newton's second law governs the dynamic response of a motor. In Equation 9.9, for example, b_m denotes the mechanical (viscous) damping constant and represents mechanical dissipation of energy. As is intuitively clear, mechanical damping torque opposes motion—hence the negative sign in the $b_m \omega_m$ term in Equation 9.9. Note, further, that the magnetic torque T_m of the motor is also dependent on speed ω_m. In particular, the back e.m.f., which is governed by ω_m, produces a magnetic field, which tends to oppose the motion of the motor rotor. This acts as a damper, and the corresponding damping constant is given by

$$b_e = -\frac{\partial T_m}{\partial \omega_m} \qquad (9.17a)$$

This parameter is termed the *electrical damping constant*. Caution should be exercised when experimentally measuring b_e. Note that in constant speed tests, the inertia torque of the rotor will be zero; there is no torque loss due to inertia. Torque measured at the motor shaft includes as well the torque reduction due to mechanical dissipation (mechanical damping) within the rotor, however. Hence the magnitude b of the slope of the speed-torque curve as obtained by steady-state tests, is equal to $b_e + b_m$, where b_m is the equivalent viscous damping constant representing mechanical dissipation at the rotor.

9.2.2.2 Linearized Experimental Model

To extract a linearized experimental model for a dc motor, consider the speed-torque curves shown in Figure 9.13(d). For each curve, the excitation voltage v_c is maintained constant. This is the voltage that is used in controlling the motor, and is termed control voltage. It can be, for example, the armature voltage, the field voltage, or the voltage that excites both armature and field windings in the case of combined excitation (e.g., shunt-wound motor). One curve in Figure 9.13(d) is obtained at control voltage v_c and the other curve is obtained at $v_c + \Delta v_c$. A tangent can be drawn at a selected point (operating point O) of a speed-torque curve. The magnitude b of the slope (which is negative) corresponds to a damping constant, which includes both electrical and mechanical damping effects. What mechanical damping effects are included in this parameter depends entirely on the nature of mechanical damping that was present during the test (primarily bearing friction). We have the *damping constant* as the magnitude of the slope at the operating point:

$$b = -\frac{\partial T_m}{\partial \omega_m}\bigg|_{v_c = \text{constant}} \qquad (9.17b)$$

Next draw a vertical line through the operating point O. The torque intercept ΔT_m between the two curves can be determined in this manner. Since a vertical line is a constant speed line, we have the *voltage gain*:

$$k_v = \frac{\partial T_m}{\partial v_c}\bigg|_{\omega_m = \text{constant}} = \frac{\Delta T_m}{\Delta v_c} \qquad (9.18)$$

Now, using the well-known relation for a total differential we have

$$\delta T_m = \frac{\partial T_m}{\partial \omega_m}\bigg|_{v_c} \delta \omega_m + \frac{\partial T_m}{\partial v_c}\bigg|_{\omega_m} \delta v_c$$

$$= -b\delta \omega_m + k_v \delta v_c \tag{9.19}$$

Equation 9.19 is the linearized model of the motor. This may be used in conjunction with the mechanical equation of the motor rotor, for the incremental motion about the operating point:

$$J_m \frac{d\delta\omega_m}{dt} = \delta T_m - \delta T_L \tag{9.20}$$

Note that Equation 9.20 is the incremental version of Equation 9.9 except that the overall damping constant of the motor (including mechanical damping) is included in Equation 9.19. The torque needed to drive the rotor inertia, however, is not included in Equation 9.19 because the steady-state curves are used in determining the parameters for this equation. The inertia term is explicitly present in Equation 9.20.

Example 9.4

Split field series-wound dc motors are sometimes used as servo actuators. A motor circuit for this arrangement, under steady-state conditions, is shown in Figure 9.14. The field windings are divided into two identical parts and supplied by a differential amplifier (such as a push-pull amplifier) such that the magnetic fields in the two winding segments oppose each other. In this manner, the difference in the two input voltage signals (i.e., an error signal) is employed in driving the motor. Split field dc motors are used in low-power applications. Determine the electrical damping constant of the motor shown in Figure 9.14.

SOLUTION
Suppose that $v_1 = \bar{v} + (\Delta v/2)$ and $v_2 = \bar{v} - (\Delta v/2)$, where $\bar{v}$ is a constant representing the average supply voltage. Hence,

$$v_1 - v_2 = \Delta v \tag{i}$$

$$v_1 + v_2 = 2\bar{v} \tag{ii}$$

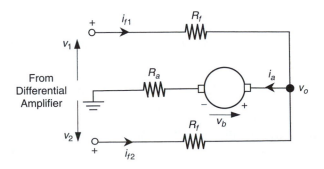

FIGURE 9.14
A split-field series-wound dc motor.

The motor is controlled using the differential voltage Δv. In a servo actuator, this differential voltage corresponds to a feedback error signal. Using the notation shown in Figure 9.14, the field current is given by (because the magnetic fields of the two stator winding segments oppose each other)

$$i_f = i_{f1} - i_{f2} \tag{iii}$$

The armature current is given by (see Figure 9.14)

$$i_a = i_{f1} + i_{f2} \tag{iv}$$

Hence, the motor magnetic torque can be expressed as

$$T_m = k i_a i_f = k(i_{f1} + i_{f2})(i_{f1} - i_{f2}) \tag{v}$$

Also, the node voltage is

$$v_o = v_1 - i_{f1} R_f = v_2 - i_{f2} R_f$$

Using this fact along with Equation i, we get

$$i_{f1} - i_{f2} = \frac{v_1 - v_2}{R_f} = \frac{\Delta v}{R_f} \tag{vi}$$

and

$$2v_o = v_1 + v_2 - R_f(i_{f1} + i_{f2}) \tag{vii}$$

But, it is clear from Figure 9.14 along with the motor back e.m.f. equation that

$$v_o = v_b + i_a R_a = k' i_f \, \omega_m + i_a R_a$$

where v_b denotes the back e.m.f. in the rotor. Hence, in view of Equation iii and Equation iv, we have

$$v_o = k'(i_{f1} - i_{f2})\omega_m + (i_{f1} + i_{f2})R_a \tag{viii}$$

Substitute Equation viii in Equation vii to eliminate v_o. We get

$$\frac{v_1 + v_2}{2} = \frac{R_f}{2}(i_{f1} + i_{f2}) + k'(i_{f1} - i_{f2})\omega_m + R_a(i_{f1} + i_{f2}) \tag{xi}$$

Substitute Equation ii and Equation vi in Equation ix. We get

$$\bar{v} = \frac{k'}{R_f} \Delta v \omega_m + \left(R_a + \frac{R_f}{2}\right)(i_{f1} + i_{f2}) \tag{x}$$

Substitute Equation vi in Equation v. We get

$$T_m = \frac{k}{R_f}(i_{f1} + i_{f2})\Delta v \tag{xi}$$

Substitute Equation xi in Equation x. We get

$$\bar{v} = \frac{k'}{R_f}\Delta v\, \omega_m + \left(R_a + \frac{R_f}{2}\right)\frac{R_f}{k\Delta v}T_m$$

or

$$T_m + \frac{kk'\Delta v^2}{R_f^2(R_a + R_f/2)}\omega_m = \frac{k\bar{v}\Delta v}{R_f(R_a + R_f/2)} \tag{9.21}$$

This is a linear relationship between T_m and ω_m. According to Equation 9.17a then, the electrical damping constant for a split-field series-wound dc motor is given by

$$b_e = \frac{kk'\Delta v^2}{R_f^2(R_a + R_f/2)} \tag{9.21a}$$

Note that the damping is zero under balanced conditions ($\Delta v = 0$). But damping increases quadratically with the differential voltage Δv.

9.3 Control of DC Motors

Both speed and torque of a dc motor may have to be controlled for proper performance in a given application of a dc motor. As we have seen, by using proper winding arrangements, dc motors can be operated over a wide range of speeds and torques. Because of this adaptability, dc motors are particularly suitable as variable-drive actuators. Historically, ac motors were employed almost exclusively in constant-speed applications, but their use in variable-speed applications was greatly limited because speed control of ac motors was found to be quite difficult, by conventional means. Since variable-speed control of a dc motor is quite convenient and straightforward, dc motors have dominated in industrial control applications for many decades.

Following a specified motion trajectory is called servoing, and servomotors (or servoactuators) are used for this purpose. The vast majority of servomotors are dc motors with feedback control of motion. Servo control is essentially a motion control problem, which involves the control of position and speed and hence their feedback. There are applications, however, that require torque control, directly or indirectly, but they usually require more sophisticated sensing and control techniques.

Control of a dc motor is accomplished by controlling either the stator field flux or the armature flux. If the armature and field windings are connected through the same circuit (see Figure 9.12), both techniques are incorporated simultaneously. Specifically, the two methods of control are,

1. Armature control
2. Field control

In armature control, the field current in the stator circuit is kept constant and the input voltage v_a to the rotor circuit is varied in order to achieve a desired performance (i.e., to reach specified values of position, speed, torque, etc.). Since v_a directly determines the motor back e.m.f., after allowance is made for the impedance drop due to resistance and leakage inductance of the armature circuit, it follows (see Equation 9.5) that armature control is particularly suitable for speed manipulation over a wide range of speeds (typically, 10 dB or more). The motor torque can be kept constant simply by keeping the armature current constant, because the field current is virtually a constant in the case of armature control (see Equation 9.4).

In field control, the armature voltage (and hence, armature current to some extent) is kept constant and the input voltage v_f to the field circuit is varied. From Equation 9.4, it can be seen that since i_a is kept more or less constant, the torque will vary in proportion to the field current i_f. Also, since the armature voltage is kept constant, the back e.m.f. will remain virtually unchanged. Hence, it follows from Equation 9.5 that the speed will be inversely proportional to i_f. This means that in field control, when the field voltage is increased the motor torque increases while the motor speed decreases, so that the output power remains somewhat constant. For this reason, field control is particularly suitable for constant power drives under varying torque-speed conditions, such as those present in winding mechanisms.

9.3.1 DC Servomotors

If the system characteristics and loading conditions are very accurately known and if the system is stable, it is possible in theory, to schedule the input signal to a motor (e.g., the armature voltage or field voltage) so as to obtain a desired response (e.g., motion trajectory or torque) from it. Parameter variations, model uncertainties, and external disturbances can produce errors that will build up (integrate) rapidly and will display unstable behavior in this case of *open-loop control* or *computed-input control* (sometimes referred to as feed-forward control). This is not acceptable in control system implementations. Feedback control is used to reduce these errors and to improve the control system performance, particularly with regard to stability, robustness, accuracy, and speed of response. In *feedback control systems*, response variables are sensed and fed back to the driver end of the system so as to reduce the response error (see Chapter 12 for details).

Servomotors are motors with motion feedback control, which are able to follow a specified motion trajectory. In a dc servomotor system, both angular position and speed might be measured (using shaft encoders, tachometers, resolvers, RVDTs, potentiometers, etc.) and compared with the desired position and speed. The error signal (= desired response – actual response) is conditioned and compensated using analog circuitry or are processed by a digital hardware processor or control computer, and is supplied to drive the servomotor toward the desired response. Both position feedback and velocity feedback are usually needed for accurate position control. For speed control, velocity feedback alone might be adequate, but position error can build up. On the other hand, if only position feedback is used, a large error in velocity is possible, even when the position error is small. Under certain conditions (e.g., high gains, large time delays), under position feedback alone, the control system may become marginally stable or even unstable. For this reason, dc servo systems historically employed tachometer feedback (velocity feedback) in addition to other types of feedback, primarily position feedback. In early generations of commercial servomotors, the motor and the tachometer (see Chapter 6) were available as a single package, within a common housing. Modern servomotors have a single built-in optical encoder (see Chapter 7) mounted on the motor shaft. This encoder is able to provide both position and speed measurements for servo control.

Motion control (position and speed control) implies control (indirect) of motor torque since it is the motor torque that causes the motion. In applications where torque itself is a primary output (e.g., metal forming operations, machining, micromanipulation, grasping, parts assembly, and tactile operations) and in situations where small motion errors could produce large unwanted forces (e.g., in parts assembly, or tracking a hard edge), direct control of motor torque would be desirable. In some applications of torque control, this is accomplished using feedback of the armature current or the field current, because the armature current and the field current determine the motor torque. The motor torque (magnetic torque), however, is not exactly equal to the load torque or the torque transmitted through the output shaft of the motor. Hence, for precise torque control, direct measurement of torque (e.g., using strain-gage, piezoelectric, or inductive sensors—see Chapter 6) would be required.

A schematic representation of an analog dc servomotor system is given in Figure 9.15. The actuator in this case is a dc motor. The sensors might include a tachometer to measure angular speed, a potentiometer to measure angular position, and a strain gage torque sensor, which is optional. More commonly, however, a single optical encoder is provided to measure both angular position and speed. The process (plant, the system that is driven) is represented by the load block in the figure. Signal-conditioning (filters, amplifiers, etc.—see Chapter 4) and compensating (lead compensation, lag compensation etc.—see Chapter 12) circuitry are represented by a single block. The power supply to the servo amplifier (and to the motor) is not shown in the figure. The motor and encoder are usually available as an integral unit, possibly (but increasingly less likely) with a tachometer as well, mounted on a common shaft. An additional position sensor (encoder, RVDT, potentiometer, resolver, etc.) may be attached to the load itself since in the presence of shaft flexibility, backlash, etc, the motor motion is not identical to the load motion.

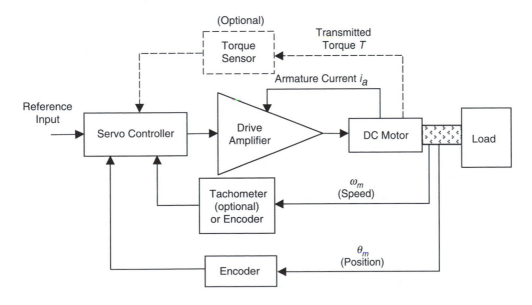

FIGURE 9.15
A dc servomotor system.

9.3.2 Armature Control

In an armature-controlled dc motor, the armature voltage v_a is used as the control input, while keeping the conditions in the field circuit constant. In particular, the field current i_f is assumed constant. Consequently, Equation 9.4 and Equation 9.5 can be written as

$$T_m = k_m i_a \tag{9.22}$$

$$v_b = k'_m \omega_m \tag{9.23}$$

The parameters k_m and k'_m are termed the *torque constant* and the *back e.m.f. constant*, respectively. Note that with consistent units, $k_m = k'_m$ in the case of ideal electrical-to-mechanical energy conversion at the motor rotor. In the Laplace domain (see Appendix A, Chapter 2, and Chapter 12), Equation 9.8 becomes

$$v_a - v_b = (L_a s + R_a)i_a \tag{9.24}$$

Note that, for convenience, time domain variables (functions of t) are used to denote their Laplace transforms (functions of s). It is understood, however, that the time functions are not identical to the Laplace functions. Also, in the Laplace domain, the mechanical Equation 9.9 becomes

$$T_m - T_L = (J_m s + b_m)\omega_m \tag{9.25}$$

where J_m and b_m denote the moment of inertia and the rotary viscous damping constant, respectively, of the motor rotor. Equations 9.22 through 9.25 are represented in the block diagram form, in Figure 9.16. Note that the speed ω_m is taken as the motor output. If the motor position θ_m is considered the output, it is obtained by passing ω_m through an integration block $1/s$. Note, further, that the load torque T_L, which is the useful (effective) torque transmitted to the load that is being driven, is an (unknown) input to the system. Usually, T_L increases with ω_m because a larger torque is necessary to drive a load at a higher speed. If a linear (and dynamic) relationship exists between T_L and ω_m at the load, a

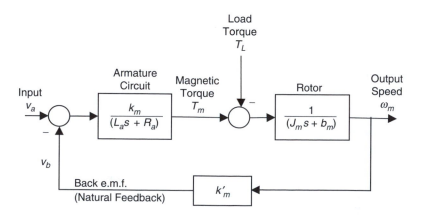

FIGURE 9.16
Open-loop block diagram for an armature-controlled dc motor.

feedback path can be completed from the output speed to the input load torque through a proper load transfer function (load block). The system shown in Figure 9.16 is not a feedback control system. The feedback path, which represents the back e.m.f., is a "natural feedback" and is characteristic of the process (dc motor); it is not an external control feedback loop.

The overall transfer relation for the system is obtained by first determining the output for one of the inputs with the other input removed, and then adding the two output components obtained in this manner, in view of the *principle of superposition*, which holds for a linear system (see Chapter 2 and Chapter 12). We get

$$\omega_m = \frac{k_m}{\Delta(s)} v_a - \frac{(L_a s + R_a)}{\Delta(s)} T_L \tag{9.26}$$

where $\Delta(s)$ is the *characteristic polynomial* of the system (see Chapter 2 and Chapter 12), given by

$$\Delta(s) = (L_a s + R_a)(J_m s + b_m) + k_m k'_m \tag{9.27}$$

This is a second-order polynomial in the Laplace variable s.

9.3.2.1 Motor Time Constants

The electrical time constant of the armature is

$$\tau_a = \frac{L_a}{R_a} \tag{9.28}$$

which is obtained from Equation 9.8 or Equation 9.24. The mechanical response of the rotor is governed by the *mechanical time constant,*

$$\tau_m = \frac{J_m}{b_m} \tag{9.29}$$

which is obtained from Equation 9.9 or Equation 9.25. Usually, τ_m is several times larger than τ_a, because the leakage inductance L_a is quite small (leakage of the flux linkage is negligible for high-quality dc motors). Hence, τ_a can be neglected in comparison to τ_m for most practical purposes. In that case, the transfer functions in Equation 9.26 become first order.

Note that the characteristic polynomial is the same for both transfer functions in Equation 9.26, regardless of the input (v_a or T_L). This should be the case because, $\Delta(s)$ determines the natural response of the system and does not depend on the system input. True time constants of the motor are obtained by first solving the characteristic equation $\Delta(s) = 0$ to determine the two roots (*poles* or *eigenvalues*), and then taking the reciprocal of the magnitudes (Note: only the real part of the two roots is used if the roots are complex). For an armature-controlled dc motor, these "true" time constants are not the same as τ_a and τ_m because of the presence of the coupling term $k_m k'_m$ in $\Delta(s)$ (see Equation 9.27). This also follows from the presence of the natural feedback path (back e.m.f.) in Figure 9.16.

Example 9.5

Determine an expression for the dominant time constant of an armature-controlled dc motor. What is the speed behavior (response) of the motor to a unit step input in armature voltage, in the absence of a mechanical load?

SOLUTION

By neglecting the electrical time constant in Equation 9.27, we have the approximate characteristic polynomial

$$\Delta(s) = R_a(J_m s + b_m) + k_m k'_m$$

This is expressed as

$$\Delta(s) = k'(\tau s + 1)$$

where τ is the overall dominant time constant of the system. It follows that the dominant time constant is given by

$$\tau = \frac{R_a J_m}{(R_a b_m + k_m k'_m)} \tag{9.30}$$

With $T_L = 0$, the motor transfer relation is

$$\omega_m = \frac{k}{(\tau s + 1)} v_a \tag{9.31a}$$

where the dc gain is

$$k = \frac{k_m}{(R_a b_m + k_m k'_m)} \tag{9.32}$$

Equation 9.31a corresponds to the system input-output differential equation:

$$\frac{d\omega_m}{dt} + \omega_m = k v_a$$

The speed response to a unit step change in v_a, with zero initial conditions, is (see chapter 2)

$$\omega_m(t) = k(1 - e^{-t/\tau}) \tag{9.33}$$

This is a nonoscillatory response. In practical situations, some oscillations will be present in the free response because, invariably, a load inertia is coupled to the motor through a shaft, which has some flexibility (i.e., it is not rigid).

9.3.2.2   *Motor Parameter Measurement*

The gain and time-constant parameters k and τ are functions of the motor parameters, as clear from Equation 9.30 and Equation 9.32. These parameters can be determined by a time domain test, where a step input is applied to the motor drive system and the response as given by Equation 9.33 is determined using either a digital oscilloscope or a data-acquisition computer. Specifically, the step response 9.33 is sketched in Figure 9.17. Note that the steady-state value of the speed is k. The slope of the response curve is obtained by differentiating Equation 9.33. Then by setting $t = 0$, we obtain the initial slope as

$$\frac{d\omega_m}{dt}(0) = \frac{k}{\tau} \tag{9.34}$$

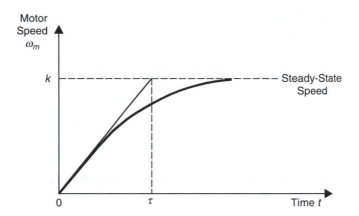

FIGURE 9.17
Open-loop step response of motor speed.

This line is drawn in Figure 9.17, which, according to Equation 9.34 intersects the steady-state level at time $t = \tau$. It follows that from an experimentally-determine step response curve it is possible to estimate the two parameters k and τ. Alternatively, a frequency-domain test can be carried out by applying a sine input and measuring the speed response, for a series of frequencies (or, by applying a transient input, measuring the speed response, and computing the ratio of the Fourier transforms of the response and the input). This gives the frequency transfer function (see Equation 9.31a)

$$G(j\omega) = \frac{k}{(\tau j\omega + 1)} \qquad (9.35)$$

The *Bode diagram* of the frequency response may be plotted as in Figure 9.18 (see Chapter 12). From the magnitude plot it is seen that

$$DC\ gain\ =\ 20\log_{10}k \qquad (9.36)$$

From either the magnitude plot or the phase plot, the *corner frequency* where the low-frequency asymptote (slope = 0 dB/decade) intersects the high-frequency asymptote (slope = −20 dB/decade), is given by

$$Corner\ frequency\ \omega_c\ =\ \frac{1}{\tau} \qquad (9.37)$$

In this manner, the frequency response plot can be used to estimate the two parameters k and τ.

Example 9.6

Analytical modeling may not be feasible for some complex engineering systems, and modeling using experimental data (i.e., *experimental modeling* or "system identification") might be the only available recourse. One approach is to use the measured response to a test input, as discussed before. In another approach, experimentally determined steady-state torque-speed characteristics are used to determine (approximately) a dynamic model

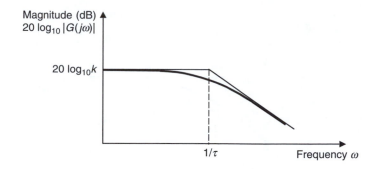

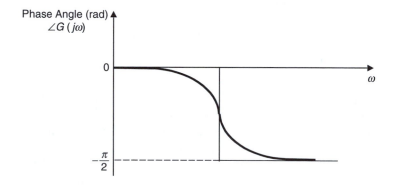

FIGURE 9.18
Open-loop frequency response of motor speed (Bode diagram).

for a motor. To illustrate this latter approach, consider an armature-controlled dc motor. Sketch steady-state speed-load torque curves using the input voltage (armature voltage) v_a as a parameter that is constant for each curve but varies from curve to curve. Obtain an equation to represent these curves. Now consider an armature-controlled dc motor driving a load of inertia J_L, which is connected directly to the rotor through a shaft that has torsional stiffness k_L. The viscous damping constant at the load is b_L. Obtain the system transfer function, with the load position θ_L as the output and armature supply voltage v_a as the input.

SOLUTION
From Equation 9.10a, the steady-state speed-torque curves for a separately excited dc motor are given by

$$T_m + \left[\frac{kk'v_f^2}{R_a R_f^2} \right] \omega_m = \left[\frac{kv_f}{R_a R_f} \right] v_a \tag{9.38}$$

Since v_f is a constant for armature-controlled motors, we can define two new constants k_m and b_e as

$$k_m = \frac{kv_f}{R_f} \tag{9.39}$$

and

$$b_e = \frac{kk'v_f^2}{R_a R_f^2} = \frac{k_m k_m'}{R_a} \tag{9.40}$$

Hence, Equation 9.38 becomes

$$T_m + b_e \omega_m = \frac{k_m}{R_a} v_a \tag{9.41}$$

Note that k_m is the torque constant defined by Equation 9.22, k_m' is the back e.m.f. constant defined by Equation 9.23, and b_e is the electrical damping constant defined by Equation 9.17a. Note, however, that because of the presence of mechanical dissipation, the torque T_{Ls} supplied to the load at steady-state (constant-speed) conditions is less than the motor magnetic torque T_m. Specifically, assuming linear viscous damping,

$$T_{Ls} = T_m - b_m \omega_m \tag{9.42}$$

It should be noted that if the motor speed is not constant, the output torque of the motor is further affected because some torque is used up in accelerating (or decelerating) the rotor inertia. Obviously, this factor does not enter into constant-speed tests. Now, by substituting Equation 9.42 in Equation 9.41, we have

$$T_{Ls} + (b_m + b_e)\omega_m = \frac{k_m}{R_a} v_a \tag{9.43}$$

Note that in constant-speed motor tests we measure T_{Ls}, not T_m. It follows from Equation 9.43 that the steady-state speed-torque curves (characteristic curves) for an armature-controlled dc motor are parallel straight lines with a negative slope of magnitude $b_m + b_e$. These curves are sketched in Figure 9.19(a). Note from Equation 9.43 that the parameters $b_m + b_e$ and k_m/R_a, can be directly extracted from an experimentally-determined characteristic curve. Once this is accomplished, Equation 9.43 is completely known and can be used for modeling the control system.

The system given in this example is shown in Figure 9.20. Suppose that θ_m denotes the motor angle. Newton's second law gives the rotor equation:

$$T_m - k_L(\theta_m - \theta_L) - b_m \dot{\theta}_m = J_m \ddot{\theta}_m \tag{i}$$

and the load equation:

$$-k_L(\theta_L - \theta_m) - b_L \dot{\theta}_L = J_L \ddot{\theta}_L \tag{ii}$$

Substituting Equation 9.41 into Equation i and Equation ii, and taking Laplace transforms, we get

$$\frac{k_m}{R_a} v_a + k_L \theta_L = \left[J_m s^2 + (b_m + b_e)s + k_L \right] \theta_m \tag{iii}$$

$$k_L \theta_m = \left(J_L s^2 + b_L s + k_L \right) \theta_L \tag{iv}$$

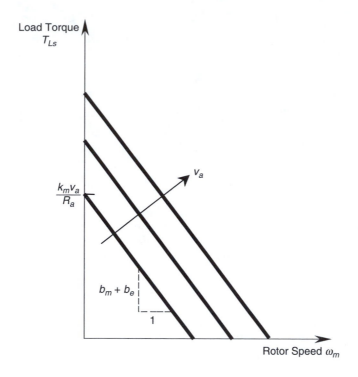

FIGURE 9.19
Steady-state speed-torque curves for an armature-controlled dc motor.

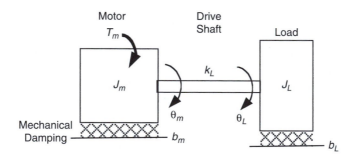

FIGURE 9.20
A motor driving an inertial load.

As usual, we use the same symbol to denote the Laplace transform as well as its time function. By substituting Equation iv in Equation iii and after straightforward algebraic manipulation, we obtain the system transfer function

$$\frac{\theta_L}{v_a} = \frac{k_L k_m / R_a}{s \left[J_m J_L s^3 + \{ J_L (b_m + b_e) + J_m b_L \} s^2 + \{ k_L (J_L + J_m) + b_L (b_m + b_e) \} s + k_L b_L + k_L (b_m + b_e) \right]} \quad (9.44)$$

Note that k_m / R_a, and $b_m + b_e$ are the experimentally determined parameters. The mechanical parameters k_L, b_L, and J_L, are assumed to be known. Notice the free integrator present in the transfer function given by Equation 9.44. This gives a pole (eigenvalue) at the origin of the s-plane ($s = 0$). It represents the *rigid-body mode* of the system, implying that the load is not externally restrained by a spring.

———————

We have seen that for some winding configurations, the speed-torque curve is not linear. Thus, the slope of a characteristic curve is not constant. Hence, an experimentally determined model would be valid only for an operating region in the neighborhood of the point where the slope was determined.

Example 9.7

A dc motor uses 2 hp under no-load conditions to maintain a constant speed of 600 rpm. The motor torque constant $k_m = 1$ V·s, the rotor moment of inertia $J_m = 0.1$ kg·m², and the armature circuit parameters are $R_a = 10$ Ω and $L_a = 0.01$ H. Determine the electrical damping constant, the mechanical damping constant, the electrical time constant of the armature circuit, the mechanical time constant of the rotor, and the true time constants of the motor.

SOLUTION

With consistent units, $k'_m = k_m$. Hence, from Equation 9.40, the electrical damping constant is

$$b_e = \frac{k_m^2}{R_a} = \frac{1}{10} = 0.1 \ \text{N·m/rad/s}$$

It is given that the power absorbed by the motor at no-load conditions is

$$2 \ \text{hp} = 2 \times 746 \ \text{W} = 1492 \ \text{W}$$

and the corresponding speed is

$$\omega_m = \frac{600}{60} \times 2\pi \ \text{rad/s} = 20\pi \ \text{rad/s}$$

This power is used against electrical and mechanical damping, at constant speed ω_m. Hence,

$$(b_m + b_e) \, \omega_m^2 = 1492$$

or

$$b_m + b_e = \frac{1492}{(20\pi)^2} = 0.38 \ \text{N·m/rad/s}$$

It follows that the mechanical damping constant is

$$b_m = 0.38 - 0.1 = 0.28 \ \text{N·m/rad/s}$$

From Equation 9.28 and Equation 9.29,

$$\tau_a = \frac{0.01}{10} = 0.001 \ \text{s}$$

$$\tau_m = \frac{0.1}{0.28} = 0.36 \ \text{s}$$

Note that τ_m is several orders larger than τ_a. In view of Equation 9.27, the characteristic polynomial of the motor transfer function can be written as

$$\Delta(s) = R_a[b_m(\tau_a s+1)(\tau_m s+1)+b_e] \tag{9.45}$$

The poles (eigenvalues) are given by $\Delta(s) = 0$ (see Chapter 2 and Chapter 12). Hence,

$$0.28(0.001s + 1)(0.36s + 1) + 0.1 = 0$$

or

$$s^2 + 1010s + 3800 = 0$$

Solving for the motor eigenvalues, we get

$$\lambda_1 = -3.8 \quad \text{and} \quad \lambda_2 = -1006$$

The two poles are real and negative. This means that any disturbance in the motor speed will die out exponentially without oscillations.

The time constants are given by the reciprocals of the magnitudes of the real parts of the eigenvalues. Hence, the true time constants are

$$\tau_1 = 1/3.8 = 0.26 \text{ s}$$

$$\tau_2 = 1/1006 = 0.001 \text{ s}$$

The smaller time constant τ_2, which derives primarily from the electrical time constant of the armature circuit, can be neglected for all practical purposes. The larger time constant τ_1 comes not only from the mechanical time constant τ_m (rotor inertia/mechanical damping constant) but also from the electrical damping constant (back e.m.f. effect) b_e. Hence, τ_1 is not equal to τ_m, even though the two are of the same order of magnitude.

9.3.3 Field Control

In field-controlled dc motors, the armature current is assumed to be kept constant, and the field voltage is used as the control input. Since i_a is assumed constant, Equation 9.4 can be written as

$$T_m = k_a i_f \tag{9.46}$$

where k_a is the electromechanical torque constant for the motor. The back e.m.f. relation and the armature circuit equation are not used in this case. Equation 9.7 and Equation 9.9 are written in the Laplace form as

$$v_f = (L_f s + R_f)i_f \tag{9.47}$$

$$T_m - T_L = (J_m s + b_m)\omega_m \tag{9.48}$$

Equations 9.46 through 9.48 can be represented by the open-loop block diagram given in Figure 9.21. Note that even though i_a is assumed constant, this is not strictly true. This should

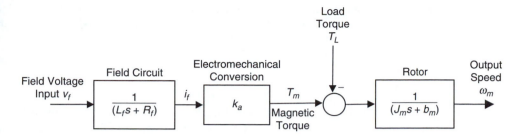

FIGURE 9.21
Open-loop block diagram for a field-controlled dc motor.

be clear from the armature circuit equation (Equation 9.8). It is the armature supply voltage v_a that is kept constant. Even though L_a can be neglected, then, i_a depends on the back e.m.f. v_b, which changes with the motor speed as well as the field current i_f. Under these conditions, the block representing k_a in Figure 9.21 is not a constant gain, and in fact it is not linear. At least, a feedback will be needed into this block from output speed. This will also add another electrical time constant, which depends on the dynamics of the armature circuit. It will also introduce a coupling effect between the mechanical dynamics (of the rotor) and the armature circuit electronics. For the present purposes, however, we assume that k_a is a constant gain.

Now, we return to Figure 9.21. Since the system is linear, the principle of superposition holds (see Chapter 2 and Chapter 12). According to this, the overall output ω_m is equal to the sum of the individual outputs due to the two inputs v_f and T_L, taken separately. It follows that the transfer relationship is given by

$$\omega_m = \frac{k_a}{(L_f s + R_f)(J_m s + b_m)} v_f - \frac{1}{(J_m s + b_m)} T_L \tag{9.49}$$

In this case, the electrical time constant originates from the field circuit and is given by

$$\tau_f = \frac{L_f}{R_f} \tag{9.50}$$

The mechanical time constant τ_m of the field-controlled motor is the same as that for the armature-controlled motor, and can be defined by Equation 9.50(a):

$$\tau_m = \frac{J_m}{b_m} \tag{9.50a}$$

The characteristic polynomial of the open-loop field-controlled motor is

$$\Delta(s) = (L_f s + R_f)(J_m s + b_m) \tag{9.51}$$

It follows that τ_f and τ_m are the true time constants of the system, unlike in an armature controlled motor. As in an armature-controlled dc motor, however, the electrical time constant is several times smaller and can be neglected in comparison to the mechanical time constant. Furthermore, as for an armature-controlled motor, the speed and the angular position of a field-controlled motor have to be measured and fed back for accurate motion control.

9.3.4 Feedback Control of DC Motors

Open-loop operation of a dc motor, as represented by Figure 9.16 (armature control) and Figure 9.21 (field control), can lead to excessive error and even instability, particularly because of the unknown load input and also due to the integration effect when position (not speed) is the desired output (as in positioning applications). Feedback control is necessary in these circumstances.

In feedback control, the motor response (position, speed, or both) is measured using an appropriate sensor and fed back into the controller, which generates the control signal for the drive hardware of the motor. An optical encoder can be used to sense both position and speed (see Chapter 7) and a tachometer may be used to measure the speed alone (see Chapter 6). The following three types of feedback control are important:

1. Velocity feedback
2. Position plus velocity feedback
3. Position feedback with a multiterm controller

9.3.4.1 Velocity Feedback Control

Velocity feedback is particularly useful in controlling the motor speed (see Chapter 12). In velocity feedback, motor speed is sensed using a device such as a tachometer or an optical encoder, and is fed back to the controller, which compares it with the desired speed, and the error is used to correct the deviation. Additional dynamic compensation (e.g., lead or lag compensation) may be needed to improve the accuracy and the effectiveness of the controller, and can be provided using either analog circuits or digital processing. The error signal is passed through the compensator in order to improve the performance of the control system.

9.3.4.2 Position Plus Velocity Feedback Control

In position control, the motor angle θ_m is the output. In this case, the open-loop system has a free integrator, and the characteristic polynomial is $s(\tau s + 1)$. This is a marginally stable system. In particular, if a slight disturbance or model error is present, it will be integrated out, which can lead to a diverging error in the motor angle. In particular, the load torque T_L is an input to the system, and is not completely known. In control systems terminology, this is a disturbance (an unknown input), which can cause unstable behavior in the open-loop system. In view of the free integrator at the position output, the resulting unstable behavior cannot be corrected using velocity feedback alone. Position feedback is needed to remedy the problem. Both position and velocity feedback are needed. The feedback gains for the position and velocity signals can be chosen so as to obtain the desired response (speed of response, overshoot limit, steady-state accuracy, etc.). Block diagram of a position plus velocity feedback control system for a dc motor is shown in Figure 9.22. The motor block is given by Figure 9.16 for an armature-controlled motor, and by Figure 9.21 for a field-control motor (Note: load torque input is integral in each of these two models). The drive unit of the motor is represented by an amplifier of gain k_a. Control system design involves selection of proper parameter values for sensors and other components in the control system.

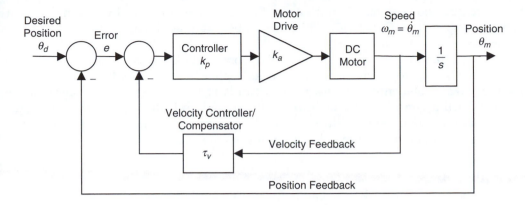

FIGURE 9.22
Position plus velocity feedback control of a dc motor.

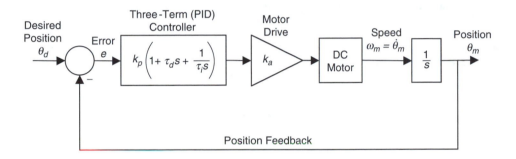

FIGURE 9.23
PID control of the position response of a dc motor.

9.3.4.3 *Position Feedback with PID Control*

A popular method of controlling a dc motor is to use just position feedback, and then reduce the error using a three-term controller having the proportional, integral and derivative (PID) actions (see Chapter 12). A block diagram for this control system is shown in Figure 9.23. Each term of the PID controller provides specific benefits. There are some undesirable side effects as well. These are discussed in Chapter 12. In particular, proportional action improves the speed of response and reduces the steady-state error, but it tends to increase the level of overshoot (i.e., less stable). Derivative action adds damping, just like velocity feedback, thereby making the system more stable (less overshoot). In doing so, it does not degrade the speed of response, however. The integral action reduces the steady-state error (typically reduces to zero), but it tends to degrade the system stability and the speed of response. A *lead compensator* provides an effect somewhat similar to the derivative action while a lag compensator provides an integrator-like effect. A compensator is used to complement the controller in realizing the required performance in the sytem. These compensators are discussed in Chapter 12.

In the control system of a dc motor (Figure 9.22 or Figure 9.23), the desired position command may be provided using a potentiometer, as a voltage signal. The measurements of position and speed also are provided as voltage signals. Specifically, in the case of an optical encoder, the pulses are detected using a digital pulse counter, and read into the digital controller (see Chapter 7). This reading has to be calibrated to be consistent with the desired position command. In the case of a tachometer, the velocity reading is generated as a voltage (see Chapter 6), which has to be calibrated then, to be consistent with the desired position signal.

It is noted that proportional plus derivative control (PPD control or PD control) with position feedback, has a similar effect as position plus velocity (speed) feedback control. But, the two are not identical because the latter adds a zero to the system transfer function, requiring further considerations in the controller design, and affecting the motor response. In particular, the zero modifies the sign and the ratio in which the two response components corresponding to the two poles contribute to the overall response.

Example 9.8

Consider the position and velocity control system of Figure 9.22. Suppose that the motor model is given by the transfer function $\frac{k_m}{(\tau_m s+1)}$. Determine the closed-loop transfer function $\frac{\theta_m}{\theta_d}$. Next consider proportional plus derivative control system (Figure 9.23, with the integral controller removed) and the same motor model. What is the corresponding closed-loop transfer function $\frac{\theta_m}{\theta_d}$? Compare these two types of control, particularly with respect to speed of response, stability (percentage overshoot), and steady-state error.

SOLUTION
From Figure 9.22 we can write

$$(\theta_d - \theta_m)k_p k_a \frac{k_m}{(\tau_m s+1)} = s\theta_m$$

Hence,

$$\frac{\theta_m}{\theta_d} = \frac{k}{\left[\tau_m s^2 + (1+\tau_v)s + k\right]} \tag{9.52}$$

where k denotes $k_p k_a k_m$.

Now from Figure 9.23, with the integral control action removed, we can write

$$(\theta_d - \theta_m)k_p(1+\tau_d s)k_a \frac{k_m}{(\tau_m s+1)} = s\theta_m$$

On simplification we get

$$\frac{\theta_m}{\theta_d} = \frac{k(1+\tau_d s)}{\left[\tau_m s^2 + (1+\tau_v)s + k\right]} \tag{9.53}$$

It is seen that the characteristic polynomials (denominators of the transfer functions) are identical, in the two cases. As a result, it is possible to place the closed-loop poles (eigenvalues) at desirable locations, in both cases.

The PPD controller, however, introduces a zero to the transfer function, as seen in the numerator of Equation 9.53. This zero can have a significant effect on the transient response of the motor. In particular, the zero contributes a time derivative term, which can be significant in the beginning. Hence, a larger overshoot (than for the position plus velocity control) will result. But, the same derivative action will cause the response to settle down quickly to the steady-state value.

The steady-state gain (or, dc gain) of both transfer functions: Equation 9.52 and Equation 9.53 is equal to 1 (which is obtained by setting $s = 0$). It follows that the steady state error is zero in both cases.

9.3.5 Phase-Locked Control

Phase-locked control is an effective approach to controlling dc motors. A block diagram of a phase-locked servo system is shown in Figure 9.24. This is a *phase* control method. The position command is generated according to the desired motion of the motor, using a signal generator (e.g., a voltage-controlled oscillator) or using digital means. This reference signal is in the form of a pulse train, which is quite analogous to the output signal of an incremental encoder (Chapter 7). The rotation of the motor (or load) is measured using an incremental encoder. The encoder pulse train forms the feedback signal. The reference signal and the feedback signal are supplied to a *phase detector*, which generates a signal representing the phase difference between the two signals and possibly some unwanted high-frequency components. The unwanted components are removed using a low-pass filter, and the resulting (error) signal is supplied to the drive amplifier of the motor. The error signal drives the motor so as to obtain the desired motion. The objective is to maintain a fixed phase difference (ideally, a zero phase difference) between the reference pulse signal and the position (encoder) pulse signal. Under these conditions, the two signals are synchronized or phase-locked together. Any deviation from the locked conditions will generate an error signal, which will bring the motor motion back in phase with the reference command. In this manner, deviations due to external disturbances, such as load changes on the motor, are also corrected.

One method of determining the phase difference of two pulse signals is by detecting the edge transitions (Chapter 7 and Chapter 10). An alternative method is to take the product of the two signals and then low-pass filter the result. To illustrate this second method, suppose that the primary (harmonic) components of the reference pulse signal and the response pulse signal are $(u_o \sin \theta_u)$ and $(y_o \sin \theta_y)$, respectively, where

$$\theta_u = \omega t + \phi_u$$

$$\theta_y = \omega t + \phi_y$$

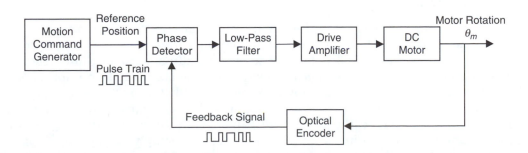

FIGURE 9.24
Schematic diagram of a phase-locked servo.

Note that ω is the frequency of the two pulse signals (assumed to be the same) and ϕ denotes the phase angle. The product signal is

$$p = u_o y_o \sin\theta_u \sin\theta_y$$

$$= \frac{1}{2} u_o y_o [\cos(\theta_u - \theta_y) - \cos(\theta_u + \theta_y)]$$

Consequently,

$$p = \frac{1}{2} u_o y_o \cos(\phi_u - \phi_y) - \frac{1}{2} u_o y_o \cos(2\omega t + \phi_u + \phi_y) \qquad (9.54)$$

Low-pass filtering will remove the high-frequency component of frequency 2ω, leaving the signal

$$e = \frac{1}{2} u_o y_o \cos(\phi_u - \phi_y) \qquad (9.55)$$

This is a nonlinear function of the phase difference $(\phi_u - \phi_y)$. By applying a $\pi/2$ phase shift to the original two signals, we also could have determined $\frac{1}{2} u_o y_o \sin(\phi_u - \phi_y)$. In this manner, the magnitude and sign of $(\phi_u - \phi_y)$ are determined. Very accurate position control can be obtained by driving this phase difference to zero. This is the objective of phase-locked control; the phase angle of the output is locked onto the phase angle of the command signal. In more sophisticated phase-locked servos, the frequency differences are also detected—for example, using pulse counting—and compensated. This is analogous to the classic proportional plus derivative (PD) control. It is clear that phase-locked servos are velocity-control devices as well, because velocity is proportional to the pulse frequency. When the two pulse signals are synchronized, the velocity error also approaches zero, subject to the available resolution of the control system components. Typically, speed error levels of ±0.002% or less are possible using phase-locked servos. Also, the overall cost of a phase-locked servo system is usually less than that of a conventional analog servo system, because less expensive solid-state devices replace bulky analog control circuitry.

9.4 Motor Driver

The driver of a dc motor is a hardware unit, which generates the necessary current to energize the windings of the motor. By controlling the current generated by the driver, the motor torque can be controlled. By receiving feedback from a motion sensor (encoder, tachometer, etc.), the angular position and the speed of the motor can be controlled. Note that when an optical encoder is provided integral with the motor unit—a typical situation—it is not necessary to use a tachometer as well, because the enoder can generate both position and speed measurements (see Chapter 7). The drive unit primarily consists of a drive amplifier, with additional circuitry and a dc power supply. In typical applications of motion control and servoing, the drive unit is a *servoamplifier* with auxiliary hardware. The driver is commanded by a control input provided by a host computer (personal computer or PC) through an interface (input/output) card. A suitable arrangement is

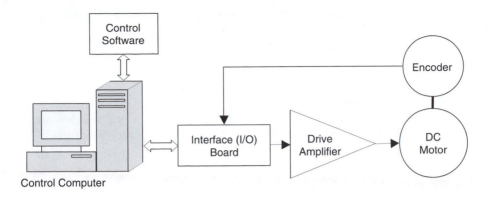

FIGURE 9.25
Components of a dc motor control system.

shown in Figure 9.25. Also, the driver parameters (e.g., amplifier gains) are software programmable and can be set by the host computer.

The control computer receives a feedback signal of the motor motion, through the interface board, and generates a control signal, which is provided to the drive amplifier, again through the interface board. Any control scheme can be programmed (say, in C language) and implemented in the control computer. In addition to typical servo control schemes such as proportional-integral-derivative (PID) and position-plus-velocity feedback, other advanced control algorithms (e.g., optimal control techniques such as linear quadratic regulator or LQR and linear quadratic Gaussian or LQG, adaptive control techniques such as model-referenced adaptive control, switching control techniques such as sliding-mode control, nonlinear control schemes such as feedback linearization technique or FLT, and intelligent control techniques such as fuzzy logic control) may be applied in this manner. If the computer does not have the processing power to carry out the control computations at the required speed (i.e., control bandwidth), a digital signal processor (DSP) may be incorporated into the computer. But, with modern computers, which can provide substantial computing power at low cost, DSPs are not needed in most applications.

9.4.1 Interface Board

The I/O card or data acquisition (DAQ) card is a hardware module with associated driver software, based in a host computer (PC), and connected through its bus (ISA bus). It forms the input-output link between the motor and the controller. It can provide many (say, eight) analog signals to drive many (eight) motors, and hence termed a *multi-axis* card. It follows that the digital-to-analog conversion (DAC) capability is built into the I/O card (e.g., 16 bit DAC including a sign bit, ±10 V output voltage range). Similarly, the analog-to-digital conversion (ADC) function is included in the I/O card (e.g., eight analog input channels with 16 bit ADC including a sign bit, ±10 V output voltage range). These input channels can be used for acquiring data from analog sensors such as tachometers, potentiometers, and strain gauges. Equally important are the encoder channels to read the pulse signals from the optical encoders mounted on the dc servomotors. Typically the encoder input channels are equal in number to the analog output channels (and the number of axes, e.g., eight). The position pulses are read using counters (e.g., 24-bit counters), and the speed is determined by the pulse rate. The rate at which the encoder pulses are counted can be quite high (e.g., 10 MHz). In addition a number of bits (e.g., 32) of digital input and output may be available through the I/O card, for use in simple digital sensing, control, and switching functions. The principles of ADC and DAC are discussed in Chapter 4. Digital counters are discussed in Chapter 10.

9.4.2 Drive Unit

The primary hardware component of the motor drive system is the drive amplifier. In typical motion control applications, these amplifiers are called servo amplifiers. Two types of drive amplifiers are commercially available:

1. Linear amplifier
2. Pulse-width-modulation (PWM) amplifier

A linear amplifier generates a voltage output, which is proportional to the control input provided to it. Since the output voltage is proportioned by dissipative means (using resistor circuitry), this is a wasteful and inefficient approach. Furthermore, fans and heat sinks have to be provided to remove the generated heat, particularly in continuous operation. To understand the inefficiency associated with a linear amplifier, suppose that the operating output range of the amplifier is 0–20 V, and that the amplifier is powered by a 20 V power supply. Under a particular operating condition, suppose that the motor is applied 10 V and draws a current of 4 A. The power used by the motor then is 10×4 W = 40 W. Still, the power supply provides 20 V at 5 A, thereby consuming 100 W. This means, 60 W of power is dissipated, and the efficiency is only 40%. The efficiency can be made close to 100% using modern PWM amplifiers, which are nondissipative devices depending on high-speed switching at constant voltage to control the power supplied to the motor, as discussed next.

Modern servo amplifiers use pulse-width modulation (PWM) to drive servomotors efficiently under variable-speed conditions, without incurring excessive power losses. Integrated microelectronic design makes them compact, accurate, and inexpensive. The components of a typical PWM drive system are shown in Figure 9.26. Other signal conditioning hardware

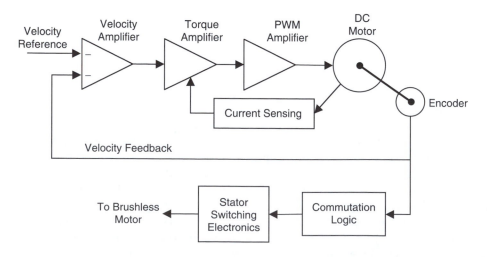

FIGURE 9.26
The main components of a PWM drive system for a dc motor.

(e.g., filters) and auxiliary components such as isolation hardware, safety devices including tripping hardware, and cooling fan are not shown in the figure. In particular, note the following components, connected in series:

1. A velocity amplifier (a differential amplifier)
2. A torque amplifier
3. A PWM amplifier

The power can come from an ac line supply, which is rectified in the drive unit to provide the necessary dc power for the electronics. Alternatively, leads may be provided for an external power supply (e.g., 15 V DC). The reference velocity signal and the feedback signal (from an encoder or a tachometer) are connected to the input leads of the velocity amplifier. The resulting difference (error signal) is conditioned and amplified by the torque amplifier to generate a current corresponding to the required torque (corresponding to the driving speed). The motor current is sensed and fed back to this amplifier, to improve the torque performance of the motor. The output from the torque amplifier is used as the modulating signal to the PWM amplifier. The reference switching frequency of a PWM amplifier is high (on the order of 25 kHz). Pulse-width modulation is accomplished by varying the duty cycle of the generated pulse signal, through switching control, as explained next. The PWM signal from the amplifier (e.g., at 10 V) is used to energize the field windings of a dc motor. A brushless dc motor needs electronic commutation. This may be accomplished using the encoder signal to time the switching on and off of the current, which flows through the stator windings.

9.4.3 Pulse Width Modulation

The final control of a dc motors is accomplished by controlling the supply voltage to either the armature circuit or the field circuit. A dissipative method of achieving this involves using a variable resistor in series with the supply source to the circuit. This method, besides being wasteful, has other disadvantages. Notably, the heat generated at the control resistor has to be removed promptly to avoid malfunction and damage due to high temperatures. A linear amplifier with a variable gain is also dissipative and inefficient. A much more desirable way to control the voltage to a dc motor is by using a solid-state switch to vary the off time of a fixed voltage level, while keeping the period (or, inverse frequency) of the pulse signal constant. Specifically, the *duty cycle* of a pulse signal is varied.

Consider the voltage pulse signal shown in Figure 9.27. The following notation is used:

T = pulse period (i.e., interval between the successive on instants)

T_o = on period (i.e., interval between on instant to the next off instant)

Then, the *duty cycle* is given by the percentage

$$d = \frac{T_o}{T} \times 100\% \qquad (9.56a)$$

Note that the voltage level v_{ref} and the pulse frequency $1/T$ are kept fixed, and what is varied is T_o. Pulse width modulation is achieved by "chopping" the reference voltage so that the average voltage is varied. As discussed later, it is easy to see that, with respect to

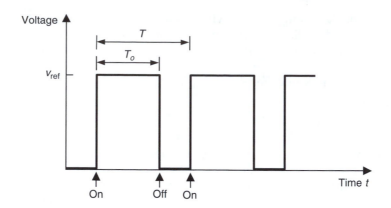

FIGURE 9.27
Duty cycle of a PWM signal.

an output pulse signal, the duty cycle is given by the ratio of average output to the peak output; specifically,

$$\text{Duty cycle} = \frac{\text{average output}}{\text{peak output}} \times 100\% \tag{9.56b}$$

Equation 9.56 also verifies that the average level of a PWM signal is proportional to the duty cycle (or the on time period T_o) of the signal. It follows that the output level (i.e., the average value) of a PWM signal can be varied simply by changing the signal-on time period (in the range 0 to T) or equivalently by changing the duty cycle (in the range 0–100%). This relationship between the average output and the duty cycle is linear. Hence a digital or software means of generating a PWM signal would be to use a straight line from 0 to the maximum signal level, spanning the period (T) of the signal. For a given output level, the straight line segment at this height, when projected on the time axis, gives the required on-time interval (T_o).

A discrete "chopper" circuit for pulse width modulation may be developed using a solid-state switch known as the *thyristor*. The thyristor is also known as a *silicon-controlled rectifier, a solid-state controlled rectifier, a semiconductor-controlled rectifier,* or simply an SCR. It is a pellet made of four layers (pnpn) of semiconductor material (e.g., silicon with a trace of dope material). It has three terminals—the anode, the cathode, and the gate—as shown in Figure 9.28(a). The anode and the cathode are connected to the circuit that carries the load current i. When the gate potential v_g is less than or equal to zero with respect to the cathode, the thyristor cannot conduct in either direction ($i = 0$). When v_g is made positive, the thyristor will conduct from anode to cathode but not in the opposite direction (i.e., it acts like a basic diode). In other words, a positive firing signal (i.e., a positive trigger voltage) v_g will close (turn on) the switch. To open (turn off) the switch again, we not only have to make v_g zero (or slightly negative) with respect to the cathode, but also the load current from the anode to the cathode has to be zero (or slightly negative). This is the natural mode of operation of a thyristor. When the supply voltage is dc, it does not drop to zero; hence, the thyristor would be unable to turn itself off. In this case, a commutating circuit that can make the voltage across the thyristor slightly negative, has to be employed. This is called forced commutation (as opposed to natural commutation) of a thyristor. Note that when a thyristor is conducting, it offers virtually no resistance, and the voltage

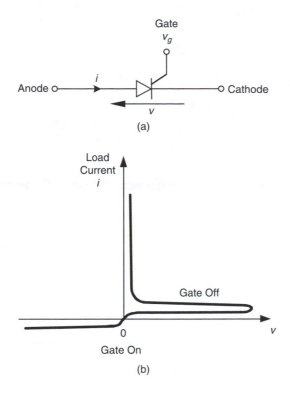

Gate
v_g

Anode o——→ i ——o Cathode

v

(a)

Load
Current
i

Gate Off

0

v

Gate On

(b)

FIGURE 9.28
(a) Symbol for a thyristor, (b) Ideal characteristic curve of a thyristor.

drop across the thyristor can be neglected for practical purposes. An idealized voltage-current characteristic curve of a thyristor is shown in Figure 9.28(b). Solid-state switching devices are lossless (or nondissipative) in nature.

A basic thyristor circuit using a dc power supply, which may be used in dc motor control, is shown in Figure 9.29(a). The dc supply voltage is v_{ref}, the PWM voltage signal supplied to the armature circuit is v_a, and the back e.m.f. in the motor is v_b. The nature of these voltage signals is shown in Figure 9.29(b). Since the supply voltage v_{ref} to the armature circuit is "chopped" in generating the PWM signal v_a, the circuit in Figure 9.29(a) is usually known as a *chopper circuit*. In armature control, the field circuit is separately excited, as shown. The armature resistance R_a is neglected to provide a qualitative explanation of the voltages appearing in various parts in the circuit; R_a should be included in a more accurate model. The inductance L_o includes the usual armature leakage inductance, self-inductance, and so forth (normally denoted by L_a), and an external inductance that is needed to avoid large fluctuations in armature current i_a, since v_a is pulsating. Alternatively, a series-wound motor in which the field inductance L_f is connected in series with the armature may be used to increase L_o. A free-wheeling diode provides a path for i_a during the off period of v_a so as to avoid large voltage buildup in the armature.

Initially, the voltage v_g applied to the gate terminal will close (turn on) the SCR, allowing v_{ref} to be applied to the armature circuit. Since v_{ref} is a dc voltage, however, the SCR will not open (turn off) by itself. Hence, a commutating circuit that is capable of applying a slightly negative voltage to the anode of the SCR is needed. The commutating circuit usually consists of a capacitor that is charged to provide the required voltage, a diode, and a thyristor.

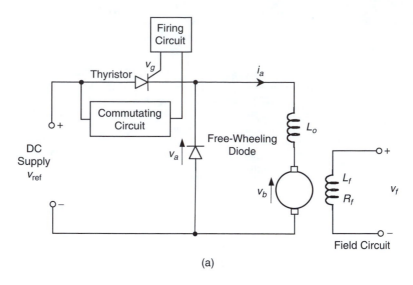

(a)

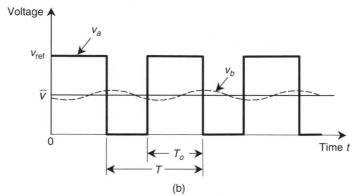

(b)

FIGURE 9.29
(a) A thyristor (SCR) control circuit (chopper) for a dc motor, (b) Circuit voltage signals.

The average voltage $\bar{v}$ supplied to the armature circuit is the average of v_a. This is given by

$$\bar{v} = \frac{T_o}{T} v_{ref} \tag{9.57}$$

where
  T_o = on time interval of the supplied voltage pulse
  T = pulse period

In other words, the reference voltage is fractioned by the duty cycle (see Equation 9.56b) to obtain the average voltage. As noted before, $\bar{v}$ can be varied either by changing T_o (called, *pulse width modulation* or PWM) or by changing T (*pulse frequency modulation,* or PFM). What is common is PWM, where the pulse frequency (or, T) is kept constant. This method of pulsing control is employed in chopper drive circuits of dc motors.

If v_a were a constant, there would not be a potential drop across the inductor L_o, and v_b would be a constant equal to v_a. It follows that the average value (i.e., the dc component) of v_b is equal to the average value of v_a, which is denoted by $\bar{v}$ in Equation 9.57. When the motor speed is properly regulated (i.e., speed fluctuations are small), and assuming that the conditions in the field circuit are steady, the back e.m.f. v_b is nearly a constant. Hence,

$$v_b = \bar{v} \tag{9.58}$$

Note that the voltage across L_o is $v_a - v_b$. It follows that

$$L_o \frac{di_a}{dt} = v_a - v_b \tag{9.59}$$

Now, in view of Equation 9.58, we can write

$$L_o \frac{di_a}{dt} = v_a - \bar{v} \tag{9.60}$$

or

$$i_a = \frac{1}{L_o} \int (v_a - \bar{v}) dt \tag{9.61}$$

Equation 9.61 tells us that the change in the armature current is proportional to the area between the v_a curve and the $\bar{v}$ line shown in Figure 9.29(b). In particular, starting from a steady-state value of i_a, the armature current will rise by

$$\Delta i_a = \frac{1}{L_o} (v_{ref} - \bar{v}) T_o$$

over the time period T_o in which the thyristor is on. Substituting Equation 9.57, we get

$$\Delta i_a = \frac{v_{ref}}{L_o T} (T - T_o) T_o \tag{9.62}$$

Then, over the time interval $(T - T_o)$ during which the thyristor is off, the armature current will drop by

$$\frac{1}{L_o} \bar{v} (T - T_o)$$

which is equal to

$$\frac{v_{ref}}{L_o T} (T - T_o) T_o$$

As a result, the armature current will go back to the initial value i_a. This cycle will repeat over the subsequent pulse cycles of duration T. It follows that the fluctuation in the armature current is given by Equation 9.62. For a given T (i.e., a given pulse frequency), this amplitude is maximum when $T_o = T/2$. Hence,

$$(\Delta i_a)_{max} = \frac{T v_{ref}}{4L_o} \tag{9.63}$$

Note that the current fluctuations can be reduced by increasing L_o and decreasing T for a given supply voltage v_{ref}. Note, further, that since $v_{ref} - \bar{v}$ is constant, it follows from Equation 9.61 that the armature current increases or decreases linearly with time.

Example 9.9

Consider the chopper circuit given in Figure 9.29(a). The chopper frequency is 200 Hz, the series inductance in the armature circuit is 50 mH, and the supply dc voltage to the chopper is 100 V. Determine the amplitude of the maximum (worst case) fluctuation in the armature current.

SOLUTION

Since we are interested in the worst case of current fluctuations, we use Equation 9.63. Then,

$$T = \frac{1}{200} \text{ sec}, \qquad L_o = 0.05 \text{ mH}, \qquad v_{ref} = 100 \text{ V}$$

Substituting values, we get

$$(\Delta i_a)_{max} = \frac{100}{200 \times 4 \times 0.05} 2.5 \text{ A}$$

High-power dc motors are usually driven by rectified ac supplies (single-phase or three-phase). In this case as well, motor control can be accomplished by using thyristor circuits. Full-wave circuits are those that use both the positive and negative parts of an ac supply voltage. A full-wave single-phase control circuit for a dc motor is shown in Figure 9.30.

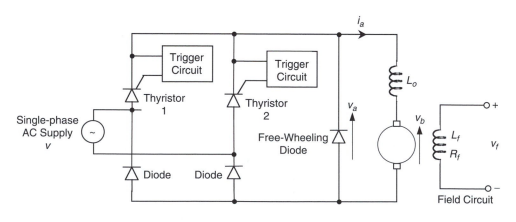

FIGURE 9.30
Full-wave single-phase control circuit for a dc motor.

It uses two thyristors. For convenience, the armature resistance R_o is not shown, although it is important in the analysis of the circuit. The inductance L_o contains the usual armature component L_a as well as an external inductor, which is connected in series with the armature to reduce surges in the armature current. Furthermore, a free-wheeling diode is used to provide a current path when the two thyristors are turned off. Two additional diodes are provided to complete the current path for each half of the supply wave period.

Various signals through the circuit are sketched in Figure 9.31. The supply voltage v is shown in Figure 9.31(a). The broken line in this figure is the voltage v_a that would result if the two thyristors were replaced by diodes. Figure 9.31(b) shows the dc voltage v_a supplied to the armature circuit and the back e.m.f. v_b across the armature; T_o is the firing time of each thyristor from the time when the voltage supplied to a thyristor begins to build from zero. Specifically, during the positive half of the supply voltage v, thyristor 1 will be triggered after time T_o, and during the negative half of the supply voltage, thyristor 2 will be triggered after time T_o. Note that in this full-wave circuit, the negative half of v also appears as positive in v_a (across the free-wheeling diode). The back e.m.f. v_b is reasonably constant, and so is the armature current i_a. The voltage across L_o is $v_a - v_b$. Hence, the armature current variation is given by the area between the two curves v_a and v_b, as is clear from Equation 9.59. As in the case of dc power supply, the motor can be

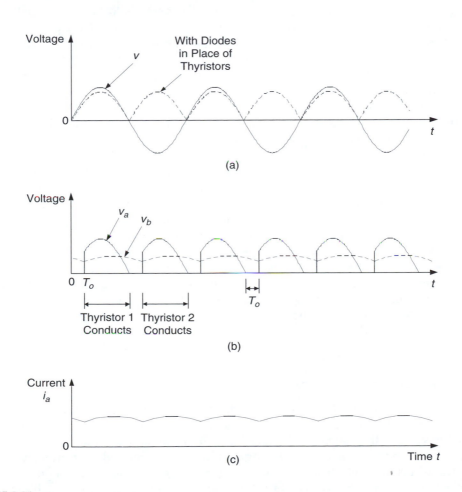

FIGURE 9.31
Voltages and armature current for the circuit in Figure 9.30: (a) Supply voltage; (b) Voltage to the armature circuit and back e.m.f.; (c) Armature current.

controlled either by varying the thyristor firing time T_o for a given supply frequency or by varying the supply frequency for a given firing time.

The two diodes in the supply part of the circuit in Figure 9.30 may be replaced by two thyristors. In that case, the two thyristors in each current path (i.e., for positive v and negative v) have to be triggered simultaneously. The operation of a three-phase control circuit for a dc motor can be analyzed by direct extension of the concepts presented here.

9.5   DC Motor Selection

DC motors, dc servomotors in particular, are suitable for applications requiring continuous operation (continuous duty) at high levels of torque and speed. Brushless permanent-magnet motors with advanced magnetic material provide high torque/mass ratio, and are preferred for continuous operation at high throughput (e.g., component insertion machines in the manufacture of printed-circuit boards, portioning and packaging machines, printing machines) and high speeds (e.g., conveyors, robotic arms), in hazardous environments (where spark generation from brushes would be dangerous), and in applications that need minimal maintenance and regular wash-down (e.g., in food processing applications). For applications that call for high torques and low speeds at high precision (e.g., inspection, sensing, product assembly), torque motors or regular motors with suitable speed reducers (e.g., harmonic drives, and gears units commonly using worm gears, etc.) may be employed.

A typical positioning application involves a "rotation stage" producing rotary motion for the load. If an application requires linear (rectilinear) motions, a "linear stage" has to be used. One option is to use a rotary motor with a rotatory-to-linear motion transmission device such a lead screw or ball screw and nut, rack and pinion, or conveyor belt. This approach introduces some degree of nonlinearity and other errors (e.g., friction, backlash). For high-precision applications, linear motor provide a better alternative. The operating principle of a linear motor is similar to that of a rotary motor, except linearly moving armatures on linear bearings or guideways are used instead of rotors mounted on rotary bearings.

When selecting a dc motor for a particular application, a matching drive unit has to be chosen as well. Due consideration must be given to the requirements (specifications) of power, speed, accuracy, resolution, size, weight, and cost, when selecting a motor and a drive system. In fact vendor catalogs give the necessary information for motors and matching drive units, thereby making the selection far more convenient. A suitable speed transmission device (harmonic drive, gear unit, lead screw and nut, etc.) may have to be chosen as well, depending on the application.

9.5.1   Motor Data and Specifications

Torque and speed are the two primary considerations in choosing a motor for a particular application. Speed-torque curves are available, in particular. The torques given in these curves are typically the maximum torques (known as peak torques), which the motor can generate at the indicated speeds. A motor should not be operated continuously at these torques (and current levels) because of the dangers of overloading, wear, and malfunction. The peak values have to be reduced (say, by 50%) in selecting a motor to match the torque requirement for continuous operation. Alternatively, the continuous torque values as given by the manufacturer, should be used in the motor selection.

Motor manufacturers' data that are usually available to users include the following:

1. Mechanical data
   - Peak torque (e.g., 65 N·m)
   - Continuous torque at zero speed or continuous stall torque (e.g., 25 N·m)
   - Frictional torque (e.g., 0.4 N·m)
   - Maximum acceleration at peak torque (e.g., 33×10^3 rad/s^2)
   - Maximum speed or no-load speed (e.g., 3000 rpm.)
   - Rated speed or speed at rated load (e.g., 2400 rpm.)
   - Rated output power (e.g., 5100 W)
   - Rotor moment of inertia (e.g., 0.002 kg·m^2)
   - Dimensions and weight (e.g., 14 cm diameter, 30 cm length, 20 kg)
   - Allowable axial load or thrust (e.g., 230 N)
   - Allowable radial load (e.g., 700 N)
   - Mechanical (viscous) damping constant (e.g., 0.12 N·m/krpm)
   - Mechanical time constant (e.g., 10 ms)
2. Electrical data
   - Electrical time constant (e.g., 2 ms)
   - Torque constant (e.g., 0.9 N·m/A for peak current or 1.2 N·m/A rms current)
   - Back e.m.f constant (e.g., 0.95 V/rad/s for peak voltage)
   - Armature/field resistance and inductance (e.g., 1.0 Ω, 2 mH)
   - Compatible drive unit data (voltage, current, etc.)
3. General data
   - Brush life and motor life (e.g., 5×10^8 revolutions at maximum speed)
   - Operating temperature and other environmental conditions (e.g., 0–40 °C)
   - Thermal resistance (e.g., 1.5 °C/W)
   - Thermal time constant (e.g., 70 min)
   - Mounting configuration

Quite commonly, motors and drive systems are chosen from what are commercially available (i.e., off the shelf). Customized production may be required, however, in highly specialized and research and development applications where the cost may not be a primary consideration. The selection process involves matching the engineering specifications for a given application with the data of commercially available motor systems.

9.5.2 Selection Considerations

When a specific application calls for large speed variations (e.g., speed tracking over a range of 10 dB or more), armature control is preferred. Note, however, that at low speeds (typically, half the rated speed), poor ventilation and associated temperature buildup can cause problems. At very high speeds, mechanical limitations and heating due to frictional dissipation become determining factors. For constant-speed applications, shunt-wound motors are preferred. Finer speed regulation may be achieved using a servo system with encoder or tachometer feedback or with phase-locked operation. For constant power applications, the series-wound or compound-wound motors are preferable over shunt-wound units. If the shortcomings of mechanical commutation and limited brush life are critical, brushless dc motors should be used.

For high-speed and transient operations of a dc motor, its mechanical time constant (or mechanical bandwidth) is an important consideration. This is limited by the moment of inertia of the rotor (armature) and the load, shaft flexibility, and the dynamics of the mounted instrumentation, such as tachometers and encoders. The mechanical bandwidth of a dc motor can be determined by simply measuring the velocity transducer signal (output) v_o for a transient drive signal (input) v_i and computing the ratio of their Fourier spectra (see Appendix A). The flat region of the resulting transfer function magnitude gives the bandwidth (or, the useful operating region) of the motor. This procedure and a result are illustrated in Figure 9.32. A better way of computing this transfer function is by the cross-spectral density method. Again, the flat region of the resulting frequency transfer function (magnitude) plot determines the mechanical bandwidth of the motor.

A simple way to determine the operating conditions of a motor is by using its torque-speed curve, as illustrated in Figure 9.33. What is normally provided by the manufacturer is the peak torque curve, which gives the maximum torque the motor (with a matching drive system) can provide at a given speed, for short periods (say, 30% duty cycle). The actual selection of a motor should be based on its continuous torque, which is the torque that the motor is able to provide continuously at a given speed, for long periods without overheating or damaging the unit. If the continuous torque curve is not provided by the manufacturer, the peak torque curve should be reduced by about 50% (or even by 70%) for matching with the specified operating requirements.

The minimum operating torque T_{min} is limited mainly by loading considerations. The minimum speed ω_{min} is determined primarily by operating temperature. These boundaries along with the continuous torque curve define the useful operating region of the particular motor (and its drive system), as indicated in Figure 9.33(a). The optimal operating points are those that fall within this segment on the continuous torque-speed curve. The upper limit on speed may be imposed by taking into account transmission limitations in addition to the continuous torque-speed capability of the motor system.

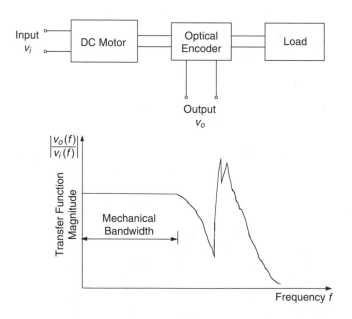

FIGURE 9.32
Determination of the mechanical bandwidth of a dc motor: (a) Test setup; (b) Test result.

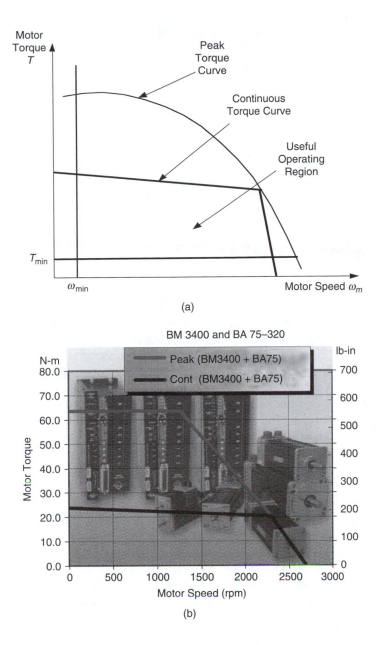

FIGURE 9.33

(a) Representation of the useful operating region for a dc motor, (b) Speed-torque characteristics of a commercial brushless dc servomotor with a matching amplifier (Aerotech, Inc., Pittsburg, PA. With permission).

9.5.3 Motor Sizing Procedure

Motor sizing is the term used to denote the procedure of matching a motor (and its drive system) to a load (i.e., "demand" of the specific application). The load may be given by a load curve, which is the speed-torque curve representing the torque requirements for operating the load at various speeds (see Figure 9.34). Clearly, greater torques are needed to drive a load at higher speeds. For a motor and a load, the acceptable operating range is the interval where the load curve overlaps with the operating region of the motor

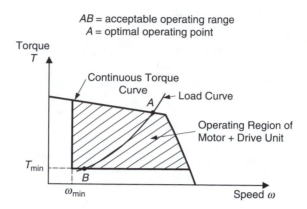

FIGURE 9.34
Sizing a motor for a given load.

(segment AB in Figure 9.34). The optimal operating point is the point where the load curve intersects with the speed-torque curve of the motor (point A in Figure 9.34).

Sizing a dc motor is similar to sizing a stepper motor, as studied in Chapter 8. The same equations may be used for computing the load torque (demand). The motor characteristic (speed-torque curve) gives the available torque, as in the case of a stepper motor. The main difference is, a stepper motor is not suitable for continuous operation for long periods and at high speeds, whereas a dc motor can perform well in such situations. In this context, a dc motor can provide high torques, as given by its "peak torque curve," for short periods, and reduced torques, as given by its "continuous torque curve" for long periods of operation. In the motor sizing procedure, then, the peak torque curve may be used for short periods of acceleration and deceleration, but the continuous torque curve (or, the peak torque curve reduced by about 50%) must be used for continuous operation for long periods.

9.5.3.1 *Inertia Matching*

The motor rotor inertia (J_m) should not be very small compared to the load inertia (J_L). This is particularly critical in high speed and highly repetitive (high throughput) applications. Typically, for high speed applications, the value of J_L/J_m may in the range 5–20. For low speed applications, J_L/J_m can be as high as 100. This assumes direct drive applications.

A gear transmission may be needed between the motor and the load in order to amplify the torque available from the motor, which also reduces the speed at which the load is driven. Then, further considerations have to be made in inertia matching. In particular, neglecting the inertial and frictional loads due to gear transmission, it can be shown that best acceleration conditions for the load are possible if (see Chapter 4 under impedance matching of mechanical devices)

$$\frac{J_L}{J_m} = r^2 \qquad\qquad (6.64)$$

where r is the step-down gear ratio (i.e., motor speed/load speed). Since J_L/r^2 is the load inertia as felt at the motor rotor, the optimal condition (Equation 9.64) is when this equivalent inertia (which moves at the same acceleration as the rotor) is equal to the rotor inertia (J_m).

9.5.3.2 *Drive Amplifier Selection*

Usually, the commercial motors come with matching drive systems. If this is not the case, some useful sizing computations can be done to assist the process of selecting a drive amplifier. As noted before, even though the control procedure becomes linear and convenient when linear amplifiers are used, it is desirable to use pulse width modulation (PWM) amplifiers in view of their high efficiency (and associated low thermal dissipation).

The required current and voltage ratings of the amplifier, for a given motor and a load, may be computed rather conveniently. The required motor torque is given by

$$T_m = J_m \alpha + T_L + T_f \tag{9.65a}$$

where
α = highest angular acceleration needed from the motor
T_L = worst-case load torque
T_f = frictional torque on the motor

If the load is a pure inertia (J_L), Equation 9.65a becomes

$$T_m = (J_m + J_L)\alpha + T_f \tag{9.65b}$$

The current required to generate this torque in the motor is given by

$$i = \frac{T_m}{k_m} \tag{9.66}$$

where k_m is the torque constant of the motor.

The voltage required to drive the motor is given by

$$v = k'_m \omega_m + Ri \tag{9.67}$$

where $k'_m = k_m$ is the back e.m.f. constant, R is the winding resistance, and ω_m is the highest operating speed of the motor in driving the load. For a PWM amplifier, the supply voltage (from a dc power supply) is computed by dividing the voltage in Equation 9.67 by the lowest duty cycle (fraction) of operation.

Example 9.10

A load of moment of inertia $J_L = 0.5 \ \text{kg} \cdot \text{m}^2$ is ramped up from rest to a steady speed of 200 rpm in 0.5 s using a dc motor and a gear unit of step down speed ratio $r = 5$. A schematic representation of the system is shown in Figure 9.35(a) and the speed profile of the load is shown in Figure 9.35(b). The load exerts a constant resistance of $T_R = 55 \ \text{N} \cdot \text{m}$ throughout the operation. The efficiency of the gear unit is $e = 0.7$. Check whether the commercial brushless dc motor and its drive unit, whose characteristics are shown in Figure 9.33(b), is suitable for this application. The moment of inertia of the motor rotor is $J_m = 0.002 \ \text{kg} \cdot \text{m}^2$.

SOLUTION
The load equation to compute the torque required from the motor is given by

$$T_m = \left(J_m + \frac{J_L}{er^2} \right) r\alpha + \frac{T_R}{er} \tag{9.68}$$

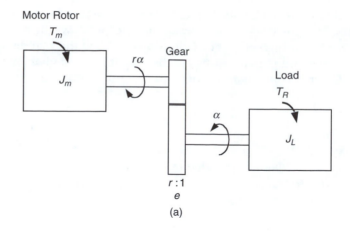

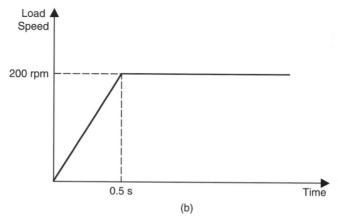

FIGURE 9.35
(a) A load driven by a dc motor through a gear transmission, (b) Speed profile of the load.

where a = load acceleration, and the remaining parameters are as defined in the example. The derivation of Equation 9.68 is straightforward. In particular, see the derivation of a similar equation for positioning table driven by a stepper motor (Chapter 8, Equation 9.50). From the given speed profile,

$$\text{Maximum load speed} = 200 \text{ rpm} = 20.94 \text{ rad/s}$$

$$\text{Load acceleration} = \frac{20.94}{0.5} \text{ rad/s}^2 = 42 \text{ rad/s}^2$$

Substitute the numerical values in Equation 9.67, under worst-case conditions, to compute the required torque from the motor. We have

$$T_m = \left(0.002 + \frac{0.05}{0.7 \times 5^2} \right) 5 \times 42 + \frac{55.0}{0.7 \times 5} \text{ N} \cdot \text{m} = 1.02 + 15.71 = 16.73 \text{ N} \cdot \text{m}$$

Under worst-case conditions, at least this much of torque would be required from the motor, operating at a speed of $200 \times 5 = 1000$ rpm. Note from Figure 9.33(b) that the load point (1000 rpm, 16.73 N · m) is sufficiently below even the continuous torque curve of the given motor (with its drive unit). Hence this motor is adequate for the task.

9.6 Induction Motors

With the widespread availability of alternating current (ac) as an economical form of power supply for operating industrial machinery and household appliances, much attention has been given to the development of ac motors. Because of the rapid progress made in this area, ac motors have managed to replace dc motors in many industrial applications until the revival of the dc motor, particularly as a servomotor in control system applications. However, ac motors are generally more attractive than conventional dc motors, in view of their robustness, lower cost, simplicity of construction, and easier maintenance, especially in heavy-duty (high-power) applications (e.g., rolling mills, presses, elevators, cranes, material handlers, and operations in paper, metal, petrochemical, cement, and other industrial plants) and in continuous constant-speed operations (e.g., conveyors, mixers, agitators, extruders, pulping machines, household and industrial appliances such as refrigerators, heating-ventilation-and-air-conditioning or HVAC devices such as pumps, compressors and fans). Many industrial applications using ac motors may involve continuous operation throughout the day for over six days a week. Also, with advances in control hardware and software and the low cost of microelectronics have led to advance controllers for ac motors, which can emulate the performance of variable-speed drives of dc motors, for example, ac servomotors that rival their dc counterpart. Some advantages of ac motors are:

1. Cost-effectiveness
2. Convenient power source (standard power grid providing single-phase and three-phase ac supplies)
3. No commutator and brush mechanisms needed in many types of ac motors
4. Low power dissipation, low rotor inertia, and light weight in some designs
5. Virtually no electric spark generation or arcing (less hazardous in chemical environments)
6. Capability of accurate constant-speed operation without needing servo control (with synchronous ac motors)
7. No drift problems in ac amplifiers in drive circuits (unlike linear dc amplifiers)
8. High reliability, robustness, easy maintenance, and long life

The primary disadvantages include

1. Low starting torque (zero starting torque in synchronous motors)
2. Need of auxiliary starting devices for ac motors that have zero starting torque
3. Difficulty of variable-speed control or servo control (unless modern sold-state and variable-frequency drives with devices with field feedback compensation are employed)

We will discuss two basic types of ac motors:

1. Induction motors (asynchronous motors)
2. Synchronous motors

9.6.1 Rotating Magnetic Field

The operation of an ac motor can be explained using the concept of a rotating magnetic field. A rotating field is generated by a set of windings uniformly distributed around a circular stator and excited by ac signals with uniform (i.e., equal) phase differences. To illustrate this, consider a standard three-phase supply. The voltage in each phase is 120° out of phase with the voltage in the next phase. The phase voltages can be represented by

$$v_1 = a\cos\omega_p t$$

$$v_2 = a\cos\left(\omega_p t - \frac{2\pi}{3}\right) \qquad (9.69)$$

$$v_3 = a\cos\left(\omega_p t - \frac{4\pi}{3}\right)$$

where ω_p is the frequency of each phase of the ac signal (i.e., the line frequency). Note that v_1 leads v_2 by $2\pi/3$ radians and v_2 leads v_3 by the same angle. Furthermore, since v_1 leads v_3 by $4\pi/3$ radians, it is correct to say that v_1 lags v_3 by $2\pi/3$ radians. In other words, v_1 leads $-v_3$ by $(\pi - 2\pi/3)$, which is equal to $\pi/3$. Now consider a group of three windings, each of which has two segments (a positive segment and a negative segment) uniformly arranged around a circle (stator), as shown in Figure 9.36, in the order $v_1, -v_3, v_2, -v_1, v_3, -v_2$.

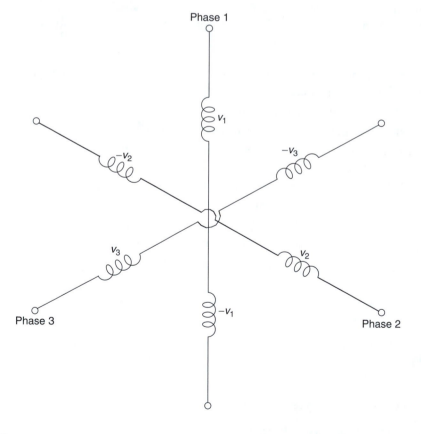

FIGURE 9.36
The generation of a rotating magnetic field using a three-phase supply and two winding sets per phase.

Note that each winding segment has a phase difference of $\pi/3$ (or 60°) from the adjacent segment. The physical (geometric) spacing of adjacent winding segments is also 60°. Now, consider the time interval $\Delta t = \pi / (3\omega_p)$. The status of $-v_3$ at the end of a time interval of Δt is identical to the status of v_1 in the beginning of the time interval. Similarly, the status of v_2 after a time Δt becomes that of $-v_3$ in the beginning, and so on. In other words, the voltage status (and hence the magnetic field status) of one segment becomes identical to that of the adjacent segment in a time interval of Δt. This means that the magnetic field generated by the winding segments appears to rotate physically around the circle (stator) at angular velocity ω_p.

It is not necessary for the three sets of three-phase windings to be distributed over the entire 360° angle of the circle. Suppose, instead, that these three sets (six segments) of windings are distributed within the first 180° of the circle, at 30° apart, and a second three sets are distributed similarly within the remaining 180°. Then, the field would appear to rotate at half the speed ($\omega_p/2$), because in this case, Δt is the time taken for the field to rotate through 30°, not 60°. It follows that the general formula for the angular speed ω_p of the rotating magnetic field generated by a set of winding segments uniformly distributed on a stator and excited by an ac supply, is

$$\omega_f = \frac{\omega_p}{n} \tag{9.70}$$

where
  ω_p = frequency of the ac signal in each phase (i.e., line frequency)
  n = number of pairs of winding sets used per phase (i.e., number of pole pairs per phase)

Note that when $n = 1$, there are two coils (+ve and −ve) for each phase (i.e., there are two poles per phase). Similarly, when $n = 2$, there are four coils for each phase. Hence, n denotes the number of "pole pairs" per phase in a stator. In this manner, the speed of the rotating magnetic field can be reduced to a fraction of the line frequency simply by adding more sets of windings. These windings occupy the stator of an ac motor. The number of phases and the number of segments wound to each phase determine the angular separation of the winding segments around the stator. For example, for the three-phase, one pole pair per phase arrangement shown in Figure 9.36, the physical separation of the winding segments is 60°. For a two-phase supply with one pole pair per phase, the physical separation is 90°, and the separation is halved to 45° if two pole pairs are used per phase. It is the rotating magnetic field, produced in this manner, which generates the driving torque by interacting with the rotor windings. The nature of this interaction determines whether a particular motor is an induction motor or a synchronous motor.

Example 9.11
Another way to interpret the concept of a rotating magnetic field is to consider the resultant field due to the individual magnetic fields in the stator windings. Consider a single set of three-phase windings arranged geometrically as in Figure 9.36. Suppose that the magnetic field due to phase 1 is denoted by $a \sin \omega_p t$. Show that the resultant magnetic field has an amplitude of $3a/2$ and that the field rotates at speed ω_p.

SOLUTION
The magnetic field vectors in the three sets of windings are shown in Figure 9.37(a). These can be resolved into two orthogonal components, as shown in Figure 9.37(b). The component

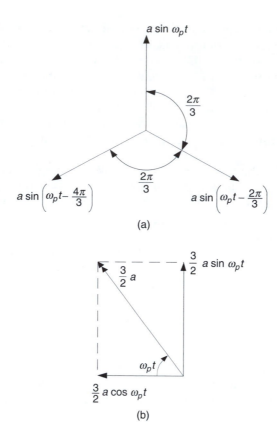

FIGURE 9.37
An alternative interpretation of a rotating magnetic field: (a) Magnetic fields of the windings; (b) Resultant magnetic field.

in the vertical direction (upwards) is

$$a\sin\omega_p t - a\sin\left(\omega_p t - \frac{2\pi}{3}\right)\cos\frac{\pi}{3} - a\sin\left(\omega_p t - \frac{4\pi}{3}\right)\cos\frac{\pi}{3}$$

$$= a\sin\omega_p t - \frac{a}{2}\left[\sin\left(\omega_p t - \frac{2\pi}{3}\right) + \sin\left(\omega_p t - \frac{4\pi}{3}\right)\right]$$

$$= a\sin\omega_p t - a\sin(\omega_p t - \pi)\cos\frac{\pi}{3}$$

$$= a\sin\omega_p t + \frac{a}{2}[\sin\omega_p t]$$

$$= \frac{3a}{2}\sin\omega_p t$$

Note that in deriving this result we have used the following trigonometric identities:

$$\sin A + \sin B = 2\sin\left(\frac{A+B}{2}\right)\cos\left(\frac{A-B}{2}\right)$$

and

$$\sin(A - \pi) = -\sin A$$

The horizontal component of the magnetic fields (which is directed to the left) is

$$a \sin\left(\omega_p t - \frac{4\pi}{3}\right)\sin\frac{\pi}{3} - a\sin\left(\omega_p t - \frac{2\pi}{3}\right)\sin\frac{\pi}{3} = \frac{\sqrt{3}}{2}a\left[\sin\left(\omega_p t - \frac{4\pi}{3}\right) - \sin\left(\omega_p t - \frac{2\pi}{3}\right)\right]$$

$$= \sqrt{3}a\cos(\omega_p t - \pi)\sin\left(-\frac{\pi}{3}\right)$$

$$= \frac{3a}{2}\cos\omega_p t$$

Here we have used the following trigonometric identities:

$$\sin A - \sin B = 2\cos\frac{A+B}{2}\sin\frac{A-B}{2}$$

$$\cos(A - \pi) = -\cos A$$

$$\sin(-A) = -\sin A$$

The resultant of the two orthogonal components is a vector of magnitude $3a/2$, making an angle $\omega_p t$ with the horizontal component, as shown in Figure 9.37(b). It follows that the resultant magnetic field has a magnitude of $3a/2$ and rotates in the clockwise direction at speed ω_p radians per second.

9.6.2 Induction Motor Characteristics

The stator windings of an *induction motor* generate a rotating magnetic field, as explained in the previous section. The rotor windings are purely secondary windings, which are not energized by an external voltage. For this reason, no commutator-brush devices are needed in induction motors (see Figure 9.38). The core of the rotor is made of ferromagnetic laminations in order to concentrate the magnetic flux and to minimize dissipation (eddy currents). The rotor windings are embedded in the axial direction on the cylindrical surface of the rotor and are interconnected in groups. The rotor windings may consist of uninsulated copper or aluminum (or any other conductor) bars (*a cage rotor*), which are fitted into slots in the end rings at the two ends of the rotor. These end rings complete the paths for electrical conduction through the rods. Alternatively, wire with one or more turns in each slot (*a wound rotor*) may be used. First, consider a stationary rotor. The rotating field in the stator intercepts the rotor windings, thereby generating an induced current due to mutual induction or transformer action (hence the name induction motor). The resulting secondary magnetic flux interacts with the primary, rotating magnetic flux, thereby producing a torque in the direction of rotation of the stator field. This torque drives the rotor. As the rotor speed increases, initially the motor torque also increases (rather moderately) because of secondary interactions between the stator circuit and the rotor circuit, even though the relative speed of the rotating field with respect to the rotor decreases, which reduces the rate of change of

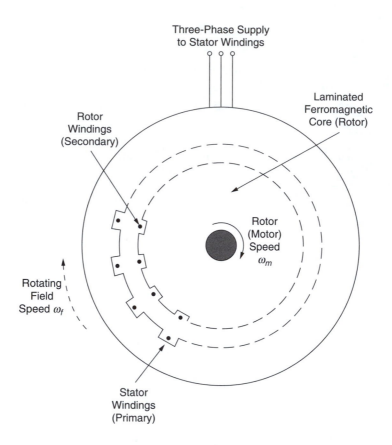

FIGURE 9.38
Schematic diagram of an induction motor.

flux linkage and hence the direct transformer action. (*Note:* The relative speed is termed the *slip rate*). Quite soon, the maximum torque will be reached. Further increase in rotor speed (i.e., a decrease in slip rate) sharply decreases the motor torque, until at synchronous speed (zero slip rate), the motor torque becomes zero. This behavior of an induction motor is illustrated by the typical characteristic curve given in Figure 9.39. From the starting torque T_s to the maximum torque (*breakdown torque*) T_{max}, the motor behavior is unstable. This can be explained as follows. An incremental increase in speed will cause an increase in torque, which will further increase the speed. Similarly, an incremental reduction in speed will bring about a reduction in torque that will further reduce the speed. The portion of the curve from T_{max} to the zero-torque (or, no-load or synchronous) condition represents the region of stable operation. Under normal operating conditions, an induction motor should operate in this region.

The fractional slip S for an induction motor is given by

$$S = \frac{\omega_f - \omega_m}{\omega_f} \tag{9.71}$$

Even when there is no external load, the synchronous operating condition (i.e., $S = 0$) is not achieved at steady state because of the presence of frictional torque, which opposes the rotor motion. When an external torque (load torque) T_L is present, under normal operating

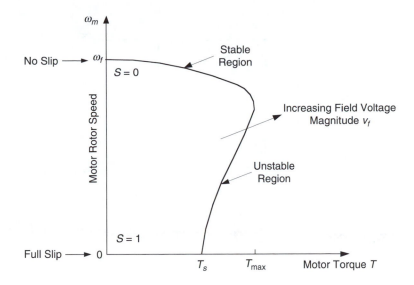

FIGURE 9.39
Torque-speed character-istic curve of an induction motor.

conditions, the slip rate will further increase so as to increase the motor torque to support this load torque. As is clear from Figure 9.39, in the stable region of the characteristic curve, the induction motor is quite insensitive to torque changes; a small change in speed would require a very large change in torque (in comparison with an equivalent dc motor). For this reason, an induction motor is relatively insensitive to load variations and can be regarded as a constant speed machine. If the rotor speed is increased beyond the synchronous speed (i.e., $S < 0$), the motor becomes a generator.

9.6.3 Torque–Speed Relationship

It is instructive to determine the torque-speed relationship for an induction motor. This relationship provides insight into possible control methods for induction motors. The equivalent circuits of the stator and the rotor for one phase of an induction motor are shown in Figure 9.40(a). The circuit parameters are

R_f = stator coil resistance
L_f = stator leakage inductance
R_c = stator core iron loss resistance (eddy current effects, etc.)
L_c = stator core (magnetizing) inductance
L_r = rotor leakage inductance
R_r = rotor coil resistance

The magnitude of the ac supply voltage for each phase of the stator windings is v_f at the line frequency ω_p. The rotor current generated by the induced e.m.f. is i_r. After allowing for the voltage drop due to stator resistance and stator leakage inductance, the voltage available for mutual induction is denoted by v. This is also the induced voltage in the secondary (rotor) windings at standstill, assuming the same number of turns. This induced voltage changes linearly with slip S, because the induced voltage is proportional to the relative velocity of the rotating field with respect to the rotor (i.e., $\omega_f - \omega_m$), as is evident

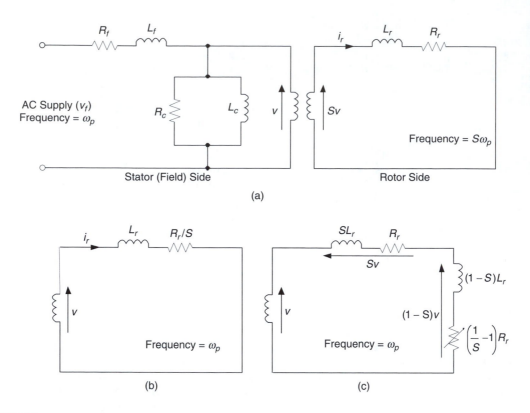

FIGURE 9.40
(a) Stator and rotor circuits for an induction motor, (b) Rotor circuit referred to the stator side, (c) Representation of available mechanical power using the rotor circuit.

from Equation 9.2. Hence, the induced voltage in the rotor windings (secondary windings) is Sv. Note, further, that at standstill (when $S = 1$), the frequency of the induced voltage in the rotor is ω_p. At synchronous speed of rotation (when $S = 0$), this frequency is zero because the magnetic field is fixed and constant relative to the rotor in this case. Now, assuming a linear variation of frequency of the induced voltage between these two extremes, we note that the frequency of the induced voltage in the rotor circuit is $S\omega_p$. These observations are indicated in Figure 9.40(a).

Using the frequency domain (complex) representation for the out-of-phase currents and voltages, the rotor current i_r in the complex form is given by

$$ i_r = \frac{Sv}{(R_r + jS\omega_p L_r)} = \frac{v}{(R_r/S + j\omega_p L_r)} \tag{9.72} $$

From Equation 9.72 it is clear that the rotor circuit can be represented by a resistance R_r/S and an inductance L_r in series and excited by voltage v at frequency ω_p. This is in fact the rotor circuit referred to the stator side, as shown in Figure 9.40(b). This circuit can be grouped into two parts, as shown in Figure 9.40(c). The inductance SL_r and resistance R_r in series, with a voltage drop Sv, are identical to the rotor circuit in Figure 9.40(a). Note that SL_r has to be used as the inductance in the new equivalent circuit segment, instead of L_r in the original rotor circuit, for the sake of circuit equivalence. The reason is simple. The new equivalent circuit operates at frequency ω_p while the original rotor circuit operates at frequency $S\omega_p$ (*Note*: impedance of an inductor is equal to the product of inductance and frequency of excitation). The second voltage drop $(1 - S)v$ in Figure 9.40(c) represents the back e.m.f. due to rotor-stator field interaction; it generates the capacity to drive an external load (mechanical power). Note here that the back e.m.f. governs the current in the rotor circuit and hence the generated torque. It follows that the available mechanical power, per phase, of an induction motor is given by $i_r^2(1/S-1)R_r$. Hence,

$$T_m\omega_m = pi_r^2\left(\frac{1}{S}-1\right)R_r \tag{9.73}$$

where
    T_m = motor torque generated in the rotor
    ω_m = rotor speed of the motor
    p = number of supply phases
    i_r = magnitude of the current in the rotor

The magnitude of the current in the rotor circuit is obtained from Equation 9.72; thus,

$$i_r = \frac{v}{\sqrt{R_r^2/S^2 + \omega_p^2 L_r^2}} \tag{9.74}$$

By substituting Equation 9.74 in Equation 9.73, we get

$$T_m = pv^2\,\frac{S(1-S)}{\omega_m}\,\frac{R_r}{\left(R_r^2 + S'\omega_p^2 L_r^2\right)} \tag{9.75}$$

From Equation 9.70 and Equation 9.71, we can express the number of pole pairs per phase of stator winding as

$$n = \frac{\omega_p}{\omega_m}(1-S) \tag{9.76}$$

Equation 9.76 is substituted in Equation 9.75; thus

$$T_m = \frac{pnv^2 S R_r}{\omega_p\left(R_r^2 + S^2\omega_p^2 L_r^2\right)} \tag{9.77}$$

If the resistance and the leakage inductance in the stator are neglected, v is approximately equal to the stator excitation voltage v_f. This gives the torque-slip relationship:

$$T_m = \frac{pnv_f^2 S R_r}{\omega_p\left(R_r^2 + S^2\omega_p^2 L_r^2\right)} \tag{9.78}$$

By using Equation 9.76, it is possible to express S in Equation 9.78 in terms of the rotor speed ω_m. This results in a torque-speed relationship, which gives the characteristic curve shown in Figure 9.39. Specifically, we employ the fact that the motor speed ω_m is related to slip through

$$S = \frac{\omega_p - n\omega_m}{\omega_p} \tag{9.79}$$

Note, further, from Equation 9.78 that the motor torque is proportional to the square of the supply voltage v_f.

Example 9.12

In the derivation of Equation 9.78, we assumed that the number of effective turns per phase in the rotor is equal to that in the stator. This assumption is generally not valid, however. Determine how the equation should be modified in the general case. Suppose that

$$r = \frac{\text{number of effective turns per phase in the rotor}}{\text{number of effective turns per phase in the stator}}$$

SOLUTION

At standstill ($S = 1$), the induced voltage in the rotor is rv and the induced current is i_r/r. Hence, the impedance in the rotor circuit is given by

$$Z_r = \frac{rv}{i_r/r} = r^2 \frac{v}{i_r} \tag{9.80}$$

or

$$Z_r = r^2 Z_{\text{req}}$$

It follows that the true rotor impedance (or resistance and inductance) simply has to be divided by r^2 to obtain the equivalent impedance. In this general case of $r \neq 1$, the resistance R_r and the inductance L_r should be replaced by $R_{\text{req}} = R_r/r^2$ and $L_{\text{req}} = L_r/r^2$, in Equation 9.78.

Example 9.13

Consider a three-phase induction motor that has one pole pair per phase. The equivalent resistance and leakage inductance in the rotor circuit are 8 Ω and 0.06 H, respectively. The motor supply voltage is 115 V in each phase, at a line frequency of 60 Hz. Compute the torque-speed curve for the motor.

SOLUTION

In this example, $R_r = 8\ \Omega$, $L_r = 0.06$ H, $v_f = 115$ V, $n = 1$, $p = 3$, and $\omega_p = 60 \times 2\pi$ rad/s. Now, using Equation 9.78 along with Equation 9.79, we can compute the torque-speed curve. The result is shown in Figure 9.41.

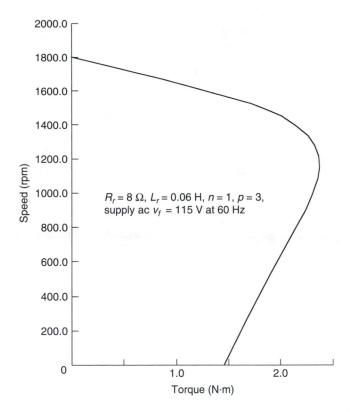

FIGURE 9.41
Torque-speed curve for an induction motor.

9.7 Induction Motor Control

DC motors are widely used in servo control applications because of their simplicity and flexible speed-torque capabilities. In particular, dc motors are easy to control and they operate accurately and efficiently over a wide range of speeds. The initial cost and the maintenance cost of a dc motor, however, are generally higher than those for a comparable ac motor. AC motors are rugged and are most common in medium- to high-power applications involving fairly constant speed operation. Of late, much effort has been invested in developing improved control methods for ac motors, and significant progress is seen in this area. Today's ac motors and their advanced drive systems with frequency control and and field feedback compensation can provide speed control that is comparable to the capabilities of dc servomotors (e.g., 1 : 20 or 26 dB range of speed variation).

Since fractional slip S determines motor speed ω_m, Equation 9.78 suggests several possibilities for controlling an induction motor. Four possible methods for induction motor control are

1. Excitation frequency control (ω_p)
2. Supply voltage control (v_f)
3. Rotor resistance control (R_r)
4. Pole changing (n)

What is given in parentheses is the parameter that is adjusted in each method of control.

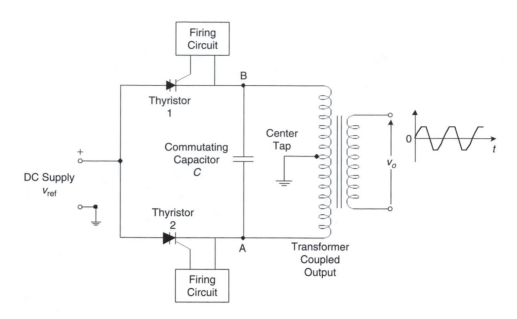

FIGURE 9.42
A single-phase inverter circuit for frequency control.

9.7.1 Excitation Frequency Control

Excitation frequency control can be accomplished using a *thyristor* circuit. As discussed under dc motor drive system, a thyristor (or SCR) is a semiconductor device, which possesses very effective, efficient, and nondissipative switching characteristics, at very high frequencies. Furthermore, thyristors can handle high voltages and power levels.

By using an *inverter* circuit, a variable-frequency ac output can be generated from a dc supply. A single-phase inverter circuit is shown in Figure 9.42. Thyristors 1 and 2 are gated by their firing circuits according to the required frequency of the output voltage v_o. The primary winding of the output transformer is center-tapped. A dc supply voltage v_{ref} is applied to the circuit as shown. If both thyristors are not conducting, the voltage across the capacitor C is zero. Now, if thyristor 1 is gated (i.e., fired), the current in the upper half of the primary winding will build to its maximum and the voltage across that half will reach v_{ref} (since the voltage drop across thyristor 1 is very small). As a result of the corresponding change in the magnetic flux, a voltage v_{ref} (approximately) will be induced in the lower half of the primary winding, complementing the voltage in the upper half. Accordingly, the voltage across the primary winding (or across the capacitor) is approximately $2v_{ref}$. Now, if thyristor 2 is fired, the voltage at point A becomes v_{ref}. Since the capacitor is already charged to $2v_{ref}$, the voltage at point B becomes $3 v_{ref}$. This means that a voltage of $2v_{ref}$ is applied across thyristor 1 in the nonconducting direction. As a result, thyristor 1 will be turned off. Then, as before, a voltage $2v_{ref}$ is generated in the primary winding, but in the opposite direction, because it is thyristor 2 that is conducting now. In this manner, an approximately rectangular pulse sequence of ac voltage v_o is generated at the circuit output. The frequency of the voltage is equal to the inverse of the firing interval between the two thyristors. A three-phase inverter can be formed by triplicating the single-phase inverter and by phasing the firing times appropriately.

Modern drive units for induction motors use pulse-width modulation (PWM) and advanced microelectronic circuitry incorporating a single monolithic integrated-circuit (IC) chip with more than 30,000 circuit elements, rather than discrete semiconductor

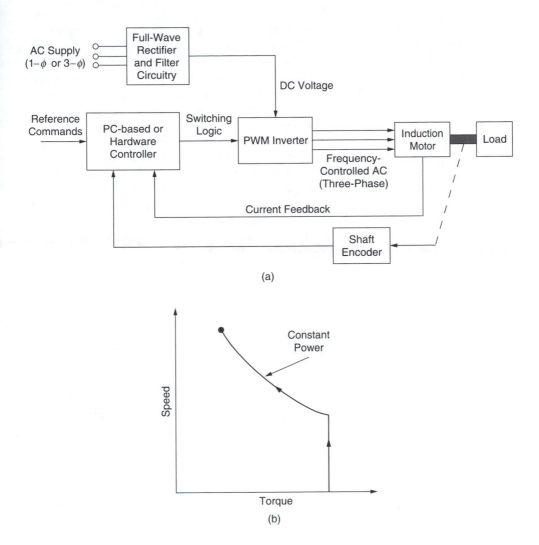

FIGURE 9.43
(a) Variable-frequency control of an induction motor, (b) A typical control strategy.

elements. The block diagram in Figure 9.43(a) shows a frequency control system for an induction motor. A standard ac supply (three-phase or single-phase) is rectified and filtered to provide the dc supply to the three-phase PWM inverter circuit. This device generates a nearly sinusoidal three-phase output at a specified frequency. Firing of the switching circuitry, for varying the frequency of the ac output, is commanded by a hardware controller. If the control requirements are simple, a variable-frequency oscillator or a voltage-to-frequency converter may be used instead. Alternatively, a digital computer (PC) based controller may be used to vary the drive frequency and to adjust other control parameters in a more flexible manner, using software. The controller may use hardware logic or software to generate the switching signal (reference frequency), while taking into account external (human-operator) commands and sensor feedback signals. A variable-frequency drive for an ac motor can effectively operate in the open-loop mode. Sensor feedback may be employed, however, for more accurate performance. Feedback signals may include shaft encoder readings (motor angle) for speed control and current (stator current, rotor current in wound rotors, dc current to PWM inverter, etc.) particularly for motor torque control. A typical control strategy is shown in Figure 9.43(b). In this case, the control processor provides a two-mode control scheme. In the initial mode, the torque is kept constant while accelerating the motor. In the next mode, the power is kept constant while further increasing the speed. Both modes of operation can be achieved through frequency control. Strategies of specified torque profiles (torque control) or specified speed profiles (speed control) can be implemented in a similar manner.

Programmable, microprocessor-based variable-frequency drives for ac motors are commercially available. One such drive is able to control the excitation frequency in the range 0.1–400 Hz with a resolution of 0.01 Hz. A three-phase ac voltage in the range 200–230 V or 380–460 V is generated by the drive, depending on the input ac voltage. AC motors with frequency control are employed in many applications, including variable-flow control of pumps, fans and blowers, industrial manipulators (robots, hoists, etc.), conveyors, elevators, process plant and factory instrumentation, and flexible operation of production machinery for flexible (variable-output) production. In particular, ac motors with frequency control and sensor feedback are able to function as servomotors.

9.7.2 Voltage Control

From Equation 9.77 it is seen that the torque of an induction motor is proportional to the square of the supply voltage. It follows that, an induction motor can be controlled by varying its supply voltage. This may be done in several ways. For example, amplitude modulation of the ac supply, using a ramp generator, will directly accomplish this objective by varying the supply amplitude. Alternatively, by introducing zero-voltage regions (i.e., blanking out) periodically (at high frequency) in the ac supply, for example, using a thyristor circuit with firing delays as in pulse width modulation, will accomplish voltage control by varying the root-mean-square (rms) value of the supply voltage. Voltage control methods are appropriate for small induction motors, but they provide poor efficiency when control over a wide speed range is required. Frequency control methods are recommended in low-power applications. An advantage of voltage control methods over frequency control methods is the lower stator copper loss.

Example 9.14

Show that the fractional slip versus motor torque characteristic of an induction motor, at steady state, may be expressed by

$$T_m = \frac{aSv_f^2}{[1+(S/S_b)^2]} \tag{9.81}$$

Identify the parameters a and S_b. Show that S_b is the slip corresponding to the breakdown torque (maximum torque) T_{max}. Obtain an expression for T_{max}.

An induction motor with parameter values $a = 4 \times 10^{-3}$ N·m/V² and $S_b = 0.2$ is driven by an ac supply that has a line frequency of 60 Hz. Stator windings have two pole pairs per phase. Initially, the line voltage is 500 V. The motor drives a mechanical load, which can be represented by an equivalent viscous damper with damping constant $b = 0.265$ N·m/rad/s. Determine the operating point (i.e., the values of torque and speed) for the system. Suppose that the supply voltage is dropped by 50% (to 250 V) using a voltage control scheme. What is the new operating point? Is this a stable operating point? In view of your answer, comment on the use of voltage control in induction motors.

SOLUTION

First, we note that Equation 9.78 can be expressed as Equation 9.81, with

$$a = \frac{pn}{\omega_p R_r} \tag{9.82}$$

and

$$S_b = \frac{R_r}{\omega_p L_r} \tag{9.83}$$

The breakdown torque is the peak torque and is defined by

$$\frac{\partial T_m}{\partial \omega_m} = 0$$

We express

$$\frac{\partial T_m}{\partial \omega_m} = \frac{\partial T_m}{\partial S} \frac{\partial S}{\partial \omega_m} = -\frac{1}{\omega_f} \frac{\partial T_m}{\partial S}$$

where, we have differentiated Equation 9.71 with respect to ω_m and substituted the result. It follows that the breakdown torque is given by

$$\frac{\partial T_m}{\partial S} = 0$$

Now, differentiate Equation 9.81 with respect to S and equate to zero. We get

$$\left[1 + \left(\frac{S}{S_b} \right)^2 \right] - S \left[2 \frac{S}{S_b^2} \right] = 0$$

or

$$1 - \left(\frac{S}{S_b} \right)^2 = 0$$

It follows that $S = S_b$ corresponds to the breakdown torque. Substituting in Equation 9.81, we have

$$T_{max} = \frac{1}{2} a S_b v_f^2 \tag{9.84}$$

Next, the speed-torque curve is computed using the given parameter values in Equation 9.81 and plotted as shown in Figure 9.44 for the two cases $v_f = 500$ V and $v_f = 250$ V. Note that with $S_b = 0.2$, we have, from Equation 9.84, $(T_{max})_1 = 100$ N·m and $v_f = 25$ N·m. These values are confirmed from the curves in Figure 9.44. The load curve is given by

$$T_m = b\omega_m$$

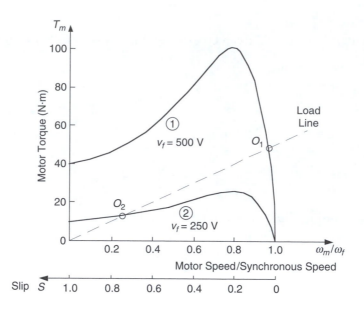

FIGURE 9.44
Speed-torque curves for induction motor voltage control.

or

$$T_m = b\omega_f \frac{\omega_m}{\omega_f}$$

Now, from Equation 9.70, the synchronous speed is computed as

$$\omega_f = \frac{60 \times 2\pi}{2} \text{ rad/s} = 188.5 \text{ rad/s}$$

Hence,

$$b\omega_f = 0.265 \times 188.5 = 50 \text{ N} \cdot \text{m}$$

This is the slope of the load line shown in Figure 9.44. The points of intersection of the load line and the motor characteristic curve are the steady-state operating points. They are;
   For case 1 ($v_f = 500$ V):

   Operating torque = 48 N · m
   Operating slip = 4%
   Operating speed = 1728 rpm

   For case 2 ($v_f = 250$ V):

   Operating torque = 12 N · m
   Operating slip = 77%
   Operating speed = 414 rpm

Note that when the supply voltage is halved, the torque drops by a factor of four and the speed drops by about 76%. But, what is worse is that the new operating point (O_2) is in the unstable region (i.e., the range from $S = S_b$ to $S = 1$) of the motor characteristic curve.

It follows that large drops in supply voltage are not feasible, and the efficiency of the motor can degrade significantly with voltage control.

9.7.3 Rotor Resistance Control

It can be seen from Equation 9.78 that an induction motor can be controlled by varying R_r. Since this is a dissipative technique, it is also a wasteful method. This was a commonly used method for induction motor control, prior to the development of more efficient variable-frequency switching circuits, digital signal processing (DSP) chips, microelectronic drive systems, and related control techniques. The rotor of an induction motor has a closed circuit (resistive-inductive), which is not connected to an external power supply, unlike in the case of a dc motor. In the wound-rotor design, windings are usually arranged and connected as polyphase groups (e.g., a delta-configuration (Δ) or star configuration (Y) in three-phase motors), just like the stator windings, but without a supply voltage. The current in the rotor circuit is generated purely by magnetic induction, but it determines the torque-speed characteristics of the motor. The motor response is controlled by changing the rotor resistance. This can be accomplished by connecting a variable resistance to each phase externally through a slip ring and brush arrangement, as schematically shown in Figure 9.45 for the three-phase star (Y) connection. Rotor resistance control has the same disadvantages as in the voltage control method. In particular, the motor efficiency drops considerably when the motor operates over a wide range of speeds. Furthermore, the energy dissipated by the control resistors will result in thermal problems. Heat sinks, fans, and other cooling methods may have to be employed, particularly for continuous operation.

9.7.4 Pole-Changing Control

The number of pole pairs per phase in the stator windings (n) is a parameter in the speed-torque Equation 9.78. It follows that changing the parameter n is an alternative method for controlling an induction motor. This can be accomplished by switching the supply

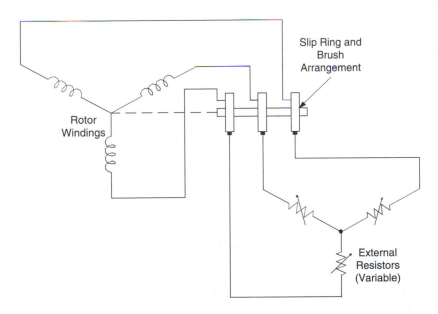

FIGURE 9.45
Rotor resistance control of an induction motor (three-phase).

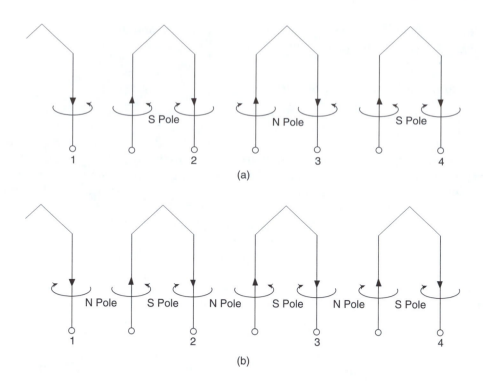

FIGURE 9.46

Pole-changing control of an induction motor (In changing from (a) to (b). The number of pole pairs per phase is doubled when the currents in stator windings 1 and 3 are reversed).

connections in the stator windings in some manner. The principle is illustrated in Figure 9.46. Consider the windings in one phase of the stator. With the coil currents as in Figure 9.46(a), the magnetic fields in the alternate pairs of adjacent coils cancel out. When the coil currents are as in Figure 9.46(b), all adjacent pairs of coils have complementary magnetic fields. As a result, the number of pole pairs per phase is doubled when the stator is switched from the configuration in Figure 9.46(a) to that in Figure 9.46(b). Note that when the stator windings are switched into a certain configuration of poles, the same switching should be done simultaneously to the rotor windings as well. This results in an additional complexity for the control circuitry in the case of a wound rotor. In a squirrel-cage rotor, a separate switching mechanism is not necessary for the rotor because it automatically reacts to configure itself according to the winding configuration in the stator. For this reason, cage rotor induction motors are better suited for pole-changing control.

9.7.5   Field Feedback Control (Flux Vector Drive)

An innovative method for controlling ac motors is through field feedback (or *flux vector*) compensation. This approach can be explained using the equivalent circuit shown in Figure 9.40(c). Note that this circuit separates the rotor-equivalent impedance into two parts—a nonproductive part and a torque-producing part—as discussed previously. There exist magnetic field vectors (or complex numbers) that correspond to these two parts of circuit impedance. As clear from Figure 9.40(c), these magnetic flux components depend on the slip S and hence the rotor speed and also on the current. In the present method of control, the magnetic field vector associated with the first part of impedance is sensed using speed measurement (from an encoder) and motor current measurement from a current-voltage transducer), and compensated for (i.e., removed through feedback) in the stator circuit. As a result, only the second part of impedance (and magnetic field vector), corresponding to the back e.m.f., will remain. Hence, the ac motor will behave quite like a dc motor that has an equivalent torque-producing back e.m.f. More sophisticated schemes of control may use a model of the motor. Flux-vector control has been commercially implemented in ac motors using customized digital signal processing (DSP) chips. Feedback of rotor current can further improve the performance of a flux-vector drive. A flux-vector drive tends to be more complex and costly than a variable-frequency drive. Need of sensory feedback introduces a further burden in this regard.

9.7.6 A Transfer-Function Model for an Induction Motor

The true dynamic behavior of an induction motor is generally nonlinear and time-varying. For small variations about an operating point, however, linear relations can be written. On this basis, a transfer-function model can be established for an induction motor, as we have done for a dc motor. The procedure described in this section uses the steady-state speed-torque relationship for an induction motor to determine the transfer-function model. The basic assumption here is that this steady-state relationship, if the inertia effects are modeled by some other means, can represent the dynamic behavior of the motor for small changes about an operating point (i.e., near steady-state) with reasonable accuracy.

Suppose that a motor rotor that has moment of inertia J_m and mechanical damping constant b_m (mainly from the bearings) is subjected to a variation δT_m in the motor torque and an associated change $\delta\omega_m$ in the rotor speed, as shown in Figure 9.47. In general, these changes may arise from a change δT_L in the load torque, and a change δv_f in supply voltage. Newton's second law gives

$$\delta T_m - \delta T_L = J_m \delta\dot{\omega}_m + b_m\omega_m \tag{9.85}$$

Now, use a linear steady-state relationship to represent the variation in motor torque as a function of the incremental change $\delta\omega_m$ in speed and a variation δv_f in the supply voltage. We get

$$\delta T_m = -b_e\delta\omega_m + k_v\delta v_f \tag{9.86}$$

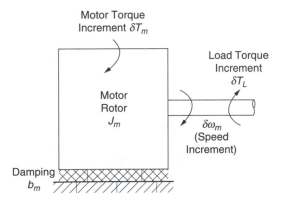

FIGURE 9.47
Incremental load model for an induction motor.

By substituting Equation 9.86 in Equation 9.85 and using the Laplace variable s, we have

$$\delta\omega_m = \frac{k_v}{[J_m s + b_m + b_e]}\delta v_f - \frac{1}{[J_m s + b_m + b_e]}\delta T_L \tag{9.87}$$

In the transfer-function Equation 9.87, note that $\delta\omega_m$ is the output, δv_f is the control input, and δT_L is an unknown (disturbance) input. The motor transfer function $\delta\omega_m/\delta v_f$ is given by

$$G_m(s) = \frac{k_v}{[J_m s + b_m + b_e]} \tag{9.88}$$

The motor time constant τ is

$$\tau = \frac{J_m}{b_m + b_e} \tag{9.89}$$

Now it remains to identify the parameters b_e (analogous to electrical damping in a dc motor) and k_v (a voltage gain parameter, as for a dc motor). To accomplish this, we use Equation 9.81, which can be written in the form

$$T_m = k(S)v_f^2 \tag{9.90}$$

where

$$k(S) = \frac{aS}{1 + (S/S_b)^2} \tag{9.91}$$

Now, using the well-known relation in differential calculus

$$\delta T_m = \frac{\partial T_m}{\partial\omega_m}\delta\omega_m + \frac{\partial T_m}{\partial v_f}\delta v_f$$

we have

$$b_e = -\frac{\partial T_m}{\partial\omega_m} \quad \text{and} \quad k_v = \frac{\partial T_m}{\partial v_f}$$

But

$$\frac{\partial T_m}{\partial\omega_m} = \frac{\partial T_m}{\partial S}\frac{\partial S}{\partial\omega_m} = -\frac{1}{\omega_f}\frac{\partial T_m}{\partial S}$$

Thus,

$$b_e = \frac{1}{\omega_f}\frac{\partial T_m}{\partial S} \tag{9.92}$$

where ω_f is the synchronous speed of the motor. Now, by differentiating Equation 9.91 with respect to S, we have

$$\frac{\partial k}{\partial S} = a\frac{1 - (S/S_b)^2}{[1 + (S/S_b)^2]^2} \tag{9.93}$$

Hence,

$$b_e = \frac{a v_f^2}{\omega_f} \frac{1-(S/S_b)^2}{[1+(S/S_b)^2]^2} \tag{9.94}$$

Next, by differentiating Equation 9.90 with respect to v_f we have

$$\frac{\partial T_m}{\partial v_f} = 2k(S)v_f$$

Accordingly, we get

$$k_v = \frac{2aSv_f}{1+(S/S_b)^2} \tag{9.95}$$

where S_b is the fractional slip at the breakdown (maximum) torque and a is a motor torque parameter, as defined by Equation 9.82. If we wish to include the effects of the electrical time constant τ_e of the motor, we may include the factor $\tau_e s + 1$ in the denominator (characteristic polynomial) on the right-hand side of Equation 9.87. Since τ_e is usually an order of magnitude smaller than τ as given by Equation 9.89, no significant improvement in accuracy results through this modification. Finally, note that the constants b_e and k_v can be obtained graphically using experimentally determined speed-torque curves for an induction motor for several values of the line voltage v_f, using a procedure similar to what we have described for a dc motor.

Example 9.15
A two-phase induction motor can serve as an ac servomotor. The field windings are identical and are placed in the stator with a geometric separation of 90°, as shown in Figure 9.48(a). One of the phases is excited by a fixed reference ac voltage $v_{ref} \cos \omega_p t$. The other phase is 90° out of phase from the reference phase; it is the control phase, with voltage amplitude v_c. The motor is controlled by varying the voltage v_c.

1. With the usual notation, obtain an expression for the motor torque T_m in terms of the rotor speed ω_m and the input voltage v_c.
2. Indicate how a transfer function model may be obtained for this ac servo
   a. Graphically, using the characteristic curves of the motor
   b. Analytically, using the relationship obtained in part 1

SOLUTION
Note that since $v_c \neq v_{ref}$, the two phases are not balanced. Hence, the resultant magnetic field vector in this two-phase induction motor does not rotate at a constant speed ω_p; as a result, the relations derived previously cannot be applied directly. The first step, then, is to decompose the field vector into two components that rotate at constant speeds. This is accomplished in Figure 9.48(b). The field component 1 is equivalent to that of an induction motor supplied with a line voltage of $(v_{ref} + v_c)/2$, and it rotates in the clockwise direction at speed ω_p. The field component 2 is equivalent to that generated with a line voltage of $(v_{ref} - v_c)/2$, and it rotates in the counterclockwise direction.

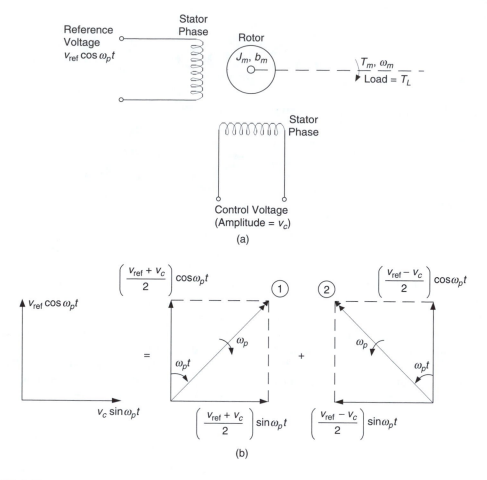

FIGURE 9.48
(a) A two-phase induction motor functioning as an ac servomotor, (b) Equivalent representation of the magnetic field vector in the stator.

Suppose that the motor rotates in the clockwise direction at speed ω_p. The slip for the equivalent system 1 is

$$S = \frac{\omega_p - \omega_m}{\omega_p}$$

and the slip for the equivalent system 2 is

$$S' = \frac{\omega_p + \omega_m}{\omega_p} = 2 - S$$

which are in opposite directions. Now, using the relationship for an induction motor with a balanced multiphase supply (Equation 9.90), we have

$$T_m = k(S)\left[\frac{v_{\text{ref}} + v_c}{2}\right]^2 - k(2-S)\left[\frac{v_{\text{ref}} - v_c}{2}\right]^2 \qquad (9.96)$$

In this derivation we have assumed that the electrical and magnetic circuits are linear, so the principle of superposition holds. The function $k(S)$ is given by the standard Equation 9.91, with a and S_b defined by Equation 9.82 and Equation 9.83, respectively. In this example, there is only one pole pair per phase ($n = 1$). Hence, the synchronous speed ω_f is equal to the line frequency ω_p. The motor speed ω_m is related to S through the usual Equation 9.87.

To obtain the transfer-function relation for operation about an operating point, we use the differential relation

$$\delta T_m = \frac{\partial T_m}{\partial \omega_m}\delta\omega_m + \frac{\partial T_m}{\partial v_c}\delta v_c$$

$$= -b_e\delta\omega_m + k_v\delta v_c$$

As derived previously, the transfer relation is

$$\delta\omega_m = \frac{k_v}{[J_m s + b_m + b_e]}\delta v_c - \frac{1}{[J_m s + b_m + b_e]}\delta T_L$$

where T_L denotes the load torque. It remains to show how to determine the parameters b_e and k_v, both graphically and analytically.

In the graphic method, we need a set of speed-torque curves for the motor, for several values of v_c in the operating range. Note that experimental measurements of motor torque contain the mechanical damping torque in the bearings. The actual electromagnetic torque of the motor is larger than the measured torque at steady state, the difference being the frictional torque. As a result, adjustments have to be made to the measured torque curve in order to get the true speed-motor torque curves. If this is done, the parameters b_e and k_v can be determined graphically as indicated in Figure 9.49. Each curve is a constant v_c curve. Hence, the magnitude of its slope gives b_e. Note that $\partial T_m/\partial v_c$ is evaluated at constant ω_m. Hence, the parameter k_v has to be determined on a vertical line (where ω_m = constant). If two curves, one for the operating value of v_c and the other for a unit increment in v_c, are available, as shown in Figure 9.49, the value of k_v is simply the vertical separation of the two curves at the operating point. If the increments in v_c are small, but not unity, the vertical separation of the two curves has to be divided by this increment (δv_c) in order to determine k_v, and the result will be more accurate.

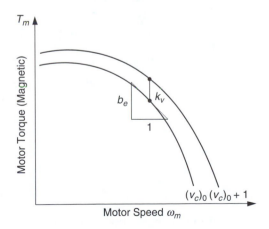

FIGURE 9.49
Graphic determination of transfer-function parameters for an induction motor.

To analytically determine b_e and k_v, we must differentiate T_m in Equation 9.96 with respect to ω_m and v_c, separately. We get

$$b_e = -\frac{\partial T_m}{\partial \omega_m} = \frac{1}{\omega_p}\left[\frac{v_{\text{ref}} + v_c}{2}\right]^2 \frac{\partial k(S)}{\partial S} - \frac{1}{\omega_p}\left[\frac{v_{\text{ref}} - v_c}{2}\right]^2 \frac{\partial k(2-S)}{\partial S} \tag{9.97}$$

where $[\partial k(S)/\partial S]$ is given by Equation 9.93. To determine $[\partial k(2-S)/\partial S]$, we note that

$$\frac{\partial k(2-S)}{\partial S} = \frac{\partial k(2-S)}{\partial(2-S)}\frac{d(2-S)}{dS} = -\frac{\partial k(2-S)}{\partial(2-S)} = -\frac{\partial k(S)}{\partial S}\bigg|_{S=2-S} \tag{9.98}$$

In other words, $[\partial k(2-S)/\partial S]$ is obtained by first replacing S by $2-S$ in the right-hand side of Equation 9.93 and then changing the sign. Finally,

$$k_v = \frac{\partial T_m}{\partial v_c} = \frac{1}{2}k(S)[v_{\text{ref}} + v_c] + \frac{1}{2}k(2-S)[v_{\text{ref}} - v_c] \tag{9.99}$$

Induction motors have the advantages of brushless operation, low maintenance, ruggedness, and low cost. They are naturally suitable for constant-speed and continuous operation applications. With modern drive systems, they are able to function well in variable-speed and servo applications as well. Applications of induction motors include household appliances, industrial instrumentation, traction devices (e.g., ground transit vehicles), machine tools (e.g., lathes and milling machines), heavy-duty factory equipment (e.g., steel rolling mills, conveyors, and centrifuges), office and business machinery and equipment in large buildings (e.g., elevator drives, compressors, fans, and HVAC systems).

9.7.7 Single-Phase AC Motors

The multiphase (polyphase) ac motors are normally employed in moderate- to high-power applications (e.g., more than 5 hp). In low-power applications (e.g., motors used in household appliances such as refrigerators, dishwashers, food processors, and hair dryers; tools such as saws, lawn mowers, and drills), single-phase ac motors are commonly used, for they have the advantages of simplicity and low cost.

The stator of a single-phase motor has only one set of drive windings (with two or more stator poles) excited by a single-phase ac supply. If the rotor is running close to the frequency of the line ac, this single phase can maintain the motor torque, operating as an induction motor. But a single phase is obviously not capable of starting the motor. To overcome this problem, a second coil that is out of phase from the first coil is used during the starting period and is turned off automatically once the operating speed is attained. The phase difference is obtained either through a difference in inductance for a given resistance in the two coils or by including a capacitor in the second coil circuit.

9.8 Synchronous Motors

Phase-locked servos and stepper motors can be considered synchronous motors because they run in synchronism with an external command signal (a pulse train) under normal operating conditions. What is described now is a different class of synchronous motors, termed ac synchronous motors. The rotor of a synchronous ac motor rotates in synchronism with the rotating magnetic field generated by the stator windings. The generation principle of this rotating field is identical to that in an induction motor. Unlike an induction motor, however, the rotor windings of a synchronous motor are energized by an external dc source. The rotor magnetic poles generated in this manner will lock themselves with the rotating magnetic field generated by the stator and will rotate at the same speed (synchronous speed). For this reason, synchronous motors are particularly suited for constant-speed applications under variable-load conditions. Synchronous motors with permanent magnet (e.g., samarium-cobalt) rotors are also commercially available, but these motors may be treated as permanent-magnet stepper motors with a harmonic command signal rather than a pulse command.

A schematic representation of the stator-rotor pair of a synchronous motor is shown in Figure 9.50. The dc voltage, which is required to energize the rotor windings, may come from several sources. An independent dc supply, an external ac supply and a rectifier, or a dc generator driven by the synchronous motor itself, are three ways of generating the dc signal.

One major drawback of the synchronous ac motor is that an auxiliary "starter" is required to start the mo*tor and bring its speed close to the synchronous speed. The reason for this is that in synchronous motors, the starting torque is virtually zero. To understand this, consider the starting conditions. The rotor is at rest and the stator field is rotating (at the synchronous speed). Consequently, there is 100% slip ($S = 1$). When, for example, an N pole of the rotating field in the stator is approaching an S pole in the rotor, the magnetic force will tend to turn the rotor in the direction opposite to the rotating field. When the same N pole of the rotating field has just passed the rotor S pole, the magnetic force will tend to pull the rotor in the same direction as the rotating field. These opposite interactions balance out, producing a zero net torque on the rotor. One method of starting a synchronous motor is by using a small dc motor. Once the synchronous motor reaches the synchronous speed, the dc motor is operated as a dc generator to supply power to the rotor windings. Alternatively, a small induction motor may be used to start the synchronous motor. A more desirable arrangement, which employs this principle, is to include several sets of induction-motor-type rotor windings (cage-type or wound-type) in the synchronous motor rotor itself. In all these cases, the supply to the rotor windings

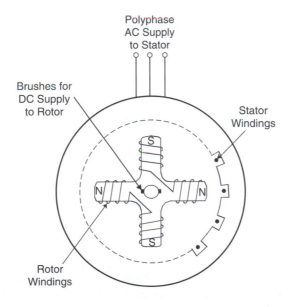

FIGURE 9.50
Schematic diagram of a stator-rotor configuration of a synchronous motor.

of the synchronous motor is disconnected during the starting conditions and is turned on only when the motor speed becomes close to the synchronous speed.

9.8.1 Control of a Synchronous Motor

Under normal operating conditions, the speed of a synchronous motor is completely determined by the frequency of the ac supply to the stator windings, because the motor speed is equal to the speed ω_f of the rotating field (see Equation 9.70). Hence, speed control can be achieved by the variable-frequency control method as described for an induction motor. In some applications of ac motors (both induction and synchronous types), clutch devices that link the motor to the driven load are used to achieve variable-speed control (e.g., using an eddy current clutch system that produces a variable coupling force through the eddy currents generated in the clutch). These dissipative techniques are quite wasteful and can considerably degrade the motor efficiency. Furthermore, heat removal methods would be needed to avoid thermal problems. Hence, they are not recommended for high-power applications where motor efficiency is a prime consideration, and for hazardous (e.g., chemical) environments.

Note that unless a permanent magnet rotor is used, a synchronous motor would require a slip ring and brush mechanism to supply the dc voltage to its rotor windings. This is a drawback that is not present in an induction motor.

The steady-state speed-torque curve of a synchronous motor, for a constant drive (line) frequency, is a straight line parallel to the torque axis. But with proper control (e.g., frequency control), an ac motor can function as a servomotor. Conventionally, a servomotor has a linear torque-speed relationship, which can be approached by an ac servomotor with a suitable drive system. Applications of synchronous ac motors include steel rolling mills, rotary cement kilns, conveyors, hoists, process compressors, recirculation pumps in hydroelectric power plants, and, more recently, servomotors and robotics. Synchronous motors are particularly suitable in high-speed, high-power applications where dc motors might not be appropriate. A synchronous motor can operate with a larger air gap between the rotor and the stator in comparison with an induction motor. This is an advantage for synchronous motors from the mechanical design point of view (e.g., bearing tolerances and rotor deflections due to thermal, static, and dynamic loads). Furthermore, rotor losses are smaller for synchronous motors than for induction motors.

9.9 Linear Actuators

Linear actuator stages are common in industrial motion applications. They may be governed by the same principles as the rotary actuators, but employing linear mechanical arrangements for the stator and the moving element, or a rotary motor with a rotary/ linear motion transmission unit. Solenoids are typically on-off (or, push-pull) type linear actuators, commonly used in relays, valve actuators, switches, and a variety of other applications. Some useful types of linear actuators are presented now.

9.9.1 Solenoid

The solenoid is a common rectilinear actuator, which consists of a coil and a soft iron core. When the coil is activated by a dc signal, the soft iron core becomes magnetized. This electromagnet can serve as an on-off (push-pull) actuator, for example to move a ferromagnetic element (moving pole or plunger). The moving element is the load, which is typically restrained by a light spring and a damping element.

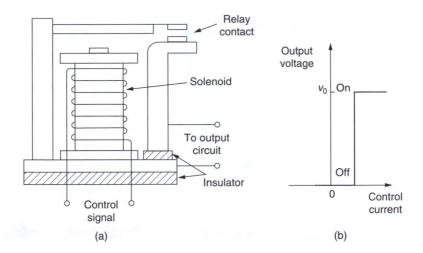

FIGURE 9.51
A solenoid operated relay: (a) Physical components; (b) Characteristic curve.

Solenoids are rugged and inexpensive devices. Common applications of solenoids include valve actuators, mechanical switches, relays, and other two-state positioning systems. An example of a solenoid-operated relay is shown in Figure 9.51.

A relay of this type may be used to turn on and off such devices as motors, heaters, and valves in industrial systems. They may be controlled by a programmable logic controller (PLC). A time-delay relay provides a delayed on-off action with an adjustable time delay, as necessary in some process applications.

The percentage on time with respect to the total on-off period is the *duty cycle* of a solenoid. A solenoid will need a sufficiently large current to move a load. There is a limit to the resulting magnetic force because the coil will eventually saturate. In order to avoid this, the ratings of the solenoid should match the needs of the load. There is another performance consideration. For long duty cycles it is necessary to maintain a current through the solenoid coil for a correspondingly long period. If the initial activating current of the solenoid is maintained over a long period, it will heat up the coil and create thermal problems. Apart from the loss of energy, this situation is undesirable because of safety issues, reduction in the coil life, and the need to have special means for cooling. A common solution is to incorporate a *hold-in circuit*, which will reduce the current through the solenoid coil shortly after it is activated. A simple hold-in circuit is shown in Figure 9.52. The resistance R_h is sufficiently large and comparable to resistance R_s of the solenoid coil. Initially, the capacitor C is fully discharged. Then the transistor is turned "on" (i.e., forward biased) and is able to conduct from the emitter (E) to collector (C). When the switch, which is normally open (denoted by NO), is turned on (i.e., closed), the dc supply sends a current through the solenoid coil (R_s), and the circuit is completed through the transistor. Since the transistor offers only a low resistance, the resulting current is large enough to actuate the solenoid. As the current flows through the circuit (while the switch is closed), the capacitor C becomes fully charged. The transistor becomes reverse-biased due to the resulting voltage of the capacitor. This turns off the transistor. Then the circuit is completed not through the transistor but through the hold-in resistor R_h. As a result, the current through the solenoid drops by a factor of $\frac{R_s}{R_s+R_h}$. This lower current is adequate to maintain the "holding" state of the solenoid without overheating it.

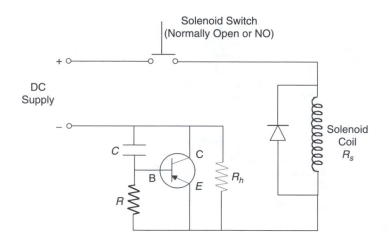

FIGURE 9.52
A hold-in circuit for a solenoid.

A *rotary solenoid* provides a rotary push-pull motion. Its principle of operation is the same as that of a linear solenoid. Another type of solenoid is the *proportional solenoid*. It is able to produce a rectilinear motion in proportion to the current through the coil. Accordingly it acts as a linear motor. Proportional solenoids are particularly useful as valve actuators in fluid power systems, for example, as actuators for spool valves in hydraulic piston-cylinder devices (hydraulic rectilinear actuators) and valve actuators for hydraulic motors (hydraulic rotary actuators).

9.9.2 Linear Motors

It is possible to obtain a rectilinear motion from a rotary electromechanical actuator (motor) by employing an auxiliary kinematic mechanism (motion transmission) such as a cam and follower, a belt and pulley, a rack and pinion, or a lead screw and nut (see Chapter 3). These devices inherently have problems of friction and backlash. Furthermore, they add inertia and flexibility (i.e., mechanical loading) to the driven load, thereby generating undesirable resonances and motion errors. Proper matching of the transmission inertia and the load inertia is essential (see Chapter 4). Particularly, the transmission inertia should be less than the load inertia, when referred to one side of the transmission mechanism. Furthermore, extra energy is needed to operate the system against the inertia and dissipation (friction) of the transmission mechanism.

For improved performance, direct rectilinear electromechanical actuators (direct drive devices) are desirable. These actuators operate according to the same principle as their rotary counterparts, except that flat stators and rectilinearly moving elements (in place of rotors) are employed. They come in different types:

1. Stepper linear actuators
2. DC linear actuators
3. AC linear actuators
4. Fluid (hydraulic and pneumatic) pistons and cylinders

In Chapter 8 we have indicated the principle of operation of a linear stepper motor. Fluids pistons and cylinders are discussed later in the present chapter. Linear electric motors are also termed *electric cylinders* and are suitable as high-precision linear stages of motion applications. For example, a dc brushless linear motor operates similar to a rotary brushless motor and using a similar drive amplifier. Advanced rare-earth magnets are used for the moving member, providing high force/mass ratio. The stator takes the form of a U-channel within which the moving member slides. Linear (sliding) bearings are standard. Since magnetic bearings can interfere with the force generating magnetic flux, air bearings are used in more sophisticated applications. The stator has the "forcer" coil for generating the drive magnetic field and Hall-effect sensors for commutation. Since conductive material will create eddy-current problems (to some degree even with a laminated construction, and to a lesser degree with powdered material), reinforced ceramic epoxy structures are used for the stator channel by leading manufacturers of linear motors. Applications of linear motors include traction devices, liquid-metal pumps, multi-axis tables, Cartesian robots, conveyor mechanisms, and servovalve actuators.

9.10 Hydraulic Actuators

The ferromagnetic material in an electric motor saturates at some level of magnetic flux density (and at a corresponding level of electric current, which generates the magnetic field). This limits the torque/mass ratio obtainable from an electric motor. Hydraulic actuators use the hydraulic power of a pressurized liquid. Since high pressures (on the order of 5000 psi) can be used, hydraulic actuators are capable of providing very high forces (and torques) at very high power levels simultaneously to several actuating locations in a flexible manner. The force limit of a hydraulic actuator can be an order of magnitude larger than that of an electromagnetic actuator. This results in higher torque/mass ratios than those available from electric motors, particularly at high levels of torque and power. This is a principal advantage of hydraulic actuators. Note that the actuator mass considered here is the mass of the final actuating element, not including auxiliary devices such as those needed to pressurize and store the fluid. Another advantage of a hydraulic actuator is that it is quite stiff when viewed from the side of the load. This is because a hydraulic medium is mechanically stiffer than an electromagnetic medium. Consequently, the control gains required in a high-power hydraulic control system would be significantly less than the gains required in a comparable electromagnetic (motor) control system. Note that the stiffness of an actuator may be measured by the slope of the speed-torque (force) curve, and is representative of the speed of response (or, bandwidth). There are other advantages of *fluid power systems*. Electric motors generate heat. In continuous operation, then, the thermal problems can be serious, and special means of heat removal will be necessary. In a fluid power system, however, any heat generated at the load can be quickly transferred to another location away from the load, by the hydraulic fluid itself, and effectively removed by means of a heat exchanger. Another advantage of fluid power systems is that they are self-lubricating and as a result, the friction in valves, cylinders, pumps, hydraulic motors, and other system components will be low and will not require external lubrication. Safety considerations will be less as well because, for example, there is no possibility of spark generation as in motors with brush mechanisms. There are several disadvantages as well. Fluid power systems are more nonlinear than electrical actuator systems. Reasons for this include valve nonlinearities, fluid friction, compressibility, thermal effects, and generally nonlinear constitutive relations. Leakage can create problems. Fluid power systems tend to be noisier than electric motors. Synchronization of multi-actuator operations may be more difficult as well. Also, when the necessary accessories are included, fluid power systems are by and large more expensive and less portable than electrical actuator systems.

Fluid power systems with analog control devices have been in use in engineering applications since the 1940s. Smaller, more sophisticated, and less costly control hardware and microprocessor-based controllers were developed in the 1980s, making fluid power control systems as sophisticated, precise, cost-effective, and versatile as electromechanical control systems. Today, miniature fluid power systems with advanced digital control and electronics, are used in numerous applications, directly competing with advanced dc and ac motion control systems. Also, logic devices based on "fluidics" or *fluid logic* devices are preferred over digital electronics in some types of industrial applications. Applications of fluid power systems include vehicle steering and braking systems, active suspension systems, material handling devices and industrial mechanical manipulators such as hoists, industrial robots, rolling mills, heavy-duty presses, actuators for aircraft control surfaces (ailerons, rudder, and elevators), excavators, actuators for opening and closing of bridge spans, tunnel boring machines, food processing machines, reaction injection molding (RIM) machines, dynamic testing machines and heavy-duty shakers for structures and components, machine tools, ship building, and dynamic props, stage backgrounds and structures in theatres and auditoriums.

9.10.1 Components of a Hydraulic Control System

A schematic diagram of a basic hydraulic control system is shown in Figure 9.53(a). A view of a practical fluid power system is shown in Figure 9.53(b). The hydraulic fluid (oil) is pressurized using a pump, which is driven by an ac motor. Typical fluids used are mineral oils or oil in water-emulsions. These fluids have the desirable properties of self-lubrication, corrosion resistance, good thermal properties and fire resistance, environmental friendliness, and low compressibility (high stiffness for good bandwidth). Note that the motor converts electrical power into mechanical power, and the pump converts mechanical power into fluid power. In terms of through and across variable pairs (see Chapter 2), these power conversions can be expressed as

$$(i,v)\xrightarrow{\ \eta_m\ }(T,\omega)\xrightarrow{\ \eta_h\ }(Q,P)$$

in the usual notation. The conversion efficiency η_m of a motor is typically very high (over 90%), whereas the efficiency η_h of a hydraulic pump is not as good (about 60%), mainly because of dissipation, leakage, and compressibility effects. Depending on the pump capacity, flow rates in the range of 1000–50,000 gallons per minute (Note: 1 gal/min = 3.8 liters/min) and pressures from 500–5000 psi (Note: 1 kPa = 0.145 psi) can be obtained. The pressure of the fluid from the pump is regulated and stabilized by a relief valve and an accumulator. A hydraulic valve provides a controlled supply of fluid into the actuator, controlling both the flow rate (including direction) and the pressure. In feedback control, this valve uses response signals (motion) sensed from the load, to achieve the desired response—hence the name *servovalve.* Usually, the servovalve is driven by an electric *valve actuator,* such as a *torque motor* or a *proportional solenoid,* which in turn is driven by the output from a *servo amplifier.* The servo amplifier receives a reference input command (corresponding to the desired position of the load) as well as a measured response of the load (in feedback). Compensation circuitry may be used in both feedback and forward paths of the control system to modify the signals so as to obtain the desired control action. The hydraulic actuator (typically a piston-cylinder device for rectilinear motions or a hydraulic motor for rotary motions) converts fluid power back into mechanical power, which is available to perform useful tasks (i.e., to drive a load). Note that some power in the fluid is lost at this stage. The low-pressure fluid at the drain of the hydraulic servovalve is filtered and returned to the reservoir, and is available to the pump.

One might argue that since the power that is required to drive the load is mechanical, it would be much more efficient to use a motor directly to drive that load. There are

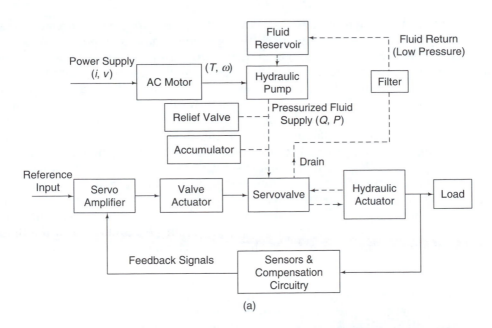

(a)

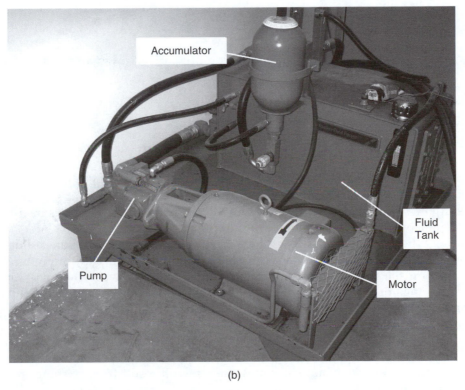

(b)

FIGURE 9.53
(a) Schematic diagram of a hydraulic control system, (b) An industrial fluid-power system.

good reasons for using hydraulic power, however. For example, ac motors are usually difficult to control, particularly under variable-load conditions. Their efficiency can drop rapidly when the speed deviates from the rated speed, particularly when voltage control is used. They need gear mechanisms for low-speed operation, with associated problems such as backlash, friction, vibration, and mechanical loading effects. Special coupling devices are also needed. Hydraulic devices usually filter out high-frequency noise, which is not the case with ac motors. Thus, hydraulic systems are ideal for high-power, high-force control applications. In high-power applications, a single high-capacity pump or several pumps can be employed to pressurize the fluid. Furthermore, in low-power applications, several servovalve and actuator systems can be operated to perform different control tasks in a distributed control environment, using the same pressurized fluid supply. In this sense, hydraulic systems are very flexible. Hydraulic systems provide excellent speed-force (or torque) capability, variable over a wide range of speeds without significantly affecting the power-conversion efficiency, because the excess high-pressure fluid is diverted to the return line. Consequently, hydraulic actuators are far more controllable than ac motors. As noted before, hydraulic actuators also have an advantage over electromagnetic actuators from the point of view of heat transfer characteristics. Specifically, the hydraulic fluid promptly carries away any heat that is generated locally and releases it through a heat exchanger at a location away from the actuator.

9.10.2   Hydraulic Pumps and Motors

The objective of a hydraulic pump is to provide pressurized oil to a hydraulic actuator. Three common types of hydraulic pumps are

1. Vane pump
2. Gear pump
3. Axial piston pump

The pump type used in a hydraulic control system is not very significant, except for the pump capacity, with respect to the control functions of the system. But since hydraulic motors can be interpreted as pumps operating in the reverse direction, it is instructional to outline the operation of these three types of pumps.

A sliding-type *vane pump is* shown schematically in Figure 9.54. The vanes slide in the interior of the housing as they rotate with the rotor of the pump. They can move within radial slots on the rotor, thereby maintaining full contact between the vanes and the housing. Springs or the pressurized hydraulic fluid itself may be used for maintaining this contact. The rotor is eccentrically mounted inside the housing. The fluid is drawn in at the inlet port as a result of the increasing volume between vane pairs as they rotate, in the first half of a rotation cycle. The oil volume trapped between two vanes is eventually compressed because of the decreasing volume of the vane compartment, in the second half of the rotation cycle. Note that a pressure rise will result from pushing the liquid volume into the high-pressure side and not allowing it to return to the low-pressure side of the pump, even when there is no significant compressibility in the liquid, when it moves from the low-pressure side to the high-pressure side. The typical operating pressure (at the outlet port) of these devices is about 2000 psi (13.8 MPa). The output pressure can be varied by adjusting the rotor eccentricity, because this alters the change in the compartment volume during a cycle. A disadvantage of any rotating device with eccentricity is the centrifugal forces that are generated even while rotating at constant speed. Dynamic balancing is needed to reduce this problem.

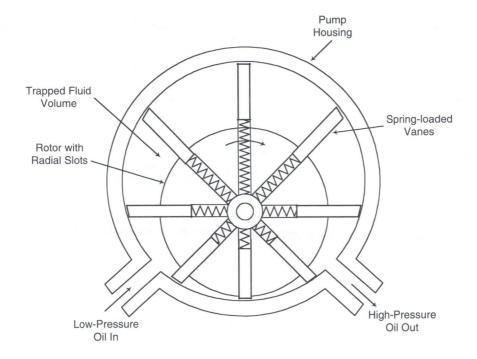

Pump Housing

Trapped Fluid Volume

Spring-loaded Vanes

Rotor with Radial Slots

Low-Pressure Oil In

High-Pressure Oil Out

FIGURE 9.54
A hydraulic vane pump.

The operation of an external-gear hydraulic pump (or, simply a *gear pump*) is illustrated in Figure 9.55. The two identical gears are externally meshed. The inlet port is facing the gear enmeshing (retracting) region. Fluid is drawn in and trapped between the pairs of teeth in each gear, in rotation. This volume of fluid is transported around by the two gear wheels into the gear meshing region, at the pump outlet. Here it undergoes an increase in pressure, as in the vane pump, as a result of forcing the fluid into the high-pressure side. Only moderate to low pressures can be realized by gear pumps (about 1000 psi or 7 MPa, maximum), because the volume changes that take place in the enmeshing and meshing regions are small (unlike in the vane pumps) and because fluid leakage between teeth and housing can be considerable. Gear pumps are robust and low-cost devices, however, and they are probably the most commonly used hydraulic pumps.

A schematic diagram of an axial piston hydraulic pump is shown in Figure 9.56. The chamber barrel is rigidly attached to the drive shaft. The two pistons themselves rotate with the chamber barrel, but since the end shoes of the pistons slide inside a slanted (skewed) slot on the stroke plate, which is stationary, the pistons simultaneously undergo a reciprocating motion as well in the axial direction. As a chamber opening reaches the inlet port of the pump housing, fluid is drawn in because of the increasing volume between the piston head and the chamber. This fluid is trapped and transported to the outlet port while undergoing compression as a result of the decreasing volume inside the chamber due to the axial motion of the piston. Fluid pressure increases in this process. High outlet pressures (4000 psi or 27.6 MPa, or more) can be achieved using piston pumps. As shown in Figure 9.56, the piston stroke can be increased by increasing the inclination angle of the stroke plate (slot). This, in turn, increases the pressure ratio of the pump. A lever mechanism is usually available to adjust the piston stroke. Piston pumps are relatively expensive.

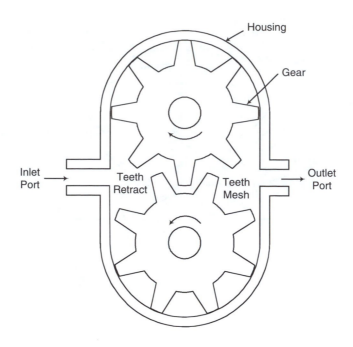

FIGURE 9.55
A hydraulic gear pump.

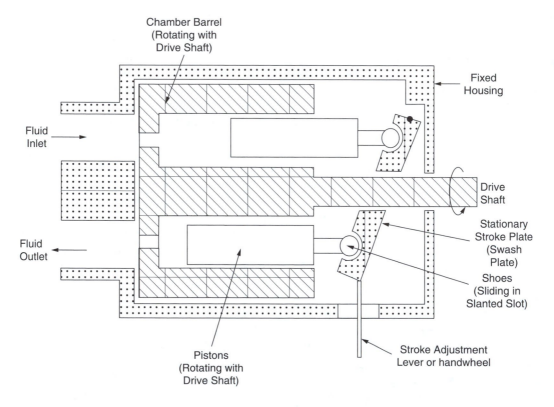

FIGURE 9.56
An axial piston hydraulic pump.

The efficiency of a hydraulic pump is given by the ratio of the output fluid power to the motor mechanical power; thus,

$$\eta_p = \frac{PQ}{\omega T} \qquad (9.100)$$

where
    P = pressure increase in the fluid
    Q = fluid (volume) flow rate
    ω = rotating speed of the pump
    T = drive torque to the pump

9.10.3   Hydraulic Valves

Fluid valves can perform three basic functions:

1. Change the flow direction
2. Change the flow rate
3. Change the fluid pressure

The valves that accomplish the first two functions are termed *flow-control valves.* The valves that regulate the fluid pressure are termed *pressure-control valves. A* simple relief valve regulates pressure, whereas the poppet valve, gate valve and globe valve are on/off flow-control valves. Some examples are shown in Figure 9.57. The directional valve (or, check valve) shown in Figure 9.57(a) allows the fluid flow in one direction and blocks it in the opposite direction. The spring provides sufficient force for the ball to return to the seat when there is no fluid flow. It does not need to sustain any fluid pressure, and hence its stiffness is relatively low. A check valve falls into the category of flow control valves. Figure 9.57(b) shows a poppet valve. It is normally in the closed position, with the ball completely seated to block the flow. When the plunger is pushed down, the ball moves with it, allowing fluid flow through the seat opening. This on/off valve is bidirectional, and may be used to permit fluid flow in either direction. The relief valve shown in Figure 9.57(c) is in the closed condition under normal conditions. The spring force, which closes the valve (by seating the ball) is adjustable. When the fluid pressure (in a container or a pipe to which the valve is connected) rises above a certain value, as governed by the spring force, the valve opens thereby letting the fluid out through vent (which may be recirculated within the system). In this manner, the pressure of the system is maintained at a nearly constant level. Typically, an accumulator is used in conjunction with a relief valve, to take up undesirable pressure fluctuations and to stabilize the system. Valves are classified by the number of flow paths present under operating conditions. For example, a four-way valve has four ways in which flow can enter and leave the valve. In high-power fluid systems, two valve stages consisting of a *pilot valve* and a main valve may be used. Here, the pilot valve is a low-capacity, low-power valve, which operates the higher-capacity main valve.

9.10.3.1   Spool Valve

Spool valves are used extensively in hydraulic servo systems. *A* schematic diagram of a four-way spool valve is shown in Figure 9.58(a). This is commonly called a *servovalve* because motion feedback is used by it to control the motion of a hydraulic actuator. The moving unit of the valve is called the *spool.* It consists of a spool rod and one or more expanded regions

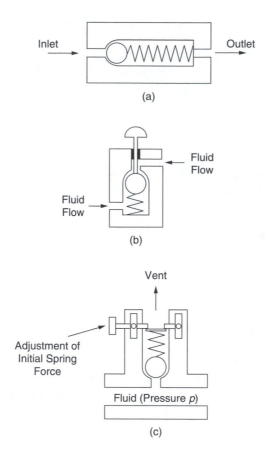

FIGURE 9.57
(a) A check valve (directional valve), (b) A poppet valve (an on/off valve), (c) A relief valve (a pressure regulating valve).

(or, lobes), which are called *lands*. Input displacement (U) applied to the spool rod, using an actuator (torque motor or proportional solenoid), regulates the flow rate (Q) to the main hydraulic actuator as well as the corresponding pressure difference (P) available to the actuator. If the land length is larger than the port width (Figure 9.58(b)), it is an *overlapped land*. This introduces a dead zone in the neighborhood of the central position of the spool, resulting in decreased sensitivity and increased stability problems. Since it is virtually impossible to exactly match the land size with the port width, the *underlapped land* configuration (Figure 9.58(c)) is commonly employed. In this case, there is a leakage flow, even in the fully closed position, which decreases the efficiency and increases the steady-state error of the hydraulic control system. For accurate operation of the valve, the leakage should not be excessive. The direct flow at various ports of the valve and the leakage flows between the lands and the valve housing should be included in a realistic analysis of a spool valve. For small displacements δU about an operating point, the following linearized equations can be written. Since the flow rate Q_2 into the actuator increases as U increases and it decreases as P_2 increases, we have

$$\delta Q_2 = k_q \delta U - k'_c \delta P_2 \tag{9.101}$$

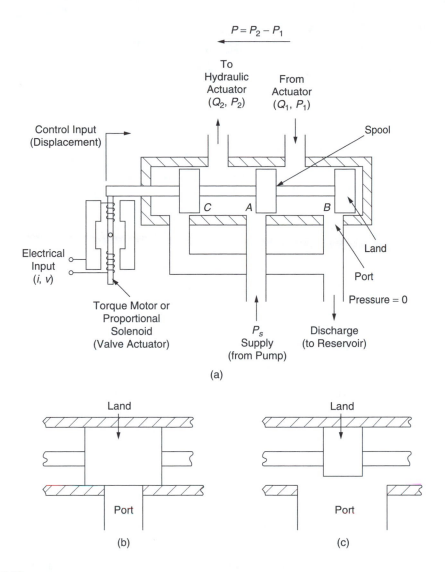

FIGURE 9.58
(a) A four-way spool valve, (b) An overlapped land, (c) An underlapped land.

Similarly, since the flow rate Q_1 from the actuator increases with both U and P_1, we have

$$\delta Q_1 = k_q \delta U + k'_c \delta P_1 \qquad (9.102)$$

The gains k_q and k'_c will be defined later.

In fact, if we disregard the compressibility of the fluid, $\delta Q_1 = \delta Q_2$, assuming that the hydraulic piston (actuator) is double-acting, with equal piston areas on the two sides of the actuator piston. We consider the general case where $Q_1 \neq Q_2$. Note, however, that the inlet port and the outlet port are assumed to have identical characteristics. By adding together Equation 9.101 and Equation 9.102 and defining an average flow rate

$$Q = \frac{Q_1 + Q_2}{2} \qquad (9.103)$$

and an equivalent flow-pressure coefficient

$$k_c = \frac{k'_c}{2} \tag{9.104}$$

we get

$$\delta Q = k_q \delta U - k_c \delta P \tag{9.105}$$

where the *flow gain* is

$$k_q = \left(\frac{\partial Q}{\partial U} \right)_P \tag{9.106}$$

and the *flow-pressure coefficient* is

$$k_c = -\left(\frac{\partial Q}{\partial P} \right)_U \tag{9.107}$$

Note further that the *pressure sensitivity* is

$$k_p = \left(\frac{\partial P}{\partial U} \right)_Q = \frac{k_q}{k_c} \tag{9.108}$$

To obtain Equation 9.108, we use the well-known result from calculus:

$$\delta Q = \left(\frac{\partial Q}{\partial U} \right)_P \delta U + \left(\frac{\partial Q}{\partial P} \right)_U \delta P$$

Since $\delta P / \delta U \to \partial P / \partial U$ as $\delta Q \to 0$, we have

$$\left(\frac{\partial P}{\partial U} \right)_Q = -\left(\frac{\partial Q}{\partial U} \right)_P \Big/ \left(\frac{\partial Q}{\partial P} \right)_U \tag{9.109}$$

Equation 9.108 directly follows from Equation 9.109.

A valve can be actuated by several methods, for example, manual operation, the use of mechanical linkages connected to the drive load, and the use of electromechanical actuators such as solenoids and torque motors (or force motors). Regular solenoids are suitable for on/off control applications, and proportional solenoids and torque motors are used in continuous control. For precise control applications, electromechanical actuation of the valve (with feedback for servo operation) is preferred.

Large valve displacements can saturate a valve because of the nonlinear nature of the flow relations at the valve ports. Several valve stages may be used to overcome this saturation problem, when controlling heavy loads. In this case, the spool motion of the first stage (pilot stage) is the input motion. It actuates the spool of the second stage, which acts as a hydraulic amplifier. The fluid supply to the main hydraulic actuator, which drives the load, is regulated by the final stage of a multistage valve.

9.10.3.2 Steady-State Valve Characteristics

Although the linearized valve Equation 9.105 is used in the analysis of hydraulic control systems, it should be noted that the flow equations of a valve are quite nonlinear. Consequently, the valve constants k_q and k_c change with the operating point. Valve constants can be determined either by experimental measurements or by using an accurate non-linear model. Now we establish a reasonably accurate nonlinear relationship relating the (average) flow rate Q through the main hydraulic actuator and the pressure difference (load pressure) P provided to the hydraulic actuator.

Assume identical rectangular ports at the supply and discharge points in Figure 9.58(a). When the valve lands are in the neutral (central) position, we set $U = 0$. We assume that the lands perfectly match the ports (i.e., no dead zone or leakage flows due to clearances). The positive direction of U is taken as shown in Figure 9.58(a). For this positive configuration, the flow directions are also indicated in the figure. The flow equations at ports A and B are

$$Q_2 = Ubc_d \sqrt{\frac{2(P_s - P_2)}{\rho}} \tag{9.110}$$

$$Q_1 = Ubc_d \sqrt{\frac{2P_1}{\rho}} \tag{9.111}$$

where
b = land width
c_d = discharge coefficient at each port
ρ = density of the hydraulic fluid
P_s = supply pressure of the hydraulic fluid

In Equation 9.111 the pressure at the discharge end is taken to be zero. For steady-state operation, we use

$$Q_1 = Q_2 = Q \tag{9.112}$$

Now, squaring Equation 9.110 and Equation 9.111 and adding, we get

$$2Q^2 = 2(Ubc_d)^2 \frac{(P_s - P)}{\rho}$$

where, the pressure difference supplied to the hydraulic actuator is denoted by

$$P = P_2 - P_1 \tag{9.113}$$

Consequently,

$$Q = Ubc_d \sqrt{\frac{P_s - P}{\rho}} \qquad \text{for } U > 0 \tag{9.114}$$

When $U < 0$, the flow direction reverses; furthermore, port A is now associated with P_1 (not P_2) and port C is associated with P_2. It follows that Equation 9.114 still holds, except that $P_2 - P_1$ is replaced by $P_1 - P_2$. Hence,

$$Q = Ubc_d \sqrt{\frac{P_s + P}{\rho}} \qquad \text{for } U < 0 \tag{9.115}$$

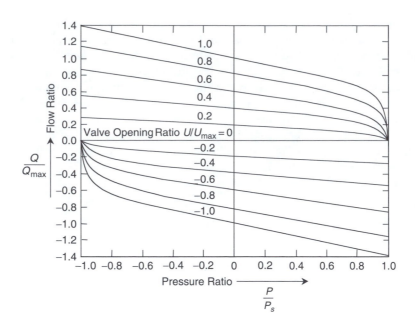

FIGURE 9.59
Steady-state characteristics for a four-way spool valve.

Combining Equation 9.114 and Equation 9.115, we have

$$Q = Ubc_d \sqrt{\frac{P_s - P\,\mathrm{sgn}(U)}{\rho}} \tag{9.116}$$

This can be written in the nondimensional form

$$\frac{Q}{Q_{\max}} = \frac{U}{U_{\max}} \sqrt{1 - \frac{P}{P_s} \mathrm{sgn}\left(\frac{U}{U_{\max}}\right)} \tag{9.117}$$

where $U_{\max}$ = maximum valve opening (> 0), and

$$Q_{\max} = U_{\max} bc_d \sqrt{\frac{P_s}{\rho}} \tag{9.118}$$

Equation 9.117 is plotted in Figure 9.59. As with the speed-torque curve for a motor, it is possible to obtain the valve constants k_q and k_c defined by Equation 9.106 and Equation 9.107, from the curves given in Figure 9.59 for various operating points. For better accuracy, hoever, experimentally determined valve characteristic curves should be used.

9.10.4   Hydraulic Primary Actuators

Rotary hydraulic actuators (hydraulic motors) operate much like the hydraulic pumps discussed earlier, except that the direction flow is reversed and the mechanical power is delivered by the shaft, rather than taken in. High-pressure fluid enters the actuator. As it

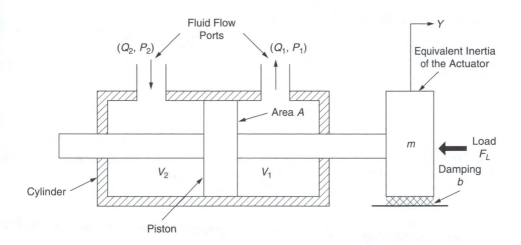

FIGURE 9.60
Double-acting piston-cylinder hydraulic actuator

passes through the hydraulic motor, the fluid power is used up in turning the rotor, and the pressure is dropped. The low-pressure fluid leaves the motor en route to the reservoir. One of the more efficient rotary hydraulic actuators is the axial piston motor, quite similar in construction to the axial piston pump shown in Figure 9.56.

The most common type of rectilinear hydraulic actuator, however, is the *hydraulic ram* (or, piston-cylinder actuator). A schematic diagram of such a device is shown in Figure 9.60. This is a *double-acting actuator* because the fluid pressure acts on both sides of the piston. If the fluid pressure is present only on one side of the piston, it is termed a *single-acting actuator.* Single-acting piston-cylinder (ram) actuators are also in common use for their simplicity and the simplicity of the other control components such as servovalves that are needed, although they have the disadvantage of asymmetry. The fluid flow at the ports of a hydraulic actuator is regulated typically by a spool valve. This valve may be operated by a pilot valve (e.g., a flapper valve).

To obtain the equations for the actuator shown in Figure 9.60, we note that the volume flow rate Q into a chamber depends primarily on two factors:

1. Increase in chamber volume
2. Increase in pressure (compressibility effect of the fluid)

When a piston of area A moves through a distance Y, the flow rate due to the increase in chamber volume is $\pm A\dot{Y}$. The positive sign denotes "into the chamber" and the negative sign denotes "out of the chamber." Now, with an increase in pressure δP, the volume of a given fluid mass would decrease by the magnitude $[-(\partial V/\partial P)\delta P]$. As a result, an equal volume of new fluid would enter the chamber. The corresponding rate of flow (into the chamber) is $[-(\partial V/\partial P)(dP/dt)]$. Since the *bulk modulus* (isothermal, or at constant temperature) is given by

$$\beta = -V\frac{\partial P}{\partial V} \tag{9.119}$$

the rate of flow due to the rate of pressure change is given by $[(V/\beta)(dP/dt)]$. Using these facts, the fluid conservation (i.e., flow continuity) equations for the two sides of the actuator chamber in Figure 9.60 can be written as

$$Q_2 = A\frac{dY}{dt} + \frac{V_2}{\beta}\frac{dP_2}{dt} \tag{9.120}$$

$$Q_1 = A\frac{dY}{dt} - \frac{V_1}{\beta}\frac{dP_1}{dt} \tag{9.121}$$

Note that for a realistic analysis, leakage flow rate terms (for leakage between piston and cylinder and between piston rod and cylinder) should be included in Equation 9.120 and Equation 9.121. For a linear analysis, these leakage flow rates can be taken as proportional to the pressure difference across the leakage path. Note, further, that V_1 and V_2 can be expressed in terms of Y, as follows:

$$V_1 + V_2 = V_o \tag{9.122}$$

$$V_1 - V_2 = V_o' + 2AY \tag{9.123}$$

where V_o and V_o' are constant volumes, which depend on the cylinder capacity and on the piston position when $Y = 0$, respectively. Now, for incremental changes about the operating point $V_1 = V_2 = V$, Equation 9.120 and Equation 9.121 can be written as

$$\delta Q_2 = A\frac{d\delta Y}{dt} + \frac{V}{\beta}\frac{d\delta P_2}{dt} \tag{9.124}$$

$$\delta Q_1 = A\frac{d\delta Y}{dt} - \frac{V}{\beta}\frac{d\delta P_1}{dt} \tag{9.125}$$

Note that the "total" Equation 9.120 and total Equation 9.121 are already linear for constant V. But, since the valve equation is nonlinear, and since V is not a constant, we should use the "incremental" Equation 9.124 and the "incremental" Equation 9.125 instead of the "total" equations, in a linear model. Adding Equation 9.124 and Equation 9.125 and dividing by 2, we get the hydraulic actuator equation

$$\delta Q = A\frac{d\delta Y}{dt} + \frac{V}{2\beta}\frac{d\delta P}{dt} \tag{9.126}$$

where

$Q = \dfrac{Q_1 + Q_2}{2}$ = average volume flow into the actuator

$P = P_2 - P_1$ = pressure difference on the piston of the actuator.

9.10.5 The Load Equation

So far, we have obtained the linearized valve Equation 9.105 and the linearized actuator actuation Equation 9.126. It remains to determine the load equation, which depends on the nature of the load that is driven by the hydraulic actuator. We may represent it by a load force F_L, as shown in Figure 9.60. Note that F_L is a dynamic term that may represent such effects as flexibility, inertia, and the dissipative effects of the load. In addition, the inertia of the moving parts of the actuator is modeled as a mass m, and the energy dissipation effects associated with these moving parts are represented by an equivalent viscous damping constant b. Accordingly, Newton's second law gives

$$m\frac{d^2Y}{dt^2} + b\frac{dY}{dt} = A(P_2 - P_1) - F_L \tag{9.127}$$

This equation is also linear already. Again, since the valve equation is nonlinear, to be consistent, we should consider incremental motions δY about an operating point. Consequently, we have

$$m\frac{d^2\delta Y}{dt^2} + b\frac{d\delta Y}{dt} = A\delta P - \delta F_L \tag{9.128}$$

where, as before

$$P = P_2 - P_1$$

If the active areas on the two sides of the piston are not equal, a net imbalance force would exist. This could lead to unstable response under some conditions.

9.11 Hydraulic Control Systems

The main components of a hydraulic control system are

1. A servovalve
2. A hydraulic actuator
3. A load
4. Feedback control elements.

We have obtained linear equations for the first three components as Equation 9.105, Equation 9.126, and Equation 9.128. Now we rewrite these equations, denoting the incremental variables about an operating point by lowercase letters.

$$\text{Valve: } q = k_q u - k_c p \tag{9.129}$$

$$\text{Hydraulic actuator: } q = A\frac{dy}{dt} + \frac{V}{2\beta}\frac{dp}{dt} \tag{9.130}$$

$$\text{Load: } m\frac{d^2y}{dt^2} + b\frac{dy}{dt} = Ap - f_L \tag{9.131}$$

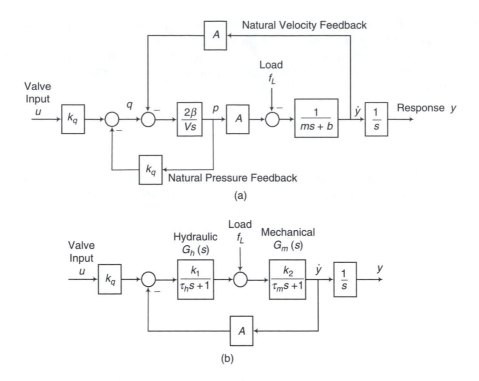

FIGURE 9.61
(a) Block diagram for an open-loop hydraulic control system, (b) An equivalent block diagram.

The feedback elements will depend on the specific feedback control method that is employed. We will revisit this aspect of a hydraulic control system later. Equations 9.129 through 9.131 can be represented by the block diagram shown in Figure 9.61(a). This is an open-loop control system because no external feedback elements have been used. Note, however, the presence of a "natural" *pressure feedback* path and a "natural" *velocity feedback* path, which are inherent to the dynamics of the open-loop system. The block diagram can be reduced to the equivalent form shown in Figure 9.61(b). To obtain this equivalent representation, combine the first two summing junctions and then obtain the equivalent transfer function for the pressure feedback loop (see Chapter 2 and Chapter 12 for further considerations of block diagram reduction). This equivalent transfer function can be obtained using the relationship for reducing a feedback control system:

$$G_h = \frac{G}{1 + GH} \tag{9.132}$$

where
  G = forward transfer function
  H = feedback transfer function

In the present case,

$$G = \frac{2\beta}{Vs}$$

and

$$H = k_c$$

Hence,

$$G_h = \frac{k_1}{\tau_h s + 1} \tag{9.133}$$

where the *pressure gain* parameter is

$$k_1 = \frac{1}{k_c} \tag{9.134}$$

and the *hydraulic time* constant is

$$\tau_h = \frac{V}{2\beta k_c} \tag{1.135}$$

The pressure gain k_1 is a measure of the load pressure p generated for a given volume flow rate q into the hydraulic actuator. The smaller the pressure coefficient k_c, the larger the pressure gain, as is clear from Equation 9.134. The hydraulic time constant increases with the volume of the actuator fluid chamber and decreases with the bulk modulus of the hydraulic fluid. This is to be expected, because the hydraulic time constant depends on the compressibility of the hydraulic fluid.

The mechanical transfer function of the hydraulic actuator is represented by

$$G_m = \frac{k_2}{\tau_m s + 1} \tag{9.136}$$

where the *mechanical time constant* is given by

$$\tau_m = \frac{m}{b} \tag{9.137}$$

and $k_2 = 1/b$. Typically, the mechanical time constant is the dominant time constant, since it is usually larger than the hydraulic time constant.

Example 9.16

A model of the automatic gage control (AGC) system of a steel rolling mill is shown in Figure 9.62. The rollers are pressed using a single-acting hydraulic actuator with valve displacement u. The rollers are displaced through y, thereby pressing the steel that is being rolled. For a given y, the rolling force F is completely known from the steel parameters.

1. Identify the inputs and the controlled variable in this control system.
2. In terms of the variables and system parameters indicated in Figure 9.62, write dynamic equations for the system, including valve nonlinearities.

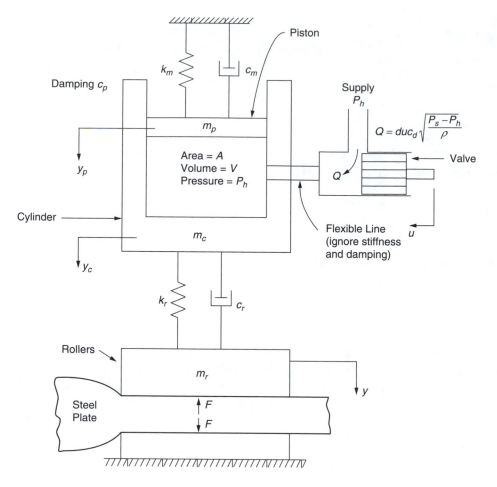

FIGURE 9.62
Automatic gage control (AGC) system of a steel rolling mill.

3. What is the order of the system? Identify the response variables.
4. Draw a block diagram for the system, clearly indicating the hydraulic actuator with valve, the mechanical structure of the mill, inputs, and the controlled variable.
5. What variables would you measure (and feed back through suitable controllers) in order to improve the performance of the control system?

SOLUTION

Part 1: *Valve* displacement u and rolling force F are inputs. Roll displacement y is the controlled variable.

Part 2: The mechanical-dynamic equations are

$$m_p\ddot{y}_p = -k_m y_p - c_m \dot{y}_p - c_p(\dot{y}_p - \dot{y}_c) - AP_h \tag{i}$$

$$m_c\ddot{y}_c = -k_r(y_c - y) - c_r(\dot{y}_c - \dot{y}) - c_p(\dot{y}_c - \dot{y}_p) + AP_h \tag{ii}$$

$$m_r\ddot{y} = -k_r(y - y_c) - c_r(\dot{y} - \dot{y}_c) - F \tag{iii}$$

The static forces balance and the displacements are measured from the corresponding equilibrium configuration, so that gravity terms do not enter into the equations.

The *hydraulic actuator equation* is derived as follows. For the valve, with the usual notation, the flow rate is given by

$$Q = buc_d\sqrt{\frac{P_s - P_h}{\rho}}$$

For the piston-cylinder,

$$Q = A(\dot{y}_c - \dot{y}_p) + \frac{V}{\beta}\frac{dP_h}{dt}$$

Hence,

$$\frac{V}{\beta}\frac{dP_h}{dt} = A(\dot{y}_c - \dot{y}_p) + buc_d\sqrt{\frac{P_s - P_h}{\rho}} \qquad \text{(iv)}$$

Part 3: There are three second-order differential equations: Equation i, Equation ii, Equation iii, and one first-order differential equation: Equation iv. Hence, the system is seventh-order. The response variables are the displacements y_p, y_c, y and the pressure P_h.

Part 4: A block diagram for the hydraulic control system of the steel rolling mill is shown in Figure 9.63.

Part 5: The hydraulic pressure P_h and the roller displacement y are the two response variables, which can be conveniently measured and used in feedback control. The rolling force F may be measured and fed forward, but this is somewhat difficult in practice.

Example 9.17
A single-stage pressure control valve is shown in Figure 9.64. The purpose of the valve is to keep the load pressure P_L constant. Volume rates of flow, pressures, and the volumes of fluid subjected to those pressures are indicated in the figure. The mass of the spool and appurtenances is m, the damping constant of the damping force acting on the moving parts is b, and the effective bulk modulus of oil is β. The accumulator volume is V_a. The flow

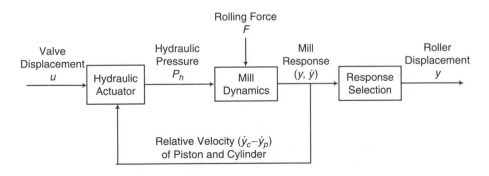

FIGURE 9.63
Block diagram for the hydraulic control system of a steel rolling mill.

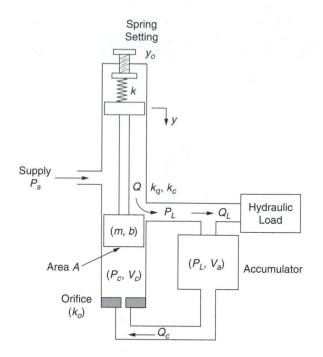

FIGURE 9.64
A single-stage pressure control valve.

into the valve chamber (volume V_c) is through an orifice. This flow may be taken as proportional to the pressure drop across the orifice, the constant of proportionality being k_o. A compressive spring of stiffness k restricts the spool motion. The initial spring force is set by adjusting the initial compression y_o of the spring.

1. Identify the reference input, the primary output, and a disturbance input for the valve system.
2. By making linearization assumptions and introducing any additional parameters that might be necessary, write equations to describe the system dynamics.
3. Set up a block diagram for the system, showing various transfer functions.

SOLUTION
Part 1:

Input setting $= y_o$
Primary response (controlled variable) $= P_L$
Disturbance input $= Q_L$

Part 2: Suppose that the valve displacement y is measured from the static equilibrium position of the system. The equation of motion for the valve spool device is

$$m\ddot{y} = -b\dot{y} - k(y - y_o) + A(P_s - P_c) \tag{i}$$

The flow through the chamber orifice is given by

$$Q_c = k_o(P_L - P_c) = -A\frac{dy}{dt} + \frac{V_c}{\beta}\frac{dP_c}{dt} \tag{ii}$$

The outflow Q from the spool port increases with y and decreases with the pressure drop $(P_L - P_s)$. Hence the linearized flow equation is

$$Q = k_q y - k_c(P_L - P_s)$$

Note that k_q and k_c are positive constants, defined previously by Equation 9.106 and Equation 9.107.

The accumulator equation is

$$Q - Q_c - Q_L = \frac{V_a}{\beta}\frac{dP_L}{dt}$$

Substituting for Q and Q_c, we have

$$k_q y - k_c(P_L - P_s) - k_o(P_L - P_c) - Q_L = \frac{V_a}{\beta}\frac{dP_L}{dt}$$

or

$$k_q y - (k_c + k_o)P_L + k_c P_s + k_o P_c - Q_L = \frac{V_a}{\beta}\frac{dP_L}{dt} \tag{iii}$$

The equations of motion are equations i through iii.

Part 3: Using equations i through iii, the block diagram shown in Figure 9.65 can be obtained. Note in particular the feedback path of load pressure P_L. This feedback is responsible for the pressure control characteristic of the valve. Another point to note is the presence of the derivative operator "s" in the block "As" as drawn in Figure 9.65. This operator is not physically realizable. In fact it is possible to recast the block diagram so that it does not contain a derivative operator.

9.11.1 Feedback Control

In Figure 9.61(a), we have identified two "natural" feedback paths that are inherent in the dynamics of the open-loop hydraulic control system. In Figure 9.61(b), we have shown the time constants associated with these natural feedback modules. Specifically, we observe the following:

1. A *pressure feedback path* and an associated *hydraulic time constant* τ_h
2. A *velocity feedback path* and an associated *mechanical time constant* τ_m

The hydraulic time constant is determined by the compressibility of the fluid. The larger the bulk modulus of the fluid, the smaller the compressibility. This results in a smaller hydraulic time constant. Furthermore, τ_h increases with the volume of the fluid in the actuator chamber, hence, this time constant is related to the capacitance of the fluid as well. The mechanical time constant has its origin in the inertia and the energy dissipation

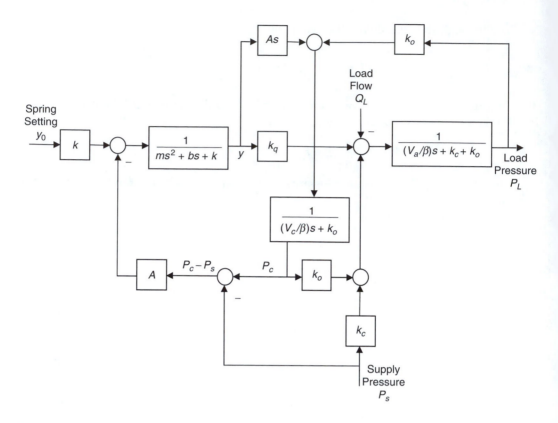

FIGURE 9.65
Block diagram for the single-stage pressure control valve.

(damping) in the moving parts of the actuator. As expected, the actuator becomes more sluggish as the inertia of the moving parts increases, resulting in an increased mechanical time constant.

These natural feedback paths usually provide a stabilizing effect to a hydraulic control system, but they are not adequate for satisfactory operation of the system. In particular, the position of the actuator is provided by an integrator (see Figure 9.61). In open-loop operation, the position response will steadily grow and will display an unstable behavior, in the presence of a slightest disturbance. Furthermore, the speed of response, which usually conflicts with stability, has to be adequate for proper performance. Consequently, it is necessary to include feedback control into the system. This is accomplished by measuring the response variables, and modifying the system inputs using them, according to some control law.

Schematic representation of a computer-controlled hydraulic system is shown in Figure 9.66. In addition to the motion (both position and speed) of the mechanical load, it is desirable to sense the pressures on the two sides of the piston of the hydraulic actuator, for feedback control. There are numerous laws of feedback control, which may be programmed into the control computer (see Chapter 12). Many of the conventional methods implement a combination of the following three basic control actions:

1. Proportional control (P)
2. Derivative control (D)
3. Integral control (I)

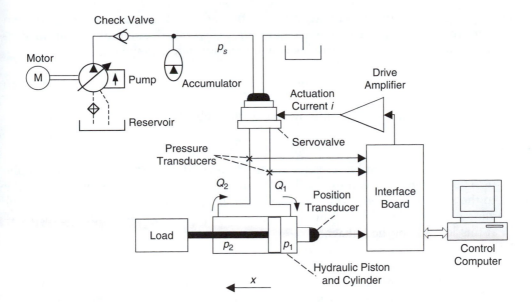

FIGURE 9.66
A computer controlled hydraulic system.

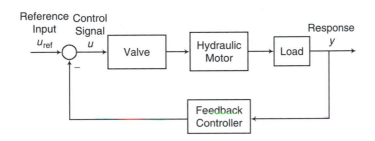

FIGURE 9.67
A closed-loop hydraulic control system.

In proportional control, the measured response (or, response error) is used directly in the control action. In derivative control, the measured response (or, the response error) is differentiated before it is used in the control action. Similarly, in integral control, the response error is integrated and used in the control action. Modification of the measured responses to obtain the control signal is done in many ways, including by electronic, digital, and mechanical means. For example, an analog hardware unit (termed a compensator or controller), which consists of electronic circuitry may be employed for this purpose. Alternatively, the measured signals, if they are analog, may be digitized and subsequently modified in a required manner through digital processing (multiplication, differentiation, integration, addition, etc.). This is the method used in digital control; either hardware control or software control may be used. What is represented in Figure 9.66 is the software approach.

Consider the feedback (closed-loop) hydraulic control system shown by the block diagram in Figure 9.67. In this case, a general controller is located in the feedback path, and its control law may be written as

$$u = u_{\text{ref}} - f(y)$$ (9.138a)

where $f(y)$ denotes the modifications made to the measured output y in order to form the control (error) signal u. The reference input u_{ref} is specified. Alternatively, if the controller is located in the forward path, as usual, the control law may be given by

$$u = f(u_{ref} - y) \tag{9.138b}$$

Mechanical components may be employed as well to obtain a robust control action.

Example 9.18

A mechanical linkage is employed as the feedback device for a servovalve of a hydraulic actuator. The arrangement is illustrated in Figure 9.68(a). The reference input is u_{ref}, the input to the servovalve is u, and the displacement (response) of the actuator piston is y. A coupling element is used to join one end of the linkage to the piston rod. The displacement at this location of the linkage is x.

Show that rigid coupling gives proportional feedback action (Figure 9.68(b)). Now, if a viscous damper (damping constant b) is used as the coupling element and if a spring (stiffness k) is used to externally restrain the coupling end of the linkage (Figure 9.68(c)), show that the resulting feedback action is a lead compensation (see Chapter 12). Next, if the damper and the spring are interchanged (Figure 9.68(d)), what is the resulting feedback control action?

SOLUTION

For all three cases of coupling, the relationship between u_{ref}, u, and x is the same. To derive this, we introduce the variable θ to denote the clockwise rotation of the linkage. With the linkage dimensions h_1 and h_2 defined as shown in Figure 9.68(a), we have

$$u = u_{ref} + h_1\theta$$

$$x = u_{ref} - h_2\theta$$

Now, by eliminating θ, we get

$$u = (r+1)u_{ref} - rx \tag{i}$$

where

$$r = h_1/h_2 \tag{ii}$$

For rigid coupling (Figure 9.68(b)),

$$y = x$$

Hence, from Equation i, we have

$$u = (r+1)u_{ref} - ry \tag{9.139}$$

Clearly, this is a proportional feedback control law.

Next, for the coupling arrangement shown in Figure 9.68(c), by equating forces in the spring and the damper, we get

$$kx = b(\dot{y} - \dot{x}) \tag{iii}$$

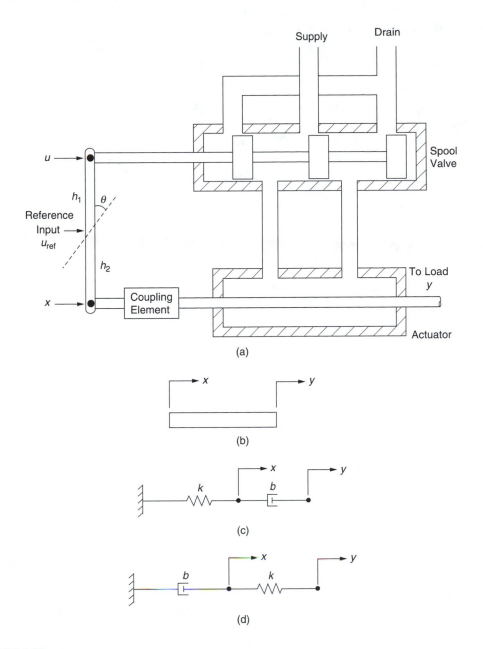

FIGURE 9.68
(a) A servovalve and actuator with mechanical feedback, (b) Rigid coupling (proportional feedback), (c) Damper-spring coupling (lead compensator), (d) Spring-damper coupling (lag compensator).

Introducing the Laplace variable s, we have the transfer-function relationship corresponding to Equation iii:

$$x = \frac{bs}{bs+k}y \tag{iv}$$

By substituting Equation iv in Equation i, we get

$$u = (r+1)u_{\text{ref}} - \frac{rbs}{bs+k}y \tag{9.140}$$

Note that the feedback transfer function

$$G_c(s) = \frac{rbs}{bs + k} \qquad (9.141)$$

is a lead compensator (as discussed in Chapter 12), because the numerator provides a pure derivative action.

Finally, for the coupling arrangement shown in Figure 9.68(d), we have

$$b\dot{x} = k(y - x) \qquad (v)$$

The corresponding transfer-function relationship is

$$x = \frac{k}{bs + k} y \qquad (vi)$$

By substituting Equation vi in Equation i, we get the transfer-function relationship for the feedback controller as

$$u = (r + 1)u_{ref} - \frac{rk}{bs + k} y \qquad (9.142)$$

In this case the feedback transfer function is

$$G_c(s) = \frac{rk}{bs + k} \qquad (9.143)$$

This is clearly a lag compensator (as discussed in Chapter 12), because the denominator dynamics of the transfer function provide the lag action and the numerator has no dynamics (independent of s).

———

Fluid power systems in general and hydraulic systems in particular are nonlinear. Nonlinearities have such origins as nonlinear physical relations of the fluid flow, compressibility, nonlinear valve characteristics, friction in the actuator (at the piston rings, which slide inside the cylinder) and the valves, unequal piston areas on the two sides of the actuator piston, and leakage. As a result accurate modeling of a fluid power system is difficult. A linear model will not represent the correct situation except near a small operating region. This situation may be addressed by using an accurate nonlinear model or a series of linear models for different operating regions. In either case, linear control laws (e.g., proportional, integral, and derivative actions) may not be adequate. This problem can be further exacerbated by factors such as parameter variations, unknown disturbances, and noise.

Many advanced control techniques have been applied to fluid power systems, in view of the limitations of such classical control techniques as PID. In one approach, observer is used to estimate velocity and friction in the actuator, and a controller is designed to compensate for friction. Adaptive control is another advanced approach used in hydraulic control systems. In model-referenced adaptive control, for example, the controller pushes the behavior of the hydraulic system towards a reference model. The reference model is designed to display the desired behavior of the physical system. Frequency-domain control techniques such as H-infinity control (H_∞ control) and quantitative feedback theory (QFT) where the system transfer function is shaped to realize a desired performance, have been studied. They are linear control techniques, which may not work perfectly when applied to a nonlinear system. Impedance control has been studied as well, with respect to hydraulic control systems. In impedance control, the objective is to realize a desired impedance function (Note: Impedance = Force/Velocity, in the frequency domain) at the output of the control system, by manipulating the controller. These advanced techniques are beyond the scope of the present introductory treatment.

9.11.2 Constant-Flow Systems

So far, we have discussed only *valve-controlled* hydraulic actuators. There are two types of valve-controlled systems:

1. Constant-pressure systems
2. Constant flow systems

Since there are four flow paths for a four-way spool valve, an analogy can be drawn between a spool-valve controlled hydraulic actuator with a Wheatstone bridge circuit (see Chapter 4), as shown in Figure 9.69. Each arm of the bridge corresponds to a flow (or, current) path. As usual, P denotes pressure, which is an across variable analogous to voltage (see Chapter 2); and Q denotes the volume flow rate, which is a through variable analogous to current. The four fluid resistors R_i represent the resistances experienced by the fluid flow in the four paths of the valve. Note that these are variable resistors whose variation is governed by the spool movement (and hence the current of the valve actuator). When the spool moves to one side of the neutral (center) position, two of the resistors (say, R_2 and R_4) change due to the port opening, and the remaining two resistors represent the leakage resistances (see Figure 9.58). The reverse is true when the spool moves in the opposite direction from the neutral position. The flow through the actuator is represented by a load resistance R_L, which is connected across the bridge.

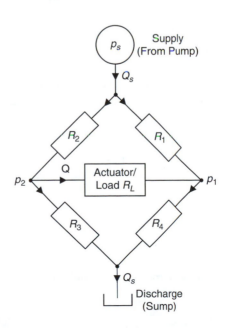

FIGURE 9.69
The bridge circuit representation of a four-way valve and an actuator load.

In our discussion thus far, we have considered only the *constant pressure system*, in which the supply pressure P_s to the servovalve is maintained constant, but the corresponding supply flow rate Q_s is variable. This system is analogous to a constant-voltage bridge (see Chapter 4). In a *constant-flow system*, the supply flow Q_s is kept constant, but the corresponding pressure P_s is variable. This system is analogous to a constant-current Wheatstone bridge. Constant-flow operation requires a constant-flow pump, which may be more economical than a variable-flow pump. But, it is easier to maintain a constant pressure level by using a pressure regulator and an accumulator. As a result, constant pressure systems are more commonly used in practical applications.

Valve-controlled hydraulic actuators are the most common type used in industrial applications. They are particularly useful when more than one actuator is powered by the same hydraulic supply. Pump-controlled actuators are gaining popularity, and are outlined next.

9.11.3 Pump-Controlled Hydraulic Actuators

Pump-controlled hydraulic drives are suitable when only one actuator is needed to drive a process. A typical configuration of a pump-controlled hydraulic drive system is shown in Figure 9.70. A variable-flow pump is driven by an electric motor (typically, an ac motor). The pump feeds a hydraulic motor, which in turn drives the load. Control is provided by the flow control of the pump. This may be accomplished in several ways—for example, by controlling the pump stroke (see Figure 9.56) or by controlling the pump speed using a frequency-controlled ac motor. Typical hydraulic drives of this type can provide positioning errors less than $1°$ at torques in the range 25–250 N·m.

9.11.4 Hydraulic Accumulators

Since hydraulic fluids are quite incompressible, one way to increase the hydraulic time constant is to use an accumulator. An accumulator is a tank, which can hold excessive fluid during pressure surges and release this fluid to the system when the pressure slacks. In this manner, pressure fluctuations can be filtered out from the hydraulic system and the pressure can be stabilized. There are two common types of hydraulic accumulators:

1. Gas-charged accumulators
2. Spring-loaded accumulators

In a gas-charged accumulator, the top half of the tank is filled with air. When high-pressure liquid enters the tank, the air compresses, making room for the incoming liquid.

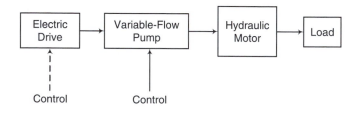

FIGURE 9.70
Configuration of a pump-controlled hydraulic drive system.

In a spring-loaded accumulator, a movable piston, restrained from the top of the tank by a spring, is used in place of air. The operation of these two types of accumulators is quite similar. By connecting the pump (hydraulic or pneumatic) to an accumulator, the flow into the servovalve can be stabilized and the excess energy can be stored for later use. This minimizes undesirable pressure pulses, vibration, and fatigue loading.

9.11.5 Pneumatic Control Systems

Pneumatic control systems operate in a manner similar to hydraulic control systems. Pneumatic pumps, servovalves, and actuators are quite similar in design to their hydraulic counterparts. The basic differences include the following:

1. The working "fluid" is air, which is far more compressible than hydraulic oils. Hence, thermal effects and compressibility should be included in any meaningful analysis.
2. The outlet of the actuator and the inlet of the pump are open to the atmosphere (No reservoir tank is needed for the working fluid).

Hydraulic systems are stiffer and usually employed in heavy-duty control tasks, whereas pneumatic systems are particularly suitable for medium to low-duty tasks (supply pressures in the range of 500 kPa to 1 MPa). Pneumatic systems are more nonlinear and less accurate than hydraulic systems. Since the working fluid is air and since regulated high-pressure air lines are available in most industrial facilities and laboratories, pneumatic systems tend to be more economical than hydraulic systems. Also, pneumatic systems are more environmentally friendly and cleaner, and the fluid leakage does not cause a hazardous condition. But, they lack the self-lubricating property of hydraulic fluids. Furthermore, atmospheric air has to be filtered and any excesses moisture removed before compressing. Heat generated in the compressor has to be removed as well.

Both hydraulic and pneumatic control loops might be present in the same control system. For example, in a manufacturing workcell, hydraulic control can be used for parts transfer, positioning, and machining operations, and pneumatic control can be used for tool change, parts grasping, switching, ejecting, and single-action cutting operations. In a fish processing machine, servo-controlled hydraulic actuators have been used for accurately positioning the cutter while pneumatic devices have been used for grasping and chopping of fish. We will not extend our analysis of hydraulic systems to include air as the working fluid. The reader may consult a book on pneumatic control for information on pneumatic actuators and valves.

9.11.6 Flapper Valves

Flapper valves, which are relatively inexpensive and operate at low-power levels, are commonly used in pneumatic control systems. This does not rule them out for hydraulic control applications, however, where they are popular in pilot valve stages. A schematic diagram of a single-jet flapper valve used in a piston-cylinder actuator is shown in Figure 9.71. If the nozzle is completely blocked by the flapper, the two pressures P_1 and P_2 will be equal, balancing the piston. As the clearance between the flapper and the nozzle increases, the pressure P_1 drops, thus creating an imbalance force on the piston of

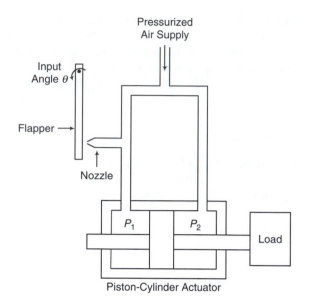

FIGURE 9.71
A pneumatic flapper valve system.

the actuator. For small displacements, a linear relationship between the flapper clearance and the imbalance force can be assumed.

Note that the operation of a flapper valve requires fluid leakage at the nozzle. This does not create problems in a pneumatic system. In a hydraulic system, however, this not only wastes power but also wastes hydraulic oil and creates a possible hazard, unless a collecting tank and a return line to the oil reservoir are employed. For more stable operation, double-jet flapper valves should be employed. In this case, the flapper is mounted symmetrically between two jets. The pressure drop is still highly sensitive to flapper motion, potentially leading to instability. To reduce instability problems, pressure feedback, using a bellows unit, can be employed.

A two-stage servovalve with a flapper stage and a spool stage is shown in Figure 9.72. Actuation of the torque motor moves the flapper. This changes the pressure in the two nozzles of the flapper, in opposite directions. The resulting pressure difference is applied across the spool, which is moved as a result, which in turn moves the actuator as in the case of a single-stage spool valve. In the system shown in Figure 9.72, there is a feedback mechanism as well, between the two stages of valve. Specifically, as the spool moves due to the flapper movement caused by the torque motor, the spool carries the flexible end of the flapper in the opposite direction to the original movement. This creates a back pressure in the opposite direction. Hence, this valve system is said to possess *force feedback* (more accurately, *pressure feedback*).

In general, a multistage servovalve uses several servovalves in series to drive a hydraulic actuator. The output of the first stage becomes the input to the second stage. As noted before, a common combination is a hydraulic flapper valve and a hydraulic spool valve, operating in series. A multistage servovalve is analogous to a multistage amplifier. The advantages of multistage servovalves are:

1. A single-stage servovalve will saturate under large displacements (loads). This may be overcomed by using several stages, with each stage being operated in

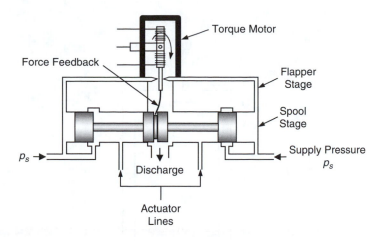

FIGURE 9.72
A two-stage servovalve with pressure feedback.

its linear region. Hence, a large operating range (load variations) is possible without introducing excessive nonlinearities, particularly saturation.

2. Each stage will filter out high-frequency noise, giving a lower overall noise-to-signal ratio.

The disadvantages are:

1. They cost more and are more complex than single-stage servovalves.
2. Because of series connection of several stages, failure of one stage will bring about failure of the overall system (a reliability problem).
3. Multiple stages will decrease the overall bandwidth of the system (i.e., lower speed of response).

Example 9.19

Draw a schematic diagram to illustrate the incorporation of pressure feedback, using a bellows, in a flapper-valve pneumatic control system. Describe the operation of this feedback control scheme, giving the advantages and disadvantages of this method of control.

SOLUTION
One possible arrangement for external pressure feedback in a flapper valve is shown in Figure 9.73. Its operation can be explained as follows: If pressure P_1 drops, the bellows will contract, thereby moving the flapper closer to the nozzle, thus increasing P_1. Hence, the bellows acts as a *mechanical feedback* device, which tends to regulate pressure disturbances. The advantages of such a device are:

1. It is a simple, robust, low-cost mechanical device.
2. It provides mechanical feedback control of pressure variations.

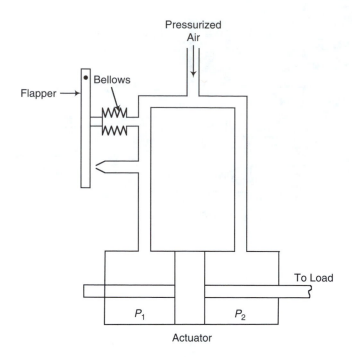

FIGURE 9.73
External pressure feedback for a flapper valve, using a bellows.

The disadvantages are:

1. It can result in a slow (i.e., low-bandwidth) system, if the inertia of the bellows is excessive.
2. It introduces a time delay, which can have a destabilizing effect, particularly at high frequencies.

9.11.7 Hydraulic Circuits

A typical hydraulic control system consists of several components such as pumps, motors, valves, piston-cylinder actuators, and accumulators, which are interconnected through piping. It is convenient to represent each component with a standard graphic symbol. The overall system can be represented by a circuit diagram where the symbols for various components are joined by lines to denote flow paths. Circuit representations of some of the many hydraulic components are shown in Figure 9.74. A few explanatory comments would be appropriate. The inward solid pointers in the motor symbols indicate that a hydraulic motor receives hydraulic energy. Similarly, the pointers in the pump symbols show that a hydraulic pump gives out hydraulic energy. In general, the arrows inside a symbol show fluid flow paths. The external spring and arrow in the relief valve symbol shows that the unit is adjustable and spring restrained. There are three basic types of hydraulic line symbols. A solid line indicates a primary hydraulic flow. A broken line with long dashes is a *pilot line*, which indicates that the outward component is controlled. For example, the broken line in the relief valve symbol indicates that the valve is controlled by pressure. A broken line with short dashes represents a drain line or leakage flow. In the spool valve symbols, P denotes the supply port (with pressure P_s) and T denotes the discharge port to the reservoir

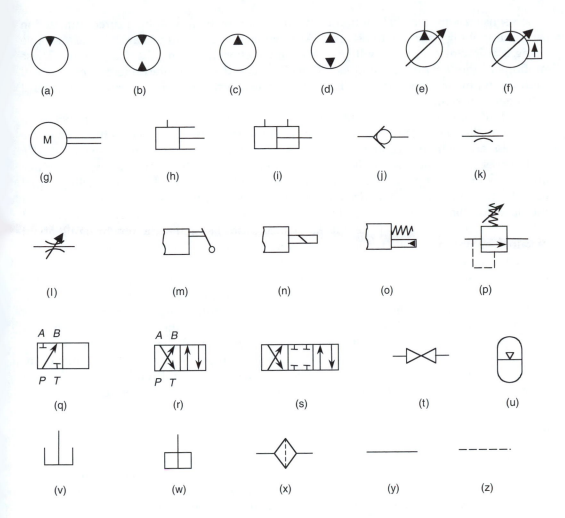

FIGURE 9.74
Standard graphic symbols used in hydraulic circuit diagrams: (a) motor; (b) reversible motor; (c) pump; (d) reversible pump; (e) variable displacement pump; (f) pressure-compensated variable displacement pump; (g) electric motor; (h) single-acting cylinder; (i) double-acting cylinder; (j) ball-and-seat check valve; (k) fixed orifice; (l) variable flow orifice; (m) manual valve; (n) solenoid-actuated valve; (o) spring centered pilot-controlled valve; (p) relief valve (adjustable and pressure-operated); (q) two-way spool valve; (r) four-way spool valve; (s) three-position four-way valve; (t) manual shut-off valve; (u) accumulator; (v) vented reservoir; (w) pressurized reservoir; (x) filter; (y) main fluid line; (z) pilot line.

(with gage zero pressure). Finally, note that ports *A* and *B* of a four-way spool valve are connected to the two ports of a double-acting hydraulic cylinder (see Figure 9.58(a)).

9.12 Fluidics

The term *fluidics* is derived probably form *fluid logic* or perhaps *fluid electronics*. In *fluidic control systems*, the basic functions such as sensing, signal conditioning, and control are accomplished by the interaction of streams of fluid (liquid or gas). Unlike in mechanical systems, no moving parts are used in fluidic devices to accomplish these tasks. Of course, when sensing a mechanical motion by a fluidic sensor there will be a direct interaction with the moving object that is sensed. Also, when actuating a mechanical load or valve using a fluidic device, there will be a direct interaction with a mechanical motion. These motions of the *input devices* and *output devices* should not be interpreted as mechanical motions within a fluidic device, but rather, mechanical motions external to the fluid interactions therein.

The fluidics technology was first introduced by the U.S. Army engineers in 1959 as a possible replacement for electronics in some control systems. The concepts themselves are quite old, and perhaps originated through electrical-hydraulic analogies (see Chapter 2); where pressure is an *across variable* analogous to voltage, and flow rate is a *through variable* analogous to current. Since electronic circuitry is widely used for tasks of sensing, signal conditioning, and control in hydraulic and pneumatic control systems, it was thought that the need for conversion between fluid-flow and electrical variables could be avoided if fluidic devices were used for these tasks in such control systems, thereby bringing about certain economic and system-performance benefits. Furthermore, fluidic devices are considered to have high reliability and can be operated in hostile environments (e.g., corrosive, radioactive, shock and vibration, high-temperature) more satisfactorily than electronic devices can. But due to rapid advances in digital electronics with associated gains in performance and versatility, and reduction in cost, the anticipated acceptance of fluidics was not actually materialized in the 1960s and 1970s. Some renewed interest in fluidics was experienced in the 1980s, with applications primarily in the aircraft, aerospace, manufacturing, and process control industries. This section provides a brief introduction to the subject of fluidics. In view of its analogy to electronics and the use in mechanical control, fluidics is a topic that is quite relevant to the subject of mechatronics.

9.12.1   Fluidic Components

Since fluidics was intended as a substitute for electronics, particularly in hydraulic and pneumatic control systems (i.e., in fluid power control systems), it is not surprising that much effort has gone into development of fluidic devices that are analogous to electronic devices. Naturally, two types of fluidic components were developed:

1. Analog fluidic components for analog systems
2. Digital fluidic components for logic circuits

Examples of analog components, which have been developed for fluidic systems are fluidic position sensor, fluidic rate sensor, fluidic accelerometer, fluidic temperature sensor, fluidic oscillator, fluidic resistor, vortex amplifier, jet-deflection amplifier, wall-attachment amplifier, fluidic summing amplifier, fluidic actuating amplifier, and fluidic modulator. A description of all such devices is beyond the scope of this book. Instead, we will describe one or two representative devices in order to introduce the nature of fluidic components.

9.12.1.1   Logic Components

Examples of digital fluidic components are switches, flip-flops, and logic gates (see Chapter 10 for electronic logic). Complex logic systems can be assembled by interconnecting these basic elements. As an example consider the fluidic AND gate shown in Figure 9.75. The control inputs u_1 and u_2 represent the "presence" or "absence" of the high-speed fluid streams applied to the corresponding ports of the device. When only one control stream is present it will pass through the drain channel aligned with it, due to the entrainment capability of the stream. When both control streams are present, there

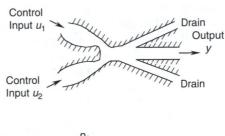

FIGURE 9.75
Fluidic AND gate.

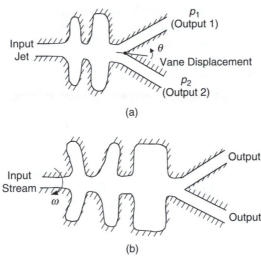

FIGURE 9.76
Fluidic motion sensors. (a) Angular displacement sensor; (b) Laminar angular speed sensor.

will be an interaction between the two streams thereby producing a sizeable output stream *y*. Hence, there is an "AND" relationship between the output *y* and the inputs u_1 and u_2.

Operation of the fluid logic components depends on the wall-attachment phenomenon (or, *coanda effect*). According to this phenomenon, a jet of fluid applied toward a wall will tend to "attach" itself to the wall. If two walls are present, the jet will be attached to one of the walls depending on the conditions at the exit of the jet and the angles, which the walls make with the jet. Hence a switching action (i.e., attachment to one wall or the other) is created. The corresponding switching state can be considered a digital output.

A measure of the capability of a digital device is its "fan-out." This is the number of similar devices that can be driven (or controlled) by the same digital component. Fluidic components have a reasonably high fan-out capability.

9.12.1.2 *Fluidic Motion Sensors*

A fluidic displacement sensor can be developed by using a mechanical vane to split a stream of incoming flow into two output streams. This is shown in Figure 9.76(a). When the vane is centrally located, the displacement is zero ($\theta = 0$). Under these conditions the pressure is the same at both output streams, the differential pressure $p_2 - p_1$ being zero. When the vane is not symmetrically located (corresponding to a nonzero displacement), the output pressures will be unequal. The differential pressure $p_2 - p_1$ will provide both magnitude and direction of the displacement θ.

A fluidic angular speed sensor is shown in Figure 9.76(b). The nozzle of the input stream is rotated at angular speed ω, which is to be measured. When $\omega = 0$ the stream will travel straight in the axial direction of the input stream. When $\omega \neq 0$ the fluid particles emitting from the nozzle will have a transverse speed (due to rotation) as well as an axial speed due to jet flow. Hence the fluid particles will be deflected from the original (axial) path. This deflection will be the cause of a pressure change at the output. Hence the output pressure change can be used as a measure of the angular speed.

There are many other ways to sense speed using a fluidic device. One type of fluidic angular speed sensor uses the *vortex flow* principle. In this sensor, the angular speed of the input device (object) is imparted on the fluid entering a vortex chamber at the periphery. In this manner a tangential speed is applied to the fluid particles, which move radially from the periphery toward the center of the vortex chamber. The resulting vortex flow will be such that the tangential speed becomes larger as the particles approach the center of the chamber. (This follows by the conservation of momentum). Consequently, a pressure drop is experienced at the output (i.e., center of the chamber). The higher the angular speed imparted to the incoming fluid, the larger the pressure drop at the output. Hence the output pressure drop can be used as a measure of angular speed.

An angular speed sensor that is particularly useful in pneumatic systems is the wobble-plate sensor. This is a flapper valve type device with two differential nozzles facing a wobble plate. The supply pressure is maintained constant. There are two output ports corresponding to the two nozzles. As the wobble plate rotates, the proximity of the plate to each nozzle will change periodically. The resulting fluctuation in the differential pressure can be used as a measure of the plate speed.

9.12.1.3 Fluidic Amplifiers

Fluidic amplifiers apply a "gain" in pressure, flow, or power to a fluidic circuit. The corresponding amplifiers are analogous to voltage, current, and power amplifiers used in electronic circuits (see Chapter 4).

Many designs of fluidic amplifiers are available. Consider the *jet-deflection amplifier* shown in Figure 9.77. When the control input pressures p_{u1} and p_{u2} are equal, the supply stream will pass through the amplifier with a symmetric flow. In this case the output pressures p_{y1} and p_{y2} will be equal. When $p_{u1} \neq p_{u2}$ the fluid stream will be deflected to one side due to the nonzero differential pressure $\Delta p_u = p_{u1} - p_{u2}$. As a result, a nonzero

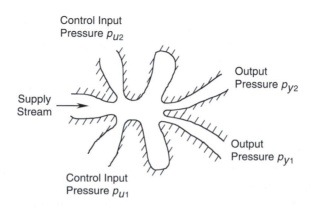

FIGURE 9.77
Jet deflection amplifier.

differential pressure $\Delta p_y = p_{y2} - p_{y1}$ will be created at the output. The *pressure gain* of the amplifier is given by

$$K_p = \frac{\Delta p_y}{\Delta p_u} \tag{9.144}$$

The gain K_p will be constant in a small operating range.

9.12.2 Fluidic Control Systems

A fluidic control system is a control system that employs fluidic components to perform one or more functions such as sensing, signal conditioning, and control. The actuator is a symmetric configuration of a hydraulic piston-cylinder (ram) device. It is controlled using a pair of spool valves. Control signals to the spool valves are generated by appropriate fluidic components. Specifically, position of the load is measured using a fluidic displacement sensor, and the speed of the load is measured using a fluidic speed sensor. These signals are properly conditioned using fluidic amplifiers, compared with a reference signal using a fluidic summing amplifier, and the error signal is used through a fluidic interface amplifier to operate the actuator valve. This type of fluidic control systems is useful in flight control, to operate the control surfaces (ailerons, rudders, and elevators) of an aircraft, particularly as a backup (e.g., for emergency maneuvering).

9.12.2.1 Interfacing Considerations

The performance of some of the early designs of fluidic control systems was disappointing because the overall control system did not function as expected while the individual fluidic components separately would function very well. Primary reason for this was the dynamic interactions between components and associated loading and impedance problems (see Chapter 4). Early designs of fluidic components, amplifiers in particular, did not have sufficient input impedances. Furthermore, output impedances were found to be higher than what was desired in order to minimize dynamic interaction problems. Much research and development effort has gone into improving the impedance characteristics of fluidic devices. Also, leakage problems are unavoidable when separate fluidic components are connected together using transmission lines. Modular laminated construction of fluidic systems has minimized these problems.

9.12.2.2 Modular Laminated Construction

Even though not to be extent that the integrated construction of electronic circuits has revolutionized the electronic technology (see Chapter 10), the modular laminated (or *integrated*) construction of fluidic systems has made a reasonable impact on the fluidic technology. The first-generation fluidic components were machined out of metal blocks or moldings. Precise duplication and quality control in mass production were difficult with these types of components. Furthermore, the components were undesirably bulky. Since individual components were joined using flexible tubing, leakage at joints presented serious problems. Since the design of a fluidic control system is often a trial-and-error process of trying out different components, system design was costly, time-consuming, and tiresome, particularly because of component costs and assembly difficulties.

Many of these problems are eliminated or reduced with the modern-day modular design of fluidic systems using component *laminates*. Individual fluidic components are precisely manufactured as thin laminates using a sophisticated stamping process. The system assembly is done by simply stacking and bonding together of various laminates (e.g., sensors, amplifiers, oscillators, resistors, modulators, vents, exhausts, drains, and gaskets) to form the required fluidic circuit. In the design stage, the stacks are clamped together without permanently bonding, and are tested. Then, design modifications can be implemented simply and quickly by replacing one or more of the laminates. Once an acceptable design is obtained, the stack is permanently bonded.

9.12.3   Applications of Fluidics

Fluidic components and systems have the advantages of small size and no moving parts, when compared with conventional mechanical systems, which typically use bulky gear systems, clutches, linkages, cables, and chains. Furthermore, fluidic systems are highly reliable and are preferable to electronic systems, in hostile environments of explosives, chemicals, extreme (high and low) temperatures, radiation, shock, vibration, and electromagnetic interference. For these reasons, fluidic control systems have received a renewed interest in aircraft and aerospace applications, particularly as backup systems. In hydraulic and pneumatic control applications, the use of fluidics in place of electronics will avoid the need for conversion between hydraulic/pneumatic signals and electrical signals, which could result in substantial cost benefits and reduction in the physical size.

Present-day fluidic components can provide high input impedances, low output impedances, and high gains comparable to those from typical electronic components. These fluidic components can provide bandwidths in the kHz range. Good dynamic range and resolution capabilities are available as well.

In addition to aircraft and aerospace flight control applications, fluidic devices are used in ground transit vehicles, heavy-duty machinery, machine tools, industrial robots, medical equipment, and process control systems. Specific examples include valve control for hydraulic/pneumatic actuated robotic joints and end effectors, windshield-wiper and windshield-washer controls for automobiles, respirator and artificial heart pump control, backup control devices for aircraft control surfaces, controllers for pneumatic power tools, control of food-packaging (e.g., bottle filling) devices, counting and timing devices for household appliances, braking systems, and spacecraft sensors. Fluidics will not replace electronics in a majority of control systems. But, there are significant advantages to using fluidics in some critical applications.

9.13 Problems

9.1 What factors generally govern (a) The electrical time constant (b) The mechanical time constant of a motor? Compare typical values for these parameters and discuss how they affect the motor response.

9.2 Write an expression for the back e.m.f. of a dc motor. Show that the armature circuit of a dc motor may be modeled by the equation

$$v_a = i_a R_a + k\phi \omega_m$$

where

   v_a = armature supply voltage
   i_a = armature current
   R_a = armature resistance
   ϕ = field flux
   ω_m = motor speed

and k is a motor constant. Suppose that $v_a = 20$ V DC. At standstill, $i_a = 20$ A. When running at a speed of 500 rpm, the armature current was found to be 15 A. If the speed is increased to 1000 rpm while maintaining the field flux constant, determine the corresponding armature current.

9.3 In equivalent circuits for dc motors, iron losses (e.g., eddy current losses) in the stator are usually neglected. A way to include these effects is shown in Figure P9.3. Iron losses in the stator poles are represented by a circuit with resistance R_e and self-inductance L_e. The mutual inductance between the field circuit and the iron loss circuit is denoted by M. Note that

$$M = k\sqrt{L_f L_e}$$

where L_f is the self-inductance in the field circuit and k denotes a coupling constant. With perfect coupling (no flux leakage between the two circuits), we have $k = 1$. But usually, k is less than 1. The circuit equations are

$$v_f = R_f i_f + L_f \frac{di_f}{dt} - M \frac{di_e}{dt}$$

$$0 = R_e i_e + L_e \frac{di_e}{dt} - M \frac{di_f}{dt}$$

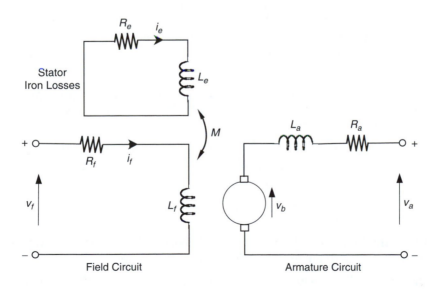

FIGURE P9.3
Equivalent circuit for a separately excited dc motor with iron losses in the stator.

The parameters and variables are defined in Figure P9.3. Obtain the transfer function for i_f/v_f. Discuss the case $k = 1$ in reference to this transfer function. In particular, show that the transfer function has a "phase lag" effect.

9.4 Explain the operation of a brushless dc motor. How does it compare with the principle of operation of a stepper motor?

9.5 Give the steady-state torque-speed relations for a dc motor with the following three types inter connections for the armature and filed windings:

   a. A shunt-wound motor

   b. A series-wound motor

   c. A compound-wound motor.

The following parameter values are given:

   $R_a = 5\ \Omega$, $R_f = 20\ \Omega$, $k = 1\ \text{N} \cdot \text{m/A}^2$, and for a compound-wound motor, $R_{f1} = R_{f2} = 10\ \Omega$. Note that

$$T_m = k i_f i_a$$

Assume that the supply voltage is 115 V. Plot the steady-state torque-speed curves for these types of winding arrangements.

Using these curves, compare the steady state performance of the three types of motors.

9.6 What is the electrical damping constant of a dc motor? Determine expressions for this constant for the three types of dc motor winding arrangements mentioned in Problem 9.5. In which case is this a constant value? Explain how the electrical damping constant could be experimentally determined. How is the "dominant" time constant of a dc motor influenced by the electrical damping constant? Discuss ways to decrease the motor time constant.

9.7 Explain why the transfer-function representation for a separately excited and armature-controlled dc motor is more accurate than that of a field-controlled motor, and still more accurate than those of shunt-wound, series-wound, or compound-wound dc motors. Give a transfer function relation (using the Laplace variable s) for a dc motor where the incremental speed $\delta\omega_m$ is the output, the incremental winding excitation voltage δv_c is the control input and the incremental load torque δT_L on the motor is a disturbance input. Assume that the parameters of the motor model are determined from experimental speed-torque curves for constant excitation voltage.

9.8 Explain the differences between full-wave circuits and half-wave circuits in thyristor (SCR) control of dc motors. For the SCR drive circuit shown in Figure 9.29(a), sketch the armature current time history.

9.9 Using sketches, describe how pulse-width modulation (PWM) effectively varies the average value of the modulated signal. Explain how one could obtain

   a. A zero average

   b. A positive average

   c. A negative average

by pulse-width modulation. Indicate how PWM is useful in the control of dc motors. List the advantages and disadvantages of pulse-width modulation.

9.10 In the chopper circuit shown in Figure 9.29(a), suppose that $L_o = 100$ mH and $v_{ref} = 200$ V. If the worst-case amplitude of the armature current is to be limited to 1 A, determine the minimum chopper frequency.

9.11 Figure P9.11 shows a schematic arrangement for driving a dc motor using a linear amplifier. The amplifier is powered by a dc power supply of regulated voltage v_s. Under a particular condition suppose that the linear amplifier drives the motor at voltage v_m and current i. Assume that the current drawn from the power supply is also i. Give an expression for the efficiency at which the linear amplifier is operating under these conditions. If $v_s = 50$ V, $v_m = 20$ V and $i = 5$ A, estimate the efficiency of operation of the linear amplifier.

9.12 For a dc motor, the starting torque and the no-load speed are known, which are denoted by T_s and ω_o, respectively. The rotor inertia is J. Determine an expression for the dominant time constant of the motor.

9.13 A schematic diagram for the servo control loop of one joint of a robotic manipulator is given in Figure P9.13.

The motion command for each joint of the robot is generated by the controller of the robot in accordance with the required trajectory. An optical (incremental) encoder is used for both position and velocity feedback in each servo loop. Note that for a six-degree-of-freedom robot there will be six such servo loops. Describe the function of each hardware component shown in the figure and explain the operation of the servo loop.

After several months of operation the motor of one joint of the robot was found to be faulty. An enthusiastic engineer quickly replaced the motor with an identical one without realizing that the encoder of the new motor was different. In particular, the original

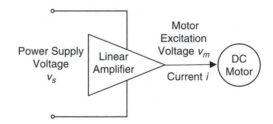

FIGURE P9.11
A linear amplifier for a dc motor.

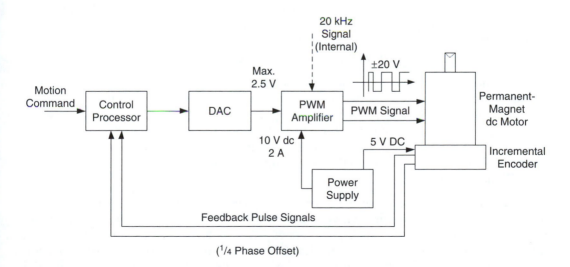

FIGURE P9.13
A servo loop of a robot.

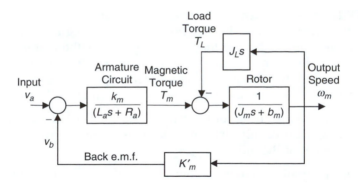

FIGURE P9.14
An armature-controlled dc motor with an inertial load.

encoder generated 200 pulses/rev whereas the new encoder generated 720 pulses/rev. When the robot was operated the engineer noticed an erratic and unstable behavior at the repaired joint. Discuss reasons for this malfunction and suggest a way to correct the situation.

9.14 Consider the block diagram in Figure 9.16, which represents a dc motor, for armature control, with the usual notation. Suppose that the load driven by the motor is a pure inertia element (e.g., a wheel or a robot arm) of moment of inertia J_L, which is directly and rigidly attached to the motor rotor.

   a. Show that, in this case, the motor block diagram may be given as in Figure P9.14. Obtain an expression for the transfer function $\frac{\omega_m}{v_a} = G_m(s)$ for the motor with the inertial load, in terms of the parameters given in Figure P9.14(a)

   b. Now neglect the leakage inductance L_a. Then, show that the transfer function in Part (a) can be expressed as $G_m(s) = \frac{k}{(\tau s+1)}$. Give expressions for τ and k in terms of the given system parameters.

   c. Suppose that the motor (with the inertial load) is to be controlled using position plus velocity feedback. The block diagram of the corresponding control system is given in Figure 9.22, where the motor transfer function $G_m(s) = (k/(\tau s + 1))$. Determine the transfer function of the (closed-loop) control system $G_{CL}(s) = \frac{\theta_m}{\theta_d}$ in terms of the given system parameters (k, k_p, τ, τ_v). Note that θ_m is the angle of rotation of the motor with inertial load, and θ_d is the desired angle of rotation.

9.15 In the joint actuators of robotic manipulators, it is necessary to minimize backlash. Discuss the reasons for this. Conventional techniques for reducing backlash in gear drives include preloading, the use of bronze bearings that automatically compensate for wear, and the use of high-strength steel and other alloys that can be machined accurately and that have minimal wear problems. Discuss the shortcomings of some of the conventional methods of backlash reduction. Discuss the operation of a joint actuator unit that has virtually no backlash problems.

9.16 The moment of inertia of the rotor of a motor (or any other rotating machine) can be determined by a run-down test. With this method, the motor is first brought up to an acceptable speed and then quickly turned off. The motor speed versus time curve is obtained during the run-down period that follows. A typical run-down curve is shown in Figure P9.16. Note that the motor decelerates because of its resisting torque T_r during this period. The slope of the run-down curve is determined at a suitable (operating) value of speed ($\bar{\omega}_m$) in Figure P9.16. Next, the motor is brought up to this speed ($\bar{\omega}_m$), and the

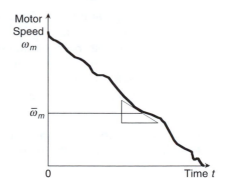

FIGURE P9.16
Data from a run-down test on an
electric motor.

torque ($\overline{T}_r$) that is needed to maintain the motor steady at this speed is obtained (either by direct measurement of torque, or by computing using field current measurement and known value for the torque constant, which is available in the data sheet of the motor). Explain how the rotor inertia J_m may be determined from this information.

9.17 In some types of (indirect-drive) robotic manipulators, joint motors are located away from the joints, and torques are transmitted to the joints through transmission devices such as gears, chains, cables, and timing belts. In some other types of (i.e., direct-drive) manipulators, joint motors are located at the joints themselves, the rotor being integral with one link and the stator being integral with the joining link. Discuss advantages and disadvantages of these two designs.

9.18 In brushless motors, commutation is achieved by switching the stator phases at the correct rotor positions (e.g., at the points of intersection of the static torque curves corresponding to the phases, for achieving maximum average static torque). We have noted that the switching points can be determined by measuring the rotor position using an incremental encoder. Incremental encoders are delicate, cannot operate at high temperatures, costly, and will increase the size and cost of the motor package. Also, precise mounting is required for proper operation. The generated signal may be subjected to electromagnetic interference (EMI) depending on the means of signal transmission. Since we need only know the switching points (i.e., continuous measurement of rotor position is not necessary), and since these points are uniquely determined by the stator magnetic field distribution, a simpler and cost effective alternative to an ecoder for detecting the switching points would be to use Hall effect sensors. Specifically, Hall effect sensors are located at switching points around the stator (forming sensor ring), and a magnet assembly is located around the rotor (In fact, the magnetic poles of the rotor can serve this purpose, without needing an additional set of poles). As the rotor rotates, a magnetic pole on the rotor will trigger an appropriate Hall effect sensor, thereby generating a switching signal (pulse) for commutation at the proper rotor position. A microelectronic switching circuit (or a switching transistor) is actuated by the corresponding pulse. Since Hall effect sensors have several disadvantages—such as hysteresis (and associated nonsymmetry of the sensor signal), low operating temperature ratings (e.g., 125°C), thermal drift problems, and noise due to stray magnetic fields and EMI—it may be more desirable to use fiber optic sensors for brushless commutation. Describe how the fiberoptic method of motor commutation works.

9.19 A brushless dc motor and a suitable drive unit are to be chosen for a continuous drive application. The load has a moment of inertia 0.016 kg · m², and faces a constant resisting torque of 35.0 N · m (excluding the inertia torque) throughout the operation.

The application involves accelerating the load from rest to a speed of 250 rpm in 0.2 sec, maintaining it at this speed for extended periods, and then decelerating to rest in 0.2 sec. A gear unit with step-down gear ratio 4 is to be used with the motor. Estimate a suitable value for the moment of inertia of the motor rotor, for a fairly optimal design. Gear efficiency is known to be 0.8. Determine a value for continuous torque and a corresponding value for operating speed on which a selection of a motor and a drive unit can be made.

9.20 Compare dc motors with ac motors in general terms. In particular, consider mechanical robustness, cost, size, maintainability, speed control capability, and implementation of complex control schemes.

9.21 Compare frequency control with voltage control in induction motor control, giving advantages and disadvantages. The steady-state slip-torque relationship of an induction motor is given by

$$T_m = \frac{aSv_f^2}{[1+(S/S_b)^2]}$$

with the parameter values $a = 1 \times 10^{-3}$ N·m/V^2 and $S_b = 0.25$. If the line voltage $v_f = 241$ V, calculate the breakdown torque. If the motor has two pole pairs per phase and if the line frequency is 60 Hz, what is the synchronous speed (in rpm)? What is the speed corresponding to the breakdown torque? If the motor drives an external load, which is modeled as a viscous damper of damping constant $b = 0.03$ N·m/rad/s, determine the operating point of the system. Now, if the supply voltage is dropped to 163 V through voltage control, what is the new operating point? Is this a stable operating point?

9.22 Consider the induction motor in Problem 9.21. Suppose that the line voltage $v_f = 200$ V and the line frequency is 60 Hz. The motor is rigidly connected to an inertial load. The combined moment of inertia of the rotor and load is $J_{eq} = 5$ kg·m^2. The combined damping constant is $b_{eq} = 0.1$ N·m/rad/s. If the system starts from rest, determine, by computer simulation, the speed time history $\omega_L(t)$ of the load (and motor rotor). (Hint: Assume that the motor is a torque source, with torque represented by the steady-state speed-torque relationship).

9.23 a. The equation of the rotor circuit of an induction motor (per phase) is given by (see Figure 9.40(a))

$$i_r = \frac{Sv}{(R_r + jS\omega_p L_r)} = \frac{v}{(R_r/S + j\omega_p L_r)}$$

which corresponds to an impedance (i.e., voltage/current, in the frequency domain) of

$$Z = R_r/S + j\omega_p L_r.$$

Show that this may be expressed as the sum of two impedance components:

$$Z = [R_r + j\omega_p SL_r] + [(1/S - 1)R_r + j\omega_p(1-S)L_r].$$

For a line frequency of ω_p, this result is equivalent to the circuit shown in Figure 9.40(c). Note that the first component of impedance corresponds to the rotor electrical loss and the second component corresponds to the useful mechanical power.

b. Consider the characteristic shape of the speed versus torque curve of an induction motor. Typically, the starting torque T_s is less than the maximum torque T_{max}, which occurs at a nonzero speed. Explain the main reason for this.

9.24 Prepare a table to compare and contrast the following types of motors:

a. Conventional dc motor with brushes

b. Brushless torque motor (dc)

c. Stepper motor

d. Induction motor

e. AC synchronous motor.

In your table, include terms such as power capability, speed controllability, speed regulation, linearity, operating bandwidth, starting torque, power supply requirements, commutation requirements, and power dissipation. Discuss a practical method for reversing the direction of rotation in each of these types of motors.

9.25 Chopper circuits are used to "chop" a dc voltage so that a dc pulse signal results. This type of signal is used in the control of dc motors, by pulse width modulation (PWM), because the pulse width for a given pulse frequency determines the mean voltage of the pulse signal. Inverter circuits are used to generate an ac voltage from a dc voltage. The switching (triggering) frequency of the inverter determines the frequency of the resulting ac signal. The inverter circuit method is used in the frequency control of ac motors. Both types of circuits use thyristor elements for switching. Either discrete circuit elements or integrated circuit (monolithic) chips may be developed for this purpose. Indicate how an ac signal may be generated by using a chopper and a high-pass filter.

9.26 Show that the root-mean-square (rms) value of a rectangular wave can be changed by phase-shifting it and adding to the original signal. What is its applicability in the control of induction motors?

9.27 The direction of the rotating magnetic field in an induction motor (or any other type of ac motor) can be reversed by changing the supply sequence of the phases to the stator poles. This is termed phase-switching. An induction motor can be decelerated quickly in this manner. This is known as "plugging" an induction motor. The slip versus torque relationship of an induction motor may be expressed as

$$T_m = k(S)v_f^2$$

Show that the same relationship holds under plugged conditions, except that k (S) has to be replaced by $-k(2 - S)$. Sketch the curves k (S), $k(2 - S)$, and $-k(2 - S)$, from $S = 0$ to $S = 2$. Using these curves, indicate the nature of the torque acting on the rotor during plugging. (*Hint:* $k(S) = (aS)/[1 + (S/S_b)^2])$.)

9.28 What is a servomotor? AC servomotors that can provide torques on the order of 100 N.m at 3000 rpm are commercially available (Note: 1 N.m = 141.6 oz.in). Describe the operation of an ac servomotor that uses a two-phase induction motor. A block diagram for an ac servo motor is shown in Figure P9.28. Describe the purpose of each component

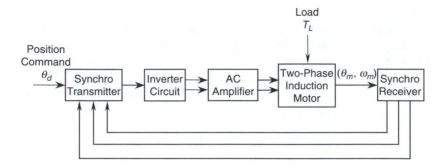

FIGURE P9.28
AC servomotor using a two-phase induction motor and a synchro transformer.

in the system and explain the operation of the overall system. What are the advantages of using an ac amplifier after the inverter circuit in comparison to using a dc amplifier before the inverter circuit?

9.29 Consider the two-phase induction motor discussed in an example in Chapter 9. Show that the motor torque T_m is a linear function of the control voltage v_c when $k(2 - S) = k(S)$. How many values of speed (or slip) satisfy this condition? Determine these values.

9.30 A magnetically levitated rail vehicle uses the principle of induction motor for traction. Magnetic levitation is used for suspension of the vehicle slightly above the emergency guide rails. Explain the operation of the traction system of this vehicle, particularly identifying the stator location and the rotor location. What kinds of sensors would be needed for the control systems for traction and levitation? What type of control strategy would you recommend for the vehicle control?

9.31 What are common techniques for controlling
   a. dc motors?
   b. ac motors?
Compare these methods with respect to speed controllability.

9.32 Describe the operation of a single-phase ac motor. List several applications of this common actuator. Is it possible to realize three-phase operation using a single-phase ac supply? Explain your answer.

9.33 Speed control of motors (ac motors as well as dc motors) can be accomplished by using solid-state switching circuitry. In one such method, a solid-state relay is activated using a switching signal generated by a microcomputer so as to turn on and off at high speed, the power into the motor drive circuit. Speed of the motor can be measured using a sensor such as tachometer or optical encoder. This signal is read by the microcomputer and is used to modify the switching signal so as to correct the motor speed. Using a schematic diagram, describe the hardware needed to implement this control scheme. Explain the operation of the control system.

9.34 In some applications, it is necessary to apply a force without creating a motion. Discuss one such application. Discuss how an induction motor could be used in such an application. What are the possible problems?

9.35 The harmonic drive-principle can be integrated with an electric motor in a particular manner in order to generate a high-torque "gear motor." Suppose that the flexispline of the harmonic drive (see Chapter 3) is made of an electromagnetic material, as the rotor of a motor. Instead of the mechanical wave generator, suppose that a rotating magnetic field is generated around the fixed spline. The magnetic attraction will cause the tooth engagement between the flexispline and the fixed spline. What type of motor principle may be used in the design of this actuator? Give an expression for the motor speed. How would one control the motor speed in this case?

9.36 List three types of hydraulic pumps and compare their performance specifications. A position servo system uses a hydraulic servo along with a synchro transformer as the feedback sensor. Draw a schematic diagram and describe the operation of the control system.

9.37 Giving typical applications and performance characteristics (bandwidth, load capacity, controllability, etc.), compare and contrast dc servos, ac servos, hydraulic servos, and pneumatic servos.

9.38 What is a multistage servovalve? Describe its operation. What are advantages of using several valve stages?

9.39 Discuss the origins of the hydraulic time constant in a hydraulic control system that consists of a four-way spool valve and a double-acting cylinder actuator. Indicate the significance of this time constant. Show that the dimensions of the right-hand-side expression in the equation: $\tau_h = \frac{V}{2\beta k_c}$ are [time].

9.40 Sometimes either a pulse width-modulated (PWM) alternating current (ac) signal or a dc signal with a superimposed constant-frequency ac signal (or, dither) is used to drive the valve actuator (torque motor) of a hydraulic actuator. What is the main reason for this? Discuss the advantages and disadvantages of this approach.

9.41 Compare and contrast valve-controlled hydraulic systems with pump-controlled hydraulic systems. Using a schematic diagram, explain the operation of a pump-controlled hydraulic motor. What are its advantages and disadvantages over a frequency-controlled ac servo?

9.42 Explain why accumulators are used in hydraulic systems. Sketch two types of hydraulic accumulators and describe their operation.

9.43 Identify and explain the components of the hydraulic system given by the circuit diagram in Figure P9.43. Describe the operation of the overall system.

9.44 If the load on the hydraulic actuator shown in Figure 9.60 consists of a rigid mass restrained by a spring, with the other end of the spring connected to a rigid wall, write equations of motion for the system. Draw a block diagram for the resulting complete system, including a four-way spool valve, and give the transfer function that corresponds to each block.

9.45 Suppose that the coupling of the feedback linkage shown in Figure 9.68(c) is modified as shown in Figure P9.45. What is the transfer function of the controller? Show that this feedback controller is a lead compensator.

9.46 The sketch in Figure P9.46 shows a half-sectional view of a flow control valve, which is intended to keep the flow to a hydraulic load constant regardless of variations of the load pressure P_3 (disturbance input).

 a. Briefly discuss the physical operation of the valve, noting that the flow will be constant if the pressure drop across the fixed area orifice is constant.

 b. Write the equations that govern the dynamics of the unit. The mass, the damping constant, and the spring constant of the valve are denoted by m, b, and k,

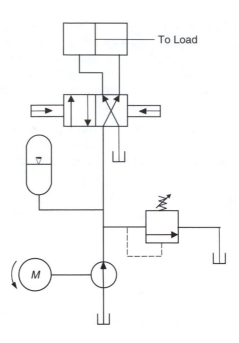

FIGURE P9.43
A hydraulic circuit diagram.

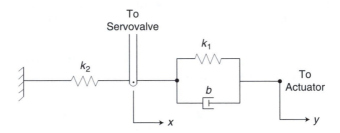

FIGURE P9.45
A mechanical coupling with lead action for a hydraulic servovalve.

respectively. The volume of oil under pressure P_2 is V, and the bulk modulus of the oil is β. Make the usual linearizing assumptions.

c. Set up a block diagram for the system from which the dynamics and stability of the valve could be studied.

9.47 A schematic diagram of a pump stroke-regulated hydraulic power supply is shown in Figure P9.47. The system uses a three-way pressure control valve of the type described in the text (see Figure 9.64). This valve controls a spring-loaded piston, which in turn regulates the pump stroke by adjusting the swash plate angle of the pump. The load pressure P_L is to be regulated. This pressure can be set by adjusting the preload x_o of the spring in the pressure control valve (y_o in Figure 9.64). The load flow Q_L enters into the hydraulic system as a disturbance input.

a. Briefly describe the operation of the control system.

b. Write the equations for the system dynamics, assuming that the pump stroke mechanism and the piston inertia can be represented by an equivalent mass m_p

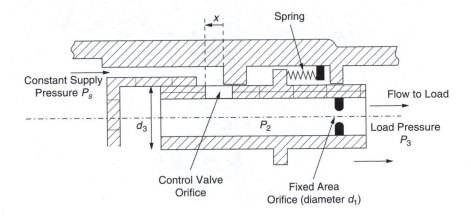

FIGURE P9.46
A flow control valve.

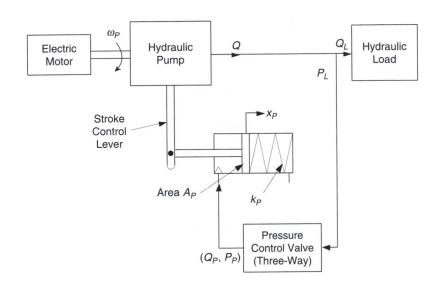

FIGURE P9.47
A pump stroke-regulated hydraulic power supply.

moving through x_p. The corresponding spring constant and damping constant are k_p and b_p, respectively. The piston area is A_p. The mass, spring constant, and damping constant of the valve are m, k, and b, respectively. The valve area is A_v and the valve spool movement is x_v. The volume of oil under pressure P_L is V_t, and the volume of oil under pressure P_p is V_o (volume of oil in the cylinder chamber). The bulk modulus of the oil is β.

c. Draw a block diagram for the system from which the behavior of the system could be investigated. Indicate the inputs and outputs.

d. If Q_p is relatively negligible, indicate which control loops can be omitted from the block diagram. Hence, derive an expression for the transfer function $x_p(s)/x_v(s)$ in terms of the system parameters.

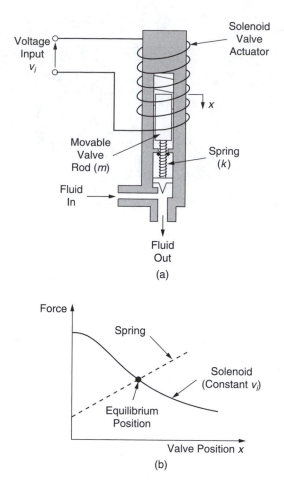

FIGURE P9.48
(a) A solenoid-actuated flow control valve, (b) Steady-state characteristics of the valve.

9.48 A schematic diagram of a solenoid-actuated flow control valve is shown in Figure P9.48(a). The downward motion x of the valve rod is resisted by a spring of stiffness k. The mass of the valve rod assembly (all moving parts) is m, and the associated equivalent viscous damping constant is b. The voltage supply to the valve actuator (proportional solenoid) is denoted by v_i. For a given voltage v_i, the solenoid force is a nonlinear (decreasing) function of the valve position x. This steady-state variation of the solenoid force (downward) and the resistive spring force (upward) are shown in Figure P9.48(b). Assuming that the inlet pressure and the outlet pressure of the fluid flow are constant, the flow rate will be determined by the valve position x. Hence, the objective of the valve actuator would be to set x using v_i.

   a. Show that for a given input voltage v_i, the resulting equilibrium position (x) of the valve is always stable.
   b. Describe how the relationship between v_i and x could be obtained
      i. under quasi-static conditions
      ii. under dynamic conditions.

9.49 What are the advantages and disadvantages of pneumatic actuators in comparison with electric motors in process control applications? A pneumatic rack-and-pinion actuator is an on/off device that is used as a rotary valve actuator. A piston or diaphragm in the actuator is moved by allowing compressed air into the valve chamber. This rectilinear motion is converted into rotary motion through a rack-and-pinion device in the actuator. Single-acting types with spring return and double-acting types are commercially available. Using a sketch, explain the operation of a piston-type single-acting rack-and-pinion actuator with a spring-restrained piston. Could the force rating, sensitivity and robustness of the device be improved by using two pistons and racks coupled with the same pinion? Explain.

9.50 Consider a pneumatic speed sensor that consists of a wobble plate and a nozzle arranged like a pneumatic flapper valve. The wobble plate is rigidly mounted at the end of a rotating shaft, so that the plane of the plate is inclined to the shaft axis. Using a sketch, explain the principle of operation of the wobble plate pneumatic speed sensor.

9.51 A two-axis hydraulic positioning mechanism is used to position the cutter of an industrial fish-cutting machine developed at the University of British Columbia, Canada. The cutter blade is pneumatically operated. The hydraulic circuit of the positioning mechanism is given in Figure P9.51.

Since the two hydraulic axes are independent, the governing equations are similar. State the nonlinear servovalve equations, hydraulic cylinder (actuator) equations and the mechanical load (cutter assembly) equations for the system. Use the following notation:

x_v = servovalve displacement.

K = valve gain (nonlinear)

P_s = supply pressure

P_1 = head-side pressure of the cylinder, with area A_1 and flow Q_1

P_2 = rod-side pressure of the cylinder, with area A_2 and flow Q_2

V_h = hydraulic volume in the cylinder chamber

β = bulk modulus of the hydraulic oil

x = actuator displacement

M = mass of the cutter assembly

F_f = frictional force against the motion of the cutter assembly

9.52 Define the following terms in relation to fluidic systems and devices:

a. Fan-in

b. Fan-out

c. Switching speed

d. Transport time

e. Load sensitivity

f. Input impedance

g. Output impedance.

9.53 A fluidic pulse generator is schematically shown in Figure P9.53(a). The fluidic switching element has a regulated supply. Typically, when the input u_2 is larger than the input u_1 to the switching element, the resulting pressure differential will turn on the output y to its high level, shown by a dotted line in Figure P9.53(b). When $u_1 > u_2$ the output y will be turned off (and its complement output $\bar{y}$ will be turned on). There is a hysteresis band for this switching process. The pulse generator consists of a switching element, a fluid

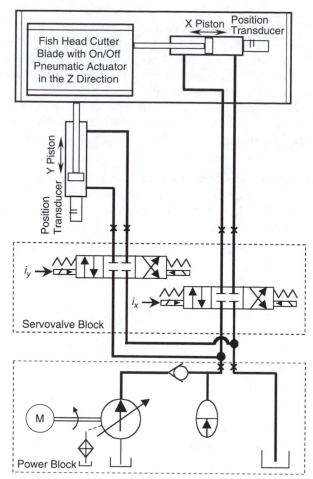

FIGURE P9.51
Two-axis hydraulic positioning system of an industrial fish cutter.

capacitor (accumulator) C and a fluid restrictor (resistor) R_1. In the feedback configuration there is also a feedback path through a second resistor R_2 as shown by the dotted line in Figure P9.53(a).

First consider the open-loop configuration (without the feedback path) of the pulse generator. When the input (pressure) signal u is applied, u_2 will immediately rise to the value of u, but u_1 will rise to the value of u only after a time delay, due to the presence of transport lag and a finite time constant (modeled using R_1 and C). Hence, the output y will be turned on (to its high level) at the switch on level of u. Subsequently, u_1 will reach u. Now if u falls to the switch off level, and so will u_2. But, due to the accumulator C, the level of u_1 will be maintained for some time. Accordingly, the output y will be switched off (i.e., y will be zero). Sketch the shape of the output signal y for the input u given in Figure P9.53(b).

Now consider the pulse generator with the feedback path through R_2. How will the output y change in this case, for the same input u?

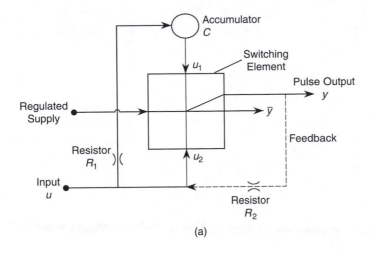

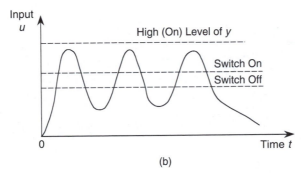

FIGURE P9.53
(a) A fluidic pulse generator, (b) An input signal.

An application of the pulse generator (with feedback) is in the pharmaceutical packaging industry. For example, consider the filling of a liquid drug into bottles and capping them. A packaging line consisting primarily of fluidic devices and fluid power devices can be designed for this purpose. Suppose that fluidic proximity sensors, fluidic amplifiers, hydraulic/pneumatic valves and actuators and other auxiliary components (including resistors and logic elements) are available. Using a sketch, briefly describe a fluidic system that can accomplish the task.

10

Digital Logic and Hardware

Digital devices, digital computers in particular, use digits (according to some code) to represent information and some form of logic to process such information. Since mechatronic systems use digital devices for information acquisition, information communication, information processing, and device control, the field of digital hardware and logic are important in Mechatronics.

In binary (or, two-state) logic a variable can take one of two discrete states: *true* (T) or *false* (F). A numerical digit can assume several values depending on the particular number system used. For example, in the familiar *decimal* number system, each digit can assume one of the ten values denoted by 0, 1, 2, ..., 9. Similarly in the *binary* number system, each digit can assume one of only two values: 0 or 1. A digital device has to process logical quantities as well as numerical data. It is convenient to use the binary number system in digital devices, not only because two-state logical variables and numerical variables can be processed in the same manner, but because components that can assume one of two states can be used as the building blocks for a digital circuit. Such two-state devices are much easier to develop than, for instance, ten-state devices which will be needed to handle the decimal system.

The particular number system itself does not introduce any error in representing and processing information, provided that sufficient memory is available to store the data. This is true because a quantity represented in one number system can be converted exactly into a unique representation in another number system. A code is needed represent such a quantity in a digital form. The result is a coded piece of information. Many types of codes and data representations are employed in digital devices and computers. Several important number systems and codes are discussed below.

10.1 Number Systems and Codes

The *base* or the *radix* (denoted by R) of a number system is the maximum number of discrete values each digit of a number can assume. This is also equal to the maximum number of different characters (symbols) that are needed to represent any number in that system. For the *decimal* system $R = 10$; for the *binary* system $R = 2$; for the *octal* system $R = 8$; and for the *hexadecimal* system $R = 16$. We are quite familiar with the decimal system. The origin of this system is perhaps linked to the fact that a human has ten fingers in her hand. Also ten is a convenient and moderate number, which is neither too large nor too small. If the base of a number system is too large that system will need as many (large) number of separate symbols, which is an inconvenience. On the other hand, if the base is too small, the number of digits needed to express a large number will be too large, which is also an inconvenience. Not withstanding this inconvenience, as justified earlier, the binary number system is what is natural for digital logic devices and digital computers.

10.1.1   Binary Representation

Since digital devices internally make use of high/low voltage levels, the presence/absence of voltage pulses, and the on/off state of bistable elements (microswitches) for data representation and processing, it is the binary representation that is natural for their internal operations. For example, this is the reason why, instructions, stored data, and addresses of memory locations in a computer are all present in the form of binary numbers, even though all such types of information are not numerical.

A binary number consists of binary digits (*bits*). Each bit can take the value 0 or 1. Typically, information is arranged in 8-bit groups called *bytes*, internally in a digital computer. A digital computer operates on one data *word* at a time, however, and the word size can be several bytes (e.g., 16 and 32 bits in microcomputers).

If a string of bits represents a binary number, then it can be converted into a decimal number in a straightforward manner. This simply amounts to a conversion from the base-2 (or radix-2) representation into the base-10 representation.

Example 10.1

Convert the binary number 10101.11 into the decimal form. Also, convert the decimal number 50.578125 into the binary form.

SOLUTION
First, note the decimal point in each number. It is clear that

$$(10101.11)_2 = 1 \times 2^4 + 0 \times 2^3 + 1 \times 2^2 + 0 \times 2^1 \times 1 \times 2^0 + 1 \times 2^{-1} + 1 \times 2^{-2}$$

By evaluating the decimal value on the right hand side, we have

$$(10101.11)_2 = (21.75)_{10}$$

The subscript in each number gives the base. Usually, the subscript is omitted when the decimal system (base 10) is used or when the base is understood by the user without any ambiguity.

In converting the given decimal number into the binary form, the integer part (50) and the fractional part (0.578125) have to be considered separately. The binary number is obtained by repeated division by 2 of the integer part, and the repeated multiplication by 2 of the fractional part. Accordingly,

2	50	Remainder
2	25	0
2	12	1
2	6	0
2	3	0
	1	1

This gives

$$(50)_{10} = (110010)_2$$

Next we perform:

$$0.578125 \times$$
$$\frac{2}{1.156250} \times$$
$$\frac{2}{0.31250} \times$$
$$\frac{2}{0.6250} \times$$
$$\frac{2}{1.250} \times$$
$$\frac{2}{0.50} \times$$
$$\frac{2}{1.0}$$

This gives

$$(0.578125)_{10} = (0.100101)_2$$

Combining these two results we have

$$(50.578125)_{10} = (110010.100101)_2$$

————

Since the decimal point should also be represented as binary information internally, some form of code or convention has to be used in representing *fractional* and *mixed* (integral + fractional) numbers within a computer. For example, the integer part may be stored in one byte and the fractional part in the adjacent byte. Alternatively, the *floating-point* representation may be used, in which the number is represented by a *mantissa* and an *exponent*. The floating-point representation is particularly suitable when the range of values that need to be handled by a digital computer (i.e., its *dynamic range*) is large, advantage being the word size that is needed to store any number in a given dynamic range would be smaller in the floating-point representation, in comparison to the *fixed-point* representation.

Because writing a number in the binary form is lengthy, and conversion between binary and decimal forms is time consuming, it is convenient to employ a base larger than 2. Such a base should also have the advantage of ease of conversion between binary and that base. Two such representations are *octal* (base-8) and *hexadecimal* or *hex* (base-16) number systems. Since the largest octal digit (7) requires 3 bits in the binary representation, conversion from binary to octal form can be done simply by grouping the integer part and the fractional part into 3-bit groups, starting from the decimal point and, then, converting each group into an octal digit. Similarly, conversion from binary to hex representation can be accomplished by using 4-bit groups.

Example 10.2

Convert $(1001010.11)_2$ into a hex number. Also convert $(6D.0F)_{16}$ into a binary number. What is the octal representation of each of these two numbers?

SOLUTION

By separating the integer part of the binary number into 4-bit groups from the decimal point to the left, and similarly separating the fractional part into 4-bit groups from the decimal point to the right, we have

0100	1010	•	1100
4	A	•	C

Observe that starting from left, the first group evaluates to decimal 4 which is hex 4 as well; the second group evaluates to decimal 10 which is hex A; and the third group evaluates to decimal 12 which is hex C. Hence, the hex representation of the given binary number is $(4A.C)_{16}$ as shown. Also, it is clear that the hex representation is much more compact than the binary representation.

Next take $(6D.0F)_{16}$ and write the four-digit binary representation for each hex digit as: 0110 1101.0000 1111 which gives the binary number $(1101101.00001111)_2$.

For the octal representation, we separate the binary number into groups of three digits stating from the decimal point, both to the left (for the integer part) and to the right (for the fractional part). We have,

$$(1001010.11)_2 = (001\ 001\ 010.110)_2 = (112.6)_8$$

Similarly,

$$(1101101.00001111)_2 = (001\ 101\ 101.000\ 011\ 110)_2 = (155.036)_8.$$

10.1.2   Negative Numbers

Consider an n-bit binary integer number. The most significant bit (msb) is the left-most bit, and a digit of 1 there has the value 2^{n-1}. The least significant bit (lsb) is the right-most bit and a digit of 1 there has the value 1. When all the n bits are assigned 1s, the value of the number is $2^n - 1$. This is the largest value of an n-bit binary number. Similarly, when all the n bits are assigned 0s, the value of the number is 0. This is the smallest value of an n-bit number. Hence an n-bit digital storage space can occupy n positive integers ranging from 0 to $2^n - 1$. The above discussion concerns positive numbers. When the number is negative, a special representation is needed to take into account the sign. Several methods are available, as mentioned below.

10.1.2.1   Signed Magnitude Representation

When negative numbers are represented we need an extra bit to represent its sign. So, to represent $-(2^n - 1)$ we need $n + 1$ bits. The most significant bit (msb) is the sign bit. In the *signed magnitude* (sm) representation, for a negative number, msb = 1; and for a positive number, msb = 0. The remaining bits represent the magnitude of the number. For example, in this representation $(11110)_{sm}$ represents decimal -14 and $(01110)_{sm}$ represents decimal $+14$. In general, in the "sm" representation with $n + 1$ bits, we can represent a number ranging from $-(2^n - 1)$ to $+(2^n - 1)$. Also, zero is not uniquely represented in this form, allowing for both a negative zero (e.g., 10000) and a positive zero (e.g., 00000).

Example 10.3
Determine $(010100)_{sm} + (111011)_{sm}$ and express the result in the sm form.

SOLUTION
$(010100)_{sm} + (111011)_{sm} = 10100 - 11011 = -(11011 - 10100) = -00111 = (100111)_{sm}$

10.1.2.2 Two's Complement Representation

Another common way to handle negative binary numbers is through the two's complement (2's) representation. Here the msb has a negative value attached to it, and the remaining bits are considered to have positive values. For a positive number, the two's complement is the same as its binary representation, with 0 as the msb.

Specifically, consider $n + 1$ bits for representing a number in two's complement. Suppose that all $n + 1$ bits are 1s. Then the value of the msb is -2^n. The value of the remaining n bits is $2^n - 1$. Hence the overall value of the number is $-2^n + (2^n - 1) = -1$. If the msb is 1 and the remaining n bits are all 0s, we have the largest (in magnitude) negative number that can be represented by $n + 1$ bits, namely, -2^n. If the msb is 0 and the remaining n bits are all 1s, we have the largest positive number that can be represented by $n + 1$ bits in the 2's complement form, namely, $2^n - 1$. Hence, in the two's complement form $n + 1$ bits can represent any number from -2^n to $+(2^n - 1)$. For example, with 5 bits, the smallest number will be $(10000)_{2's} = -16$ and the largest number will be $(01111)_{2's} = +15$. Also, in the two's complement method, zero is uniquely represented (there is no negative zero).

Given a negative binary number, its 2's complement representation can be formed by the following method.

> **Step 1**: Switch the bit values (i.e., change 1s to 0s and 0s to 1s)
>
> **Step 2**: Append a "1" as the new msb
>
> **Step 3**: Add 1 to the resulting binary number

Example 10.4

Consider the negative number $-(1101)$, which is in the binary form. It has a decimal value of -13. Determine its 2's complement representation.

SOLUTION

> **Step 1**: 0010
>
> **Step 2**: 10010
>
> **Step 3**: 10011

CHECK $(10011)_{2's} = -2^4 + 2 + 1 = -13$.

The two's complement representation has an interesting property called *sign extension*. Specifically, any number of leading 1s can be appended to a negative 2's complement number (assuming that an adequate number of bits are available for storage) and this does not change its value. Similarly any number of leading zeros can be appended to a positive 2's complement number and this does not change its value. The latter is more obvious than the former. As examples, $(1011)_{2's}$, $(11011)_{2's}$, $(111011)_{2's}$, $(1111011)_{2's}$,... etc. all represent the same negative number -5. Similarly, $(0110)_{2's}$, $(00110)_{2's}$, $(000110)_{2's}$, $(0000110)_{2's}$,... etc. represent the same positive number $+6$.

Example 10.5

Carry out the following operations and express the results in the 2's complement form. Check the results using the decimal values of the numbers.

> i. $(10001)_{2's} + (01011)_{2's}$
>
> ii. $(11001)_{2's} - (00011)_{2's}$

iii. $(10001)_{2's} + (10010)_{2's}$

iv. $(01110)_{2's} + (01101)_{2's}$

SOLUTION

i.

10001

01011
___

11100
___

Hence we have the result $(11100)_{2's}$. By the property of sign extension, this number has the same value as $(100)_{2's}$, which is decimal $-2^2 = -4$. Now, since $(10001)_{2's}$ has the decimal value $-2^4 + 1 = -15$ and $(01011)_{2's}$ has the decimal value $+(2^3 + 2 + 1) = +11$ we have $-15 + 11 = -4$, which checks out.

ii. Note that $-(00011)_{2's} = -(0011)_2 = (11101)_{2's}$

Next, we perform the binary addition:

11001

11101
___

110110
___

Now, $(110110)_{2's} = (10110)_{2's}$ in view of the sign extension property. We should use $(10110)_{2's}$ as the answer because the given numbers have 5 bits and the result also should use no more than 5 bits.

CHECK $(11001)_{2's} = (1001)_{2's} = -2^3 + 1 = -7$

$(00011)_{2's} = 2 + 1 = 3$

Now $-7 - 3 = -10$

This checks out because $(10110)_{2's} = -2^4 + 2^2 + 2 = -10$.

iii.

Perform the addition

10001

10010
___

100011
___

Since the given numbers have 5 bits, the answer also should have the same number of bits. This means, the leading 1 in the above result has to be omitted. This gives $(00011)_{2's}$, which is not correct.

CHECK $(10001)_{2's} = -2^4 + 1 = -15$ and $(10010)_{2's} = -2^4 + 2 = -14$. Hence the answer should be $-15 - 14 = -29$. This number cannot be represented using 5 bits (We know that in the 2's complement form, 5 bits can represent an integer from -2^4 to $+2^4 - 1$, or from -16 to $+15$, and -29 is outside this range). The obtained answer (after truncating the msb to retain only 5 bits) is $(00011)_{2's} = +3$. This error is called an *underflow*. Specifically since there are not enough bits to represent the correct negative answer, an incorrect number is produced.

NOTE   This problem can be resolved by using 6 bits and through sign extension. Then we have $(110001)_{2's} + (110010)_{2's} = (1100011)_{2's}$. Now remove the msb, in view of the sign-extension property. This gives the six-bit result $(100011)_{2's}$, which is the correct answer since $-2^5 + 2 + 1 = -32 + 2 + 1 = -29$.

iv. Perform the addition

$$
\begin{array}{r}
01110 \\
01101 \\
\hline
11011
\end{array}
$$

This gives $(11011)_{2's}$, which is not correct. Specifically, $(11011)_{2's} = (1011)_{2's} = -2^3 + 2 + 1 = -5$. But, the correct answer cannot be negative since we are adding two positive numbers. In fact $(01110)_{2's} = 2^3 + 2^2 + 2 = 14$ and $(01101)_{2's} = 2^3 + 2^2 + 1 = 13$, and hence the answer should be 27. Since 5 bits are not adequate to represent this correct positive number, an incorrect number is produced. This error is called an *overflow*.

NOTE This problem can be resolved by using 6 bits and through sign extension. Then we have $(001110)_{2's} + (001101)_{2's} = (011011)_{2's} = 2^4 + 2^3 + 2 + 1 = 16 + 8 + 2 + 1 = 27$ which is the correct answer.

10.1.2.3 One's Complement

The one's complement representation is somewhat similar to the two's complement representation, and it too is useful in dealing with negative numbers. In 1's complement form the msb is the sign bit; 0 represents a positive number and 1 represents a negative number as usual. The remaining bits give the magnitude of the number. For a positive number, the magnitude is represented in the regular binary form. For a negative number, the magnitude is represented by the complement of the regular binary form (i.e., 0s are changed to 1s and 1s are changed to 0s). Note that for positive numbers the 1's complement form, 2's complement form, and the signed magnitude form are identical. But for negative numbers these three representations are different. The property of sign extension holds for the 1's complement representation as well.

For example consider decimal 9, which has the binary form 1001. Then

$$9 = (01001)_{1's} = (001001)_{1's} = (0001001)_{1's} \text{ etc.}$$

$$-9 = (10110)_{1's} = (110110)_{1's} = (1110110)_{1's} \text{ etc.}$$

With $n + 1$ bits, in the 1's complement form, we can represent an integer from $-(2^n - 1)$ to $+(2^n - 1)$. Note that unlike in the two's complement method, zero is not uniquely represented; there is a negative zero (e.g., 11111) as well as a positive zero (e.g., 00000).

10.1.3 Binary Multiplication and Division

Binary arithmetic is quite analogous to decimal arithmetic. The main difference is the number of distinct digits available (two for binary and ten for decimal). The operations and the rules are the same. In binary addition, as clear from the examples discussed before, only 0 and 1 are used as the primary digits. When the value of a summation becomes 2, it is taken to the next higher order place and added as 1 there (i.e., a carry of 1). Binary subtraction is done by sign reversal and addition.

Binary multiplication too is straightforward, as for decimal multiplication. For example, $(1011)_2 \times (101)_2$ is carried out as follows:

$$
\begin{array}{r}
01011 \times \\
101 \\
\hline
1011 \\
0000 \\
1011 \\
\hline
110111
\end{array}
$$

It follows that the answer is $(110111)_2$, and it requires 6 bits of storage space.

CHECK $(11)_{10} \times (5)_{10} = (55)_{10} = (110111)_2$

Binary division is done in the same way as the decimal division. For example $(1101101)_2 \div (101)_2$ is carried out as follows:

$$
\begin{array}{r}
10101 \bullet 1100 \\
\hline
101\,|\,\overline{1101101} \\
\underline{101} \\
111 \\
\underline{101} \\
1001 \\
\underline{101} \\
1000 \\
\underline{101} \\
110 \\
\underline{101} \\
1000 \\
\underline{101} \\
11
\end{array}
$$

Hence the answer is $(10101.11001100....)_2$

CHECK $(109)_{10} \div (5)_{10} = (21.8)_{10} = (10101.11001100...)_2$

It should be clear that as long as the rules for the arithmetical operations are provided and sufficient memory (storage space) is available, a digital computer can follow the instructions very fast to produce accurate results.

10.1.4   Binary Gray Codes

The binary representation of numbers, as discussed above, is known as the *natural binary code* or *straight binary code* because it comes directly from the standard representation of numbers with respect to an arbitrary base. A practical drawback of this natural binary code is that more than one bit switching would be needed when advancing from one number to the next number, consecutively. Since bit switching might not occur simultaneously in actual hardware, there will be some ambiguity concerning the actual value during switching, depending on which bit is switched first. A particular disadvantage of this characteristic of the straight binary code has been discussed in Chapter 7, under absolute encoders. In Table 10.1 a binary gray code is given, which uses a 4-bit word. Note that according to this code, only one bit is switched when the value of an integer number changes by unity. Many such binary gray codes could be developed depending on the bit pattern used to represent the starting number (typically zero). In Table 10.1 zero is represented by 0111 in a 4-bit word. This resulted because in that example we have chosen to make the the normal binary code and the binary gray code identical (1111) for the largest number (15). Alternatively, we could have chosen to make the normal binary code and the binary gray code identical for the smallest number (0). Then, we will end up with a different gray code but, still, the property that only one bit is switched when changing the value of a number by unity will be maintained. For example, for a 4-bit word, the gray code would be

0000, 0001, 0011, 0010,....etc.

starting from zero. We can obtain this particular code from the natural binary code simply by scanning the straight binary word from MSB to LSB (i.e., from left to right) and

TABLE 10.1

A Gray Code Example

Straight Binary Code (MSB → LSB)	Gray code
0 0 0 0	0 1 1 1
0 0 0 1	0 1 1 0
0 0 1 0	0 1 0 0
0 0 1 1	0 1 0 1
0 1 0 0	0 0 0 1
0 1 0 1	0 0 0 0
0 1 1 0	0 0 1 0
0 1 1 1	0 0 1 1
1 0 0 0	1 0 1 1
1 0 0 1	1 0 1 0
1 0 1 0	1 0 0 0
1 0 1 1	1 0 0 1
1 1 0 0	1 1 0 1
1 1 0 1	1 1 0 0
1 1 1 0	1 1 1 0
1 1 1 1	1 1 1 1

complementing the bit next to each 1-bit that is encountered. Since, the straight binary code for 15 is 1111, then, the binary gray code becomes 1000 in this case.

10.1.5   Binary Coded Decimal (BCD)

Instead of converting a decimal number directly to its binary form, one could convert each decimal digit separately to its binary form. Here, we are coding each decimal digit and not the entire number. We call it the *binary coded decimal* representation. Note that the binary representation of the largest decimal digit (9) is 1001. Hence, in BCD each decimal digit will require 4 bits. A drawback of BCD, however, is that the codes 1010 through 1111 are not used (wasted) because they do not represent separate decimal digits. The main advantage is the ease of conversion, similar to conversion between binary and hex.

As an example let us determine the BCD representation of decimal 79. In this case, using the fact that

$$(7)_{10} = (0111)_2 \text{ and } (9)_{10} = (1001)_2$$

we have

$$(79)_{10} = (01111001)_{BCD}$$

Note that this result is quite different from the binary representation of 79, which is found to be $(01001111)_2$.

From the input/output point of view, it is convenient to use BCD for arithmetic manipulation because in view of its familiarity we usually prefer the decimal representation in the real world, outside a digital computer. But, within a computer, arithmetic manipulation of BCD numbers will require a different (and more complex) set of rules from those used for binary numbers. Even with this disadvantage and inefficient use of memory, BCD might be preferred in input/output oriented applications such as point-of-sale terminals where processing itself is not complex, but the ease of conversion from and to the decimal form is a major advantage. BCD is also commonly used in digital

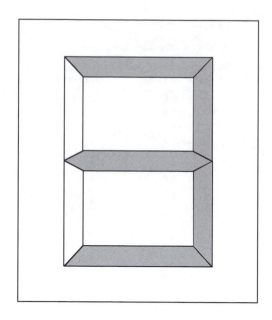

FIGURE 10.1
A seven-segment LED display for a decimal digit (digit 3 is illuminated).

displays such as the seven-segment numeric LED display (see Figure 10.1) where each decimal digit is illuminated separately by energizing a suitable set of the seven light emitting diode (LED) segments.

A number computed by a microcomputer would be in binary representation, but it has to be converted into a proper form according to some code, depending on whether it is displayed as a decimal number on an LED (light-emitting diode) display, or printed on paper, or used as a control signal to control a process.

10.1.6   ASCII (Askey) Code

Information received by a digital computer from a peripheral device such as keyboard (an input device) may include nonnumeric data (e.g., letters in the alphabet and special characters and control commands) as well as numeric data. Furthermore, numbers that appear in a text file (e.g., in a letter or report) take a different meaning internally within a computer, from numerical data that will be processed (e.g., multiply, subtract) by the computer. Similarly, information transmitted from a computer to an output device (e.g., display) also can take these different forms. Usually, information enters and leaves a keyboard as a sequence of pulses (in bit serial manner) and the information is handled by a computer in a bit parallel manner (as bytes or words). The standard code used for the information transfer between a computer and a keyboard is the *United States of America Standard Code for Information Exchange* (US-ASCII), which is usually abbreviated as the ASCII (pronounced "Askey") code. The ASCII code is given in Table 10.2. Note that decimal numbers, lowercase (simple) and upper-case (capital) letters of the alphabet, special keyboard characters, and control commands can be transmitted as binary information, using ASCII. Seven bits are needed for ASCII. But one byte (8 bits) is usually employed, with the eighth bit used as a *parity bit* for checking whether errors have entered the transmitted data (*odd parity* or *even parity* check).

TABLE 10.2

The ASCII Code

Character	Code	Character	Code	Character	Code
NUL	000 0000	+	010 1011	V	101 0110
SOH	000 0001	'	010 1100	W	101 0111
STX	000 0010	−	010 1101	X	101 1000
ETX	000 0011	.	010 1110	Y	101 1001
EOT	000 0100	/	010 1111	Z	101 1010
ENO	000 0101	0	011 0000	[	101 1011
ACK	000 0110	1	011 0001	\	101 1100
BEL	000 0111	2	011 0010	]	101 1101
BS	000 1000	3	011 0011	^	101 1110
HT	000 1001	4	011 0100	−	101 1111
LF	000 1010	5	011 0101	`	110 0000
VT	000 1011	6	011 0110	a	110 0001
FF	000 1100	7	011 0111	b	110 0010
CR	000 1101	8	011 1000	c	110 0011
SO	000 1110	9	011 1001	d	110 0100
SI	000 1111	:	011 1010	e	110 0101
DLE	001 0000	;	011 1011	f	110 0110
DC1	001 0001	<	011 1100	g	110 0111
DC2	001 0010	=	011 1101	h	110 1000
DC3	001 0011	>	011 1110	i	110 1001
DC4	001 0100	?	011 1111	j	110 1010
NAK	001 0101	@	100 0000	k	110 1011
SYN	001 0110	A	100 0001	l	110 1100
ETB	001 0111	B	100 0010	m	110 1101
CAN	001 1000	C	100 0011	n	110 1110
EM	001 1001	D	100 0100	o	110 1111
SUB	001 1010	E	100 0101	p	111 0000
ESC	001 1011	F	100 0110	q	111 1001
FS	001 1100	G	100 0111	r	111 1010
GS	001 1101	H	100 1000	s	111 1011
RS	001 1110	I	100 1001	t	111 1100
US	001 1111	J	100 1010	u	110 0101
SP	010 0000	K	100 1011	v	110 0110
!	010 0001	L	100 1100	w	110 0111
"	010 0010	M	100 1101	x	110 1000
#	010 0011	N	100 1110	y	110 1001
$	010 0100	O	100 1111	z	110 1010
%	010 0101	P	101 0000	{	110 1011
&	010 0110	Q	101 0001	|	110 1100
'	010 0111	R	101 0010	}	110 1101
(	010 1000	S	101 0011	~	110 1110
)	010 1001	T	101 0100	DEL	110 1111
*	010 1010	U	101 0101		

Note: Explanation of Control Characters

NUL	Null, or all zeros	ENQ	Enquiry	LF	Line feed
SOH	Start of heading	ACK	Acknowledge	VT	Vertical tabulation
STX	Start of text	BELL	Bell	FF	Form feed
ETX	End of text	BS	Backspace	CR	Carriage return
EOT	End of transmission	HT	Horizontal tabulation	SO	Shift out
SI	Shift in	NAK	Negative acknowledge	ESC	Escape
DLE	Data link escape	SYN	Synchronous idle	FS	File separator
DCI	Device control 1	ETB	End of transmission block	GS	Group separator
DC2	Device control 2	CAN	Cancel	RS	Record separator
DC3	Device control 3	EM	End of medium	US	Unit separator
DC4	Device control 4	SUB	Substitute	SP	Space

10.2 Logic and Boolean Algebra

Digital circuits can perform logical and binary operations at very high speeds and very accurately. Similarly, they can perform numerical operations, particularly when implemented in digital computers. Logical and numerical operations are directly useful in all types of mechatronic systems. An understanding of the concepts of binary logic is important in the analysis and design of digital logic circuits. Crisp sets and binary logic are analogous. Furthermore, Boolean algebra is useful in the representation and analysis of sets and binary logic, both. This section presents an introduction to crisp sets, conventional (binary) logic, and Boolean algebra, and highlights the isomorphism that exists between them.

10.2.1   Sets

A crisp set is a collection of elements within a crisp boundary. Since there cannot be any elements on the boundary, this is called a "crisp"set. A set of this type may be graphically represented by a *Venn diagram*, as shown in Figure 10.2(a). Here A denotes a set. The

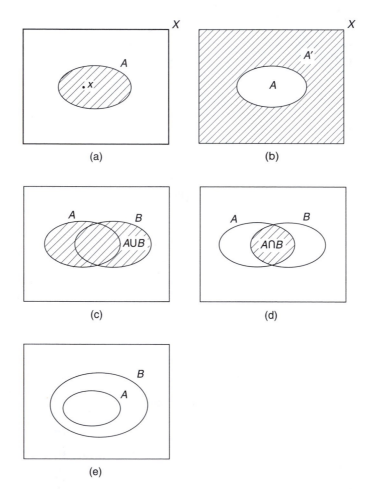

FIGURE 10.2
Some concepts of crisp sets: (a) Membership, (b) Complement, (c) Union, (d) Intersection, (e) Subset (proper).

universal set (or, universe of discourse), as denoted by X, is the largest set that one could consider in a particular problem domain. It contains all possible elements, and there cannot be any elements outside it. The null set is the empty set, and is denoted by $\varnothing$. It does not have any elements. If an element x is inside set A, then x is called a member of A. This is represented by the notation $x \in A$. Of course, by definition, $x \in X$. If x is outside A (i.e., x is not a member of A) we write $x \notin A$. An example of a set is the group of people who live in Cincinnati who are over 50 years old. As another example,

$$A = \{a_1, a_2, ..., a_n\}$$

represents the set containing the n elements $a_1, a_2, ..., a_n$. As yet another example, the set of numbers greater than 50 may be denoted by

$$A = \{x \mid x > 50\}$$

Here, the expression that follows the symbol "|" gives the condition (or property) to which the elements of the set are subjected.

10.2.1.1 Operations of Sets

The basic operations involving sets are complement, union, and intersection.

Complement: A complement of a set A is the set formed by all the elements outside A. It is denoted by A' and is shown by the shaded area in Figure 10.2(b). In particular, the universal set X is the complement of the null set $\varnothing$, and vice versa.

Union: The union of two sets A and B is the set formed by all the elements in A and B. This is denoted by $A \cup B$, and is shown by the shaded area in Figure 10.2(c).

Intersection: The intersection of two sets A and B is the set formed by all the elements that are common to both A and B. This is denoted by $A \cap B$, and is shown by the shaded area in Figure 10.2(d).

Subset: A set A is a subset of another set B if all the elements in A are common to (i.e., contained in) B. This is denoted by $A \subseteq B$. This is equivalent to $B \supseteq A$. If A is a subset of B but B is not a subset of A, then A is said to be a "proper subset" of B, and is denoted by $A \subset B$. This is equivalent to $B \supset A$, and is shown by Figure 10.2(e).
   The above definitions assume that the sets involved belong to the same universe (X).

10.2.2 Logic

Conventional logic deals with statements called "propositions." In binary (or, *two-valued*) logic, a proposition can assume one of only two truth values: true (T), false (F). An example of a proposition would be "John is over 50 years old." Now consider the following propositions:

1. Charcoal is white
2. Snow is cold
3. Temperature is above 60°C

Here proposition 1 has the truth value F, and proposition 2 has the truth value T. But, for proposition 3 the truth value depends on the actual value of the temperature: if it is above 60°C the truth value is T and otherwise it is F.

TABLE 10.3

Truth Tables of Some Logical Connectives

(a) Negation (*NOT*)	
A	~*A*
T	*F*
F	*T*

(b) Disjunction (*OR*)		
A	*B*	*A* V *B*
T	*T*	*T*
T	*F*	*T*
F	*T*	*T*
F	*F*	*F*

(c) Conjunction (*AND*)		
A	*B*	*A* ∧ *B*
T	*T*	*T*
T	*F*	*F*
F	*T*	*F*
F	*F*	*F*

(d) Implication (*IF-THEN*)		
A	*B*	*A* → *B*
T	*T*	*T*
T	*F*	*F*
F	*T*	*T*
F	*F*	*T*

Propositions may be connected/modified by logical connectives such as *AND, OR, NOT,* and *IMPLIES.* These basic logical operations are defined below.

Negation: The negation of a proposition A is "*NOT A*" and may be denoted as ~A (also $\overline{A}$). It is clear that when A is *TRUE*, then *NOT A* is *FALSE* and vice versa. These properties of negation may be expressed by a *truth table* as shown in Table 10.3(a). Note that a truth table gives the truth values of a combined proposition in terms of the truth values of its individual components.

As an example, consider the proposition "John is over 50 years old." If John's age is actually over 50 years, then this proposition is true. Otherwise it is false. Now consider the set of people who are more than 50 years old. If John is a member of this set, then the above proposition is true. If John is a member of the complement of the set, the above proposition is false. Then the negated proposition "John is not over 50 years old" becomes true. This shows that there is a correspondence between "complement" in set theory and negation in logic.

Disjunction: The disjunction of the two propositions A and B is "*A OR B*" and is denoted by the symbol A V B. Its truth table is given in Table 10.3(b). In this case, the combined proposition is true if at least one of its constituents is true. Note that, this is not the "Exclusive *OR*" where "*A OR B*" is false also when both A and B are true, and is true only when either A is true or B is true. It is easy to see that disjunction in logic corresponds to "union" in sets.

Conjunction: The conjunction of two propositions A and B is "*A AND B*" and is denoted by a symbol A ∧ B. Its truth table is given in Table 10.3(c). In this case the combined proposition is true if and only if both constituents are true. There is correspondence between conjunction in logic and "intersection" in set theory.

Implication: Consider two propositions A and B. The statement "*A implies B*" is the same as "*IF A THEN B.*" This may be denoted by $A \rightarrow B$. Note that if both A and B are true then $A \rightarrow B$ is true. If A is false, the statement "When A is true then B is also true" is not violated regardless of whether B is true or false. But if A is true and B is false, then clearly the statement $A \rightarrow B$ is false. These facts are represented by the truth Table 10.3(d).

Example 10.6

Consider two propositions A and B. Form the truth table of the combined proposition (*NOT A*) *OR B*. Show that, in accordance with two-valued crisp logic, this proposition is identical to the statement "*IF A THEN B.*"

SOLUTION
The truth of the combined proposition is given below.

A	$\sim A$	B	$\sim A \vee B$
T	F	T	T
T	F	F	F
F	T	T	T
F	T	F	T

Here we have used the truth Table 10.3(a) and Table 10.3(b). Note that the result is identical to Table 10.3(d). This equivalence is commonly exploited in logic associated with knowledge-based decision making.

In logic, knowledge is represented by propositions. A simple proposition does not usually make a knowledge base. Many propositions connected by logical connectives may be needed. Knowledge may be processed through *reasoning*, by the application of various laws of logic including an appropriate *rule of inference*, subjected to a given set of data (measurements, observations, external commands, previous decisions, etc.) to arrive at new inferences or decisions.

10.2.2.1 Correspondence Between Sets and Logic

As indicated before, there is an isomorphism between set theory and logic. For crisp sets and conventional two-valued logic, this correspondence is summarized in Table 10.4. Boolean algebra is the algebra of two-valued logic. The two values used are 1 (corresponding to *true*) and 0 (corresponding to *false*). Accordingly there is also a correspondence between the operations of Boolean algebra and logic, as indicated in Table 10.4. The two values in Boolean algebra may represent any type of two states (e.g., *on* or *off* state; *presence* or *absence* state; *high* or *low* state; two voltage levels in *transistor-to-transistor logic*) in practical applications.

10.2.3 Boolean Algebra

Boolean algebra is useful in the analysis and design of digital logic circuits. Laws of Boolean algebra follow from the characteristics of two-valued logic. Some basic laws are given in Table 10.5. Some of these are obvious and the others may be verified using truth tables.

TABLE 10.4

Isomorphism Between Set Theory, Logic, and Boolean Algebra

Set Theory Concept	Set Theory Notation	Binary Logic Concept	Binary Logic Notation	Boolean Algebra Notation
Universal set	X	(Always) T	(Always) T	(Always) 1
Null set	$\varnothing$	(Always) F	(Always) F	(Always) 0
Complement	A'	Negation (*NOT*)	$\sim A$	$\overline{A}$
Union	$A \cup B$	Disjunction (*OR*)	$A \vee B$	$A + B$
Intersection	$A \cap B$	Conjunction (*AND*)	$A \wedge B$	$A \cdot B$
Subset	$A \subseteq B$	Implication (If-Then)	$A \rightarrow B$	$A \leq B$

TABLE 10.5

Some Laws of Boolean Algebra and Two-Valued Logic

Name of Property/Law	Statement of Property/Law
Commutativity	$a + b = b + a$
	$a \cdot b = b \cdot a$
Associativity	$(a + b) + c = a + (b + c)$
	$(a \cdot b) \cdot c = a \cdot (b \cdot c)$
Distributivity	$a \cdot (b + c) = (a \cdot b) + (a \cdot c)$
	$a + (b \cdot c) = (a + b) \cdot (a + c)$
Absorption	$a + (a \cdot b) = a$
	$a \cdot (a + b) = a$
Idempotency	$a + a = a$
	$a \cdot a = a$
Exclusion	$a + \bar{a} = 1$
	$a \cdot \bar{a} = 0$
DeMorgan' laws	$\overline{a \cdot b} = \bar{a} + \bar{b}$
	$\overline{a + b} = \bar{a} \cdot \bar{b}$
Boundary conditions	$a + 1 = 1$
	$a \cdot 1 = a$
	$a + 0 = a$
	$a \cdot 0 = 0$

The notation used for the basic operations is given in Table 10.4. For "Exclusive OR" (XOR), the Boolean notation is $\oplus$. It should be remembered that when the order of performing the operations is not explicitly indicated (e.g., through the use of parentheses) the "NOT" operations are performed before the "AND" operations; and the "AND" operations are performed before the "OR" operations.

Example 10.7

Using Boolean algebra verify the following "Absorption Properties" of logic:

$$A \text{ OR } A \text{ AND } B = A$$

$$A \text{ AND } (A \text{ OR } B) = A$$

Using one of these results and other Boolean properties/laws simplify the Boolean expression

$$\overline{(a + \bar{b}) \cdot (c + \bar{d})} \cdot c.$$

NOTE   AND operations are carried out before OR operations.

SOLUTION

$$a + a \cdot b = a \cdot 1 + a \cdot b = a \cdot (1 + b) = a \cdot 1 = a$$

(Using boundary conditions and distributivity, Table 10.5)

$$a \cdot (a + b) = a \cdot a + a \cdot b = a + a \cdot b = a \cdot (1 + b) = a \cdot 1 = a$$

(Again, using Table 10.5)

Now, using the second result above, we have $(c + \bar{d}) \cdot c = c$. Hence,

$$\overline{(a + \bar{b}) \cdot (c + \bar{d}) \cdot c} = \overline{(a + \bar{b}) \cdot c} = \overline{a + \bar{b}} + \bar{c} \quad \text{(De Morgan's law and 2nd result above)}$$

$$= \bar{a} \cdot b + c \quad \text{(De Morgan's law)}.$$

Example 10.8

Using Boolean algebra verify the following "Simplification Properties" of logic:

$$A \text{ OR NOT } A \text{ AND } B = A \text{ OR } B$$

$$A \text{ AND (NOT } A \text{ OR } B) = A \text{ AND } B$$

NOTE NOT operations are performed before AND; AND operations are performed before OR.

SOLUTION
First property:

$$a + \bar{a} \cdot b = a + a \cdot b + \bar{a} \cdot b \quad \text{(since } a + a \cdot b = a)$$

$$= a + (a + \bar{a}) \cdot b = a + b \quad \text{(since } a + \bar{a} = 1)$$

Second property:

$$a \cdot (\bar{a} + b) = a \cdot \bar{a} + a \cdot b$$

$$= a \cdot b \quad \text{(since } a \cdot \bar{a} = 0)$$

10.2.3.1 Sum and Product Forms

Conversion of a Boolean expression from the "sum" form to the "product" form, and vice versa, may be accomplished by using De Morgan's laws and the fact that performance of two negations will result in the original expression.

First start with $\overline{a + b} = \bar{a} \cdot \bar{b}$ and negate both sides. We have $a + b = \overline{\bar{a} \cdot \bar{b}}$.

Similarly, start with $\overline{a \cdot b} = \bar{a} + \bar{b}$ and negate both sides. We have $a \cdot b = \overline{\bar{a} + \bar{b}}$.

These results are useful in the analysis and design of logic circuits.

Example 10.9

Prepare a truth table to represent the Boolean relation $a = x \cdot \bar{y} + z$. Using this table express a in a

  i. Sum-of-product form
  ii. Product-of-sum form

Verify that your answers are equivalent to the original expression for a.

SOLUTION

The truth table is prepared by first listing all possible combinations of Boolean values for
(x, y, z) systematically as rows of the table, and then forming the columns for the required
expressions according to laws of logic. This procedure results in the truth table below.

x	y	z	$\bar{y}$	$x \cdot \bar{y}$	a
0	0	0	1	0	0
0	0	1	1	0	1
0	1	0	0	0	0
0	1	1	0	0	1
1	0	0	1	1	1
1	0	1	1	1	1
1	1	0	0	0	0
1	1	1	0	0	1

Now note from the last column of the truth table that a is formed by the combinations for
(x "and" y "and" z) in row 2, "or" row 4, or row 5, or row 6, or row 8. This fact may be
expressed in the Boolean form (Note: "and" is "·" and "or" is "+") as:

$$a = \bar{x} \cdot \bar{y} \cdot z + \bar{x} \cdot y \cdot z + x \cdot \bar{y} \cdot \bar{z} + x \cdot \bar{y} \cdot z + x \cdot y \cdot z \tag{i}$$

This is the sum-of-product form, as required in Part (i) of the question.

Next note from the last column of the truth table that $\bar{a}$ is formed by the combinations
for (x and y and z) in row 1, or row 3, or row 7. This fact may be expressed in the Boolean
form as:

$$\bar{a} = \bar{x} \cdot \bar{y} \cdot \bar{z} + \bar{x} \cdot y \cdot \bar{z} + x \cdot y \cdot \bar{z}$$

Now negate both sides and successively apply De Morgan's law. We get

$$a = \overline{\bar{x} \cdot \bar{y} \cdot \bar{z} + \bar{x} \cdot y \cdot \bar{z} + x \cdot y \cdot \bar{z}} = \overline{\bar{x} \cdot \bar{y} \cdot \bar{z}} \cdot \overline{\bar{x} \cdot y \cdot \bar{z}} \cdot \overline{x \cdot y \cdot \bar{z}}$$

$$= (x + y + z) \cdot (x + \bar{y} + z) \cdot (\bar{x} + \bar{y} + z) \tag{ii}$$

This is the product-of-the sum form, as required in Part (ii) of the question.

Now we verify Equation i by applying the rules of Boolean algebra to the result.
Specifically,

$$a = \bar{x} \cdot \bar{y} \cdot z + \bar{x} \cdot y \cdot z + x \cdot \bar{y} \cdot \bar{z} + x \cdot \bar{y} \cdot z + x \cdot y \cdot z = \bar{x} \cdot \bar{y} \cdot z + \bar{x} \cdot y \cdot z + x \cdot \bar{y} \cdot (\bar{z} + z) + x \cdot y \cdot z$$

$$= \bar{x} \cdot \bar{y} \cdot z + \bar{x} \cdot y \cdot z + x \cdot \bar{y} + x \cdot y \cdot z = \bar{x} \cdot \bar{y} \cdot z + \bar{x} \cdot y \cdot z + x \cdot \bar{y} \cdot (1 + z) + x \cdot y \cdot z$$

(Using properties: $\bar{a} + a = 1$; $a \cdot (1 + b) = a$)

$$= x \cdot \bar{y} + \bar{x} \cdot \bar{y} \cdot z + \bar{x} \cdot y \cdot z + x \cdot \bar{y} \cdot z + x \cdot y \cdot z = x \cdot \bar{y} + (\bar{x} \cdot \bar{y} + \bar{x} \cdot y + x \cdot \bar{y} + x \cdot y) \cdot z$$

$$= x \cdot \bar{y} + (\bar{x} \cdot (\bar{y} + y) + x \cdot (\bar{y} + y)) \cdot z = x \cdot \bar{y} + (\bar{x} + x) \cdot z \quad \text{(Using } \bar{a} + a = 1)$$

$$= x \cdot \bar{y} + z$$

Next we verify Equation ii. Specifically, consider the first two terms of the RHS product of Equation ii:

$$(x+y+z)\cdot(x+\bar{y}+z) = x\cdot x + x\cdot\bar{y} + x\cdot z + y\cdot x + y\cdot\bar{y} + y\cdot z + z\cdot x + z\cdot\bar{y} + z\cdot z$$

$$= x + x\cdot\bar{y} + x\cdot z + y\cdot x + y\cdot z + z\cdot x + z\cdot\bar{y} + z \quad \text{(Using } a\cdot a = a \text{ and } a\cdot\bar{a} = 0\text{)}$$

$$= x + x\cdot(\bar{y}+z+y+z) + y\cdot z + z\cdot\bar{y} + z = x + y\cdot z + z\cdot\bar{y} + z \quad \text{(Using } a + a\cdot b = a\text{)}$$

$$= x + z\cdot(y+\bar{y}) + z = x + z \quad \text{(Using } a + a\cdot b = a\text{)}$$

Hence,

$$(x+y+z)\cdot(x+\bar{y}+z)\cdot(\bar{x}+\bar{y}+z) = (x+z)\cdot(\bar{x}+\bar{y}+z)$$

$$= x\cdot\bar{x} + x\cdot\bar{y} + x\cdot z + z\cdot\bar{x} + z\cdot\bar{y} + z\cdot z$$

$$= x\cdot\bar{y} + x\cdot z + z\cdot\bar{x} + z\cdot\bar{y} + z = x\cdot\bar{y} + z\cdot(x+\bar{x}+\bar{y}) + z$$

$$\text{(Using } a\cdot\bar{a} = 0; \quad a\cdot a = 1\text{)}$$

$$= x\cdot\bar{y} + z \quad \text{(Using } a\cdot b + a = a\text{)}$$

10.3 Combinational Logic Circuits

A digital circuit converts digital inputs into digital outputs. There are two types of logic devices, classified as either *combinational logic* or *sequential logic*. Combinational logic devices are "static" where the present inputs completely (and uniquely) determine the present outputs, without using any past information (history) or memory. In contrary, the output of a sequential logic device depends on the past values of the inputs as well as the present values. In other words they depend on the time sequence of the input data, and hence some form of *memory* would be needed. The present section deals with the realization of combinational logic circuits. Sequential logic realizations will be discussed in a later section.

A digital device may have to process both logical quantities and numerical data. The purpose of a digital circuit might be to turn on or off a device depending some logical conditions, for example, turn the light on or off in a room depending whether there are people in it or not. Such an implementation can be done using logic circuitry, which performs logical operations. In some other application, a digital circuit might have to perform numerical computations on a measured signal (available in digital form) and then generate a control signal (in digital form). A digital device can perform such numerical functions as well, using the binary number system where each digit can assume one of only two values: 0 or 1. Because the logic state of each output line of a digital circuit corresponds to binary 0 or 1, a combination of several output lines can form a numerical output as well (in binary form). It is convenient to use the binary number system in digital devices, not only because logical variables and numerical variables can be processed in the same manner, but because components that can assume one of two states can be used as the building blocks for digital circuitry. Such two-state devices are much easier to develop than, for instance, ten-state devices based on the decimal system. In the present chapter we are primarily concerned with the use of digital circuits to carry out logic operations. But the extension of these concepts for numerical computation (using the binary number system) should be clear.

10.3.1 Logic Gates

Logic gates are the basic circuit elements found in IC circuits that are used in digital systems. A logic gate has one or more logical inputs and only one logical output. Each input line or output line of the gate can have one of two states: *true* (represented by the binary digit 1) and *false* (represented by the binary digit 0). It is easy to see how a combination of switches can be used to form a logic gate. For example, two switches connected in series can serve as an AND gate because current passes through the circuit only when both switches are closed. Output of the gate is the state of the circuit. Hence, by denoting the closed-circuit state as true (1) and the open-circuit state as false (0), we can obtain a truth table for the AND gate, as shown in Figure 10.3(a). Note that the gate has two inputs, which are the states of the two switches. Similarly, an OR gate can be formed by connecting two switches in parallel, as shown in Figure 10.3(b). In this OR gate, the output state is considered true (1) also when both inputs are true simultaneously. Alternatively, in an EXCLUSIVE OR (XOR), the output state is taken as false (0) when both inputs are true, and the output is true when only one of the inputs is true. Mechanical switches and relays are not suitable for logic gates in digital circuits. Solid-state switches are the preferred variety. Semiconductor elements such as diodes and transistors can function as solid-state switches. In digital systems, these elements are present in the IC form and not in their discrete component form.

The three basic logic gates are AND, OR, and NOT. The three gates NAND, XOR, and NOR can be constructed from the first three gates, but all six of these gates may be

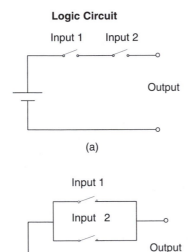

Logic Circuit

Truth Table

(a)

Input 1	Input 2	Output
0	0	0
0	1	0
1	0	0
1	1	1

(b)

Input 1	Input 2	Output
0	0	0
0	1	1
1	0	1
1	1	1

FIGURE 10.3
Examples of basic logic gates formed using switches: (a) AND gate and its truth table; (b) OR gate and its truth table.

Symbol	Truth Table

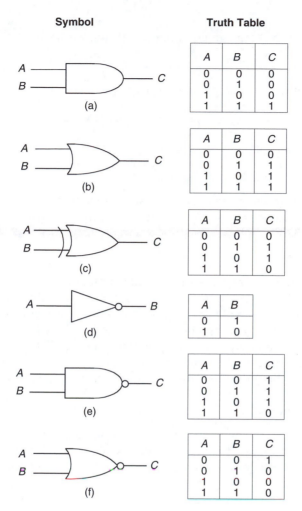

	A	B	C
(a)	0	0	0
	0	1	0
	1	0	0
	1	1	1

	A	B	C
(b)	0	0	0
	0	1	1
	1	0	1
	1	1	1

	A	B	C
(c)	0	0	0
	0	1	1
	1	0	1
	1	1	0

	A	B
(d)	0	1
	1	0

	A	B	C
(e)	0	0	1
	0	1	1
	1	0	1
	1	1	0

	A	B	C
(f)	0	0	1
	0	1	0
	1	0	0
	1	1	0

FIGURE 10.4
American National Standards Institute (ANSI) symbols and truth tables for the six basic logic gates: (a) AND; (b) OR; (c) XOR; (d) NOT; (e) NAND; (f) NOR.

considered basic. American National Standards Institute (ANSI) symbols for these six logic gates are shown in Figure 10.4. A circle at the end of a logic signal line represents an inverter (negation). This is a simplification of the complete symbol of a NOT gate (triangle and a circle, as shown in Figure 10.4(d)). The truth tables give the state of the output of each gate for various possible states of the inputs.

Example 10.10

An OR gate obeys the Boolean operation $a + b$. For an XOR, denoted by $a \oplus b$, the same relation holds, except when both $a = 1$ and $b = 1$, in which case the output becomes 0. It follows that XOR obeys the Boolean relation $c = a \oplus b = (a + b) \cdot \overline{(a \cdot b)}$. This follows from the fact that the term $\overline{(a \cdot b)} = 0$ when both $a = 1$ and $b = 1$, and it is equal to 1 otherwise. Obtain a Boolean relation for XOR using the truth table in Figure 10.4(c), and show that this is equivalent to the above result. Give a digital circuit using AND, OR, and NOT gates only to realize XOR.

SOLUTION

From the truth table of XOR, as given in Figure 10.4(c), we have

$$c = \bar{a} \cdot b + a \cdot \bar{b}$$

The expression given in the example may be expanded as follows:

$$c = (a+b) \cdot \overline{(a \cdot b)} = (a+b) \cdot (\bar{a} + \bar{b}) \quad \text{(De Morgan)}$$

$$= a \cdot \bar{a} + a \cdot \bar{b} + b \cdot \bar{a} + b \cdot \bar{b}$$

$$= a \cdot \bar{b} + b \cdot \bar{a} \quad (\text{since } x \cdot \bar{x} = 0)$$

This result is identical to what we obtained from the truth table.

A realization of the XOR gate is done by following the governing Boolean relation, as shown in Figure 10.5.

It can be shown that NAND gate is functionally complete. That is, the three basic gates AND, OR, and NOT can be implemented with NAND gates alone. For example, Figure 10.6 shows a realization of OR using NAND gates alone. Similarly, it can be shown that NOR gate is functionally complete.

A gate symbol can have many (more than two) input lines. Then, the logic combination that produces the output should be interpreted accordingly. For example, if an AND gate has three inputs, then the output is true only when all three inputs are true.

10.3.2   IC Logic Families

A logic family is identified by the nature of the solid-state elements (resistors, diodes, bipolar transistors, MOSFETS, etc.) used in the integrated circuit. Examples include *Resistor-Transistor Logic* (RTL), *Diode-Transistor Logic* (DTL), *Emitter-Coupled Logic* (ECL), *Transistor-Transistor*

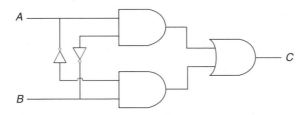

FIGURE 10.5
A realization of XOR using AND, OR, and NOT gates.

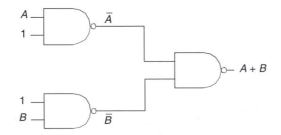

FIGURE 10.6
A realization of OR using NAND gates alone.

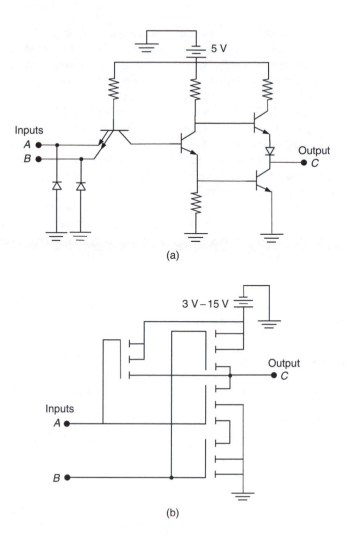

(a)

(b)

FIGURE 10.7
Typical logic gates from TTL and CMOS logic families: (a) A TTL NAND gate; (b) A CMOS NAND gate.

Logic (TTL), and *Complementary-Symmetry Metal-Oxide Semiconductor Logic* (CMOS). Currently, TTL is the most popular logic family, but CMOS is fast becoming popular for its lower power dissipation (particularly useful in battery powered applications), smaller size (higher circuit density on chip), and better noise immunity properties. But they are less robust and more easily damaged by electrical fields than TTL devices. Each logic family uses logic gates as basic elements. Two examples are shown in Figure 10.7. A TTL NAND gate is shown in Figure 10.7(a). In this case, *A* and *B* are the inputs and *C* is the output. It can be shown that this circuit element satisfies the truth table given in Figure 10.4. Specifically, the output will be at a low voltage level (logic 0 or logical FALSE) only when both inputs are simultaneously at a high voltage level (logic 1 or logical TRUE).

For an output signal of a TTL device, the low value corresponds to a voltage in the range 0–0.4 V, and high value corresponds a voltage in the range 2.4–5 V. For an input signal of a TTL device, the low value corresponds to a voltage in the range 0–0.8 V, and the high value corresponds a voltage greater than 2 V. These voltage ranges provide a high degree of immunity to noise because logic 0 and logic 1 are separated by about 2 V. Furthermore, even if a noise level of 0.4 V is added to an output of a logic device, its logic state will remain the same at the input of another logic device. Note that Figure 10.7(a) shows only two input lines, but more than two inputs could be connected to the emitter of the input transistor, if so desired. The power supply voltage is usually limited to the range 4.75 – 5.25 V in a TTL circuit. A 5 V power supply is shown in Figure 10.7(a).

A NAND gate in the C-MOS logic family is shown in Figure 10.7(b). Several advantages of C-MOS over TTL were mentioned before. A further advantage of CMOS logic circuits is that they can use any power supply in the voltage range of 3–15 V. For example, a CMOS circuit can operate with a 12 V automobile battery, without having to use special circuitry to reduce the supply voltage, which is the case for a TTL circuit. The logic levels of CMOS circuits change with the supply voltage.

Sometimes, two or more logic families might be used in the same digital device (e.g., computer). When connecting two IC units that belong to different logic families (e.g., one is CMOS and the other TTL), it is necessary to make sure, as when interconnecting any two devices, that the two units are compatible. In particular, operating voltages of the two units have to be compatible and operating currents of the two units have to be compatible as well. In short, the two units should have matching impedances. Otherwise, interface circuitry (e.g., *level shifters* for voltage compatibility and *buffers* for current compatibility) has to be used between the two IC units that are interconnected.

10.3.3   Design of Logic Circuits

The main steps in realizing a logic circuit for a practical application are given below.

1. Identify the inputs and outputs of the circuit, for the particular application.
2. State the logic that connects the inputs and outputs and express it as a Boolean relation.
3. Using the basic logic gates, sketch the realization that will satisfy the Boolean relation.

In a practical realizations of logic circuitry it is important to minimize the cost and complexity. Accordingly, a minimal realization, which uses the least number of basic gates would be preferred. This topic will be addressed under Karnaugh maps. Also, it is desirable to use least number of types of logic gates (e.g., all NOR, all NAND). We now illustrate the design/realization of combinational logic circuits for several practical applications.

10.3.3.1   Multiplexer Circuit

A digital multiplexer selects one digital input channel from a group and connects it to the output channel (i.e., reads the input channel). Consider the case of a 2-input multiplexer. The inputs are denoted by x and y. The output is a. The control signal for input selection is c. The logic function of the circuit may be expressed as follows:

$$a = x \quad \text{if } c = 0$$

$$a = y \quad \text{if } c = 1$$

This logic may be translated into the following Boolean relation:

$$a = x \cdot \bar{c} + y \cdot c$$

Its realization is shown in Figure 10.8.

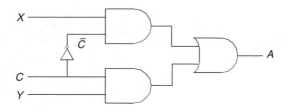

FIGURE 10.8
A two-input digital multiplexer.

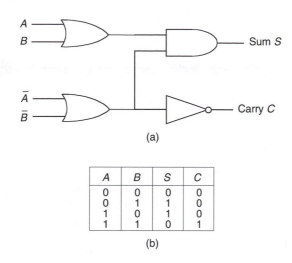

(a)

A	B	S	C
0	0	0	0
0	1	1	0
1	0	1	0
1	1	0	1

(b)

FIGURE 10.9
A half-adder: (a) Logic circuit; (b) Truth table.

10.3.3.2 Adder Circuits

Logic gates are used in the *arithmetic and logic unit* (ALU) of a microprocessor to perform various processing operations. A basic arithmetic operation that is performed by the ALU is *addition*. A simple logic circuit, which performs the addition of two (input) binary digits (bits) is shown in Figure 10.9(a). This is called a *half adder*. Note that the circuit contains two OR gates, one AND gate, and one NOT gate. The added two digits are denoted by *A* and *B* (the inputs to the logic circuit). The sum of the two digits is denoted by *S*, and the carry to the next higher place is denoted by *C*. The truth table given in Figure 10.9(b) agrees with binary addition of two bits. Note that $\bar{A}$ is the complement of *A*. A *full adder* circuit performs binary addition on three (input) bits.

Example 10.11

An electronic switch that uses digital logic is to be developed for switching the lights on and off in an art gallery. The switch has to be turned on at 7:00 p.m. and turned off at 6:00 a.m. Also if there are people in the gallery and the light that enters the gallery from outside is inadequate, the lights have to be turned on regardless of the time of the day. Assume that a sensor detects the presence of people in the gallery and produces a logic state of high (1). Also, there is sensor that detects the level of light entering the gallery from outside and generates a logic state of high (1) when the light level is inadequate. Furthermore a digital clock produces logic high state when the time is between 6:00 a.m. to 7:00 p.m. Design a logic circuit that uses NAND gates only, to operate the switch.

SOLUTION

The logic state of the people sensor is denoted by p, that of the light sensor is denoted by l, and that of the time sensor is dented by t. The output of the logic circuit is denoted by s, which is high (binary 1) when the switch has to be on. The logic of operation is, the switch has to be on when there are people in the gallery "and" the light entering from outside is not adequate, "or" if the time is not between 6:00 a.m. to 7:00 p.m. This statement translates into the Boolean relation

$$s = p \cdot l + \bar{t}$$

Now in order to determine a realization with NAND gates only, we proceed as follows:

$$s = p \cdot l + \bar{t} = \overline{\overline{p \cdot l + \bar{t}}} = \overline{\overline{p \cdot l} \cdot t} \quad \text{(by double negtion and application of De Morgan)}$$

It is seen that this relation can be realized using two NAND gates, as shown in Figure 10.10.

10.3.4 Active-Low Signals

Logic devices often use *npn* transistors (or, *n*-channel transistors) at their output terminals. This is because these transistors can handle larger currents and can operate at higher switching rates than the *pnp* transistors (or, *p*-channel transistors) can. To operate devices like motors and solenoids, reasonably high current levels are needed, and hence switching devices based on *npn* transistors are more desirable. A logic device based on this principle is shown in Figure 10.11. Note that the base B is the input to the transistor and the collector C represents the output. The emitter E is grounded. The collector has to be internally

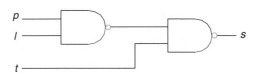

FIGURE 10.10
A logic circuit for operating lights in an art gallery.

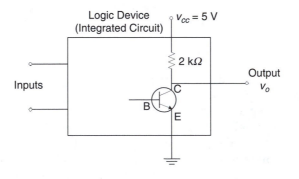

FIGURE 10.11
A logic device (*npn*) with active-low output.

connected to a suitable resistor and a power supply v_{cc} (typically, 5 V). If the collector is not internally connected in this manner, we have an *open-collector* output, which is not desirable. In that case, the collector has to be connected externally through a *pull-up resistor* to the power source v_{cc}. This resistance has to be sufficiently high (e.g., 2 kΩ). Otherwise, the resulting high currents will increase the power dissipation, and furthermore the output voltage (which has to be small for logic-0 state) will increase.

The *npn* transistor in Figure 10.11 operates like a semiconductor switch. This can be explained as follows. When the input voltage to the transistor (i.e., base voltage with respect to the emitter, v_{be}) is small (<0.7 V), there will not be any current through the collector (and hence through R_c). Then the transistor is in its *off* state, and the output v_o will be close to v_{cc}. This is a high-voltage state for the output. The output impedance of the transistor is high (several kΩ) in this state. When the input to the transistor is greater than 0.7 V, the transistor is forward biased, and is turned *on*. A finite current will conduct through the collector, thereby saturating the transistor, and resulting in a significant voltage drop across R_c. As a result, the output voltage v_o will be close to zero (about 0.2 V). This is a low-voltage state for the output. The output impedance of the transistor is low in this state. In summary, when the transistor is turned off (logic 0), the output signal is high; and when the transistor is turned on (logic 1), the transistor is saturated, and the output signal is low (yet providing a reasonable amount of current to drive a device). In other words, the logic device has an *active-low* output.

A TTL logic device (as shown in Figure 10.11, for example) has some disadvantages over a CMOS logic device in view of the faster switching speed, higher fan-out capability (i.e., the number of devices it can drive at the output), and increased immunity to noise, of the latter. To eliminate these disadvantages, a logic gates should have a small output impedances regardless of the output at logic-0 or logic-1. This is essential so that at fast switching rates the output transistor will be capable of both supplying and sinking large currents. This is not the case for the circuit in Figure 10.11. Here when the input voltage (at the base) is small (logic-0), the transistor will not conduct (off), as indicated before, and the output will be at a high voltage (logic-1). In this state, the output impedance will be high (e.g., 1.4 kΩ). When the input (base) voltage high enough (logic-1), the transistor will be turned on and the voltage at the collector will drop to a small value (logic-0). Then the output impedance (resistance between the collector of the saturated transistor and ground) will be small.

In the developments of previous sections we have assumed *active-high* signals, where the high voltage level represents logic 1 (logical TRUE) and the low voltage level represents logic 0 (logical FALSE). This is the default case. In other words if "active-low" (denoted by *signal.L*) is not specified we assume "active-high" signals (denoted by *signal.H*). Hence, a signal denoted by A is identical to $A.H$. But, as indicated before, active-low signal usage is appropriate in some practical applications. Note that *signal.L* can be converted into *signal.H*, and vice versa, by simply adding an inverter (NOT gate). For example, if A denotes an active-high signal, then $A = A \cdot H$; $\overline{A} = A \cdot L$; $\overline{A \cdot L} = A$.

Example 10.12

Using NAND gates and NOT gates (inverters) only, develop a logic circuit to realize

$$C = A \oplus B$$

assuming that

a. Both input signals (A, B) are available in the active-high form and the output C is also needed in the active-high form.

b. Signals $A \cdot L$, $B \cdot L$ are available and the desired output from the circuit is $C \cdot L$.

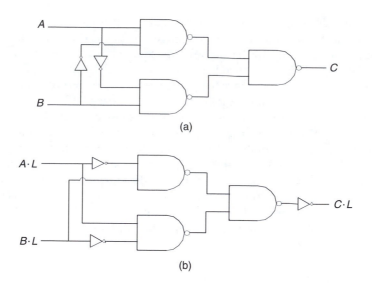

(a)

(b)

FIGURE 10.12
Implementation of XOR using NAND gates using: (a) Active-high signals; (b) Active-low signals.

SOLUTION
Since $A \oplus B = A \cdot \overline{B} + \overline{A} \cdot B$ we have

$$C = \overline{\overline{A \cdot \overline{B} + \overline{A} \cdot B}} = \overline{\overline{A \cdot \overline{B}} \cdot \overline{\overline{A} \cdot B}} \text{ (double negation and De Morgan)}$$

This is implemented in the active-high form in Figure 10.12(a), and in the active-low form in Figure 10.12(b).

Compare these implementations with what is given in Figure 10.5.

10.3.5   Minimal Realization

Boolean expressions of logic functions may contain some redundancy. Their implementation in such a form will require a larger number of logic gates, and will be less desirable (with respect to cost, complexity, reliability, etc.). Procedures area available to reduce (and in fact minimize) logic functions before they are implemented. In particular, various reduction/simplification operations of Boolean algebra, as given in Table 10.5 and illustrated in various examples in the previous sections, may be employed. The use of *Karnaugh maps* (or, *K-maps*) is a popular and convenient procedure, which is explained next.

10.3.5.1   Karnaugh Map Method

Logic function minimization using K-maps is based on the property $A + \overline{A} = 1$ and its extensions such as $AB + A\overline{B} = A$, and $\overline{A}\overline{B} + \overline{A}B + AB + A\overline{B} = 1$. We have seen that any Boolean (logic) function may be expressed as a truth table with respect to its variables, in a 1-dimensional array. A K-map is a convenient version of a truth table where a matrix-type map is used, which is more convenient when the number of variables is large. Each row or column of the matrix represents a variable group (typically, a pair). The variable groups are arranged such that only one of the variables is complemented from one group to the adjacent group (i.e., *adjacency* is preserved; hence there is only one bit-switching between two adjacent groups). The reduction procedure is summarized below:

1. Separate the variables of the logic function into two groups, one group representing the rows of a matrix and the other representing the columns of the matrix, as follows: If there are only two variables, each group will contain one variable; if there are three variables, one group will have two variables and the other will have the remaining one; if there are four variables, each group will have two variables, and so on.

2. Mark the rows and columns of the matrix with various combinations of logic states for the two groups such that adjacent rows or adjacent columns correspond to only one bit switching (i.e., complementation of just one variable).

3. Mark the cells of the matrix with 1s corresponding to the terms of the logic function (the unmarked cells have 0s).

4. Identify all rectangular blocks that are filled (i.e., with no 0s), starting from blocks with the largest number of cells and ending at the remaining single cells, as follows: blocks with an even number of rows and columns; blocks with one row and an even number of columns; blocks with one column and an even number of rows; single cells. In this procedure, allow for *wrap-around* (i.e., the first and the last rows are linked, and the first and the last columns are linked, in a cylindrical manner). Note that a subset of a union of already marked blocks should not be marked as a new block.

5. Reduce each block into a minimal form by using the reduction formulas of Boolean algebra.

This procedure may be explained using illustrative examples.

Example 10.13
Reduce the logic function $\bar{A} + A \cdot B + \bar{B} \cdot C + A \cdot B \cdot C$ by preparing a Karnaugh map for the function.

SOLUTION
The K-map is drawn for each term of the given function as given below (this step is given as an explanation of the procedure, and is skipped in future examples), and then combined to form the overall K-map.

$\bar{A}$:

	$\bar{C}$	C
$\bar{A}\bar{B}$	1	1
$\bar{A}B$	1	1
AB		
$A\bar{B}$		

(B brackets rows $\bar{A}B$ and AB; A brackets rows AB and $A\bar{B}$)

$A \cdot B$:

	$\bar{C}$	C
$\bar{A}\bar{B}$		
$\bar{A}B$		
AB	1	1
$A\bar{B}$		

B { $\bar{A}B$, AB } A { AB, $A\bar{B}$ }

$\bar{B} \cdot C$:

	$\bar{C}$	C
$\bar{A}\bar{B}$		1
$\bar{A}B$		
AB		
$A\bar{B}$		1

$A \cdot B \cdot C$:

	$\bar{C}$	C
$\bar{A}\bar{B}$		
$\bar{A}B$		
AB		1
$A\bar{B}$		

$\bar{A} + A \cdot B + \bar{B} \cdot C + A \cdot B \cdot C$:

	$\bar{C}$	C
$\bar{A}\bar{B}$	1	1
$\bar{A}B$	1	1
AB	1	1
$A\bar{B}$		1

Two 2×2 blocks each containing four filled cells, and one 4×1 block also containing four filled cells have been identified. They exhaust all the filled cells in the K-map. Now we can determine the reduced function for each block. Specifically, the top square block corresponds to $\overline{A}$. The bottom square block corresponds to B. Finally, the right column with four filled cells corresponds to C. It follows that the reduced function is

$$\overline{A} + B + C$$

This result may be verified by applying the reduction property: $a + \overline{a} \cdot b = a + b$ to the original function.

Example 10.14

Consider the logic function $A \cdot (\overline{B} + \overline{C}) + \overline{A} \cdot \overline{B} \cdot C + A \cdot B \cdot C \cdot \overline{D}$. Using a Karnaugh map reduce this function and express it in the

   i. Sum of product form
  ii. Product of sum form

SOLUTION
The given expression may be expanded as

$$Y = A \cdot \overline{B} + A \cdot \overline{C} + \overline{A} \cdot \overline{B} \cdot C + A \cdot B \cdot C \cdot \overline{D}.$$

Its K-map is shown below.

Y:

	$\overline{C}\overline{D}$	$\overline{C}D$	CD	$C\overline{D}$
$\overline{A}\overline{B}$			1	1
$\overline{A}B$				
AB	1	1		1
$A\overline{B}$	1	1	1	1

B brackets rows $\overline{A}B$ and AB; A brackets rows AB and $A\overline{B}$.

   i. First we partition the K-map, as shown below, to form four 4-element blocks:
      Accordingly, we write the reduced form

$$Y = A \cdot \overline{B} + A \cdot \overline{C} + \overline{B} \cdot C + A \cdot \overline{D}.$$

      This is in the minimal, sum of product form. Note that in arriving at this expression, the K-map was partitioned into four blocks each having four elements. The first term corresponds to the bottom 1×4 row. The second term corresponds to the 2×2 block at the bottom left corner. The third term corresponds to the

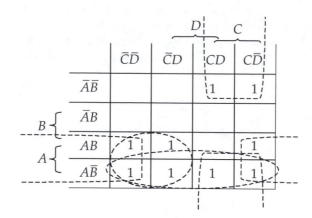

wrapped-around 2×2 block formed by the last two elements in the bottom row and the last two elements of the top row. The last term corresponds to the wrapped-around 2×2 block formed by the last two elements of the first column and the last two elements of the last column.

ii. The 0-cells in the K-map above gives the truth-value terms for $\overline{Y}$. Hence, we can form the K-map for $\overline{Y}$ as given below.

$\overline{Y}$:

	$\overline{C}\overline{D}$	$\overline{C}D$	CD	$C\overline{D}$
$\overline{A}\overline{B}$	1	1		
$\overline{A}B$	1	1	1	1
AB			1	
$A\overline{B}$				

This K-map can be partitioned into two 4-cell blocks and one 2-cell block, giving the minimal expression:

$$\overline{Y} = \overline{A} \cdot B + \overline{A} \cdot \overline{C} + B \cdot C \cdot D$$

Now negate (complement) this expression to get

$$Y = \overline{\overline{A} \cdot B + \overline{A} \cdot \overline{C} + B \cdot C \cdot D} = \overline{\overline{A} \cdot B} \cdot \overline{\overline{A} \cdot \overline{C}} \cdot \overline{B \cdot C \cdot D} = \overline{\overline{A} \cdot B} \cdot \overline{\overline{A} \cdot \overline{C}} \cdot \overline{B \cdot C \cdot D}$$

$$= (A + \overline{B}) \cdot (A + C) \cdot (\overline{B} + \overline{C} + \overline{D})$$

This is the minimal function in the product of sum form.

10.4 Sequential Logic Devices

The logic circuits considered in the previous section do not have feedback paths. Hence, when the inputs are removed, the outputs also disappear. Such logic circuits (called combinational logic circuits) are "static" (algebraic) in nature. If an output of a logic circuit is fed back, it is possible to maintain an output logic state even after the inputs are removed. Furthermore, in this case the output of the circuit will depend on inputs applied earlier and, consequently, on the time sequence of the previous inputs/outputs. Such feedback logic circuits are called *sequential logic circuits*. They are "dynamic" devices. Their present output depends on the past history and timing of the inputs.

A clock signal is available as the time reference for operation of a sequential logic circuit. If the operation is synchronized with the clock signal and the actions are triggered by it, we have *synchronous* operation. If the actions are triggered by the inputs (which are generally nonperiodic) and not by the clock signals, we have *asynchronous* operation. Edge-triggered devices use the transitions in clock pulses for triggering. Positive *edge-triggered* devices use the rising edge (i.e., transition from 0–1) of a clock pulse for triggering. Negative edge-triggered devices use the falling edge (i.e., transition from 1–0) of a clock pulse for triggering. This nomenclature is illustrated in Figure 10.13. Edge trigger is needed in synchronous operation. For asynchronous operation, *level-trigger* is used where triggering does not occur in a periodic manner but rather depends on the level of the triggering signal.

Examples of sequential logic devices include flip-flops, latches, shift registers, counters, and trigger devices with memory. In fact the property of "memory" is an important consideration in sequential logic devices. Furthermore, complex digital systems like *microprocessors* and *state machines* are formed using basic sequential circuits such as flip-flops as the building blocks. Flip-flops are called bistable devices because they can assume two and only two stable output states (0 or 1). There are many types of flip-flops. A "latch" is a flip-flop that is able to latch onto a binary state. "Up-counters" generate a binary number sequence where each number is generated by incrementing the previous number by 1. Similarly, "down-counters" generate a binary number sequence in the consecutive descending order. Integrated circuits (IC) are used to perform important functions such as data storage (memory) and processing (ALU) in a microcomputer. Solid state memory devices (e.g., buffers, registers) and the logic elements that perform data processing (e.g., division) in microcomputers employ sequential logic ICs. Several basic devices, which fall into the category of sequential logic circuits, are discussed next.

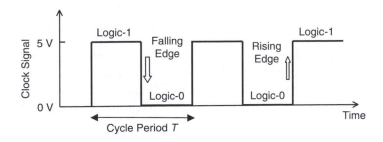

FIGURE 10.13
A clock signal for edge-triggered synchronous operation of a logic device.

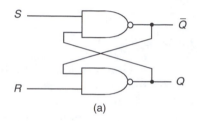

(a)

Applied Input (Data)		Resulting Output		Explanation
S	R	Q	$\bar{Q}$	
0	0	1	1	Not acceptable
0	1	0	1	Independent of previous output
1	0	1	0	
1	1	Q_o	$\bar{Q}_o$	Retains state (same as previous output)

(b)

S	Q
R	$\bar{Q}$

(c)

FIGURE 10.14
RS flip-flop formed by using NAND gates: (a) Logic circuit; (b) Truth table; (c) ANSI symbol.

10.4.1   RS Flip-Flop

An RS flip-flop (or, *reset-set flip-flop*) is shown in Figure 10.14(a). This circuit uses two cross-coupled NAND gates. Two cross-coupled NOR gates may be used as well to form an RS flip-flop. The truth table (*function table*) for an RS flip-flop, as given in Figure 10.14(b), is obtained from the truth table of a NAND gate. Under normal operating conditions, Q and $\bar{Q}$ are "double-rail" outputs, and the latter output is the complement of the former. But it is seen from Figure 10.14(b) that when both inputs are 0, both outputs become 1. This condition violates the requirement, and is unacceptable. Furthermore, it is seen from the truth table that when one input is the complement of the other input, the output state simply becomes equal to the input state, irrespective of the previous state. In other words the input data are directly transferred to the output (i.e., a "data transfer"). In this mode, when the output becomes 0 (when the input is 0), it is called a "reset" operation, and when the output becomes 1 (when the input is 1), it is called a "set" operation. When both inputs to the RS flip-flop are 1, the resulting output state becomes identical to the previous output state (denoted by Q_o). Hence, the device has memory. Note that the resulting output, as given in Figure 10.14(b), is the final "settled" output, in cases where it is different from the previous output. The ANSI symbol for an RS flip-flop is given in Figure 10.14(c). The property that the current output of a flip-flop depends on the previous output implies that the unit has memory (i.e., it can remember the state history). Hence, a flip-flop is used as a basic element in semiconductor memory units. Since the RS flip-flop can assume two stable states (01 and 10), it is a *bistable* device. A device such as an oscillator, which does not have a stable state, is known as an *astable* device.

10.4.2   Latch

A latch circuit is able to retain the previous output state when triggered. Triggering may be done using either an enable signal (asynchronous operation) or a clock signal (synchronous operation). Figure 10.15(a) shows a latch circuit formed using an RS flip-flop. Its truth

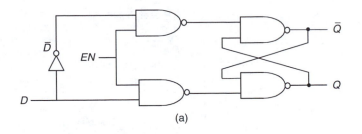

(a)

Input	Enable	Resulting Output		Explanation
D	EN	Q	$\bar{Q}$	
0	1	0	1	Equal to input
1	1	1	0	
Any	0	Q_o	$\bar{Q}_o$	Latches to previous output

(b)

(c)

FIGURE 10.15
A latch circuit formed using an RS flip-flop and NAND gates: (a) Logic circuit; (b) Truth table; (c) Symbol.

table is given in Figure 10.15(b), which can be easily verified using the truth tables of the RS flip-flop (Figure 10.14(b)), the NAND gate, and the NOT gate. The symbol of the latch is shown in Figure 10.15(c).

10.4.3 JK Flip-Flop

In a JK flip-flop, the input data are denoted by J and K (similar to R and S in an RS flip-flop). But, unlike an RS flip-flop, a JS flip-flop does not have an unacceptable state of input data. Furthermore, a JS flip-flop can provide a "toggle" action where the output is the complement of the previous output. A toggle action is important in "counting" operations (A toggle will change the count by 1. Both up count and down count are possible). Note that an RS flip-flop does not provide a toggle action.

The ANSI symbol of a JK flip-flop is shown in Figure 10.16. As before, Q is the output that results due to the input data J and K. Again, $\bar{Q}$ denotes the complement of Q, and is available as an output. The action of a JK flip-flop can be triggered either by external signals (specifically, a "Preset" signal P and a "Clear" signal C) or by the clock signal CK. The former is the asynchronous operation and the latter the synchronous operation. Typically, active-low signals are needed for P and C (i.e., logic-0 activates the operation) and the falling edge of a clock pulse is used for triggering of the device in the synchronous operation. The truth table for asynchronous operation of a JS flip-flop is given in Table 10.6. Note that the values the complement signals $\bar{P}$ and $\bar{C}$ are listed to emphasize the active-low operation. The circles that end these signal lines in Figure 10.16, denoting "NOT" operations, further indicate the use of active-low signals. The state of $\bar{P}=0, \bar{C}=0$ is not used in a JK flip-flop. When $\bar{P}=0, \bar{C}=1$, the output Q is "set" to 1 (which is the value of P). Similarly, when $\bar{P}=1, \bar{C}=0$, the output Q is "cleared" (i.e., becomes 0, which is the value of $\bar{C}$). When $\bar{P}=1, \bar{C}=1$, the JK flip-flop goes into synchronous operation where

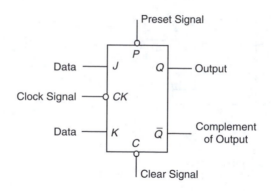

FIGURE 10.16
The symbol of a JK flip-flop.

TABLE 10.6

Truth Table for Asynchronous
Operation of a JK Flip-Flop

$\bar{P}$	$\bar{C}$	Q	$\bar{Q}$
0	0	Not used	
0	1	1	0
1	0	0	1
1	1	Synchronous operation	

TABLE 10.7

Truth Table of a JK Flip-Flop in Synchronous (Clocked) Operation

Clock Signal	Applied Input		Resulting Output		
CK	J	K	Q	$\bar{Q}$	Explanation
↓	0	0	Q_o	$\bar{Q}_o$	Retains state (same as previous output; i.e., memory)
↓	0	1	0	1	Data transfer independent of previous output (clear, set)
↓	1	0	1	0	
↓	1	1	$\bar{Q}_o$	Q	Toggles state (complement of previous output; i.e., counting)

triggering is done by the clock signal *CK*. The circle ending the line of the clock signal in Figure 10.16 indicates that the device is triggered by the falling edge of a clock pulse.

The function table (truth table) for a JK flip-flop, in clocked (synchronous) operation is given in Table 10.7. Note that falling-edge triggering is used (denoted by ↓). As usual, Q_o denotes the previous output (i.e., the output prior to trigger).

Since a JK flip-flop has the capabilities of memory and toggle, it has many applications (e.g., counters and other types of flip-flops such as D flip-flop and T flip-flop). In designing a circuit for a particular application, one may start with the desired output sequence from the circuit and proceed backwards to develop a suitable circuit that will achieve this output. In this approach to design, what is particularly useful is the "Excitation Table" of the flip-flop. This table gives the required input (data) for a given combination of previous output (before trigger) and new output (after trigger). The excitation table for a JK flip-flop is given in Table 10.8.

TABLE 10.8

Excitation Table of a JK Flip-Flop

Q_n	Q_{n+1}	J	K
0	0	0	Any
0	1	1	Any
1	0	Any	1
1	1	Any	0

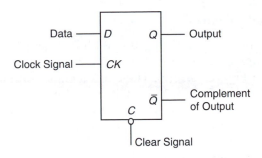

FIGURE 10.17
The symbol of a D flip-flop.

TABLE 10.9

Function Table (truth table) of a D Flip-Flop

Clock Signal	Data Input	Resulting Output		Explanation
CK	D	Q	$\bar{Q}$	
↑	0	0	1	Data transfer independent
↑	1	1	0	of previous output (clear, set)

TABLE 10.10

Excitation Table of a D Flip-Flop

Q_n	Q_{n+1}	D
0	0	0
0	1	1
1	0	0
1	1	0

10.4.4 D Flip-Flop

A D flip-flop (or, *data flip-flop*) has a single input (or, data) D. Its output Q is equal to the input data value. A D flip-flop may be interpreted as a special case of a JK flip-flop with $J = D$ and $K = \bar{D}$ (Hence, there is no need to have the input K). The symbol of a D flip-flop ·is shown in Figure 10.17. The clock signal is used in synchronous operation. Typically, rising-edge trigger is used, as shown in Figure 10.17 (Note: there is no circle at the end of the CK line entering the flip-flop). Asynchronous operation is possible with the "Clear" signal C. In Figure 10.17, an active-low signal is used for C, as typical. Specifically, when $C = 0$, the output is cleared (i.e., $Q = 0$) regardless of the input data. The function table (truth table) of a clocked D flip-flop is given in Table 10.9. The excitation table is given in Table 10.10.

10.4.4.1   Shift Register

A shift register shifts the stored data in a word, one bit at a time to the right, as triggered by incoming clock pulses. A set of D flip-flops connected in series, with the first flip-flop receiving a 0 as the input data and the subsequent flip-flops receiving the output of the previous flip-flop as their input data, can serve as a shift register. Edge-trigger using a clock signal will cause the originally stored (output) data of the flip-flops to be shifted to the right, one at a time. At the end of the shifting sequence all the bits will be 0 (i.e., data will be cleared).

10.4.5   T Flip-Flop and Counters

A T flip-flop (or, *toggle flip-flop*) is a JK flip-flop with $J = 1$ and $K = 1$. Also, it is assumed that $\overline{P} = 1$. Hence the signal lines for J, K, and P are not marked on the symbol of the T flip-flop, as shown in Figure 10.18(a). The clock signal is in fact the data signal of a T flip-flop, and is denoted by T. When $\overline{C} = 1$ it should be clear from the function table for a JK flip-flop (Table 10.6 and Table 10.7) that synchronous (i.e., clocked) toggle operation takes place in the T flip-flop. The output can be cleared ($Q = 0$) simply by setting $\overline{C} = 0$ (asynchronous mode).

In the device shown in Figure 10.18(a), the falling edge (i.e., transition from 1–0) of T results in a *toggle* (i.e., changing from 0–1 or 1–0) of the output Q. This property enables a group of T flip-flops to function as either a binary counter or a frequency divider (by multiples of 2). A 3-bit binary counter using T flip-flops is shown in Figure 10.18(b). The corresponding timing diagram is shown in Figure 10.18(C). The outputs Q_0, Q_1, and Q_2 form the 3-bit word $[Q_2 Q_1 Q_0]_2$. Bit changing occurs at every clock period, triggered by the falling edge of a clock pulse. From the timing diagram it is clear that the 3-bit output changes in the sequence 000, 001, 010, 011, 100, 101, 110, 111 at the clock frequency. Hence we have a 3-bit up-counter. Also, from the timing diagram it is clear that the frequency of Q_0 is 1/2 the clock frequency, the frequency of Q_1 is 1/4 the clock frequency, and the frequency of Q_2 is 1/8 the clock frequency. Hence, these outputs can also serve as frequency-divider signals.

In a counter that uses the straight binary code (as in Figure 10.18(c)) there can be more than one bit switching at a time. For example, when advancing from 001–010 there will be two bit switches, and when advancing from 011–100 there will be 3-bit switches. Due to signal delays in the hardware, these switches will not happen simultaneously. As a result, an intermediate value, which will be different from the correct value, will result. For example, when advancing from 011–100, if the middle bit is switched first, the intermediate result will be 1, which is different from the correct result (decimal 4). Such a condition is called a *hazard*. It can be avoided by using a code other than the straight binary code; specifically, a gray code as discussed before in the present chapter and also in Chapter 7.

Example 10.15

A 3-bit binary counter is to be designed using three D flip-flops and basic logic gates. In order to avoid hazards, a gray code is used, where the required sequence is 000, 001, 011, 010, 110, 111, 101, 100. Note that in this sequence only one bit will switch for each increment in the count. Develop a suitable circuit for this counter.

SOLUTION

The output from the three flip-flops is denoted by $[Q_2 Q_1 Q_0]$. The required output sequence may be represented either by a timing diagram or as a table. The data inputs to the three flip-flops are denoted by $[D_2 D_1 D_0]$. The values of the input data, which are necessary to

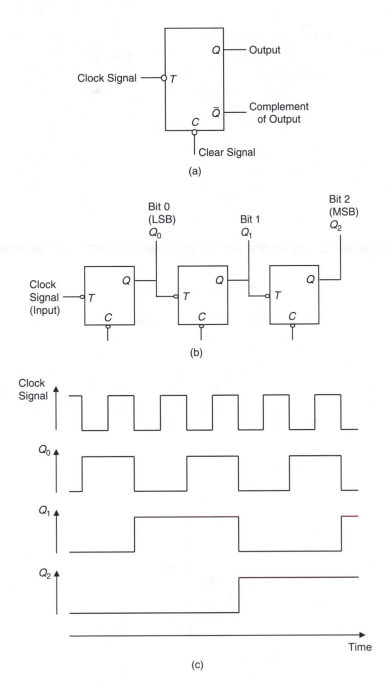

FIGURE 10.18
(a) The symbol of a T flip-flop. (b) A 3-bit binary up-counter using T flip-flops. (c) The timing diagram for the up-counter.

achieve the required changes in the three bits at each clock trigger may be conveniently determined using the excitation table for a D flip-flop (Table 10.10). This information is given in Table 10.11.

From Table 10.11 (truth table) we can express the necessary input data at clock trigger in terms of the output that has to be present at (or, just before) clock trigger in order to

TABLE 10.11

Input-Output Values for the Gray-Code Counter

Output just Before Clock Trigger			Output Just After Clock Trigger			Input Data at Clock Trigger		
Q_2	Q_1	Q_0	Q_2	Q_1	Q_0	D_2	D_1	D_0
0	0	0	0	0	1	0	0	1
0	0	1	0	1	1	0	1	1
0	1	1	0	1	0	0	1	0
0	1	0	1	1	0	1	1	0
1	1	0	1	1	1	1	1	1
1	1	1	1	0	1	1	0	1
1	0	1	1	0	0	1	0	0
1	0	0	0	0	0	0	0	0

achieve the required results. Specifically we have,

$$D_0 = \overline{Q}_0\overline{Q}_1\overline{Q}_2 + Q_0\overline{Q}_1\overline{Q}_2 + \overline{Q}_0Q_1Q_2 + Q_0Q_1Q_2$$

$$D_1 = Q_0\overline{Q}_1\overline{Q}_2 + Q_0Q_1\overline{Q}_2 + \overline{Q}_0Q_1\overline{Q}_2 + \overline{Q}_0Q_1Q_2$$

$$D_2 = \overline{Q}_0Q_1\overline{Q}_2 + \overline{Q}_0Q_1Q_2 + Q_0Q_1Q_2 + Q_0\overline{Q}_1Q_2$$

Note that what is related is the output Q that is present "prior to" clock trigger, and the necessary input data D which will generate the required output after clock trigger. In other words, the above three relations give the necessary *feedback paths* in the circuit. We now simplify (minimize) these expressions using Karnaugh maps, as follows:

D_0:

	$\overline{Q}_2$	Q_2
$\overline{Q}_0\overline{Q}_1$	1	
$\overline{Q}_0Q_1$		1
Q_0Q_1		1
$Q_0\overline{Q}_1$	1	

D_1:

	$\overline{Q}_2$	Q_2
$\overline{Q}_0\overline{Q}_1$		
$\overline{Q}_0Q_1$	1	1
Q_0Q_1	1	
$Q_0\overline{Q}_1$	1	

D_2:

	$\bar{Q}_2$	Q_2
$\bar{Q}_0\bar{Q}_1$		
$\bar{Q}_0Q_1$	1	1
Q_0Q_1		1
$Q_0\bar{Q}_1$		1

Q_1 brackets rows $\bar{Q}_0Q_1$ and Q_0Q_1; Q_0 brackets rows Q_0Q_1 and $Q_0\bar{Q}_1$.

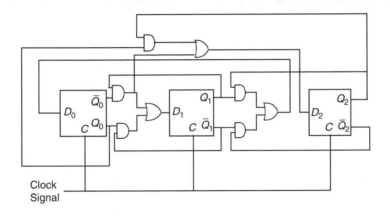

FIGURE 10.19
A gray-code counter using D flip-flops.

The minimized expressions are:

$$D_0 = Q_1Q_2 + \overline{Q}_1\overline{Q}_2$$

$$D_1 = \overline{Q}_0Q_1 + Q_0\overline{Q}_2$$

$$D_2 = \overline{Q}_0Q_1 + Q_0Q_2$$

The circuit shown in Figure 10.19 is now developed using these minimal expressions.

10.4.6 Schmitt Trigger

Switching or triggering elements are useful in both analog and digital circuitry. If switching is ideal, slightest noise in a signal can create undesirable chatter near the switching region, producing erroneous results. Hence, it is desirable to have some hysteresis in the switching element so that the "switch-on" signal level is slightly higher than the "switch-off" signal level, thereby the switch becoming insensitive to noise. Schmitt trigger is a solid-state switch, which has this type of desirable hysteresis. A bipolar transistor circuit for Schmitt trigger is shown in Figure 10.20(a). The input/output characteristic of the circuit under *quasi-static* conditions (i.e., assuming that the input varies very slowly) is shown in Figure 10.20(b).

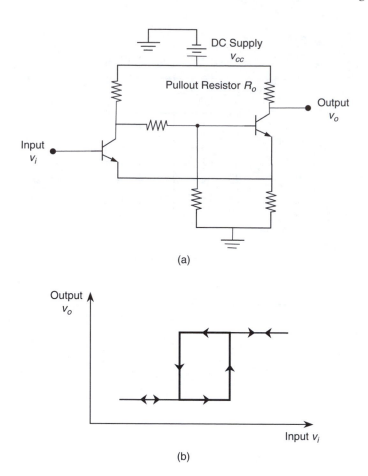

(a)

(b)

FIGURE 10.20
Schmitt trigger; (a) A circuit using bipolar transistors; (b) Quasi-static characteristic curve.

Typically six Schmitt triggers are included in a single IC package. The symbol for a logic gate with hysteresis is its usual ANSI symbol, with a hysteresis curve marked inside the symbol.

10.5 Practical Considerations of IC Chips

Logic circuitry in digital devices and in various parts of a microcomputer (microprocessor, memory, interface hardware, etc.) is usually present in the integrated circuit (IC) form. Elements such as logic gates and bistable circuits such as flip-flops are building blocks for an IC, but these elements are present in the monolithic form and not in the discrete form. In this section, let us examine the manufacture and several practical considerations of digital IC chips. This insight is useful, even though not always essential, to a mechatronics engineer.

In a digital system, data storage and processing are performed using logic circuitry. These circuits consist of logic gate elements. The logic gates are arranged in an IC circuit in a suitable manner during the manufacture of the IC, so that the IC could accomplish the required digital function. A typical IC chip contains a large number of microelectronic circuit elements manufactured on a small slice of silicon (smaller than a finger nail). Each circuit element consists of basic semiconductor elements such as diodes and transistors. These semiconductor elements are not present as conventional *discrete components* with leads and wiring for interconnection, but rather in the *monolithic form*. The entire integrated circuit is produced through a delicate manufacturing process and it is not possible to remove or replace, for example, a transistor in the IC without destroying the entire integrated circuit.

10.5.1 IC Chip Production

Unlike discrete electrical circuits, integrated circuits (microelectronic circuits) are not manufactured by starting with discrete semiconductor elements and connecting them together to form a circuit. Such a "discrete" production method would not only be expensive and time consuming but would also result in much larger chip size and poor reliability due to the large number of circuit connections that would be needed. The circuit density of an IC chip used in digital systems can be expressed in terms of the number of elementary logic circuits (logic gates) present on the chip. Four classifications are common, as given in Table 10.12. Note that the numbers given are approximate and represent the order of magnitude only. The important thing to note, however, is that it is practically impossible to produce a VLSI chip (e.g., a RAM memory chip containing 100,000 logic gates) using discrete transistors.

Integrated circuits are produced by various methods. In all such methods, the required circuit is first designed on paper or computer screen and then the entire circuit is produced on a semiconductor chip (silicon, GaAs, etc.) by several steps of a delicate and carefully controlled manufacturing process. To give an idea of the various steps involved, consider the production of a monolithic IC chip using *planar diffusion technology*. A flat piece of doped silicon crystal that is polished and cleaned is used as the substrate on the surface of which the IC is formed. Suppose that the substrate has a layer of *n*-type silicon. The method of forming the IC is somewhat similar to that for manufacturing a discrete bipolar transistor.

The surface of the silicon wafer is oxidized to form a coating of silicon dioxide (insulation). Then, a chemical coating that is sensitive to light is applied on the oxidized surface. Next, a light beam is focused onto the coating through a mask. The mask has a window structure (openings) which conforms to the required (designed) circuit pattern. Areas exposed to light become insoluble to a chemical solution (developer) that is used to wash off the unexposed chemical. Next, hydrofluoric acid is used to remove the silicon dioxide layer underneath the window pattern produced by removing the unexposed coating. The element is next placed in a heated diffusion furnace, and a stream of gas containing an acceptor-type dopant is applied. This forms a pattern of *p*-type regions underneath the windows. Next, as in the case of a discrete transistor, it is possible to form

TABLE 10.12

IC Chip Classification According to Circuit Density

IC Chip Type	Number of Basic Logic Gates Present
Small-scale integration (SSI)	Less than 10
Medium-scale integration (MSI)	10–1000
Large-scale integration (LSI)	1000–10,000
Very-large-scale integration (VLSI)	10,000 to over 1×10^6

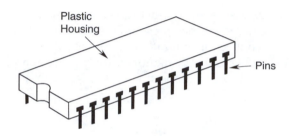

FIGURE 10.21
A dual-in-line packaged (DIP) IC.

a pattern of n-type regions over this by a systematic process of oxidation, chemical coating, masked exposure to light, developing, and diffusion using a donor-type dopant. Finally, electrical connections have to be made between various semiconductor elements formed this way on the silicon crystal. To accomplish this, the same process of photo masking is used first to open up areas that need connection. Then aluminum vapor is used to deposit aluminum on these exposed areas, thereby making electrical connection. There is no guarantee that an IC element manufactured in this manner will function correctly. Hence, the element has to be tested for proper operation, preferably before packaging.

10.5.2 Chip Packaging

Once an IC is formed on a silicon wafer, the wafer has to be cut into the correct shape and then packaged. The packaging material could be plastic or ceramic. Typically, a packaged chip is rectangular in shape with pins protruding out. Some of these pins serve as leads for information signals (data in and out) and others as power leads.

A popular packaging geometry is shown in Figure 10.21. This is known as the *dual-in-line package* (DIP). The pins of a DIP protrude down from the two longitudinal edges of the housing. Typically, up to 48 pins can be provided in a single package. An out-line package has lead pins protruding outward from the edges in plane with the underside of the chip. About 64 pins can be provided with such a package. With increased complexity of IC chips, efforts are continuing to increase the number of pins per package.

The assembly of an IC chip on a printed-circuit board (PC board) is usually a simple matter of pushing the pins of the chip through a corresponding set of holes (socket). Typically, soldering would be necessary to hold the chip in place and make electrical connections. In addition to IC chips, a printed-circuit (PC) board may contain a few discrete elements such as resistors, capacitors, and additional discrete diodes and transistors, and auxiliary devices such as switches and perhaps a socket for a ribbon cable, all permanently interconnected in a required manner usingconductive lines deposited on the PC board.

10.5.3 Applications

Solid state logic devices are common in mechatronics applications. Examples include the control of dc motors using chopper circuits, the control of ac motors using variable-frequency drives, intelligent instruments with embedded control. GaAs appears superior to silicon for high frequency and high-speed devices. Electron mobility in GaAs field effect transistors (FET) is an order of magnitude higher than in Silicon FETs. Also the power dissipation (active power) is lower for a GaAs device. Typical data rates are: Silicon: <1000 Mb/s; GaAs: 500–2500 Mb/s; CMOS: over 1500 Mb/s. It is advantageous to put logic gates and power-control devices on the same chip, in view of lower cost, improved reliability, less wiring, and lower number of components.

10.6 Problems

10.1 Discuss the reasons for using binary codes to represent information within a digital computer.

    i. Convert the following binary numbers to the decimal form:

        a. 10110

        b. 0.101

        c. 1101.1101

    ii. Convert the following decimal numbers to the binary form:

        a. 29

        b. 0.5625

        c. 10.3125

    Why is it convenient to handle binary numbers in the hexadecimal form? Write the hex values of the six numbers given above.

10.2 Convert $(18347.319)_{10}$ into the octal representation, then to binary, and from that to the hex representation.

10.3 Perform the following two's complement operations and express the results in the two's complement form:

    i. $(10010)_{2's} + (00110)_{2,s}$

    ii. $(10010)_{2's} - (00110)_{2,s}$

    iii. $(10011)_{2's} + (10100)_{2's}$

    iv. $(01101)_{2's} + (01011)_{2's}$

    Check your results using the equivalent decimal numbers.

10.4 Explain the property of "sign extension" as applied to the coding of signed numbers for use in digital computers.

    Carry out the following one's complement operations and give the answers in the one's complement form:

    i. $(01010)_{1's} + (10010)_{1's}$

    ii. $(01010)_{1's} - (10010)_{1's}$

    iii. $(10101)_{1's} + (10010)_{1's}$

Check your answers using decimal numbers.

10.5 Carry out the following binary arithmetic computations and express the results in binary. Check your answers using decimal computations.

    i. $(1101)_2 \times (110)_2$

    ii. $(11001101)_2 \div (1011)_2$

10.6 Codes are employed in digital systems to represent information such as numerical values and text (or characters). Consider the following four codes:

    a. Natural binary code

    b. A binary gray code

    c. Binary coded decimal (BCD)

    d. American standard code for information interchange (ASCII)

State which of these codes are used to represent numerical values, which ones are used to represent characters, and which ones for both numerical values and characters. Give one or more applications of these four codes in digital systems.

Using a 4-bit word, write the straight binary and gray codes for numbers from 0–15. For the starting value (zero), make the straight binary and gray codes identical. Write the BCD representation for the decimal values from 0–15.

10.7 i. Convert $(456.128)_{10}$ into the bcd form.

ii. Convert $(10000110.10010010)_{bcd}$ into the decimal form.

10.8 Demonstrate how you would apply De Morgan's law to expressions containing either products (logical AND) or sums (logical OR) of more than two terms. Illustrate your approach for

   i. $\overline{a+b+c}$

   ii. $\overline{a \cdot b \cdot c}$

   Simplify the Boolean expression

$$\overline{(\overline{a}+b) \cdot a + \overline{(b+c) \cdot \overline{c}}}$$

   NOTE: AND operations are carried out before OR operations.

10.9 Prepare a truth table to represent the Boolean relation $a = x \cdot y + \overline{z}$. Using this table express a in:

   i. A sum-of-product form

   ii. A product-of-sum form

   Verify that your answers are equivalent to the original expression for a.

10.10 The NOR gate is said to be functionally complete. Explain the meaning of this statement. Give a realization of AND using NOR gates alone.

10.11 A "2–4 decoder" is a digital device whose input is a 2-bit binary word. The output has 4 lines, one of which is activated high (1) depending on the input value. Note that the input is one of the four (decimal) values 0, 1, 2, and 3 (or, binary 00, 01, 10, 11). Realize a logic circuit for this decoder using AND gates and NOT gates only.

10.12 Illustrate how the Boolean function $\overline{a} \cdot b + c \cdot d$ may be implemented using

   i. NAND gates only

   ii. NOR gates only

10.13 What are advantages of CMOS logic family over TTL logic family? How could the two logic families be combined in the same digital system?

A TTL logic gate is shown in Figure P10.13. The inputs to the gate are A and B, and the output is C. What logic operation does this gate provide? Explain the principle of operation of this logic gate.

10.14 A basic security system for a home operates as follows:

When the system is activated in the "home" mode, only an opening of a door or window will sound the alarm. When the system is activated in the "away" mode, any motion inside the house or opening of a door/window will sound the alarm.

The activation of the system is done manually by using a binary switch with logic-1 representing "home" and logic-0 representing "away." This logic state is denoted by H.

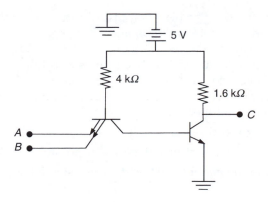

FIGURE P11.13
A basic TTL logic gatte.

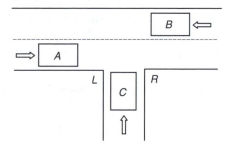

FIGURE P10.16
Turn signals for a road intersection.

The detection of a motion by the motion sensor is denoted by M. The opening of a door or a window is denoted by O. Activation of the alarm is denoted by A.

  a. Assuming that the logic signals H, M, O, and A are all active-high, develop a logic circuit using OR gates and NOT gates only for this alarm system.

  b. Assuming that the signals M and O are available in the active-low form (i.e., M.L and O.L) and the output signal A is also needed in the active-low form (i.e., A.L), modify the circuit in part (a) to realize the system (using OR and NOT gates only).

10.15 For each of the following logic functions, form the Karnaugh map and using it obtain a minimal logic function that is equivalent:

  a. $\overline{A} \cdot B + A \cdot \overline{B} + \overline{A} \cdot C + A \cdot B \cdot \overline{C} + \overline{A} \cdot \overline{B} \cdot \overline{C}$

  b. $A \cdot B + \overline{C} \cdot D + C \cdot \overline{D} + \overline{A} \cdot C \cdot D + A \cdot \overline{B} \cdot C \cdot D + A \cdot \overline{B} \cdot \overline{C} \cdot \overline{D}$.

10.16 A one-way road meets a two-way road, as shown in Figure P10.16. The vehicles on the one-way road are allowed to make right turns and left turns only (i.e., it is a dead end). There are sensors to detect vehicles at positions A, B, and C, which generate high signals (logic-1) when vehicles are present and low signals (logic-0) when vehicles are not present. Logic devices are to be developed to operate the right-turn signal (with active-high output R)

and the left-turn signal (with active-high output L) in the one-way road. Inputs to these devices are the logic signals denoted by A, B, and C, as generated by the vehicle-detection sensors.

The logic governing the operation of the two signals is as follows: The right-turn signal is on when there are no vehicles at A and there are vehicles at C. The right-turn signal is on as well, when there are no vehicles at A, B, and C. The left turn signal is on when there are no vehicles at A and B and there are vehicles at C.

   a. Express the logic governing the two devices.

   b. Using a Karnaugh map, minimize the logic, if possible.

   c. Give circuits for implementing the two devices using NOR gates and NOT gates only.

10.17 a. How do sequential logic devices differ from combinational logic devices?

   b. How does the asynchronous operation of a logic device differ from the synchronous operation?

   c. Give a circuit for an RS flip-flop using NOR gates only. Extend this to an asynchronous latch circuit, using NOR gates and NOT gates.

10.18 a. Compare and contrast the following three types of flip-flops: JK flip-flop, D flip-flop, and T flip-flop. Indicate how the latter two flip-flops can be derived as special cases of a JK flip-flop. What is a practical use of the data "toggle" capability of a T flip-flop?

   b. Develop a 3-bit binary counter using D flip-flops. A counting sequence of straight binary 3-bit words (000, 001, 010, 011, 100, 101, 110, 111) is needed.

10.19 Outline the production process of a typical IC chip. Suppose a digital control circuit is assembled using discrete elements such as bipolar junction transistors, diodes, capacitors, and resistors, instead of their monolithic versions. What are shortcomings of such a controller in comparison to a single-board controller that uses (monolithic) IC chips?

10.20 Explain the acronyms IC, PC board, SSI, MSI, LSI, and VLSI. Give a classification for IC devices based on the logic gate density. Into what category would you put a modern 32-bit microprocessor chip?

10.21 A basic characteristic of a digital system is that many hardware components of the system are able to store and/or transfer binary data. Since a two-state element is needed to represent a binary digit (bit), this type of digital hardware should physically possess the two-state characteristic. Briefly state the two physical states associated with elements in the following types of hardware:

   a. TTL circuit

   b. Nonvolatile MOSFET memory

   c. Magnetic bubble memory

   d. Optical fiber

   e. CCD (Charge-coupled device)

   f. Microcomputer diskette

   g. Microcomputer hard disk

   h. EAROM (Electrically alterable read only memory)

10.22 a. What is a combinational logic circuit and what is a sequential logic circuit? Describe the use of a flip-flop or a latch as a basic element in semiconductor memory.

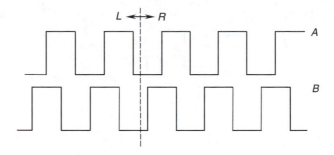

FIGURE P10.22
Two pulse sequences which are 90° out-of-phase as generated by a motion encoder.

Prepare a truth table similar to what is shown in Figure 10.14(b), for an RS flip-flop that uses two cross-coupled NOR gates.

b. An optical encoder (a motion sensor, which senses displacement in steps and produces pulses correspondingly) generates two pulse sequences, which are 90° out of phase. In one direction of motion (denoted by L), one pulse sequence (denoted by A) leads the other (denoted by B) by 90°, and in the opposite direction of motion (denoted by R) this pulse sequence (A) lags the other (B) by 90°. This situation is shown in Figure P10.22. Suppose that the two sequences are read into a two-bit register, the high voltage level of each signal being represented by a 1-bit and the low (zero) voltage level being represented by a 0-bit. Verify that in the L direction of motion, the register value will change according to the sequence:

$$A \quad 0 \quad 1 \quad 1 \quad 0 \quad 0 \quad \ldots$$
$$B \quad 0 \quad 0 \quad 1 \quad 1 \quad 0 \quad \ldots$$

and in the R direction of motion it will be:

$$A \quad 0 \quad 0 \quad 1 \quad 1 \quad 0 \quad \ldots$$
$$B \quad 0 \quad 1 \quad 1 \quad 0 \quad 0 \quad \ldots$$

Explain a simple way to detect the direction of motion by checking the binary value in the register. Discuss a way to physically implement this direction-detection method.

10.23 What is the main advantage of including hysteresis in a switching element? What is a Schmitt trigger? Describe the operation of the Schmitt trigger circuit, which employs two bipolar junction transistors as in Figure 10.20(a). Even though a Schmitt trigger can be constructed using discrete elements in this manner, this device is commercially available in the monolithic form as a single IC chip. Suppose a sinusoidal signal is applied to a Schmitt trigger, as schematically shown in Figure P10.23. Discuss the shape of the output signal. What is the output frequency?

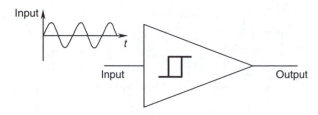

FIGURE P10.23
A Schmitt trigger excited by a sinusoidal signal.

10.24 Typically in a digital controller, the arithmetic operations ADD and SUBTRACT are performed using hardware. More complex operations including MULTIPLY and DIVIDE could be implemented either by hardware or by software. The software operations could be more than an order of magnitude slower. Compare the use of hardware-intensive controllers (or hard-wired controllers) with software-intensive controllers in terms of factors such as speed of processing, flexibility, possibility of using complex control algorithms, and controller cost. Would you classify a ROM-based controller that does not have programmable memory, as a hard-wired controller or as a software-intensive controller?

11

Microprocessors and PLCs

Digital hardware can perform a variety of functions in a mechatronic system. Signal acquisition and processing, system monitoring and control, switching, and information display are such functions. Simple functions may be implemented using logic hardware, as discussed in Chapter 10. More complex tasks will require programmable digital devices, or in a broader sense, digital computers. In particular microprocessors are commonly used in mechatronic applications. For example, in digital control of a mechatronic system, a digital computer serves as the controller. The computer uses external input commands along with measured responses of the system to generate suitable control signals. These control signals are used to drive the system. In microprocessor-based control, one or more microprocessors and related hardware (memory, storage, input/output devices, etc.) are used as the controller. If many mechatronic devices have to be operated in a coordinated manner, for example in a plant setting, a programmable logic controller or PLC can be used to control their operation. This chapter deals with programmable digital devices, particularly microprocessors and PLCs.

11.1 Digital Computer

The heart of a computer is a set of printed circuit (PC) boards (or PC cards or PC modules). Of course, other components such as various data storage devices including compact disk (CD) drives, a display unit (e.g., liquid crystal display or LCD screen), keyboard, mouse, and printer might be needed for the general operation of the computer. A PC board is typically an assembly of integrated-circuit (IC) chips interconnected together and, perhaps, with discrete components (discrete capacitors, discrete diodes, discrete transistors, discrete resistors, etc.). An IC chip is simply a packaged integrated circuit. It can represent a microprocessor, a memory unit, or some other circuit used in a digital system, or even an analog circuit.

The minimum hardware requirement for a digital computer is a *central processing unit* (CPU) for processing digital data, a *memory* unit for storing data, and input/output circuitry (I/O) for sending information into the computer and obtaining information back from the computer. Early types of digital computers used discrete semiconductor elements in their logic circuitry within the CPU, and magnetic core memory for storing information. Disadvantages such as large size, high power consumption (and heat generation), slow data processing speed, and low data handling capacity lead to extensive research and development activities that have resulted in miniaturization of the digital computer through the use of microelectronics, which has radically improved the functionality, capacity, and speed while reducing the cost.

The use of very large scale integration (VLSI) chips not only for data processing (Microprocessor) but also for memory (RAM, ROM, PROM, etc.) and interfacing (I/O) hardware, was responsible for the advent of the microcomputer in the early 1970s. Since then, great strides have been made in the microprocessor technology. Today, microprocessor has become an integral part of many mechatronic devices along with such components as smart sensors, digital signal processors, intelligent controllers, and microelectromechanical systems (MEMS).

11.1.1   Microcomputer Organization

A microcomputer consists of at least a microprocessor, memory unit, and input/output hardware. In a typical microcomputer, these devices interact through multiline electrical communication paths (e.g., conductors or cables) known as buses. Three types of buses can be identified: The data needed at various locations in the microcomputer, are carried by a *data bus*. The address (digital name) of a memory location or hardware component that needs to be accessed by the microprocessor, is carried by an *address bus*. The commands issued by the microprocessor that controls the operation of the microcomputer system, are carried by a control bus. A schematic representation of a computer system of this type is given in Figure 11.1. Note that more than one memory unit may exist in a computer. Furthermore, input/output hardware can include an interface board (interface card) consisting of many hardware units including sample-and-hold circuits (S/H), analog to digital converters (ADC), digital to analog converters (DAC), and *multiplexers* (MUX), as discussed in Chapter 4. Peripheral devices such as a hard drive for mass storage of data, display screen, printer, mouse, and keyboard are not exclusively shown in Figure 11.1.

11.1.1.1   Software

A microprocessor performs its intended functions using *instructions* and *data*, both provided through *programs* and external commands (say, using a keyboard or mouse or touchscreen). Instructions (programs) provided by the vendor (manufacturer) or developed by the user and stored in computer memory are known as software. In particular, vendor-provided software that is stored in a *read-only-memory* (ROM) unit and that cannot be generally altered by the user, are called *firmware*. Also, for specialized functions, some programs may be permanently converted into digital hardware. These hardware implementations have the advantage of fast speed, lower cost (in mass production), and convenience of operation, but

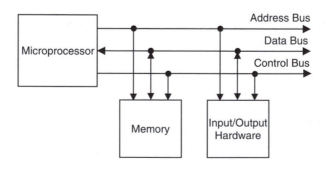

FIGURE 11.1
Basic structure of a microcomputer.

lack flexibility because the program is permanently placed in the logic microcircuits and cannot be modified or reprogrammed.

11.1.1.2   Operation

All information (*data, addresses, control instructions, operation codes*, etc.) within a microcomputer is present in the binary form as, for example, voltage levels, voltage pulses, or on/off states of digital logic elements, as discussed in Chapter 10. Consider the operation of the microcomputer system shown in Figure 11.1. For example, suppose that the microprocessor wants to read a piece of data from memory, for the purpose of numerical processing (addition, multiplication, etc.). The microprocessor knows (it should) the address of the particular location in the memory at which the piece of data is stored. It will place this address code (as voltage pulses, for example) on the address bus and also will place the control signal corresponding to the instruction "Read from memory," on the control bus. In response, the memory will place the piece of data on the data bus, from which the microprocessor will read the data into one of its registers. Next, consider storing in memory a piece of data processed by the microprocessor. First, the microprocessor will place the piece of data (perhaps currently located in its *accumulator* or *data register*) on the data bus, and will place on the address bus the address code of the location in the memory at which the piece of data is to be stored. Finally, the microprocessor will place the "Write in memory" control signal on the control bus. This will store the piece of data at the proper address location in the memory. Note, however, that this process will not automatically erase the piece of data generated inside the microprocessor (which is present in the accumulator or the data register). Data input from other (peripheral) devices and data output to other devices can also be handled in the same manner.

Now consider the task of data processing by a microprocessor. A computer program to perform the processing task is stored in the memory. The particular step of the program (i.e., *instruction*) that is being executed, is selected according to the value in the *program counter*. This value is the *instruction address*. Suppose that the task, according to the computer program, is to add the data stored in the memory address location B to the data at the address location A, and store the result at the address location C. First instruction address (picked up according to the program counter) will identify an instruction in the program, which will "instruct" the microprocessor to read the data at the memory location A into the microprocessor accumulator. Consequently, this instruction address location (in the memory where program is stored) will contain the *operation code* for "Read from memory" and the address of the memory location A. Similarly, the second instruction address location will contain the operation code for "Addition" and the address of the memory location B. This will result in addition of the data at location B into what is already in the accumulator of the microprocessor. The result will remain in the accumulator. The third instruction address location will contain the operation code for "Store in memory" and the address of the memory location C. This will cause the data in the accumulator to be placed at location C of the memory.

11.1.2   Microprocessor

The brain of a microcomputer system is the microprocessor unit. This is a programmed "artificial brain." It processes data (numerical and logical) according to instructions (computer program) supplied to the computer. Conventionally, a microprocessor chip contains the processor only. But some modern chips may contain elements of user-accessible memory and input/output hardware in addition to a microprocessor. In the latter case,

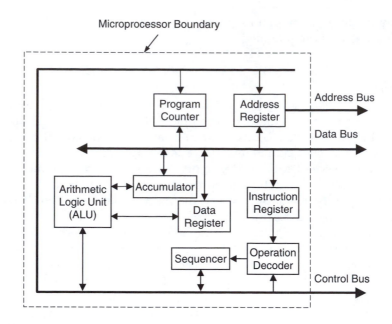

FIGURE 11.2
Schematic representation of a microprocessor.

we have the so-called "computer on a chip." Let us briefly discuss the architecture of the microprocessor portion alone. Memory and input/output hardware will be discussed subsequently.

A simplified structure of a microprocessor unit is shown in Figure 11.2. We have already mentioned several components included in this schematic representation. Carefully note the interaction among various components shown in the figure. A brief description of the function of each component is given next.

Arithmetic Logic Unit (ALU): The ALU of a microprocessor can perform arithmetic operations such as *addition, subtraction, multiplication,* and *division,* and logic operations such as *AND, OR, NOT, NAND,* and *NOR*. Depending on the number of *primitive operations* available with the particular microprocessor, some or all of these operations could be performed by the ALU in response to the corresponding operation codes. For instance, if multiplication is not a primitive operation for a particular microprocessor, it can be performed by several ways such as repeated addition, or addition and shift. Similarly, subtraction can be performed by using addition (e.g., by the two's complement method, as discussed in Chapter 10). It follows that the most primitive ALU may contain logic circuitry for addition (i.e., *adder*) only.

Typically, data (i.e., *operands*) to be processed by the ALU are read (for instance, from memory) into the *accumulator* and the *data register* (see Figure 11.2). The operation code corresponding to the operation to be performed by the ALU is placed in the *instruction register*. Note that the operation code is available from the computer program, which is stored in the memory. This *"opcode"* is in the program step (instruction), which is addressed according to the contents of the program counter. The operation code is decoded by the *operation decoder*. A control signal is next supplied to the ALU by the *sequencer* in order to perform the particular operation. Once the operation is performed, the result is stored in the accumulator and may be subsequently stored in the memory or sent to an output device through the data bus.

Program Counter: During execution, the computer program is stored in the memory. Each instruction (or program step) is stored at a specific memory location with an associated address. The program counter is a storage area within the microprocessor, which contains the address of the instruction (program step) to be executed next. When this address is placed on the address bus, the memory will respond by placing the data (program step) contained in that memory location on the data bus. The address in the program counter will advance (increment) sequentially so as to execute the program in a sequential manner unless instructions such as jump or branch are encountered. In the case of a branch instruction, the contents of the program counter will be changed to reflect the program step to which the execution is branched off.

Address Register: The address of a memory location or peripheral device that needs to be accessed by the microprocessor is contained in a temporary register known as the address register. When this address is placed on the address bus, the data in the corresponding memory location will be placed on the data bus, or the addressed hardware unit will be "enabled," perhaps placing some data on the data bus.

Accumulator and Data Register: These are registers in the microprocessor, which can temporarily hold data, to be processed by the arithmetic logic unit. Typically, when a program step is addressed, data to be processed in that instruction are placed on these registers. Furthermore, results produced by the ALU are temporarily stored in these registers, to be placed on the data bus, for example, for subsequent storage in the memory.

Instruction Register: Instructions such as an operation code contained in a program step (stored in the memory and accessed by the microprocessor) are temporarily stored in an instruction register, awaiting decoding.

Operation Decoder: The operation decoder interprets an operation code stored in the instruction register. The decoded information is then passed on to a sequencer to be sent to the ALU. In this manner, the ALU finds out which operation is to be performed on the data (i.e., operands) that are stored in the data registers.

Sequencer: It is usually not possible for the ALU to process data immediately after the operation decoder interprets the opcode. The sequencer controls the sequence in which the ALU performs its operations. The sequencer may receive control signals from peripheral devices or a "Ready for action" signal from the ALU.

11.1.3   Memory

A microprocessor must have access to the data that it processes, and the instructions (computer program), which inform it how the data should be processed. Computer programs (in the software form) and data are stored in the computer memory for access during processing. Memory is different from mass storage media such as disks and "memory sticks," which are off-line devices. During operation, the microcomputer may transfer data (and programs) stored in a mass storage medium into the memory, and may transfer processed data currently stored in the memory back to the mass storage medium. It is the contents of the memory that can be directly accessed by the microprocessor.

11.1.3.1   RAM, ROM, PROM, EPROM, and EAROM

There are many categories of memory. Read and write memory (popularly termed random access memory) is denoted by RAM. The user can read the contents of a RAM and also write (store) information (data, programs) into a RAM. Hence, user generated software is typically stored in a RAM chip.

Read only memory is denoted by ROM. The user can read contents of a ROM but he cannot modify the contents; the user cannot write into a ROM. It does not lose its contents when the power is removed. Typically, the permanent software provided by the manufacturer (i.e., firmware) is stored in a ROM.

Sometimes, the user (not the manufacturer) may wish to store information permanently in memory. A programmable read only memory or PROM is used for this purpose. With a special writing device, user can write into a blank PROM once, but the stored contents cannot be modified thereafter.

One may wish to have permanent memories that can be modified occasionally, under special circumstances. Erasable programmable read only memory (EPROM) is useful in that case. The contents of an EPROM can be erased using specialized equipment (e.g., by applying ultraviolet light for 30 min). Then, new information can be stored again on the EPROM (permanently, until erased using the special equipment).

Electrically alterable PROM is denoted by EAPROM. This is also denoted by EEPROM (electrically erasable PROM). This memory can be used permanently as a ROM, but it can be modified occasionally by using electrical currents (rather than ultraviolet light) of fairly high voltage produced by special circuits. Advantages of EAPROM include the fact that carefully controlled modification of memory contents (say, erasure of a selected portion of the contents) is possible and that the memory chip does not have to be removed from the PC board for modification. Disadvantages include the slowness of the modification process and the complexity of the electrical circuitry that would be required in the process.

11.1.3.2　Bits, Bytes, and Words

Information within a microcomputer is stored, processed, and transferred as *binary digits* or bits. As discussed in Chapter 10, one bit of information can be represented by the presence or absence of an electric pulse, high or low level of a voltage signal or, basically, the state (on/off) of a bistable (two-state) element. Data storage in computer memory may be achieved using a grid of bistable elements. A memory chip that can store 1024 bits (i.e., 2^{10} bits) is a 1 Kb memory chip. Note that RAM chips of hundreds of megabits (Mb) in capacity (Note: 1 Mb = 1024 Kb) are commonly available, and the cost is decreasing as well. This is a rapidly growing area.

A basic microcomputer handles information grouped into 8 bits simultaneously. Such a group of 8 bits is usually termed a *byte*. (Note: 4 bits is termed a *nibble*). In computer terminology, a collection of 1024 bytes is a one *killobyte* or simply 1 K. Also, 1 *megabyte* (or 1 M) is equal to 1024 K.

The number of bits that can be manipulated by a computer in one operation is called the *word size* of the computer. One word of data is stored in each register of a microprocessor and it can be processed by the ALU in a single operation. This usually is also the size of data that can be placed on the data bus of the microcomputer. Today the 32 bit microcomputers (i.e., data word = 32 bits) are commonly available.

Volatile Memory: If the contents of a memory are automatically erased when the power source of the computer is turned off, the memory is said to be volatile. Many types of RAM chips have volatile memories. ROM chips should maintain their contents under power-off conditions as well. Hence, they should have *nonvolatile* memories.

11.1.3.3　Physical Form of Memory

Any bistable element can store a bit of data. It follows that a variety of physical devices can serve as computer memory. The density of data storage (which governs the physical size), speed of data transfer, cost, and compatibility with the overall microcomputer system are the deciding factors in the development of microcomputer memory.

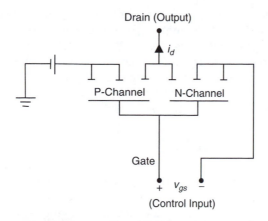

FIGURE 11.3
CMOS (Complementary Symmetry MOSFET) memory element.

Semiconductor memory is widely used in microcomputers, but magnetic bubble memory technology is used as well. A *semiconductor memory* is simply an IC chip consisting of many bistable circuit elements (e.g., flip-flops), each element representing a bit of data. The technique used to store data in a memory chip depends on the particular type of semiconductor memory.

One type of memory chip uses metal oxide semiconductor field effect transistor (MOSFET) elements (see Figure 11.3) to store bits of data. The level of electric charge between the gate lead and the channel lead is used to represent a bit. A level of charge above a threshold value will represent 1 and a charge level below the threshold value will represent 0. Hence, a 1-bit can be stored by energizing the gate of the MOSFET at the correct memory location. Since this charge gradually leaks off, it has to be periodically refreshed (say, at a frequency of 1 kHz) in order to maintain its value. Hence, this type of RAM is called *dynamic RAM* or, *DRAM*), and is implemented using dynamic circuitry. This type of memory is volatile and cannot be used as ROM.

A *static* memory, unlike dynamic memory, does not have to be periodically refreshed. A common static RAM uses flip-flop elements. As discussed in Chapter 10, a flip-flop element consists of logic gates made of semiconductor elements, and it will maintain its output state even after the inputs are removed. A static RAM of this type tends to be more expensive than a dynamic RAM, because more complex circuitry is involved. An advantage is, however, that a static RAM does not require circuitry for continuously refreshing the inputs.

A semiconductor memory that is commonly used as EPROM employs MOSFET elements to permanently store electrical charges. Each MOSFET has an additional (floating) gate, which is insulated (using a silicon dioxide layer) from the rest of the transistor. When a high voltage (e.g., 25 V) is applied between the regular gate lead and the drain lead, the floating gate will be charged, but this charge will be trapped by the insulation when the voltage is removed. Hence, a 1-bit will be retained permanently at that memory location. The chip can be erased by exposing it to ultraviolet light for about a half an hour. This process makes the silicon dioxide insulation of the floating gate temporarily conducting, thereby discharging the gate. New data can be stored again in the chip, by charging the floating gate as before.

A type of permanent memory chip that cannot be erased and reused is made of semiconductor elements linked using nichrome wire. Initially, the circuit is such that all bits are set to 0. To store data (i.e., 1-bits at appropriate locations), an electrical current is passed through a properly selected set of nichrome links in the chip. The current will fuse the links converting a selected set of elements to 1s. Note that the process cannot be reversed. Hence, these types of chips may be used as ROM or PROM but not as EPROM.

Another type of memory that is slower than semiconductor memory is the *magnetic bubble memory*. A bubble memory consists of a thin film of magnetic material. Very small (microscopic) cylindrical domains (bubbles) of magnetism are created on the film by a delicate method. Each bubble represents a 1-bit of data. This type of memory is nonvolatile, yet erasable.

Memory Access: Each memory location has a unique address so that it could be accessed by the microprocessor. The address of a memory location is certainly not the same as the information (data) stored at that location. Typically, each memory location can store one data word. Each memory location can be connected to the data bus through a *data buffer* using a control signal. A specific memory location to be connected to the data bus is chosen depending on the address on the address bus. This address is decoded and the corresponding memory location is activated. Depending on the control signal, the contents temporarily stored in the buffer are then stored at the memory location, or vice versa. A schematic representation of this scheme of memory access (read/write) is shown in Figure 11.4. Note that the buffer has three states (*read, write, no action*). The access scheme can be extended to the case where there are more than one memory chip to be accessed. In this case, first the "chip select" control is activated and information on the address bus is used to choose the proper memory chip. Next, write control or read control is activated as in the single chip case to perform the appropriate operation.

Consider the static RAM chip shown in Figure 11.5. This is a $32K \times 8$ semiconductor (CMOS) RAM. Specifically, it has 32K memory locations, each holding 8 bits of memory. Since $32K = 32 \times 2^{10} = 2^{15}$, we need a 15-bit address to identify all 32K locations of memory. These address bits are denoted by A0 to A14 in Figure 11.5. When a particular (8-bit) memory location is addressed using the corresponding combination of 15-bit address, the contents of the memory (8 bits) can be accessed by the eight I/O pins denoted by I/O1 to I/O8. In particular, when the pin WE is activated (using logic 0; because "active low" as discussed in chapter 10) the eight I/O pins can be used to send 8 bits of data to be stored at the addressed memory location. The "overbar" in $\overline{WE}$ denotes "active low." When the lead OE is activated (using logic 0, because "active low"), the contents (8 bits) of the addressed

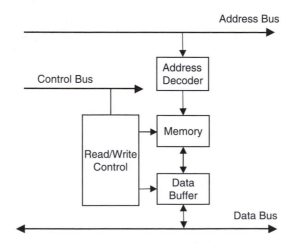

FIGURE 11.4
Memory access scheme.

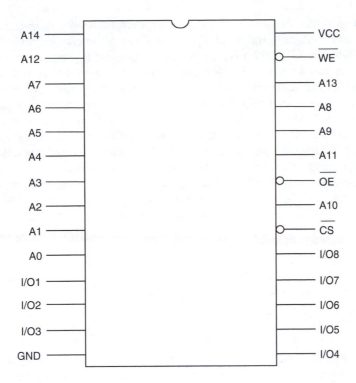

A14	VCC
A12	$\overline{\text{WE}}$
A7	A13
A6	A8
A5	A9
A4	A11
A3	$\overline{\text{OE}}$
A2	A10
A1	$\overline{\text{CS}}$
A0	I/O8
I/O1	I/O7
I/O2	I/O6
I/O3	I/O5
GND	I/O4

A0 – A14 : Address Pins
I/O1– I/O8 : Data Input/Output Pins
$\overline{\text{CS}}$: Chip Select (Active Low) Pin
$\overline{\text{WE}}$: Write Enable (Active Low) Pin
$\overline{\text{OE}}$: Output Enable (Active Low) Pin
VCC : Power Supply Pin
GND : Ground Pin

FIGURE 11.5
A 32K × 8 RAM chip.

memory location can be read from the eight I/O leads. The particular memory chip is selected by the processor by sending a logic low signal to the lead CS, which is the "chip select" lead (which is active low).

Example 11.1

Consider a RAM chip having 256 memory locations. Discuss different ways this memory can be structured and accessed using the address. What is the minimum word size of the memory address that is required?

SOLUTION
One possible way of arranging the 256 locations of memory is as a single row having 256 locations. In this case an 8-bit address would be needed (Note: $2^8 = 256$). All 8 bits of the address word are used simultaneously to define the 256 locations along the row. Alternatively, the memory could be arranged as a matrix of two rows, each row having 128 locations. In this latter case 1 bit of the address is used to denote the row number. For example, value 0

of this bit could represent first row and, then, value 1 would represent the second row. Now, 7 bits are necessary to denote the memory location along the chosen row (Note: $2^7 = 128$). Here again, an 8-bit word is required as the memory address. Yet, another possibility is a memory matrix having 4 rows and 64 locations on each row. In this case 2 bits are required to represent the row number, and 6 bits ($2^6 = 64$) are necessary to define the location along each row. Again, we need an 8-bit word. Next, consider a matrix of 8 rows, each row having 32 memory locations. In this case 3 bits of the address word have to be allocated for defining the row number and 5 bits ($2^5 = 32$) are needed for defining the location along each row. Once again, an 8-bit address would be needed. These four memory arrangements are the optimal structures, and in each case an 8-bit address is required. Other structures are not optimal. For example, consider a structure having 3 rows. In this case 86 memory locations are needed on each row. Since 2 bits are needed to define the row number and 7 bits for defining the memory location on each row, we need a 9-bit address for this structure.

11.1.3.4   *Memory Card Design*

Several memory chips can be placed in a memory card, to form a memory of larger capacity, as needed. Consider an $mK \times p$ memory chip. This has mK memory locations, each location being able to hold p bits of data. Suppose that $m = 2^r$. Since $1K = 2^{10}$, it is noted that there are 2^{10+r} memory locations, and as a result, $10 + r$ bits are needed to address these locations.

When several memory chips are placed in a memory card, then, an address is needed to select a specific chip and a further address is needed for a memory location in that chip. A decoder may be used to convert the chip selecting address to the required $\overline{\text{CS}}$ logic of the selected chip. The I/O will then match the addressed memory location of the selected chip.

Example 11.2

Suppose that a required number of 2K × 8 EPROM chips and 2K × 8 RAM chips are available. Construct a memory card containing 4K × 8 EPROM and 4K × 8 RAM.

SOLUTION

We have to use two 2K × 8 EPROM chips and two 2K × 8 RAM chips in order to meet the requirement. A 2-bit address is needed to select one of these four chips. A further 11 bits are needed to select a memory location in the selected chip. Let us denote these 13 bits by $A_0, A_1, \ldots, A_{12}$. The bits A_{11} and A_{12} are used to select a chip and the bits $A_0, A_1, \ldots, A_{10}$ are used to address a memory location of the selected chip. This arrangement is shown in Figure 11.6.

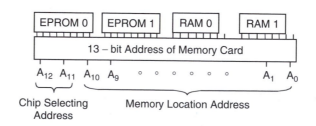

FIGURE 11.6
A memory card containing two EPROM chips and two RAM chips.

TABLE 11.1

Chip Selecting Logic

		$\overline{\text{CS}}$ of				
A_{12}	A_{11}	**EPROM 0**	**EPROM 1**	**RAM 0**	**RAM 1**	**Select**
0	0	0	1	1	1	EPROM 0
0	1	1	0	1	1	EPROM 1
1	0	1	1	0	1	RAM 0
1	1	1	1	1	0	RAM 1

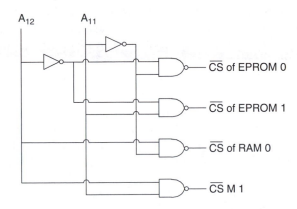

FIGURE 11.7
A 2 to 4 decoder for chip selection.

The values of A_{11} and A_{12} have to be chosen such the when a particular chip is addressed, its $\overline{\text{CS}}$ is set to 0 while the $\overline{\text{CS}}$ of the other chips are all set to 1. This is the case for "active low" logic, as discussed in Chapter 10. A "2 to 4 decoder" may be used for this purpose, as indicated in Table 11.1. A logic gate realization of this decoder is shown in Figure 11.7.

11.1.4 Input/Output Hardware

Returning to the microcomputer arrangement shown in Figure 11.1, note that so far we have discussed the microprocessor and memory in some detail. In this section, we will examine the input/output hardware and related aspects. Several pertinent issues have been discussed in Chapter 4. Buses and information networks will be discussed in a subsequent section.

Input/output hardware forms the link between a microcomputer and a *peripheral device* such as a printer, display, keyboard, or a system being monitored or controlled (i.e., a plant). It is virtually impossible and often useless to directly connect a peripheral device to the microprocessor or the memory. Some of the reasons for this are as follows:

1. If devices are directly connected to a microprocessor, it will not know from which devices or memory location the data are coming from (during input), and also will not be able to "select" a device to supply data (during output).

2. Data may be received from several peripheral devices simultaneously. If a peripheral device does not wait until the microcomputer is ready to accept data

from that device, the system cannot function properly. In monitoring and control of a mechatronic system, often, many response variables from the system may have to be read and more than one control signal may have to be generated. The two items mentioned in Item 1 above are applicable here as well.

3. Data from the external device might be *analog signals*, which have to be converted into the digital form before entering into the microcomputer.

4. Speed of data processing of the microcomputer will be different (often faster) than the speed at which data are handled (printed, displayed, or generated) by a peripheral device. Hence, *synchronization* (or *proportioning*) of data rates would be necessary, particularly in real-time monitoring and control.

5. Some computations (e.g., fast Fourier transform) require blocks of data, not just one data sample. Hence a means of *buffering* the data would be needed. Furthermore, a peripheral device might generate data in a *bit-serial* manner, as a sequence of pulses, whereas, a microprocessor processes words (e.g., 16 or 32 bits) of data, and the memory stores words of data as well. A means for this serial-parallel conversion is needed.

6. Voltage levels and currents of signals handled by external devices are often different from what are compatible with a microcomputer. Furthermore, the *impedance* of a peripheral device has to be matched with that of the microcomputer (see Chapter 4). Otherwise signal distortion due to loading will result.

The required input/output hardware (or *interface hardware*) will depend on many factors including the characteristics of the peripheral device that is connected to the microcomputer (e.g., analog or digital data, data rate, signal level, serial or parallel data), the nature of application (e.g., automatic control, real-time process monitoring, data acquisition and logging for off-line processing), and the number of input signal channels and output signal channels. Often, the function of interface hardware can be interpreted as signal modification. For this reason, we have already discussed some types of input/output hardware in Chapter 4. They are briefly mentioned again in this section, for the sake of completeness.

The input/output process can take place in many ways. Three main ways of data transfer through interface hardware are as follows:

1. Programmed input/output
2. Interrupt input/output
3. Direct memory access (DMA)

The three methods are illustrated in Figure 11.8.

Programmed I/O: In this case data transfer takes place under the control of a program running in the microcomputer. This is illustrated in Figure 11.8(a). First, the program selects the proper peripheral device to be accessed and the microprocessor sends a control signal to activate the corresponding interface hardware and, perhaps, to inform whether it is a data input (i.e., read) operation or a data output (i.e., write) operation. In an input operation, data are transferred from the peripheral device into the data register (or accumulator) of the microprocessor, which might eventually be stored in the computer memory. In an output operation, data are transferred from the data register of the microprocessor to the peripheral device. Since there might be many peripheral devices or data channels interfaced with a microcomputer, there should be a way to pick the proper device or channel for data transfer. A common method used for this, in programmed I/O, is known as *Memory Mapped I/O*. This method is illustrated in Figure 11.9. In this case, each data

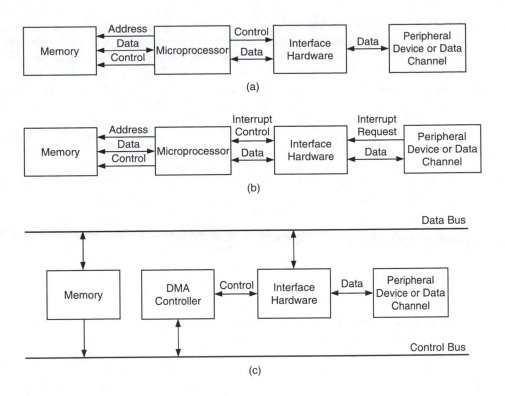

FIGURE 11.8
Three methods of input/output data transfer: (a) Programmed I/O; (b) Interrupt I/O; (c) Direct memory access (DMA).

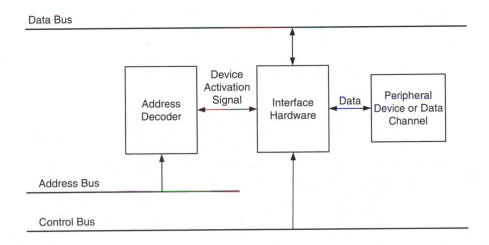

FIGURE 11.9
Memory mapped I/O.

channel (or peripheral device) is treated by the microprocessor as a memory location, and is assigned an address. Clearly this address must be unique and not an address already used for a memory location. To select a data channel, the microprocessor places the corresponding address on the address bus. The address decoder interprets the address and activates the proper I/O line. Also, the microprocessor sends a control signal to instruct whether it is an input operation or an output operation, unless automatically known, depending on the type of peripheral device (or channel) chosen. In the case of an output operation, the microprocessor places the data on the data bus. The peripheral device then picks up this data. In an input operation, the interface hardware places the data on the data bus and the microprocessor picks up that data. The microprocessor should have a way to find out whether a peripheral device is ready to accept data from the computer or whether a device has data ready to be read into the computer. One method of doing this is by *polling*. In this method, the microprocessor (when not engaged in an important processing activity) will periodically scan the I/O channels to see whether a particular channel (or device) is ready for an input/output activity. This readiness may be indicated by a single-bit *status register* (e.g., 1-bit denoting "ready" and 0-bit denoting "not ready"). Some scheme of *priority assignment* would be needed to select a channel when more than one channel are ready. Polling is a slow and somewhat wasteful technique of device selection. It has to scan all channels in each cycle of polling and it uses up the microprocessor even when the input/output channels are not ready. A method of selecting an I/O device, which does not have these disadvantages is the use of *interrupts*, as described below.

Interrupt I/O: This method of input/output data transfer is illustrated in Figure 11.8(b). In this method, a peripheral device, when ready for an input/output activity, sends an *interrupt request* signal to the microprocessor. The microprocessor suspends its current activity (after completing the execution of the instruction that it is currently executing) and performs the I/O activity and, then, returns to the original (interrupted) activity. Like in the case of polling, some priority assignment method is needed to handle situations where there are several interrupt requests simultaneously. For example, the microcomputer can have more than one interrupt request line, which are occupied according to some priority (e.g., on the first-come-first-served basis). The microprocessor services the interrupt requests one at a time, according to that priority. The basic steps of servicing an interrupt request are as follows: first, the microprocessor completes execution of the current instruction; second, it saves the contents of its registers in the microcomputer memory; third, it sends an *interrupt acknowledge* signal through a control line to the peripheral device. This initiates the data transfer process. Finally, the microprocessor sends a *service complete* signal, which will disable the particular I/O line, and will resume the interrupted activity.

Direct Memory Access (DMA) Method: This method of input/output data transfer is illustrated in Figure 11.8(c). Unlike the two previous methods, this method needs very little microprocessor activity, and it is suitable for fast transfer of bulk data. In this method, I/O data transfer takes place directly between the microcomputer memory and the peripheral device through a data bus, under control of a *DMA Controller*. The DMA Controller temporarily suspends the operation of the microprocessor, generates the control signals that are needed to perform the data transfer operation, and reactivates the microprocessor at the completion of the data transfer. This is known as *cycle stealing*, because the DMA Controller steals an operating cycle of the microprocessor. In fact, during the data transfer operation, the DMA Controller acts as a toned-down version of the main microprocessor.

   Some of the basic components found on an interface board are shown in Figure 11.10. Let us briefly describe these components. Additional details are given in Chapter 4.

Analog Signal Conditioning Hardware: Noisy analog signals have to be filtered. If the voltage level is not compatible with the connected hardware, it has to be adjusted using proper amplifier circuitry. If the impedances of the interconnected devices are not matched, signal distortion will result. Hence, impedance matching circuitry might be needed for

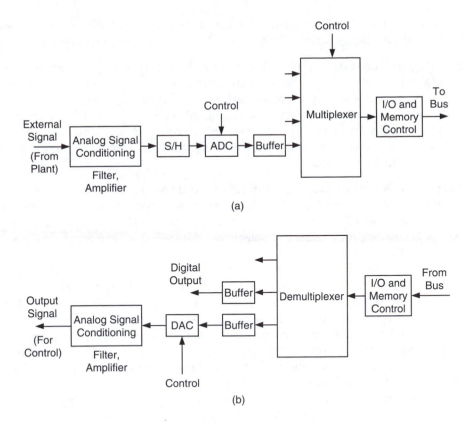

FIGURE 11.10
Basic interface hardware: (a) Data input components; (b) Data output components.

accurate data transfer. Filters, amplifiers, and related circuitry used for signal conditioning and impedance modification have been discussed in Chapter 4.

Sample and Hold (S/H) Circuit: Input signals and output signal of a computer (e.g., measured response signals and control/actuation signals) are transient in nature. They continuously change with time. The analog to digital conversion process takes a finite length of time. If the analog signal is continuously supplied to the analog to digital converter (ADC), there will be ambiguity as to what data sample is to be converted into the digital form. Hence, it is necessary to sample a data value from an analog signal and maintain that value constant until the ADC completes the conversion process. This is accomplished by means of a sample and hold circuit. The device consists basically of a solid-state switch (such as a MOSFET) and a capacitor connected to a high-input-impedance amplifier such as voltage follower. The switch is closed instantly (in the *sample mode*) by a control (timing) signal at each sampling instant. This charges the capacitor up to the level of the input voltage. When the switch is opened, (in the *hold mode*) the voltage level of the capacitor output will be maintained constant because the capacitor is unable to discharge quickly due to high input impedance of the amplifier. Hence, a virtually constant signal is supplied to the ADC. The sampling rate is controlled by the timing signal (pulse sequence) applied to the solid-state switch.

Digital to Analog Converter (DAC): Data processed or generated by a microcomputer are supplied to the interface hardware as digital data. Since typically, analog signals are needed for monitoring, control and recording purposes, it is important to convert the digital outputs into analog form. DAC is a circuit that accomplishes this task. Several types of DAC modules are commercially available, as discussed in Chapter 4.

Analog to Digital Converter (ADC): Data read in by a microcomputer have to be in a binary digital form that is compatible with the data buses and registers of the microcomputer. An ADC module accomplishes the task of converting each constant data sample provided by a sample and hold circuit into a digital word that could be read in by the microcomputer. Many types of ADCs operate by making use of internal DACs. Hence, ADCs are usually several times more expensive than DACs. Details of the operation of typical ADCs are given in Chapter 4.

Multiplexer and Demultiplexer: A multiplexer is used to "time share" expensive signal modification hardware among several data channels. Suppose that there are several analog data channels to be connected to a microcomputer. One way of accomplishing this is to use an analog multiplexer to select one channel at a time and to connect it to a time-shared unit of interface hardware. The time-shared hardware may consist of, for example, analog filters and amplifiers, sample and hold circuit, ADC, and data buffer. Hence, only one interface hardware unit will be needed for all channels. An analog multiplexer is typically a bank of switches. It uses solid-state switches such as MOSFET units operated using electrical control signals (pulses).

If the analog to digital conversion hardware is inexpensive, then each channel can have its own signal modification circuitry that produces a digital output. In this case we have a set of digital data channels. Connection of these data channels to a common microcomputer can be accomplished using a *digital multiplexer*. This may take the form of an integrated circuit (IC) chip that has several digital input pins and one digital output pin (in the bit-serial case), along with suitable timing and control pins. But any form of I/O arrangement where several digital data sources are connected to a common data bus may be considered as a digital multiplexer. Various methods such as polling, interrupt request, and memory-mapped input described earlier may be used for connecting one digital data source at a time, to the data bus.

A demultiplexer performs the reverse function of a multiplexer. For example, when more than one control signal are needed to control a machine, the control computer has to generate all these signals. In that case, the computer has to sequentially connect its output data register to the machine control channels. Demultiplexing is useful as well when several loads are driven by the same signal. Note that demultiplexer is also a switching device. Both analog and digital demultiplexing may be used. In digital demultiplexing over a large number of data channels, the "channel address method" may be used to connect one channel at a time to the data bus, as in the case of digital multiplexing.

11.1.4.1   Data Buffer

A storage register at input or output of a computer that temporarily holds data is called data buffer. The reasons for using a data buffer at computer interface, include the following:

1. Data outside the computer may be generated or accepted in the bit-serial manner (one bit at a time), whereas data bus and registers of the computer will hold data as bytes or words. A temporary storage (buffer) is necessary during the serial/parallel conversion at I/O.
2. Data transmitted into a computer might be generated (by a process) as bursts or impulses, with very high frequency content. The computer might not be able to read in such high-speed data as soon as they are generated.

3. Data generated by a computer might have to be stored temporarily until a peripheral device is ready to accept them.

4. Computer might not be ready to accept incoming data because it is executing an instruction or some crucial task.

The physical form of a buffer is typically identical to a computer memory. By being able to hold highly transient types of data (such as bursts) and by temporarily relieving the computer or peripheral device from receiving data when the receiving device is not ready, a data buffer functions as a "buffer" or smoothing device in the true sense of the word.

A buffer should have three states of operation; *data in, data out,* and *no action* (or *high-impedance*). Hence, it is known as a *tri-state buffer.* During the data-in state, data will be stored into the buffer but its contents cannot be read out. During the data-out state, data in the buffer can be read but data cannot be stored in. During the high-impedance state the buffer is inactive and is not connected to a data source or receiver. Buffer receives a control signal, which determines the state of the buffer.

Data received by a buffer are usually momentary pulses. Hence, the buffer should be able to instantaneously capture each data pulse and store it (as a data bit) until the buffer contents are read subsequently. Buffer consists of *data latch* elements, which can accomplish this. A latch is typically a *flip-flop* circuit, which can assume one of two states (representing 0 or 1), as discussed in Chapter 10.

11.1.4.2 Handshaking Operation

During data input, it is desirable for the microcomputer to be notified when its peripheral device has placed data in the buffer of its I/O port. Similarly, the peripheral device should be notified by the microcomputer when the data in the I/O port buffer have been read by the microcomputer. This type of two-way signalling is known as *handshaking.* For example, a voltage level could be set high when the buffer is full and subsequently set low when the buffer is read by the microcomputer. This type of handshaking signal for data input is shown in Figure 11.11(a). Similar handshaking will take place at data output as well. For example, the output handshaking signal is set high when the microprocessor has stored processed data in the output buffer. This signal will be set low when the peripheral device receives the contents of the buffer. The associated handshaking signal is shown in Figure 11.11(b).

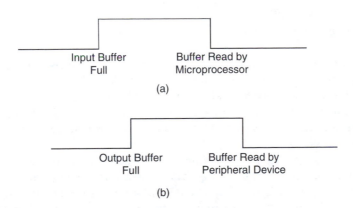

(a)

(b)

FIGURE 11.11
A handshaking signal: (a) for data input; (b) for data output.

The handshaking function could be accomplished by using a *status register* as well. A status register consists of a data byte or word. One bit location of the register is assigned to each I/O port. Then the bit value can be used in the handshaking operation. For example, during data input at I/O port n, the bit at the nth location of the status register is set to 1 when the data buffer is full. The bit value at that location is set to 0 again when the microprocessor has completed reading the contents of that port.

Clock, Counter, Timer: Clocks or timing signals are crucial to the operation of a micro-computer. Clock signal is typically a *pulse sequence* at a known constant frequency, generated, for example, by a quartz crystal oscillator. Many clock signals are used in a microcomputer. In particular, the microprocessor clock is a high frequency (e.g., hundreds of MHz) pulse signal that is used to time, coordinate, and synchronize various activities of the micropro-cessor. In addition, clock signals are used to time or count events, synchronize input/output data transfer, and for generating control signals and interrupts in the operation of sample-and-hold circuits, ADC, DAC, multiplexers and other interface hardware components. For example, analog to digital conversion can be controlled (or synchronized with other activ-ities) by triggering the ADC using a clock signal. The frequency of an ADC trigger signal could be in the MHz range. Variable clock frequencies can be obtained by using program-mable timers. Note that timing of input/output operations is particularly crucial in real-time microcomputer applications such as digital control and on-line monitoring of mecha-tronic systems.

11.1.4.3   Serial/Parallel Interface

Data generated or received by a peripheral device such as keyboard, display screen, or printer typically exist in the serial form. Furthermore, in long distance, networked communication, data might be carried by single (or double) telephone lines. A *modem* (modulator-demodulator) device at the computer interface will do the conversion between telephone-line signals and digital logic signals for this purpose. A computer, on the other hand, receives data as bytes (8 bits) or words (e.g., 16 bits) that are compatible with its buses and registers. Hence, conversion between serial data (a sequence of bits) and parallel data (say bytes of data) is an important consideration for a standard interface.

When data are in the serial form, there should be a set of rules to identify and properly interpret the data. This set of rules is known as a *communication protocol*. Two types of serial data communicating protocols are available. They are:

1. Synchronous data transfer
2. Asynchronous data transfer

In *synchronous* data transfer, a data unit (or character) must be transmitted at a fixed rate (frequency) regardless of whether data are available. In particular, when there are no data to transmit (while waiting for data), *synchronizing characters*, which are empty char-acters, have to be transmitted at the fixed data frequency in order to fill the data gaps. The data source and data receiver must properly synchronize their own data rates, for correct data interpretation.

In *asynchronous* data transfer, a data character is transmitted only when data are avail-able. In this case, data are not transmitted at a constant frequency. Hence, each data unit has to be identified by a start signal (*start bits*) and an end signal (*end bits*). This is known as *framing*. A handshaking protocol may be necessary for asynchronous data transfer between a source and a receiver so as to signal the receiver when the data are ready and to signal the source when the data have been received.

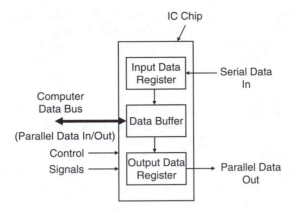

FIGURE 11.12
Basic hardware structure for a serial-parallel interface.

Data transfer is timed and synchronized using a clock signal, known as serial data clock. A data bit could be read at each *rising edge* or each *falling edge* of the clock pulses. The data clock of the source has to be synchronized with the data clock of the receiver. Serial data transfer rate is measured in bits/s. This is known as the *baud rate*. For example, 9600 Bd implies 9600 bits/s and corresponds to a data transfer rate (or data clock) of 9.6 kHz.

The structure of a typical IC chip for serial-parallel data transfer is shown in Figure 11.12. Incoming serial data are stored bit by bit in the *input data register* (typical size = 8 bits). When the register is full, the data block is transferred to the *data buffer*, and the input register is now ready to accept new data. The data in the data buffer could be placed on the data bus of the microcomputer for reading by the microprocessor. All these actions are properly timed and controlled using control signals. For serial output of parallel data from the microprocessor, the data unit present on the data bus is first stored in the data buffer of the interface IC chip. This data block is then transferred into the *output data register*, and the data buffer is now ready to receive a new slice of data from the data bus. Then each data bit in the output data register is read serially by the receiver. Again, these actions are properly timed and controlled. The time available for the microprocessor to read data is improved by including a data buffer. To explain this, note that if a data buffer is not present, the computer has two options for reading the data. It can read one bit at a time, serially, as the data are coming in, or it can read the contents of the input register in a single step. In the latter case the incoming data stream has to be halted (or wasted) while the computer is reading the contents of the input register. If a data buffer is present, however, the contents of the input register are immediately transferred into the buffer once the input register is full. Then the input register is ready to accept new data while the computer is reading the buffer. Hence, no data are wasted or lost. Wasting of data actually corresponds to slowing down of the average speed of data input. It follows that the data input rate is improved by buffering. Furthermore, it is more efficient to read an entire data buffer than to read data serially, one bit at a time. As a result, there is a further improvement in the data input rate due to buffering. Data buffer size can range from a few data words to millions of words. The computer reads the buffer on the first-come, first-read basis.

Transmitted data are subjected to errors, for example, due to noise signals entering the transmission path. One way to detect erroneous data transmission is by using a *parity bit*.

In this case an additional bit known as the parity bit is included in each transmitted character (set of bits). If *even parity* is used, the parity bit is set in such a manner that the total number of 1-bits in a transmitted data unit is even. Then, if the number of 1-bits was found to be odd at the receiver, it means that the receiving data are in error. Similarly, if *odd parity* is used, the parity bit is set such that the total number of 1-bits in each transmitted data unit is odd.

A standard asynchronous device that does the serial/parallel data conversion within interface hardware is known as *universal asynchronous receiver-transmitter* (UART). Various standards such as RS 232-C, RS 422-A, and RS 423-A have been developed by industry for serial interface devices. For example, the interface standard known as the RS 232-C applies to serial transmissions of data bits (i.e., bit serial transmission) between two devices. This standard is provided by the Electrical Industries Association (EIA) and it specifies interface characteristics such as voltages, currents and impedances, associated with interface signals, pin configuration and pin definitions of the connector (25-pin connector is standard), and interface hardware requirements, particularly for data timing and control. The RS 232-C standard applies to both synchronous and asynchronous bit-serial data transmission. Only two wires are needed to interconnect two devices for bit-serial data transmission. A standard 25-pin connector used for interface hardware that follows the RS 232-C standard is known as the 232-C connector.

If data receiver and data source both handle data in parallel form (say, bytes of 8 bits at a time), then a parallel/parallel interface unit could be used. This is typically less complex than a serial/parallel interface unit because data bus structure could be used to transfer bytes of data between a receiver and a source. Both devices are connected to a common interface bus and the data transfer can take place by using standard methods such as *addressing*, *interrupt requesting*, and *handshaking*. A standard interface of this type is the IEEE 488-1975 Instrumentation Bus, which is also known as the *general purpose interface bus* (GPIB). It agrees with the specifications provided in a standard known as the IEEE 488 Standard. This standard applies to bit-parallel, byte-serial data transmission between devices, and, is particularly useful in mechatronic applications where, typically, many sensors/transducers and actuators have to be connected to a single control micro-computer. Such standards are necessary because we often need to interconnect devices from different manufacturers.

11.1.4.4 *Operation Codes and Mnemonics*

Within a microcomputer, the *instructions* for manipulating data and performing various *operations* have to be given in the binary form. The code used to represent a computer operation or instruction is known as the *operation code* or simply *opcode*. Internally, the opcode is used in the binary form. This corresponds to the machine code or *machine language* of the microcomputer. Externally, however, it is inconvenient to use the binary representation for each code. Hence, in programming in a low level language such as the *assembly language*, a meaningful abbreviation is used to represent each opcode. These abbreviations are known as mnemonics. For example, the operation code for division of one number by another might be 0011011 in the machine code, and the mnemonic of this operation code might be DIV.

A program written in the assembly language (or generally any other programming language) has to be converted into the machine language for storing it in the computer memory and subsequent use by the processor. The software program, which performs this conversion, is known as an *assembler* if the original program is in the assembly language (low level), or as a *compiler* if the original program is in a high level language such as FORTRAN or PASCAL or C.

11.1.5   Operation Cycle of a Microcomputer

A microcomputer cannot perform a task unless it is completely and correctly instructed how to perform that task. In other words, the task has to be "programmed". The computer program contains the necessary instructions to perform the task, step by step. Each operation step of a microcomputer requires an operation code along with *data* (or *operands*) on which the operation is performed. Hence, the associated program step (or instruction) should contain two addresses, the address of the operation code and the address of the data location in the memory of the microcomputer. Operation codes are usually permanently stored in read only memory (ROM) and the data are usually stored in read and write memory (RAM).

Suppose that a computer program is written correctly and is converted into the machine language by the assembler or compiler software. This machine code program (in the binary form) is first stored in RAM of the microcomputer. In other words, the program is "loaded" into the microcomputer. Note that a machine language program could be stored in a peripheral medium such as hard disk and loaded into RAM only when the program is executed. This economizes the valuable RAM space. Each program step has an address associated with it. This address gives the RAM location where that program step is stored. Next, the microprocessor of the computer should follow the instructions given in each of these program steps.

Program execution commences by some command (say, by an external signal from the keyboard or trigger from a response variable of a mechatronic system). First, the address of the first program step is read into the *program counter* of the microprocessor. This address is then placed on the *address bus*. The contents of that address location in memory are returned to the microprocessor through the *data bus*. The contents are both the address of an opcode and the address of the data on which the opcode would operate. The microprocessor first places the address of the opcode on the address bus. Typically, this would be an address of a ROM location. In response, the contents of the ROM location (opcode) are returned to the microprocessor through the data bus. This is automatically stored in the *instruction register* of the microprocessor. Next, the microprocessor places on the address bus the address of the data (operand). Usually, this would be an address of a RAM location. In response, the contents of the RAM location (data) are returned to the microprocessor through the data bus. This data will be stored in the *accumulator* (a *data register*) within the microprocessor. Data are "fetched" in this manner by the microprocessor. Subsequently, the operation code stored in the instruction register is decoded by the *operation decoder* and presented to the *sequencer* (see Figure 11.2), which in turn sequences the proper operation. The actual operation takes place by means of the processing hardware of the arithmetic and logic unit (ALU) of the microprocessor. Once this program step is executed, the program counter is advanced by one. This new number in the program counter corresponds to the address of the next program step. Consequently, the microprocessor will execute the next program step as before. Throughout the operation, control signals will be sent through the *control bus* to signal various activities and states of the microcomputer. This was implied but not specifically mentioned in the foregoing discussion.

Branching: A microprocessor executes program steps one by one in the order in which they are written (i.e., sequentially), unless instructed by the present program step to branch (or *jump*) to a step other than the one next to it. Normally, the program counter is incremented by 1 after executing each program step, as mentioned before. If a branch statement is encountered, the contents of the program counter have to be changed to give the address of the program step that would be executed next. Hence, a branch statement should contain that address. When a branch statement in the program is encountered, the microprocessor first stores the present contents of the program counter in a memory location (known as the stack), and then loads into the program counter the address of the program step to which the execution would be branched. Once the program steps following the branch statement are executed, the program will encounter a "return" statement. On encountering this statement, the microprocessor will load the address stored in the stack back into the program counter and continue operation as usual.

Operating System: Commercial computers and workstations are provided with *firmware* (permanent software provided by the computer manufacturer or vendor). A necessary firmware is the *operating system* of the computer. Management functions such as program scheduling for execution, allocation of hardware resources such as I/O devices, and management of data files are performed by the operating system. The operating system of a microcomputer is usually stored on a mass storage device such as hard disk and loaded into RAM when the computer is turned on. Simple and dedicated microprocessor devices usually do not need an operating system because their tasks are well defined and restricted.

11.1.6 Programming and Languages

A microprocessor executes the programs stored in the memory of the microcomputer. These programs are present in the form of machine language (in binary code) of the microcomputer. It is rather difficult to write a program in a machine code, however. Hence, many computer languages that are more convenient for programming by humans have been developed over the years. Some of these languages are *low level*, in the sense that they are closely related to a particular machine language and are typically machine specific, and hence not interchangeable with different types of microcomputers (i.e., not portable). Some other types of languages are high level and generic. A high level language may be used with any microcomputer provided that an appropriate *compiler*, which is a program that can convert the high level program into the machine language of the computer, is available with the computer.

It is usually much easier to write a program in a high-level language. Debugging (error removal) is also easier. But, since a high-level program is neither machine specific nor customized, it can be inefficient and the execution speed is slower than that for a program written in a low-level machine-dependent language.

11.1.6.1 Assembly Language Programming

A program in assembly language has a one-to-one correspondence to its machine language program. Specifically, each program step (or line) of a machine language program will consist of three items as follows:

Instruction Address   Operation Code   Data Address

The main difference is that the machine language uses the binary representation for each item, and the assembly language uses a *mnemonic* for the operation code and a variable name to denote the data address. Instruction addresses are not required in an assembly program because they are automatically generated during the assembly process. A program known as *assembler* is used to convert an assembly language program into the machine language. It will properly interpret the mnemonics for opcodes and also will assign memory addresses for data variable names.

Assembly language is machine dependent and, hence, it will vary from one microcomputer to the next. In order to program a microcomputer in the assembly language, the *instruction set* of the microprocessor should be known. The mnemonics for the opcodes are given in the instruction set.

A program written in assembly language is a *source program*. This is converted into machine language by an *assembler program* and the resulting binary information (*object module* or *object code*) is placed in the computer memory by a *loader* program. Errors in the assembly-language program are also checked and indicated during the assembly process. Only an error-free assembly program will generate an object module, to be stored in computer memory.

In many microcomputers and all commercial computers and workstations, the assembly process takes place in the same computer on which the program (object module) is finally executed. For some special applications (e.g., dedicated digital control), and in the case of toned down versions of microcomputers, it is possible to perform the assembly process in a different (larger and more versatile) computer and then load the resulting object module into the microcomputer on which the program is executed (i.e., the *target computer*). This is known as *cross assembly,* and the machine that generated the object module is known as the cross assembler. The target computer can then work like a dedicated piece of hardware for efficiently performing the programmed task at high speed (e.g., real-time control of a mechatronic system).

11.1.6.2 High-Level Languages

Due to the one-to-one correspondence between machine language and assembly language, it is complex and tiresome to program in assembly language. Furthermore, an assembly language program cannot be easily transferred (ported) from one microcomputer to another. Hence, computer programs are frequently written in a high-level language.

High-level languages also employ mnemonics for operations and variable names for data addresses. But many operations are generally combined into one program statement, and hence there is no one-to-one correspondence between a high-level-language and its machine-code program. A program known as the compiler is used to convert a high-level-language program (the source program) into a machine-code program (the object code).

Popular high-level languages include C, FORTRAN, PASCAL, and BASIC. A mechatronics engineer should be able to program using at least one of these languages. Unlike the first three languages, BASIC (Beginner's All-purpose Symbolic Instruction Coding) uses an interpreter-compiler. In this case the entire source program is not compiled in a single step. Instead, each BASIC program line (statement) is compiled and executed before proceeding to the next statement. This process is obviously slower than the regular compiling, but is more convenient and advantageous in on-line and interactive computations.

11.1.7 Real-Time Processing

Computation requirements, particularly those pertaining to speed and efficiency, are usually more stringent for real-time processing. In mechatronic applications, control signals have to be generated in real time. In other words, data processing cycles have to be synchronized with a real-time clock. Processing efficiency is a major consideration in real-time computing. For example, in digital control, processing speed has to increase proportionately with the sampling rate of the measured data. *Shannon's sampling theorem* (see Chapter 5) indicates that the sampling rate should be at least twice (and preferably 5 or 10 times) the maximum frequency component of interest in the measured signal. It follows that there is a definite need to use faster computers, better algorithms, efficient programs, and perhaps digital signal processors (DSP) or parallel processing and hardware implementations, in computer control of mechatronic systems.

Input/output (I/O) operations as well are crucial in a real-time computing environment. In many situations, I/O operations are performed by computer programs (software) known as *I/O drivers*. In this case an I/O operation is handled simply by calling a particular I/O driver subroutine by the main program. The programming of an I/O driver should be done with utmost care, preferably by experienced programmers in consultation with hardware specialists. Typically, the driver software is provided with commercial I/O hardware. The main advantage is that when an I/O driver is employed the programmer of the main program does not have to worry about the actual I/O hardware interface.

Operating system requirements are quite stringent in real-time computing. In particular, priorities have to be assigned for various tasks, demands, and needs. For example, *interrupt handling* methods should be clearly defined, or polling could be used instead, to service I/O channels. Resource allocation includes memory allocation, and this too is quite crucial in real-time systems. Also, alarm handling and operation under abnormal conditions (e.g., safety and security aspects) should be considered. Real-time operating systems or real-time kernels for general operating systems are commercially available for use in real-time computer applications.

Another consideration that is of utmost importance in real time computing is the issue of whether *software driven devices* or *hardware logic devices* should be used. For example, in direct digital control the control signals have to be computed within a control cycle. This time period is limited by factors such as ADC sampling period for analog sensors, pulse period for encoders, and step period (time for one step at maximum speed) for stepping motors. The software-based approach has advantages such as flexibility and the ability to implement quite complex control strategies. The hardware-based approach (a hardware controller) has advantages such as fast operating speed, simplicity, and low cost (in mass production). Depending on the task at hand, a compromise could be reached; a hybrid system consisting of software-based devices and hardware logic devices could be used. For example, simple tasks such as timing, counting and sequencing could be implemented by hardware logic circuitry and control algorithms could be software programmed on a control computer.

11.2 Programmable Logic Controllers

A programmable logic controller (PLC) is essentially a digital-computer-like system that can properly sequence a complex task, consisting of many discrete operations and involving several devices, which need to be carried out in a sequential manner. PLCs are rugged computers typically used in factories and process plants, to connect input devices such as switches to output devices such as valves, at high speed at appropriate times in a task, operation as governed by a program. Internally, a PLC performs basic computer functions such as logic, sequencing, timing, and counting. It can carry out simpler computations and control tasks such as proportional-integral-derivative (PID) control (see Chapter 12). Such control operations are called *continuous-state control*, where process variables are continuously monitored and made to stay very close to desired values. There is another important class of controls, known as *discrete-state control*, where the control objective is for the process to follow a required sequence of states (or steps). In each state, however, some form of continuous-state control might be operated, but it is not quite relevant to the discrete-state control task. Programmable logic controllers are particularly intended for accomplishing discrete-state control tasks.

There are many mechatronic systems and industrial tasks that involve the execution of a sequence of steps, depending on the state of some elements in the system and on some external input states. For example, consider an operation of turbine blade manufacture. The discrete steps in this operation might be:

1. Move the cylindrical steel billets into furnace
2. Heat the billets
3. When a billet is properly heated, move it to the forging machine and fixture it
4. Forge the billet into shape
5. Perform surface finishing operations to get the required aerofoil shape
6. When the surface finish is satisfactory, machine the blade root

This entire task involves a sequence of events where each event depends on the completion of the previous event. In addition, it may be necessary for each event to start and end at specified time instants. Such *time sequencing* would be important for coordinating the operation with other activities, and perhaps for proper execution of each operation step. For example, activities of the parts handling robot have to be coordinated with the schedules of the forging machine and milling machine. Furthermore, the billets will have to be heated for a specified time, and machining operation cannot be rushed without compromising product quality, tool failure rate, safety, etc. The task of each step in the discrete sequence might be carried out under continuous-state control. For example, the milling machine would operate using several direct digital control (DDC) loops (say, PID control loops), but discrete-state control is not concerned with this except for the starting point and the end point of each task.

A process operation might consist of a set of two-state (on-off) actions. A PLC can handle the sequencing of these actions in a proper order and at correct times. Examples of such tasks include sequencing the production line operations, starting a complex process plant, and activating the local controllers in a distributed control environment. In the early days of industrial control solenoid-operated electromechanical relays, mechanical timers, and drum controllers were used to sequence such operations. An advantage of using a PLC is that the devices in a plant can be permanently wired, and the plant operation can be modified or restructured by software means (by properly programming the PLC) without requiring hardware modifications and reconnection.

A programmable logic controller operates according to some "logic" sequence programmed into it. Connected to a PLC are a set of input devices (e.g., pushbuttons, limit switches, and analog sensors such as RTD temperature sensors, diaphragm-type pressure sensors, piezoelectric accelerometers, and strain-gauge load sensors) and a set of output devices (e.g., actuators such as dc motors, solenoids, and hydraulic rams, warning signal indicators such as lights, alphanumeric LED displays and bells, valves, and continuous control elements such as PID controllers). Each such device is assumed to be a two-state device (taking the logical value 0 or 1). Now, depending on the condition of each input device and according to the programmed-in logic, the PLC will activate the proper state (e.g., on or off) of each output device. Hence, the PLC performs a switching function. Unlike the older generation of sequencing controllers, in the case of a PLC, the logic that determines the state of each output device is processed using software, and not by hardware elements such as hardware relays. Hardware switching takes place at the output port, however, for turning on or off the output devices controlled by the PLC.

11.2.1   PLC Hardware

As noted before, a PLC is a digital computer that is dedicated to perform discrete-state control tasks. A typical PLC consists of a microprocessor, RAM and ROM memory units, and interface hardware, all interconnected through a suitable bus structure. In addition, there will be a keyboard, a display screen, and other common peripherals. A basic PLC system can be expanded by adding expansion modules (memory, I/O modules, etc.) into the system rack.

A PLC can be programmed using a keyboard or touch-screen. An already developed program could be transferred into the PLC memory from another computer or a peripheral mass-storage medium such as hard disk. The primary function of a PLC is to switch (energize or de-energize) the output devices connected to it, in a proper sequence, depending on the states of the input devices and according to the logic dictated by the program. A schematic representation of a PLC is shown in Figure 11.13.

In addition to turning on and off the discrete output components in a correct sequence at proper times, a PLC can perform other useful operations. In particular, it can perform simple arithmetic operations such as addition, subtraction, multiplication, and division on input data. It is also capable of performing counting and timing operations, usually as part of its normal functional requirements. Conversion between binary and binary-coded decimal (BCD) might be required for displaying digits on an LED panel (see Chapter 10), and for interfacing the PLC with other digital hardware (e.g., digital input devices and digital output devices). For example, a PLC can be programmed to make a temperature measurement and a load measurement, display them on an LED panel, make some computations on these (input) values, and provide a warning signal (output) depending on the result.

The capabilities of a PLC can be determined by such parameters as the number of input devices (e.g., 16) and the number of output devices (e.g., 12) which it can handle, the number

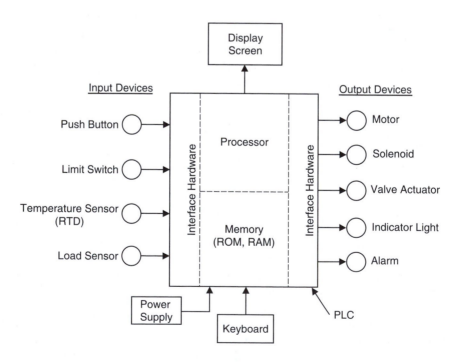

FIGURE 11.13
Schematic representation of a PLC.

of program steps (e.g., 2000), and the speed at which a program can be executed (e.g., 1 M steps/s). Other factors such as the size and the nature of memory and the nature of timers and counters in the PLC, signal voltage levels, and choices of outputs, are all important factors.

11.2.2 Ladder Diagrams

Just like a computer, a PLC has to be programmed in order to perform a required sequencing task. The PLC will check the states of the input devices and then according to the logic of the program (*ladder logic*) it will change the states of (i.e., switch on or off) the output devices that it controls. When many input devices and output devices are involved and when the sequencing logic is complex, the program gets complex. Then a systematic approach is needed in order to efficiently program a PLC. A *ladder diagram* is a graphic representation of sequencing logic of a task to be executed by the PLC. Once a ladder diagram is developed, it can be directly entered into the PLC memory using keyboard commands. This is the conventional way to program a PLC. Ladder logic can be entered into a PLC using a mouse or loaded as a software program from a personal computer or workstation as well.

Historically, ladder diagrams were used to represent the interconnection of hardware components in a mechanical sequencing controller. Then, each symbol in a ladder diagram represented a physical component such as electromechanical relay contact. In PLCs, however, switching logic is available in the software form. Hence, only the input devices and the output devices shown in a ladder diagram are actual physical devices. The relay-contact symbols used to represent sequencing logic in a ladder diagram for a modern PLC are not actual physical relays but represent programmed logic states, in the software form.

A ladder diagram is termed so because it looks like a ladder having two side rails joined by a series of rungs. Each rung can be considered as a circuit that is viewed from the left end (left rail) to the right end (right rail). If the symbol located at the right end of the rung corresponds to an *output device*, then the state of the device (i.e., whether energized or deenergized) is determined according to the logic dictated by the previous segment of the same rung (i.e., the segment from the left end up to the output device symbol). If it is an *input device*, then its state will determine the state of the logic (i.e., 1 or 0) of the previous portion of the same rung. As far as the PLC is concerned, every such device (input devices and output devices) is a two-state device, with logical 1 representing the "on state" or *energized state*, and logical 0 representing the "off state" or *de-energized state*. Hence, the state of each device can be represented simply by a relay contact or switch. If this hypothetical relay contact (or switch) is *normally open* (NO), then the switch will close when the device associated with that switch is energized. If the hypothetical relay contact (or switch) is *normally closed* (NC), then the switch will open when the device associated with that switch is energized. These relay contacts or switches are not physically present, but programmed as shown in a ladder diagram, so as to provide the proper logic needed to sequence the output devices according to a set of rules and depending on the states of the input devices.

Let us consider the symbols used in a ladder diagram. Strictly speaking, only four symbols are sufficient to draw a ladder diagram. They are

1. Symbol for a normally open (NO) relay contact.
2. Symbol for a normally closed (NC) relay contact.
3. A circle representing a physical (input or output) device.
4. A rectangle is used to provide special instruction such as "end of the program."

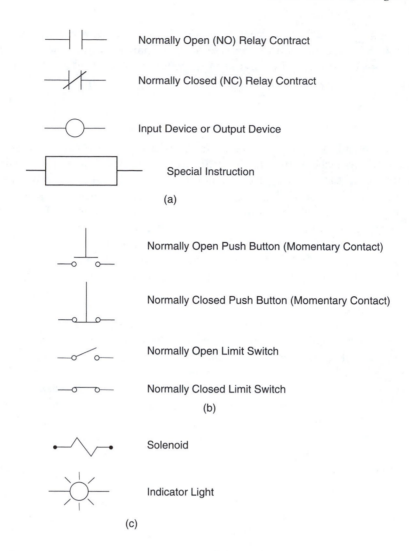

FIGURE 11.14
Ladder diagram symbols: (a) Basic symbols; (b) Symbols for typical input devices; (c) Symbols for two output devices.

These symbols are shown in Figure 11.14(a). But, for clarity, more symbols are used so that the physical nature of the input device or output device is explicitly indicated by the symbol. Furthermore, if the hypothetical (software) switch (or relay contact) represents an input device which itself is a switch (say, a pushbutton or limit switch), then the specific symbol for that input device may be used in the logic circuit instead of using the generic "relay contact" symbols shown in Figure 11.14(a). This is a good practice, for example, if the software switch is used only a very few times (once or twice) in the logic circuit part of the ladder diagram. Several of such more specific symbols are given in Figures 11.14(b) and Figure 11.14(c). Specifically, symbols for two input devices (pushbutton and limit switch) are shown in Figure 11.14(b) and two output devices (solenoid and indicator lamp) are shown in Figure 11.14(c). A special instruction in the ladder diagram may be given in a box.

A very useful element that is often employed in ladder diagrams is the *timer*, denoted by TM or TR (*time relay*). This is not explicitly shown in Figure 11.14. A timer is an output device that could be either a hardware timer, or a software timer that is programmed in. The function of a timer is to introduce a time delay. Specifically, when a timer is energized

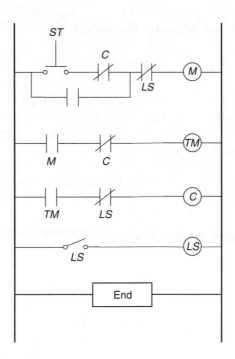

FIGURE 11.15
A ladder diagram example.

(physically in the case of a hardware timer and hypothetically in the case of a software timer), the logic switch corresponding to the timer is activated (closed) after a specified time delay. The amount of delay is specified either by the physical device used or by programming.

Example 11.3
Consider the ladder diagram shown in Figure 11.15. The symbols M and C denote a motor and a clutch, respectively. Write the program logic in the Boolean form and explain the operation of the programmed sequence of activities.

SOLUTION
In the first rung of the ladder diagram, the motor is activated according to the Boolean logic relation

$$M = \overline{LS} \bullet (M + ST \bullet \overline{C}) \qquad \text{(i)}$$

Specifically, when the right hand side (RHS) of Equation i is true, the motor is activated. Note that a series connection is an AND operation and a parallel connection is an OR operation. Also, AND operation is denoted by ($\bullet$), OR operation by (+), and NOT operation by the overbar (̄). The logical variables M, LS, ST, and C can take values 1 (for true) or 0 (for false). In essence then Equation i is a logical "if-then" statement—when the RHS becomes true the LHS is activated.

In the second rung, the timer is activated according to the relation

$$TM = M \bullet \overline{C} \qquad \text{(ii)}$$

In the third rung, the clutch is activated according to the relation

$$C = TM \bullet \overline{LS} \qquad\qquad\qquad (iii)$$

The fourth rung simply says that LS represents a limit switch (an input device) that is normally open and will close when externally activated. This does not yield an additional logic relationship since LS is an external input. The fifth rung is an "instruction" statement, which indicates the end of the ladder program.

Usually, we first identify the steps in a discrete control sequence and then develop a ladder diagram that satisfies them. We shall follow the opposite process here. Specifically, we shall list the steps of a discrete control cycle on the basis of the ladder diagram shown in Figure 11.15.

Initial Conditions: Since all the input and output devices are off, we have the logical values

$$M = 0,\ TM = 0,\ C = 0,\quad \text{and}\quad LS = 0$$

(Note: Accordingly, we have $\overline{M} = \overline{TM} = \overline{C} = \overline{LS} = 1$)

Step 1: Press the start button. This makes $ST = 1$ momentarily (Note: eventually the push button will return to the open position: $ST = 0$, but this is of no consequence until the tasks are completed, as we shall see later). Now since $\overline{C} = 1$ and $\overline{LS} = 1$, the first rung of the ladder (Equation i) will generate the logic condition

$$M = 1$$

Hence, the PLC will turn on the motor. (Note: in a classic electromechanical sequencer, which actually uses physical relay contacts, what actually happens in the first rung of the ladder is that when the pushbutton is pressed, a current path is generated into the motor starter. This will not only start the motor but also will close all "normally open" relay contacts denoted by M (or, open all "normally closed" relay contacts denoted by M).

Step 2: Since $M = 1$ and $\overline{C} = 1$ now, the second rung of the ladder (Equation ii) will turn on the timer (denoted by TM). After the time delay associated with the timer is elapsed, we have the logic state

$$TM = 1$$

Accordingly, the normally open logic relay contacts denoted by TM will be closed in the ladder diagram. Note that there is only one normally open TM relay contact in Figure 11.15, which is in rung 3.

Step 3: Once the time delay specified by the timer has elapsed, we have $TM = 1$ and also $\overline{LS} = 1$. Then, according to the third rung of the ladder (Equation iii), the following logic state is generated

$$C = 1$$

Consequently, the PLC will engage the clutch (denoted by C), thereby driving the load connected to the motor through the clutch. Since $\overline{C} = 0$ now, according to Equation ii, we have $TM = 0$. Hence, the timer will be turned off. Then, after a specified time has elapsed, according to rung 3, the logic state $C = 0$ and the clutch will be disengaged. Consequently, since $\overline{C} = 1$ according to rung 2, the time is turned on. This cycle of operations (clutch engaged, time elapsed, clutch disengaged, time elapsed) will continue while the motor is running, until the limit switch is closed in rung 4.

Step 4: Suppose the limit switch is activated (momentarily) in rung 4. This opens the normally closed limit switch, and generates the logic condition $\overline{LS} = 0$ (or, $LS = 1$). Hence, from Equation i we get

$$M = 0$$

and the motor is turned off. Also, from Equation ii we get

$$TM = 0$$

and the timer is turned off. Furthermore, from Equation iii we get

$$C = 0$$

and the clutch is disengaged. It follows that when the limit switch is physically activated (rung 4), the PLC simultaneously turns off the motor, disengages the clutch, and turns off the timer. Also, since the limit switch is only activated momentarily, it will return to its normally closed state $\overline{LS} = 1$ (or, $LS = 0$). It follows that the systems has returned to the initial conditions as given before. Since $ST = 0$, the motor cannot start until the start button is momentarily pressed again, as in Step 1.

When the program corresponding to a ladder diagram runs, the PLC scans the ladder from top to bottom at very high speed (governed by the microprocessor instruction cycle time and the number of instructions needed to process the logic relation in each rung). The output device on each rung will be turned on or off according to the logic relation associated with that rung, depending on the states of the logic relay contacts on that rung. Of course, if the logic state of an output device does not change in successive scans, no action is taken on that device.

11.2.3 Programming a PLC

The standard way to program a PLC is by using ladder diagrams. First, the physical process is thoroughly studied and understood and then the sequence of events that should be accomplished is listed. Next, a ladder diagram is drawn for the sequence. It is then a straightforward task to enter the ladder diagram into the PLC.

For example, a PLC will allow a certain maximum number of rungs and a maximum number of logic switch elements per rung. This forms a matrix structure of logic elements. Furthermore, each row of the matrix will have an associated output device or input device, as the last element. Then, programming the ladder diagram will be a matter of entering the elements of this matrix structure and joining the elements in the manner dictated by the ladder diagram. This can be accomplished by a keyboard (mnemonic programming) or a mouse (graphic programming). Also, PLC programs can be developed using a personal computer or a workstation and then loaded into the PLC.

11.3 Data Acquisition and Control

Acquisition and utilization of data are crucial for a mechatronic system in all stages of design, development, testing, operation, and maintenance. In particular, sensor-based data acquisition is the basis of feedback control. Furthermore, input signals are measured for feedforward control; output signals are measured for system monitoring, tuning and supervisory control; and input-output signal pairs are measured for testing, experimental modeling and evaluation of a mechatronic system. A typical mechatronic system operates as a self-contained unit and involves direct data acquisition. There are situations, however, where a group of mechatronic systems have to operate in a coordinated manner, as in a mechatronic workcell in a factory (or process plant) setting. Also it may be necessary to monitor a mechatronic system remotely and to share common resources between several applications. Networked control is needed then. Several important issues of computer-assisted data acquisition are discussed next.

11.3.1 Buses and Local Area Networks

An important item that pertains to the computer architecture shown in Figure 11.1 is the *information bus*. A bus is a communication pathway that serves a common purpose of information transfer. For example, the physical medium of a bus may consist of 24 electrical lines available in the form of a ribbon cable, a twisted pair of wire, a fiber-optic network with a fiber count of 48, or even a wireless (e.g., microwave, radio wave, infrared) link. Each line ribbon cable carries one bit of information at a time, using a voltage level or pulse. Hence, number of electrical lines in a ribbon cable is determined by the bit size of the information word carried by that medium.

Buses are communication links of a microcomputer. As discussed before in this chapter, a typical bus structure of a microcomputer consists of an address bus, a data bus, and a control bus. A data bus carries data to or from components connected to the bus. Hence, it is a *bidirectional bus* in general. An address bus carries an address of a memory location or device or data channel connected to the data bus. Microprocessor places an address on the address bus. When conditions are right (e.g., when the addressee is ready), the addressed memory location or the device gets connected to the data bus so that the microprocessor can either send data to that location or receive data from the location. Since an address is issued by a microprocessor, the address bus is generally *unidirectional*. Commands and control instructions can be given by the microprocessor through a *control bus*. For example, when a memory address is placed on the address bus, it will still be necessary to inform the memory whether it is a "write into" operation or a "read from" operation. Furthermore, in handshaking operations and interrupt operations, the micro-processor has to let the other device know whether it is ready to take an action and that it has completed taking the action. Also, trigger signals and activity initiation signals might be needed for devices such as ADC, DAC, and multiplexer. This type of information is transmitted through a control bus. Typically, control bus is unidirectional, from the micro-processor to the other devices. But, in operations such as handshaking, a bidirectional control bus would be needed. In short, a control bus carries rules (or *protocols*) according to which information on the other buses, such as data bus and address bus, is handled.

A computer system typically has many users who would need to use the resources simultaneously and, perhaps, would wish to communicate with each other. A *distributed computing environment* is needed to accomplish these tasks. As an example of another distributed computing environment, consider a *distributed control system*. It will have several localized microprocessors performing *direct digital control* functions at those locations, in a coordinated manner through a higher level (*supervisory*) computer. In such situations, there might be several devices such as personal computers, workstations, PLCs, and various peripheral devices and equipment, which would need to communicate with each other. This may be accomplished by interconnecting these devices, using a *communication network*.

When high data rates (e.g., 100 Mb/s) are required and when data transmission is limited to relatively short distances (e.g., 2 km or less), a *local area network* (LAN) is a very popular communications network. Components are connected at *nodes* of an LAN. Many types of structure (*topology*) are possible. Three possibilities are shown in Figure 11.16. A centralized topology is the *star structure* shown in Figure 11.16(a). In this case several nodes are directly connected to a host computer. In the structure shown, communication between nodes has to take place through the host. An alternative network is the *bus structure* shown in Figure 11.16(b). In this case all the nodes are connected to a common bus. This is somewhat faster than the centralized star structure because data do not have to go through a host. The *ring structure* shown in Figure 11.16(c) is simple. But, it has poor reliability because failure at one node can paralyze the operation of the entire LAN.

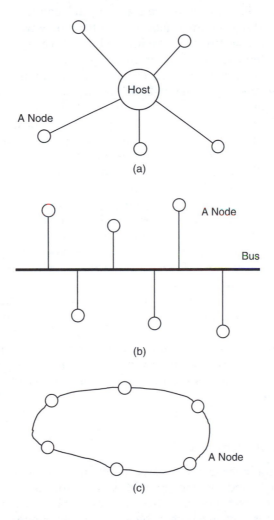

FIGURE 11.16
Typical local area network (LAN) topologies: (a) Star structure; (b) Bus structure; (c) Ring structure.

11.3.2   Data Acquisition

Sensors in a mechatronic system are by and large analog in nature. They cannot be directly linked to the system computer without proper interface hardware and software. This is generally the case even when the transducers are pulse-generating (e.g., an incremental encoder, as discussed in Chapter 7). A digital instrument or an instrument with a digital interface may be directly connected to a computer only if it is compatible with the computer bus. Lack of compatibility among proprietary equipment is the main reason for connectivity problems.

A typical computer-controlled mechatronic system uses a plug-in interface board called *data acquisition board* or DAQ, installed in an expansion slot of the computer, for data acquisition and control. This I/O card is a hardware module with associated driver software, based in a host computer (PC), and connected through its bus (e.g., ISA bus). It will be multifunctional in general, and will accommodate both analog and digital data. In a motion control application, for example, it forms the input-output link between the motor and the controller. It can provide many (say, eight) analog signals to drive many (eight) motors, and hence termed a multi-axis card. The digital-to-analog conversion (DAC) capability is built into the I/O card (e.g., 16 bit DAC including a sign bit, ±10 V output voltage range). Similarly, the analog-to-digital conversion (ADC) function is included in the I/O card (e.g., eight analog input channels with 16 bit ADC including a sign bit, ±10 V output voltage range). These input channels can be used for analog sensors such as tachometers, potentiometers, and strain gauges. Equally important are the encoder channels to read the pulse signals from the optical encoders mounted on the dc servomotors. Typically the encoder input channels are equal in number to the analog output channels (and the number of axes, e.g., eight). The position pulses are read using counters (e.g., 24-bit counters), and the speed is determined by the associated pulse rate. The rate at which the encoder pulses are counted can be quite high (e.g., 10 MHz). In addition, a number of bits (e.g., 32) of digital input and output may be available through the I/O card, for use in simple digital sensing, control, and switching functions.

Direct interfacing of sensors with the control computer is the common architecture of data acquisition in a mechatronic system (see Figure 11.17). This arrangement has many advantages. It is fast (e.g., 20,000 bits/s for an RS-232 serial link, 20 MB/s for parallel GPIB) with no delay and providing real-time data acquisition, it is self-contained and self sufficient, the transmitted information is safe and secure, possibilities of noise entering the transmission system are rare, and it is supported by many and competitive choices of hardware and driver software. Direct interfacing has disadvantages as well. For example, the distance between the data source (sensor) and the receiver (computer) is limited (e.g., 15 m for an RS-232 link; 20 m for a GPIB), remote monitoring and supervision would not be feasible, cannot accommodate distributed control of multiple mechatronic systems in an industrial plant setting, and resource sharing (e.g., processing power of many interconnected computers, public information systems and libraries, web resources and the Internet) is not possible on line. Networked systems provide a solution to these problems.

11.3.3   Communication Networks

A distributed mechatronic system will have many users who would need to use the resources simultaneously and, perhaps, would wish to communicate with each other as well. Furthermore, different types of devices from a variety of suppliers with different specifications, data types and levels may have to be interconnected. A communication network with switching nodes and multiple routes is needed for this purpose.

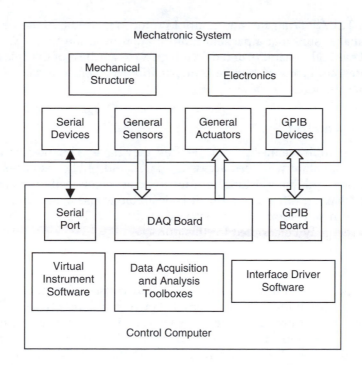

FIGURE 11.17
Direct data acquisition.

11.3.3.1 Protocols

Information is transmitted through a communication network as units of information called "packets." An information packet will contain not only the specific piece of data generated by the source but other information such as the type and form of information, and the addresses of the source and the receiver. The procedures followed when transmitting information are governed by the specific communication protocol. A communication protocol is a set of rules according to which information is transmitted on a bus so that errors, data loss, and delays are reduced and that the network is shared by many users in a fair manner. A protocol may be implemented according a model. For example, the open system interconnection (OSI) model of the International Standards Organization (ISO) has the following modules (layers):

Application layer (source and receiver formats)

Presentation layer (form in which information is presented to the lower layer)

Session layer (establishment of an information transfer session between the source and the receiver)

Transport layer (information exchange between the source and the receiver)

Network layer (information path through the network)

Data link layer (handling of data through one link of the path)

Physical layer (transmission medium)

This communication protocol stack consists of seven layers. The function of each layer is indicated in parentheses.

The protocol TCP/IP (Transmission Control Protocol) is used by Ethernet and Internet. This protocol makes sure that a packet of information from the source will arrive at the receiver (i.e., it has high reliability and data integrity), regardless of the delay. The protocol UDP (User Datagram Protocol) transmits information at a constant rate, but data may be lost during transmission.

11.3.4 Networked Plant

When a plant uses multiple mechatronic devices, they have to be interconnected through a network. Also, the plant will need access to shared and public resources and means of remote monitoring and supervision. In order to achieve connectivity between different types of devices having different origins it is desirable to use a standardized bus that is supported by all major suppliers of the needed devices. The Foundation Fieldbus or Industrial Ethernet may be adopted for this purpose.

11.3.4.1 Fieldbus

Fieldbus is a standardized bus for a plant, which may consist of an interconnected system of mechatronic devices. It provides connectivity between different types of devices having different origins. Also, it provides access to shared and public resources. Furthermore, it can provide means of remote monitoring and supervision. The fieldbus uses a four-layer communication protocol stack, as shown in Figure 11.18. This is similar to the seven-layer ISO OSI model, with the layers 3 to 6 removed, and a new layer (User Layer) added at the top. In Figure 11.18, the function of each layer is described in a manner analogous to the communication method that uses an ordinary letter in a public postal system.

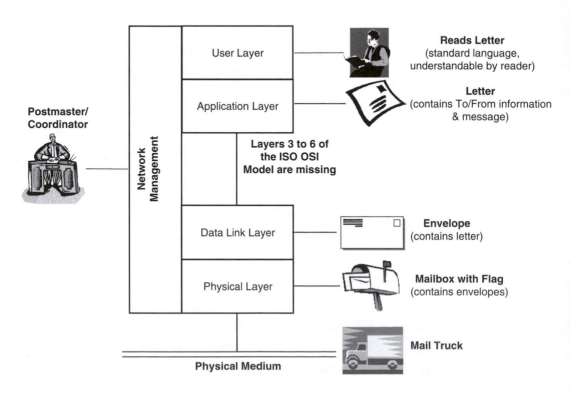

FIGURE 11.18
Postal analogy of fieldbus.

A suitable architecture for networking a mechatronic plant is shown in Figure 11.19. The industrial plant in this case consists of many intelligent mechatronic devices (IMD), one or more programmable logic controllers (PLC) and a distributed control system (DSC) or a supervisory controller. The IMDs will have direct I/O communication with their own components and also will process connectivity through the plant network. Similarly, a PLC may have direct connectivity with a group of devices as well as networked connectivity with other devices. The DSC will supervise, manage, coordinate and control the overall plant.

11.3.5 A Networked Application

A machine which we have developed for head removal of salmon is shown in Figure 11.20. The conveyor, driven by an ac motor, indexes the fish in an intermittent manner. Image of each fish, obtained using a CCD camera, is processed to determine the geometric features, which in turn establish the proper cutting location. A two-axis hydraulic drive positions the cutter accordingly, and the cutting blade is operated using a pneumatic actuator. Position sensing of the hydraulic manipulator is done using linear magnetostrictive

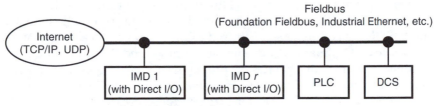

FIGURE. 11.19
A networked mechatronic plant.

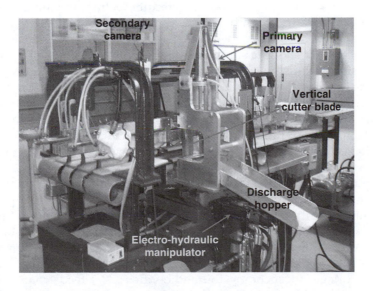

FIGURE 11.20
An intelligent iron butcher.

displacement transducers, which have a resolution of 0.025 mm when used with a 12-bit analog-to-digital converter. A set of six gage-pressure transducers are installed to measure the fluid pressure in the head and rod sides of each hydraulic cylinder, and also in the supply lines. A high-level imaging system determines the cutting quality, according to which adjustments may be made on-line, to the parameters of the control system so as to improve the process performance. The control system has a hierarchical structure with conventional direct control at the component level (low level) and an intelligent monitoring and supervisory control system at an upper level.

The primary vision module of the machine is responsible for fast and accurate detection of the gill position of a fish, on the basis of an image of the fish as captured by the primary CCD camera. This module is located in the machine host and comprised of a CCD camera for image grabbing, an ultrasonic sensor for thickness measurement of fish, a trigger switch for detecting a fish on the conveyor, GPB-1 image processing board for image analysis, and a PCL-I/O board for digital data communication with the control computer of the electro-hydraulic manipulator. This vision module is capable of reliably detecting and computing the cutting location of a fish in approximately 300–400 ms. The secondary vision module is responsible for acquisition and processing of visual information pertaining to the quality of the processed fish that leaves the cutter assembly. This module functions as an intelligent sensor in providing high-level information feedback into the control computer. The hardware associated with this module are a CCD camera at the exit end for grabbing images of processed fish, and a GPB-1 image processing board for visual data analysis. The CCD camera acquires images of processed fish under the direct control of the host computer, which determines the proper instance to trigger the camera by timing the duration it takes for the cutting operation to complete. The image is then transferred to the image buffer in the GPB board for further processing. In this case (i.e., high-level vision), however, image processing is accomplished to extract high level information such as the quality of processed fish.

11.3.5.1   Network Infrastructure

With the objective of web-based monitoring and control of intelligent mechatronic system, we have developed a universal network architecture, both hardware and software. The developed infrastructure has been designed to perform optimally with a Fast Ethernet (100Base-T) backbone where each network device only needs a low cost Network Interface Card (NIC). Figure 11.21 shows a simplified hardware architecture, which networks two mechatronic systems, namely a fish processing machine and an industrial robot. Each machine is directly connected to its individual control server, which handles networked communication between the process and the web-server, data acquisition, sending of control signals to the process, and the execution of low level control laws. The control server of the fish-processing machine contains several data acquisition boards that have analog-to-digital conversion (ADC), digital-to-analog conversion (DAC), digital I/O, and frame grabbers for image processing.

Video cameras and microphones are placed at strategic locations to capture live audio and video signals allowing the remote user to view and listen to the process facility, and to communicate with local research personnel. The camera selected in the present application is the Panasonic Model KXDP702 color camera with built-in pan, tilt and 21 × zoom, which can be controlled through a standard RS-232C communication protocol. Multiple cameras can be connected in daisy-chained manner, to the video-streaming server. For capturing and encoding the audio-video (AV) feed from the camera, the Winnov Videum 1000 PCI board is installed in the video-streaming server. It can capture video signals at a maximum resolution of 640 × 480 pixels at 30 fps (frames per second),

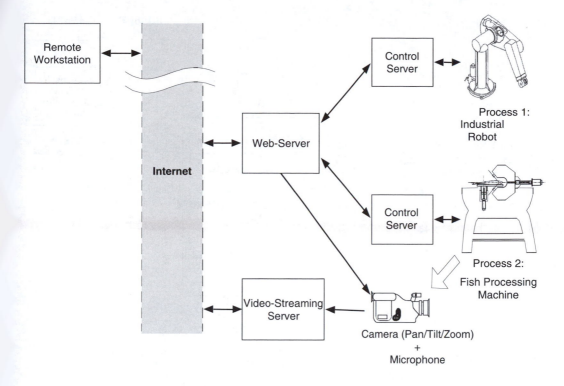

FIGURE 11.21
Network hardware architecture.

with a hardware compression that significantly reduces computational overheads of the video-streaming server. Each of the AV capture boards can only support one AV input. Hence multiple boards have to be installed.

Figure 11.22 shows the interaction of the system components and associated information flow. The control servers may run different types of operating systems. In particular, the control server of the fish-processing machine runs Microsoft Windows NT 4.0 with Service Pack 6. In order to facilitate real-time processing and control in this NT-based system, Venturcom's Real-Time Extension (RTX) software is installed on the control server, allowing direct access to legacy data acquisition boards. This provides high-performance deterministic, real-time and non-real-time processing within the control server. The web-server runs on Microsoft Windows 2000 operating system with Apache Web Server. Microsoft Windows 2000 Professional is used on the video-steaming server due to specific requirements of the AV capture device.

The control server handles controlling and data acquisition. It sends data and receives commands from the web-server through an Ethernet network. Up-to-date system responses and states are transmitted in real time to the web-server. All external communications with a remote VPS are handled by the web-server. Depending the type of service that is requested, both UDP and TCP are implemented for client-server communication. Java™'s scalability, portability, and platform independence allow remote monitoring and control applications (applets) to run on any web-browser, eliminating the need to develop custom communication protocol software. The communication between the web-server and a remote virtual project station (VPS) is achieved by using Java™, whose programs are compiled to platform-independent codes, and dispersed on the web-server as Java™ applets. When a user is logged on to the web-server through a

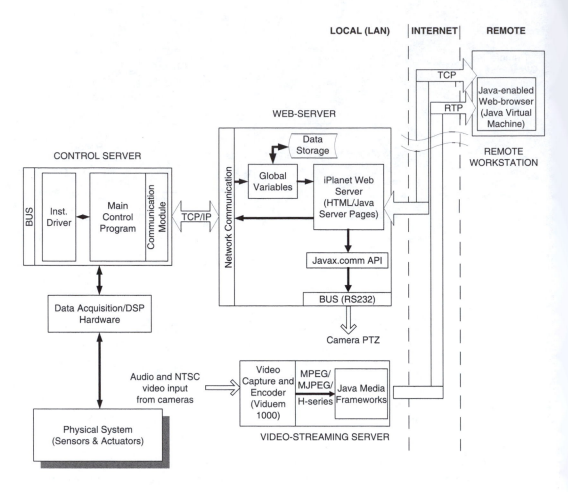

FIGURE 11.22
Component interaction and information flow.

VPS, the Java™ applets are automatically loaded into the VPS (client) and a temporary communication socket is created between the VPS and the web-server. This client-server communication is established for the entire session until the user logs out or the network connection is terminated.

Color cameras and microphones are connected to audio-visual (AV) capturing boards, and are installed in the video-streaming server to provide live audio and video feedback to the VPS. In line with the objective of platform independency, the current implementation has adopted Java™ Media Framework by Sun to implement AV streaming to the Internet. It uses Real-Time Protocol (RTP) to transmit live AV feeds resulting in minimal latency (live to view in less than 1 second) and low transmission bandwidth. Java™ Media Framework can be easily embedded into Java™ applets and provides the option to distribute the AV streams through multicasting, unicasting or video conferencing. Java™ Media Framework can also work together with Darwin Streaming Server, an open source streaming solution from Apple Computer, Inc., to broadcast AV streams to the Internet. Camera control (pan, tilt and zoom) is achieved through serial RS-232C communication interface by using Java™ Communications API (Application Program Interface) package, which adds serial support for Java™.

11.4 Problems

11.1 What are the essential hardware components in a basic microcomputer? Using a schematic diagram show the organization of such a basic computer.

Explain the acronyms RAM, ROM, PROM, EPROM, and EAPROM (or, EEPROM). What is the difference between:

a. Volatile memory and non-volatile memory?

b. Static memory and dynamic memory?

c. Semi-conductor memory and magnetic bubble memory?

Which of these types of memory cannot be used as a ROM?

11.2 Explain the following methods of data transfer between a microcomputer and a peripheral device:

a. Programmed I/O

b. Interrupt I/O

c. DMA

Which method would you use in data acquisition for real-time control? Into which of the three categories would you put memory-mapped I/O?

What is a handshaking operation in I/O data transfer? Explain the role of a status register in a handshaking operation.

11.3 Give reasons why interface hardware would be needed in connecting a device such as sensor or actuator to a control computer. Compare and contrast the following pairs of terms:

a. Interrupt and polling

b. Synchronous data transmissions and asynchronous data transmission

11.4 Explain what roles the following factors play in real-time control using a microcomputer:

a. Word size of the microprocessor

b. Machine cycle time of the microprocessor

c. Instruction cycle time of the microcomputer system

d. Number of instructions in the machine language program of the control algorithm

What are advantages and disadvantages of expanding the instruction set of a microprocessor?

The frequency of the main clock of a microprocessor is known to be 5 MHz. One machine cycle takes two clock periods, and two machine cycles are needed for one instruction. The microprocessor is used in a real-time mechatronic control application. The input hardware of the controller includes a multiplexer and a bank of 10-bit ADC units. The ADC cycle time is 5 ms and the multiplexer takes 0.5 ms for one channel switch.

a. How many input channels can the multiplexer handle optimally?

b. What is the instruction rate in MIPS (million instructions per second)?

c. Estimate the maximum program size for a control algorithm in K (1024 bytes), if three bytes (one byte for the opcode and two bytes for the address field) are used for one instruction

d. Estimate the control bandwidth

11.5 Consider the following three possible applications of a personal computer:

a. Data logging

b. Supervisory control

c. Direct digital control (DDC)

For each application assign a "high," "medium," or "low" weighting for the computer, with respect to the following functional requirements:

i. Ability to handle high data rates

ii. Responsibility for making high-level decisions

11.6 Using a suitable schematic diagram, identify the basic interface hardware components in a microcomputer for real-time control, and explain their functions. List several factors that determine the

a. ADC rate (rate at which data are supplied into the computer)

b. DAC rate (rate at which data are sent out from the computer)

c. Control frequency (rate at which control signal is updated)

of a real-time control computer. Explain why an input register and a buffer in series (*double buffering*) would be needed to improve the ADC rate.

A microprocessor-based real-time control loop of a mechatronic device has a single ADC with a double buffer, a control microcomputer, and a single DAC. The ADC rate at which the measured feedback signal is sampled into the input buffer is 10k words/s. The microprocessor can read data from the input buffer at a rate of 1M words/s. The computer can write data into the output register at the same rate. A typical control computation cycle involves reading the contents of the input buffer, performing a "system identification" (i.e., computation of a dynamic model) using this data, computation of the new control signal, and loading the control signal into the output register to be picked up by the DAC of the control channel. The computations alone require a processing time of 1 ms during which time the microprocessor is not available for reading data from the input buffer or for sending data to the DAC.

a. Estimate the required minimum size of the input buffer.

b. What is the best control frequency that can be provided by this controller?

11.7 What are the two main functions of the hardware used in controlling a stepper motor? Using a schematic diagram, explain a microprocessor-based feedback control system for a stepper motor.

A stepper motor controller uses a 12-bit counter chip (a hardware counter). If one step of the motor corresponds to a load movement of 0.2 mm, what is the maximum load movement (stroke) that could be controlled?

11.8 Explain the functions of a cross-assembler and a cross-compiler. List advantages and disadvantages of using assembly language or machine language to program a control algorithm into a microcomputer in comparison to using a problem-oriented high-level language such as C, FORTRAN, or PASCAL. Specifically, consider the memory requirements for storing the control program, speed of program execution, machine dependency of the programming language, ease of programming, and cost. From the point of view of cost per unit, which language level would you prefer if very large numbers of controller units are expected to be sold?

11.9 What is the main function of an input/output adapter chip? Consider an I/O chip having 40 pins. Pin allocation includes a power supply pin, a microprocessor clock signal pin, a reset pin, interrupt request pins, a read/write control pin, chip select pins, and register select pins. Do you think that this I/O chip:

  a. Can handle bidirectional data (i.e., data transfer in both directions) in bytes?

  b. Can handle interrupts?

  c. Has the handshaking capability?

11.10 Machine language program that is stored in a microcomputer memory is simply a set of "instructions." Consider an instruction that is 3 bytes long. State what information it might carry. Instruction cycle time will vary from instruction to instruction depending on the information carried by an instruction. What are the factors that contribute to determining the instruction cycle time?

Explain why the accuracy and the speed of a microcomputer can be increased by increasing the word size of the microprocessor and by addressing one word, and not one byte, at a time. How many memory locations (words in a word-addressable machine) can be addressed using a 16-bit address?

11.11 Shannon's sampling theorem tells us that the largest useful frequency component that is retained when a signal is sampled is half the sampling rate. In practice, however, we wish to make the sampling rate at least 5 times larger than the maximum frequency of interest. In a servo-controller of a robotic manipulator, suppose that the maximum frequency of interest in a measured motion signal (angular position and angular speed) is 50 Hz. What is an acceptable ADC conversion rate to be used in an associated digital controller? Suppose that 12 channels of data (six joints with a position channel and speed channel for each joint) are sampled and read into the digital controller. The control microprocessor needs 10 μs to read a data sample from one channel. Assuming 3-byte instructions on the average, the average instruction cycle time (fetch and execute), is known to be 6 μs. Estimate the program size (number of instructions) of the control algorithm.

11.12 Consider the ladder logic given in Figure P11.12. Show that this corresponds to a "latch," where the output remains actuated even when the input is deactivated. Suggest a way to make this latch more practical.

11.13 Discuss the differences between discrete-state control and continuous-state control. What is the basic function of a PLC? Give an industrial application where one of the tasks in a discrete-state control sequence is to enable or disable a continuous-state controller.

A spray-painting station of an automotive production line is schematically shown in Figure P11.13. In this simplified representation, a single robot is used. The event sequence to be controlled by the PLC is as follows:

  1. Check limit switch *LS1* for the initiating the position of the painting robot

  2. Activate a time delay (typically using a software timer) of 10 sec

  3. Start the line conveyor drive *CD*

  4. Wait for the second limit switch (*LS2*) signal to detect a car

  5. Stop the conveyor

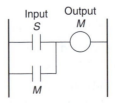

FIGURE P11.12
A nonpractical latch.

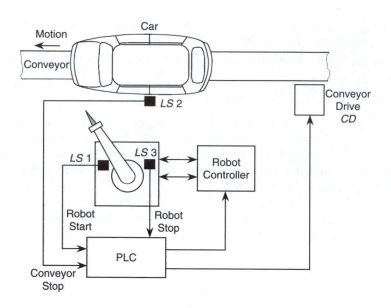

FIGURE P11.13
A robot-automated spray-painting station.

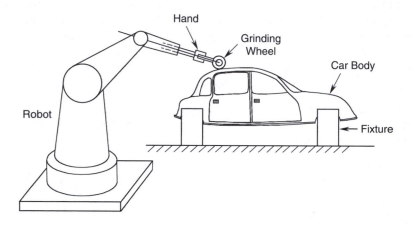

FIGURE P11.14
Weld-seam grinding of a car body by an industrial robot.

6. Activate the robot controller for painting (*RP*)
7. Wait for the "painting completed" limit switch (*LS3*) signal
8. Activate the robot controller for robot initiation (*RI*)
9. Go to Step 1.

A pushbutton (momentary-contact) is provided for manual start and emergency stop of the entire system.

   a. Draw a ladder diagram to represent this discrete-state control sequence.

   b. Using the given ladder logic, write Boolean logic equations for generation of a suitable control program.

11.14 Consider an industrial robot that performs a grinding operation on a weld seam of a car body. A schematic diagram of this arrangement is shown in Figure P11.14. The trajectory of the seam, with reference to some coordinate frame, is stored in the robot controller. Fixturing of the workpiece (car body) is not perfect, however, and the robot has to first determine the true position of the workpiece before grinding the seam. Tool motion during grinding will include trajectory following (seam tracking) by the grinding wheel. Since the required tolerance in position normal to the seam is very tight, force control also would be necessary in this normal direction. Classify these three control tasks (i.e., identification of orientation, position control in trajectory following, and force control) according to

1. Control bandwidth (speed of control)
2. Complexity of control or of the decision making algorithms

Discuss the implementation of these tasks in terms of instrumentation (sensors, data acquisition, etc.) and controller (hardware, algorithms, etc.).

The entire system in Figure P11.14 can be divided into three parts (subsystems) as

1. Robot
2. End effector (or robot hand) with tool (grinding wheel)
3. Workpiece (car body)

On a coordinate plane with natural frequency (ω_n) and damping ratio (ζ) as the two axes, mark relative locations of these three subsystems, for a typical industrial robot system. How many combinations of such relative locations are possible, in general, for these three subsystems? Select three such combinations and compare them, giving attention to implementation (instrumentation, control, etc.) of the overall control system and the system performance (accuracy, speed, control demand, etc.).

12

Control Systems

The purpose of a controller is to make a *plant* (i.e., the system to be controlled) behave in a desired manner. The overall system that includes the plant and the *controller* is called the *control system*. Any mechatronic system is a control system. The system can be quite complex and may be subjected to known and unknown excitations (inputs), as in the case of an aircraft (see Figure 12.1). The system may have many responses (outputs) as well. Modern control techniques are suitable for controlling complex multi-input-multi-output (MIMO) systems of this type. Control engineers should be able to identify or select components for a control system, model and analyze individual components or the overall system, design the control system, and choose parameter values so as to perform the intended functions of the particular system in accordance with some specifications. Component identification, analysis, selection, matching and interfacing, and system design and tuning (i.e., adjusting parameters to obtain the required response) are essential tasks in the instrumentation, design and development of a methatronic control system. In this chapter we present the fundamentals of control systems. Techniques that are widely used in the study and practice of control engineering are discussed. Modeling issues related to control systems are given in Chapter 2. Sensing, actuation, and instrumentation issues are discussed in several previous chapters. The present chapter concentrates on the analysis, performance evaluation, and design of control systems. Both time domain techniques and frequency domain techniques are studied. The emphasis is on linear time-invariant (LTI) single-input-single-output (SISO) systems.

12.1 Control Engineering

A schematic diagram of a control system is shown in Figure 12.2. The physical dynamic system (e.g., a mechanical system) whose response (e.g., speed) needs to be controlled is called the *plant* or *process*. The device that generates the signal (or, command) according to some scheme (or, control law) and controls the response of the plant, is called the *controller*. The plant and the controller are the two essential components of a *control system*. Usually the plant has to be monitored and its response needs to be measured using *sensors* and *transducers*, for feeding back into the controller. Then, the controller compares the sensed signal with a desired response as specified externally, and uses the error to generate a proper control signal. This is a *feedback control system*. Let us examine the generalized control system represented by the block diagram in Figure 12.3. We have identified several discrete blocks, depending on various functions that take place in a typical control system. In a practical mechatronic control system, this type of clear demarcation of components might be difficult; one piece of hardware might perform several functions, or more than one distinct unit of equipment might be associated with one function. Nevertheless, Figure 12.3 is useful in understanding the architecture of a general control system. This is an analog

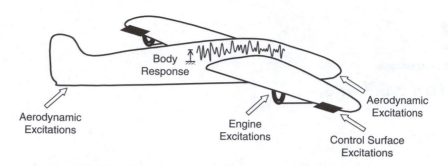

FIGURE 12.1
Aircraft is a complex multi-input-multi-output (MIMO) control system.

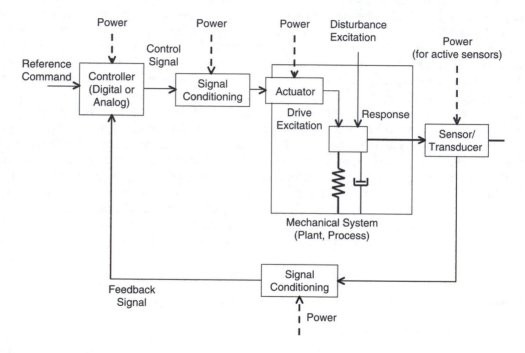

FIGURE 12.2
Schematic diagram of a feedback control system.

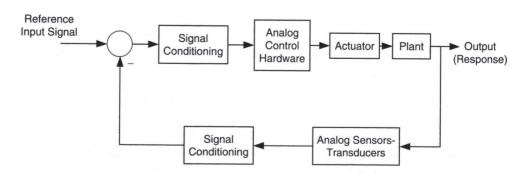

FIGURE 12.3
Components of a typical analog control system.

control system because the associated signals depend on the continuous time variable; no signal sampling or data encoding is involved in the system. As noted before, *plant* is the system or "process" that we are interested in controlling. By *control*, we mean making the system respond in a desired manner. To be able to accomplish this, we must have access to the *drive system* or *actuator* of the plant. We apply certain *command signals*, or inputs, to the *controller* and expect the plant to behave in a desirable manner.

An important factor that we must consider in any practical control system is noise, including external disturbances. Noise may represent actual contamination of signals or the presence of other unknowns, uncertainties, and errors, such as parameter variations and modeling errors. Furthermore, weak signals will have to be amplified, and the form of a signal might have to be modified at various points of interaction. Identification of the hardware components (perhaps commercially available off-the-shelf items) corresponding to each functional block in Figure 12.3 is one of the first steps of instrumentation of a control system. For example, in process control applications off-the shelf analog proportional-integral-derivative (PID) controllers are commonly used. These controllers for process control applications have knobs or dials for control parameter settings—that is, proportional band or gain, reset rate (in repeats of the proportional action per unit time), and rate time constant. The control bandwidth (frequency range of operation) of these control devices is specified (see Chapter 5). Various control modes-such as on/off, proportional, integral, and derivative, or their combinations are provided by the same control box.

Actuating devices (actuators) include dc motors, ac motors, stepper motors, solenoids, valves, and relays (see Chapter 8 and Chapter 9), which are also commercially available to various specifications. Potentiometers, differential transformers, resolvers, synchros, strain gauges, tachometers, piezoelectric devices, pressure gauges, thermocouples, thermistors, and resistance temperature detectors (RTDs) are examples of sensors used to measure process response for monitoring performance and possible feedback (see Chapter 6 and Chapter 7). Charge amplifiers, lock-in amplifiers, power amplifiers, switching amplifiers, linear amplifiers, tracking filters, low-pass filters, high-pass filters, and notch filters are some of the signal-conditioning devices used in analog control systems (see Chapter 4). Additional components, such power supplies and surge-protection units, are often needed in control, but they are not indicated in Figure 12.3 because they are only indirectly related to control functions. Relays and other switching and transmission devices, and modulators and demodulators may also be included.

12.1.1 Control System Architectures

In *open-loop control*, we do not use current information on *system response* to determine the control signals. In other words, there is no feedback. The structure of an open-loop control system is shown in Figure 12.4. On the other hand, in a *feedback control system*, as shown in Figure 12.2 and Figure 12.3, the control loop has to be closed, making measurements of the system response and employing that information to generate control signals so as to correct any output errors. Hence, feedback control is also known as *closed-loop control*.

FIGURE 12.4
An open-loop control system.

12.1.1.1   *Feedforward Control*

Many control systems have inputs that do not participate in feedback control. In other words, these inputs are not compared with feedback (measurement) signals to generate control signals. Some of these inputs might be important variables in the plant (process) itself. Others might be undesirable inputs, such as external disturbances, which are unwanted yet unavoidable. Performance of a control system can generally be improved by measuring these (unknown) inputs and somehow using the information to generate control signals. In *feedforward control*, unknown "inputs" are measured and that information, along with desired inputs, is used to generate control signals that can reduce errors due to these unknown inputs or variations in them. The reason for calling this method "feedforward control" stems from the fact that the associated measurement and control (and compensation) concern input (not outputs) and take place in the forward path (not the feedback path) of the control system. In feedback control, on the other hand unknown "outputs" are measured and compared with known (desired) inputs to generate control signals. Both feedback and feedforward schemes may be used in the same control system.

A block diagram of a typical control loop that uses feedforward control is shown in Figure 12.5. In this system, in addition to feedback control, a feedforward control scheme is used to reduce the effects of a disturbance input that enters the plant. The disturbance input is measured and fed into the controller. The controller uses this information to modify the control action so as to compensate for the effect of the disturbance input. See Appendix A and Chapter 2, Chapter 5, and Chapter 9 for concepts of transfer functions and control system representation in the Laplace domain.

12.1.1.2   *Terminology*

Some useful terminology introduced in the above discussion is summarized below.

Plant or Process: System to be controlled

Inputs: Excitations (known, unknown) to the system

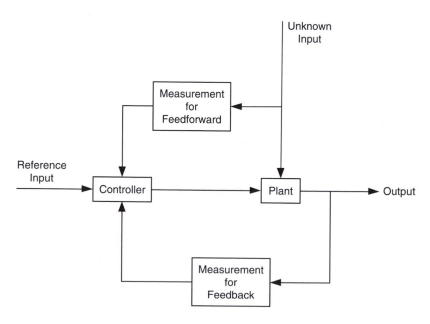

FIGURE 12.5
A system with feedback and feedforward control.

Outputs: Responses of the system

Sensors: Measure system variables (excitations, responses, etc.)

Actuators: Drive various parts of the system

Controller: Device that generates control signal to drive the plant toward a desired response

Control Law: Relation or scheme according to which control signal is generated

Control System: Plant and controller, at least (May include sensors, signal conditioning, etc, as well)

Feedback Control: Control signal is determined according to plant "response"

Open-loop Control: No feedback of plant response to controller

Feed-forward Control: Control signal is determined according to plant "excitations"

Example 12.1

As a practical example, consider the natural gas home heating system shown in Figure 12.6(a). A simplified block diagram of the system is shown in Figure 12.6(b). In conventional feedback control, the room temperature is measured and its deviation from the desired temperature (set point) is used to adjust the natural gas flow into the furnace. On/off control through a thermostat, is used in most such applications. Even if proportional or three-mode (proportional-integral-derivative) control is employed, it is not easy to steadily maintain the room temperature at the desired value if there are large changes in other (unknown) inputs to the system, such as water flow rate through the furnace, temperature of water entering the furnace, and outdoor temperature. Better results can be obtained by measuring these "disturbance" inputs and using that information in generating the control action. This is then feedforward control. Note that in the absence of feedforward control, any changes in the inputs w_1, w_2, and w_3 in Figure 12.6 would be detected only through their effect on the response signal (room temperature), in feedback. Hence, the subsequent corrective action can considerably lag behind the cause (changes in w_i). This delay will lead to large errors and possible instability problems. With feedforward control, information on the disturbance input w_i will be available to the controller immediately, thereby speeding up the control action and also improving the response accuracy. Faster action and improved accuracy are two very desirable effects of feedforward control.

In some applications, control inputs are computed using the desired outputs and accurate dynamic models for the plants, and these computed inputs are used for control purposes. This is the "inverse model" (or "inverse dynamics") approach because the input is computed by substituting the output into a model (i.e., the inverse model). This is a popular way for controlling robotic manipulators, for example. This method is also known as feedforward control. To avoid confusion, however, it is appropriate to denote this method as *computed-input control*.

Example 12.2

Consider the system shown by the block diagram in Figure 12.7(a). Note that

$G_p(s)$ = plant transfer function

$G_c(s)$ = controller transfer function

$H(s)$ = feedback transfer function

$G_f(s)$ = feedforward compensation transfer function

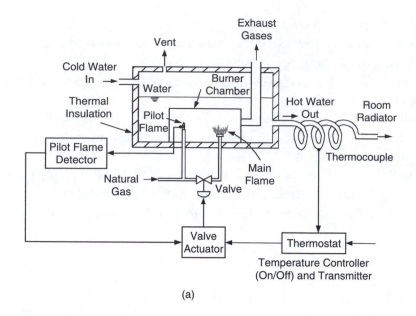

FIGURE 12.6
(a) A natural gas home heating system; (b) A block diagram representation of the system.

The disturbance input w is measured, compensated using G_f, and fed into the controller, along with the driving input u.

a. Obtain the transfer function relationship between the output y and the driving input u in the absence of the disturbance input w.

b. Show that in the absence of u, the block diagram can be drawn as in Figure 12.7(b). Obtain the transfer relationship between y and w in this case.

c. From parts (a) and (b), write an expression for y in terms of u and w, when both these inputs are present

d. Show that the effect of disturbance is fully compensated if the feedforward compensator is selected according to the law

$$G_f(s) = \frac{1}{G_c(s)}$$

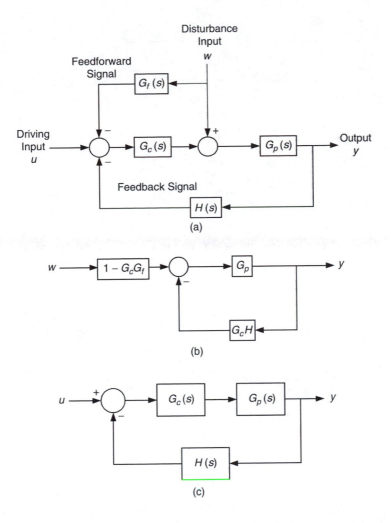

FIGURE 12.7
(a) A block diagram for a system with feedforward control; (b) Reduced form in the absence of the driving input; (c) The block diagram when the disturbance is removed; (d) Steps of block diagram reduction in the absence of the main input.

SOLUTION

a. When $w = 0$, we have the system shown in Figure 12.7(c). Then, from the usual block diagram analysis (see Chapter 2), we have the transfer function

$$\frac{y}{u} = \frac{G_c G_p}{1 + G_c G_p H}$$

b. When $u = 0$ the block diagram may be reduced using the approach described in Chapter 2, in three steps as shown in Figure 12.7(d). After Step 2, by reducing only the left segment of the block diagram we get Figure 12.7(b). From the last step in Figure 12.7(d), we have the transfer function

$$\frac{y}{w} = \frac{(1 - G_f G_c)G_p}{1 + G_p G_c H}$$

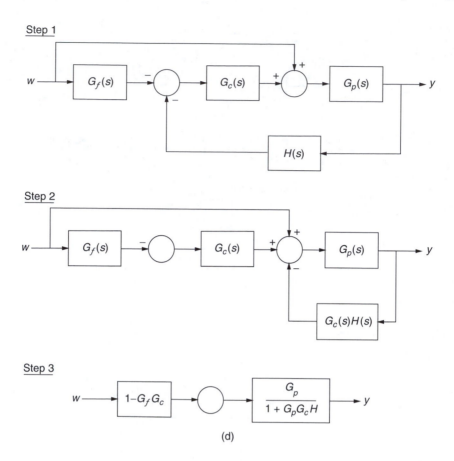

FIGURE 12.7
(Continued)

c. Since the system is linear, the *principle of superposition* applies. Accordingly, the overall transfer function relation when both inputs u and w are present, is given by

$$y = \frac{G_c G_p}{(1+G_c G_p H)} u + \frac{(1-G_f G_c)G_p}{1+G_p G_c H} w$$

d. When $G_f G_c = 1$, the second term of the RHS of the equation in Part (c) vanishes; that is,

$$\frac{(1-G_f G_c)G_p}{1+G_p G_c H} = 0$$

Hence the effect of the disturbance (w), on the output, is fully compensated.

12.1.2 Instrumentation and Design

As we have noted, the characteristic constituents of a control system are:

- The *plant,* or the dynamic system to be controlled
- Signal *measurement* devices for system evaluation (monitoring) and for feedback and feedforward control
- The *drive system* that actuates the plant

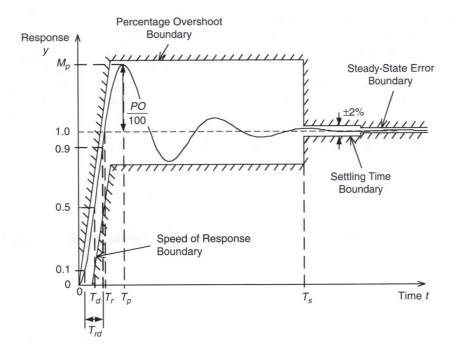

FIGURE 12.8
Conventional performance specifications used in the time-domain design of control systems.

- *Signal conditioning* by filtering and amplification and *signal modification* by modulation, demodulation, ADC, DAC, and so forth, to convert the signals into an appropriate form
- The *controller* that generates appropriate drive signals for the plant in order to realize a desired performance

Each function or operation within a control system can be associated with one or more physical devices, components, or pieces of equipment, and one hardware unit may accomplish several of the control system functions. By *instrumentation,* in the present context, we mean the identification of these various instruments or hardware components with respect to their functions, operation and interaction with each other, and the proper selection and interfacing of these components for a given application—in short, "instrumenting" a control system.

By *design,* we mean the process of selecting suitable equipment to accomplish various functions in the control system; developing the system architecture; matching and interfacing these devices; and selecting the parameter values, depending on the system characteristics, in order to achieve the desired objectives (performance) of the overall control system (i.e., to meet design specifications), preferably in an optimal manner and according to some performance criterion. Design may be included as an objective of the instrumentation. In particular, there can be many designs that meet a given set of performance requirements. Identification of key design parameters, modeling of various components, and analysis are often useful in the design process. Modeling (both analytical and experimental) is important in analyzing, designing, and evaluation of a control system. Chapter 2 discusses various issues of modeling of a mechatronic system.

12.1.3 History of Control Engineering

The demand for servomechanisms in military applications during World War II provided much incentive and many resources for the growth of control technology. Early efforts were devoted to the development of analog controllers, which are electronic devices or circuits that generate proper drive signals for a plant (process). Parallel advances were necessary in actuation devices such as motors, solenoids, and valves, which are the driving means of a plant. For feedback control and feedforward control, further developments in sensors and transducers became essential. Innovations and improvements were necessary as well in the devices for signal processing, conditioning, and modification. With added sophistication in control systems, it was soon apparent that analog control techniques had serious limitations. In particular, linear assumptions had been used to develop controllers even for highly nonlinear plants. Furthermore, complex and costly circuitry was often needed to generate even simple control signals. Consequently, most analog controllers were limited to on/off and proportional-integral-derivative (PID) actions, and lead and lag compensation networks were employed to compensate for weaknesses in such simple control actions.

The digital computer, first developed for large number-crunching jobs, was employed as a controller in complex control systems in the 1950s and 1960s. Originally, cost constraints restricted its use primarily to aerospace applications, which required the manipulation of large amounts of data (complex models, several hundred signals, and thousands of system parameters) for control and which did not face serious cost restraints. Real-time control requires fast computation (see Chapter 11), and the required speed of computation is determined by the control bandwidth (or the speed of control) and parameters (e.g., time constants, natural frequencies, and damping constants) of the process (plant) that is being controlled. For instance, prelaunch monitoring and control of a space vehicle would require digital data acquisition at very high sampling rates (e.g., 50,000 samples/second). As a result of a favorable decline of computation cost (both hardware and software) in subsequent years, widespread application of digital computers as control devices (i.e., digital control) has become feasible. Dramatic developments in large-scale integration (LSI) technology and microprocessors (see Chapter 11) in the 1970s resulted in very significant drops of digital processing costs, which made digital control a very attractive alternative to analog control. Today, digital control has become an integral part of numerous systems and applications, including machine tools, robotic manipulators, automobiles, aircraft autopilots, nuclear power plants, traffic control systems, chemical process plants, and practically all mechatronic systems. Both software-based digital controllers (Chapter 11) and faster yet less flexible hardware-based digital controllers (Chapter 10), which employ digital circuits to perform control functions, are commonly used now. Landmark developments in the history of control engineering are listed below.

300 B.C.	Greece (Float valves and regulators for liquid level control)
1770	James Watt (Steam engine; Governor for speed control)
1868	James Maxwell (Cambridge University, Theory of governors)
1877	E.J. Routh (Stability criterion)
1893	A.M. Lyapunov (Soviet Union, Stability theory, basis of state space formulation)
1927	H.S. Black and H.W. Bode (AT&T Bell Labs, Electronic feedback amplifier)
1930	Norbert Wiener (MIT, Theory of stochastic processes)
1932	H. Nyquist (AT&T Bell Labs, Stability criterion from Nyquist gain/phase plot)
1936	A. Callender, D.R. Hartee, and A. Porter (England, PID Control)
1948	Claude Shannon (MIT, Mathematical Theory of Communication)
1948	W.R. Evans (Root locus method)

1940s   Theory and applications of servomechanisms, cybernetics, and control (MIT, Bell Labs, etc.)

1959   H.M. Paynter (MIT, Bond graph techniques for system modeling)

1960s   Rapid developments in State-space techniques, Optimal control, Space applications (R. Bellman and R.E. Kalman in USA, L.S. Pontryagin in USSR, NASA)

1965   Theory of fuzzy sets and fuzzy logic (L.A. Zadeh)

1970s   Intelligent control; Developments of neural networks and soft computing; Widespread developments of robotics and industrial automation (North America, Japan, Europe)

1980s   Robust control; Widespread applications of robotics, and flexible automation

1990s   Increased application of smart products; Developments in Mechatronics, MEMS, and nanotechnology

Future challenges will concern developments, innovations, and applications in such domains as microelectromechanical systems (MEMS); nanotechnology; embedded, distributed, and integrated sensors, actuators, and controllers; intelligent multiagent systems; smart and adaptive structures; and intelligent vehicle-highway systems.

12.2   Control System Performance

A good control system should possess the following performance characteristics:

1. Sufficiently stable response (*stability*). Specifically, the response of the system to an initial-condition excitation should decay back to the initial steady state (asymptotic stability). The response to a bounded input should be bounded (bounded-input-bounded-output—BIBO stability).

2. Sufficiently fast response (*speed of response* or *bandwidth*). The system should react quickly to a control input.

3. Low sensitivity to noise, external disturbances, modeling errors and parameter variations (*sensitivity* and *robustness*).

4. High sensitivity to control inputs (*input sensitivity*).

5. Low error, for example, tracking error and steady-state error (*accuracy*).

6. Reduced coupling among system variables (*cross sensitivity* or *dynamic coupling*).

Table 12.1 summarizes typical performance requirements for a control system. Some of these are time-domain parameters and the others are frequency-domain parameters. We will address only the time-domain parameters at present. Frequency domain considerations will be discussed in subsequent sections.

Some performance requirements might be conflicting. For example, fast response is often achieved by increasing the system gain, but increased gain increases the actuation signal, which has a tendency to destabilize a control system. Note further that what is given above are qualitative descriptions for "good" performance. In designing a control system, however, these descriptions have to be specified in a quantitative manner. The nature of the design specifications used depends considerably on the particular design technique that is employed.

TABLE 12.1

Performance Specifications for a Control System

Attribute	Desired Value	Purpose	Specifications
Stability level	High	The response does not grow without limit and decays to the desired value	Percentage overshoot, settling time, pole (eigenvalue) locations, time constants, phase and gain margins, damping ratios
Speed of response	Fast	The plant responds quickly to inputs	Rise time, peak time, delay time, natural frequencies, resonant frequencies, bandwidth
Steady-state error	Low	The offset from the desired response is negligible	Error tolerance for a step input
Robustness	High	(Insensitivity to signal noise, model error, parameter variation etc.	Tolerance levels on input noise, measurement error, model error, and parameter variation
Dynamic interaction	Low	One input affects only one output	Cross-sensitivity, cross-transfer function magnitudes

12.2.1   Performance Specification in Time-Domain

Speed of response and degree of stability are two commonly used specifications in the conventional time-domain design of a control system. These two types of specifications are conflicting requirements in general. In addition, steady-state error is also commonly specified. Speed of response can be increased by increasing the gain of the control system. This, in turn, can result in reduced steady-state error. Furthermore, steady-date error requirement can often be satisfied by employing integral control. For these reasons, first we will treat speed of response and degree of stability as the requirements for performance specification in time domain, tacitly assuming that there is no steady state error. The steady-state error requirement will be treated separately.

Performance specifications in the time-domain are usually given in terms of the response of an oscillatory (under-damped) system to a unit step input, as shown in Figure 12.8. First, assuming that the steady-state error is zero, note that the response will eventually settle at the steady-state value of unity. Then the following performance specifications can be stipulated (Also see Chapter 5):

Peak Time (T_p): Time at which the response reaches its first peak value.

Rise Time (T_r): Time at which the response passes through the steady-state value (i.e., in the normalized case 1.0) for the first time.

Modified Rise Time (T_{rd}): Time taken for the response to rise from 0.1 to 0.9.

Delay Time (T_d): Time taken for the response to reach 0.5 for the first time.

Two-Percent Settling Time (T_s): Time taken for the response to settle within ±2% of the steady-state value (i.e., between 0.98 and 1.02).

Peak Magnitude (M_p): Response value at the peak time.

Percentage Overshoot (PO): This is defined as

$$\text{PO} = \frac{(\text{peak magnitude} - \text{steady state value})}{\text{steady state value}} \times 100\% \qquad (12.1a)$$

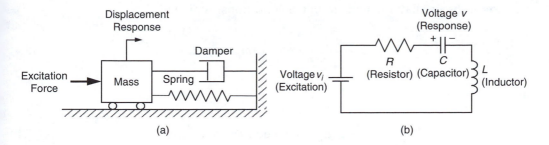

FIGURE 12.9
A damped simple oscillator: (a) Mechanical; (b) Electrical.

In the present case of unity (i.e., normalized) steady-state value, this may be expressed as

$$PO = 100(M_p - 1)\%$$ (12.1b)

Note that T_r, T_p, T_{rd}, and T_d are "primarily" measures of the speed of response whereas T_s, M_p, and PO are "primarily" measures of the level of stability. Note further that T_r, T_p, M_p, and PO are not defined for nonoscillatory responses. Simple expressions for these time-domain design specifications may be obtained, assuming that the system is approximated by a simple oscillator.

Specifications on the slope of the step-response curve (speed of response), percentage overshoot (stability), settling time (stability), and steady-state error can also be represented as boundaries to the step response curve. This representation of conventional time-domain specifications is shown in Figure 12.8.

12.2.2 Simple Oscillator

A damped simple oscillator (mechanical or electrical, as shown in Figure 12.9) may be expressed by the input-output differential equation (see Chapter 2):

$$\frac{d^2y}{dt^2} + 2\zeta\omega_n\frac{dy}{dt} + \omega_n^2 y = \omega_n^2 u$$ (12.2)

where
   u = input (normalized)
   y = output (normalized)
   ω_n = undamped natural frequency
   ζ = damping ratio

The corresponding transfer function is given by (see Chapter 2, Chapter 5, and Appendix A)

$$\frac{Y(s)}{U(s)} = \frac{\omega_n^2}{\left(s^2 + 2\zeta\omega_n s + \omega_n^2\right)}$$ (12.3)

Suppose that a unit step input (i.e., $U(s) = \frac{1}{s}$) is applied to the system. We can show that (say, using the Laplace transform technique — See Chapter 2 and Appendix A) the resulting response of the oscillator, with zero initial conditions, is given by

$$y(t) = 1 - \frac{1}{\sqrt{1-\zeta^2}} e^{-\zeta\omega_n t} \sin(\omega_d t + \phi) \quad \text{for } \zeta < 1 \tag{12.4}$$

where

$$\omega_d = \sqrt{1-\zeta^2}\,\omega_n = \text{damped natural frequency} \tag{12.5}$$

$$\cos\phi = \zeta; \ \sin\phi = \sqrt{1-\zeta^2} \tag{12.6}$$

Now, let us obtain expressions for some of the design specifications that were defined earlier (also see Chapter 5). The response given by Equation 12.4 is of the form shown in Figure 12.8. Clearly, the first peak of the response occurs at the end of the first (damped) half cycle. It follows that the *peak time* is given by

$$T_p = \frac{\pi}{\omega_d} \tag{12.7}$$

The same result may be obtained by differentiating Equation 12.4, setting the result equal to zero, and solving for the first peak.

The *peak magnitude* M_p and the *percentage overshoot* PO are obtained by substituting Equation 12.7 into Equation 12.4; thus,

$$M_p = 1 + \exp(-\zeta\omega_n T_p) \tag{12.8}$$

$$PO = 100 \exp(-\zeta\omega_n T_p) \tag{12.9a}$$

Note that in obtaining this result we have used the fact that $\sin\phi = \sqrt{1-\zeta^2}$. Alternatively, by substituting Equation 12.5 and Equation 12.7 into Equation 12.9a we get

$$PO = 100 \exp\left(-\pi\zeta/\sqrt{1-\zeta^2}\right) \tag{12.9b}$$

The *settling time* is determined by the exponential decay envelope of Equation 12.4. Specifically the 2% settling time is given by

$$\exp(-\zeta\omega_n T_s) = 0.02\sqrt{1-\zeta^2} \tag{12.10}$$

For small damping ratios, T_s is approximately equal to $4/(\zeta\omega_n)$. This should be clear from the fact that $\exp(-4) \approx 0.02$. Note further that the *poles (eigenvalues)* of the system, as given by the roots of the *characteristic equation*

$$s^2 + 2\zeta\omega_n s + \omega_n^2 = 0 \tag{12.11}$$

are

$$p_1, p_2 = -\zeta\omega_n \pm j\omega_d \tag{12.12}$$

It follows that the *time constant* of the system (inverse of the real part of the dominant pole) is

$$\tau = \frac{1}{\zeta\omega_n} \qquad (12.13)$$

Hence, an approximate expression for the 2% settling time is

$$T_s = 4\tau \qquad (12.14)$$

Rise time is obtained by substituting $y = 1$ in Equation 12.14 and solving for t. This gives

$$\sin(\omega_d T_r + \phi) = 0$$

or

$$T_r = \frac{\pi - \phi}{\omega_d} \qquad (12.15)$$

in which the *phase angle* ϕ is directly related to the damping ratio, through Equation 12.6.

The expressions for the performance specifications, as obtained using the simple oscillator approximation, are summarized in Table 12.2. For a higher order system, the damping ratio and the natural frequency that are needed to evaluate these expressions may be obtained from the "dominant" complex pole pair of the system, when applicable. In conventional time-domain design, relative stability specification is usually provided by a limit on the percentage overshoot. This can be related to damping ratio (ζ) using a design curve. For the simple oscillator approximation, Equation 12.9b is used to calculate ζ when PO is specified. This relationship is plotted in Figure 12.10.

TABLE 12.2

Analytical Expressions for Time-Domain Performance Specifications (Simple Oscillator Approximation)

Performance Specification Parameter	Analytical Expression(Exact for a Simple Oscillator)
Peak time T_p	π/ω_d
Rise time T_r	$(\pi - \phi)/\omega_d$
Time constant τ	$1/\zeta\omega_n$
2% Settling time T_s	$-\left(\ln 0.02\sqrt{1-\zeta^2}\right)\tau \approx 4\tau$
Peak magnitude M_p	$1+\exp\left(-\dfrac{\pi\zeta}{\sqrt{1-\zeta^2}}\right)$
Percentage overshoot PO	$100\exp\left(-\dfrac{\pi\zeta}{\sqrt{1-\zeta^2}}\right)$

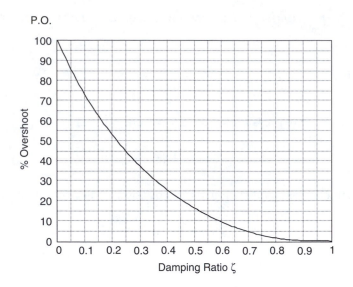

FIGURE 12.10
Damping specification in terms of percentage overshoot using a simple oscillator model.

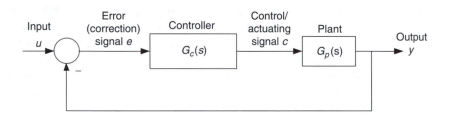

FIGURE 12.11
A feedback control system with unity feedback.

12.3 Control Schemes

By control, we mean making a plant (process, machine, mechatronic system, etc.) respond to inputs in a desired manner. In a *regulator*-type control system the objective is to maintain the output at a desired (constant) value. In a *servomechanism*-type control system the objective is for the output to follow (i.e., track) a desired trajectory (i.e., a specified time response or a path with respect to time). In the design of a control system the desired behavior in response is achieved by meeting a set of performance *specifications*, usually expressed with respect to the attributes given in Table 12.1.

If the plant is stable and is completely and accurately known, and if the inputs to the plant can be precisely generated (by the controller) and applied, accurate control might be possible even without feedback control. In this case a measurement system is not needed (or at least not needed for feedback) and thus we have an *open-loop control* system. More often, however, this is not the case, and feedback control will be necessary. Consider the feedback control system with a single input u, a single output y, and a unity feedback (i.e., $H = 1$ in Figure 12.7(c)), as shown in Figure 12.11.

In selecting a proper controller $G_c(s)$ for an application, the prime factors to be considered are:

1. Precision and complexity of the required response
2. Difficulty of controlling the plant (process) $G_p(s)$

Although the simplest controller that would produce the desired result should be selected, in some instances the required response cannot be realized even with a sophisticated controller. Selection of the controller may be approached from several angles:

1. Frequency domain or process transfer function (reaction curve) analysis
2. Time domain system (model) analysis
3. Previous experience and human expertise (knowledge-based)
4. Experimental investigation

A *control law* is a relationship between the controller output (or, plant input) and the controller input (or, plant output or response error, in the case of feedback control). Common control modes are:

On-off (bang-bang) control

Proportional (P) control

Proportional control combined with reset (integral—I) and/or rate (derivative—D) action (i.e., multi-mode or multi-term control).

Control laws for commonly used control actions are given in Table 12.3. Some advantages and disadvantages of each control action are also indicated. Compare this information with what is given in Table 12.1. Many control systems employ *three-mode controllers* or *three-term controllers* (i.e., PID controllers), which can provide the combined action of proportional (P), integral (I) and derivative (D) modes. The control law for proportional plus integral plus derivative (PID) control is given by

$$c = k_p \left(e + \tau_d \dot{e} + \frac{1}{\tau_i} \int e\, dt \right) \tag{12.16a}$$

or in the transfer function form

$$\frac{c}{e} = k_p \left(1 + \tau_d s + \frac{1}{\tau_i s} \right) \tag{12.16b}$$

TABLE 12.3

Comparison of Some Common Control Actions

Control Action	Control Law	Advantages	Disadvantages
On-Off	$\dfrac{c_{max}}{2}[\text{sgn}(e)+1]$	Simple I, expensive	Continuous chatter, Mechanical problems, Poor accuracy
Proportional	$k_p e$	Simple, Fast response	Offset error (steady state error), Poor stability
Reset (Integral)	$\dfrac{1}{\tau_i}\int e\, dt$	Eliminates offset, Filters out noise	Low bandwidth (slow response), Reset windup, Instability problems
Rate (Derivative)	$\tau_d \dfrac{de}{dt}$	High bandwidth (Fast response), Improves stability	Insensitive to dc error, Allows high-frequency noise, Amplifies noise, Difficult analog implementation

in which
  e = error signal (controller input)
  c = control/actuating signal (controller output or plant input)
  k_p = proportional gain
  τ_d = derivative time constant
  τ_i = integral time constant

Another parameter, which is frequently used in process control, is the *integral rate*. This is defined as

$$r_i = \frac{1}{\tau_i} \tag{12.17}$$

The parameters k_p, τ_d, and τ_i or r_i may be adjusted in controller tuning. The proportional action provides the necessary speed of response and adequate signal level to drive a plant. Besides, increased proportional action has the tendency to reduce steady-state error. A shortcoming of increased proportional action is the degradation of stability, as will be studied later. Derivative action (or rate action) provides stability that is necessary for satisfactory performance of a control system. In the time domain this is explained by the fact that the derivative action tends to oppose sudden changes (large rates) in the system response. Derivative control has its shortcomings, however. If the error signal that drives the controller is constant, for example, derivative action will be zero and it has no effect on the system response. In particular, as a consequence, derivative control cannot reduce steady-state error in a system. Also, derivative control increases the system bandwidth, which has the desirable effect of increasing the speed of response (and tracking capability) of control system but has the drawback of allowing and amplifying high-frequency disturbance inputs and noise components. Hence, derivative action is not practically implemented in its pure analytic form, but rather as a lead circuit, and together with proportional (and possibly integral action).

Example 12.3
Consider an actuator having transfer function

$$G_p = \frac{1}{s(0.5s + 1)}$$

Design a position feedback controller and a tacho-feedback controller (i.e., position plus velocity feedback controller) that will meet the design specifications $T_p = 0.1$ and PO = 25%.

SOLUTION
Note that the block diagram for this problem can be derived from Figure 12.7(c) with $G_c(s) = k$. For position feedback, $H = 1$; and for position plus velocity feedback control, $H = 1 + \tau_v s$. The transfer functions for the two cases can be derived using the general result for the closed-loop transfer function

$$\frac{y}{u} = \frac{G_c G_p}{1 + G_c G_p H} \tag{i}$$

as obtained for Figure 12.7(c).

a. Position Feedback:

From (i) we get

$$\text{Closed-loop TF} = \frac{kG_p(s)}{1+kG_p(s)} = \frac{k}{s(0.5s+1)+k} = \frac{2k}{s^2+2s+2k} \tag{ii}$$

where k is the gain of the proportional controller in the forward path. Note that, in the present case only one parameter (k) is available for specifying two performance requirements. Hence, it is unlikely that both specifications can be met. To check this note from the denominator (characteristic polynomial) of Equation ii that $\zeta\omega_n = 1$ and $\omega_n^2 = 2k$. Hence,

$$\zeta = \frac{1}{\sqrt{2k}} \tag{iii}$$

and

$$\omega_d = \sqrt{\omega_n^2 - (\zeta\omega_n)^2} = \sqrt{2k-1} \tag{iv}$$

Now, for a given T_p we can compute ω_d using the result in Table 12.2; k using Equation iv; ζ using Equation iii; and finally PO using Table 12.2. Alternatively, for a given PO, we can determine ζ using Table 12.2; k using Equation iii; ω_d using Equation iv; and finally T_p using Table 12.8. These two sets of results are given in the following Table.

Results for Position Control

T_p	ω_d	k	ζ	PO
0.1	10π	494.0	0.032	90.5%
1.39	2.264	3.063	0.404	25%

Note that, for $T_p = 0.1$ we have PO = 90.5%. For PO = 25% we have $T_p = 1.39$. Hence both requirements cannot be met with the single design parameter k, as expected.

b. Tacho-Feedback:

Tachometer is a velocity sensor (see Chapter 6). Customarily, tacho feedback uses feedback of both position and velocity, as in the case of encoder feedback. Hence, as noted before, $H = 1 + \tau_v s$, and

$$\text{Closed-loop TF} = \frac{kG_p(s)}{1+kG_p(s)\times(\tau_v s+1)} = \frac{k}{s(0.5s+1)+k(\tau_v s+1)}$$

$$= \frac{2k}{s^2+2(1+k\tau_v)s+2k}$$

where k is the proportional gain and τ_v is the velocity feedback parameter (*tacho gain*). By comparing with the simple oscillator characteristic polynomial (i.e., denominator of the transfer function), we note

$$\omega_n^2 = 2k \tag{v}$$

$$\zeta\omega_n = 1 + k\tau_v \tag{vi}$$

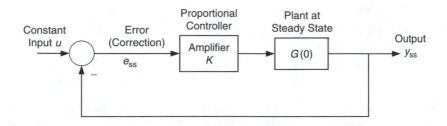

FIGURE 12.12
Presence of offset error in a type-zero system under proportional control.

Since two parameters (k and τ_v) are available to meet the two specifications, it is likely that the goal can be achieved. The computation steps are given below:

As before, for $T_p = 0.1$ we have $\omega_d = 10\pi$.

Also, for PO = 25% we have $\zeta = 0.404$.

Hence, $\omega_n = \omega_d / \sqrt{1 - \zeta^2} = 10.93\pi$ and $\zeta\omega_n = 0.404 \times 10.93\pi = 13.873$.

Then we use Equation v to compute k.

Substitute in Equation vi to compute τ_v.

We get $k = 590$ and $\tau_v = 0.022$.

12.3.1   Integral Control and Steady State Error

It is easy to explain the presence of an *offset* (i.e., *steady-state error*) when proportional control is used for a system having finite dc gain (i.e., in a "type-zero" system, which will be further discussed in a later section). Consider a unity feedback system under proportional control, with proportional gain K_p. Since at steady state (i.e., at zero frequency), the system transfer function can be represented by its dc gain, the steady-state behavior of this system can be represented by the block diagram in Figure 12.12, with a constant input u. Note that y_{ss} is the steady-state response of the system and e_{ss} is the steady state error. It should be clear that e_{ss} could not be zero because then there would be no actuating (driving) signal for the plant and, consequently, the plant output would be zero, thereby violating the zero error ($u - y_{ss} = 0$) condition. Hence, an offset error will always be present in this control system. To obtain an expression for the offset, note from Figure 12.12 that

$$e_{ss} = u - y_{ss} \tag{12.18}$$

and

$$y_{ss} = KG(0)e_{ss}$$

By straightforward substitution we get

$$e_{ss} = \left[\frac{1}{1 + KG(0)}\right]u \tag{12.19}$$

This result confirms the fact that a steady state error is always present in the present system, with P control. An offset could result from an external disturbance such as a change in load as well.

12.3.2 Final Value Theorem

The result given by Equation 12.19 can be formally obtained using the final value theorem (FVT), which states: The steady-state value of a signal $x(t)$ is given by

$$x_{ss} = \lim_{s \to 0} sX(s) \tag{12.20}$$

in which $X(s)$ is the Laplace transform of $x(t)$.

NOTE In determining the steady-state value of a response (or, of an error) the nature of the input in the beginning is not important. What matters is the nature of the input for a sufficiently large time period at the end, inclusive of the settling time.

Accordingly, in this context there is no difference between a constant input, a step input, or any input that reaches a constant value at steady state. It follows that in the system shown in Figure 12.12, the constant input can be treated as a "step input." Hence, in the Laplace domain, the input is given by $\frac{u}{s}$. Then, by the final value theorem, the steady-state error is given by

$$e_{ss} = \lim_{s \to 0} s \left[\frac{1}{1 + K_p G(s)} \right] \frac{u}{s} = \left[\frac{1}{1 + K_p G(0)} \right] u$$

which is identical to the previous result given by Equation 12.19.

See Appendix A and Chapter 2 for an introduction to Laplace transform techniques and transfer functions.

12.3.3 Manual Reset

When there is an offset (i.e., a steady state error), one way to make the actual steady state value equal to the desired value would be to change the set point (i.e., the input value) in proportion to the desired change. This is known as *manual reset*. Note that this method does not remove the presence of an offset but rather changes the output value to the desired value. Note further that the percentage overshoot given in Table 12.3 is obtained with respect to the steady state value, which is equal to the steady input value if there is no offset. These concepts can be illustrated by a simple example.

Example 12.4

A feedback control system of a machine tool is shown in Figure 12.13.

a. Determine the steady-state value y_{ss} of the response and the steady state error e_{ss}.
b. Determine the percentage overshoot of the system with respect to the set point value.

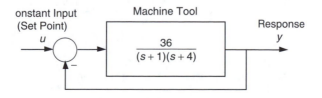

FIGURE 12.13
A feedback control system for a machine tool.

c. What should be the set point value in order to obtained a desired steady state response value of y_o?

SOLUTION

a. Forward transfer function (machine tool) is

$$G(s) = \frac{36}{(s+1)(s+4)}$$

Note that

$$G(0) = 9$$

For a constant u, the steady state response value is,

$$y_{ss} = \frac{G(0)}{1+G(0)} u = \frac{9}{1+9} u = 0.9u \qquad\qquad\qquad (i)$$

Steady state error

$$e_{ss} = \frac{1}{1+G(0)} u = \frac{1}{1+9} u = 0.1u$$

b. Closed-loop transfer function $= \dfrac{G(s)}{1+G(s)} = \dfrac{36}{(s+1)(s+4)+36} = \dfrac{36}{s^2+5s+40}$

From the characteristic polynomial (denominator of the TF) of the closed-loop system we have $2\zeta\omega_n = 5$ and $\omega_n^2 = 40$. Hence,

$$\zeta = \frac{5}{2\sqrt{40}} = 0.395.$$

Then, using the formula in Table 12.2, we have, PO = 25.9%. This value is computed with respect to the steady state value. In order to determine the % overshoot w.r.t. the set point value, we proceed as follows:

$$\text{Peak magnitude} \quad M_p = y_{ss}\left[1+\exp\left(-\frac{\pi\zeta}{\sqrt{1-\zeta^2}}\right)\right] = 0.9u \times (1+0.259)$$

$$\text{PO with respect to the set point } u = \frac{M_p - u}{u} \times 100 = \frac{0.9u[1+0.259]-u}{u} \times 100$$

$$= (0.9[1+0.259]-1) \times 100 = 13.3\%$$

c. From Equation i it is seen that in order to get $y_{ss} = y_o$, we must use a set point of

$$u = \frac{y_o}{0.9} = 1.11 y_o.$$

12.3.4   Automatic Reset (Integral Control)

Another way to zero out the steady-state error given by Equation 12.19 is to make $G(0)$ infinity. This can be achieved by introducing an integral term $(1/s)$ in the forward path of the control system (because $1/s \to \infty$ when $s = 0$). This is known as *integral control* or *reset control* or *automatic reset*. An alternative way to arrive at the same conclusion is by noting that an integrator can provide a constant output even when the error input is zero, because the value of the integrated (accumulated) error signal will be maintained even after the error goes to zero. Integral control is known as *reset control* because it can reduce the offset to zero, and can counteract external disturbances including load changes. A further advantage of integral control is its low-pass-filter action, which filters out high-frequency noise. Since integral action cannot respond to sudden changes (rates) quickly, it has a destabilizing effect, however. For this reason integral control is practically implemented in conjunction with proportional control, in the form of proportional plus integral (PI) control (a two-term controller), or also including derivative (D) control as PID control (*a three-term controller*).

12.3.4.1   Reset Windup

The integral action integrates the error to generate the control action. Suppose that the sign of the error signal remains the same over a sufficiently long period of time. Then, the error signal can integrate into a very large control signal. As a result, it can saturate the device, which performs the control action (e.g., a valve actuator can be in the fully-open or fully-closed position). Beyond that point, a change in the control action has no effect on the process. Effectively, the controller has locked in one position while the control signal may keep growing. This "winding up" of the control signal due to error integration, without causing a further control action, is known as *reset windup*.

Due to reset windup, not only will the controller take a longer time to bring the error to zero, but when the error reaches zero, the integrated error signal will not be zero and the control actuator will remain saturated. As a result, the response will be pushed with the maximum control force, beyond the zero error value (desired value). It will take a further period of time to unsaturate the control actuator and adjust the control action in order to push the response back to the desired value. Consequently, large oscillations (and stability problems) can result.

Reset windup problems can be prevented by stopping the reset action before the control actuator becomes saturated. A simple approach is to first determine two limiting values (positive and negative) for the control signal beyond which the control actuator becomes saturated (or very nonlinear). Then, a simple logic implementation (hardware or software) can be used to compare the actual control signal with the limiting values, and deactivate the integral action if the control signal is outside the limits. This is termed *reset inhibit* or *reset windup inhibit*.

12.3.5   System Type and Error Constants

Characteristics of a system can be determined by applying a known input (*test input*) and studying the resulting response or the resulting error of the response. For a given input, system error will depend on the nature of the system. It follows that, error, particularly the *steady-state error*, to a standard test input, may be used as a parameter that characterizes a control system. This is the basis of the definition of error constants. Before studying this topic we should explain the term "system type."

Consider the general feedback control system shown in Figure 12.14. The forward transfer function is $G(s)$ and the feedback transfer function is $H(s)$. Since it is the *loop transfer function*

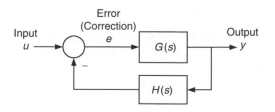

FIGURE 12.14
A general feedback control system.

GH that primarily determines the dynamic behavior of the closed-loop system, we may assume without loss of generality that the controller elements and any compensators as well as the plant (process) are in the forward path, and their dynamics are jointly represented by $G(s)$. The feedback transfer function $H(s)$ is considered to have the necessary transducers and associated signal conditioning for the sensor that measures the system output *y*. Then it is proper to assume that the dc gain of $H(s)$ to be unity (i.e., $H(0) = 1$) because the sensor would be properly calibrated to make a correct measurement, at least under static conditions.

Customarily, the controller input signal

$$e = u - Hy \tag{12.21}$$

is termed *error signal* (even though the true error is $y - u$), because input *u* is "compared" with the feedback signal *Hy* and the difference is used to actuate the corrective measures (i.e., control action) so as to obtain the desired response. Note further that the signals *u* and *Hy* should have consistent units. For example in a velocity servo, the input command might be a speed setting on a dial (a physical angle in degrees, for example) and the output would be a velocity (in m/s, for example). But if the velocity is measured using a tachometer, the feedback signal, as typically is the case, would be a voltage. The velocity setting has to be properly converted into a voltage, for example by using a calibrated potentiometer as the input device, so that it could be compared with the feedback voltage signal for control purposes. Throughout the present discussion we assume consistent units for signals that are compared, and unity dc gain for the feedback transfer function.

12.3.5.1   Definition of System Type

Assuming that the *feedback transfer function H(s)* has unity dc gain, the system type is defined as the number of free integrators present in the forward transfer function $G(s)$. For example, if there are no free integrators, it is a *type-zero system*. If there is only one free integrator, it is a *type-1 system*, and so on.

Obviously, the system type is a system property. Also, the steady-state error to a test input is also a system property. We will see that these two properties are related. Furthermore, system type is a measure of *robustness* with regard to steady-state error, in the presence of variations in system parameters.

12.3.6   Error Constants

Steady-state error of a system, when actuated by a "normalized" test signal such as a *unit step, unit ramp,* or *unit parabola,* may be considered a system property. Such system properties may be expressed as error constants. Note, however, that error constants can be defined only if the associated steady-state error is finite. That, in turn, will depend on the number of integrators in $G(s)$, which is known as the system type (or *type number*).

To formally define the three types of error constants associated with unit step input, unit ramp input, and unit parabolic input, consider once again the general system shown in Figure 12.14. The error (correction) signal e is given by Equation 12.21. Furthermore, it is clear from Figure 12.14 that

$$y = Ge \tag{12.22}$$

Note that, for convenience, the same lower case letter is used to denote a time signal and its Laplace transform even though they are two entirely different variables. By substituting Equation 12.22 into Equation 12.21 we get

$$e = \left[\frac{1}{1+GH}\right] u \tag{12.23}$$

By applying the final value theorem to Equation 12.23, the steady-state error may be expressed as

$$e_{ss} = \lim_{s \to 0} \left[\frac{su(s)}{1+GH(s)}\right] = \lim_{s \to 0} \left[\frac{su(s)}{1+G(s)H(0)}\right]$$

Since we assume unity dc gain for the feedback transfer function, we have $H(0) = 1$. Hence,

$$e_{ss} = \lim_{s \to 0} \left[\frac{su(s)}{1+G(s)}\right] \tag{12.24}$$

We will use this result to define three commonly used error constants.

12.3.6.1 *Position Error Constant K_p*

Consider a *unit step input* defined as

$$u(t) = 1 \quad \text{for } t \geq 0$$
$$= 0 \quad \text{for } t < 0 \tag{12.25a}$$

Its Laplace transform is given by

$$u(s) = \frac{1}{s} \tag{12.25b}$$

Then from Equation 12.24 we have

$$e_{ss} = \frac{1}{1+G(0)} \tag{12.26}$$

Now $G(0)$ will be finite only if the system is type zero. In that case the dc gain of $G(s)$ is defined (i.e., finite), and is denoted by K_p. This is termed position error constant because, for a position control system, a step input can be interpreted as a constant position input. Thus,

$$e_{ss} = \frac{1}{1 + K_p}$$ (12.27)

in which, the position error constant,

$$K_p = \lim_{s \to 0} G(s) = G(0)$$ (12.28)

It is seen from Equation 12.26 that for systems of type-1 or higher, the steady-state error to a step input would be zero, because $G(0) \to \infty$ in those cases.

12.3.6.2　Velocity Error Constant K_v

Consider a *unit ramp input* defined as

$$u(t) = t \quad \text{for } t \geq 0$$
$$= 0 \quad \text{for } t < 0$$ (12.29a)

Its Laplace transform is given by

$$u(s) = \frac{1}{s^2}$$ (12.29b)

Then from Equation 12.24 we get

$$e_{ss} = \lim_{s \to 0} \left[\frac{1}{s + sG(s)} \right]$$

or

$$e_{ss} = \frac{1}{\lim_{s \to 0} sG(s)}$$ (12.30)

Now note that:

For a type-0 system $\lim_{s \to 0} sG(s) = 0$
For a type-1 system $\lim_{s \to 0} sG(s) = \text{constant}$
For a type-2 system $\lim_{s \to 0} sG(s) = \infty$

It follows from Equation 12.30 that for a type-0 system $e_{ss} \to \infty$ and for a type-2 system $e_{ss} \to 0$. The steady-state error for a unit ramp is a non-zero constant only for a type-1 system, and is given by

$$e_{ss} = \frac{1}{K_v}$$ (12.31)

in which

$$K_v = \lim_{s \to 0} sG(s) \tag{12.32}$$

The constant K_v is termed *velocity error constant* because for a position control system, a ramp position input is a constant velocity input.

12.3.6.3 Acceleration Error Constant K_a

Consider a *unit parabolic input* defined as a

$$u(t) = \frac{t^2}{2} \quad \text{for } t \geq 0$$
$$= 0 \quad \text{for } t < 0 \tag{12.33a}$$

Its Laplace transform is given by

$$u(s) = \frac{1}{s^3} \tag{12.33b}$$

Then from Equation 12.24 we have

$$e_{ss} = \frac{1}{\lim_{s \to 0} s^2 G(s)} \tag{12.34}$$

It is now clear that for a type-0 system or type-1 system this steady-state error goes to infinity. For a type-2 system the steady-state error to a unit parabolic input is finite, however, and is given by

$$e_{ss} = \frac{1}{K_a} \tag{12.35}$$

in which

$$K_a = \lim_{s \to 0} s^2 G(s) \tag{12.36}$$

The constant K_a is termed *acceleration error constant* because, for a position control system, a parabolic position input is a constant acceleration input.

These results on steady-state error to the test inputs are summarized in Table 12.4. Note that for control loops with one or more free integrators (i.e., system type-1 or higher) the steady-state error to a step input would be zero. This explains why integral control is used to eliminate the offset error in systems under steady inputs, as noted previously.

TABLE 12.4

Steady-State Error for Standard Test Inputs

Input	$u(t)$ $t \geq 0$	$u(s)$	Steady-state error		
			Type 0 System	Type 1 System	Type 2 System
Unit Step	1	$\dfrac{1}{s}$	$\dfrac{1}{1+K_p}$	0	0
Unit Ramp	t	$\dfrac{1}{s^2}$	∞	$\dfrac{1}{K_v}$	0
Unit Parabola	$\dfrac{t^2}{2}$	$\dfrac{1}{s^3}$	∞	∞	$\dfrac{1}{K_a}$

Example 12.5

Synchro transformer is a feedback sensor that is used in control applications. It consists of a transmitter whose rotor is connected to the input member, and a receiver whose rotor is connected to the output member of the system to be controlled. The stator field windings and rotor field windings are Y-connected together. The transmitter rotor has a single set of windings and is activated by an external ac source. Then the signal generated in the rotor windings of the receiver is a measure of the position error (difference between the transmitter rotor angle and the receiver rotor angle). Consider a position servo that uses a synchro transformer. When the transmitter rotor turns at a constant speed of 120 rpm, the position error was found to be 2°. Estimate the velocity error constant of the control system. The loop gain of the control system is 50. Determine the required loop gain in order to obtain a steady-state error of 1° at the same input speed.

SOLUTION

$$\text{Input speed} = 120 \text{ rpm } = \frac{120}{60} \times 2\pi \, \text{rad/s}$$

$$\text{Steady-state error} = 2° = 2 \times \frac{\pi}{180} \, \text{rad}$$

Steady-state error for a unit ramp input (i.e., a constant speed of 1 rad/s at the input)

$$e_{ss} = 2 \times \frac{\pi}{180} \bigg/ \frac{120 \times 2\pi}{60} \, \text{rad/rad/s} = \frac{1}{360} \, \text{s}$$

From Equation 12.31, the velocity error constant is obtained as

$$K_v = 360 \text{ s}^{-1}$$

Since K_v is directly proportional to the loop gain of a control system, required loop gain for a specified steady-state error can be determined with the knowledge of K_v. In the present example, in view of Equation 12.31, the K_v required to obtain a steady-state error of 1° at 120 rpm is double the value for a steady-state error of 2°. The corresponding loop gain is 100. Note that the units of loop gain are determined by the particular control system, and is not known for this example even though the units of K_v are known to be s^{-1}.

12.3.7 System Type as a Robustness Property

For a type-0 system, the steady-state error to a unit step input may be used as a design specification. The gain of the control loop (loop gain) can be chosen on that basis. For a system of type-1 or higher, this approach is not quite appropriate, however, because the steady state error would be zero no matter what the loop gain is. An appropriate design specification for a type-1 system (i.e., a system having a single integrator in the loop) would be the steady-state error under a unit ramp input. Alternatively, the velocity error constant K_v may be used as a design specification. Similarly, for a type-2 system, the steady-state error under a unit parabolic input, or alternatively the acceleration error constant K_a may be used as a design specification. Note further that the system type number is a measure of the *robustness* of a system. To understand this, we recall the fact that for a type-0 system, steady-state error to a unit step input depends on plant gain and, hence, error will vary due to variations in the gain parameter. For a system of type-1 or higher, steady-state error to a step input will remain zero even in the presence of variations in the gain parameter (a robust situation). For a system of type-2 or higher, steady-state error to a ramp input as well as step input will remain zero in the presence of gain variations. It follows that as the type number increases, the system robustness with respect to steady-state error to a polynomial input (step, ramp, parabola, etc.), improves.

12.3.8 Performance Specification Using S Plane

The poles of a damped oscillator are given by Equation 12.12. The s-plane is defined by a horizontal axis corresponding to the real part of s and a vertical axis corresponding to the imaginary part of s. Then, the pole location on the s-plane is defined by the real part $-\zeta\omega_n$ and the imaginary part ω_d of the two roots. Note that the magnitude of the real part is the reciprocal of the *time constant* τ (Equation 12.13), and ω_d is the *damped natural frequency* (Equation 12.5). Now recall the expressions for the performance specifications as given in Table 12.2. It is seen that a "constant settling time line" corresponds to a "constant time constant line" (i.e., a vertical line on the s-plane). Also, a "constant peak time line" corresponds to a "constant ω_d line" (i.e., a horizontal line on the s-plane). Next, a "constant percentage overshoot line" corresponds to a "constant damping ratio line." But we know that this is a radial line, cosine of whose angle with reference to the negative real axis is equal to the damping ratio ζ (see Equation 12.6). These lines are shown in Figures 12.15(a) through 12.15(c). Since a satisfactory design is expressed by an inequality constraint on each of the design parameters, we have indicated the acceptable design region in each case.

To obtain an appropriate measure for steady-state error we recall that for a type-zero system, the steady-state error to a step input decreases with increasing loop gain. Furthermore, it is also known that the undamped natural frequency ω_n increases with the loop gain. It follows that for a system with variable gain, a "constant steady-state error line" is in fact a "constant ω_n line." This is a circle on the s-plane, with radius ω_n and centered at the origin of the coordinate system. This line is shown in Figure 12.15(d).

A composite design boundary and a design region (corresponding to a combined design specification) can be obtained by simply overlaying the four regions given in Figures 12.15(a) through 12.15(d). This is shown in Figure 12.15(e). In Figure 12.15 we have disregarded the right half of the s-plane, at the outset, because this region corresponds to an *unstable* system.

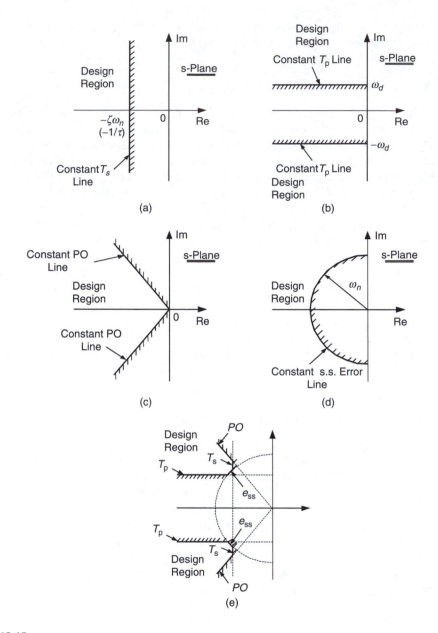

FIGURE 12.15
Performance specification on the s-plane: (a) Settling time; (b) Peak time; (c) Percentage overshoot; (d) Steady-state error; (e) Combined specification.

Example 12.6

A control system is required to have a percentage overshoot of less than 10% and a settling time of less than 0.8 sec. Indicate this design specification as a region on the s-plane.

SOLUTION
A 10% overshot means (see Table 12.2)

$$0.1 = \exp\left(-\frac{\pi\zeta}{\sqrt{1-\zeta^2}}\right)$$

Hence

$$\zeta = 0.60 = \cos\phi$$

or

$$\phi = 53°$$

Next, T_s of 0.8 s means (see Table 12.2)

$$T_s = 4\tau = \frac{4}{\zeta\omega_n} = 0.8$$

or

$$\zeta\omega_n = 5.0$$

For the given specifications we require $\phi \le 53°$ and $\zeta\omega_n \ge 5.0$. The corresponding region on the s-plane is given by the shaded area in Figure 12.16.

Example 12.7
A simplified model of a control system for a paper winding mechanism used in a paper plant is shown in Figure 12.17. The model includes the time constants of the drive motor

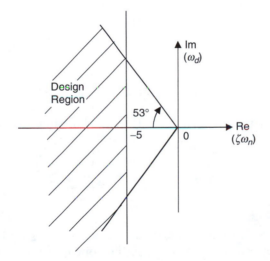

FIGURE 12.16
Design specification on the s-plane.

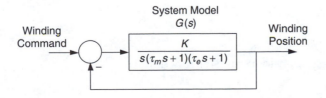

FIGURE 12.17
A simplified control system for a paper winding mechanism.

with load, and the gain of the control amplifier and other circuitry. For a preliminary design it is required to determine the system parameters K, τ_m, and τ_e so as to meet the following design specifications:

a. 2% Settling time = 4/3 s
b. Peak Time = $\pi/4$

Determine the minimum value of steady-state error e_{ss} to a unit ramp input and, hence, the maximum value of the velocity error constant K_v that is possible for this control system.

Select a value for K so that the closed-loop system behaves like a simple oscillator. Determine the corresponding values of e_{ss} and K_v and system parameters τ_m and τ_e. Estimate the percentage overshoot of the resulting closed-loop system.

SOLUTION

First we note that with a settling time of $T_s = 4/3$, the closed-loop system must satisfy (see Table 12.2):

$$\frac{4}{\zeta\omega_n} = \frac{4}{3} \quad \text{or} \quad \zeta\omega_n = 3$$

Also, with a peak time of $T_p = \frac{\pi}{4}$ we have

$$\frac{\pi}{\omega_d} = \frac{\pi}{4} \quad \text{or} \quad \omega_d = 4$$

Accordingly, we have

$$\omega_n = 5 \quad \text{and} \quad \zeta = 0.6$$

Since the system is third-order, the closed-loop system should have a transfer function given by

$$\tilde{G}(s) = \frac{\tilde{K}}{(s+p)\left(s^2 + 2\zeta\omega_n s + \omega_n^2\right)} \tag{i}$$

with p sufficiently large so that the expressions used for T_s and T_p would hold with reasonable accuracy. Specifically, for large p, the contribution of this pole (Ce^{-pt}) to the overall response of the system, will decay very fast. Once this happens, there is no effect from p, and the system will behave like a simple oscillator. With the values computed above, we have

$$\tilde{G}(s) = \frac{\tilde{K}}{(s+p)(s^2 + 6s + 25)} \tag{ii}$$

The corresponding open-loop transfer function is

$$G(s) = \frac{\tilde{G}(s)}{1 - \tilde{G}(s)} = \frac{\tilde{K}}{(s+p)(s^2 + 6s + 25) - \tilde{K}}$$

Since this transfer function has a free integrator (open-loop system is Type-1), as given in Figure 12.17, we must have the constant terms in the denominator cancel out; thus

$$\tilde{K} = 25p \tag{iii}$$

Hence,

$$G(s) = \frac{\tilde{K}}{s[s^2 + (p+6)s + 6p + 25]} \tag{iv}$$

The velocity error constant (Equation 12.32)

$$K_v = \lim_{s \to 0} sG(s) = \frac{\tilde{K}}{6p + 25}$$

Now using Equation iii we have

$$K_v = \frac{\tilde{K}}{6\tilde{K}/25 + 25} \tag{v}$$

This is a monotonically increasing function for positive values of $\tilde{K}$. It follows that the maximum value of K_v is obtained when $\tilde{K} \to \infty$. Hence,

$$K_{v_{max}} = \frac{1}{6/25} = \frac{25}{6}$$

Corresponding steady-state error for a unit ramp input (see Table 12.4)

$$e_{ss_{min}} = \frac{6}{25} = 0.24$$

As noted before, if the closed-loop system is to behave like a simple oscillator, the real pole has to move far to the left of the complex pole pair so that the effect of the real pole could be neglected. A rule of thumb is to use a factor of 10 as an adequate distance. Since $\zeta\omega_n = 3$, we pick

$$p = 30$$

From Equation iii then $\tilde{K} = 25 \times 30 = 750$. Substituting in Equation v, we get

$$K_v = \frac{750}{6 \times 750/25 + 25} = 3.66$$

Hence,

$$e_{ss} = \frac{1}{3.66} = 0.25$$

Note that this value is quite close to the minimum possible value (0.24), and is quite satisfactory. From Equation iv

$$G(s) = \frac{750}{s(s^2 + 36s + 205)} = \frac{750}{s(s+7.09)(s+28.91)} = \frac{3.66}{s(0.141s + 1)(0.0346s + 1)}$$

So we have

$$K = 3.66, \quad \tau_m = 0.141, \quad \tau_e = 0.0346$$

Furthermore, since $\zeta = 0.6$ for the closed-loop system, the percentage overshoot is (Table 12.2)

$$PO = 100\exp\left(-\pi \times 0.6/\sqrt{1 - 0.36}\right)$$

or

$$PO = 9.5\%$$

12.3.9   Control System Sensitivity

Accuracy of a control system is affected by parameter changes in the control system components and by the influence of external disturbances. It follows that analyzing the sensitivity of a feedback control system to parameter changes and to external disturbances is important.

Consider the block diagram of a typical feedback control system, as shown in Figure 12.18. In the usual notation:

$G_p(s)$ = transfer function of the plant (or the system to be controlled)

$G_c(s)$ = transfer function of the controller (including compensators)

$H(s)$ = transfer function of the output feedback system (including the measurement system)

u = system input command

u_d = external distrubance input

y = system output

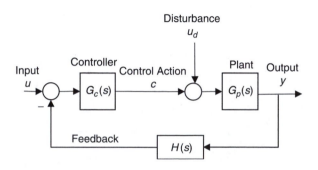

FIGURE 12.18
Block diagram representation of a feedback control system.

Since what we have is a linear system (as necessary in the transfer function representation), the *principle of superposition* is valid. In particular, if we know the outputs corresponding to the two inputs applied separately, the output when both inputs are applied simultaneously is given by the sum of the individual outputs.

First set $u_d = 0$

Then it is straightforward to obtain the input-output relationship:

$$y = \left[\frac{G_c G_p}{1 + G_c G_p H} \right] u \tag{i}$$

Next set $u = 0$

Then we obtain the input-output relationship:

$$y = \left[\frac{G_p}{1 + G_c G_p H} \right] u_d \tag{ii}$$

By applying the principle of superposition on Equation i and Equation ii, we obtain the overall input-output relationship:

$$y = \left[\frac{G_c G_p}{1 + G_c G_p H} \right] u + \left[\frac{G_p}{1 + G_c G_p H} \right] u_d \tag{12.37}$$

The closed-loop transfer function $\tilde{G}$ is given by y/u, with $u_d = 0$; thus,

$$\tilde{G} = \frac{G_c G_p}{[1 + G_c G_p H]} \tag{12.38}$$

The sensitivity of the system to a change in some parameter k may be expressed as the ratio of the change in output to the change in the parameter, that is, $\Delta y / \Delta k$. In the nondimensional form, this sensitivity is given by

$$S_k = \frac{k}{y} \frac{\Delta y}{\Delta k} \tag{12.39}$$

Since $y = \tilde{G} u$, with $u_d = 0$, it follows that for a given input u,

$$\frac{\Delta y}{y} = \frac{\Delta \tilde{G}}{\tilde{G}}$$

Consequently, Equation 12.39 may be expressed as

$$S_k = \frac{k}{\tilde{G}} \frac{\Delta \tilde{G}}{\Delta k} \tag{12.40}$$

or, in the limit,

$$S_k = \frac{k}{\tilde{G}} \frac{\partial \tilde{G}}{\partial k} \qquad (12.41)$$

Now, by applying Equation 12.41 to Equation 12.38, we are able to determine expressions for the control system sensitivity to changes in various components in the control system. Specifically, by straightforward partial differentiation of Equation 12.38, separately with respect to G_p, G_c, and H, we get

$$S_{Gp} = \frac{1}{[1 + G_c G_p H]} \qquad (12.42)$$

$$S_{Gc} = \frac{1}{[1 + G_c G_p H]} \qquad (12.43)$$

$$S_H = -\frac{G_c G_p H}{[1 + G_c G_p H]} \qquad (12.44)$$

It is clear from these three relations that as the static gain (or, dc gain) of the loop (i.e., $G_c G_p H$, with $s = 0$) is increased, the sensitivity of the control system to changes in the plant and the controller decreases, but the sensitivity to changes in the feedback (measurement) system approaches (negative) unity. Furthermore, it is clear from Equation 12.37 that the effect of the disturbance input can be reduced by increasing the static gain of $G_c H$. By combining these observations, the following design criteria can be stipulated for a feedback control system:

1. Make the measurement system (H) very accurate and stable.
2. Increase the loop gain (i.e., gain of $G_c G_p H$) to reduce the sensitivity of the control system to changes in the plant and controller.
3. Increase the gain of $G_c H$ to reduce the influence of external disturbances.

In practical situations, the plant G_p is usually fixed and cannot be modified. Furthermore, once an accurate measurement system is chosen, H is essentially fixed. Hence, most of the design freedom is available with respect to G_c only. It is virtually impossible to achieve all the design requirements simply by increasing the gain of G_c. The dynamics (i.e., the entire transfer function) of G_c (not just the gain value at $s = 0$) also have to be properly designed in order to obtain the desired performance of a control system.

Example 12.8

Consider the cruise control system given by the block diagram in Figure 12.19. The vehicle travels up a constant incline with a constant speed setting for the cruise controller.

   a. For a speed setting of $u = u_o$ and a constant road inclination of $u_o = u_{do}$ derive an expressions for the steady state values y_{ss} of the speed and e_{ss} of the speed error. Express your answers in terms of K, K_c, u_o and u_{do}.
   b. At what minimum percentage grade would the vehicle stall? Use steady-state conditions, and express your answer in terms of the speed setting u_o and controller gain K_c.

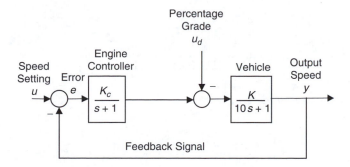

FIGURE 12.19
A cruise control system.

c. Suggest a way to reduce e_{ss}.

d. If $u_o = 4$, $u_{do} = 2$ and $K = 2$, determine the value of K_c such that $e_{ss} = 0.1$.

SOLUTION

a. For $u_d = 0$:

$$y = \frac{\frac{K_c K}{(s+1)(10s+1)}}{\left[1 + \frac{K_c K}{(s+1)(10s+1)}\right]} u = \frac{K_c K}{[(s+1)(10s+1) + K_c K]} u$$

For $u = 0$

$$y = \frac{\frac{K}{(10s+1)}}{\left[1 + \frac{K_v K}{(s+1)(10s+1)}\right]} (-u_d) = -\frac{K(s+1)}{[(s+1)(10s+1) + K_c K]} u_d$$

Hence, with both u and u_d present, using the principle of superposition (linear system):

$$y = \frac{K_c K}{[(s+1)(10s+1) + K_c K]} u - \frac{K(s+1)}{[(s+1)(10s+1) + K_c K]} u_d \qquad (i)$$

If the inputs are constant at steady state, the corresponding steady-state output does not depend on the nature of the inputs under starting conditions. Hence, in this problem what matters is the fact that the inputs and the output are constant at steady state. Hence, without loss of generality, we can assume the inputs to be step functions.

NOTE Even if we assume a different starting shape for the inputs, we should get the same answer for the steady state output, for the same steady-state input values. But the mathematics of getting that answer would be more complex.

Now, using final value theorem, at steady state:

$$y_{ss} = \lim_{s \to 0} \left[\frac{K_c K}{[(s+1)(10s+1) + K_c K]} \frac{u_o}{s} s - \frac{K(s+1)}{[(s+1)(10s+1) + K_c K]} \frac{u_{do}}{s} s \right]$$

or

$$y_{ss} = \frac{K_c K}{(1+K_c K)} u_o - \frac{K}{(1+K_c K)} u_{do} \quad \text{(ii)}$$

Hence, the steady-state error,

$$e_{ss} = u_o - y_{ss} = u_o - \frac{K_c K}{(1+K_c K)} u_o + \frac{K}{(1+K_c K)} u_{do}$$

or,

$$e_{ss} = \frac{1}{(1+K_c K)} u_o + \frac{K}{(1+K_c K)} u_{do} \quad \text{(iii)}$$

b. Stalling condition is $y_{ss} = 0$. Hence, from Equation ii we get

$$u_{do} = K_c u_o$$

c. Since $\underline{K}$ is usually fixed (a plant parameter) and cannot be adjusted, we should increase K_c to reduce e_{ss}.
d. We are given $u_o = 4$, $u_{do} = 2$, $K = 2$, $e_{ss} = 0.1$

Substitute in (iii):

$$0.1 = \frac{1}{(1+2K_c)} \times 4 + \frac{2}{(1+2K_c)} \times 2$$

Hence, $1 + 2K_c = 80$, or $K_c = 39.5$.

12.4   Stability and Routh–Hurwitz Criterion

In this section we will formally study stability of a linear time-invariant (LTI) system, in the time domain. The natural response (homogeneous solution) of a system (differential equation) is determined by the eigenvalues (poles) of the system. Hence, stability is determined by the system poles.

12.4.1   Natural Response

Consider the linear, time-invariant system, discussed in Chapter 2:

$$a_n \frac{d^n y}{dt^n} + a_{n-1} \frac{d^{n-1} y}{dt^{n-1}} + \cdots + a_0 y = b_m \frac{d^m u}{dt^m} + b_{m-1} \frac{d^{m-1} u}{dt^{m-1}} + \cdots + b_0 u \quad (12.45)$$

where u is the input and y is the output. When there is no input, we have the *homogeneous equation*

$$a_n \frac{d^n y}{dt^n} + a_{n-1} \frac{d^{n-1} y}{dt^{n-1}} + \cdots + a_0 y = 0 \tag{12.46}$$

It is well known that the solution to this equation is of the exponential form

$$y = C e^{\lambda t} \tag{12.47}$$

which can be verified by substituting Equation 12.47 into Equation 12.46. Then, on canceling the common factor $C e^{\lambda t}$ since $C e^{\lambda t}$ is not zero for a general t, we get

$$a_n \lambda^n + a_{n-1} \lambda^{n-1} + \cdots + a_0 = 0 \tag{12.48a}$$

If we use s instead of λ in Equation 12.47 we have

$$a_n s^n + a_{n-1} s^{n-1} + \cdots + a_0 = 0 \tag{12.48}$$

Equation 12.48 is called the *characteristic equation* of the system given by Equation 12.45 and its roots are called *eigenvalues* or *poles* of the system. In general there will be n roots to Equation 12.48, and let us denote them by $\lambda_1, \lambda_2, \ldots, \lambda_n$. The solution Equation 12.47 will be formed by combining the contributions of all these roots. Hence, the general solution to Equation 12.46, assuming that there are no repeated roots among $\lambda_1, \lambda_2, \ldots, \lambda_n$, is

$$y_h = C_1 e^{\lambda_1 t} + C_2 e^{\lambda_2 t} + \cdots + C_n e^{\lambda_n t} \tag{12.49}$$

This is the homogeneous solution of Equation 12.45, and it represents the natural response of the system (it does not depend on the input). The constants $C_1, C_2, \ldots, C_n$ are unknowns (integration constants), which have to be determined by using n initial conditions (say, $y(0), y'(0), \ldots, y^{n-1}(0)$) of the system. If two roots are equal, then the corresponding constants are not independent and can be combined in Equation 12.49. In that case a factor t has to be included into one of the corresponding terms, for example, if $\lambda_1 = \lambda_2$ without loss of generality, we have

$$y_h = C_1 e^{\lambda_1 t} + C_2 t e^{\lambda_1 t} + \cdots + C_n e^{\lambda_n t} \tag{12.50}$$

Next let us see how the natural response depends on the nature of the poles. First, since the system coefficients (parameters) $a_0, a_1, \ldots, a_n$ are all real, any complex roots of Equation 12.48 should occur in complex conjugates (i.e., $\lambda_r + j\lambda_i$ and $\lambda_r - j\lambda_i$, where λ_r and λ_i are the real and the imaginary parts, respectively, of a complex root). The contribution of a complex pole to the natural (homogeneous) response of the system is of the form

$$e^{(\lambda_r + j\lambda_i)t} = e^{\lambda_r t}(\cos \lambda_i t + j \sin \lambda_i t) \tag{12.51}$$

Hence, it is clear that if a root (pole) has an imaginary part, then that pole produces an oscillatory response. Also, if the real part is negative, it generates an exponential decay and if the real part is positive, it generates an exponential growth (i.e., an unstable response). These observations are summarized in Table 12.5 and further illustrated in Figure 12.20. Note that poles are the roots of Equation 12.48 and hence they can be marked on the s-plane.

TABLE 12.5

Dependence of Natural Response on System Poles

Pole		Nature of Response	
Real	Negative	Transient (nonoscillatory)	Decaying (stable)
	Positive		Growing (unstable)
Imaginary		Oscillatory with constant amplitude	Steady (marginally stable)
Complex	negative real part	Oscillatory with varying amplitude	Decaying (stable)
	positive real part		Growing (unstable)
Zero value		Constant	

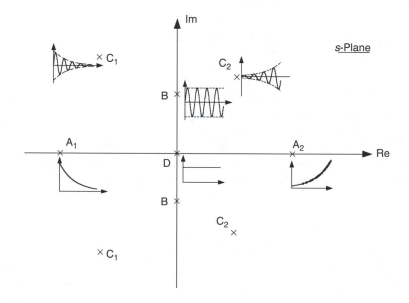

FIGURE 12.20
Pole location on the *s*-plane and the corresponding response.

12.4.2   Routh–Hurwitz Criterion

The Routh test or Routh-Hurwitz stability criterion is a simple way to determine whether a system is stable (i.e., whether none of the poles have positive real parts) by simply examining the characteristic polynomial, without actually solving for the roots. If the system is unstable, the Routh test also tells us how many poles are on the right half plane (RHP), that is, the number of unstable poles. First, a Routh array has to be formed in order to perform the test.

12.4.2.1   Routh Array

The characteristic equation (i.e., the denominator of the transfer function equated to zero) of an *n*th-order system can be expressed as Equation 12.48. This has *n* roots, which are the poles of the system. It is possible to determine the stability of the system without actually finding these *n* roots, by forming a Routh array, as follows:

	First Column	Second Column	Third Column	...	
S^n	a_n	a_{n-2}	a_{n-4}	...	← First row
S^{n-1}	a_{n-1}	a_{n-3}	a_{n-5}	...	← Second row
S^{n-2}	b_1	b_2	b_3	...	← Third row
S^{n-3}	c_1	c_2	c_3	...	← Fourth row
				...	
				...	
S^0	h_1				← Last row

The first two rows are completed first, using the coefficients $a_n, a_{n-1}, \ldots, a_1, a_0$ of the characteristic polynomial, as shown. Note the use of alternate coefficients in these two rows. Each subsequent row is computed from the elements of the two rows immediately above it, by *cross-multiplying* the elements of those two rows. For example,

$$b_1 = \frac{a_{n-1}a_{n-2} - a_n a_{n-3}}{a_{n-1}}$$

$$b_2 = \frac{a_{n-1}a_{n-4} - a_n a_{n-5}}{a_{n-1}}, \text{ etc.}$$

$$c_1 = \frac{b_1 a_{n-3} - a_{n-1} b_2}{b_1}$$

$$c_2 = \frac{b_1 a_{n-5} - a_{n-1} b_3}{b_1}$$

and so on.

The Routh-Hurwitz stability criterion states that for the system to be stable, the following two conditions must be satisfied:

1. All the coefficients $(a_0, a_1, \ldots, a_n)$ of the characteristic polynomial must be positive (i.e., must have the same sign).
2. All the elements in the first column of the Routh array must be positive (i.e., same sign).

If the system is unstable, the number of unstable poles is given by the number of successive sign changes in elements of the first column of Routh array.

Example 12.9

Consider a system whose closed-loop transfer function is

$$G(s) = \frac{2}{(s^3 - s^2 + 2s + 1)}$$

Without even completing a Routh array, it is seen that the system is unstable—from Condition 1 of the Routh test (because, a negative coefficient is present in the characteristic polynomial).

Example 12.10

Consider a system having the (closed loop) transfer function

$$G(s) = \frac{2(s+5)}{3s^3 + s^2 + 4s + 2}$$

Its Routh array is formed by examining the characteristic equation

$$3s^3 + s^2 + 4s + 2 = 0$$

The Routh array is

$$
\begin{array}{ccc}
s^3 & 3 & 4 \\
s^2 & 1 & 2 \\
s^1 & b_1 & 0 \\
s^0 & c_1 & 0
\end{array}
$$

where

$$b_1 = \frac{1 \times 4 - 3 \times 2}{1} = -2$$

$$c_1 = \frac{b_1 \times 2 - 1 \times 0}{b_1} = 2$$

The first column of the array has a negative value, indicating that the system is unstable. Furthermore, since there are "two" sign changes (positive to negative and then back to positive) in the first column, there are two unstable poles in this system.

12.4.2.2 Auxiliary Equation

A Routh array may have a row consisting of zero elements only. This usually indicates a *marginally stable* system (i.e., a pair of purely imaginary poles). The polynomial equation formed by the row which immediately precedes the row with zero elements is called the "auxiliary equation." The roots of this equation, will give the values of these marginally stable poles.

Example 12.11

Consider a plant

$$G(s) = \frac{1}{s(s+1)}$$

and a feedback controller

$$H(s) = \frac{K(s+5)}{(s+3)}$$

Its closed-loop characteristic polynomial ($1 + GH = 0$) is

$$1 + \frac{K(s+5)}{s(s+1)(s+3)} = 0$$

or

$$s(s+1)(s+3) + K(s+5) = 0$$

or

$$s^3 + 4s^2 + (3+K)s + 5K = 0$$

The Routh array is:

$$
\begin{array}{c c c}
s^3 & 1 & 3+K \\
s^2 & 4 & 5K \\
s^1 & \dfrac{12-K}{4} & 0 \\
s^0 & 5K &
\end{array}
$$

Note that when $K = 12$, the 3rd row (corresponding to s^1) of the Routh array will have all zero elements. The polynomial equation corresponding to the previous row (s^2) is

$$4s^2 + 5K = 0$$

With $K = 12$, we have the *auxiliary equation*

$$4s^2 + 5 \times 12 = 0 \quad \text{or} \quad s^2 + 15 = 0$$

whose roots are $s = \pm j\sqrt{15}$. Hence when $K = 12$ we have a marginally stable closed-loop system, with the two marginally stable poles given by $\pm j\sqrt{15}$. The third pole can be determined by comparing coefficients as shown below. This root has to be real (because there is only one root left apart from the complex roots; complex roots occur in conjugate pairs). Call it p. Then, with $K = 12$, the characteristic equation may be expressed as

$$s^3 + 4s^2 + 15s + 60 = (s-p)(s^2+15) = 0$$

By comparing coefficients

$$60 = -15p \quad \text{or} \quad p = -4$$

Hence the real pole is at -4 and is stable. Note: In the above Routh array, the last row cannot be completed when the previous row has all zeros, because of the needed division by zero in that case.

12.4.2.3 Zero Coefficient Problem

If the first element in a particular row of a Routh array is zero, a division by zero will be needed in computing the next row. This will create an ambiguity as to the sign of the next element. This problem can be avoided by replacing the zero element by a small positive element ε, and then completing the array as usual.

Example 12.12

Consider a system whose characteristic equation is

$$s^4 + 5s^3 + 5s^2 + 25 + 10 = 0$$

Let us study the stability of the system. Routh array is:

$$
\begin{array}{cccc}
s^4 & 1 & 5 & 10 \\
s^3 & 5 & 25 & 0 \\
s^2 & \varepsilon & 10 & 0 \\
s^1 & \dfrac{25\varepsilon - 50}{\varepsilon} & 0 & \\
s^0 & 10 & &
\end{array}
$$

Note that the first element of the 3rd row (s^2) should be $\frac{5 \times 5 - 25 \times 1}{5} = 0$. But we have represented it by ε, which is positive and will tend to zero. Then, the first element of the 4th row (s^1) becomes $\frac{25\varepsilon - 50}{\varepsilon}$. Since ε is very small, 25ε is also very small. Hence, the numerator of this quantity is negative, but the denominator (ε) is positive. Hence, this element is negative (and large). This indicates two sign changes in the 1st column, and hence the system has two unstable poles.

In applying Routh-Hurwitz stability criterion, we may start with the system differential equation. Consider the following example.

Example 12.13

Consider the differential equation

$$
3\frac{d^3 y}{dt^3} + 2\frac{d^2 y}{dt^2} + \frac{dy}{dt} + y = 2\frac{du}{dt} + u
$$

in which u is the system input and y is the system output. To obtain the transfer function: change d/dt to s. The result of substituting s for d/dt (and s^2 for d^2/dt^2) in the system differential equation is

$$
(3s^2 + 2s^2 + s + 1)y = (2s + 1)u
$$

The system transfer function (output/input) is

$$
\frac{y}{u} = \frac{(2s + 1)}{(3s^3 + 2s^2 + s + 1)}
$$

Now, Routh array for the characteristic polynomial is constructed as:

$$
\begin{array}{ccc}
s^3 & 3 & 1 \\
s^2 & 2 & 1 \\
s^1 & -1/2 & 0 \\
s^0 & 1 & 0
\end{array}
$$

This first column has two sign changes. Hence the system is unstable, with two unstable poles.

12.4.3 Relative Stability

Consider a stable system. The pole that is closest to the imaginary axis is the *dominant pole*, because the natural response from remaining poles will decay to zero faster, leaving behind the natural response of this pole. It should be clear that the distance of the dominant pole from the imaginary axis is a measure of the "level of stability" or "degree of stability" or "stability margin" or "relative stability" of the system. In other words, if we shift the dominant pole closer to the imaginary axis, the system becomes less stable (i.e., the "relative stability" of the resulting system is lower).

The stability margin of a system can be determined by the Routh test. Specifically, consider a stable system. All its poles will be on the LHP. Now, if we shift all the poles to the right by a known amount, the resulting system will be less stable. The stability of the shifted system can be established using the Routh test. If we continue this process of pole shifting in small steps, and repeatedly apply the Routh test, until the resulting system just goes unstable, then the total distance by which the poles have been shifted to the right provides a measure of the stability margin (or, relative stability) of the original system. The larger the amount by which the poles have to be shifted to the right in order to reach marginal stability the greater the stability margin of the original system.

Example 12.14

A system has the characteristic equation

$$s^3 + 6s^2 + 11s + 36 = 0$$

a. Using Routh-Hurwitz criterion determine the number of unstable poles in the system.

b. Now move all the poles of the given system to the right of the s-plane by the real value 1. (i.e., add 1 to every pole). Now how many poles are on the right-half plane?

NOTE You should answer this question *without* actually finding the poles (i.e., without solving the characteristic equation).

SOLUTION

a. Characteristic equation: $s^3 + 6s^2 + 11s + 36 = 0$

Routh Array:

$$
\begin{array}{ccc}
s^3 & 1 & 11 \\
s^2 & 6 & 36 \\
s^1 & \dfrac{6 \times 11 - 1 \times 36}{6} = 5 & 0 \\
s^0 & 36 &
\end{array}
$$

Since the entries of the first column are all positive, there are no unstable poles. The system is stable.

b. Denote the shifted poles by $\tilde{s}$

We have $\tilde{s} = s + 1$ or $s = \tilde{s} - 1$

Substitute this in the original characteristic equation. The characteristic equation of the system with shifted poles is:

$$(\tilde{s} - 1)^3 + 6(\tilde{s} - 1)^2 + 11(\tilde{s} - 1) + 36 = 0$$

or

$$\tilde{s}^3 - 3\tilde{s}^2 + 3\tilde{s} - 1 + 6\tilde{s}^2 - 12\tilde{s} + 6 + 11\tilde{s} - 11 + 36 = 0$$

or

$$\tilde{s}^3 + 3\tilde{s}^2 + 2\tilde{s} + 30 = 0$$

Routh Array:

$\tilde{s}^3$	1	2
$\tilde{s}^2$	3	30
$\tilde{s}^1$	$\dfrac{3\times2 - 1\times30}{3} = -8$	0
$\tilde{s}^0$	30	

There are two sign changes in the first column. Hence, there are two unstable poles.

12.5 Root Locus Method

Root locus is the locus of (or the continuous path traced by) the closed-loop poles (i.e., roots of the closed-loop characteristic equation) of a system, as one parameter of the system (typically the loop gain) is varied. Specifically, the root locus shows how the locations of the poles of a closed-loop system change when some parameter of the loop transfer function is varied. Hence, it indicates the stability of the closed-loop system as a function of the variable parameter. The method was first published by W. R. Evans in 1948 but is used even today as a powerful tool for analysis and design of control systems. Since in the root locus method we use the system transfer function (specifically, the *loop transfer function*) and the associated "algebraic" approach to draw the root locus, we may consider this method as a frequency domain (strictly, Laplace domain) technique. On the other hand, since the closed-loop poles are directly related to the system response (in the time domain) and also to time domain design specifications such as settling time, peak time, and percentage overshoot, the root locus method can be considered as a time domain technique as well. For these reasons we shall discuss the concepts of root locus without classifying the method into either the time domain or the Laplace domain. The root locus method relies on first locating the open-loop poles (strictly speaking, "loop" poles) on the complex s-plane.

Consider the feedback control structure, as shown in Figure 12.21. The overall transfer function of this system (i.e., the closed-loop transfer function) is

$$\frac{Y(s)}{U(s)} = \frac{G(s)}{1 + G(s)H(s)} \tag{12.52}$$

The stability of the closed loop system is completely determined by the poles (not zeros) of the closed-loop transfer function, Equation 12.52. These poles are obtained by solving the characteristic equation (the equation of the denominator polynomial)

$$1 + G(s)H(s) = 0 \tag{12.53a}$$

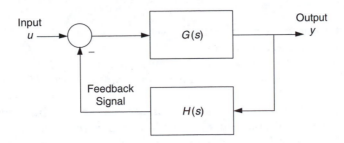

FIGURE 12.21
A feedback control system.

It follows that the closed-loop poles are (and hence, the stability of the closed loop system is) completely determined by the *loop transfer function* $G(s)H(s)$. It is immediately clear that the roots of Equation 12.53a depend on both poles and zeros of $G(s)H(s)$. Hence, stability of a closed loop system depends on both poles and zeros of the loop transfer function.

12.5.1 Rules for Plotting Root Locus

In theory, root locus can be plotted by successively varying any one parameter of the system and solving the corresponding closed-loop characteristic equation. This is in fact the method used in most computer programs for root locus analysis. But in many situations an accurate plot is not needed. In preliminary studies it is adequate to roughly sketch the root locus, with the exact numerical values being computed only for several critical parameter values. In this section we will summarize the rules for sketching a root locus. The principle behind each rule will be explained but the results are not fully.

Since the loop transfer function $G(s)H(s)$ completely determines the closed-loop characteristic equation given by Equation 12.53a, it is the transfer function GH that is analyzed in plotting a root locus. Equation 12.53a can be written in several useful and equivalent forms, as indicated now. First, we have

$$GH = -1 \qquad (12.53b)$$

Next, since GH can be expressed as a ratio of two *monic polynomials* (i.e., polynomials with coefficient of whose highest order term equal to unity) $N(s)$ and $D(s)$, we can write Equation 12.53b as

$$K \frac{N(s)}{D(s)} = -1 \qquad (12.53c)$$

in which
  $N(s) =$ numerator polynomial (monic) of the loop transfer function
  $D(s) =$ denominator polynomial (monic) of the loop transfer function
  $K =$ loop gain

Now in Equation 12.53c the polynomials by factorizing, we can write

$$K \frac{(s-z_1)(s-z_2)\cdots(s-z_m)}{(s-p_1)(s-p_2)\cdots(s-p_n)} = -1 \qquad (12.53d)$$

in which

z_i = a zero of the loop transfer function
p_i = a pole of the loop transfer function
m = order of the numerator polynomial = number of zeros of GH
n = order of the denominator polynomial = number of poles of GH

For physically realizable systems we have $m \le n$ (see Chapter 2). Equation 12.53c is in the "ratio-of-polynomials form" and Equation 12.53d is in the "pole-zero form." Now we will list the main rules for sketching a root locus, and subsequently explain each rule. Note that it is just one equation, the characteristic Equation 12.53 of the closed-loop system, which generates all these rules.

12.5.1.1 Complex Numbers

As we have observed, a pole or a zero of a system can be complex. Hence before stating the rules for sketching a root locus, we need to understand some basic mathematics of complex numbers (Also see Appendix A and Chapter 2). A complex number has a real part and an imaginary part. This can be represented by a vector (or a directed line) on the two-dimensional plane formed by a real axis and an orthogonal imaginary axis (or, the s-plane), as shown in Figure 12.22(a). In particular, a complex number $\underline{r}$ can be expressed as

$$\underline{r} = r \cos \theta + jr \sin \theta = re^{j\theta} \tag{i}$$

where r denoted by $|\underline{r}|$, is the *magnitude* of the complex number (or vector) $\underline{r}$, and θ denoted by $\angle \underline{r}$ is the *phase angle* of the complex number (or vector) $\underline{r}$. Now consider two complex numbers

$$\underline{r_1} = r_1 \cos \theta_1 + jr_1 \sin \theta_1 = r_1 e^{j\theta_1} \tag{ii}$$

$$\underline{r_2} = r_2 \cos \theta_2 + jr_2 \sin \theta_2 = r_2 e^{j\theta_2} \tag{iii}$$

Their product is

$$\underline{r_1 r_2} = (r_1 \cos \theta_1 + jr_1 \sin \theta_1)(r_2 \cos \theta_2 + jr_2 \sin \theta_2)$$

$$= r_1 r_2 [(\cos \theta_1 \cos \theta_2 - \sin \theta_1 \sin \theta_2) + j(\cos \theta_1 \sin \theta_2 + \sin \theta_1 \cos \theta_2)] = r_1 r_2 e^{j(\theta_1 + \theta_1)} \tag{iv}$$

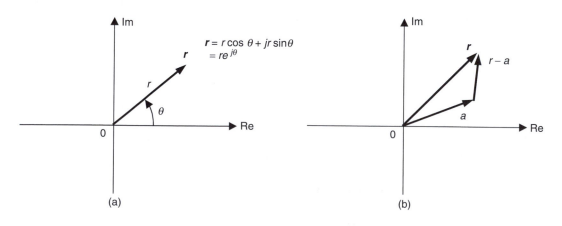

FIGURE 12.22
(a) Complex number represented as a 2-D vector; (b) The complex number subtraction.

and the quotient is

$$r_1/r_2 = (r_1 \cos\theta_1 + jr_1 \sin\theta_1)/(r_2 \cos\theta_2 + jr_2 \sin\theta_2)$$

Multiply the numerator and the denominator by $(r_2 \cos\theta_2 - jr_2 \sin\theta_2)$ and simplify. We get

$$\frac{r_1}{r_2} = \frac{r_1}{r_2}(\cos\theta_1 + j\sin\theta_1)(\cos\theta_2 - j\sin\theta_2)$$

$$\text{(v)}$$

$$= \frac{r_1}{r_2}[(\cos\theta_1\cos\theta_2 + \sin\theta_1\sin\theta_2) + j(\sin\theta_1\cos\theta_2 - \cos\theta_1\sin\theta_2)] = \frac{r_1}{r_2}e^{j(\theta_1-\theta_1)}$$

Hence, In a product of complex numbers, the magnitudes multiply and the phase angles add. In a quotient of complex numbers, the magnitudes divide and the phase angles subtract.

Next consider two complex numbers $\underline{r}$ and $\underline{a}$, as shown in Figure 12.22(b). The complex number $\underline{r}$-$\underline{a}$ is given by the vector line starting from the head of $\underline{a}$ and ending at the head of $\underline{r}$, as shown. This can be confirmed by vector addition using triangle of vectors since

$$\underline{a} + \underline{r} - \underline{a} = \underline{r} \qquad \text{(vi)}$$

12.5.1.2 Root Locus Rules

Rule 1 (*Number of Branches*): Root locus has n branches. They start at the n poles of the loop transfer function. Out of these, m branches terminate at the zeros of GH and the remaining $(n-m)$ branches go to infinity, tangential to a set of $n-m$ lines called *asymptotes*.

Rule 2 (*Magnitude and Phase Conditions*): The *magnitude condition* is

$$K \frac{\displaystyle\prod_{i=1}^{m}|s - z_i|}{\displaystyle\prod_{i=1}^{n}|s - p_i|} = 1 \qquad (12.54a)$$

The *phase angle condition* is

$$\sum_{i=1}^{n} \angle(s - p_i) - \sum_{i=1}^{m} \angle(s - z_i) = \pi + 2r\pi \qquad (12.54b)$$

$$r = 0, \pm 1, \pm 2,\ldots$$

Rule 3 (*Root Locus on Real Axis*): Pick any point on the real axis. If (# poles − # zeros) of GH to the right of the point is odd, the point lies on the root locus.

Rule 4 (*Asymptote Angles*): The $n - m$ asymptotes form angles

$$\theta_r = \frac{\pi + 2\pi r}{n - m}$$

$$r = 0, \pm 1, \pm 2,\ldots \qquad (12.55)$$

with the positive real axis of the s-plane.

Rule 5 (Break Points): *Break-in points* and *breakaway points* of root locus (where two or more branches intersect) correspond to the points of repeated (multiple) poles of the closed-loop system. These points are determined by differentiating the characteristic equation, substituting for K, and solving for s; thus

$$N(s)\frac{dD}{ds} - D(s)\frac{dN}{ds} = 0 \tag{12.56}$$

Rule 6 (Intersection with Imaginary Axis): If the root locus intersects the imaginary axis, the points of intersection are given by setting $s = jw$ and solving the resulting characteristic equation,

$$D(jw) + KN(jw) = 0 \tag{12.57}$$

which gives two equations (one for the real terms and the other for the imaginary terms).

Alternatively, the Routh-Hurwitz criterion and the auxiliary equation for marginal stability, may be used to determine these points and the corresponding gain value.

Rule 7 (Angles of Approach and Departure): The *departure angle* α of root locus, from a GH pole, is obtained using

$$\alpha + \angle \text{ at other poles} - \angle \text{ at zeros} = \pi + 2r\pi \tag{12.58a}$$

The *approach angle* α to a GH zero is obtained using

$$\angle \text{ at poles} - \alpha - \angle \text{ at other zeros} = \pi + 2r\pi \tag{12.58b}$$

Note that an angle mentioned in Rule 7 is measured by drawing a line "to" the point of interest on the root locus "from" the considered pole or zero of GH and determining the angle of that line measured from the positive real axis (or, horizontal line drawn to the right at the considered pole or zero).

Rule 8 (Intersection of Asymptotes with Real Axis): Asymptotes meet the real axis at the *centroid* about the imaginary axis, of the poles and zeros of GH. Each pole is considered to have a weight of +1 and each zero a weight of –1.

12.5.1.3   Explanation of the Rules

Rule 1 comes from the fact that when we set, $K = 0$ the closed loop poles are identical to the poles of GH, and when we set $K \to \infty$, the closed-loop poles are either equal to the zeros of GH or are at infinity. This fact is confirmed by Equation 12.53d, which may be written as

$$-K(s - z_1)(s - z_2) \cdots (s - z_m) = (s - p_1)(s - p_2) \cdots (s - p_n)$$

or

$$(s - z_1)(s - z_2) \cdots (s - z_m) = -(s - p_1)(s - p_2) \cdots (s - p_n)/K$$

Rule 2 as well is obtained directly from Equation 12.53d noting the fact that the magnitude of –1 is 1 and the phase angle of –1 is $\pi + 2r\pi$, where r is any integer.

Rule 3 is a direct result of the phase angle condition given by Equation 12.54. Unless Rule 3 is satisfied, the right hand side of Equation (12.54b) would be an even multiple of π (which is incorrect) rather than an odd multiple of π (correct).

Rule 4 is also a result of the phase angle condition, Equation 12.54b. But this time we use the fact that at infinity (i.e., $s \to \infty$), a root locus and its asymptote are identical, and the fact that when s is at infinity the finite values z_i and p_i can be neglected in Equation 12.54b.

To understand Rule 5, consider the characteristic polynomial of a closed-loop system that has two identical poles. It will have a factor $(s - p)^2$ where p is the double pole of the closed-loop system. It follows that if we differentiate the characteristic polynomial with respect to s, there still will remain a common factor $(s - p)$. This means that, at a double pole, both the characteristic polynomial and its first derivative will be equal to zero. Similarly, at a triple pole (three identical poles), the characteristic polynomial and its first and second derivatives will vanish, and so on.

Rule 6 is clear from common sense because s is purely imaginary on the imaginary axis. Furthermore, purely imaginary poles are *marginally stable* poles. If all the remaining poles are stable, we have a marginally stable system.

Rule 7 is also a direct consequence of the phase angle condition, Equation 12.54b. Specifically, consider a *GH* pole p_i and a point (s) on the root locus very close to this pole. Then the angle $\angle(s - p_i)$ is in fact the angle of departure of the root locus from that pole p_i (see complex number subtraction discussed previously). A similar argument can be made for the angle of approach to a *GH* zero z_i.

To establish Rule 8 using intuitive notions, consider Equation 12.53d, which can be written in the form

$$(s - p_1)(s - p_2) \cdots (s - p_n) = -K(s - z_1)(s - z_2) \cdots (s - z_m) \tag{12.53e}$$

Now define,

$$\bar{s} = \text{point at which asymptotes meet the real axis.}$$

Then, the vector $s - \bar{s}$ defines an asymptote line, where s denotes any general point on the asymptote.

Note that by definition, at infinity, a root locus branch that goes to infinity and its asymptote line become identical. As $s \to \infty$ on root locus, we notice that the vector lines $(s - p_i)$, $(s - z_j)$, $(s - \bar{s})$ all appear to be identical (for any finite p_i and z_j) and they appear to come from the same point $\bar{s}$ on the real axis. Hence, Equation 12.53e becomes

$$(s - \bar{s})^n = -K(s - \bar{s})^m \quad \text{as } s \to \infty \text{ on root locus} \tag{12.59}$$

Substitute Equation 12.59 in Equation 12.53e. We get

$$(s - p_1)(s - p_2) \cdots (s - p_n) = (s - \bar{s})^{n-m}(s - z_1)(s - z_2) \cdots (s - z_m) \quad \text{as } s \to \infty \text{ on root locus} \tag{12.60}$$

Now, by equating the coefficients of s^{n-1} in Equation 12.60 we have

$$\sum_{i=1}^{n} p_i = (n - m)\bar{s} + \sum_{i=1}^{m} z_i$$

or

$$\bar{s} = \frac{1}{(n-m)} \left[\sum_{i=1}^{n} p_i - \sum_{i=1}^{m} z_i \right] \tag{12.61}$$

Equation 12.61 is in fact Rule 8. Remember here that we need to consider only the real parts of p_i and z_i because the imaginary parts occur as conjugate pairs in complex poles or zeros, and they cancel out in the summation.

12.5.1.4   Steps of Sketching Root Locus

Now we list the basic steps of the normal procedure that is followed in sketching a root locus.

Step 1: Identify the loop transfer function and the parameter (gain K) to be varied in the root locus.

Step 2: Mark the poles of GH with the symbol ($\times$) and the zeros of GH with the symbol (o) on the s-plane.

Step 3: Using Rule 3 sketch the root locus segments on the real axis.

Step 4: Compute the asymptote angles using Rule 4 and the asymptote origin using Rule 8, and draw the asymptotes.

Step 5: Using Rule 5, determine the break points, if any.

Step 6: Using Rule 7 compute the departure angles and approach angles, if necessary.

Step 7: Using Rule 6, determine the points of intersection with the imaginary axis, if any.

Step 8: Complete the root locus by appropriately joining the points and segments that have been determined in the previous steps.

Example 12.15

Newton's second law suggests that a very approximate model for an unconstrained rigid body is a double integrator. Aircraft, satellites, guideway vehicles, and robotic manipulators, in their direction of motion, can be crudely approximated by this model. Consider position feedback control of such a system. A compensator is used in the forward path of the control system as shown in Figure 12.23. The controller gain K includes the gains in other components such as plant, sensors, and compensator. Also, the parameters p and z, which determine the compensator pole and zero, are assumed positive, which is the case in practice. We are interested in studying the behavior of the control system for different values of the controller gain. This is easily accomplished by sketching the root locus of the system. Show that:

    a. there is only *one* break point   if $p < z$.
    b. there is only *one* break point   if $z < p < 9z$.
    c. there are only *two* break points   if $p = 9z$.
    d. there are *three* break points   if $p > 9z$.

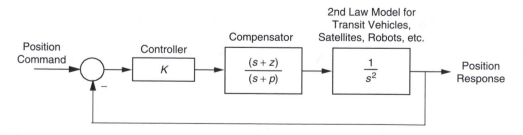

FIGURE 12.23
Feedback control of unconstrained motion of a mechatronic device.

Sketch the root locus for each of the four cases mentioned above and discuss stability of the closed-loop system in these cases.

SOLUTION
The loop transfer function for this example is

$$GH = \frac{K(s+z)}{s^2(s+p)} \tag{i}$$

and the parameter that is varied in the root locus is the loop gain K. Before sketching the root locus, let us examine the possible break points, using Rule 5. Since

$$N(s) = s + z \tag{ii}$$

and

$$D(s) = s^2(s+p) \tag{iii}$$

we have the condition for break points,

$$(s+z)(3s^2 + 2sp) - s^2(s+p) = 0$$

which simplifies to

$$s[2s^2 + (p+3z)s + 2pz] = 0 \tag{iv}$$

We observe from Equation iv that the point $s = 0$ is always a break point. The remaining break points are obtained by solving

$$2s^2 + (p+3z)s + 2pz = 0 \tag{v}$$

which gives

$$s = -\frac{1}{4}(p+3z) \pm \frac{1}{4}\sqrt{(p+3z)^2 - 16pz} \tag{vi}$$

The number of possible break points will depend on the sign of the *discriminant*—the expression under the square root sign in Equation vi:

$$\Delta = (p+3z)^2 - 16pz$$

This expression can be expanded and factorized as

$$\Delta = (p-z)(p-9z) \tag{vii}$$

Five cases can be identified for examining the sign of Δ. But since the case $p = z$ provides a "pole-zero cancellation" in the compensator, it is not considered here. The remaining four cases correspond to those stated in the problem. First, note from Equation vi that the two roots are always negative because the square root term is always less than $(p + 3z)$. Now, let us examine the four possible cases.

Case 1: $p < z$

In this case we have $\Delta > 0$ and we get two real (and negative) roots from Equation vi. But we can show that these two points do not lie on the root locus and, hence, are not valid break points. First note from Step 3 of root locus sketching that the root locus segment on the real axis extends from $-p$ to $-z$ to only. Now in the present case we have

$$z = rp \quad \text{with } r > 1 \tag{viii}$$

Then from the coefficient of s^0 in Equation v we note that the product of the roots is $pz = rp^2$. If one of these roots is to the right of $-p$ then, the other root will be to the left of $-rp$ (i.e., to the left of $-z$) and, so, both roots fall outside the root locus segment. Therefore, it is adequate to show that one root of Equation v falls outside the root locus segment on the real axis, because, the other root also will then fall outside the root locus, in the present case. Now substitute Equation viii in Equation vi with

$$r = 1 + \delta, \quad \delta > 0 \tag{ix}$$

Then, we have

$$s = -p - \frac{3\delta p}{4} \pm \frac{1}{4}\sqrt{\delta(8 + 9\delta)p} \tag{x}$$

If we look at the +ve square root we get a root that is to the right of $-p$. Hence, the other root will be to the left of $-z$ and both roots are not acceptable as break points. In this case the only valid break point is $s = 0$. The asymptote angles are $90°$ and $-90°$ (Step 4) and the asymptote origin is

$$\bar{s} = \frac{-p + z}{(3 - 1)} = (r - 1)\frac{p}{2} = \delta\frac{p}{2} > 0 \tag{xi}$$

on the positive real axis. Hence, the asymptotes are on the right half plane. The root locus for this case is sketched in Figure 12.24(a). Note that the closed-loop system is unstable for all values of gain K because two branches of the root locus are entirely on the right half plane. This is to be expected because the plant (double integrator) is marginally stable (unstable) and the compensator is a *lag compensator* ($p < z$), which has a destabilizing effect on systems.

Case 2: $z < p < 9z$

In this case, Δ as given by Equation vii, will be negative. Hence, the roots of Equation v will be complex. It follows that the only possible break point in this case is $s = 0$, as in the previous case. The asymptote angles are $90°$ and $-90°$ as before. Also, with $z = rp$ where $r < 1$, we have

$$\bar{s} = -(1 - r)\frac{p}{2} < 0 \tag{xii}$$

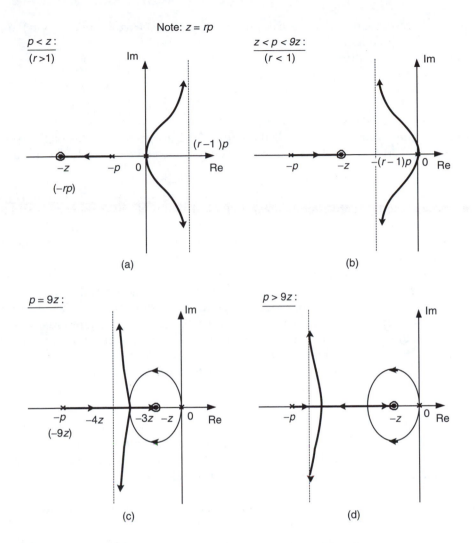

FIGURE 12.24
Root loci for the system in Figure 12.23: (a) $p < z$; (b) $z < p < 9z$; (c) $p = 9z$; (d) $p > 9z$.

indicating that the asymptotes are on the left half plane. Note that $\bar{s}$ could lie anywhere from 0 to $-\frac{p}{2}$, and could be on or outside the root locus segment on the real axis. What is shown in Figure 12.24(b) is the situation when $\bar{s}$ is to the right of $-z$. The system is stable in the present case, which is to be expected because $z < p$ corresponds to a *lead compensator*.

Case 3: $p = 9z$
In this case, $\Delta = 0$ and we have two identical roots for Equation v at

$$s = -\frac{1}{4}(p + 3z) = -3z$$

Since this point falls within the root locus segment on the real axis (i.e., between $-z$ and $-p$), this is an acceptable break point. Furthermore, if the break point condition gives two identical roots, the characteristic equation has to have three identical roots (at $s = -3z$).

Hence, in this case, there are two break points, one at $s = 0$ (two identical roots) and the other at $s = -3z$ (three identical roots).

The origin of the asymptotes is

$$\bar{s} = \frac{-p + z}{(3 - 1)} = \frac{-9z + z}{2} = -4z$$

The corresponding root locus is sketched in Figure 12.24(c). In this case as well the system is stable for all values of K, which is to be expected because we have a lead compensator.

Case 4: $p > 9z$

In this case we note from Equation vii that $\Delta > 0$. Hence, we have two real roots for Equation v. We can show using the same arguments as in Case 1 that both these roots are valid break points, both roots falling between $-z$ and $-p$. It follows that there are three break points in this case. The root locus for this case is shown in Figure 12.24(d). Once again the system is stable, for we are using a lead compensator.

This is a good example to caution that one should not rush to sketch a root locus without using complete information. In particular, one should obtain the asymptote origin and the break points before sketching a root locus. Depending on the nature of this information, the root loci can vary significantly even when the system transfer functions are very similar.

Another observation we can make from the present example is that a dominant zero on the left hand plane (LHP) will tend to attract the root locus to the LHP (a stabilizing effect or "lead" action), and a dominant pole will tend to push the root locus away from the LHP (a destabilizing effect or "lag" action).

Example 12.16

A dc servomotor uses a proportional feedback controller along with a low-pass filter to eliminate signal noise. A block diagram of the control system is shown in Figure 12.25. The component transfer functions are as given in the diagram.

a. Sketch the root locus for the closed-loop system indicating the numerical values for possible break points and asymptotes.

b. Determine the value of gain K when the closed-loop system has two equal poles. What is the value of these poles?

c. Determine the range of K for which the closed-loop system is stable.

d. At what frequency will the system naturally oscillate when it is marginally stable?

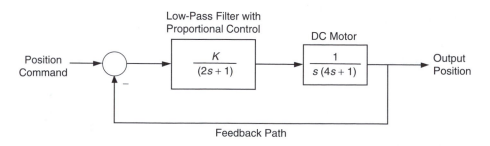

FIGURE 12.25
Block diagram of a dc servomotor.

SOLUTION

a. The loop transfer function is

$$GH = \frac{K}{s(2s+1)(4s+1)}$$

There are three loop poles, at 0, $-\frac{1}{4}$ and $-\frac{1}{2}$

There are no loops zeros. Three asymptotes with asymptote angles $= \frac{\pi \pm 2r\pi}{3} = \pm 60^\circ$ and 180°

Centroid of the poles, $\bar{s}$, is given by $3\bar{s} = 0 - \frac{1}{4} - \frac{1}{2}$

$$\Rightarrow \bar{s} = -\frac{1}{4}$$

Closed-loop characteristic equation:

$$s(2s+1)(4s+1) + K = 0$$

$$\text{or } 8s^3 + 6s^2 + s + K = 0 \tag{i}$$

Differentiate: $24s^2 + 12s + 1 = 0 \Rightarrow s = -\frac{1}{4} \pm \frac{1}{4\sqrt{3}}$

The correct break point $= -\frac{1}{4} + \frac{1}{4\sqrt{3}}$

The root locus is sketched in Figure 12.26

b. As obtained in Part (a), the repeated roots are given by the break point

$$s = -\frac{1}{4} + \frac{1}{4\sqrt{3}} = -0.106$$

Substitute in (i): $K = -8(-0.106)^3 - 6(-0.106)^2 - (-0.106) = 0.048$

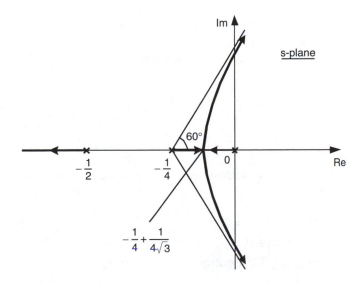

FIGURE 12.26
The root locus of the system in Figure 12.25.

c. Use Routh-Hurwitz criterion.

Routh Array:

$$
\begin{array}{ccc}
s^3 & 8 & 1 \\
s^2 & 6 & K \\
s^1 & \dfrac{6-8K}{6} & 0 \\
s^0 & K &
\end{array}
$$

For stability we need and $K > 0$ and $6 - 8K > 0$

Hence, the stability region is given by: $0 < K < \frac{3}{4}$

d. Method 1:

From Routh array, the row s^1 becomes null when $K = \frac{3}{4}$. The corresponding auxiliary equation is given by the previous row:

$$6s^2 + K = 0 \quad \Rightarrow \quad 6s^2 + 3/4 = 0$$

The corresponding imaginary roots are $s = \pm j\frac{1}{2\sqrt{2}}$

These are the points of intersection of the root locus with the imaginary axis, and they correspond the natural frequency of oscillation at marginal stability.

Method 2:

Substitute $s = jw$ in the closed-loop characteristic Equation i for marginal stability. We get

$$8(j\omega)^3 + 6(j\omega)^2 + j\omega + K = 0$$

$$\Rightarrow \quad -8j\omega^3 - 6\omega^2 + j\omega + K = 0$$

$$\Rightarrow \quad -6\omega^2 + K = 0 \quad \text{and} \quad -8\omega^3 + \omega = 0$$

$$\Rightarrow \quad \omega^2 = \frac{1}{8} \quad \text{and} \quad K = 6\omega^2$$

$$\Rightarrow \quad \omega = \frac{1}{2\sqrt{2}} \quad \text{and} \quad K = \frac{3}{4}$$

Example 12.17

Consider the feedback control system shown in Figure 12.27. The following three types of control may be used:

a. Proportional (P) Control: $G_c = K$
b. Proportional + Derivative (PD) Control: $G_c = K(1 + s)$
c. Proportional + Integral (PI) Control: $G_c = K(1 + 1/s)$

Sketch the root loci for these three cases and compare the behavior of the controlled systems.

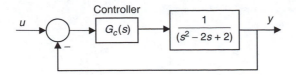

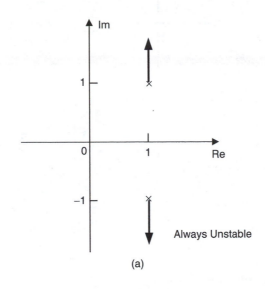

FIGURE 12.27
A feedback control system.

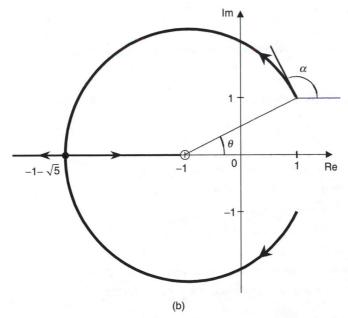

FIGURE 12.28
(a) Root locus for the system with P control; (b) Root locus for the system with PD control; (c) Root locus for the system with PI control.

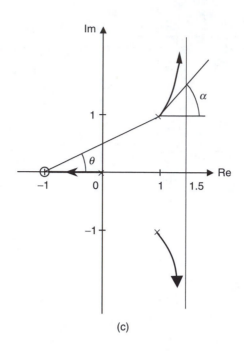

(c)

FIGURE 12.28
(Continued).

SOLUTION

a. The loop transfer function $GH = \frac{K}{(s^2-2s+2)}$

The loop poles are at $s = 1 \pm j$

There are no zeros.

Hence, there are two asymptotes with asymptote angle $\pm 90°$

The pole centroid $\bar{s} = \frac{1 \times 2}{2} = 1$

NOTE Even though obvious, the departure angle α at the pole $1 + j$ can be determined by:

$$\alpha + 90° = 180° \Rightarrow \alpha = 90°$$

The root locus is sketched in Figure 12.28(a).

It is seen that the system is always unstable.

b. The loop transfer function $GH = \frac{K(1+s)}{(s^2-2s+2)}$

The loop poles are at $s = 1 \pm j$

There is a zero at -1.

Hence, there is one asymptote with asymptote angle $180°$

The departure angle α at the pole $1 + j$ is determine by:

$$\alpha + 90° - \theta = 180° \quad \text{where} \quad \tan\theta = \frac{1}{2} \quad \text{or} \quad \theta = 26.6° \Rightarrow \alpha = 116.6°$$

Break Points:

$$N(s) = (1 + s) \quad \text{and} \quad D(s) = s^2 - 2s + 2$$

Hence, the break point is given by

$$(1 + s)(2s - 2) - (s^2 - 2s + 2) = 0 \Rightarrow s^2 + 2s - 4 = 0 \Rightarrow s = -1 \pm \sqrt{5}$$

The correct break point must be on the root locus (i.e., < -1). Hence we pick

$$s = -1 - \sqrt{5}$$

The root locus is sketched in Figure 12.28(b).

It is seen that the system becomes stable beyond a certain gain value. This is due to the inclusion of derivative control (or, a zero on the LHP), which has a stabilizing effect. The gain value and the frequency of marginal stability can be determined as usual. In the present case this is a relatively simple exercise. The closed-loop characteristic equation is

$$\frac{K(1+s)}{(s^2 - 2s + 2)} = -1 \quad \Rightarrow \quad (s^2 - 2s + 2) + K(1 + s) = 0$$

$$\Rightarrow \quad s^2 + (K - 2)s + K + 2 = 0$$

Hence, for stability, we must have $K > 2$. The gain for marginal stability is $K = 2$. The corresponding closed-loop poles are $s = \pm j2$

c. The loop transfer function $GH = \frac{K(1+s)}{s(s^2 - 2s + 2)}$

The loop poles are at $s = 0$ and $1 \pm j$

There is a zero at -1.

Hence, there are two asymptotes at asymptote angles $= \pm 90°$

The pole centroid $\bar{s} = \bar{s} = \frac{1 \times 2 - 1 \times (-1)}{3 - 1} = \frac{3}{2}$

The departure angle α at the pole $1 + j$ is determined by:

$$\alpha + 90° + 45° - \theta = 180° \quad \text{where} \quad \tan \theta = \frac{1}{2} \quad \text{or} \quad \theta = 26.6° \Rightarrow \quad \alpha = 71.6°$$

The root locus is sketched in Figure 12.28(c).

It is seen that the system is always unstable. In fact, the system is more unstable than with P control alone. This is due to the presence of the integral (I) action in the controller, which has a destabilizing effect.

12.6   Frequency Domain Analysis

The concept of transfer function and frequency domain models have been discussed in Chapter 2. Some related concepts of Fourier transform are outlined in Appendix A. Furthermore, transfer functions, in the Laplace domain, have been used extensively in the previous sections of the present chapter. Now we will specifically use the concept of *frequency transfer function* (or, *frequency response function*), where the independent variable is frequency ω radians/s or f cycles/s (or hertz), to develop some useful techniques in the analysis and design of control systems. First we will review some fundamentals of frequency transfer function, which we have already covered in Chapter 2.

12.6.1 Response to a Harmonic Input

Consider a linear, time-invariant system, given in the time domain (linear ordinary differential equation with constant coefficients) by

$$a_n \frac{d^n y}{dt^n} + a_{n-1} \frac{d^{n-1}y}{dt^{n-1}} + \cdots + a_0 y = b_m \frac{d^m u}{dt^m} + b_{m-1} \frac{d^{m-1}u}{dt^{m-1}} + \cdots + b_0 u \qquad (12.62a)$$

The system parameters (coefficients) $a_0, a_1, \ldots, a_n$ and $b_0, b_1, \ldots, b_m$ are constant (time invariant) by definition. The differential Equation 12.62a is an *input-output model* where

u = system input

y = system output

The order of the system is n. The system transfer function (i.e., output/input ratio, in the Laplace domain), is

$$G(s) = \frac{Y(s)}{U(s)} = \frac{b_m s^m + b_{m-1}s^{m-1} + \cdots + b_0}{a_n s^n + a_{n-1}s^{n-1} + \cdots + a_0} \qquad (12.62b)$$

Note that when we know Equation 12.62a we can immediately write down Equation 12.62b, and vice versa. This confirms that the two representations Equation 12.62a and Equation 12.62b are completely equivalent.

Suppose that a sinusoidal input of amplitude u_0 and frequency ω is applied to the system given by Equation 12.62a. This input may be represented in the "complex" form

$$u = u_o e^{j\omega t} = u_o (\cos \omega t + j \sin \omega t) \qquad (12.63)$$

Actually what we are applying to the system is the real part of the right hand side of Equation 12.63. But in view of the relative ease of manipulating an exponential function in comparison to a sinusoidal function, we use Equation 12.63 and then at the end take the real part of the result. The simplicity of analysis by using the exponential function stems particularly from the fact that $d/dt\, e^{st} = se^{st}$ and hence, after differentiation, the original exponential function remains (albeit with a multiplication factor). This is easier than, say, using $d/dt\, \sin \omega t = \omega \cos \omega t$ and $d/dt\, \cos \omega t = -\omega \sin \omega t$ where the function type changes on differentiation.

It is reasonable to assume (and, in fact, it can be verified through experiments and practical observations of real systems) that when a harmonic excitation is applied to a system (strictly, to a "linear" system) the response after a while becomes harmonic as well, oscillating at the same frequency (Note: If experiments are carried out to observe this property, the system has to be "stable" as well). The amplitude of the response, however, will not be the same as that of the input, in general. Hence, the steady-state harmonic response of Equation 12.62a may be expressed as

$$y = y_o e^{j\omega t} \qquad (12.64)$$

We will show later that not only the amplitude but also the "phase" of the output will be different from that of the input. Now substitute Equation 12.63 and Equation 12.64 into Equation 12.62a and cancel the common term $e^{j\omega t}$ which is not zero for a general value of t. Then we have

$$y_o = \left[\frac{b_m(j\omega)^m + b_{m-1}(j\omega)^{m-1} + \cdots + b_0}{a_n(j\omega)^n + a_{n-1}(j\omega)^{n-1} + \cdots + a_0} \right] u_o \tag{12.65a}$$

In view of Equation 12.62b, we note that what is in brackets is the transfer function $G(s)$ with s substituted by $j\omega$. This is called the *frequency transfer function* (or, *frequency response function* in the terminology of mechanical vibration). Hence,

$$y_o = G(s)|_{s=j\omega} u_o \tag{12.65b}$$

or

$$y_o = G(j\omega)u_o \tag{12.65c}$$

The amplitude u_0 of the input is clearly a real value. Also, $G(j\omega)$ is a complex number in general, which has a real part and an imaginary, or, a magnitude and a phase angle. Suppose that this magnitude is M and the phase angle is ϕ. Then

$$\text{Magnitude of } G(j\omega) = |G(j\omega)| = M \tag{12.66a}$$

$$\text{Phase angle of } G(j\omega) = \angle G(j\omega) = \phi \tag{12.66b}$$

Furthermore,

$$G(j\omega) = M\cos\phi + jM\sin\phi = Me^{j\phi} \tag{12.66c}$$

Substitute Equation 12.66c into Equation 12.65c, and use Equation 12.64. We get

$$y = u_o Me^{j(\omega t+\phi)} \tag{12.67}$$

What the result Equation 12.67 means, in view of Equation 12.63, is that for an input of $u_0 \cos\omega t$ the output will be $u_o M \cos(\omega t + \phi)$ and similarly for an input of $u_o \sin \omega t$ the output will be $u_o M \sin(\omega t + \phi)$. In summary, when a harmonic input is applied to the system having transfer function $G(s)$;

1. The output will be magnified by magnitude $|G(j\omega)|$
2. The output will have a *phase lead* equal to $\angle G (j\omega)$, with respect to the input. In fact $\angle G(j\omega)$ is typically a negative phase lead. Hence the output usually lags the input.

Example 12.18

Consider the mass-spring system shown in Figure 12.29(a).

By Newton's second law, the equation of motion of the system with a forcing input u is given by

$$m\ddot{y} + ky = u \tag{i}$$

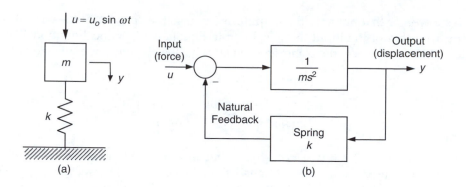

FIGURE 12.29
(a) An undamped simple oscillator; (b) Block diagram representation with a natural feedback.

where u denotes the input force applied to the mass m, and y denotes the displacement output (response) of the mass. The spring constant (stiffness) is k. The overall system transfer function of the system (i.e., y/u in the Laplace domain) is

$$\tilde{G}(s) = \frac{1}{ms^2 + k} \tag{ii}$$

Suppose that the forcing function is harmonic, as given by

$$u = u_o \sin \omega t \tag{iii}$$

Note that

$$\tilde{G}(j\omega) = \frac{1}{k - \omega^2 m} \tag{iv}$$

Hence,

$$\text{Magnitude } M = |G(j\omega)| = \frac{1}{|k - \omega^2 m|} \tag{v}$$

and

$$\text{Phase } \phi = \angle G(j\omega) = 0 \quad \text{for } \omega < \sqrt{k/m}$$

$$= -\pi \quad \text{for } \omega > \sqrt{k/m} \tag{vi}$$

The harmonic response of the system (in steady state) is

$$y = u_o M \sin(\omega t + \phi) \tag{vii}$$

It is clear from Equation vi that when $\omega < \sqrt{k/m}$ the response will be in phase with excitation, and when $\omega > \sqrt{k/m}$ the response will be 180° out of phase with the input. The phase will switch from 0° to 180° at the excitation frequency value $\omega = \sqrt{k/m}$.

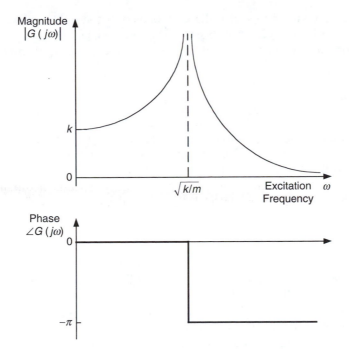

FIGURE 12.30
The Magnitude and phase plots of an undamped simple oscillator.

Furthermore, the response amplitude will increase with the excitation frequency, up to the frequency $\omega = \sqrt{k/m}$. Beyond that, the response amplitude will decrease with the excitation frequency. In particular, at $\omega = \sqrt{k/m}$ the response amplitude will be infinity. This is called a *resonance* and corresponding frequency is called *resonant frequency*. These results are shown in the plots of magnitude $|G(j\omega)|$ and phase lead $\angle G(j\omega)$ in Figure 12.30.

NOTE An important observation can be made about this system. By examining the system Equation i, we can draw a block diagram for the system as in Figure 12.29(b). The system has a natural feedback as a result of the spring, as shown (which can be established by writing Equation i in the form $u - ky = m\ddot{y}$, which is the form used in drawing the block diagram). According to this model then, the forward transfer function is

$$G(s) = \frac{1}{ms^2} \tag{viii}$$

and the feedback transfer function is

$$H(s) = k \tag{ix}$$

The overall (closed-loop) transfer function is

$$\tilde{G}(s) = \frac{G}{1+GH} = \frac{1}{ms^2 + k} \tag{x}$$

which is the same as Equation ii, as expected.
   The loop transfer function is

$$GH = \frac{k}{ms^2} \tag{xi}$$

It is *GH* that should be used to sketch the root locus of the simple oscillator. In view of the double pole at the origin ($s = 0$) we notice that the root locus falls entirely on the imaginary axis for any k (or, for any m or any k/m). Hence, the system is always marginally stable.

12.6.2 Marginal Stability

If a dynamic system oscillates steadily in the absence of a steady external excitation, that represents a condition of *marginal stability*. The "distance" to such a state of marginal stability is a measure of the level of stability, and is called a *stability margin*. We have already discussed this topic. In the present section we will formally develop the concepts of marginal stability and stability margin using a frequency transfer function model. First let us address the issue of marginal stability in qualitative terms and then develop an analytical basis.

12.6.2.1   *The 1,0 Condition*

Consider a feedback control system represented by the block diagram in Figure 12.31. We have assumed unity feedback, but this can be generalized later. In fact, without loss of generality we can interpret *G* in Figure 12.31 as the loop transfer function *GH*, because a system with a general feedback transfer function *H* can be reduced to a unity feedback system, through block diagram reduction, by placing *GH* as the forward transfer function. Suppose that the open-loop transfer function $G(s)$ is such that at a specific frequency, ω, we have:

1. Magnitude $|G(j\omega)| = 1$
2. Phase angle $\angle G(j\omega) = -\pi$ (12.68a)

Then, if an error signal of frequency ω is injected into the loop (due to noise, initial excitation, etc.) its amplitude will not change in passing through $G(s)$, but the phase angle will reduce by π. Hence the output signal y will have the same amplitude as the error signal e, but y will "lag" e by π. Since y is fed back into the loop with a negative feedback (Note: –1 corresponds to a further phase lag of π) the overall phase lag of the feedback signal, when reaching the forward path of the loop, will be 2π. Since $2\pi = 0$ in terms of phase angles, the feedback signal will have the same amplitude as the forward signal (i.e., gain = 1) and the same phase angle as the forward signal (i.e., phase = 0). This is called the "(1,0) *condition*." Under this condition, it is clear that even in the absence of an external input u, a harmonic signal of a specific frequency ω can sustain in the loop without growing or decaying. This is a state of steady oscillation. If such a condition of steady oscillation

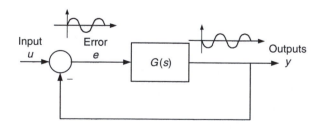

FIGURE 12.31
A feedback control system with unity feedback.

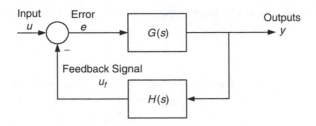

FIGURE 12.32
A control system with non-unity feedback.

is "naturally" possible, in the absence of a steady external input, the system is said to be marginally stable.

NOTE the specific frequency ω at which this condition of steady oscillation would be feasible, is itself a property of the system, and depends on system parameters. This the so-called "natural frequency" of the system. In the simple oscillator example, which we discussed earlier, this frequency is $\sqrt{k/m}$.

Next, consider a system with non-unity feedback with a feedback transfer function of $H(s)$ as shown in Figure 12.32. In applying the (1, 0) condition for marginal stability, what matters is the overall gain and phase shift in the entire loop. Hence, we need to consider the overall loop transfer function $G(s)H(s)$ and not the individual components. The (1, 0) condition for *marginal stability*, at a specific "natural" frequency ω, is

1. Magnitude $|G(j\omega)H(j\omega)| = 1$
2. Phase angle $\angle G(j\omega)H(j\omega) = -\pi$ (12.68b)

With respect to the closed-loop system shown in Figure 12.32, the following nomenclature should be recalled:

$G(s)$ = open-loop transfer function (or, forward transfer function)

$H(s)$ = feedback transfer function

$G(s)H(s)$ = loop transfer function

$\dfrac{G}{1+GH} = \tilde{G}$ = closed-loop transfer function.

NOTE The characteristic equation of the closed-loop system $\tilde{G}$ is $GH + 1 = 0$. Hence, the stability of a closed-loop system is completely determined by the loop transfer function GH. Specifically, we need to study the magnitude and the phase of $GH(j\omega)$, as we will elaborate later. In the special case of unity feedback ($H = 1$) we need to study $G(j\omega)$. For convenience, in these studies we denote GH simply by G, keeping in mind that then G represents GH, the loop transfer function. Let us now introduce some graphical representations of a transfer function ($G(j\omega)$) that are particularly useful in stability studies.

12.6.3 Bode Diagram

Bode plot of a transfer function $G(s)$ constitutes the following pair of curves:

Magnitude $|G(f)|$ versus frequency f

Phase angle $\angle G(f)$ versus frequency f

Also, see Chapter 2, Chapter 5 and Appendix A, for related fundamentals and applications.

Example 12.19

We will sketch the Bode plot of the transfer function of an armature controlled dc motor. It can be shown that the transfer function [Output speed/Input voltage] is given by

$$G(s) = \frac{K}{(\tau s + 1)} \tag{12.69a}$$

where
K = gain parameter (depends on motor constants, armature resistance, and damping)
τ = time constant (depends on motor inertia, motor constants, armature resistance, and damping)

This is a first order system. The frequency transfer function corresponding to Equation 12.69a is

$$G(j\omega) = \frac{K}{(\tau j\omega + 1)} \tag{12.69b}$$

or

$$G(j2\pi f) = \frac{K}{(\tau j2\pi f + 1)} \tag{12.69c}$$

The complex functions $G(j\omega)$ and $G(j2\pi f)$ may be denoted by $G(\omega)$ and $G(f)$, respectively, for notational convenience (even though contrary to strict mathematical meanings).

The Bode plot for Equation 12.69c is shown in Figure 12.33. Observe, in particular, the two *asymptotes* to the magnitude curve. These are obtained as follows:

First we define a critical frequency as:

$$f_b = \frac{1}{2\pi\tau} \tag{12.70}$$

When $f \ll f_b$

$$G(f) \approx K \tag{12.71a}$$

The corresponding magnitude value is K (or $20 \log_{10} K$ dB). This asymptote is a horizontal line as shown in Figure 12.33. The corresponding phase angle is zero.

When $f \gg f_b$

$$G(f) \approx \frac{K}{\tau j2\pi f}$$

The magnitude of this function is $K/(\tau 2\pi f)$. It monotonically decreases with frequency. If decibel scale (i.e., $20 \log_{10}(\ )$ dB) is used for the magnitude axis and decade scale (i.e., multiples of 10) for the frequency axis, the slope of this asymptote is -20 dB/decade.

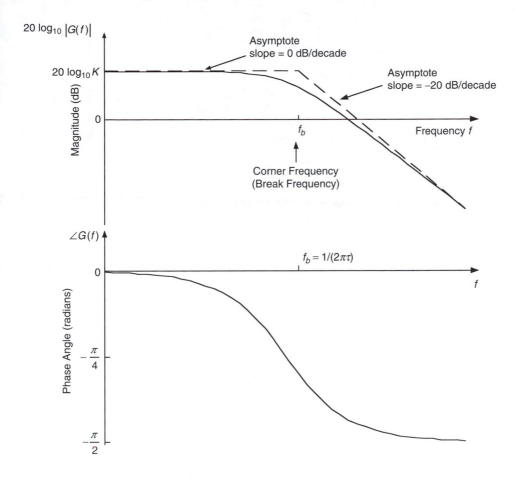

FIGURE 12.33
Bode diagram of a dc motor transfer function.

The two asymptotes intersect at $f = f_b$. This frequency is known as the *break frequency* (or *corner frequency*). Since a significant magnitude attenuation takes place for input signal frequencies greater than f_b and in view of the fast decay of the natural response for large f_b, it is appropriate to consider f_b, given by Equation 12.70, as a measure of *bandwidth* for a dc motor.

For the normalized case of $K = 1$ the bode diagrams of the transfer function Equation 12.69 is shown in Figure 12.34, where angular frequency (ω rad/s) is used instead of cyclic frequency (f Hz) in Figure 12.33. Suppose that a sinusoidal signal is used as the input test signal to the plant (dc motor). As the input frequency is raised, it is found that the output amplitude decreases and the phase-lag increases.

The x-axis (frequency axis) of the Bode plot is marked in units of frequency, which may be incremented by factors of 2 (*octaves*) or factors of 10 (*decades*). The y-axis (magnitude axis or gain axis) of the Bode magnitude curve is nondimensional, and is scaled in decibels (dB).

The advantage of the log scale for magnitude is that the Bode diagram for a product of several transfer functions can be obtained by simply adding the Bode plots for the individual transfer functions. In this manner, the Bode plot of a complex system can be conveniently obtained with the knowledge of the Bode plots of its components (plant, controller, actuator, sensor, etc.) The advantages of using a log scale for frequency are the fact that a wide range of frequencies can be accommodated in a limited plotting area, and

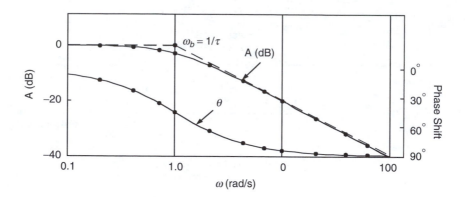

FIGURE 12.34
Bode diagram (plot of amplitude versus frequency and phase versus frequency of a first order system.

that asymptotes to the magnitude curve become straight lines with slopes differing by fixed increments (by ±20 dB/decade if decibel scale is used for magnitude and decade scale is used for frequency).

12.6.4   Phase and Gain Margins

In the conventional frequency-domain design of control systems, stability requirement is specified using a phase margin or a gain margin. First, we will revisit the concept of marginal stability to introduce these two stability margins. Next, we will present the Nyquist stability criterion, which will lead to formal definitions of the two stability margins.

We have observed that stability of a closed-loop system is completely determined by the loop transfer function GH. Consider the feedback control system shown in Figure 12.32. The characteristic equation of the closed-loop system is $G(s)H(s) = -1$. The system is marginally stable if one pair of roots of the characteristic equation is purely imaginary (i.e., $\pm j\omega$) while the remaining roots are not unstable. Hence the condition for marginal stability is, there exists a frequency ω such that

$$G(j\omega)H(j\omega) = -1 \tag{12.72}$$

In fact the two conditions given by Equation 12.68b are exactly equivalent to this single complex equation, because −1 has a magnitude of 1 and a phase angle of −π. It follows that if the plot of the loop transfer function in the complex plane (i.e., Imaginary $GH(j\omega)$ versus Real $GH(j\omega)$), as ω changes, passes through the point $(-1,0)$, then the control system is *marginally stable*.

12.6.4.1   Gain Margin

Suppose that at frequency ω, the phase $\phi = -180°$ but the gain (magnitude) M is less than unity (i.e., $M < 1$). Then, if the external input u is disconnected, the amplitude of the feedback signal will steadily decay. This, of course, corresponds to a stable system. The smaller the value of M, the more stable the system. Hence, a stability margin known as the gain margin g_m can be defined as

$$g_m = \frac{1}{|G(j\omega)H(j\omega)|} \tag{12.73a}$$

at a frequency ω where $\angle G(j\omega)H(j\omega) = -180°$.

If the gain of the transfer function is increased by a factor of g_m at this frequency, then the marginal stability conditions (Equations 12.68b) will be satisfied. Hence, g_m is the margin by which system gain could be increased before the system becomes just unstable. It follows that the larger the g_m, the better the degree of stability. It is convenient to express g_m in decibels (dB) because the transfer function magnitude (in the frequency domain) is usually expressed in dB (particularly in Bode diagrams). Then,

$$g_m = -20 \log_{10} |G(j\omega)H(j\omega)| \qquad (12.73b)$$

at the frequency ω where $\angle G(j\omega)H(j\omega) = -180°$.

12.6.4.2 Phase Margin

Suppose that there exists some frequency ω at which the magnitude (gain) of the loop transfer function is $M = 1$ (i.e., 0 dB) but the phase ϕ lies between 0 and $-180°$. This frequency ω_c is called the *crossing frequency* or *crossover frequency*, because it corresponds to the point where the gain (magnitude) curve crosses the 0 dB line. Since the phase angle typically decreases with frequency, there will be a higher frequency at which the phase is $-180°$ but the gain (magnitude) will be less than unity at this frequency (because the loop transfer function magnitude usually decreases with increasing frequency, at high frequencies). This again corresponds to a stable system. The amount by which the phase of the loop transfer function at gain $= 0$ dB, could be decreased (a lag) until it reaches the $-180°$ value, is termed phase margin (ϕ_m). Specifically,

$$\phi_m = 180° + \angle G(j\omega)H(j\omega) \qquad (12.74)$$

at the frequency ω where $|G(j\omega)H(j\omega)| = 1$. The larger the phase margin, the more stable the system.

In summary, gain margin tells us the amount (margin) by which the gain could be increased at a phase of $-180°$, before the system becomes marginally stable; and phase margin tells us the amount (margin) by which the phase could be "decreased" at unity gain, before the system becomes marginally unstable. A more rigorous development of the concepts of gain margin and phase margin requires a knowledge of *Nyquist stability criterion*.

12.6.5 Nyquist Plot

The Bode plot requires two curves—one for gain and one for phase. These two curves can be represented as a single curve by using a so-called *polar plot* with a real axis and an imaginary axis (Figure 12.35(a)). A polar plot is a way to represent both magnitude and phase of a rotating vector, with a single curve. When phase $= 0°$, the vector (which represents the complex number) points to the right; when phase $= 90°$, the vector points up; and when phase $= -180°$, the vector points to the left, etc.

The solid curve in Figure 12.35(a) represents the path of the tip of the directed line (two-dimensional vector) representing the frequency transfer function (i.e., complex transfer function in the frequency domain) as the frequency varies. Any one point of the Nyquist curve represents both the *amplitude* (distance to origin) and the *phase* (angle measured from the positive real axis), at one given frequency. Thus all the information in the two curves of a Bode plot (Figure 12.35(b)) is represented by this single curve called *Nyquist plot* (or *polar plot* or *argand plot*).

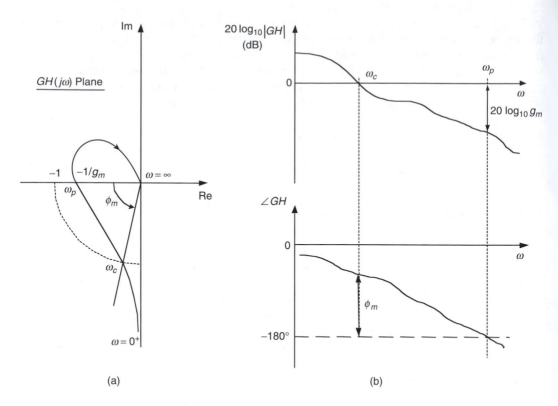

FIGURE 12.35
Definition of gain margin and phase margin: (a) Using Nyquist diagram; (b) Using Bode diagram.

The marginal stability condition is: (a) gain = 1 or 0 dB and (b) phase lag = 180° at some specific (natural or resonant) frequency. A gain of 1 is represented by a vector of unit length. Its tip traces a unit circle with its center at the origin of the coordinate frame (see the dotted circle in Figure 12.35(a)). A phase of 180°. Corresponds to a horizontal vector pointing to the left from the origin (i.e., the negative real axis). The intersection of the Nyquist plot with the unity-gain circle gives the *phase* at 0 dB point; the length of the vector (distance from origin) of the point where the Nyquist plot intersects the negative side of the real axis gives the gain at the critical 180° phase point.

Now formal definitions for the two stability margins are given using Nyquist diagram and Bode diagram. Gain margin g_m and phase margin ϕ_m are defined in Figure 12.35. Consider a stable closed-loop system with transfer function $\tilde{G}$. First we plot the Nyquist diagram for the loop transfer function GH as, for example, shown in Figure 12.35(a). The factor by which the Nyquist curve should be expanded (say, by increasing gain) in order to make the system marginally stable (i.e., to pass through the point (-1, 0)) measures the relative stability (or, stability margin) of the closed-loop system.

Gain Margin (GM): Gain margin g_m is the reciprocal of the magnitude of the loop transfer function $GH(j\omega)$ at a frequency (ω) where the phase angle of the loop transfer function is −180°. It follows that the larger the g_m, the larger the separation of the Nyquist curve from the critical point (−1) and the better the closed-loop stability.

Phase Margin (PM): Phase margin ϕ_m is the sum of 180° and the phase angle (in degrees) of the loop transfer function $GH(j\omega)$ at a frequency (ω) where the magnitude of $GH(j\omega)$ is unity (or 0 dB).

The relative stability (stability margin) of a control system can be improved by adding a compensator so as to increase the phase margin (PM) and the gain margin (GM). Since, in general, the gain margin of a system automatically improves when the phase margin of the system is improved, in design specifications it is adequate to consider only one of these two stability margins, say the phase margin. For a closed-loop system that can be approximated by a simple oscillator, a reasonably accurate relationship for damping ratio (of the closed-loop system) is

$$\zeta = 0.01\phi_m \qquad (12.75)$$

in which ϕ_m is the phase margin in degrees, as determined from the loop transfer function. This relationship is acceptable in the damping range $0 \le \zeta \le 0.6$, and it provides a slightly conservative estimate for damping ratio in terms of phase margin.

Gain margin and phase margin may not be defined for some loop transfer functions. Specifically, if the magnitude curve does not cross the 0 dB line, the PM of the system is not defined. Similarly, if the phase angle curve does not cross the ($-180°$) line, the GM is not defined. In some such cases it is still possible and quite valid to use the concepts of PM and GM.

For example, if the magnitude curve of the loop transfer function (GH) stays less than 0 dB throughout the entire frequency range of operation (operating bandwidth) of the control system, then the control system is considered stable (amplitude stabilization). In this case a positive gain margin can be defined using the gain at the frequency where the phase angle is closest to $-180°$ (or, alternatively, at the high frequency end of the operating bandwidth no matter what the phase angle at that end might be). Similarly, if the magnitude curve of GH remains greater than 0 dB throughout the operating bandwidth, the system is considered unstable. Then a negative GM can be defined as before.

If the phase angle of GH remains between 0 and $-180°$ within the entire operating bandwidth of the system, the closed-loop control system is considered stable, assuming that the magnitude of GH is less than 0 dB at the high-frequency end of the operating bandwidth. In this case a positive GM and a positive PM can be defined at the high-frequency end.

If there are multiple crossings of the 0 dB line, a unique PM is not defined. Similarly, if there are multiple crossings of the ($-180°$) line, a unique GM is not defined. In these cases, a single PM or a single GM may be defined by taking the worst case (i.e., the smallest of the stability margins) or by considering a limited operating bandwidth that contains only a single crossing. A system is considered stable only if *both* the phase margin and gain margin are positive.

Example 12.20
Figure 12.36 and Figure 12.37 show Bode and Nyquist plots (of the loop transfer functions GH) for two systems. The one on the left is stable because the phase lag is less than 180° at the critical 0 dB (i.e., where gain = 1) point, and the gain is less than 1 at the critical phase lag point (i.e., where phase lag = 180°). Note that the amount by which the gain is less than 0 dB at the phase-crossover ($-180°$) point is the *gain margin*. Similarly, the amount by which the phase lag is less than 180° at the gain-crossover point (0 dB) is the *phase margin*.

Example 12.21
The open-loop transfer function of a control system with unity feedback (i.e., the loop transfer function), is given by

$$G(s) = \frac{(s+3)}{(s^2 + 4s + 16)}$$

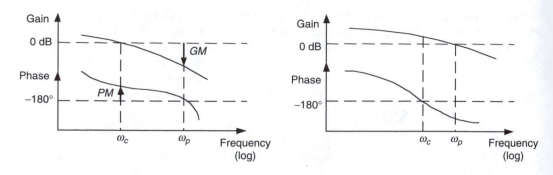

FIGURE 12.36
Bode plots of a stable system (left), and an unstable system (right). GM = gain margin, PM = phase margin.

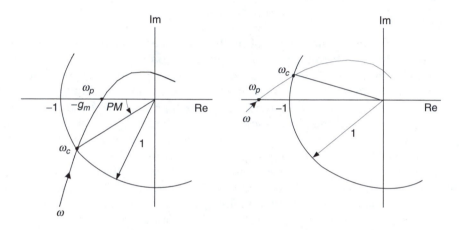

FIGURE 12.37
Nyquist plots of a stable system (left), and an unstable system (right).

a. Tabulate the magnitude $|G(j\omega)|$ and phase angle $\angle G(j\omega)$ values for about 6 points of frequency in the range $\omega = 0$ to $\omega = 5$.
b. Plot the Nyquist diagram for G.
c. Plot the Bode diagram for G, and indicate the asymptotes.
d. If the open-loop system (G) is given the sinusoidal input

$$u = 2\cos 2t$$

what is the output at steady state?
e. Explain, using the Nyquist plot, why the closed-loop system $\tilde{G}$ given by

$$\tilde{G} = \frac{G}{1+G}$$

is stable.

SOLUTION

By setting $s = j\omega$ we get the frequency transfer function

$$G(j\omega) = \frac{j\omega + 3}{16 - \omega^2 + 4j\omega}$$

Hence

$$|G(j\omega)| = \sqrt{\frac{3^2 + \omega^2}{(16 - \omega^2)^2 + 16\omega^2}}$$

and

$$\angle G(j\omega) = \tan^{-1}\frac{\omega}{3} - \tan^{-1}\frac{4\omega}{16 - \omega^2} \qquad \text{for } \omega < 4$$

$$= \tan^{-1}\frac{\omega}{3} - \pi + \tan^{-1}\frac{4\omega}{\omega^2 - 16} \qquad \text{for } \omega > 4$$

a.

Frequency ω	0	1	2	3	4	5	∞
Magnitude $|G(j\omega)|$	3/16	0.204	0.25	0.305	0.3125	0.266	0
(dB)	(−14.5)	(−13.8)	(−12)	(−10.3)	(−10.1)	(−11.5)	(−∞)
Phase $\angle G(j\omega)$ (degrees)	0	3.5	0	−14.7	−36.8	−55.2	−90

b. The Nyquist curve is shown in Figure 12.38.

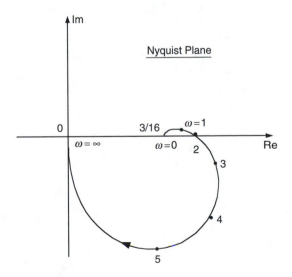

FIGURE 12.38
Nyquist curve of the example.

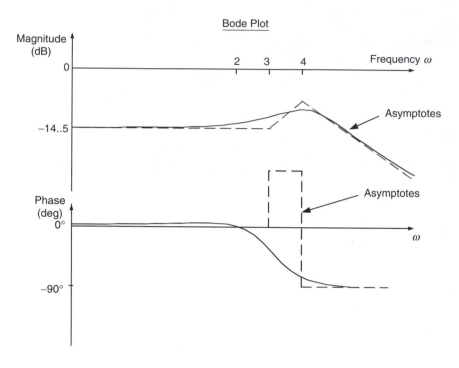

FIGURE 12.39
Bode diagram of the example.

c. The Bode plot is shown in Figure 12.39. Note that for small frequencies $w \ll 3$, the magnitude is approximately constant and equal to the static gain −14.5 dB and the phase angle is approximately zero. At $\omega = 3$, the first break point occurs due to the zero in the loop TF. A slope of + 20 dB/decade is added to the magnitude asymptote and a phase "lead" of 90° is added to the phase asymptote, as shown. At $\omega = 4$, the second break pint occurs, due to the second order (simple oscillator) term in the loop transfer function. This adds slope of −40 dB/decade to the magnitude asymptote (resulting in a net slope of −20 dB/decade) and a phase angle of -180° to the phase asymptote (resulting in a net phase angle of −90°, which is a phase lag of −90°), as shown in Figure 12.39.

d. For the open-loop system,

$$G(j\omega) = \frac{j\omega + 3}{(16 - \omega^2 + 4j\omega)}$$

At $\omega = 2$:

$$|G(j\omega)| = 0.25$$

$$\angle G(j\omega) = 0° = 0 \text{ rad}$$

Hence, the steady-state response for an input of $u = 2\cos 2t$ is

$$y = 2 \times 0.25 \cos (2t + 0)$$

or

$$y = 0.5 \cos 2t$$

e. It is noted from the Nyquist plot that as the frequency increases, the magnitude (of the loop transfer function) remains less than 1 and the phase "lag" never increases beyond 90°, in the entire frequency range (from 0 to ∞). Hence, the system is stable. Even though the phase margin and the gain margin are not numerically defined for this system, both may be taken positive in qualitative terms. The Nyquist plot remains entirely to the right of the critical point −1. Hence there is no possibility of the closed-loop system ever becoming even marginally stable (i.e., Nyquist plot will not cross the critical point).

12.6.6 Slope Relationship for Bode Magnitude Curve

H. W. Bode obtained an equation that relates to phase angle, the slope of the Bode magnitude (gain) curve. This approximate relationship is valid for a system whose loop transfer function does not contain poles or zeros on the right hand plane (i.e., for a *minimum-phase system*). The approximate relationship is

$$\phi = r \times 90° \tag{12.76}$$

in which

ϕ = phase angle (in degrees) at frequency ω

r = normalized slope of the asymptote to Bode magnitude curve at ω, (dB/decade/ 20dB/decade)

Note that r is obtained by determining the slope in decibels per decade and dividing the result by 20.

Example 12.22

Consider an underdamped simple oscillator (see Figure 12.9 and Equation 12.3), which has the frequency transfer function

$$G(j\omega) = \frac{K}{(\omega_n^2 - \omega^2 + 2j\zeta\omega_n\omega)} \qquad 0 < \zeta < 1 \tag{i}$$

Note that "underdamped" means the damping ratio ζ is less than 1. Bode curve pair for this system is shown in Figure 12.40. The break point for the asymptotes is the undamped natural frequency ω_n (as used in the previous example as well, see Figure 12.39). One asymptote is drawn for $\omega \ll \omega_n$ in which case the frequency transfer function Equation i can be approximated by the static gain (i.e., the zero-frequency magnitude)

$$G(j\omega) \approx \frac{K}{\omega_n^2} \tag{ii}$$

The magnitude of this transfer function is a constant and hence the slope is zero ($r = 0$). The phase angle is zero as well, in this region.

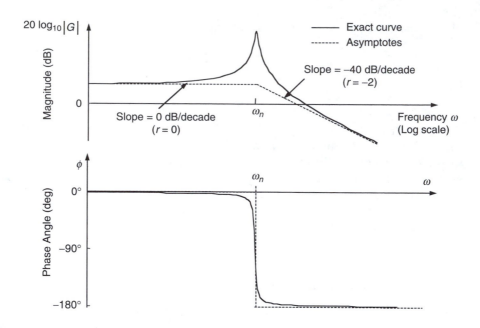

FIGURE 12.40
An example to illustrate the slope relationship for Bode magnitude curve.

The second asymptote is drawn for $\omega \gg \omega_n$. Then the frequency transfer function Equation i can be approximated by

$$G(j\omega) \approx -\frac{K}{\omega^2}$$

In this region the magnitude in decibels is

$$20\log_{10}\left(\frac{K}{K_o}\right) - 40\log_{10}\left(\frac{\omega}{\omega_o}\right)\mathrm{dB}$$

Note that K and ω are nondimensionalized because mathematically it is not proper to obtain the logarithm of a dimensional quantity. The important thing to note, however, is that when the frequency changes by 1 decade (i.e., when $\omega/\omega_o = 10$), the magnitude changes by -40 dB. Hence, the slope of this asymptote is -40 dB/decade. It follows that $r = -2$. Note that the phase angle is $(-180°)$ in this region. These observations verify Bode's approximate relationship for the slope of the magnitude curve (Equation 12.76).

12.6.7 Nyquist Stability Criterion

Let us revisit the feedback control system shown in Figure 12.32. The closed loop transfer function is given by

$$\tilde{G}(s) = \frac{G}{1+GH} \qquad\qquad (12.77)$$

As noted previously, system stability is determined by the poles (eigenvalues) of $\tilde{G}$. These are the roots of the characteristic equation $1 + G(s)H(s) = 0$. If all the poles are located on the left hand s-plane (i.e., if the real parts of the roots are all negative), the closed-loop system is stable. If there is at least one pole of $\tilde{G}$ on the right hand s-plane (RHP), the system is unstable. If there is a pole on the imaginary axis of s-plane (including, of course, the origin), the closed-loop system is considered marginally stable, provided that there are no poles on the right half of the s-plane.

Nyquist stability criterion follows from *Cauchy's theorem* on complex mapping. Using this criterion, the stability of a closed-loop system can be determined simply by sketching the Nyquist diagram of the corresponding loop transfer function GH. As discussed before, to obtain the Nyquist plot, first set $s = j\omega$ in $GH(s)$ and plot the resulting function using the imaginary part as the y-coordinate and the real part as the x-coordinate, while varying the frequency parameter ω from $-\infty$ to $+\infty$. This is explained in Figure 12.41.

Consider the area to the right of the imaginary axis on the s-plane, which is enclosed by the closed contour as ω, is varied from $-\infty$ to $+\infty$, and the contour is completed in the clockwise (right-handed) sense at infinity. As shown in Figure 12.41, the corresponding mapping of $GH(j\omega)$ on the GH-plane is also a closed contour, which is the Nyquist diagram.

NOTE Changing j to $-j$ in a function of $j\omega$ amounts to the same thing as changing ω to $-\omega$. Furthermore, changing j to $-j$ corresponds to changing the sign of the imaginary part of the complex function (i.e., *complex conjugation*), which forms a mirror image about the real axis. It follows that the Nyquist plot for negative frequencies is the mirror image of that for positive frequencies, about the real axis. Consequently, it is only necessary to plot the Nyquist diagram for positive frequencies.

The Nyquist stability criterion states that

$$\tilde{p} - p = N \tag{12.78}$$

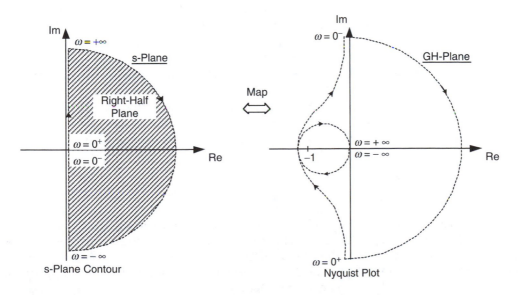

FIGURE 12.41
Generation of a Nyquist diagram.

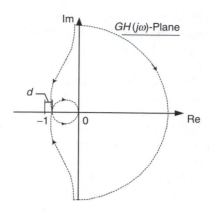

FIGURE 12.42
Gain margin represented on Nyquist plane.

in which
    $\tilde{p}$ = number of unstable poles in the closed-loop transfer function $\tilde{G}$
    p = number of unstable poles in the loop transfer function GH
    N = number of clockwise encirclements of point –1 on the real axis by the Nyquist plot

For stability of the closed-loop system we need $\tilde{p} = 0$. Hence, we should have $N = -p$. For example, in Figure 12.41 we have two clockwise encirclements of the point –1. Hence $N = 2$. From Equation 12.78 we have $\tilde{p} > 0$ and the closed-loop system will always be unstable (even if GH is stable, i.e., $p = 0$). Note: p and $\tilde{p}$ are nonnegative integers while the integer N can be positive, negative, or zero.

The gain of GH can be changed (usually decreased) to shrink the Nyquist plot so that the point (–1) is no longer encircled, however. This situation is shown in Figure 12.42. In this case $N = 0$. Hence, it follows from Equation 12.78 that if GH is stable ($p = 0$) to begin with, the closed-loop system will remain stable. This indicates how, in many cases, an unstable system can be stabilized by decreasing the gain, and vice versa. The *degree of stability* (or *stability margin*) can be measured by the distance d through which the Nyquist plot must be expanded along the real axis in order to have an encirclement of the point –1 (see Figure 12.42). The concepts of gain margin and phase margin are based on this fact.

Caution should be exercised when there are poles of GH on the imaginary axis of the s-plane. In this case, by properly choosing the contour on the s-plane, these marginally stable poles of GH can be excluded from the right half of the s-plane. Consequently, they are not counted in p of Equation 12.78.

Example 12.23

Two examples are given in Figure 12.43. There are no GH poles inside the chosen contour on the s-plane in Figure 12.43(a), hence, $p = 0$. The corresponding Nyquist plot has two encirclements around –1. Hence $N = 2$. This gives $\tilde{p} = 2$, and the closed-loop system has two unstable poles. Similarly, for the system shown in Figure 12.43(b) we have $p = 0$ and $N = 0$. Hence $\tilde{p} = 0$, and the closed-loop system is stable.

Now let us examine the possibility of stabilizing the system shown in Figure 12.43(a) by using a proportional plus derivative (PPD or PD) controller. As we know, the controller

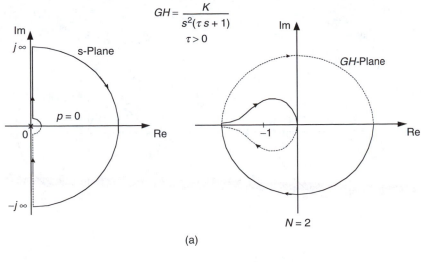

(a)

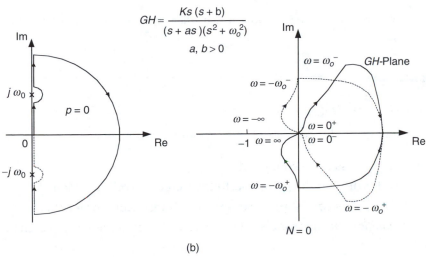

(b)

FIGURE 12.43
Application of Nyquist stability criterion: (a) Unstable system; (b) Stable system.

is given by $K(\tau_d s + 1)$. The corresponding loop transfer function is

$$GH = \frac{K(\tau_d s + 1)}{s^2(\tau s + 1)}$$

Two possible cases are considered.

Case 1 ($\tau_d > \tau$):
In this case the derivative (lead) action of the PPD controller overcomes the integral (lag) action of the non-zero pole of the plant. We observe from Figure 12.44(a) that the closed-loop system is stabilized in this case.

Case 2 ($\tau_d < \tau$):
In this case the lead action of the PPD controller cannot overcome the lag effect of the non-zero pole of the plant. We see from Figure 12.44(b) that the closed-loop system remains unstable in this case.

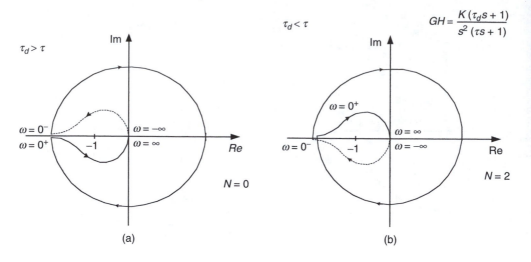

FIGURE 12.44
Stabilization using a lead compensator: (a) Adequate lead action; (b) Inadequate lead action.

12.7 Controller Design

A control system generally, consists of five types of components:

- *Plant*, the process to be controlled
- *Actuators* for driving the plant and for introducing control actions
- *Measuring devices* consisting of sensors and transducers for feedback control
- *Signal modification units* for conditioning and changing the form of signals in the control system
- *Controllers* for generating control actions for the plant

Designing a control system may involve development, selection, modification, addition, removal, and relocation of these components, and furthermore, tuning or the selection of parameter values for one or more of the components in the control system. We have already encountered several concepts and examples of designing a controller of a control system, so as to satisfy a set of performance specifications for the system. These concepts are further examined and enhanced now.

The performance of a control system can be improved in many ways. Three common approaches are as follows:

1. Redesign or modify the plant
2. Substitute, add, modify, or relocate sensors, actuators and associated hardware
3. Introduce a new controller or add a compensator for an existing controller

Plant redesign includes modification of the general structure of the plant as well as adjustment of parameters. Plant redesign is usually a costly approach to design improvement, particularly when structural modifications are involved. Furthermore, there are serious limitations to the degree of plant modification that is feasible and the level of performance improvement that could be achieved by this method both from practical and economical points of view. This approach is, in general, unsuitable when quick and short-term solutions are called for.

Adding or improving sensors and actuators normally result in improvements in the flexibility and versatility of a control system. This is a very practical method of design improvement. At least in theory, system observability controllability and reliability can be improved by adding sensors. For example, state-space design techniques (i.e., those based on state-space models — see Chapter 2) often provide superior controller designs when all state variables in the plant are measurable. This is often not feasible, however, because all responses might not be available for measurement and the increased sensor cost might not justify the expected performance improvement. System controllability can be improved as well by adding new actuators or improving the existing ones. Often, plant power require-ments alone may call for improved actuators. As in the case of sensors, increased cost and system complexity are two major drawbacks of this approach.

Controller improvement (redesign) is usually a rather economical and convenient method of design implementation. This may be accomplished either by analog means or by using a digital controller. In the former case, appropriately designed analog circuitry (analog controllers and compensators) is added at suitable locations in the control system. Modified control action that is generated in this manner can "compensate" for system weaknesses, thereby improving the overall performance. Proportional, integral, and derivative (PID) control actions and lead and lag compensation methods are commonly employed in indus-trial control systems.

12.7.1 Design Specifications

Typical step of designing a control system are given below:

- Establish specifications for system performance (performance specifications or design specifications). These may be provided by the customer or have to be developed by the control engineer based on the system objectives.

- Analyze and/or test the plant (or the original system in the case of design modifications). This step may involve system modeling.

- Select suitable components and determine the parameter values to meet the performance specifications. Commonly available (off the shelf) components should be used whenever possible. The needed parameter values may be avail-able from the product data sheets provided by the manufacturers or vendors.

- Analyze and/or test the overall system to evaluate and verify performance. Testing should involve the users (e.g., personnel from an industrial facility) of the designed control system when feasible, and should be done under realistic operating conditions.

- Repeat the design iteration if the specifications are not satisfied.

The design process is greatly influenced by the nature of the performance specifications used. Control systems can be designed by using either time-domain techniques or frequency-domain techniques. Many of the conventional methods of control system design are frequency-domain techniques and are particularly convenient when designing single-input, single-output (SISO) systems. State-space design techniques primarily use time-domain concepts. These latter techniques are useful with multi-input, multi-output (MIMO) or multivariable systems. Since poles and zeros of a transfer function determine the time response of the corresponding system, it is difficult to classify some design techniques into time and frequency domains. For example, pole assignment using state-space techniques is considered a time-domain approach whereas root locus design may be considered a frequency domain (actually, Laplace domain) approach. Design specifications may concern such attributes of a control system as: stability, bandwidth, sensitivity and robustness, input sensitivity, and accuracy. In designing a control system, these attributes or descriptions have to be specified in quantitative terms. The nature of the design specifications used depends considerably on the particular design technique that is employed. As discussed before, performance specifications can be made in both time domain and frequency domain.

Time-Domain Design Techniques:

1. Conventional design of proportional, derivative and integral controllers (*Specifications*: percentage overshoot, rise time, delay time, peak time, settling time, time constant, damping ratio, and steady-state error)

2. Optimal control using state-space approach (*Design specifications*: expressed as a performance function that will be optimized in the design process, e.g., final time, weighted quadratic integral of response variables, inputs, and error variables)

3. Modal control and pole assignment using state-space approach (*Design specifications*: required pole locations and mode shape vectors)

Frequency-Domain Design Techniques:

1. Bandwidth design (*Specifications*: resonant peak, bandwidth, resonant frequency)

2. Bode and Nyquist design (*Specifications*: phase margin, gain margin, steady-state error, gain crossover frequency, slope of the transfer-function magnitude at gain crossover)

3. Ziegler-Nichols tuning (*Specification*: e.g., a decay ratio of 4 in the closed loop response)

4. Root locus design (*Specifications*: pole locations, or other parameters such as error constants, gain, natural frequency, damping ratio, settling time, peak time, and percentage overshoot that can be expressed in terms of pole locations).

12.7.2   Conventional Time-Domain Design

Speed of response, degree of stability, and steady-state error are the three specifications that are most commonly used in conventional time-domain design of a control system. Speed of response can be increased by increasing the control system gain. This, in turn, can result in reduced steady-state error. Furthermore, steady-date error requirement can often be satisfied by employing integral control. For these reasons, one can treat speed of response and degree of stability as the only requirements in conventional time-domain design, and treat the steady-state error requirement separately.

Time domain design of a proportional controller and a position plus velocity servo has been discussed in a previous section. The approach has to be modified for the design of a proportional plus derivative (PD or PPD) controller. This issue is treated next.

12.7.2.1   *Proportional Plus Derivative Controller Design*

Actuators with proportional plus derivative (PPD) error control are commonly used as position servos. The two main parameters that can be adjusted in a PPD control element are the control gain K and the derivative time constant τ_d. Values for these two parameters

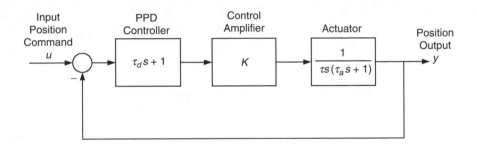

FIGURE 12.45
Block diagram of a PPD position servo system.

can be chosen to provide specified levels of stability and speed of response in a control system that employs a PPD servo.

The time domain design problem is quite straightforward for proportional control and position plus velocity (tachometer) feedback control, as discussed previously. This is so because, in both cases, the closed-loop transfer function does not contain finite zeros. With PPD error control, however, a finite zero enters into the closed-loop transfer function, making the design problem more difficult.

A common practice in the classical time domain design of PPD controllers is to use the same design equations or curves as for position plus velocity feedback. This approach, however, results in large errors when the finite zero that is introduced by the PPD controller becomes dominant. Hence a modification to the previous design approach is needed, which is discussed now.

Consider the PPD servo system represented by the block diagram in Figure 12.45. The derivative time constant of the PPD controller is denoted by τ_d. The closed loop transfer function of the system is

$$G(s) = \frac{K(\tau_d s + 1)}{\tau\tau_a s^2 + (\tau + K\tau_d)s + K} \tag{12.79a}$$

which is of the form

$$G(s) = \frac{\omega_n^2(\tau_d s + 1)}{\left(s^2 + 2\zeta\omega_n s + \omega_n^2\right)} \tag{12.79b}$$

The undamped natural frequency ω_n of the closed loop system is given by

$$\omega_n^2 = \frac{K}{\tau\tau_m} \tag{12.80}$$

and the damping ratio ζ is given by

$$2\zeta\omega_n = \frac{\tau + K\tau_d}{\tau\tau_m} \tag{12.81}$$

Equation 12.79 has a finite zero (at $s = -1/\tau_d$). It follows that (by using the principle of superposition, which is valid for linear systems) the step response of the PPD servo is given by

$$y = y^* + \tau_d \frac{dy^*}{dt} \tag{12.82}$$

in which $y*$ is the simple oscillator step response as given by (also see Chapter 2)

$$y* = 1 - \frac{\exp(-\zeta\omega_n t)}{\sqrt{1-\zeta^2}} \sin(\omega_d t + \phi) \tag{12.83a}$$

By substituting Equation 12.83a into Equation 12.82 we obtain

$$y = 1 - \frac{\omega_n \tau_d}{\sin\eta} \exp(-\zeta\omega_n t)\sin(\omega_d t + \phi + \eta) \tag{12.83}$$

in which

$$\eta = \tan^{-1}\left[\frac{\omega_d \tau_d}{1 - \zeta\omega_n \tau_d}\right] \tag{12.84}$$

Peak time T_p is determined by the condition $\frac{dy}{dt} = 0$. This gives

$$\tan(\omega_d T_p + \phi + \eta) = \frac{\omega_d}{\zeta\omega_n} = \tan\phi$$

or

$$\eta = \pi - \omega_d T_p \tag{12.85}$$

By substituting Equation 12.85 into Equation 12.83 we get the peak response (at $t = T_p$),

$$M_p = 1 + \frac{\omega_d \tau_d}{\sin\omega_d T_p} \exp(-\zeta\omega_n T_p) \tag{12.86}$$

Design Equations: There are four equations that govern the design of a PPD position servo. They can be obtained directly from the theory outlined above. With straightforward manipulation of the previous equations we get:

$$\frac{1}{\tau_d} = \zeta\omega_n - \frac{\omega_d}{\tan\omega_d T_p} \tag{12.87}$$

$$\frac{K}{\tau\tau_a} = (\zeta\omega_n)^2 + \omega_d^2 \tag{12.88}$$

$$\frac{1}{\tau_a} + \frac{K\tau_d}{\tau\tau_a} = 2\zeta\omega_n \tag{12.89}$$

Percentage overshoot

$$PO = \frac{100\omega_d \tau_d}{\sin \omega_d T_p} \exp(-\zeta\omega_n T_p) \qquad (12.90)$$

These results can be expressed graphically and used in the design process. Alternatively, the equations can be solved using a nonlinear equation solver, after substituting the specifications and known parameter values.

Example 12.24

Consider PPD control of a dc motor represented in Figure 12.45, with mechanical time constant $\tau_a = 0.5$ s and the gain parameter $\tau = 1$. Suppose that the design specifications are $T_p = 0.09$ s. and PO = 12%. The plant transfer function for this example is

$$G_p(s) = \frac{1}{s(0.5s+1)}$$

It can be verified that a PPD controller with parameters $K = 277$ and $\tau_d = 0.078$ will meet these design specifications.

12.8   Compensator Design in the Frequency Domain

Once the components such as actuators, sensors, and transducers are chosen for a control system and the control strategy (e.g., proportional feedback control) is decided upon, the design of the control system can be accomplished by determining the parameter values for the controller (and possibly for other components) that will bring about the desired performance. If the design specifications cannot be met by adjusting the system parameters, then new components (called *compensators*) have to be added to the control system. This process is known as control system compensation.

   Lead compensation and lag compensation are the most commonly used methods of compensation in the conventional frequency domain design of control systems. These compensators improve the system performance by modifying the frequency response of the original control system. Both types of compensators can provide improved stability. The way this improvement is brought about by a lag compensator is not quite the same as the way it is achieved by a lead compensator, however. Lead compensation improves the speed of response (or bandwidth) as well, of the control system. Lag compensation improves the low-frequency performance (steady-state accuracy, in particular). Unfortunately, lag compensation has the disadvantage of decreasing the system bandwidth. In general, combined lead-lag compensation, and perhaps several stages of this, would be needed for achieving large improvements in performance. Now we will study the behavior of lead compensation and lag compensation and how these compensators may be designed into a control system.

   To get the loop transfer function of the compensated system, the compensator transfer function is multiplied by the loop transfer function *GH* of the uncompensated system. It follows that the compensator element may be added at any point in the control loop. It may be included either in the forward path or in the feedback path of the loop. In either case the analysis and the design procedures are the same.

The design of a lead compensator or a lag compensator consists of two basic steps. They are:

a. Select the system gain to meet the steady state accuracy specification. This will result in improved speed of response as well.

b. Choose the zero and the pole of the compensator to meet the phase margin specification (for relative stability).

The compensator design amounts to the selection of appropriate parameters (gains, poles, and zeros) for the compensator elements. Bode diagrams are particularly useful in the frequency-domain design of compensators. A particular advantage is the fact that the Bode plot for the compensated system is obtained by simply adding the Bode plot for the compensator to the Bode plot of the original uncompensated system. This follows from the fact that the phase angles of a transfer function product are additive, and the magnitudes are additive as well when a log scale (or decibel scale) is used.

12.8.1 Lead Compensation

Lead compensation is a conventional frequency-domain design approach, which employs the derivative action (lead action) of a compensator circuit (or software) to improve stability by directly increasing the phase margin. Also, lead compensation increases the system bandwidth, thereby improving the speed of response. The speed of response may be further increased and the steady state accuracy improved as well by increasing the dc gain (zero-frequency gain) of the control loop. Lead compensation primarily modifies the high-frequency region of the system bandwidth, thereby improving transient characteristics of the control system.

The transfer function of a lead compensator is expressed as

$$G_d(s) = \left[\frac{\tau s + 1}{\alpha \tau s + 1} \right] \quad 0 < \alpha < 1 \tag{12.91}$$

Nyquist and Bode diagrams for this compensator are shown in Figure 12.46. Note that the maximum phase (lead) angle is obtained at a point within the frequency range $[1/\tau, 1/(\alpha\tau)]$, and is given by

$$\Delta\phi_m = \sin^{-1}\left[\frac{1-\alpha}{1+\alpha} \right] \tag{12.92}$$

This is obtained by drawing a tangent line to the Nyquist plot (which is a semicircle) as shown in Figure 12.46(a). It can be shown that the corresponding frequency is

$$\omega_c = \frac{1}{\sqrt{\alpha}\tau} \tag{12.93}$$

and the transfer-function magnitude is $1/\sqrt{\alpha}$. In decibels this magnitude can be expressed as

$$M_c = 20\log_{10}\left(\frac{1}{\sqrt{\alpha}} \right) \text{dB} \tag{12.94}$$

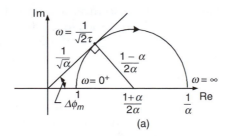

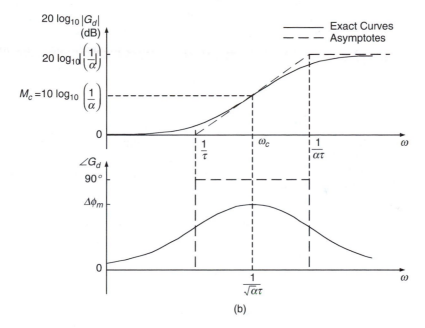

FIGURE 12.46
Frequency transfer function of a lead compensator: (a) Nyquist plot; (b) Bode plot.

To verify these results note that in Figure 12.46(a), the transfer-function magnitude at the maximum phase angle is given by the length of the tangent to the Nyquist curve:

$$\sqrt{\left(\frac{1+\alpha}{2\alpha}\right)^2 - \left(\frac{1-\alpha}{2\alpha}\right)^2} = \frac{1}{\sqrt{\alpha}}$$

This verifies Equation 12.94. Now if we substitute $s = j/(\sqrt{\alpha}\tau)$ in Equation 12.91 we get

$$G_d(s)\Big|_{s=\frac{j}{\sqrt{\alpha}\tau}} = \frac{1}{\sqrt{\alpha}}\frac{(j+\sqrt{\alpha})}{(\sqrt{\alpha}j+1)}$$

This has a magnitude of $1/\sqrt{\alpha}$, thus verifying Equation 12.93.

Notice from Figure 12.46(b) that when a lead compensator is added to a system, its gain is amplified, particularly at high frequencies. This increases the crossing frequency. The phase angle of the uncompensated system at this new crossing frequency is typically more negative than that at the original crossing frequency. This means more phase lead (than would be required if the compensator had not produced a magnitude increase) has to be provided by the lead compensator in order to meet a specified phase margin. Usually, a correction angle $\delta\phi$ of a few degrees is added to the required phase lead of the compensator (given by the difference: specified phase margin—phase margin of the uncompensated system). The actual correction that is required depends on the rate at which the phase angle of the original system changes in the vicinity of the crossing frequency, but it is not precisely known beforehand.

Final value theorem dictates that the lead compensator given by Equation 12.91 does not affect the steady state accuracy. The reason is that the compensator has a unity dc gain (i.e., magnitude at zero frequency is 1), and it is the dc gain that determines the steady state response of a transfer function. Note further that since the crossover frequency increases by lead compensation, the system bandwidth also increases.

Using the results outlined above, an iterative procedure for lead compensator design can be stated. The procedure is considered optimal because, in each iteration the maximum phase lead that is offered by the compensator is utilized.

12.8.1.1 Design Steps for a Lead Compensator

The main steps of an iterative procedure for designing a lead compensator are listed now. Note that just one iteration is adequate for most purposes. Each step in the design procedure is explained at the end of the listing. The control system configuration considered is shown in Figure 12.47. *Design Specifications*: PM_{spec}, e_{ss}

Step 1: Compute the loop gain K needed to meet the steady-state error specification. Obtain the Bode curves for the uncompensated system with this gain included in the loop. Determine the phase margin PM_o.

Step 2: Compute the required phase margin improvement

$$\Delta\phi_m = PM_{spec} - PM_o + \delta\phi \tag{12.95}$$

Step 3: Compute the lead compensator parameter

$$\alpha = \frac{1-\sin\Delta\phi_m}{1+\sin\Delta\phi_m} \tag{12.96}$$

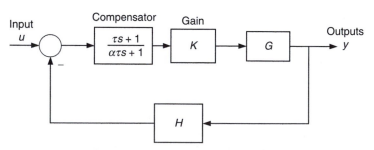

G = uncompensated forward transfer function
H = feedback transfer function
KGH = uncompensated loop transfer function

FIGURE 12.47
Forward (cascade) compensation of a control system.

Step 4: Compute the compensator gain at the maximum phase lead

$$M_c = 10\log_{10}\left(\frac{1}{\alpha}\right)\,\text{dB}$$

Step 5: From the uncompensated Bode gain curve, determine the frequency ω_c at which gain is $(-M_c)$ dB.

Step 6: Compute the remaining compensator parameter

$$\tau = \frac{1}{\sqrt{\alpha}\omega_c} \tag{12.98}$$

Step 7: Compute the Bode curves for the system including the compensator and determine the *PM* of the compensated system.

Step 8: If $|PM_{\text{spec}} - PM| \le \delta\phi_o$ Stop. If the number of iterations exceeds the limit, stop. Otherwise, increase $\delta\phi$ by $(PM_{\text{spec}} - PM)$ and Go To Step 2.

Typically, a phase margin specification PM_{spec} and a steady-state error specification e_{ss} are specified in the compensator design. Gain K that is required to meet the e_{ss} (for a step input) is computed in Step 1. Note that if the uncompensated loop has a free integrator (i.e., it is a Type 1 system), the steady-state error for a step input will be zero. Then, there is no need to change the system gain, unless the steady-state error specification is based on a rate input such as a ramp or parabola. The steady state response to a unit step input $(1/s)$ is given by

$$y_{ss} = \lim_{s\to 0} \frac{s}{s}\tilde{G}(s) = \frac{KG(0)}{1 + KG(0)H(0)} = \frac{K}{1+K} \tag{12.99}$$

This follows from the *final value theorem*. This assumes that the transfer functions $G(s)$ and $H(s)$ do not contain free integrators and have unity dc gains. The steady-state error is given by

$$e_{ss} = 1 - y_{ss} = 1 - \frac{K}{1+K} = \frac{1}{1+K} \tag{12.100}$$

The system gain computed in this manner, is added to the uncompensated system. Note that if the uncompensated loop has one or more free integrators, the original gain is unchanged. The phase margin PM_o of the uncompensated system (with gain K included) is determined in Step 1.

In Step 2 the required increase in phase margin is computed. A typical starting value for the correction angle $\delta\phi$ is $5°$. It will be changed in subsequent iterations. One of the compensator parameters (α) is computed in Step 3. Equation 12.96 follows from Equation 12.92. Note that if $\Delta\phi_m$ is excessive, a single compensator may not be adequate. For example, if the typical value of 1 decade is used as the separation between the zero and the pole of the lead compensator, then $\alpha = 0.1$ and

$$\Delta\phi_m = \sin^{-1}\left[\frac{1-0.1}{1+0.1}\right] = 50°$$

A phase margin improvement of better than 50° would be very demanding on a single compensator. In general, a single compensator should not be used to obtain a phase increase of more than 70°.

Equation 12.97, which is used in Step 4 to compute the gain at maximum phase lead of the compensator, follows directly from Equation 12.94. Frequency ω_c will be the crossing frequency of the compensated system. Manual determination of ω_c can be done simply by noting the frequency at which the gain is $-M_c$ on the uncompensated Bode gain plot. For computer determination of this quantity, one has to add M_c to the uncompensated gain value and then obtain the crossing frequency of this modified gain curve.

The remaining compensator parameter τ is computed using Equation 12.98 in Step 6. This equation is the same as Equation 12.93. Note that ω_c is equal to the geometric mean of the compensator zero and compensator pole. In Step 7, the compensator transfer function is included in the loop transfer function; the new Bode curves are computed; and the new phase margin *PM* is determined. The absolute error in the new phase margin is computed in Step 8. If this is less than a prechosen error tolerance $\delta\phi_o$, the design is concluded. Otherwise, the phase margin correction $\delta\phi$ is increased by an amount that is equal to the current error, and the design is repeated. An error tolerance of 1° is usually adequate.

If a single lead compensator is unable to provide the necessary gain increase, two or more compensators in cascade should be used. In that case the design could be done with one compensator at a time. The first compensator is designed for its optimum performance. It is then included in the loop transfer function and the second compensator is designed for this modified loop transfer function, and so on. The computer-aided design procedure that was outlined previously can be easily extended to this case of designing higher order compensators in sequence.

When a lead compensator is added, the resulting crossing frequency ω_c of the compensated system will be larger than that of the uncompensated system. This should be obvious from the gain curve in Figure 12.46(b). This means that the bandwidth of the compensated system is larger. This has the favorable effect of increasing the speed of response of the control system. Unfortunately, a lead compensator is also a high-pass filter. This means that the compensated system can allow higher-frequency noise to distort the signals in the control system, which is a shortcoming of lead compensation.

12.8.2   Lag Compensation

If a control system has more than adequate bandwidth, a lag compensator can be used to simultaneously improve both the steady-state accuracy and the stability of the system. Since a lag compensator adds a phase lag to the loop, it actually has a destabilizing effect. But since the crossing frequency (hence, the system bandwidth) is reduced by a lag compensator, the phase lag will be lower in the neighborhood of the new crossing frequency than the phase lag near the old crossing frequency, thereby increasing the phase margin and improving system stability. It follows that even though a lead compensator and a lag compensator both improve system stability, the way they accomplish this is quite different. Also, a lag compensator inherently improves the low-frequency behavior, steady-state accuracy in particular, of the system. Furthermore, since a lag compensator is essentially a low-pass filter, it has the added advantage of filtering out high-frequency noise.

A lag compensator is given by the transfer function

$$G_g(s) = \left[\frac{\tau s + 1}{\beta \tau s + 1}\right] \quad \beta > 1 \qquad (12.101)$$

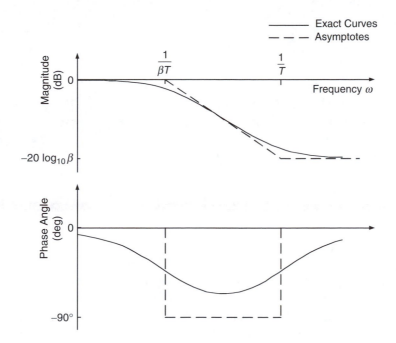

FIGURE 12.48
Bode diagram of a lag compensator.

Its Bode diagram is shown in Figure 12.48. Since this compensator adds a negative slope (in the frequency range $1/(\beta\tau)$ to $1/\tau$) to the original loop, the crossing frequency will decrease. If the phase margin at this new crossing frequency is adequate to meet the *PM* specification, the lag compensator will provide the required stability. Otherwise, a lead compensator should be cascaded to further improve the stability of the system.

Phase lag contribution from a lag compensator is primarily limited to the frequency interval $1/(\beta\tau)$ to $1/\tau$. This range should be shifted far enough to the left of the crossing frequency of the compensated system so that the phase lag of the compensator has a negligible effect on the phase-margin potential of the original system. One way to accomplish this would be to make $1/\tau$ a small fraction (typically 0.1) of the required crossing frequency. Using these considerations, a design procedure for a lag compensator is outlined below.

12.8.2.1 Design Steps for a Lag Compensator

Main steps of designing a lag compensator are given in this section. The considered control structure is the same as that shown in Figure 12.47. Design Specifications: PM_{spec}, e_{ss}

Step 1: Compute the loop gain K that is needed to meet the steady-state error specification. Obtain the Bode curves for the uncompensated system with this gain included in the loop.

Step 2: Compute the phase angle required at the new crossing frequency,

$$\phi = PM_{spec} - 180° + \delta\phi \tag{12.102}$$

Step 3: From the uncompensated Bode phase curve determine the frequency ω_c where the phase angle is ϕ.

***Step* 4:** Compute the lag compensator parameter

$$\tau = \frac{10}{\omega_c} \tag{12.103}$$

***Step* 5:** From the uncompensated Bode gain curve determine the magnitude M_c dB at ω_c.

***Step* 6:** Compute the remaining compensator parameter β as follows:

$$a = \frac{M_c}{10} + \log_{10} 101 \tag{12.104}$$

$$\beta = 0.1\sqrt{10^a - 1} \tag{13.105}$$

***Step* 7:** Compute the Bode curves for the system including the lag compensator and determine the *PM* of the compensated system.

***Step* 8:** If $|PM_{\text{spec}} - PM| \leq \delta\phi_o$ Stop. If the number of iterations exceeds the limit, stop. Otherwise increase $\delta\phi$ by $(PM_{\text{spec}} - PM)$ and Go To Step 2.

Most of the steps given above are self-explanatory. In Step 2, a correction angle $\delta\phi$ is used to account for the phase lag contributed by the lag compensator at frequency ω_c. In the first iteration, a correction angle of $5°$ is typically used. The required correction angle is small provided that the magnitude of the compensator zero $(1/\tau)$ is quite small compared to the crossing frequency (ω_c). This is guaranteed in Step 4, through Equation 12.103. The magnitude of the uncompensated system (M_c) at ω_c (see Step 5) has to be exactly cancelled by the magnitude of the lag compensator at ω_c, in order to force ω_c to be the crossing frequency of the compensated system. The necessary condition is

$$-M_c = 20\log\frac{|\tau j\omega_c + 1|}{|\beta\tau j\omega_c + 1|} \tag{12.106}$$

By substituting Equation 12.103 into Equation 12.106 we get

$$M_c = 20\log_{10}\frac{|10\beta j + 1|}{|10 j + 1|} \tag{13.107}$$

Equation 12.104 and Equation 12.105 given in Step 6 are obtained directly from Equation 12.107. An approximate relation is obtained by neglecting 1 compared to 10; thus,

$$\beta = 10^{M_c/20} \tag{13.108}$$

Usually, acceptable results are obtained in just one design iteration. In computer-aided design routines a maximum number of design iterations (typically 5) should be specified. Then, if the design does not converge, the design computation is stopped when the maximum number of iterations is exceeded, and the best design among the several iterations performed is presented as the final design. Note that if the phase margin potential (i.e., the difference: $180°-$ minimum phase lag angle of the uncompensated system at a frequency above the required bandwidth) is smaller than the phase margin specification, the lag compensator is unable to meet the phase margin specification.

In some designs it may be necessary to modify both the low frequency region and the high frequency region of the loop transfer function in order to simultaneously reduce the steady-state error to a desired level; improve slow transients; filter out high-frequency noise; increase the speed of response; and improve the input-tracking capability. This may be achieved by using one or more lead-lag compensator stages in cascade.

For a good compensator design, the slope of the loop gain curve at the crossover frequency should be approximately equal to (−20 dB/decade). Often, this condition is given as a design specification. It can be shown that if the slope at crossover is substantially smaller (algebraically) than (−20 dB/decade), it is quite difficult to accurately meet a phase-margin specification. To illustrate this, consider an underdamped oscillator. In this example, the slope at crossover is (−40 dB/decade) which is considerably smaller than the required (−20 dB/decade), and the phase angle changes rapidly from 0° to −180° in the neighborhood of the natural frequency ω_n. Now, recall the fact that we need to include a correction angle $\delta\phi$ in the design of a lead compensator because the phase lag angle of the uncompensated system at the compensated (new) crossover frequency is different from (usually larger than) that at the uncompensated (old) crossover frequency. Also, a correction angle $\delta\phi$ has to be included in the design of a lag compensator because the compensator adds a small phase lag in the neighborhood of the compensated crossover frequency. Should the phase angle of the uncompensated system change rapidly (which is the case when the slope of the gain curve is (40 dB/decade or smaller), the design would be very sensitive to the phase angle correction $\delta\phi$. Then it would be very difficult to meet the *PM* specification, usually resulting in either a substantially overdesigned compensator or a substantially underdesigned compensator. For example, for a lightly damped simple oscillator, if the compensated crossover frequency is less than ω_n (which will be the case with a lag compensator), the uncompensated loop itself will provide a phase margin of nearly 180° (an overdesigned case). On the other hand, if the compensated crossover frequency is greater than ω_n, the uncompensated loop will have a very small phase margin (approximately zero). Hence, a lead compensator will have to provide the entire requirement of design phase margin, which is not usually possible with a single compensator (an underdesigned case).

12.8.3 Design Specifications in Compensator Design

More than a phase margin and a steady-state error might be specified in the design of a compensator. For example, a settling time and a bandwidth of the closed-loop system might also be specified. Since settling time is related to stability, one approach to meet this specification would be to first design the compensator to meet the phase margin specification and subsequently check to see whether the design satisfies the settling time specification. If not, the phase margin specification should be increased and the compensator redesigned on that basis. Similarly, the bandwidth of the designed closed-loop system should also be checked. If it is not satisfied, the system gain should be increased and the compensator redesigned.

As noted before (see Equation 12.75) it is possible to relate a phase margin specification (ϕ_m) to a damping ratio specification, using the simple oscillator approximation. First note that the loop transfer function

$$GH(s) = \frac{\omega_n^2}{s(s + 2\zeta\omega_n)} \tag{12.109a}$$

with unity feedback ($H = 1$), gives the closed-loop transfer function ($G/(1 + GH)$):

$$\frac{\omega_n^2}{s^2 + 2\zeta\omega_n s + \omega_n^2} \tag{12.109b}$$

which represents the simple oscillator equation. The crossover frequency (corresponding to $|GH(j\omega)| = 1|$) is given by

$$\frac{\omega_n^2}{\omega\sqrt{\omega^2 + 4\zeta^2\omega_n^2}} = 1 \qquad (12.110a)$$

whose positive solution for ω is

$$\omega_c = a\omega_n \qquad (12.110b)$$

in which

$$a = \sqrt{\sqrt{4\zeta^2 + 1} - 2\zeta^2} \qquad (12.111)$$

Now note from Equation 12.109a that

$$\angle GH(j\omega) = -90° - \tan^{-1}\frac{\omega}{2\zeta\omega_n} \qquad (12.112)$$

Hence, the phase margin of the system is

$$\phi_m = 180° - 90° - \tan^{-1}\frac{\omega_c}{2\zeta\omega_n} = 90° - \tan^{-1}\frac{\omega_c}{2\zeta\omega_n} = \tan^{-1}\frac{2\zeta\omega_n}{\omega_c}$$

Now, in view of Equation 12.111 we have

$$\phi_m = \tan^{-1}\frac{2\zeta}{a} \text{ deg} \qquad (12.113)$$

Equation 12.113 along with Equation 12.111 provides a relationship for the specifications of *PM* and ζ. In particular for small ζ, if we neglect terms compared to unity, Equation 12.111 can be approximated by $a = 1$. Then Equation 12.113 can be approximated as

$$\phi_m = 2\zeta \text{ radians} = 2\zeta \times \frac{180}{\pi} \text{ deg}$$

or, approximately

$$\phi_m = 100\zeta \text{ degrees} \qquad (12.114)$$

Example 12.25

A speed control system is shown by the block diagram in Figure 12.49. The motor is driven by control circuitry, approximated in this example by an amplifier of gain *K*. The signal *y* from the speed sensor is conditioned by a low-pass filter and compared with the speed command *u*. The resulting error signal is fed into the control amplifier. The controller may be tuned by adjusting the gain *K*. Since the required performance was not achieved by

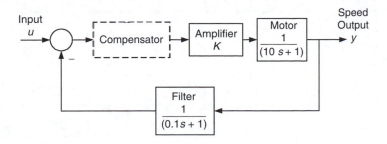

FIGURE 12.49
Compensator design for a velocity servo.

this adjustment alone, it was decided to add a compensator network into the forward path of the control loop. The design specifications are:

1. Steady-state accuracy of 99.9% for a step input
2. Percentage overshoot of 10%

Design:

a. A lead compensator
b. A lag compensator

to meet these design specifications.

SOLUTION
From the time-domain considerations using a damped oscillator model, a percentage overshoot of PO = 10 corresponds to a damping ratio of $\zeta = 0.6$. Then in view of Equation 12.114 we have the equivalent phase margin specification

$$PM_{spec} = 60°.$$

Next, note that the dc gain of the filter is $H(0) = 1$. Accordingly, we use Equation 12.100 to determine the gain that satisfies the s.s. error specification; thus

$$\frac{1}{1+K} = \frac{0.1}{100}$$

or

$$K = 999$$

With this gain, the transfer function of the uncompensated loop is

$$GH = \frac{999}{(10s+1)(0.1s+1)}$$

Bode curve pair for this transfer function is shown in Figure 12.50.

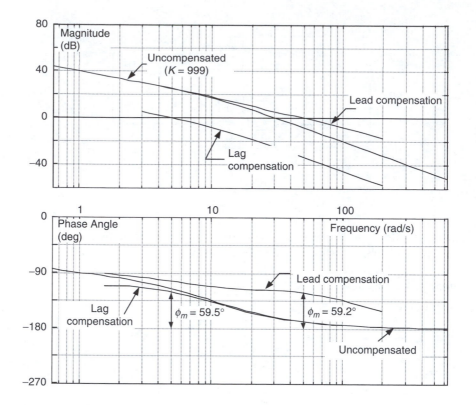

FIGURE 12.50
Bode curves for the compensator design example.

a. Lead Compensation: Here we follow the design steps as given earlier.

Step **1:** The Bode plot with $K = 999$ gives a phase margin of

$$PM_o = 18° \text{ at } 31 \text{ rad/s (4.9 Hz)}$$

Step **2:** Using a correction angle of 6° we have from Equation 12.95

$$\Delta\phi_m = 60 - 18 + 6 = 48°$$

(This correction angle was actually obtained after one design iteration starting with a correction angle of 1°.)

Step **3:** From Equation 12.96 we obtain one of the lead compensator parameters:

$$\alpha = \frac{1 - \sin 48°}{1 + \sin 48°} = 0.15$$

Step **4:** From Equation 12.97

$$M_c = 10 \log_{10}\left(\frac{1}{0.15}\right) = 8.2 \text{ dB}$$

Step 5: The frequency corresponding to an uncompensated magnitude of –8.2 dB is obtained from the uncompensated Bode plot. This value (new crossing frequency) is

$$\omega_c = 51 \text{ rad/s } (8.12 \text{ Hz})$$

Step 6: From Equation 12.98, the second parameter of the lead compensator is obtained as

$$\tau = \frac{1}{\sqrt{0.15 \times 51}} = 0.051 \text{ s}$$

It follows that the transfer function of the lead compensator is

$$G_d(s) = \left[\frac{0.051s + 1}{0.008s + 1} \right]$$

The Bode curve pair for the compensated loop transfer function is shown in Figure 12.50. It is seen that the phase margin specification has been satisfied. Also note the increased crossover frequency (and hence increased bandwidth).

b. Lag Compensation: By following the steps outlined earlier for lag compensator design, we can obtain a lag compensator that satisfies the design specifications.

Step 1: As before we have $K = 999$.

Step 2: Using a correction angle of 5° (this was obtained after one computer iteration) we have from Equation 12.102

$$\phi = 60 - 180 + 5 = -115°$$

Step 3: From the uncompensated Bode curves, the frequency corresponding to a phase angle of –115° is

$$\omega_c = 5 \text{ rad/s } (0.79 \text{ Hz})$$

Note that this step also verifies that a lag compensator can meet the given design specification (i.e., adequate phase margin is present in the uncompensated system), provided that a low bandwidth (less than 1 Hz) is acceptable.

Step 4: One parameter of the lag compensator is obtained from Equation 12.103,

$$\tau = \frac{10}{5} = 2.0 \text{ s}$$

Step 5: The magnitude of the uncompensated system at ω_c is (from Bode curve)

$$M_c = 25.2 \text{ dB}$$

Step 6: From Equation 12.104

$$a = \frac{25.2}{10} + \log_{10} 101 = 4.52$$

From Equation 12.105 the second parameter of the lag compensator is obtained; thus

$$\beta = 0.1\sqrt{10^{4.52} - 1} = 18.3$$

It follows that the transfer function of the lag compensator is

$$G_g(s) = \left[\frac{2.0s + 1}{36.6s + 1}\right]$$

Note from the Bode curves for the lag-compensated system (Figure 12.50) that the phase margin specification has been satisfied. Note further that the crossover frequency has been decreased substantially (from 31 to 5 rad/s) indicating a large reduction in the bandwidth of the control system.

12.8.4 Destabilizing Effect of Time Delays

Time delays, which are inherently present in control systems, can have a destabilizing effect on the system response. Time delays can result from various causes including transport lags in systems such as chemical processes (e.g., flow changes under transient pressure conditions and temperature changes due to mixing of fluids), measurement delays due to large time constants in sensors, and dynamic delays in mechanical systems with high inertia and damping.

A block diagram representation of a time delay is shown in Figure 12.51. Here a signal $x(t)$ undergoes a pure delay by time τ. Since the Laplace transform of the delayed signal is given by

$$\mathcal{L}x(t - \tau) = \exp(-\tau s)\mathcal{L}x(t) \tag{12.115}$$

it is clear that the transfer function for a pure delay is

$$G(s)_{\text{delay}} = \exp(-\tau s) \tag{12.116}$$

In the frequency domain ($s = j\omega$) the magnitude of this transfer function is unity, and the phase angle is negative and monotonically decreasing with frequency ω; thus

$$\left|G(j\omega)\right|_{\text{delay}} = 1 \tag{12.117}$$

$$\angle G(j\omega)_{\text{delay}} = -\tau\omega \tag{12.118}$$

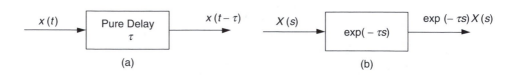

(a)   (b)

FIGURE 12.51
Representation of a time delay: (a) Time domain representation; (b) Transfer function.

t follows that due to a pure delay, the system magnitude is unchanged, but the phase angle is decreased. Consequently, the phase margin and gain margin of the system are reduced; a destabilizing effect. Note further that the condition gets worse as the frequency increases. It follows that a system operating at high frequencies is more likely to become unstable due to time delays. In control system design, specified phase margin should allow for the time delays that are present in various components in the control loop.

12.9 Controller Tuning

Ziegler-Nichols tuning is a procedure commonly used to set parameters of PID controllers (*three-mode controllers* or *three-term controllers*) in industrial control systems. It uses rules of thumb based on practical experience and experimental observation, with common types of control systems. We now present this tuning method as it provides quite satisfactory results even though it lacks theoretical rigor.

12.9.1 Ziegler-Nichols Tuning

Adjustment of controller parameters to obtain an acceptable system response is known as controller tuning. Selection of parameter values for a controller is an important step in control system design. This of course assumes that the design has progressed to the extent that everything about the control system (e.g., control system structure, process parameters, controller type) is known except for the parameter values of the controller. Even if the controller parameters are known for the initial design of the controller system, they may have to be further adjusted (or, "tuned") during operation, as further information on the system performance becomes available and as the operating conditions change.

Controller tuning can be accomplished either by analysis of the control system or by testing. Many mechatronic systems are complex and nonlinear with noisy signals and unknown and time-varying parameter values. Controller tuning by analysis becomes a difficult task for such systems, and consequently controller tuning by testing becomes useful. In their original work Ziegler and Nichols proposed two empirical methods for tuning three mode (PID) controllers:

Reaction curve method

Ultimate response method

Both methods of tuning are expected to provide approximately a *quarter decay ratio* (i.e., amplitude decays by a factor of 4 in each cycle) in the closed-loop system response. On the basis of a simple oscillator model, damping ratio may be expressed by the approximate relationship

$$\zeta = \frac{1}{2\pi r} \ln \frac{A_i}{A_{i+r}}$$
(12.119)

in which
  A_i = response amplitude in the *i*th cycle
  A_{i+r} = response amplitude in the $(i + r)$th cycle

We note from Equation 12.119 that the quarter decay ratio corresponds to

$$\zeta = \frac{1}{2\pi} \ln 4 \approx 0.22 \tag{12.120}$$

or a phase margin of approximately 22°. These are rough estimates, however, because their derivation is based on the simple oscillator model.

12.9.1.1   Reaction Curve Method

In this method, first the open loop response of the plant alone (without any feedback and control) to a *unit step* input is determined. This response is known as the *reaction curve*. Note that if the step input that is used in the test is not unity, the response curve has to be appropriately scaled (i.e., divided by the magnitude of the step input) in order to obtain the reaction curve. We assume that the process is *self-regulating*, for open-loop test purposes, implying that it is stable and its (open-loop) step response eventually settles to a steady state value, even though this assumption is actually not needed in Ziegler-Nichols tuning. The reaction curve of many processes that are self-regulating has the well-known S-shape as represented in Figure 12.52. Note the parameters identified in the figure. The *lag time L* is also known as *dead time* or *delay time*; K is the steady state value of the process variable (process response) for a "unit" step of process demand (process input); T is termed cycle time; and R is the maximum slope of the process reaction curve. These parameters alone completely determine a first-order process with a time delay, as given by the transfer function

$$G_p(s) = \frac{Ke^{-Ls}}{(Ts+1)} \tag{12.121}$$

This is a self-regulating plant because it has a stable pole at $-1/T$. If instead the pole is at the origin of the s-plane (integrator), we have a non-self-regulating plant. Higher-order

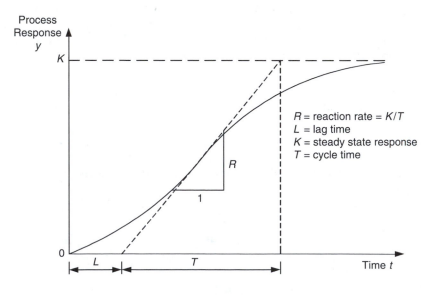

FIGURE 12.52
The reaction curve (Response to a unit step input) of an idealized self-regulating process.

TABLE 12.6

Ziegler Nichols Controller Settings

Controller	Parameter	Reaction Curve Method	Ultimate Response Method
P	K_p	$\dfrac{1}{RL}$	$0.5 K_u$
PI	K_p	$\dfrac{0.9}{RL}$	$0.45 K_u$
	τ_i	$3.3L$	$0.83 P_u$
PID	K_p	$\dfrac{1.2}{RL}$	$0.6 K_u$
	τ_i	$2L$	$0.5 P_u$
	τ_d	$0.5L$	$0.125 P_u$

processes have to be approximated by Equation 12.121. Once the parameters L and R are obtained from the experimentally-determined reaction curve for the open loop process, the controller parameter values for proportional (P), proportional plus integral (PI) and proportional, integral, derivative (PID) controllers are determined according to the Ziegler-Nichols method as tabulated in the column named reaction curve method in Table 12.6.

In conducting the step response test, the process should be first maintained steady at normal operating conditions, with the feedback transmitter disconnected and the controller set to "manual." The corresponding process load and other conditions should be kept constant during the test. Step test is conducted by changing the set point by 5% of the full range and recording the response until the steady state is reached. The response curve should be divided by the value of the step in order to get the reaction curve. Since hysteresis effects are usually present, it is a good practice to apply a step change in the reverse direction and determine the corresponding reaction curve and then take the average of the two curves. Accuracy can be improved by conducting several step tests in each direction and taking the average of all measured reaction curves.

The Ziegler-Nichols method is applicable even if the process is non-self-regulating, because the method does not directly depend on the steady state value K or cycle time T (it depends on the slope K/T). Ziegler-Nichols method is particularly suitable for non-self-regulating processes. In fact, the settings have to be modified when K is of the order of LR, which is the self-regulating case.

12.9.1.2   Ultimate Response Method

In this method the closed-loop system with proportional control alone (i.e., integral and derivative control actions disconnected) is tested to determine the *ultimate gain* K_u and the corresponding period of oscillation (*ultimate period*) P_u of the process response. Ultimate gain is the controller gain at which the closed-loop system is marginally stable; that is, when the system response continuously cycles without noticeable growth or decay. The Ziegler-Nichols controller settings are then determined using K_u and P_u, as given under the ultimate response method in Table 12.6.

In conducting the test, first the process is connected with the controller in the feedback control mode with the integral rate r_i (i.e., inverse of the integral time constant τ_i) and the derivative time τ_d set to zero. Then, the process conditions are maintained at normal

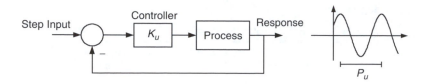

FIGURE 12.53
Test parameters in the ultimate response method of Ziegler–Nichols tuning.

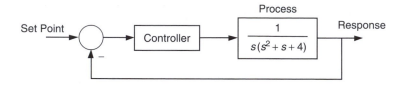

FIGURE 12.54
An example for the ultimate response method of Ziegler-Nichols tuning.

operating values. Next, the proportional gain is set to a small value and maintained there until the conditions are steady. Then, a step input change (typically 5% of the full scale) is applied and the process response is noted. It should decay quickly with this low value of K_p. The proportional gain is increased in sufficiently large steps and the test repeated to roughly estimate the value of K_p that makes the system marginally stable. Once a rough estimate is found, the test should be repeated in that neighborhood using small changes in proportional gain, in order to obtain a more accurate value for the ultimate gain. With the proportional gain set at this value, the process response (closed-loop) to a step input is recorded and from this data, the ultimate period of oscillations is determined. This step is explained in Figure 12.53.

Example 12.26

Consider the feedback control system shown in Figure 12.54. The process transfer function is

$$G_p(s) = \frac{1}{s(s^2 + s + 4)}$$

With proportional control, determine the proportional gain that will make the system marginally stable. What is the period of oscillations for that condition? Give suitable settings for a

- a. Proportional (P) controller
- b. PI controller
- c. PID controller

SOLUTION
With proportional control (gain = K_p), the closed-loop characteristic equation is

$$1 + \frac{K_p}{s(s^2 + s + 4)} = 0$$

or

$$s^3 + s^2 + 4s + K_p = 0$$

To find the condition for marginal stability, we use Routh-Hurwitz method. First, form the Routh array:

	Column 1	Column 2
s^3	1	4
s^2	1	K_p
s^1	$4 - K_p$	0
s^0	K_p	

For stability, the terms in the first column should be all positive. Hence we must have $4 - K_p > 0$ and $K_p > 0$. Thus, the stability region is $0 < K_p < 4$. Accordingly, the gain for marginal stability (ultimate gain) is

$$K_u = 4$$

The corresponding characteristic equation is

$$s^3 + s^2 + 4s + 4 = 0$$

which factorizes into

$$(s+1)(s^2 + 4) = 0$$

NOTE $(s^2 + 4 = 0)$ is the *auxiliary equation*, corresponding to the row s^2 of the Routh array with $(K_p = K_u = 4)$

The oscillatory root pair is

$$\pm j\omega_n = \pm j2$$

Hence, the frequency of oscillations is

$$\omega_n = 2 \text{ rad/s}$$

The period of oscillations (ultimate period) is

$$P_u = \frac{2\pi}{\omega_n} = \frac{2\pi}{2} = \pi \text{ sec}$$

Now, from Table 12.6, we can determine the controller settings.

a. *Proportional control*

$$K_p = 0.5 \times 4 = 2$$

b. *PI control*

$$K_p = 0.45 \times 4 = 1.8$$

$$\tau_i = 0.83\pi = 2.61 \text{ sec}$$

c. *PID control*

$$K_p = 0.6 \times 4 = 2.4$$

$$\tau_i = 0.5\pi = 1.57 \text{ sec}$$

$$\tau_d = 0.125\pi = 0.393 \text{ sec}$$

12.10 Design Using Root Locus

As discussed in Section 12.5, Root locus is the locus of the closed-loop poles as one parameter of the system (typically the loop gain) is varied. A set of design specifications can be met by locating the closed-loop poles inside the corresponding design region on the s-plane. Then the design process will involve the selection of parameters such as control gain, compensator poles and compensator zeros, so as to place the closed-loop poles in the proper design region. This can be accomplished by the root locus method.

12.10.1 Design Steps

Once the design region on the s-plane is chosen, the next step is to check whether the dominant branch (i.e., branch closest to $s = 0$) of the root locus (for the closed-loop system) passes through that region. If it does, the corresponding value of the root locus variable (typically the loop gain) is computed using the magnitude condition (see Rule 2 given for sketching a root locus). If the steady-state error requirement is already included in the design specification (on the s-plane) as a limit on ω_n, then we do not need to proceed further. Otherwise, the applicable error constant (K_p, K_v, or K_a) should be computed using the design value of the loop gain, to check whether the steady-state accuracy is adequate. If the design requirements are not met with the existing control loop, a compensator with dynamics (i.e., one having s terms), such as a lead compensator or a lag compensator, should be added to the loop and the compensator parameters should be chosen to satisfy the design specifications. To summarize, the root locus design steps are:

Step **1:** Represent the design specifications as a region on the s-plane.

Step **2:** Plot the root locus to check whether at least one branch passes through the design region while the other branches pass through regions to the left of the design region, for the same parameter values.

Step **3:** If the system does not satisfy the requirement in Step 2, add a compensator to the control loop and adjust the compensator parameters to achieve the requirement. If it does, then compute the corresponding root locus parameter (typically the loop gain) using the magnitude condition.

NOTE It is the dominant poles of the closed-loop system that should fall inside the design region (the remaining poles being to the left of the region).

The magnitude condition is an important equation in the root locus design. In this regard, a useful result can be obtained by using the closed-loop characteristic equation as given before,

$$K\frac{(s-z_1)(s-z_2)\cdots(s-z_m)}{(s-p_1)(s-p_2)\cdots(s-p_n)} = -1 \qquad (12.53d)$$

Since the coefficient of the second highest power (i.e., coefficient of s^{n-1}) of a monic characteristic polynomial (i.e., a polynomial with the coefficient of s^n equal to 1) is equal to the sum of the roots except for a sign change, we observe from Equation 12.53d the following fact, which can be given as a rule:

RULE 9 If $m < n - 1$, then the (sum of the closed loop poles) = (sum of the *GH* poles) and this sum is independent of *K*.

12.10.2 Lead Compensation

The lead compensator design using the root locus method is illustrated now. Essentially, we follow the three steps for root locus design.

Lead compensator design in the frequency domain using Bode diagram was discussed previously. The objective of the method was to determine the zero ($-z$) and the pole ($-p$) of the lead compensator transfer function

$$G_c = \frac{p}{z}\left[\frac{s+z}{s+p}\right] \qquad z<p \qquad (12.122)$$

so that the design specifications are satisfied. The steps that are usually followed in the root locus method to design a lead compensator are given below.

Step 1: Select a closed loop pole pair (complex conjugates) that meets the design specifications. This should be the dominant pole pair of the closed-loop system.

Step 2: Locate the compensator zero ($-z$) vertically below the specified closed-loop pole.

Step 3: Locate the compensator pole ($-p$), to the left of so as to satisfy the angle condition of the root locus (Rule 2 given under the root locus method, Section 12.5)

Step 4: Compute the root locus parameter (usually gain *K*) at the design pole location, using the magnitude condition (Rule 2).

In Step 1, the design pole pair of the closed-loop system is located in the design region of the *s*-plane, as discussed earlier. In Step 2, if there is a *GH* pole at the location where the compensator zero is to be located, we should locate the compensator zero sufficiently to the left of that location so that the compensator would not drastically alter the dynamic characteristics of the uncompensated system, by producing a closed-loop pole that dominates over the design poles. Step 3 makes sure that the root locus passes through the design point, and Step 4 provides the root locus parameter value at the design point.

Example 12.27

Sheet steel, which many major industries such as automobile industry and household appliance industry depend on, is obtained by either hot rolling or cold rolling of thick plates or slabs of steel castings. The steel is passed through a pair of work rolls, which are driven by heavy duty a motor. The thickness of rolled steel depends on the roll

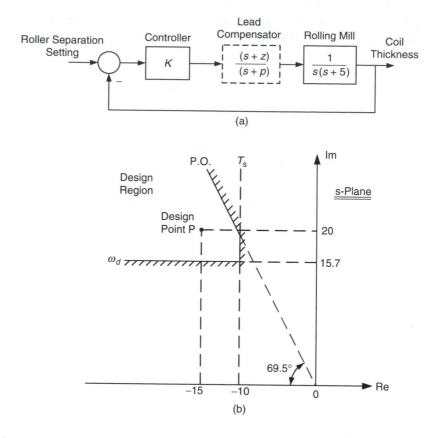

FIGURE 12.55
(a) Block diagram for the control system of a steel rolling mill; (b) Selection of a design point (*P*) on the *s*-plane.

separation, which is adjusted by a hydraulic actuator (ram). Open-loop adjustment is not satisfactory for reasons such as roll deformation, flexibility of rolled steel, and mill stretch. The thickness of the output steel coil is measured, and the roller separation is corrected accordingly, using feedback control. A simplified model for this control loop is shown in Figure 12.55(a). Show that simple proportional control is inadequate for simultaneously meeting the following three specifications:

a. A peak time < 0.2 s
b. 2% settling time < 0.4 s
c. Percentage overshoot < 10

Determine the parameters for a suitable lead compensator that satisfies these three control specifications. Compute the velocity error constant of the resulting system and the steady-state error to a unit ramp input.

SOLUTION
Using the formulas given in Table 12.2 we find that

$T_p = 0.2$ s corresponds to $\omega_d = 15.7$ rad/s
$T_s = 0.4$ s corresponds to $\zeta\omega_n = 10$ rad/s
PO = 10% corresponds to $\zeta = 0.35$ or $\cos^{-1} \zeta = 69.5°$

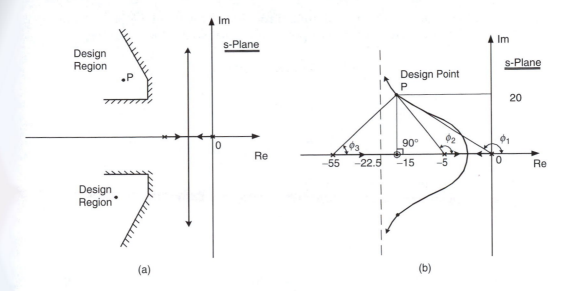

FIGURE 12.56
(a) Design region and the root locus of the uncompensated system; (b) Root locus of the compensated system.

These design boundaries are drawn in Figure 12.55(b) and an acceptable design region is determined. Any point in this region would be satisfactory. Usually, the design point in this region that is closest to the origin of the s-plane is chosen. But since a steady-state error specification is not available, it is not known whether the design gain is satisfactory. As a compromise, the design point P (corresponding to $\omega_d = 20$, $\zeta\omega_n = 15$, $\omega_n = 25$, and $\zeta = 0.6$) is chosen as shown in Figure 12.55(b). The root locus of the uncompensated system is sketched in Figure 12.56(a). Since it does not pass through the design region as superimposed in Figure 12.56(a), it is concluded that proportional control without compensation cannot meet the specifications.

Next, the compensator zero is located vertically below P. This gives

$$z = -15$$

We know that the compensator pole lies to the left of this point. Now we are able to sketch the root locus for the compensated system, as in Figure 12.56(b). Note that we have not yet determined all the parameters that are marked in this figure. We see from the figure that the design pole pair are the dominant poles of the closed-loop system, the third pole being located to their left (perhaps close to $-p$).

From geometry we can compute the following angles:

$$\text{Angle at the GH pole (0): } \phi_1 = 180° - \tan^{-1}\frac{20}{15} = 126.87°$$

$$\text{Angle at the GH pole (-5): } \phi_2 = 180° - \tan^{-1}\frac{20}{10} = 116.57°$$

$$\text{Angle at the GH pole (-15): } = 90°$$

Hence, from the angle condition of plotting a root locus, for P to be a point on the root locus we must have the angle at the GH pole $(-p)$, ϕ_3, satisfying the condition:

$$\phi_3 + 116.57 + 126.87 - 90 = 180$$

Hence

$$\phi_3 = 26.56° = \tan^{-1} \frac{20}{(p - 15)}$$

From this we have the compensator pole;

$$p = 55$$

The compensator transfer function is

$$G_c(s) = \frac{(s + 10)}{(s + 55)}$$

The loop gain K at the design point is obtained using the magnitude condition (Rule 2 of root locus method). From geometry of Figure 12.56(b) we obtain

Distance from P to $0 = \sqrt{20^2 + 15^2} = 25$
Distance from P to $-5 = \sqrt{20^2 + 10^2} = 22.36$
Distance from P to $-15 = 20$
Distance from P to $-55 = \sqrt{20^2 + 40^2} = 44.72$

Hence the magnitude condition gives

$$\frac{K \times 20}{25 \times 22.36 \times 44.72} = 1$$

or

$$K = 1250$$

Now using Equation 12.32 the velocity error constant for the control system is computed as

$$K_v = \lim_{s \to 0} \frac{s \times 1250(s + 15)}{s(s + 5)(s + 40)} = 93.75$$

Hence, the steady-state error to a unit ramp input is

$$e_{ss} = \frac{1}{93.75} = 0.011$$

This error is very small (compared to 1), and the design is concluded to be satisfactory. Note that since the compensator pole is far to the left of the system poles, the real pole of the closed-loop system will not dominate and the compensated system will behave like a second order system.

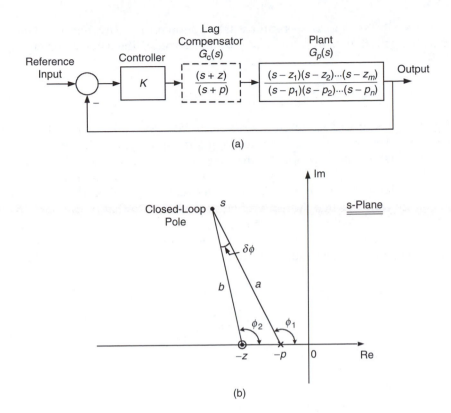

FIGURE 12.57
(a) Block diagram of a system with a lag compensator; (b) Relationship of pole and zero of a lag compensator to dominant closed-loop pole (s) of uncompensated system.

12.10.3 Lag Compensation

We know that lag compensation improves the behavior in low frequency operation (particularly, steady-state accuracy) of a control system. Accordingly, lag compensation is recommended if the uncompensated system has good transient response (i.e., satisfactory moderate-to-high-frequency performance) but has poor steady state accuracy.

To explain the principle of lag compensation by the root locus method, consider the control system shown in Figure 12.57(a). For a controller with gain value $K = K_o$ (i.e., the value of the root locus parameter), the characteristic equation of the system without lag compensator $G_c(s)$ is

$$1 + K_o G_p(s) = 0 \tag{12.123}$$

The roots of this characteristic equation give closed-loop poles. Suppose that the dominant pole (one closest to the origin of the s-plane) obtained this way is shown as point s in Figure 12.57(b). At point s, the angle condition and the magnitude condition are satisfied; thus

$$\angle(s - p_1) + \cdots + \angle(s - p_n) - \angle(s - z_1) - \cdots - \angle(s - z_m) = \pi + 2r\pi \tag{12.124}$$

$$\frac{|s - p_1| \cdots |s - p_n|}{|s - z_1| \cdots |s - z_m|} = K_o \tag{12.125}$$

Now let us include the lag compensator $G_c(s)$ as shown in Figure 12.57(a). Then for the same controller gain K_o (i.e., the root locus parameter value), the closed-loop characteristic equation is

$$1 + K_o G_c(s) G_p(s) = 0 \qquad (12.126)$$

Generally, the roots of this equation are different from the roots of Equation 12.123. In particular, now there may not be a root at point s in Figure 12.57(b). But if we select the compensator $G_c(s)$ in an appropriate manner, we can make sure that the roots of Equation 12.123 are "very close" to the roots of Equation 12.126, so that the transient performance of the system is not significantly affected by compensation. In particular, the root locus of the compensated system will be very similar to that of the uncompensated system except in the neighborhood of the compensator pole and zero. To examine the conditions that should be satisfied by the compensator to achieve this, let us write the angle condition and the magnitude condition corresponding to Equation 12.126; thus

$$\angle(s - p_1) + \cdots + \angle(s - p_n) - \angle(s - z_1) - \cdots - \angle(s - z_m) + \delta\phi = \pi + 2r\pi \qquad (12.127)$$

$$\frac{|s - p_1| \cdots |s - p_n|}{|s - z_1| \cdots |s - z_m|} \cdot \frac{a}{b} = K_o \qquad (12.128)$$

in which

$$\delta\phi = \angle(s + p) - \angle(s + z) = \phi_1 - \phi_2 \qquad (12.129a)$$

$$a = |s + p| \qquad (12.129b)$$

$$b = |s + z| \qquad (12.129c)$$

It follows that if $\delta\phi$ is very small, the Equation 12.129a approximates to Equation 12.124. Similarly, if $a \approx b$, then Equation 12.128 approximates to Equation 12.125.

Now we can conclude that the requirements for the closed-loop poles of the uncompensated system to be "very close" to the closed-loop poles of the compensated system, for the same value of controller gain K_o, are

$$\delta\phi \approx 0 \qquad (12.130)$$

and

$$a \approx b \qquad (12.131)$$

It follows that the lag compensator must satisfy these two requirements. But note that the dc gain of the compensated loop is z/p times the dc gain of the uncompensated loop, thus

$$(\text{DC Gain})_{comp} = \frac{z}{p}(\text{DC Gain})_{uncomp} \qquad (12.132)$$

Since $z > p$ for a lag compensator, this means that the dc gain has increased or, in other words, the error constant has increased. This is the reason for decreased steady-state error due to lag compensation.

A disadvantage of lag compensation can be easily pointed out. Note that the number of closed-loop poles has increased by one due to compensation. One of these poles (the one that is not close to the uncompensated closed-loop poles) is very close to the compensator pole and zero. Since compensator pole and zero have to be chosen very close to the origin (and close together) in order to satisfy the conditions given by Equation 12.130 and Equation 12.131, it follows that the dominant closed-loop pole now is the one created by the compensator. This is obviously a slow pole producing a slowly decaying transient, even though the magnitude of this transient is usually small. Hence, the settling time will be increased to some extent due to lag compensation.

Example 12.28

The motor and the load of a position servo system are represented by the plant transfer function

$$G_p(s) = \frac{1}{s(s+4)}$$

The controller is represented by a pure gain K along with unity feedback. The system is shown in Figure 12.58(a). Sketch the root locus and show that this servo system cannot simultaneously meet the following performance specifications:

a. Percentage overshoot of 4.3
b. Velocity error constant of 10

Design a lag compensator to meet these specifications, within a margin of error. Estimate this margin of error. Sketch the root locus of the compensated system.

SOLUTION

The loop transfer function of the uncompensated system is $\frac{s}{s(s+4)}$. The root locus is sketched in Figure 12.58(b).

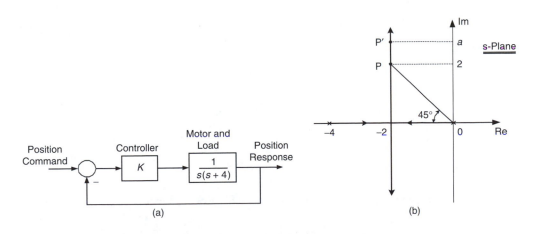

(a) (b)

FIGURE 12.58
(a) Block diagram of an uncompensated position servo system, (b) Root locus of the uncompensated system.

A PO of 4.3% corresponds to a damping ratio of $\zeta = 0.707$ (this can be verified using the equation given in Table 12.2). The corresponding closed-loop pole location is shown as point P in Figure 12.58(b). The controller gain for these operating conditions is obtained by using the magnitude conditions; thus

$$K = \sqrt{2^2 + 2^2} \times \sqrt{2^2 + 2^2} = 8$$

The corresponding velocity error constant (Equation 12.32) is

$$K_v = \lim_{s \to 0} s \times \frac{s}{s(s+4)} = \frac{8}{4} = 2$$

This is less than the specified value of 10. We can meet the K_v specification simply by increasing the controller gain to $K = 40$. Then, however, the operating point moves to P' in Figure 12.58(b), where the closed-loop poles are $-2 \pm ja$. Now the value of "a" is determined using the magnitude condition, thus

$$\sqrt{a^2 + 2^2} \times \sqrt{a^2 + 2^2} = 40$$

or

$$a = 6$$

The corresponding undamped natural frequency $= \sqrt{6^2 + 2^2} = \sqrt{40}$ and the damping ratio $\zeta = \frac{2}{\sqrt{40}} = 0.316$. The percentage overshoot with this ζ is found to be 35.1%, which is much higher than the specified value. It follows that the specifications cannot be met by adjusting the controller gain alone.

Next, we add a lag compensator to the control loop. Since we want to keep the pole near point P in Figure 12.58(b), the pole and the zero of the compensator transfer function

$$G_c(s) = \frac{(s+z)}{(z+p)} \tag{12.133}$$

should be chosen to be much closer to origin than the operating point P. This is accomplished by making z equal to 10% of the real part of the operating pole. So, we have

$$z = 0.2$$

The velocity error constant with the compensator added, is

$$K_v = K \frac{z}{p} \frac{1}{4} = \frac{8 \times 0.2}{p \times 4}$$

We have to make this value equal to 10 (the specification). Hence, we have

$$\frac{8 \times 0.2}{p \times 4} = 10$$

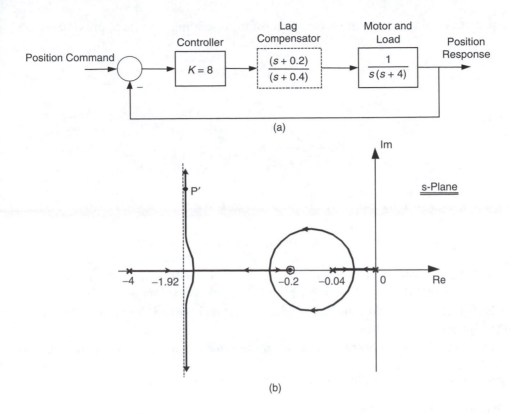

FIGURE 12.59
(a) Block diagram of the lag-compensated position servo; (b) Root locus of the compensated system (Operating point = P').

or

$$p = 0.04$$

The lag compensator transfer function is

$$G_c(s) = \frac{(s+0.2)}{(s+0.04)}$$

The compensated system is shown in Figure 12.59(a) and its root locus is sketched in Figure 12.59(b). The operating point (with $K = 8$) of the system is shown as P' in Figure 12.59(b). The damping ratio at this point will be slightly different from the required 0.707, even though the velocity error constant is exactly met ($K_v = 10$). The error in damping ratio (or percentage overshoot) can be estimated by estimating the angle error and magnitude error near the operating point, as introduced by the compensator. Since the operating point (P') is approximately at $-2 + j2$, the angle error

$$\delta\phi \approx \angle(-2 + j2 + 0.04) - \angle(-2 + j2 + 0.2)$$

$$\approx 134.4° - 132.0° = 2.4°$$

The percentage error (since the vector angle of the operating point is approximately 45°) is

$$\frac{2.4°}{45°} \times 100 = 5.3\%$$

The magnitude error

$$= \frac{|-2 + j2 + 0.04|}{|-2 + j2 + 0.2|} - 1 = 0.041 = 4.1\%$$

It follows that the error will be about 5%.

Usual steps of designing a lag compensator by the root locus method are given below.

Step **1:** Determine an appropriate operating point for the uncompensated system so as to satisfy all performance specifications, except the error constant specification.

Step **2:** Select the compensator zero at about 0.1 × real part of the closed-loop operating pole of the uncompensated system.

Step **3:** Select the compensator pole to meet the error constant (steady-state error) specification.

Step **4:** Check for the margin of error introduced by the compensator.

All these steps should be clear from the previous design example.

12.11   Digital Control

In a digital control system, a digital device is used as the controller. The digital controller may be a *hardware* device that uses permanent logic circuitry to generate control signals (see Chapter 10). Such a device is termed a hardware controller. It does not have programmable memory. This type of controller is not flexible in the sense that the control algorithm cannot be modified without replacing hardware and furthermore, implementation of complex control algorithms by this hardware-based method can become difficult and expensive. But the method is typically very fast from the point of view of speed of generating the control signals, and is suitable for simple dedicated controllers. In mass production, hardware controllers are inexpensive. In the present treatment we will primarily consider *software-based* digital controllers, where a digital computer serves as the controller (see Chapter 11). A controller of this type has programmable memory devices in addition to a central processor. The control algorithm is stored in the computer memory in machine code (in binary code) and is used by the processor in real time to generate the control signals, perhaps on the basis of measured outputs from the plant (which is the case in feedback control) and other types of data. This, along with associated input/output hardware and driver software, forms the digital controller. The control algorithm in such a controller can be modified simply by reprogramming, without the need for hardware changes. A convenient way to analyze and design digital control systems is by the z-transform method. The theory behind this method will be presented and such issues as stability analysis and controller/compensator design by the z-transform method will be described.

12.11.1 Digital Control Using *Z*-Transform

In computer-based control systems, a suitable control algorithm has to be programmed into the memory of the control computer. A digital controller is functionally similar to its analog counterpart except that the input data to the controller and the output data from the controller are in the digital form. The control law can be represented by a set of *difference equations*. These difference equations relate the discrete output signals generated by controller and the discrete input signals provided into the controller. The problem of developing a digital controller can be interpreted as the formulation of appropriate difference equations that are able to generate the required control signals. Just as an analog controller may be represented by a set of analog transfer functions, a digital controller can be represented by a set of *discrete transfer functions*. These discrete transfer functions, in turn, can be transformed into a set of difference equations.

Once a control law is available in the analog form, as a transfer function, the corresponding digital control law may be realized by determining the discrete transfer function that is equivalent to the analog transfer function. This approach is useful when, for example, it is required to update (modernize) a well-established analog control system by replacing its analog compensator circuitry with a digital controller/compensator. In this case, the (Laplace) transfer function of the analog compensator can be obtained by testing or analysis (or both) of the compensator. The objective then would be to develop a difference equation to represent the analog compensator. This is a basic problem in digital control, and is conveniently handled by the z-transform method. This approach is developed in the present section.

A discrete transfer function necessarily depends on the sampling period T used to convert analog signals into discrete data (sampled data). Digital control action approaches the corresponding analog control action when T approaches zero. Faster sampling rates provide better accuracy and less aliasing error (see Chapter 5), but demand smaller processing cycle times, which, in turn, call for efficient processors and improved control algorithms for a given level of control complexity. Faster sampling rates are more demanding on the interface hardware as well. A large word size is needed to accurately represent data. By increasing the word size (number of bits per word), the *dynamic range* and the *resolution* of the represented data can be improved and the *quantization error* decreased. Even though the processing cycle time will generally increase by increasing the word size, on average there is also a speed advantage to increasing the word size of a computer. The larger the program size (number of instructions per program), the greater the memory requirements and, furthermore, the slower the associated control cycle for a given control computer. It follows that sampling rate, processing cycle time, data word size, and memory requirements are interrelated and crucial parameters in digital control.

12.11.1.1 The Z-Transform

Consider an infinite sequence of data

$$\{x_k\} = \{\cdots x_{-k}, x_{-k+1}, \ldots, x_0, x_1, \ldots, x_k, x_{k+1}, \ldots\} \tag{12.134}$$

This sequence can be represented by a polynomial function of the complex variable z; thus

$$X(z) = \sum_{k=-\infty}^{\infty} x_k z^{-k} \tag{12.135}$$

Here $X(z)$ is termed the z-transform of the sequence $\{x_k\}$. This relationship may be expressed using the z-transform operator "Z" as

$$Z\{x_k\} = X(z) \tag{12.136}$$

Note from Equation 12.135 that $\{x_k\}$ uniquely determines $X(z)$ and vice versa. Since $X(z)$ is a continuous polynomial function of z it is convenient to use the z-transform instead of the sequence which it represents, in analyses that involve sequences of data. In digital control systems in particular, inputs and outputs of a digital controller are such data sequences, which are defined at discrete time points. Hence z-transform techniques are very useful in the design of digital controllers and compensators. Note that generally, for the summation in Equation 12.135 to converge, the magnitude of z has to be restricted to at least $|z| < 1$.

In a digital control system, the controller reads sampled values of a continuous signal $x(t)$. Assuming that the sampling period T is constant, the corresponding discrete data values are given by

$$x_k = x(k \cdot T) \tag{12.137}$$

Typically the signal is zero for negative values of time; hence $x_k = 0$, for $k < 0$. But we will retain the full sequence including the negative portion of Equation 12.134 for the sake of analytical convenience.

Example 12.29

Consider a unit step signal given by

$$\mathcal{U}(t) = 1 \quad \text{for } t \geq 0$$
$$= 0 \quad \text{for } t < 0$$

Suppose that this signal is sampled at sampling period T. The corresponding data sequence is

$$\{\mathcal{U}_k\} = \{0, 0, \dots, 0, 0, 1, 1, \dots, 1, 1, \dots\}$$

Determine the z-transform of this sequence.

SOLUTION

By definition, the z-transform is given by

$$\mathcal{U}(z) = \sum_{0}^{\infty} z^{-k}$$

By summation of the series, this can be expressed in the closed form

$$\mathcal{U}(z) = \frac{1}{(1 - z^{-1})}$$

or

$$U(z) = \frac{z}{(z-1)}$$

Example 12.30

Consider the unit ramp signal given by

$$x(t) = t \quad \text{for } t \geq 0$$

$$= 0 \quad \text{for } t < 0$$

What is the corresponding z-transform if the signal is sampled at period T?

SOLUTION

By definition, the z-transform of the sampled data is expressed as

$$X(z) = \sum_{k=0}^{\infty} kTz^{-k}$$

$$= T\sum_{k=0}^{\infty} z^{-k}kz^{-k}$$

By using a well-known result in summation of series, this can be expressed in the closed form

$$X(z) = \frac{Tz}{(z-1)^2}$$

A z-transform depends on the sampling period T in general. Table 12.7 lists z-transforms corresponding to a selected set of time signals, sampled at T.

12.11.2 Difference Equations

In the context of dynamic systems, difference equations are discrete-time models. Consider, in particular, a single-input single-output (SISO) system. The input to a discrete-time model of this system is the sequence $\{u_k\}$ and the output from the model is the sequence $\{y_k\}$, as represented in Figure 12.60.

An nth order linear dynamic system can be modeled in the continuous-time case by the nth order linear ordinary differential equation

$$\bar{a}_n \frac{d^n y}{dt^n} + \bar{a}_{n-1} \frac{d^{n-1} y}{dt^{n-1}} + \cdots + \bar{a}_0 y = \bar{b}_m \frac{d^m u}{dt^m} + \bar{b}_{m-1} \frac{d^{m-1} u}{dt^{m-1}} + \cdots + \bar{b}_0 u \qquad (12.138)$$

in which
$u(t) = $ input
$y = $ output

TABLE 12.7

Some Useful z-Transforms

Time Signal $x(t)$	z-Transform $X(z)$
Unit impulse $\delta(t)$	$1/T$
Unit pulse	1
Unit step $u(t)$	$\dfrac{z}{z-1}$
Unit ramp t	$\dfrac{Tz}{(z-1)^2}$
$\exp(-at)$	$\dfrac{z}{z-\exp(-aT)}$
$\sin \omega t$	$\dfrac{z\sin\omega T}{z^2 - 2z\cos\omega T + 1}$
$\cos \omega t$	$\dfrac{z(z-\cos\omega T)}{z^2 - 2z\cos\omega T + 1}$
$\exp(-at)\sin \omega t$	$\dfrac{z\exp(-aT)\sin\omega T}{z^2 - 2z\exp(-aT)\cos\omega T + \exp(-2aT)}$
$\exp(-at)\cos \omega t$	$\dfrac{z^2 - z\exp(-aT)\cos \omega T}{z^2 - 2z\exp(-aT)\cos \omega T + \exp(-2aT)}$
$t\exp(-at)$	$\dfrac{Tz\exp(-aT)}{[z-\exp(-aT)]^2}$

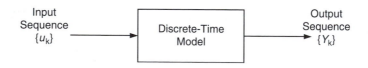

FIGURE 12.60
Block Diagram representation of a single-input single-output discrete model.

The corresponding discrete-time model may be expressed by the nth order linear difference equation

$$a_0 y_k + a_1 y_{k-1} + \cdots + a_n y_{k-n} = b_0 u_k + b_1 u_{k-1} + \cdots + b_m u_{k-m} \qquad (12.139)$$

NOTE The coefficients of the difference equation and the coefficients of the original differential equation are not identical. It is clear from Equation 12.139 that, provided the input sequence $\{u_k\}$ is known, the output sequence $\{y_k\}$ can be computed starting with the first n values of the sequence, which should be known. These initial n values are the *initial conditions*, which are required to determine the complete solution of a difference equation. In general, the model parameters a_i and b_i in Equation 12.139 depend on the sampling period T.

2.11.3   Discrete Transfer Functions

For a time-invariant (i.e., constant-parameter) linear system, the coefficients $\bar{a}_i$ and $\bar{b}_i$ in Equation 12.138 are constants. Then, the system transfer function is given by

$$G(s) = \frac{\bar{b}_0 + \bar{b}_1 s + \cdots + \bar{b}_m s^m}{\bar{a}_0 + \bar{a}_1 s + \cdots + \bar{a}_n s^n} \tag{12.140}$$

This analog transfer function is obtained, in theory, by applying the Laplace transformation, with zero initial conditions, to Equation 12.138. In an analogous manner, by applying the z-transform to Equation 12.139, the corresponding discrete transfer function is obtained. To show this approach, multiply Equation 12.139 by z^{-k} and sum over $k(-\infty, \infty)$. This gives

$$a_0 \sum y_k z^{-k} + a_1 \sum y_{k-1} z^{-k} + \cdots + a_n \sum y_{k-n} z^{-k} = b_0 \sum u_k z^{-k} + b_1 \sum u_{k-1} z^{-k} + \cdots + b_m \sum u_{k-m} z^{-k}$$

which can be rewritten in the form

$$a_0 \sum y_k z^{-k} + a_1 z^{-1} \sum y_{k-1} z^{-(k-1)} + \cdots + a_n z^{-n} \sum y_{k-n} z^{-(k-n)} =$$
$$b_0 \sum u_k z^{-k} + b_1 z^{-1} \sum u_{k-1} z^{-(k-1)} + \cdots + b_m z^{-m} \sum u_{k-m} z^{-(k-m)} \tag{i}$$

Since all summations on the left hand side of Equation i run through $-\infty$ to $+\infty$ each of them is equal to $Y(z)$ as evident from Equation 12.135. Similarly, each summation on the right hand side of Equation i is equal to $U(z)$. It follows that Equation i can be written as

$$(a_0 + a_1 z^{-1} + \cdots + a_n z^{-n}) Y(z) = (b_0 + b_1 z^{-1} + \cdots + b_m z^{-m}) U(z)$$

Hence, the discrete transfer function is given by

$$G(z) = \frac{Y(z)}{U(z)} = \frac{b_0 + b_1 z^{-1} + \cdots + b_m z^{-m}}{a_0 + a_1 z^{-1} + \cdots + a_n z^{-n}} \tag{12.141}$$

As mentioned earlier, the parameters a_i and b_i depend on the sampling period T, in general. Equations 12.139 and 12.141 represent *discrete models* and Equations 12.138 and 12.140 represent continuous (analog) models for a dynamic system.

12.11.4   Time Delay

It should be intuitively clear from the preceding development that z^{-1} can be interpreted as an operator representing a time delay by one sampling period. Specifically,

$$\mathcal{Z}\{x_{k-1}\} = z^{-1} X(z) \tag{12.142}$$

This result can be verified directly from the definition of the z-transform, Equation 12.135. In general, for a delay of r sampling periods we have

$$\mathcal{Z}(x_{k-r}) = z^{-r} X(z) \tag{12.143}$$

This should be compared with the property of Laplace transformation that allows us t interpret the Laplace variable s as the time-derivative operator.

A time delay by T in the continuous-time case can be represented by the transfer functio $\exp(-Ts)$, as can be verified using the definition of Laplace transform. This establishes "correspondence" between the z-transform and the Laplace traform, through the relatio

$$z^{-1} = \exp(-Ts)$$

or

$$z = \exp(Ts) \tag{12.144}$$

Caution should be exercised in using Equation 12.144. This equation provides a "mapping between the s-domain and the z-domain. Specifically, suppose that we have the tw functions $G_1(s)$ and $G_2(z)$. As s varies in some manner (say, along some contour) on the s plane, there is a corresponding variation of $G_1(s)$ in the $G_1(s)$-plane. Then, z varies on th z-plane according to the mapping Equation 12.144 and $G_2(z)$ varies on the $G_2(z)$-plan according to this variation of z. Note carefully that, nowhere did we imply that G_1 and G are the same functions. Hence, it is clear that $G_2(z)$ is "not" obtained by simply substitutin Equation 12.144 in to $G_1(s)$. Hence, one should not attempt to obtain the discrete transfe function corresponding to a continuous (analog) transfer function by substituting Equatio 12.144 into the analog transfer function.

Now we will establish another important property of the z-transform. Consider a signa $x(t)$. Its Laplace transform is denoted by $X(s)$. If we sample $x(t)$ at period T, we have the dat sequence $\{x_k\}$. The corresponding z-transform is denoted by $X(z)$. Note that $X(s)$ and $X(z$ are convenient notations. We know that they represent two entirely different functions; $X(z$ is not obtained from $X(s)$ by substituting z for s. If we delay $x(t)$ by an integer multiple c T, say rT, the resulting signal is $x(t - rT)$. Its Laplace transform is given by $\exp(-rTs)X(s)$.

Now if we sample this delayed signal $x(t - rT)$ at sampling period T, we get the delaye sequence $\{x_{k-r}\}$. From Equation 12.143 we note that the corresponding z-transform is $z^{-r}X(z$ Thus we can state the following general result.

General Result: Consider a signal $x(t)$ whose Laplace transform is $X(s)$. Consider also signal $y(t)$ whose Laplace transform can be expressed as

$$Y(s) = f(\exp(Ts)) \cdot X(s)$$

in which $f(\ )$ is a polynomial function of $\exp(Ts)$. Then the z-transform of the sequenc $\{y_k\}$, obtained by sampling $y(t)$ at period T, is given by

$$Y(z) = f(z) \cdot X(z) \tag{12.145}$$

in which $X(z)$ is the z-transform of the sequence $\{x_k\}$ that is obtained by sampling $x(t)$ a period T.

This result and several other properties of the z-transform are given in Table 12.8.

12.11.5   The $s - z$ Mapping

Equation 12.144 represents a mapping between the complex z-plane and the comple: s-plane. This is one of the most important relationships in z-transform analysis. Let u further discuss the nature of this mapping.

TABLE 12.8

Some Useful Properties of the z-Transform

Item	z-Transform Result
$x(t)$	$X(z)$
$a_1 x_1(t) + a_2 x_2(t)$	$a_1 X_1(z) + a_2 X_2(z)$
$x(t - rT)$	$z^{-r} X(z)$
$Y(x) = f(\exp(Ts))\, X(s)$	$Y(z) = f(z)X(z)$
Final value theorem	$x_{ss} = \lim\limits_{z \to 1}(z - 1)X(z)$

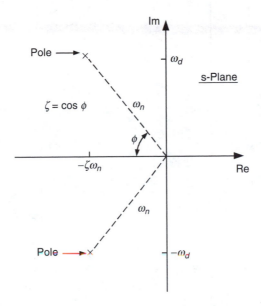

FIGURE 12.61

Representation of complex-conjugate poles on the s-plane.

Consider the case of complex poles

$$s = -\zeta\omega_n \pm j\omega_d \tag{12.146}$$

These poles correspond to, for instance, a simple oscillator with undamped natural frequency ω_n, damped natural frequency ω_d, and damping ratio ζ. An s-plane representation of these complex–conjugate poles is given in Figure 12.61. By substituting Equation 12.146 into Equation 12.144 we get the corresponding pole locations on the z-plane:

$$z = \exp\left(T(-\zeta\omega_n \pm j\omega_d)\right) \tag{12.147}$$

The magnitude of these two poles (on the z-plane) is

$$|z| = \exp(-T\zeta\omega_n) \tag{12.148}$$

and the phase angle is

$$\angle z = \pm T\omega_d \tag{12.149}$$

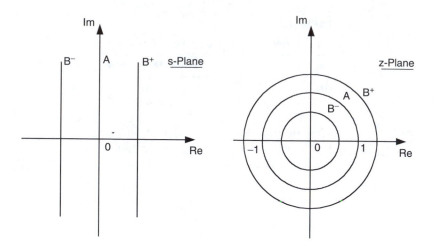

FIGURE12.62
Constant $\zeta\omega_n$ lines.

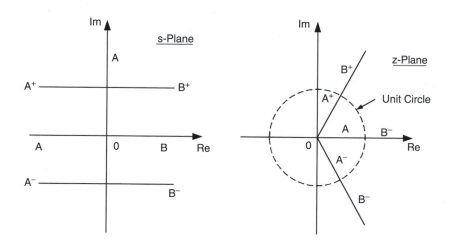

FIGURE 12.63
Constant ω_d lines.

Let us discuss several special mappings given by these (magnitude and phase) relationships.

1. **Constant $\zeta\omega_n$ lines:** From Equation 12.148 it follows that when $\zeta\omega_n$ is constant, $|z|$ is also constant. Hence, constant $\zeta\omega_n$ lines on the s-plane (i.e., lines parallel to the imaginary axis) map onto circles centered at origin on the z-plane, as shown in Figure 12.62.

   Note in particular that when $\zeta\omega_n = 0$ (line A), we have $|z| = 1$, a *unit circle*. The left hand side of the s-plane corresponds to "inside" of the unit circle and the right hand side of the s-plane corresponds to "outside" of the unit circle on the z-plane.

2. **Constant ω_d lines:** It should be clear from Equation 12.149 that constant ω_d lines on the s-plane (i.e., lines parallel to the real axis) map onto constant phase angle lines on the z-plane. This is shown in Figure 12.63. Each line on the s-plane can

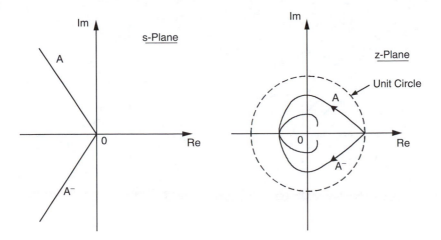

FIGURE 12.64
Constant ζ lines are spirals on the z-plane.

be divided into a part on the left hand plane and a part on the right hand plane. Correspondingly, on the z-plane the line is divided into a part within the unit circle and part outside the unit circle.

The fact that a mapping from the s-plane to the z-plane is a *many-to-one mapping* should be clear from Equation 12.149. Note in particular that all the lines given by

$$T\omega_d = 2\pi r + c \tag{12.150}$$

on the s-plane, for integer r and constant c, are mapped onto the same line on the z-plane. This is because any integer change of r corresponds to a phase change by 2π on the z-plane, which returns the line to its original location.

3. **Constant ζ lines:** It can be shown that constant ζ lines on the s-plane (i.e., straight lines through the origin) map onto spirals on the z-plane. This situation is sketched in Figure 12.64.

12.11.6 Stability of Discrete Models

Under the heading of constant $\zeta\omega_n$ lines, in the previous section, we noted that the left-half s-plane is mapped onto the inside of the unit circle on z-plane, as shown in Figure 12.65. It follows that stable poles in a discrete transfer function are those within the unit circle on the z-plane.

12.11.7 Discrete Final Value Theorem

In the continuous-time case, we have the familiar final value theorem (FVT)

$$x_{ss} = \lim_{s \to 0} sX(s) \tag{12.151}$$

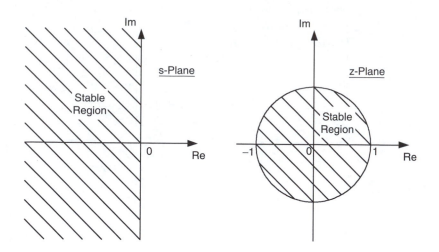

FIGURE 12.65
Stability region for discrete-time models.

in which x_{ss} is the steady state value of the signal $x(t)$;

$$x_{ss} = \lim_{t \to \infty} x(t) \tag{12.152}$$

and $X(s)$ is the Laplace transform of $x(t)$. It is assumed that a steady state value exists for the considered signal.

To establish the discrete-time counterpart of the final value theorem, let us return to the definition of the z-transform—Equation 12.135. Assume that the steady state value given by

$$x_{ss} = \lim_{k \to \infty} x_k \tag{12.153}$$

exists for the discrete-time signal (sequence) $\{x_k\}$. Now consider the product

$$(z-1)X(z) = (z-1)\sum x_k z^{-k}$$

If the sequence $\{x_k\}$ converges to a steady state value for a sufficiently large value of $k = N$, we can assume that

$$x_N \approx x_{N+r} \quad \text{for all } r > 0$$

Hence, we can write

$$(z-1)X(z) \approx (z-1)\sum_{-\infty}^{N-1} x_k z^{-k} + (z-1)x_N \sum_{N}^{\infty} z^{-k}$$

Since the first summation on the right hand side of this relation is finite, its product with $(z-1)$ vanishes as $z \to 1$. The second summation on the RHS can be written as

$$\sum_{N}^{\infty} z^{-k} = z^{-N} \sum_{0}^{\infty} z^{-k} = \frac{z^{-N}}{1-z^{-1}} = \frac{z^{-N+1}}{z-1}$$

its product with $(z - 1)$ approaches unity as $z \to 1$. Hence,

$$\lim_{z \to 1}(z-1)X(z) \approx x_N$$

Exact equality is obtained as $N \to \infty$; thus

$$x_{ss} = \lim_{z \to 1}(z-1)X(z) \tag{12.154}$$

Equation 12.154 is the discrete-time final value theorem. This is listed in Table 12.8, along with other important properties of z-transform.

12.11.8   Pulse Response Function

It is well known (see Chapter 2) that if $g(t)$ is the impulse response function (i.e., response to a unit impulse input $\delta(t)$) of a system, then its Laplace transform $G(s)$ is the transfer function of the system. We write

$$\mathcal{L}g(t) = G(s) \tag{12.155}$$

The response $y(t)$ to a general input $u(t)$ is given by the *convolution integral*; thus

$$y(t) = \int g(t-\tau)u(\tau)d\tau = \int g(\tau)u\,(t-\tau)d\tau \tag{12.156}$$

The limits of integration may be chosen depending on the nonzero regions of the two functions $g(t)$ and $u(t)$.

Now consider the unit pulse input (given at $k = 0$), which is defined by the sequence

$$\{\ldots, 0, 0, \ldots, 0, 1, 0, \ldots, 0, 0, \ldots\}$$

Note that the only non-zero sample value in this sequence is the "1" given at $k = 0$. The z-transform of this sequence is unity, as clear from Equation 12.135. This is given as the first entry in Table 12.7. From Equation 12.141 it follows that the z-transform of the pulse response sequence $\{g_k\}$ is the discrete transfer function $G(z)$ thus

$$\mathcal{Z}(g_k) = G(z) \tag{12.157}$$

The *discrete convolution* given below may be used to obtain the response to a general input, once the pulse response is known:

$$y_k = \sum_r g_{k-r}u_r = \sum_r g_r u_{k-r} \tag{12.158}$$

Equation 12.158 may be verified by direct substitution of Equation 12.135 into Equation 12.141.

12.11.8.1 Unit Pulse and Unit Impulse

The unit pulse is a pulse of unity height (magnitude), applied at $t = 0$. Pulse width is the sampling period T. Note that the area of this pulse is T and not unity. Since the sample value of the unit pulse is 1 at the first sample and zero thereafter, as shown earlier, the z-transform of the unit pulse is 1.

A distinction between the unit pulse and the unit impulse should be recognized. The unit impulse $\delta(t)$ has an infinite height at $t = 0$ and its area is unity. To sample such a signal, in practice, a very small sampling period T has to be used. Then, in the discrete approximation, the unit impulse is assumed to extend over the entire sample period of the first sample. The magnitude of the first sample has to be $1/T$ so that the area under the signal is unity, and this is inconsistent with the definition of the unit impulse. It follows that the z-transform of the unit impulse signal (whose Laplace transform is s) is given by $1/T$. These two important observations are listed as the first two entries of Table 12.7.

12.11.9 Digital Compensation

In sections 12.8 and 12.10 we noticed that analog compensator design could be interpreted as the development of a transfer function $G(s)$ that would modify the control signal so as to generate a desired system response. Signal modification by an analog transfer function is schematically shown in Figure 12.66(a). Let us consider the possibility of using a digital device to accomplish the same task. This is termed *digital compensation*.

In a software-based digital compensator, a digital computer reads a sequence of data $\{u_k\}$ obtained by sampling the true continuous-time signal $u(t)$, and produces a sequence of output data $\{y_k^*\}$ according to the *compensation algorithm* stored inside the computer. This algorithm may be expressed as an appropriate difference equation. Suppose that the discrete transfer function corresponding to this difference equation is $G(z)$. This process is schematically shown in Figure 12.66(b) or equivalently, in Figure 12.66(c). Our objective in digital compensation is to establish a $G(z)$ that corresponds to an ideal analog compensator $G(s)$ such that the error between $\{y_k\}$ and $\{y_k^*\}$ is negligible. Note that $\{y_k\}$ is obtained by sampling the ideal (analog) compensator output $y(t)$ according to $y_k = y(k \cdot T)$. It should

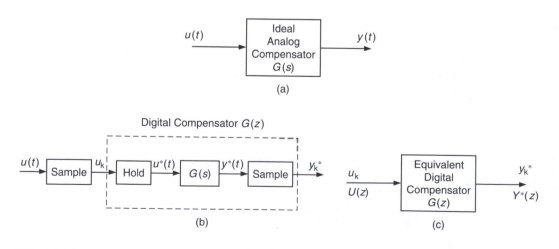

FIGURE 12.66

Digital compensation: (a) An ideal analog compensator; (b) Schematic representation of digital compensation; (c) Equivalent digital compensator.

be intuitively clear that if $T = 0$ (i.e., infinite sampling rate) the output sequence $\{y_k^*\}$ would be identical to $\{y_k\}$, in theory. We know, however, that due to practical limitations it is impossible to achieve such a realization. The error $(y_k^* - y_k)$ depends primarily on the sampling period T, and the holding method used for each input sample during this period.

The continuous and discrete equivalence is achieved by modeling the discrete compensation process as follows: First sample the input signal $u(t)$ at sampling period T and then hold (or extrapolate) the sampled data to generate an equivalent continuous signal $u^*(t)$. Next, pass this through the ideal compensator $G(s)$. The resulting output is $y^*(t)$, which when sampled at period T, produces the sampled data sequence $\{y_k^*\}$. Clearly in general, $y^*(t)$ is different from the ideal $y(t)$, due to the error between $u(t)$ and $u^*(t)$ as a result of the "sample and hold" operation (with non-zero T). Hence, the error between the digital compensator and the analog compensator is caused essentially by this discrepancy between $u(t)$ and $u^*(t)$.

12.11.9.1   Hold Operation

The purpose of a hold operation is to extrapolate a data sequence (sampled data points) to obtain a piecewise continuous signal. The type of hold is determined by the extrapolation scheme that is used. If the data value is held constant until the next data value has arrived (i.e., extrapolation by a zero-th order polynomial) we have a *zero-order hold*. Only the current data value is used in this extrapolation. This is typically the hold method used in analog to digital conversion (ADC), since a constant data value is needed during the analog to digital conversion process (see Chapter 4). If the current data value and the previous data value are used to extrapolate to the next data value using a straight line (i.e., extrapolation by a first order polynomial), it is a *first order hold*. If the current data sample and the two previous data samples are used to extrapolate to the next data sample by a quadratic curve (i.e., extrapolation by a second order polynomial), we have a *second-order hold*, and so on. These various types of data hold are illustrated in Figure 12.67.

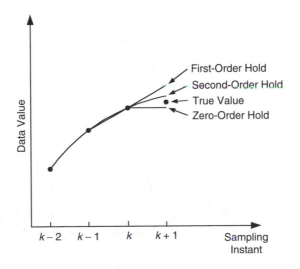

FIGURE 12.67
Several types of data hold.

12.11.9.2 *Discrete Compensator*

We now return to the schematic diagram in Figure 12.66(b). In the present analysis we will employ a zero-order hold, as this is what is commonly used in ADC. This may not be accurate enough, however, in some applications a unless sufficiently small sampling period T is employed.

The zero-order hold converts the sampled data sequence $\{u_k\}$ into a continuous series of pulses. Consider, in general, the pulse corresponding to the data sample u_k. The Laplace transform of a pulse of unit height and width T that is originating at time $t = \tau$, is given by subtracting a unit step signal at $t = \tau + T$ from a unit step signal at $t = \tau$; thus

$$\Delta_\tau(s) = \mathcal{L}u(t - \tau) - \mathcal{L}u(t - \tau - T)$$

$$= \frac{1}{s}\exp(-\tau s) - \frac{1}{s}\exp(-(\tau + T)s)$$

or

$$\Delta_\tau(s) = \frac{1}{s}\exp(-\tau s)[1 - \exp(-Ts)] \tag{12.159}$$

Hence, the Laplace transform of a pulse of magnitude u_k given at time $t = kT$ is

$$u_k\Delta_k(s) = \frac{u_k}{s}\exp(-kTs)[1 - \exp(-Ts)] \tag{12.160}$$

Note that this result is obtained by using $\tau = kT$ in Equation 12.159. It follows that, in Figure 12.66(b), the Laplace transform of the sampled and extrapolated (S/H) signal $u^*(t)$ is given by

$$U^*(s) = \sum_k \frac{u_k}{s}\exp(-kTs)[1 - \exp(-Ts)]$$

or

$$U^*(s) = \frac{1}{s}[1 - \exp(-Ts)]\sum_k u_k\exp(-kTs) \tag{13.161}$$

The corresponding output signal $y^*(t)$ has the Laplace transform

$$Y^*(s) = \left\{[1 - \exp(-Ts)]\sum_k u_k\exp(-kTs)\right\}\frac{G(s)}{s} \tag{12.162}$$

At this stage we use Equation 12.145 to obtain $Y^*(z)$; thus

$$Y^*(z) = \left\{[1 - z^{-1}]\sum_k u_k z^{-k}\right\}Z\left[\frac{G(s)}{s}\right]$$

in which $Z[\ ]$ denotes the z-transform of the sequence generated by a continuous signal whose Laplace transform is given in $[\ ]$. Now using the definition of $U(z)$—Equation 12.135, we get

$$Y^*(z) = (1 - z^{-1})U(z)Z\left[\frac{G(s)}{s}\right] \tag{12.163}$$

it follows that the discrete transfer function $G(z)$ that is equivalent to the analog transfer function $G(s)$ is given by (see Figure 12.66(c)),

$$G(z) = (1 - z^{-1}) Z \left[\frac{G(s)}{s} \right] \qquad (12.164)$$

For the sake of emphasis, let us repeat what we have mentioned earlier. $G(z)$ is "not" obtained by replacing s in $G(s)$ according to the mapping Equation 12.144. Furthermore, $G(z)$ is "not" obtained by replacing s by z in $G(s)$.

Equation 12.164 provides an important result, which may be directly employed to design digital compensators. The main steps of this procedure are given below.

Step 1: Design an analog compensator $G(s)$ to meet a given set of design specifications.

Step 2: Using Equation 12.164 obtain the discrete transfer function $G(z)$ of the corresponding digital compensator.

Step 3: Obtain the difference equation corresponding to $G(z)$.

The resulting difference equation can be programmed into a digital computer, or implemented in digital hardware, to form the digital compensator.

Example 12.31
Consider the lead compensator

$$G_c(s) = \left[\frac{0.051s + 1}{0.008s + 1} \right]$$

which was previously obtained for a motor speed control design problem. This is of the form

$$G_c(s) = \left[\frac{bs + 1}{as + 1} \right]$$

Derive a digital compensator corresponding to this analog compensator.

SOLUTION
To use Equation 12.164 we first determine the partial fractions:

$$\frac{(bs + 1)}{s(as + 1)} = \frac{A}{s} + \frac{B}{(as + 1)}$$

This gives the constants A and B. The time signals corresponding to these partial fractions are known from Laplace transform tables. Specifically,

$$\mathcal{L}^{-1} \frac{1}{s} = \text{unit step function}$$

$$\mathcal{L}^{-1} \left[\frac{1}{s + 1/a} \right] = \exp(-t/a)$$

The corresponding discrete sequences have the z-transforms (see Table 12.7):

$$\frac{z}{z-1} \quad \text{and} \quad \frac{z}{z-\exp(-T/a)}$$

Hence, we have from Equation 12.164,

$$G_c(z) = (1-z^{-1})\left[\frac{Az}{(z-1)} + \frac{B}{a}\frac{z}{(z-\exp(-T/a))}\right]$$

$$= A + \frac{B}{a}\frac{(z-1)}{(z-\exp(-T/a))}$$

This is written as

$$G_c(z) = K\frac{(z-\beta)}{(z-\alpha)} = K\frac{(1-\beta z^{-1})}{(1-\alpha z^{-1})}$$

In view of Equations 12.139 through 12.141 we can write the corresponding difference equation as

$$y_k - \alpha y_{k-1} = K(u_k - \beta u_{k-1})$$

This is the difference equation that should be programmed into the computer, or implemented in digital hardware, for digital compensation.

Example 12.32

Using Equation 12.164 develop a discrete-time integrator and a discrete-time differentiator. Note that since a zero-order hold is assumed, the results might not be very accurate.

SOLUTION
For the continuous integrator we have

$$G(s) = \frac{1}{s}$$

To use Equation 12.164 we have to determine

$$\mathcal{Z}\left[\frac{G(s)}{s}\right] = \mathcal{Z}\left[\frac{1}{s^2}\right]$$

Note that

$$\mathcal{L}^{-1}\frac{1}{s^2} = t$$

Hence, from Table 12.7,

$$Z\left[\frac{1}{s^2}\right] = \frac{Tz}{(z-1)^2}$$

Then from Equation 12.164 we get the discrete transfer function for the integrator as

$$G(z) = (1-z^{-1})\frac{Tz}{(z-1)^2} = \frac{T}{(z-1)} = \frac{Tz^{-1}}{(1-z^{-1})}$$

In view of equations 12.139 and 12.141 we have the corresponding difference equation

$$y_k - y_{k-1} = Tu_{k-1}$$

or

$$y_k = y_{k-1} + Tu_{k-1}$$

This is the familiar *forward rectangular rule* of integration. This difference equation can be programmed into a control computer for use as a simple integration algorithm.

Next, since the continuous differentiator is given by

$$G(s) = s$$

we have from Equation 12.164, the corresponding z-transform representation as

$$G(z) = (1-z^{-1})z\left(\frac{s}{s}\right)$$

$$= (1-z^{-1})z(1)$$

We know that 1 is the Laplace transform of the unit impulse $\delta(t)$. Hence, from Table 12.7, its z-transform is given by $1/T$. Accordingly, the discrete differentiator is given by

$$G(z) = (1-z^{-1})\frac{1}{T}$$

If the input to the differentiator is $u(t)$ and the differentiated output is $y(t)$, the difference equation of this discrete differentiator is given by

$$y_k = \frac{1}{T}(u_k - u_{k-1})$$

This is the familiar *backward difference rule* for differentiation. The results of the present example may be used to develop an algorithm or digital hardware for a *digital PID controller*.

12.11.9.3   *Direct Synthesis of Digital Compensators*

The method of digital compensation (and control) as described above is an "indirect" method in the sense that first an analog compensator or controller is developed and then it is approximated by a discrete transfer function using the z-transform method. Finally the corresponding difference equation is programmed into a digital device. An alternative method (a direct synthesis method) starts with a discrete transfer function $\tilde{G}(z)$ of a closed-loop system that responds with a desired response $\{y_k\}$ when a known input $\{u_k\}$ is applied, as indicated in Figure 12.68(a). For example, $\tilde{G}(z)$ might be the discrete transfer function of a simple oscillator with specified values for damping ratio, natural frequency, and dc gain. Now, since the discrete transfer function $G_p(z)$ of the process (plant) to be controlled is assumed to be known, it is a straightforward algebraic exercise to determine a compensator transfer function $G_c(z)$ that will produce $\tilde{G}(z)$ for a given feedback structure. For example, if the unity feedback control structure, as shown in Figure 12.68(b), is used we notice that

$$\tilde{G}(z) = \frac{G_c(z)G_p(z)}{[1+G_c(z)G_p(z)]} \tag{12.165}$$

Now by straightforward algebraic manipulation, we get an expression for the desired compensator/controller; thus

$$G_c(z) = \frac{\tilde{G}(z)}{G_p(z)[1-\tilde{G}(z)]} \tag{12.166}$$

Since both $\tilde{G}(z)$ and $G_p(z)$ are known, we can directly determine $G_c(z)$. Note, however, that the discrete transfer function $G_c(z)$ that is obtained in this manner may not be physically realizable, and even when physically realizable it may not be well behaved or may be too complex for physical implementation. Hence, physical implementation of this compensator may not be feasible or direct programming of the difference equation corresponding to the synthesized $G_c(z)$ into the control computer may not always produce the desired response. To illustrate this difficulty, note that when a discrete transfer function $G_c(z)$ is expanded as a series in powers of z^{-1}, we cannot have negative powers (i.e., we cannot have positive powers of z in the series expansion of $G_c(z)$). Otherwise, the corresponding difference equation will require future input values to determine the present output value, thereby violating the *causality* requirement for a "physically realizable" dynamic system. Hence, it is clear from Equation 12.166 that the specification $\tilde{G}(z)$ for a direct design will be limited by the nature of the plant transfer function $G_p(z)$. In particular, it can be shown that for $G_c(z)$ to be

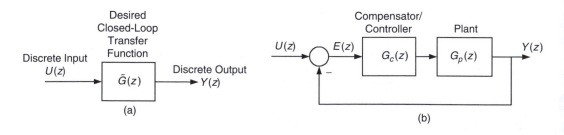

FIGURE 12.68
A direct synthesis method for digital compensators and controllers: (a) Desired system; (b) Unity feedback control structure.

physically realizable (i.e., for not to violate the causality requirement) it is required that the lowest power of z^{-1} in the series expansion of the plant transfer function $\tilde{G}(z)$ be greater than or equal to the lowest power of z^{-1} in the series expansion of $G_p(z)$.

It should be understood that, in the discrete transfer function block diagrams shown in Figure 12.68(a) and Figure 12.68(b), the signal paths carry discrete sequences such as $\{u_k\}$ and $\{y_k\}$, and not continuous signals. In particular, in Figure 12.68(b), the error (correction) signal is in fact $\{e_k\}$ and is given by

$$e_k = u_k - y_k \tag{12.167}$$

and hence

$$E(z) = U(z) - Y(z) \tag{12.168}$$

The closed loop Equation 12.165 comes from Equation 12.168. If some of the signal paths were analog, Equation 12.165 would not hold exactly. For example, suppose that $G_c(z)$ and $G_p(z)$ correspond to the continuous transfer functions $G_c(s)$ and $G_p(s)$, respectively. But, $\tilde{G}(z)$ would not be exactly equal to the discrete transfer function corresponding to the analog closed-loop transfer function

$$\frac{G_c(s)G_p(s)}{[1 + G_c(s)G_p(s)]}$$

12.11.10 Stability Analysis Using Bilinear Transformation

Nyquist stability criterion and associated concepts of gain margin and phase margin cannot be directly extended to discrete transfer functions. The reason for this is straightforward: the stability region on the s-plane is the left hand plane whereas the stability region on the z-plane is the unit circle area. To overcome this difficulty, another transformation that maps a unit circle on to the left hand side of a coordinate plane is used. One such transformation is the *bilinear transformation* given by

$$w = \frac{z-1}{z+1} \tag{12.169}$$

In this case the unit circle on the z-plane ($|z| \leq 1$) is mapped onto the left hand side of the w-plane (i.e., $\mathrm{Re}(w) \leq 0$). This is illustrated in Figure 12.69. Its relationship to the s-plane is given by the mapping Equation 12.144. Specifically,

$$w = \frac{\exp(Ts) - 1}{\exp(Ts) + 1} \tag{12.170}$$

On the imaginary axis (frequency axis) of the s-plane we have $s = j\omega$. On the w-plane, this corresponds to the line,

$$w = \frac{\exp(Tj\omega) - 1}{\exp(Tj\omega) + 1} = j\tan\left(\frac{\omega T}{2}\right) \tag{12.171}$$

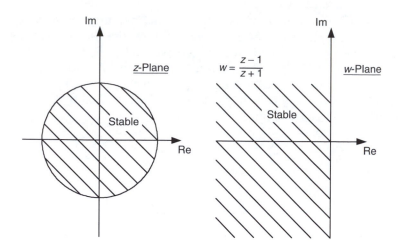

FIGURE 12.69
Bilinear transformation.

This is clearly the imaginary axis of the w-plane. It follows that the frequency axis of the s-plane corresponds to the imaginary axis of the w-plane. Consequently, we can define a new frequency variable ω^* for the w-plane. In view of Equation 12.171, ω^* is related to the true frequency ω through

$$\omega^* = \tan\left(\frac{\omega T}{2}\right) \tag{12.172}$$

This is a monotonic, but nonlinear, relationship. As a result of this monotonic relationship between the s-plane and the w-plane, we are able to use the same concepts of *relative stability* (*margin, phase margin*, etc.) for a discrete transfer function, when expressed in terms of the w variable (i.e., for a transfer function $G(w)$). Similarly, the well-known *Routh-Hurwitz stability criterion* can be applied to the denominator polynomial of $G(w)$, in order to determine the stability of a discrete-time system.

12.11.11 Computer Implementation

Digital control is particularly preferred when the control algorithms are complex. The algorithm for a *three-point controller* (PID controller), for example, is quite simple and straightforward. Even though a PID controller can be easily implemented by analog means, or even by a *hardware digital controller*, one may decide to employ a simple microprocessor as the controller in each proportional-integral-derivative (PID) loop of a control system. The microprocessor approach has the advantages of low cost, small size, and flexibility (see Chapter 10 and Chapter 11). In particular, integration with a higher-level supervisory controller in a distributed-control environment would be rather convenient when microprocessor-based loop controllers are employed. Also integration of PID loops with more complex control schemes such as linearizing control, adaptive control, and fuzzy-neural control would be simplified when the software-based digital control approach is used. Digital implementation of lead and lag compensators can be slightly more difficult than the implementation of three-point controllers.

12.12   Problems

12.1 a. What is an open-loop control system and what is a feedback control system? Give one example for each case.

b. A simple mass-spring-damper system (simple oscillator) is excited by an external force $f(t)$. Its displacement response y (see Figure P12.1(a)) is given by the differential equation

$$m\ddot{y} + b\dot{y} + ky = f(t)$$

A block diagram representation of this system is shown in Figure P12.1(b). Is this a feedback control system? Explain and justify your answer.

12.2 You are asked to design a control system to turn on lights in an art gallery at night, provided that there are people inside the gallery. Explain a suitable control system, identifying the open-loop and feedback functions, if any, and describing the control system components.

12.3 Into what classification of control system components (actuators, signal modification devices, controllers, and measuring devices) would you put the following?

a. Stepping motor

b. Proportional-plus-integration circuit

c. Power amplifier

d. ADC

e. DAC

f. Optical incremental encoder

g. Process computer

h. FFT analyzer

i. Digital signal processor (DSP)

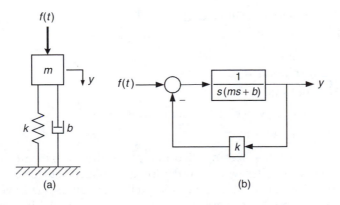

(a)   (b)

FIGURE P12.1

(a) A mechanical system representing a simple oscillator, (b) A block diagram representation of the simple oscillator.

12.4 a. Discuss possible sources of error that can make open-loop control or feedforward control meaningless in some applications.

b. How would you correct the situation?

12.5 Compare analog control and direct digital control for motion control in high-speed applications of industrial manipulators. Give some advantages and disadvantages of each control method for this application.

12.6 A soft-drink bottling plant uses an automated bottle-filling system. Describe the operation of such a system, indicating various components in the control system and their functions. Typical components would include a conveyor belt; a motor for the conveyor, with start/stop controls; a measuring cylinder, with an inlet valve, an exit valve, and level sensors; valve actuators; and an alignment sensor for the bottle and the measuring cylinder.

12.7 Consider the natural gas home heating system shown Figure 12.6. Describe the functions of various components in the system and classify them into the function groups: controller, actuator, sensor, and signal modification device. Explain the operation of the overall system and suggest possible improvements to obtain more stable and accurate temperature control.

12.8 In each of the following examples, indicate at least one (unknown) input that should be measured and used for feedforward control to improve the accuracy of the control system.

a. A servo system for positioning a mechanical load. The servo motor is a field-controlled dc motor, with position feedback using a potentiometer and velocity feedback using a tachometer.

b. An electric heating system for a pipeline carrying a liquid. The exit temperature of the liquid is measured using a thermocouple and is used to adjust the power of the heater.

c. A room heating system. Room temperature is measured and compared with the set point. If it is low, a valve of a steam radiator is opened; if it is high, the valve is shut.

d. An assembly robot that grips a delicate part to pick it up without damaging the part.

e. A welding robot that tracks the seam of a part to be welded.

12.9 A typical input variable is identified for each of the following examples of dynamic systems. Give at least one output variable for each system.

a. Human body: neuroelectric pulses

b. Company: information

c. Power plant: fuel rate

d. Automobile: steering wheel movement

e. Robot: voltage to joint motor.

12.10 Hierarchical control has been applied in many industries, including steel mills, oil refineries, chemical plants, glass works, and automated manufacturing. Most applications have been limited to two or three levels of hierarchy, however. The lower levels usually consist of tight servo loops, with bandwidths on the order of 1 kHz. The upper levels typically control production planning and scheduling events measured in units of days or weeks.

A five-level hierarchy for a flexible manufacturing facility is as follows: The lowest level (level 1) handles servo control of robotic manipulator joints and machine tool degrees of freedom. The second level performs activities such as coordinate transformation in machine tools, which are required in generating control commands for various servo loops. The third level converts task commands into motion trajectories (of manipulator end effector, machine tool bit, etc.) expressed in world coordinates. The fourth level converts complex and general task commands into simple task commands. The top level (level 5) performs supervisory control tasks for various machine tools and material-handling devices, including coordination, scheduling, and definition of basic moves. Suppose that this facility is used as a flexible manufacturing workcell for turbine blade production. Estimate the event duration at the highest level and the control bandwidth (in hertz) at the lowest level for this type of application.

12.11 According to some observers in the process control industry, early brands of analog control hardware had a product life of about 20 years. New hardware controllers can become obsolete in a couple of years, even before their development costs are recovered. As a control instrumentation engineer responsible for developing an off-the-shelf controller for a mechatronic application, what features would you incorporate into the controller in order to correct this problem to a great extent?

12.12 The programmable logic controller (PLC) is a sequential control device, which can sequentially and repeatedly activate a series of output devices (e.g., motors, valves, alarms, signal lights) on the basis of the states of a series of input devices (e.g., switches, two-state sensors). Show how a programmable controller and a vision system consisting of a solid-state camera and a simple image processor (say, with an edge-detection algorithm) could be used for sorting fruits on the basis of quality and size for packaging and pricing.

12.13 It is well known that the block diagram in Figure P12.13(a) represents a dc motor, for armature control, with the usual notation. Suppose that the load driven by the motor is a pure inertia element (e.g., a wheel or a robot arm) of moment of inertia J_L, which is directly and rigidly attached to the motor rotor.

   a. Obtain an expression for the transfer function $\frac{\omega_m}{v_a} = G_m(s)$ for the motor with the inertial load, in terms of the parameters given in Figure P12.13(a), and J_L.

   b. Now neglect the leakage inductance L_a. Then, show that the transfer function in Part (a) can be expressed as $G_m(s) = \frac{k}{(\tau s + 1)}$. Give expressions for τ and k in terms of the given system parameters.

   c. Suppose that the motor (with the inertial load) is to be controlled using position plus velocity feedback. The block diagram of the corresponding control system is given in Figure P12.13(b), where $G_m(s) = \frac{k}{(\tau s + 1)}$. Determine the transfer function of the (closed-loop) control system $G_{CL}(s) = \frac{\theta_m}{\theta_d}$ in terms of the given system parameters (k, k_p, τ, τ_v). Note that θ_m is the angle of rotation of the motor with inertial load, and θ_d is the desired angle of rotation.

12.14 Consider a thermal process whose output temperature T is related to the heat input as W follows:

$$T = 0.5W$$

The units of T are °C and the units of W are watts. The heat generated by the proportional controller of the process is given by

$$W = 10(T_o - T) + 1000$$

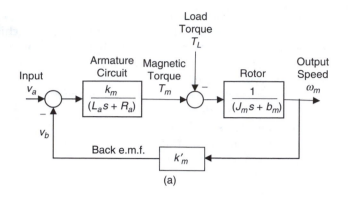

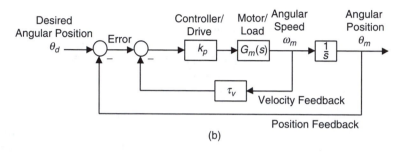

FIGURE P12.13
(a) Block diagram of a dc motor for armature control, (b) Motor control with feedback of position and velocity.

in which T_o denotes the temperature set point. Determine the offset for the following three set points:

Case **1:** $T_o = 500°$

Case **2:** $T_o = 200°$

Case **3:** $T_o = 800°$

12.15 A linearized thermal plant is represented by $T = gc$
where
   T = temperature (response) of the plant
   c = heat input to the plant
   g = transfer function (a constant gain)

A proportional controller is implemented on the plant, to achieve temperature regulation, and it is represented by

$$c = ke + \Delta c$$

where
   $e = T_o - T$ = error signal into the controller
   T_o = temperature set point (desired plant temperature)
   k = controller gain
   Δc = controller output when the error is zero

Under the given conditions suppose that there is no offset.

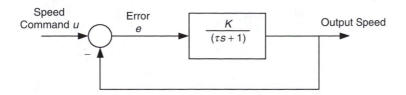

FIGURE P12.16
A speed control system.

    a. Express Δc in terms of T_o and g.
    b. If the plant transfer function changes to g' what is the resulting offset $T_o - T$?
    c. If the set point is changed to T'_o for the original plant (g) what is the resulting offset $T'_o - T$?

12.16 A block diagram of a speed-control system is shown in Figure P12.16. Here the controlled process (say, an inertia with a damper) has a transfer function given by

$$\frac{1}{\tau s + 1}$$

and the controller is a proportional controller with gain K. Hence, the combined transfer function of the process and the controller is $\frac{K}{\tau s+1}$. If the process is given a unit step input, find the final steady-state value of the output. What is the resulting offset? Show that this decreases when K is increased.

12.17 Fraction (or percentage) of the full scale of controller input that corresponds to the full operating range of the final control element (controller output) is *termed proportional band*. Hence,

$$PB = \frac{\Delta e}{\Delta E} \times 100\%$$

in which
    PB = proportional band (percent)
    ΔE = full scale of controller input
    Δe = range of controller input corresponding to the full range of final control element.

Obtain an equation relating PB to the proportional gain K_p of the controller. Use the additional parameter, Δc = full range of the final control element. Note that PB is dimensionless but K_p has physical units because Δc and Δe do not have the same units in general.
    A feedback control system for a water heater is shown in Figure P12.17. A proportional controller with the control law

$$W = 20e + 100$$

is used, where W denotes the heat transfer rate into the tank in watts and e denotes the temperature error in °C. The tank characteristic for a given fixed rate of water flow is known to be

$$W = 20T$$

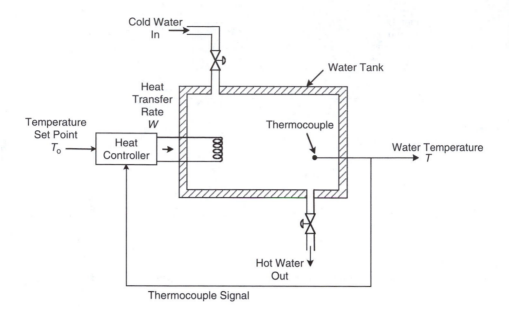

FIGURE P12.17
Feedback control of a water heater.

where T is the temperature (°C) of the hot water leaving the tank. The full scale (span) of the controller is 500°C.

a. What is the proportional gain and what is the proportional band of the controller?
b. What is the set-point value for which there is no offset error?
c. If the set point is 40°C determine the offset.
d. If the set point is 60°C determine the offset.
e. If the proportional gain is increased to 80 watts/°C, what is the proportional band and what is the offset in part (c)?
f. Suppose the water flow rate is increased (load increased) so that the process law changes to $W = 25T$. Determine the set point corresponding to zero offset noting that this is different from the answer for part (b). Also, determine the offset in this case when the set point is 50°C.

12.18 Consider six control systems whose loop transfer functions are given by:

$$(a) \ \frac{1}{(s^2 + 2s + 17)(s + 5)} \qquad\qquad (d) \ \frac{10(s + 2)}{(s^2 + 2s + 101)}$$

$$(b) \ \frac{10(s + 2)}{(s^2 + 2s + 17)(s + 5)} \qquad\qquad (e) \ \frac{1}{s(s + 2)}$$

$$(c) \ \frac{10}{(s^2 + 2s + 101)} \qquad\qquad (f) \ \frac{s}{(s^2 + 2s + 101)}$$

Compute the additional gain (multiplication) k needed in each case to meet a steady-state error specification of 5% for a step input.

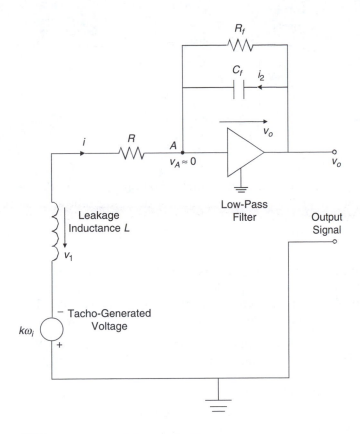

FIGURE P12.19
An approximate model for a tachometer-filter combination.

12.19 A tachometer is a device that is commonly used to measure speed, both rotatory and translatory. It consists of a coil which moves in a magnetic field. When the tachometer is connected to the object whose speed is to be sensed, the coil moves with the object and a voltage is induced in the coil. In the ideal case, the generated voltage is proportional to the speed. Accordingly, the output voltage of the tachometer serves as a measure of the speed of the object. High frequency noise that may be present in the tachometer signal can be removed using a low-pass filter.

Figure P12.19 shows a circuit, which may be used to model the tachometer-filter combination. The angular speed of the object is ω_i and the tachometer gain is k. The leakage inductance in the tachometer is demoted by L and the coil resistance (possibly combined with the input resistance of the filter) is denoted by R. The low-pass filter has an operational amplifier with a feedback capacitance C_f and a feedback resistor R_f. Since the operational amplifier has a very high gain (typically 10^5–10^9) and the output signal v_o is not large, the voltage at the input node A of the op-amp is approximately zero. It follows that v_o is also the voltage across the capacitor.

   a. Comment on why the speed of response and the settling time are important in this application. Give two ways of specifying each of these two performance parameters.

   b. Using voltage v_o across the capacitor C_f and the current i through the inductor L as the state variables and v_o itself as the output variable, develop a state space model for the circuit. Obtain the matrices A, B, C, and D for the model.

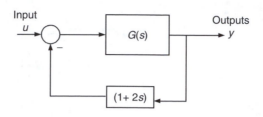

FIGURE P12.20
A control system with tacho feedback.

    c. Obtain the input-output differential equation of the model and express the undamped natural frequency ω_n and the damping ratio ζ in terms of L, R, R_f, and C_f. What is the output of the circuit at steady state? Show that the filter gain k_f is given by R_f/R and discuss ways of improving the overall amplification of the system.

    d. Suppose that the percentage overshoot of the system is maintained at or below 5% and the peak time at or below 1 ms. Also it is known that $L = 5.0$ mH and $C_f = 10.0$ μF. Determine numerical values for R and R_f that will satisfy the given performance specifications.

12.20 A control system with tacho feedback is represented by the block diagram in Figure P12.20. The following facts are known about the control system:

1. It is a third order system but behaves almost like a second order system.
2. Its 2% settling time is 1 second, for a step input.
3. Its peak time is $\pi/3$ seconds, for a step input.
4. Its steady-state error, to a step input, is zero.

    a. Completely determine the third order forward transfer function $G(s)$.

    b. Estimate the damping ratio of the closed-loop system.

12.21 A satellite-tracking system (typically a position control system) having the plant transfer function

$$G_p(s) = \frac{1}{s(2s+1)}$$

is controlled by a control amplifier with *compensator*, having the combined transfer function

$$G_c(s) = \frac{K(s+1)}{(\tau s+1)}$$

and unity feedback ($H = 1$). A block diagram of the control system is shown in Figure P12.21.
    a. Write the closed-loop characteristic equation (in polynomial form).
    b. Using the Routh-Hurwitz criterion for stability, determine the conditions that should be satisfied by the compensation parameter τ and the controller gain K in order to maintain stability in the closed-loop system.
    c. Sketch this stability region using K as the horizontal axis and τ as the vertical axis.
    d. When $K = 5$ and $\tau = 3$ find the poles (i.e., eigenvalues or roots) of the closed-loop system. What is the natural frequency of the system for these parameters values?

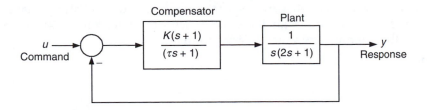

FIGURE P12.21
Block diagram of a satellite tracking system.

12.22 Consider the six transfer functions:

a. $\dfrac{1}{(s^2 + 2s + 17)(s + 5)}$

b. $\dfrac{10(s + 2)}{(s^2 + 2s + 17)(s + 5)}$

c. $\dfrac{10}{(s^2 + 2s + 2)}$

d. $\dfrac{10(s + 2)}{(s^2 + 2s + 2)}$

e. $\dfrac{s}{(s^2 + 2s + 2)}$

f. $\dfrac{1}{s(s + 2)}$

Suppose these transfer functions are the plant transfer functions of five control systems with proportional feedback control. If the loop gain is variable, sketch the root loci of the five systems and discuss their stability.

12.23 The loop transfer function of a feedback control system is given by

$$GH = \frac{K}{s(s + 1)(s + 2)}$$

a. Sketch the root locus of the closed-loop system by first determining the:
   i. Location and the angles of the asymptotes
   ii. Break points
   iii. Points at which the root locus intersects with the imaginary axis, and the corresponding gain value
b. Fully justifying your answer, state whether the system is stable for $K = 10$.
c. Suppose that a zero at −3 is introduced to the control loop so that

$$GH = \frac{K(s + 3)}{s(s + 1)(s + 2)}$$

Sketch the root locus of the new system.

12.24 Explain why the variable parameter in a root locus does not always have to be the loop gain. Sketch the root loci for the systems with the following closed-loop transfer functions, as the unknown parameter varies from 0 to ∞:

a. $\dfrac{2(s+z)}{(s^2+s+2)}$

b. $\dfrac{2(s+4)}{(\tau s+10)}$

c. $\dfrac{1}{s^2+(2a+3)s+3a+1}$

12.25 Sketch the Bode magnitudes plots (asymptotes only, if the exact curve needs numerical computation) of the following common system elements:

a. Derivative Controller (τs)
b. Integral Controller ($1/(\tau s)$)
c. First Order Simple Lag Network ($1/(\tau s+1)$)
d. Proportional Plus Derivative Controller ($\tau s+1$).

Also, sketch the polar plots (Nyquist plots) for these elements.

12.26 The open-loop transfer function of a control system is given by

$$G(s)=\frac{s+1}{(s^2+s+4)}$$

a. With this transfer function, if the loop is closed through a unity feedback, determine the phase margin. You should use direct computation rather than a graphical approach.
b. If an input of $u=3\cos 2t$ is applied to the open-loop system, determine the output y under steady conditions.

12.27 Consider the proportional plus derivative (PPD) servo system shown in Figure 12.45. The actuator transfer function is given by

$$G_p(s)=\frac{1}{s(0.5s+1)}$$

By hand calculation complete the following table:

$\zeta\omega_n T_p$	2.5	2.5	2.5	2.5	2.2	2.1	2.0	2.0	2.0
$\omega_d T_p$	0.9π	0.75π	0.6π	0.55π	0.55π	0.25π	0.55π	0.2π	0.1π
T_p									
PO									

Design a PPD controller that approximately meets $T_p=0.09$ and PO = 10%.

12.28 Compare position feedback servo, tacho-feedback servo, and PPD servo with particular reference to design flexibility, ease of design, and cost. Consider an actuator with transfer function

$$G_p=\frac{1}{s(0.5s+1)}$$

Design a position feedback controller and a tacho-feedback controller that will meet the design specifications $T_p = 0.09$ and PO = 10%.

12.29 Consider the six control systems in Problem 12.18. Plot Bode curves for the systems with modified gain values to meet the steady-state error specification of 5% for a step input. Determine the gain margins and phase margins.

If you were asked to pick one of these systems to design a single lead compensator or a lag compensator so that the compensated system would have a phase margin of exactly 60°, which system would you pick? Discuss your answer indicating why you did not pick the remaining five systems.

12.30 A control system was found to have poor accuracy at low frequencies, poor speed of response at high frequencies, and a low stability margin in the operating bandwidth. Discuss what type of compensation you would recommend in order to improve the performance of this control system.

12.31 Consider the problem of tracking an aircraft using a radar device that has a velocity error constant of 10 s⁻¹. If the airplane flies at a speed of 2000 km/hr at an altitude of 10 km, estimate the angular position error of the radar antenna that tracks the aircraft.

12.32 The feedback transmitter of the temperature control system of a heating process was disconnected and the set point was changed by 10°C manually with the controller dynamics inactive. The steady-state response was found to be 80°C. The recorded response provided the following values:

$$\text{Lag time } L = 0.5 \text{ min}$$

$$\text{Cycle time } T = 2.0 \text{ min}$$

Determine suitable settings for the PID controller used in this temperature control system.

12.33 Explain a situation when the reaction curve method of controller tuning is preferred and a situation when the ultimate response method of controller tuning is preferred. Consider a process with the transfer function

$$G_p(s) = \frac{R}{s} e^{-Ls}$$

The control system has unity feedback. Determine suitable parameter settings for a three-mode (PID) controller.

12.34 List three advantages and three advantages of open-loop control. Note that the disadvantages will correspond to advantages of feedback control. A process plant is represented by the block diagram shown in Figure P12.34(a).

a. What is the transfer function (y/u) of the plant?

b. Determine the undamped natural frequency and the damping ratio of the plant.

A constant-gain feedback controller with gains K_1 and K_2 is added to the plant as shown in Figure P12.34(b).

c. Determine the new transfer function of the overall control system.

d. Determine the values of K_1 and K_2 that will keep the undamped natural frequency at the plant value (of part(b)) but will make the system critically damped.

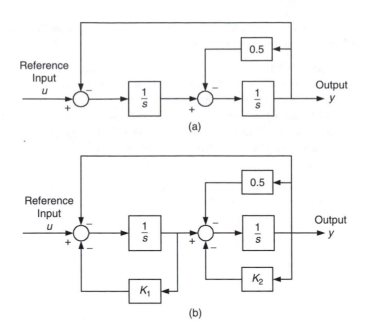

FIGURE P12.34
(a) Simulation block diagram of a plant, (b) Plant with a feedback controller.

 e. Determine the values of K_1 and K_2 such that the overall control system has a natural frequency of $\sqrt{2}$ and a 2% settling time of 4 seconds.

12.35 A control system has an unstable plant given by the transfer function

$$G_p(s) = \frac{1}{(s^2 - s + 1)}$$

By sketching root locus, discuss whether the plant can be stabilized using
 a. proportional (P) feedback control
 b. proportional plus derivative (PPD) control
 c. proportional plus integral (PI) control
Are these observations intuitively clear?

12.36 A control loop used for controlling the roll motion of an aircraft is shown by the block diagram in Figure P12.36. The following specifications must be met by the control system:
 a. A peak time $\leq \pi/2$ sec
 b. 2% settling time ≤ 2 sec
 c. Percentage overshoot $\leq 3\%$
Show that the settling time specification is redundant.
  Using root locus design, determine a set of control parameters (K, z, p) that will satisfy these specifications. What is the acceleration error constant of the system?

12.37 Both frequency domain Bode method and the root locus method can be used for designing lead and lag compensators for control systems. Which method would you prefer in each of the following cases?

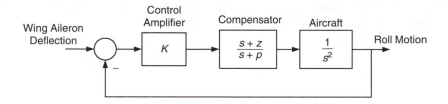

FIGURE P12.36
Roll motion control loop of an aircraft.

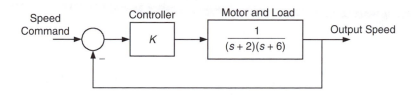

FIGURE P12.38
Block diagram of an uncompensated speed control system.

a. Compensator design for a system with large time delays.

b. Lead Compensator design for a stable system with negligible time delay.

c. Lag Compensator design for an unstable system with negligible time delay.

    i. Even though a lag compensator has a destabilizing effect in general, it can be used to stabilize some unstable systems. Using sketches of Bode diagrams for a typical situation, explain why this is true. What is a major shortcoming of this method of compensation?

    ii. Specifications on peak time, settling time, and percentage overshoot are commonly used in controller design that employs the root locus method. Using sketches of design regions on the s-plane, give an example for a situation where all three specifications are necessary and an example where one specification is redundant.

12.38 A speed control system is shown in Figure P12.38. Show that the specifications:

a. Percentage overshoot ≤ 5%

b. Steady state error ≤ 2% to a step input

cannot be simultaneously satisfied by this system. You may use a sketch of the root locus of the system for this purpose.

  Design a suitable lag compensator to meet both specifications given by the equality conditions (i.e., PO = 5% and s.s. error = 2%). Estimate the margin of error in the (actual) PO of the compensated control system. Sketch the root locus of the compensated system.

12.39 A microprocessor-based loop tuner uses the response signal for a test input to compute in "almost" real-time, controller settings for P, PI, and PID control. Typical tuning specifications include one of the following:

a. Response with minimum overshoot for a step input

b. Response with 10% overshoot for a step input

c. Response with quarter decay ratio

Using schematic diagrams, illustrate how such a tuner is physically connected to a process control loop. Describe the main steps of control loop tuning.

12.40 What is a self-regulating process? Strictly speaking, Ziegler-Nichols controller settings are applicable to non-self-regulating plants that may be approximated by

$$G_p(s) = \frac{R}{s} e^{-Ls}$$

In the self-regulating case the plant transfer function is approximated by

$$G_p(s) = \frac{K}{(Ts+1)} e^{-Ls}$$

In this latter case an index of self regulation can be defined as

$$m = \frac{LR}{K}$$

When $m = 0$ (i.e., K is large and T is large, but the reaction rate R is finite), we have a non-self-regulating case. Explain how the Ziegler-Nichols controller settings in the reaction curve method should be modified to include the parameter m.

12.41 Sketch an operational amplifier circuit for a PI controller and one for a lag-lead compensator. In each case derive the circuit transfer function.

12.42 Describe the operation of the cruise control loop of an automobile, indicating the *input*, the output, and a *disturbance input* for the control loop. Discuss how the effect of a disturbance input can be reduced using feedforward control.
   Synthesis of feedforward compensators is an important problem in control system design. Consider the control system shown in Figure P12.42. Derive the transfer function relating the disturbance input u_d and the plant output y. If you have the complete freedom to select any transfer function for the feedforward compensator G_f, what would be your choice?

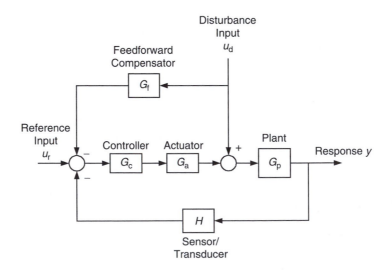

FIGURE P12.42
A feedback control system with feedforward compensation.

If the process bandwidth is known to be very low and if G_f is a pure gain, suggest a suitable value for this gain.

Suppose a unit step input is applied to the system in Figure P12.42. For what value of step disturbance u_d will the output y be zero at steady state?

12.43 The transfer function of a field controlled dc motor is given by

$$G(s) = \frac{K_m}{s(Js+b)(Ls+R)}$$

for open-loop control, with the usual notation. The following parameter values are given:

$J = 10 \text{ kg} \cdot \text{m}^2$

$L = 1$ Henry

$b = 0.1 \text{ N} \cdot \text{m/rad/s}$

$R = 10$ Ohms

Calculate the electrical time constant and the mechanical time constant for the motor. Plot the open-loop poles on the s-plane. Obtain an approximate second-order transfer function for the motor.

A proportional (position) feedback control system for the motor is shown in Figure P12.43. What is the closed-loop transfer function? Express the undamped natural frequency and damping ratio of the closed-loop system in terms of K, K_m, J, R and b.

12.44 A third-order closed-loop system with unity feedback is known to have the following characteristics:

a. It behaves like a second-order system
b. Its 2% settling time is 4 sec
c. Its peak time is π sec
d. Its steady-state value for a unit step input is $= 1$

Determine the corresponding third-order open loop system (i.e., when the feedback is disconnected). Explain why the steady-state error for a step input is zero for this closed-loop system.

12.45 An interesting issue of force feedback control in robotic manipulators is discussed in the literature by Eppinger and Seering of Massachusetts Institute of Technology. Consider the two models representing a robotic manipulator that interacts with a workpiece, as shown in Figure P12.45(a) and Figure P12.45(b). In (a) robot is modeled as a rigid body (without flexibility) connected to ground through a viscous damper, and the workpiece is modeled as a mass-spring-damper system. In this case only the rigid body mode of the

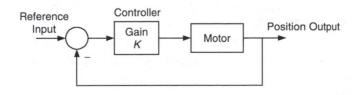

FIGURE P12.43
Block diagram of a position servo.

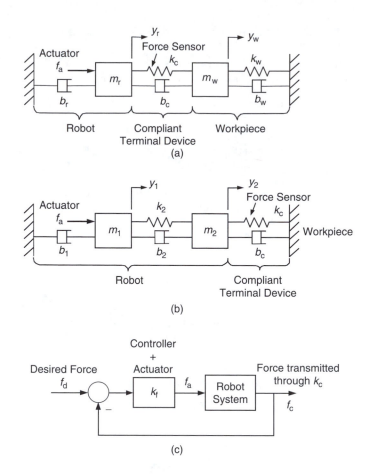

FIGURE P12.45

(a) A model for robot-workpiece interaction, (b) An alternative model, (c) A simple force feedback scheme.

robot is modeled. The robot interacts with the workpiece through a compliant device (e.g., remote center compliance—RCC device or a robot hand), which has an effective stiffness and damping. In (b) the robot model has flexibility, and the workpiece is modeled as a clamped rigid body, which cannot move. Note that in this second case a flexible (vibrating) mode as well as a rigid body mode of the robot are modeled. The interaction between the robot and the workpiece is represented the same way as in case (a). In both cases, the employed force feedback control strategy is to sense the force f_c transmitted through the compliant terminal device (the force in spring k_c), compare it with a desired force f_d, and use the error to generate the actuator force f_a. The controller (with driving actuator) is represented by a simple gain k_f. This gain is adjusted in designing or tuning the feedback control system shown in Figure P12.45(c).

a. Derive the dynamic equations for the two systems and obtain the closed-loop characteristic equations.

b. Give complete block diagrams for the two cases showing all the transfer functions.

c. Using the controller gain k_f as the variable parameter, sketch the root loci for the two cases.

d. Discuss stability of the two feedback control systems. In particular, discuss how stability may be affected by the location of the force sensor. (Note that the two models are analytically identical except for the force sensor location.)

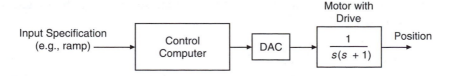

FIGURE P12.46
Open-loop digital control of a dc motor (unstable).

12.46 A computer-based open-loop control system for a dc motor is shown by the block diagram in Figure P12.46. In response to an input command, the control computer generates digital values corresponding to a desired continuous input. These digital values are converted into the analog form and supplied to the motor drive circuit in real time. The motor and its drive circuitry can be modeled by the transfer function

$$G(s) = \frac{1}{s(s+1)}$$

Suppose that the conversion time T of the DAC is $T = 1$ s.

a. Using the z-transform approach, obtain a difference equation that will approximate the motor response.

b. Assuming that the continuous input is a unit ramp, compute the first 4 discrete output values (at sampling period $T = 1$ s) using the difference equations, with zero initial values.

c. For a unit ramp input, determine an expression for $Y(z)$—the z-transform of the discrete output values. By long division, obtain the first four discrete output values and compare them with the results in part (b).

d. Suppose the input is a unit step. Using the difference equation, with zero initial conditions, compute the first few output samples. Show that the response behaves in an unstable manner. What is the main source of this instability (nature of the system or recursive algorithm)?

12.47 Consider a continuous-time system given by the transfer function

$$G(s) = \frac{1}{(s^2 + 1.2s + 1)}$$

a. Determine the undamped natural frequency and damping ratio of the system.

b. Write the input-output differential equation for this system.

c. Write an expression for the system response $y(t)$ for a unit step input. Plot this response.

d. From the plotted time response determine the % overshoot (PO) and the peak time (T_p) for the system.

e. Determine the discrete transfer function $G(z)$, which relates the discrete input and discrete output data.

f. Write the difference equation corresponding to this discrete transfer function.

    g. Using this difference equation, compute the time response of the system for a unit step input,

       i. With a sampling period $T = 1$ s.

      ii. With a sampling period $T = 0.5$ s.

Plot these two curves on the same paper (same scale) along with the exact response curve obtained in part (c). Compare the three curves. Discuss any discrepancies. (Compare, in particular, PO and T_p for the three curves).

12.48 What are advantages of digital control over analog control? A lead compensator was designed for a control system using the classical analog approach. Its transfer function was found to be

$$G_c(s) = \frac{(2s+1)}{(s+1)}$$

Assuming a sampling period of $T = 1$, obtain a difference equation for digital implementation of this lead compensator.

12.49 Consider a system whose transfer function is $G(s)$. The corresponding discrete transfer function $G(z)$ is obtained using

$$G(z) = (1-z^{-1})\, Z\left[\frac{G(s)}{s}\right]$$

in which the operator Z denotes z-transformation. Note that the response of the system computed using the difference equation given by $G(z)$ is identical to the response of the original system $G(s)$, if the analog input is first sampled, then each sample is held using a zero-order hold circuit, then the resulting analog signal is applied to the original analog system $G(s)$ and, finally, the resulting response is sampled at the same sampling frequency as the input. Using this fact, explain why the discrete transfer function corresponding to the product $G_1(s)G_2(s)$ is not identical to the product $G_1(z)G_2(z)$ in which

    $G_1(z)$ = discrete transfer function corresponding to $G_1(s)$

    $G_2(z)$ = discrete transfer function corresponding to $G_2(s)$

If $G_1(s)$ and $G_2(s)$ are known, would you prefer the former process (i.e., convert the product $G_1(s)G_2(s)$ or the latter process (i.e., convert $G_1(s)$ and $G_2(s)$ separately and take the product) in order to obtain a difference equation for the product transfer function $G_1(s)G_2(s)$, for computer implementation?

12.50 Schematic representation of signal processing associated with a vibration-test system is shown in Figure P12.50. An acceleration signal $x(t)$ of the test object is measured using an accelerometer/charge amplifier combination and sampled into the signal-processing computer at sampling period T. This discrete acceleration sequence $\{x_k\}$ is then integrated using parabolic integration, to form a velocity sequence $\{y_k\}$. Show that the parabolic integration algorithm is given by

$$y_{k+1} = y_k + \frac{T}{12}[5x_{k+1} + 8x_k - x_{k-1}]$$

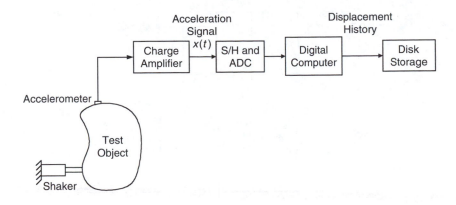

FIGURE P12.50
A vibration test system.

Note that in the parabolic integration algorithm, the discrete data are interpolated (every three points x_{k-1}, x_k, x_{k+1}) using a quadratic curve (a parabola). What is the discrete transfer function $G(z)$ corresponding to this difference equation? Using this, derive an algorithm for double integration of $x(t)$, to obtain the displacement response. Show that the associated discrete transfer function is marginally stable. What is the practical implication of this observation? Suggest a way to improve this situation.

12.51 Consider the analog feedback control system shown in Figure P12.51(a). If we sample the error (correction) signal as shown in Figure P12.51(b), this is equivalent to sampling both input signal and feedback signal. If the output signal is also sampled, then the signal paths can be completely represented by discrete sequences (only the holding operation is needed to convert the discrete sequences to sample and hold outputs). Hence, the discrete transfer function of the system shown in Figure P12.51(c) is identical to that of the system shown in Figure P12.51(b). Note that $G(z)$ and $H(z)$ are, respectively, the discrete transfer functions corresponding to $G(s)$ and $H(s)$. What is the closed loop discrete transfer function $\tilde{G}(z)$ corresponding to Figure P12.51(b) or Figure P12.51(c)? Now suppose that the input and the output are sampled (and not the feedback signal). In this case we have the system shown in Figure P12.51(d). What is the closed loop discrete transfer function in this case? Verify that this is not identical to, but could be approximated by, the previous case.

Which discrete model (Figure P12.51(c) or Figure P12.51(d)) would be more accurate in the following two situations:

a. The forward path of the control loop has a digital controller, and a digital transducer (say, optical encoder) is used to measure the response signal for feedback.

b. The forward path of the control loop has a digital controller (computer) but the output sensor used to obtain the feedback signal is analog.

Is any one of the two discrete models shown in Figure P12.51 exactly equivalent to the case where the forward path and the feedback path are completely analog except that a digital transducer (with a zero-order hold) is used to measure the output signal for feedback?

a. Analog system

b. Sampling the error and the output

c. Sampling the input and the output.

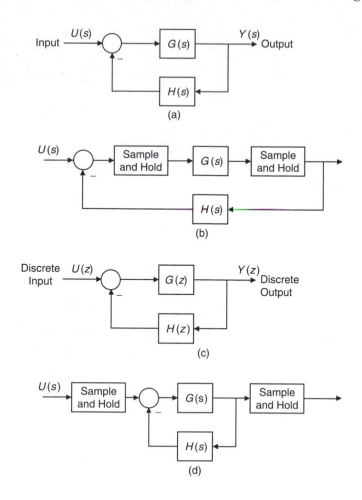

FIGURE P12.51
Discrete transfer function models of a closed loop analog system. A discrete system equivalent to (b)

12.52 Consider the general control loop shown in Figure P12.52. The case of unity feedback (i.e., feedback transfer function $H = 1$) is shown. A loop with nonunity feedback $H(s)$ can be reduced to this equivalent form by moving $H(s)$ to the forward path of the loop and dividing the reference input by $H(s)$, since $U - HY = (U/H - Y)H$. Hence, the unity feedback case shown in Figure P12.52 can be considered the general case. The controller, which has the analog transfer function $G_c(s)$, converts the error (actually correction) signal $e(t)$ into the control signal $c(t)$. Consider the following three types of controllers, which are commonly used in industrial control systems:

a. Proportional plus integral (PI) control given by

$$G_c(s) = K_p\left(1 + \frac{1}{\tau_i s}\right)$$

b. Proportional plus derivative (PPD) control given by

$$G_c(s) = K_p(1 + \tau_d s)$$

c. Proportional plus integral plus derivative (PID) control given by

$$G_c(s) = K_p\left(1 + \frac{1}{\tau_i s} + \tau_d s\right)$$

3y approximating the integration

$$i(t) = \int e(t)dt$$

by the forward rectangular rule:

$$i_k = i_{k-1} + Te_{k-1}$$

and the differentiation

$$d(t) = \frac{d}{dt}e(t)$$

by the backward difference rule:

$$d_k = \frac{1}{T}(e_k - e_{k-1})$$

obtain difference equations (i.e., digital control algorithms) for the three types of controllers PI, PPD, and PID. Note that the discrete control sequence $\{c_i\}$ can be computed from the discrete error (correction) sequence $\{e_i\}$ using these difference equations.

Give the discrete transfer functions (functions of z) corresponding to these three difference equations. Compare them with the functions obtained by converting the continuous transfer functions $G_c(s)$ into the corresponding discrete transfer functions $G_c(z)$ using the standard z-transform method.

12.53 Consider a discrete transfer function $G_c(z)$. It may be expressed as a ratio of rational polynomials in z^{-1}. Suppose that the numerator polynomial is

$$N(z^{-1}) = b_m z^{-m} + b_{m+1} z^{-(m+1)} + \cdots$$

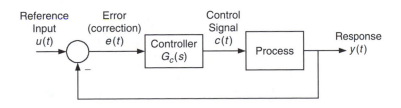

FIGURE P12.52
A general control loop.

and the denominator polynomial is

$$D(z^{-1}) = a_n z^{-n} + a_{n+1} z^{-(n+1)} + \cdots$$

in which m and n are nonnegative integers. Show that for the causality requirement of $G_c(z)$ to be satisfied (i.e., for $G_c(z)$ to be physically realizable) we must have $m \geq n$.

Now consider the direct synthesis of digital controllers and compensators using the z-transform approach. Note that $G_p(z)$ is fixed by the given system and $\tilde{G}(z)$ is defined by the specified (required) performance of the closed-loop system. Of course, both $G_c(z)$ and $\tilde{G}(z)$ have to be realizable. Suppose that the series expansion (obtained, say, by long division) of $G_p(z)$ in powers of z^{-1} has n as the lowest power, and a similar series expansion of $\tilde{G}(z)$ has m as the lowest power. Show that for $G_c(z)$ to be physically realizable, we must have $m \geq n$.

12.54 A *deadbeat controller* is a digital controller that provides a system response, which will settle down to the steady-state value in a finite (a very few) number of sample periods. Note that the settling time of a system can be determined by the free response of the system to a unit pulse. Consider a closed-loop system $\tilde{G}(z)$ that includes a deadbeat controller. Suppose that $\tilde{G}(z)$ is expanded as a series in powers of z^{-1}. Discuss characteristics of this series such that $\tilde{G}(z)$ represents a deadbeat control system.

12.55 One way to classify controllers is to consider their sophistication and physical complexity separately. For instance, we can use an x-y plane with the x-axis denoting the physical complexity and the y-axis denoting the controller sophistication. In this graphical representation, simple open-loop on-off controllers (say, opening and closing a valve) would have a very low controller sophistication value and an artificial-intelligence (AI)-based "intelligent" controller would have a high controller sophistication value. Also, a passive device is considered to have less physical complexity than an active device. Hence, a passive spring-operated device (e.g., a relief valve) would occupy a position very close to the origin of the x-y plane and an intelligent mechatronic device (e.g., sophisticated robot) would occupy a position diagonally far from the origin. Consider five control devices of your choice. Mark the locations that you expect them to occupy (in relative terms) on this classification plane.

13

Case Studies in Mechatronics

Mechatronics is an interdisciplinary engineering field, which involves a synergistic integration of several areas such as mechanical engineering, electronic engineering, control engineering, and computer engineering. Similarly, the design and development of a mechatronic system will require an integrated approach to deal with the subsystems and subprocesses of a mixed system, specifically, an electromechanical system. As reiterated throughout the book, the subsystems of a mechatronic system should not be designed or developed independently without addressing the system integration, subsystem interactions and matching, and the intended operation of the overall system. Such an integrated approach will make a mechatronic design more optimal than a conventional design. In this chapter, some important issues in the design and development of a mechatronic product are highlighted. As illustrative examples, several case studies of practical mechatronic systems are provided.

13.1 Design of a Mechatronic System

A mechatronic system may be treated as a control system, consisting of a plant (which is the process, machine, device, or system to be controlled), actuators, sensors, interfacing and communication structures, signal modification devices, and controllers and compensators. The function of the mechatronic system is primarily centered at the plant. Actuators (see Chapter 3, Chapter 8, and Chapter 9), sensors (see Chapter 6 and Chapter 7), and signal modification devices (see Chapter 4) might be integral with the plant itself, or might be needed as components that are external to the plant, for proper operation of the overall mechatronic system. Controller (see Chapter 12) is an essential part of a mechatronic system. It generates control signals to the actuators in order to operate (drive) the plant in a desired manner. Sensed signals might be used for system monitoring and feedforward control, in addition to feedback control. Both digital hardware devices (Chapter 10) and software-based computer devices (Chapter 11) may be needed in a mechatronic system. These various components may not be present as physically autonomous units in a mechatronic system in general, even though they may be separately identified, from a functional point of view. In particular, as pointed out before, an actuator and a sensor might be an integral part of the plant itself.

As an example, consider a robotic manipulator. The joint motors are usually considered as a part of the manipulator because, from the perspective of robot dynamics it is virtually impossible to uncouple the actuators from the main structure of the robot. Specifically, the torque transmitted to a manipulator link will depend on the (magnetic) torque of the motor at that joint as well as on the motor motion (speed and acceleration). Furthermore, magnetic torque will depend on the back e.m.f. in the rotor, which in turn will be determined by the motor motion (speed and acceleration). The transmitted torque will determine the link motion (displacement, speed, and acceleration), which is directly related to the motor motion (say, through a gear ratio). In the presence of such dynamic coupling, it is not proper to treat the actuator as a component external to the plant. But the sensors (e.g., tachometers and encoders) at the joints can be treated separately from the plant because their dynamic coupling with the manipulator structure (and mechanical and electrical loading—see Chapter 4) is usually negligible.

Design of a mechatronic system can be interpreted as the process of integrating (physical/functional) components such as actuators, sensors, signal modification devices, interfacing and communicating structures, and controllers with a plant so that the plant in the overall mechatronic system will respond to inputs (or, commands) in a desired manner (see Chapter 5). From this point of view design is an essential procedure in the instrumentation of a mechatronic system. The instrumentation will include the design of a component structure (including addition and removal of components and interconnecting them into various structural forms and locations), selection of components (giving consideration to types, ratings, and capacities), interfacing various components (perhaps through signal modification devices, properly considering impedances, loading, signal types and signal levels), adding controllers and compensators (including the selection of a control structure), implementing control algorithms, and tuning (selecting and adjusting the unknown parameters of) the overall mechatronic system. Many of these instrumentation tasks are also design tasks.

13.1.1 Intelligent Mechatronic Devices

A mechatronic system generally has some degree of "intelligence" built into it. An *intelligent mechatronic system* (IMS) is a system that can exhibit one or more intelligent characteristics of a human. As much as neurons themselves in a brain are not intelligent but certain behaviors that are effected by those neurons are, the basic physical elements of a mechatronic system are not necessarily intelligent but the system can be programmed to behave in an intelligent manner. An intelligent mechatronic device embodies machine intelligence. An IMS, however, may take a broader meaning than an *intelligent computer*. The term may be used to represent any electromechanical process, plant, system, device, or machinery that possesses machine intelligence. Sensors, actuators, and controllers will be integral components of such a system and will work cooperatively in making the behavior of the system intelligent. Sensing with understanding or "feeling" what is sensed is known as *sensory perception*, and this is very important for intelligent behavior. Humans use vision, smell, hearing, and touch (tactile sensing) in the context of their intelligent behavior. Intelligent mechatronic systems too should possess some degree of sensory perception. The "mind" of an IMS is represented by machine intelligence. For proper functioning of an IMS it should have effective communication links between various components. An IMS may consist of an electromechanical structure for carrying out the intended functions of the system. Computers that can be programmed to perform "intelligent" tasks such as playing chess or understanding a natural language are known to employ AI techniques for those purposes, and may be classified as intelligent computers. When integrated with a dynamic electromechanical structure such as robotic hands and visual, sonic, chemical, and tactile interfaces, they may be considered as intelligent mechatronic systems. By taking these various requirements into consideration a general-purpose structure of an intelligent mechatronic device is given in Figure 13.1.

In broad terms, an IMS may be viewed to consist of a *knowledge system* and a *structural system*. The knowledge system effects and manages intelligent behavior of the system, loosely analogous to the brain, and consists of various knowledge sources and reasoning strategies. The structural system consists of physical hardware and devices that are necessary to perform the system objectives yet do not necessarily need a knowledge system for their individual functions. Sensors, actuators, controllers (nonintelligent), communication

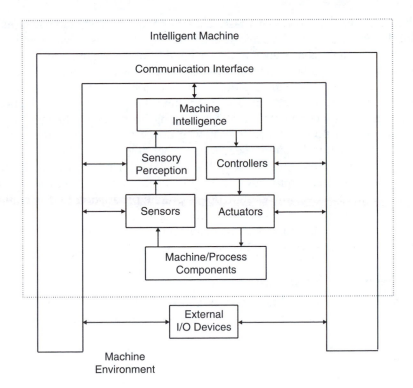

FIGURE 13.1
An intelligent mechatronic device.

interfaces, mechanical devices, and other physical components fall into this category. The broad division of the structure of an IMS, as mentioned above, is primarily functional rather than physical. In particular the knowledge system may be distributed throughout the system, and individual components by themselves may be interpreted as being "intelligent" as well (e.g., intelligent sensors, intelligent controllers, intelligent multi-agent systems). It needs to be emphasized that an actual implementation of an IMS will be domain specific, and much more detail than what is alluded to in Figure 13.1 may have to be incorporated into the system structure. Even from the viewpoint of system efficiency, domain-specific and special purpose implementations are preferred over general purpose mechatronic systems. Advances in digital electronics (see Chapter 10), technologies of semiconductor processing, and micro-electromechanical systems (MEMS) have set the stage for the integration of intelligence into sensors, actuators, and controllers. The physical segregation between these devices may well be lost in due time as it becomes possible to perform diversified functionalities such as sensing, conditioning (filtering, amplification, processing, modification, etc.), transmission of signals, and intelligent control all within the same physical device. Due to the absence of adequate analytical models, sensing assumes an increased importance in the operation and control of intelligent mechatronic systems.

Smart mechatronic devices will exhibit an increased presence and significance in a wide variety of applications. The trend in the applications has been towards mechatronic technologies where intelligence is embedded at the component level, particularly in sensors and actuators, and distributed throughout the system. Application areas such as industrial automation, service sector, and mass transportation have a significant potential for using intelligent mechatronics, incorporating advanced sensor technology and intelligent control. Tasks involved may include handling, cleaning, machining, joining, assembly, inspection, repair, packaging, product dispensing, automated transit, ride quality control, and vehicle entraining.

In industrial plants, for example, many tasks are still not automated, and use human labor. I is important that intelligent mechatronic systems perform their tasks with minimal interven tion of humans, maintain consistency and repeatability of operation, and cope with distur bances and unexpected variations in the machine, its operating environment, and performance objectives. In essence, these systems should be autonomous and should have the capability to accommodate rapid reconfiguration and adaptation. For example, a production machine should be able to quickly cope with variations ranging from design changes for an existing product to the introduction of an entirely new product line. The required flexibility and autonomous operation, translate into a need for a higher degree of intelligence in the support ing devices. This will require proper integration of such devices as sensors, actuators, and controllers, which themselves may have to be "intelligent" and, furthermore, appropriately distributed throughout the system. Design, development, production, and operation of intel ligent mechatronic systems, which integrate technologies of sensing, actuation, signal condi tioning, interfacing, communication, and intelligent control, have been possible today through ongoing research and development in the field of intelligent mechatronic systems.

13.1.1.1   Hierarchical Architecture

A hierarchical structure can facilitate efficient control and communication in an intelligent mechatronic system. A three-level hierarchy is shown in Figure 13.2. The bottom level

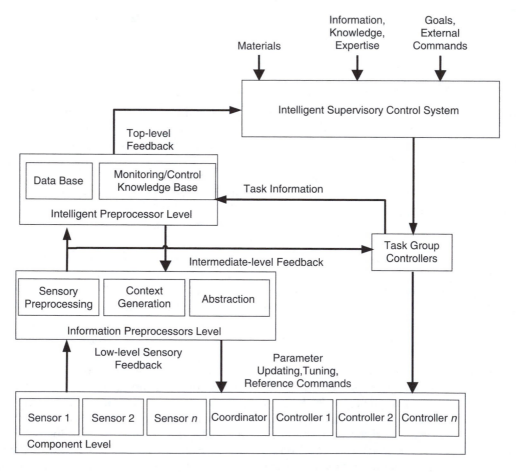

FIGURE 13.2
A hierarchical control/communications structure for an intelligent mechatronic system.

consists of electromechanical components with component-level sensing and control. Actuation and direct feedback control are carried out at this level. The intermediate level uses intelligent preprocessors for abstraction of the information generated by the component-level sensors. The sensors and their intelligent preprocessors together perform tasks of intelligent sensing. State of performance of the system components may be evaluated by this means, and component tuning and component-group control may be carried out as a result. The top level of the hierarchy performs task-level activities including planning, scheduling, monitoring of the system performance, and overall supervisory control. Resources such as materials and expertise may be provided at this level and a human-machine interface would be available. Knowledge-based decision making is carried out at both intermediate and top levels. The resolution of the information that is involved will generally decrease as the hierarchical level increases, while the level of "intelligence" that would be needed in decision-making will increase.

Within the overall system, the communication protocol provides a standard interface between various components such as sensors, actuators, signal conditioners, and controllers, and also with the system environment (see Chapter 11). The protocol will not only allow highly flexible implementations, but will also enable the system to use distributed intelligence to perform preprocessing and information understanding. The communication protocol should be based on an application-level standard. In essence, it should outline what components can communicate with each other and with the environment, without defining the physical data link and network levels. The communication protocol should allow for different component types and different data abstractions to be interchanged within the same framework. It should also allow for information from geographically removed locations to be communicated to the control and communication system of the IMS.

13.1.1.2 Blackboard Architecture

A suitable implementation architecture for an intelligent mechatronic system would be a blackboard architecture, which is a cooperative problem-solving architecture. It consists of a global data base called *blackboard*, several intelligent modules called *knowledge sources*, and a *control unit*, which manages the operation of the system. The main feature of the blackboard architecture is the common data region (blackboard), which is shared by and visible to the entire system. In particular, the data base is shared by the knowledge sources. A blackboard-based system has the flexibility of accommodating different types of knowledge sources, with associated methods of knowledge representation and processing. Unlike in a hierarchical architecture, the knowledge sources are not arranged in a hierarchical manner and will cooperate as equal partners (specialists) in making a knowledge-based decision. The knowledge sources interact with the shared data region under the supervision of the control unit.

When the data in the blackboard change, which corresponds to a change in the context (data condition), the knowledge sources are triggered in an opportunistic manner and an appropriate decision is made. That decision could then result in further changes to the blackboard data and subsequent triggering of other knowledge sources. Data may be changed by external means (e.g., through the user interface) as well as by knowledge-source actions. External data entering the system go directly to the blackboard. Also, the user interface is linked to the blackboard. The operation of a blackboard-based system is controlled by its control unit. Specifically, when an updating of data occurs in the blackboard, the control unit triggers the appropriate knowledge source, which results in the execution of some reasoning procedure and possibly, generation of new information. The blackboard architecture is generally fast, and particularly suitable for real-time applications.

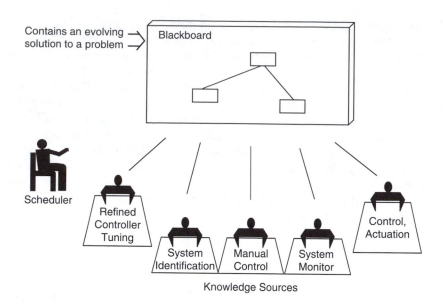

FIGURE 13.3
A blackboard architecture for an IMS.

A blackboard may consist of more than one layer (hierarchical level), each layer consisting of a subsystem which itself having a blackboard-system architecture. In this manner, a hierarchical structure can be organized for the system. Hybrid systems, consisting of subsystems having production and frame-based architectures, can be developed as well.

Figure 13.3 shows a blackboard application of controller tuning for a mechatronic system, as implemented by T.H. Lee and his co-workers at the National University of Singapore. A knowledge-based controller is implemented in a hierarchical architecture. The hardware consists of a computer workstation connected through an IEEE-488 parallel bus to a front-end dedicated controller board, which is built around the Intel 8096 microcontroller. This board constitutes the lower layer of the hierarchy where actual real-time control is performed. The microcontroller computes and sends the control signal to the plant at every sampling interval. At the upper level of the hierarchy is the workstation where the knowledge-based system is implemented. The workstation runs on UNIX and has three processes running concurrently. The knowledge-based process oversees and supervises the microcontroller on line. This involves starting and stopping of particular algorithms, calculating algorithm parameters, analyzing results from identification algorithms, bumpless transfer from one algorithm to another, and reacting correctly on alarms from the monitoring algorithm. The user-interface process provides man-machine interaction for the operator, while the IEEE-488 communication process exchanges information with the microcontroller board.

The knowledge-based system is implemented using a commercial expert system shell (NEXPERT Object), which is a hybrid shell that has rule-based and frame-based knowledge representation. The knowledge-based process is implemented in a blackboard architecture and has six knowledge sources as depicted in Figure 13.3. The manual control knowledge source supervises the manual control operation and is also responsible for gathering open-loop process information like open-loop gain and process noise during manual control. The system identification knowledge source supervises the on-off (relay) control operation, gathers process information such as input-output data, ultimate gain and ultimate period (see Chapter 12), and carries out experimental modeling (model identification). The refined controller-tuning knowledge source supervises the fine-tuning of the automatic controller. The system-monitor knowledge source examines plant response information. It contains heuristics to-determine the achievable performance of the plant and the controller, and decides whether the system performance is acceptable. The control-and-actuation knowledge source oversees the control and actuation tasks for the plant. The scheduler knowledge source takes overall charge of the blackboard. It contains rules that keep track of changes on the blackboard and decides which knowledge source to activate.

13.1.1.3 Technology Needs

Even though a considerable effort has gone into the development of intelligent mechatronic systems that somewhat mimic humans in their actions, the present generation of intelligent systems do not claim to possess all the capabilities of human intelligence; for example, common sense, display of emotions, and inventiveness. Significant advances have been made, however, in machine implementation of characteristics of intelligence such as sensory perception, pattern recognition, knowledge acquisition and learning, inference from incomplete information, inference from qualitative or approximate information, ability to handle unfamiliar situations, adaptability to deal with new yet related situations, and inductive reasoning. Much research and development would be needed in these areas, pertaining to techniques, hardware, and software before a machine could reliably and consistently possess the level of intelligence of say, a dog.

13.1.2 General Design Procedure

In some situations of mechatronic system design, the plant is already available. For example, the plant could be a vehicle structure or a building, which has been provided to the mechatronic engineer. In some other situations, the design of the mechatronic system will include the development of the plant as well. An example would be the design development of a robotic manipulator. In either case, the first step in the design of a mechatronic system is to develop a model for the plant (see Chapter 2). Both mechanical engineering components and electrical/electronic engineering components should be included in a mixed system model of the plant. If the prototype plant (or even a scaled physical model) is available, then an experimental model could be developed by testing and analyzing the test data (i.e., by *model identification*). An analytical model could be developed regardless of whether the plant is available in the physical form or not. Initially, such a model should be fairly simple. It need not contain all the details of the physical system, including all types of nonlinearities, inputs, outputs, degrees of freedom, coupling, flexibilities, distributed parameters, and other physical details. A linear time-invariant model with one input and one output is often adequate at this stage, even for relatively complex mechatronic systems. The model parameter values are automatically determined in experimental modeling. If analytical modeling is used, however, one has to decide on the numerical values for the plant parameters at this stage. If the plant is not yet built, the designer has the freedom to modify these values in the subsequent design iterations. If the plant is physically available, the numerical values used should be changed, either if they were found to be inaccurate in a later stage or if the model was upgraded using new information.

Once a plant model is available, the next step would be to decide on the design specifications for the mechatronic system. These specifications are based on the required performance. Specifications can be made in a variety of forms including those that we have discussed in Chapter 5 and Chapter 12 (e.g., rise time, peak time, percentage overshoot, steady-state error, settling time, bandwidth, gain margin, phase margin, time constants, damping ratios, optimal value of a performance index, pole and zero locations). Then, the model should be analyzed to check whether the given design specifications are satisfied. If the specifications are satisfied (which is rather unlikely), then, if the plant is not available, it could be built according to the model and tested to verify its performance. If the prototype plant behaves as required, the design need not advance any further.

If the plant does not satisfy the design specifications, at this stage we should check to see whether some modifications to the plant could yield the required results. These may include structural modifications (e.g., fixtures, couplings, transmissions) and replacement of existing actuators (e.g., actuators of different type or capacity) and associated hardware (e.g., power supplies, amplifiers, and drive circuitry). If the plant modifications could realize the desired performance, the design process should stop here.

If the plant cannot meet the design specifications even with direct modifications, the next step would be to plan a new control structure for the mechatronic system. Such a structure may include the plant, additional actuators, feedback and feedforward control paths, additional sensors for control purposes, controller/compensator locations and possible signal modification devices. Initially, this structure should be kept as simple as possible. Specifically, no new components should be added to the structure of the mechatronic system unless it is intuitively clear that this could lead to satisfying at least some of the design specifications. The development of this simple control structure will include the selection of actuators, sensors, and signal modification devices according to their type, capacity, and compatibility. Proper consideration should be given to device ratings and the overall cost/benefit aspects. The controllers should be kept very simple. For example, simple constant-gain feedback control might be attempted at first (see Chapter 12). Then, the resulting structure should be analyzed to see whether it could satisfy the design specifications. This will involve the selection of parameters (e.g., calibration constants, gains, time constants, impedances, signal levels, and transmission ratios) to meet the design specifications. Then, it should be verified that the parameter values obtained in this manner would be consistent with components that are commercially available or that could be built conveniently and economically. If this is feasible, then the mechatronic system should developed and tested to verify the performance.

If the design specifications are still not met, simple modifications to the control components and algorithms should be attempted. This may include the introduction of simple compensators (e.g., lead/lag compensators) and moderate improvements to the control method (e.g., PID control, pole placement, linear quadratic regulator—LQR). If further improvements are needed to satisfy the performance requirements, more sophisticated control techniques and algorithms should be implemented. If the problems persist, we may want to return to the previous step and change the system components or even modify the control system structure. If the performance is still not satisfactory, we should seriously consider replacing or redesigning the plant itself and repeating the design steps outlined above.

In summary, a general procedure for designing a mechatronic system may be given by the following basic steps:

Step 1: Develop an experimental or analytical model of the plant (along with the parameter values), taking into account the mixed-system nature of a mechatronic plant.

Step 2: Obtain (or, develop) the design specifications for the mechatronic system and check whether the plant satisfies them. If satisfied, stop.

Step 3: Modify the plant (e.g., structural modifications and actuator replacements) while properly matching the components, and test the resulting system. If the specifications are satisfied, stop.

Step 4: Develop a mechatronic system structure. Select appropriate components (e.g., actuators, sensors, signal modification devices, interface and communication devices, and simple controllers). Integrate the system by appropriately matching the interconnected components, possibly for optimization of the performance. Test the system. If the specifications are satisfied, stop.

Step 5: Add compensators, and upgrade the controllers and control algorithms. Test the system. If the specifications are satisfied, stop. Otherwise, go to Step 4. If the specifications are not satisfied after a few design iterations, modify or replace the plant and repeat the design steps.

Note that these design steps are quite general and will cover a large class of mechatronic systems. But, when more specific details are available on the mechatronic system, the engineer will be able to modify or improve these steps depending on the specific needs.

13.1.2.1 Development of an IMS

Development of an intelligent mechatronic system (IMS) will require a parallel development of the knowledge system and the structural system of the IMS. It may be the case that the structural system (a nonintelligent system) is already available. Still, modifications might be necessary in the structural system in order to accommodate the needs of the intelligent system, for example, new sensors, transducers, and communication links for information acquisition. In any event, the development of an IMS is typically a multidisciplinary task often involving the collaboration of a group of professionals such as engineers (electrical, mechanical, etc.), domain experts, computer scientists, programmers, software engineers, and technicians. The success of a project of this nature will depend on proper planning of the necessary activities. The involved tasks will be domain specific and depend on many factors, particularly the objectives and specifications of the mechatronic system itself. The main steps would be:

1. Conceptual development
2. System design
3. Prototyping
4. Testing and refinement
5. Technology transfer and commercialisation

It should be noted that, generally these are not sequential and independent activities; furthermore, several iterations of multiple steps may be required before satisfactory completion of any single step.

Conceptual development will usually evolve from a specific application need. A general concept needs to be expanded into an implementation model of some detail. A preliminary study of feasibility, costs, and benefits should be made at this stage. The interdisciplinary groups that would form the project team should be actively consulted and their input should be incorporated into the process of concept development. Major obstacles and criticism that may arise from both prospective developers and users of the technology should be seriously addressed in this stage. The prospects of abandoning the project altogether due to valid reasons such as infeasibilty, time constraints, and cost-benefit factors should not be overlooked.

Once a satisfactory conceptual model has been developed, and the system goals are found to be realistic and feasible, the next logical step of development would be the system design. Here the conceptual model is refined and sufficient practical details for implementation of the IMS are identified. The structural design may follow traditional practices. Commer cially available components and tools should be identified. In the context of an IMS, carefu attention needs to be given for design of the knowledge system and the human-machine interface. System architecture, types of knowledge that are required, appropriate tech niques of knowledge representation, reasoning strategies, and related tools of knowledge engineering should be identified at this stage. Considerations in the context of user inter face would include graphic displays; interactive use; input types including visual, vocal, and other sensory inputs; voice recognition; and natural language processing. These con siderations will be application specific to a large degree and will depend on what tech nologies are available and feasible. A detail design of the overall system should be finalized after obtaining input from the entire project team, and the cost-benefit analysis should be refined. At this stage the financial sponsors and managers of the project as well as the developers and users of the technology should be convinced of the project outcomes.

Prototype development may be carried out in two stages. First a research prototype may be developed in laboratory, for the proof of concept. For this prototype it is not necessary to adhere strictly to industry standards and specifications. Through the experience gained in this manner a practical prototype should be developed next, in close collaboration with industrial users. Actual operating conditions and performance specifications should be carefully taken into account for this prototype. Industry-standard components and tools should be used whenever possible.

Testing of the practical prototype should be done under normal operating conditions and preferably at an actual industrial site. During this exercise, prospective operators and users of the IMS should be involved cooperatively with the project team. This should be used as a good opportunity to educate, train, and generally prepare the users and operators for the new technology. Unnecessary fears and prejudices can kill a practical implementation of advanced technology. Familiarization with the technology is the most effective way to overcome such difficulties. The shortcomings of the IMS should be identified through thor ough testing and performance evaluation where all interested parties should be involved. The IMS should be refined (and possibly redesigned) to overcome any shortcomings.

Once the developed system is tested, refined, and confirmed to satisfy the required performance specifications, the processes of technology transfer to industry, and commer cialization, could begin. An approved business plan, necessary infrastructure, and funds should be in place for commercial development and marketing. According to the existing practice, engineers, scientists, and technicians provide minimal input into these activities. This situation is not desirable and needs to be greatly improved. The lawyers, financiers, and business managers should work closely with the technical team during the processes of production planning and commercialization. Typically the industrial sponsors of the project will have the right of first refusal of the developed technology. The process of technology transfer would normally begin during the stage of prototype testing and con tinue through commercial development and marketing. A suitable plan and infrastructure for product maintenance and upgrading are musts for sustaining any commercial product.

13.2   Robotics Case Study

Robots are mechatronic devices (see Chapter 3). They may be employed either individ ually or within workcells, to carry out industrial tasks, service functions, and household chores. Regardless of the application, design, development, and selection of a robot for a specific task have to be carried out by giving careful attention to practical, economic, and social considerations. Issues such as process requirements, commercial availability, cost and economic realities, time constraints of implementation and production, and human factors have to be taken into consideration in the process of robotization.

13.2.1   General Considerations

Applications of robots may be classified into the following broad categories of activity:

  a. Point-to-point motion
  b. Trajectory following
  c. Local/fine manipulation

A specific task may need a combination of two or more of these activities. Examples are: mixing and dispensing of drugs (activities a and c); assisting a disabled person to walk (activities b and c); vacuum cleaning a floor (activity b); loading and unloading of parts to and from machine tools (activity a); tool replacement (activities a and c); spot welding (activity a); seam welding (activity b); die casting (activity c); sealant application (activity b); packaging (activities a and c); and product inspection (activities a and c). There are many benefits of using robots in practical applications. They include improved task flexibility, elimination of low-quality human labor, increased productivity, improved product quality, consistency and repeatabilty of production, improved utilization of capital equipment, hazard reduction, robustness to external economic factors (e.g., inflation), round-the-clock and on-demand operation, better inventory management, better production planning, increased production competitiveness and flexibility, improved work environment, and overall improvement of the quality of life. Prior to robotization of a task, however, it is necessary to evaluate many factors such as appropriateness, feasibility, time constraints of production, costs, and benefits. In an industrial application, for example, it is advisable to carry out the following studies:

  1. **Evaluate the plant for tasks of potential application of robotics.** Here, the level and nature of automation (hard versus flexible automation), needed level of production flexibility (e.g., parts on demand), degree of structure in the operation and plant environment (for highly structured and fixed processes, hard automation without robots might be more appropriate), work environment (e.g., potential hazards, user friendliness), desired production rates and volumes, and the existence of similar proven applications (on site or elsewhere) should be considered.

  2. **Determine the features of robots needed for the specific application.** Important considerations in this context include payload, operating speed/bandwidth, accuracy (repeatability, precision, resolution, etc.), work envelope (reachability), method of actuation (dc or ac servomotors, stepper motors, hydraulic, pneumatic), desired robotic structure (degrees of freedom, revolute or prismatic joints, and in what combination), end-effector requirements (gripper, hand, tools, sensors, and the characteristics of objects which are to be handled), method of control (servo, adaptive, hierarchical, etc.), instrumentation requirements (sensory, monitoring, data acquisition, fixturing, control, and coordination needs), operating environment (moisture, chemicals, fire hazards, dust, temperature, etc.), and commercial availability of the desired types of robots.

  3. **Study the robot installation requirements and their consequences in plant operation.** Relevant considerations will include necessary utilities (dc or ac, single-phase or three-phase power, compressed air, etc.), installation timing (down time of operations, product demands, etc.), other machinery within workcells

(interfacing, communication, networking, control, etc.), plant layout (integration with raw material, product and tool flow plans, services, operator interaction and interfaces, safety, etc.), and personnel (training, programming, maintenance, operation, etc.).

4. **Perform an economic analysis.** This is a cost-benefit evaluation, taking into account such considerations as availability of off-the-shelf units, capital investment, down time, efficiency, wastage reduction, labor reduction, operating costs, and the rate of return on investment. Specifically, cost of purchase and installation of the robot, useful life, maintenance, service, and utility costs, depreciation, and money cost should be considered in the analysis. It may be estimated, for example, that the approximate hourly cost of labor has increased rather exponentially from $6.00 in 1975 to $35.00 in 2004, whereas the approximate hourly cost of operation of a robot has only increased from $4.00 to $8.00 during the same period. The former cost continues to rise and the latter is leveling off.

5. **Address human relations considerations.** Here, the relationship between workers and management staff, loss of employment due to automation, massive loss of employment due to plant closures (without automation), inefficiency, worker retraining, and union representation should be considered.

13.2.1.1   Economic Analysis

Economic analysis involves a cost-benefit evaluation, and will require computation of the payback period. Figures on the initial investment, number of people replaced by the robot and corresponding wage savings, productivity increase due to robotization, inflation rate, corporate tax rate, and operating expenses such as utilities, maintenance, and insurance, would be required for an analysis of this type. The procedure that is given in the next case study may be applied in the present case study as well.

13.2.2   Robot Selection

Selection of the "best" robot for a given task is a crucial step in robotic application. Clearly, the term "best" is used within the set of constraints that govern the problem; for example, cost, timing of installation and operation, and availability of hardware and personnel. For precision tasks (e.g., manufacture of products of fine tolerance), it is important to make sure that the accuracy specifications (including repeatability and motion resolution) can be met. In such applications, structural integrity and strength (e.g., robot stiffness) also would be prime considerations. Other aspects such as the controller and its architecture (e.g., open and user-programmable at low level), compatibility and ease of communication with other machinery and their controllers in coordinated operation (e.g., networking considerations such as the use of fieldbus, communication protocols, etc. as applied to factories, plants, and workcells), necessary end effectors, tools, fixtures, and instrumentation, and ease of programming and operator friendliness would be important. Special requirements (e.g., clean room tasks) may require custom modifications and component/ system qualification (i.e., analysis and/or testing to evaluate and determine the suitability of the device or system for the specific application). A typical set of steps that would be followed in selecting an industrial robot is given below:

1. **Define the tasks, which are to be carried out by the robot.** First a verbal description of the task sequence would be appropriate (e.g., pick a bulb from the conveyor, inspect for faults, decide category, place in the appropriate bin). Next, motion sequence should be defined in a quantitative/analytical form, for

example, giving time sequences and/or trajectories for the set of actions. Also, define the time sequences for nonmotion tasks (e.g., grasp the object, release the object, wait for part). Error tolerances should be specified as well.

2. **Develop robot specifications for the tasks.** For example, work envelope, speed limits and cycle time, force and payload capability, repeatability, and motion resolution, have to be specified for the robot.

3. **Identify the necessary mechanical structure for the robot.** The required number of degrees of freedom, type of joint combination (e.g., SCARA robot with three revolute joints and one prismatic joint), required end-effector motions (e.g., complete or partial rotations, strokes of linear motions), and lengths of robot links have to be decided.

4. **Identify the sensor and actuator preferences.** The nature of the desired drive system for each joint (e.g., backlash-free transmission or harmonic drive with ac servomotor, direct-drive joint) and the associated sensory preferences (e.g., incremental optical encoders, resolvers, joint torque sensors) have to be identified as completely as possible.

5. **Identify the end-effector requirements.** Depending on the expected tasks, a variety of end effectors might be needed. Simple grasping operations may need basic two-finger grippers. More sophisticated tasks would require robotic hands, tools, and custom devices with adequate dexterity, motion resolution, force resolution, and sensors (e.g., tactile, wrist torque and force, optical, and ultrasonic sensors).

6. **Identify the control and programming requirements.** Decide whether high-level task programming alone is adequate or whether low-level direct programming of the joint controllers would be needed. Also, depending on the expected operators and programmers, decide upon the desired difficulty level of programming. In addition, communication and interfacing needs and compatibility with other interacting devices and tools have to be established; for example, when the robot is expected to be an integral part of a larger system such as a workcell.

7. **Identify the user interface needs.** This is somewhat related to Item 6 above. A graphic user interface (GUI) that suits the specific application has to be considered. Generally, it has to be user friendly, with simple input/output means such as touch screens, voice activation, and hand-written commanding being particularly useful when technologically nonsophisticated users are involved.

8. **Decide on a budget, and contact suppliers.** If a suitable robot is commercially available, within budget and the required time frame, the selection would be straightforward. Otherwise, several iterations of robot specification and identification would be needed, each time relaxing/modifying some of the specifications.

13.2.2.1 *Commercial Robots*

The task of selecting a robot is greatly simplified if the required specifications can be matched to those of a commercially available robot. A typical set of commercial robots and some useful specifications attributed to them are listed in Table 13.1. Since an end effector has to be chosen separately and does not usually come with the robot, the payload that is indicated in the table, includes the weight of the end effector. The *repeatability* of a robot specifies how accurately a robot can reach a commanded point in space. Repeatability error may result from such factors as Coulomb friction, backlash, and poor control, and is one of many factors that determine the overall *accuracy* of a robot. The accuracy itself will depend on such considerations as the speed of operation, payload, and the specific trajectory of motion. The *resolution* of a robot, like repeatability, is a lower bound for

TABLE 13.1

Data for Several Commercial Robots

Robot	Mechanical Structure	Drive System	Payload (kg)	End-Effector Speed, with Payload (m/s)	Repeatability (mm)
PUMA 560	6-axis revolute	Geared dc servomotors	2.3	0.5	0.10
SEIKO RT5000	4-axis cylindrical	Gear/rack/belt/ harmonic drive, dc servomotors	5.0	2.0	0.04
SCORA-ER 14	4-axis SCARA (3 revolute + 1 prismatic)	Harmonic-drive, dc servomotors	2.0	1.9	0.05
Adept-3	4-axis SCARA (3R + 1P)	Direct-drive, dc servomotors	25.0	0.7	0.05
Pana Rob HR-50	4-axis SCARA (3R+1P)	ac servomotors with speed reducers	5.0	1.2	0.05
Staubli RX 130	6-axis revolute	ac servomotors with speed reducers/gears	12.0	1.5	0.025
CRS Robotics A 465	6-axis revolute	dc servomotors with harmonic drives and belts	3.0	1.0	0.05
GMF Robotics M-300	4-axis cylindrical	dc servomotors	100.0	Moderate	1.0
Hitachi A4030	4-axis SCARA	dc servomotors with speed reducers	10.0	1.5	0.05

accuracy, and represents the smallest motion increment that can be executed. Again, resolution may depend on factors such as friction, backlash, unknown disturbances, the resolution of digital motion transducers such as encoders, the bit size of a control command, and for a robot that uses stepper motors, the step size of incremental motion.

Often, what is given in product specifications of commercial robots is the no-load speed. A more meaningful specification is the *cycle time* for a specified pick-and-place cycle, when carrying the specified payload. Specifically, by assuming a triangular speed profile, where the robot steadily accelerates from rest after the "pick" to reach the peak speed, and then steadily decelerates to the "place" point, the peak velocity v_{peak} is given by

$$v_{peak} = \frac{2\Delta\ell}{\Delta t} \tag{13.1}$$

where

Δt = cycle time

$\Delta\ell$ = pick-and-place distance

The corresponding acceleration (and deceleration) is given by

$$a = \frac{2v_{peak}}{\Delta t} \tag{13.2}$$

or

$$a = \frac{4\Delta\ell}{\Delta t^2} \tag{13.3}$$

If the payload is M, the force f_e exerted at the end effector for steady acceleration to peak speed, is given by

$$f_e = Ma \tag{13.4}$$

or

$$f_e = \frac{4M\Delta\ell}{\Delta t^2} \tag{13.5}$$

The capabilities of the robot, as specified by the cycle time, payload, peak operating speed, and the force at peak speed and at steady acceleration have to be matched with the requirements of the robotic task.

As an example, consider a robot that is able to execute a pick-and-place operation over a distance of 0.5 m in less than 1.0 s, carrying a payload of 30 kg. Suppose that it has a resolution of ± 0.01 mm and a repeatability of ± 0.05 mm. We can determine the maximum operating speed of the robot when carrying a payload of 30 kg. Also, we can determine the maximum force that can be exerted by the end effector of the robot at its peak operating speed.

Here, $\Delta\ell = 0.5$ m and $\Delta t = 1.05$. Hence, from Equation 13.1 we have

$$v_{peak} = \frac{2 \times 0.5}{1.0} = 1.0 \text{ m/s}$$

which is the maximum operating speed with the payload. From Equation 13.2,

$$a = \frac{2 \times 1.0}{1.0} = 2.0 \text{ m/s}^2$$

From Equation 13.4,

$$f_e = 30 \times 2.0 = 60.0 \text{ N}$$

which is the maximum force that can be exerted by the end effector.

13.2.2.2 Robotic Workcells

A robotic workcell is a group of machinery (generically termed *machine tools*) such as material removal devices (e.g., computer-numerical-control or CNC mills, lathes, drills, borers, cutters), material handling equipment (e.g., conveyors, gantry mechanisms, positioning tables, automated guided vehicles or AGVs) along with one or more robots, working together to achieve a common task objective, under the supervision and control of a *cell host* computer. Each machine tool will be controlled by its own *machine-tool controller*, but will communicate with, and coordinated by the cell host. The *cell supervisor* is a high-level program that runs the cell host. A schematic representation of the concept of workcells is shown in Figure 13.4. A production/manufacturing system or a process plant or a factory may consist of two or more workcells, which communicate with each other through a local area network (LAN, see Chapter 11) under the supervision of a *system control computer*. Note that the machine tools and robots operate under their own controllers, running their own programs (e.g., CNC parts programs and robot motion/operation

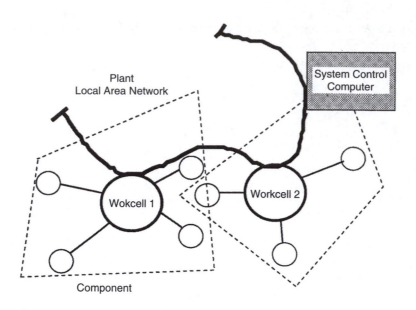

FIGURE 13.4
A flexible production/manufacturing system consisting of robotic workcells.

sequence programs) on command, coordination, and supervision by the cell host. Parts and material movement within a workcell is guided by the cell host. During a parts run, very little communication is needed with the system control computer.

Autonomous operation is a desired feature for a robotic workcell. In this context, a workcell should be able to carry out its task without relying on external assistance. In particular, unmanned and automated operation, flexibility, and capabilities of self recon-figuration, self-repair, learning, and adaptation would be desirable. Smart sensors, intel-ligent controllers, programmable machine tools, and effective communication and control structures would be needed as well. Networking and communication protocol consider-ations (see Chapter 11) are paramount here.

The design philosophy for robotic workcells is based primarily on flexibility and auto-nomy. To achieve flexibility, the use of programmable and modular components along with communication and control architectures that allow fast restructuring would be necessary. Accordingly, robotic workcells fall into the category of *flexible automation*. For autonomous operation, sensors to obtain the necessary information on each component of the workcell and its environment, and intelligent control systems that can handle unfamiliar and unexpected conditions, along with capabilities of self-reconfiguration and repair would be needed. Structured communication and control, particularly a hierarchical control architecture, where the required information and control signals can be related to a specific and well-defined layer and where lower layers may be modified with little effect on the upper layers, would be desirable. Allowance should be made for ease of upgrading and expansion of the workcell. Standard and programmable components that are com-patible with the plant network and the associated communication protocols, and whose operation can be defined in terms of input-output characteristics, would help simplify workcell integration, component replacement, and workcell modification. The following guidelines may be followed in the design and development of a robotic workcell:

1. Identify the workcell tasks and process requirements (production rates, toler-ances, etc.). Reach a compromise between product flexibility and production rate.

2. Identify the machine tool and robot requirements, limiting if possible, to either existing or commercially available units.

3. Initial workcell need not be fully autonomous. Integrate humans if feasible.

4. Develop a workcell architecture. Use a simple networking topology and communication protocols, and existing computer technology and hardware.

5. Identify the necessary accessories such as grippers, fixtures, sensors, and instrumentation. Use simple and off-the-shelf components with simple interfacing requirements where possible.

6. Modify and enhance existing components and controllers (e.g., by adding sensors, processors, memory, software) as required and feasible.

7. Identify the critical components of the workcell and consider the possibility of incorporating either software or hardware redundancy.

When designing a flexible production system having multiple workcells, an analysis has to be made to determine the workload demand on each component under normal operating conditions. Then, in selecting the workcell components, care has to be exercised to ensure that the component capacity is greater than or equal to the workload demand, and to reach a somewhat optimal balance between these two levels. A dynamic restructuring system with proper monitoring, control, and rescheduling capabilities, will be able to accomplish this. If there is excess capacity in each component, it may be possible to share common components within or between workcells, thereby releasing appropriate components that operate well below their capacity. Also, if component overloading is present, a similar procedure may be used to shed the overload into a partner component that operates below capacity.

13.2.3 Robot Design and Development

The main steps of the design and development of a customized robot may be given as follows:

1. Arrive at kinematic (motion) and dynamic (force/torque-motion) specifications for the range of tasks to be carried out by the robot. Modeling and analysis will be required.

2. Determine the geometric requirements (e.g., degrees of freedom, choice of revolute and/or prismatic joints, lengths of links, motion resolution and accuracy) for the robot, based on the task kinematic specifications. Modeling and analysis will be required.

3. Determine the geometric requirements for the end effector. Modeling and analysis will be required.

4. Determine the dynamic (forces and torque) requirements for the end effector and the robotic joints, based on the dynamic specifications. Modeling and analysis will be required.

5. Select actuators (type, load, motion, and power capacities) from commercially available units to match the kinematic and dynamic requirements. Analysis and design will be required.

6. If available direct-drive actuators cannot match the requirements, select motion transmission units (perhaps available in integral from with the actuators) to meet the requirements. Analysis and design will be required.

7. Select matching drive systems and power supplies for the actuators.

8. Select the digital control platform for the robot. This may include the control computer, input/output hardware and software, user interface, and other communication needs.

9. Apart from the sensors provided with the actuators (e.g., encoders, tachometers, resolvers) determine what other sensors are necessary for the tasks (e.g., proximity sensors, ultrasonic, optical and vision sensors) and select them from commercially available units, to meet the requirements (of task, accuracy, resolution, bandwidth, etc.).

10. Decide upon a preliminary design for the robot and carry out a model analysis/simulation exercise to validate the design. Fine tune the design (some of the previous steps may have to repeated here) to meet the specifications.

11. Acquire/build the components for the robotic system. Assemble/integrate the robotic system.

12. Test the robot and compare with computer simulations. Carry out further improvements based on the test results.

13.2.3.1   Prototype Robot

We were given the task of developing a laboratory robot at the University of British Columbia, for use in research and development, particularly in space robotics. The developed manipulator can be used to assess, through real-time experiments, the effectiveness of a variety of control procedures for their possible application to space-based systems. Robotic manipulators play an important role in space exploration because of the harsh environment in which they have to operate and the challenges associated with it. Their tasks include capture and release of spacecraft including satellites, maneuver of payloads, and support of extra-vehicular activities (EVA). One example is the mobile servicing system (MSS), which is an example of Canada's contribution to the International Space Station project.

After a preliminary study, we decided to develop a nonconventional robot consisting of multiple modules connected in series, each module consisting of a revolute (slewing) joint and a prismatic (deploying) joint. The particular robotic design is termed Multimodule Deployable Manipulator System (MDMS). A robot of this type offers several useful characteristics with respect to the dynamics and control, over the conventional manipulator designs that involve only revolute joints:

- Reduced inertial coupling and simpler kinematics, for the same number of joints
- Better capability to overcome obstacles
- Reduced number of singular positions for a given number of joints
- Simpler decision making during task execution

We have developed a detail, nonlinear, dynamic model for the robot, based on the task requirements; carried out extensive analysis and computer simulation; designed and developed the robot; implemented a variety of control schemes; and extensively tested the prototype robot.

13.2.3.2   Robot Design

In the process of developing the present 4-module manipulator, we first developed a 2-module variable geometry manipulator (VGM). Through the working experience with this initial prototype, the new four-module system was developed so as to achieve the

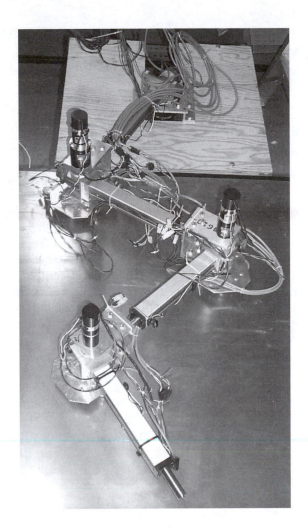

FIGURE 13.5
The MDMS prototype.

following: a higher level of rigidity in the joint connections; reduced size, weight and inertia; reduced machine shop time; and ease of construction, assembly and maintenance. In addition to these criteria, the mechanical and electrical components were chosen using a mechatronic approach, based on their performance characteristics such as power, speed, accuracy, reliability, and robustness. A view of the developed manipulator is shown in Figure 13.5. Starting from the top of the picture, mounted underneath the wooden board is a dc motor for shoulder joint motion. Each of the subsequent elbow joints is also equipped with a dc motor for slewing motion. In between the elbow joints are the deploying links. Each link, which can vary its length, provides not only an extra degree of freedom, but also increased versatility and maneuverability to this robotic arm. In addition, each elbow joint is provided with a rolling support comprising three ball transfers, which run smoothly on a bench with a steel surface. The prototype has the following main features and specifications:

- It is a planar, eight-axis robotic manipulator with four modules, each consisting of one slewing link and one deploying link

- It uses rolling supports on a flat surface to compensate for gravity
- Maximum extension of each deploying link is 15 cm ($\approx$ 6 in)
- Maximum acceleration of 0.08 m/s^2 at a design payload of 5 kg
- Maximum slew speed = 60°/s
- Maximum deployment speed = 4 cm/s
- PC-based control system on which a variety of control strategies could be implemented.

The 15 cm extension provides sufficient change in length for demonstrating the characteristically improved performance due to prismatic joints. The rolling supports is an important feature because in the fully extended configuration the manipulator induces a high bending load at each unit, particularly at the shoulder joint. The workspace of the prototype manipulator is a circle of approximately 4.5 m in diameter.

Two categories of components are employed in the construction of the manipulator: those commercially available (supplied by outside manufacturers/vendors) and those designed and machined in-house. The selection of the components in the former category is focused here. The latter type of components include machined parts like the base, motor mounts and joint supports, and they are made of aluminum 6061 for its low mass density (ρ_{Al} = 2710 kg/m^3), strength (yield strength = 255 MPa) and high machineability.

13.2.3.3   Actuator Selection/Sizing

The manipulator was built employing four harmonic-drive dc servomotors from HD Systems, Inc. for the revolute links, and four Pulse Power I (PPI) linear actuators from Dynact, Inc. for the deploying links. Electromagnetic actuators were chosen as opposed to hydraulic or pneumatic actuators, because of their simplicity, availability, lower cost, and ease of integration and control. These actuators are particularly suitable for carrying out light duty tasks in laboratory experimentation using the robot. Our older robot prototype consisting of two modules (i.e., the variable-geometry manipulator or VGM) uses dc servomotors with conventional gear transmissions. The resulting problems include increased weight, noise, and gear backlash, which contribute to the nonlinearity of the system, making the control problem more challenging. Low backlash gear units are 25–50% more expensive. Direct drive servomotor eliminate the backlash and noise problems. However, the size and weight of the units available to us did not meet the present design requirements. With harmonic-drive actuators (see Chapter 3), which are virtually backlash free and available in compact packages, provide high positional accuracy and stiffness. Precise motion control can be performed using them. Furthermore, these actuators have a large torque capacity and are 10–30% less expensive than the conventional geared servomotor units.

As in the 2-module VGM, the new prototype manipulator also uses a ball-screw mechanism for its prismatic (deploying) links. In view of several attractive features, direct drive linear servomotors were considered for these links, but their main drawback is the heavy magnetic assembly, which did not satisfy our design criteria. Also they generally have a lower thrust-force to weight (size) ratio when compared to ball-screw actuators. Ball-screw actuators with drive nuts that have very low backlash are available as integral packages from Dynact Incorporated in Orchard Park, NY. Each actuator includes a dc servomotor, a ball-screw mechanism, a deploying shaft and an optical encoder for sensing and feedback of the angular position. The machine shop time was greatly reduced as well with these integral packages. The actuator housing and the extensible shaft are made of formed and machined aircraft-quality aluminum, and are rugged and lightweight. In addition, the actuators incorporate magnetic reed switches to protect them from over-travel in both "extend" and "retract" positions. The drive screws of the linear actuators are available in either ACME or ball thread form. Ball thread type screws were selected in the present prototype because of their high efficiency, high duty cycle, low misalignment problems, low friction, long design life, and high speed and load capabilities (see Chapter 3). Optional nuts with lower backlash can be chosen as well, for the ball screw.

The particular commercial models for the revolute link actuators were selected based on the torque and speed requirements (see Chapter 8 and Chapter 9). The computed design torque consists of three parts: actuating torque for the driven load, resistance torque in the bearings, and the resistance torque contributed by the ball-transfer rolling supports. It is required that the revolute joint motion is capable of achieving a speed of 60°/s. The computed design torque requirements for the four slewing joint actuators, starting from the shoulder joint, are listed in Table 13.2. With these torque values and the speed requirement, harmonic-drive actuators were selected by referring to their torque-speed curves as provided by the manufacturer (see Chapter 8 and Chapter 9), and they are listed in Table 13.3.

For the deploying links, the design thrust consists of the driving force for the payload and the force to overcome the friction between the actuated load and the workspace surface. The maximum required thrust was computed to be 92.0 N. Pulse Power I (PPI) linear actuators, which correspond to the smallest direct (in-line) drive model available from Dynact, were selected. With a standard motor, the actuator can produce a maximum thrust of 550.0 N. In order to reduce the motor mass, the smaller, nonstandard motors were used instead. DC servomotors, model 14201 from Pittman that are smaller and lighter but still meet the torque-speed requirements, were chosen to drive the linear actuators.

TABLE 13.2

Computed Design Torque
for Revolute Joints

Slewing Joint Motor No.	Design Torque (N·m)
1	30
2	13
3	4
4	0.5

TABLE 13.3

Selected Revolute Joint Actuators and Their Characteristics

	Joint #1	Joint #2	Joint #3	Joint #4
Model	RFS-20-3007	RH-14C-3002	RH-11C-3001	RH-8C-3006
Rated voltage (V)	75	24	24	24
Rated current (A)	1.9	1.8	1.3	0.8
Peak current (A)	4.8	4.1	2.1	1.1
Rated output torque (N·m)	24	5.9	3.9	2.0
Rated output speed (rpm)	30	30	30	30
Max. continuous stall torque (N·m)	28	7.8	4.4	2.3
Peak torque (N·m)	84	20	7.8	3.5
Motor positioning accuracy (arc-min)	1.0	2.0	2.0	2.5
Diameter (mm)	85	50	40	33
Length (mm)	216	148	125	107
Mass (kg)	3.6	0.78	0.51	0.32
Inertia (kg·m²)	1.2	0.082	0.043	0.015
Transmission ratio	100	100	100	100

TABLE 13.4

Characteristics of a Deploying Joint Motor

Pittman Model 14201 DC Servomotor	
Rated voltage (V)	24.0
Rated current (A)	1.1
Peak current (A)	8.6
Rated output torque (N·m)	0.0244
Rated output speed (rpm)	3700
Max. continuous stall torque (N·m)	0.07
Peak torque (N·m)	0.5
Diameter (mm)	54
Length (mm)	75
Mass (kg)	1.3
Inertia (kg·m²)	1.15E-5
Transmission ratio	1

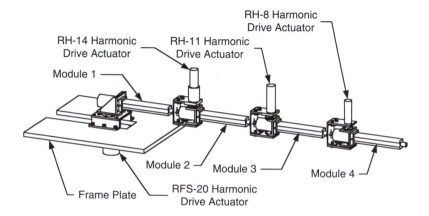

FIGURE 13.6
The CAD model of the MDMS prototype.

They are 50% lighter than the PPI standard motors. Some characteristics of this servomotor unit are given in Table 13.4.

13.2.3.4   Final Design

Having sized and selected the actuators for the MDMS, the remaining components, mounting brackets and connectors were designed with the aid of a computer-aided design (CAD) software package. The assembled CAD model is shown in Figure 13.6. The base plate is the wooden board in Figure 13.5. The modules 1–4 are the PPI actuators. The whole manipulator is mounted on a base plate attached to a steel frame. The modules 2–4 have the same type of connection to the adjacent modules. Hence, the system can be easily reduced to a two- or three-module manipulator for specific experimental studies. The mount plate for the shoulder joint actuator also features four jacking screws for leveling the manipulator system. Overall, the new prototype manipulator design has improved on the rigidity and compactness over its forerunner, the VGM.

A detailed drawing of one of the revolute joints is given in Figure 13.7. The joint is constructed with two C-shape channels connected through ball bearings and machined shaft-like connectors. The module connector integrates the joint bracket to the end of the deploying shaft of the previous module through a taper pin. The connectors and the

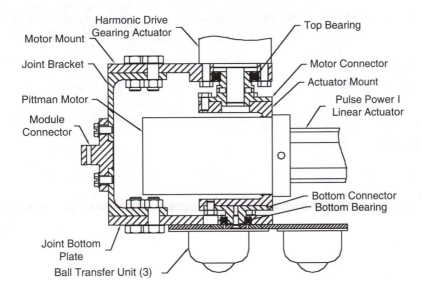

FIGURE 13.7
Details of a revolute joint.

channels are machined out of aluminum. The size and the weight of the joint are minimized through careful design practices. For example, the size of the outer channel was minimized by allowing just enough clearance for the linear actuator motor and the electrical wiring (not shown in the figure) for required movement. Furthermore, the length of the channel flange was minimized; hence, flange thickness need not be too large in order to realize a rigid structure. With the PPI linear actuator mounted on the inside smaller channel, the slewing link rotates with respect to the outer channel. Two deep groove ball bearings are installed, one each at the top and bottom plates, to support the slewing link. The motor connector is attached to the harmonic drive gear actuator and is secured in place with two setscrews. Two setscrews are used to prevent any possible slack during operation. Rolling support with ball transfer units is attached to the bottom plate of the joint.

13.2.3.5 Amplifiers and Power Supplies

All the actuators are driven by brush type pulse-width modulated (PWM) servo amplifiers and power supplies from advanced motion controls (AMC) in Camarillo, CA. The amplifiers and power supplies were selected to match the actuators. For example, the harmonic drive actuator at the base has the following ratings:

- Rated voltage = 75 V
- Rated current = 1.9 A
- Peak current = 4.8 A
- Rated output torque = 24 N · m
- Rated output speed = 30 rpm
- Maximum continuous stall torque = 28 N · m
- Peak torque = 84 N · m
- Motor positioning accuracy = 1.0 arc-min
- Transmission ratio (harmonic drive) = 1 : 100
- Optical encoder resolution = 500 P/rev

The AMC PWM amplifier model 12A8 selected to drive this actuator, has the following ratings:

- DC supply voltage = 20–80 V
- Peak current (over a max. period of 2 sec., internally limited) = ±12 A
- Max. continuous current (internally limited) = ±6 A
- Switching frequency = 36 kHz
- Bandwidth = 2.5 kHz

Both the dc supply voltage range and the peak current meet the motor needs. The continuous current of amplifier (±6 A) is greater than the rated current of the motor. Moreover, the operating bandwidth of the amplifier, at 2.5 kHz, is far greater than what is needed for a typical robotic task (e.g., 125 Hz). The power supply, AMC model PS16L30, has the following ratings:

- Supply voltage = 120 VA
- Rated output voltage = 30 VDC
- Nominal output current = 53 A

The output voltage of the power supply matches the range of operation of the amplifier. Also, the continuous output current of the power supply (53 A) is sufficient even for eight motors (a total of 14 A) or eight amplifiers (a total of 48 A). Besides, when driving the actuator based on a command signal from the control computer, each amplifier is able to receive end-of-stroke signals from the magnetic reed switches (limit switches, see Chapter 7), which are mounted on the linear actuators, in order to disconnect the power to the linear actuators.

13.2.3.6   Control System

The control system is schematically shown in Figure 13.8. Controlled motion of the prototype manipulator system is carried out using optical encoders as feedback sensors, which come integral with the actuators, a data acquisition board, and an IBM-PC compatible computer. To implement different control algorithms on this robotic manipulator system, an open architecture real-time control system has been established using an 8-axis ISA bus servo I/O card from Servo To Go, Inc. in Indianapolis, IN. This data acquisition board features the following functionalities:

- 8 channels of encoder input of up to 10 MHz input rate
- 8 channels of 13-bit analog output of +10 V to –10 V range
- Interval timers capable of interrupting the PC
- Timer interval programmable to 10 minutes in 25 μs increments
- Board base address determined automatically without a configuration file
- IRQ number is software selectable; that is, no board jumper is required.

For real-time control, the 8-axis card was originally operated under QNX real-time operating system. The necessary drivers and function library are available in C language from quality real-time systems (QRTS) in Falls Church, VA. Controller programs have been written in C language. The control program also serves as the manager for data exchange and the coordinator of the following functions: setting the sampling rate, acquiring encoder signals, and sending out command signals to the amplifiers through the digital-to-analogue

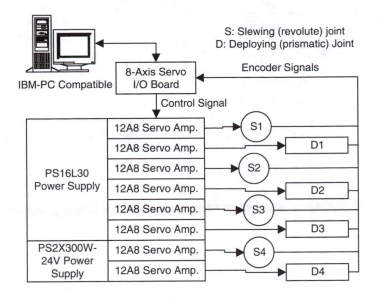

FIGURE 13.8
The robot control system.

conversion (DAC) channels. The MDMS prototype uses a single host computer for the development and implementation of its control system. This enables easier debugging and faster modification of program code, while maintaining the performance of a dedicated real-time controller. This provided a cost effective option, particularly since no co-processors or DSPs were employed. Originally, real-time control of MDMS was achieved using QNX real-time operating system. Even though it has many attractive and multi-tasking features, the range of supported hardware configurations was not as extensive as for a popular operating system such as Windows, which was found to me more user friendly. Even though Windows NT has been used in real-time applications, it was not designed for "hard" real-time applications. Any events that are given the highest priority setting are still subject to unpredictable delays due to lower level processes running in Windows NT. Windows NT is better suited for operations that only require precision in the 100 ms range. VenturCom's real-time extension (RTX) for Windows NT adds real-time capabilities down to the sub-millisecond range. RTX provides real-time capabilities by adding a new subsystem known as RTSS to the Windows NT architecture. It allows users to schedule events ahead of all Windows NT schedules and to create threads that are not subject to time-sliced sharing of the processor. RTX enables users to take advantage of the sophisticated graphics user interface (GUI) and connectivity options of Windows and the high-performance, reliability and determinism of a real-time operating system at the same time, on the same computer. RTX provides a set of real-time functions, allowing the developer to program a real-time controller in C/C++. These functions are similar to those available in the Win32 API, but they allow the user to set thread priority at any one of the 128 levels provided by RTX, and particularly higher priority than Windows NT schedules.

A PID controller was implanted on the MDMS, particularly using the features of the clock and timer provided by RTX. The control program was developed with two threads: the timer function and the main function. The timer function was scheduled to execute every sampling period and was given priority over the main function. The main function waits until the ESC key is hit or the robot task is completed and then zeroes the digital-to-analog

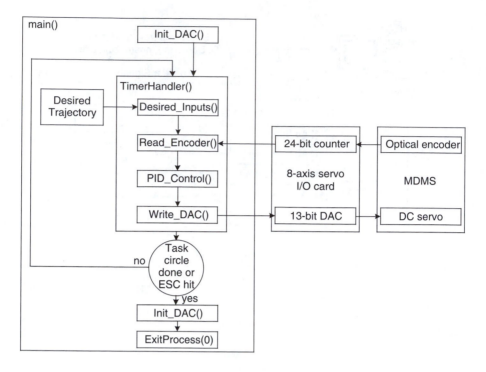

FIGURE 13.9
PID control of the MDMS.

converters (DACs) and kill the process. The timer function repeats the following steps at each sampling period:

- Determine the desired output from a function or file
- Read the actual output from the encoders (motion sensors at robot joints)
- Employ a PID control law to determine the input to the joint motors
- Write the control input value to the DACs

Rather than using the Windows drivers for the servo card, simple read and write functions were developed in RTX. This improved performance, by ensuring that each function call has priority over other Windows NT schedules. This was not strictly necessary for the PID controller, but would be useful for implementing more computationally intensive controllers in the future. Figure 13.9 shows a schematic diagram of the control systems with the PID control program. The control action at time instant n is computed by (see Chapter 12)

$$u_n = K_p \left[e_n + \frac{T_d(e_n - e_{n-1})}{T_s} + \frac{1}{T_i} \left\{ \frac{(e_n + e_{n-1})T_s}{2} + PInt \right\} \right] \tag{13.6}$$

where

T_s = sampling time
e_n = response error
u_n = control action

$$K_p = \text{proportional gain}$$
$$K_i = \text{integral gain}$$
$$K_d = \text{derivative gain}$$
$$T_i = K_p/K_i = \text{integral action time}$$
$$T_d = K_d/K_p = \text{derivative action time}$$

The value of *PInt* is initialized at step 0. In addition to the standard PID control, other sophisticated control schemes such as predictive control have been implemented and successfully tested using the MDMS.

13.3 Iron Butcher Case Study

The Iron Butcher, which is commonly used in the fish processing industry for head cutting of fish, is known to be inefficient and wasteful, and the resulting product quality may be unacceptable for high-end markets. We have developed two improved designs of the Iron Butcher machines for fish cutting. The two machines are similar, except for the cutter design and the types of actuators used. The development of one machine is presented here. The machine, which uses a variety of sensors, actuators, and hardware for component interface and control, operates with the help of a dedicated supervisory control system. A layered architecture has been used for the system. It has several knowledge-based modules for carrying out tasks such as machine monitoring, controller tuning, machine conditioning, and product quality assessment.

13.3.1 Technology Needs

Fish processing, a multi-billion dollar industry in North America alone, by and large employs outdated technology. Wastage of the useful fish meat during processing, which reaches an average of about 5%, is increasingly becoming a matter of concern for reasons such as dwindling fish stocks and increasing production costs. Due to the seasonal nature of the industry, it is not cost effective to maintain a highly skilled labor force to carry out fish processing tasks and to operate and service the processing machinery. Due to the rising cost of fish products and also diverse tastes and needs of the consumer, the issues of product quality and products-on-demand are gaining prominence. To address these needs and concerns, the technology of fish processing should be upgraded so that the required tasks could be carried out in a more efficient and flexible manner.

A machine termed the Iron Butcher, which was originally designed in the beginning of the last century and has not undergone any major changes, is widely used in the fish processing industry for carrying out the head cutting operation of various species of salmon. The sketch given in Figure 13.10 illustrates the operating principle of the Iron Butcher. The fish are manually fed onto the moving conveyor, at one end of the machine. A pair of pins is provided at the feeding end of the machine, with respect to which the fish should be manually positioned across the conveyor. But, since the typical feeding rate is 2 fish/s, accurate manual positioning is infeasible. Instead, a mechanical indexing mechanism is employed, for automatic positioning. The main component of this mechanism is an indexer foot, which drops onto the fish body. The indexer permits finer adjustment of the lateral position of a fish with respect to the cutter blade. The indexer mechanism moves diagonally along the conveyor, on a carriage guideway, and is driven by the same actuator as for the conveyor. Accordingly, the indexer moves at the same speed in the direction of

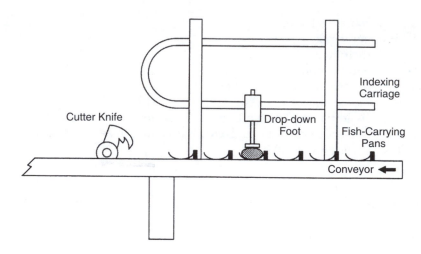

FIGURE 13.10
The industrial Iron Butcher.

motion of the conveyor. In addition, the indexer also has a lateral motion toward the cutter edge, in view of the overall diagonal movement. During this lateral motion, the indexer foot slides over the fish body until it engages with the collarbone at the gill plate of the fish. Subsequently, due to this engagement, the indexer foot pushes the entire fish laterally toward the cutter blade, thereby accomplishing the required positioning action. Just before the fish reaches the cutter, the indexer foot lifts off and starts the return motion along the carriage guideway. The rotary knife of the cutter, then, lops off the head of the fish.

The straightforward indexing mechanism of the Iron Butcher, even though mechanically robust, can result in two common types of positioning error. An "over-feed error" results if the indexer foot becomes engaged into some location of the fish body prior to reaching the collarbone. This could result due to an external damage or some other structural nonuniformity on the fish body. Consequently, the fish would be pushed too much in the lateral direction, with respect to the cutter. The subsequent cut would be wasteful, removing a chunk of valuable meat with the head. An "under-feed error" results if the indexer foot slides over the collarbone and the gill plate, or pushes too much into a soft or damaged gill plate. In this case, a portion of the head would be retained with the fish body after the cut. Such fish needs to be manually trimmed in a subsequent operation, and this would represent a reduction in the throughput rate and a wastage of labor. Another shortcoming of the Iron Butcher is the poor quality of cut even with accurate positioning. The primary contributing factors are clear. The fish move continuously, not intermittently, with the conveyor, and are not stationary during the cutting process. Also, the high-inertia rotary cutter has a pointed outer end, which first hooks the fish and then completes a guillotine-type cut. The combined effect of the cutter inertia, hooking engagement, and the motion of the fish during cutting is an irregular cut with excessive stressing and deformation of the fish body. This results in a lower product quality.

Motivated by the potential for improvement of this conventional machine, specifically with regard to waste reduction, productivity and production flexibility, a project was undertaken by us to develop an innovative machine for fish cutting. The machine has been designed, integrated, tested, and refined. The test results have shown satisfactory performance. The new Iron Butcher possesses the following important features:

1. High cutting accuracy; obtained using mechanical fixtures, positioners, tools and associated sensors, actuators, and controllers, which have been properly designed and integrated into the machine.

2. Improved product quality; achieved through high-accuracy cutting and also through mechanical designs that do not result in product damage during handling and processing, along with a quality assessment and supervisory control system, which monitors the machine performance, determines the product quality, and automatically makes corrective adjustments.

3. Increased productivity and efficiency; attained through accurate operation and low wastage, with a reduction in both downtime and the need for reprocessing of poorly processed fish.

4. Flexible automation; requiring fewer workers for operation and maintenance than the number needed for a conventional machine, which is possible due to the capabilities of self-monitoring, tuning, reorganization, and somewhat autonomous operation, as a result of advanced components, instrumentation, and the intelligent and hierarchical supervisory control system of the machine.

13.3.2   Machine Features

A view of the new machine for fish cutting is shown in Figure 13.11(a). The hardware structure of the machine is schematically shown in Figure 13.11(b). A brief description of the machine is given now. The fish are placed one by one on the conveyor at the feeding end of the machine. Subsequently, as a fish passes through the primary imaging station, an image of the fish is generated by a CCD camera and captured by a SHARP GPB image processor. This is the primary imaging module of the machine. Simultaneously, the thickness of the fish in the head region is sensed through an ultrasonic position sensor (see Chapter 6). The image is processed by the GPB card, which occupies one slot of the

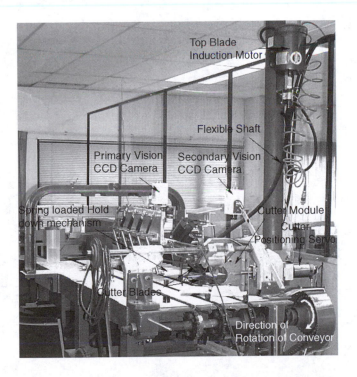

FIGURE 13.11(a)
The "Intelligent Iron Butcher."

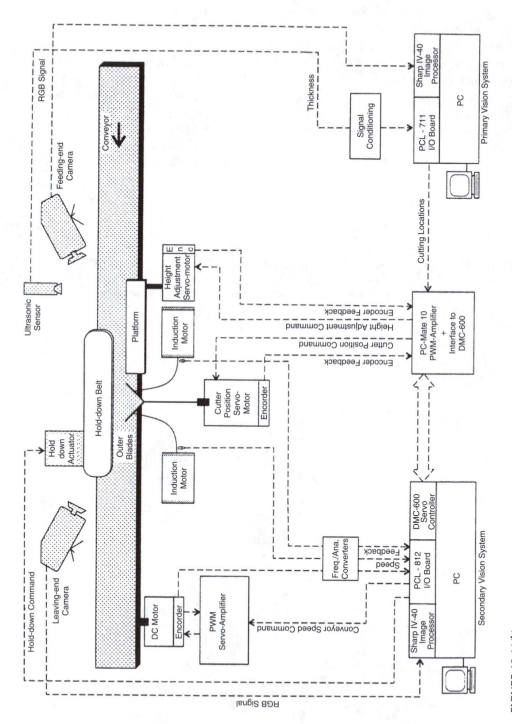

FIGURE 13.11(b)
The hardware architecture of the machine.

primary control computer, a dedicated personal computer (PC). The main image processing steps are filtering, enhancing, thresholding and labeling of the image objects. In this manner, the gill plate and correspondingly, the collarbone, of the fish are identified and gauged. This information is then used to determine the desired position of the cutter. As the fish leaves the primary imaging module it enters a positioning platform whose purpose is to deliver each fish symmetrically into the cutter unit. The desired vertical position of the delivery platform is determined according to the thickness of the particular fish, as measured by an ultrasonic sensor and read into the control computer through a PCL-711 Input/Output board. The platform is driven by a dc servomotor (see Chapter 9) through a lead-screw-and-nut arrangement (see Chapter 3). The desired position is the reference with which the actual position, as measured by the optical encoder (see Chapter 7) of the servomotor, is compared to generate the servo signal for platform control. The cutter unit consists of a pair of circular blades arranged in a V-configuration and driven by two remotely placed, 3-phase induction motors (see Chapter 9), which are connected to the blades through long flexible shafts (see Chapter 3). The V-configuration allows the cutter to reach into the head region below the collarbone and recover the useful meat that is present in this region. In a manner similar to the delivery platform, the cutter is positioned by a dc servomotor, through a lead-screw and nut arrangement. The two servomotors, one for the delivery platform and the other for the cutter, are controlled by means of a GALIL DMC-600 two-axis controller, which receives reference commands from the control computer and generates input signals for the two pulse-width-modulated (PWM) amplifiers (see Chapter 9) which drive the motors.

There is a hold-down mechanism for fish along the conveyor, starting from the primary imaging station and ending at the cutting station. This device consists of a matrix of spring-loaded rollers, which can apply a desired distribution of forces on the fish body so that each fish would be properly restrained. The hold-down mechanism will ensure that the configuration of a fish, as measured by the imaging system, would not change as it reaches the cutter and also will provide an adequate holding force during the cutting process itself. Clearly, the restraining forces should not crush or damage the fish. The hold-down force is controlled by the supervisory control system of the machine. A secondary CCD camera has been mounted on the product-exit side of the cutting station. The images generated by this camera are captured and processed by a dedicated SHARP GPB card that is mounted within another personal computer. This computer, which hosts the secondary vision system, also serves as the supervisory-control computer. A direct means of monitoring and assessing cutting quality is provided through the secondary imaging module. The conveyor is driven by a variable-speed dc motor through a PWM amplifier, according to a speed command given by the supervisory control computer. This computer reads the optical encoders of the conveyor motor and the cutter induction motors, through frequency-to-analog conversion circuitry and a PCL-812 Input/Output board; and from this information the conveyor speed and the loads at the two cutter blades are determined. The responses of the two dc servo systems together with the cutter loads represent additional data for on-line assessment of the machine performance; primarily, the cutting quality. The conveyor speed provides useful operational data. The integrated hardware configuration of the prototype machine is shown in Figure 13.12.

The typical throughput rate of the machine is 2 fish/s, which allows a maximum of 500 ms for a group of tasks associated with the processing of each fish. Sensing, sensory processing, platform positioning, cutter positioning, and cutting are the primary tasks that are important in this regard, and should be carried out in the given sequence. It is clear that the platform and the cutter could be positioned for the next fish only after the current fish has been cut. Sensing and sensory processing may be done while another fish is being cut. It follows that a maximum duration of 500 ms is available for simultaneous positioning

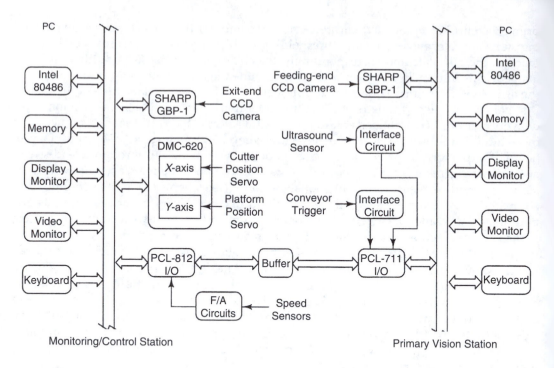

FIGURE 13.12
Hardware configuration of the new Iron Butcher.

of the platform and the cutter, and for the actual cutting process. In the present prototype, sensing and sensory processing take less than 400 ms. Also, positioning of the platform and the cutter takes less than 100 ms, which allows up to 400 ms for the cutting operation itself. This means, the maximum width of a fish can be up to 80% of the inter-fish spacing of the fish on the conveyor.

13.3.3   Hardware Development

Starting with a conceptual design for the new machine, the overall design and development were carried out, which involved component design and selection so as to meet the system specifications. Some components were developed in our laboratory, and several off-the-shelf components were acquired as well, depending on the availability, cost, and compatibility with the design specifications and system requirements. The present section outlines the development activity related to system hardware. In particular, the conveyor system, the cutter assembly and actuators, motion sensors, position controllers, and system interfacing hardware are discussed. The primary imaging system for position control and the secondary imaging system for supervisory control will be described in other sections.

13.3.3.1   Conveyor System

It was established that the throughput rate and the structural strength of an existing Iron Butcher were quite adequate for the new prototype. Accordingly, the conveyor and the support frame of an existing machine were adapted, with the addition of a new motor controller and other essential modifications. The fish conveyor consists of a set of three chains, which loop over the machine surface, with spacing lugs attached to them. The lug spacing along each chain is 22.8 cm (9 in) allowing for fish of widths up to 18 cm. The dc notor with belt drive and the variable-speed controller, in the conveyor system meet the ollowing specifications:

Throughput rate	= 2 fish/s
Linear speed	= 45.6 cm/s
Drive shaft speed	= 1 rev/s
Speed fluctuations	= 5% or less
Step response	= a rise time of less than 1 s

Speed regulation capability is more important for the conveyor, than the ability of fast speed control. It follows that the rise time need not be as low as that of, say, the cutter position controller. The conveyor motor is a separately-excited, 115 V, 1.5 hp, dc motor. The armature current rating of 12A is more than adequate even under highly transient motion conditions of high motor load. A speed control loop as shown in Figure 13.13, was implemented, as the speed control capability is an integral part of the supervisory controller of the overall machine. This aspect will be discussed in a separate section. A pulse-width-modulated (PWM) amplifier, model ESA-10/75, provided by GALIL and having a voltage rating of 25–75 VDC, a current rating of 10 A continuous and 25 A peak, and a gain adjustable in the range of 0.3–0.8 A/V is employed in the control loop. The conveyor speed is sensed using an optical encoder mounted on the conveyor-drive shaft, which provides a feedback signal for the system-monitoring and supervisory-control computer. The computer in turn generates a 12 bit digital command for the speed controller, which is converted into an analog signal in the range of 0–5 VDC by the PCL-812 Input/Output interface board within the computer. This analog signal is fed into the PWM amplifier, which converts the voltage to a pulse-width-modulated current for driving the conveyor motor.

13.3.3.2 Cutter Assembly and Actuators

The cutter assembly consists of the cutter unit itself and a platform, which ensures symmetric delivery of a fish into the cutter unit, as sketched in Figure 13.14. The two cutter blades of rotary type, arranged in a V-configuration to improve meat recovery, are driven through long flexible shafts by remotely placed 3-phase induction motors.

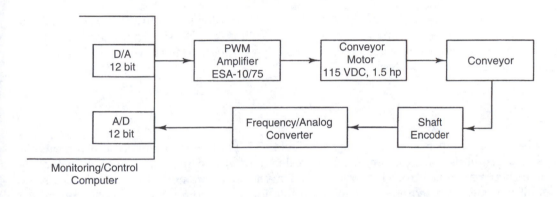

FIGURE 13.13
Speed control loop of the conveyor system.

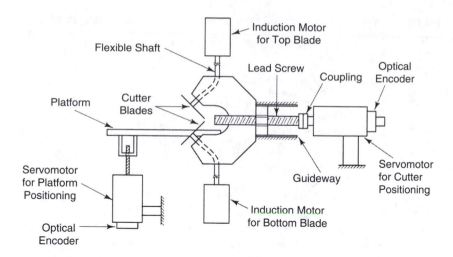

FIGURE 13.14
The cutter assembly.

The cutter unit is laterally positioned (*x*-axis) and the delivery platform is vertically positioned (*y*-axis) using two independent dc servomotors. The specifications of these two degrees of freedom are given below:

x-axis motion:

Time between cuts	= 500 ms
Cutting duration	= 300–400 ms
Time available for positioning	= 100 ms
Total moving mass	= 9 kg
Maximum positioning stroke	= 25.4 mm
Positioning accuracy	= ±0.5 mm

y-axis motion:

Time available for positioning	= 100 ms
Total moving mass (against gravity)	= 2 kg
Maximum positioning stroke	= 12.5 mm
Positioning accuracy	= ±0.5 mm

Lead-screw-and-nut pairs are used for linking the motor shafts to the moving units. In addition, a flexible coupling is incorporated into the *x*-axis, as shown in Figure 13.14. The motor and the lead-screw mechanism are chosen through an iterative use of the equation (see Chapter 8 and Chapter 9),

$$T = \left[J + \frac{mr^2}{e} \right] \frac{a}{r} = \frac{r}{e} \cdot F_R \qquad (13.7)$$

where, the parameters are as defined in Table 13.5. For a worst-case scenario, a triangular speed profile is used to compute the acceleration (*a*) and the peak speed such that the specified maximum stroke is traveled in 100 ms. Next, a commercially available lead-screw unit that meets the required positioning resolution, and a compatible dc motor, are chosen, and the associated motor torque (*T*) is computed using Equation 13.7. Then the worst-case torque-speed values, as computed in this manner, are checked against the torque-speed

TABLE 13.5

Servomotor Parameters for the Cutter Assembly

Parameter	x-axis	y-axis
Moved mass (m)	9 kg	2 kg
Rotational inertia (J)	1.9×10^{-4} kg$\cdot$m^2	2.6×10^{-5} kg$\cdot$m^2
Lead-screw pitch (r)	2.02×10^{-3} m/rad	1.01×10^{-3} m/rads
Efficiency (e)	75%	75%
Acceleration (a)	10.16 m$\cdot$s^{-2}	5.08 m.s^{-2}
Resisting force (F_R)	0.1 kg	2.0 kg
Max. torque (T)	2.4 N$\cdot$m	0.32 N$\cdot$m
Speed (s)	2400 rpm	2400 rpm

characteristics of the particular motor as given in the motor data sheet (see Chapter 9). Through an iterative procedure, which follows these steps, the motors and the lead-screws as specified in Table 13.5, were chosen for the cutter assembly. Parameters of the other chosen components of the cutter assembly are given below.

x-axis:

  i. A dc permanent magnet motor, Model 500/1000B GALIL, Voltage = 64 VDC, Peak torque = 5.1 N$\cdot$m, Continuous torque = 1.11 N$\cdot$m, Moment of inertia = 1.9 $\times$ 10^{-4} kg$\cdot$m^2, and Maximum speed = 3750 rpm.

  ii. A backlash-free lead-screw arrangement with 1/2 in (1.27 cm) pitch.

  iii. Two circular blades, each 6 in (15.24 cm) in diameter, arranged in a V-configuration and mounted on the cutter structure.

  iv. Two high-speed induction motors: 3 phase, 230 V, 3600 rpm, 1.5 hp, and two flexible shafts for individually coupling the induction motors to the cutter blades.

y-axis:

  i. A dc permanent magnet motor, Model 50/1000 GALIL, Voltage = 32 VDC, Peak torque = 1.45 N$\cdot$m, Continuous torque = 0.21 N$\cdot$m, Moment of inertia = 2.6 $\times$ 10^{-5} kg$\cdot$m^2, Maximum speed = 3750 rpm.

  ii. A backlash-free lead-screw arrangement with 1/4 in (0.635 cm) pitch.

13.3.3.3   *Motion Sensors*

The two imaging systems, which also fall within the category of machine sensors, will be described in separate sections. In the present section the remaining motion sensors will be considered. Of particular interest are the optical encoders (see Chapter 7) of the conveyor, cutter blades and the positioning axes of the cutter assembly, and the ultrasonic sensor (see Chapter 6) for measuring the fish thickness.

Figure 13.15 shows the hardware of motion sensing for the cutter blades and the conveyor. Reflective type optical sensors are used as shaft encoders for the induction motors of the cutter blades (Figure 13.15(a)). An incremental optical encoder (see Chapter 7) mounted on the conveyor shaft provides motion information of the conveyor (Figure 13.15(b)). All three sensors generate frequency signals proportional to the corresponding shaft speeds. Three separate frequency-to-analog (F/A) converters (see Chapter 4), each designed based on the LM2917 monolithic frequency-to-voltage converter (Figure 13.15(c)), facilitate simultaneous reading of the three speed channels. The linear operating range and the gain of each speed transducer are adjustable through the external components R and C, as shown

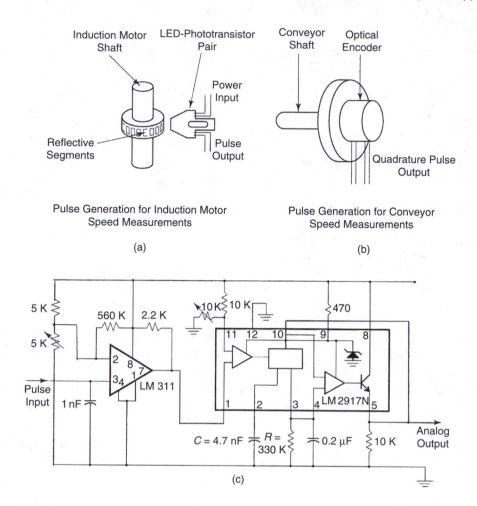

FIGURE 13.15

Speed sensing hardware: (a) Optical encoder for cutter blade load sensing; (b) Optical encoder for sensing conveyor speed; (c) Circuit of the frequency-to-analog converter for an encoder.

TABLE 13.6

Transducer Parameters of the Conveyor and the Cutter Blade

Component	Conveyor	Induction Motor
Transducer type	Incremental encoder	Optical object sensor
Counts/rev	512	8
Displacement/rev	45.7 cm/rev	n/a
A/D conversion	12 bit	12 bit
A/D output	±10 V	±10 V
A/D resolution	4.88 mV	4.88 mV
F/A gain	12.5 mV/Hz	12.5 mV/Hz

in Figure 13.15(c). The induction motor speed also provides the associated slip and hence the motor torque (see Chapter 9). The parameters of the sensory system are given in Table 13.6. The resolution of the conveyor speed measurement, which corresponds to a single count of the encoder, is given by

$$45.72 \frac{\text{cm}}{\text{rev}} \times \frac{1 \text{ rev/s}}{512 \text{ Hz}} \times \frac{1 \text{ Hz}}{12.5 \text{ mV}} \times 4.88 \text{ mV} = 0.035 \text{ cm/s}$$

since the nominal speed of the conveyor is 25 cm/s, this resolution is better than 1%. Similarly, the speed resolution of the induction-motor sensors is given by

$$\frac{1 \text{ rev/s}}{8 \text{ Hz}} \times \frac{1 \text{ Hz}}{12.1 \text{ mV}} \times 4.88 \text{ mV} = 0.0503 \text{ rev/s} = 3.02 \text{ rpm}.$$

At a nominal speed of 3600 rpm, this represents a resolution better than 0.1%. Also, this translates into a 2% resolution for load sensing of the induction motors.

The parameters of the sensory systems of the cutter position (x-axis) servo and the platform position (y-axis) servo are given in Table 13.7. The servo controllers are of DMC-620 type, with an accuracy of ± 1 quadrature count. It follows that the positioning resolution of the x-axis drive is given by

$$\frac{1}{2} \text{ in/rev} \times \frac{1}{4000} \text{ rev/count} \times 25.4 \text{ mm/in} = 0.003 \text{ mm/count}.$$

Similarly, for the y-axis, it is 0.0015 mm/count. It follows that both axes have a positioning resolution that is well in excess of the required accuracy of 0.5 mm. The resolution alone will not guarantee this level of accuracy, however.

Vertical positioning of the platform, which delivers fish into the cutter, is based on the thickness of the fish. Since a single CCD camera, mounted vertically perpendicular to the conveyor surface, is used as the primary imaging sensor, it cannot provide the thickness information of a fish. An ultrasonic displacement sensor, mounted alongside the camera is used to obtain the necessary thickness information. It provides an analog voltage signal proportional to the distance from the sensor to the top surface of the fish in the gill region. This reading is activated simultaneously with the camera trigger. Since the distance from the conveyor bed to the sensor head is fixed and accurately known, the thickness of a fish is easily computed. Table 13.8 provides relevant parameters of the ultrasonic sensor.

TABLE 13.7

Sensory Parameters of the Positioning Servo Loops

Parameter	x-axis	y-axis
Encoder Type	Incremental, Quadrature	Incremental, Quadrature
Output	1000 pulses/rev	1000 pulses/rev
Counts	4000 counts/rev	4000 counts/rev
Accuracy	± 1 count	± 1 count
Lead screw	2 rev/inch (0.787 rev/cm)	4 rev/inch (1.575 rev/cm)

TABLE 13.8

Parameters of the Ultrasonic Sensor

Parameter	Value
Type	Analog ultrasonic proximity sensor
Range	10–75 cm, adjusted to 24–32 cm
Input power	10-30 VDC, 50 mA
Output	0-5 VDC
Response speed	50 ms
Bandwidth	20 Hz
Repeat accuracy	1% of full scale

13.3.3.4   *Position Controllers and Interfacing Hardware*

The desired positioning command for the cutter (*x*-axis) is provided by the primary imag processing system while the command for the fish-delivery platform (*y*-axis) is provide by the ultrasonic sensor. The associated sensory information is captured by the primar computer, through a SHARP GPB image processing card and a PCL-711 I/O interfacing board, respectively, as indicated in Figure 13.11(b). The direct control is achieved by mean of a DMC-620, two-axis servomotor controller. These modules and a PCL-812 I/O inter facing board are linked with the PC bus of the secondary computer. The positioning commands from the primary computer are received by the DMC servo controller. The drive signals for the two servomotors are generated by the controller and are sent to the PWM amplifiers of the two motors.

The DMC controller card consists of a digital-control compensator (see Chapter 12), a command interpreter, and a communication interface. It utilizes the encoder signals from the two motors and the desired positions from the primary computer, to complete a control loop as shown in Figure 13.16. The digital compensator has the *z*-transformed (discrete) transfer function

$$D(z) = K\frac{z-q}{z-p} + \frac{K_i}{z-1} \tag{13.8}$$

as discussed in Chapter 12 and, *FA* in Figure 13.16 is a simple feedforward gain, which boosts the output to the motor by an amount proportional to a commanded acceleration. The transfer function $D(z)$ is fully programmable in the sense that the characteristic parameters K, p, q, K_i, and *FA* are adjustable on the fly, through the PC-bus communication interface (see Chapter 11). Similarly the torque limit block (*TL*) has an adjustable deadband and a saturation limit.

The DMC controller can operate in various modes of servo control including indepen-dent, profiled positioning of each axis; incremental positioning (commands are interpreted with respect to the current position); jogging (constant speed operation); and coordinated motion of the two axes (synchronized operation of the two axes). In the present application, independent profiled motion of the two servomotors is used. In this mode, the acceleration (*AC*), the slew speed (*SP*) and the end position (absolute-*PA* or relative-*PR*) for each axis are specified in advance. With a motion command of begin (*BG*), the controller generates either a trapezoidal or triangular velocity profile and a position trajectory. A new position output is generated every 1 ms. The motion is deemed complete when the target position has been reached by the controller.

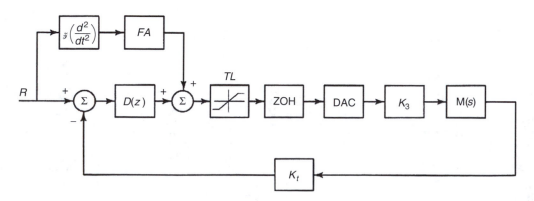

FIGURE 13.16
Control loop implementation in the DMC controller.

Communication between the DMC and the host PC is in the form of ASCII characters see Chapter 11), where data are sent and received via READ and WRITE registers of the controller card, which map into the I/O space of the host computer. A C-language-callable, high-level library of subroutines have been developed for the purpose of the present application, particularly to handle the communication between the controller and the machine. The commands that are required to position the cutter servo and the platform servo are collected into a single program file and downloaded into the DMC in the beginning of the machine operation. During each cycle of operation, this program reads the position commands that are sent to it from the primary image processing subsystem (including the ultrasonic sensor) and then converts them into the respective DMC position commands. Once the position commands are issued, the program jumps back to the starting address and waits for the next command from the primary imaging subsystem.

Simultaneously to the motion of the servomotors, the secondary (monitoring and supervisory control) computer (see Figure 13.11(b)) keeps track of the position response of the two motors through its own communication interface (PCL 812) to the controller. This computer also intervenes the compensator transfer functions at a supervisory level in order to update or tune the parameters of the compensators when such an action is being demanded by the intelligent supervisory controller.

The PCL is a high speed, multi-function, data acquisition card for PC compatible computers (also see Chapter 11). The principal features of interest are its 12-bit successive-approximation A/D converter (or ADC—see Chapter 4) with a maximum sampling rate of 30 kHz, two 12-bit D/A (or, DAC—see Chapter 4) output channels, and 16 TTL compatible digital-input and 16 digital-output channels. Also a programmable timer/counter (see Chapter 10) enables precise trigger control for A/D data conversion. Two such data acquisition cards (711 and 812) serve the prototype machine, one for the primary image processing subsystem and the other for the control/monitoring subsystem, as indicated in Figure 13.11(b).

The cutting locations are determined by the primary image processing system and are communicated to the DMC controller via the digital output ports of the 711 I/O card. The hand-shaking mechanism for this data transfer (*Send, Receive,* and *Acknowledge* functions—see Chapter 11) is automatically handled by the program that is downloaded to the DMC controller. The analog output of the ultrasonic sensor is also read by means of the A/D ports of the same I/O card in the primary imaging subsystem.

The analog inputs corresponding to the three speed signals from the conveyor, and the top and bottom blades of the cutter are digitized by means of the ADC at a rate of 30 kHz, through the 812 I/O card in the supervisory computer. Each sample of the blade speed comprises the average of 100 readings. For the conveyor, 20 readings are averaged for each sample. The averaging process is particularly important since it smoothes out ripples that are inherently present in the output signal of a frequency-to-analog converter (see Chapter 4). At this rate of conversion, a sampling frequency of 100 Hz is easily achieved for all three-speed channels. The analog output that drives the conveyor motor is provided through the D/A converter of the PCL-812. A 12-bit D/A conversion to a maximum of 10 VDC yields a resolution of 4.8 mV, which is quite adequate for the control of the conveyor motor.

The two servomotors of the two positioning axes are coupled to the servo-motor controller through a matching pair of amplifiers. Both amplifiers are of the PWM type (see Chapter 9), working on a primary line voltage of 115 VAC at 60 Hz. The governing characteristics of the amplifiers are as follows:

x-axis:

Output = 64 VDC, 7 A continuous/10 A peak, Gain = 1.0 A/V,
Maximum signal input =10 VDC.

y-axis:

$$\text{Output} = 32 \text{ VDC}, 4 \text{ A continuous}/5 \text{ A peak, Gain} = 0.5 \text{ A/V},$$
$$\text{Maximum signal input} = 10 \text{ VDC}.$$

13.3.4 Image Processing for Cutter Positioning

In the new machine, accurate positioning of the cutter is important for maximizing the meat recovery. The approach used in the machine is to image the head region of each fish using a (primary) CCD camera and process the image so as to detect the gill plate. Since the collarbone is known to lie underneath the gill edge, the desired cutter position can be measured in this manner. A GPB-1 image processing board, which is located in a single slot of the primary computer of the machine, is used for this purpose. The GPB-1 is interfaced to the host processor through a block of 64 I/O addresses. It provides programmability through a library of extensive image processing subroutines. Figure 13.17 schematically illustrates the internal arrangement of a GPB-1 board. From the viewpoint of the user, the two most important parts of the board are the core and the memory banks. The on-board memory is divided into 4 image banks and each bank contains 3 planes, giving a total of 12 memory banks. Each bank of memory can hold a full camera image from one color channel of red (R), green (G), or blue (B). A special display buffer, consisting of three more memory banks corresponding to R, G, and B channels, facilitates displaying of a full-color composite image. Image processing is accomplished through the core of the GPB board, which reads an image from the memory, processes it according to the function specified in the image-processing program, and then writes the result into a new memory bank.

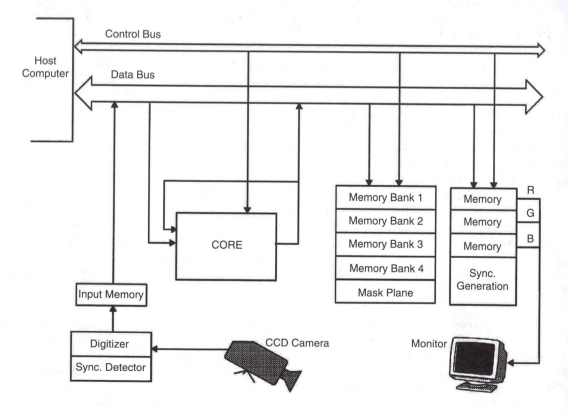

FIGURE 13.17
Functional diagram of a GPB-1 image processing board.

The size of each bank of image memory is 512 columns by 512 rows (pixels) by 8 bits (gray-level resolution). At any time, a subset of this matrix may be defined as the region of interest in order to limit the area of an image that would be processed, thereby reducing the processing time. The processing speed of the core is 40 ns/pixel, or equivalently, 700 MIPS.

The image source can be one of four video (NTSC), composite signals (RS-170); provided either from the CCD cameras or directly from the data bus of the host. The latter method allows the host machine to display images that have been processed and saved from earlier sessions. Camera inputs provide on-line (live) imaging capabilities. An image in a memory bank can be moved, under software control, to a special display memory, which will convert the image into a standard VGA format for displaying on a video monitor.

The prototype fish-processing machine that has been developed by us is equipped with two GPB-1 processors, one for primary image processing and the other for visual data monitoring for product-quality assessment and supervisory control. A dedicated VGA monitor is attached to the secondary GPB-1 processor (Figure 13.11(b)) so that the user can visually examine the images of the processed fish. Capturing an image from the camera takes about 33 ms, and the ensuing processing time depends on the required details of image understanding. In the case of quality evaluation through vision, the captured image undergoes smoothing, thresholding, filtering, and labeling before some useful information on the quality is extracted. These image processing operations and further preprocessing take close to 1 sec. At the present level of development, a cycle time in the order of few seconds is considered adequate for monitoring purposes since any improvement to this figure is likely to increase the cost of additional processing power and hence make the machine economically unattractive.

Several methods have been developed by us for measurement of the gill position, which is required in cutter control. The technique employed in the new fish-processing machine is termed the binary projection method, and is outlined below.

The method consists of the following four main steps:

i. Reverse the intensity of the image; specifically,

$$I'(x,y) = 255 - I(x,y), \quad x = 0, \ldots, i, \ldots, 512; \quad y = 0, \ldots, j, \ldots, 512 \tag{13.9}$$

where, $I(x, y)$ is the intensity value at a pixel coordinate point (x, y).

ii. Perform the Sobel operation along the x direction. This is a directional filtering operation, which will enhance the edges along the y direction and will suppress edges along the x direction. Consequently the gill plate edge (in the y direction) will be emphasized.

iii. Threshold the image and project it to the x axis using:

$$h(k) = \sum_{j=0}^{512} PI(k), \quad k = 0, \ldots, i, \ldots, 512 \tag{13.10}$$

where, $h(k)$ represents the projected signature, and PI is the pixel intensity value. Thresholding involves generation of a binary image, where the intensity values above the threshold ae considered "1" and those below the threshold are considered "0." In particular, "1" represents "white" and "0" represents "black."

iv. Detect the peak point of the projected binary image to find the gill position; thus,

$$x_g = \max h(k), \quad k = 0, \ldots, i, \ldots, 512 \tag{13.11}$$

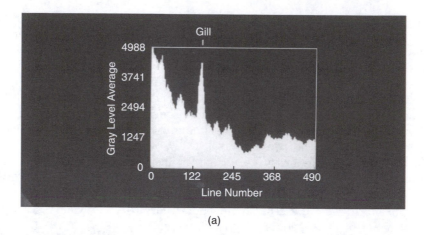

(a)

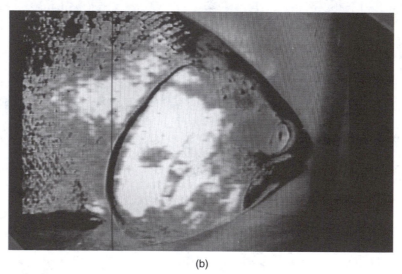

(b)

FIGURE 13.18
Cutting position measurement through image processing: (a) A projected image intensity signature; (b) Detected gill (cutter) position (denoted by the line).

A projected intensity signature for the head region of a fish, as obtained by this method, is shown in Figure 13.18(a). The detected gill position on a fish, and accordingly, the desired cutter position, is shown in Figure 13.18(b).

For proper performance of the cutter control, it is necessary to synchronize the image capture and processing with the conveyor motion. In the prototype machine, this is accomplished by using a magnetic switch (or, a limit switch — see Chapter 7), which is activated as a fish reaches the imaging region. Since the signal from the magnetic switch is noisy, the circuit shown in Figure 13.19 is used, which employs two serially-connected, retriggerable, monostable, multivibrators. The first multivibrator is triggered by the first falling edge of the switch signal (see Chapter 10), which will mask any other signal transitions for a specified period of time (typically 300 ms, for a throughput rate of 2 fish/s), which is adjustable using the variable resistor $R3$ in Figure 13.19. The second multivibrator is triggered by the subsequent rising edge output of the first multivibrator. A clean pulse is generated in this manner (see Chapter 7 and Chapter 10). This pulse is read by the primary image-processing computer and used for synchronization purposes. Specifically, the pulse is polled by the image processing program. An active pulse will cause the system

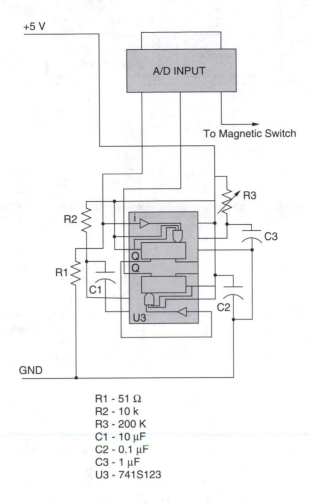

R1 - 51 Ω
R2 - 10 k
R3 - 200 K
C1 - 10 µF
C2 - 0.1 µF
C3 - 1 µF
U3 - 741S123

FIGURE 13.19
The synchronization circuit for object detection and image processing.

to capture an image, carry out the necessary processing for gill measurement, and transmit the corresponding cutter position to the DMC controller.

13.3.5 Supervisory Control System

The overall control system of the fish-cutting machine has a variety of functions which may be organized according to an appropriate criterion. The criteria that are appropriate here will include the required speed of control (control bandwidth), the level of detail of the information that is handled (information resolution), crispness of the knowledge that is involved (fuzzy resolution), and the level of intelligence that is needed to carry out a particular function. It may not be necessary to incorporate all such criteria in a particular system since there exist at least intuitive relationships among these criteria. This is the case particularly if the control system is organized into a hierarchical architecture.

A fully automated fish-cutting machine will incorporate a multitude of components that have to be controlled and properly coordinated in order to obtain satisfactory overall performance. The plant operation will involve many different types of actions such as sensing, image processing, direct control, performance monitoring, and component coordination. Furthermore, an intelligent agent operating in a supervisory capacity will be

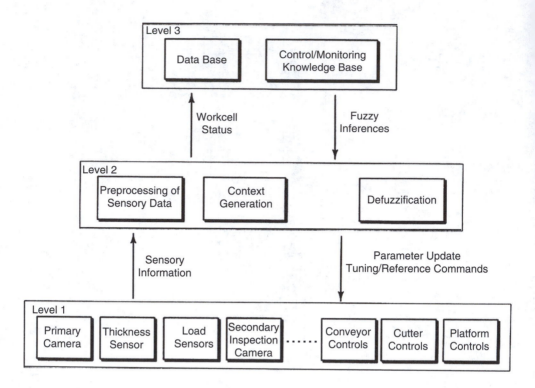

FIGURE 13.20

Hierarchical supervisory control system of the machine.

able to perform such high-level assignments as goal description, decomposition of a goal into subtasks, and sub-task allocation. A multi-layered architecture may be developed for the machine by considering the nature and requirements of these various functions. In the present application the three-level architecture that is schematically shown in Figure 13.20 has been implemented. This structure may be considered as a supervisory control system. The lowest layer consists of high-resolution sensors, process actuators, and associated direct controllers. The intermediate layers primarily carry out information preprocessing, determination of the machine status, and command transfer for operating and modifying the performance of the system components. The top layer carries out system monitoring and high-level supervisory control activities.

The bottom layer of the control system of Figure 13.20 carries out direct control of system components in a conventional sense. Specifically, the component responses are sensed at high resolution, control commands are generated on this basis at high speed, according to hard (i.e., crisp) algorithms, and the components are directly actuated using the control commands. These high-resolution, high-bandwidth, direct control operations do not need a high degree of intelligence for their execution. Specifically, in our prototype machine, the following sensory information is acquired in the bottom layer:

1. Position and speed information of the positioning servo systems and the conveyor, from the respective optical encoders.
2. Cutter load information from the induction motors, which drive the cutter blades.
3. Visual images from the position sensing CCD camera (primary).
4. Visual images from the task monitoring CCD camera (secondary).
5. Thickness information from the ultrasound height sensor.

The associated raw signals are characterized by the following properties:

1. Large quantities of data points are collected.
2. Data points are generated in a rapid succession.
3. The sensory signals have a high resolution and tend to be quite precise, in the absence of noise.

For example, an image may contain about 250 kilobytes of information, and may be captured every 0.5 s. This will represent an information rate of 500 kB/s for this optical sensor alone. Within the bottom layer, some sensory signals would be directly used, without "intelligent" preprocessing, for device control. This is the case with the encoder signals from the servomotors and the ultrasonic thickness sensor, which are directly used in feedback for positioning control.

The top layer of the hierarchical system shown in Figure 13.20 is considered the most intelligent level of the machine. It is responsible for monitoring the overall performance of the system, including the product quality, production efficiency, and the throughput rate. On this basis, it makes decisions not only on product quality (grade) but also for appropriate adjustments and modifications to the system so as to improve the machine performance and to maintain within specification the operating conditions of various components in the machine. The nature of these objectives is such that they are typically qualitative and descriptive. As such, it is difficult to transcribe them into a set of quantitative specifications. The required actions to achieve these goals are typically based on knowledge, experience, and expertise of human operators and skilled workers. Consequently, they may be available as a set of linguistic statements or protocols and not as crisp algorithms. Furthermore, the available information for decision making at this top level, may be imprecise, qualitative, and incomplete as well. It follows that a high level of intelligence would be needed at the top level to interpret the available information on the system status and to take appropriate "knowledge-based" actions for proper operation of the machine. In the machine, the knowledge base is represented as a set of fuzzy-logic rules. This is in fact a collection of expert instructions on how to judge the product quality and what actions are needed to achieve a required performance from the machine under various operating conditions. The decision-making actions are carried out by matching the machine status with the knowledge base, using fuzzy logic, and particularly the composition-based inference.

The high resolution information from the crisp sensors in the bottom layer of the machine have to be converted into a form that is compatible with the knowledge base in the top layer before a matching process could be executed. The intermediate layer of the hierarchy of Figure 13.20 accomplishes this task of information abstraction, where the sensory information from the bottom layer is preprocessed, interpreted, and represented for use in inference making in the top layer. Each stream of sensory data from the bottom layer is filtered using a dedicated processor, which identifies trends and patterns of a set of representative parameters in these signals. The operations performed by the preprocessors may include averaging, statistical computations such as standard deviation and correlation, peak detection, pattern recognition, and computation of performance attributes such as rise time, damping ratio, and settling time of a machine component (see Chapter 5 and Chapter 12). Next, these parameters are evaluated according to appropriate performance criteria and transcribed into fuzzy quantities that are compatible with the fuzzy condition descriptors of the top-level knowledge base.

Next we will describe the procedures that are employed in processing the low-level sensory signals from the machine components so as to establish the performance status of the machine, and in particular, the quality of the processed fish. The specific sensory information that is processed in this manner includes visual images from the quality-monitoring

CCD camera (secondary), servomotor position signals from the optical encoders, and the load profiles of the cutter blades.

13.3.5.1 Image Preprocessing

The main purpose of the secondary image preprocessor of the machine is to extract context information from raw images, which will eventually lead to the extraction of high-level information such as the quality of processing and the quality (or, grade) of the end product. The problem of assessing the quality of a fish product is mostly aesthetic in nature. It differs from the apparently similar problem of evaluating the quality of manufacture of an industrial component such as a turbine blade. In the latter case, the quality could be gauged through direct measurement of dimensions and to a lesser extent by the detection of surface defects. The quality of processed fish, in contrast, is not amenable to simple quantification based on a few dimensions, shapes, or surface defects. At least in theory, however, it could be considered as a complex combination of such features. In the sequel, several of these features are considered as qualitative or imprecise estimates that approximately indicate quality, and are consequently parameterized by means of fuzzy linguistic variables.

The top-level knowledge base of the hierarchical system of the machine will derive two context variables from the visual information; namely, the accuracy of cut and the quality of cut. The accuracy of cut can be considered deterministic since it is determined by the final error values of the cutter positioning system and the platform positioning system. In particular, the cutter position error is given by the measured difference between the gill position as gauged by the primary vision system and the actual position of cut as determined by the cutter-positioning sensor. The quality of cut is assessed based on the following geometric features:

1. Depth of the solid region of the cut section
2. Smoothness of the boundary contour of the cut section
3. Smoothness of the cut surface.

13.3.5.1.1 Image Acquisition and Processing

A well-designed lighting system is crucial in the acquisition of a good image for processing. The machine employs a structured lighting system custom-designed to flush out unnecessary details from the background as well as such details of the fish body as scales and dirt, while retaining salient features of the cutting region. Figure 13.21(a) and Figure 13.21(b) show such images of two separately processed fish; the former illustrating a typical cut of acceptable quality and the latter a cut of rather unacceptable quality. The image is smoothed by using a 3×3 binomial convolution kernal in order to reduce the effects of noise and irregularities. Since smoothing could wipe out some cutting irregularities that are of interest, convolution with a smoothing kernal has been applied only twice over the entire image. Application of the smoothing kernal, shown in Table 13.9, results in the transformation given by

$$F(i,j) = \frac{1}{16}[S(i-1,j-1) + 2S(i,j-1) + S(i+1,j-1) + S(i-1,j) + 4S(i,j)$$

$$+ 2S(i+1,j) + (i-1,j+1) + 2S(i,j+1) + S(i+1,j+1)] \tag{13.12}$$

where, $S(i,j)$ is the (i,j)th pixel of the original source image and $F(i,j)$ is that of the filtered (smoothed) image. The image is then thresholded at a fixed grey level of mid-range (128)

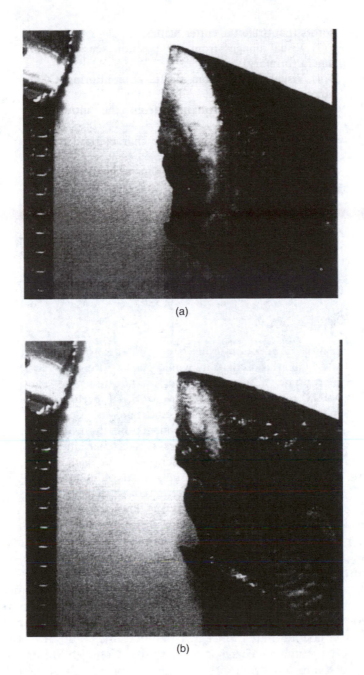

(a)

(b)

FIGURE 13.21
Camera images of processed fish: (a) Acceptable process quality; (b) Unacceptable process quality

to accentuate the body of the fish and the cutting region. Sobel filtering given by

$$L(i,j) = \Big| F(i-1, j-1) + 2F(i, j-1) + F(i+1, j-1) - F(i-1, j+1)$$

$$- 2F(i, j+1) - F(i+1, j+1)\Big| + \Big| F(i-1, j-1) + 2F(i-1, j) + F(i-1, j+1)$$

$$- F(i+1, j-1) - 2F(i+1, j) - F(i+1, j+1)\Big| \tag{13.13}$$

TABLE 13.9

Kernals used in the Convolution Operations:
(a) For Smoothing. (b) For Sobel filtering

(a)			(b)						
1	2	1	1	2	1		1	0	−1
2	4	2	0	0	0	+	2	0	−2
1	2	1	−1	−2	−1		1	0	−1

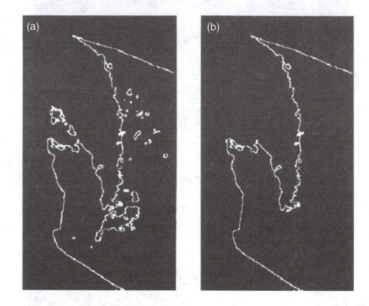

FIGURE 13.22
(a) Segmented binary image after processing, for an acceptable cut, (b) Extracted boundary contour of cut for the image.

with the kernel in Table 13.9, is applied to the thresholded image in order to highlight the edge features in both directions X and Y of the image. The regions of interest of the two images of fish after performing these two operations are shown in Figure 13.22 and Figure 13.23. Segmentation and labeling are then performed over the area of interest shown in Figure 13.22(a) and Figure 13.23(a), for the two images. The boundary contours of cut, shown in Figure 13.22(b) and Figure 13.23(b), respectively, are easily extracted as they represent the object with the maximum area in each segmented image. In order to recognize and compute the quality parameters corresponding to the images in Figure 13.22 and Figure 13.23, the image data are converted into "arc-length" profiles as shown in Figure 13.24. The arc length is measured along the boundary contour of each cut, starting from the lower left corner of the image. For the purpose of measurement, the X and Y image pixels are considered as the basis of the Cartesian coordinate system. The distance between adjacent pixels in both horizontal and vertical directions is reckoned as the unit length and in diagonal directions as $\sqrt{2}$ of the unit length.

The resulting arc-length profiles retain some of the angular properties of the original boundary contours. The data for the arc lengths can be readily derived from the chain codes for the corresponding contours of Figure 13.22(b) and Figure 13.23(b). Although the arc-lengths in Figure 13.24 are profiled against the Y-pixel coordinates, a similar set of profiles (not shown) could be obtained for the X-pixel coordinates simultaneously, without an additional computational burden.

FIGURE 13.23
(a) Segmented binary image after processing, for an unacceptable cut, (b) Extracted boundary contour of cut for the image.

13.3.5.1.2 Indicative Features of Quality

The arc-length profiles are made up of somewhat linear curve segments where certain points of interest on the boundary contour of cut could be easily identified. For the purpose of assessing quality, these features of interest are: Point a, which signifies the start of the boundary contour; Point b—the start of the solid region of cut; and Point c—the end of the solid region of cut; Point f — the end of the boundary contour. The characteristic arc-length profile of a good, symmetric cut is shown in Figure 13.24 (a) while Figure 13.24 (b) corresponds to an asymmetric and jagged cut. An algorithm which searches for gross maxima and minima is used to recognize the points of interests (points a, b, c and f), within each curve. These points give the X and Y coordinates of the salient points of the boundary contour of cut. Now, the following quality indicative indices can be defined:

$$\text{Index of depth of the solid region of cut } D_c = \frac{Y_{bc}}{Y_{af}} \tag{13.14}$$

where
Y_{bc} = depth of the solid region of cut from b to c, in pixel coordinates
Y_{af} = width of fish from a to f in pixel coordinates

$$\text{Index of smoohness of the boundary contour of cut } S_c = \frac{Y_{bc}}{l_{bc}} \tag{13.15}$$

where
l_{bc} = arc length from b to c

A larger value for D_c signifies a better cut while a larger value for S_c signifies a smoother cut. The smoothness of the surface over a cutting region is strongly indicated by the number of labeled image objects in the region of interest, as clear from Figure 13.22 and Figure 13.23.

$$\text{Index of smoothness of the cut surface } S_r = N_l \tag{13.16}$$

where, N_l = number of labelled image objects within the region of interest.

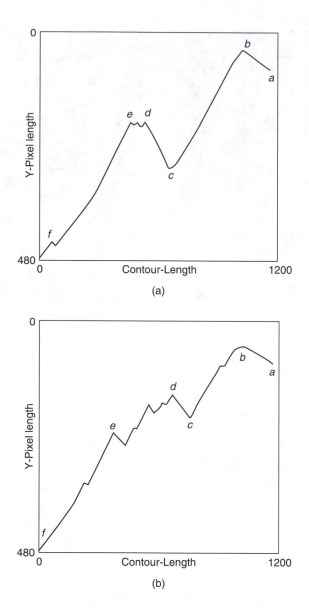

FIGURE 13.24
Arc-length profiles for processed fish: (a) Acceptable cut; (b) Unacceptable cut.

Figure 13.23(a) contains roughly twice as many image objects as Figure 13.22 (a). This signifies, for example, the roughness of the cut region of the two images of fish in Figure 13.21 (a) and (b). Ideally, under good operating conditions of the machine, the left side of a boundary contour of cut should be free of debris and loose objects. Otherwise, an incomplete cut would be implied and readily observed on its image.

13.3.5.1.3 *Building the Visual Context Database*

Computation of quality-indicative indices is the first step of preprocessing that is carried out on the raw images of fish. The next step is to assign proper membership functions for the fuzzy sets characterizing the relative importance of each indexing parameter and building a fuzzy linguistic context taking the quality-indicative indices as its variables.

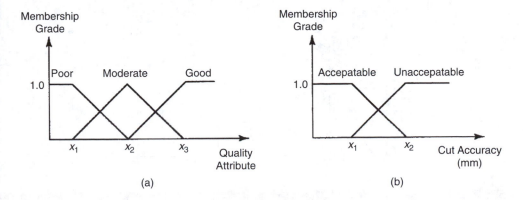

FIGURE 13.25
Context membership functions: (a) For visual quality; (b) For cutting accuracy.

Altogether there are four fuzzy context variables, three obtained from the quality-indicative indices and one from cut accuracy, in the visual context database. The fuzzy context for the quality-indicative features of D_c, S_c, and S_r is generated by using a uniform membership function for all three variables as shown in Figure 13.25(a). The resulting fuzzy variables are *Cut_depth*, *Cut_contour*, and *Cut_surface*, respectively. All three variables have a primary fuzzy term set (fuzzy states) of *Poor*, *Moderate*, and *Good* which is defined over a universe of discourse given by the real interval [0, 1.0]. The fuzzy variable *Cut_accuracy* is generated from the value of the cutting accuracy using the membership function of Figure 13.25(b) and has a primary term set of *Acceptable* and *Unacceptable*, which is defined over a millimeter scale.

13.3.5.2 Servomotor Response Preprocessing

As indicated in Figure 13.11(b), the control and monitoring computer receives feedback from the two servomotors that drive the cutter assembly and the vertical positioning platform, in the form of position information via the GALIL DMC-620 two-axis controller. This information is filtered by the servomotor preprocessor and the following performance parameters, as applicable, are extracted:

1. Rise time (time to reach 95% of the step input)
2. Damping ratio
3. Damped natural frequency (if underdamped)
4. Overshoot (if underdamped)
5. Offset (steady state error)

Figure 13.26 shows the significance of these parameters for two typical cases of responses where one is underdamped and the other is overdamped. Since the required processing involves determination of maxima and minima, it is imperative that the stream of position data be filtered to remove any effect of noise and local disturbances. Consequently, the following second order binomial-filter function is used to smooth the servo-response data prior to further processing:

$$f(n) = \frac{1}{4}[x(n-1) + 2x(n) + x(n+1)] \tag{13.17}$$

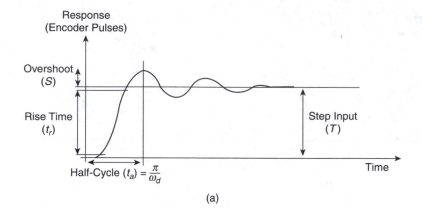

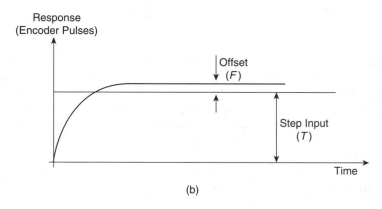

FIGURE 13.26
Typical parameters of servomotor response: (a) Underdamped case; (b) Overdamped case.

The next step of processing is to locate the turning points of the response curves of Figure 13.26; specifically, let us denote them as the points a (first peak), b (first vally) and c (second peak), and the corresponding time instants t_a, t_b, and t_c. This is done by means of an algorithm that searches for zero crossings of the first derivative of the step response. The rise time (T_r) is then considered to be the time it takes for the response to reach 95% of the step input. The overshoot (S_p) is calculated at the first peak of response, as a percentage of the step input. The steady-state error, which is computed as the percentage offset (F_p), is determined by comparing the average of the last third of the response with the step input. With reference to the servomotor response in Figure 13.26, the following performance parameters can be estimated:

$$\text{Rise time } (T_r) = t_r \tag{13.18}$$

$$\text{Percentage overshoot } (S_p) = \left(\frac{S}{T}\right) \times 100\% \tag{13.19}$$

$$\text{Damped natural frequency } (\omega_d) = \frac{\pi}{t_a} \tag{13.20}$$

$$\text{Damping ratio } (\zeta) = \sqrt{\frac{\ln^2(S/T)}{\pi^2 + \ln^2(S/T)}} \tag{13.21}$$

In Equations 13.18 through 13.21, the dominant mode of the servomotor response is assumed to be second order—an assumption that is quite justifiable for the servomotors of the prototype machine. Equation 13.21 is obtained by rearranging the terms of the more familiar equation for the overshoot of a second order system, in terms of its damping ratio (see Chapter 12); thus,

$$\text{Overshoot } (S/T) = \exp - \left(\frac{\pi \zeta}{\sqrt{1 - \zeta^2}} \right) \qquad (13.22)$$

In the next step of data preprocessing, the performance parameters computed by Equations 13.18 through 13.21 are compared with those of a reference model. This model is considered as exhibiting the desired response, which is the performance specification, and is typically not a model of the actual physical system. For the servomotor system of the prototype machine, the following second order plant, with appropriate parameters is used as the reference model:

$$G_r = \frac{\omega_n^2}{\left(s^2 + 2\zeta\omega_n s + \omega_n^2 \right)} + G_d \qquad (13.23)$$

where, ω_n and ζ are, respectively, the undamped natural frequency and the damping ratio. The term G_d represents external disturbances and can be regarded as a fixed offset for all practical purposes.

The knowledge base for servomotor tuning contains a set of rules prescribing what tuning actions should be taken when the actual response of the system deviates from the specifications. In the case of an intelligent tuner, this deviation is expressed in qualitative terms, signifying the experience of an expert. Accordingly, the context database of servomotor status uses fuzzy linguistic variables symbolizing the qualitative estimates of how the performance can deviate from the desired conditions. This is accomplished for rise time, percentage overshoot, and offset, first by comparing each actual performance parameter with the corresponding parameter of the reference model according to

$$\text{Index of deviation} = 1 - \frac{\text{attribute of model}}{\text{attribute of servomotor}} \qquad (13.24)$$

and then using the resulting index to fuzzify each performance attribute into one of the five primary fuzzy sets (fuzzy states). For the performance parameters damped natural frequency and damping ratio, the index used is given by

$$\text{Index of deviation} = 1 - \frac{\text{attribute of servomotor}}{\text{attribute of model}} \qquad (13.25)$$

Each index of deviation corresponds to one antecedent variable in the rule-base and is defined such that when the actual performance is within specification, the index is zero, and for the worst-case performance the index approaches unity. These numerical values are then fuzzified into membership functions with five primary fuzzy states, as shown in Figure 13.27. These data processing operations eventually result in a set of five fuzzy condition variables standing for the performance attributes as given in row 1 of Table 13.10, and a set of five primary fuzzy states (row 2) that each variable can assume. The database

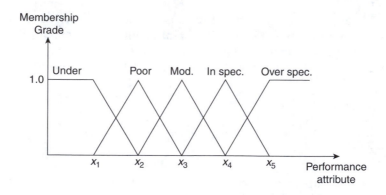

FIGURE 13.27
Multistate membership function for each servomotor performance index.

TABLE 13.10

Context Database for Servomotor Performance

Context Database of Performance Attributes	rise_time, damped_natural_frequency, damping_ratio, overshoot, offset
Primary Term Set for Each Performance Deviation	over_specification, in_specification, moderate, poor, unsatisfactory

in fact contains two sets of such variables, one set for the cutter positioning servomotor and the other for the platform positioning servomotor.

13.3.5.3 Cutter Load Preprocessing

The purpose of the cutter-load-profile preprocessor is to extract useful high-level information pertaining to the performance of the cutting action itself. To this end, the speeds of the top and the bottom blades of the cutter unit are monitored through speed sensors and associated circuitry. The goal is to eventually derive from these data, such context information as peak and average load on the cutter motors, symmetry of feeding of fish into the cutter, and the presence of any undesirable lateral movement (slip) of fish while cutting takes place.

The speeds of the induction motors, which drive the cutter blades, are first converted into the corresponding *slip* values and then to the percentage nominal loads, using

$$\text{Measured slip } (s_m) = \frac{\text{synchronous speed} - \text{measured shaft speed}}{\text{synchronous speed}} \qquad (13.26)$$

where, the synchronous speed for an induction motor is 3600 rpm, for a 60 Hz ac and a motor with one pole pair per phase (see Chapter 9).

$$\text{Percent load} (L_p) = \frac{\text{measured slip } (s_m)}{\text{slip at full load } (s_f)} \times 100\% \qquad (13.27)$$

Equation 13.27 assumes that the motor operates in the linear region of its load-slip curve (see Chapter 9), as indicated in Figure 13.28. This assumption holds true in general for all commercially available induction motors of type 'D', such as those used in the machine.

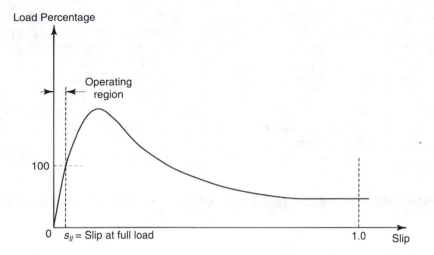

FIGURE 13.28
Load percentage versus slip for the induction motors.

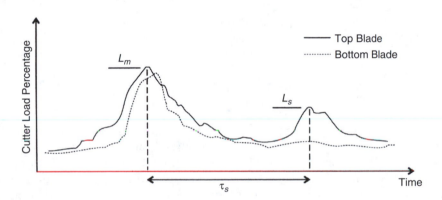

FIGURE 13.29
Typical load profiles of the two cutter blades.

A typical load profile obtained after these preliminary data processing operations are carried out is shown in Figure 13.29, where the load on each of the two cutter blades during a typical cutting operation is plotted with respect to time. The load-profile data contain interference or noise created by the frequency-to-analog conversion, which usually appears as easily-visible ripples on the load profile. The fourth order binomial filter given by

$$f(n) = \frac{1}{16}\left[x(n-2) + 4x(n-1) + 6x(n) + 4x(n+1) + x(n+2)\right] \tag{13.28}$$

is used in order to screen out the effect of these ripples before further processing. Now, the following parameters can be conveniently measured from the two individual profiles: peak load, average load, magnitude of secondary peaks, if present, and their times of occurrence relative to the first peak. The average load of each curve over a single cutting cycle signifies the average cutting load on the induction motors. The difference in magnitudes between the two peak loads of the top and the bottom blades is typically an indication of the degree of asymmetry in feeding fish into the cutter. The presence of secondary peaks may indicate slipping of fish during cutting, and hence, improper holding by the gripper. A secondary peak that is located closely to the primary peak is a common feature in a majority of cut profiles, indicating the load variations when the blades are cutting through bones rather than meat. This feature, however, can be set apart form the undesirable secondary peaks due to improper holding simply by timing the secondary peaks with respect to the primary one and then ignoring the ones that are sufficiently close to the primary peak. It follows that the timing of secondary peaks is crucial in recognizing the presence of improper holding.

For the purpose of context generation, the nondimensional indices given by

$$\text{Cut load percentage } (L_{pc}) = \frac{1}{2}(L_{mt} + L_{mb}) \tag{13.29}$$

$$\text{Index of asymmetry } (I_a) = \frac{|L_{mt} - L_{mb}|}{\frac{1}{2}(L_{mt} + L_{mb})} \tag{13.30}$$

$$\text{Secondary peak load index } (I_{ls}) = \frac{2L_s}{(L_{mt} + L_{mb})} \tag{13.31}$$

$$\text{Index of secondary-peak time } (I_{ts}) = \frac{\tau_s}{T} \tag{13.32}$$

are computed using the parameters illustrated in Figure 13.29. The subscripts t and b denote the "top" and the "bottom" blades, respectively, and $1/T$ is the feed rate of fish on the conveyor. The indices for the secondary peak load (I_{ls}) and the time of secondary peak load (I_{ts}) are computed for the two blades, separately. These indices are fuzzified using the membership function shown in Figure 13.30, and then the following rule-base

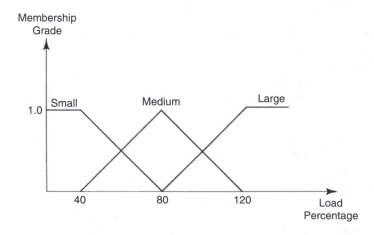

FIGURE 13.30
Membership function for the cutter load context.

submodule is used in conjunction with the compositional rule of inference, to generate the membership functions of *Slip_at_top_blade* and *Slip_at_bottom_blade*:

- **If** secondary peak load (I_{ts}) is present **and** secondary peak load index (I_{ls}) is medium, **then** *slip* is moderate.
- **Else if** secondary peak load (I_{ts}) is present **and** secondary peak load index (I_{ls}) is high **then** *slip* is large.
- **Else** *slip* is small.

The knowledge base of the machine supervisory system has an antecedent variable representing the slip at the blades. In order to generate the context of this composite slip, the two membership states of *Slip_at_top_blade* and *Slip_at_bottom_blade* are composed further by the following rule-base module:

- **If** slip at top blade is large **and** slip at bottom blade is large **then** *composite_slip* is large.
- **Else if** slip at top blade is moderate **or** slip at bottom blade is moderate **then** *composite_slip* is moderate.
- **Else if** slip at top blade is small **and** slip at bottom blade is small **then** *composite_slip* is small.

These preprocessing operations eventually result in a set of three fuzzy-status variables, which stand for the cutter performance as depicted in column 1 of Table 13.11. Column 2 shows the primary fuzzy terms (fuzzy states) that each variable can assume.

13.3.5.4 Conveyor Speed Preprocessing

The purpose of the preprocessor for conveyor speed profile is to extract high-level information pertaining to the performance of the drive unit of the conveyor during operation of the machine. Since the conveyor speed is measured directly through dedicated hardware, the degree of preprocessing required on this channel of information is minimal. The following two items of context data are extracted from the conveyor speed samples:

$$\text{Average speed } (V_{av}) = \frac{1}{n} \sum_{k=1}^{n} v_{kt} \tag{13.33}$$

$$\text{Speed fluctuations } (V_\Delta) = \frac{\max_{k=1}^{n}(v_{kt}) - \min_{k=1}^{n}(v_{kt})}{V_{av}} \tag{13.34}$$

which are acquired during each cycle of fish processing; where, $T = nt$ is the cycle time of processing. Table 13.12 shows the resulting fuzzy variables, which represent the status

TABLE 13.11

Context Database for Cutter Performance

Context Variable of Cutter Performance	Primary Term Set for Each Variable
Percent_cut_load	*Small, medium, large*
Index_of_asymmetry	*Small, medium, large*
Composite_slip	*Small, moderate, large*

TABLE 13.12

Context Database for Conveyor Performance

Context of Conveyor Performance	Primary Term Set
Conveyor_speed	*Low, medium, high*
Speed_fluctuations	*Small, medium, large*

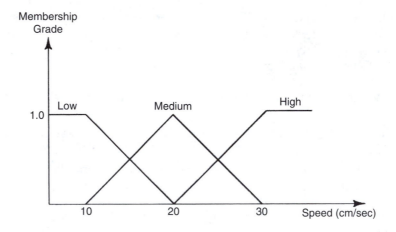

FIGURE 13.31
Membership functions for conveyor performance.

of the conveyor. The input space is partitioned into primary fuzzy sets that are named in column 2 of Table 13.12, and shown in Figure 13.31.

13.3.5.5   Servo Tuning

For tuning under the control environment of DMC-600, a tuning rule-base was developed by applying control-engineering knowledge to the machine in an on-line basis. The antecedent variables, (i.e., the condition or status of the servomotor) and the fuzzy values that each variable can assume are given in Table 13.13. The consequent (i.e., action) variables represent the design attributes of the controller that have to be tuned in order to achieve a desired performance. These design attributes relate to the frequency domain specifications of a lead compensator (see Chapter 12) as illustrated in Figure 13.32, along with a pure integral controller in parallel. Each of these consequent variables, as listed in Table 13.13, can assume one of five possible fuzzy states that are defined over a normalized output space [−1.0, +1.0], as shown in Figure 13.33. The tuning rule-base is given in Table 13.14. Those consequent variables that do not appear in some of the rules in this rule-base must be considered as not contributing to tuning of the servomotor controller. In other words, they all take the fuzzy value *ze* (zero) in the output space. The design attributes are updated by the incremental values that the rule-base infers. The resulting values are then mapped to the controller parameters of the frequency domain transfer function of a lead compensator

$$G(j\omega) = K\left(\frac{1+A\cdot j\omega}{1+\alpha A\cdot j\omega}\right), \quad 0 < \alpha < 1 \tag{13.35}$$

TABLE 13.13

Consequent Variables for Servomotor Tuning

Variable	Description	Notation
ω_{cog}	Frequency at maximum phase lead of compensator	FRCOG
ϕ_{cof}	Maximum phase lead of compensator	PHCOF
G_{cof}	Compensator gain at maximum phase lead	CNCOF
ω_{log}	Integrator frequency for a gain of G_{cof}	LFCOG

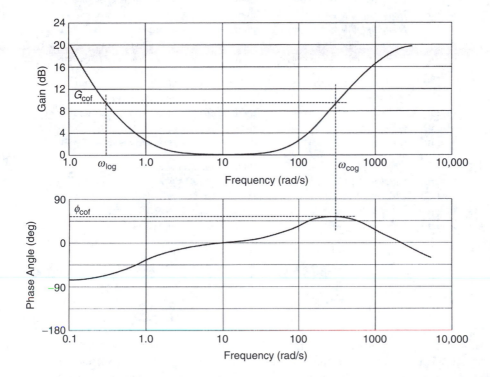

FIGURE 13.32

Frequency domain specifications (design parameters) of a lead compensator as applicable to the rule-base for servomotor tuning.

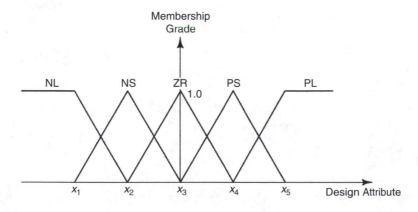

FIGURE 13.33

Output space partitioning for the rule-base of servomotor tuning.

TABLE 13.14

The Rule-Base for Servomotor Tuning

If	*Rise_time* is *Unsatis*	then	*frcog* is *nl* & *phcof* is *nl* & *gncof* is *pl*
else if	*Rise_time* is *Poor*	then	*frcog* is *ns* & *phcof* is *ns* & *gncof* is *ps*
else if	*Rise_time* is *Moderate*	then	*frcog* is *ns* & *phcof* is *ns* & *gncof* is *ps*
else if	*Rise_time* is *In_spec*	then	*frcog* is *ze* & *phcof* is *ze* & *gncof* is *ze*
else if	*Rise_time* is *Over_spec.*	then	*frcog* is *ps* & *phcof* is *ps* & *gncof* is *ns*
If	*Damping_ratio* is *Unsatis.*	then	*frcog* is *nl* & *phcof* is *pl* & *gncof* is *pl*
else if	*Damping_ratio* is *Poor*	then	*frcog* is *ns* & *phcof* is *ps* & *gncof* is *ps*
else if	*Damping_ratio* is *Moderate*	then	*frcog* is *ns* & *phcof* is *ps* & *gncof* is *ps*
else if	*Damping_ratio* is *In_spec.*	then	*frcog* is *ze* & *phcof* is *ze* & *gncof* is *ze*
else if	*Damping_ratio* is *Over_spec.*	then	*frcog* is *ps* & *phcof* is *ns* & *gncof* is *ns*
If	*Overshoot* is *Unsatis*	then	*frcog* is *nl* & *phcof* is *pl* & *gncof* is *pl*
else if	*Overshoot* is *Poor*	then	*frcog* is *ns* & *phcof* is *ps* & *gncof* is *ps*
else if	*Overshoot* is *Moderate*	then	*frcog* is *ns* & *phcof* is *ps* & *gncof* is *ps*
else if	*Overshoot* is *In_spec.*	then	*frcog* is *ze* & *phcof* is *ze* & *gncof* is *ze*
else if	*Overshoot* is *Over_spec.*	then	*frocg* is *ps* & *phcof* is *ns* & *gncof* is *ns*
If	*Offset* is *Unsatis.*	then	*gncof* is *pl* & *lfcog* is *pl*
else if	*Offset* is *Poor*	then	*gncof* is *pl* & *lfcog* is *ps*
else if	*Offset* is *Moderate*	then	*gncof* is *ps* & *lfcog* is *ps*
else if	*Offset* is *In_spec.*	then	*gncof* is *ps* & *lfcog* is *ze*
else if	*Offset* is *Over_spec.*	then	*gncof* is *ze* & *lfcog* is *ns*
If	*Dmp_nat_freq.* is *Unsatis.*	then	*gncof* is *pl*
else if	*Dmp_nat_freq.* is *Poor*	then	*gncof* is *ps*
else if	*Dmp_nat_freq.* is *Moderate*	then	*gncof* is *ps*
else if	*Dmp_nat_freq.* is *In_spec.*	then	*gncof* is *ze*
else if	*Dmp_nat_freq.* is *Over_spec.*	then	*gncof* is *ns*

where, K is the gain parameter of the compensator, while A and αA are time constants associated with the lead compensator. The following equations complete the mapping procedure from design attributes to the controller parameters (see Chapter 12):

$$\alpha = \frac{1 - \sin(\phi_{cof})}{1 + \sin(\phi_{cof})} \tag{13.36}$$

$$A = \frac{1}{\omega_{cog}\sqrt{\alpha}} \tag{13.37}$$

$$K = \sqrt{\alpha}\left|G_{cof}\right| \tag{13.38}$$

according to the definitions given in Table 13.13. In addition, the controller has an integrator branch with the transfer function,

$$G_i(j\omega) = \frac{K_i}{j\omega} \tag{13.39}$$

According to Table 13.13, the gain parameter K_i is given by,

$$K_i = \omega_{log}\left|G_{cof}\right| \tag{13.40}$$

The above procedure of parameter updating is repeated for both the cutter servomotor nd the platform servomotor. The rule-base for servomotor tuning operates independently n the separate context databases of the two servomotors, producing inferences for tuning ach servomotor independently.

3.3.5.6   *Product Quality Assessment*

Quality of the finished product from the machine is assessed primarily on the basis of visual context that is determined through preprocessing of image data from the secondary CCD camera. Specifically, the features that indicate the quality are (a) Accuracy of cut (*Accuracy*), (b) Depth of the solid region of the cut section (*Depth*), (c) Smoothness of the boundary contour of the cut section (*Contour*), and (d) Smoothness of the cut surface (*Surface*). The inference of the associated rule-base is a single fuzzy variable indicating the quality of the finished product, as one of two possible fuzzy outcomes of either *Acceptable* or *Unacceptable*, as represented by the membership functions in Figure 13.34. The rule-base has been devised to mimic the judgment of a human expert who assesses the quality of fish products by visual inspection. The rule-base module utilized in the machine for this purpose is shown in Table 13.15. In this rule-base, the term *Any* in the antecedent variables

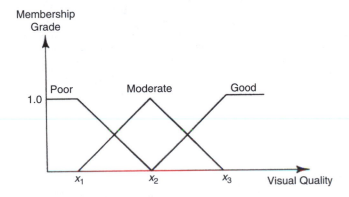

FIGURE 13.34
Output space partitioning for the quality assessment rule-base.

TABLE 13.15

Rule-Base for Assessment of Product Quality

- IF *Accuracy* is *Unacceptable* and *Depth* is *Any* and *Contour* is *Any* and *Surface* is *Any* THEN *Quality* is *Unacceptable*.
- IF *Accuracy* is *Acceptable* and *Depth* is *Good* and *Contour* is *Good* and *Surface* is *Good* THEN *Quality* is *Acceptable*.
- IF *Accuracy* is *Acceptable* and *Depth* is *Good* and *Contour* is *Moderate* and *Surface* is *Good* THEN *Quality* is *Acceptable*.
- IF *Accuracy* is *Acceptable* and *Depth* is *Moderate* and *Contour* is *Moderate* and *Surface* is *Moderate* THEN *Quality* if *Acceptable*.
- IF *Accuracy* is *Acceptable* and *Depth* is *Poor* and *Contour* is *Any* and *Surface* is *Any* THEN *Quality* is *Unacceptable*.
- IF *Accuracy* is *Acceptable* and *Depth* is *Any* and *Contour* is *Poor* and *Surface* is *Any* THEN *Quality* is *Unacceptable*.
- IF *Accuracy* is *Acceptable* and *Depth* is *Any* and *Contour* is *Any* and *Surface* is *Poor* THEN *Quality* is *Unacceptable*.

refers to the conjunction of all possible values in that variable. For example, *Depth* is *Any* equivalent to the conjunction of *Depth* is *Poor* or *Depth* is *Moderate* or *Depth* is *Good*.

13.3.5.7 Machine Tuning

The rule-base for machine tuning operates on the context derived from the low-level sensory measurements of conveyor speed, cutter load, and information on poor holding (slip) at the cutter. This database consists of the following fuzzy antecedent variables: (a) percentage cutter load (*Cutload*), (b) composite slip at cutter (*Slip*), (c) average conveyor speed (*Speed*), and (d) conveyor speed fluctuations (*Fluctuations*). The consequent or action variables are the tuning inferences: (a) the change in conveyor speed (*Speed change*), and (b) the degree of grasping (*Hold-down*). It should be noted that while the inference on conveyor speed is incremental, the inference on grasping, that is, the hold-down force, is absolute. In addition, the absolute speed of the conveyor is limited by a pair of upper and lower bounds that are set through the defuzzification algorithm. Table 13.16 shows the rule-base module that has been adopted for tuning of the machine. Although this table depicts the rule-base as a multi-input-multi-output (MIMO) system, the machine implements the two inferences individually in the form of two independent multi-input-single-output (MISO) systems.

TABLE 13.16

Rule-base for Machine Tuning

- IF *Cutload* is *Low* and *Slip* is *Small* and *Fluctuations* is *Not Large* and *Speed* is *Low* THEN *Speed change* is *Pos. Large* and *Hold-down* is *Soft*.
- IF *Cutload* is *Low* and *Slip* is *Small* and *Fluctuations* is *Not Large* and *Speed* is *Medium* THEN *Speed change* is *Pos. Small* and *Hold-down* is *Soft*.
- IF *Cutload* is *Low* and *Slip* is *Small* and *Fluctuations* is *Not Large* and *Speed* is *High* THEN *Speed change* is *No Change* and *Hold-down* is *Soft*.
- IF *Cutload* is *Medium* and *Slip* is *Small* and *Fluctuations* is *Not Large* and *Speed* is *Low* THEN *Speed change* is *Pos. Small* and *Hold-down* is *Soft*.
- IF *Cutload* is *Medium* and *Slip* is *Small* and *Fluctuations* is *Not Large* and *Speed* is *High* THEN *Speed change* is *Pos. Small* and *Hold-down* is *Soft*.
- IF *Cutload* is *Medium* and *Slip* is *Small* and *Fluctuations* is *Not Large* and *Speed* is *High* THEN *Speed change* is *No Change* and *Hold-down* is *Soft*.
- IF *Cutload* is *High* and *Slip* is *Small* and *Fluctuations* is *Not Large* and *Speed* is *Low* THEN *Speed change* is *No Change* and *Hold-down* is *Moderate*.
- IF *Cutload* is *High* and *Slip* is *Small* and *Fluctuations* is *Not Large* and *Speed* is *Medium* THEN *Speed change* is *Neg Small* and *Hold-down* is *Moderate*.
- IF *Cutload* is *High* and *Slip* is *Small* and *Fluctuations* is *Not Large* and *Speed* is *High* THEN *Speed change* is *Neg Large* and *Hold-down* is *Moderate*.
- IF *Cutload* is *High* and *Slip* is *Small* and *Fluctuations* is *Not Large* and *Speed* is *Medium* THEN *Speed change* is *Neg Small* and *Hold-down* is *Moderate*.
- IF *Cutload* is *Any* and *Slip* is *Moderate* and *Fluctuations* is *Not Large* and *Speed* is *Medium* THEN *Speed change* is *Neg Small* and *Hold-down* is *Moderate*.
- IF *Cutload* is *Any* and *Slip* is *Moderate* and *Fluctuations* is *Not Large* and *Speed* is *Large* THEN *Speed change* is *Neg Large* and *Hold-down* is *Moderate*.
- IF *Cutload* is *Any* and *Slip* is *Large* and *Fluctuations* is *Not Large* and *Speed* is *Any* THEN *Speed change* is *Neg Large* and *Hold-down* is *Tight*.
- IF *Cutload* is *Any* and *Slip* is *Any* and *Fluctuations* is *Moderate* and *Speed* is *Medium* THEN *Speed change* is *No Change* and *Hold-down* is *Moderate*.
- IF *Cutload* is *Any* and *Slip* is *Any* and *Fluctuations* is *Moderate* and *Speed* is *Large* THEN *Speed change* is *Neg Small* and *Hold-down* is *Moderate*.
- IF *Cutload* is *Any* and *Slip* is *Any* and *Fluctuation* is *Large* and *Speed* is *Any* THEN *Speed change* is *Neg Small* and *Hold-down* is *Tight*.

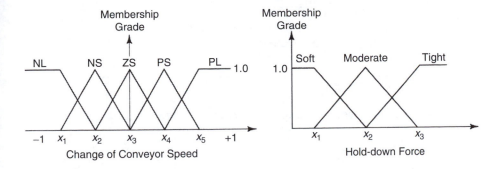

FIGURE 13.35
Output space partitioning for the rule-base of machine tuning: (a) For conveyor speed; (b) For hold-down force.

Figure 13.35 shows the partitioning of the two fuzzy output spaces corresponding to the respective inferences of change of conveyor speed and hold-down force.

13.3.5.8 System Modules

The hardware indicated in Figure 13.11(b) has been integrated into a complete machine, as shown in Figure 13.11(a). Software for both low-level and high-level functions of the system have been developed and implemented. Associated functions involve sensor reading and generation of low-level control actions, component synchronization, signal preprocessing, supervisory decision making, and execution of tuning actions.

The control hierarchy has been realized in several C-language-programmed modules. Selection of the language C for realization of the machine software was particularly influenced by factors such as:

- Availability of low-level machine instructions in "C" facilitates easy interfacing with the low-level controllers.
- Fast speed of execution and the availability of support for a graphical user interface.

The modular development of software closely resembles the architecture of the control hierarchy such that every module is identified with a particular function in the hierarchy. Each module was realized separately in "C" and after debugging and testing, all the modules were compiled and linked together to build a single executable file, MACHINE.EXE. Figure 13.36 illustrates the functions assigned to different software modules and their layout in relation to the data flow between them. The functional description of each system module is given below.

Sensor (Input) Module: Reading of the input ports of the data acquisition board, recording of the position responses of the servomotors, and capturing of images of processed fish are carried out in this module. In order to reduce the effect of noise and ripple, all analog-to-digital-conversion (A/D) ports are read many times in rapid succession and the average is taken. All data are then converted into proper engineering units and transmitted to the preprocessor module.

Preprocessor Module: This module contains the software required to interpret the sensor readings. Specifically, it transcribes crisp, sensory information into a linguistic form as required by the fuzzy knowledge base, in a compatible form.

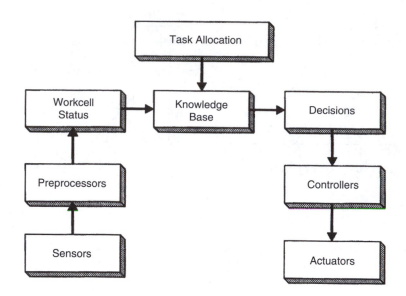

FIGURE 13.36
Software modules of the machine.

Knowledge Base: This module contains the rule-set and the fuzzy relational matrix derived from it. Since the relational matrix could be computed off line, the module simply stores the matrix in the form of a multidimensional array with each dimension corresponding to either an antecedent or a consequent variable. The size of this array is determined by the resolution of each fuzzy variable (the number of fuzzy states) and the total number of such variables.

Decisional Module: Application of the compositional rule of inference is carried out in this module. The center of area method is applied to obtain a crisp value and the output is sent to the actuator (output) module.

Actuator (Output) Module: The crisp values of the consequent variables are scaled before writing them into the output ports of the digital-to-analog converter (D/A) within the PCL-812 board of the machine such that the generated analog signals are directly interfaced with the corresponding servo amplifiers. The passing of tuning commands to the DMC-600 controller is also handled in this module.

The overall control of variable passing between modules and sequencing of various actions is coordinated by a main module, which incorporates all the modules into a single program. At run time, the user can monitor the machine through a menu-driven user interface, which facilitates interactive monitoring of each condition variable of the machine. Also, data may be stored for off-line analysis and further evaluation.

13.3.5.9   User Interface of the Machine

In the present implementation, the complexity of operation and the large quantity of information available for monitoring necessitate that the machine employ a user friendly human-machine interface. The hierarchical structure of the control system itself provides a good basis for a hierarchically-layered, menu-driven, user interface. The *Main Menu* which is the top level of the user interface, provides a graphical representation that is analogous to the hierarchy of the machine. At any level of the interface hierarchy, the user can select and open up an item at a lower level or in the level of the hierarchy in order to monitor or interrogate a particular level or a device in the machine.

3.3.5.10 Machine Tuning Example

Performance of the machine tuner is illustrated now, using an example. First, several cutter load profiles were obtained through experimentation. Then, by the use of the cutter-load preprocessor, a context database corresponding to each load profile was generated. Next, the rule-base for machine tuning was used to obtain the inferences pertaining to the conveyor speed and hold-down force.

Figures 13.37(a) through 13.37(d) show four cutter load profiles corresponding to actual fish cutting experiments carried out using the machine. Each curve in Figure 13.37 was obtained under different operating conditions of holding force and conveyor speed, as indicated. The numerical parameters resulting from preprocessing of these cutter load profiles are given in Table 13.17. The corresponding fuzzy linguistic context, as generated by the cutter-load preprocessor, is shown in Figures 13.38(a) through 13.38(d).

The rule-base for machine tuning has two fuzzy inferences; namely, the change of conveyor speed and the magnitude of hold-down (grasping) force. Figures 13.39(a) through 13.39(d) illustrate the corresponding inferences when the knowledge-based system operates on the context shown in Figure 13.38. Figure 13.37(a) shows a case where the conveyor speed is quite high, as can be gauged from the peak load that is higher than normal. Also, in this case, it is clear that the fish has not been delivered symmetrically into the cutter, since there is a mismatch of peak loads corresponding to the top and the bottom load profiles. Furthermore, it is seen that the two peak values do not occur simultaneously because of the slight offset with an overlap, that has been provided in the placement of the two circular blades, so as to achieve a complete cut. The load profile of the top blade (solid curve) also shows a slight degree of slipping as can be seen from the higher than normal trailing region of the load profile. The fuzzy linguistic context obtained after preprocessing the load profile confirms those visual observations, as seen in Figure 13.38(a). The inferences corresponding to this context, shown is Figure 13.39(a), have recommended a large reduction in conveyor speed and a moderate increase in hold-down force.

Figure 13.37(b) and Figure 13.37(c) illustrate two cases where slipping occurs towards the tail end of the cutter load profiles. Here as well, the fuzzy linguistic context shown in Figure 13.38(b) and Figure 13.38(c) captures the information as expected. Consequently, the knowledge-based system recommends tight holding and a reduced speed, as can be seen from Figure 13.39(b) and Figure 13.39(c).

Figure 13.37(d) shows a satisfactory cut at a moderate conveyor speed. In this case, the two closely placed peaks visible on the load profile of the top blade are not due to slipping of fish, and the preprocessor correctly interprets this situation. The corresponding inferences suggest a soft holding force and a slight increase in conveyor speed, to take into account the fact that the capacity of the cutter motors is underutilized in this case.

13.3.6 Economic Analysis

Economic analysis involves a cost-benefit evaluation, and will require computation of the payback period. Figures on the initial investment, number of people replaced by a new machine and corresponding wage savings, productivity increase due to new technology, inflation rate, corporate tax rate, and operating expenses such as those for utilities, maintenance,

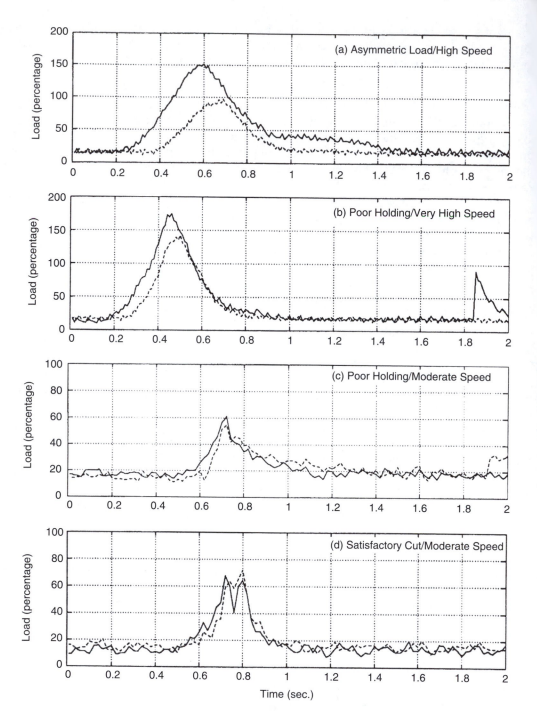

FIGURE 13.37

Cutter load profiles (Solid curves: Top blade; Dashed curves: Bottom blade).

and insurance, are required for an analysis of this type. Through our knowledge of the fish processing industry and details of the new machine, the following representative data are used:

$$\text{Initial investment on a machine} = \$100,000$$
$$\text{Company expenses for machine start-up} = \$10,000$$

TABLE 13.17

Preprocessed Cutter Load Information

Load Profile	Mean Load%	Asymmetry %	I^{ry} Peak (%)/Time (s)		II^{ry} Peak (%)/Time (s)	
			Top	Bottom	Top	Bottom
(a)	118	47	146/0.58	91/0.67	33/1.22	None
(b)	149	22	166/0.44	133/0.48	71/1.85	None
(c)	55	11	58/0.70	52/0.70	None	32/1.93
(d)	65	7.6	63/0.72	68/0.78	62/0.79	None

First Year Revenues:

Wage savings due to replaced workers = $20,000
Increased revenues due to higher productivity = $50,000

It has been assumed that only four workers would be needed per new machine whereas six workers are needed for an Iron Butcher. A processing season of 3 months is assumed as well. Productivity increase given here is primarily from the increased meat recovery.

First year Operating Expenses = $10,000
(maintenance, utilities, insurance, etc.)

Economic Parameters:

Inflation rate = 5%
Corporate tax rate = 25%
Tax credit on capital expenditure = 10%
With a planning horizon of 5 years,
   assume a depreciation rate = 20%

Next, a cash flow table is completed, as shown in Table 13.18, for the data given here.

Internal Rate of Return:

Taking into consideration that the risk of payback is greater for the later years from the time of initial investment, the applicable equation is

$$C = \sum_{i=1}^{n} \frac{S_i}{(1+r)^n} \qquad (13.41)$$

in which
  C = initial cost of new technology
  S_i = net cash savings in the ith year
  n = design life (planning horizon or period of amortization)
  r = internal rate of return.

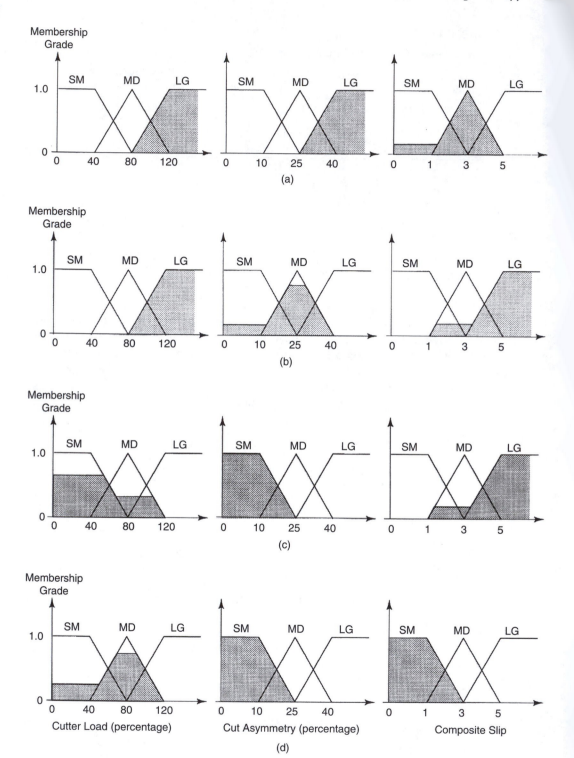

FIGURE 13.38
Fuzzy linguistic context corresponding to the experimental cutter load profiles in Figure 13.37 (SM — Small, MD — Moderate, LG — Large).

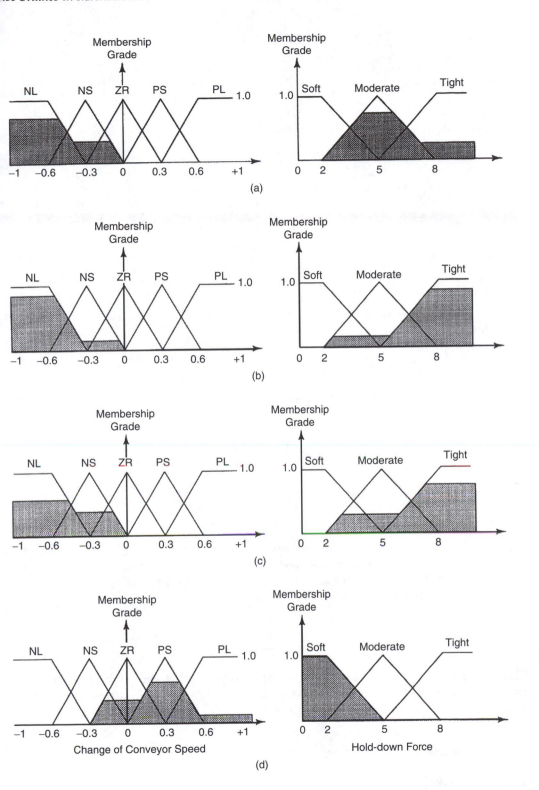

FIGURE 13.39

Fuzzy linguistic inferences of the machine tuner.

TABLE 13.18

Cash Flow Table ($) for a Machine Installation

	Initial	Year 1	Year 2	Year 3	Year 4	Year 5
Capital investment	−100,000	10,000				
Tax credit (10%)						
Machine start-up cost	−10,000					
Tax gain (25%)		2,500				
Machine depreciation (20%)		20,000	20,000	20,000	20,000	20,000
Tax loss (25%)		−5,000	−5,000	−5,000	−5,000	−5,000
Wage savings (5% inflation)		20,000	21,000	22,050	23,153	24,310
Tax loss (25%)		−5,000	−5,250	−5,513	−5,788	−6,078
Productivity increase (5% inflation)		50,000	52,500	55,125	57,881	60,775
Tax loss (25%)		−12,500	−13,125	−13,781	−14,470	−15,194
Operating expenses (5% inflation)		−10,000	−10,500	−11,025	−11,576	−12,155
Tax gain (25%)		2,500	2,625	2,756	2,894	3,039
Net savings	−110,000	72,500	62,250	64,612	67,094	69,697

Then, from the cash flow table (Table 13.18) for one machine installation, and using Equation 13.41 we have

$$110,000 = \frac{72,500}{(1+r)} + \frac{62,250}{(1+r)^2} + \frac{64,612}{(1+r)^3} + \frac{67,094}{(1+r)^4} + \frac{69,697}{(1+r)^5}$$

The solution for r may be obtained iteratively. An approximate value for r is obtained by using

$$r_{apr} = \frac{1}{Cn} \sum_{i=1}^{n} S_i \qquad (13.42)$$

Hence,

$$r_{apr} = \frac{72,500 + 62,250 + 64,612 + 67,094 + 69,697}{110,000 \times 5}$$

or, $r_{apr} = 0.6$. A more accurate value for r would be 0.5465. Now, using $r = 0.55$, the payback period, defined as $1/r$, is approximately 1.8 years. A payback period of less than 2 years, as predicted in this analysis, is quite acceptable for the fish processing industry.

A simplified version of Equation 13.41 is given by

$$C = S \sum_{i=1}^{n} \frac{1}{(1+r)^n} \qquad (13.43)$$

This approximate equation is commonly used in predicting the payback period, but it cannot be fully justified, as the net savings figure is not constant during the accounting life of a machine, as clear from Table 13.18.

13.4 Projects

You are a mechatronic engineer who has been assigned the task of designing, developing, and instrumenting a mechatronic system. Giving the necessary details, describe the steps of system design, integration, testing, and application of the projects outlined below.

The design should include the structural system, electronic system, hardware, and software, including sensors, actuators, motion transmission devices, power sources, controllers, signal conditioning and interfacing needs. Off-the shelf components may be used where available and appropriate. The given specifications may be modified and additional specifications may be established, when necessary. Provide data for the main components of the system. Using suitable diagrams and sketches, describe how the overall system (plant, sensors, actuators, controllers, signal modification devices, etc.) is interconnected (integrated). Explain how the system operates (e.g., what is the purpose of the system, what commands are provided to the system, what responses and signals are generated by the system and sensed, how the control signals are generated and according to what criteria, how the plant is actuated, and what type of plant response would be achieved under proper operation). Study how well the required performance specifications are satisfied by the designed system.

Project 1: Automated Glue Dispensing System

Application of an adhesive is useful in many industrial processes; for example automotive, wood product, and building. Consider a system for two-dimensional application of glue. The system has to sense the application area, position the dispensing gun, and operate it, which will include simultaneous dispensing of glue and moving the dispensing head accurately with respect to the glued object (e.g., window). Speed of the system will be governed by the constraints of the glue gun (including the properties of the glue) and the moving system. The following preliminary specifications are given.

> Maximum area of application: 1 m × 1 m
>
> Speed of the dispensing unit = 20 cm/s
>
> Positioning accuracy = ±0.5 mm
>
> Glue application accuracy (uniformity) = ±0.2 mm
>
> Glue dispensing pressure = 100 psi (690 kPa)

The design should include the glue dispensing system as well. Also, describe in detail an application of your design.

Project 2: Material Testing Machine

Testing of material (e.g., tensile/compressive, bending, torsional, fatigue, impact) is important in product development, monitoring, and qualification. Testing of biological material may be needed in medical, agricultural, and food-product applications. Design a machine for *in vitro* (outside the body, in an artificial environment) testing of spine segments obtained from a human cadaver. The spine segment is mounted on a form base, and various load profiles (forces and moments) are applied. The resulting motions (displacements and rotations) are measured and recoded for further analysis. Some of the test requirements are given below.

> Moment increment (for a ramp test) = 1 N · m
>
> Moment range = −15 N · m to +15 N · m
>
> Speed = 4 increments/s
>
> Moment step (for a step test) = ±10 N · m
>
> Accuracy = 2%

Project 3: Active Orthosis

Powered prosthetic devices are increasingly used to assist deformed, disabled or injured upper and lower limbs of humans. An orthosis is a fully integrated prosthetic device attached to a human body, and it assumes that the limb is not missing and the sensation of the limb is not completely lost. An active device (as opposed to a passive device) will require a power source. Sensations of temperature, pressure, and texture (e.g., tactile sensing) are particularly important in human functions. Design an active upper-limb orthosis. The head and the shoulder may be used to control the device, and two-state, multi-state, or continuous commands may be provided. Functionality, reliability, convenience and comfort, speed, accuracy, cost, and appearance are important considerations in the design. The assisted functions and movements may include those of upper arm, elbow, forearm, wrist, and fingers. You may establish the necessary design specifications for the orthotic device through self-testing, experience, and literature search.

Project 4: Railway Car Braking System

In braking a train, the braking forces have to be applied to the wheels rapidly, systematically and under control. Derailments should be avoided and the braking operation should not be damaging to the train and its occupants. Under normal conditions, occupant discomfort should be minimized, and braking should be done while minimizing discomfort to the passengers. Consider a multi-car light-rail system (e.g., an elevated guideway transit system or a subway system). Assume that a hydraulic system is used to apply the braking force to the wheels through brake shoes, and these forces can be quite high (e.g., 3×10^4 N). Establish suitable design specifications. Note that passenger comfort limits are available as applied acceleration (g) levels as a function of excitation frequency, for different exposure times of excitation. Design a braking system that includes antilock braking features. Train speed, wheel-rail conditions, weather conditions, and the nature of the stop (normal or emergency) should be factored into the control system.

Project 5: Machine Tool Control System

Productivity, product quality, machine life, tool life, and safety will improve through proper control of machine tools. Consider a standard vertical milling machine consisting of a positioning (x-y) table and a vertical spindle assembly, which carries the toolbit. The following parameters values and specifications are available.

Mass of the positioning table: 250 kg

Mass of the spindle assembly: 50 kg

Maximum mass of workpiece: 50 kg

Positioning accuracy: 0.01 mm

Maximum speed of the positioning table: 0.2 m/s

Maximum acceleration of the positioning table: 1.0 m/s²

Maximum cutting force: 2000 N

Operating bandwidth of the milling machine: 100 Hz.

Assume that dc motors and ball screws are used to drive the positioning table. The servo rise time may be taken as 50 ms. Design a suitable control system for tool positioning and machining. Examine whether/how the design should be modified depending on the cutting (workpiece) material (e.g., steel, aluminum, other metals and alloys, plastic, wood, rubber).

Project 6: Welding Robot

Considerations of productivity, flexibility, hazards, and cost have provided the motivation for using robots for industrial welding applications. Automotive industry is a good example. Both seam welding and spot welding may be carried out by robots. Design an arc-welding industrial robot for a production line of an industrial plant. Select a specific industrial application and on that basis, establish a set of specifications for the robot. The design should consider pertinent aspects of kinematics, dynamics, mechanics, electronics, control, and system integration and plant networking. The design should involve detection of the welded part prior to positioning the welding torch. In particular, consider a six degree-of-freedom robot with three prismatic joints and three revolute joints. The prismatic joints are primarily used for gross positioning of the end effector (the welding torch) and the revolute joints are primarily used for fine manipulation (e.g., orientation and proximity adjustments) of the welding torch with respect to the welded part. Some preliminary design specifications are given below.

Maximum speed of a prismatic joint: 1.0 m/s

Maximum speed of a revolute joint: 2.0 rad/s

Maximum linear acceleration: 1 g

Linear positioning accuracy: ±0.1 mm

Angular positioning accuracy: ±1°

Payload (including the welding torch): 15 kg

Work envelope: hemisphere with 1.5 m radius

You may use ac servomotors.

Project 7: Wood Strander

A strander is a machine for producing wood chips (flakes) in the manufacture of strand boards (chipboards). A schematic diagram of a strander is shown in Figure P13.7. The lumber is intermittently delivered into the cutting ring by a conveyor. The feeder pushes the lumber against the rotating ring whose blades chip the lumber.

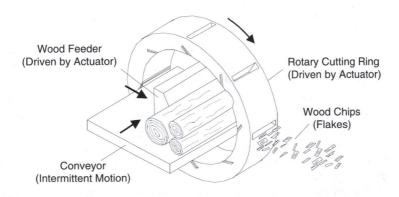

FIGURE P13.7
A strander for strand board manufacture.

Design a strander that can chip logs of maximum diameter 0.4 m, in 0.5 m segmen (which corresponds to the axial length of the cutting ring). The following preliminar specifications are provided.

Cutting ring diameter: 3.0 m

Number of cutting knives: 88

Cutting ring speed: 900 rpm

Feed velocity: 10 cm/s

Average cutting force: 5000 N

Average chip thickness: 0.3 cm

Chip thickness accuracy: 1%

Project 8: Automated Mining Shovel

A mining shovel is similar to an excavator used in earth removal operations. It has a boom to hold the stick, which carries the shoveling bucket. The stick is moved and manipulated by means of a cable winch mechanism. It is proposed to automate the shoveling process, by incorporating appropriate sensors, actuators, controllers, and other hardware. The following partial specifications are available.

Shovel (bucket or dipper) capacity: 40 m^3

Boom length: 15 m

Stick length: 7 m

Hoist speed: 2 m/s

Stick speed: 1 m/s

Stick rotation speed: 5 rpm

Dipper rotation speed: 10 rpm

Cycle time: 10 sec

Project 9: Can Filling Machine

Packaging of measured portions of food products is an important operation in the food processing industry. In particular, "portion control" is necessary. Consider an automated can-filling operation. It optimally portions and fills fish into cans. The overall system includes the stages of prefilling, filling, and postfilling. An integrated approach to optimal grouping and cutting, according to a weight-based portioning criterion, is central to the advantages of the approach. The system that is considered here incorporates optimal grouping and cutting of fish using robotic devices in order to minimize the deviation of a fish portion from the target weight of a can. Devices have to be designed and developed for mechanical handling and cutting of fish, with associated sensing and control systems. Automated can filling, and sensor-based postfilling inspection and integrated correction/ repair, under the supervision and control of a high-level control system, are needed. The associated technology includes fast and accurate estimation of the weight distribution of each fish; a portion-optimization method; handling, conveying, and cutting devices; advanced sensor technology; and multi-layered intelligent control. An important feature of the system is the weight-based optimization of the fill weight so as to minimize over-fills and under-fills of canning. The typical throughput rate is 5 cans/s.

roject 10: Fish Marking Machine

Marking of juvenile fish in hatcheries, before releasing into lakes and rivers, is an activity that is very valuable in fishery management. Samples of grown fish could be subsequently harvested and examined to collect data, which would be useful for many purposes such as predicting fish stocks, determining migration patterns of fish, and ascertaining the survival ratio of hatchery fish. A simple presence/absence type mark can provide a straightforward identification means for hatchery fish, at high speed. This project concerns the design and development of an automated machine for spray marking of fish. The machine consists of four main modules: the feeding unit; the conveying unit; the spray marking unit; and the pigment recirculation unit. A fluorescent pigment may be used for mass marking, which is fast and inexpensive, the associated fish survival was excellent, and a mark retention of 130 days or more is possible. A commercial spray gun is adopted, which uses high-pressure air to embed microscopic fluorescent granules into the epidermis of fish. A pigment emulsion in water is used instead of the dry powdered pigment. A conveyor system is used to transport live juvenile fish into the spray marker, which dispenses the pigment mixed with water, through a nozzle with the aid of compressed air. An agitator is used to continuously mix the container of pigment suspended in water, in order to reduce clogging of the nozzle and ensure the uniformity of marking. The marks are detected in the field by examining samples of fish under a low-power ultraviolet (UV) lamp. During examination under a UV light source, the spray-marked areas on the fish body, which contain the fluorescent particles, will shine brightly in a specific color such as red, green, or orange in the visible spectrum and these marks can be very easily detected through the naked eye or by optical means. By this method, fish can be marked at a rate of approximately 15000/h using about one pound (0.45 kg) of fluorescent pigment per 7000 fish. The length of the fish ranges from 38 mm to 52 mm. Although the particles of the fluorescent pigment spray impinge on fish body with a reasonable momentum, only a fraction of the pigment particles actually embed into the scales of fish. Much of the sprayed emulsion is collected in a container attached to the underside of the machine at the exit, and is recycled, providing a degree of environmental friendliness. Water in the emulsion increases the momentum of the pigment spray during marking, thereby increasing the mark retention. Also an emulsion facilitates the use of the system under damp conditions, thereby reducing nozzle clogging. An air pressure of 120 psi is used by the spray gun. Spraying a target (fish) positioned 30 cm from the nozzle resulted in a 8.5 cm spray region diameter and a 5 cm marking region diameter. The conveyor belt is roughly 15 cm wide. According to typical experimental conditions, one spray gun is able to cover only a 5 cm width. A triple-gun system is needed in the design, to produce the required marking area on the conveyor.

Project 11: Machine for Grading Herring Roe

Quality is crucial for manufactured products such as automotive parts and high-end fish products such as herring roe. Vision-based systems are used in industry for inspection, quality assessment, and grading of products. Processed herring roe, which is considered a delicacy, has a lucrative market in countries like Japan. Size, shape, color, texture, and firmness are important in determining the overall quality of a skein of roe. A Grade 1 product may command double the price of a Grade 2 product. Accurate grading is quite important in this respect. Grading of herring roe is done mainly by manual labor at present. In view of associated difficulties such as speed and maintaining a uniform product quality, machine grading has received much attention. Design a grading machine that employs intelligent sensor integration and fusion, for herring roe. In the machine, the roe skeins

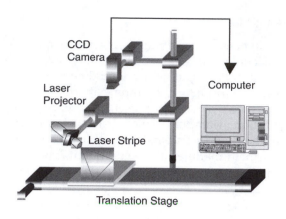

FIGURE P13.11
Laser-based sensing system.

are arranged in a single file at the feeder, and sent through the sensory system. A camera-based sensory system, as schematically shown in Figure P13.11, may be used for high-speed sensing of size, shape and color. Ultrasound or other methods may be used for sensing texture and firmness. Roe firmness, geometric features, associated weight estimates, and color are incorporated in an intelligent sensor fusion system, to arrive at a grading decision. Typical throughput for the machine is 5 skeins/s. The dimension of a skein of roe will not exceed 20 cm × 5 cm. Consider the experimental arrangement shown in Figure P13.11. Images are captured continuously through a PULNIX 6701 progressive scan camera equipped with a 12.5 mm, *f*.1.4 Cosmicar TV lens, by a PCI frame grabber board manufactured by Matrox Genesis. Single and multiple linear laser stripes generated from the LASIRIS laser diode structured-light projector (30 mW, 670 nm), are projected onto the object (herring roe). Microsoft Visual Basic 6.0 programming language is used as an object oriented environment to develop the human machine interface. Visual C++ 6.0 programming language is used to develop the algorithm for quality assessment, as a Dynamic Link Library (DLL) in a powerful and high-speed PC computer.

Project 12: Hydraulic Control System

Component sizing is an important consideration in the design of a hydraulic control system. You are asked to design a hydraulic system for a radar positioning drive. Specifically, you must

a. Select a suitable hydraulic motor and suitable gearing to drive the load (radar).
b. Select a suitable pump for continuous hydraulic power supply.
c. Design a speed transmission unit (e.g., gear) for the motor-to-load (radar) coupling.
d. Determine the inlet pressure at the hydraulic motor.
e. Determine the pump outlet pressure.

The following data are given:

Load inertia = 2000 kg · m²
Maximum load speed = 1 rad/s
Maximum load acceleration = 10 rad/s²

Wind torque = 2000 N · m

Distance from pump to hydraulic motor = 4 m

Size of pipeline (steel) = 1.25 cm O.D. and 1.0 mm thickness

Maximum supply pressure of fluid = 20,000 kPa

Hydraulic power loss in pipeline = 5%

Pump leakage = motor leakage = 5%

Motor efficiency = gear box efficiency = 95%

Assume that the hydraulic fluid is MIL-H-5606.

Design the following:

1. A pump-controlled system, in which the pump directly supplies a controlled flow to the hydraulic motor.

2. A valve-controlled system, in which a valve is used between the pump and the hydraulic motor, to supply a controlled flow to the motor (assume a 5% valve leakage).

You may use a mechanical engineering handbook to obtain the specifications for the pump and the hydraulic motor (usually the same specs are given for both pumps and motors) and to estimate the pressure loss in the steel piping carrying MIL-H-5606 oil. Commercially available sensors, transmission units, motors, pumps, valves, and power supplies may be used.

Comment: Consider the problem of servovalve selection for a hydraulic drive. The first step is to choose a suitable hydraulic actuator (ram or motor) that meets the load requirements. This establishes the load flow Q_L and the load pressure P_L at the operating speed of the load. The supply pressure P_s is also known. Note that under no-load conditions, the pressure drop across the servovalve is P_s, and under normal operating conditions, it is $P_s - P_L$. In manufacturers' specifications, the rated flow of a servovalve is given at some specified pressure drop (e.g., 7000 kPa). Since the flow is proportional to the square root of the pressure drop, we can determine the required flow rating for the valve (typically, by increasing the computed value of the rated flow by 10% to allow for leakage, fluctuations in load, etc.). This rated flow is one factor that governs the choice of a servovalve. The second factor is the valve bandwidth, which should be several times larger than the primary resonant frequency of the load, for proper control. Typical information available from servovalve manufacturers includes the frequency corresponding to the 90° phase lag point of the valve response. This frequency may be used as a measure of the valve bandwidth. When selecting a servovalve for a valve-controlled radar drive, the first step would be to obtain a catalog with a data sheet from a well-known servovalve manufacturer. Assume that the resonant frequency of the radar system is 10 Hz.

Appendix A

Transform Techniques

Many people use "transforms" without even knowing it. A "transform" is simply a number, variable, or function in a different form. For example, since $10^2 = 100$, you can use the exponent (2) to represent the number 100. Doing this for all numbers (i.e., using their exponent to the base 10), results in a "table of logarithms." One can perform mathematical computations using only "logarithms." The logarithm transforms all numbers into their exponential equivalents; a table of such transforms (i.e., a log table) enables a user to quickly transform any number into its exponent, do the computations using exponents (where, a product becomes an addition and a division becomes a subtraction), and transform the result back (i.e., inverse logarithm) into the original form. It is seen that the computations have become simpler by using logarithms, but at the cost of the time and effort needed for transformation and inverse transformation.

Other common transforms include the Laplace transform, Fourier transform, and Z transform. In particular, the Laplace transform provides a simple, algebraic way to solve (i.e., integrate) a linear differential equation. Most functions that we use are of the form t^n, $\sin\omega t$, or e^t, or some combination of them. Thus, in the expression

$$y = f(t)$$

the function y is quite likely a power, a sine, or an exponential function. Also, often, we have to work with derivatives and integrals of these functions, and differential equations containing these functions. These tasks can be greatly simplified by the use of the *Laplace transform*.

Concepts of frequency-response analysis originate from the nature of the response of a dynamic system to a sinusoidal (i.e., harmonic) excitation. These concepts can be generalized because the time-domain analysis, where the independent variable is time (t) and the frequency-domain analysis, where the independent variable is frequency (ω) are linked through the *Fourier transformation*. Analytically, it is more general and versatile to use the Laplace transformation, where the independent variable is the Laplace variable (s) which is complex (nonreal). This is true because analytical Laplace transforms may exist even for time functions that do not have "analytical" Fourier transforms. But with compatible definitions, the Fourier transform results can be obtained form the Laplace transform results simply by setting $s = j\omega$. In the present appendix we will formally introduce the Laplace transformation and the Fourier transformation, and will illustrate how these techniques are useful in the analysis of mechatronic systems. The preference of one domain over another will depend on such factors as the nature of the excitation input, the type of the available analytical model, the time duration of interest, and the quantities that need to be determined.

A.1 Laplace Transform

The Laplace transformation relates the time domain to the *Laplace domain* (also called *s-domain* or complex frequency domain). The Laplace transform $Y(s)$ of a piecewise-continuous function or signal $y(t)$ is given, by definition, as

$$Y(s) = \int_0^\infty y(t)\exp(-st)\,dt \tag{A.1}$$

and is denoted using the Laplace operator $\mathcal{L}$, as

$$Y(s) = \mathcal{L}y(t) \tag{A.1a}$$

Here, s is a complex independent variable known as the *Laplace variable*, defined by

$$s = \sigma + j\omega \tag{A.2}$$

where σ is a real-valued constant that will make the transform Equation A.1 finite, ω is simply frequency, and $j = \sqrt{-1}$. The real value (σ) can be chosen sufficiently large so that the integral in Equation A.1 is finite even when the integral of the signal itself (i.e., $\int y(t)\,dt$) is not finite. This is the reason why, for example, Laplace transform is better behaved than Fourier transform, which will be defined later, from the analytical point of view. The symbol s can be considered to be a constant, when integrating with respect to t, in Equation A.1.

The inverse relation (i.e., obtaining y from its Laplace transform) is

$$y(t) = \frac{1}{2\pi j} \int_{\sigma-j\omega}^{\sigma+j\omega} Y(s)\exp(st)\,ds \tag{A.3}$$

and is denoted using the inverse Laplace operator $\mathcal{L}^{-1}$, as

$$y(t) = \mathcal{L}^{-1}Y(s \tag{A.3a}$$

The integration in Equation A.3 is performed along a vertical line parallel to the imaginary (vertical) axis, located at σ from the origin in the complex Laplace plane (s-plane). For a given piecewise-continuous function $y(t)$, the Laplace transform exists if the integral in Equation A.1 converges. A sufficient condition for this is

$$\int_0^\infty |y(t)|\exp(-\sigma t)\,dt < \infty \tag{A.4}$$

Convergence is guaranteed by choosing a sufficiently large and positive σ. This property is an advantage of the Laplace transformation over the Fourier transformation.

A.1.1 Laplace Transforms of Some Common Functions

Now we determine the Laplace transform of some useful functions using the definition Equation A.1. Usually, however, we use Laplace transform tables to obtain these results.

A.1.1.1 Laplace Transform of a Constant

Suppose our function $y(t)$ is a constant, B. Then the Laplace transform is

$$\mathcal{L}(B) = Y(s) = \int_0^\infty B e^{-st} dt$$

$$= B \frac{e^{-st}}{-s} \Big|_0^\infty = \frac{B}{s}$$

A.1.1.2 Laplace Transform of the Exponential

If $y(t)$ is e^{at}, its Laplace transform is

$$\mathcal{L}(e^{at}) = \int_0^\infty e^{-st} e^{at} dt$$

$$= \int_0^\infty e^{(a-s)t} dt$$

$$= \frac{1}{(a-s)} e^{(a-s)t} \Big|_0^\infty = \frac{1}{s-a}$$

NOTE If $y(t)$ is e^{-at}, it is obvious that the Laplace transform is

$$\mathcal{L}(e^{-at}) = \int_0^\infty e^{-st} e^{-at} dt$$

$$= \int_0^\infty e^{-(a+s)t} dt$$

$$= \frac{-1}{(a-s)} e^{-(a+s)t} \Big|_0^\infty = \frac{1}{s+a}$$

This result can be obtained from the previous result simply by replacing a with $-a$.

A.1.1.3 Laplace Transform of Sine and Cosine

In the following, the letter $j = \sqrt{-1}$, which is imaginary unit. If $y(t)$ is $\sin\omega t$, the Laplace transform is

$$\mathcal{L}(\sin\omega t) = \int_0^\infty e^{-st} (\sin\omega t) dt$$

Consider the identities:

$$e^{j\omega t} = \cos\omega t + j\sin\omega t$$

$$e^{-j\omega t} = \cos\omega t - j\sin\omega t$$

If we add and subtract these two equations, respectively, we obtain the expressions for the sine and the cosine in terms of $e^{j\omega t}$ and $e^{-j\omega t}$:

$$\cos \omega t = \frac{1}{2}(e^{j\omega t} + e^{-j\omega t})$$

$$\sin \omega t = \frac{1}{2i}(e^{j\omega t} - e^{-j\omega t})$$

$$\mathcal{L}(\cos \omega t) = \frac{1}{2}\mathcal{L}(e^{j\omega t}) + \frac{1}{2}\mathcal{L}(e^{-j\omega t})$$

$$\mathcal{L}(\sin \omega t) = \frac{1}{2}\mathcal{L}(e^{j\omega t}) - \frac{1}{2}\mathcal{L}(e^{-j\omega t})$$

We have just seen that

$$\mathcal{L}(e^{at}) = \frac{1}{s-a}; \quad \mathcal{L}(e^{-at}) = \frac{1}{s+a}$$

Hence,

$$\mathcal{L}(e^{j\omega t}) = \frac{1}{s-j\omega t}; \quad \mathcal{L}(e^{-j\omega t}) = \frac{1}{s+j\omega t}$$

Substituting these expressions, we get

$$\mathcal{L}(\cos \omega t) = \frac{1}{2}\left[\frac{1}{s-j\omega}\right] + \frac{1}{2}\left[\frac{1}{s+j\omega}\right]$$

$$= \frac{1}{2}\left[\frac{s+j\omega}{s^2 - (j\omega)^2} + \frac{s-j\omega}{s^2 - (j\omega)^2}\right]$$

$$= \frac{s}{s^2 + \omega^2}$$

$$\mathcal{L}(\sin \omega t) = \frac{1}{2j}\mathcal{L}(e^{j\omega t} - e^{-j\omega t})$$

$$= \frac{1}{2j}\left[\frac{1}{s-j\omega}\right] - \frac{1}{2j}\left[\frac{1}{s+j\omega}\right]$$

$$= \frac{1}{2j}\left[\frac{s+j\omega}{s^2 - (j\omega)^2} + \frac{s-j\omega}{s^2 - (j\omega)^2}\right]$$

$$= \frac{1}{2j}\left[\frac{2j\omega}{s^2 + \omega^2}\right]$$

$$= \frac{\omega}{s^2 + \omega^2}$$

A.1.1.4   Transform of a Derivative

Let us transform a derivative of a function. Specifically, the derivative of a function y of is denoted by $\dot{y} = dy / dt$. Its Lapalace transform is given by

$$\mathcal{L}(\dot{y}) = \int_0^\infty e^{-st} \dot{y} dt = \int_0^\infty e^{-st} \frac{dy}{dt} dt \tag{A.5}$$

Now we integrate by parts, to eliminate the derivative within the integrand.

A.1.1.4.1   Integration by Parts

From calculus we know that $d(uv) = udv + vdu$
  By integrating we get $uv = \int udv + \int vdu$
Hence,

$$\int udv = uv - vdu \tag{A.6}$$

This is known as integration by parts.
  In Equation A.5, let

$$u = e^{-st} \quad \text{and} \quad v = y$$

Then,

$$dv = dy = \frac{dy}{dt} dt = \dot{y} dt$$

$$du = \frac{du}{dt} dt = -se^{-st} dt$$

Substitute in Equation A.5 to integrate by parts:

$$\mathcal{L}(\dot{y}) = \int_0^\infty e^{-st} dy$$

$$= \int udv = uv - \int vdu$$

$$= e^{-st} y(t) \Big|_0^\infty - \int_0^\infty -se^{-st} y(t) dt$$

$$= -y(0) + s\mathcal{L}[y(t)]$$

$$= s\mathcal{L}(y) - y(0)$$

where $y(0)$ = initial value of y. This says that the Laplace transform of a first derivative $\dot{y}$, equals s times the Laplace transform of the function y minus the initial value of the function (the initial condition).

NOTE We can determine the Laplace transforms of the second and higher derivatives by repeated application this result, for the first derivative. For example, the transform of the second derivative is given by

$$\mathcal{L}[\ddot{y}(t)] = \mathcal{L}\left[\frac{d\dot{y}(t)}{dt}\right] = s\mathcal{L}[\dot{y}(t)] - \dot{y}(0) = s\{s\mathcal{L}[y(t)] - y(0)\} - \dot{y}(0)$$

Hence

$$\mathcal{L}[\ddot{y}(t)] = s^2\,\mathcal{L}[y(t)] - sy(0) - \dot{y}(0)$$

A.1.2 Table of Laplace Transforms

Table A.1 shows the Laplace transforms of some common functions. Specifically, the table lists functions as $y(t)$, and their Laplace transforms (on the right) as $Y(s)$ or $\mathcal{L}\,y(t)$. If one is given a function, one can get its Laplace transform from the table. Conversely, if one is given the transform, one can get the function from the table.

TABLE A.1

Laplace Transform Pairs

$y(t) = \mathcal{L}^{-1}[Y(s)]$	$\mathcal{L}[y(t)] = Y(s)$
B	B/s
e^{-at}	$\dfrac{1}{s+a}$
e^{at}	$\dfrac{1}{s-a}$
$\sinh at$	$\dfrac{a}{s^2 - a^2}$
$\cosh at$	$\dfrac{s}{s^2 - a^2}$
$\sin \omega t$	$\dfrac{\omega}{s^2 + \omega^2}$
$\cos \omega t$	$\dfrac{s}{s^2 + \omega^2}$
$e^{-at}\sin \omega t$	$\dfrac{\omega}{(s+a)^2 + \omega^2}$
$e^{-at}\cos \omega t$	$\dfrac{s+a}{(s+a)^2 + \omega^2}$
Ramp t	$\dfrac{1}{s^2}$
$e^{-at}(1 - at)$	$\dfrac{s}{(s+a)^2}$
$y(t)$	$Y(s)$
$\dfrac{dy}{dt} = \dot{y}$	$sY(s) - y(0)$

(Continued)

TABLE A.1

(Continued)

$y(t) = \mathcal{L}^{-1}[Y(s)]$	$\mathcal{L}[y(t)] = Y(s)$
$\dfrac{d^2 y}{dt^2} = \ddot{y}$	$s^2 Y(s) - sy(0) - \dot{y}(0)$
$\dfrac{d^3 y}{dt^3} = \dddot{y}$	$s^3 Y(s) - s^2 y(0) - s\dot{y}(0) - \ddot{y}(0)$
$\int_a^c y(t)dt$	$\dfrac{1}{s}Y(s) - \dfrac{1}{s}\int_0^a y(t)dt$
$af(t) + bg(t)$	$aF(s) + bG(s)$
Unit step $U(t) = 1$ for $t \geq 0$ $\quad\quad\quad = 0$ otherwise	$\dfrac{1}{s}$
Delayed step $cU(t-b)$	$\dfrac{c}{s}e^{-bs}$
Pulse $c[U(t) - U(t-b)]$	$c\left(\dfrac{1-e^{-bs}}{s}\right)$
Impulse function $\delta(t)$	1
Delayed impulse $\delta(t-b) = \dot{U}(t-b)$	e^{-bs}
Sine pulse	$\left(\dfrac{\omega}{s^2 + \omega^2}\right)\left(1 + e^{-(\pi s/\omega)}\right)$

Some general properties and results of the Laplace transform are given in Table A.2. In particular, note that, with zero initial conditions, differentiation can be interpreted as multiplication by s. Also, integration can be interpreted as division by s.

A.2 Response Analysis

The Laplace transform method can be used in the response analysis of dynamic systems, mechatronic and control systems in particular. We will give examples for the approach.

Example A.1

The capacitor-charge equation of the RC circuit shown in Figure A.1 is

$$e = iR + v \tag{i}$$

TABLE A.2

Important Laplace Transform Relations

$\mathcal{L}^{-1}\, F(s) = f(t)$	$\mathcal{L}f(t) = F(s)$
$\dfrac{1}{2\pi j}\int_{\sigma-j\infty}^{\sigma+j\infty} F(s)\exp(st)ds$	$\int_0^\infty f(t)\exp(-st)dt$
$k_1 f_1(t) + k_2 f_2(t)$	$k_1\, F_1(s) + k_2\, F_2(s)$
$\exp(-at)f(t)$	$F(s+a)$
$f(t-\tau)$	$\exp(-\tau s)F(s)$
$f^{(n)}(t) = \dfrac{d^n f(t)}{dt^n}$	$s^n F(s) - s^{n-1} f(0^+) - s^{n-2} f^1(0^+)$ $-\cdots - f^{n-1}(0^+)$
$\int_{-\infty}^{t} f(t)dt$	$\dfrac{F(s)}{s} + \dfrac{\int_{-\infty}^{0} f(t)dt}{s}$
t^n	$\dfrac{n!}{s^{n+1}}$
$t^n e^{-at}$	$\dfrac{n!}{(s+a)^{n+1}}$

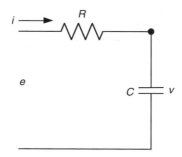

FIGURE A.1
An RC circuit with applied voltage e and voltage v across capacitor.

For the capacitor,

$$i = C\frac{dv}{dt} \tag{ii}$$

Substitute Equation ii in Equation i to get the circuit equation:

$$e = RC\frac{dv}{dt} + v \tag{iii}$$

Take the Laplace transform of each term in Equation iii, with all initial conditions = 0:

$$E(s) = RCsV(s) + V(s)$$

The transfer function expressed as the output/input ratio (in the transform form) is:

$$\frac{V(s)}{E(s)} = \frac{V(s)}{sRCV(s) + V(s)} = \frac{1}{sRC+1} = \frac{1}{\tau s+1} \tag{iv}$$

where $\tau = RC$ = time constant.

The actual response can now be found from Table A.1 for a given input E. The first step is to get the transform into proper form (as in line 2):

$$\frac{1}{\tau s+1} = \frac{1/\tau}{s+(1/\tau)} = \frac{a}{s+a} = a\left(\frac{1}{s+a}\right)$$

where $a = 1/\tau$. Suppose that input (excitation) e is a unit impulse. Its Laplace transform (see Table A.1) is $E = 1$. Then from Equation iv,

$$V(s) = \frac{1}{\tau s+1}$$

From line 2 of Table A.1, the response is

$$v = ae^{-at} = \frac{1}{\tau}e^{-t/\tau} = \frac{1}{RC}e^{-t/RC}$$

A common transfer function for an overdamped second-order system (e.g., one with two RC circuit components of Figure A.1) would be

$$\frac{V(s)}{E(s)} = \frac{1}{(1+\tau_1 s)(1+\tau_2 s)}$$

This can be expressed as "partial fractions" in the from

$$\frac{A}{1+\tau_1 s} + \frac{B}{1+\tau_2 s}$$

and solved in the usual manner.

Example A.2

The transfer function of a thermal system is given by

$$G(s) = \frac{2}{(s+1)(s+3)}$$

If a unit step input is applied to the system, with zero initial conditions, what is the resulting response?

SOLUTION
Input

$$U(s) = \frac{1}{s} \quad \text{(for a unit step)}$$

Since

$$\frac{Y(s)}{U(s)} = \frac{2}{(s+1)(s+3)}$$

we have the output (response)

$$Y(s) = \frac{2}{s(s+1)(s+3)}$$

Its inverse Laplace transform gives the time response. For this, first convert the expression into partial fractions as

$$\frac{2}{s(s+1)(s+3)} = \frac{A}{s} + \frac{B}{(s+1)} + \frac{C}{(s+3)} \tag{i}$$

The unknown A is determined by multiplying Equation i throughout by s and then setting $s = 0$. We get

$$A = \frac{2}{(0+1)(0+3)} = \frac{2}{3}$$

Similarly, B is obtained by multiplying Equation i throughout by $(s + 1)$ and then setting $s = -1$. We get

$$B = \frac{2}{(-1)(-1+3)} = -1$$

Next, C is obtained by multiplying Equation i throughout by $(s + 3)$ and then setting $s = -3$. We get

$$C = \frac{2}{(-3)(-3+1)} = \frac{1}{3}$$

Hence,

$$Y(s) = \frac{2}{3s} - \frac{1}{(s+1)} + \frac{1}{3(s+3)}$$

Take the inverse transform using line 2 of Table A.1.

$$y(t) = \frac{2}{3} - e^{-t} + \frac{1}{3}e^{-3t}$$

Example A.3

The transfer function of a damped simple oscillator is known to be of the form

$$\frac{Y(s)}{U(s)} = \frac{\omega_n^2}{\left(s^2 + 2\zeta\omega_n s + \omega_n^2\right)}$$

where

   ω_n = Undamped natural frequency

   ζ = Damping ratio

Suppose that a unit step input (i.e., $U(s) = \frac{1}{s}$) is applied to the system. Using Laplace transform tables determine the resulting response, with zero initial conditions.

SOLUTION

$$Y(s) = \frac{1}{s} \cdot \frac{\omega_n^2}{\left(s^2 + 2\zeta\omega_n s + \omega_n^2\right)}$$

The corresponding partial fractions are of the form

$$Y(s) = \frac{A}{s} + \frac{Bs + C}{\left(s^2 + 2\zeta\omega_n s + \omega_n^2\right)} = \frac{\omega_n^2}{s\left(s^2 + 2\zeta\omega_n s + \omega_n^2\right)} \tag{i}$$

We need to determine A, B, and C.

Multiply Equation i throughout by s and set $s = 0$. We get

$$A = 1$$

Next note that the roots of the characteristic equation

$$s^2 + 2\zeta\omega_n s + \omega_n^2 = 0$$

are

$$s = -\zeta\omega_n \pm \sqrt{\zeta^2 - 1}\,\omega_n = -\zeta\omega_n \pm j\omega_d$$

These are the poles of the system and are complex conjugates. Two equations for B and C are obtained by multiplying Equation i by $s + \zeta\omega_n - \sqrt{\zeta^2 - 1}\,\omega_n$ and setting $s = -\zeta\omega_n + \sqrt{\zeta^2 - 1}\,\omega_n$ and by multiplying Equation i by $s + \zeta\omega_n + \sqrt{\zeta^2 - 1}\,\omega_n$ and setting $s = -\zeta\omega_n - \sqrt{\zeta^2 - 1}\,\omega_n$. From them, we obtain $B = -1$ and $C = 0$. Consequently,

$$Y(s) = \frac{1}{s} - \frac{s}{\left(s^2 + 2\zeta\omega_n s + \omega_n^2\right)}$$

$$= \frac{1}{s} - \frac{s + \zeta\omega_n}{\left[(s + \zeta\omega_n)^2 + \omega_d^2\right]} + \frac{\zeta}{\sqrt{1 - \zeta^2}} \cdot \frac{\omega_d}{\left[(s + \zeta\omega_n)^2 + \omega_d^2\right]}$$

where, $\omega_d = \sqrt{1 - \zeta^2}\,\omega_n$ = damped natural frequency.

Now use Table A.1 to obtain the inverse Laplae transform:

$$y_{step}(t) = 1 - e^{-\zeta\omega_n t}\cos\omega_d t - \frac{\zeta}{\sqrt{1-\zeta^2}}e^{-\zeta\omega_n t}\sin\omega_d t$$

$$= 1 - \frac{e^{-\zeta\omega_n t}}{\sqrt{1-\zeta^2}}[\sin\phi\cos\omega_d t + \cos\phi\sin\omega_d t]$$

$$= 1 - \frac{e^{-\zeta\omega_n t}}{\sqrt{1-\zeta^2}}\sin(\omega_d t + \phi)$$

where, $\cos\phi = \zeta$ = damping ratio; $\sin\phi = \sqrt{1-\zeta^2}$. This result is identical to what was given in Chapter 2.

Example A.4

The open-loop response of a plant to a unit impulse input, with zero ICs, was found to be $2e^{-t}\sin t$. What is the transfer function of the plant?

SOLUTION

By linearity, since a unit impulse is the derivative of a unit step, the response to a unit impulse is given by the derivative of the result given in the previous example; thus

$$y_{impulse}(t) = \frac{\zeta\omega_n}{\sqrt{1-\zeta^2}}e^{-\zeta\omega_n t}\sin(\omega_d t + \phi) - \frac{\omega_d}{\sqrt{1-\zeta^2}}e^{-\zeta\omega_n t}\cos(\omega_d t + \phi)$$

$$= \frac{\omega_n}{\sqrt{1-\zeta^2}}e^{-\zeta\omega_n t}[\cos\phi\sin(\omega_d t + \phi) - \sin\phi\cos(\omega_d t + \phi)]$$

or,

$$y_{impulse}(t) = \frac{\omega_n}{\sqrt{1-\zeta^2}}e^{-\zeta\omega_n t}\sin\omega_d t$$

Compare this with the given expression. We have

$$\frac{\omega_n}{\sqrt{1-\zeta^2}} = 2; \ \zeta\omega_n = 1; \ \omega_d = 1$$

But,

$$\omega_n^2 = (\zeta\omega_n)^2 + \omega_d^2 = 1 + 1 = 2$$

Hence

$$\omega_n = \sqrt{2}$$

Hence

$$\zeta = \frac{1}{\sqrt{2}}$$

The system transfer function is

$$\frac{\omega_n^2}{\left(s^2 + 2\zeta\omega_n s + \omega_n^2\right)} = \frac{2}{s^2 + 2s + 2}$$

Example A.5

Express the Laplace transformed expression

$$X(s) = \frac{s^3 + 5s^2 + 9s + 7}{(s+1)(s+2)}$$

as partial fractions. From the result, determine the inverse Laplace function $x(t)$.

SOLUTION

$$X(s) = s + 2 + \frac{2}{s+1} - \frac{1}{s+2}$$

From Table A.1, we get the inverse Laplace transform $x(t)$

$$x(t) = \frac{d}{dt}\delta(t) + 2\delta(t) + 2e^{-t} - e^{-2t}$$

where $\delta(t)$ = unit impulse function.

A.3 Transfer Function

By the use of Laplace transformation, a *convolution integral* equation can be converted into an algebraic relationship. To illustrate this, consider the convolution integral which gives the response $y(t)$ of a dynamic system to an excitation input $u(t)$, with zero ICs, as discussed in Chapter 2. By definition of Laplace transform as given by Equation A.1, the Laplace transform of the response may be expressed as

$$Y(s) = \int_0^\infty \int_0^\infty h(\tau)u(t-\tau)d\tau \exp(-st)dt \tag{A.7}$$

Note that $h(t)$ is the *impulse response function* of the system. Since the integration with respect to t is performed while keeping τ constant, we have $dt = d(t-\tau)$. Consequently,

$$Y(s) = \int_{-\tau}^\infty u(t-\tau)\exp[-s(t-\tau)]\,d(t-\tau) \int_0^\infty h(\tau)\exp(-s\tau)d\tau$$

The lower limit of the first integration can be made equal to zero, in view of the fact that $u(t) = 0$ for $t < 0$. Again, by using the definition of Laplace transformation, the foregoing relation can be expressed as

$$Y(s) = H(s)U(s) \tag{A.8}$$

in which

$$H(s) = \mathscr{L}h(t) = \int_0^\infty h(t)\exp(-st)dt \tag{A.9}$$

Note that, by definition, the transfer function of a system, denoted by $H(s)$, is given by Equation A.8. More specifically, system transfer function is given by the ratio of the Laplace-transformed output and the Laplace-transformed input, with zero initial conditions. In view of Equation A.9, it is clear that the system transfer function can be expressed as the Laplace transform of the impulse-response function of the system. Transfer function of a linear and constant-parameter system is a unique function that completely represents the system. A physically realizable, linear, constant-parameter system possesses a unique transfer function, even if the Laplace transforms of a particular input and the corresponding output do not exist. This is clear from the fact that the transfer function is a system model and does not depend on the system input itself.

NOTE The transfer function is also commonly denoted by $G(s)$. But in the present context we use $H(s)$ in view of its relation to $h(t)$.

Consider the nth-order linear, constant-parameter dynamic system given by

$$a_n \frac{d^n y}{dt^n} + a_{n-1} \frac{d^{n-1}y}{dt^{n-1}} + \cdots + a_0 y = b_0 u + b_1 \frac{du(t)}{dt} + \cdots + b_m \frac{d^m u(t)}{dt^m} \tag{A.10}$$

For a physically realizable system, $m \leq n$. By applying Laplace transformation and then integrating by parts, it may be verified that

$$\mathscr{L}\frac{d^k f(t)}{dt^k} = s^k F(s) - s^{k-1}f(0) - s^{k-2}\frac{df(0)}{dt} - \cdots + \frac{d^{k-1}f(0)}{dt^{k-1}} \tag{A.11}$$

By definition, the initial conditions are set to zero in obtaining the transfer function. This results in the transfer function

$$H(s) = \frac{b_0 + b_1 s + \cdots + b_m s^m}{a_0 + a_1 s + \cdots + a_n s^n} \tag{A.12}$$

for $m \leq n$. Note that Equation A.12 contains all the information that is contained in Equation A.10. Consequently, transfer function is an analytical model of a system. The transfer function may be employed to determine the total response of a system for a given input, even though it is defined in terms of the response under zero initial conditions. This is quite logical because the analytical model of a system is independent of the initial conditions of the system.

A.4 Fourier Transform

The Fourier transform $Y(f)$ of a signal $y(t)$ relates the time domain to the frequency domain. Specifically,

$$Y(f) = \int_{-\infty}^{+\infty} y(t)\exp(-j2\pi ft)dt$$

(A.13)

$$= \int_{-\infty}^{+\infty} y(t)e^{-\omega t}dt$$

Using the Fourier operator "$\mathscr{F}$" terminology:

$$Y(f) = \mathscr{F}\, y(t)$$

(A.14)

Note that if $y(t) = 0$ for $t < 0$, as in the conventional definition of system excitations and responses, the Fourier transform is obtained from the Laplace transform by simply changing the variable according to $s = j2\pi f$ or $s = j\omega$. The Fourier is a special case of the Laplace, where, in Equation A.2, $\sigma = 0$:

$$Y(f) = Y(s)|_{s=j2\pi f}$$

(A.15)

or

$$Y(\omega) = Y(s)|_{s=j\omega}$$

(A.16)

The (complex) function $Y(f)$ is also termed the (continuous) *Fourier spectrum* of the (real) signal $y(t)$. The inverse transform is given by:

$$y(t) = \int_{-\infty}^{+\infty} Y(f)\exp(j2\pi ft)df$$

(A.17)

or,

$$y(t) = \mathscr{F}^{-1}Y(f)$$

Note that according to the definition given by Equation A.13, the Fourier spectrum $Y(f)$ is defined for the entire frequency range $f(-\infty, +\infty)$ which includes negative values. This is termed the *two-sided spectrum*. Since, in practical applications it is not possible to have "negative frequencies," the *one-sided spectrum* is usually defined only for the frequency range $f(0, \infty)$.

In order that a two-sided spectrum have the same amount of *power* as a one-sided spectrum, it is necessary to make the one-sided spectrum double the two-sided spectrum for $f > 0$.

If the signal is not sufficiently transient (fast-decaying or damped), the infinite integral given by Equation A.13 might not exist, but the corresponding Laplace transform might still exist.

A.4.1 Frequency-Response Function (Frequency Transfer Function)

The Fourier integral transform of the impulse-response function is given by

$$H(f) = \int_{-\infty}^{\infty} h(t) \exp(-j2\pi ft) \, dt \qquad \text{(A.18}$$

where f is the *cyclic frequency* (measured in cycles/s or hertz). This is known as the frequency-response function (or, frequency transfer function) of a system. Fourier transform operation is denoted as $\mathscr{F} h(t) = H(f)$. In view of the fact that $h(t) = 0$ for $t < 0$, the lower limit of integration in Equation A.18 could be made zero. Then, from Equation A.9, it is clear that $H(f)$ is obtained simply by setting $s = j2\pi f$ in $H(s)$. Hence, strictly speaking, we should use the notation $H(j2\pi f)$ and not $H(f)$. But for the notational simplicity we denote $H(j2\pi f)$ by $H(f)$. Furthermore, since the angular frequency $\omega = 2\pi f$, we can express the frequency response function by $H(j\omega)$, or simply by $H(\omega)$ for the notational convenience. It should be noted that the frequency-response function, like the (Laplace) transfer function, is a complete representation of a linear, constant-parameter system. In view of the fact that both $u(t) = 0$ and $y(t) = 0$ for $t < 0$, we can write the Fourier transforms of the input and the output of a system directly by setting $s = j2\pi f = j\omega$ in the corresponding Laplace transforms. Then, from Equation A.8, we have

$$Y(f) = H(f) \, U(f) \qquad \text{(A.19)}$$

NOTE Sometimes for notational convenience, the same lowercase letters are used to represent the Laplace and Fourier transforms as well as the original time-domain variables.

If the Fourier integral transform of a function exists, then its Laplace transform also exists. The converse is not generally true, however, because of poor convergence of the Fourier integral in comparison to the Laplace integral. This arises from the fact that the factor $\exp(-\sigma t)$ is not present in the Fourier integral. For a physically realizable, linear, constant-parameter system, $H(f)$ exists even if $U(f)$ and $Y(f)$ do not exist for a particular input. The experimental determination of $H(f)$, however, requires system stability. For the nth-order system given by Equation A.10, the frequency-response function is determined by setting $s = j2\pi f$ in Equation A.12, as

$$H(f) = \frac{b_0 + b_1 j2\pi f + \cdots + b_m (j2\pi f)^m}{a_0 + a_1 j2\pi f + \cdots + a_n (j2\pi f)^n} \qquad \text{(A.20)}$$

This, generally, is a complex function of f, which has a magnitude denoted by $|H(f)|$ and a phase angle denoted by $\angle H(f)$.

A.5 The s-plane

We have noted that the Laplace variable s is a complex variable, with a real part and an imaginary part. Hence, to represent it we will need two axes at right angles to each other— the real axis and the imaginary axis. These two axes from a plane, which is called the s-plane. Any general value of s (or, any variation or trace of s) may be marked on the s-plane.

A.5.1   An Interpretation of Laplace and Fourier Transforms

In the Laplace transformation of a function $f(t)$ we multiply the function by e^{-st} and integrate with respect to t. This process may be interpreted as determining the "components" $\mathscr{F}(s)$ of $f(t)$ in the "direction" e^{-st} where s is a complex variable. All such components $\mathscr{F}(s)$ should be equivalent to the original function $f(t)$.

In the Fourier transformation of $f(t)$ we multiply it by $e^{-j\omega t}$ and integrate with respect to t. This is the same as setting $s = j\omega$. Hence, the Fourier transform of $f(t)$ is $F(j\omega)$. Furthermore, $F(j\omega)$ represents the components of $f(t)$ that are in the direction of $e^{-j\omega t}$. Since $e^{-j\omega t} = \cos \omega t - j \sin \omega t$, in the Fourier transformation what we do is to determine the sinusoidal components of frequency ω, of a time function $f(t)$. Since s is complex, $F(s)$ is also complex and so is $F(j\omega)$. Hence they all will have a real part and an imaginary part.

A.5.2   Application in Circuit Analysis

The fact that $\sin \omega t$ and $\cos \omega t$ are 90° out of phase is further confirmed in view of

$$e^{jwt} = \cos \omega t + j \sin \omega t \tag{A.21}$$

Consider the R-L-C circuit shown in Figure A.2. For the capacitor, the current (i) and the voltage (v) are related through

$$i = C \frac{dv}{dt} \tag{A.22}$$

If the voltage $v = v_o \sin \omega t$, the current $i = v_o \, \omega C \cos \omega t$. Note that the magnitude of v/i is $\frac{1}{\omega C}$ (or, $\frac{1}{2\pi f C}$ where $\omega = 2\pi f$; f is the cyclic frequency and ω is the angular frequency). But v and i are out of phase by 90°. In fact, in the case of a capacitor, i leads v by 90°. The equivalent circuit resistance of a capacitance is called *reactance*, and is given by

$$X_C = \frac{1}{2\pi j f C} \tag{A.23}$$

$$= \frac{1}{j\omega Cj} \tag{A.24}$$

Note that this parameter changes with the frequency.

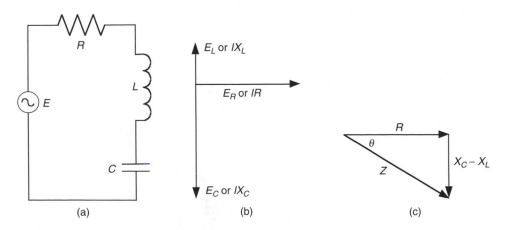

FIGURE A.2
(a) Series RLC circuit; (b) Phases of voltage drops; (c) Impedance triangle.

We cannot add the reactance of the capacitor and the resistance of the resisto algebraically; we must add them vectorially because the voltages across a capacito and resistor in series are not in phase. Also, the resistance in a resistor does not chang with frequency. In a series circuit, as in Figure A.2, the current is identical in eac element, but the voltages differ in both amplitude and phase; in a parallel circuit, th voltages are identical, but the currents differ in amplitude and phase.

Similarly, for an inductor

$$v = L\frac{di}{dt}$$ (A.25)

The corresponding reactance is

$$X_L = j\omega L = 2\pi jfL$$ (A.26)

If the voltage (E) across R in Figure A.2(a) is in the direction shown in Figure A.2(b) (i.e., pointing to the right), then the voltage across the inductor L must point upwards (90° leading) and the voltage across the capacitor C must point down (90° lagging). Since the current (I) is identical in each component of a series circuit, we see the directions of IR, IX_L and IX_C as in Figure A.2(b), giving the impedance triangle shown in Figure A.2(c).

To express these reactances in the s domain, we simply substitute s for $j\omega$ in the following, as clear from the Laplace versions of Equation A.22 and A.25:

$$X_C = \frac{1}{sC}$$

$$X_L = sL$$

The series impedance of the RLC circuit can be expressed as

$$Z = R + sL + \frac{1}{sC}$$

In this discussion, note the use of $\sqrt{-1}$ or j, to indicate a 90° phase change.

Appendix B

Software Tools

Modeling, analysis, design, data acquisition, and control are important activities within the field of Mechatronics. Computer software tools and environments are available for effectively carrying out, both at the learning level and at the professional application level. Several such environments and tools are commercially available. A selected few, which are particularly useful for the tasks related to the present book are outlined here.

MATLAB[1] is an interactive computer environment with a high-level language and tools for scientific and technical computation, modeling and simulation, design, and control of dynamic systems. SIMULINK[1] is a graphical environment for modeling, simulation, and analysis of dynamic systems, and is available as an extension to MATLAB. LabVIEW[1] is graphical programming language and a program development environment for data acquisition, processing, display, and instrument control.

B.1 Simulink

Perhaps the most convenient computer-based approach to simulation of a dynamic model (see Chapter 2) is by using a graphic environment that uses block diagrams. Several such environments are commercially available. One that is widely used is SIMULINK, which is an extension to MATLAB. It provides a graphical environment for modeling, simulating, and analyzing linear and nonlinear dynamic systems. Its use is quite convenient. First a suitable block diagram model of the system is developed on the computer screen, and stored. The SIMULINK environment provides almost any block that is used in a typical block diagram. These include transfer functions, integrators, gains, summing junctions, inputs (i.e., source blocks), and outputs (i.e., graph blocks or scope blocks). Such a block may be selected and inserted into the workspace as many times as needed, by clicking and dragging using the mouse. These blocks may be connected as required, using directed lines. A block may be opened by clicking on it, and the parameter values and text may be inserted or modified as needed. Once the simulation block diagram is generated in this manner, it may be run and the response may be observed through an output block (graph block or scope block). Since Simulink is integrated with MATLAB, data can be easily transferred between programs within various tools and applications.

B.1.1 Starting Simulink

First enter the MATLAB environment. You will see the MATLAB command prompt ≫. To start SIMULINK, enter the command: simulink. Alternatively, you may click on the "Simulink" button at the top of the MATLAB command window.

---

[1] MATLAB and SIMULINK are registered trademarks and products of The MathWorks, Inc. LabVIEW is a product of National Instruments, Inc.

The Simulink library browser window should now appear on the screen. Most of the blocks needed for modeling basic systems can be found in the subfolders of the main Simulink folder.

B.1.2 Basic Elements

There are two types of elements in SIMULINK: **blocks** and **lines**. Blocks are used to generate (or input), modify, combine, output, and display signals. Lines are used to transfer signals from one block to another.

B.1.2.1 Blocks

The subfolders below the SIMULINK folder show the general classes of blocks available for use. They are

- Continuous: Linear, continuous-time system elements (integrators, transfer functions, state-space models, etc.)
- Discrete: Linear, discrete-time system elements (integrators, transfer functions, state-space models, etc.)
- Functions and tables: User-defined functions and tables for interpolating function values
- Math: Mathematical operators (sum, gain, dot product, etc.)
- Nonlinear: Nonlinear operators (Coulomb/viscous friction, switches, relays, etc.)
- Signals and systems: Blocks for controlling/monitoring signals and for creating subsystems
- Sinks: For output or display of signals (displays, scopes, graphs, etc.)
- Sources: To generate various types of signals (step, ramp, sinusoidal, etc.)

Blocks may have zero or more input terminals and zero or more output terminals.

B.1.2.2 Lines

A directed line segment transmits signals in the direction indicated by its arrow. Typically, a line must transmit signals from the output terminal of one block to the input terminal of another block. One exception to this is, a line may be used to tap off the signal from another line. In this manner, the tapped original signal can be sent to other (one or more) destination blocks. However, a line can never inject a signal into another line; combining (or, summing) of signals has to be done by using a summing junction. A signal can be either a scalar signal (single signal) or a vector signal (several signals in parallel). The lines used to transmit scalar signals and vector signals are identical; whether it is a scalar or vector is determined by the blocks connected by the line.

B.1.3 Building an Application

To build a system for simulation, first bring up a "new model" window for creating the block diagram. To do this, click on the "New Model" button in the toolbar of the SIMULINK library browser. Initially the window will be blank. Then, build the system using the following three steps.

1. Gather Blocks: From the Simulink library browser, collect the blocks you need in your model. This can be done by simply clicking on a required block and dragging it into your workspace.

2. Modify the Blocks: SIMULINK allows you to modify the blocks in your model so that they accurately reflect the characteristics of your system. Double-click on the block to be modified. You can modify the parameters of the block in the "Block Parameters" window. SIMULINK gives a brief explanation of the function of the block in the top portion of this window.

3. Connect the Blocks: The block diagram must accurately reflect the system to be modeled. The selected SIMULINK blocks have to be properly connected by lines, to realize the correct block diagram. Draw the necessary lines for signal paths by dragging the mouse from the starting point of a signal (i.e., output terminal of a block) to the terminating point of the signal (i.e., input terminal of another block). SIMULINK converts the mouse pointer into a "crosshair" when it is close to an output terminal, to begin drawing a line, and the pointer will become a "double crosshair" when it is close enough to be snapped to an input terminal. When drawing a line, the path you follow is not important. The lines will route themselves automatically. The terminals points are what matter. Once the blocks are connected, they can be moved around for neater appearance. A block can be simply clicked and dragged to its desired location (the signal lines will remain connected and will reroute themselves).

It may be necessary to branch a signal and transmit it to more than one input terminal. To do this, first place the mouse cursor at the location where the signal is to be branched (tapped). Then, using either the CTRL key in conjunction with the left mouse button or just the right mouse button, drag the new line to its intended destination.

B.1.4   Running a Simulation

Once the model is constructed, you are ready to simulate the system. To do this, go to the **Simulation** menu and click on **Start**, or just click on the "Start/Pause Simulation" button in the model window toolbar (this will look like the "Play" button on a VCR). The simulation will be carried out and the necessary signals will be generated.

B.1.4.1   General Tips

1. You can save your model by selecting **Save** from the file menu and clicking the **OK** button (you should give a name to a file).

2. The results of a simulation can be sent to the MATLAB window by the use of the "**to workshop**" icon from the **Sinks** window.

3. Use the **Demux** (i.e., demultiplexing) icon to convert a vector into several scalar lines. The **Mux** icon takes several scalar inputs and multiplexes them into a vector. This is useful, for example, when transferring the results from a simulation to the MATLAB workspace.

4. A sign of a **Sum** icon may be changed by double clicking on the icon and manually changing the sign. The number of inputs to a **Sum** icon may be changed by double clicking on the icon and correctly setting the number of inputs in the window.

5. Be sure to set the integration parameters in the simulation menu. In particular, the default minimum and maximum step sizes must be changed to the required values. They should be around 1/100 to 1/10 of the dominant (i.e., slowest) time constant of your system.

Example B.1

Consider the example of robotic sewing system, as studied in Chapter 2 (Figure 2.28). To carry out a imulation using SIMULINK, we use the following parameter values:

$m_c = 0.6$ kg

$k_c = 100$ N/m

$b_c = 0.3$ N/m/s

$m_h = 1$ kg

$b_h = 1$ N/m/s

$k_r = 200$ N/m

$b_r = 1$ N/m/s

$J_r = 2$ kg·m²

$r = 0.05$ m

The matrices of the linear state-space model are obtained as:

$$A = \begin{bmatrix} -0.00125 & -0.025 & 0 & 0 & 0 \\ 10 & 0 & -200 & 0 & 0 \\ 0 & 1 & -1.3 & 1 & 0.3 \\ 0 & 0 & -100 & 0 & 100 \\ 0 & 0 & 0.5 & -1.67 & 0.5 \end{bmatrix}, \quad B = \begin{bmatrix} 0.5 & 0 \\ 0 & 0 \\ 0 & 0 \\ 0 & 0 \\ 0 & 1.67 \end{bmatrix},$$

$$C = \begin{bmatrix} 0 & 0 & 0 & 1 & 0 \\ 1 & 0 & 0 & 0 & 0 \end{bmatrix}, \quad D = \begin{bmatrix} 0 & 0 \\ 0 & 0 \end{bmatrix}$$

The SIMULINK model may be built, as shown in Figure B.1(a).
The response of the system to two impulse inputs is shown in Figure B.1(b).

Example B.2

Consider the application of the superposition method to the time domain input–output model given by:

$$\dddot{y} + 13\ddot{y} + 56\dot{y} + 80y = \ddot{u} + 6\ddot{u} + 11\dot{u} + 6u$$

as discussed in Chapter 2. We use the simulation block diagram shown in Figure 2.58 to build the SIMULINK model, as given in Figure B.2(a). The system response to an impulse input is shown in Figure B.2(b).

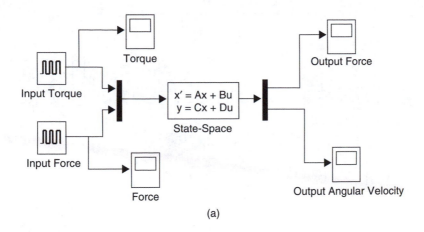

(a)

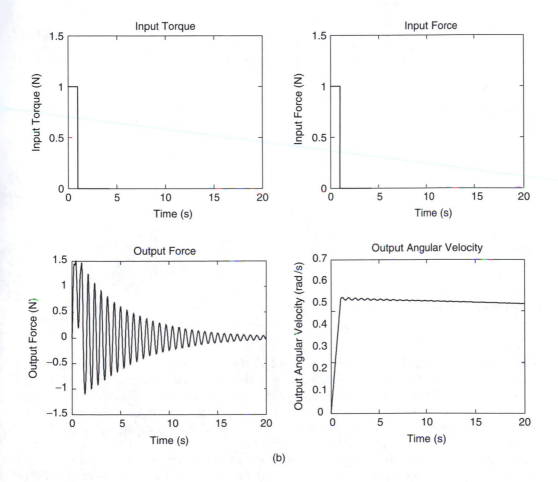

(b)

FIGURE B.1
(a) SIMULINK model of a robotic sewing machine; (b) Simulation results.

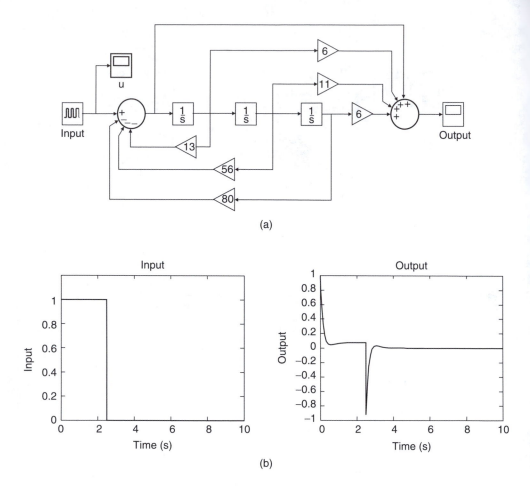

(a)

(b)

FIGURE B.2
(a) SIMULINK model of the simulation block diagram; (b) System response.

B.2 Stateflow

Stateflow is a tool that is particularly useful in the modeling and simulation of event-driven systems. It is available with MATLAB and SIMULINK, and may be used integral with Simulink. This interactive graphical tool is useful in modeling, simulation, design, and control of complex systems, mechatronic systems in particular, containing embedded control (see Chapter 12), logic (e.g., digital logic and state machines; see Chapter 10), and supervisory and hierarchical systems (see Chapter 13).

Stateflow is an interactive tool for visual/graphical modeling and animated simulation of finite state machines. Finite state machines are those having a finite number of states. A common example is a logic device or a digital circuit having two states (binary 0 or 1; on-off; True/False, etc.), as discussed in Chapter 10. A more general example is a discrete state system similar to one controlled by a programmable logic controller (or PLC), as discussed in Chapter 11. Stateflow uses control flow diagrams and state-transition diagrams all in the same Stateflow chart. In this manner it enables graphical representation of hierarchical and parallel states, and event-driven transitions between them. Since a "state" itself does not *r*ovide conditions or constraints under which the state can change to another state, state *t*ansition conditions, as given by a chart or a diagram, are very important in simulating *th*e associated system. A user-written C code can be integrated into a Stateflow chart, and *a* custom C function can be called from a state or transition action. Stateflow supports *v*ector and matrix data for use with Simulink.

In a Simulink model, a Stateflow chart is represented as a separate block called a *"*Stateflow block," which can exchange data, signals, and events with other blocks of the *m*odel. The collection of Stateflow blocks in a Simulink model is called a "Stateflow *m*achine." Stateflow can control the execution of a Simulink block using a function-call *t*rigger. Some important procedures of Stateflow are given next, with a simple example of *c*reating and running a power switch model.

B.2.1   Create a Simulink Model

Using the following steps, create a Simulink model containing a Stateflow block, and label the block.

1. At the MATLAB prompt, enter sfnew. Stateflow displays an untitled Simulink model window with an untitled Stateflow block.
2. Label the Stateflow block in the new untitled model by clicking in the text area and replacing the text "Untitled" with "On_Off," as in Figure B.3(a).

B.2.2   Create a Stateflow Diagram

Create a Stateflow diagram using the graphics editor, as in the following steps:

1. Double-click the Stateflow block in the Simulink model window to invoke the graphics editor window (Figure B.3(b)).
2. Select the "**State**" tool button in the drawing toolbar.
3. Move the cursor into the drawing area and left-click to place the state.
4. Position the cursor over that state, click the right mouse button, and drag to another location in the drawing area to make a copy of the state.
5. Click the "?" character within each state to enter the appropriate state label.
   Label each state with the title "On" or "Off". The Stateflow diagram should look as in Figure B.3(c).
6. Draw a transition starting from the right side of state On to the top of state Off, as follows:
   a. Place the cursor at a straight portion of the right border of the On state.
   b. When the cursor changes to crosshairs, click-drag the mouse to the top border of the Off state.
   c. When the transition snaps to the border of the Off state, release the mouse button.
7. Select the **Junction** button in the drawing toolbar. Move the cursor into the drawing area and click to place the junction, as in Figure B.3(c).
8. Draw a transition segment from the state Off to the junction. Transitions exist only between one state and another. However, transitions can be made up of transition segments that define alternate flow paths with junctions.

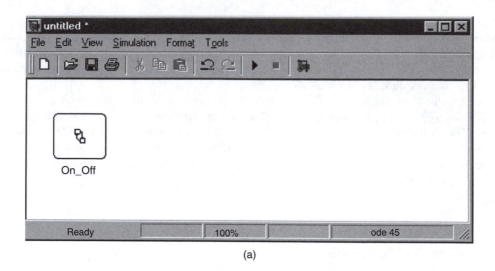

(a)

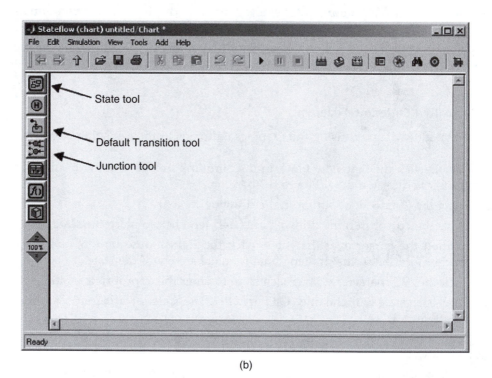

(b)

FIGURE B.3

(a) State flow block; (b) Graphics editor window; (c) Stateflow diagram with a junction.; (d) Completed Stateflow diagram with event triggers; (e) Submenu for defining an input event; (f) Screen containing Stateflow diagram with input port; (g) Stateflow diagram with input data temp; (h) Property dialog submenu for new data; (i) Simulink model with Stateflow diagram.

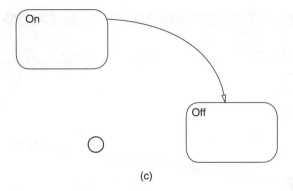

(c)

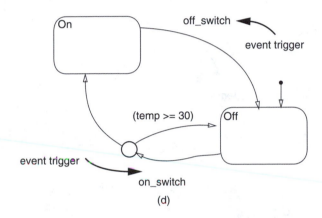

(d)

Event event			_ □ X
Name: event			
Parent: (chart) On_Off/On_Off			
Scope: Input from Simulink ▼ **Index:** 1 ▼ **Trigger:** Rising Edg ▼			
Debugger breakpoints: ☐ Start of broadcast ☐ End of broadcast			
Description:			
Document Link:			
ID# 1394	OK	Cancel Help	Apply

(e)

FIGURE B.3
(Continued)

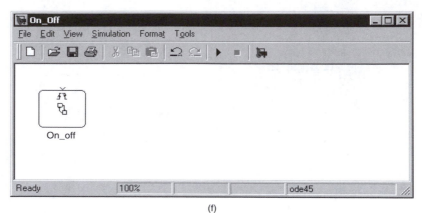

(f)

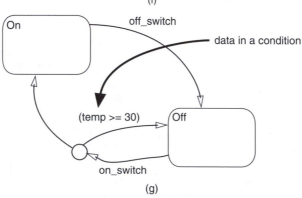

(g)

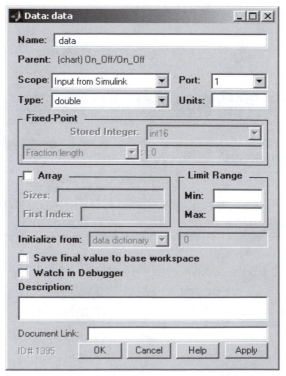

(h)

FIGURE B.3
(Continued)

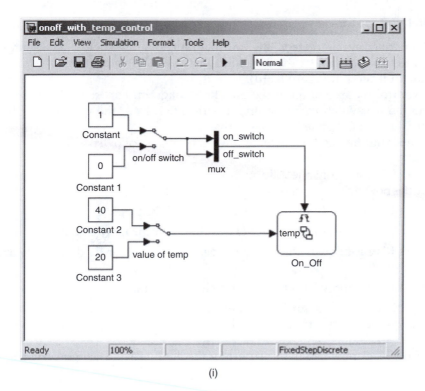

(i)

FIGURE B.3
(Continued)

9. Draw a transition segment from the junction to the state On.

10. Draw a transition segment from the junction to the state Off.

11. Label the transition from the state On to the state Off as follows:

   a. Click on the transition to select it.

   b. Click on the "?" character that appears alongside the transition. A blinking cursor appears.

   c. Enter the label text "off_switch."

12. Label the transition segment from the state Off to the junction with the text on_switch.

13. Label the transition segment from the junction to the state Off with the text "[temp >= 30]."

14. Add a default transition to the state Off with the following steps:

   a. Select the **Default Transition** tool in the drawing toolbar.

   b. Move the mouse cursor to the top border of the Off state.

   c. Click and release the mouse when the transition arrowhead of the cursor snaps straight up and down to the border of the Off state.

The completed Stateflow diagram should appear as in Figure B.3(d).

B.2.3 Define Input Events

Before simulating the completed Stateflow diagram, you must define the events on_switch and off_switch, which trigger the transitions between On state and Off state. These event triggers are indicated in Figure B.3(d).

Events control the execution of the Stateflow diagram. For example, if the state On is active and the on_switch event occurs, the transition from the On state to the Off state takes place and the Off state becomes active.

Define the input events for the Stateflow diagram, as follows:

1. Select "**Event**" from the **Add** menu of the diagram editor. The property dialog for the new event appears as in Figure B.3(e).
2. In the resulting submenu, select **Input from Simulink**.
3. Enter on_switch in the **Name** field of the **Event properties** dialog box.
4. Select **Rising Edge** as the **Trigger** type. Leave all other fields with their default values.
5. Select **OK** to apply the changes and close the window.
6. Repeat steps 1 through 5 to define the event off_switch, except enter **Falling Edge** as the **Trigger** type in step 4.

The Stateflow block now has an input port for the defined events, as shown in Figure B.3(f).

B.2.4 Define Input Data

Since the transition segment between the junction and the state Off has a condition based on the value of the Stateflow data temp, you must define temp in the diagram, as in Figure B.3(g).

To define the input data temp for your Stateflow diagram, do the following:

1. Select **Data** from the **Add** menu of the diagram editor. The property dialog for the new data appears as in Figure B.3(h).
2. In the resulting submenu, select **Input from Simulink**.
3. Enter temp in the **Name** field of the properties dialog. Leave the remaining fields with their default values.
4. Select **OK** to apply the changes and close the window.

B.2.5 Define the Stateflow Interface

In the Simulink model, make connections between the Stateflow block and other blocks, to provide sources for the events and data. The resulting model will appear as in Figure B.3(i). To construct this model, follow the general procedure of building a Simulink model.

B.2.6 Define Simulink Parameters

1. Choose **Simulation Parameters** from the **Simulation** menu of the Simulink model window and set the simulation **Stop time** to 100s.
2. Select **OK** to apply the changes and close the dialog box.
3. Select **Save** from the **File** menu to save the model.

B.2.7 Parse the Stateflow Diagram

Parsing the Stateflow diagram ensures that the notations you specified are valid and correct. To parse a Stateflow diagram, choose **Parse Diagram** from the **Tools** menu of the graphics editor. Informational messages are displayed in the MATLAB Command Window. Any error messages are displayed in red. If no messages appear with a red button, the parse operation is successful.

B.2.8 Run a Simulation

The following steps illustrate how to run a simulation:

1. Double-click the On_off Stateflow block to display the Stateflow diagram.
2. Select **Open Simulation Target** from the graphics editor **Tools** menu.
3. Select **Coder Options** on the **Simulation Target Builder** dialog box.
4. Ensure that the check box to **Enable Debugging/Animation** is selected.
5. Select **Debug** from the graphics editor **Tools** menu.
6. Ensure that the **Enabled** radio button in the **Animation** section is selected.
7. Choose **Start** from the diagram editor **Simulation** menu to start a simulation of the model. Notice that the background of the Stateflow diagram editor becomes darker. Then the highlighting of the default transition into the state Off, followed by the state Off itself, will occur.
8. Toggle the on/off manual switch (double click it) between its inputs, 0 and 1. This sends on_switch and off_switch events to the Stateflow chart. These events toggle the active state of the chart from Off to On and back to Off again. Both on_switch and off_switch events are defined as **Input from Simulink** events. This means that the events are input to the chart from Simulink, in this case, from a manual switch. When the input to the on/off switch rises from 0 to 1, it sends an on_switch event to the On_Off chart because on_switch is defined as a rising trigger. Similarly, when the input to the on/off switch falls from 1 to 0, it sends an off_switch event to the chart because the off_switch event is defined as a falling trigger.
9. Toggle the value of the temp switch to change the value of the data temp from 20 to 40.
10. Toggle the on/off switch again and notice the difference. Because of the different value for temp, the transition from the Off state to the On state no longer takes place. The transition from Off to On passes through a junction, which has an alternate path back to Off. Because the alternate path has a condition, it gets priority. In this case, the condition evaluates to true (nonzero) and the alternate path is taken.
11. Choose **Stop** from the graphics editor **Simulation** menu to stop a simulation. Once the simulation stops, Stateflow resets the model to be editable.

B.3 MATLAB

MATLAB interactive computer environment is very useful in computational activities in Mechatronics. Computations involving scalars, vectors, and matrices can be carried out and the results can be graphically displayed and printed. MATLAB toolboxes are available for performing specific tasks in a particular area of study such as control systems, fuzz logic, neural network, data acquisition, image processing, signal processing, system iden tification, optimization, model predictive control, robust control, and statistics. User guides Web-based help, and on-line help are available from the parent company, MathWorks Inc. and various other sources. What is given here is a brief introduction to get started in MATLAB for tasks that are particularly related to Control Systems and Mechatronics.

B.3.1 Computations

Mathematical computations can be done by using the MATLAB command window. Simply type in the computations against the MATLAB prompt ">>" as illustrated next.

B.3.2 Arithmetic

An example of a simple computation using MATLAB is given below.

```
>>x=2; y=-3;
>> z =x^2-x*y+4
z=-14
```

In the first line we have assigned values 2 and 3 to two variables x and y. In the next line, the value of an algebraic function of these two variables is indicated. Then, MATLAB provides the answer as 14. Note that if you place a ";" at the end of the line, the answer will not be printed/displayed.

Table B.1 gives the symbols for common arithmetic operations used in MATLAB. Following example shows the solution of the quadratic equation $ax^2 + bx + c = 0$

```
>>  a=2;b=3c=4;
>>  x=(-b+sqrt(b^2-4*a*c))/(2*a)
x  =  -0.7500+1.1990i
```

The answer is complex, where i denotes $\sqrt{-1}$. Note that the function sqrt() is used, which provides the positive root only. Some useful mathematical functions are given in Table B.2. Note that MATLAB is case sensitive.

B.3.3 Arrays

An array may be specified by giving the start value, increment, and the end value limit. An example is given below.

```
>>x=(.9:-.1:0.42)
x=0.9000  0.8000  0.7000  0.6000  0.5000
```

TABLE B.1

MATLAB Arithmetic Operations

Symbol	Operation
+	Addition
−	Subtraction
*	Multiplication
/	Division
^	Power

TABLE B.2

Useful Mathematical Functions in MATLAB

Function	Description
abs()	Absolute value/magnitude
acos()	Arc-cosine (inverse cosine)
acosh()	Arc-hyperbolic-cosine
asin()	Arc-sine
atan()	Arc-tan
cos()	Cosine
cosh()	Hyperbolic cosine
exp()	Exponential function
imag()	Imaginary part of a complex number
log()	Natural logarithm
log10()	Log to base 10 (common log)
real()	Real part of a complex number
sign()	Signum function
sin()	Sine
sqrt()	Positive square root
tan()	Tan function

The entire array may be manipulated. For example, all the elements are multiplied by π as below:

```
>>x=x*pi
x=2.8274 2.5133 2.1991 1.8850 1.5708
```

The second and the fifth elements are obtained by:

```
>>x (2 5)
ans = 2.5133 1.5708
```

Next we form a new array y using x, and then plot the two arrays, as shown in Figure B.4.

```
>>y=sin(x);
>>plot(x,y)
```

A polynomial may be represented as an array of its coefficients. For example, the quadratic equation $ax^2 + bx + c = 0$ as given before, with $a = 2$, $b = 3$, and $c = 4$, may be solved using the function "roots" as below.

```
>>p=[2 3 4];
>>roots(p)
ans=-0.7500+1.1990i
-0.7500-1.1990i
```

The answer is the same as what we obtained before.

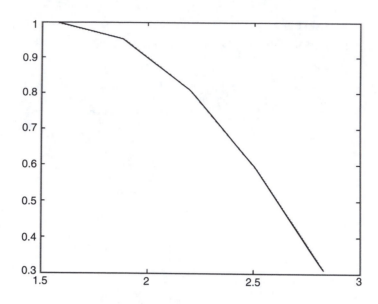

FIGURE B.4
A plot using MATLAB.

TABLE B.3

Some Relational Operations

Operator	Description
<	Less than
<=	Less than or equal to
>	Greater than
>=	Greater than or equal to
= =	Equal to
~=	Not equal to

TABLE B.4

Basic Logical Operations

Operator	Description
&	AND
|	OR
~	NOT

B.3.4 Relational and Logical Operations

Useful relational operations in MATLAB are given in Table B.3. Basic logical operations are given in Table B.4. Consider the following example.

```
>>x=(0:0.25:1)*pi
x=0  0.7854  1.5708  2.3562  3.1416
>>cos(x)>0
ans=  1  1  1  0  0
>>(cos(x)>0)&(sin(x)>0)
ans=  0  1  1  0  0
```

In this example, first an array is computed. Then the cosine of each element is computed. Next it is checked whether the elements are positive (A truth value of 1 is sent out if true and a truth value of 0 if false). Finally the "AND" operation is used to check whether the corresponding elements of the two arrays are positive.

B.3.5 Linear Algebra

MATLAB can perform various computations with vectors and matrices. Some basic illustrations are given here.

A vector or a matrix may be specified by assigning values to its elements. Consider the following example.

```
>> b=[1.5 -2];
>> A=[2 1;-1 1];
>> b=b'
b = 1.5000 -2.0000
>> x=inv(A)*b
x = 1.1667 -0.8333
```

In this example, first a 2nd order row vector and a 2×2 matrix are defined. Next the row vector is transposed to get a column vector. Finally the matrix-vector equation $Ax = b$ is solved according to $x = A^{-1}b$. The determinant and the eigenvalues of A are determined by:

```
>> det(A)
ans=3
>> eig(A)
ans = 1.5000 + 0.8660i
1.5000 - 0.8660i
```

Both eigenvectors and eigenvalues of A are computed as:

```
>>[V,P]=eig(A)
V = 0.7071 0.7071
-0.3536 + 0.6124i -0.3536 -0.6124i
P = 1.5000 + 0.8660i 0
0 1.5000 -0.8660i
```

Here the symbol V is used to denote the matrix of eigenvectors. The symbol P is used to denote the diagonal matrix whose diagonal elements are the eigenvalues.

Useful matrix operations in MATLAB are given in Table B.5 and several matrix functions are given in Table B.6.

B.3.6 M-Files

The MATLAB commands have to be keyed in on the command window, one by one. When several commands are needed to carry out a task, the required effort can be tedious.

TABLE B.5

Some Matrix Operations in MATLAB

Operation	Description
+	Addition
−	Subtraction
*	Multiplication
/	Division
^	Power
'	Transpose

TABLE B.6

Useful Matrix Functions in MATLAB

Function	Description
det()	Determinant
inv()	Inverse
eig()	Eigenvalues
[,]=eig()	Eigenvectors and eigenvalues

Instead, the necessary commands can be placed in a text file, edited as appropriate (using text editor), which MATLAB can use to execute the complete task. Such a file is called an M-file. The file name must have the extension "m" in the form *filename.m*. A toolbox is a collection of such files, for use in a particular application area (e.g., control systems, fuzzy logic). Then, by keying in the M-file name at the MATLAB command prompt, the file will be executed. The necessary data values for executing the file have to be assigned beforehand.

B.4   Control Systems Toolbox

There are several toolboxes with MATLAB, which can be used to analyze, compute, simulate, and design control problems. Both time-domain representations and frequency-domain representations can be used. Also, both classical and modern control problems can be handled. The application is illustrated here through several conventional control problems discussed in Chapter 12.

B.4.1   Compensator Design Example

Consider again the design problem given in Chapter 12, Figure 12.49. The MATLAB SISO (single-input-single-output) Design Tool is used here to solve this problem.

B.4.1.1   Building the System Model

Build the transfer function model of the Motor and Filter, in the MATLAB workspace, as follows:

```
Motor_G = tf([999], [10 1]);
Filter_H = tf([1], [0.1 1]);
```

To Open the SISO Design Tool, type

```
sisotool
```

at the MATLAB prompt (>>).

B.4.1.2 Importing Model into SISO Design Tool

Select *Import Model* under the *File* menu. This opens the *Import System Data* dialog box, as shown in Figure B.5(a).
Use the following steps to import the motor and filter models:

1. Select Motor_G under *SISO Models*
2. Place it into the *G* Field under *Design Model* by pressing the right arrow button to the left of *G*.
3. Similarly import the filter model.
4. Press **OK**

Now the main window of the SISO Design Tool, will show the root locus and Bode plots of open loop transfer function *GH* (see Figure B.5(b)). As given in the figure, the phase margin is 18.2°, which occurs at 30.8 rad/s (4.9 Hz).

The closed-loop step response, without compensation, is obtained by selecting *Tools→Loop responses→closed-loop step* from the main menu. The response is shown in Figure B.5(c). It is noted that the phase margin is not adequate, which explains the oscillations and the long settling time. Also the *P.O.* is about 140%, which is considerably higher than the desired one (10%) and is not acceptable.

B.4.1.3 Adding Lead and Lag Compensators

To add a lead compensator, right click the mouse in the white space of the Bode magnitude plot, choose *Add Pole/Zero* and then *lead* in the right-click menu for the open-loop Bode diagram. Move the zero and the pole of the lead compensator to get a desired phase margin of about 60°.

To add a lag compensator, choose *Add Pole/Zero* and then *lag* in the right-click menu for the open-loop Bode diagram. Move the zero and the pole of the lag compensator to get a desired phase angle of about –115° at the crossing frequency, which corresponds to a phase margin of 180° – 115° = 65°.

With the added lead and leg compensators, the root locus and Bode plots of the system are shown in Figure B.5(d). The closed-loop step response of the system is shown in Figure B.5(e).

B.4.2 PID Control with Ziegler–Nichols Tuning

Consider the example shown in Chapter 12, Figure 12.54. The SISO Design Tool is used. First build the transfer function model of the given system (call it Mill).

    Mill_G = tf([1], [1 1 4 0]);
    Filter_H = tf([1], [1]);

As before, import the system model into the SISO Design Tool.

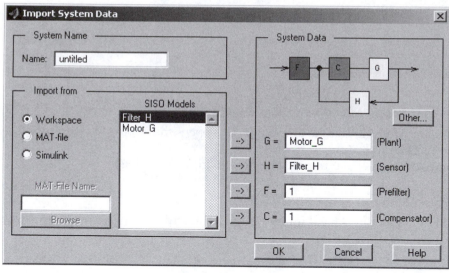

(a)

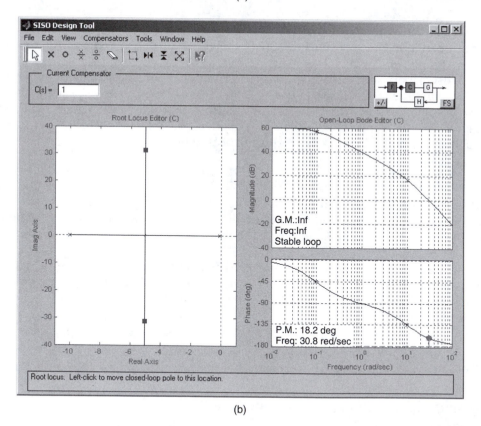

(b)

FIGURE B.5

(a) Importing the model into the SISO Design Tool; (b) Root locus and Bode plots for the motor model; (c) Closed-loop step response of the motor system without compensation; (d) Root locus and Bode plots of the compensated system; (e) Closed-loop step response of the compensated system.

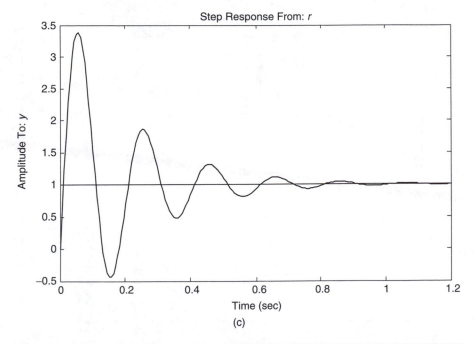

(c)

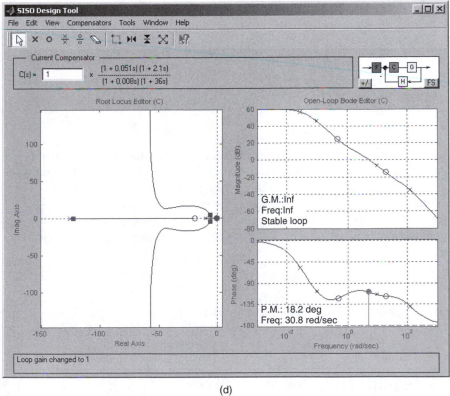

(d)

FIGURE B.5
(Continued)

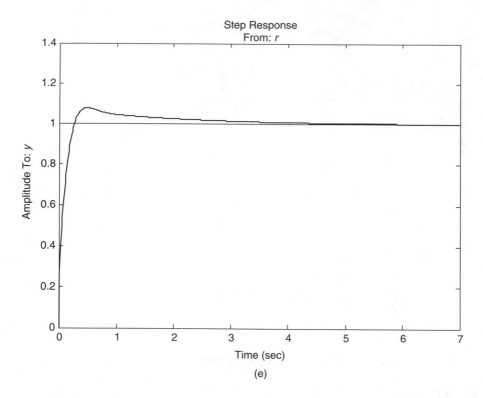

FIGURE B.5
(Continued)

B.4.2.1   Proportional Control

Even without using the Routh-Hurwitz method, we can change the gain setting by trial and error to obtain the proportional gain that will make the system marginally stable. As seen in Figure B.6(a), when $K = 4$, the Gain Margin is just below 0 dB, which makes the system unstable. The response of the system is shown in Figure B.6(b).

Referring to the Ziegler-Nichols controller settings, as given in Table 12.6, we can obtain the proper proportional gain as $K_p = 0.5 \times 4 = 2$. The corresponding system response is shown in Figure B.6(c).

B.4.2.2   PI Control

Note that the period of oscillations (ultimate period) is

$$P_u = \frac{2\pi}{\omega_n} = \frac{2\pi}{2} = \pi \text{ sec}$$

Hence, from the Ziegler-Nichols settings given in Table 12.6, we have for a PI controller,

$$K_p = 0.45 \times 4 = 1.8$$

$$\tau_i = 0.83\pi = 2.61 \text{ sec}$$

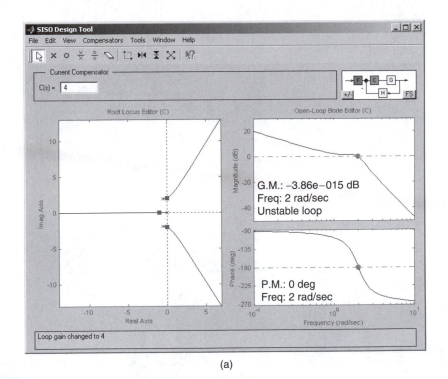

(a)

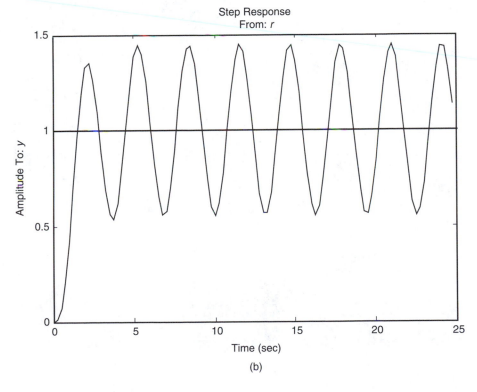

(b)

FIGURE B.6

(a) Root locus and Bode plots of the system with proportional gain $K_p = 4$; (b) Step response of the closed-loop system with $K_p = 4$; (c) Step response of the closed-loop system with $K_p = 2$; (d) Bode plot of the system with PI control; (e) Step response of the system with PI control; (f) Bode plot of the system with PID control; (g) Step response of the system with PID control.

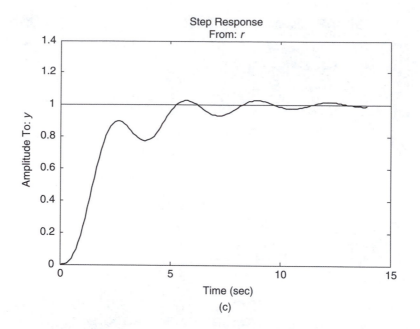

(c)

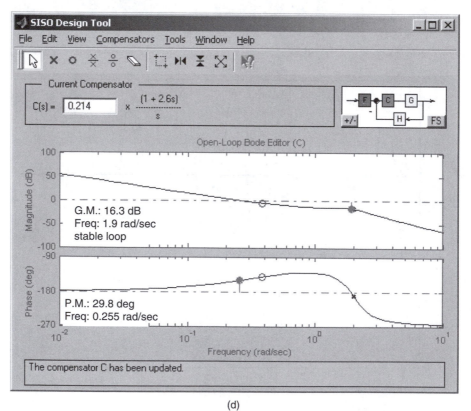

(d)

FIGURE B.6
(Continued)

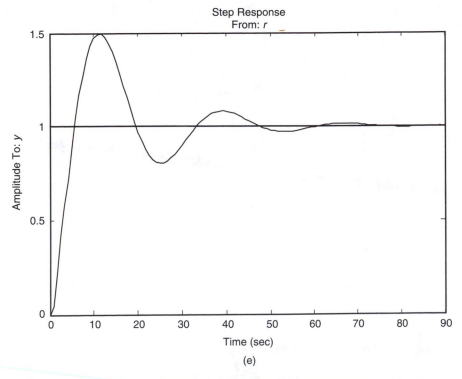

(e)

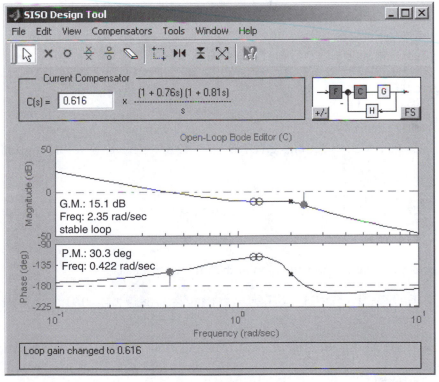

(f)

FIGURE B.6
(Continued)

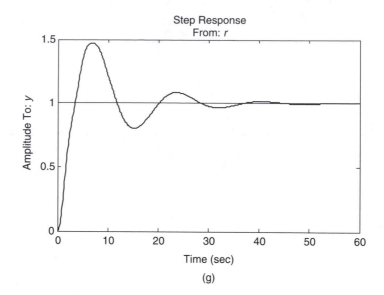

FIGURE B.6
(Countued)

Hence, the PI controller transfer function is

$$K_p\left(1+\frac{1}{\tau_i s}\right)=\frac{K_p\tau_i s+K_p}{\tau_i s}=\frac{4.68s+1.8}{2.61s}=0.214\frac{2.6s+1}{s}$$

Insert this controller into C in the SISO Design Tool.

The corresponding system Bode plot and the step response are shown in Figure B.6(d) and Figure B.6(e), respectively.

B.4.2.3   PID Control

From the Ziegler-Nichols settings given in Table 12.6, we have for a PID controller,

$$K_p = 0.6 \times 4 = 2.4$$

$$\tau_i = 0.5\pi = 1.57\text{sec}$$

$$\tau_i = 0.125\pi = 0.393 \text{ sec}$$

The corresponding transfer function of the PID controller is

$$K_p\left(1+\frac{1}{\tau_i s}+\tau_d s\right)=\frac{K_p\tau_i\tau_d s^2+K_p\tau_i s+K_p}{\tau_i s}=\frac{1.48s^2+3.768s+2.4}{1.57s}=\frac{0.94s^2+2.4s+1.53}{s}$$

Use the MATLAB function **roots** to calculate the roots of the numerator polynomial.

```
R  =  roots([0.94  2.4  1.53]);
R  =  -1.3217  -1.2315
```

Hence, the transfer function of the PID controller is

$$\frac{(s+1.32)(s+1.23)}{s} = 0.616\frac{(0.76s+1)(0.81s+1)}{s}$$

Insert this controller into C of the SISO Design Tool. The corresponding Bode plot and the step response of the controlled system are shown in Figure B.6(f) and Figure B.6(g).

3.4.3 Root Locus Design Example

Consider the example given in Chapter 12, Figure 12.55. Again, the SISO Design Tool is used. First build the transfer function model for the rolling mill with no filter:

    Mill_G = tf([1], [1 5 0]);
    Filter_H = tf([1], [1]);

Then, as before, import the system model into the SISO Design Tool. The root locus and the step response of the closed-loop system are shown in Figure B.7(a) and Figure B.7(b). From Figure B.7(b), it is seen that the peak time and the 2% settling time do not meet the design specifications. To add a lead compensator, right click in the white space of the root locus plot, choose **Add Pole/Zero** and then **lead** in the right-click menu. Left click on the root locus plot where we want to add the lead compensator.

Now we have to adjust the pole and zero of the lead compensator and the loop gain so that the root locus passes through the design region. To speed up the design process, turn on the grid setting for the root locus plot. The radial lines are constant damping ratio lines and the semicircular curves are constant undamped natural frequency lines (see Chapter 12).

On the root locus plot, drag the pole and zero of the lead compensator (pink cross or circle symbol on the plot) so that the root locus moves toward the design region. Left click and move the closed-loop pole (small pink-color square box) to adjust the loop gain. As you drag the closed-loop pole along the locus, the current location of that pole, and the system damping ratio and natural frequency will be shown at the bottom of the graph.

Drag the closed-loop pole into the design region. The resulting lead compensator, the loop gain, and the corresponding root locus are shown in Figure B.7(c). The step response of the compensated closed-loop system is shown in Figure B.7(d).

B.5 LabVIEW

LabVIEW or Laboratory Virtual Engineering Workbench is a product of National Instruments. It is a software development environment for data acquisition, instrument control, image acquisition, motion control, and presentation. LabVIEW is a complied graphical environment, which allows the user to create programs graphically through wired icons similar to creating a flowchart.

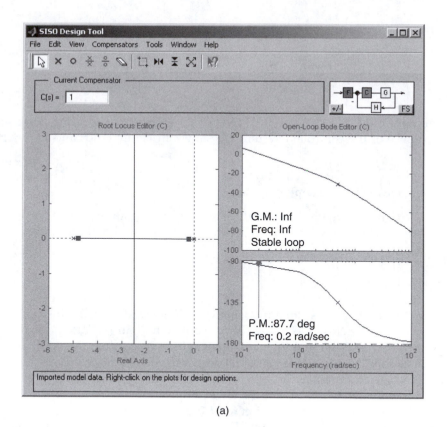

(a)

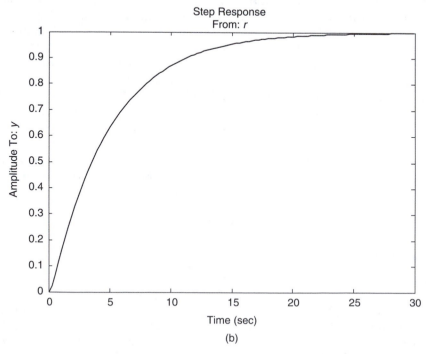

(b)

FIGURE B.7

(a) Root locus and Bode plots of the rolling mill system without compensation.; (b) Step response of the closed-loop system without compensation; (c) Root locus of the compensated system; (d) Step response of the compensated closed-loop system.

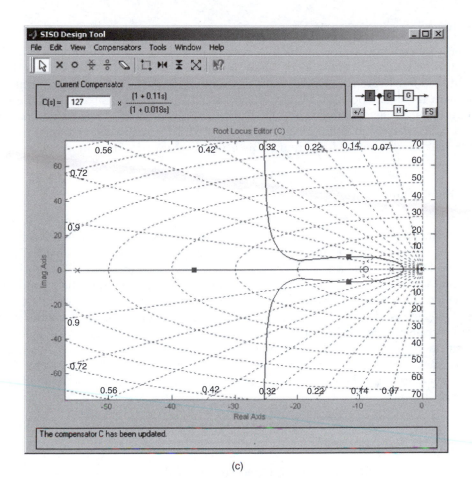

(c)

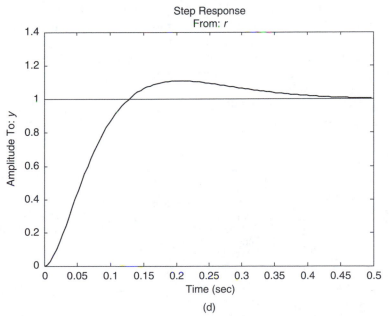

(d)

FIGURE B.7
(Continued)

B.5.1 Working with LabVIEW

As a software centered system, LabVIEW resides in a desktop computer, laptop or PXI a an application where it acts as a set of virtual instruments (VIs), providing the functionalit of traditional hardware instruments such as oscilloscopes. Comparing to physical instru ments with fixed functions, LabVIEW VIs are flexible and can easily be reconfigured t different applications. It is able to interface with various hardware devices (see Chapter 11 such as GPIB, data acquisition modules, distributed I/O, image acquisition, and motior control, making it a modular solution. This utility is shown in Figure B.8.

B.5.2 Front Panel

Upon launching LabVIEW, you will be able to create or open an existing VI where the layout of the graphical user interface can be designed. Figure B.9 shows the front panel of the simple alarm slide control (alarmsld.lib) example included with LabVIEW suite of examples. This is the first phase in developing a VI. Buttons, indicators, I/O and dialogs are placed appropriately. These control components are selected from the "Controls Pal- ette," which contains a list of pre-built library or user-customized components.

A component is selected from the controls palette by left clicking the mouse on the particular control icon, and can be placed on the front panel by left clicking again. Then the component can be resized, reshaped or moved to any desired position. A component property such as visibility, format, precision, labels, data range, or action can be changed by right-clicking, with the cursor placed anywhere on the selected component, to bring up the pop-up menu.

B.5.3 Block Diagrams

After designing the graphical user interface (GUI) in the front panel, the VI has to be programmed graphically through the block diagram window in order to implement the intended functionality of the VI. The block diagram window can be brought forward by clicking on the "Window" pull menu and selecting "Show Diagram." For every control component created on the front panel, there is a corresponding terminal automatically created in the block diagram window. Figure B.10 shows the block diagram for the alarm slide control example provided with LabVIEW.

The terminal is labeled automatically according to the data type of each control. For example, the stop button has a terminal labeled TF, which is a Boolean type. The vertical

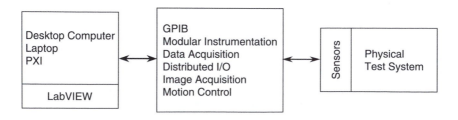

FIGURE B.8
Modular solution of LabVIEW.

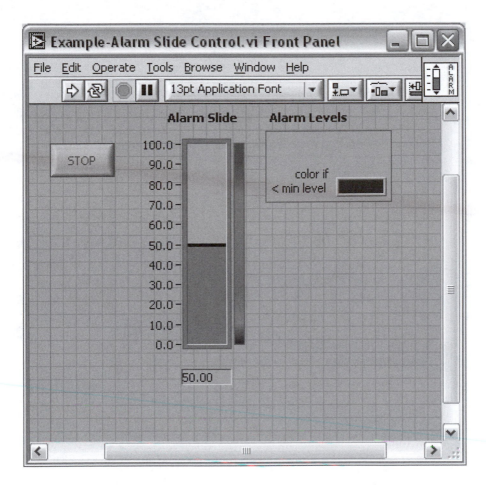

FIGURE B.9
Front panel of the alarm slide control example.

level indicator has a DBL type terminal, indicating double-precision number. Other common controls with a DBL terminal include various numeric indicators, sliders, and graphs.

LabVIEW uses the G-programming language to implement the functionality of a VI. It provides an extensive library of basic conditional and looping structures, mathematical operators, Boolean operators, comparison operators, and more advanced analysis and conditioning tools, provided through the Functions Palette. A function may be placed on the block diagram window similar to how a control component is placed on the front panel. Depending on the required flow of execution, they are then wired together using the connect wire tool in the tools palette. In order to wire two terminals together, first click on the connect wire icon in the tools palette, then move the cursor to the input/output hotspot of one terminal, left click to make the connection, and then move the cursor to the output/input hotspot of the other terminal, and left click again to complete the connection. The corresponding control component on the front panel can be selected by double clicking on the terminal block.

The general flow of execution is to first acquire the data, then analyze, followed by the presentation of results. The terminals and functional components are wired in such a way that data flows from the sources (e.g., data acquisition) to the sinks (e.g., presentation).

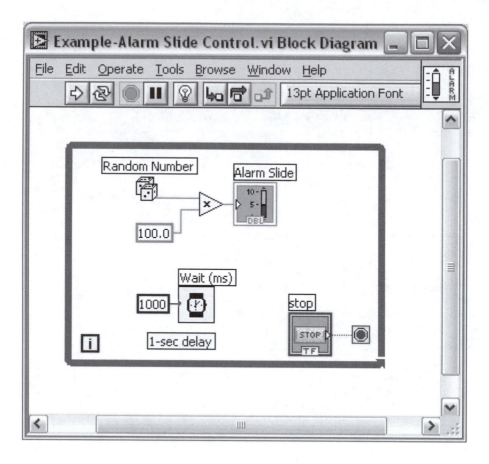

FIGURE B.10
Block diagram of the alarm slide control example.

LabVIEW executes its G-programming code in data flow manner, executing an icon as data becomes available to it through connecting wires.

The dice terminal is a random number generator and its output is multiplied by a constant using the multiplier operator (see Figure B.10). The multiplication result is connected to the input of the alarm slide, which will show up as the level in the vertical indicator on the front panel during VI execution. The gray box surrounding the terminals is the while loop in which all the flow within the gray box will run continuously until the loop is terminated by the stop button with the corresponding Boolean terminal. When the stop terminal is true, the while loop terminates upon reading a "false" through the "not" operator. The wait terminal (watch icon) controls the speed of the while loop. The wait terminal input is given in milliseconds. In the figure, the loop runs at an interval of 1 sec since a constant of 1000 is wired to the wait terminal. In order to run the VI, left click on the arrow icon on the top rows of icons or click on "Operate" and then select "Run." No compilation is required.

Note the remove broken wire command found in the edit pull-down menu. This command cleans up the block diagram of any unwanted or incomplete wiring. The debugging pop-up window that appears when an erroneous VI is executed is very helpful in troubleshooting of the VI. Double-clicking on the items in the errors list will automatically highlight the problematic areas or wires or terminals in the diagram.

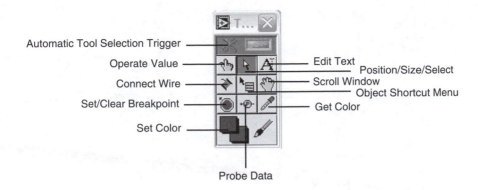

FIGURE B.11
LabVIEW tools palette.

B.5.4   Tools Palette

LabVIEW has three main floating palettes for creating VIs. They are the tools palette, controls palette, and functions palette. The tools palette, shown in Figure B.11, is the general editing palette with tools for editing components in the front panel and block diagram panel, modifying the position, shape and size of components, labeling, wiring of terminals in the block diagram panel, debugging, and coloring. When manipulating the front panel and the block diagram panel, note which tool icon is selected. For example, the values of a control or terminal cannot be selected or edited when the positioning icon is selected.

B.5.5   Controls Palette

Figure B.12 shows the controls palette, which contains the prebuilt and user-defined controls to create a graphical user interface. This palette will be available when the front panel is selected. If it is not showing, click on the "Window" pull-down menu and select the "Show Controls Palette" option. The figure shows the main group of top-level components available in its prebuilt library. Clicking on the appropriate top-level icons will bring up the subpalettes of the available controls and indicators. To go back to the top-level icons, click on the up arrow icon on the top-left of the controls palette.

B.5.6   Functions Palette

When the block diagram panel is selected, the functions palette is shown as in Figure B.13, enabling you to program the VI. The functions palette contains a complete library of necessary operations for developing the functionality of the VI. Similar to the controls palette, the top-level icons show the grouping of different sub-functions available for the programmer. Several commonly used groups are indicated below.

- Structures: The structures icon consists of the usual programming language sequences, conditional statements and conditional loops. These structures are in the form of boxes where the terminals within the boxes are executed when the statements or loops are invoked. In addition, there is a formula node where custom text-based formulas can be included if you prefer the traditional text-based equations. There are also variable declaration nodes where local and global variables can be declared.

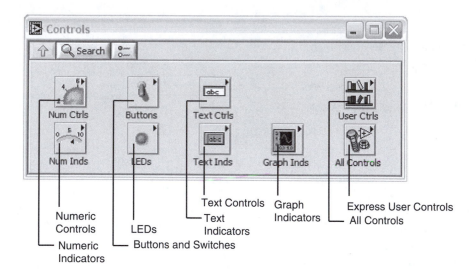

FIGURE B.12
LabVIEW controls palette.

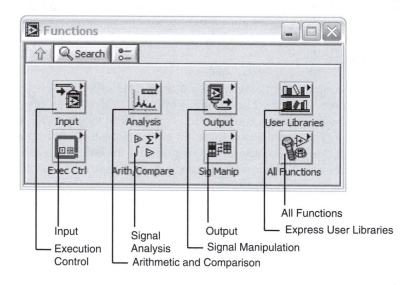

FIGURE B.13
LabVIEW functions palette.

- Numeric: The elementary operators such as summation, subtraction, multiplication, division, and power, are grouped under this icon.
- Boolean: This icon contains the Boolean operators required for logic manipulation.
- Array: The array grouping consists of tools for array manipulation.

- Comparison: Operators for numerical comparison, which provide Boolean outputs, are found under this icon.
- Analyze: This icon more advanced analysis tools such as FFT spectrum, power spectrum, filters, triggering, and waveform generation.
- Mathematics: Under this icon, the tools for mathematical manipulation such as calculus, statistics and probability, linear algebra, optimization, and numeric functions are found.

References and Further Reading

This book has relied on many publications, directly and indirectly, in its development and evolution. Many of these publications are based on the work of the author and his co-workers. Also, there are some excellent books the reader may refer to for further information and knowledge. Some selected publications are listed below.

Books

Auslander, D.M. and Kemp f, C.J., *Mechatronics Mechanical System Interfacing*, Prentice Hall, Upper Saddle River, NJ, 1996.

Bolton, W., *Mechatronics*, 2nd ed., Longman, Essex, England, 1999.

Chen, B.M., Lee, T.H., and Venkataramenan, V., *Hard Disk Drive Servo Systems*, Springer-Verlag, London, England, 2002.

De Silva, C.W., *Intelligent Control : Fuzzy Logic Applications*, CRC Press, Boca Raton, FL, 1995.

De Silva, C.W., *Vibration Fundamentals and Practice*, CRC Press, Boca Raton, FL, 2000.

De Silva, C.W. Ed., *Intelligent Machines: Myths and Realities*, CRC Press, Boca Raton, FL, 2000.

Histand, M.B. and Alciatore, D.G., *Introduction to Mechatronics and Measurement Systems*, WCB McGraw-Hill, New York, NY, 1999.

Jain, L. and de Silva, C.W. Eds., *Intelligent Adaptive Control: Industrial Applications*, CRC Press, Boca Raton, FL, 1999.

Karray, F. and de Silva, C.W., *Soft Computing Techniques and Their Applications*, Pearson, U.K., 2004.

Necsulescu, D., *Mechatronics*, Prentice Hall, Upper Saddle Rivver, NJ, 2002.

Shetty, D. and Kolk, R.A., *Mechatronics System Design*, PWS Publishing Co., Boston, MA, 1997.

Tan, K.K., Lee, T.H., Dou, H, and Huang, S., *Precision Motion Control*, Springer-Verlag, London, England, 2001.

Sections of Books

Cao, Y. and de Silva, C.W., Adaptive control of a manipulator with flexible joints using neural networks, *Mechatronics and Machine Vision: Future Trends*, Billingsley J. Ed., Research Studies Press LTD., Baldock, Hertfordshire, England, pp. 71–78, 2003.

De Silva, C.W., Considerations of hierarchical fuzzy control, *Theoretical Aspects of Fuzzy Control*, Nguyen, H.T., Sugeno, M., Tong, R.M., and R.R. Yager Eds., John Wiley & Sons, New York, 1995.

De Silva, C.W., Intelligent restructuring of automated production systems, *Fuzzy Logic and Its Applications to Engineering, Information Sciences, and Intelligent Systems*, Bien, Z., and Min, K.C. Eds., Kluwer Academic Publishers, Boston, MA, 1995.

De Silva, C.W., Electronic components, *Encyclopedia of Electrical and Electronics Engineering*, Webster, J.G., Ed., John Wiley & Sons, New York, Vol. 6, pp. 577–594, 1999.

De Silva, C.W., Sensors for control, *Encyclopedia of Physical Science and Technology*, 3rd ed., Meyers, R.A., Ed., Academic Press, San Diego, CA, Vol. 14, pp. 609–650, 2001.

De Silva, C.W., Wong, K.H., and Modi, V.J., Development of a novel multi-module manipulator system: dynamic model and prototype design, *Mechatronics and Machine Vision: Current Practice*, Bradbeer, R.S. and Billingsley J., Eds., Research Studies Press, Ltd., Hertfordshire, Baldock, England, pp. 161–168, 2002.

Zhang, J. and de Silva, C.W., Intelligent hierarchical control of a space-based deployable manipulato: *Mechatronics and Machine Vision: Future Trends*, Billingsley J. Ed., Research Studies Press LTD Baldock, Hertfordshire, England, pp. 61–70, 2003.

Journal Papers

Croft, E.A., de Silva, C.W., and Kurnianto, S., Sensor technology integration in an intelligent machine for herring Roe grading, *IEEE/ASME Transactions on Mechatronics*, Vol. 1, No. 3, pp. 204–215, September 1996.

Chen, Y., Wang, X.G., Sun, C., Devine, F., and de Silva, C.W., Active vibration control with state feedback in woodcutting, *Journal of Vibration and Control*, Vol. 9, No. 6, pp. 645–664, 2003.

De Silva, C.W., Price, T.E., and Kanade, T., A torque sensor for direct-drive manipulators, *Journal of Engineering for Industry*, Trans. ASME, Vol. 109, No. 2, pp. 122–127, May 1987.

De Silva, C.W., Design of PPD controllers for position servos, *Journal of Dynamic Systems, Measurement, and Control*, Trans. ASME, Vol. 112, No. 3, pp. 519–523, September 1990.

De Silva, C.W., Singh, M., and Zaldonis, J., Improvement of response spectrum specifications in dynamic testing, *Journal of Engineering for Industry*, Trans. ASME, Vol. 112, No. 4, pp. 384–387, November 1990.

De Silva, C.W., Trajectory design for robotic manipulators in space applications, *Journal of Guidance, Control, and Dynamics*, Vol. 14, No. 3, pp. 670–674, June 1991.

De Silva, C.W., An analytical framework for knowledge-based tuning of servo controllers, *International Journal of Engineering Applications of Artificial Intelligence*, Vol. 4, No. 3, pp. 177–189, 1991.

De Silva, C.W., Schultz, M., and Dolejsi, E., Kinematic analysis and design of a continuously-variable transmission, *Mechanism and Machine Theory*, Vol. 29, No. 1, pp. 149–167, January 1994.

De Silva, C.W. and Lee, T.H., Knowledge-based intelligent control, *Measurements and Control*, Vol. 28, No. 2, pp. 102–113, April 1994.

De Silva, C.W. and Lee, T.H., fuzzy logic in process control, *Measurements and Control*, Vol. 28, No. 3, pp. 114–124, June 1994.

De Silva, C.W., A criterion for knowledge base decoupling in fuzzy-logic control systems, *IEEE Transactions on Systems, Man, and Cybernetics*, Vol. 24, No. 10, pp. 1548–1552, October 1994.

De Silva, C.W. and Gu, J., An intelligent system for dynamic sharing of workcell components in process automation, *International Journal of Engineering Applications of Artificial Intelligence*, Vol. 7, No. 5, pp. 571–586, 1994.

De Silva, C.W., Applications of fuzzy logic in the control of robotic manipulators, *Fuzzy Sets and Systems*, Vol. 70, No. 2–3, pp. 223–234, 1995.

De Silva, C.W., Gamage, L.B., and Gosine, R.G., An intelligent firmness sensor for an automated herring roe grader, *International Journal of Intelligent Automation and Soft Computing*, Vol. 1, No. 1, pp. 99–114, 1995.

De Silva, C.W. and Wickramarachchi, N., An innovative machine for automated cutting of fish, *IEEE/ASME Transactions on Mechatronics*, Vol. 2, No. 2, pp. 86–98, 1997.

De Silva, C.W., Intelligent control of robotic systems, *International Journal of Robotics and Autonomous Systems*, Vol. 21, pp. 221–237, 1997.

De Silva, C.W., The role of soft computing in intelligent machines, *Philosophical Transactions of the Royal Society*, Series A, UK, Vol. 361, No. 1809, pp. 1749–1780, 2003 (By Invitation).

Goulet, J.F., de Silva, C.W., Modi, V.J., and Misra, A.K., Hierarchical knowledge-based control of a deployable orbiting manipulator, *Acta Astronautica*, Vol. 50, No. 3, pp. 139–148, 2002.

Hu, B.G., Gosine, R.G., Cao, L.X., and de Silva, C.W., Application of a fuzzy classification technique in computer grading of fish products, *IEEE Transactions on Fuzzy Systems*, Vol. 6, No. 1, pp. 144–152, February 1998.

Lee, T.H., Yue, P.K., and de Silva, C.W., Neural networks improve control, *Measurements and Control*, Vol. 28, No. 4, pp. 148–153, September 1994.

Lee, M.F.R, de Silva, C.W., Croft, E.A., and Wu, Q.M.J., Machine vision system for curved surface inspection, *Machine Vision and Applications*, Vol. 12, pp. 177–188, 2000.

Omar, F.K. and de Silva, C.W., Optimal portion control of natural objects with application in automated cannery processing of fish, *Journal of Food Engineering*, Vol. 46, pp. 31–41, 2000.

Kahbari, R. and de Silva, C.W., Comparison of two inference methods for P-type fuzzy logic control through experimental investigation using a hydraulic manipulator, *International Journal Engineering Applications of Artificial Intelligence*, Vol. 14, No. 6, pp. 763–784, 2001.

Stanley, K., Wu, Q.M.J., de Silva, C.W., and Gruver, W., Modular neural-visual servoing with image compression input, *Journal of Intelligent and Fuzzy Systems*, Vol. 10, No. 1, pp. 1–11, 2001.

Tafazoli, S., de Silva, C.W., and Lawrence, P.D., Tracking control of an electrohydraulic manipulator in the presence of friction, *IEEE Transactions on Control Systems Technology*, Vol. 6, No. 3, pp. 401–411, May 1998.

Tang, P.L., Poo, A.N., and de Silva, C.W., Knowledge-based extension of model-referenced adaptive control with application to an industrial process, *Journal of Intelligent and Fuzzy Systems*, Vol. 10, No. 3,4, pp. 159–183, 2001.

Tang, P.L., de Silva, C.W., and Poo, A.N., Intelligent adaptive control of an industrial fish cutting machine, *Transactions of the South African Institute of Electrical Engineers*, Vol. 93, No. 2, pp. 60–72, June 2002.

Wu, Q.M. and de Silva, C.W., Dynamic switching of fuzzy resolution in knowledge-based self-tuning control, *Journal of Intelligent and Fuzzy Systems*, Vol. 4, No. 1, pp. 75–87, 1996.

Wu, Q.M.J., Lee, M.F.R., and de Silva, C.W., An imaging system with structured lighting for on-line generic sensing of three-dimensional objects, *Sensor Review*, Vol. 22, No. 1, pp. 46–50, 2002.

Yan, G.K.C., de Silva, C.W., and Wang, G.X., Experimental modeling and intelligent control of a wood-drying Kiln, *International Journal of Adaptive Control and Signal Processing*, Vol. 15, pp. 787–814, 2001.

Zhou, Y. and de Silva, C.W., Adaptive control of an industrial robot retrofitted with an open-architecture controller, *Journal of Dynamic Systems, Measurement, and Control*; Trans. ASME, Vol. 118, No. 1, pp. 143–150, March 1996.

Conference Papers

De Silva, C.W. and Riahi, N., System Integration and Image Processing Techniques in a Fish Processing Workcell, *ASME Winter Annual Meeting*, Atlanta, GA, DSC-Vol. 30, pp. 59–64, December 1991.

De Silva, C.W. and Saliba, M., Instrumentation Issues in the Handling of Fish for Automated Processing, *Proceedings of the IEEE Industrial Electronics Conference-IECON '92*, San Diego, CA, Vol. 2, pp. 789–794, Nov. 1992.

De Silva, C.W. and Wang, Y., Use of Kinematic Redundancy for Design Optimization in Space-Robot Systems, *Proceedings of the 1993 American Control Conference*, San Francisco, California, Vol. 2, pp. 1806–1810, June 1993.

De Silva, C.W., Dong, C., Gosine, R., and Yang, J., Design of an Automated Fish Marking Machine, *Proceedings of the IEEE International Conference on Systems, Man, and Cybernetics*, San Antonio, TX, Vol. 1, pp. 171–176, Oct. 1994.

De Silva, C.W. and Gu, J.H., Use of Mechanical Impedance in Sensing Object-End Effector Interaction, *Proceedings of the IEEE International Symposium on Industrial Electronics*, Warsaw, Poland, Vol. 1, pp. 60–65, June 1996.

De Silva, C.W., Robotization and Its Benefits, *Proceedings of the IASTED International Conference on Intelligent Systems and Control*, Halifax, Nova Scotia, pp. 197–200, June 1998.

De Silva, C.W., University-Industry Collaborative Research: Observations, Criticism, and Suggestions through Personal Experience, *Proceedings of The IEEE International Conference on Control Theory and Applications*, Pretoria, South Africa, pp. 557–563, December 12–14, 2001.

De Silva, C.W., Wong, K.H., and Modi, V.J., Development of a Novel Multi-module Manipulator System: Dynamic Model and Prototype Design, *The Nineth IEEE Conference on Mechatronics and Machine Vision in Practice*, Chang Mai, Thailand, pp. 69–74, September 2002.

De Silva, C.W., Wong, K.H., and Modi, V.J., Design Development of a Prototype Multi-module Manipulator, *Proceedings of the 2003 IEEE/ASME International Conference on Advanced Intelligent Mechatronics (AIM 2003)*, Kobe, Japan, pp. 1384–1389, July 2003.

De Silva, C.W., McCourt, R., and Ohmiya, M., Control of a Multi-Module Deployable Manipulator Using RTX, *Proceedings of the 2003 IEEE Pacific Rim Conference on Communications, Computers and Signal Processing (PACRIM '03)*, Victoria, BC, pp. 864–867, August 2003.

De Silva, C.W., Sensing and Information Acquisition for Intelligent Mechatronic Systems, *Scienc* *and Technology of Information Acquisition and Their Applications, Proceedings of the Symposium o* *Information Acquisition*, Chinese Academy of Sciences, Hefei, China, pp. 9–18, November 200?

Dong, C., de Silva, C.W., and Gosine, R.G., Particle Retention Measurement Using Image Processin; for a Spray Marking System, *Proceedings of the IEEE Pacific Rim Conference on Communication: Computers and Signal Processing*, Victoria, B.C., Vol. 2, pp. 790–793, May 1993.

Gamage, L.B. and de Silva, C.W., Rule-Based Contour Generation for Robotic Processes, *Proceeding of IEEE TENCON*, Melbourne, Australia, pp. 649–653, Nov. 1992.

McCourt, R. and de Silva, C.W., Application of Predictive Control for Autonomous Satellite Capture Using a Deployable Manipulator System, *Proceedings of the ASME International Mechanica. Engineering Congress*, Washington, DC, CD ROM Paper IMECE2003-43718, November 2003.

Saliba, M. and de Silva, C.W., An Innovative Robotic Gripper for Grasping and Handling, *Proceedings of the IEEE International Conference on Industrial Electronics, Control, and Instrumentation*, Kobe, Japan, IEEE 91CH2976-9, pp. 975–979, October 1991.

Tang, P.L. and de Silva, C.W., Ethernet-Based Intelligent Switching of Controllers for Performance Improvement in an Industrial Plant, *Proceedings of the ASME International Mechanical Engineering Congress*, Washington, DC, CD ROM Paper IMECE2003-42281, November 2003.

Index

A

T